Engineering Mechanics
STATICS AND DYNAMICS

Engineering Mechanics
STATICS AND DYNAMICS

I. C. JONG

Professor of Mechanical Engineering
University of Arkansas
Fayetteville, Arkansas

B. G. ROGERS

Professor of Civil Engineering
Lamar University
Beaumont, Texas

Saunders College Publishing

A Division of
Holt, Rinehart and Winston, Inc.
Philadelphia   Chicago   Fort Worth   San Francisco   Montreal   Toronto   London   Sydney   Tokyo

Text typeface: Times Roman
Compositor: The Clarinda Company
Acquisitions Editor: Barbara Gingery
Managing Editor: Carol Field
Project Editor: Marc Sherman
Manager of Art and Design: Carol Bleistine
Art and Design Coordinator: Doris Bruey
Text Designer: Arlene Putterman
Cover Designer: Lawrence R. Didona
Text Artwork: Tech-Graphics
Director of EDP: Tim Frelick
Production Manager: Bob Butler
Marketing Manager: Denise Watrobsky

Cover: © COMSTOCK INC./ Brooklyn Bridge

Printed in the United States of America

ENGINEERING MECHANICS: STATICS AND DYNAMICS

ISBN 0–03–026314–X

Library of Congress Catalog Card Number: 90-053086

0 1 2 3   032   9 8 7 6 5 4 3 2 1

THIS BOOK IS PRINTED ON **ACID-FREE, RECYCLED** PAPER

RELATIONS FOR CONVERSIONS OF UNITS[†]

Acceleration:	1 ft/s^2	= 0.3048 m/s^{2*}
	1 in./s^2	= 25.4 mm/s^{2*}
Area:	1 ft^2	= 0.092903 m^2
	1 in^2	= 645.16 mm^{2*}
Density:	1 lbm/ft^3	= 16.018 kg/m^3
Energy:	1 Btu	= 1055.1 J
	1 ft·lb	= 1.3558 J
Force:	1 kip	= 4.4482 kN
	1 lb	= 4.4482 N
Impulse:	1 lb·s	= 4.4482 N·s
Length:	1 ft	= 0.3048 m*
	1 in.	= 25.4 mm*
	1 mi (= 5280 ft)	= 1.6093 km
	1 NM	= 1852 m*
Mass:	1 lbm	= 0.45359237 kg*
	1 slug	= 14.594 kg
	1 ton (= 2000 lbm)	= 907.18 kg
Moment of Force:	1 lb·ft	= 1.3558 N·m
	1 lb·in.	= 0.11298 N·m
Moment of Inertia, area:	1 in^4	= 416231 mm^4
Moment of Inertia, mass:	1 slug·ft^2 (= 1 lb·ft·s^2)	= 1.3558 kg·m^2
Momentum, linear:	1 lb·s	= 4.4482 N·s
Momentum, angular:	1 lb·ft·s	= 1.3558 kg·m^2/s
Power:	1 ft·lb/s	= 1.3558 W
	1 hp (= 550 ft·lb/s)	= 745.70 W
Pressure, stress:	1 lb/ft^2	= 47.880 Pa
	1 psi (= 1 lb/in^2)	= 6.8948 kPa
Velocity:	1 ft/s	= 0.3048 m/s*
	1 mph (= 1 mi/h)	= 0.44704 m/s*
	1 kn (= 1NM/h)	= 0.51444 m/s
Volume:	1 ft^3	= 0.028317 m^3
	1 in^3	= 16.387 cm^3
	1 gal (= 231 in^3)	= 3.7854 L
Work:	1 ft·lb	= 1.3558 J

[†]Refer to Table 1.1 on p. 9.

*Exact value

Preface

This combined text of statics and dynamics is designed for the first courses in engineering mechanics, which are usually offered in the sophomore or junior year in engineering curricula. In the development of the text, students are assumed to have a background in algebra, geometry, trigonometry, and basic differential and integral calculus. Nevertheless, students with a prior knowledge in college physics normally have a higher degree of readiness for learning engineering mechanics.

Mechanics has long been recognized as mainly a deductive science; however, the learning process is largely inductive. In the text, simple topics and problems precede those that are more complex and advanced. For example, the computation of moments of forces using scalar algebra is presented before that using vector algebra, explanations of the rudiments of vector analysis are given before their applications in mechanics, and exercise problems are arranged in order of increasing difficulty. Overall, the text is written to give a clear and up-to-date presentation of the theory and application of engineering mechanics; it is aimed at helping engineering students develop an ability to apply well-established principles to analyze and solve problems in a logical and effective manner.

Organization and Features

As shown in the contents overview, the first ten chapters cover statics, and the remaining ten chapters cover dynamics. The text has been especially organized to help students understand and retain the concepts presented. A more definitive approach is used in developing the text. Each chapter is divided into two or three definitive parts with descriptive headings, and each part contains an appropriate number of sections. The *key concepts* to be covered are listed at the beginning of each chapter, and the list is followed by an introductory paragraph. In addition to sets of traditional problems, concept-reinforcing *developmental exercises* are provided, and concluding remarks as well as a set of review problems are given at the end of each chapter.

Appendices A and B include materials for a *review of basic mathematics* from algebra through calculus, Appendix C verifies the vector characteristics of infinitesimal angular displacements, while Appendices D and E provide general *computer programs* for solving linear and nonlinear equations in mechanics. Yet, omission of computer programs will not prejudice the understanding of the text.

The list of *key concepts* at the beginning of each chapter serves as a summary of major topics presented. By identifying the key concepts in a given chapter, this list helps students focus their efforts and increase their efficiency of study. The introductory paragraph, which follows the list, gives an overview of the subject matter and the relevance of the topics to be covered.

The numerous *developmental exercises* interspersed throughout the text are intended for use as learning exercises. Users may either use or skip them as they see fit. These exercises are designed to (a) prompt students to describe and reinforce their understanding of the concepts, laws, and principles they are learning; and (b) provide opportunities for students to practice problem-solving skills by working through exercises that contain cue questions or similarities to preceding examples.

It is believed that *examples* depicting real-life situations are a useful means of providing motivation for students to learn and to develop the skills needed to reduce actual physical situations to models so that basic principles can be applied. Thus, the text includes many examples that illustrate real-life, as well as theoretical, situations. Each example contains a sample problem and a solution to that problem. The solution stresses the interpretation of the situation and the setting up of equations. Whenever intermediate steps in the solution are routine in nature, they are omitted to avoid distraction from the main solution.

The sets of *problems* provided in each chapter are also designed to create real-life, as well as theoretical, situations. Students can practice applying the principles presented in the text to these problems. Further, the text offers an abundance of problems, which provide the instructor with choices of assignments. Approximately two thirds of the problems have answers given in the back of the book. To alert users to a problem without a given answer, an asterisk (∗) is placed after the problem number.

The last section in each chapter is set aside to give *concluding remarks* on the subject matter in that chapter. Generally, these remarks cover the highlights and special features of the chapter. At the end of each chapter, a set of *review problems* is given. These problems generally illustrate the key concepts listed at the beginning of the chapter. Besides providing additional homework problems, they may be used by students as a test of the knowledge and skills they are expected to have acquired upon completing the study of that chapter.

Systems of Units

In recognition of current trends and practices in the United States, both the SI units *(Le Système International d'Unités)* and the U.S. customary units are used. Chapter 1 offers a discussion of the conversion between the two systems of units. Such a conversion is also discussed briefly in Chapter 12. Approximately 60 percent of the problems are stated in SI units, and the remainder retain U.S. customary units.

Optional Sections and Order of Coverage

To allow choices of coverage and to serve a wider group of users, the text includes more than a few optional sections. Each of these sections is marked with a star (★) before the section number for distinction from the other sections that form the core of the basic mechanics course. Optional sections may be skipped without affecting the understanding of the core of the course.

At the discretion of the instructor, some of the materials may be covered in a different sequence without losing continuity. In *Statics,* one may cover the concepts of forces and moments in Chapters 2 and 4 prior to covering the concepts of equilibrium in Chapters 3 and 5; and moments of inertia in Chapter 7 may be covered toward the end of the course, such as after Chapter 9. In *Dynamics*, one may cover the kinematics in Chapters 11 and 15 prior to covering the kinetics of particles and rigid bodies.

Computer Programs and Diskettes

Although students may use a *calculator* as a tool to solve linear simultaneous equations and find roots of nonlinear equations, they will often benefit further by using a *computer*, which is nowadays utilized in a wide variety of situations. Therefore, some computer programs are included. These programs are written in BASIC and will run on IBM PC and compatibles. Students using a computer should understand the essential concepts involved in solving the problem and be prepared to exercise good judgement in interpreting the output from the computer. The short programs appearing in some sections of the text are simple application programs, which serve as illustrations. The program in Appendix D is a general program for solving a set of *n* linear simultaneous equations in *n* unknowns, while the program in Appendix E is a general program for finding roots of a nonlinear or transcendental equation. Several problems that are amenable to computer solutions are identified in the text by footnotes, hints, or suggestions.

All programs listed in the text and the supplementary software, as described in Appendix F, are stored on computer diskettes, which may be obtained from bookstores or the publisher. The authors wish to thank Professor Lynn D. Wills of the University of Arkansas for his contribution in preparing and programming the supplementary software.

Unique Topic Coverage

The statics and dynamics texts include a number of presentations, illustrations, and emphases not usually found in similar texts. In particular, note the following:

1. Contrasting *relativistic viewpoints* on length, time, and mass are presented in Sec. 1.14. These viewpoints provide glimpses of the difference between engineering mechanics and modern physics.

2. The *computer solution of simultaneous equations,* encountered in Example 3.8, is illustrated in Appendix D. Opportunities to practice this skill are provided in some problems (e.g., Probs. 3.43, 3.44, 3.45, 3.57, 8.24, 8.25, 14.35, 14.36, etc.).

3. The *principle of moments,* synthesized from the concept of equivalent systems of forces, is identified as the key principle for locating the resultant of parallel forces, population center, centroid, geographic center, center of mass, and center of gravity in Sec. 6.2. It is a unifying principle which simplifies the study of the topics mentioned.

4. *Useful formulas* are established in Example 6.7 for locating the centroid of any plane area using triangular finite elements (cf. Probs. 6.40, 6.41, 6.42, and 6.96). Their application to any composite plane area furnishes an illustration of the finite element approach.

5. *Useful formulas* are established in Example 7.11 and the associated developmental exercises D7.28 and D7.29 for finding the moments of inertia and the product of inertia of any plane area using triangular finite elements (cf. Probs. 7.67 through 7.78, and 7.161). Their application to any composite plane area furnishes another illustration of the finite element approach.

6. *Effects of distributed moments* are taken into account in Eq. 8.9 and illustrated in Example 8.21. Without that equation, the slope in the bending-moment diagram for a beam with distributed moments would be inexplicable.

7. The *computer solution of a transcendental equation,* encountered in Example 8.24, is illustrated in Appendix E. Opportunities to practice using a computer to solve transcendental and nonlinear equations are provided in some problems (e.g., Probs. 8.113, 9.114, 10.24, 10.25, 10.26, 10.34, 13.57, 13.58, 14.52, 18.51, etc.).

8. The distinction between *compatible virtual displacement* and *constrained virtual displacement* is made in Secs. 10.4 and 13.13 and illustrated in Examples 10.7 through 10.10, as well as Examples 13.16, 14.14, 17.8, 17.9, 18.10, and 18.11.

9. *Infinitesimal angular displacements* are shown to be vectors in Appendix C. It establishes the background against which angular velocities and angular accelerations can readily be discerned as vectors.

10. The *displacement center* for finding virtual displacements is introduced in Sec. 10.6 and illustrated in Examples 10.6, 10.10, and 17.8. This concept provides an alternative to the traditional approach using variational calculus. (The displacement center is located in a manner similar to the velocity center.)

11. A *quantitative definition for potential energy* is stated in Secs. 10.9 and 13.8. This definition makes potential energy as easy to understand and compute as work and kinetic energy.

12. *Taylor's series expansion* is employed to establish a *higher-order criterion for stability* in Sec. 10.14. It explicitly shows why the system is stable when the first nonzero derivative of the potential energy is of *even* order and has a *positive* value.

13. The *inertia force* of a body is interpreted in Sec. 12.3 as the reaction of the body to the action of the resultant force on the body. This interpretation helps demystify inertia forces for dynamics students.

14. The *rigor in the definition of work,* when the force does not act continually on the same particle, is emphasized in Secs. 10.2, 13.2, and 17.1. It brings home to students why the friction force exerted by the ground on a wheel rolling without slipping does no work on the wheel.

15. *Extensions and generalizations of the principle of virtual work* are presented in Secs. 13.13, 14.10, 17.7, and 18.7 as complements to traditional methods. The virtual work approach often obviates the necessity of solving simultaneous equations in kinetics as well as statics.

16. The *corollary to the theorem of acceleration center* of a rigid body is emphasized in Sec. 15.9 and used in Chapters 15 and 16. This concept greatly reduces the algebraic work required in a traditional solution. Moreover, it alerts students to the frequent mistake of taking a velocity center as an acceleration center.

Acknowledgments

The authors thank their colleagues Jim H. Akin, Charles W. Crook, and Charles R. Fagg for their thoughtful suggestions, and department heads, William F. Schmidt (University of Arkansas) and Enno Koehn (Lamar University), for their encouragement. They appreciate the numerous helpful suggestions and comments made on drafts of the text by the reviewers:

Paul Chan, New Jersey Institute of Technology
Sahib Chehl, Southern University
Rick Gill, University of Idaho
Richard Golembiewski, Milwaukee School of Engineering
Reese Goodwin, Brigham Young University
David Hartman, Northern Arizona University
Robert Howland, Notre Dame University
Richard Longman, Columbia University
Scott Martin, Youngstown State University
Luis Clemente Mesquita, University of Nebraska
Saeed Niku, California State Polytechnic University
Semih Oktay, University of Maryland
Rick Reimer, University of California, Berkeley
Eugene Stolarik, University of Minnesota
Robert Swaim, Oklahoma State University
William Van Arsdale, University of Houston
Carl Vilmann, Michigan Technological University

Special thanks are extended to Barbara Gingery (Acquisitions Editor), Marc Sherman (Project Editor), and Saunders College Publishing for their effort and support during the various stages of the project.

I. C. Jong
B. G. Rogers

December 1990

A Note for Students

As students, your goal here is learning mechanics as well as developing a high level of analytical capability. It is going to require some of your *time* and *effort*. However, many people find that it may be more efficiently attained by noting the following hints:

- **Study methodically.** As you begin each new chapter, try first to gain an overview. This may be achieved by doing the following: (a) thumb through the chapter to look it over, (b) go over the key concepts at the beginning of the chapter, (c) read the introduction following the key concepts, and (d) turn to the last section to read the concluding remarks. Then, read rapidly the sections as assigned or desired, and later return for an in-depth study.

- **Draw diagrams.** It is important to realize early on that engineering mechanics is a physical science which heavily involves *geometry*. Pertinent diagrams or sketches are valuable aids in studying the state of rest or motion of physical systems. There is no substitute for the perception and insight gained from the diagrams or sketches drawn.

- **Work problems.** Work the problems assigned or suggested by your instructor. This is not only an effective way to learn the subject but also a way to show to yourself how well you have learned it.

- **Ask questions.** Maintain your inquisitive mind in the study. Most instructors and teaching assistants have a sincere interest in helping students deepen their understanding of the new material.

The authors sincerely hope that the above hints will prove to be helpful, and that you will see the logic, beauty, and relevance of mechanics as you progress in your learning.

List of Symbols

$\mathbf{a}$	Acceleration
a	Constant; radius; distance; axis; semimajor axis
$\bar{\mathbf{a}}$	Acceleration of mass center
$\mathbf{a}_A, \mathbf{a}_B, \ldots$	Accelerations of $A, B, \ldots$
$\mathbf{a}_{B/A}$	Acceleration of B in a nonrotating reference frame translating with A; acceleration of B relative to A
$\mathbf{a}_{B/Axyz}$	Acceleration of B in the moving reference frame $Axyz$
$\mathbf{a}_c$	Constant acceleration
$\mathbf{A, B, C}, \ldots$	Vectors; forces; reactions from supports and connections
$A, B, C, \ldots$	Points; magnitudes
$A', B', C', \ldots$	Points; magnitudes
A	Area
$\overline{AB}, \overline{AC}, \ldots$	Line segments; distances
$\mathbf{Aimp}_A$	Angular impulse about A
b	Width; distance; axis; semiminor axis
c	Constant; coefficient of viscous damping
C	Centroid; constant; compression; velocity center
d	Distance
d_s	Shortest distance
$\mathbf{e}_t, \mathbf{e}_n$	Unit vectors in tangential and normal directions
$\mathbf{e}_r, \mathbf{e}_\theta$	Unit vectors in radial and transverse directions
e	Base of natural logarithms; coefficient of restitution
f	Scalar function; cyclic frequency
$\mathbf{F}$	Force; friction force
$\mathbf{F}_k$	Kinetic friction force
$\mathbf{F}_m$	Maximum friction force
$\mathbf{F}_\parallel$	Component of $\mathbf{F}$ parallel to a specific axis
$\mathbf{F}_\perp$	Component of $\mathbf{F}$ perpendicular to a specific axis
g	Gravitational acceleration
G	Constant of gravitation; center of gravity; mass center
h	Height; sag of cable; angular momentum per unit mass
$\mathbf{H}_G, \mathbf{H}_O$	Angular momentum about mass center G, angular momentum about fixed point O
$\mathbf{i, j, k}$	Unit vectors along Cartesian coordinate axes

$I, I_x, I_y, \ldots$	Moments of inertia
$I_{xy}, I_{uv}, \ldots$	Products of inertia
$\bar{I}$	Centroidal moment of inertia of an area; central moment of inertia of a body
$\mathbf{Imp}_{1 \to 2}$	Linear impulse exerted during $t_1 \leq t \leq t_2$
J	Polar moment of inertia
k	Spring modulus; constant
k_x, k_y, k_O	Radii of gyration
$\bar{k}$	Centroidal radius of gyration of an area; central radius of gyration of a body
K	Torsional spring modulus; constant
$\mathbf{L}$	Linear momentum
L	Length; span
m	Mass of a body
$\tilde{m}$	Distributed torque; moment per unit length
$\mathbf{M}$	Moment; torque
M	Magnitude of moment or torque; mass of earth
M_{AB}	Moment about axis AB
$\mathbf{M}_O$	Moment about point O
$\mathbf{M}_O^R$	Resultant moment about point O
$\mathbf{M}_\parallel$	Component of $\mathbf{M}$ parallel to a specific axis
$\mathbf{M}_\perp$	Component of $\mathbf{M}$ perpendicular to a specific axis
n	Normal axis; normal direction
$\mathbf{N}$	Normal component of reaction
O	Origin of coordinates; point
p	Pressure; distance; pitch of a screw
P_{cr}	Critical load for buckling
$\mathbf{q}$	Displacement
$\mathbf{q}_\parallel$	Component of $\mathbf{q}$ parallel to a specific force
Q	Time rate of volumetric flow; magnitude; point
$\dot{\mathbf{Q}}$	Time derivative of $\mathbf{Q}$ in a fixed reference frame (e.g., $OXYZ$)
$(\dot{\mathbf{Q}})_{Axyz}$	Time derivative of Q in the moving reference frame $Axyz$
$\mathbf{r}$	Position vector
$\mathbf{r}_{B/A}$	Position vector of B relative to A
$\mathbf{r}_\parallel$	Component of $\mathbf{r}$ parallel to a specific axis
$\mathbf{r}_\perp$	Component of $\mathbf{r}$ perpendicular to a specific axis
r	Radius; distance; polar coordinate
r_f	Radius of circle of friction
$\mathbf{R}$	Resultant force; resultant vector; reaction
R	Radius; radius of earth; magnitude
s	Length of arc; length of cable; axis
t	Time; thickness; tangential axis; tangential direction
T	Tension; concentrated moment; kinetic energy
$\mathbf{u}$	Velocity
u	rectangular coordinate or axis; polar coordinate; speed
U	Work
$U_{1 \to 2}$	Work of a force acting on a body from positions 1 to 2

$\mathbf{v}$	Velocity
v	rectangular coordinate or axis; speed; rectilinear velocity
$\bar{\mathbf{v}}$	Velocity of mass center
$\mathbf{v}_A, \mathbf{v}_B, \ldots$	Velocities of A, B, . . .
$\mathbf{v}_{B/A}$	Velocity of B in a nonrotating reference frame translating with A; velocity of B relative to A
$\mathbf{v}_{B/Axyz}$	Velocity of B in the moving reference frame $Axyz$
V	Volume; shear; potential energy
V_a, V_e, V_g	Applied, elastic, and gravitational potential energies
w	Distributed load; force per unit length
$\mathbf{W}$	Weight force
W	Weight
x, y, z	Rectangular coordinates or axes; distances
$\bar{x}, \bar{y}, \bar{z}$	Rectangular coordinates of centroid or center of mass
x', y', z'	Centroidal axes; central axes; auxiliary rectangular coordinate axes
x'', y'', z''	Auxiliary rectangular coordinate axes
X	Amplitude of vibration
Z	Acceleration center
$\boldsymbol{\alpha}$	Angular acceleration
α, β, γ	Angles
β	Angle of contact in belt friction
γ	Specific weight
γ_A, γ_L	Weight per unit area, weight per unit length
δ	Deflection; logarithmic decrement
δ_{st}	Static deflection
$\delta\mathbf{r}$	Virtual displacement vector
$\delta x, \delta y$	Virtual displacements
$\delta\theta$	Angular virtual displacement
$\Delta\theta$	Angular displacement
δU	Virtual work
$\delta U'$	Generalized virtual work
ε	Eccentricity of conic section or of orbit
ζ	Variable analogous to z; axis; damping factor
η	Variable analogous to y; axis; efficiency
θ	Angle; angular coordinate; polar coordinate; Eulerian angle (for nutation)
$\theta_1, \theta_2, \ldots$	Angles; generalized coordinates
θ_A	Directional angle of $\mathbf{A}$; angle at point A
$\theta_x, \theta_y, \theta_z$	Direction angles of a vector
Θ	Angular amplitude of vibration
$\boldsymbol{\lambda}$	Unit vector along a line
μ_k	Coefficient of kinetic friction
μ_s	Coefficient of static friction
ξ	Variable analogous to x; axis
ρ	Mass density; radius of curvature
ρ_A, ρ_L	Mass per unit area, mass per unit length
τ	Variable analogous to t; period of orbit; period of vibration

ϕ	Angle; Eulerian angle (for precession); phase angle (in free vibration)
ϕ_k	Angle of kinetic friction
ϕ_s	Angle of static friction
φ	Phase angle (in forced vibration)
ψ	Angle; Eulerian angle (for spin)
$\boldsymbol{\omega}$	Angular velocity
ω	Angular speed; circular frequency
$\boldsymbol{\Omega}$	Angular velocity of reference frame
*	Asterisk, marking a problem with no answer given
★	Star, marking an optional section
†	Dagger, marking a footnote
⇑	Hollow arrow, representing a moment or torque
⇒	Equipollent to
◄	Answer pointer

Contents Overview

STATICS

1. Fundamentals
2. Forces
3. Equilibrium of Particles
4. Moments of Forces
5. Equilibrium of Rigid Bodies
6. First Moments: Centroids and Centers of Gravity
7. Second Moments: Moments of Inertia[†]
8. Structures
9. Friction
10. Virtual Work

DYNAMICS

11. Kinematics of Particles
12. Kinetics of Particles: Force and Acceleration
13. Kinetics of Particles: Work and Energy
14. Kinetics of Particles: Impulse and Momentum
15. Plane Kinematics of Rigid Bodies
16. Plane Kinetics of Rigid Bodies: Force and Acceleration
17. Plane Kinetics of Rigid Bodies: Work and Energy
18. Plane Kinetics of Rigid Bodies: Impulse and Momentum
19. Motion of Rigid Bodies in Three Dimensions
20. Vibrations

APPENDICES

A. Prior Basic Mathematical Skills
B. Review of Basic Mathematics
C. Infinitesimal Angular Displacements
D. Simultaneous Equations Solver
E. Digital Root Finder
F. Supplementary Software

[†]Chapter 7 may be covered after Chap. 9 without affecting continuity.

Contents

Preface   v

A Note for Students   x

List of Symbols   xi

Chapter 1
Fundamentals   1

1.1   Prior Basic Mathematical Skills   1

FUNDAMENTAL CONCEPTS   2

1.2   Scalar and Vector Quantities   2
1.3   Representation of a Vector   3
1.4   Characteristics of a Force   3
1.5   Mechanics   4
1.6   Idealizations in Mechanics   5
1.7   Dimensions: Physical Quantities   7
1.8   Units and Prefixes   8
1.9   Significant Digits and Rounding Rule   10
1.10 Numerical Accuracy and Conversion of Units   11

FUNDAMENTAL LAWS   13

1.11 Newton's Laws of Motion and Law of Gravitation   13
1.12 Parallelogram Law and Rigid-Body Principle   15
1.13 Mass Versus Weight and Force   16
1.14 Concluding Remarks   20

Chapter 2
Forces   23

FORCES IN A PLANE   23

2.1   Graphical Representation   23
2.2   Addition of Forces: Parallelogram Law   25
2.3   Resolution of Forces: Parallelogram Law   27
2.4   Analytical Representation   33
2.5   Addition and Resolution of Forces: Vector Algebra   35

FORCES IN SPACE   41
2.6   Representation of a Force in Space   41
2.7   Position Vector and Unit Vector in Space   42
2.8   Addition and Resolution of Forces in Space   44
2.9   Concluding Remarks   48

Chapter 3
Equilibrium of Particles   51

FUNDAMENTALS AND FREE-BODY DIAGRAM   51
3.1   Inertial Reference Frame   51
3.2   Concurrent Force Systems   52
3.3   Reaction, Internal Force, and External Force   53
3.4   Cordlike Elements, Pulleys, and Connections   54
3.5   Linear Springs   55
3.6   Rollers and Smooth Supports   56
3.7   Free-Body Diagram   57
3.8   Guides for Drawing a Good Free-Body Diagram   59
EQUILIBRIUM IN A PLANE   62
3.9   Equilibrium of Particles in a Plane   62
3.10   Scalar Method of Solution   64
3.11   Analytical Vector Method of Solution   69
3.12   Semigraphical Method of Solution   69
EQUILIBRIUM IN SPACE   73
3.13   Equilibrium of Particles in Space   73
3.14   Method of Solution: Equilibrium in Space   76
3.15   Concluding Remarks   79

Chapter 4
Moments of Forces   82

SCALAR FORMULATION   82
4.1   Moment of a Force   82
4.2   Moment Vectors and the Right-Hand Rule   84
4.3   Varignon's Theorem: Forces in a Plane   86
4.4   Moment of a Couple: Torque   89
4.5   Moments about Coordinate Axes   91
APPLICATION OF CROSS PRODUCTS   97
4.6   Trihedrals   97
4.7   Cross Product of Two Vectors   98
4.8   Properties of Cross Products   99
4.9   Analytical Expression of the Cross Product   101
4.10   Moment of a Force about a Point by Cross Product   103
4.11   Varignon's Theorem: Forces in Space   105
4.12   Moment of a Couple by Cross Product   107
4.13   Equivalent Couples and Addition of Moments of Couples   108
APPLICATION OF DOT PRODUCTS   112
4.14   Dot Product and Its Properties   112
4.15   Scalar Triple Product   114

4.16   Moment of a Force about an Axis   115
★4.17   Shortest Distance between an Axis and a Force   118
4.18   Concluding Remarks   122

Chapter 5
Equilibrium of Rigid Bodies   124

EQUIVALENT SYSTEMS OF FORCES   125
5.1   Resolution of a Given Force into a Force-Moment System at Another Point   125
5.2   Reduction of a System of Forces to a Force-Moment System   126
5.3   Equivalent and Equipollent Systems of Forces   128
★5.4   Reduction of a System of Forces to a Wrench   130
EQUILIBRIUM IN A PLANE   134
5.5   Reactions from Supports and Connections in a Plane   134
5.6   Equilibrium of a Rigid Body in a Plane   137
5.7   Equilibrium of Two- and Three-Force Bodies   141
5.8   Statical Indeterminacy and Constraint Conditions   143
5.9   Possible Sets of Independent Scalar Equilibrium Equations   145
5.10   Method of Solution: Equilibrium in a Plane   147
EQUILIBRIUM IN SPACE   153
5.11   Reactions from Supports and Connections in Space   153
5.12   Maximum Number of Statically Determinate Unknowns   156
5.13   Method of Solution: Equilibrium in Space   159
5.14   Concluding Remarks   172

Chapter 6
First Moments:
Centroids and Centers of Gravity   176

6.1   Dimensional Homogeneity in Equations   176
6.2   Resultant of Parallel Forces: Principle of Moments   177
LINES AND SLENDER MEMBERS   179
6.3   Centroids of Lines   179
6.4   Composite Lines   182
6.5   Centers of Gravity of Slender Members   184
AREAS AND LAMINAS   188
6.6   Orders of Differential Area Elements   188
6.7   Centroids of Areas   189
6.8   Composite Areas   194
6.9   Centers of Gravity of Laminas   197
6.10   Distributed Loads on Beams   202
6.11   Theorems of Pappus-Guldinus   204
VOLUMES AND BODIES   208
6.12   Centroids of Volumes   208
6.13   Composite Volumes and Center of Pressure   209
6.14   Centers of Mass and Centers of Gravity of Bodies   212
6.15   Concluding Remarks   220

Chapter 7
Second Moments: Moments of Inertia   223

AREA MOMENTS OF INERTIA[†]   224

7.1   Situations Involving Area Moments of Inertia   224
7.2   Moments of Inertia of an Area   225
7.3   Radius of Gyration of an Area   231
7.4   Parallel-Axis Theorem: Area Moments of Inertia   234
7.5   Product of Inertia of an Area   236
7.6   Parallel-Axis Theorem: Area Products of Inertia   239
7.7   Composite Areas   241
★7.8   Rotation of Axes   250
★7.9   Representation by Mohr's Circle   251
★7.10   Principal Moments and Axes of Inertia   254

MASS MOMENTS OF INERTIA   258

7.11   Moments of Inertia of a Mass   258
7.12   Radius of Gyration of a Mass   264
7.13   Parallel-Axis Theorem: Mass Moments of Inertia   266
7.14   Composite Bodies   268
7.15   Concluding Remarks   274

Chapter 8
Structures   278

TRUSSES   278

8.1   Trusses and Their Uses   278
8.2   Usual Assumptions for Trusses   280
8.3   Rigid Truss, Plane Truss, and Simple Truss   281
8.4   Compound and Complex Trusses   283
8.5   Forces in Truss Members   284
8.6   Zero-Force Members at T and V Joints   285
8.7   Method of Joints for Plane Trusses   286
8.8   Method of Sections for Plane Trusses   291
8.9   X Joints and Successive Method   297
★8.10   Space Trusses   299
★8.11   Method of Joints for Simple Space Trusses   300

FRAMES AND MACHINES   307

8.12   Structures Containing Multiforce Members   307
8.13   Force System Acting on a Detached Member   309
8.14   Method of Solution for Frames and Machines in a Plane   311

BEAMS AND CABLES   323

★8.15   Internal Forces and Moments in Multiforce Members   323
★8.16   Shears and Bending Moments in Beams   325
★8.17   Shear and Bending-Moment Diagrams   328
★8.18   Cables and the Governing Differential Equation   336
★8.19   Parabolic Cable   340
★8.20   Catenary Cable   343
8.21   Concluding Remarks   348

[†]Also called *second moments of areas*.

Chapter 9
Friction   351

FRICTION BETWEEN RIGID BODIES   351

9.1   State of Sliding Surfaces   351
9.2   Friction Force   352
9.3   Laws of Dry Friction   353
9.4   Angles of Friction   355
9.5   Types of Friction Problems   356
9.6   Method of Solution for Type I Problems   357
9.7   Method of Solution for Type II Problems   359
9.8   Method of Solution for Type III Problems   368
★9.9   Dry Friction in Square-Threaded Screws   374
★9.10   Axle Friction in Journal Bearings   379
★9.11   Disk Friction in Thrust Bearings   380

BELT FRICTION   386

9.12   Flat Belts   386
9.13   V Belts   390
9.14   Concluding Remarks   396

Chapter 10
Virtual Work   400

PRINCIPLE OF VIRTUAL WORK   400

10.1   Displacements   400
10.2   Work of a Force   401
10.3   Work of a Couple   406
10.4   Virtual Displacements   409
10.5   Components of a Linear Virtual Displacement   412
10.6   Differential Calculus and Displacement Center   413
10.7   Principle of Virtual Work   417
10.8   Applications to Equilibrium Problems   418

POTENTIAL ENERGY AND STABILITY   430

★10.9   Conservative System and Potential Energy   430
★10.10   Gravitational Potential Energy   431
★10.11   Elastic Potential Energy   433
★10.12   Applied Potential Energy   434
★10.13   Principle of Potential Energy   436
★10.14   Stability of Equilibrium   439
10.15   Concluding Remarks   446

Chapter 11
Kinematics of Particles   451

11.1   Types of Motion of a Particle   451

RECTILINEAR MOTION   452

11.2   Kinematic Quantities   452
11.3   Position, Displacement, and Velocity   452
11.4   Acceleration and Jerk   455
11.5   Determination of Rectilinear Motion   457
11.6   Relative Rectilinear Motion   465

11.7   Dependent Rectilinear Motions   467
★11.8   Graphical Solution   470
CURVILINEAR MOTION   477
11.9   Position Vector, Velocity, and Acceleration   477
11.10   Derivatives of Vector Functions   481
11.11   Rectangular Components   482
11.12   Free Flight of a Projectile   485
11.13   Tangential and Normal Components   492
11.14   Radial and Transverse Components   496
11.15   Cylindrical Components   498
11.16   Concluding Remarks   503

Chapter 12
Kinetics of Particles:
Force and Acceleration   507

NEWTON'S SECOND LAW   507
12.1   Method of Force and Acceleration   507
12.2   Consistent Systems of Kinetic Units   509
12.3   Dynamic Equilibrium: Inertia Force   511
12.4   Rectilinear Motion of a Particle   512
12.5   Curvilinear Motion of a Particle   520
12.6   Systems of Particles   525
12.7   Principle of Motion of the Mass Center   527
CENTRAL-FORCE MOTION   534
12.8   Gravitational Force   534
12.9   Motion under a Central Force   535
12.10   Governing Differential Equation   538
12.11   Trajectories of Spacecraft   539
★12.12   Kepler's Laws of Planetary Motion   545
12.13   Concluding Remarks   550

Chapter 13
Kinetics of Particles: Work and Energy   552

WORK AND KINETIC ENERGY   552
13.1   Work of a Force   552
13.2   Power and the Subtlety in Defining Work   557
13.3   Work of a Gravitational Force   560
13.4   Work of a Spring Force   561
13.5   Kinetic Energy of a Particle   564
13.6   Principle of Work and Energy   565
13.7   System of Particles   569
CONSERVATION OF ENERGY AND SPECIAL TOPIC   576
13.8   Conservative System and Potential Energy   576
13.9   Gravitational Potential Energy   577
13.10   Elastic Potential Energy   580
13.11   Applied Potential Energy   581
13.12   Conservation of Energy   583
★13.13   Virtual Work in Kinetics: Force and Acceleration   587
13.14   Concluding Remarks   592

Chapter 14
Kinetics of Particles:
Impulse and Momentum   594

IMPULSE AND MOMENTUM   594

14.1   Linear Impulse and Linear Momentum   594
14.2   Principle of Impulse and Momentum   596
14.3   Angular Impulse and Angular Momentum   599
14.4   System of Particles   602
14.5   Impulsive Motion   610
14.6   Central-Force Motion   616

IMPACT AND SPECIAL TOPIC   620

14.7   Impact   620
14.8   Direct Central Impact   621
14.9   Oblique Central Impact   624
★14.10   Generalized Virtual Work: Impulse and Momentum   626

VARIABLE SYSTEMS   630

★14.11   Variable Systems of Particles   630
★14.12   Systems with Steady Mass Flow   631
★14.13   Systems with Variable Mass: Motion of Rockets   636
14.14   Concluding Remarks   645

Chapter 15
Plane Kinematics of Rigid Bodies   649

15.1   Types of Plane Motion of a Rigid Body   649

USE OF NONROTATING REFERENCE FRAMES   651

15.2   Translation   651
15.3   Rotation   652
15.4   Linear and Angular Motions   654
15.5   General Plane Motion: Chasles' Theorem   660
15.6   Velocities in Relative Motion   662
15.7   Velocity Center   669
15.8   Accelerations in Relative Motion   676
15.9   Acceleration Center   683
15.10   Parametric Method   688

USE OF ROTATING REFERENCE FRAMES   694

15.11   Time Derivatives of Rotating Unit Vectors   694
15.12   Time Derivatives of a Vector in Two Reference Frames   696
15.13   Velocities in Different Reference Frames   698
15.14   Accelerations in Different Reference Frames   702
15.15   Interpretations for Coriolis Acceleration   705
15.16   Concluding Remarks   711

Chapter 16
Plane Kinetics of Rigid Bodies:
Force and Acceleration   715

MASS MOMENTS OF INERTIA   716

16.1   Moments of Inertia of a Mass   716
16.2   Radius of Gyration of a Mass   720

16.3   Parallel-Axis Theorem: Mass Moments of Inertia   722
16.4   Composite Bodies   724

FORCE AND ACCELERATION   731
16.5   Effective Force-Moment System on a Rigid Body   731
16.6   Center of Percussion   735
16.7   Method of Force and Acceleration   737
16.8   Interpretations for Mass Moments of Inertia   743
16.9   Constrained General Plane Motion   750
16.10   Concluding Remarks   760

Chapter 17
Plane Kinetics of Rigid Bodies:
Work and Energy   765

WORK AND KINETIC ENERGY   765
17.1   Work of a Force Acting on a Rigid Body   765
17.2   Work of a Couple Acting on a Rigid Body   767
17.3   Kinetic Energy of a Rigid Body in Plane Motion   768
17.4   Principle of Work and Energy   770

CONSERVATION OF ENERGY AND SPECIAL TOPIC   782
17.5   Conservation of Energy   782
17.6   Power   785
★17.7   Virtual Work in Kinetics: Force and Acceleration   785
17.8   Concluding Remarks   794

Chapter 18
Plane Kinetics of Rigid Bodies:
Impulse and Momentum   797

IMPULSE AND MOMENTUM   797
18.1   Linear and Angular Impulses on a Rigid Body   797
18.2   Momentum of a Rigid Body in Plane Motion   799
18.3   Principle of Impulse and Momentum   803
18.4   System of Rigid Bodies   806
18.5   Conservation of Momentum   809

IMPULSIVE MOTION AND SPECIAL TOPIC   817
18.6   Impulsive Motion of Rigid Bodies   817
★18.7   Generalized Virtual Work: Impulse and Momentum   822
18.8   Concluding Remarks   830

Chapter 19
Motion of Rigid Bodies in Three Dimensions   834

KINEMATICS OF RIGID BODIES IN SPACE   834
★19.1   Time Derivatives of a Vector in Two Reference Frames   834
★19.2   Velocities in Different Reference Frames   837
★19.3   Accelerations in Different Reference Frames   845

KINETICS OF RIGID BODIES IN SPACE   857
★19.4   Moments and Products of Inertia of a Mass   857
★19.5   Rotation of Axes: Principal Axes of Inertia   861
★19.6   Momentum of a Rigid Body   866

★19.7   Kinetic Energy of a Rigid Body   872
★19.8   Equations of Motion for a Rigid Body   878
★19.9   Gyroscopic Motion: Steady Precession   888
★19.10   Concluding Remarks   898

Chapter 20
Vibrations   901

FREE VIBRATIONS   901
★20.1   Undamped Free Vibrations   901
★20.2   Energy Method   910
★20.3   Damped Free Vibrations   917
★20.4   Logarithmic Decrement   922
FORCED VIBRATIONS   925
★20.5   Harmonic Excitation   925
★20.6   Transmissibility   929
★20.7   Concluding Remarks   933

APPENDIX A
PRIOR BASIC MATHEMATICS   A–1

APPENDIX B
REVIEW OF BASIC MATHEMATICS   A–7
        B1   Algebra   A–7
        B2   Geometry   A–10
        B3   Trigonometry   A–11
        B4   Analytic Geometry   A–13
        B5   Calculus   A–16

APPENDIX C
INFINITESIMAL ANGULAR DISPLACEMENTS   A–20

APPENDIX D
SIMULTANEOUS EQUATIONS SOLVER   A–22

APPENDIX E
DIGITAL ROOT FINDER   A–25

APPENDIX F
SUPPLEMENTARY SOFTWARE   A–27

ANSWERS TO SELECTED DEVELOPMENTAL
EXERCISES AND PROBLEMS   ANS–1

INDEX   I–1

Fundamentals

Chapter 1

KEY CONCEPTS

- Scalar, vector, force, mechanics, statics, equilibrium, dynamics, kinematics, kinetics, particle, rigid body, continuum, concentrated force, dimension, length, mass, time, weight, apparent weight, unit, slug, and newton.
- Characteristics of a force and idealizations in mechanics.
- Dimensions and units in the SI and U.S. customary system.
- Rounding rule, numerical accuracy, and conversion of units.
- Newton's laws of motion and law of gravitation, the parallelogram law, and the rigid-body principle.

To form a basis for the study of elementary mechanics, this chapter presents fundamental concepts, definitions, laws, and principles. The fundamental importance of the concept of force is obvious since the study of effects of forces on physical bodies at rest or in motion is, in truth, the central effort of mechanicists. Particles, rigid bodies, continua, and concentrated forces are all idealized concepts which have practical significance in the solution of mechanics problems. Dimensions and units are important to engineers since they work with numerical quantities which are usually used in conjunction with certain units. The distinction between *mass* and *weight* is emphasized because some beginners in mechanics frequently confuse the mass with the weight of a body. As a guide, some rules and concepts related to numerical accuracy are given and discussed. Furthermore, problems are provided to let students practice converting units from the SI to the U.S. customary system, and vice versa.

1.1 Prior Basic Mathematical Skills[†]

The effective study of mechanics as presented in this text requires certain mathematical skills. A set of problems reflecting some of such skills is presented in App. A, which may be used for self-assessment. The mathematical

---

[†]Skills in computer usage and programming are not required for a basic course of study; however, such skills will be helpful if the use of computers is incorporated to enrich the course of study. As mentioned in the Preface, some computer programs are included in the text.

background needed in answering the problems in App. A is tacitly assumed and will generally not be explained in the development of the text material. For convenience, a short review of basic mathematics is presented in App. B, which includes basic concepts and useful topics in algebra, geometry, trigonometry, analytic geometry, and calculus.

FUNDAMENTAL CONCEPTS

1.2 Scalar and Vector Quantities

A *scalar* may be defined as a quantity which has magnitude and possibly sense, but no orientation. The sense of a scalar is denoted by a positive or a negative sign before the magnitude of the scalar. Examples of scalar quantities include length, time, mass, temperature, cost, humidity, work, energy, and power.

A *vector* may be defined as a directed quantity which has magnitude, sense, and orientation, and which obeys the laws for addition of vectors. The sense and orientation of a vector constitute its direction. The result of the addition of two vectors must be independent of the order of addition of these two vectors. Concepts and operations involving vectors will be defined and illustrated later. Typical vector quantities are force, moment of force, linear displacement, velocity, acceleration, impulse, and momentum.

In tensor analysis, whose general theory is not needed in the present study, both scalars and vectors are regarded as special cases of tensors: A scalar is defined as a tensor of the zeroth order, and a vector is defined as a tensor of the first order. Furthermore, specially defined moments of inertia[†] as well as other quantities are classified as tensors of the second order.[††] Operations involving scalars and vectors will be employed for the effective study of the topics in this text.

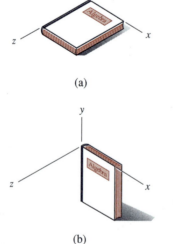

(a)

(b)

Fig. D1.3

Developmental Exercises

D1.1[†††]   Define a scalar and give three examples of scalars.

D1.2   Define a vector and give three examples of vectors.

D1.3   A book is shown in Fig. D1.3(a). First, rotate the book about the x axis through 90° as shown in Fig. D1.3(b). This finite rotation is said to have a direction parallel to the x axis and a magnitude of 90°. From this new position, rotate the book about the y axis through 90° and note the orientation of the book. (i) If you first rotated the book in Fig. D1.3(a) about the y axis through 90°, then rotated it about the x axis through 90°, would you obtain the same orientation for

---

[†]Cf. the first footnote on p. 237.

[††]Second-order tensors are sometimes called *bisors*. There are higher-order tensors defined and used in advanced mechanics.

[†††]The developmental exercises (D's) may be *used* or *skipped* as explained in the Preface. Whenever a developmental exercise asks for the definition of a term or description of a law or principle, students are encouraged to answer in their *own words,* which reflect their understanding of the concept involved. Answers to selected developmental exercises are given at the end of the text.

the book as that obtained earlier? (ii) Can we classify finite rotations (which have directions and magnitudes) of a body as vectors? (Cf. App. C.)

D1.4   Is it proper to say that all physical quantities can be classified as either scalar or vector quantities?

1.3   Representation of a Vector

In a typeset text, a vector quantity is often represented by a **boldface** letter, while its magnitude is simply represented by the *italic* version of the same letter. If not typeset, a vector is often represented by a letter with a *short arrow* drawn above it (e.g., $\vec{F}$), while its magnitude is simply represented by that letter (e.g., F).

At times, a vector is graphically represented by a directed line segment. The *magnitude* of the vector is usually denoted by a lightface letter or by a numerical value near the directed line segment; alternatively, the length of the directed line segment drawn to scale may be used to denote the magnitude. The *direction* of the vector is specified by the orientation of the line segment and by the arrowhead indicating the sense at the tip of the segment. For instance, a vector **V** in the *xy* plane may be represented as shown in Fig. 1.1, where the sense indicates that **V** is directed from the point *A* toward the point *B*. If the sense of **V** is reversed, we obtain $-$**V**. Furthermore, a vector may be represented by its analytical expression, which will be covered in Chap. 2.

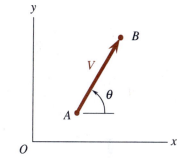

Fig. 1.1

Developmental Exercises

D1.5   Which are vector quantities in the equation written as $\mathbf{F} = m\mathbf{a}$?

D1.6   In Fig. 1.1, what part of the vector **V** does each of the following represent? (a) The letter V, (b) the angle θ, (c) the arrowhead at the point *B*.

1.4   Characteristics of a Force

A *force* may be defined as an action in the form of a push, a pull, or a shear exerted by one body on another body where the action changes or tends to change the state of motion of the body on which it acts. A force is a vector. The force exerted by one body on another body *by contact* is called a *contact force* or *surface force*. The force exerted by one body on another body *at a distance* is called a *body force*. Gravitational forces and magnetic forces are examples of body forces. In some cases, a force applied to a body will cause the body to change its direction of motion, or its speed, or both. In other cases, the application of a force to a body at rest will cause the other forces (e.g., the reaction forces from its supports) acting on the same body to change without causing the body to move.

The properties which distinguish one force from a different force are called the characteristics of a force. The *characteristics of a force* are (a) its point of application, (b) its magnitude, and (c) its direction (orientation and sense). The rotational effect produced by a given force acting at a point on

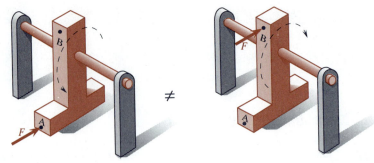

Fig. 1.2

a body is generally different from that produced by another force with the same magnitude and direction (as the given force) but acting at a different point on the same body. This may be illustrated in Fig. 1.2, where the curved arrow indicates the rotational effect of the force **F** on the rigid body.

In mathematics, any two vectors which have the same magnitude and direction are considered to be equal, regardless of whether or not they have the same point of application or the same line of action. Force vectors are not so. Efforts to differentiate such situations have led to the practice of referring to vectors in mathematics as *free vectors* and to force vectors as *bound* (or *localized*) *vectors* in some circumstances and as *sliding vectors* in other circumstances (cf. Sec. 1.12). Note that there are some free vectors in mechanics (cf. Secs. 4.4 and 4.12).

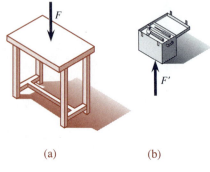

(a)   (b)

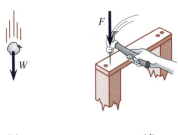

(c)   (d)

Fig. D1.9

Developmental Exercises

D1.7   Define a force and state the two ways in which a body may exert a force on another body.

D1.8   What are the three characteristics of a force?

D1.9   Which of the following forces are contact forces? (a) The force exerted by a toolbox on a workbench, (b) the force exerted by the workbench on the toolbox, (c) the force exerted by the earth on an apple as it falls to the ground, (d) the force exerted by a hammer on a nail.

D1.10   A force **F** is applied to the point A as shown. (a) What are the coordinates of the point of application? (b) What is the direction of the force? (c) Is a force vector a free vector?

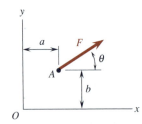

Fig. D1.10

1.5   Mechanics

Mechanics may be defined as the physical science which describes and predicts the conditions of rest or motion of bodies under the action of force systems. In other words, where there is motion or force, there is mechanics. In engineering, mechanics is generally based on Newton's laws (cf. Sec. 1.11) and is often called *newtonian mechanics,* after the English physical scientist and mathematician Sir Isaac Newton (1642–1727). Conditions involving speeds of bodies close to the speed of light (about 300×10^6 meters per second) and conditions requiring consideration of bodies with extremely small mass and size (such as subatomic particles with sizes in the order of 10^{-12} meter and smaller) cannot be adequately described by Newton's laws; these extreme conditions are treated in *relativistic mechanics* and *quantum mechanics*. However, for a vast range of problems between these extremes, Newton's laws give accurate results and are far simpler to apply.

A force system acting on a body is said to be *balanced* if it has no tendency to change the state of rest or motion of the body in any way. If a body is in *equilibrium,* the force system acting on it must be *balanced.* Furthermore, a body in a state of *equilibrium* must be at rest or moving at a constant speed in a straight line. *Statics* is the branch of mechanics that describes and predicts the conditions of bodies under the action of balanced force systems. Most problems in statics concern bodies at rest.

Dynamics is the branch of mechanics that describes and predicts the conditions of bodies under the action of unbalanced force systems. For the purpose of study, dynamics is often divided into kinematics and kinetics. *Kinematics* deals with the study of motion of bodies without considering the cause of motion. *Kinetics* relates the motion of a body to the force system causing the motion; it usually contains some kinematics.

Developmental Exercises

D1.11   Define the terms: (a) mechanics, (b) newtonian mechanics, (c) statics, (d) dynamics.

D1.12   Give an example of a condition which cannot be adequately described by Newton's laws.

D1.13   State the condition in which a body may be said to be in equilibrium.

D1.14   Name the two parts into which dynamics is often divided for study.

D1.15   How does kinetics differ from kinematics?

1.6   Idealizations in Mechanics

A *particle* is an idealized body which may have finite or negligible mass[†] and whose size and shape can be disregarded without introducing appreciable errors in the description and prediction of its state of rest or motion. It is sometimes viewed as a mass point or a corpuscle.

In *statics,* if the lines of action of all the forces acting on a body intersect at a common point, the body (regardless of the amount and distribution of its mass) may be treated as a particle located at that common point. In *kinematics,* if only the translational motion (in which the body maintains the same orientation) of a body is of concern, the body is treated as a particle. In *kinetics,* if the motion of a body caused by the action of an unbalanced force system is exclusively translational, the body is often treated as a particle, which must have a finite mass.

[†]The equilibrium of a particle with negligible mass (e.g., a knot or small connector) is studied in statics.

A *rigid body* is an idealized body composed of a large number of particles, all of which always remain at fixed distances from each other. In reality, no physical body under the action of forces can remain absolutely rigid. In practice, a body may be regarded as rigid when its deformation is too small to affect the desired accuracy of the analysis. Most bodies in this text are assumed to be rigid bodies. Elastic springs and cordlike elements, such as cords, strings, cables, wires, ropes, belts, and chains, are, of course, nonrigid bodies.

A *continuum* is an idealized body whose matter is assumed to be totally continuous, not porous. The continuum theory is especially practical when one is interested only in the macroscopic manifestations of the matter and nothing else. It can be said that, for many practical purposes in engineering, we are not considering what happens at the molecular level. In this text, each body is assumed to be a continuum; i.e., the molecular structure of matter is disregarded. Note that a continuum may be in the form of a rigid body or a deformable body.

A *concentrated force* (also called a *point force*) is an idealized force assumed to act at a point of a continuum. Although the force exerted by the earth on a body is actually distributed throughout the body, we frequently idealize this situation by assuming the total force exerted by the earth on a body to be a single concentrated force acting at a specific point, which is called the *center of gravity* of the body. It is clear that a contact force exerted on a body by another body is actually distributed over the area of contact between the two bodies. If the area of contact is relatively small, the contact force between the two bodies may be considered as a concentrated force. However, a distributed force system acting on a rigid body is often replaced by a concentrated force to simplify the analysis and solution of a problem.

The assumptions and simplifications leading to the formulation and frequent utilization of the concepts of *particle, rigid body, continuum,* and *concentrated force* are some of the most important *idealizations in mechanics.* Without them, the study of mechanics problems in engineering would be unnecessarily difficult in many cases.

Developmental Exercises

D1.16   Define a particle.

D1.17   Which of the following bodies in the circumstances described may be considered as particles? (a) The determination of the tensile forces holding a *crate and its connecting ring* in equilibrium, (b) the determination of the unknown forces acting on a *beam* in equilibrium, (c) the determination of the maximum altitude of a *satellite,* (d) the determination of the safe speed for a *car* traveling over a vertical curve, (e) the determination of the force holding up one end of a *car.*

D1.18   Define (a) a rigid body, (b) a continuum, (c) a concentrated force.

D1.19   Describe when a body may be regarded as a rigid body.

D1.20   A load of sandbags is uniformly placed on a beam. The reactions from

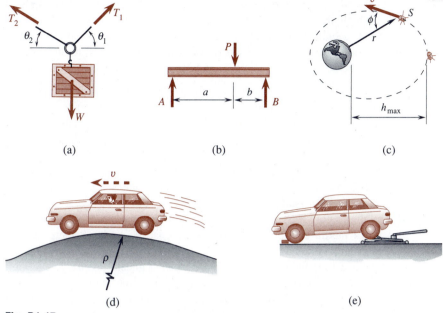

Fig. D1.17

the supports at A and B are to be determined. Can the distributed weight force of the sandbags be replaced by a single concentrated force?

D1.21 What concepts represent some of the most important idealizations in mechanics?

Fig. D1.20

1.7 Dimensions: Physical Quantities

A *dimension* is a qualitative description of a physical quantity which can be quantified by a certain standard of measure, called *unit*.[†] Examples of dimensions include length, mass, time, force, velocity, area, volume, and others. In 1971, the 14th General Conference of Weights and Measures, an international body, selected *length, mass, time, electric current, thermodynamic temperature, amount of matter,* and *luminous intensity* as the *seven base dimensions* for practical measurement in the International System of Units, abbreviated SI from the French *Le Système International d'Unités,* which is an *absolute* system of units. However, we need to use only *length, mass,* and *time* out of the foregoing seven as the base dimensions in elementary mechanics. A dimension which is developed in terms of base dimensions is called a *derived dimension*; e.g., velocity and volume are derived dimensions. The U.S. customary system of units is a *gravitational* system of units, which uses *length, force,* and *time* as base dimensions in elementary mechanics. Since the United States is in a transition period to

---

[†]Units are discussed in Sec. 1.8. The term *dimension* is often referred to as *quantity* in other literature. Elsewhere, *dimension* is used to indicate the measurement in width, length, thickness, or others.

adopt the SI, U.S. engineers will need to be familiar with both systems of units, which are therefore used in this text.

The base dimensions length, mass, and time in the SI are all scalars and may be described as follows, where the definitions given were adopted by the General Conference of Weights and Measures in the years indicated:[†]

- The dimension *length* is a concept for measuring quantitatively the linear extent of the size or position of an object. The standard of length is defined in terms of the distance traveled by light during a specified fraction of a second. Precisely, *one meter* is defined as the length of the path traveled by light in vacuum in 1/299 792 458 of a second.[††] (Adopted in 1983.)[†††] This means that the speed of light is now *defined* to be exactly 299 792 458 meters per second. In the United States, *one foot* is legally equal to 0.3048 meter.

- The dimension *mass* is a concept for describing quantitatively the translational inertia of a body. Besides, the mass of a body characterizes the mutual gravitational attraction with another body. The standard of mass is defined in terms of the mass of a certain cylinder. Precisely, *one kilogram* is defined as the amount of mass equal to the mass of the *international prototype kilogram,* which is a platinum-iridium cylinder kept at the International Bureau of Weights and Measures at Sèvres, near Paris, France. (Adopted in 1889.) Note that the mass of 1 liter of water is approximately equal to 1 kilogram. In the United States, *one pound-mass is* legally equal to 0.453 592 37 kilogram.

- The dimension *time* is a concept for ordering the flow, or for measuring the duration, of events. The standard of time is defined in terms of a certain number of periods of atomic radiation. Precisely, *one second* is defined as the duration of 9 192 631 770 periods of the radiation corresponding to the transition between the two hyperfine levels of the ground state of the cesium-133 atom. (Adopted in 1967.)

Developmental Exercises

D1.22   Define (a) dimension, (b) derived dimension.

D1.23   How many physical quantities have been selected as base dimensions for practical measurement in the SI?

D1.24   Which SI base dimensions are used in elementary mechanics?

D1.25   Explain briefly the dimensions length, mass, and time.

D1.26   Give three examples of derived dimensions in the SI.

1.8   Units and Prefixes

A *unit* of a dimension is a standard of measure for the quantitative description of the dimension. The primary unit of a base dimension is called a *base*

[†]The relativistic viewpoints of length, mass, and time are briefly described in Sec. 1.14.

[††]When it is desirable to separate the digits in a number into groups of three, the SI practice is to use spaces, not commas, to do so.

[†††]*One meter* was previously defined in 1960 to be equal to 1 650 763.73 vacuum wavelengths of the orange-red radiation emitted under specified conditions by the krypton-86 atom. This definition is now superseded by the 1983 definition of the speed-of-light meter.

unit. The SI uses the *meter* (m), *kilogram* (kg), and *second* (s) as the base units for the base dimensions length, mass, and time, respectively. Furthermore, the base units for the base dimensions electric current, thermodynamic temperature, amount of matter, and luminous intensity are ampere (A), kelvin (K), mole (mol), and candela (cd), respectively. In the U.S. customary system of units, the *foot* (ft), *pound* (lb), and *second* (s) are the base units for length, force, and time, respectively. The unit of a derived dimension is called a *derived unit* and is generally a compound unit, unless it is given a special name; e.g.,[†]

$$1 \text{ kg} \cdot \text{m/s}^2 = 1 \text{ newton} = 1 \text{ N} \quad \text{(for force)}$$

$$1 \text{ N/m}^2 = 1 \text{ pascal} = 1 \text{ Pa} \quad \text{(for pressure or stress)}$$

$$1000 \text{ cm}^3 = 10^{-3} \text{ m}^3 = 1 \text{ liter} = 1 \text{ L} \quad \text{(for volume)}$$

The abbreviations of SI units and other units commonly used in mechanics are listed in Table 1.1. As a rule, symbols for the units are not pluralized, the practice with respect to upper and lower case modes is to be followed, and a period is used only with the abbreviation for inch. However, when the inch is raised to a power, the period may be omitted.

Table 1.1 Units and Their Abbreviations

Name	Symbol	Name	Symbol	Name	Symbol
meter	m	liter	L	mile	mi
kilogram	kg	minute	min	pound	lb
second	s	hour	h	pound-mass	lbm
newton	N	day	d	gallon	gal
joule	J	radian	rad	barrel	bbl
hertz	Hz	degree	°	nautical mile	NM
watt	W	inch	in.	knot	kn
pascal	Pa	foot	ft	acre	a

Table 1.2 SI Prefixes

Factor	Prefix	Symbol	Factor	Prefix	Symbol
10^{18}	exa	E	10^{-1}	deci*	d
10^{15}	peta	P	10^{-2}	centi*	c
10^{12}	tera	T	10^{-3}	milli	m
10^{9}	giga	G	10^{-6}	micro	μ
10^{6}	mega	M	10^{-9}	nano	n
10^{3}	kilo	k	10^{-12}	pico	p
10^{2}	hecto*	h	10^{-15}	femto	f
10^{1}	deka*	da	10^{-18}	atto	a

*The use of these prefixes should be avoided, except for measuring area, volume, body and clothing, and others where numbers would be otherwise awkward.

Generally, each dimension has a *primary unit* and several *secondary units.* In forming secondary units in the SI, the *prefixes* in Table 1.2 are used. For example, 1 gigapascal = 10^9 pascals (or 1 GPa = 10^9 Pa), 1 kilonewton = 10^3 newtons (or 1 kN = 10^3 N), and 1 millimeter = 10^{-3}

[†]The primary unit of a derived dimension is *not* called a base unit. Cf. *ASME Guide SI-1,* 9th ed., 1982.

meter (or 1 mm $= 10^{-3}$ m). Note that *gram* (g) is to be considered a *secondary unit* because kilogram (kg) is selected as a base unit, where $1 \text{ g} = 10^{-3}$ kg.

Developmental Exercises

D1.27 (a) Define the terms: unit, base unit, derived unit. (b) Write the abbreviations for newton, kilogram, pound, and pound-mass.

D1.28 Which of the following units are base units (a) in the SI, (b) in the U.S. customary system? Units: N, lb, lbm, slug, g, kg, m, mm, km, Pa, ft, mi, s, h, gal, and L.

D1.29 Express the following quantities in base units in the SI or U.S. customary system: (a) 322 lbm, (b) 0.00524 kN, (c) 1452 g, (d) 30 μs.

1.9 Significant Digits and Rounding Rule

Any nonzero digit in a number is a significant digit; and the digit 0 is also a significant digit except when it is used to fix the decimal point or to fill the places of discarded digits. To avoid ambiguity, we define the *significant digits* of a number as the digits beginning with the first nonzero digit on the left and ending with the last digit on the right.[†]

The *scientific notation* and the *engineering notation* for numerical values are powers-of-ten notations in which each number is expressed as a product consisting of a number multiplied by 10 raised to an appropriate power. For instance, 1.9312128×10^7 and 3.048×10^{-1} are scientific notations, while 19.312128×10^6 and 304.8×10^{-3} (where each exponent is a multiple of ± 3) are engineering notations for the numbers 19312128 and 0.3048, respectively. These notations are particularly useful in expressing very large and very small numbers which have just a few significant digits. (Note that 19.312128 Mm $=$ 12000 mi and 304.8 mm $=$ 1 ft.)

From time to time, it is necessary to *round off* or simply *round* a number. The rule for rounding a number is summarized in the following table:

Table 1.3 Rounding Rule*

When the First Digit Dropped Is:	The Last Digit Retained Is:
Smaller than 5	Unchanged
Greater than 5 or 5 followed by one or more nonzero digits	Increased by 1
5 which is not followed by any nonzero digit	Unchanged if it is even Increased by 1 if it is odd

*For discussion of the rule summarized here, see J. B. Scarborough, *Numerical Mathematical Analysis,* 6th ed. (Baltimore, Md: The Johns Hopkins Press, 1966), pp. 2–3. Note that when the first digit dropped is 5 which is not followed by any nonzero digit, the last digit retained is such that it becomes, or remains, an *even* digit. The errors due to rounding are generally minimized when the rule is followed consistently.

[†]The number of significant digits of a number has also been defined as the number of digits of the number beginning with the first nonzero digit on the left and ending with the last nonzero digit on the right.

Developmental Exercises

D1.30   State the number of significant digits in each of the following numbers: (a) 41.500, (b) 41.5021, (c) 42.50, (d) 42.501, (e) 0.006245, (f) 1360000, (g) 5.218×10^6.

D1.31   Round the numbers in D1.30 to *three* significant digits.

D1.32   Round the numbers in D1.30 to *two* significant digits.

1.10   Numerical Accuracy and Conversion of Units

The accuracy of the result in an answer should be closely related to the accuracy of the given data. Let us here assume that the last significant digit of each given number is an estimated digit. When a given number is *added to* or *subtracted from* another given number, the best possible value of the resulting number will contain no significant digits after any of the decimal places where the estimated digits of these two given numbers are located. For instance, we write

$$1.2 + 9 = 10 \qquad 9.81 - 20.3 = -10.5$$

$$12.34 \times 10^3 + 3.41 = 12.34 \times 10^3$$

$$12.34 \times 10^3 - 341 \times 10^6 = -341 \times 10^6$$

When a given number is *multiplied* or *divided by* another given number, the best possible value of the resulting number may contain one more digit than, but at least as many digits as, the number of significant digits of the less accurate given number. For instance, we write

$$1.2(9) = 11 \qquad 9.81/2.34 = 4.19$$

$$(12.34 \times 10^3)(3.41) = 42.1 \times 10^3$$

$$(12.34 \times 10^3)/(3.41 \times 10^6) = 3.62 \times 10^{-3}$$

In the practice of engineering, the data are seldom known with an accuracy greater than 0.2%. We note that the relative error between 9.97 and 9.99 is about the same as that between 1.001 and 1.003 (about 0.2% in both cases). Unless otherwise specified, it will generally be assumed and understood in the *given data* and in the *reporting of numerical final answers* in this text that all numbers whose first significant digit is 1 have a degree of accuracy of at least four significant digits, and other numbers have a degree of accuracy of at least three significant digits. For example, a force of 12 N is to be read 12.00 N, a length of 1.7 m is to be read 1.700 m, and a force of 50 N is to be read 50.0 N. Nevertheless, it is suggested that, if possible, *all intermediate computations be carried through with more significant digits than wanted or justified in the final results.* As a rule of thumb, *a numerical final answer is generally reported to four significant digits if its first significant digit is 1; otherwise, it is generally reported to three significant digits.*[†]

[†]An exception is given to the case where a computed number is added to or subtracted from an exact number (e.g., 360°). The accuracy in that rule of thumb is known as the *slide rule accuracy*. A slide rule is an analog calculator, which was widely used by engineers before the advent of electronic digital calculators in the early 1970s.

The practice of reporting final results according to the number of significant digits may have no more practical value than according to the number of digits after the decimal point. For instance, reporting the reaction components from the support at A as $A_x = 1.234$ kN and $A_y = 78.9$ kN has no more practical value than reporting them as $A_x = 1.2$ kN and $A_y = 78.9$ kN for design and general purposes. However, the former practice allows readers to check the accuracy of numerical results more closely.

From time to time, engineers may need to convert units from the U.S. customary system to the SI, and vice versa. The following conversions are basic and well-known to engineers:

1 mi = 5280 ft	1 yd = 3 ft	1 ft = 12 in.
1 ft = 0.3048 m	1 d = 24 h	1 h = 60 min
1 min = 60 s	π rad = 180°	1 ton = 2000 lbm
1 kip = 1000 lb	1 lb = 16 oz	1 hp = 550 ft · lb/s

Developmental Exercises

D1.33 Assuming that the last significant digit of each number is a doubtful digit, compute the values of the following: (a) $32.2 + 1.234$, (b) $32.2 - 1.234$, (c) $5.35(3.14)$, (d) $12.67/25.4$, (e) $26.7 \times 10^3 - 21.9$, (f) $3.92/(322 \times 10^3)$, (g) $\sqrt{32.2}$, (h) $\ln 9.81$, (i) $e^{0.234}$, (j) $\sin 125.5°$, (k) $\cos 125.5°$.

D1.34 When you use an electronic calculator in your solution of a problem, should you report your final numerical answer to as many significant digits as shown on the calculator?

EXAMPLE 1.1

The angle of contact between a belt and a drum is $\beta = 36.9°$. Express β in radians.

Solution. We know that π rad = 180°. Thus, we write

$$\beta = 36.9° = 36.9° \left(\frac{\pi \text{ rad}}{180°} \right)$$

$$\beta = 0.644 \text{ rad} \blacktriangleleft$$

REMARK. Note that the final answer is given to only three significant digits because the first significant digit of β is other than 1.

Developmental Exercise

D1.35 Express the following quantities in SI base units: (a) 2 in., (b) 6 yd, (c) 1000 ft³.

EXAMPLE 1.2

The efficiency of an automobile is often measured by its fuel economy (or gas mileage). Making use of 1 gal = 231 in^3 and 1 m^3 = 1000 L, determine the fuel economy of 40 mi/gal of a compact car in km/L.

Solution. Note that the given gas mileage of 40 mi/gal is to be read 40.0 mi/gal. Using the basic conversions (cf. p. 12) as well as those given in the problem, we write

$$40 \text{ mi/gal} = \frac{40 \text{ mi}}{\text{gal}} \cdot \frac{5280 \text{ ft}}{1 \text{ mi}} \cdot \frac{0.3048 \text{ m}}{1 \text{ ft}} \cdot \frac{1 \text{ km}}{1000 \text{ m}} \cdot \frac{1 \text{ gal}}{231 \text{ in}^3}$$

$$\cdot \frac{(12)^3 \text{ in}^3}{1 \text{ ft}^3} \cdot \frac{1 \text{ ft}^3}{(0.3048)^3 \text{ m}^3} \cdot \frac{1 \text{ m}^3}{1000 \text{ L}}$$

$$= \frac{40(5280)(0.3048)(12)^3}{231(0.3048)^3(10)^6} \frac{\text{km}}{\text{L}}$$

$$40 \text{ mi/gal} = 17.01 \text{ km/L} \quad \blacktriangleleft$$

REMARK. Note that the final answer is given to four significant digits in this case because its first significant digit is 1.

PROBLEMS[†]

1.1 A motorist traveling on a highway notes that the posted speed limit is 90 km/h. Determine the speed limit in (a) mi/h, (b) ft/s.

1.2 The rate of flow of oil in a pipe is 90 bbl/min. Making use of 1 bbl = 42 gal, 1 gal = 231 in^3 and 1 m^3 = 1000 L, determine the rate of flow of oil in L/s.[††]

1.3* A meteorologist reports a rainfall of 1 in. Making use of 1 a = 43560 ft^2, 1 gal = 231 in^3, and 1 m^3 = 1000 L, determine the rainfall on a residential lot of one half acre in (a) liters, (b) gallons.

1.4 A submarine is cruising at a speed of 45 kn. Making use of 1 kn = 1 NM/h and 1 NM = 1852 m, determine the speed of the submarine in ft/s.

FUNDAMENTAL LAWS

1.11 Newton's Laws of Motion and Law of Gravitation[†††]

Newton's laws of motion are Newton's first law, Newton's second law, and Newton's third law. *Newton's first law* states that every particle remains at

---

[†]Answers are provided at the end of the text for all problems except those marked with an asterisk.

[††]It may be noted that 1 barrel = 1 bbl = 42 gal of petroleum or 31 gal of fermented beverage.

[†††]Newton's work, entitled *Philosophiæ Naturalis Principia Mathematica,* was written in Latin and published in 1687; it was revised in 1713 and 1726. For a modern translation, see Florian Cajori, trans., *Sir Isaac Newton's Mathematical Principles of Natural Philosophy and His System of the World* (Berkeley: University of California Press, 1934).

rest, or in motion with constant velocity, unless an unbalanced force acts on it. In other words, when the resultant force acting on the particle is zero, the particle will remain at rest if it is originally at rest, and the particle will move with constant speed in a straight line if it is originally in motion. Thus, Newton's first law is also called the *law of inertia*. Note that the *inertia* of a particle is an *intrinsic property* of the particle to preserve its status quo. The *mass* of a particle is the quantitative measure of the *inertia* of the particle.

In modern terminology, *Newton's second law* states that the derivative of the momentum of a particle with respect to the time t is proportional to the resultant force **F** acting on the particle, where the momentum is equal to the product of the mass m and the velocity **v** of the particle. The constant of proportionality can be made unity by appropriate choices of units so that Newton's second law becomes

$$\mathbf{F} = \frac{d}{dt}(m\mathbf{v}) \tag{1.1}$$

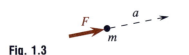

Fig. 1.3

We know that the acceleration **a** of the particle is the time derivative of its velocity **v**; i.e., $\mathbf{a} = d\mathbf{v}/dt$. For a particle with constant mass m as shown in Fig. 1.3, Eq. (1.1) yields

$$\mathbf{F} = m\mathbf{a} \tag{1.2}$$

Besides being applicable to a body with a constant mass m, Newton's second law in the form of Eq. (1.1) is found to be directly applicable to a body gaining or losing mass due to the change in its velocity such as those encountered in relativistic mechanics (cf. Sec. 1.14). However, Newton's second law in the form of either Eq. (1.1) or Eq. (1.2) is *not* directly applicable to a variable body gaining or losing mass through absorbing or expelling particles such as those encountered in the motion of a chain being pulled up from a pile or the motion of a rocket (cf. Sec. 14.13). Thus, it is a *misconception* to think that Newton's second law in the form of Eq. (1.1) is intended for direct application to a body where the particles it contains are continuously changing.

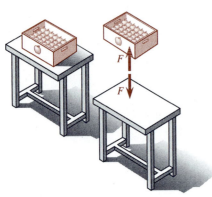

Fig. 1.4

By Newton's second law, we know that when the resultant of all forces acting on the particle is zero, the acceleration of the particle must also be zero. Since zero acceleration means constant velocity and constant velocity includes zero velocity (i.e., at rest), we see that Newton's first law can be regarded as a *special case* of Newton's second law where the acceleration of the particle is zero. On the other hand, Newton's first law may well be considered as the *definition of a reference frame* in space (called the *inertial reference frame*) for which the second law is then valid.[†] In statics, Newton's second law is mainly used to relate weight to mass (cf. Sec. 1.13). In dynamics, Newton's second law is extensively used.

Newton's third law states that to every action there is a collinear reaction having the same magnitude and being opposite in sense. This is to say that the mutual actions of two bodies upon each other are always equal in magnitude and directed in opposite senses along a common line, as shown in Figs. 1.4 and 1.5. From this law, we note that any action must always be

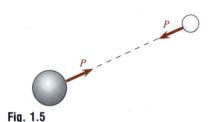

Fig. 1.5

[†]Cf. Sec. 12.1.

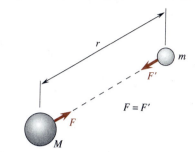

accompanied by a corresponding reaction. Newton's third law is very important in both statics and dynamics.

Newton's law of gravitation states that two particles are mutually attracted by a pair of opposite forces that are directed along the line joining the particles and have the same magnitude which is directly proportional to the product of their masses and inversely proportional to the square of the distance between them.[†] For two particles with masses M and m separated by a distance r, as shown in Fig. 1.6, the magnitude of the attractive force may be written as

$$F = G\frac{Mm}{r^2} \qquad (1.3)$$

Fig. 1.6

where G is the *constant of gravitation*. Experiments show that

$$G = 66.726 \times 10^{-12}\ \frac{m^3}{kg \cdot s^2} \qquad (1.4)^{[††]}$$

This law has little application in statics; it is mainly used in dynamics in studying the motions of spacecraft, satellites, and planets.

Developmental Exercises

D1.36   Describe (a) Newton's first law, (b) Newton's second law, (c) Newton's third law, (d) Newton's law of gravitation.

D1.37   Is it correct to infer from Newton's third law that "forces always occur in pairs"?

1.12   Parallelogram Law and Rigid-Body Principle

The *parallelogram law* states that the combined effect of two forces acting on a particle is equivalent to the effect of a single force, called their *resultant,* which acts on the particle and is equal to the diagonal of the parallelogram whose consecutive sides intersecting at that diagonal are equal to those two forces.[†††] This is the basic law which defines the addition of two vectors in physical space. The parallelogram law may be illustrated with a graphical equation as shown in Fig. 1.7, which means that

$$\mathbf{P} + \mathbf{Q} = \mathbf{R} \qquad (1.5)$$

where $\mathbf{R}$ is the *resultant* of $\mathbf{P}$ and $\mathbf{Q}$. The *resultant* of more than two forces may be obtained by repeated applications of Eq. (1.5).

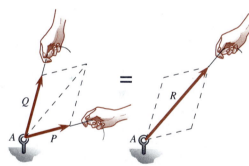

Fig. 1.7

[†]The statement of this law is a synthesis from *Book Three, System of the World* in Newton's *Principia* (Cf. Cajori, trans. *op. cit.*). In certain astronomical studies, this law is replaced by the law of gravitation as covered in the *general* theory of relativity (1915) of Albert Einstein (1879–1955). Cf. Sec. 1.14.

[††]The accepted value of G as of 1982 is $G = (66.726 \pm 0.005) \times 10^{-12}\ m^3/(kg \cdot s^2)$.

[†††]Although fundamental in nature, the parallelogram law was treated by Newton as *Corollary II* to his laws of motion. (Cf. Cajori, trans. *op. cit.*, pp. 13–15.) For Varignon's contribution to this law, cf. Sec. 4.18.

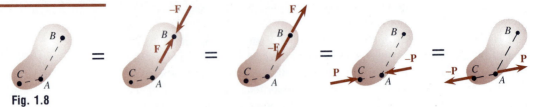

Fig. 1.8

The *rigid-body principle* states that any two opposite forces acting on a rigid body will have no effect on the condition of rest or motion of the rigid body if they have the same magnitude and the same line of action. This principle is illustrated in Fig. 1.8 and is employed in deriving the so-called *principle of transmissibility* in Fig. 1.9, which may be stated as follows: The effect of a force on the condition of rest or motion of a rigid body will remain unchanged if the force is moved to act at another point on its line of action. Thus, a force acting on a rigid body is sometimes called a *sliding vector*.

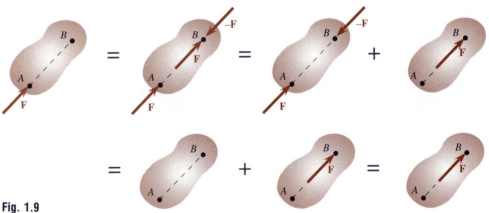

Fig. 1.9

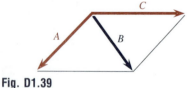

Fig. D1.39

Developmental Exercises

D1.38 Describe (a) the parallelogram law, (b) the rigid-body principle, (c) the principle of transmissibility.

D1.39 Three forces **A**, **B**, and **C** form the sides and a diagonal of a parallelogram as shown. (a) Is $\mathbf{A} + \mathbf{B} = \mathbf{C}$? (b) Is $\mathbf{B} + \mathbf{C} = \mathbf{A}$? (c) Is $\mathbf{C} + \mathbf{A} = \mathbf{B}$? (d) Is $B = A + C$?

1.13   Mass Versus Weight and Force

The *mass* of a body is a quantitative measure of the *inertia of the body* with respect to translational motion. In newtonian mechanics, the mass of a body is a positive scalar quantity and does not vary with the orientation, location, or state of motion of the body. However, the inertia of a body with respect to rotational motion is dependent on not only the amount of mass but also on the way the mass is distributed with respect to the axis of rotation of the body.[†] The *weight force* **W** of a body is the *gravitational force* exerted by the earth on the body and is always directed vertically downward toward the earth.[††] A weight force is a body force. The *weight W* of a body is the magnitude of the weight force **W** of the body. Such a weight is a *gravitational weight*. By Newton's second law, the weight of a body with mass m may be written as

$$W = mg \qquad (1.6)$$

where g is the gravitational acceleration of the body. The value of g is known to vary with the location of the body.[†††] The *standard gravitational acceleration* is defined to be exactly equal to 9.80665 m/s^2. However, for practical purposes, we will use

$$g = 9.81 \text{ m/s}^2 \quad \text{or} \quad g = 32.2 \text{ ft/s}^2 \qquad (1.7)$$

as the gravitational acceleration of all bodies near the surface of the earth. These approximate values of g are well-known to engineers.

On the other hand, people feel their "weight" by the contact forces exerted on them from their supports (e.g., seats, floors, and the ground). If their supports exert no contact forces on them, they feel "weightless." Such a "weight" felt by people should be classified as *apparent weight* (or contact weight). Obviously, the apparent weight of a body in motion is generally different from the gravitational weight of the body at the same location. In mechanics, the term *weight* is to be understood as the *gravitational weight,* not the apparent weight. Note that the mass of a one-kilogram gold bullion remains the same, no matter if we place it on the surfaces of the Moon, Mars, or Jupiter. However, the weight of this bullion will be different at these different locations. Remember that we use kilogram (kg) as a unit of mass, not of weight or force, in the SI.

Newton's second law and the concept of weight force as described earlier are useful in defining the units of force and mass. *One newton* (N) is defined as the net force required to give a body with a mass of 1 kg an acceleration of 1 m/s^2; i.e.,

$$1 \text{ N} = 1 \text{ kg} \cdot \text{m/s}^2 \qquad (1.8)$$

The SI is termed an *absolute system* since mass is taken as an absolute or base quantity. The U.S. customary system is termed a *gravitational system* since force as measured from gravity is taken as a base quantity.

One pound (lb) of force is defined as the weight of a body with one pound-mass (1 lbm) measured under the standard gravitational acceleration, which is 9.80665 m/s^2. Thus, by Newton's second law, we write

$$1 \text{ lb} = 1 \text{ lbm} \cdot 9.80665 \text{ m/s}^2 \qquad 1 \text{ lb} = \frac{9.80665}{0.3048} \text{ lbm} \cdot \text{ft/s}^2 \qquad (1.9)$$

[†]Cf. Sec. 7.11.

[††]The weight force of a body on the Moon or Mars is, of course, taken to be the gravitational force exerted by the Moon or Mars on the body, as appropriate.

[†††]Considering that the earth is a rotating oblate spheroid with flattening at the poles, the International Gravity Formula gives $g = 9.78049 (1 + 0.0052884 \sin^2 \phi - 0.0000059 \sin^2 2\phi)$, where ϕ is the latitude and g is measured in m/s^2, at the sea level. Thus, g varies from 9.780 m/s^2 at the equator to 9.832 m/s^2 at the poles.

Notice that the abbreviations for *pound-mass* and *pound* are lbm and lb, respectively; the former is a unit of mass, while the latter is a unit of force. *One slug* is defined as the amount of mass of a body which will be accelerated at the rate of 1 ft/s^2 by a net force of 1 lb; i.e.,

$$1 \text{ lb} = 1 \text{ slug} \cdot \text{ft/s}^2 \qquad (1.10)$$

In the U.S. customary system, the *slug* is the *primary unit* of mass, while the *pound-mass* is a *secondary unit*. We see from Eqs. (1.9) and (1.10) that

$$1 \text{ slug} = 9.80665/0.3048 \text{ lbm} \qquad 1 \text{ slug} \approx 32.2 \text{ lbm} \qquad (1.11)$$

Developmental Exercises

D1.40 How does the *weight* of a body differ from its *mass*?

D1.41 Astronauts in a space shuttle orbiting around the earth experience no contact force from their surroundings and are reported as being "weightless." Which of the following is zero: (a) their gravitational weight or (b) their apparent weight?

EXAMPLE 1.3

Express 1 slug in terms of the base units ft, lb, and s.

Solution. From Newton's second law and the definition of 1 slug, we write

$$1 \text{ lb} = 1 \text{ slug} \cdot (1 \text{ ft/s}^2)$$

$$1 \text{ slug} = 1 \ \frac{\text{lb} \cdot \text{s}^2}{\text{ft}} \quad \blacktriangleleft$$

Developmental Exercises

D1.42 Define the units: (a) 1 N, (b) 1 lbm, (c) 1 lb, (d) 1 slug.

D1.43 Which of the units in D1.42 are for (i) force, (ii) mass?

EXAMPLE 1.4

Making use of the exact relations that 1 lbm = 0.45359237 kg and 1 slug = 9.80665/0.3048 lbm, express 1 newton in terms of pounds to four significant digits.

Solution. From Newton's second law and the definitions of units, we write

$$1 \text{ N} = 1 \text{ kg} \cdot (1 \text{ m/s}^2)$$

$$= 1 \text{ kg} \cdot \frac{1 \text{ m}}{s^2} \cdot \frac{1 \text{ ft}}{0.3048 \text{ m}} \cdot \frac{1 \text{ lbm}}{0.45359237 \text{ kg}} \cdot \frac{1 \text{ slug}}{9.80665/0.3048 \text{ lbm}}$$

$$= \frac{1}{0.45359237(9.80665)} \frac{\text{slug} \cdot \text{ft}}{s^2} = 0.224809 \text{ slug} \cdot \text{ft/s}^2$$

$$1 \text{ N} = 0.2248 \text{ lb} \blacktriangleleft$$

Developmental Exercises

D1.44 Express the value of G in Eq. (1.4) in base units in the U.S. customary system.

D1.45 A cart weighing 981 N is being held on a ramp as shown. Which of the following will give the correct mass of the cart?

(a) $\dfrac{981 \sin 30°}{9.81}$ kg. (b) $\dfrac{981 \cos 30°}{9.81}$ kg. (c) $\dfrac{981}{9.81}$ kg. (d) 981(9.81) kg.

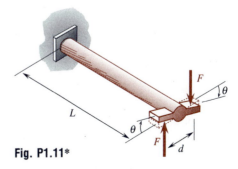

Fig. D1.45

PROBLEMS[†]

1.5 A newborn baby in a hospital weighs 7 pounds and 5 ounces. Making use of 1 lbm = 0.4536 kg, determine the baby's mass in kg.

1.6 Making use of 1 lbm = 0.4536 kg, express a gas pressure of 120 kPa in psi (pounds per square inch).

1.7 The allowable compressive stress for a structural member is 20 ksi (kips per square inch). Making use of 1 lbm = 0.4536 kg, determine the allowable compressive stress in MPa.

1.8* Making use of 1 lbm = 0.4536 kg, determine the acceleration in ft/s^2 of a 10-Mg space vehicle under the action of a 10-lb thrust force exerted by a course correction rocket.

1.9 The gravitational acceleration on the surface of the moon is approximately 5.35 ft/s^2. Making use of 1 lbm = 0.4536 kg, determine the weight of an instrument of 10 slugs on the surface of the moon in (a) pounds, (b) newtons.

1.10 The deflection δ of the free end of a cantilever beam, as shown, is given by $\delta = PL^3/(3EI)$, where P is the weight of the 10-kg crate, L = 2 m, E = 200 GPa, and $I = 5 \times 10^4$ mm^4. Determine the value of δ in inches.

1.11* The angle of twist (in radians) of a circular shaft, as shown, is given by $\theta = TL/(JG)$, where $T = Fd$, F = 300 lb, d = 2 in., L = 300 mm, $J = 10^4$ mm^4, and $G = 11.5 \times 10^6$ psi. Determine the value of θ in degrees.

Fig. P1.10

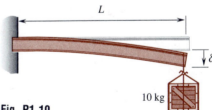

Fig. P1.11*

[†]Recall that 1 pascal = 1 Pa = 1 N/m^2 in solving some of the problems.

1.14   Concluding Remarks

The concepts of force, equilibrium, and idealization, as well as the parallelogram law, the rigid-body principle, and Newton's first and third laws, are extensively used in the study of statics. In dynamics, Newton's second law and law of gravitation are additionally used in the study. Note that the applicability of Newton's third law is *not* limited to bodies in contact. In explaining the third law in his *Principia,* Newton stated that "this law takes place also in attractions . . . ."[†] Thus, Newton's third law should be understood to be applicable to any two bodies interacting with each other either *by contact* or *at a distance.*

In relativistic mechanics, the measurements of dimensions are affected by the relative motion of the observer with respect to the object. For example, the concepts of length, time, and mass are not absolute in relativistic mechanics as they are in newtonian mechanics. For enrichment, a brief glance at some of the contrasting relativistic viewpoints would perhaps be of interest, although they are not needed for understanding the remainder of this text. To describe them, we first let the speed of light be denoted by c and consider two coordinate systems S and S' where S' moves with a constant velocity $\mathbf{v}$ with respect to S. By Einstein's special theory of relativity (1905), we have the following:

1. A rigid body of length L_0 as measured in the direction of $\mathbf{v}$ in the system S, in which the rigid body is at rest, measures only L as observed from the system S' where $L = L_0(1 - v^2/c^2)^{1/2}$; and reciprocally an identical *length contraction* occurs for a rigid body at rest in S' as observed from S.
2. The time interval t_0 measured between two events that occur at the same place in the system S is found to be t as observed from the system S' where $t = t_0/(1 - v^2/c^2)^{1/2}$; and reciprocally an identical *time dilation* (or retardation) occurs for a time interval measured in S' as observed from S.
3. A body of mass m_0 as measured in the system S, in which the body is at rest, manifests a mass m as observed from the system S' where $m = m_0/(1 - v^2/c^2)^{1/2}$; and reciprocally an identical *mass increase* occurs for a body at rest in S' as observed from S. Thus, Newton's second law is to be written as $\mathbf{F} = d(m\mathbf{v})/dt$ and the total energy of the particle is $E = mc^2$. It can be shown that $mc^2 = m_0c^2 + \frac{1}{2}m_0v^2$ when $v \ll c$.

By Einstein's general theory of relativity (1915), which deals with the law of gravitation in the large scale structure of the universe, gravity is looked upon not merely as action at a distance but rather as some modification of space. According to Einstein, mass is the cause of the *curvature* of the four-dimensional space-time, which is not flat. In the universe, everything (including light) must experience gravitation. Einstein's law of gravitation controls a geometrical quantity called *curvature* in contrast to Newton's law of gravitation, which controls a mechanical quantity called *force.*

[†]Cf. Cajori, trans. *op. cit.,* pp. 13–14.

From the relativistic viewpoints, we see that the fundamental principles formulated by Newton do have limitations. However, the relativistic effects are unobservably small in most engineering practice where $v \ll c$, and newtonian mechanics still remains the basis of today's engineering sciences.

REVIEW PROBLEMS

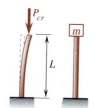

Fig. P1.12

1.12 The critical load P_{cr} for the buckling of a column fixed at one end as shown is given by $P_{cr} = (\pi^2 EI)/(4L^2)$. For $E = 10 \times 10^6$ lb/in², $I = 10^3$ mm⁴, and $L = 1.5$ m, determine the largest mass m in kg of a block which may be placed on the top of the column without causing the column to buckle.

1.13 The magnitude v_{esc} of the escape velocity of a space vehicle at a distance r from the center of the earth is equal to $[(2gR^2)/r]^{1/2}$ where g is the standard gravitational acceleration and R is the radius of the earth. Knowing that the circumference of the earth is 40 Mm and the space vehicle is 300 miles above the surface of the earth, determine the value of v_{esc} in miles per hour.

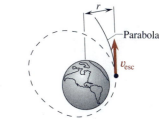

Fig. P1.13

1.14* It is known that $GM = gR^2$ where G is the constant of gravitation, M is the mass of the earth, and g and R are the same as those defined in Prob. 1.13. Using the value of G as given in Eq. (1.4), determine the mass of the earth in pounds-mass.

Choose the correct or best item to complete the sentence in each of the following:

1.15 A scalar may be defined as a quantity which has (a) magnitude only, (b) magnitude and possibly sense, (c) magnitude and direction, (d) sense, but no orientation, (e) magnitude and possibly orientation.

1.16 A vector may be defined as a directed quantity which has (a) sense and orientation, and which obeys the laws for addition of vectors; (b) magnitude and orientation, and which obeys the principle of addition; (c) magnitude and direction, and which obeys the laws for addition of vectors; (d) magnitude and sense, and which obeys the principle of addition; (e) magnitude, orientation, and sense.

1.17 In statics, if the lines of action of the forces acting on a body intersect at a common point, the body may be treated as a (a) rigid body, (b) particle, (c) beam, (d) elastic body, (e) continuum.

1.18 The dimension which is a concept for measuring quantitatively the linear extent of the size or position of an object is (a) a base dimension, (b) mass, (c) time, (d) length, (e) a unit.

1.19 The base unit for mass in the SI is (a) kilogram, (b) slug, (c) gram, (d) newton, (e) pound-mass.

1.20 Rounding the number 0.03065 to three significant digits, we write (a) 0.0306, (b) 0.03, (c) 0.0307, (d) 3.065×10^{-2}, (e) 0.031.

1.21 It can be shown that 1 slug is equivalent to (a) 1 ft · lb/s², (b) 1 lb · s²/ft, (c) 1 ft · s²/lb, (d) 32.2 lb, (e) 9.81 kg.

1.22 One newton is defined as (a) the weight of a 1-kg mass, (b) the weight of a 1-slug mass, (c) the net force required to accelerate a mass of 1 kg at the rate of 9.80665 m/s², (d) the net force required to accelerate a mass of 1 slug at the rate of 1 ft/s², (e) the net force required to accelerate a mass of 1 kg at the rate of 1 m/s².

1.23 The mass of a body is (a) equal to the weight of the body, (b) the magnitude of the attractive force exerted by the earth on the body, (c) the magnitude of the inertia force of the body, (d) a qualitative measure of the rotational inertia of the body, (e) a quantitative measure of the translational inertia of the body.

1.24 When a net force of 1 lb acts on a body of mass m, the acceleration of the body is found to be 1 ft/s^2. The value of m is (a) 1 newton, (b) 1 kilogram, (c) 1 pascal, (d) 1 slug, (e) 1 pound-mass.

Forces

Chapter 2

KEY CONCEPTS

- Directional angle, slope triangle, zero vector, couple, triangle rule, polygon rule, unit vector, vector components, scalar components, magnitude of a vector, analytical expression of a vector, direction angles, and position vector.
- Addition and resolution of forces in a plane using the parallelogram law.
- Addition and resolution of forces in a plane using vector algebra.
- Addition and resolution of forces in space using vector algebra.

A knowledge of how forces are added and resolved is basic to the study of the conditions of bodies which are acted on by forces. This chapter presents fundamental studies on the addition and resolution of forces in a plane and of forces in space. Two basic ways are used in representing forces. The *graphical* representation of forces is a basic step in applying the parallelogram law to the addition and resolution of forces, while the *analytical* representation of forces is a basic step in using vector algebra for adding and resolving forces. The method using vector algebra is shown to be consistent with the parallelogram law. The distinction between the vector components and the scalar components of a vector is stressed. Furthermore, the generation of a set of simultaneous scalar equations from a vector equation is illustrated in the method using vector algebra. The representations and operations pertaining to force vectors are applicable to other vector quantities.

FORCES IN A PLANE[†]

2.1 Graphical Representation

A force in a plane may be represented graphically by a segment of its line of action plus an arrowhead at the tip of the line segment to indicate the sense of the force. Using an appropriate scale, we may let the magnitude of

---

[†]A plane is a *two-dimensional space*. Thus, *forces in a plane* are forces in a two-dimensional space, and are, at times, referred to as *two-dimensional forces*. This seems to be an appropriate terminology. However, some engineers and mathematicians object to attaching the term "dimension" to a single force or vector in space because dimension is a property of space, not of a force or vector. (Some suggest the use of an *n*-component force or vector when it contains *n* components.)

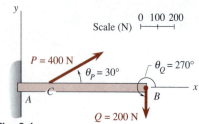

Fig. 2.1

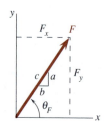

Fig. 2.2

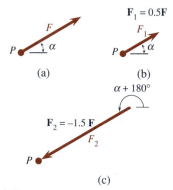

Fig. 2.3

Fig. D2.4

the force be represented by the length of that line segment. Alternatively, we may denote the magnitude of a force with an *italic* letter or a numerical value near the directed line segment used to represent the force. When a force lies in the xy plane, the counterclockwise angle measured from the positive direction of the x axis to the positive direction of the force is used to specify the orientation and sense of the force and is called the *directional angle* of the force. Figure 2.1 illustrates the graphical representations of a 400-N force **P** at C and a 200-N force **Q** at B, where θ_P and θ_Q, as indicated, are the *directional angles* of **P** and **Q**.

The *orientation* of a force in the xy plane may also be indicated with a *slope triangle*, which is a small right triangle having sides a, b, and hypotenuse c, as shown in Fig. 2.2. Since $c = (a^2 + b^2)^{1/2}$, the value of c is often not indicated on the slope triangle. By proportion, we may write the projections of **F** as follows:

$$F_x = \frac{b}{c} F \qquad F_y = \frac{a}{c} F \qquad (2.1)$$

We see that the *directional angle* θ_F of **F** can be found from

$$\tan \theta_F = \frac{a}{b} \qquad (2.2)$$

Depending on the direction of **F**, the directional angle θ_F may have a value ranging from 0 to 360° (or 0 to 2π rad).

The *product* $c\mathbf{F}$ of a scalar c and a force **F** represents another force which has a magnitude equal to the product of the magnitude of **F** and the absolute value of c, and which points in the direction of **F** if c is positive, but in the opposite direction of **F** if c is negative. These concepts are illustrated in Fig. 2.3. If $c = 0$, then $c\mathbf{F}$ is a *zero* (or *null*) *vector* **0**. A *zero vector* has a magnitude of zero and an indeterminate direction. If $c = -1$, then $c\mathbf{F} = (-1)\mathbf{F} = -\mathbf{F}$, which is the *negative vector* of **F**. When **F** and $-\mathbf{F}$ represent forces at different points of a rigid body and have different lines of action, they are said to form a *couple*. The sum of the forces in a couple is

$$\mathbf{F} + (-\mathbf{F}) = \mathbf{0} \qquad (2.3)$$

However, the effect of a couple is not simply zero, but is a twisting or turning action, which is discussed in Chap. 4.

Developmental Exercises

D2.1 Define (a) the directional angle of a force in the xy plane, (b) a zero vector, (c) a couple.

D2.2 Represent graphically (a) a 50-N force **F** at the point $A(1, 2)$ m with $\theta_F = 210°$, (b) a 30-N force **T** at the point $B(2, 1)$ m with $\theta_T = 60°$.

D2.3 A 50-lb force **F** with $\theta_F = 60°$ acts at the point $(2, 0)$ ft. Sketch the following: (a) **F**, (b) $\mathbf{F}_1 = 2\mathbf{F}$, (c) $\mathbf{F}_2 = -0.5\mathbf{F}$.

D2.4 Two forces **P** and **Q** are applied to an anchor as shown, where slope triangles are indicated. Determine the directional angles θ_P and θ_Q of **P** and **Q**, respectively.

2.2 Addition of Forces: Parallelogram Law

Forces must be added in a way consistent with the parallelogram law, or its extended rules: the *triangle rule* and the *polygon rule*. The *triangle rule* states that when two vectors are drawn to scale and in tip-to-tail fashion, the vector connecting, and directed from, the tail of the first vector to the tip of the second one gives the resultant **R** of those two vectors. This rule may be dubbed the "half-of-a-parallelogram law." The triangle rule, as illustrated in Fig. 2.4, shows that

$$\mathbf{P} + \mathbf{Q} = \mathbf{Q} + \mathbf{P} \qquad (2.4)$$

Thus, *vector addition is commutative*.

The *polygon rule* states that when several vectors are drawn to scale and in tip-to-tail fashion, the vector connecting, and directed from, the tail of the first vector to the tip of the last one gives the resultant of those several vectors. This rule can be shown to be valid through repeated applications of the triangle rule. The polygon rule, as illustrated in Fig. 2.5, shows that

$$\mathbf{P} + \mathbf{Q} + \mathbf{S} = (\mathbf{P} + \mathbf{Q}) + \mathbf{S} = \mathbf{P} + (\mathbf{Q} + \mathbf{S}) \qquad (2.5)$$

Thus, *vector addition is associative*.

Generally, the parallelogram law alone gives the magnitude and direction, but not the point of application, of the sum of the forces. Figure 2.6 shows that the sum of the two forces **P** and **Q** may graphically be obtained by placing the tails of the two force vectors at the same point, say *A*, and forming a parallelogram with these two force vectors as consecutive sides.

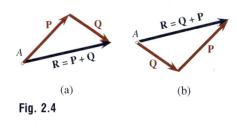

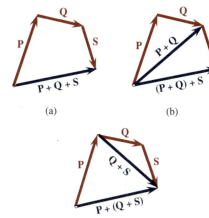

Fig. 2.4

Fig. 2.5

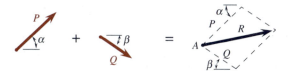

Fig. 2.6

The sum is denoted by **R** as shown, and is equal to the *diagonal vector* whose tail is located at the point *A*. This sum is the *resultant force*, or simply the *resultant*, of the two forces.[†] When the two forces **P** and **Q** act on a particle, it is evident that the particle itself is also the *point of application* of the resultant force **R** as shown in Fig. 2.7, where R and θ_R of **R** may analytically be determined.

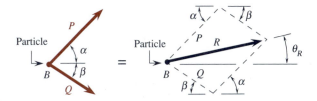

Fig. 2.7

[†]Note that a *resultant force* is not an actually applied force. Nevertheless, it is a vector representing an *equivalent force*.

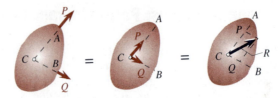

Fig. 2.8

When the two forces **P** and **Q** act on a rigid body, we see from the principle of transmissibility that the line of action of the resultant force **R** must contain the *point of intersection* C of the lines of action of **P** and **Q** as shown in Fig. 2.8. To locate the resultant force of two or more parallel forces, we need the concepts of *moment* and *equivalence*, which are presented in Chaps. 4 and 5.

The subtraction of the force **Q** from the force **P** may be written as

$$\mathbf{P} - \mathbf{Q} = \mathbf{P} + (-\mathbf{Q}) \qquad (2.6)$$

Thus, the *subtraction* of **Q** from **P** is the same as the *addition* of **P** and the negative vector of **Q**. The parallelogram law may, therefore, be used to find the *difference of two forces* as shown in Fig. 2.9.

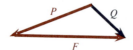

Fig. 2.9

Developmental Exercises

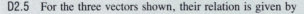

D2.5   For the three vectors shown, their relation is given by

(a) **F** + **Q** = **P**.   (b) **P** + **Q** = **F**.   (c) **P** + **F** = **Q**.
(d) **F** − **Q** = **P**.   (e) **F** − **P** = **Q**.

D2.6   The three vectors shown are drawn to scale. Their resultant **R** is given by

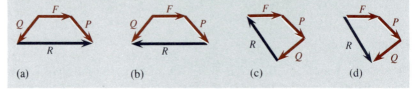

(a)   (b)   (c)   (d)

Fig. D2.5

Fig. D2.6

EXAMPLE 2.1

A 200-N force **P** and a 400-N force **Q** are applied to a small boat stuck in a shallow stream as shown. The resultant force of **P** and **Q** is **R**. Determine the magnitude R and the directional angle θ_R of **R**.

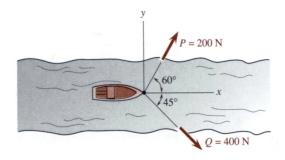

Solution.   The parallelogram for the forces and the relevant angles are shown. Using the laws of cosines and sines, we write

$$R^2 = (200)^2 + (400)^2 - 2(200)(400) \cos 75° \qquad R = 398 \text{ N} \blacktriangleleft$$

$$\frac{200}{\sin \alpha} = \frac{R}{\sin 75°} \qquad \alpha = 29.0°$$

$$\theta_R = 270° + 45° + \alpha \qquad \theta_R = 344.0° \blacktriangleleft$$

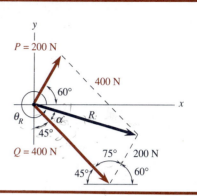

EXAMPLE 2.2

The boat in Example 2.1 remains stuck. A decision is made to get the boat out by reversing the sense of **Q** such that the resultant is **D** = **P** − **Q**. Determine the magnitude D and the directional angle θ_D of **D**.

Solution.   The parallelogram for finding **D** is shown. Using the laws of cosines and sines, we write

$$D^2 = (200)^2 + (400)^2 - 2(200)(400) \cos 105° \qquad D = 491 \text{ N} \blacktriangleleft$$

$$\frac{200}{\sin \alpha} = \frac{D}{\sin 105°} \qquad \alpha = 23.2°$$

$$\theta_D = 180° - 45° - \alpha \qquad \theta_D = 111.8° \blacktriangleleft$$

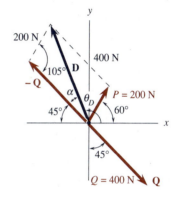

Developmental Exercise

D2.7   Describe the procedure for (a) adding two given forces, (b) subtracting a given force from another given force.

2.3   Resolution of Forces: Parallelogram Law

The reverse of the process of adding two forces to obtain an equivalent single force is the process of resolving a single force into two component forces which, together, have the same effect as that of the single force. The *resolution* as well as the *addition* of forces must conform to the *parallelogram law*, or its extended rules: the *triangle rule* and the *polygon rule*, described in Sec. 2.2.

As illustrations, let us consider the following three different specifications and problems where *a given single force* **F** *is resolved into two component forces* **P** *and* **Q**:

(1) *The magnitudes of* **P** *and* **Q** *are specified and* P + Q > F, *the directions of* **P** *and* **Q** *are to be determined.*

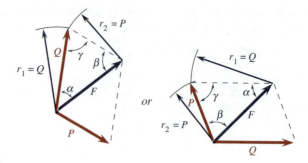

Fig. 2.10

The force triangle can be constructed by drawing two arcs of radii P and Q with centers at the tail and tip of $\mathbf{F}$ as shown in Fig. 2.10. In this case, F, P, and Q are known quantities; the senses of $\mathbf{P}$ and $\mathbf{Q}$ are readily determined from the triangle rule or the parallelogram law. The main task is the determination of the angles α and β. To solve the problem analytically, we may first use the law of cosines to find the angle α:

$$P^2 = F^2 + Q^2 - 2FQ\cos\alpha$$

$$\alpha = \cos^{-1}[(F^2 + Q^2 - P^2)/(2FQ)] \tag{2.7}$$

Then, using the law of sines, we have

$$\frac{P}{\sin\alpha} = \frac{Q}{\sin\beta}$$

$$\beta = \sin^{-1}[(Q\sin\alpha)/P] \tag{2.8}$$

The angles α and β define the directions of $\mathbf{P}$ and $\mathbf{Q}$ with respect to the given force $\mathbf{F}$ as shown in Fig. 2.10.[†] Note that it is possible to have *two different solutions* for the directions of $\mathbf{P}$ and $\mathbf{Q}$ relative to the direction of $\mathbf{F}$ if the magnitudes only are specified.

(2) *The orientations of $\mathbf{P}$ and $\mathbf{Q}$ are specified, the magnitudes and senses of $\mathbf{P}$ and $\mathbf{Q}$ are to be determined.*

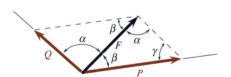

Fig. 2.11

The parallelogram may be constructed by drawing lines from the tip of $\mathbf{F}$, parallel to the specified orientations of $\mathbf{P}$ and $\mathbf{Q}$, as shown in Fig. 2.11. In this case, F, α, and β are known quantities; the senses of $\mathbf{P}$ and $\mathbf{Q}$ are readily determined from the parallelogram law once the parallelogram is constructed as shown. The main task is the determination of the magnitudes P and Q. Since the sum of α, β, and γ must be 180°, we have $\gamma = 180° - \alpha - \beta$. By the law of sines, we write

$$\frac{F}{\sin\gamma} = \frac{P}{\sin\alpha} = \frac{Q}{\sin\beta}$$

$$P = \frac{F\sin\alpha}{\sin\gamma} = \frac{F\sin\alpha}{\sin(\alpha + \beta)} \qquad Q = \frac{F\sin\beta}{\sin\gamma} = \frac{F\sin\beta}{\sin(\alpha + \beta)} \tag{2.9}$$

[†]The same solutions for α and β may alternatively be obtained by constructing right triangles without using the law of cosines and the law of sines. This is illustrated in Example 2.8.

(3) *The force* **P** *is specified, the force* **Q** *is to be determined.*
Drawing **F** and **P** from the same point, we immediately see from the relation
F = **P** + **Q** and the triangle rule that the closing side of the force triangle
must be the force **Q** as shown in Fig. 2.12. In this case, F, P, and β are
known quantities; the sense of **Q** is readily determined from the triangle rule
as shown. The main task is the determination of Q and α (or γ). Using the
laws of cosines and sines, we write

Fig. 2.12

$$Q^2 = F^2 + P^2 - 2FP \cos\beta$$

$$Q = (F^2 + P^2 - 2FP \cos\beta)^{1/2} \qquad (2.10)$$

$$\frac{Q}{\sin\beta} = \frac{P}{\sin\alpha} = \frac{F}{\sin\gamma}$$

$$\alpha = \sin^{-1}\left(\frac{P \sin\beta}{Q}\right) \qquad \gamma = \sin^{-1}\left(\frac{F \sin\beta}{Q}\right) \qquad (2.11)$$

EXAMPLE 2.3

A man and his son attach a piece of rope to their disabled car in the garage
and pull with a force **F** in the x direction, but the rope breaks at a tension
of 400 N just as the car starts to move. The man then attaches one piece of
the rope to the car and pulls with a force **Q** of 300 N, while his son uses
the other piece of rope and pulls with a force **P** of 200 N as shown. It is
known that $\alpha < \beta$, and the resultant of **P** and **Q** is to be the same as the
previous 400-N force **F**. Determine the directional angles θ_P and θ_Q of **P**
and **Q**.

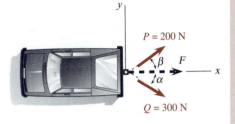

Solution. Since $\alpha < \beta$ is specified, there is only one solution. The parallelogram for the forces is constructed as shown, where $\theta_P = \beta$. By the
law of cosines, we write

$$(300)^2 = (400)^2 + (200)^2 - 2(400)(200) \cos\beta$$

$$\beta = 46.6° \qquad\qquad \theta_P = 46.6° \blacktriangleleft$$

Applying the law of sines, we have

$$\frac{200}{\sin\alpha} = \frac{300}{\sin\beta} \qquad \alpha = 29.0°$$

$$\theta_Q = 360° - \alpha \qquad \theta_Q = 331.0° \blacktriangleleft$$

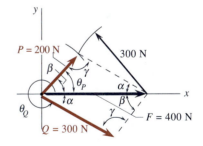

Developmental Exercise

D2.8 What fundamental law must we observe in resolving a force into two
component forces?

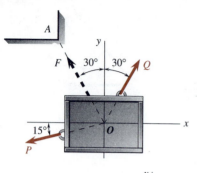

EXAMPLE 2.4

A crate is to be moved in the direction of $\overrightarrow{OA}$ as shown. It is known that a 50-kip force **F** in this direction is required but obstructions prevent a direct application of such a force. Consequently, the forces **P** and **Q** are applied instead as shown. It is desired that the resultant of **P** and **Q** be equivalent to **F**. Determine the magnitudes of **P** and **Q**.

Solution. The parallelogram may be constructed by drawing lines parallel to the specified orientations of **P** and **Q** from the tip of **F** as shown, where the angles indicated are determined from geometry. Referring to the figure and using the law of sines, we write

$$\frac{50}{\sin 45°} = \frac{P}{\sin 60°} = \frac{Q}{\sin 75°}$$

$$P = 61.2 \text{ kips} \qquad Q = 68.3 \text{ kips} \blacktriangleleft$$

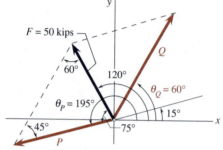

Developmental Exercise

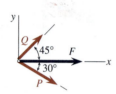

Fig. D2.9

D2.9 A 100-lb force **F** is to be resolved into two forces **P** and **Q** in the directions as shown. Determine their magnitudes P and Q.

EXAMPLE 2.5

A 300-lb force **F** is to be resolved into a 400-lb force **P** along the direction of the line OA and another force **Q**. Determine the magnitude Q and the directional angle θ_Q of **Q**.

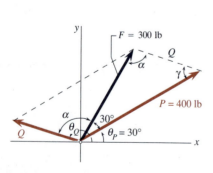

Solution. The force **Q** may be obtained from the parallelogram law as shown. By the laws of cosines and sines, we write

$$Q^2 = (300)^2 + (400)^2 - 2(300)(400)\cos 30°$$

$$Q = 205.31 \qquad\qquad Q = 205 \text{ lb} \blacktriangleleft$$

$$\frac{Q}{\sin 30°} = \frac{400}{\sin \alpha} \qquad \alpha = 76.9° \text{ or } 103.1°$$

Since the projection of **P** onto **F** is

$$(400 \text{ lb}) \cos 30° = 346 \text{ lb} > 300 \text{ lb} = F$$

we conclude that $\alpha > 90°$. We choose $\alpha = 103.1°$ and write

$$\theta_Q = 60° + \alpha \qquad\qquad \theta_Q = 163.1° \blacktriangleleft$$

Developmental Exercise

D2.10   A 30-lb force **F** is to be resolved into a 20-lb force **P** and another force **Q** as shown. Determine the magnitude Q and the directional angle θ_Q of **Q**.

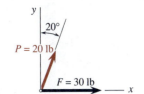

Fig. D2.10

PROBLEMS

2.1   A 300-lb force **A** and a 280-lb force **B** act on a hook as shown. If their resultant is **C**, determine the magnitude C and the directional angle θ_C of **C**.

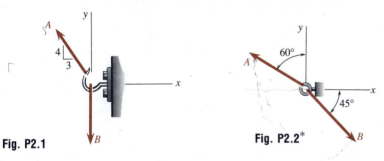

Fig. P2.1   **Fig. P2.2***

2.2*   A 400-N force **A** and a 600-N force **B** act on an eyebolt as shown. If their resultant is **C**, determine the magnitude C and the directional angle θ_C of **C**.

2.3   The hydraulic cylinder AB exerts at B a force **F** of 600 N in the direction of $\overrightarrow{AB}$ against the adjustable platform BC as shown. Determine the magnitudes P and Q and the directional angles θ_P and θ_Q of **P** and **Q**, which are the components of **F** parallel and normal to BC, respectively.

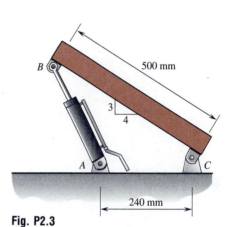

Fig. P2.3

2.4   A 200-lb force **A** and a 600-lb force **B** act on the top of a post as shown. If **D** = **A** − **B**, determine the magnitude D and the directional angle θ_D of **D**.

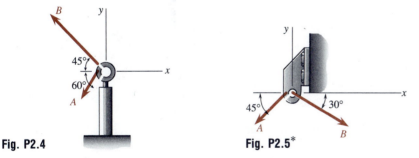

Fig. P2.4   **Fig. P2.5***

Fig. P2.6

2.5*   A 200-N force **A** and a 300-N force **B** act on a bracket as shown. If **D** = **A** − **B**, determine the magnitude D and the directional angle θ_D of **D**.

2.6   A man exerts a force **F** of 150 N on the handles of a wheelbarrow as shown. Resolve **F** into two forces **P** and **Q**, which are parallel and normal to the incline, respectively, and determine the values of P, Q, θ_P, and θ_Q.

2.7   The 400-N force **F** as shown is to be resolved into two forces **A** and **B** where $A = 350$ N, $B = 600$ N, and $90° < \theta_B < 150°$. Determine the directional angles θ_A and θ_B.

2.8*   Solve Prob. 2.7 if $A = 300$ N, $B = 600$ N, and $0 < \theta_A < 150°$.

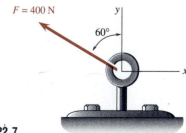

Fig. P2.7

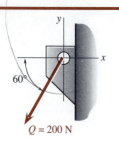

Fig. P2.9

2.9   The force **Q** as shown is to be resolved into two forces **A** and **B** where **A** is parallel to the x axis and **B** is parallel to a line whose slope is −1. Determine A, B, and θ_B.

2.10* Solve Prob. 2.9 if **A** is parallel to the y axis.

2.11   A tractor is being transported by a helicopter to a remote location. If the tensile force **T₁** in the cable AB as shown is 4 kips and the resultant of the tensile forces **T₁** and **T₂** is vertical, determine T_2.

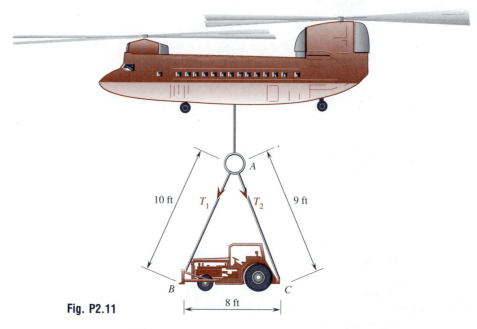

Fig. P2.11

2.12   A 400-N force **F₁** and a 500-N force **F₂** act on a hook as shown. If $\theta_1 = 45°$ and the resultant of **F₁** and **F₂** is vertical, determine θ_2.

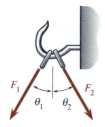

Fig. P2.12 and P2.13*

2.13* Two forces **F₁** and **F₂** act on a hook as shown, where $\theta_1 = 30°$ and $F_1 = 200$ lb. If the resultant of **F₁** and **F₂** acts vertically downward and has a magnitude of 500 lb, determine F_2 and θ_2.

2.14   A 600-N force **S** acts on an anchor as shown. If **S** = **A** + **B**, where A = 500 N and $\theta_A = 350°$, determine B and θ_B.

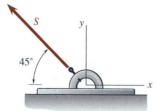

Fig. P2.14

2.15* Solve Prob. 2.14 if $A = 500$ N and $\theta_A = 170°$.

2.16   A beam BC is being lowered into position by a motor crane. At the instant shown, $\theta_1 = 35°$ and the tensions in the cables AB and AC are 800 lb and 1000 lb, respectively. Furthermore, the resultant **R** of the tensile forces in AB and AC is known to act vertically. Determine θ_2 and R.

Fig. P2.16

2.4   Analytical Representation

Unit vectors are vectors of magnitude equal to 1. The unit vectors directed along or parallel to the x and y axes, as shown in Fig. 2.13, are denoted by **i** and **j**, respectively, and are referred to as *cartesian unit vectors*,[†] which have fixed directions in the xy plane. A *general unit vector* is a unit vector which is not parallel to any of the coordinate axes. In Fig. 2.13, the line AB is not parallel to any of the coordinate axes. Therefore, the unit vector **λ** directed along the line AB is a general unit vector.

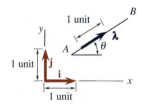

Fig. 2.13

When a vector is resolved into two vectors which are perpendicular to each other, the two vectors are called the *rectangular vector components*, or

---

[†]René Du Perron Descartes (born 1596 in France, died 1650 in Sweden) published in 1637 his magnum opus, *Géométrie*, in which he applied algebra to geometry and introduced the rectangular coordinate system for studying geometric problems.

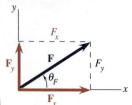

Fig. 2.14

more briefly the *vector components*, of the original vector. A vector in the xy plane is often resolved into two components parallel to the x and y axes. We may denote the vector components of **F** parallel to the x and y axes as $\mathbf{F}_x$ and $\mathbf{F}_y$, respectively, as shown in Fig. 2.14. We write

$$\mathbf{F} = \mathbf{F}_x + \mathbf{F}_y \tag{2.12}$$

Furthermore, the vector components $\mathbf{F}_x$ and $\mathbf{F}_y$ of **F** may be expressed in terms of scalars and the unit vectors **i** and **j** as follows:

$$\mathbf{F}_x = F_x \mathbf{i} \qquad \mathbf{F}_y = F_y \mathbf{j} \tag{2.13}$$

where the scalar quantities F_x and F_y are called the *rectangular scalar components*, or more briefly the *scalar components*, of **F**. The scalar components F_x and F_y are also simply called the *x component* and the *y component* of **F**, respectively; they may be positive or negative, depending on whether or not the senses of the vector components $\mathbf{F}_x$ and $\mathbf{F}_y$ are the same as the senses of **i** and **j**, respectively. Of course, either of the values of F_x and F_y may be zero.

In our study, the term *analytical expression* of a vector will be used to refer to the expression of the vector written in terms of scalars and cartesian unit vectors. The analytical expression of a vector is the so-called *cartesian vector form* of the vector. We may write the *analytical expression* of a force **F** in the xy plane as

$$\mathbf{F} = F_x \mathbf{i} + F_y \mathbf{j} \tag{2.14}$$

The *magnitude F* of **F** is equal to the *absolute value* $|\mathbf{F}|$, a nonnegative scalar. Applying the Pythagorean theorem, we write

$$F = (F_x^2 + F_y^2)^{1/2} \tag{2.15}$$

Letting $\boldsymbol{\lambda}_F$ be the unit vector in the direction of **F**, we write

$$\mathbf{F} = F\boldsymbol{\lambda}_F \tag{2.16}$$

$$\boldsymbol{\lambda}_F = \cos\theta_F \mathbf{i} + \sin\theta_F \mathbf{j} \tag{2.17}$$

Note that the magnitude of $\boldsymbol{\lambda}_F$ is

$$|\boldsymbol{\lambda}_F| = (\cos^2\theta_F + \sin^2\theta_F)^{1/2} = 1 \tag{2.18}$$

Moreover, the scalar components F_x and F_y of **F** may be expressed in terms of F and θ_F as follows:

$$F_x = F\cos\theta_F \qquad F_y = F\sin\theta_F \tag{2.19}$$

The signs of F_x and F_y are, of course, dependent on the quadrant of θ_F.

Developmental Exercises

D2.11 Discuss the differences and similarities among the terms: (a) unit vector, (b) cartesian unit vector, (c) general unit vector.

D2.12 Define the terms: (a) vector components, (b) scalar components, (c) analytical expression of a vector.

EXAMPLE 2.6

For the forces **P** and **Q** in Example 2.1, write (a) the analytical expressions of the unit vectors λ_P and λ_Q in the directions of **P** and **Q**, (b) the analytical expressions of **P** and **Q**.

Solution.   The directional angles of **P**, **Q**, λ_P, and λ_Q may be indicated as shown. By Eq. (2.17) and then Eq. (2.16), we write

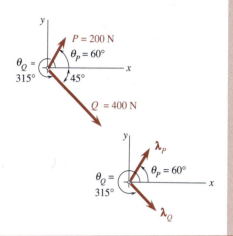

$$\lambda_P = \cos\theta_P\mathbf{i} + \sin\theta_P\mathbf{j} = \cos 60°\mathbf{i} + \sin 60°\mathbf{j}$$

$$\lambda_Q = \cos\theta_Q\mathbf{i} + \sin\theta_Q\mathbf{j} = \cos 315°\mathbf{i} + \sin 315°\mathbf{j}$$

$$\lambda_P = 0.5\mathbf{i} + 0.866\mathbf{j} \qquad \lambda_Q = 0.707\mathbf{i} - 0.707\mathbf{j} \blacktriangleleft$$

$$\mathbf{P} = P\lambda_P = 200(0.5\mathbf{i} + 0.886\mathbf{j})$$

$$\mathbf{Q} = Q\lambda_Q = 400(0.707\mathbf{i} - 0.707\mathbf{j})$$

$$\mathbf{P} = 100\mathbf{i} + 173.2\mathbf{j}\ \text{N} \qquad \mathbf{Q} = 283\mathbf{i} - 283\mathbf{j}\ \text{N} \blacktriangleleft$$

Developmental Exercises

D2.13   A 100-N force **G** and a 130-N force **H** are applied to an eyebolt as shown. Determine (a) the vector components of **G** and **H**, (b) the scalar components of **G** and **H**, (c) the magnitude of H_x.

D2.14   A stump is pulled with a 250-N force **P** as shown. (a) Write the analytical expression of **P**. (b) What is the directional angle of **P**?

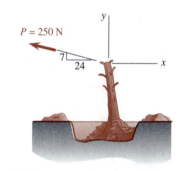

Fig. D2.13

2.5   Addition and Resolution of Forces: Vector Algebra

If two forces **P** and **Q** in a plane are analytically expressed as

$$\mathbf{P} = P_x\mathbf{i} + P_y\mathbf{j} \qquad \mathbf{Q} = Q_x\mathbf{i} + Q_y\mathbf{j}$$

then, by the associative property of the vector addition, we may write the *addition* of these two forces as

$$\mathbf{P} + \mathbf{Q} = (P_x + Q_x)\mathbf{i} + (P_y + Q_y)\mathbf{j} \qquad (2.20)$$

Note that this result is consistent with the parallelogram law for the addition of two forces as sketched in Fig. 2.15. The *subtraction* of **Q** from **P** may

Fig. D2.14

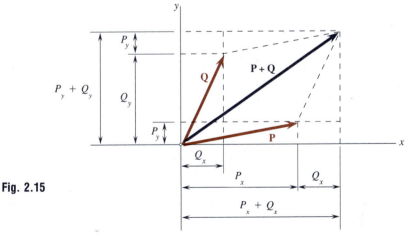

Fig. 2.15

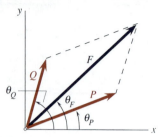

Fig. 2.16

similarly be written as

$$\mathbf{P} - \mathbf{Q} = (P_x - Q_x)\mathbf{i} + (P_y - Q_y)\mathbf{j} \qquad (2.21)$$

The analytical expressions of forces may likewise be used to perform the addition or subtraction of more than two forces.

The *resolution* of a given force **F** in a plane into two component forces **P** and **Q** may *analytically* be formulated in terms of their magnitudes F, P, and Q, and their directional angles θ_F, θ_P, and θ_Q as shown in Fig. 2.16, where

$$\mathbf{F} = \mathbf{P} + \mathbf{Q} \qquad (2.22)$$

In terms of magnitudes and directional angles, we write

$$F \cos\theta_F \mathbf{i} + F \sin\theta_F \mathbf{j} = (P \cos\theta_P + Q \cos\theta_Q)\mathbf{i}$$
$$+ (P \sin\theta_P + Q \sin\theta_Q)\mathbf{j}$$

For this vector equation to be true, we must require that the coefficients of the **i**'s and **j**'s on both sides of the equal sign be equal; i.e., the ensuing set of simultaneous scalar equations is

$$F \cos\theta_F = P \cos\theta_P + Q \cos\theta_Q$$
$$F \sin\theta_F = P \sin\theta_P + Q \sin\theta_Q \qquad (2.23)$$

In these two simultaneous equations, F and θ_F are given or known; therefore, if any two of the four quantities P, Q, θ_P, and θ_Q are specified or known, the other two may be determined by solving the two simultaneous equations in Eqs. (2.23). Furthermore, if any two vectors in Eq. (2.22) are known, the third vector can be found using vector algebra.

EXAMPLE 2.7

Determine the resultant force **R** of the five forces $\mathbf{F}_1$, $\mathbf{F}_2$, $\mathbf{F}_3$, $\mathbf{F}_4$, and $\mathbf{F}_5$ acting on an eyebolt as shown.

Solution. This problem can be solved by using *vector algebra* as follows:

$$\mathbf{F}_1 = 500\boldsymbol{\lambda}_1 = 500\mathbf{i} \qquad \mathbf{F}_2 = 200\boldsymbol{\lambda}_2 = 200(\cos 30°\mathbf{i} + \sin 30°\mathbf{j})$$

$$\mathbf{F}_3 = 100\boldsymbol{\lambda}_3 = 100\mathbf{j} \qquad \mathbf{F}_4 = 340\boldsymbol{\lambda}_4 = 340\left(-\frac{8}{17}\mathbf{i} - \frac{15}{17}\mathbf{j}\right)$$

$$\mathbf{F}_5 = 350\boldsymbol{\lambda}_5 = 350\left(\frac{3}{5}\mathbf{i} - \frac{4}{5}\mathbf{j}\right) \qquad \mathbf{R} = \mathbf{F}_1 + \mathbf{F}_2 + \mathbf{F}_3 + \mathbf{F}_4 + \mathbf{F}_5$$

$$\mathbf{R} = \left[500 + 200 \cos 30° + 340\left(-\frac{8}{17}\right) + 350\left(\frac{3}{5}\right)\right]\mathbf{i}$$

$$+ \left[200 \sin 30° + 100 + 340\left(-\frac{15}{17}\right) + 350\left(-\frac{4}{5}\right)\right]\mathbf{j}$$

$$\mathbf{R} = 723\mathbf{i} - 380\mathbf{j} \text{ lb} \qquad \blacktriangleleft$$

Developmental Exercise

D2.15   Refer to Example 2.7. Determine the resultant force **P** of the forces **F**$_3$, **F**$_4$, and **F**$_5$.

EXAMPLE 2.8

Work Example 2.3 by constructing right triangles and using the *Pythagorean theorem*, instead of the laws of cosines and sines.

Solution.   The parallelogram for resolving **F** into **P** and **Q** was shown in Example 2.3. The top portion of this parallelogram may be divided into two right triangles where their perpendicular sides are related to the scalar components of **P** and **Q** as indicated.

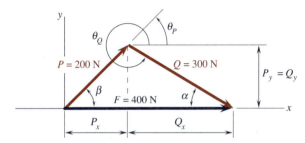

We are to find the directional angles θ_P and θ_Q. By the Pythagorean theorem and the figure shown, we write

$$(200)^2 = P_x^2 + P_y^2 \qquad (300)^2 = Q_x^2 + Q_y^2$$

$$P_y = Q_y \qquad P_x + Q_x = 400$$

To solve the above four simultaneous equations, we write

$$(200)^2 - (300)^2 = (P_x^2 + P_y^2) - (Q_x^2 + Q_y^2) = P_x^2 - Q_x^2$$

$$= (P_x + Q_x)(P_x - Q_x) = 400(P_x - Q_x)$$

Thus, $P_x - Q_x = -125$. This and the preceding equations readily yield

$$P_x = 137.50 \qquad P_y = 145.24 \qquad Q_x = 262.50 \qquad Q_y = 145.24$$

Referring to the figure shown, we write

$$\tan \beta = \frac{145.24}{137.50} \qquad \beta = 46.6° = \theta_P$$

$$\tan \alpha = \frac{145.24}{262.50} \qquad \alpha = 29.0°$$

$$\theta_Q = 360° - \alpha \qquad \theta_P = 46.6° \qquad \theta_Q = 331.0° \blacktriangleleft$$

Developmental Exercise

D2.16   A 10-kip force **F** is to be resolved into an 8-kip force **P** and a 9-kip force **Q**. Determine the angle γ as shown by constructing right triangles and using the Pythagorean theorem, instead of the laws of cosines and sines.

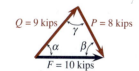

Fig. D2.16

EXAMPLE 2.9

Using *vector algebra*, work Example 2.4.

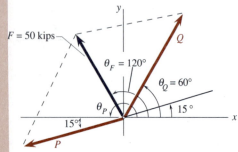

Solution. The parallelogram for resolving **F** into **P** and **Q** is shown, where the magnitudes P and Q are to be determined. Referring to the figure, we write

$$\mathbf{F} = 50\boldsymbol{\lambda}_F = 50(\cos 120°\mathbf{i} + \sin 120°\mathbf{j})$$

$$\mathbf{P} = P\boldsymbol{\lambda}_P = P(\cos 195°\mathbf{i} + \sin 195°\mathbf{j})$$

$$\mathbf{Q} = Q\boldsymbol{\lambda}_Q = Q(\cos 60°\mathbf{i} + \sin 60°\mathbf{j})$$

Since $\mathbf{F} = \mathbf{P} + \mathbf{Q}$, we write

$$50 \cos 120°\mathbf{i} + 50 \sin 120°\mathbf{j}$$
$$= (P \cos 195° + Q \cos 60°)\mathbf{i} + (P \sin 195° + Q \sin 60°)\mathbf{j}$$

Equating the coefficients of **i**'s and **j**'s, we have

$$50 \cos 120° = (\cos 195°)P + (\cos 60°)Q$$

$$50 \sin 120° = (\sin 195°)P + (\sin 60°)Q$$

Solution of the above two equations yields

$$P = 61.2 \text{ kips} \qquad Q = 68.3 \text{ kips} \quad \blacktriangleleft$$

EXAMPLE 2.10

Using *vector algebra*, work Example 2.5.

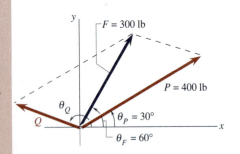

Solution. The parallelogram for resolving **F** into **P** and **Q**, as specified, is shown, where the magnitude Q and the directional angle θ_Q are to be determined. Referring to the figure, we write

$$\mathbf{F} = 300\boldsymbol{\lambda}_F = 300 \cos 60°\mathbf{i} + 300 \sin 60°\mathbf{j}$$

$$\mathbf{P} = 400\boldsymbol{\lambda}_P = 400 \cos 30°\mathbf{i} + 400 \sin 30°\mathbf{j}$$

Since $\mathbf{F} = \mathbf{P} + \mathbf{Q}$, we write

$$\mathbf{Q} = \mathbf{F} - \mathbf{P} = (300 \cos 60° - 400 \cos 30°)\mathbf{i}$$
$$+ (300 \sin 60° - 400 \sin 30°)\mathbf{j}$$
$$= -196.4\mathbf{i} + 59.8\mathbf{j}$$

$$Q^2 = (-196.4)^2 + (59.8)^2 \qquad Q = 205 \text{ lb} \quad \blacktriangleleft$$

$$\tan \theta_Q = 59.8/(-196.4) \qquad \theta_Q = 163.1° \quad \blacktriangleleft$$

Developmental Exercise

D2.17 It is known that $(2 + 3x)\mathbf{i} + 37\mathbf{j} = -4\mathbf{i} + (2x + y)\mathbf{j}$. Determine the values of x and y.

PROBLEMS

2.17 Knowing that $\mathbf{F} = 80\mathbf{i} + 150\mathbf{j}$ N and $\mathbf{S} = 90\mathbf{i} - 400\mathbf{j}$ N, determine their magnitudes F and S and directional angles θ_F and θ_S.

2.18* Knowing that $\mathbf{P} = -240\mathbf{i} + 70\mathbf{j}$ lb and $\mathbf{Q} = -22\mathbf{i} - 120\mathbf{j}$ lb, determine their magnitudes P and Q and directional angles θ_P and θ_Q.

2.19 The cable of a motor crane is loaded with a 3.4-kip force $\mathbf{F}$ as shown. Determine the magnitudes of the forces $\mathbf{P}$ and $\mathbf{Q}$ which are the components of $\mathbf{F}$ parallel and normal to the boom AB, respectively.

Fig. P2.19

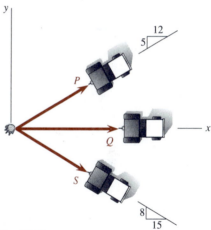

Fig. P2.20

2.20 A heavy stump is being dislodged with three tractors as shown, where the forces exerted by the tractors are $P = 6.5$ kN, $Q = 5$ kN, and $S = 8.5$ kN. Determine the resultant $\mathbf{R}$ of these three forces.

2.21* The two forces $\mathbf{P}$ and $\mathbf{Q}$ applied to a hook as shown have a resultant $\mathbf{R} = -100\mathbf{i} - 600\mathbf{j}$ N. If $\mathbf{P}$ has an x component $\mathbf{P}_x = -200\mathbf{i}$ N and $\mathbf{Q}$ has a y component $\mathbf{Q}_y = -300\mathbf{j}$ N, determine $\mathbf{P}$ and $\mathbf{Q}$.

2.22 Determine the resultant $\mathbf{R}$ of the three forces which act on the riveted bracket as shown.

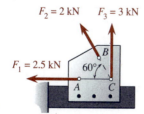

Fig. P2.22

Fig. P2.21*

2.23* Determine the resultant **R** of the four axial forces in the truss members that are fastened to a gusset plate at a hinge support as shown.

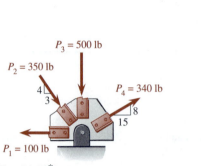

Fig. P2.23*

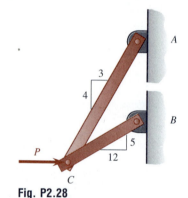

Fig. P2.24

2.24 The resultant of the four forces acting on the anchor shown is known to be **R** = 559**i** + 788**j** N. Determine the force **Q**$_3$.

2.25 The horizontal component of the resultant **R** of the four forces acting on the crate shown is **R**$_x$ = 1160**i** N. If F_1 = 260 N, F_2 = 500 N, F_4 = 760 N, and θ_3 = tan^{-1} (5/12), determine F_3.

2.26* The crate shown is acted on by four forces, where F_1 = 260 N, F_2 = 500 N, and F_3 = 50 N. If the resultant of the four forces is **R** = 160**i** + 750**j** N, determine θ_3 and F_4.

2.27 A small boat at A is being towed with two forces **F**$_1$ and **F**$_2$ as shown. If F_1 = 60 lb and the resultant of **F**$_1$ and **F**$_2$ is directed from A to B, determine F_2.

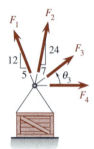

Fig. P2.25 and P2.26*

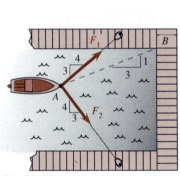

Fig. P2.27

Fig. P2.28

2.28 When a 330-N force **P** is applied to the structure at the joint C as shown, the forces **F**$_{AC}$ and **F**$_{BC}$ developed in the members AC and BC are known to direct along the axes of the respective members, and the resultant of **F**$_{AC}$ and **F**$_{BC}$ is equal to $-$**P**. Determine the magnitudes of **F**$_{AC}$ and **F**$_{BC}$.

2.29* through 2.42* Using *vector algebra,* solve the following problems:

2.29*	Prob. 2.1.	2.30*	Prob. 2.2*.
2.31*	Prob. 2.3.	2.32*	Prob. 2.4.
2.33*	Prob. 2.5*.	2.34*	Prob. 2.6.
2.35*	Prob. 2.9.	2.36*	Prob. 2.10*.
2.37*	Prob. 2.11.	2.38*	Prob. 2.12.
2.39*	Prob. 2.13*.	2.40*	Prob. 2.14.
2.41*	Prob. 2.15*.	2.42*	Prob. 2.16.

FORCES IN SPACE[†]

2.6   Representation of a Force in Space

The unit vectors directed along or parallel to the x, y, and z axes of a fixed rectangular coordinate system are denoted by $\mathbf{i}$, $\mathbf{j}$, and $\mathbf{k}$, respectively, which form a set of *cartesian unit vectors* in space. Each of these three unit vectors is a *constant vector* and is perpendicular to the other two as shown in Fig. 2.17.

A force $\mathbf{F}$ in space is often resolved into three components $\mathbf{F}_x$, $\mathbf{F}_y$, and $\mathbf{F}_z$ parallel respectively to the x, y, and z rectangular coordinate axes as shown in Fig. 2.18, where

$$\mathbf{F} = \mathbf{F}_y + \mathbf{F}_h = \mathbf{F}_y + (\mathbf{F}_x + \mathbf{F}_z)$$

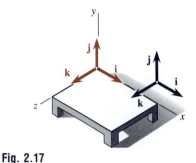

Fig. 2.17

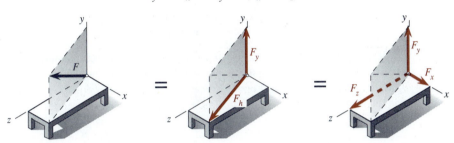

Fig. 2.18

Since vector addition is associative and commutative, we write

$$\mathbf{F} = \mathbf{F}_x + \mathbf{F}_y + \mathbf{F}_z \qquad (2.24)$$

where $\mathbf{F}_x$, $\mathbf{F}_y$, and $\mathbf{F}_z$ are called the *rectangular vector components*, or more briefly the *vector components*, of $\mathbf{F}$. Expressing them as products of appropriate scalars and the unit vectors $\mathbf{i}$, $\mathbf{j}$, and $\mathbf{k}$, we write

$$\mathbf{F}_x = F_x\mathbf{i} \qquad \mathbf{F}_y = F_y\mathbf{j} \qquad \mathbf{F}_z = F_z\mathbf{k}$$

$$\mathbf{F} = F_x\mathbf{i} + F_y\mathbf{j} + F_z\mathbf{k} \qquad (2.25)$$

where F_x, F_y, and F_z are called the *rectangular scalar components*, or more briefly the *scalar components*, of $\mathbf{F}$. Furthermore, F_x, F_y, and F_z are also called the *x component*, the *y component*, and the *z component* of $\mathbf{F}$, respectively. Note that Eq. (2.25) shows the *analytical expression of the vector* $\mathbf{F}$ *in space*. Referring to Fig. 2.18 and applying the Pythagorean theorem, we see that the *magnitude* (or *absolute value*) of $\mathbf{F}$ is

$$F = (F_y^2 + F_h^2)^{1/2} = [F_y^2 + (F_x^2 + F_z^2)]^{1/2}$$

$$F = (F_x^2 + F_y^2 + F_z^2)^{1/2} \qquad (2.26)$$

The nonnegative angles measured from the positive directions of the x, y, and z coordinate axes to the positive direction of $\mathbf{F}$ are referred to as the *direction angles* of $\mathbf{F}$ and are denoted by θ_x, θ_y, and θ_z, respectively, as

---

[†]Unless otherwise stated, the term space will be understood to imply *three-dimensional space*. *Forces in space* are, at times, called *three-dimensional forces*. However, some object to the latter terminology. Cf. the footnote on p. 23.

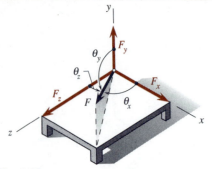

Fig. 2.19

shown in Fig. 2.19.[†] Depending on the signs of the scalar components, the direction angles may have values ranging from 0 to 180° (or from 0 to π rad).

The cosines of the direction angles are called the *direction cosines*. Referring to Fig. 2.19, we write

$$F_x = F\cos\theta_x \qquad F_y = F\cos\theta_y \qquad F_z = F\cos\theta_z \quad (2.27)$$

Thus, the scalar components F_x, F_y, and F_z of **F** are, in fact, the *projections* of **F** onto the x, y, and z axes, respectively. If a scalar component or a projection of the vector onto an axis is positive, zero, or negative, then the corresponding direction angle must be less than, equal to, or greater than 90°, respectively. Using Eqs. (2.26) and (2.27), we write

$$F^2 = F_x^2 + F_y^2 + F_z^2 = F^2(\cos^2\theta_x + \cos^2\theta_y + \cos^2\theta_z)$$

$$\cos^2\theta_x + \cos^2\theta_y + \cos^2\theta_z = 1 \quad (2.28)$$

Equation (2.28) is called the *constraint equation* on the direction angles. We see from this equation that only two of the three direction angles θ_x, θ_y, and θ_z are independent; the third one must satisfy the constraint equation, which states that the sum of the squares of the three direction cosines equals 1.

Developmental Exercises

D2.18   What are the scalar components of the vectors (a) **F** = 3**i** − 4**j** − 12**k** kN, (b) **r** = 40**i** − 9**k** m?

D2.19   What is the magnitude of the force **F** = 4**i** + 4**j** − 7**k** kips?

D2.20   A force vector **F** is shown. (a) Which direction angles are less than 90°? (b) Is there any direction angle which is greater than 180°? (c) Is there any direction angle which is negative? (d) Is there any direction cosine which is negative?

D2.21   For **F** = 8**j** − 15**k** lb, what are (a) the projections of **F** onto the coordinate axes, (b) the direction angles of **F**?

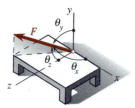

Fig. D2.20

2.7   Position Vector and Unit Vector in Space

The *position vector* of a point P is defined as the vector drawn from the origin O of the coordinate system to the point P. It has a magnitude equal to the distance between O and P and is directed from O to P.

As shown in Fig. 2.20, the position vector **r** of the point $P(x, y, z)$ is

$$\mathbf{r} = \overrightarrow{OP} \quad (2.29)$$

The projections of **r** onto the coordinate axes are equal to x, y, and z, respectively. Thus, we may express **r** as

$$\mathbf{r} = x\mathbf{i} + y\mathbf{j} + z\mathbf{k} \quad (2.30)$$

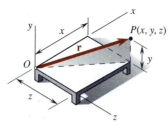

Fig. 2.20

[†]Note that the term *direction angle* is associated with a vector in three-dimensional space, while the term *directional angle*, as defined in Sec. 2.1, is associated with a vector in a plane.

The name *position vector* of the point P for $\mathbf{r}$ is quite fitting because $\mathbf{r}$ contains the information which completely defines the position of P.

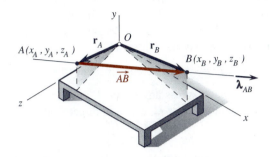

Fig. 2.21

The position vectors $\mathbf{r}_A$ and $\mathbf{r}_B$ of the points $A(x_A, y_A, z_A)$ and $B(x_B, y_B, z_B)$ in Fig. 2.21 may be expressed as

$$\mathbf{r}_A = x_A\mathbf{i} + y_A\mathbf{j} + z_A\mathbf{k} \qquad \mathbf{r}_B = x_B\mathbf{i} + y_B\mathbf{j} + z_B\mathbf{k}$$

The vector which is directed from A to B is denoted by $\overrightarrow{AB}$. The vector $\overrightarrow{AB}$ is called *the position vector of B relative to A*. As shown in Fig. 2.21, the vectors $\mathbf{r}_A$, $\overrightarrow{AB}$, and $\mathbf{r}_B$ form a vector triangle OAB. Applying the *triangle rule* in Sec. 2.2, we write

$$\mathbf{r}_A + \overrightarrow{AB} = \mathbf{r}_B$$

$$\overrightarrow{AB} = \mathbf{r}_B - \mathbf{r}_A = (x_B\mathbf{i} + y_B\mathbf{j} + z_B\mathbf{k}) - (x_A\mathbf{i} + y_A\mathbf{j} + z_A\mathbf{k})$$

$$\overrightarrow{AB} = (x_B - x_A)\mathbf{i} + (y_B - y_A)\mathbf{j} + (z_B - z_A)\mathbf{k} \qquad (2.31)$$

The magnitude of $\overrightarrow{AB}$ is given by the absolute value

$$|\overrightarrow{AB}| = [(x_B - x_A)^2 + (y_B - y_A)^2 + (z_B - z_A)^2]^{1/2} \qquad (2.32)$$

Thus, the unit vector $\boldsymbol{\lambda}_{AB}$ directed from $A(x_A, y_A, z_A)$ to $B(x_B, y_B, z_B)$ may be computed from the formula

$$\boldsymbol{\lambda}_{AB} = \frac{\overrightarrow{AB}}{|\overrightarrow{AB}|} \qquad (2.33)$$

where $\overrightarrow{AB}$ and $|\overrightarrow{AB}|$ are given by Eqs. (2.31) and (2.32).

In Fig. 2.22, a unit vector $\boldsymbol{\lambda}$ is shown, where the direction angles θ_x and θ_y are known, but θ_z is unknown. In this case, the cosine of θ_z may first be

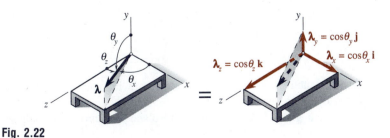

Fig. 2.22

obtained from Eq. (2.28) as

$$\cos\theta_z = \pm(1 - \cos^2\theta_x - \cos^2\theta_y)^{1/2} \qquad (2.34)$$

where the upper sign corresponds to $0 \leq \theta_z \leq 90°$, and the lower sign corresponds to $90° < \theta_z \leq 180°$. If the range of values of θ_z is not specified, there will be two possible values for θ_z. In any case, the analytical expression of the unit vector $\boldsymbol{\lambda}$ may be written as

$$\boldsymbol{\lambda} = \cos\theta_x\mathbf{i} + \cos\theta_y\mathbf{j} + \cos\theta_z\mathbf{k} \qquad (2.35)$$

Furthermore, we may express any vector $\mathbf{F}$ in space as the product of its magnitude F and the unit vector $\boldsymbol{\lambda}_F$ which is directed along $\mathbf{F}$; i.e.,

$$\mathbf{F} = F\boldsymbol{\lambda}_F \qquad (2.36)$$

Equation (2.36) is a useful *formula* for obtaining the analytical expression of any vector $\mathbf{F}$ when F and $\boldsymbol{\lambda}_F$ are known. On the other hand, if the analytical expression of $\mathbf{F}$ is known, we may obtain the analytical expression of $\boldsymbol{\lambda}_F$ by dividing F into $\mathbf{F}$.

Developmental Exercises

D2.22   A point P is located at $(6, -4, 12)$ m. Determine (a) the analytical expression of the position vector of P, (b) the distance from the origin to P.

D2.23   A vector directed from A to B is shown. (a) Write the analytical expression of $\overrightarrow{AB}$. (b) Determine the length of $\overrightarrow{AB}$. (c) Write the analytical expression of the unit vector $\boldsymbol{\lambda}_{AB}$.

D2.24   A 100-lb force $\mathbf{F}$ has direction angles $\theta_x = 45°$, $90° < \theta_y \leq 180°$, and $\theta_z = 60°$. Determine the analytical expressions of (a) the unit vector $\boldsymbol{\lambda}_F$, (b) the force $\mathbf{F}$.

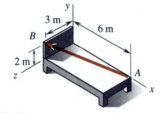

Fig. D2.23

2.8   Addition and Resolution of Forces in Space

Forces in space are seldom added or resolved by direct application of the parallelogram law. Such forces are more effectively handled by using vector algebra. Suppose that the analytical expressions of the forces $\mathbf{P}$ and $\mathbf{Q}$ are

$$\mathbf{P} = P_x\mathbf{i} + P_y\mathbf{j} + P_z\mathbf{k} \qquad \mathbf{Q} = Q_x\mathbf{i} + Q_y\mathbf{j} + Q_z\mathbf{k}$$

Then, by the associative property of the vector addition, we write

$$\mathbf{P} + \mathbf{Q} = (P_x + Q_x)\mathbf{i} + (P_y + Q_y)\mathbf{j} + (P_z + Q_z)\mathbf{k} \qquad (2.37)$$

The subtraction of $\mathbf{Q}$ from $\mathbf{P}$ may similarly be written as

$$\mathbf{P} - \mathbf{Q} = (P_x - Q_x)\mathbf{i} + (P_y - Q_y)\mathbf{j} + (P_z - Q_z)\mathbf{k} \qquad (2.38)$$

The addition or subtraction of three or more forces in space may likewise be performed by using vector algebra.

Suppose that a given force $\mathbf{F}$ is to be resolved into three noncoplanar forces **A**, **B**, and **C** with *unknown magnitudes* but *known directions*. We write

$$\mathbf{F} = \mathbf{A} + \mathbf{B} + \mathbf{C} \tag{2.39}$$

$$F_x\mathbf{i} + F_y\mathbf{j} + F_z\mathbf{k} = A\boldsymbol{\lambda}_A + B\boldsymbol{\lambda}_B + C\boldsymbol{\lambda}_C \tag{2.40}$$

where $\boldsymbol{\lambda}_A$, $\boldsymbol{\lambda}_B$, and $\boldsymbol{\lambda}_C$ are known and may be expressed in terms of the unit vectors **i**, **j**, and **k**. We may rearrange and equate the coefficients of **i**'s, **j**'s, and **k**'s on both sides of Eq. (2.40) to yield a set of three simultaneous scalar equations with the magnitudes A, B, and C as the three unknowns, which may then be solved. (Cf. Example 2.13.)

EXAMPLE 2.11

The guy wire AB, as shown, has a tension of 2 kips. Determine the analytical expressions of (a) the unit vector $\boldsymbol{\lambda}_{AB}$ directed along AB, (b) the tensile force $\mathbf{T}_{AB}$ exerted by the guy wire on the utility pole.

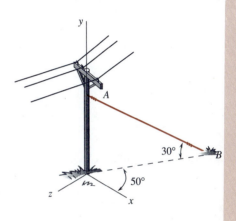

Solution. Let the unit vector $\boldsymbol{\lambda}_{AB}$ as shown be first resolved into a vertical component $\boldsymbol{\lambda}_y$ and a horizontal component $\boldsymbol{\lambda}_h$. The horizontal component $\boldsymbol{\lambda}_h$ is then projected onto the zx plane and resolved into two rectangular components $\boldsymbol{\lambda}_x$ and $\boldsymbol{\lambda}_z$ which lie along the x and z axes, respectively. We see from the figure that

$$|\boldsymbol{\lambda}_y| = |\boldsymbol{\lambda}_{AB}| (\cos 60°) = (1)(\cos 60°) = \cos 60°$$

$$|\boldsymbol{\lambda}_h| = |\boldsymbol{\lambda}_{AB}| (\sin 60°) = (1)(\sin 60°) = \sin 60°$$

$$|\boldsymbol{\lambda}_x| = |\boldsymbol{\lambda}_h| (\cos 50°) = \sin 60° \cos 50°$$

$$|\boldsymbol{\lambda}_z| = |\boldsymbol{\lambda}_h| (\sin 50°) = \sin 60° \sin 50°$$

By Eq. (2.36), we write

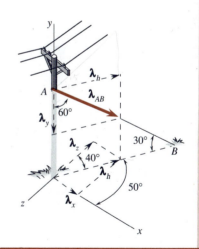

$$\boldsymbol{\lambda}_y = |\boldsymbol{\lambda}_y|(-\mathbf{j}) = -\cos 60°\mathbf{j}$$

$$\boldsymbol{\lambda}_h = \boldsymbol{\lambda}_x + \boldsymbol{\lambda}_z = |\boldsymbol{\lambda}_x|\mathbf{i} + |\boldsymbol{\lambda}_z|(-\mathbf{k}) = \sin 60° \cos 50°\mathbf{i} - \sin 60° \sin 50°\mathbf{k}$$

$$\boldsymbol{\lambda}_{AB} = \boldsymbol{\lambda}_y + \boldsymbol{\lambda}_h = -\cos 60°\mathbf{j} + (\sin 60° \cos 50°\mathbf{i} - \sin 60° \sin 50°\mathbf{k})$$

$$\boldsymbol{\lambda}_{AB} = 0.557\mathbf{i} - 0.5\mathbf{j} - 0.663\mathbf{k} \blacktriangleleft$$

Check. $(0.557)^2 + (0.500)^2 + (0.663)^2 \approx 1.000$

Since $T_{AB} = 2$ kips, we apply Eq. (2.36) to write

$$\mathbf{T}_{AB} = T_{AB}\boldsymbol{\lambda}_{AB} = 2(0.5567\mathbf{i} - 0.5\mathbf{j} - 0.6634\mathbf{k})$$

$$\mathbf{T}_{AB} = 1.113\mathbf{i} - \mathbf{j} - 1.327\mathbf{k} \text{ kips} \blacktriangleleft$$

Developmental Exercise

D2.25 Write the analytical expressions of (a) $\boldsymbol{\lambda}_F$ if $\mathbf{F} = 400\mathbf{i} + 700\mathbf{j} - 400\mathbf{k}$ lb, (b) the 260-lb force **P** if $\boldsymbol{\lambda}_P = \frac{1}{13}(3\mathbf{i} - 12\mathbf{j} + 4\mathbf{k})$.

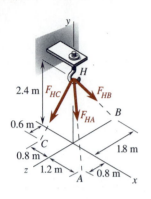

2.4 m F_{HC} F_{HB}

0.6 m F_{HA} B

C

0.8 m   1.8 m

z 1.2 m   0.8 m

A 0.8 m x

EXAMPLE 2.12

Three forces $\mathbf{F}_{HA}$, $\mathbf{F}_{HB}$, and $\mathbf{F}_{HC}$ act on a hook at H as shown. If the magnitudes of these forces are $F_{HA} = 420$ N, $F_{HB} = 500$ N, and $F_{HC} = 390$ N, determine the resultant force $\mathbf{R}$ acting on the hook.

Solution. From the geometry and the data given, we write

$$\overrightarrow{HA} = 1.2\mathbf{i} - 2.4\mathbf{j} + 0.8\mathbf{k} \qquad \boldsymbol{\lambda}_{HA} = \tfrac{1}{7}(3\mathbf{i} - 6\mathbf{j} + 2\mathbf{k})$$

$$\overrightarrow{HB} = -2.4\mathbf{j} - 1.8\mathbf{k} \qquad \boldsymbol{\lambda}_{HB} = -\tfrac{1}{5}(4\mathbf{j} + 3\mathbf{k})$$

$$\overrightarrow{HC} = -0.8\mathbf{i} - 2.4\mathbf{j} + 0.6\mathbf{k} \qquad \boldsymbol{\lambda}_{HC} = \tfrac{1}{13}(-4\mathbf{i} - 12\mathbf{j} + 3\mathbf{k})$$

$$\mathbf{F}_{HA} = F_{HA}\boldsymbol{\lambda}_{HA} = 420\boldsymbol{\lambda}_{HA} \qquad \mathbf{F}_{HA} = 60(3\mathbf{i} - 6\mathbf{j} + 2\mathbf{k})$$

$$\mathbf{F}_{HB} = F_{HB}\boldsymbol{\lambda}_{HB} = 500\boldsymbol{\lambda}_{HB} \qquad \mathbf{F}_{HB} = -100(4\mathbf{j} + 3\mathbf{k})$$

$$\mathbf{F}_{HC} = F_{HC}\boldsymbol{\lambda}_{HC} = 390\boldsymbol{\lambda}_{HC} \qquad \mathbf{F}_{HC} = 30(-4\mathbf{i} - 12\mathbf{j} + 3\mathbf{k})$$

Thus, we have

$$\mathbf{R} = \mathbf{F}_{HA} + \mathbf{F}_{HB} + \mathbf{F}_{HC} \qquad \mathbf{R} = 60\mathbf{i} - 1120\mathbf{j} - 90\mathbf{k} \text{ N} \blacktriangleleft$$

Developmental Exercise

D2.26 If the forces $\mathbf{F}$ and $\mathbf{P}$ in D2.25 act at the origin O, determine their resultant $\mathbf{R}$.

EXAMPLE 2.13

If the resultant of the three forces $\mathbf{F}_{HA}$, $\mathbf{F}_{HB}$, and $\mathbf{F}_{HC}$ in Example 2.12 is $\mathbf{R} = -1000\mathbf{j}$ N, determine the magnitude of each force.

Solution. From the solution in Example 2.12, we have

$$\mathbf{F}_{HA} = F_{HA}\boldsymbol{\lambda}_{HA} = \frac{F_{HA}}{7}(3\mathbf{i} - 6\mathbf{j} + 2\mathbf{k})$$

$$\mathbf{F}_{HB} = F_{HB}\boldsymbol{\lambda}_{HB} = -\frac{F_{HB}}{5}(4\mathbf{j} + 3\mathbf{k})$$

$$\mathbf{F}_{HC} = F_{HC}\boldsymbol{\lambda}_{HC} = \frac{F_{HC}}{13}(-4\mathbf{i} - 12\mathbf{j} + 3\mathbf{k})$$

Since $\mathbf{F}_{HA} + \mathbf{F}_{HB} + \mathbf{F}_{HC} = \mathbf{R} = -1000\mathbf{j}$, we write

$$\left(\frac{3}{7}F_{HA} - \frac{4}{13}F_{HC}\right)\mathbf{i} + \left(-\frac{6}{7}F_{HA} - \frac{4}{5}F_{HB} - \frac{12}{13}F_{HC}\right)\mathbf{j}$$

$$+ \left(\frac{2}{7}F_{HA} - \frac{3}{5}F_{HB} + \frac{3}{13}F_{HC}\right)\mathbf{k} = -1000\mathbf{j}$$

Equating the coefficients of **i**'s, **j**'s, and **k**'s, we obtain

$$\frac{3}{7}F_{HA} - \frac{4}{13}F_{HC} = 0 \tag{1}$$

$$-\frac{6}{7}F_{HA} - \frac{4}{5}F_{HB} - \frac{12}{13}F_{HC} = -1000 \tag{2}$$

$$\frac{2}{7}F_{HA} - \frac{3}{5}F_{HB} + \frac{3}{13}F_{HC} = 0 \tag{3}$$

which may be solved to yield[†]

$$F_{HA} = 339 \text{ N} \qquad F_{HB} = 343 \text{ N} \qquad F_{HC} = 472 \text{ N} \blacktriangleleft$$

PROBLEMS

2.43 A force $\mathbf{F} = 3\mathbf{i} - 12\mathbf{j} + 4\mathbf{k}$ kN acts at the origin O. Determine (a) the scalar components of $\mathbf{F}$, (b) the magnitude of the horizontal component $\mathbf{F}_h$, (c) the unit vector $\boldsymbol{\lambda}_F$ directed along $\mathbf{F}$, (d) the direction angles of $\mathbf{F}$.

2.44* through **2.49** Let $\boldsymbol{\lambda}_{AB}$ be a unit vector directed along $\overrightarrow{AB}$ as shown. Determine (a) $\boldsymbol{\lambda}_{AB}$, (b) the magnitude of the horizontal projection of $\boldsymbol{\lambda}_{AB}$.

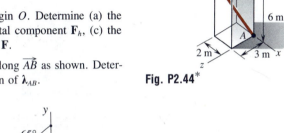

Fig. P2.44*

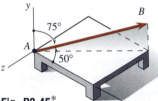

Fig. P2.45*

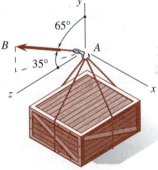

Fig. P2.46

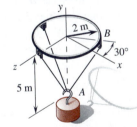

Fig. P2.47

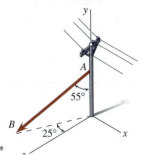

Fig. P2.48*

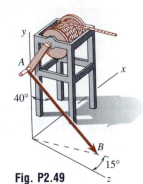

Fig. P2.49

[†]Cf. Sec. B1.3 of App. B.

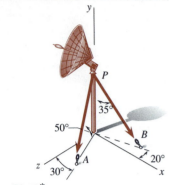

Fig. P2.50*

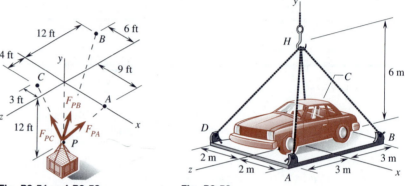

Fig. P2.51 and P2.52   **Fig. P2.53**

2.50* Two guy wires are attached to a pole supporting a dish antenna as shown. If the tensions in the guy wires are $T_{PA} = 400$ N and $T_{PB} = 600$ N, determine their resultant force **R** at P.

2.51 Three forces $\mathbf{F}_{PA}$, $\mathbf{F}_{PB}$, and $\mathbf{F}_{PC}$ are applied to a crate at P as shown. If $F_{PA} = 100$ lb, $F_{PB} = 150$ lb, and $F_{PC} = 195$ lb, determine their resultant force **R** at P.

2.52 Three forces $\mathbf{F}_{PA}$, $\mathbf{F}_{PB}$, and $\mathbf{F}_{PC}$ are applied to a crate at P as shown. If their resultant is $340\mathbf{j}$ lb, determine the magnitude of each force.

2.53 A car is suspended by four cables HA, HB, HC, and HD as shown. If the tension in each cable is 4.9 kN, determine their resultant force **R** at H.

2.9   Concluding Remarks

We saw in this chapter that force vectors may be represented *graphically* or *analytically*. These two basic representations are put to use in presenting the two traditional methods for adding and resolving forces. The first method uses the *parallelogram law*, while the second uses *vector algebra*. In the first method, the laws of cosines and sines along with the parallelogram law are extensively utilized. In the second method, the skill to write the analytical expressions of vectors is a basic requirement. We have chosen to use the vector algebra in handling forces in space because it is generally rather awkward to add or resolve forces in space by the direct application of the parallelogram law.

The knowledge of how forces are added and resolved is fundamental in mechanics. It is a prerequisite for studying the conditions of any physical bodies under the action of forces. All representations and operations of force vectors discussed in this chapter are also applicable to other vector quantities, such as those in dynamics.

REVIEW PROBLEMS

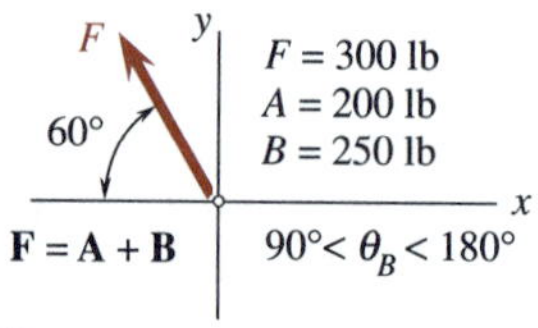

2.54 If $\mathbf{P} = -10\mathbf{i} + 4\mathbf{j}$ N, $\mathbf{Q} = 6\mathbf{i} - 2\mathbf{j}$ N, and $\mathbf{D} = \mathbf{P} - \mathbf{Q}$, determine P, Q, D, θ_P, θ_Q, and θ_D.

2.55 The force **F** is to be resolved into the forces **A** and **B** as indicated. Using the *parallelogram law*, determine θ_A and θ_B.

$F = 300$ lb
$A = 200$ lb
$B = 250$ lb

$\mathbf{F} = \mathbf{A} + \mathbf{B}$ $90° < \theta_B < 180°$

Fig. P2.55

2.56 The force **Q** is to be resolved into the forces **S** and **T** where **S** is parallel to the y axis and **T** is parallel to the line *AB* as indicated. Using *vector algebra*, determine the magnitudes *S* and *T*.

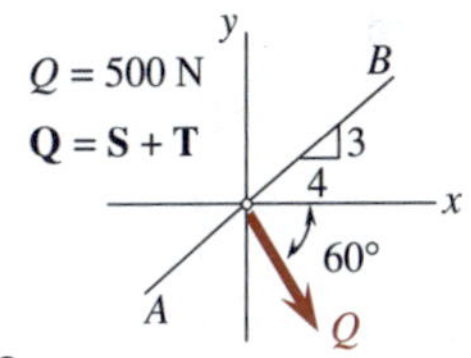

2.57 Three mooring lines are attached to the bow of a ship as shown. If the tensions in the lines are $F_{AB} = 900$ N, $F_{AC} = 1200$ N, and $F_{AD} = 1400$ N, determine the magnitude of their resultant force **R** at *A*.

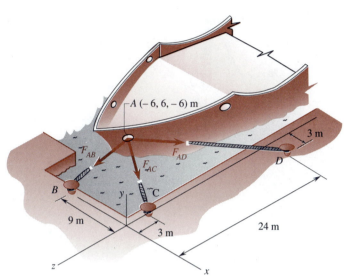

Fig. P2.56

Fig. P2.57

2.58 Three forces act on an eyebolt at *D* as shown, where $F_{DA} = 150$ lb, $F_{DB} = 180$ lb, and $F_{DC} = 140$ lb. Determine their resultant force **R** at *D*.

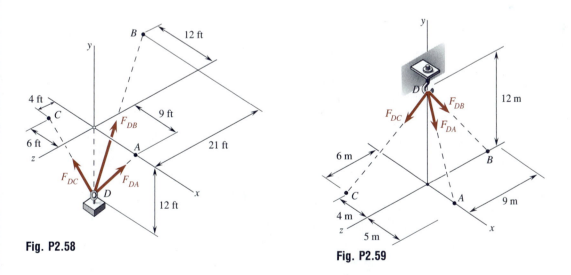

Fig. P2.58

Fig. P2.59

2.59 Three forces act on a hook at *D* as shown. If their resultant force at *D* is **R** = -370**j** N, determine the magnitude of each force.

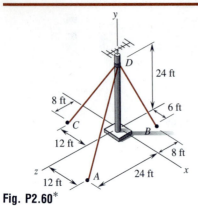

Fig. P2.60*

2.60* The pole for an antenna is held by three guy wires as shown. If the resultant force exerted by these wires on the pole at D is $\mathbf{R} = -168\mathbf{j}$ lb, determine the tension in each wire.

2.61 A transmission tower OD is guyed with three cables as shown. If the resultant force exerted by these cables on the tower at D is $\mathbf{R} = -22.9\mathbf{j}$ kN, determine the tension in each cable.

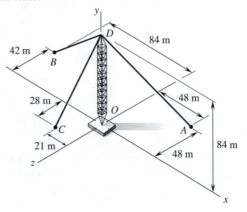

Fig. P2.61

Equilibrium of Particles

Chapter 3

KEY CONCEPTS

- Inertial reference frame, concurrent force system, reaction, internal force, external force, linear spring, space diagram, free-body diagram, and equilibrium of a particle.
- Usual assumptions and properties related to cordlike elements, pulleys, linear springs, rollers, and smooth supports.
- Drawing a good free-body diagram.
- Scalar method, analytical vector method, and semigraphical method for the solution of equilibrium problems of particles in a plane, where the force systems are concurrent and coplanar.
- Method of solution for equilibrium problems of particles in space, where the force systems are concurrent but noncoplanar.

A system of particles in equilibrium in a plane or in space is characterized by the feature that each particle in the system is subjected to a balanced system of forces whose lines of action all intersect at a common point. This chapter covers the equilibrium of systems with a single particle and systems with two or more connected particles. The fundamental concepts of inertial reference frame, reaction, internal force, external force, and free-body diagram are discussed in detail. In particular, helpful guides for drawing a good free-body diagram are given and emphasized. The scalar method, the analytical vector method, and the semigraphical method for solving equilibrium problems of particles in a plane are described and illustrated. Furthermore, a general procedure for solving equilibrium problems of particles in space is presented.

FUNDAMENTALS AND FREE-BODY DIAGRAM

3.1 Inertial Reference Frame

To describe accurately the state of rest or motion of particles or rigid bodies, a proper reference frame must be used. An *inertial reference frame* is a frame of reference in which Newton's laws of motion are valid or highly

Fig. 3.1

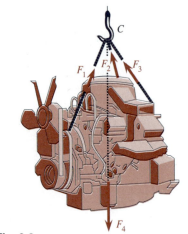

Fig. 3.2

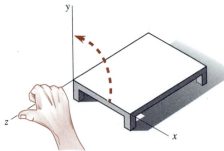

Fig. 3.3

Fig. 3.4

accurate.[†] Therefore, an inertial reference frame is also called a *newtonian reference frame*. Experiments indicate that a nonaccelerating reference frame, which is either stationary or translating uniformly relative to the sun,[††] is generally valid as an inertial reference frame.

A frame of reference in which Newton's laws of motion are only accurate enough for practical purposes is called an *approximate inertial reference frame*. Since the earth rotates about its own axis and revolves around the sun, any reference frame fixed to the earth is in an accelerated motion relative to the sun. However, in most situations, a reference frame fixed on the earth is an approximate inertial reference frame because the errors introduced are generally negligible.

In situations such as the studies of motion of satellites, spacecraft, intercontinental missiles, ocean currents, and hurricanes, where the duration and distance of motion are relatively large, a reference frame fixed on the surface of the earth will become a *poor* approximate inertial reference frame because the errors introduced can be significant. In such situations, *a nonrotating reference frame attached to the center of the earth* serves better as an approximate inertial reference frame.[†††]

For practical purposes, a satisfactory approximate inertial reference frame will simply be referred to as an *inertial reference frame*. In situations under consideration, if *no* coordinate system is shown, it will be assumed that fixed rectangular coordinate systems, oriented as shown in Fig. 3.1 for systems in a plane and in Fig. 3.2 for systems in space, are being used. Note that a set of x, y, and z axes, as shown in Fig. 3.2, is said to form a *right-handed rectangular coordinate system* if when the thumb of the right hand extends along the z axis the other four fingers of the right hand can curl only in the direction in which the x axis could be rotated through 90° to coincide with the y axis.

Developmental Exercise

D3.1   What is an inertial reference frame?

3.2   Concurrent Force Systems

A *concurrent force system* is a system of forces whose lines of action intersect at a common point, which is called the *point of concurrence*. For instance, the point C in Fig. 3.3 is the point of concurrence of the concurrent forces $\mathbf{F}_1$, $\mathbf{F}_2$, $\mathbf{F}_3$, and $\mathbf{F}_4$. A *coplanar concurrent force system* is a concurrent force system where the lines of action of the forces all lie in the same plane. A *collinear force system* is a concurrent force system consisting of forces which have the same line of action, as illustrated in Fig. 3.4.

[†]A *reference frame* is a frame (or body) of reference in which one or more coordinate systems may be embedded. For conciseness, the phrase "in a reference frame" is, from time to time, used to mean "with respect to a reference frame."

[††]More accurately, to the center of mass of the solar system.

[†††]The magnitude of the acceleration of the center of the earth relative to the sun is about 6 mm/s².

Developmental Exercise

D3.2 Define (a) a concurrent force system, (b) a coplanar concurrent force system.

3.3 Reaction, Internal Force, and External Force

We know that an *action* of a body is an exertion of the body on another body. A *reaction* is an induced opposing action; it is a counteraction. Thus, we may regard the action as the cause and the reaction as the effect. *Reactions* from the supports or connections of a body are the *constraining forces* acting on the body to oppose the movements of the body relative to the supports or connections. In both statics and dynamics, an action exerted at a location is always accompanied, at the same location, by a collinear reaction, which is equal in magnitude but opposite in direction to the action.

The relation between action and reaction is described by *Newton's third law* and is illustrated in Fig. 3.5 for a cart of weight W_C supported by a rope and an incline, and in Fig. 3.6 for a girl of weight W_G in a swing. Note that the *weight force* of a body is the attractive force exerted at a distance by the earth on the body. At the same time, the *weight force* of a

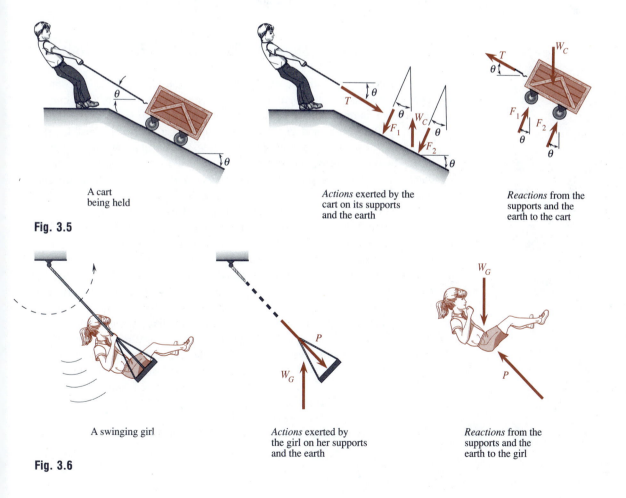

A cart
being held

Actions exerted by the
cart on its supports
and the earth

Reactions from the
supports and the
earth to the cart

Fig. 3.5

A swinging girl

Actions exerted by
the girl on her supports
and the earth

Reactions from the
supports and the
earth to the girl

Fig. 3.6

body may be viewed as a *reaction* from the earth to the attractive force exerted on it by the body.[†]

An *internal force* to a body (which may be a system of connected bodies) is a force of interaction existing *within* the body under consideration. By Newton's third law, internal forces must occur in collinear pairs which are equal in magnitude and opposite in direction; i.e., the sum of internal forces in a system must be zero. An *external force* to a body is a force exerted on the body by another body, which is *outside* the body. Notice that the weight force of a body is an external force to the body.

Developmental Exercises

D3.3 Ascertain whether each of the following sentences gives a true or false statement about the *weight force* of a body. (a) It is the attractive force exerted by the earth on the body. (b) It is equal in magnitude but opposite in direction to the attractive force exerted by the body on the earth. (c) It is the action exerted by the body on the earth. (d) It is the reaction from the earth to the body. (e) It is equal to the mass of the body.

D3.4 Is the coupling force in a two-car train an internal force or an external force to (a) the train, (b) either car?

D3.5 Define the terms: (a) reaction, (b) internal force, (c) external force.

3.4 Cordlike Elements, Pulleys, and Connections

In engineering applications, we frequently use cordlike elements to connect different bodies and exert pulling forces along straight lines or over curved surfaces. Examples of *cordlike elements* include *cords*, *strings*, *ropes*, *wires*, *cables*, *belts*, and *chains*. Unless otherwise specified, cordlike elements will be assumed to be *perfectly flexible*, *inextensible*, and *of negligible mass*. In other words, the internal forces developed in cordlike elements under the action of applied loads are always *tensile forces*, which are collinear with or tangent to their respective cordlike elements.

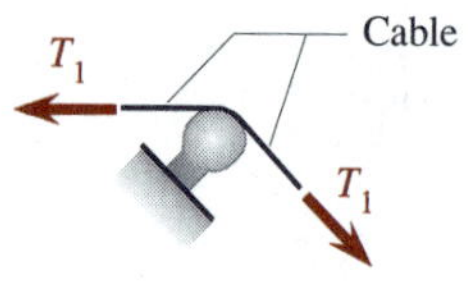

In statics, the letter T with one or two appropriate subscripts (e.g., T_1, T_{AB}) is often used to represent the *magnitude* of the tensile force developed in a cordlike element. This magnitude of the tensile force is usually referred to as the *tension* in the cordlike element. Although the order of two subscripts may indicate the sense of the direction of a vector, it should be remembered that the magnitude of a vector is a scalar quantity. Thus, $\mathbf{T}_{AB} = -\mathbf{T}_{BA}$ but $T_{AB} = T_{BA}$.

From the assumptions stated earlier, it is clear that the *magnitude* of the tensile force is the *same* everywhere in the *same* cordlike element which passes around smooth pegs or pulleys (or sheaves) with frictionless bearings. This property, which is illustrated in Fig. 3.7 for a smooth peg and in Fig. 3.8 for a pulley with frictionless bearing, may be proved using the concepts of rigid-body equilibrium presented in Chap. 5. Unless otherwise specified, *pegs and pulleys will be assumed to be frictionless and the weights of pulleys will be assumed to be negligible*. The words "smooth" and "friction-

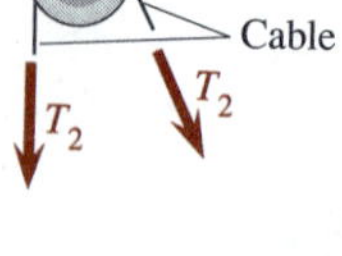

Fig. 3.7

Fig. 3.8

[†]Cf. Newton's law of gravitation in Sec. 1.11.

less'' will be assumed to have the same meaning and hence may be used interchangeably.

Two or more cordlike elements may be fastened together to form a *knot* or be connected to a *small ring* or *connector* (whose weight is taken as negligible) in supporting a load. The tensile forces in *different* cordlike elements, which are connected by a knot or a ring, are generally *different* in magnitude as well as direction. This property is illustrated in Fig. 3.9 for a *knot connection* and in Fig. 3.10 for a *ring connection*.

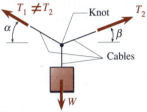

Fig. 3.9

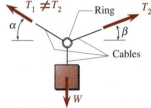

Fig. 3.10

Developmental Exercises

D3.6   What are the usual assumptions and properties related to (a) cordlike elements, (b) pegs, (c) pulleys?

D3.7   A tow truck A is pulling a car B out of a ditch as shown. What is the sense of the reaction (a) from the cable to the car B, (b) from the cable to the truck A?

D3.8   A crate suspended by a system of pulleys is held with a force $\mathbf{P}$ as shown. (a) Is the tension T_{AB} in the cable from A to B equal to P? (b) Is T_{AB} equal to the tension T_{CD} in the cable from C to D? (c) Is $T_{CD} = T_{EF}$? (d) Is $T_{AB} = T_{GH}$?

D3.9   In the system shown, which of the following pairs of sections of string have tensile forces which are equal in magnitude? (a) AB and BC, (b) BC and CD, (c) CD and DE, (d) CD and DF, (e) DE and DF.

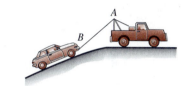

Fig. D3.7

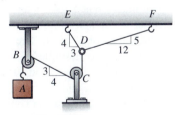

Fig. D3.9

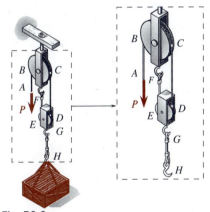

Fig. D3.8

3.5   Linear Springs

Springs are common elastic elements in mechanical systems. In a broader sense, any material which is capable of developing elastic restoring forces may be considered as a spring. A *linear spring* develops a tensile or compressive force whose magnitude is proportional to the magnitude of the change in length (i.e., elongation or contraction) of the spring. Otherwise, the spring is called a *nonlinear spring*. Unless otherwise noted, all springs are taken as linear and have negligible mass.

The *free length* (or *natural length*) of a spring is the length of a spring in its undeformed state.[†] For a spring of free length L experiencing an elongation x as shown in Fig. 3.11(a), the magnitude F of the tensile force developed is proportional to x and is depicted in Fig. 3.11(b). Mathematically, we write

$$F = kx \qquad (3.1)$$

where the constant k is called the *spring modulus* (or *spring constant*). The dimensions of k are *force/length*. Note in Eq. (3.1) that F represents a compression if x is the contraction (or shortening) of the spring.

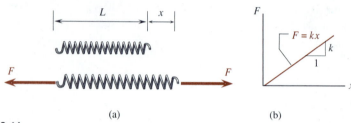

(a)   (b)

Fig. 3.11

Developmental Exercise

D3.10   The spring AB has a spring modulus of 2 kN/m. In the position shown, determine the magnitude F of the spring force if the free length of the spring is (a) 0.4 m, (b) 0.3 m.

3.6   Rollers and Smooth Supports

A *roller* is a support which allows the point of support to move freely parallel to the surface on which the roller rests. For instance, a caster or an unpowered wheel is a roller support. Unless otherwise stated, rollers are assumed to have negligible mass. When *rollers* are used to support a body, the reaction from each roller to the body must be normal (i.e., perpendicular) to the surface on which the roller rests. The normal reactions at the rollers are sometimes combined into a single resultant force to allow the treatment of the body as a particle, as shown in Fig. 3.12, where a cart of weight W is held in equilibrium in (a), and the two reactions from the rollers in (b) are combined into one in (c). However, in some situations (e.g., beam problems in Chap. 5), this treatment may cease to be appropriate.

A friction force on a surface is a tangential force acting on the surface. When the effect of the friction force between a body and its support is negligibly small compared with the effects of other forces, the support may be taken as smooth. If a body rests on a *smooth support*, the reactions from the support to the body must be normal to the surface of the support and are generally distributed over the surface of contact. However, the distributed

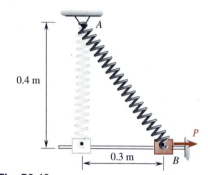

0.4 m

0.3 m

Fig. D3.10

[†]This term is also called the *unstretched length* (or *undeformed length*) of the spring by others.

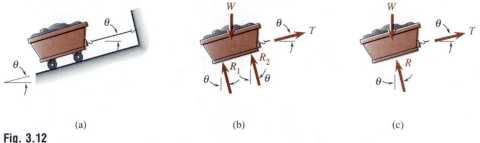

(a)   (b)   (c)

Fig. 3.12

reactions are sometimes combined into a single resultant force to allow the treatment of the body as a particle, as shown in Fig. 3.13, where a block of weight W resting on a smooth incline is held in equilibrium in (a), and the distributed reactions from the incline in (b) are combined into one in (c). Note that $\mathbf{R}$, $\mathbf{T}$, and $\mathbf{W}$ form a concurrent force system as indicated.

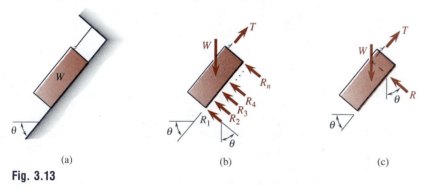

(a)   (b)   (c)

Fig. 3.13

Developmental Exercises

D3.11   How is the reaction from a roller support directed?

D3.12   When may a support be taken as smooth?

3.7   Free-Body Diagram

A *space diagram* is a *sketch* showing the *physical and geometrical conditions* of a system. In other words, a space diagram is a graphical description of the system. It generally shows the shape and size of the system, the weights, the externally applied loads, the connections, and the supports of the system.

The *free-body diagram* of a body is a *sketch* in which we *free* (or isolate) the *body* by removing its supports and connections in the space diagram and then *show* on this freed body all the pertinent weight forces, the externally applied loads, and the reactions from its supports and connections existing just before their removal. The ''freed body'' is often simply called the ''free body.'' A *free-body diagram* may be viewed as an *engineer's ledger* which accounts the system of forces that bring about the state of rest or motion of the body under consideration. A simplified but practical definition of the free-body diagram may be written in the form of an equation as follows:

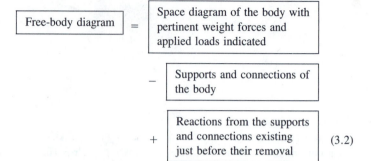

$$\boxed{\text{Free-body diagram}} = \boxed{\begin{array}{l}\text{Space diagram of the body with}\\ \text{pertinent weight forces and}\\ \text{applied loads indicated}\end{array}}$$

$$- \boxed{\begin{array}{l}\text{Supports and connections of}\\ \text{the body}\end{array}}$$

$$+ \boxed{\begin{array}{l}\text{Reactions from the supports}\\ \text{and connections existing}\\ \text{just before their removal}\end{array}} \qquad (3.2)$$

The importance of *drawing free-body diagrams* cannot be overemphasized for mechanics students. Those who develop the habit of drawing appropriate free-body diagrams usually find their study of mechanics considerably better organized, more at their command, and easier.

EXAMPLE 3.1

A crate of weight W is supported by three ropes in a ring connection as shown. Draw the free-body diagrams of (a) the crate and the ring combination, (b) the crate, (c) the ring.

Solution.   The free-body diagrams are drawn as shown.

REMARK. Note that (1) all forces developed in the ropes are tensile forces, (2) the tensile force in the rope CD is not shown in case (a) because it is an internal force to the crate and the ring combination, (3) the tensile force in the rope CD becomes an external force to the free body of just the crate or just the ring and is shown in cases (b) and (c), (4) when ropes are cut, portions of them are usually shown on the free body to help us identify the reactions from the connections existing just before their removal.

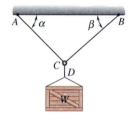

Space diagram

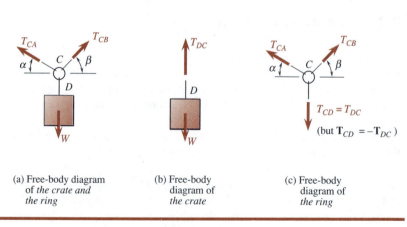

(a) Free-body diagram of *the crate and the ring*

(b) Free-body diagram of *the crate*

(c) Free-body diagram of *the ring*

Developmental Exercises

D3.13 Define the terms: (a) space diagram, (b) free-body diagram.

D3.14 Refer to the free-body diagrams drawn in Example 3.1. Is each of the "freed particles" shown as a *point* or as a *body with a shape* as depicted in the space diagram?

3.8 Guides for Drawing a Good Free-Body Diagram

A *good* free-body diagram is a sketch which *correctly* and *effectively* exhibits the system of forces that brings about the state of rest or motion of the body under consideration. The following are helpful guides for drawing a good free-body diagram:

1. The *body to be freed* (or isolated) for consideration may be the entire system or any portion of the system; however, it is important to make a clear decision as to which portion of the system is to be freed.
2. The *free body* drawn should have no external supports or connections.
3. Any *adopted coordinate system* whose axes are not in the horizontal and vertical directions should be shown.
4. Appropriate *dimensions* (including slopes or angles) or the *coordinates* of certain key points, which are needed in defining the configuration of the force system, should be indicated.
5. Each *applied load* is externally applied to act on the body and should be indicated with an arrow and labeled either with its known magnitude or with a letter when it is not known.
6. The *weight force* of the free body is an external force exerted by the earth on the body and should be indicated with a *vertical* downward arrow and labeled like an applied load if the weight is not negligible.
7. The *actions exerted by the body* on its supports and connections existing just before their removal *should not be indicated* on the free body in the diagram. (It should be noted that such actions are exerted *by*, not *on*, the body. Only those forces which are exerted *on* the body by other bodies are to be indicated in the free-body diagram.)
8. The *reactions* (or components of the reactions) from the supports and connections existing just before their removal should be indicated with arrows and labeled like the applied loads. (This is obviously essential in making the free body exhibit the same condition of rest or motion as that of the body shown in the space diagram.)
9. The *sense* of an unknown force, when not reasonably obvious, may be assumed and corrected later if the value of the variable representing the magnitude of the force is subsequently found to be negative.[†]
10. The *forces in the cords or members which are uncut* in the free body are internal forces to the free body and *should not be shown*.

[†]If the value of an unknown quantity is found to be negative in the solution, the magnitude of this unknown quantity is simply equal to the absolute value of the solved value, but the true sense of this unknown quantity is opposite to that initially assumed. However, this practice may cease to be valid when the unknown quantity is a *friction force* at a location where slipping *impends* or *occurs*. The sense of a friction force is explained in Sec. 9.2.

Developmental Exercises

D3.15 Is it true that the weight force of a body is an internal force to the body and should be excluded from the free-body diagram of the body?

D3.16 Which of the following should be indicated in the free-body diagram of a body? (a) The actions exerted by this body on its surrounding bodies, (b) the reactions from the surrounding bodies to the actions exerted by this body, (c) the forces in the cords or members which are uncut in the free body.

EXAMPLE 3.2

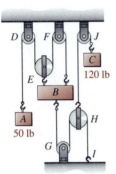

Three cylinders *A*, *B*, and *C* suspended by a system of cables and pulleys are in equilibrium as shown. The weights of the cylinders *A* and *C* are indicated. Denoting the weight of the cylinder *B* by W_B, draw the free-body diagrams of (a) the cylinder *A*, (b) the cylinder *C*, (c) the cylinder *B* and the pulley *E* combination.

Solution. By the usual assumptions, we note that (1) the weights of the pulleys are negligible, (2) the magnitude of the tensile force is the same everywhere in the same cable which passes around pulleys, (3) we may use T_1, T_2, and T_3 to denote the tensions in the cables *ADEFB*, *HJC*, and *BGHI*, respectively. Thus, we may draw the required free-body diagrams as shown.

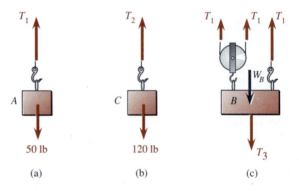

Developmental Exercises

D3.17 Draw the free-body diagram of the pulley *H* in Example 3.2.

D3.18 Draw the free-body diagram of the ring *A* in Example 3.3.

EXAMPLE 3.3

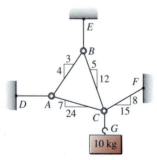

A 10-kg block is hung with a system of wires and small rings, as shown. Draw the free-body diagrams of (a) the block, (b) the ring *B*, (c) the ring *C*.

Solution. Since all forces in the wires are tensile forces, we may draw the required free-body diagrams as shown.[†]

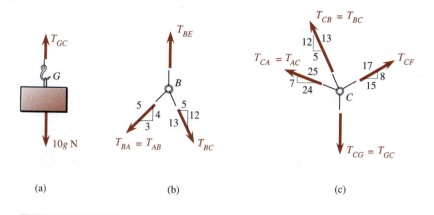

(a) (b) (c)

PROBLEMS

3.1 through 3.6* A crate is supported by three cables in a ring connection as shown. Draw the free-body diagrams of (a) the crate, (b) the ring, (c) the crate and the ring combination.

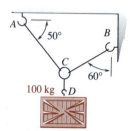

Fig. P3.1 and P3.17

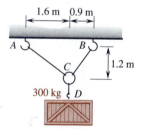

Fig. P3.2* and P3.18

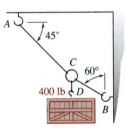

Fig. P3.3* and P3.19*

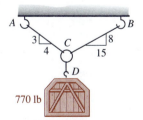

Fig. P3.4 and P3.20

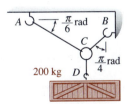

Fig. P3.5* and P3.21

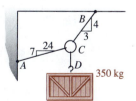

Fig. P3.6* and P3.22*

[†]Note in this text that *g* which is not separated from a number by a space, nor stands alone, is *not* the abbreviation for the unit *gram*, but represents the value of 9.81 (m/s²) in problems with SI units.

Fig. P3.7*

Fig. P3.8*

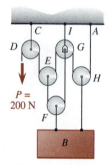

Fig. P3.11*, P3.14*, and P3.27*

3.7* A man exerts a horizontal force **P** to hold a 60-lb ice block in equilibrium on a ramp as shown. Neglecting the friction force between the ice block and the ramp, draw the free-body diagram of the ice block.

3.8* A boy standing on a skateboard on a steep hill is held in equilibrium by a force **P** as shown. The boy and his skateboard have a mass of 40 kg. Draw the free-body diagram of the boy and his skateboard.

3.9 through 3.11* A block of mass m_B is being held in equilibrium by the force **P** through a system of ropes and pulleys as shown. (a) How many different ropes are there in the system? (b) Using T_i ($i = 1, 2, \ldots, n$) to represent the tensions in different ropes, draw the free-body diagrams of (i) the block B, (ii) each pulley, separately.

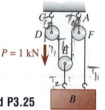

Fig. P3.9, P3.12, and P3.25

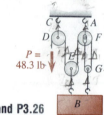

Fig. P3.10*, P3.13*, and P3.26

3.12 Draw the free-body diagram of the block B and the pulley E combination.

3.13* Draw the free-body diagram of the block B and the pulleys E and G combination.

3.14* Draw the free-body diagram of the block B and the pulleys E, F, and H combination.

3.15* and 3.16* The package at G is supported by wires and rings as shown. Draw the free-body diagrams of (a) the package, (b) each ring, separately.

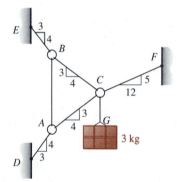

Fig. P3.15* and P3.43

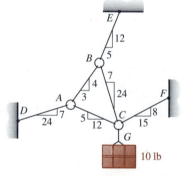

Fig. P3.16* and P3.44*

EQUILIBRIUM IN A PLANE[†]

3.9 Equilibrium of Particles in a Plane

If a body is originally in equilibrium in the space diagram, the *free body* of this body must also be in equilibrium under the action of the forces indicated in the free-body diagram. Since all pairs of collinear forces having equal

---

[†]*Equilibrium in a plane* is equilibrium in a two-dimensional space, and is also termed *equilibrium in two dimensions*.

magnitudes and opposite senses are balanced themselves, they can have no net external effect on a particle or a rigid body on which they act. Such forces may be considered as *trivial forces*. The *nontrivial forces* acting on a particle must be *concurrent forces*, regardless of the amount and distribution of mass of the body which is being treated as a particle.[†]

The *equilibrium of a particle* is the state of the particle in which the particle is at rest or moving with a constant velocity in an inertial reference frame. We know from Newton's second law that the acceleration of the particle is zero if the resultant force acting on it is zero. Since zero acceleration means constant (including zero) velocity, which is a manifestation of equilibrium, we see that the *necessary and sufficient condition* for the equilibrium of a single particle is that the resultant force **R** of all known and unknown forces $\mathbf{F}_1, \mathbf{F}_2, \ldots, \mathbf{F}_n$ indicated in the free-body diagram of the particle equals zero. Since $\mathbf{R} = \Sigma\mathbf{F}$, the *necessary and sufficient condition* for equilibrium of a particle is

$$\Sigma\mathbf{F} = \mathbf{0} \qquad (3.3)$$

which is also called the *vector equation of equilibrium* of a particle.

An *equilibrium problem of particles in a plane* is characterized by the feature that at least one set of coplanar, concurrent, but noncollinear unknown forces acting on a particle will be encountered in the solution; however, no set of noncoplanar unknown forces will be encountered in that solution. For each particle in equilibrium in the *xy* plane, a set of *scalar equations of equilibrium* may be written from Eq. (3.3) as

$$\Sigma F_x = 0 \qquad \Sigma F_y = 0 \qquad (3.4)$$

It is to be noted that the *x* and *y* axes for Eqs. (3.4) may be chosen to be any two orthogonal axes in the plane of the forces. This is illustrated in Example 3.4, which serves as a forerunner of the scalar method of solution presented in Sec. 3.10.

Forces in some equilibrium problems involving pulleys are parallel, rather than concurrent, forces. However, many of such problems may simply be solved by applying one of the two equations in Eqs. (3.4). For this reason, some of them are included in this chapter.

EXAMPLE 3.4

A girl is sitting in an old automobile tire which is suspended as shown. If the girl and the tire together have a mass of 60 kg, determine the tension T_{CA} in the rope *CA*.

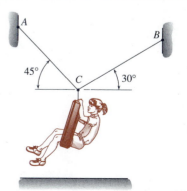

[†]Cf. Example 3.1 and the concept of particle discussed in Sec. 1.6.

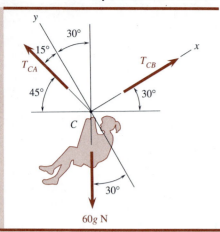

Solution. By choosing the x axis to be along the unknown tensile force $\mathbf{T}_{CB}$ in the free-body diagram shown, we can solve for T_{CA} from a single equilibrium equation as follows:

$$+\nwarrow \Sigma F_y = 0: \quad T_{CA} \cos 15° - 60g \cos 30° = 0$$

$$T_{CA} = 528 \text{ N} \blacktriangleleft$$

Developmental Exercises

D3.19 Solve for the tension T_{CB} in Example 3.4 in a similar manner.

D3.20 Is the resultant force acting on the free body necessarily equal to zero if the body in the space diagram is in equilibrium?

D3.21 What is the characterizing feature in an equilibrium problem of particles in a plane?

D3.22 What is the necessary and sufficient condition for the equilibrium of a particle?

3.10   Scalar Method of Solution[†]

A *scalar method of solution* places emphasis on the scalar components of vectors and makes use of scalar equations, rather than vector equations. Although certain problems can best be solved by employing a special set of steps, the following procedure is generally helpful in *solving equilibrium problems of particles in a plane*:

1. *Analyze the problem situation* carefully and make preliminary computations from the given geometric and material properties if needed.
2. *Indicate the chosen xy rectangular coordinate axes* if they are not in the horizontal and vertical directions.
3. *Draw correctly the relevant free-body diagram* of a particle where no more than two unknowns will be involved. [This is desirable because no more than two independent equilibrium equations can be obtained in step (4) below.] If this step is impossible, go to step (7) below.
4. *Sum to zero the scalar components of all forces* in the x and y directions, respectively, to obtain a set of two scalar equations.
5. *Solve for the unknowns* in the equations obtained in step (4).
6. *Repeat steps (2) through (5)* for other particles until the desired unknown or unknowns have been determined.

[†]Students are encouraged to be creative and to treat *any* method outlined in this text as only a "guide" in solving a class of problems. The guide is intended to help students get started. There is no need to follow the suggested procedure in a rigid fashion.

7. If the system consists of *n* particles (some of which may have negligible weight) and *the free-body diagram drawn for any of the n particles in the system involves more than two unknowns*, then we draw the free-body diagram of each particle in the system. We may write two scalar equilibrium equations for each particle as described in step (4) and obtain a total of 2*n* simultaneous equations. If the total number of unknowns equals 2*n*, the unknowns may be determined.

8. *Add appropriate units to the final answers.*

EXAMPLE 3.5

Two identical springs *AB* and *BC* are connected by a hook at *B* and stretched horizontally between two posts with a tension of 200 N. A 35-kg boy grabs the hook at *B* and hangs from it as shown. When equilibrium exists, the deflection $\overline{BB'}$ of the hook is found to be 41 mm. Determine the spring modulus *k* of each spring.

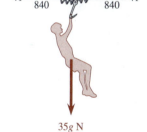

Solution. By virtue of symmetry (or by applying $\Sigma F_x = 0$), we know that the tensions in the two springs must be the same. For simplicity in the solution, let the spring modulus of each spring be *k* N/mm, the free length of each spring be *L* mm, the elongation of each spring in the horizontal position be x_1 mm, the elongation of each spring in the inclined position be x_2 mm, and the magnitude of the tension in each spring in the inclined position be T_2. Applying the Pythagorean theorem, we write

$$\overline{AB'} = [(840)^2 + (41)^2]^{1/2} = 841$$

The initial tension in each spring is $T_1 = 200$ N. By the property of linear springs, we write

$$T_1 = kx_1 = 200 \qquad L = \overline{AB} - x_1 = 840 - \frac{200}{k}$$

$$x_2 = \overline{AB'} - L = 1 + \frac{200}{k} \qquad T_2 = kx_2 = k + 200$$

From the free-body diagram of the boy and the hook combination, we write

$$+\uparrow \ \Sigma F_y = 0: \qquad 2T_2\left(\frac{41}{841}\right) - 35g = 0$$

Substituting the expression of T_2 into this equilibrium equation and noting that $g = 9.81$ (m/s^2), we write

$$2(k + 200)\left(\frac{41}{841}\right) - 35g = 0 \qquad k = 3321 \ \text{N/mm}$$

$$k = 3.32 \ \text{MN/m} \ \blacktriangleleft^\dagger$$

Developmental Exercise

D3.23   Solve for *k* in Example 3.5 if the mass of the boy is 40 kg.

---

†In writing the final answer, it is recommended that SI prefixes be used only in the numerator, except the base unit kilogram.

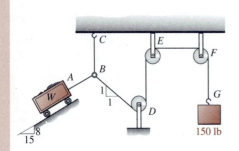

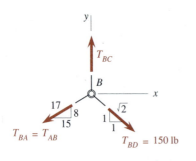

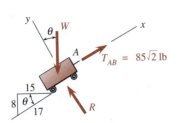

EXAMPLE 3.6

A cart and a block are connected and supported by a system of cables, pulleys, rollers, and a small ring as shown. The weight of the block at G is 150 lb. Determine the weight W of the cart at A consistent with equilibrium.

Solution. We will not start with the free-body diagram of the cart at A because it involves three unknowns: the weight W, the magnitude T_{AB} of the tension in the cable AB, and the magnitude R of the resultant of the roller reactions. Since the cable $BDEFG$ supports the 150-lb block at G and passes around pulleys D, E, and F, it is clear that the tension in the cable BD must be 150 lb. Thus, the free-body diagram of the ring at B involves only two unknown magnitudes T_{AB} and T_{BC} of the tensile forces in the cables AB and BC as shown. For equilibrium, we write

$$\xrightarrow{+} \Sigma F_x = 0: \quad \frac{1}{\sqrt{2}}(150) - \frac{15}{17}T_{BA} = 0 \qquad T_{BA} = 85\sqrt{2}\ \text{lb}$$

Having determined T_{BA}, we are now ready to consider the cart at A as shown. Note in the free-body diagram of the cart at A that the angle between the vertical line of action of the weight force $\mathbf{W}$ and the y axis is θ and from the slope triangle that $\sin\theta = 8/17$. For equilibrium, we write

$$+\nearrow \Sigma F_x = 0: \quad 85\sqrt{2} - W\sin\theta = 0$$

$$W = 255\ \text{lb} \quad \blacktriangleleft$$

Developmental Exercise

D3.24   Solve for the tension T_{BC} in Example 3.6.

EXAMPLE 3.7

Refer to Example 3.2. Determine the weight W_B of the block B necessary to hold the system in equilibrium.

Solution.   The relevant free-body diagrams for the solution of the problem are first drawn as shown. For equilibrium of *the block A*, we write

$$+\uparrow \Sigma F_y = 0: \quad T_1 - 50 = 0 \qquad T_1 = 50\ \text{lb}$$

For equilibrium of *the block C*, we write

$$+\uparrow \Sigma F_y = 0: \quad T_2 - 120 = 0 \qquad T_2 = 120\ \text{lb}$$

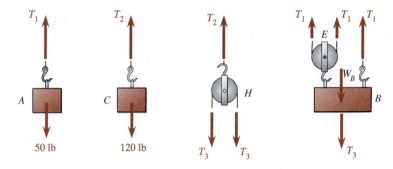

For equilibrium of *the pulley H*, we write

$$+\uparrow \ \Sigma F_y = 0: \quad T_2 - T_3 - T_3 = 0 \qquad T_3 = \tfrac{1}{2} T_2 = 60 \text{ lb}$$

For equilibrium of *the block B and the pulley E*, we write

$$+\uparrow \ \Sigma F_y = 0: \quad T_1 + T_1 + T_1 - W_B - T_3 = 0$$

$$W_B = 3T_1 - T_3 = 90 \qquad W_B = 90 \text{ lb} \blacktriangleleft$$

EXAMPLE 3.8

Refer to Example 3.3. Determine the tension in each wire of the system.

Solution. The relevant free-body diagrams for the solution of the problem are first drawn as shown.

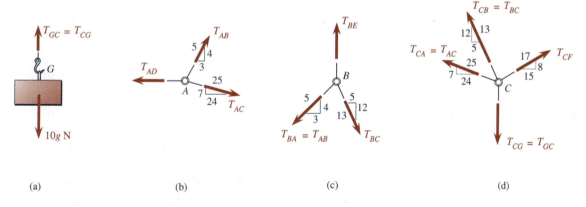

(a) (b) (c) (d)

For equilibrium of *the block at G*, we write

$$+\uparrow \ \Sigma F_y = 0: \quad T_{CG} - 10g = 0 \qquad T_{CG} = 98.1 \text{ N} \blacktriangleleft$$

Since the free-body diagram drawn for any of the three small rings involves three (not two) unknowns, we should solve the rest of this problem according to step (7) in Sec. 3.10. From the three free-body diagrams of the rings A, B, and C, we see a total of six unknowns: T_{AB}, T_{AC}, T_{AD}, T_{BC}, T_{BE}, and T_{CF}. Moreover, from each of these three free-body diagrams, we may write two independent force equilibrium equations and obtain a total of six independent equations in six unknowns. Thus, the six unknowns may be determined. For equilibrium of *the rings* as shown in their free-body diagrams, we write

Ring A

$$\xrightarrow{+} \Sigma F_x = 0: \qquad \frac{3}{5} T_{AB} + \frac{24}{25} T_{AC} - T_{AD} = 0 \qquad (1)$$

$$+ \uparrow \Sigma F_y = 0: \qquad \frac{4}{5} T_{AB} - \frac{7}{25} T_{AC} = 0 \qquad (2)$$

Ring B

$$\xrightarrow{+} \Sigma F_x = 0: \qquad -\frac{3}{5} T_{AB} + \frac{5}{13} T_{BC} = 0 \qquad (3)$$

$$+ \uparrow \Sigma F_y = 0: \qquad -\frac{4}{5} T_{AB} - \frac{12}{13} T_{BC} + T_{BE} = 0 \qquad (4)$$

Ring C

$$\xrightarrow{+} \Sigma F_x = 0: \qquad -\frac{24}{25} T_{AC} - \frac{5}{13} T_{BC} + \frac{15}{17} T_{CF} = 0 \qquad (5)$$

$$+ \uparrow \Sigma F_y = 0: \qquad \frac{7}{25} T_{AC} + \frac{12}{13} T_{BC} + \frac{8}{17} T_{CF} - 10g = 0 \qquad (6)$$

The preceding six equations may be written in the form of a matrix equation as follows:[†]

$$
\begin{bmatrix}
\frac{3}{5} & \frac{24}{25} & -1 & 0 & 0 & 0 \\
\frac{4}{5} & -\frac{7}{25} & 0 & 0 & 0 & 0 \\
-\frac{3}{5} & 0 & 0 & \frac{5}{13} & 0 & 0 \\
-\frac{4}{5} & 0 & 0 & -\frac{12}{13} & 1 & 0 \\
0 & -\frac{24}{25} & 0 & -\frac{5}{13} & 0 & \frac{15}{17} \\
0 & \frac{7}{25} & 0 & \frac{12}{13} & 0 & \frac{8}{17}
\end{bmatrix}
\begin{bmatrix}
T_{AB} \\
T_{AC} \\
T_{AD} \\
T_{BC} \\
T_{BE} \\
T_{CF}
\end{bmatrix}
=
\begin{bmatrix}
0 \\
0 \\
0 \\
0 \\
0 \\
10g
\end{bmatrix}
$$

Note that *the elements in the rows of the coefficient matrix correspond to coefficients of the unknowns as arranged in the unknown column matrix.* The solution of these simultaneous equations yields:

$$T_{AB} = 24.4 \text{ N} \qquad T_{AD} = 81.5 \text{ N} \qquad T_{BE} = 54.6 \text{ N} \quad \blacktriangleleft$$

$$T_{AC} = 69.7 \text{ N} \qquad T_{BC} = 38.0 \text{ N} \qquad T_{CF} = 92.4 \text{ N} \quad \blacktriangleleft$$

[†]The matrix representation of simultaneous equations is employed to facilitate the solution with a computer as demonstrated in App. D. No other background in matrix algebra is required for understanding the rest of this text.

Developmental Exercises

D3.25   What is the maximum number of independent equilibrium equations we can write for a particle in equilibrium in a plane?

D3.26   If a system of particles in equilibrium in a plane is to be analyzed and we cannot find a single particle whose free-body diagram involves no more than two unknowns, what should we do?

3.11   Analytical Vector Method of Solution

The *analytical vector method of solution* for equilibrium problems of particles in a plane may be formulated by amending the procedure given in Sec. 3.10 in such a way that step (4) is replaced by the following two *alternative steps*:

4a. *Add the analytical expressions of all forces* shown in the free-body diagram of the particle to obtain the resultant force in the *xy* plane in terms of known and unknown quantities.

4b. *Set each of the two scalar components* of the resultant force, obtained in step (4a), *equal to zero* to obtain a set of two scalar equations. (This step follows from the requirement that each of the scalar components of a zero vector must be equal to zero. For a body in equilibrium, the resultant force acting on it is, of course, a zero vector.)

The use of vector analysis frequently leads to more concise formulations and derivations of the fundamental principles of mechanics. This method is illustrated in Example 3.9.

Developmental Exercise

D3.27   In solving an equilibrium problem of particles in a plane by the analytical vector method, why do we set the two scalar components of the resultant force equal to zero?

3.12   Semigraphical Method of Solution

When a particle is in equilibrium in a plane, the resultant force of all coplanar concurrent forces shown in the free-body diagram of the particle must be equal to zero. By the polygon rule in Sec. 2.2, we can readily conclude that a *closed figure* or *polygon* will always be obtained when all the forces shown in the free-body diagram of the particle in equilibrium are drawn to scale and arranged in tip-to-tail fashion.

By graphics, the unknowns associated with an equilibrium problem may, of course, be determined by direct measurement from the force polygon drawn. However, the *unknowns* may also be *determined from the force polygon by using geometry and trigonometry*. This latter approach is called a *semigraphical method of solution*; it is useful in solving simple equilibrium problems of particles in a plane, for which the force polygons are in the form of triangles.

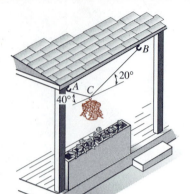

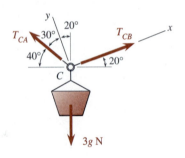

EXAMPLE 3.9

A 3-kg flowerpot is suspended as shown. Determine the tensions T_{CA} and T_{CB} in the wires CA and CB by (a) the scalar method, (b) the analytical vector method, (c) the semigraphical method.

Solution. Regardless of the method to be used, the free-body diagram of the flowerpot and the ring combination is drawn as shown. Note that the x axis is chosen to be collinear with the unknown force $\mathbf{T}_{CB}$ to simplify the algebraic work in the solution.

(a) *Scalar method.* For equilibrium of the free body as shown, we write

$$+\nwarrow \ \Sigma F_y = 0: \quad T_{CA} \cos 30° - 3g \cos 20° = 0$$

$$+\nearrow \ \Sigma F_x = 0: \quad T_{CB} - T_{CA} \sin 30° - 3g \sin 20° = 0$$

Solving these equations, we obtain

$$T_{CA} = 31.9 \text{ N} \qquad T_{CB} = 26.0 \text{ N} \ \blacktriangleleft$$

(Note the *algebraic advantage* in applying $\Sigma F_y = 0$ before $\Sigma F_x = 0$ in the above solution.)

(b) *Analytical vector method.* From the free-body diagram, we write

$$\mathbf{T}_{CB} = T_{CB}\boldsymbol{\lambda}_{CB} = T_{CB}\mathbf{i}$$

$$\mathbf{T}_{CA} = T_{CA}\boldsymbol{\lambda}_{CA} = T_{CA}(-\sin 30°\mathbf{i} + \cos 30°\mathbf{j})$$

$$\mathbf{W} = W\boldsymbol{\lambda}_W = 3g(-\sin 20°\mathbf{i} - \cos 20°\mathbf{j})$$

For equilibrium, we write

$$\Sigma\mathbf{F} = \mathbf{T}_{CB} + \mathbf{T}_{CA} + \mathbf{W} = \mathbf{0}:$$

$$(T_{CB} - T_{CA}\sin 30° - 3g\sin 20°)\mathbf{i} + (T_{CA}\cos 30° - 3g\cos 20°)\mathbf{j} = \mathbf{0}$$

Equating the coefficients of $\mathbf{i}$ and $\mathbf{j}$ to zero, we have

$$T_{CB} - T_{CA}\sin 30° - 3g\sin 20° = 0$$

$$T_{CA}\cos 30° - 3g\cos 20° = 0$$

Solving these equations, we obtain

$$T_{CA} = 31.9 \text{ N} \qquad T_{CB} = 26.0 \text{ N} \ \blacktriangleleft$$

(c) *Semigraphical method.* Referring to the free-body diagram in (a), we first draw the $3g$-N weight force. The tensile forces $\mathbf{T}_{CA}$ and $\mathbf{T}_{CB}$ have known directions and are, therefore, drawn as shown to form a force triangle. Indicating the angles in the triangle and applying the law of sines, we write

$$\frac{3g}{\sin 60°} = \frac{T_{CA}}{\sin 70°} = \frac{T_{CB}}{\sin 50°}$$

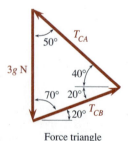

Force triangle

$$T_{CA} = 31.9 \text{ N} \qquad T_{CB} = 26.0 \text{ N} \ \blacktriangleleft$$

Finally, we may check the answers by summing horizontal and vertical scalar components of the resultant force as follows:

$$\Sigma F_{horiz} = T_{CB} \cos 20° - T_{CA} \cos 40°$$

$$= 26.0 \cos 20° - 31.9 \cos 40° \approx 0$$

$$\Sigma F_{vert} = T_{CB} \sin 20° + T_{CA} \sin 40° - 3g$$

$$= 26.0 \sin 20° + 31.9 \sin 40° - 3g \approx 0$$

PROBLEMS

3.17 through 3.22* *(The figures for these problems are specified along with those for Probs. 3.1 through 3.6* on p. 61.)* For the system shown, determine the tension in the cables *CA* and *CB*.

3.23 Determine the magnitude of **P** in Prob. 3.7*.

3.24* Determine the tension in the rope in Prob. 3.8*.

3.25 through 3.27* *(The figures for these problems are specified along with those for Probs. 3.9 through 3.11* on p. 62.)* For the system shown, determine the mass m_B of the block *B* consistent with equilibrium.

3.28 A welded joint is in equilibrium as shown. Determine the magnitudes of the forces **P** and **Q**.

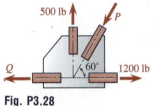

Fig. P3.28

Fig. P3.29

3.29 An instrument on a base is to be isolated from floor vibrations as shown. If the instrument and base together have a mass of 2 kg and $\theta = 60°$, find the magnitudes S_1 and S_2 of the forces in the springs when the instrument is in equilibrium.

3.30 A 30-kg boy grabs a wire and hangs from it as shown. If $\theta = 8°$ when the boy is in equilibrium, determine the tension T in the wire.

3.31 A 40-kg cart rests on an incline and is held by a cable which passes over a smooth pulley and has one end fastened to a scale as shown. Determine the angle θ which the incline makes with the horizontal if the scale indicates that the tension in the cable is 300 N.

Fig. P3.30

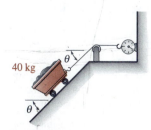

Fig. P3.31

3.32 A crate of newly cured concrete cylinders in a laboratory is lifted with a force **P** as shown. Determine the magnitude of **P** when the crate is held in equilibrium.

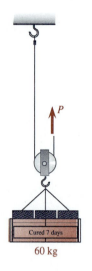

Fig. P3.32   (a)   (b)   (c)   (d)

3.33* A collar C sliding on a smooth horizontal rod is connected to an anchor at A by a spring which has a modulus of 5 kN/m. When the collar is directly above the anchor, the tension in the spring is 100 N. Determine the magnitude of the force **P** required to hold the collar in equilibrium as shown if $a = 50$ mm.

3.34 A 140-kg crate is hung on the hook of a crane as shown. If the length of the cable sling is 1.98 m and the crate is in equilibrium, determine the tension in the cable sling.

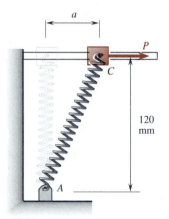

Fig. P3.33*

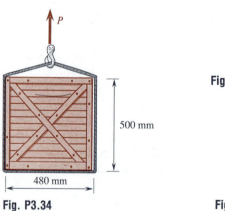

Fig. P3.34

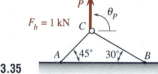

Fig. P3.35

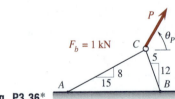

Fig. P3.36*

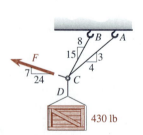

Fig. P3.37 and P3.38

3.35 and 3.36* The breaking strength of each of the two wires AC and BC is F_b as indicated. Determine the maximum magnitude of **P** and its directional angle θ_P for which neither wire will break.

3.37 Determine the tension in each cable if $F = 250$ lb and the crate shown is in equilibrium.

3.38 Determine the range of values of F for which no cable will become slack and the crate shown remains in equilibrium.

3.39 through 3.41* Determine the mass of the body B necessary to hold the system shown in equilibrium.

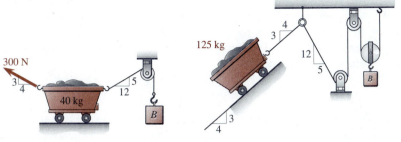

Fig. P3.39

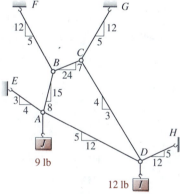

Fig. P3.40

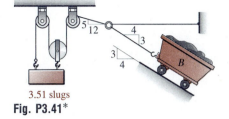

Fig. P3.41*

3.42 The tension in the cable CE as indicated by the scale is 630 N. Determine the masses of the carts A and B consistent with equilibrium.

3.43 and 3.44* *(The figures for these problems are specified along with those for Probs. 3.15* and 3.16* on p. 62.)* Determine the tension in each wire.[†]

3.45 Two bodies are hung as shown. Determine the tension in each wire.[†]

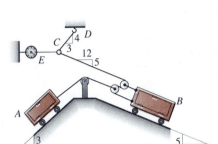

Fig. P3.42

Fig. P3.45

EQUILIBRIUM IN SPACE[††]

3.13 Equilibrium of Particles in Space

The equilibrium of a particle has been discussed in Sec. 3.9. An *equilibrium problem of particles in space* is characterized by the feature that at least one set of *noncoplanar but concurrent* unknown forces acting on a particle will be encountered in the solution. The *necessary and sufficient condition* for equilibrium of a particle in space is still given by the *vector equation of*

[†]Cf. App. D for computer solution of simultaneous equations.

[††]*Equilibrium in space* is equilibrium in three-dimensional space, and is also termed *equilibrium in three dimensions*.

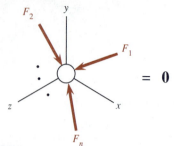

Fig. 3.14

equilibrium in Eq. (3.3); i.e.,

$$\Sigma \mathbf{F} = \mathbf{0} \tag{3.3}$$
(repeated)

where $\Sigma \mathbf{F}$ represents the summation (or the resultant) of all the known and unknown noncoplanar but concurrent forces $\mathbf{F}_1, \mathbf{F}_2, \ldots, \mathbf{F}_n$ which appear in the free-body diagram of the particle as illustrated in Fig. 3.14.

For each particle in equilibrium in space, a set of *scalar equations of equilibrium* may be written from Eq. (3.3) as

$$\Sigma F_x = 0 \qquad \Sigma F_y = 0 \qquad \Sigma F_z = 0 \tag{3.5}$$

The unknowns in the problem are solved from the simultaneous scalar equations in Eqs. (3.5). This is illustrated in Example 3.10, which serves as a forerunner of the method of solution presented in Sec. 3.14.

EXAMPLE 3.10

A 13-Mg crate containing a piece of equipment is being held by three cranes with cables joined at the ring D as shown. Determine the tensions in the cables DA, DB, and DC.

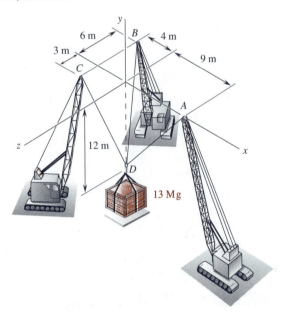

Solution. We first draw the free-body diagram of the crate and the ring combination as shown. From this, we write

$$\overrightarrow{DA} = 9\mathbf{i} + 12\mathbf{j} \qquad\qquad |\overrightarrow{DA}| = 15$$

$$\lambda_{DA} = \frac{1}{5}\,(3\mathbf{i} + 4\mathbf{j})$$

$$\overrightarrow{DB} = -4\mathbf{i} + 12\mathbf{j} - 6\mathbf{k} \qquad\qquad |\overrightarrow{DB}| = 14$$

$$\lambda_{DB} = \frac{1}{7}(-2\mathbf{i} + 6\mathbf{j} - 3\mathbf{k})$$

$$\overrightarrow{DC} = -4\mathbf{i} + 12\mathbf{j} + 3\mathbf{k} \qquad |\overrightarrow{DC}| = 13$$

$$\lambda_{DC} = \frac{1}{13}(-4\mathbf{i} + 12\mathbf{j} + 3\mathbf{k})$$

$$\mathbf{T}_{DA} = T_{DA}\lambda_{DA} = \frac{T_{DA}}{5}(3\mathbf{i} + 4\mathbf{j})$$

$$\mathbf{T}_{DB} = T_{DB}\lambda_{DB} = \frac{T_{DB}}{7}(-2\mathbf{i} + 6\mathbf{j} - 3\mathbf{k})$$

$$\mathbf{T}_{DC} = T_{DC}\lambda_{DC} = \frac{T_{DC}}{13}(-4\mathbf{i} + 12\mathbf{j} + 3\mathbf{k})$$

$$\mathbf{W} = W\lambda_W = 13g(-\mathbf{j}) = -13g\mathbf{j}$$

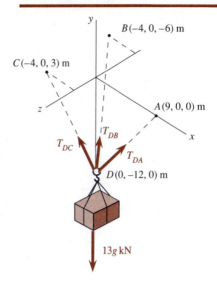

For equilibrium, we write

$$\Sigma\mathbf{F} = \mathbf{T}_{DA} + \mathbf{T}_{DB} + \mathbf{T}_{DC} + \mathbf{W} = \mathbf{0}:$$

$$\left(\frac{3}{5}T_{DA} - \frac{2}{7}T_{DB} - \frac{4}{13}T_{DC}\right)\mathbf{i}$$

$$+ \left(\frac{4}{5}T_{DA} + \frac{6}{7}T_{DB} + \frac{12}{13}T_{DC} - 13g\right)\mathbf{j}$$

$$+ \left(-\frac{3}{7}T_{DB} + \frac{3}{13}T_{DC}\right)\mathbf{k} = \mathbf{0}$$

which entails that

$$\frac{3}{5}T_{DA} - \frac{2}{7}T_{DB} - \frac{4}{13}T_{DC} = 0 \qquad (1)$$

$$\frac{4}{5}T_{DA} + \frac{6}{7}T_{DB} + \frac{12}{13}T_{DC} - 13g = 0 \qquad (2)$$

$$-\frac{3}{7}T_{DB} + \frac{3}{13}T_{DC} = 0 \qquad (3)$$

Solving these three equations, we get

$$T_{DA} = 49.0 \text{ kN} \qquad T_{DB} = 34.3 \text{ kN} \qquad T_{DC} = 63.8 \text{ kN} \blacktriangleleft$$

Developmental Exercise

D3.28 What characterizing feature does an equilibrium problem of particles in space have?

3.14 Method of Solution: Equilibrium in Space

It is true that certain problems can best be solved by employing a special set of steps. However, the following procedure may serve as a guide in *solving equilibrium problems of particles in space*:

1. *Draw the relevant free-body diagram* of a particle where no more than three unknowns will be involved. [This is desirable because no more than three independent equilibrium equations can be obtained in step (4) below.] If this step is impossible, go to step (7) below.
2. *Write the analytical expression of each force*, which may contain known and unknown quantities.
3. *Add the analytical expressions of all of the forces* acting on the particle to obtain the analytical expression of the resultant force in terms of known and unknown quantities.
4. *Set each of the three scalar components* of the resultant force, obtained in step (3), *equal to zero* to obtain a set of three scalar equations.
5. *Solve for the unknowns* in the equations obtained in step (4).
6. *Repeat steps (1) through (5)* for other particles until the desired unknown or unknowns have been determined.
7. If the system consists of *n* particles and *the free-body diagram drawn for any of the n particles in the system involves more than three unknowns*, then we draw the free-body diagram of each particle in the system. We may write three force equilibrium equations for each particle as described in steps (2) through (4) and obtain a total of 3*n* simultaneous equations. If the total number of unknowns equals 3*n*, the unknowns may be determined.

EXAMPLE 3.11

Collars *A* and *B*, each weighing 450 lb, are connected by the wire *AB* and may slide freely on the smooth rod having the shape shown. If the collars are in equilibrium in the position shown, determine the magnitude of the applied force **P**.

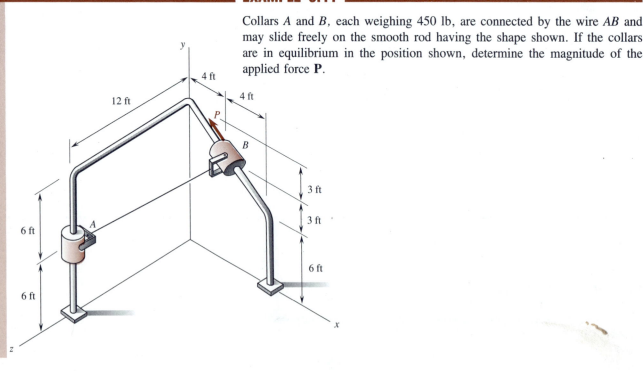

Solution. Since the collars may slide freely on the smooth rod, the reactive force from the rod to a collar must lie somewhere in the plane perpendicular to the portion of the rod at the collar. The free-body diagram of the collar A should be considered first because it contains only three unknowns: the reaction components A_x and A_z and the tension T_{AB}. Since the coordinates of the collars are $A(0, 6, 12)$ ft and $B(4, 9, 0)$ ft, we write

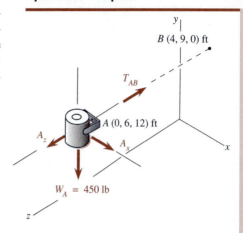

$$\overrightarrow{AB} = 4\mathbf{i} + 3\mathbf{j} - 12\mathbf{k} \qquad \boldsymbol{\lambda}_{AB} = \frac{1}{13}(4\mathbf{i} + 3\mathbf{j} - 12\mathbf{k})$$

$$\mathbf{T}_{AB} = T_{AB}\boldsymbol{\lambda}_{AB} = \frac{T_{AB}}{13}(4\mathbf{i} + 3\mathbf{j} - 12\mathbf{k})$$

$$\mathbf{A} = A_x\mathbf{i} + A_z\mathbf{k} \qquad \mathbf{W}_A = -450\mathbf{j}$$

For equilibrium of *the collar A*, we write

$$\mathbf{R}_A = \Sigma\mathbf{F} = \mathbf{T}_{AB} + \mathbf{A} + \mathbf{W}_A = \mathbf{0}:$$

$$\left(\frac{4}{13}T_{AB} + A_x\right)\mathbf{i} + \left(\frac{3}{13}T_{AB} - 450\right)\mathbf{j} + \left(-\frac{12}{13}T_{AB} + A_z\right)\mathbf{k} = \mathbf{0}$$

Equating the coefficient of $\mathbf{j}$ to zero, we obtain $T_{AB} = 1950$ lb. Thus,

$$\mathbf{T}_{AB} = 600\mathbf{i} + 450\mathbf{j} - 1800\mathbf{k} \text{ lb}$$

The reactive force from the rod to the collar B may be resolved into two rectangular components $\mathbf{B}_{xy}$ and $\mathbf{B}_z$, where $\mathbf{B}_{xy}$ lies in the xy plane and $\mathbf{B}_z$ is parallel to the z axis. Thus, the forces acting on the free body of the collar B are

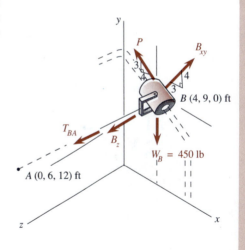

$$\mathbf{T}_{BA} = -\mathbf{T}_{AB} = -600\mathbf{i} - 450\mathbf{j} + 1800\mathbf{k}$$

$$\mathbf{B} = B_{xy}\left(\frac{3}{5}\mathbf{i} + \frac{4}{5}\mathbf{j}\right) + B_z\mathbf{k}$$

$$\mathbf{P} = P\left(-\frac{4}{5}\mathbf{i} + \frac{3}{5}\mathbf{j}\right) \qquad \mathbf{W}_B = -450\mathbf{j}$$

For equilibrium of *the collar B*, we write

$$\mathbf{R}_B = \Sigma\mathbf{F} = \mathbf{T}_{BA} + \mathbf{B} + \mathbf{P} + \mathbf{W}_B = \mathbf{0}:$$

$$\left(-600 + \frac{3}{5}B_{xy} - \frac{4}{5}P\right)\mathbf{i} + \left(-450 + \frac{4}{5}B_{xy} + \frac{3}{5}P - 450\right)\mathbf{j}$$

$$+ (1800 + B_z)\mathbf{k} = \mathbf{0}$$

Solving the three ensuing scalar equations from this vector equation, we obtain

$$B_{xy} = 1080 \text{ lb} \qquad B_z = -1800 \text{ lb} \qquad P = 60 \text{ lb} \blacktriangleleft$$

Note that the negative value for B_z above indicates that the correct sense of $\mathbf{B}_z$ is opposite to that assumed in the free-body diagram.

Developmental Exercises

D3.29 Determine the components of reaction from the rod to the collar A in Example 3.11.

D3.30 If a system of particles in equilibrium in space is to be analyzed and we cannot find a single particle whose free-body diagram involves no more than three unknowns, what should we do?

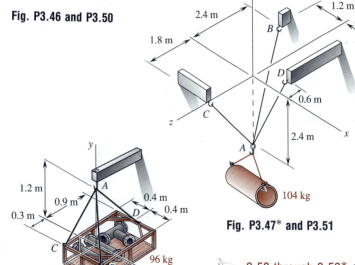

Fig. P3.46 and P3.50

PROBLEMS

3.46 through 3.49* A body is suspended as shown. Write the analytical expression of the resultant force **R** of the force system acting on the free body, which is obtained by cutting through the supporting cables AB, AC, and AD, in terms of the unknown cable tensions.

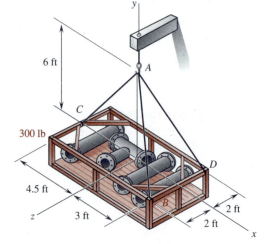

Fig. P3.47* and P3.51

Fig. P3.48 and P3.52

Fig. P3.49* and P3.53*

3.50 through 3.53* Determine the tension in the cable AB shown.

3.54 A 12-kg box is suspended as shown. Determine the applied force **P** which will induce cable tensions $T_{AB} = 70$ N, $T_{AC} = 45$ N, and $T_{AD} = 0$.

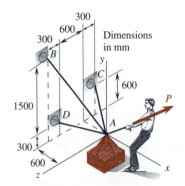

Fig. P3.54

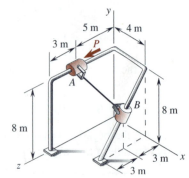

Fig. P3.55

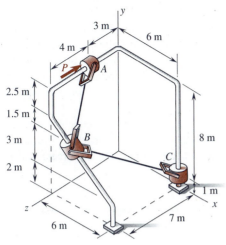

Fig. P3.56

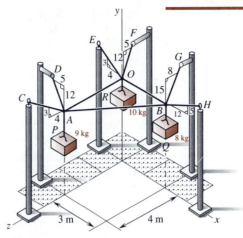

Fig. P3.57

3.55 and 3.56   Each of the smooth collars connected as shown has a mass of 140 kg. Determine the magnitude of the horizontal force **P** which must be applied to the collar *A* to maintain equilibrium.

3.57   Three blocks are hung by a system of wires as shown. Determine the tension in each of the wires.[†]

3.15   Concluding Remarks

The central fact in the equilibrium of a particle is the equality to zero of the resultant force of all forces acting on the free body of the particle. The free-body diagram of a body is a sketch which exhibits the system of applied external forces and reactions that act on the body. From such a sketch, we can have a clear geometric view of the situation in which the relations between the known and unknown quantities in the system are to be established. Therefore, *the drawing of free-body diagrams is an important step in solving equilibrium problems*.

   The equilibrium of a body acted on by a pair of collinear forces which are equal in magnitude and opposite in sense is simple and virtually trivial. Thus, the first nontrivial fundamental study of equilibrium is the study of the equilibrium of particles in a plane. Three methods of solution for equilibrium problems of particles in a plane have been presented in this chapter: the scalar method, the analytical vector method, and the semigraphical method. In dealing with forces in space, it is often awkward to apply the parallelogram law directly. Thus, we choose to use the analytical vector method in our formulation, analysis, and solution of equilibrium problems of particles in space.

   It has been shown in the examples that the solution of some equilibrium problems involving several connected particles may lead to a sizable set of simultaneous equations whose solution could present a challenge. However, with the widespread availability of personal as well as mainframe computers to most engineers, such a challenge can readily be met and overcome. One

[†]Cf. App. D for computer solution of simultaneous equations.

may find the BASIC computer program in App. D useful. This program contains "directions" and "caution" in its beginning REMark section and automatically prints out the value of the determinant of the coefficient matrix as well as the checking of the solution by back substitution in the output. These features are particularly useful in helping beginners of statics watch out for the mistake of including "dependent" equations in formulating the solution.

Equilibrium problems of rigid bodies acted on by nonconcurrent force systems are not within the province of equilibrium of particles. The study of such problems requires other concepts (e.g., moments of forces) and is treated in later chapters.

REVIEW PROBLEMS

3.58   Determine the magnitude of the force **P** required to hold the 100-kg crate A in equilibrium as shown.

3.59   Determine the weight of the cart A if the system shown is in equilibrium.

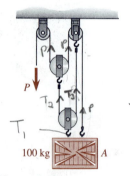

Fig. P3.58

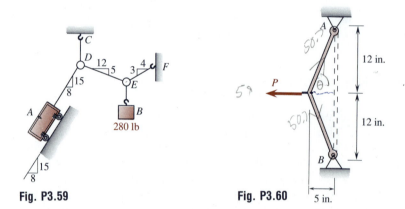

Fig. P3.59   **Fig. P3.60**   5 in.

3.60   An elastic cord is held in equilibrium by a 59-lb force **P** as shown. When the cord is stretched directly between the supports A and B, the tension is 50.7 lb. Determine for the cord (a) the spring modulus k, (b) the free length L.

3.61*   A 50-kg cart is held in equilibrium as shown. Determine the tension T_{AC} in the cable AC.

3.62*   Two cylinders are hung by a system of wires and small rings as shown. Determine the tension in each wire.[†]

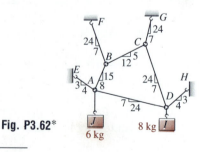

Fig. P3.62*   J 6 kg   8 kg I

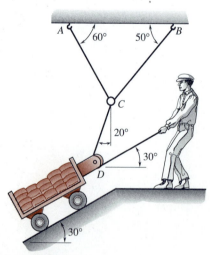

Fig. P3.61*

[†]Cf. App. D for computer solution of simultaneous equations.

3.63 Three cables *DA*, *DB*, and *DC* are used to tie down a balloon at *D* as shown. If the balloon exerts an upward force of 640 N at *D*, determine the tension in each cable.

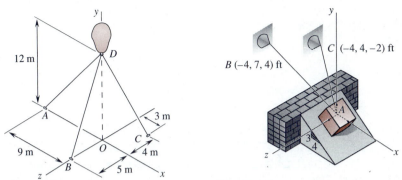

Fig. P3.63   **Fig. P3.64***

3.64* A 310-lb block on a smooth incline is held in equilibrium by cables *AB* and *AC* as shown. Determine the tensions T_{AB} and T_{AC} in the cables.

3.65 Two smooth collars *A* and *B*, each having a weight of 36 lb, are connected by a cable. A force **P** is applied to the collar *A* as shown to maintain equilibrium. Determine the magnitude *P*.

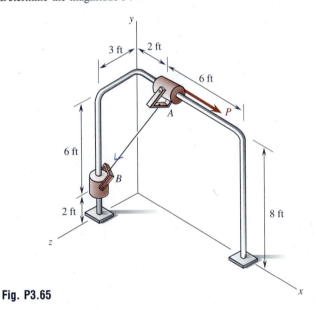

Fig. P3.65

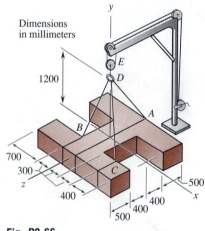

Fig. P3.66

3.66 A 2.6-Mg H-shaped slab is suspended by cables and pulleys as shown. Determine the tension T_{DC} in the cable *DC*.

Chapter 4

Moments of Forces

KEY CONCEPTS

- Moment of a force about an axis, moment of a force about a point, moment center, moment arm, right-hand rule for moment vectors, Varignon's theorem, torque, cross product, dot product, and scalar triple product.
- Moment of a system of forces about a point.
- Moment of a couple and the addition of moments of couples.
- Moment of a system of forces about an axis.
- Shortest distance between an axis and a force.

This chapter presents the concepts and methods of computation of moments of forces as well as their applications. The moment of a force acting on a body about an axis is a measure of the effectiveness or tendency of the force to rotate the body about that axis. The turning, twisting, or bending done to a body is an effect directly related to the moments of forces acting on the body about the axis with respect to which the body is turned, twisted, or bent. The computation of moments of forces using scalar algebra is presented first so that the physical implication of the moment of a force may clearly be comprehended in the early stage of the development. Operations involving cross and dot products of vectors are later presented and utilized as an effective tool to handle moments of forces and their applications in three-dimensional space. A good understanding of moments of forces is a necessary step toward a meaningful and successful study of the conditions of rigid bodies subjected to the action of force systems.

SCALAR FORMULATION

4.1 Moment of a Force

Besides the tendency to move a body in the direction of its action, a force acting on a body has also a tendency to rotate the body about an axis, which does not intersect the line of action of the force and is not parallel to the force. The *moment of a force about an axis* is an action reflecting the effectiveness or tendency of the force to turn (or rotate) a body about that axis; such a turning action should be well understood early by all students of mechanics. Observations show that the moment of a force about an axis is *zero* if any of the following conditions is true: (a) the line of action of the force *intersects the axis*, (b) the force is *parallel to the axis*, (c) the force itself is *zero*.

Let us consider a rigid slab acted on by a force $\mathbf{F}$ in the plane of the slab as shown in Fig. 4.1. The effectiveness or tendency of $\mathbf{F}$ to rotate the slab about the axis AB normal to the plane of the slab is, clearly, proportional to the magnitude F of the force $\mathbf{F}$ and the shortest distance d_s between the axis AB and the line of action of $\mathbf{F}$. The axis about which the moment of a force is computed is referred to as the *moment axis*. Thus, the *magnitude* of the moment $\mathbf{M}_{AB}$ of $\mathbf{F}$ about the moment axis AB is given by

$$M_{AB} = d_s F \qquad (4.1)$$

Note that the shortest distance d_s between the line of action of the force $\mathbf{F}$ and the moment axis AB, as indicated in Fig. 4.1, is referred to as the *moment arm* of $\mathbf{F}$ about the axis AB.

To distinguish a moment vector from a force vector, we use a *hollow arrow* to represent the moment $\mathbf{M}_{AB}$ as shown in Fig. 4.1.[†] The *direction* of $\mathbf{M}_{AB}$ is defined to be the same as that of the extended thumb of the right hand when the other fingers of the right hand are curled in the direction of the tendency to rotate. This definition is elaborated upon in Sec. 4.2.

The phrase *moment of a force about a point* is frequently used in situations where the forces and the points under consideration all lie in a plane. This phrase is commonly understood to mean "moment of a force about an axis which passes through the point and is perpendicular to the plane containing the point and the line of action of the force." In Fig. 4.1, the axis AB is perpendicular to the plane of the slab and intersects it at the point C. The point about which the moment of a force is computed is referred to as the *moment center* of the force. Thus, the moment $\mathbf{M}_C$ of the force $\mathbf{F}$ about the moment center C is the same as the moment $\mathbf{M}_{AB}$ of $\mathbf{F}$ about the moment axis AB. Besides representation by a hollow arrow like $\mathbf{M}_{AB}$, the moment $\mathbf{M}_C$ is often represented graphically by a *curved arrow* pointing in the direction of the tendency of the moment to rotate about the moment center C as shown in Fig. 4.1. The *magnitude* of $\mathbf{M}_C$ is, of course, identical with that defined in Eq. (4.1); i.e.,

$$M_C = d_s F \qquad (4.2)$$

In this case, d_s is also referred to as the *moment arm* of $\mathbf{F}$ about C.

The moments of coplanar forces about any given point in the plane of the forces are vectors all perpendicular to the plane of the forces and are hence parallel vectors. The senses of these parallel moment vectors point outward from the plane if they are counterclockwise moments, and inward to the plane if they are clockwise moments. Thus, the directions of these parallel moment vectors may be accounted for by letting *counterclockwise* moments be *positive* and *clockwise* moments be *negative*. Such a sign

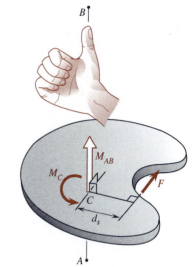

Fig. 4.1

[†]A moment vector is also represented in other literature by a straight arrow with a curved arrow around it; e.g., $\psi = \Uparrow$.

convention for parallel moment vectors is generally used, although it may be defined the other way around. The addition (and hence the computation of the resultant moment) of parallel moment vectors may, therefore, be accomplished with *scalar algebra*.

From Eqs. (4.1) and (4.2), it is obvious that the moment of a force has the dimensions *length × force*. The primary units for the moment of a force are newton-meters (N·m) in the SI and pound-feet (lb·ft) in the U.S. customary system.

Developmental Exercises

D4.1   Define the terms: (a) moment of a force about an axis, (b) moment axis, (c) moment arm, (d) moment of a force about a point, (e) moment center.

D4.2   What are the conditions under which the moment of a force about an axis is equal to zero?

D4.3   How is the moment of a force represented in a figure so that it can readily be distinguished from a force?

D4.4   What are the primary units for the moment of a force in (a) the SI, (b) the U.S. customary system?

4.2   Moment Vectors and the Right-Hand Rule

In general, the *moment* $\mathbf{M}_Q$ *of a force* $\mathbf{F}$ *about a point* Q in space is defined as the *moment* $\mathbf{M}_{QB}$ *of* $\mathbf{F}$ *about the axis* QB which passes through Q and is perpendicular to the plane containing Q and the line of action of $\mathbf{F}$; i.e., $\mathbf{M}_Q = \mathbf{M}_{QB}$. The geometry involved in this definition is illustrated in Fig. 4.2, where the point P is the point of application of $\mathbf{F}$, the point A is any convenient point on the line of action of $\mathbf{F}$, $\mathbf{r}_{QA}$ is the position vector from Q to A, θ is the angle between the positive directions of $\mathbf{r}_{QA}$ and $\mathbf{F}$, and d_s is the shortest distance between Q and the line of action of $\mathbf{F}$. Note that $\mathbf{M}_Q$ is

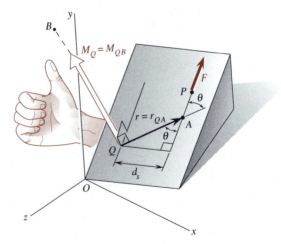

Fig. 4.2

perpendicular to both $\mathbf{r}_{QA}$ and $\mathbf{F}$. The *magnitude* of $\mathbf{M}_Q$ may be written as

$$M_Q = d_s F = (r \sin\theta)F \qquad (4.3)$$

The *sense* of $\mathbf{M}_Q$ is defined to point from the knuckle to the nail of the extended thumb of the right hand when the other fingers of the right hand are curled in the direction of the tendency of the moment of the force to rotate about the axis QB. This is called the *right-hand rule for moment of a force about a point*. Physically, the *magnitude* of $\mathbf{M}_Q$ measures the effectiveness or tendency of the force $\mathbf{F}$ to impart to a rigid body a rotational motion about an axis which passes through Q and is parallel to the direction of $\mathbf{M}_Q$.

If a shaft is *turned* by a *single force* $\mathbf{F}$, the shaft will, at the same time, be bent by $\mathbf{F}$. The turning action of $\mathbf{F}$ about the shaft is the moment of $\mathbf{F}$ about the shaft. The illustration in Fig. 4.3 shows that the moment of the force $\mathbf{F}$ at A about the shaft CD is $d_s F$ pointing to the left along the shaft, while the moment of the force $\mathbf{F}'$ at B about the shaft CD is $d_s' F'$ pointing to the right along the shaft. Since these two moments are parallel to each other, we may say that the resultant moment of $\mathbf{F}$ and $\mathbf{F}'$ about the shaft CD is either $(d_s F - d_s' F')$ pointing to the left or $(d_s' F' - d_s F)$ pointing to the right along the shaft.

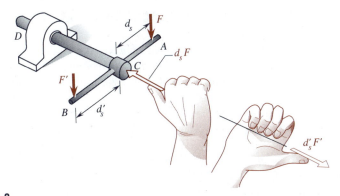

Fig. 4.3

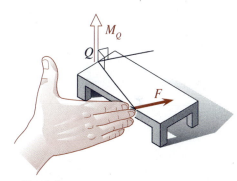

Fig. 4.4

The right-hand rule for moment vectors described above may equivalently be stated in one of the following two additional versions:

1. When the extended four fingers of the right hand point along the line of action of the force $\mathbf{F}$ and the palm of the right hand faces the point Q, the extended thumb of the right hand will point in the direction of the moment vector $\mathbf{M}_Q$ of $\mathbf{F}$ about the moment center Q. This is illustrated in Fig. 4.4.
2. If the orientation of the axis of a right-hand threaded screw at the point Q is the same as the orientation of the moment vector $\mathbf{M}_Q$ which acts on (or twists) the screw head at the point Q, the screw will advance along its axis in the same sense as that of $\mathbf{M}_Q$. This is illustrated in Fig. 4.5.

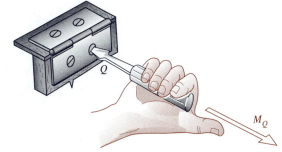

Fig. 4.5

The above equivalent versions of the right-hand rule for moment vectors are very handy because they may readily help us decide which way we should apply our moment to a right-hand threaded screw, bolt, lug nut, oil filter, etc. when we want to tighten or loosen it.

Developmental Exercises

D4.5   Describe two versions of the right-hand rule for moment vectors.

D4.6   A wrench is attached to an ordinary bolt and pushed by a horizontal force **F** as shown. (a) Which way does the bolt advance or tend to advance? (b) What is the physical implication of the magnitude of the moment **M**$_Q$ of **F** about Q?

Fig. D4.6

4.3   Varignon's Theorem: Forces in a Plane

The moment of a force is often determined indirectly from the moments of its components by using the so-called *Varignon's theorem*, after the French mathematician Pierre Varignon (1654–1722).[†] *Varignon's theorem* states that *the moment of a force about any point is equal to the sum of the moments of its components about the same point*. This theorem is here proved for a force which is resolved into only two components at a point.[††]

Suppose that the force **R** acting at the point O is resolved into two components **A** and **B** as shown in Fig. 4.6. The moment arms of **A**, **B**, and **R** drawn from the moment center P are a, b, and r as shown. Letting the y axis be directed along the line OP and projecting **A**, **B**, and **R** onto the x axis, we note that $\overline{CE} = \overline{OD}$. We find from the geometry in this figure that the moment of **R** about the moment center P is

$$rR = (\overline{OP}\sin\gamma)R = (R\sin\gamma)\overline{OP} = \overline{OE}(\overline{OP}) = (\overline{OC} + \overline{CE})\overline{OP}$$
$$= (\overline{OC} + \overline{OD})\overline{OP} = (A\sin\alpha + B\sin\beta)\overline{OP}$$
$$= (\overline{OP}\sin\alpha)A + (\overline{OP}\sin\beta)B = aA + bB$$

This proves that the moment of **R** about any point P is equal to the sum of the moments of its two components **A** and **B** about the same point P. In fact, Varignon's theorem applies equally well to the case where a force has three or more components, since the number of components can always be reduced to two by direct combination. We shall note in Sec. 6.2 that Varignon's theorem is a special case of the *principle of moments*.

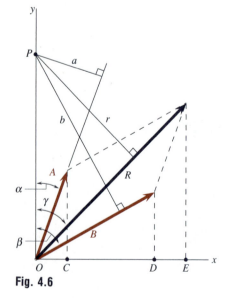

Fig. 4.6

[†]Varignon's major contribution to mechanics was his book *Projet d'uné nouvelle méchanique*, published in 1687 (the same year as Newton's *Principia*) and dedicated to the French Academy of Sciences. Cf. Sec. 4.18 for additional comments.

[††]Varignon's theorem for forces in space is presented in Sec. 4.11.

For a *moment vector* perpendicular to the *xy* plane, we often use a curved arrow in the *xy* plane to show its direction according to the right-hand rule for moment vectors; this is a *geometric notation* for moment vectors. For instance, if $\mathbf{M}_A = 200\mathbf{k}$ N·m and $\mathbf{M}_B = -300\mathbf{k}$ N·m, we may write

$$\mathbf{M}_A = 200 \text{ N·m} \circlearrowright \qquad \mathbf{M}_B = 300 \text{ N·m} \circlearrowleft$$

Furthermore, a vector quantity in a plane is often expressed by indicating its magnitude with a number (or a symbol) and its direction with an arrow sign at a slope or an angle. Such a notation is a *geometric* or *polar notation* for vectors in a plane. For instance, if $\mathbf{r} = -8\mathbf{i} + 15\mathbf{j}$ m, $\mathbf{F} = 3\mathbf{i} + 4\mathbf{j}$ kN, $\mathbf{P} = 5\mathbf{i} - 6\mathbf{j}$ kN, and $\mathbf{Q} = -50\mathbf{i} - 120\mathbf{j}$ lb, we may write

$$\mathbf{r} = 17 \text{ m} \quad \text{or} \quad \mathbf{r} = 17 \text{ m} \measuredangle 61.9°$$

$$\mathbf{F} = 5 \text{ kN} \quad \text{or} \quad \mathbf{F} = 5 \text{ kN} \measuredangle 53.1°$$

$$\mathbf{P} = 7.81 \text{ kN} \quad \text{or} \quad \mathbf{P} = 7.81 \text{ kN} \measuredangle 50.2°$$

$$\mathbf{Q} = 130 \text{ lb} \quad \text{or} \quad \mathbf{Q} = 130 \text{ lb} \measuredangle 67.4°$$

It is often preferable to work with the slope triangles as given if the corresponding angles are not explicitly required.

EXAMPLE 4.1

A 500-N force $\mathbf{F}$ is applied to a bracket fixed on a wall as shown. Determine the moment $\mathbf{M}_A$ of $\mathbf{F}$ about the point A.

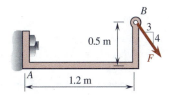

Solution. The system is coplanar. We may take advantage of its simplicity and solve it in four different ways for illustrative purposes.

(a) *Using shortest distance*. The moment $\mathbf{M}_A$ may be determined by first calculating its *magnitude* $d_s F$ and then using the right-hand rule to ascertain its *direction*. From the sketch shown, we write

$$\theta_1 = \tan^{-1}\left(\frac{0.5}{1.2}\right) = 22.62° \qquad \theta_2 = \tan^{-1}\left(\frac{4}{3}\right) = 53.13°$$

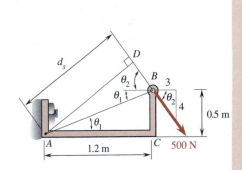

$$\overline{AB} = [(1.2)^2 + (0.5)^2]^{1/2} = 1.3 \qquad d_s = \overline{AB} \sin(\theta_1 + \theta_2) = 1.26$$

$$M_A = -d_s F = -1.26(500) \qquad \mathbf{M_A = 630 \text{ N·m}} \circlearrowright \blacktriangleleft$$

Note that the negative value for $\mathbf{M}_A$ indicates that $\mathbf{M}_A$ is clockwise.

(b) *Using Varignon's theorem at B*. We first resolve the 500-N force $\mathbf{F}$ at B into two components as shown. By Varignon's theorem and the right-hand rule, we write

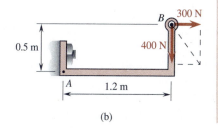

$$M_A = -0.5(300) - 1.2(400) \qquad \mathbf{M_A = 630 \text{ N·m}} \circlearrowright \blacktriangleleft$$

(b)

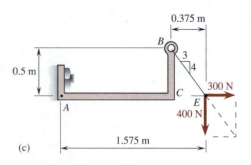

(c)

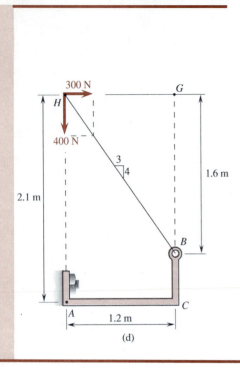

(d)

(c) *Using Varignon's theorem at E.* We apply the principle of transmissibility to transmit **F** to E and then resolve it there into two components as shown. Note that $\overline{CE} = (3/4)(0.5)$ m $= 0.375$ m. By Varignon's theorem and the right-hand rule, we write

$$M_A = 0(300) - 1.575(400) \qquad \mathbf{M}_A = 630 \text{ N} \cdot \text{m} \;\downdownarrows \blacktriangleleft$$

(d) *Using Varignon's theorem at H.* We apply the principle of transmissibility to transmit **F** to H and then resolve it there into two components as shown. Note that $\overline{BG} = (4/3)(1.2)$ m $= 1.6$ m. By Varignon's theorem and the right-hand rule, we write

$$M_A = -2.1(300) + 0(400) \qquad \mathbf{M}_A = 630 \text{ N} \cdot \text{m} \;\downdownarrows \blacktriangleleft$$

Developmental Exercises

D4.7 Describe Varignon's theorem.

D4.8 Express the following vectors in geometric notations: (a) $\mathbf{F} = 24\mathbf{i} - 7\mathbf{j}$ kN, (b) $\mathbf{r} = -200\mathbf{i} + 300\mathbf{j}$ m, (c) $\mathbf{M}_A = 4\mathbf{k}$ kN·m, (d) $\mathbf{C} = -700\mathbf{k}$ lb·ft.

EXAMPLE 4.2

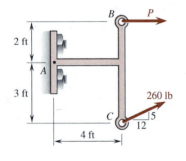

The two forces acting on the bracket as shown have the same unit. Determine the resultant moment $\mathbf{M}_A^R$ of these two forces about the point A in terms of known and unknown quantities.

Solution. We first resolve the 260-lb force at C into two components as shown. By Varignon's theorem and the right-hand rule, we write

$$M_A^R = -2P + 3(240) + 4(100) = 1120 - 2P$$

$$\mathbf{M}_A^R = (1120 - 2P) \text{ lb} \cdot \text{ft} \;\downdownarrows \blacktriangleleft$$

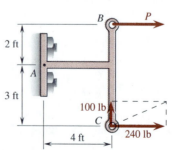

Developmental Exercises

D4.9   Determine the moment $\mathbf{M}_A$ of the force shown (see right) about A.

D4.10   Determine the resultant moment $\mathbf{M}_A^R$ of the two forces shown (below) about A.

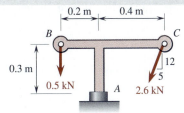

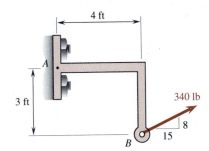

Fig. D4.10

Fig. D4.9

4.4   Moment of a Couple: Torque

A *couple* is a pair of parallel forces equal in magnitude, opposite in direction, and having different lines of action, such as $\mathbf{F}$ and $-\mathbf{F}$ shown in Fig. 4.7. Since $\mathbf{F} + (-\mathbf{F}) = \mathbf{0}$, a couple can have no translational (i.e., pushing or pulling) effect on a body. However, a couple does have a moment and a purely torsional (i.e., twisting) effect on a body. The *moment of a couple* about any point is defined as the sum of the moments of the two forces in the couple about that point.

Fig. 4.7

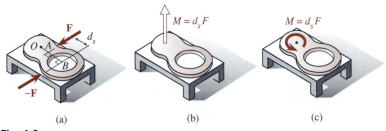

(a)   (b)   (c)

Fig. 4.8

Since only a couple can exert a purely torsional action on a shaft, the *moment of a couple* is often called the *torque of the couple*, or simply a *torque.*[†] Let a rigid slab be acted on by a couple consisting of $\mathbf{F}$ and $-\mathbf{F}$ in the plane of the slab as shown in Fig. 4.8(a). Then, the *magnitude* of the moment $\mathbf{M}_O$ of the couple about any chosen point O, as shown, is

$$M_O = \overline{OB}(F) - \overline{OA}(F) = (\overline{OB} - \overline{OA})(F) = \overline{AB}(F) = d_s F$$

where d_s is the shortest distance between the lines of action of $\mathbf{F}$ and $-\mathbf{F}$. Thus, the value of M_O contains no dimension which relates the moment center O to the couple. It follows that the moment of a couple has the same value for *all* moment centers and is, therefore, a *free vector*. It is logical that we drop the subscript O in $\mathbf{M}_O$ and write the *magnitude* of the moment of the couple as

$$M = d_s F \qquad\qquad (4.4)$$

[†]The term *torque* is usually used to denote the moment of a couple and the moment about the axis of a shaft.

The moment **M** of a couple consisting of **F** and $-\mathbf{F}$ is a vector perpendicular to the plane containing **F** and $-\mathbf{F}$. This vector points in the *direction* of the extended thumb of the right hand when the other fingers of the right hand are curled in the direction of the tendency of the couple to rotate. The moment **M** is illustrated in Fig. 4.8(b) by a *hollow arrow* and in Fig. 4.8(c) by a *curved arrow* pointing in the direction of the tendency of the couple to rotate.

In coplanar systems, torques of couples tending to rotate counterclockwise and clockwise are usually represented as illustrated in Fig. 4.9. Note that the *torque*, or *moment of a couple*, is a *vector* and is an *attribute* of the couple.[†]

         Counterclockwise torque   Clockwise torque

Fig. 4.9

EXAMPLE 4.3

A force system is applied to a cantilever beam as shown. Determine the resultant moment $\mathbf{M}_A^R$ of the force system about the point A.

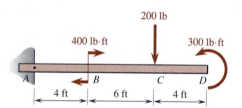

Solution. We note that the force system shown consists of a force and two torques, which are free vectors. Thus, referring to the given space diagram, we write

$$M_A^R = -400 - 10(200) + 300 \qquad M_A^R = -2100 \text{ lb} \cdot \text{ft}$$

$$\mathbf{M}_A^R = 2.1 \text{ kip} \cdot \text{ft} \, \circlearrowright \quad \blacktriangleleft$$

REMARK. In the above computation, only the 200-lb force is to be multiplied by its moment arm of 10 ft; *no* moment arms are used in computing the moments of the 400-lb · ft and 300-lb · ft torques about the point A.

[†]Since a couple is composed of two opposite forces which are noncollinear and have equal magnitudes, some mechanicists feel that the *moment of the couple*, rather than the *couple*, is a vector. Perhaps, a source contributing to the dispute is the fact that the terms *moment* and *couple* have been taken as interchangeable. For further discussion of this and other basic concepts in mechanics, see T. R. Kane, "Teaching of Mechanics to Undergraduates," *The International Journal of Mechanical Engineering Education (IJMEE)*, Vol. 6, No. 4, 1978, pp. 187–189.

Developmental Exercises

D4.11   Define (a) a couple, (b) the moment of a couple, (c) a torque.

D4.12   Why is the moment of a couple a free vector?

D4.13   How are couples represented in coplanar systems?

D4.14   In the system shown, A_x and A_y are in newtons and M_A is in newton-meters. Determine the resultant moment of the force system about (a) the point A, (b) the point B.

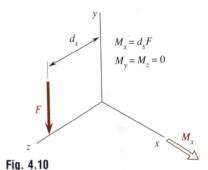

Fig. D4.14

4.5   Moments about Coordinate Axes

When the line of action of a force $\mathbf{F}$ is parallel to, or intersects, a coordinate axis, the moment of $\mathbf{F}$ about the coordinate axis is zero. On the other hand, when the line of action of $\mathbf{F}$ is perpendicular to a coordinate axis, the moment of $\mathbf{F}$ about that axis has a magnitude equal to $d_s F$ where F is the magnitude of $\mathbf{F}$ and d_s is the shortest distance between that axis and the line of action of $\mathbf{F}$. This is illustrated in Fig. 4.10. If a moment points to the positive direction of the coordinate axis, it is a positive moment; otherwise, it is a negative moment.

If the force $\mathbf{F}$ is neither parallel nor perpendicular to a coordinate axis, we may first resolve $\mathbf{F}$ into rectangular components parallel to the coordinate axes. The moment of $\mathbf{F}$ about an axis is, by Varignon's theorem, equal to the sum of the moments of the rectangular components of $\mathbf{F}$ about that axis. Note that only those components which are perpendicular to the axis may contribute moments about that axis.

$$M_x = d_s F$$
$$M_y = M_z = 0$$

Fig. 4.10

EXAMPLE 4.4

A 350-lb force $\mathbf{F}$ is applied to a plate along the direction of the line AB as shown. Determine the moments M_x, M_y, and M_z of $\mathbf{F}$ about the coordinate axes x, y, and z, respectively.

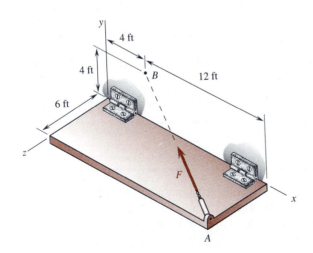

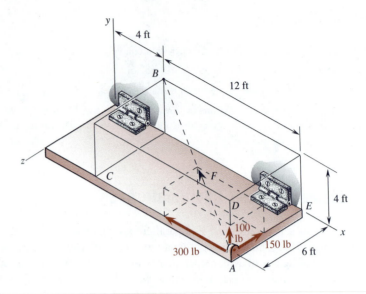

Solution. From the plate shown above, we have

$$\overline{AB} = [(12)^2 + (4)^2 + (6)^2]^{1/2} = 14$$

The scalar components of the 350-lb force **F** are simply the projections of **F** onto the directions parallel to the *xyz* axes. By proportions of the two parallelepipeds as sketched, we write

$$F_x = -\frac{\overline{AC}}{\overline{AB}} F = -\frac{12}{14}(350) = -300$$

$$F_y = \frac{\overline{AD}}{\overline{AB}} F = \frac{4}{14}(350) = 100$$

$$F_z = -\frac{\overline{AE}}{\overline{AB}} F = -\frac{6}{14}(350) = -150$$

Thus, **F** is resolved into three rectangular components at *A* as indicated. Applying Varignon's theorem and the right-hand rule, we write

$$M_x = 0 - 6(100) + 0 \qquad\qquad M_x = -600 \text{ lb} \cdot \text{ft} \quad \blacktriangleleft$$

$$M_y = -6(300) + 0 + 16(150) \qquad M_y = 600 \text{ lb} \cdot \text{ft} \quad \blacktriangleleft$$

$$M_z = 0 + 16(100) + 0 \qquad\qquad M_z = 1600 \text{ lb} \cdot \text{ft} \quad \blacktriangleleft$$

Note that $\mathbf{M}_x$ points to the negative direction of the *x* axis, while $\mathbf{M}_y$ and $\mathbf{M}_z$ point to the positive directions of the *y* and *z* axes, respectively.

EXAMPLE 4.5

The free body of a plate is acted on by a force system as shown, where the unknown forces are expressed in kilonewtons and the unknown moments in kilonewton-meters. Determine the resultant moments M_x^R, M_y^R, and M_z^R of the force system about the *x*, *y*, and *z* axes, respectively.

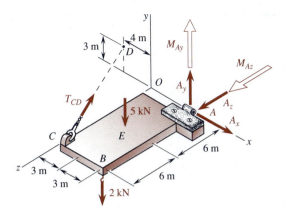

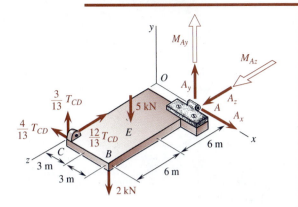

Solution. Similar to the solution in Example 4.4, we first resolve the force $\mathbf{T}_{CD}$ into three rectangular components at C as shown. By Varignon's theorem and the right-hand rule, we write

$$M_x^R = -12\left(\frac{3}{13}T_{CD}\right) + 6(5) + 12(2)$$

$$M_x^R = -\frac{36}{13}T_{CD} + 54 \quad \text{kN}\cdot\text{m} \quad \blacktriangleleft$$

$$M_y^R = M_{Ay} - 6A_z - 12\left(\frac{4}{13}T_{CD}\right)$$

$$M_y^R = M_{Ay} - 6A_z - \frac{48}{13}T_{CD} \quad \text{kN}\cdot\text{m} \quad \blacktriangleleft$$

$$M_z^R = M_{Az} + 6A_y - 3(5) - 6(2)$$
$$M_z^R = M_{Az} + 6A_y - 27 \quad \text{kN}\cdot\text{m} \quad \blacktriangleleft$$

Developmental Exercises

D4.15 Determine the moments M_x, M_y, and M_z of the 180-lb force about the x, y, and z axes as shown.

D4.16 Refer to Example 4.5. Determine the moments M_{AB} and M_{CB} of the force system about the axes along AB and CB, respectively.

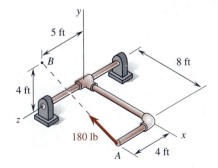

Fig. D4.15

PROBLEMS

4.1 through 4.4　A force **F** acts on the body as shown. Determine the moment **M**$_O$ of **F** about the point O.

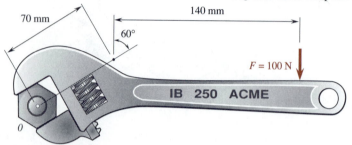

70 mm　　140 mm　　60°　　$F = 100\ N$　　IB　250　ACME　　0

Fig. P4.1

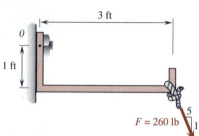

3 ft　　0　　1 ft　　$F = 260\ lb$　　5　12

Fig. P4.2

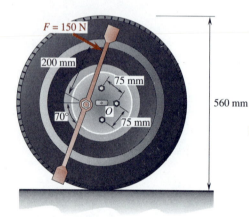

$F = 150\ N$　　200 mm　　75 mm　　70°　　O　　75 mm　　560 mm

Fig. P4.3*

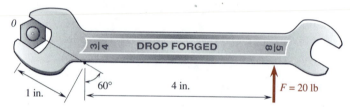

0　　DROP FORGED　　1 in.　　60°　　4 in.　　$F = 20\ lb$

Fig. P4.4

4.5 through 4.7*　Determine the resultant moment **M**$_A^R$ of the two forces about the point A as shown.

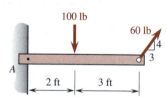

100 lb　　60 lb　　4　　3　　A　　2 ft　　3 ft

Fig. P4.5

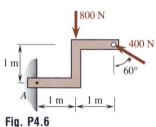

800 N　　400 N　　1 m　　60°　　A　　1 m　　1 m

Fig. P4.6

2 m　　250 N　　1 m　　A　　3 m　　340 N　　15　　8

Fig. P4.7*

3 kN　　6 kN　　A_x　　A　　B　　A_y　　2 m　　2 m　　2 m　　P　　60°

Fig. P4.8 and P4.11

4.8 through 4.10　A free body is acted on by a system of five forces as shown. The unknown forces, as labeled, and the known forces have the same unit. Determine the resultant moment **M**$_A^R$ of the force system about the point A.

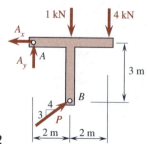

1 kN　　4 kN　　A_x　　A　　A_y　　3 m　　B　　4　　3　　P　　2 m　　2 m

Fig. P4.9* and P4.12

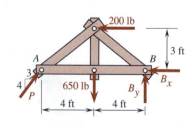

200 lb　　A　　B　　3 ft　　3　　4　　P　　650 lb　　B_y　　B_x　　4 ft　　4 ft

Fig. P4.10 and P4.13*

4.11 through 4.13* A free body is acted on by a system of five forces as shown. The unknown forces, as labeled, and the known forces have the same unit. Determine the resultant moment $\mathbf{M}_B^R$ of the force system about the point B.

4.14 For the system shown, determine (a) the angle θ for which the moment $\mathbf{M}_A$ of the 200-N force about the point A has a maximum magnitude, (b) the maximum magnitude of $\mathbf{M}_A$.

4.15* The resultant moment of the two forces about the point A as shown is zero. Determine the magnitude of $\mathbf{P}$.

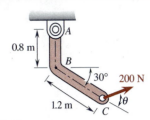

Fig. P4.14

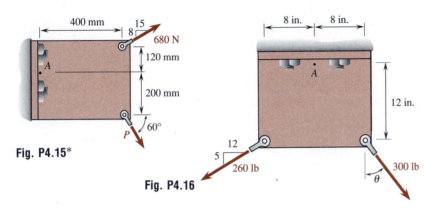

Fig. P4.15*

Fig. P4.16

4.16 The resultant moment of the two forces about the point A as shown is zero. Determine the angle θ.

4.17 through 4.19* A force system is applied to a body cantilevered at one end as shown. Determine the resultant moment $\mathbf{M}_A^R$ of the force system about the end at A.

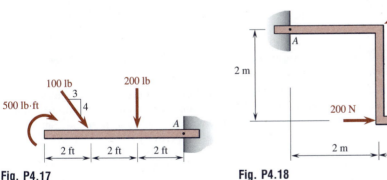

Fig. P4.17   **Fig. P4.18**   **Fig. P4.19***

4.20 through 4.22* A free body is acted on by a force system as shown, where the unknown and known forces have the same unit. Determine the resultant moment $\mathbf{M}_A^R$ of the force system about the point A.

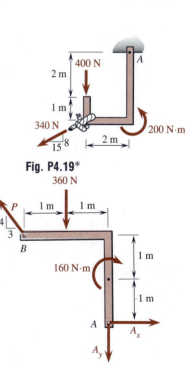

Fig. P4.20 and P4.23*

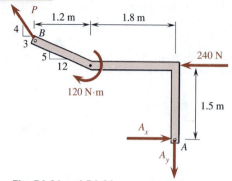

Fig. P4.21 and P4.24

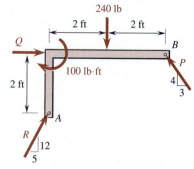

Fig. P4.22* and P4.25

4.23* through 4.25   A free body is acted on by a force system as shown, where the unknown and known forces have the same unit. Determine the resultant moment $\mathbf{M}_B^R$ of the force system about the point B.

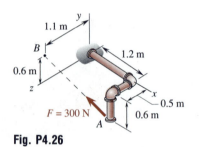

Fig. P4.26

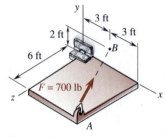

Fig. P4.27*

Fig. P4.28

4.26 through 4.28   Determine the moments M_x, M_y, and M_z of the force $\mathbf{F}$ shown about the coordinate axes x, y, and z, respectively.

4.29 through 4.32   A free body is acted on by a force system as shown, where the unknown and known forces have the same unit, and the unknown and known moments have the same units. Determine the resultant moments M_x^R, M_y^R, and M_z^R of the force system about the x, y, and z axes, respectively.

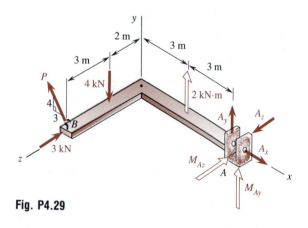

Fig. P4.29

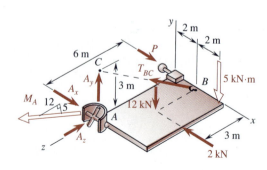

Fig. P4.30

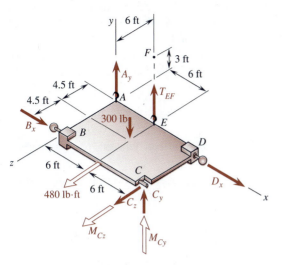

Fig. P4.31*

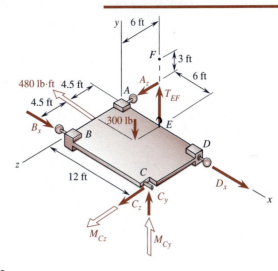

Fig. P4.32

APPLICATION OF CROSS PRODUCTS

Note. Those who are familiar with the rudiments of cross products as presented in Secs. 4.6 through 4.9 may proceed directly to Sec. 4.10.

4.6   Trihedrals

A *trihedral* is a figure formed by three noncoplanar line segments which intersect at a common point. This is illustrated by the three line segments $\overline{OA}$, $\overline{OB}$, and $\overline{OC}$ as shown in Fig. 4.11. A *directed trihedral* is a trihedral formed by three directed line segments (or three vectors). The trihedral formed by the x, y, and z axes of a rectangular coordinate system, as shown in Fig. 4.12, is a directed trihedral and is called the *coordinate trihedral*.

The three concurrent vectors **A**, **B**, **C** are said to form a *right-handed trihedral* if, when the thumb of the right hand extends from the point of concurrence along the first vector **A**, the other fingers of the right hand can curl only in the direction in which the second vector **B** could be rotated through an angle of less than 180° to coincide with the third vector **C**. This is illustrated in Fig. 4.13(a). If the vectors **A**, **B**, **C** (in the order **A**, **B**, **C**)

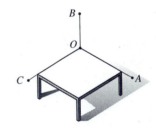

Fig. 4.11

Fig. 4.12

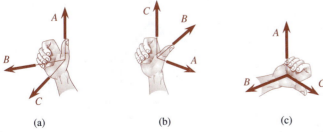

     (a)   (b)   (c)

Fig. 4.13

form a right-handed trihedral, then it must be true that these vectors in the order **B**, **C**, **A** as well as in the order **C**, **A**, **B** also form a right-handed trihedral, as shown in parts (b) and (c) of Fig. 4.13. A trihedral which is not right-handed is called a left-handed trihedral. In general, coordinate trihedrals used in mechanics are right-handed trihedrals.

Developmental Exercises

D4.17   Define the terms: (a) trihedral, (b) coordinate trihedral, (c) right-handed trihedral.

D4.18   Which of the sets of vectors in the order **A**, **B**, **C** form right-handed trihedrals?

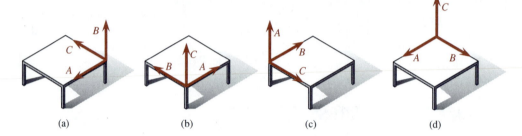

Fig. D4.18

(a)   (b)   (c)   (d)

4.7   Cross Product of Two Vectors

Two basic ways for multiplying two vectors are employed in this text. The first way is presented here and the other in Sec. 4.14. The *cross product* of two vectors **P** and **Q** is defined as a vector **V** which has the following characteristics: the *magnitude* of **V** is equal to the product of the magnitudes of **P** and **Q** and the sine of the angle θ between them; the *orientation* of **V** is perpendicular to the plane of **P** and **Q**; and the *sense* of **V** is such that the three vectors **V**, **P**, **Q** (in this order) form a right-handed trihedral. Without losing generality, the cross product of **P** and **Q** may be illustrated as shown in Fig. 4.14. As in vector analysis, we write the above cross product as

$$\mathbf{V} = \mathbf{P} \times \mathbf{Q} \tag{4.5}$$

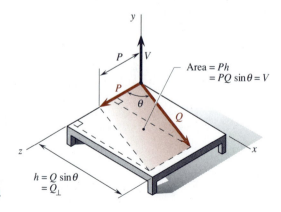

Area $= Ph$
$= PQ \sin\theta = V$

$h = Q \sin\theta$
$= Q_\perp$

Fig. 4.14

which is read "**V** equals **P** cross **Q**." Because of the nature of the product, the cross product is also called the *vector product*.

By definition, the magnitude of **P** × **Q** in Eq. (4.5) is given by

$$|\mathbf{P} \times \mathbf{Q}| = PQ \sin\theta \tag{4.6}$$

where θ is the angle between the positive directions of **P** and **Q** and $0 \le \theta \le 180°$. Writing the magnitude of **P** × **Q** as $P(Q \sin\theta)$, we readily obtain from Fig. 4.14 a simple *geometric interpretation* that the magnitude of the cross product of **P** and **Q** is equal to the area of the parallelogram, or twice the area of the triangle formed by the vectors **P** and **Q** as the adjacent sides. On the other hand, if **P** is a unit vector (i.e., $P = 1$), then $|\mathbf{P} \times \mathbf{Q}| = P(Q \sin\theta) = Q \sin\theta = h =$ the magnitude of the *perpendicular component* $\mathbf{Q}_\perp$ of **Q** with respect to **P**. Furthermore, when **P** × **Q** = **0**, it implies one or more of the following cases: (a) **P** = **Q** (where $P = Q$ and $\theta = 0$), (b) **P** is parallel to **Q** (where $\theta = 0$ or 180°), (c) **P** is a zero vector (i.e., $P = 0$), (d) **Q** is a zero vector (i.e., $Q = 0$).

Applying the definition of the cross product, we may readily obtain the cross products of the various possible pairs of the cartesian unit vectors **i**, **j**, and **k**, as shown in Fig. 4.15. These cross products are

$$\begin{array}{lll}
\mathbf{i} \times \mathbf{i} = \mathbf{0} & \mathbf{i} \times \mathbf{j} = \mathbf{k} & \mathbf{i} \times \mathbf{k} = -\mathbf{j} \\
\mathbf{j} \times \mathbf{i} = -\mathbf{k} & \mathbf{j} \times \mathbf{j} = \mathbf{0} & \mathbf{j} \times \mathbf{k} = \mathbf{i} \\
\mathbf{k} \times \mathbf{i} = \mathbf{j} & \mathbf{k} \times \mathbf{j} = -\mathbf{i} & \mathbf{k} \times \mathbf{k} = \mathbf{0}
\end{array} \tag{4.7}$$

It may be noted from the identities in Eqs. (4.7) that the cross product of any two of the three unit vectors **i**, **j**, **k** is equal to positive or negative of the remaining third unit vector depending on whether or not they, in their order, form a right-handed trihedral. A useful mnemonic device for this rule may be constructed as shown in Fig. 4.16.

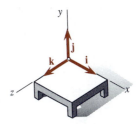

Fig. 4.15

Fig. 4.16

Developmental Exercises

D4.19 What is the value of **P** × (s**P**) where s is a scalar quantity?

D4.20 A 300-lb force **Q** is shown. For $\mathbf{P} = \overrightarrow{OA} = 5\mathbf{i}$ ft and $\mathbf{V} = \mathbf{P} \times \mathbf{Q}$, determine (a) the angle θ between **P** and **Q**, (b) the magnitude V, (c) the unit vector which points in the same direction as **V**, and (d) the analytical expression of **V**.

D4.21 Discuss the cases in which **P** × **Q** = **0**.

D4.22 Find the cross products of the various possible pairs of the three cartesian unit vectors **i**, **j**, **k**.

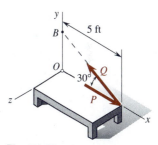

Fig. D4.20

4.8 Properties of Cross Products

By definition, it is obvious that the cross products **P** × **Q** and **Q** × **P** have the same magnitude and orientation but opposite senses as shown in Fig. 4.17; i.e.,

$$\mathbf{P} \times \mathbf{Q} = -\mathbf{Q} \times \mathbf{P} \tag{4.8}$$

This shows that *cross products are not commutative*.

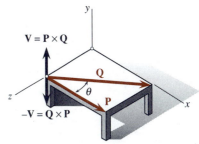

Fig. 4.17

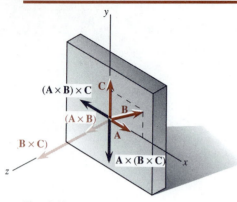

Fig. 4.18

Suppose that the vectors **A**, **B**, and **C** all lie in the *xy* plane as shown in Fig. 4.18. We readily see that both **A** × **B** and **B** × **C** are directed along the *z* axis, (**A** × **B**) × **C** is directed along the negative *x* axis, but **A** × (**B** × **C**) is directed along the negative *y* axis. Thus, we note that, in general

$$\mathbf{A} \times (\mathbf{B} \times \mathbf{C}) \neq (\mathbf{A} \times \mathbf{B}) \times \mathbf{C} \tag{4.9}$$

This shows that *cross products are not associative*.

Suppose that the vectors **Q**, **Q**′, and **P** are coplanar, the tails of **Q**, **Q**′, and **P** are coincident, and the line joining the tips of **Q** and **Q**′ is parallel to **P**, as shown in Fig. 4.19. From the definition and geometric interpretation of the cross product of two vectors in Sec. 4.7, we readily see that

$$\mathbf{P} \times \mathbf{Q} = \mathbf{P} \times \mathbf{Q}' \tag{4.10}$$

This is true because *both* **P** × **Q** *and* **P** × **Q**′ *have the same magnitude and the same direction and are hence equivalent.*

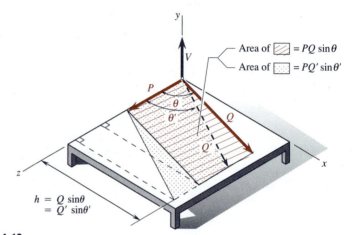

Fig. 4.19

Knowing the equivalent relation expressed in Eq. (4.10), we may proceed to prove that *cross products are distributive* over vector addition; i.e.,

$$\mathbf{A} \times (\mathbf{P} + \mathbf{Q}) = \mathbf{A} \times \mathbf{P} + \mathbf{A} \times \mathbf{Q} \tag{4.11}$$

We first, without losing generality, assume that **A** is directed along the *y* axis as shown in Fig. 4.20. Let the sum of **P** and **Q** be denoted by **R**; i.e., **R** = **P** + **Q**. Dropping perpendiculars from the tips of **P**, **Q**, and **R** onto the *zx* plane, we obtain **P**′, **Q**′, and **R**′ as their projections on the *zx* plane as shown in Fig. 4.20(a). As the projection of a parallelogram onto another plane is also a parallelogram, we note that **R**′ = **P**′ + **Q**′. Furthermore, the quadrilateral with **A** × **P**′ and **A** × **Q**′ as sides and **A** × **R**′ as diagonal is also a parallelogram in the *zx* plane as shown in Fig. 4.20(b); i.e.,

$$\mathbf{A} \times \mathbf{R}' = \mathbf{A} \times \mathbf{P}' + \mathbf{A} \times \mathbf{Q}'$$

Applying Eq. (4.10), we see that **A** × **R**′, **A** × **P**′, and **A** × **Q**′ may, respectively, be replaced by **A** × **R**, **A** × **P**, and **A** × **Q**. Thus, we have

$$\mathbf{A} \times \mathbf{R} = \mathbf{A} \times \mathbf{P} + \mathbf{A} \times \mathbf{Q}$$

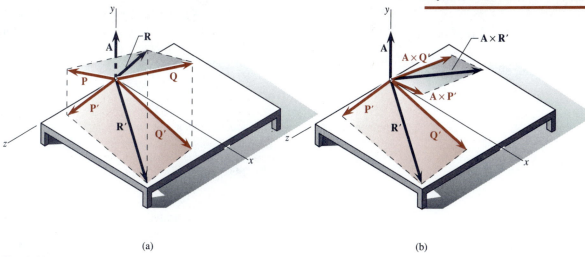

(a)   (b)

Fig. 4.20

which, upon the substitution of **R** by **P** + **Q**, leads to Eq. (4.11). This completes the proof. Moreover, applying Eqs. (4.8) and (4.11), we can readily show that

$$(\mathbf{P} + \mathbf{Q}) \times \mathbf{A} = \mathbf{P} \times \mathbf{A} + \mathbf{Q} \times \mathbf{A} \qquad (4.12)$$

Developmental Exercises

D4.23 It is known that $\mathbf{A} \times \mathbf{B} = 3\mathbf{i} - 4\mathbf{j} + 12\mathbf{k}$. What is the analytical expression of $\mathbf{B} \times \mathbf{A}$?

D4.24 Compute (a) $\mathbf{j} \times (\mathbf{j} \times \mathbf{k})$, (b) $(\mathbf{j} \times \mathbf{j}) \times \mathbf{k}$.

D4.25 What are the conditions that $\mathbf{Q}'$ has to satisfy if $\mathbf{P} \times \mathbf{Q} = \mathbf{P} \times \mathbf{Q}'$?

D4.26 Applying the distributive property of cross products, compute (a) $3\mathbf{j} \times (2\mathbf{i} + \mathbf{j} - \mathbf{k})$, (b) $(4\mathbf{i} - \mathbf{j}) \times (3\mathbf{k})$, (c) $(\mathbf{i} - \mathbf{j}) \times (\mathbf{j} - \mathbf{k})$.

4.9 Analytical Expression of the Cross Product

If the analytical expressions of **P** and **Q** are given as

$$\mathbf{P} = P_x\mathbf{i} + P_y\mathbf{j} + P_z\mathbf{k} \qquad \mathbf{Q} = Q_x\mathbf{i} + Q_y\mathbf{j} + Q_z\mathbf{k}$$

we may use the distributive property of cross products as well as the identities in Eqs. (4.7) to write

$$\mathbf{V} = \mathbf{P} \times \mathbf{Q}$$

$$= P_x\mathbf{i} \times (Q_x\mathbf{i} + Q_y\mathbf{j} + Q_z\mathbf{k}) + P_y\mathbf{j} \times (Q_x\mathbf{i} + Q_y\mathbf{j} + Q_z\mathbf{k})$$

$$+ P_z\mathbf{k} \times (Q_x\mathbf{i} + Q_y\mathbf{j} + Q_z\mathbf{k})$$

$$\mathbf{V} = (P_yQ_z - P_zQ_y)\mathbf{i} + (P_zQ_x - P_xQ_z)\mathbf{j} + (P_xQ_y - P_yQ_x)\mathbf{k} \qquad (4.13)$$

which is the analytical expression of **P** cross **Q**. The scalar components of this cross product are thus found to be

$$V_x = P_y Q_z - P_z Q_y$$

$$V_y = P_z Q_x - P_x Q_z \qquad (4.14)$$

$$V_z = P_x Q_y - P_y Q_x$$

Frequently, we make use of a determinant and write[†]

$$\mathbf{P} \times \mathbf{Q} = \begin{vmatrix} \mathbf{i} & \mathbf{j} & \mathbf{k} \\ P_x & P_y & P_z \\ Q_x & Q_y & Q_z \end{vmatrix} \qquad (4.15)$$

which, when expanded, yields the same analytical expression for **V** as that in Eq. (4.13). The use of a computer to compute the cross product of two vectors is illustrated in Fig. 4.21.

```
10 REM    CROSS PRODUCT OF TWO VECTORS, P X Q = V        (File Name: CROSSPD)
20 REM       Enter the scalar components of P and Q as data in Line 100.
30 REM     To output to the printer, change PRINT to LPRINT in the program.
40 READ PX,PY,PZ,QX,QY,QZ:VX=PY*QZ-PZ*QY:VY=PZ*QX-PX*QZ:VZ=PX*QY-PY*QX:PRINT
50 PRINT"(";PX;"i";:W=PY:GOSUB 80:PRINT"j";:W=PZ:GOSUB 80:PRINT"k ) X (";
60 PRINT QX;"i";:W=QY:GOSUB 80:PRINT"j";:W=QZ:GOSUB 80:PRINT"k )":PRINT"=";
70 PRINT VX;"i";:W=VY:GOSUB 80:PRINT"j";:W=VZ:GOSUB 80:PRINT"k":END
80 S$=" +": IF W<0 THEN W=-W: S$=" -"
90 PRINT S$;W;: RETURN
100 DATA 12,-12,6,16,-12,0

( 12 i - 12 j + 6 k ) X ( 16 i - 12 j + 0 k )
= 72 i + 96 j + 48 k
```

Fig. 4.21 (Cf. the cross product $\overrightarrow{AC} \times \overrightarrow{AB}$ in the solution in Example 4.6)

EXAMPLE 4.6

A pyramid is shown. Making use of the cross product, determine (a) the area A_{ABC} of the side ABC, (b) the angle α as indicated, (c) the outward unit vector $\boldsymbol{\lambda}_{ABC}$, which is perpendicular to the side ABC.

Solution.   From the given sketch, we write

$$\overrightarrow{AC} = 12\mathbf{i} - 12\mathbf{j} + 6\mathbf{k} \qquad |\overrightarrow{AC}| = 18$$

$$\overrightarrow{AB} = 16\mathbf{i} - 12\mathbf{j} \qquad |\overrightarrow{AB}| = 20$$

$$\overrightarrow{AC} \times \overrightarrow{AB} = \begin{vmatrix} \mathbf{i} & \mathbf{j} & \mathbf{k} \\ 12 & -12 & 6 \\ 16 & -12 & 0 \end{vmatrix}$$

$$= \mathbf{i} \begin{vmatrix} -12 & 6 \\ -12 & 0 \end{vmatrix} - \mathbf{j} \begin{vmatrix} 12 & 6 \\ 16 & 0 \end{vmatrix} + \mathbf{k} \begin{vmatrix} 12 & -12 \\ 16 & -12 \end{vmatrix}$$

$$= 72\mathbf{i} + 96\mathbf{j} + 48\mathbf{k} \qquad |\overrightarrow{AC} \times \overrightarrow{AB}| = 129.24$$

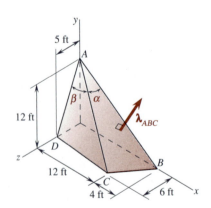

[†]Cf. Sec. B1.4 of App. B for the expansion of a determinant.

From the *geometric interpretation* and the *definition* of cross product, we write

$$A_{ABC} = \frac{1}{2} \left| \overrightarrow{AC} \times \overrightarrow{AB} \right| = \frac{1}{2} (129.24) \qquad A_{ABC} = 64.6 \text{ ft}^2 \quad \blacktriangleleft$$

$$\left| \overrightarrow{AC} \times \overrightarrow{AB} \right| = \left| \overrightarrow{AC} \right| \left| \overrightarrow{AB} \right| \sin\alpha \qquad 129.24 = 18(20) \sin\alpha$$

$$\alpha = 21.0° \quad \blacktriangleleft$$

We know that λ_{ABC} must be parallel to the cross product $\overrightarrow{AC} \times \overrightarrow{AB}$. We write

$$\lambda_{ABC} = \frac{\overrightarrow{AC} \times \overrightarrow{AB}}{\left| \overrightarrow{AC} \times \overrightarrow{AB} \right|}$$

$$\lambda_{ABC} = 0.557\mathbf{i} + 0.743\mathbf{j} + 0.371\mathbf{k} \quad \blacktriangleleft$$

Developmental Exercise

D4.27   Refer to the pyramid in Example 4.6 and the program in Fig. 4.21. Determine (a) the area A_{ACD} of the side ACD, (b) the angle β between the edges AC and AD, (c) the outward unit vector λ_{ACD} which is perpendicular to the side ACD.

4.10   Moment of a Force about a Point by Cross Product

We recall that the moment of a force **F**, acting at the point P, about a point Q in space is a vector $\mathbf{M}_Q$ as shown in Fig. 4.2, which is repeated here for convenience of reference. The *magnitude* of $\mathbf{M}_Q$ is equal to d_sF where F is the magnitude of **F** and d_s is the shortest distance from the point Q to the

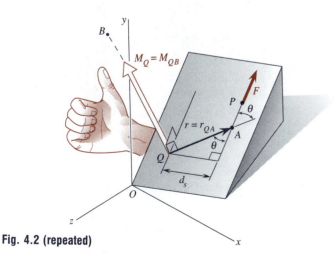

Fig. 4.2 (repeated)

line of action of the force $\mathbf{F}$ as shown. The *orientation* of $\mathbf{M}_Q$ is perpendicular to the plane containing the point Q and the line of action of $\mathbf{F}$. Let the position vector from the point Q to *any* convenient point A on the line of action of $\mathbf{F}$ be denoted by $\mathbf{r}$ as shown. It should be noted carefully that $\mathbf{r}$ is directed *from Q to A*; i.e.,

$$\mathbf{r} = \mathbf{r}_{QA} \qquad (\mathbf{r} \neq \mathbf{r}_{AQ}) \tag{4.16}$$

The *sense* of $\mathbf{M}_Q$ is defined to be such that $\mathbf{r}$, $\mathbf{F}$, and $\mathbf{M}_Q$ in order form a right-handed trihedral. From Fig. 4.2 and the above definition, we have

$$d_s = r \sin\theta \qquad \boxed{M_Q = d_s F} \tag{4.17}$$

Making use of Eq. (4.17) and the cross product defined in Sec. 4.7, we may, therefore, write the moment vector $\mathbf{M}_Q$ as

$$\boxed{\mathbf{M}_Q = \mathbf{r} \times \mathbf{F}} \tag{4.18}$$

Equation (4.18) is the vector formula for computing the moment of a force $\mathbf{F}$ about a point Q. The *resultant moment of a system of forces about a point* is, of course, equal to the vector sum of the moments of the individual forces in the system about that point. If $\mathbf{F}$ is a force of magnitude one, we have, from Eq. (4.17), $M_Q = d_s$. This means that *the shortest distance between a point and a line* is numerically equal to the magnitude of the moment of a unit force, directed along that line, about that point.

EXAMPLE 4.7

A 130-N thrust $\mathbf{F}$ is generated at the point $B(0.5, 0.6, 0)$ m of a pipe and points toward the joint at $C(1.7, 1.1, 0)$ m as shown. Determine (a) the moment $\mathbf{M}_A$ of $\mathbf{F}$ about the joint at $A(0.3, -0.2, 1)$ m, (b) the shortest distance d_s from A to the line of action of $\mathbf{F}$.

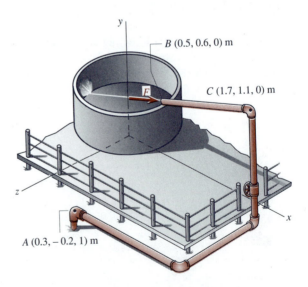

B (0.5, 0.6, 0) m
C (1.7, 1.1, 0) m
A (0.3, −0.2, 1) m

Solution.   For determining the moment $\mathbf{M}_A$, we write

$$\vec{BC} = 1.2\mathbf{i} + 0.5\mathbf{j} \qquad |\vec{BC}| = 1.3 \qquad \lambda_{BC} = \frac{\vec{BC}}{|\vec{BC}|} = \frac{1}{13}(12\mathbf{i} + 5\mathbf{j})$$

$$F = 130 \qquad \mathbf{F} = F\boldsymbol{\lambda}_{BC} = 120\mathbf{i} + 50\mathbf{j} \qquad \mathbf{r} = \overrightarrow{AB} = 0.2\mathbf{i} + 0.8\mathbf{j} - \mathbf{k}$$

$$\mathbf{M}_A = \mathbf{r} \times \mathbf{F} = \begin{vmatrix} \mathbf{i} & \mathbf{j} & \mathbf{k} \\ 0.2 & 0.8 & -1 \\ 120 & 50 & 0 \end{vmatrix}$$

$$\mathbf{M}_A = 50\mathbf{i} - 120\mathbf{j} - 86\mathbf{k} \text{ N·m} \blacktriangleleft$$

Since $M_A = d_s F$, we write

$$[(50)^2 + (120)^2 + (86)^2]^{1/2} = d_s(130) \qquad d_s = 1.199 \text{ m} \blacktriangleleft$$

Developmental Exercises

D4.28 Define the magnitude, the orientation, and the sense of the moment $\mathbf{M}_Q$ of a force $\mathbf{F}$ about a point Q.

D4.29 A 150-N force $\mathbf{P}$ is shown below. (a) Write the analytical expression of $\mathbf{P}$. (b) Is the moment $\mathbf{M}_A$ of $\mathbf{P}$ about the point A given by $\mathbf{r}_{AB} \times \mathbf{P}$ or $\mathbf{r}_{BA} \times \mathbf{P}$? (c) Determine $\mathbf{M}_A$. (d) Is $\mathbf{M}_A = \mathbf{r}_{AD} \times \mathbf{P}$? (e) Determine the shortest distance d_s from A to the line of action of $\mathbf{P}$.

D4.30 Refer to D4.29. In addition to the force $\mathbf{P}$ shown, suppose that another force $\mathbf{Q} = 100\mathbf{k}$ N is acting at the point C. Determine the resultant moment of $\mathbf{P}$ and $\mathbf{Q}$ about the point A.

D4.31 Using the concept of moment of a force, describe the general procedure to find the shortest distance between a point and a line.

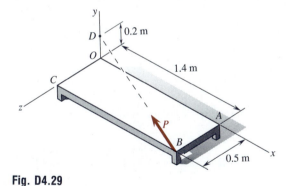

Fig. D4.29

4.11 Varignon's Theorem: Forces in Space

In general, *Varignon's theorem* may be stated as follows: *the moment $\mathbf{M}_P$, about any point P, of the resultant $\mathbf{R}$ of two or more concurrent forces $\mathbf{F}_1$, $\mathbf{F}_2$, . . . , $\mathbf{F}_n$ acting at the point C is equal to the sum of the moments $\mathbf{M}_1$, $\mathbf{M}_2$, . . . , $\mathbf{M}_n$ of the various forces $\mathbf{F}_1$, $\mathbf{F}_2$, . . . , $\mathbf{F}_n$ at C about the same point P.* To prove this theorem, let the mentioned system be first illustrated as shown in Fig. 4.22. Then, applying the distributive property of cross

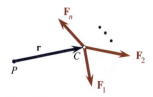

Fig. 4.22

products, we write

$$\mathbf{M}_P = \mathbf{r} \times \mathbf{R} = \mathbf{r} \times (\mathbf{F}_1 + \mathbf{F}_2 + \cdots + \mathbf{F}_n)$$
$$= \mathbf{r} \times \mathbf{F}_1 + \mathbf{r} \times \mathbf{F}_2 + \cdots + \mathbf{r} \times \mathbf{F}_n$$

Since $\mathbf{M}_1 = \mathbf{r} \times \mathbf{F}_1$, $\mathbf{M}_2 = \mathbf{r} \times \mathbf{F}_2, \ldots,$ and $\mathbf{M}_n = \mathbf{r} \times \mathbf{F}_n$, we have

$$\mathbf{M}_P = \mathbf{M}_1 + \mathbf{M}_2 + \cdots + \mathbf{M}_n \qquad (4.19)$$

This completes the proof.

EXAMPLE 4.8

A guyed pole AO is shown. The tensions in the cables AB and AC are T_{AB} = 27 kN and T_{AC} = 28 kN. Determine the range of values of the tension T_{AD} in the cable AD for which the magnitude of the resultant moment developed at the base O of the pole will not exceed 200 kN·m.

Solution.　Referring to the figure shown, we write

$$\mathbf{T}_{AB} = 27\boldsymbol{\lambda}_{AB} = \frac{27}{3} (2\mathbf{i} - 2\mathbf{j} - \mathbf{k})$$

$$\mathbf{T}_{AC} = 28\boldsymbol{\lambda}_{AC} = \frac{28}{7} (-2\mathbf{i} - 6\mathbf{j} - 3\mathbf{k})$$

$$\mathbf{T}_{AD} = T_{AD}\boldsymbol{\lambda}_{AD} = \frac{T_{AD}}{5} (-4\mathbf{j} + 3\mathbf{k})$$

The resultant force $\mathbf{R}$ at A is

$$\mathbf{R} = \mathbf{T}_{AB} + \mathbf{T}_{AC} + \mathbf{T}_{AD}$$
$$= 10\mathbf{i} - \left(42 + \frac{4}{5} T_{AD}\right)\mathbf{j} + \left(-21 + \frac{3}{5} T_{AD}\right)\mathbf{k}$$

By Varignon's theorem, the resultant moment $\mathbf{M}_O$ developed at the base O is

$$\mathbf{M}_O = \mathbf{r}_{OA} \times \mathbf{R} = 12\mathbf{j} \times \mathbf{R} = \left(\frac{36}{5} T_{AD} - 252\right)\mathbf{i} - 120\mathbf{k}$$

Setting the magnitude of $\mathbf{M}_O$ equal to 200 kN·m, we write

$$\left[\left(-252 + \frac{36}{5} T_{AD}\right)^2 + (120)^2\right]^{1/2} = 200$$

Squaring both sides of this equation and solving the resulting quadratic equation, we obtain T_{AD} = 12.78 or 57.2. For $M_O \leq$ 200 kN·m, we must have

$$12.78 \text{ kN} \leq T_{AD} \leq 57.2 \text{ kN} \blacktriangleleft$$

Developmental Exercise

D4.32   Three forces act at the free end of a cantilevered bent rod as shown. Determine the resultant moment of these forces about the fixed end at O.

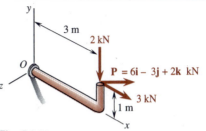

Fig. D4.32

4.12   Moment of a Couple by Cross Product

We know that the moment of a couple about a point in space is the sum of the moments of the two forces in the couple about that point. Thus, the moment $\mathbf{M}_O$ of the couple $\mathbf{F}$ and $-\mathbf{F}$ about the point O in space, as shown in Fig. 4.23, is

$$\mathbf{M}_O = \mathbf{r}_A \times \mathbf{F} + \mathbf{r}_B \times (-\mathbf{F}) = (\mathbf{r}_A - \mathbf{r}_B) \times \mathbf{F}$$

$$\mathbf{r} = \mathbf{r}_A - \mathbf{r}_B$$

$$\mathbf{M}_O = \mathbf{r} \times \mathbf{F} \qquad (4.20)$$

where the points A and B may be any points on the lines of action of $\mathbf{F}$ and $-\mathbf{F}$, respectively. Since $\mathbf{r}$ is actually any position vector *from* any point on the line of action of $-\mathbf{F}$ *to* any point on the line of action of $\mathbf{F}$, the moment

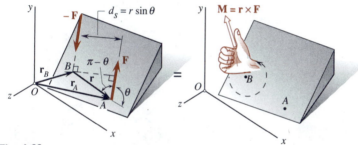

Fig. 4.23

$\mathbf{M}_O$ of the couple is, in fact, independent of the choice of the point O. This confirms that the moment of a couple is a *free vector*. Dropping the subscript O in $\mathbf{M}_O$, we may write this free vector as

$$\mathbf{M} = \mathbf{r} \times \mathbf{F} \qquad (4.21)$$

In using Eq. (4.21), note that *either* of the two forces in a couple can be designated as $\mathbf{F}$ and the other as $-\mathbf{F}$, even though the analytical expression of the force designated as $\mathbf{F}$ may contain negative terms. However, $\mathbf{r}$ must be drawn *from* the line of action of the force designated as $-\mathbf{F}$ *to* the line of action of the force designated as $\mathbf{F}$.

The moment $\mathbf{M}$ of a couple consisting of $\mathbf{F}$ and $-\mathbf{F}$ is a vector perpendicular to the plane containing $\mathbf{F}$, $-\mathbf{F}$, and $\mathbf{r}$. This moment points in the *direction* of the extended thumb of the right hand when the other four fingers of the right hand are curled in the direction of the tendency of the couple to rotate. The *magnitude* of $\mathbf{M}$ is

$$M = rF \sin\theta = (r \sin\theta)F \qquad M = d_s F \qquad (4.22)$$

where d_s is the shortest distance between the lines of action of $\mathbf{F}$ and $-\mathbf{F}$.

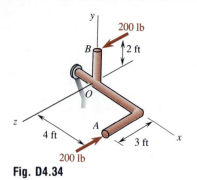

Fig. D4.34

Developmental Exercises

D4.33 The formula $\mathbf{M} = \mathbf{r} \times \mathbf{F}$ is to be used to compute the moment of a couple. (a) Which of the two forces in the couple should be designated as $\mathbf{F}$ and which as $-\mathbf{F}$? (b) Is $\mathbf{r}$ drawn from $\mathbf{F}$ to $-\mathbf{F}$ or from $-\mathbf{F}$ to $\mathbf{F}$?

D4.34 A couple consists of two 200-lb forces as shown. Compute the moments of the couple about the points (a) A, (b) B, (c) O, (d) any point P.

4.13 Equivalent Couples and Addition of Moments of Couples

We know that a couple is characterized by its moment $\mathbf{M}$. Two couples are *equivalent* if they have the *same moment* $\mathbf{M}$. The couples illustrated in Fig. 4.24 are equivalent because they have the same moment, $120\mathbf{j}$ N·m. All equivalent couples must lie in parallel planes and have the same effect of rotation on a rigid body. The plane of the couple in Fig. 4.24 may be any plane perpendicular to $\mathbf{M}$.

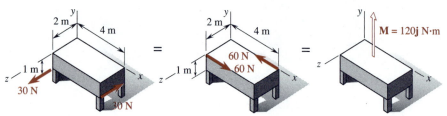

Fig. 4.24

Couples are added according to the parallelogram law (in the same way as forces or any other vectors are added), as illustrated in Fig. 4.25. Furthermore, the *resultant moment* $\mathbf{M}$ of any number of couples is equal to the vector sum of the moments, $\mathbf{M}_1, \mathbf{M}_2, \ldots, \mathbf{M}_n$, of these couples; i.e.,

$$\mathbf{M} = \mathbf{M}_1 + \mathbf{M}_2 + \cdots + \mathbf{M}_n \tag{4.23}$$

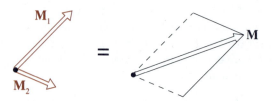

Fig. 4.25

EXAMPLE 4.9

A system of three couples is shown. Determine the analytical expressions of
(a) the moment $\mathbf{M}_1$ produced by the pair of 100-N forces, (b) the moment
$\mathbf{M}_2$ produced by the pair of 200-N forces, (c) the moment $\mathbf{M}_3$ produced by
the pair of 150-N forces, (d) the resultant moment $\mathbf{M}$ of $\mathbf{M}_1$, $\mathbf{M}_2$, and $\mathbf{M}_3$.

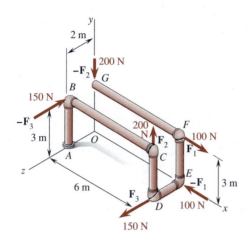

Solution.   Using Eqs. (4.21) and (4.23), we write

$$\mathbf{M}_1 = \mathbf{r}_{EF} \times \mathbf{F}_1 = 3\mathbf{j} \times (100\mathbf{i})$$

$$\mathbf{M}_1 = -300\mathbf{k} \ \text{N} \cdot \text{m} \blacktriangleleft$$

$$\mathbf{M}_2 = \mathbf{r}_{GC} \times \mathbf{F}_2 = (6\mathbf{i} + 2\mathbf{k}) \times (200\mathbf{j})$$

$$\mathbf{M}_2 = -400\mathbf{i} + 1200\mathbf{k} \ \text{N} \cdot \text{m} \blacktriangleleft$$

$$\mathbf{M}_3 = \mathbf{r}_{BD} \times \mathbf{F}_3 = (6\mathbf{i} - 3\mathbf{j}) \times (150\mathbf{k})$$

$$\mathbf{M}_3 = -450\mathbf{i} - 900\mathbf{j} \ \text{N} \cdot \text{m} \blacktriangleleft$$

$$\mathbf{M} = \mathbf{M}_1 + \mathbf{M}_2 + \mathbf{M}_3$$

$$\mathbf{M} = -850\mathbf{i} - 900\mathbf{j} + 900\mathbf{k} \ \text{N} \cdot \text{m} \blacktriangleleft$$

Developmental Exercise

D4.35   For the system of forces shown, determine the analytical expressions of
(a) the moment $\mathbf{M}_1$ produced by the pair of 2-kN forces, (b) the moment $\mathbf{M}_2$
produced by the pair of 4-kN forces, (c) the resultant moment $\mathbf{M}$ of $\mathbf{M}_1$ and $\mathbf{M}_2$.

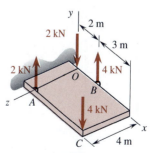

Fig. D4.35

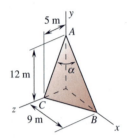

Fig. P4.35 and P4.38

PROBLEMS

4.33 If $\mathbf{P} = 2\mathbf{i} - \mathbf{j} + 3\mathbf{k}$ and $\mathbf{Q} = \mathbf{i} - 2\mathbf{j} + \mathbf{k}$, determine (a) $\mathbf{P} \times \mathbf{Q}$, (b) $\mathbf{Q} \times \mathbf{P}$, (c) $\mathbf{P} \times (\mathbf{P} - \mathbf{Q})$, (d) $|\mathbf{P} \times \mathbf{Q}|$, (e) the angle between $\mathbf{P}$ and $\mathbf{Q}$.

4.34* If $\mathbf{S} = \mathbf{i} - 3\mathbf{j} - 2\mathbf{k}$ and $\mathbf{T} = 2\mathbf{i} + \mathbf{j} + \mathbf{k}$, determine (a) $(\mathbf{T} + \mathbf{S}) \times \mathbf{T}$, (b) $(\mathbf{T} + \mathbf{S}) \times \mathbf{S}$, (c) $|\mathbf{S} \times \mathbf{T}|$, (d) the angle between $\mathbf{S}$ and $\mathbf{T}$.

4.35 through 4.37* Determine the area A_{ABC} of the side ABC of the pyramid shown.

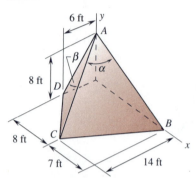

Fig. P4.36, P4.39, and P4.41

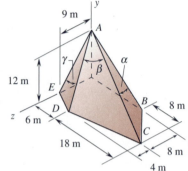

Fig. P4.37*, P4.40, P4.42, and P4.43*

4.38 through 4.40 Determine the angle α indicated on the pyramid shown.

4.41 and 4.42 For the pyramid shown, determine (a) the area A_{ACD} of the side ACD, (b) the angle β as indicated.

4.43* For the pyramid shown, determine (a) the area A_{ADE} of the side ADE, (b) the angle γ as indicated.

4.44 and 4.45 For the system shown, determine the moment $\mathbf{M}_A$ of $\mathbf{F}$ about the point A in five different ways.

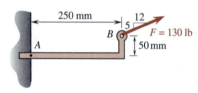

Fig. P4.44

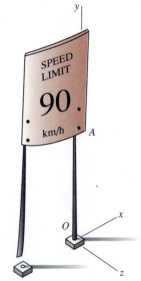

Fig. P4.46

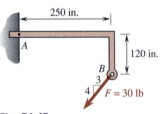

Fig. P4.45

4.46 A car sideswipes a highway sign and breaks one of its supports, leaving the broken and twisted sign supported only by the post OA as shown. A wind produces a resultant force of $100\mathbf{i} - 400\mathbf{k}$ N acting at the center $C(-0.4, 2, -0.1)$ m of the sign. Determine the moment $\mathbf{M}_O$ of the wind force about the point O.

4.47 and 4.48* A system of four forces acts on a cantilevered bent rod as shown. Determine the moment about the fixed end A of (a) $\mathbf{P}$, (b) the entire system of four forces.

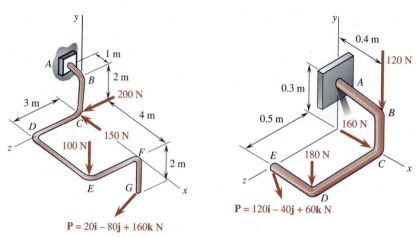

Fig. P4.47

Fig. P4.48*

4.49 and 4.50   Find the analytical expression of the resultant moment $\mathbf{M}_A^R$, in terms of known and unknown quantities, of the force system acting on the free body about the point A as shown.

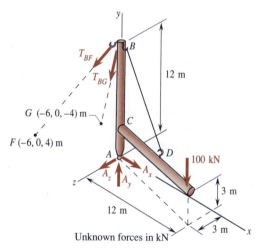

Fig. P4.49 and P4.51

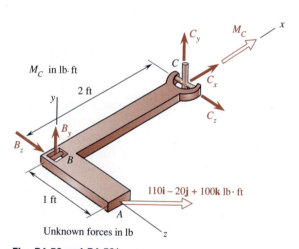

Fig. P4.50 and P4.52*

4.51 and 4.52*   Find the analytical expression of the resultant moment $\mathbf{M}_B^R$ of the force system acting on the free body about the point B as shown.

4.53 through 4.56*   A force system consisting of couples is shown. Determine the analytical expressions of (a) the moment $\mathbf{M}_i$ ($i = 1, 2, \ldots$) of each individual couple, (b) the resultant moment $\mathbf{M}$ of all couples, (c) the resultant moment $\mathbf{M}_A$ of the force system about the support at A.

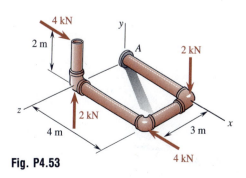

Fig. P4.53

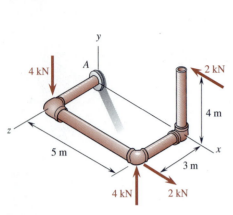

Fig. P4.54

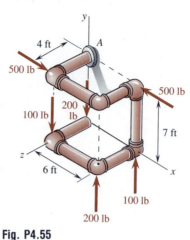

Fig. P4.55

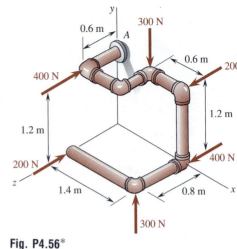

Fig. P4.56*

APPLICATION OF DOT PRODUCTS

Note. Those who are familiar with the rudiments of dot products as presented in Secs. 4.14 and 4.15 may proceed directly to Sec. 4.16.

4.14   Dot Product and Its Properties

The *dot product* of two vectors **P** and **Q**, as shown in Fig. 4.26, is defined as a scalar which is equal to the product of the magnitudes of **P** and **Q** and the cosine of the angle θ between the positive directions of **P** and **Q**. Usually, we take $0 \le \theta \le 180°$, although this is not necessary because $\cos(360° - \theta) = \cos\theta$. As in vector analysis, the dot product of **P** and **Q** is denoted by **P** · **Q**, which is read "**P** dot **Q**." Thus, we write

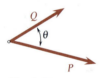

Fig. 4.26

$$\mathbf{P} \cdot \mathbf{Q} = PQ \cos\theta \tag{4.24}$$

$$PQ \cos\theta = QP \cos\theta \qquad \mathbf{P} \cdot \mathbf{Q} = \mathbf{Q} \cdot \mathbf{P} \tag{4.25}$$

Equations (4.25) show that the *dot product of two vectors is commutative*. Because of the nature of the product, the dot product is also called the *scalar product*; sometimes, it is also called the *inner product*.

From the above definition and Fig. 4.15, it can readily be shown that

$$\mathbf{i} \cdot \mathbf{i} = 1 \qquad \mathbf{j} \cdot \mathbf{j} = 1 \qquad \mathbf{k} \cdot \mathbf{k} = 1$$
$$\mathbf{i} \cdot \mathbf{j} = 0 \qquad \mathbf{j} \cdot \mathbf{k} = 0 \qquad \mathbf{k} \cdot \mathbf{i} = 0 \tag{4.26}$$

Furthermore, when **P** · **Q** = 0, it implies one or more of the following cases: (a) **P** and **Q** are perpendicular to each other, (b) **P** is a zero vector, (c) **Q** is a zero vector.

From the definition of the dot product, we see that the projection of **Q** on **P**, as shown in Fig. 4.27, may be written as

Fig. 4.27

Projection of **Q** *on* **P** $= Q \cos\theta = (1)Q \cos\theta = \lambda_P \cdot \mathbf{Q}$

where λ_P is a unit vector in the direction of **P**. Similarly, the projection of **P** on **Q** may be written as

$$\textit{Projection of } \mathbf{P} \textit{ on } \mathbf{Q} = P \cos\theta = (1)P \cos\theta = \lambda_Q \cdot \mathbf{P}$$

where λ_Q is a unit vector in the direction of **Q**.

For the vectors **A**, **B**, **C**, **B** + **C**, and the unit vector λ_A as shown in Fig. 4.28, we see that the projections on **A** have the following relationship:

$$\lambda_A \cdot (\mathbf{B} + \mathbf{C}) = \lambda_A \cdot \mathbf{B} + \lambda_A \cdot \mathbf{C} \qquad (4.27)$$

Multiplying both sides of Eq. (4.27) by A and noting that $\mathbf{A} = A\lambda_A$, we write

$$A\lambda_A \cdot (\mathbf{B} + \mathbf{C}) = A\lambda_A \cdot \mathbf{B} + A\lambda_A \cdot \mathbf{C}$$

$$\mathbf{A} \cdot (\mathbf{B} + \mathbf{C}) = \mathbf{A} \cdot \mathbf{B} + \mathbf{A} \cdot \mathbf{C} \qquad (4.28)$$

which reveals that *dot products are distributive*.

Using the above distributive property and Eqs. (4.26), we write

$$\mathbf{P} \cdot \mathbf{Q} = (P_x\mathbf{i} + P_y\mathbf{j} + P_z\mathbf{k}) \cdot (Q_x\mathbf{i} + Q_y\mathbf{j} + Q_z\mathbf{k})$$

$$\mathbf{P} \cdot \mathbf{Q} = P_xQ_x + P_yQ_y + P_zQ_z \qquad (4.29)$$

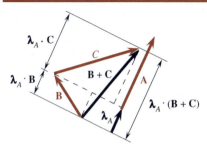

Fig. 4.28

Developmental Exercises

D4.36 Vectors **A**, **B**, **C**, and **D** as shown have magnitudes of 4, 2, 3, and 5, respectively. Compute (a) $\mathbf{A} \cdot \mathbf{B}$, (b) $\mathbf{B} \cdot \mathbf{A}$, (c) $\mathbf{A} \cdot \mathbf{C}$, (d) $\mathbf{B} \cdot \mathbf{C}$, (e) $\mathbf{A} \cdot \mathbf{A}$, (f) $\mathbf{B} \cdot \mathbf{D}$.

D4.37 Vectors **P** and **Q** have magnitudes of 10 and 20, respectively, and the angle θ between them is 30°. Determine (a) $\mathbf{P} \cdot \mathbf{Q}$, (b) the projection of **P** on **Q**, (c) the projection of **Q** on **P**.

D4.38 If $\mathbf{P} = 2\mathbf{i} - \mathbf{j} + 3\mathbf{k}$ and $\mathbf{Q} = \mathbf{i} - 2\mathbf{j} + \mathbf{k}$, determine $\mathbf{P} \cdot \mathbf{Q}$.

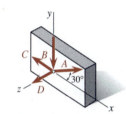

Fig. D4.36

EXAMPLE 4.10

Refer to the pyramid in Example 4.6. Making use of the cross product and the dot product, determine the interior angle $\theta_{ABC/zx}$ between the side ABC and the zx plane.

Solution. The unit vector λ_{ABC} perpendicular to the side ABC has been found, by using the cross product, in Example 4.6 as

$$\lambda_{ABC} = 0.557\mathbf{i} + 0.743\mathbf{j} + 0.371\mathbf{k}$$

The unit vector perpendicular to the zx plane is **j**. The interior angle $\theta_{ABC/zx}$ is congruent to the angle between λ_{ABC} and **j**. Applying the dot product, we write

$$\cos\theta_{ABC/zx} = 1 \cdot 1 \cdot \cos\theta_{ABC/zx} = \lambda_{ABC} \cdot \mathbf{j} = 0.743$$

$$\theta_{ABC/zx} = 42.0° \blacktriangleleft$$

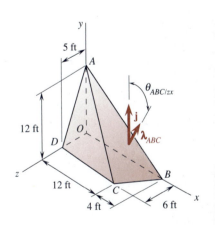

Developmental Exercises

D4.39 In the figure shown, determine, as appropriate, the analytical expression or value of (a) λ_{OA}, (b) λ_{OB}, (c) $\lambda_{OA} \cdot \lambda_{OB}$, (d) $\cos\theta$, (e) θ, (f) $\lambda_{OA} \cdot \overrightarrow{OB}$, (g) the projection of $\overrightarrow{OB}$ on $\overrightarrow{OA}$.

D4.40 Refer to Example 4.10. Determine the interior angle $\theta_{ABC/ACD}$ between the side *ABC* and the side *ACD*.

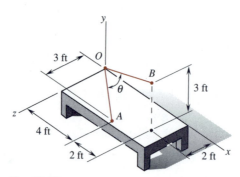

Fig. D4.39

4.15 Scalar Triple Product

The dot product of **A** with the cross product of **P** and **Q** expressed as **A** · (**P** × **Q**) is called the *scalar triple product* of **A**, **P**, and **Q**. In terms of scalar components, we write

$$\mathbf{A} \cdot (\mathbf{P} \times \mathbf{Q}) = A_x(P_yQ_z - P_zQ_y) + A_y(P_zQ_x - P_xQ_z) \\ + A_z(P_xQ_y - P_yQ_x) \tag{4.30}$$

In the form of a determinant, we write

$$\mathbf{A} \cdot (\mathbf{P} \times \mathbf{Q}) = \begin{vmatrix} A_x & A_y & A_z \\ P_x & P_y & P_z \\ Q_x & Q_y & Q_z \end{vmatrix} \tag{4.31}$$

which is more easily remembered. Using the rules governing the interchanging of rows in a determinant (cf. Sec. B1.4 of App. B) and the commutation of dot products, we can readily verify that

$$\mathbf{A} \cdot (\mathbf{P} \times \mathbf{Q}) = \mathbf{P} \cdot (\mathbf{Q} \times \mathbf{A}) = \mathbf{Q} \cdot (\mathbf{A} \times \mathbf{P}) = (\mathbf{A} \times \mathbf{P}) \cdot \mathbf{Q}$$

Thus, we see that[†]

$$\mathbf{A} \cdot (\mathbf{P} \times \mathbf{Q}) = (\mathbf{A} \times \mathbf{P}) \cdot \mathbf{Q} \tag{4.32}$$

where · and × can be interchanged if the cross product is computed before the dot product and the order of the vectors is maintained.

[†]Since (**A** · **P**) × **Q** is not defined and has no meaning, the parentheses in Eq. (4.32) may be omitted without causing ambiguity.

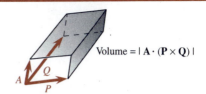

The scalar triple product of **A**, **P**, and **Q** is sometimes called the *mixed triple product* or *box product* and may be denoted as [**A P Q**]. Geometrically, the absolute value of the scalar triple product is equal to the volume of the parallelepiped, as illustrated in Fig. 4.29, which has the three vectors as its adjacent edges. Of course, if any two of the three vectors **A**, **P**, and **Q** are parallel to each other, or **A**, **P**, and **Q** are coplanar, then the scalar triple product [**A P Q**] must be zero. The use of a computer to compute a scalar triple product is illustrated in Fig. 4.30.

Fig. 4.29

```
10 REM    SCALAR TRIPLE PRODUCT, A . (P X Q) = S      (File Name: STRIPPD)
20 REM    Enter the scalar components of A, P, and Q as data in Line 100.
30 REM    To output to the printer, change PRINT to LPRINT in the program.
40 READ AX,AY,AZ,PX,PY,PZ,QX,QY,QZ:S1=AX*(PY*QZ-PZ*QY)+AY*(PZ*QX-PX*QZ)
50 S=S1+AZ*(PX*QY-PY*QX): PRINT:PRINT"The components of A, P, and Q are"
60 PRINT AX,AY,AZ:PRINT PX,PY,PZ:PRINT QX,QY,QZ:PRINT"Their scalar ";
70 PRINT "triple product is ";S;"."
100 DATA -12,0,5,-4,4,0,-2,1,2

The components of A, P, and Q are
-12          0          5
-4           4          0
-2           1          2
Their scalar triple product is -76 .
```

Fig. 4.30   (Cf. the scalar triple product in the solution in Example 4.12.)

Developmental Exercise

D4.41   Let **A** = **i** + 2**j** ft, **P** = 3**i** ft, and **Q** = −**k** ft be the adjacent edges of a parallelepiped. (a) Evaluate **A** · (**P** × **Q**). (b) What is the volume of this parallelepiped? (c) Is (**A** · **P**) × **Q** defined? (d) Using the program in Fig. 4.30, verify the answer to (a).

4.16   Moment of a Force about an Axis

The *moment of a force about an axis* is equal to the projection of the moment vector of the force, computed about any point on the axis, onto that axis. Suppose that the line of action of a force **F** passes through the point *A*, and a point *D* is on the axis *BC* as shown on the next page in Fig. 4.31(a). Then, the *moment* M_{BC} of **F** about the axis *BC* is the projection of its moment $\mathbf{M}_D$ onto the axis *BC* as shown in Fig. 4.31(b). We write

$$M_{BC} = \boldsymbol{\lambda}_{BC} \cdot \mathbf{M}_D \qquad \boxed{M_{BC} = \boldsymbol{\lambda}_{BC} \cdot (\mathbf{r} \times \mathbf{F})} \qquad (4.33)$$

where

$$\boldsymbol{\lambda}_{BC} = \frac{\overrightarrow{BC}}{|\overrightarrow{BC}|} = \text{unit vector pointing in the direction of } \overrightarrow{BC}.$$

r = a position vector from any point on the axis *BC* (e.g., the point *D*) to any convenient point on the line of action of **F** (e.g., the point *A*).

The vector form of the moment M_{BC} may be written as

$$\mathbf{M}_{BC} = [\boldsymbol{\lambda}_{BC} \cdot (\mathbf{r} \times \mathbf{F})]\boldsymbol{\lambda}_{BC} \qquad (4.34)$$

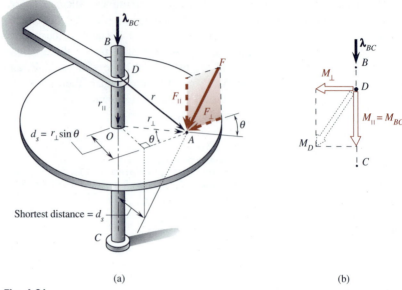

(a)   (b)

Fig. 4.31

If M_x, M_y, and M_z are the rectangular scalar components of the moment $\mathbf{M}_O$ of a force about the origin O, as shown in Fig. 4.32, then

$$\mathbf{M}_O = M_x\mathbf{i} + M_y\mathbf{j} + M_z\mathbf{k} \qquad (4.35)$$

From Eqs. (4.35) and (4.33), we note that the rectangular scalar components M_x, M_y, and M_z are in fact the moment of the force about the x, y, and z axes, respectively.

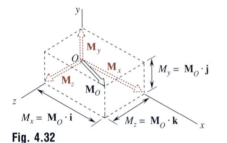

Fig. 4.32

EXAMPLE 4.11

A free body is acted on by a force system as shown, where the unknown and known forces have the same unit, and $\mathbf{B}_h$ lies in the horizontal plane and is perpendicular to $\overline{AB}$. Determine (a) the resultant moment $\mathbf{M}_A^R$ of the force system about the point A, (b) the resultant moment M_{AB}^R of the force system about the axis passing through the points A and B.

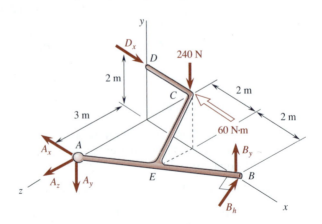

Solution. Since $\mathbf{B}_h$ is perpendicular to $\overline{AB}$, we see from the top view of the free body, as shown, that

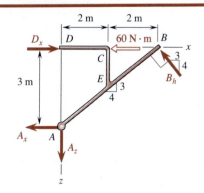

$$\mathbf{B}_h = \frac{B_h}{5}(-3\mathbf{i} - 4\mathbf{k})$$

The resultant force of $\mathbf{B}_y$ and $\mathbf{B}_h$ is

$$\mathbf{B} = -\frac{3}{5}B_h\mathbf{i} + B_y\mathbf{j} - \frac{4}{5}B_h\mathbf{k}$$

The forces at the points C and D are

$$\mathbf{C} = -240\mathbf{j} \qquad \mathbf{D} = D_x\mathbf{i}$$

The relevant position vectors for computing $\mathbf{M}_A^R$ are

$$\mathbf{r}_{AB} = \overrightarrow{AB} = 4\mathbf{i} - 3\mathbf{k}$$
$$\mathbf{r}_{AC} = \overrightarrow{AC} = 2\mathbf{i} + 2\mathbf{j} - 3\mathbf{k}$$
$$\mathbf{r}_{AD} = \overrightarrow{AD} = 2\mathbf{j} - 3\mathbf{k}$$

The moment at the point C is $\mathbf{M}_C = -60\mathbf{i}$, which is a free vector. Noting that the forces at the point A have no contribution toward $\mathbf{M}_A^R$, we write

$$\mathbf{M}_A^R = \mathbf{M}_C + \mathbf{r}_{AB} \times \mathbf{B} + \mathbf{r}_{AC} \times \mathbf{C} + \mathbf{r}_{AD} \times \mathbf{D}$$

$$= -60\mathbf{i} + \begin{vmatrix} \mathbf{i} & \mathbf{j} & \mathbf{k} \\ 4 & 0 & -3 \\ -\frac{3}{5}B_h & B_y & -\frac{4}{5}B_h \end{vmatrix} + \begin{vmatrix} \mathbf{i} & \mathbf{j} & \mathbf{k} \\ 2 & 2 & -3 \\ 0 & -240 & 0 \end{vmatrix}$$

$$+ \begin{vmatrix} \mathbf{i} & \mathbf{j} & \mathbf{k} \\ 0 & 2 & -3 \\ D_x & 0 & 0 \end{vmatrix}$$

$$\mathbf{M}_A^R = (3B_y - 780)\mathbf{i} + (5B_h - 3D_x)\mathbf{j}$$
$$+ (4B_y - 2D_x - 480)\mathbf{k} \quad \text{N·m} \quad \blacktriangleleft$$

The unit vector in the direction of $\overrightarrow{AB}$ is

$$\boldsymbol{\lambda}_{AB} = \frac{\overrightarrow{AB}}{|\overrightarrow{AB}|} = \frac{1}{5}(4\mathbf{i} - 3\mathbf{k})$$

The resultant moment M_{AB}^R is the projection of $\mathbf{M}_A^R$ onto $\overline{AB}$. Thus, we write

$$M_{AB}^R = \boldsymbol{\lambda}_{AB} \cdot \mathbf{M}_A^R \qquad M_{AB}^R = 1.2D_x - 336 \quad \text{N·m} \quad \blacktriangleleft$$

Developmental Exercise

D4.42 Describe the procedure for finding the moment of a force about a given axis.

★4.17   Shortest Distance between an Axis and a Force[†]

Physically, the *moment* M_{BC} about the axis BC of a force $\mathbf{F}$ acting on a body measures the effectiveness or tendency of $\mathbf{F}$ to rotate the body about that axis. If M_{BC} is *positive*, it means that the sense of the effectiveness or tendency to rotate the body about the axis BC is, according to the right-hand rule, the same as that of $\boldsymbol{\lambda}_{BC}$; *otherwise*, it is opposite to the sense of $\boldsymbol{\lambda}_{BC}$.

By resolving each of the vectors $\mathbf{r}$ and $\mathbf{F}$ into two rectangular components, one parallel to the axis BC and the other perpendicular to the axis BC as shown in Fig. 4.31, we have

$$\mathbf{r} = \mathbf{r}_{\parallel} + \mathbf{r}_{\perp} \qquad \mathbf{F} = \mathbf{F}_{\parallel} + \mathbf{F}_{\perp} \tag{4.36}$$

where the subscripts $_{\parallel}$ and $_{\perp}$ denote parallel and perpendicular components, respectively. Noting that both $\mathbf{r}_{\parallel}$ and $\mathbf{F}_{\parallel}$ are parallel to $\boldsymbol{\lambda}_{BC}$ in Fig. 4.31 and applying Eqs. (4.36) and the conditions for a scalar triple product to have zero value in Sec. 4.15, we may write the moment M_{BC} as

$$\begin{aligned}
M_{BC} &= \boldsymbol{\lambda}_{BC} \cdot (\mathbf{r} \times \mathbf{F}) = \boldsymbol{\lambda}_{BC} \cdot [(\mathbf{r}_{\parallel} + \mathbf{r}_{\perp}) \times (\mathbf{F}_{\parallel} + \mathbf{F}_{\perp})] \\
&= \boldsymbol{\lambda}_{BC} \cdot (\mathbf{r}_{\parallel} \times \mathbf{F}_{\parallel}) + \boldsymbol{\lambda}_{BC} \cdot (\mathbf{r}_{\parallel} \times \mathbf{F}_{\perp}) + \boldsymbol{\lambda}_{BC} \cdot (\mathbf{r}_{\perp} \times \mathbf{F}_{\parallel}) \\
&\quad + \boldsymbol{\lambda}_{BC} \cdot (\mathbf{r}_{\perp} \times \mathbf{F}_{\perp}) \\
&= 0 + 0 + 0 + \boldsymbol{\lambda}_{BC} \cdot (\mathbf{r}_{\perp} \times \mathbf{F}_{\perp})
\end{aligned}$$

$$M_{BC} = \boldsymbol{\lambda}_{BC} \cdot (\mathbf{r}_{\perp} \times \mathbf{F}_{\perp}) \tag{4.37}$$

Since both $\mathbf{r}_{\perp}$ and $\mathbf{F}_{\perp}$ are perpendicular to the axis BC, the direction of $\mathbf{r}_{\perp} \times \mathbf{F}_{\perp}$ must be parallel to $\boldsymbol{\lambda}_{BC}$. Thus, from Fig. 4.31 and the definition of the cross product, we write

$$\mathbf{r}_{\perp} \times \mathbf{F}_{\perp} = (r_{\perp} F_{\perp} \sin\theta)(\pm \boldsymbol{\lambda}_{BC})$$

$$|M_{BC}| = |\boldsymbol{\lambda}_{BC} \cdot (r_{\perp} F_{\perp} \sin\theta)(\pm \boldsymbol{\lambda}_{BC})| = r_{\perp} F_{\perp} \sin\theta |\pm \boldsymbol{\lambda}_{BC} \cdot \boldsymbol{\lambda}_{BC}|$$

$$= (r_{\perp} \sin\theta) F_{\perp} = d_s F_{\perp}$$

$$\boxed{|M_{BC}| = d_s F_{\perp}} \tag{4.38}$$

where d_s is the *shortest distance* between the axis BC and the line of action of $\mathbf{F}$.

EXAMPLE 4.12

A T-shaped lever is supported by bearings at B and C and leans against a wall at A. The lever is found to have an impending rotation about the axis BC when a 50.7-lb force $\mathbf{F}$ is applied to it as shown. Determine (a) the moment M_{BC} of $\mathbf{F}$ about the axis BC when rotation impends, (b) the analytical expression of $\mathbf{M}_{BC}$, (c) the shortest distance d_s between the axis BC and the line AD.

[†]A five-point star preceding a section number indicates that the section is an optional section as described in the Preface.

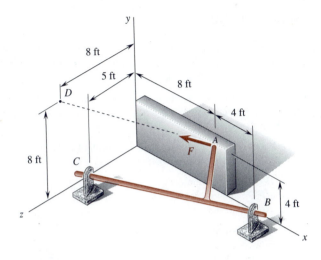

Solution. For determining M_{BC}, we may choose $\mathbf{r}$ to be $\overrightarrow{BA}$ and write

$$\mathbf{r} = \overrightarrow{BA} = -4\mathbf{i} + 4\mathbf{j}$$

$$\overrightarrow{BC} = -12\mathbf{i} + 5\mathbf{k} \qquad \boldsymbol{\lambda}_{BC} = \frac{1}{13}(-12\mathbf{i} + 5\mathbf{k})$$

$$\overrightarrow{AD} = -8\mathbf{i} + 4\mathbf{j} + 8\mathbf{k} \qquad \boldsymbol{\lambda}_{AD} = \frac{1}{3}(-2\mathbf{i} + \mathbf{j} + 2\mathbf{k})$$

$$\mathbf{F} = 50.7\boldsymbol{\lambda}_{AD} = 16.9(-2\mathbf{i} + \mathbf{j} + 2\mathbf{k})$$

$$M_{BC} = \boldsymbol{\lambda}_{BC} \cdot (\mathbf{r} \times \mathbf{F}) = \frac{16.9}{13} \begin{vmatrix} -12 & 0 & 5 \\ -4 & 4 & 0 \\ -2 & 1 & 2 \end{vmatrix} = 1.3\,(-76)$$

$$M_{BC} = -98.8 \text{ lb·ft} \quad \blacktriangleleft$$

The negative sign in the value of M_{BC} indicates that the sense of the moment of $\mathbf{F}$ about the axis BC is opposite to the sense of $\boldsymbol{\lambda}_{BC}$. This may be confirmed by studying the figure and applying the right-hand rule. For the vector form of the moment, we write

$$\mathbf{M}_{BC} = M_{BC}\boldsymbol{\lambda}_{BC} = \frac{-98.8}{13}(-12\mathbf{i} + 5\mathbf{k})$$

$$\mathbf{M}_{BC} = 91.2\mathbf{i} - 38\mathbf{k} \quad \text{lb·ft} \quad \blacktriangleleft$$

For finding the shortest distance d_s, we need to compute $F_\perp$. Placing $\boldsymbol{\lambda}_{BC}$ at A and letting the angle between $\mathbf{F}$ and $\boldsymbol{\lambda}_{BC}$ be θ, we write

$$F_\perp = F \sin\theta = 1 \cdot F \cdot \sin\theta = |\boldsymbol{\lambda}_{BC} \times \mathbf{F}|$$

$$= |1.3(-5\mathbf{i} + 14\mathbf{j} - 12\mathbf{k})| = 24.84$$

By Eq. (4.38), we have

$$|-98.8| = d_s (24.84) \qquad d_s = 3.98 \text{ ft} \quad \blacktriangleleft$$

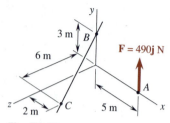

Fig. D4.45

Developmental Exercises

D4.43 The moment of a force $\mathbf{F}$ about an axis BC is 100 lb·ft. If the component of $\mathbf{F}$ perpendicular to BC is $F_\perp = 20$ lb, determine the shortest distance d_s between the axis BC and the line of action of $\mathbf{F}$.

D4.44 The moments of a force $\mathbf{F}$ about the x, y, and z axes are 450 lb·ft, -700 lb·ft, and 500 lb·ft, respectively. What is the analytical expression of the moment $\mathbf{M}_O$ of $\mathbf{F}$ about the origin O?

D4.45 A force $\mathbf{F}$ and an axis BC are shown. Determine (a) the unit vector $\boldsymbol{\lambda}_{BC}$, (b) the moment M_{BC} of $\mathbf{F}$ about the axis BC, (c) $\mathbf{M}_{BC}$.

D4.46 Refer to D4.45. Determine (a) $\boldsymbol{\lambda}_{BC} \times \mathbf{F}$, (b) $|\boldsymbol{\lambda}_{BC} \times \mathbf{F}|$, (c) the perpendicular component $F_\perp$ of $\mathbf{F}$ with respect to the axis BC, (d) the shortest distance d_s between the axis BC and the line of action of $\mathbf{F}$.

D4.47 Which of the following statements are true?
(a) The moment of a couple is a free vector.
(b) The moment of a couple about a point in space is the sum of the moments of the two forces in the couple about that point.
(c) The moment of a couple about any point has the same value.
(d) The moment of a couple about any axis has the same value.
(e) The moment of a couple is a vector perpendicular to the plane containing the two forces in the couple.

PROBLEMS

4.57 Refer to the pyramid in Prob. 4.36. Determine (a) the interior angle $\theta_{ABC/xy}$ between the side ABC and the xy plane, (b) the interior angle $\theta_{ABC/ACD}$ between the side ABC and the side ACD.

4.58* Refer to the pyramid in Prob. 4.37*. Determine (a) the interior angle $\theta_{ABC/ACD}$ between the side ABC and the side ACD, (b) the interior angle $\theta_{ACD/ADE}$ between the side ACD and the side ADE.

4.59* and 4.60 For the system shown, determine (a) the moment $\mathbf{M}_O$ of $\mathbf{F}$ about the point O, (b) the projection of $\mathbf{M}_O$ onto the axis OB, (c) the moment M_{OB} of $\mathbf{F}$ about the axis OB, (d) $\mathbf{M}_{OB}$.

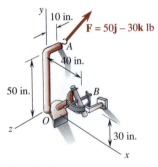

Fig. P4.59*

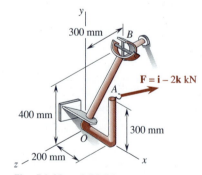

Fig. P4.60 and P4.61

4.61 Using the scalar triple product, determine the moment M_{OB} of $\mathbf{F}$ about the axis OB as shown.

4.62 The moment of the force **P** acting at the point $A(5, 10, 20)$ mm about the z axis, as shown, shall not exceed the limit of 0.24 N·m. Determine the allowable magnitude of **P** which acts in the direction of

$$\lambda_P = -\frac{5}{13}\mathbf{i} + \frac{12}{13}\mathbf{k}$$

4.63* Solve Prob. 4.62 if **P** acts in the direction of

$$\lambda_P = -\frac{3}{5}\mathbf{i} + \frac{4}{5}\mathbf{k}$$

4.64 The faucet at the end of the pipe DF is to be replaced. In order to loosen the joint at the faucet and at the same time to prevent the pipe DF from rotating, a plumber uses two pipe wrenches EG and FH and exerts two 300-N forces parallel to the yz plane as shown. All threads at the joints are right-handed. (a) Compute the moment M_{BC} of the two 300-N forces. (b) Determine whether the two 300-N forces exerted by the plumber tend to tighten or loosen the joint between the pipe BC and the elbow B.

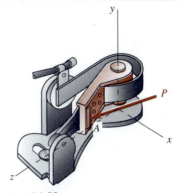

Fig. P4.62

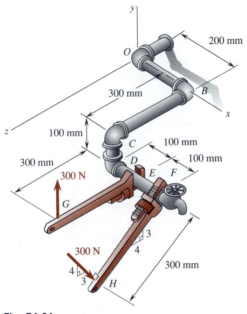

Fig. P4.64

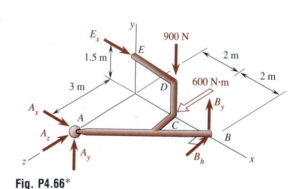

Fig. P4.66*

4.65 Refer to Prob. 4.64. Employing the concept of moments, determine the shortest distance d_s between the x axis and the line of action of the 300-N force acting at H.

4.66* The unknown and known forces in the force system shown have the same unit. Determine (a) the resultant moment $\mathbf{M}_A^R$ of the force system about the point A, (b) the resultant moment M_{AB}^R of the force system about the axis AB.

4.67 A laser beam emanates from the point P through the point Q, while another laser beam emanates from the point R through the point S. Employing the concept of moments, determine the shortest distance d_s between these two laser beams.

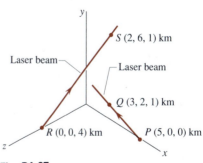

Fig. P4.67

4.18   Concluding Remarks

Three basic cases of moments of forces have been treated in this chapter. They are (1) the moment of a force about an axis, (2) the moment of a force about a point, and (3) the moment of a couple, which is a free vector having a purely torsional effect on a body. Note that the *moment* of a couple is an *attribute* of the couple and is often called the *torque of the couple*, or simply a *torque*. The primary objective of this chapter is to instill in engineering students a good understanding of moments of forces and their applications. A basic knowledge of moments of forces is a prerequisite to the study of the conditions of rigid bodies under the action of forces. The several equivalent versions of the right-hand rule for moment vectors, as presented in Sec. 4.2, should be very helpful in gaining a good understanding of the physical implications of the moments of forces.

Varignon's theorem as presented in Secs. 4.3 and 4.11 is, indeed, a direct consequence of the *parallelogram law* and the *distributive property of the cross product*. Historically, we note that the technique of composing forces by the *parallelogram law* had undergone more than a century of development before it was simultaneously published in 1687 by Newton in his *Philosophiæ Naturalis Principia Mathematica* and by Varignon in his *Projet d'uné nouvelle méchanique*. Although Varignon's *Projet* is not to be compared with Newton's *Principia*, Varignon did contribute his share to mechanics.

In general, any moment acting on a right cross section of a longitudinal member (e.g., a shaft, beam, column, or post) may be resolved into two rectangular components: one perpendicular and the other parallel to the right cross section. We recognize that the moment acting perpendicular to the right cross section of a longitudinal member is called a *torque,* which is a moment parallel to the axis of the member. The torque is a measure of the effectiveness or tendency of an action to rotate or twist a longitudinal member about its axis. However, the moment acting parallel to a right cross section of a longitudinal member is a measure of the effectiveness or tendency of an action to bend or flex the member at that cross section and is called the *bending moment.* An elementary study of bending moments in beams is presented in Chap. 8. A more in-depth study of bending moments is usually found in texts on mechanics of materials.

REVIEW PROBLEMS

4.68   A free body is acted on by a force system as shown, where unknown and known forces have the same unit. Determine the resultant moment M_A^R of the force system about the point A.

4.69   The end D of a pipeline is being adjusted with a 270-N force **F** as shown. Determine (a) the moment $\mathbf{M}_A$ of **F** about the joint A, (b) the moment M_{AB} of **F** about the axis of the pipe AB, (c) whether the action of **F** tends to tighten or loosen the joint A where the threads are right-handed, (d) the shortest distance between the point A and the line of action of **F**, (e) the shortest distance between the line containing the axis AB and the line of action of **F**.

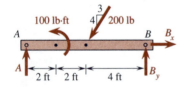

Fig. P4.68

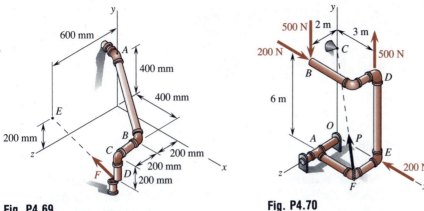

Fig. P4.69

Fig. P4.70

4.70 A system of forces consists of a 210-N force **P** and two couples as shown. Determine (a) the resultant moment **M** of the two couples, (b) the moment M_z of **P** and the two couples about the z axis.

4.71 A free body is acted on by a force system as shown, where the unknown and known forces have the same unit. Determine (a) the resultant moment M_A^R of the force system about the point A, (b) the resultant moment M_{AB}^R of the force system about the axis passing through the points A and B.

4.72* A laser beam emanates from the point P through the point Q, while another laser beam emanates from the point R through the point S. Employing the concept of moments, determine the shortest distance d_s between these two laser beams.

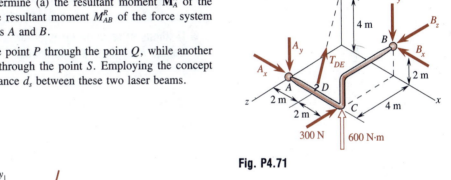

Fig. P4.71

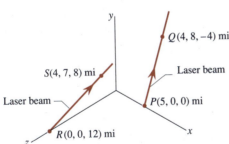

Fig. P4.72*

Chapter 5

Equilibrium of Rigid Bodies

KEY CONCEPTS

- Wrench, equivalent systems of forces, equipollent systems of forces, equilibrium of a rigid body, two-force body, three-force body, statically indeterminate, partially constrained, and improperly constrained.
- Reduction of a system of forces to a force-moment system at a chosen point.
- Reactions from supports and connections for rigid bodies in a plane and in space, where the force systems are nonconcurrent.
- Maximum number of statically determinate unknowns.
- Methods of solution for equilibrium problems of simple rigid bodies in a plane and in space.

When a body has no translational acceleration, the resultant force of the force system shown in the free-body diagram of the body must be equal to zero. Similarly, if a body has no rotational acceleration, the resultant moment, about a point, of that force system must be equal to zero. In general, a rigid body may be caused to have either translational or rotational acceleration, or both. Thus, the equilibrium of a rigid body requires that the aforementioned resultant force and resultant moment be both equal to zero. This chapter is focused on the basic study of equilibrium problems of *simple* rigid bodies acted on by forces in a plane and by forces in space, where the solution of each problem generally entails the drawing of only *one* free-body diagram. Equilibrium problems which require more advanced considerations are treated in later chapters.

5.1 Resolution of a Given
Force into a Force-Moment
System at Another Point   125

EQUIVALENT SYSTEMS OF FORCES

5.1 Resolution of a Given Force into a Force-Moment System at Another Point†

The action of a force **F** applied at a given point A of a rigid body is *equivalent to the combined action* of the force **F** applied at another point B of the rigid body and an applied moment $\mathbf{M}_B$ which is equal to the moment of **F** about B when **F** was applied at A. This may be verified step by step as shown in Fig. 5.1, where we first apply the *rigid-body principle*†† and then employ the fact that the action of a couple can be replaced by the action of the moment of the couple.

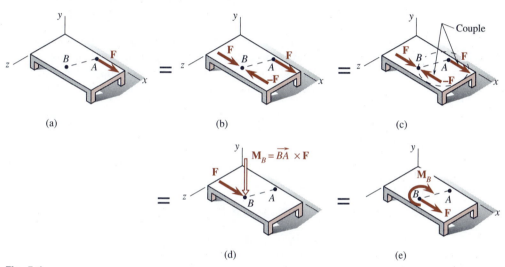

(a)   (b)   (c)

(d)   (e)

Fig. 5.1

Note that the moment $\mathbf{M}_B$ may be represented graphically by a hollow arrow or a curved arrow, as shown in Fig. 5.1. The moment $\mathbf{M}_B$ is given by

$$\mathbf{M}_B = \overrightarrow{BA} \times \mathbf{F} \tag{5.1}$$

We note from the definition of the cross product that the moment $\mathbf{M}_B$ is perpendicular to $\overrightarrow{BA}$ and **F**. The moment $\mathbf{M}_B$ is actually a free vector, but it is, for convenience, shown to act at the point B. Furthermore, it can be shown that the new force system in Fig. 5.1(e) and the original force in Fig. 5.1(a) do have the *same resultant moment* $\mathbf{M}_P$ about any point P as well as the *same resultant force* **F**.

---

†A *force-moment system* is also called a *force-couple system*. Cf. the footnote on p. 90.

††Cf. Sec. 1.12.

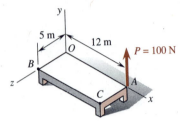

Fig. D5.2

Developmental Exercises

D5.1 The moment of the force $\mathbf{F}$ at the point A in Fig. 5.1(a) about any point P is $\mathbf{M}_P = \overrightarrow{PA} \times \mathbf{F}$, while the moment of the force-moment system in Fig. 5.1(e) about the point P is $\mathbf{M}'_P = \overrightarrow{PB} \times \mathbf{F} + \mathbf{M}_B$, where $\mathbf{M}_B = \overrightarrow{BA} \times \mathbf{F}$. Show that $\mathbf{M}_P = \mathbf{M}'_P$.

D5.2 A 100-N vertical force $\mathbf{P}$ acts at the point A as shown. (a) Resolve $\mathbf{P}$ into a force-moment system $\mathbf{F}_O$ and $\mathbf{M}_O$ acting at the origin O. (b) Resolve $\mathbf{P}$ into a force-moment system $\mathbf{F}_B$ and $\mathbf{M}_B$ acting at the point B. (c) Does each of the force-moment systems obtained in (a) and (b) produce a moment about the point C equal to $\overrightarrow{CA} \times \mathbf{P}$?

5.2 Reduction of a System of Forces to a Force-Moment System

It follows from Sec. 5.1 that a system of forces $\mathbf{F}_1, \mathbf{F}_2, \ldots, \mathbf{F}_n$ acting on a rigid body may be moved from their respective points of application $A_1, A_2, \ldots, A_n$ to a chosen point O on the rigid body, as shown in Fig. 5.2(a), if the moments $\mathbf{M}_1, \mathbf{M}_2, \ldots, \mathbf{M}_n$ equal to the moments of the forces $\mathbf{F}_1, \mathbf{F}_2, \ldots, \mathbf{F}_n$ about O are also added to act on the rigid body, as shown in Fig. 5.2(b), where

$$\mathbf{M}_i = \overrightarrow{OA_i} \times \mathbf{F}_i \qquad (i = 1, 2, \ldots n) \qquad (5.2)$$

The forces $\mathbf{F}_1, \mathbf{F}_2, \ldots, \mathbf{F}_n$ and the moments $\mathbf{M}_1, \mathbf{M}_2, \ldots, \mathbf{M}_n$ acting at O in Fig. 5.2(b) may, in turn, be combined to yield a single resultant force $\mathbf{R}$ and a single resultant moment $\mathbf{M}_O^R$, as shown in Fig. 5.2(c), where

$$\mathbf{R} = \mathbf{F}_1 + \mathbf{F}_2 + \cdots + \mathbf{F}_n$$

$$\mathbf{M}_O^R = \mathbf{M}_1 + \mathbf{M}_2 + \cdots + \mathbf{M}_n \qquad (5.3)$$

Thus, any system of forces may be reduced to a force-moment system as defined by Eqs. (5.2) and (5.3) and illustrated in Fig. 5.2.

A force-moment system consisting of a force $\mathbf{F}$ and a moment $\mathbf{M}$ which are parallel to each other is, by tradition, called a *wrench* (perhaps, a *screw-*

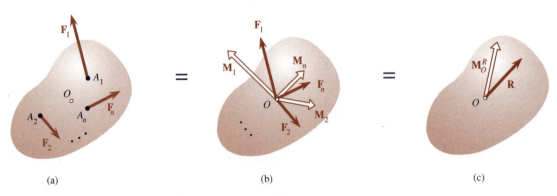

(a)   (b)   (c)

Fig. 5.2

5.2 Reduction of a
System of Forces to a
Force-Moment System

127

driver is a better name).[†] The action of a wrench is physically equivalent to a push or pull along an axis plus a twist about that axis. The line of action of **F** is called the *axis of the wrench*, and the ratio M/F is called the *pitch of the wrench*. If **F** and **M** of the wrench are of the same sense, the wrench is a *positive wrench*; otherwise, the wrench is a *negative wrench*. This is illustrated in Fig. 5.3.

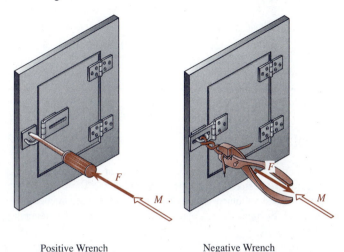

Positive Wrench   Negative Wrench

Fig. 5.3

When the moment **M** is zero, the wrench degenerates into a force **F**; when the force **F** is zero, the wrench degenerates into a moment **M**. A *zero* (or *null*) *wrench* is a condition in which both the force **F** and the moment **M** are zero. It will be shown in Sec. 5.4 that any force system acting on a rigid body can be reduced to a wrench, or one of its degenerate cases, at an appropriate location of the rigid body (or its massless extension).

_____

[†]The combined action of a force **F** and a moment **M** parallel to **F** is more closely simulated by the action transmitted by a *screwdriver* or a pair of *pliers*, as illustrated in Fig. 5.3, than the action transmitted by a common *wrench*, which usually turns, but does not push or pull.

EXAMPLE 5.1 _____

Reduce the system of forces shown to a force-moment system consisting of a force **R** and a moment $\mathbf{M}_O^R$ at the origin O.

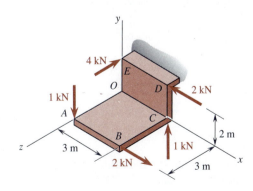

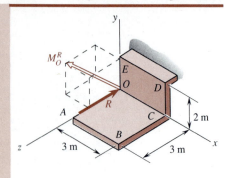

Solution.   For the forces at the points A, B, C, D, and E, as shown, we write

$$\mathbf{R} = \Sigma \mathbf{F} = -\mathbf{j} + 2\mathbf{i} + \mathbf{j} - 2\mathbf{i} - 4\mathbf{k}$$

$$\mathbf{R} = -4\mathbf{k} \text{ kN} \blacktriangleleft$$

$$\mathbf{M}_O^R = \Sigma(\mathbf{r} \times \mathbf{F})$$
$$= 3\mathbf{k} \times (-\mathbf{j}) + (3\mathbf{i} + 3\mathbf{k}) \times (2\mathbf{i}) + 3\mathbf{i} \times \mathbf{j}$$
$$+ (3\mathbf{i} + 2\mathbf{j}) \times (-2\mathbf{i}) + 2\mathbf{j} \times (-4\mathbf{k})$$

$$\mathbf{M}_O^R = -5\mathbf{i} + 6\mathbf{j} + 7\mathbf{k} \text{ kN} \cdot \text{m} \blacktriangleleft$$

Developmental Exercises

D5.3   The forces $\mathbf{F}_1 = 3\mathbf{j}$ kN and $\mathbf{F}_2 = 4\mathbf{i}$ kN act at the points $A_1(3, 0, 0)$ m and $A_2(0, 0, 2)$ m, respectively. They are to be reduced to a force $\mathbf{R}$ acting at the point $B(1, 0, 1)$ m and a moment $\mathbf{M}_B^R$. Determine (a) $\mathbf{R}$, (b) the moment $\mathbf{M}_1$ equal to the moment of $\mathbf{F}_1$ about B, (c) the moment $\mathbf{M}_2$ equal to the moment of $\mathbf{F}_2$ about B, (d) $\mathbf{M}_B^R$.

D5.4   Define the terms: (a) wrench, (b) axis of the wrench, (c) pitch of the wrench, (d) positive wrench, (e) negative wrench, (f) zero wrench.

D5.5   A negative wrench consists of a force $\mathbf{F} = 2\mathbf{i} - \mathbf{j} - 2\mathbf{k}$ lb and a moment $\mathbf{M}$ of 15 lb·ft. What is the analytical expression of $\mathbf{M}$?

5.3   Equivalent and Equipollent Systems of Forces

We have learned in Sec. 5.2 that any system of forces acting on a rigid body may be reduced to a force-moment system at a given point. The action of the original system of forces is shown to be equivalent to the action of such a force-moment system. Therefore, *equivalent systems of forces* are force systems which act on a rigid body and are reducible to the same force-moment system at a given point, and either system may be referred to as a *replacement* of the other.

If the unprimed system of forces $\mathbf{F}_1$, $\mathbf{F}_2$, . . . , $\mathbf{F}_n$ is equivalent to the primed system of forces $\mathbf{F}_1'$, $\mathbf{F}_2'$, . . . , $\mathbf{F}_m'$ then their resultant forces $\mathbf{R}$ and $\mathbf{R}'$ and their resultant moments $\mathbf{M}_P^R$ and $(\mathbf{M}_P^R)'$ about a given point P, as shown in Fig. 5.4, must respectively be equal; i.e.,

$$\mathbf{R} = \mathbf{R}' \qquad \mathbf{M}_P^R = (\mathbf{M}_P^R)' \qquad (5.4)$$

$$\mathbf{R} = \mathbf{F}_1 + \mathbf{F}_2 + \cdots + \mathbf{F}_n \qquad (5.5a)$$

$$\mathbf{R}' = \mathbf{F}_1' + \mathbf{F}_2' + \cdots + \mathbf{F}_m' \qquad (5.5b)$$

$$\mathbf{M}_P^R = \mathbf{M}_1 + \mathbf{M}_2 + \cdots + \mathbf{M}_n \qquad (5.6a)$$

$$(\mathbf{M}_P^R)' = \mathbf{M}_1' + \mathbf{M}_2' + \cdots + \mathbf{M}_m' \qquad (5.6b)$$

where $\mathbf{M}_1$, $\mathbf{M}_2$, . . . , $\mathbf{M}_n$ are the moments of $\mathbf{F}_1$, $\mathbf{F}_2$, . . . , $\mathbf{F}_n$ about the given point P, respectively; and $\mathbf{M}_1'$, $\mathbf{M}_2'$, . . . , $\mathbf{M}_m'$ are the moments of $\mathbf{F}_1'$, $\mathbf{F}_2'$, . . . , $\mathbf{F}_m'$ about P, respectively. Equations (5.4) represent the

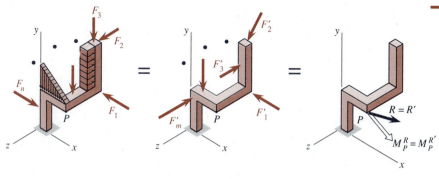

Fig. 5.4

necessary and sufficient conditions for equivalent systems of forces acting on a rigid body.

If two systems of forces satisfy Eqs. (5.4) but act on a set of nonrigidly connected bodies, where different forces may act on the various individual bodies, then they are called *equipollent systems of forces*, which may have *different effects* on the individual nonrigidly connected bodies. This is illustrated for systems containing only one force in Fig. 5.5, where = indicates *equivalence* and ⇒ indicates *equipollence*. The scalar version of Eqs. (5.4) may be written as

$$\Sigma F_x = \Sigma F'_x \qquad \Sigma F_y = \Sigma F'_y \qquad \Sigma F_z = \Sigma F'_z$$
$$\Sigma(M_P)_x = \Sigma(M_P)'_x \quad \Sigma(M_P)_y = \Sigma(M_P)'_y \quad \Sigma(M_P)_z = \Sigma(M_P)'_z \tag{5.7}$$

If the forces in the system all lie in the *xy* plane, Eqs. (5.7) reduce to

$$\boxed{\Sigma F_x = \Sigma F'_x} \quad \boxed{\Sigma F_y = \Sigma F'_y} \quad \boxed{\Sigma M_P = \Sigma M'_P} \tag{5.8}$$

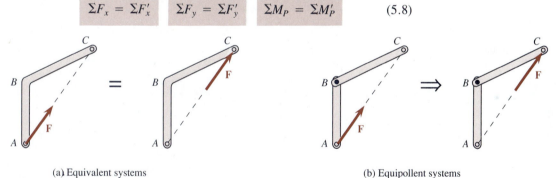

(a) Equivalent systems   (b) Equipollent systems   **Fig. 5.5**

EXAMPLE 5.2

Suppose that there are two equivalent systems of forces as shown. Determine the magnitudes of the forces **F** and **Q** and the moment **M**.

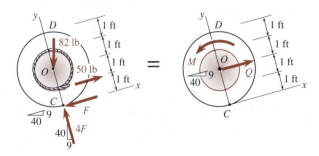

Solution.   Since the hypotenuse of the slope triangle is 41, we write

$$+\nwarrow \ \Sigma F_y \ = \ +\nwarrow \ \Sigma F'_y: \qquad 4F \ - \ 82\left(\frac{40}{41}\right) \ = \ 0$$

$$+\nearrow \ \Sigma F_x \ = \ +\nearrow \ \Sigma F'_x: \qquad -F \ + \ 50 \ - \ 82\left(\frac{9}{41}\right) \ = \ Q$$

$$+\,\circlearrowleft \ \Sigma M_C \ = \ +\,\circlearrowleft \ \Sigma M'_C: \qquad -1(50) \ + \ 2(82)\left(\frac{9}{41}\right) \ = \ M \ - \ 2Q$$

Solving these three equations, we obtain

$$F \ = \ 20 \text{ lb} \qquad Q \ = \ 12 \text{ lb} \qquad M \ = \ 10 \text{ lb} \cdot \text{ft} \quad \blacktriangleleft$$

For a check on these answers, we write

$$+\,\circlearrowleft \ \Sigma M_D \ = \ +\,\circlearrowleft \ \Sigma M'_D:$$

$$-2(82)\left(\frac{9}{41}\right) \ + \ 3(50) \ - \ 4(20) \ = \ 10 \ + \ 2(12)$$

Developmental Exercises

D5.6   Define the terms: (a) equivalent systems of forces, (b) equipollent systems of forces.

D5.7   Which of the six force systems shown are equivalent?

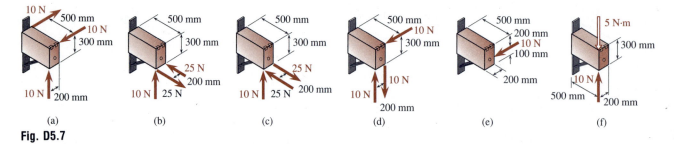

Fig. D5.7

★5.4   Reduction of a System of Forces to a Wrench

We saw in Sec. 5.2 that a system of forces may be reduced to a force-moment system consisting of a resultant force $\mathbf{R}$ acting at a chosen point O and a resultant moment $\mathbf{M}_O^R$ about O as illustrated in Fig. 5.6(a). If $\mathbf{M}_O^R$ is not parallel to $\mathbf{R}$ and neither of them is zero, we may resolve $\mathbf{M}_O^R$ into two rectangular components: $\mathbf{M}_\parallel$ and $\mathbf{M}_\perp$, where $\mathbf{M}_\parallel \parallel \mathbf{R}$ and $\mathbf{M}_\perp \perp \mathbf{R}$, as shown in Fig. 5.6(b). Letting $\boldsymbol{\lambda}_R = \mathbf{R}/R$, we write

$$\mathbf{M}_O^R \ = \ \mathbf{M}_\parallel \ + \ \mathbf{M}_\perp \qquad \mathbf{M}_\parallel \ = \ (\boldsymbol{\lambda}_R \cdot \mathbf{M}_O^R)\boldsymbol{\lambda}_R \qquad \mathbf{M}_\perp \ = \ \mathbf{M}_O^R \ - \ \mathbf{M}_\parallel \qquad (5.9)$$

By Sec. 5.1, we may equivalently replace $\mathbf{R}$ acting at the point O with $\mathbf{R}$ acting at a point P plus a moment $\mathbf{M}_P$ such that

$$\mathbf{M}_P \ = \ \overrightarrow{PO} \times \mathbf{R} \ = \ -\mathbf{M}_\perp \qquad (5.10)$$

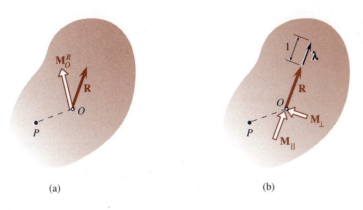
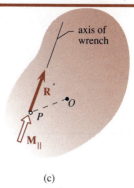

(a) (b) (c)

Fig. 5.6

Then, $\mathbf{M}_P$ cancels $\mathbf{M}_\perp$ and the original system of forces is reduced to a *wrench* consisting of $\mathbf{M}_\parallel$ and $\mathbf{R}$ acting at the point P of the rigid body (or its massless extension), as shown in Fig. 5.6(c). If $\mathbf{M}_\parallel = \mathbf{0}$, then the system of forces is equivalent to a single force $\mathbf{R}$ at P.

EXAMPLE 5.3

Reduce the system of forces in Example 5.1 to a wrench.

Solution. From the solution in Example 5.1, we know that the given system of forces is equivalent to a force-moment system $\mathbf{R}$ and $\mathbf{M}_O^R$ at the origin O, where

$$\mathbf{R} = -4\mathbf{k} \text{ kN} \qquad \mathbf{M}_O^R = -5\mathbf{i} + 6\mathbf{j} + 7\mathbf{k} \text{ kN} \cdot \text{m}$$

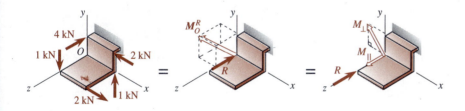

As illustrated, we first resolve $\mathbf{M}_O^R$ into $\mathbf{M}_\parallel$ and $\mathbf{M}_\perp$ as follows:

$$\lambda_R = \frac{\mathbf{R}}{R} = -\mathbf{k} \qquad \mathbf{M}_\parallel = (\lambda_R \cdot \mathbf{M}_O^R)\lambda_R = 7\mathbf{k}$$

$$\mathbf{M}_\perp = \mathbf{M}_O^R - \mathbf{M}_\parallel = -5\mathbf{i} + 6\mathbf{j}$$

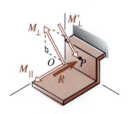

Next, we move the resultant force $\mathbf{R}$ to the point $P(x, y, z)$ and add a moment equal to $\overrightarrow{PO} \times \mathbf{R}$ to keep the effect of the force system on the body unchanged. If the point P is chosen so that the added moment is $\mathbf{M}'_\perp = -\mathbf{M}_\perp$ as shown, then a wrench is obtained. Thus, we write

$$\overrightarrow{PO} \times \mathbf{R} = -\mathbf{M}_\perp: \quad (-x\mathbf{i} - y\mathbf{j} - z\mathbf{k}) \times (-4\mathbf{k}) = -(-5\mathbf{i} + 6\mathbf{j})$$

$$4y\mathbf{i} - 4x\mathbf{j} = 5\mathbf{i} - 6\mathbf{j}$$

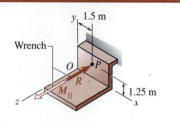

The ensuing scalar equations are

$$4y = 5 \quad \text{and} \quad -4x = -6$$

which yield $x = 1.5$, $y = 1.25$, and z can have any value (including zero). Thus, we may say that the wrench acts through the point $P(1.5, 1.25, 0)$ m and consists of

$$\mathbf{R} = -4\mathbf{k} \text{ kN} \quad \text{and} \quad \mathbf{M}_\parallel = 7\mathbf{k} \text{ kN} \cdot \text{m} \blacktriangleleft$$

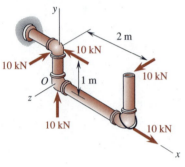

Fig. D5.8

Developmental Exercise

D5.8 The force system shown is to be reduced first to a force-moment system at the origin O and then to a wrench at a point P. Determine (a) the resultant force $\mathbf{R}$, (b) the resultant moment $\mathbf{M}_O^R$ about O, (c) λ_R, (d)$\mathbf{M}_\parallel$, (e) $\mathbf{M}_\perp$, (f) $\overrightarrow{PO}$ from $\overrightarrow{PO} \times \mathbf{R} = -\mathbf{M}_\perp$, (g) the coordinates of P.

PROBLEMS

5.1 and 5.2 Resolve the force shown into a force-moment system at the origin O.

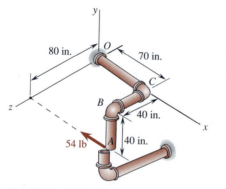

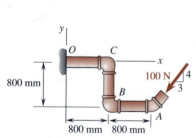

Fig. P5.1 and P5.3

Fig. P5.2 and P5.4*

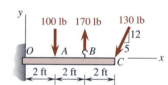

Fig. P5.5, P5.9, and P5.21

5.3 and 5.4* Resolve the force shown into a force-moment system at the point C.

5.5 through 5.8* For the system of forces shown, determine its equivalent force-moment system at the origin O.

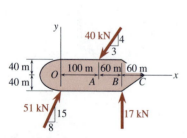

Fig. P5.6, P5.10, and P5.22

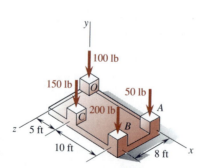

Fig. P5.7, P5.11, and P5.23

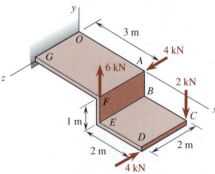

Fig. P5.8*, P5.12*, and P5.24*

5.9 through 5.12* For the system of forces shown, determine its equivalent force-moment system at the point A.

5.13 and 5.14* Four equivalent force systems are shown. Determine the magnitudes of **P**, **Q**, and **M** in the force systems (b), (c), and (d).

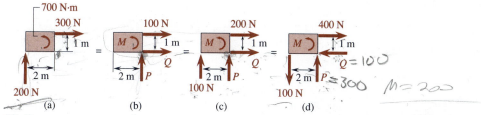

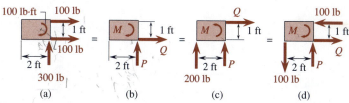

Fig. P5.13

Fig. P5.14*

5.15 Suppose that there are two equivalent systems of forces as shown. Determine the magnitudes of the forces **P** and **Q** and the moment **M**.

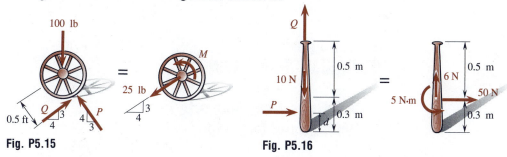

Fig. P5.15

Fig. P5.16

5.16 The two force systems shown are known to be equivalent. Determine the magnitudes of the forces **P** and **Q** and the distance d.

5.17 Three men have attempted twice to free their boat from a sand bar as shown in the left and right figures. If the effects of their efforts to free the boat during these two attempts are equivalent and $A_y = 100$ N, $C = 150$ N, determine the values of A_x, B, P_y, P_z, Q, and S.

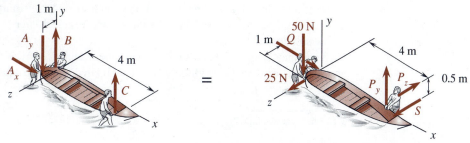

Fig. P5.17

5.18* Solve Prob. 5.17 if $A_y = 80$ N, $C = 125$ N.

5.19 and 5.20* Two equivalent force systems are shown. Determine (a) the magnitudes of **P**, **Q**, and **S**, (b) the wrench equivalent to either of the force systems.

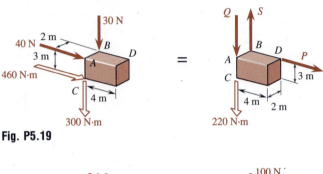

Fig. P5.19

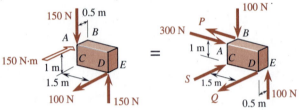

Fig. P5.20*

5.21 through 5.24* *(The figures for these problems are shown and specified along with those for Probs. 5.5 through 5.8* on p. 132.)* Reduce the system of forces shown to either a wrench or a single force.

EQUILIBRIUM IN A PLANE

5.5   Reactions from Supports and Connections in a Plane

Reactions from the supports or connections of a rigid body are the *constraining forces and moments* acting on the body at the points of support or connection. Each *reaction* must reflect the *restriction of movement* imposed on the body at the point of support or connection. Therefore, the number of components of reaction must always correspond to the number of ways in which the support or connection is *capable of restricting* the movement of the body at the point of support or connection. If the sense of a component of reaction is initially not known from the conditions of constraint and loading, it may first be assumed in the analysis and checked later in the subsequent solution.[†]

Note that, unless otherwise indicated, the weights of rigid bodies are usually assumed to be negligibly small compared with the applied loads.

[†]Cf. the footnote on p. 59.

Furthermore, the forces acting on a rigid body in *equilibrium in a plane* are *coplanar* but generally *nonconcurrent*. A rigid body acted on by coplanar but nonconcurrent forces is referred to as a *rigid body in a plane*. Likewise, supports or connections of a rigid body in a plane are referred to as *supports or connections in a plane*. The reactions from supports and connections in a plane may be classified into the following three types:

TYPE I: REACTIONS WITH ONE UNKNOWN.   This type of reaction is equivalent to a force with either unknown magnitude but known line of action or known magnitude but unknown line of action in the plane of the system. Supports and connections producing this type of reaction include *rollers*, *rockers*, *smooth surfaces*, *collars on smooth rods*, *pins in smooth slots*, *sliders in smooth slots*, *cordlike elements*, and *short links*. Each of these supports and connections *allows rotation about the point of support but restricts translation in only one direction*. The unknown magnitude of the reaction is usually represented by an appropriate letter as shown in Fig. 5.7.

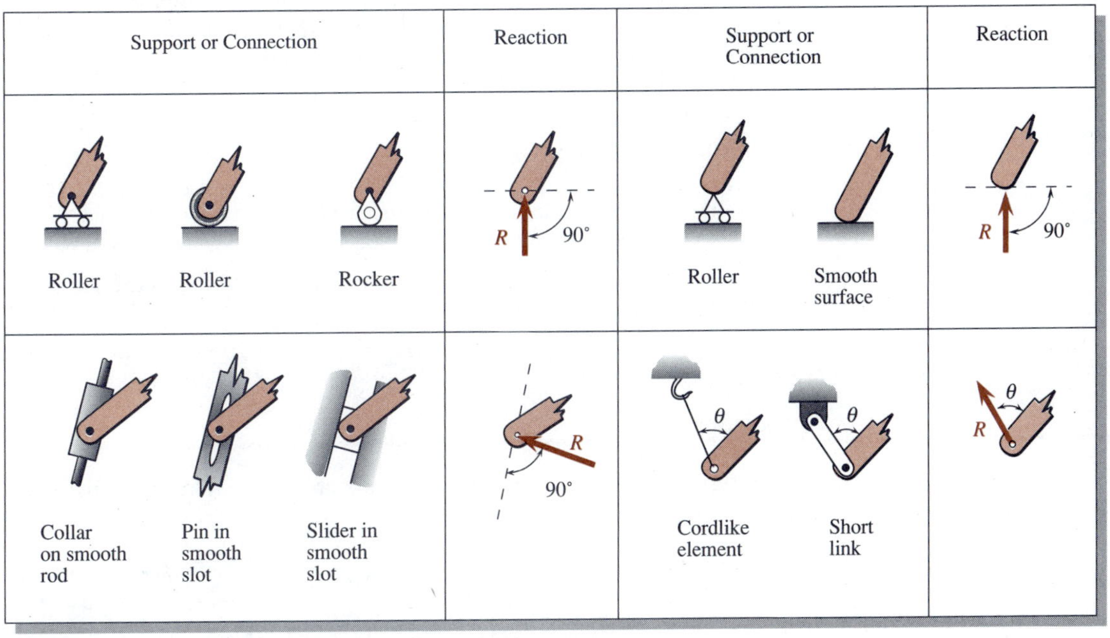

Fig. 5.7 Reactions with one unknown.

TYPE II: REACTIONS WITH TWO UNKNOWNS.   This type of reaction is equivalent to a force with two unknown rectangular components, or equivalently, a force with an unknown magnitude and an unknown direction in the plane of the system. Supports and connections producing this type of reaction include *knife edges*, *rough surfaces*, *smooth corners*, *hinges*, *pins*, and *pivots*. Each of these supports and connections *allows rotation about the point of support but restricts translation in two directions*. Therefore, this type of reaction is usually represented by *two rectangular force components* as shown in Fig. 5.8. Its resultant form is a force at an angle.

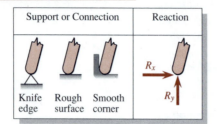

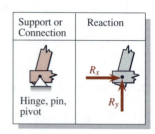

Fig. 5.8 Reactions with two unknowns.

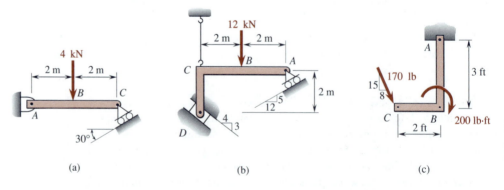

Fig. 5.9 Reactions with three unknowns.

TYPE III: REACTIONS WITH THREE UNKNOWNS. This type of reaction is equivalent to a force with two unknown rectangular components (or equivalently, a force with an unknown magnitude and an unknown direction) plus a moment with an unknown magnitude but a known orientation which is perpendicular to the plane of the system. This type of reaction is caused by a *fixed support* which is also called a *clamped* or *built-in support*, such as the support of a cantilever beam. A fixed support *restricts both translation and rotation* at the location of the support. In this case, the rotation is resisted by the moment developed at the fixed support. Thus, we usually represent this type of reaction by *two rectangular force components and a reaction moment* as shown in Fig. 5.9.

Developmental Exercises

D5.9 (a) Describe reactions from supports and connections for a rigid body. (b) What is a rigid body in a plane? (c) What are supports or connections in a plane?

D5.10 Give two examples of supports or connections that produce reactions with (a) one unknown, (b) two unknowns.

EXAMPLE 5.4

Three rigid bodies in equilibrium in a plane are shown. Draw the free-body diagram of each body.

(a)

(b)

(c)

Solution. Recalling the guides for drawing a good free-body diagram as stated in Sec. 3.8 and recognizing the types of reactions from the supports, we may draw the free-body diagram of each rigid body as shown. Note that

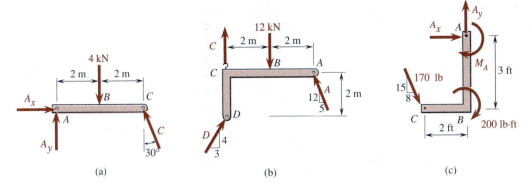

(a)   (b)   (c)

we use the letters naming the points of supports, with appropriate subscripts
as needed, to label the magnitudes of the reaction forces. A reaction moment
is labeled with the letter M, and the letter naming the point of support is
added to M as a subscript. Moreover, note that the senses of the unknown
reactions are assumed at this stage of the study.[†]

Developmental Exercises

D5.11 What are the possible ways in which we may represent the reaction from
a fixed support for a rigid body in a plane?

D5.12 Draw the free-body diagram of each of the rigid bodies shown.

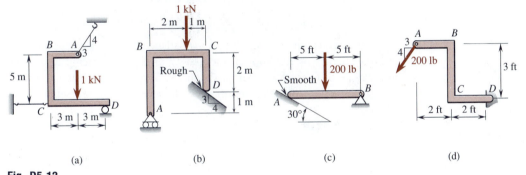

(a)   (b)   (c)   (d)

Fig. D5.12

5.6 Equilibrium of a Rigid Body in a Plane

The *equilibrium of a rigid body* is a state of the rigid body in which *all
particles* of the body are at rest or moving uniformly along straight lines in
an inertial reference frame (cf. Sec. 3.1). The equilibrium of a rigid body is
maintained if it is subjected to a balanced system of external forces. In this
balanced system, the resultant force **R** and resultant moment $\mathbf{M}_P^R$ of the force
system $\mathbf{F}_1, \mathbf{F}_2, \ldots, \mathbf{F}_n$ about an arbirary point P are both equal to zero

---

[†]Continued in Example 5.5.

as illustrated in Fig. 5.10,[†] where $\mathbf{R} = \Sigma\mathbf{F}$ and $\mathbf{M}_P^R = \Sigma\mathbf{M}_P$. Therefore, the *necessary conditions* for equilibrium of a rigid body are

$$\Sigma\mathbf{F} = 0 \qquad \Sigma\mathbf{M}_P = 0 \qquad (5.11)$$

As an example, note that the force system in the free-body diagram of a circular disk rotating frictionlessly and uniformly about its axis of symmetry will satisfy Eqs. (5.11). However, *all particles* away from the axis of this disk are moving along circular paths with accelerations directed inward toward that axis. By the preceding definition, this rotating disk is *not* in equilibrium. Thus, Eqs. (5.11) do *not* constitute the *sufficient conditions* for equilibrium of a rigid body in the kinematic sense.[††]

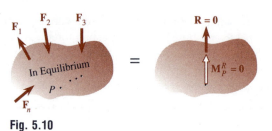

Fig. 5.10

An *equilibrium problem of a rigid body in a plane* is characterized by the feature that a set of *coplanar but nonconcurrent* unknown forces will be encountered in the solution; however, no noncoplanar unknown forces will be encountered. For a rigid body in equilibrium in the *xy* plane, the *scalar equations of equilibrium* may be written from Eqs. (5.11) as

$$\Sigma F_x = 0 \qquad \Sigma F_y = 0 \qquad \Sigma M_P = 0 \qquad (5.12)$$

Although equations of equilibrium are satisfied by the force systems acting on all bodies in equilibrium, the quality or degree of being in equilibrium may be different for different situations. This fact is illustrated in Fig. 5.11 for three pin-supported uniform rods in a vertical plane. Suppose that

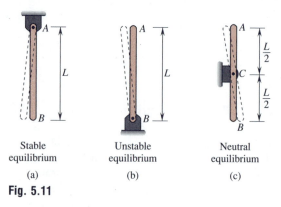

Fig. 5.11

[†]This means that the system of forces in the free-body diagram of a rigid body in equilibrium is equivalent to a *zero wrench*.

[††]In some texts, *equilibrium* was defined in kinetic (or force) terms, rather than in kinematic (or motion) terms. Consequently, Eqs. (5.11) were taken as the necessary and sufficient conditions for rigid-body equilibrium.

each rod is slightly displaced from its upright position of equilibrium and then released. It is obvious that the rod in Fig. 5.11(a) will move back toward its original position, the rod in Fig. 5.11(b) will move farther away from its original position, and the rod in Fig. 5.11(c) will stay in its new position. These three rods as shown are said to be in *stable*, *unstable*, and *neutral equilibrium*, respectively.

The *instability* exhibited by the rod in Fig. 5.11(b) is triggered by a small external disturbance given to the equilibrium configuration of the rod. This type of instability is different from the nonequilibrium of a rigid body due to insufficient or improper constraints on the rigid body. Recognitions of partially and improperly constrained rigid bodies are discussed in Secs. 5.8 and 5.12.

Developmental Exercises

D5.13 Define the equilibrium of a rigid body.

D5.14 What is the characterizing feature in an equilibrium problem of a rigid body in a plane?

EXAMPLE 5.5

Applying the equations of equilibrium, determine the reactions from the supports for each rigid body shown in Example 5.4. (The example here serves only as a preliminary study. Cf. Sec. 5.10.)

Solution. The free-body diagram of each rigid body is drawn as shown.

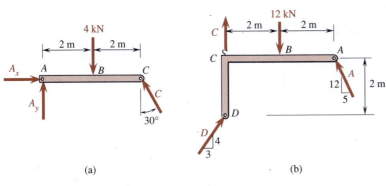

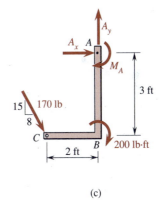

(a)   (b)   (c)

(1) Referring to the free-body diagram in Fig. (a), we write

$$+\circlearrowleft \ \Sigma M_A = 0: \ 4(C\cos 30°) - 2(4) = 0 \qquad C = 2.3094$$

$$\xrightarrow{+} \ \Sigma F_x = 0: \ A_x - C\sin 30° = 0 \qquad A_x = 1.155$$

$$+\uparrow \ \Sigma F_y = 0: \ A_y + C\cos 30° - 4 = 0 \qquad A_y = 2$$

$$\mathbf{A}_x = 1.555 \ \text{kN} \rightarrow \qquad \mathbf{A}_y = 2 \ \text{kN} \uparrow \qquad \mathbf{C} = 2.31 \ \text{kN} \ \measuredangle \ 60° \ \blacktriangleleft^{\dagger}$$

[†]We may alternatively combine the components $\mathbf{A}_x$ and $\mathbf{A}_y$ and report the reaction in its resultant form as $\mathbf{A} = 2.31$ kN $\measuredangle$ 60° or in its analytical expression as $\mathbf{A} = 1.155\mathbf{i} + 2\mathbf{j}$ kN.

(2) Referring to the free-body diagram in Fig. (b), we write

$$+\circlearrowleft \Sigma M_D = 0: \quad 2\left(\frac{5}{13}A\right) + 4\left(\frac{12}{13}A\right) - 2(12) = 0$$

$$\xrightarrow{+} \Sigma F_x = 0: \quad \frac{3}{5}D - \frac{5}{13}A = 0$$

$$+\uparrow \Sigma F_y = 0: \quad \frac{4}{5}D + C + \frac{12}{13}A - 12 = 0$$

These three equations yield $A = 5.379$, $D = 3.448$, and $C = 4.276$. We write

$$\mathbf{A} = 5.38 \text{ kN} \ \measuredangle\ 67.4° \quad \mathbf{C} = 4.28 \text{ kN} \uparrow \ \blacktriangleleft$$

$$\mathbf{D} = 3.45 \text{ kN} \ \measuredangle\ 53.1° \ \blacktriangleleft$$

(3) Referring to the free-body diagram in Fig. (c), we write

$$+\circlearrowleft \Sigma M_A = 0: \quad -M_A - 200 + 3\left(\frac{8}{17}\right)(170) + 2\left(\frac{15}{17}\right)(170) = 0$$

$$\xrightarrow{+} \Sigma F_x = 0: \quad A_x + \frac{8}{17}(170) = 0$$

$$+\uparrow \Sigma F_y = 0: \quad A_y - \frac{15}{17}(170) = 0$$

These three equations yield $M_A = 340$, $A_x = -80$, and $A_y = 150$. Since A_x is found to be negative, we write

$$\mathbf{A}_x = 80 \text{ lb} \leftarrow \quad \mathbf{A}_y = 150 \text{ lb} \uparrow \quad \mathbf{M}_A = 340 \text{ lb·ft} \ \circlearrowright \ \blacktriangleleft$$

Developmental Exercises

D5.15 The frames shown were at rest. Upon the action of the applied loads, which will still be in equilibrium?

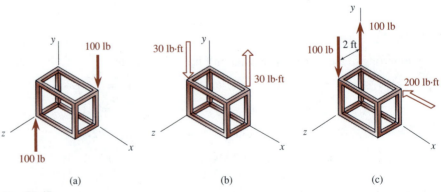

(a)   (b)   (c)

Fig. D5.15

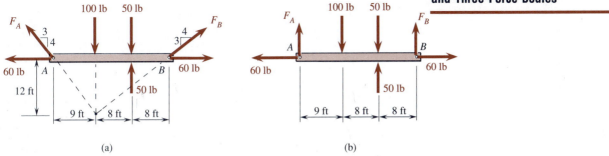

(a)   (b)

Fig. D5.16 and D5.17

D5.16   If equilibrium exists and F_A and F_B are to be determined, identify the type of equilibrium problem for each case as shown.

D5.17   The body AB is in equilibrium. Determine the unknown F_A and F_B in each case as shown.

D5.18   Three cylindrical pipes are in equilibrium on the supports shown. Identify the type of equilibrium in each case.

(a)   (b)   (c)

Fig. D5.18

5.7   Equilibrium of Two- and Three-Force Bodies

We know that a set of concurrent forces can always be combined to form a single resultant force at the point of concurrence. A *two-force body* is a rigid body acted on by a force system composed of, or reducible to, two forces at two different points. For instance, the body AB in Fig. 5.12(a) is a two-force body, whose free-body diagram is shown in Fig. 5.12(b) and (c). *When a two-force body is in equilibrium, the forces acting on it must be (i) directed along the line joining their points of application, (ii) opposite in direction, and (iii) equal in magnitude.* These three properties are illustrated in Fig. 5.12(c). The first property must be true; otherwise, a summation of moments of those two forces about either one of their points of application will not be zero. The second and third properties must be true; otherwise, a summation of forces in the direction of either one of those two forces will not be zero.

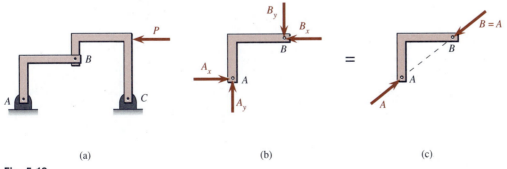

(a)   (b)   (c)

Fig. 5.12

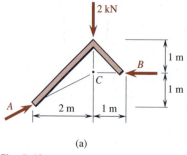

(a)

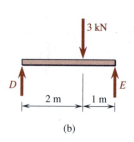

(b)

Fig. 5.13

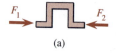

(a)

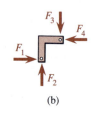

(b)

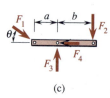

(c)

Fig. D5.20

A *three-force body* is a rigid body acted on by a force system composed of, or reducible to, three forces at three different points. *If a three-force body is in equilibrium, the lines of action of the three forces must be (1) coplanar, and (2) either concurrent or parallel to each other.* These properties are illustrated in Fig. 5.13. Their truth can be seen from the following reasoning: (1') if those three forces were concurrent but noncoplanar, the summation of forces in a direction perpendicular to a plane containing two of the three forces would not be zero; (2') if those three forces were coplanar but neither concurrent nor parallel, the summation of moments about the point of intersection of the lines of action of two of those three forces would not be zero.

Developmental Exercises

D5.19   When equilibrium exists, describe the properties of the forces acting on (a) a two-force body, (b) a three-force body.

D5.20   Ascertain whether each of the bodies shown is a two-force body or a three-force body.

D5.21   Name the two-force body as well as the three-force body, if any, in each of the systems shown.

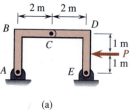

(a)

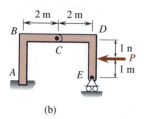

(b)

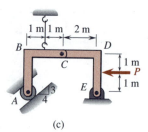

(c)

Fig. D5.21

5.8 Statical Indeterminacy and Constraint Conditions

An unknown, or a problem, associated with a body in equilibrium is said to be *statically determinate* if it can analytically be determined, or solved, solely from the equations of equilibrium and appropriate geometric relations in the system, without using any material properties of the system. Otherwise, it is said to be *statically indeterminate*. Meantime, a rigid body in equilibrium is said to be *statically indeterminate* if its free-body diagram contains one or more statically indeterminate unknowns. In general, the number of unknowns (in the free-body diagram of the body) in excess of the number of independent scalar equilibrium equations available is called the *degree of statical indeterminacy* of the body. For illustrations, see Fig. 5.14.

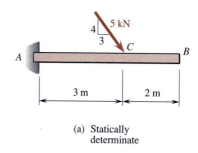

(a) Statically determinate

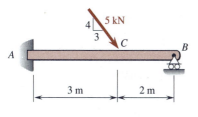

(b) Statically indeterminate to the first degree

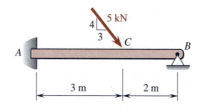

(c) Statically indeterminate to the second degree

Fig. 5.14

A rigid body is said to be *partially constrained* for the loading if (i) there is at least one prospective equilibrium equation which cannot be satisfied by all of the forces and moments in the free-body diagram of the body, and (ii) the total number of unknowns is *smaller* than the total number of prospective independent equilibrium equations to be applied. A partially constrained rigid body is not in equilibrium because the loading is unbalanced and will either accelerate the body into *motion* or cause its supports and connections to *deform* or *rupture*.

A rigid body is said to be *improperly constrained* for the loading if (i) there is at least one prospective equilibrium equation which cannot be satisfied by all of the forces and moments in the free-body diagram of the body, and (ii) the total number of unknowns is *equal to* or *greater than* the total number of prospective independent equilibrium equations to be applied. The unbalanced loading will either accelerate the improperly constrained body into *motion* or cause its supports and connections to *deform* or *rupture*.

Developmental Exercise

D5.22 State the degree of statical indeterminacy of each of the beams shown.

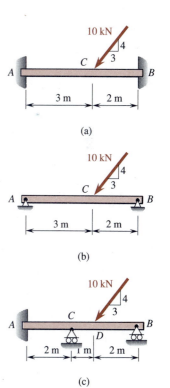

Fig. D5.22

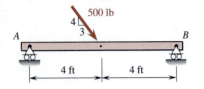

EXAMPLE 5.6

Verify that the beam *AB* is partially constrained for the loading as shown.

Solution.   The free-body diagram of the beam reveals that there are *three* prospective independent equilibrium equations, $\Sigma F_x = 0$, $\Sigma F_y = 0$, and $\Sigma M_A = 0$, to be applied. We note that (i) the prospective equilibrium equation $\Sigma F_x = 0$, cannot be satisfied, and (ii) there are *two* unknown reactions, *A* and *B*. The beam *AB* is, therefore, *partially constrained* for the loading. This beam will be accelerated to *move* to the right.

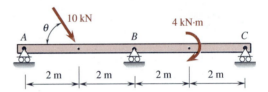

EXAMPLE 5.7

Determine whether or not the beam *ABC* is improperly constrained for the loading as shown if (a) $\theta = 60°$, (b) $\theta = 90°$.

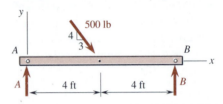

Solution.   This problem may be solved as follows:

Case (a) $\theta = 60°$. From the free-body diagram drawn below for this case, we see that there are *three* prospective independent equilibrium equations, $\Sigma F_x = 0$, $\Sigma F_y = 0$, and $\Sigma M_A = 0$, to be applied. We find (i) $\Sigma F_x \neq 0$; and (ii) there are *three* unknowns, *A*, *B*, and *C*. Thus, if $\theta = 60°$, the beam *ABC* is *improperly constrained* for the loading. This beam will be accelerated to *move* to the right.

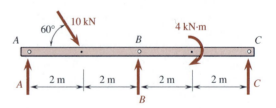

Case (b) $\theta = 90°$. From the free-body diagram drawn for this case, we see that there are only *two* prospective independent equilibrium equations, $\Sigma F_y = 0$ and $\Sigma M_A = 0$, to be applied. We find the following: (i) *no* prospective equilibrium equation which cannot be satisfied; and (ii) there are *three*

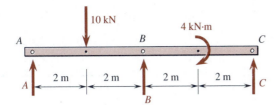

unknowns, A, B, and C. Thus, if $\theta = 90°$, the beam ABC is *not* improperly constrained for the loading. Rather, this beam is *statically indeterminate to the first degree* and may remain *in equilibrium*.

Developmental Exercises

D5.23 A rigid body is shown. (a) Draw the free-body diagram of the body. (b) Is $\Sigma M_B = 0$? (c) Is the body in equilibrium? (d) Is the body *partially* or *improperly* constrained for the loading?

D5.24 Ascertain whether the beam AB is *partially* or *improperly* constrained for the loading as shown. Give reasons for your answer.

D5.25 A plate $ABCD$ is shown. (a) Draw the free-body diagram of the plate. (b) Is the plate *partially* or *improperly* constrained for the loading? (c) Give reasons for your answer to part (b).

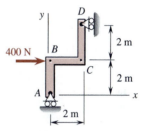

Fig. D5.23

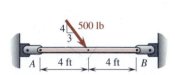

Fig. D5.24

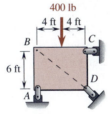

Fig. D5.25

5.9 Possible Sets of Independent Scalar Equilibrium Equations

When a rigid body is in equilibrium, we know that Eqs. (5.11) must be satisfied. Consequently, there exist *two necessary conditions for equilibrium of any body*, which are very useful concepts for solving equilibrium problems and may be stated as follows:

1. *The sum of the components of forces* (in the free-body diagram) *in any desirable direction must be zero*.
2. *The sum of the moments of forces and couples*, if any (in the free-body diagram), *about any desirable point or axis must be zero*.

By repeated applications of one or both of the above *two necessary conditions for equilibrium of any body*, it is possible to solve a statically deter-

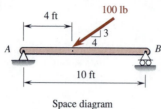

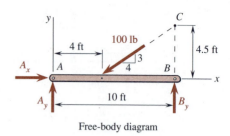

Space diagram   Free-body diagram

Fig. 5.15

minate problem with *different sets of independent scalar equilibrium equations*. For instance, let us consider the equilibrium of the simple beam in Fig. 5.15. We see that there are three possible sets of independent scalar equilibrium equations for solving the three unknown reaction components: A_x, A_y, and B. They are listed in Table 5.1.

Table 5.1   Possible Sets of Independent Scalar Equilibrium Equations for the Beam in Fig. 5.15

Set	1	2	3	Unknowns Determined
Equations of equilibrium	$\Sigma M_A = 0$	$\Sigma M_A = 0$	$\Sigma M_A = 0$	$B_y = 24$ lb
	$\Sigma F_y = 0$	$\Sigma M_B = 0$	$\Sigma M_B = 0$	$A_y = 36$ lb
	$\Sigma F_x = 0$	$\Sigma F_x = 0$	$\Sigma M_C = 0$	$A_x = 80$ lb

Note that *in each set of independent scalar equilibrium equations, the number of equations cannot exceed that obtained by applying Eqs. (5.11).*[†] Furthermore, *each such set should include a moment equilibrium equation if the unknown force components are nonconcurrent.* This is necessary because only moment equilibrium equations can account for the effects of moments and the positions of the forces. In this respect, a moment equilibrium equation is *more efficacious* than a force equilibrium equation.

Developmental Exercise

D5.26   A beam is supported and loaded as shown. (a) How many unknown reaction components are there in the free-body diagram of the beam? (b) How many independent scalar equilibrium equations are there for the solution? (c) What will $\Sigma M_A = 0$ yield? (d) What will $\Sigma M_B = 0$ yield? (e) Does $\Sigma F_y = 0$ yield any additional nontrivial information? (f) Does $\Sigma M_C = 0$ yield any additional nontrivial information? (g) Can we ever solve for A_x and B_x by repeated applications of the two necessary conditions for equilibrium stated in Sec. 5.9? (h) Is the beam statically determinate?

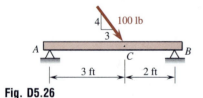

Fig. D5.26

[†]For instance, the *maximum number* of independent scalar equilibrium equations for *a* rigid body is equal to *three* if the body is in equilibrium in a plane; it is equal to *six* if the body is in equilibrium in space.

5.10   Method of Solution: Equilibrium in a Plane

Equilibrium problems of rigid bodies which are statically determinate and relatively simple are considered in this chapter. Those that are more complex are treated in Chaps. 8, 9, and 10. We know that certain problems can best be solved by employing a special set of steps. However, the following procedure may serve as a guide in *solving equilibrium problems of simple rigid bodies in a plane*:

1. *Identify two-force bodies*, if any, in the system.
2. *Indicate the chosen xy rectangular coordinate axes* if they are not in the horizontal and vertical directions.
3. *Draw a good free-body diagram* of the rigid body. (Cf. Sec. 3.8.)
4. *Sum to zero the moments of all forces and couples about a chosen point* (preferably, a point where the lines of action of two unknown forces intersect) to obtain a scalar moment equilibrium equation. If this equation contains only one unknown, it is often solved right away.
5. *Sum to zero the scalar components of all forces in the x and y directions, respectively*, to obtain two scalar force equilibrium equations. However, only one independent equation will be obtained here if the unknown forces are parallel forces.
6. *Solve for the unknowns* contained in all equations obtained.

Note that the total number of unknowns and the total number of independent scalar equilibrium equations must match if *all* unknowns are to be determined. If only a partial solution of a rigid-body equilibrium problem is desired, we may appropriately apply the *two necessary conditions for equilibrium of any body* stated in Sec. 5.9.

EXAMPLE 5.8

A 600-lb crate is suspended as shown. Determine the reaction from the support at *A*.

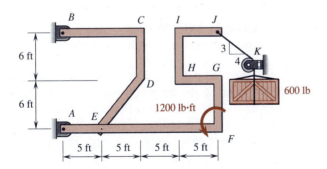

Solution.   We first note that the member *BCDE* is a two-force body. Thus, the reaction from the support at *B* must be directed along the line containing the points *E* and *B*. Since the system is in equilibrium, the tension in the cable fastened at *J* and passing around the pulley at *K* must be

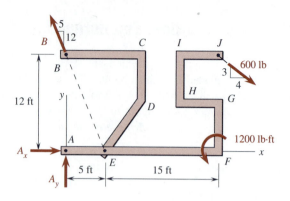

equal to the weight of the block.[†] The free-body diagram of the entire frame may be drawn as shown. In this diagram, there are three unknowns: A_x, A_y, and B. Since we need to determine only A_x and A_y, we write

$$+\circlearrowright \Sigma M_B = 0: \quad 12A_x - 20\left(\frac{3}{5}\right)(600) + 1200 = 0$$

$$+\circlearrowright \Sigma M_E = 0: \quad -5A_y - 12\left(\frac{4}{5}\right)(600) - 15\left(\frac{3}{5}\right)(600) + 1200 = 0$$

These two equations yield $A_x = 500$ and $A_y = -1992$. Since A_y is negative, the true sense of $\mathbf{A}_y$ is opposite to that assumed in the free-body diagram. Thus, we write

$$\mathbf{A}_x = 500 \text{ lb} \rightarrow \qquad \mathbf{A}_y = 1992 \text{ lb} \downarrow \quad \blacktriangleleft$$

Developmental Exercise

D5.27   Determine the reaction from the support at B in Example 5.8.

EXAMPLE 5.9

A frame $ABCD$ supports a 330-lb crate E with the cable and pulley system as shown. Determine the reaction from the support at D.

Solution.   The cable and pulley system acts on the frame at B. To determine this action, we draw the free-body diagram of the pulley at B and readily note that $B_x = T$ from $\Sigma F_x = 0$ and $B_y = T$ from $\Sigma F_y = 0$, where T is the tension in the cable. Referring to the free-body diagram of the crate E together with the lowest pulley as shown, we write

$$+\uparrow \Sigma F_y = 0: \quad 3T - 330 = 0 \qquad B_x = B_y = T = 110 \text{ lb}$$

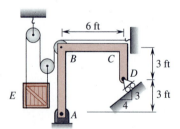

[†]In dynamics, the tension in a cable is generally *not* equal to the weight of the body to which the cable is connected.

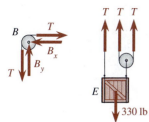

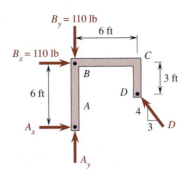

We next draw the free-body diagram of the frame $ABCD$ as shown, where the forces at B are consistent with Newton's third law. We write

$$+\circlearrowleft \Sigma M_A = 0: \quad -6(110) + 6\left(\frac{4}{5}D\right) + 3\left(\frac{3}{5}D\right) = 0$$

$$\mathbf{D} = 100 \text{ lb} \ \measuredangle \ 53.1° \ \blacktriangleleft$$

Developmental Exercises

D5.28   If a rigid-body equilibrium problem requires only a partial solution, what should we do?

D5.29   Refer to Example 5.9. Determine $\mathbf{A}_x$ and $\mathbf{A}_y$.

EXAMPLE 5.10

Using the concept of rigid-body equilibrium, determine the tensions in the wires AD, BE, and CF of the system in Example 3.3.

Solution.   For ease of reference, the space diagram for Example 3.3 is repeated here. We first draw the free-body diagram of the block at G. Referring to the free-body diagram of the block, we write

$$+\uparrow \Sigma F_y = 0: \quad T_{CG} - 10g = 0 \qquad T_{CG} = 10g \text{ N}$$

Next, we draw the free-body diagram of the body ABC obtained by cutting the wires AD, BE, CF, and CG. To facilitate the solution, the horizontal and vertical projections of the wire AC are assumed to be $24L$ and $7L$ units long, respectively, where L is a constant. The length of $\overline{BH}$ may be determined from the fact that

$$\frac{3}{4}\overline{BH} + \frac{5}{12}(\overline{BH} + 7L) = 24L$$

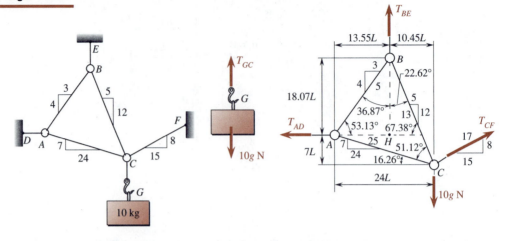

which yields $\overline{BH} = 18.07L$. In the free-body diagram of the body ABC, there are three unknowns: T_{AD}, T_{BE}, and T_{CF}. Treating ABC as a rigid body in equilibrium, we write

$$+\circlearrowleft \Sigma M_H = 0: \quad 7L\left(\frac{15}{17}\right)T_{CF} + 10.45L\left(\frac{8}{17}\right)T_{CF} - 10.45L(10g) = 0$$

$$T_{CF} = 9.419g \qquad T_{CF} = 92.4 \text{ N} \blacktriangleleft$$

$$\xrightarrow{+} \Sigma F_x = 0: \quad -T_{AD} + \frac{15}{17}T_{CF} = 0 \qquad T_{AD} = 81.5 \text{ N} \blacktriangleleft$$

$$+\uparrow \Sigma F_y = 0: \quad T_{BE} + \frac{8}{17}T_{CF} - 10g = 0 \qquad T_{BE} = 54.6 \text{ N} \blacktriangleleft$$

Note that these answers agree with those obtained in Example 3.8.

Developmental Exercise

D5.30   Refer to Example 5.10. Verify the dimensions indicated for ABC.

PROBLEMS

5.25 through 5.27* Draw the free-body diagram of the body AB as shown. *(Postpone the determination of reactions until Probs. 5.34 through 5.36.)*

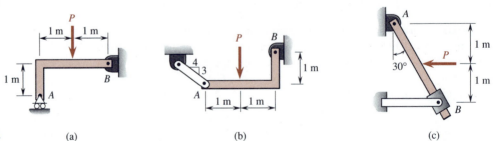

Fig. P5.25 and P5.34   (a)   (b)   (c)

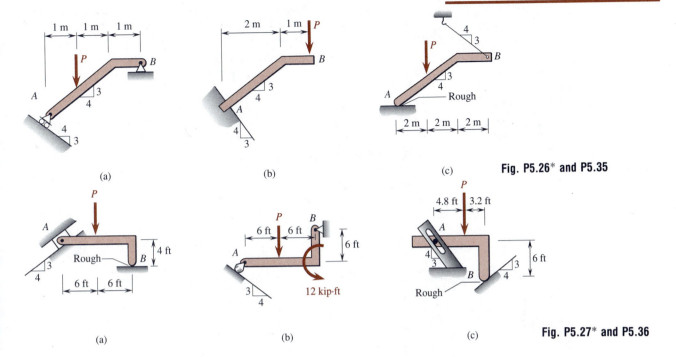

Fig. P5.26* and P5.35

Fig. P5.27* and P5.36

5.28 through 5.30* For P = 120 lb, determine the reactions from the supports of the body AB as shown.

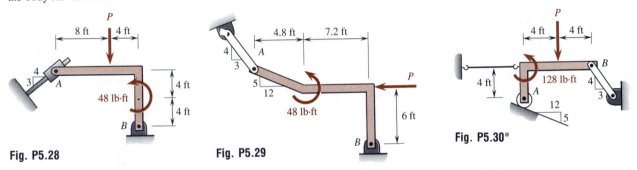

Fig. P5.28

Fig. P5.29

Fig. P5.30*

5.31 through 5.33* For P = 8 kN, determine the reactions from the supports at A and C as shown.

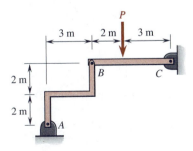

Fig. P5.31

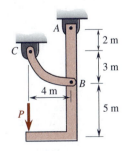

Fig. P5.32

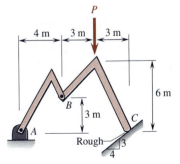

Fig. P5.33*

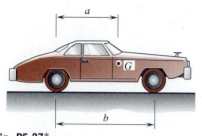

Fig. P5.37*

5.34 and 5.35   *(The figures for these problems are shown and specified along with those for Probs. 5.25 and 5.26*.)* For $P = 6$ kN, determine the reactions from the supports of the body AB as shown.

5.36   *(The figures for this problem are shown and specified along with those for Prob. 5.27*.)* For $P = 600$ lb, determine the reactions from the supports of the body AB as shown.

5.37*   A car is parked on a driveway. It has a wheelbase b and a total weight W which acts at a distance a forward of the rear axle as shown. Derive the expression of the reaction **R** to the pair of rear wheels in terms of a, b, and W.

5.38 and 5.39   For the frame shown, determine the reactions from the supports at A and D.

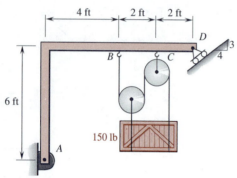

Fig. P5.38

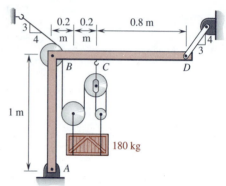

Fig. P5.39

5.40   A truck carries a load of coal at a constant speed up a grade as shown. The resultant weight force of the load is 61 kN acting through the point G_1 and the resultant weight force of the empty truck is 30.5 kN acting through the point G_2. Determine the reaction from the ground to the pair of rear driving wheels.

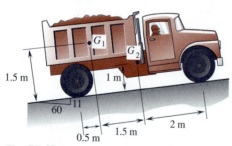

Fig. P5.40

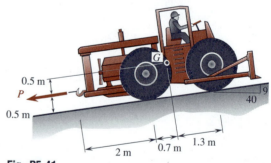

Fig. P5.41

5.41   A 4.1-Mg rubber-tired tractor is pulling with constant speed a load **P** up a grade as shown. The resultant weight force of the tractor acts through the point G, and the friction force exerted by the ground on each of the four wheels is three-fourths of the normal reaction to the wheel. Determine (a) the magnitude of **P**, (b) the reaction from the ground to the pair of front wheels.

5.42 When no counterweight is provided, the resultant weight force of the 10-Mg hoisting crane acts at the point G as shown. If the crane is to pick up a load of 5 Mg in the position shown, what is the minimum amount of mass m_C of the counterweight that must be provided at the point C?

5.43 and 5.44* A plane truss, which may here be considered as a rigid body, is loaded and supported as shown. Determine the reaction from the support at A.

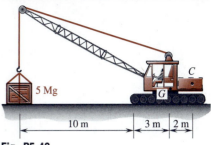

Fig. P5.42

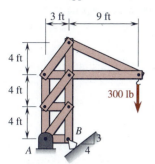

Fig. P5.43

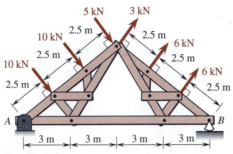

Fig. P5.44*

5.45 A model airplane is being displayed as shown. (a) Using a single moment equilibrium equation, determine the tension T_{CF} in the wire CF. (b) Determine the tensions in the other wires.

5.46 The 100-lb resultant weight force of the rod AB acts through the point G and the rod is in equilibrium as shown. Determine (a) the value of θ which the rod makes with the horizontal, (b) the tension T_{AD} in the wire AD.

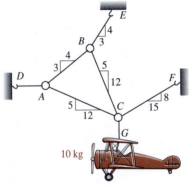

Fig. P5.45

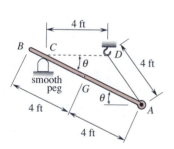

Fig. P5.46

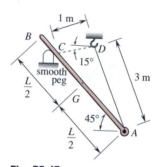

Fig. P5.47

5.47 The resultant weight force of the 10-kg rod AB acts through the point G and the rod is in equilibrium as shown. Determine (a) the length L of the rod, (b) the tension T_{AD} in the wire AD.

EQUILIBRIUM IN SPACE

5.11 Reactions from Supports and Connections in Space

When a rigid body is in *equilibrium in space*, the force system acting on it is generally *noncoplanar* and *nonconcurrent*. A rigid body acted on by non-coplanar and nonconcurrent forces is referred to as a *rigid body in space*.

Likewise, supports or connections of a rigid body in space are referred to as *supports* or *connections in space*. We recall from Sec. 5.5 that *reactions* from supports or connections of a rigid body are the *constraining forces and moments* acting on the body at the points of support or connection. Each *reaction* must reflect the *restriction of movement* imposed on the body at the point of support or connection. Consequently, the number of unknowns associated with the reaction from a support or connection in space may range from *one* to *six*, which are illustrated and labeled in Figs. 5.16 through 5.21.

TYPE I: REACTIONS WITH ONE UNKNOWN.   See illustrations in Fig. 5.16.

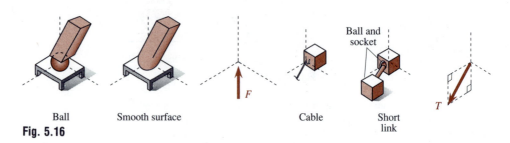

Ball   Smooth surface   Cable   Short
                                                                     link

Fig. 5.16

TYPE II: REACTIONS WITH TWO UNKNOWNS.   See illustrations in Fig. 5.17.

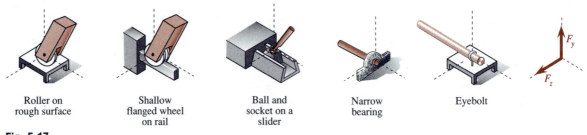

Roller on   Shallow   Ball and   Narrow   Eyebolt
rough surface   flanged wheel   socket on a   bearing
                 on rail   slider

Fig. 5.17

TYPE III: REACTIONS WITH THREE UNKNOWNS.   See illustrations in Fig. 5.18.

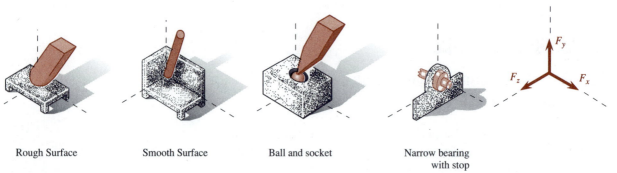

Rough Surface   Smooth Surface   Ball and socket   Narrow bearing
                                                                  with stop

Fig. 5.18

TYPE IV: REACTIONS WITH FOUR UNKNOWNS.   See illustrations in Fig. 5.19.

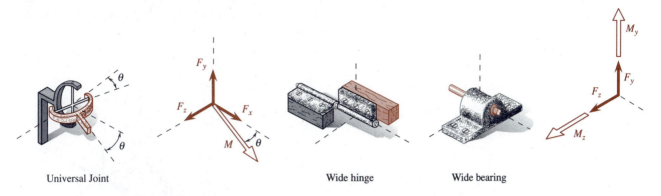

Universal Joint   Wide hinge   Wide bearing

Fig. 5.19   Note that the reaction moment developed in a universal joint is perpendicular to the *crosspiece* (or ''spider'') in the joint.

TYPE V: REACTIONS WITH FIVE UNKNOWNS.   See illustrations in Fig. 5.20.

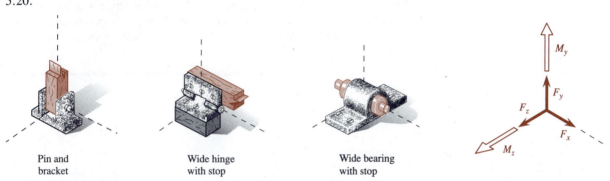

Pin and   Wide hinge   Wide bearing
bracket   with stop   with stop

Fig. 5.20

TYPE VI: REACTIONS WITH SIX UNKNOWNS.   See the illustration in Fig. 5.21. Of course, the supports or connections in Figs. 5.16 through 5.21 may be modified to give a certain type of reaction as illustrated in Fig. 5.22.

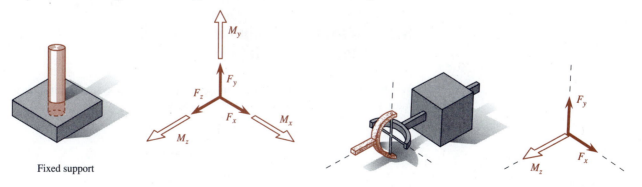

Fixed support

Fig. 5.21   **Fig. 5.22**

EXAMPLE 5.11

A 70-kg plate is supported and loaded as shown (below left). Assume that the resultant weight force of the plate acts through the point G. Draw the free-body diagram of the plate.

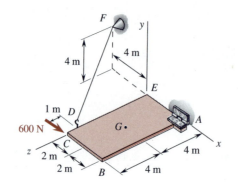
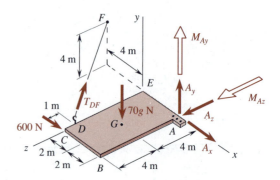

Solution.   We note that the support at A is a wide hinge with stop and the support at D is a cable. The reactions from these two supports are, respectively, type V and type I reactions. Thus, the free-body diagram of the plate is drawn as shown (above right).[†]

Fig. D5.32

Fig. D5.33

Developmental Exercises

D5.31 (a) What is a rigid body in space? (b) What are supports or connections in space?

D5.32 A horizontal shaft is supported by two *narrow* bearings as shown. If one of the bearings is removed, will the shaft remain horizontal?

D5.33 A horizontal shaft supported by a single bearing remains horizontal before and after a force is applied as shown. Should the bearing be considered as *narrow* or *wide*?

5.12 Maximum Number of Statically Determinate Unknowns

When a rigid body is in equilibrium in space, the force system acting on it must satisfy Eqs. (5.11), which may also be written as

$$(\Sigma F_x)\mathbf{i} + (\Sigma F_y)\mathbf{j} + (\Sigma F_z)\mathbf{k} = \mathbf{0}$$
$$(\Sigma M_P)_x\mathbf{i} + (\Sigma M_P)_y\mathbf{j} + (\Sigma M_P)_z\mathbf{k} = \mathbf{0}$$

$$(5.13)$$

[†]Continued in Example 5.18.

These two vector equations of equilibrium are equivalent to the following
six *scalar equations of equilibrium*:

$$\Sigma F_x = 0 \qquad \Sigma F_y = 0 \qquad \Sigma F_z = 0$$

$$(\Sigma M_P)_x = 0 \qquad (\Sigma M_P)_y = 0 \qquad (\Sigma M_P)_z = 0$$

$$(5.14)$$

The first three of Eqs. (5.14) are referred to as the *force equilibrium
equations,* while the other three are referred to as the *moment equilibrium
equations.* We recall that an independent equilibrium equation cannot be
obtained through a linear combination of other equilibrium equations. If the
total number of unknowns associated with the equilibrium of a rigid body
in space is greater than the total number of independent scalar equilibrium
equations available, the rigid body is *statically indeterminate* and *some* or
all of the unknowns *cannot* be determined solely from those equations.

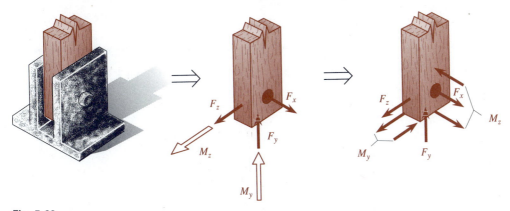

Fig. 5.23

For ascertaining the total number of statically determinate unknowns,
each unknown component of *reaction moment* may be replaced by an equivalent unknown *couple* where the parallel forces are set at a unit distance
from each other as illustrated in Fig. 5.23. The number of statically determinate unknowns in an equilibrium problem of a *simple* rigid body in space
is dependent on the configuration of the noncoplanar and nonconcurrent unknown forces in the free-body diagram of the rigid body. It is important to
recognize that the *maximum number of statically determinate unknowns* allowed by that free-body diagram is equal to *six or less* as described below:

1. The maximum number of statically determinate unknowns allowed is
   equal to *three* if the unknown forces are all parallel to each other. In this
   case, two force equilibrium equations and one moment equilibrium equation are trivial. (Cf. Example 5.14.)
2. It is equal to *four* if the unknown forces are all parallel to a plane and
   all intersect a common straight line. In this case, one force equilibrium
   equation and one moment equilibrium equation are trivial. (Cf. Example
   5.15.)
3. It is equal to *five* if the unknown forces are all parallel to a plane but are
   not all parallel to a line and do not all intersect a common straight line.
   In this case, one force equilibrium equation is trivial. (Cf. Example
   5.16.)

4. It is also equal to *five* if the unknown forces either are parallel to or intersect a common straight line but are not all parallel to a plane. In this case, one moment equilibrium equation is trivial. (Cf. Example 5.17.)
5. It is equal to *six* if the unknown forces are not all parallel to a plane nor all intersect a common straight line. (Cf. Example 5.18.)

If the total number of unknowns in the free-body diagram exceeds that described in the list above, the equilibrium problem is *statically indeterminate*. However, it may still be possible to determine *some*, but not all, of the unknowns by using the *two necessary conditions for equilibrium of any body* as stated in Sec. 5.9. (Cf. Example 5.19.) Furthermore, note that the definitions for the *degree of statical indeterminacy*, the *partially constrained rigid body*, and the *improperly constrained rigid body* in Sec. 5.8 continue to be applicable to rigid bodies in space.

EXAMPLE 5.12

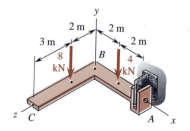

Verify that the bent bar *ABC* is partially constrained for the loading as shown at left.

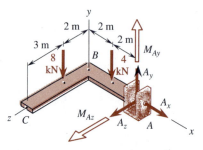

Solution.   From the free-body diagram shown above, we see that there are *six* prospective independent equilibrium equations, $\Sigma F_x = 0$, $\Sigma F_y = 0$, $\Sigma F_z = 0$, $\Sigma M_x = 0$, $\Sigma M_y = 0$, and $\Sigma M_z = 0$, to be applied. We find that (i) $\Sigma M_x \neq 0$; and (ii) there are *five* unknowns, A_x, A_y, A_z, M_{Ay}, and M_{Az}. Thus, the bent bar *ABC* is *partially constrained* for the loading, which will cause it to rotate about the *x* axis.

EXAMPLE 5.13

Determine whether the bent rod *ABC* is partially or improperly constrained for the loading as shown.

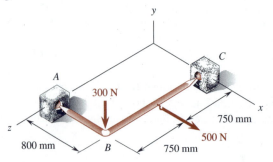

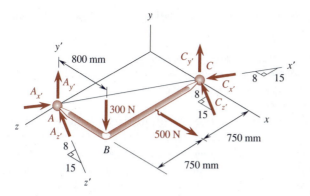

Solution. Introducing the $x'y'z'$ coordinate system in the free-body diagram as shown, we see that there are *six* prospective independent equilibrium equations, $\Sigma F_{x'} = 0$, $\Sigma F_{y'} = 0$, $\Sigma F_{z'} = 0$, $\Sigma M_{x'} = 0$, $\Sigma M_{y'} = 0$, and $\Sigma M_{z'} = 0$, to be applied. We find the following: (i) $\Sigma M_{x'} \neq 0$; and (ii) there are *six* unknowns, $A_{x'}$, $A_{y'}$, $A_{z'}$, $C_{x'}$, $C_{y'}$, and $C_{z'}$. Thus, the bent rod *ABC* is *improperly constrained* for the loading, which will cause the bent rod to rotate about the x' axis.

Developmental Exercises

D5.34 For each of the bent bars shown, indicate (a) the total number of unknowns in the free-body diagram of the bent bar, (b) the total number of independent scalar equilibrium equations available, (c) the statical indeterminacy.

D5.35 For ascertaining the total number of statically determinate unknowns, how is each unknown component of reaction moment replaced?

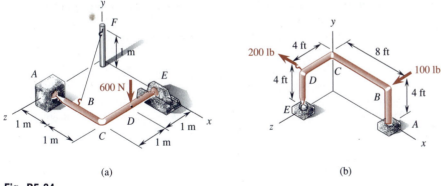

(a)   (b)

Fig. D5.34

5.13 Method of Solution: Equilibrium in Space

The characterizing feature of an *equilibrium problem of a rigid body in space* is that a set of *noncoplanar* and *nonconcurrent* unknown forces acting on the rigid body will be encountered in the solution. Although certain problems can best be solved by a special set of steps, the following procedure may serve as a guide in *solving equilibrium problems of simple rigid bodies in space*:

1. *Identify two-force bodies*, if any, in the system.
2. *Draw correctly the free-body diagram* of the rigid body.
3. *Write the analytical expression* of each external force and moment acting on the free body.
4. *Add the analytical expressions* of all forces acting on the free body to obtain the resultant force in terms of known and unknown quantities.
5. *Set each of the scalar components of the resultant force*, obtained in step (4), *equal to zero* to obtain the force equilibrium equations.
6. *Compute the resultant moment* of the force system about a chosen point (preferably, a point where several of the lines of action of the unknown forces intersect). This resultant moment is now expressed in terms of known and unknown quantities. Sometimes, a partial solution is facilitated if the resultant moment is taken about an axis.
7. *Set each of the scalar components of the resultant moment*, obtained in step (6), *equal to zero* to obtain the moment equilibrium equations.
8. *Solve for the unknowns* in the equilibrium equations obtained. Note that the total number of unknowns and the total number of independent scalar equilibrium equations must match if all unknowns are statically determinate.

If only a *partial solution* of a rigid-body equilibrium problem is desired, we may appropriately apply the *two necessary conditions for equilibrium of any body* stated in Sec. 5.9. Of course, a *complete solution* can also be achieved by obtaining several *partial solutions*.

EXAMPLE 5.14

A horizontal rectangular plate is supported by three vertical cables as shown. The weight force of the plate and the loads on the plate are equivalent to a 1000-lb force acting at the point D as shown. Are the three tensions in the cables statically determinate? If so, determine their values.

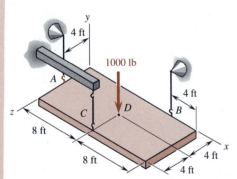

Solution. The free-body diagram of the plate may be drawn as shown. We note that the unknown reaction forces acting on the free body are all parallel to each other as well as to the y axis (or the xy and yz planes). Therefore, $(\Sigma M_D)_y = 0$, $\Sigma F_x = 0$, and $\Sigma F_z = 0$ are all trivial equations.

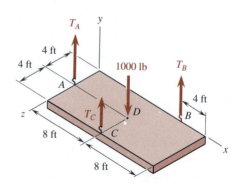

We expect to have *three* independent equilibrium equations available for the solution of the *three* unknowns: T_A, T_B, and T_C. [Cf. case (1) in Sec. 5.12.] Thus, the three tensions are *statically determinate*. Referring to the free-body diagram, we write

$$\mathbf{T}_A = T_A\mathbf{j} \qquad \mathbf{T}_B = T_B\mathbf{j} \qquad \mathbf{T}_C = T_C\mathbf{j} \qquad \mathbf{D} = -1000\mathbf{j}$$
$$\overrightarrow{DA} = -8\mathbf{i} \qquad \overrightarrow{DB} = 4\mathbf{i} - 4\mathbf{k} \qquad \overrightarrow{DC} = 4\mathbf{k}$$

For equilibrium, we write

$$\Sigma\mathbf{F} = \mathbf{T}_A + \mathbf{T}_B + \mathbf{T}_C + \mathbf{D} = \mathbf{0}:$$
$$(T_A + T_B + T_C - 1000)\mathbf{j} = \mathbf{0} \tag{1}$$
$$\Sigma\mathbf{M}_D = \overrightarrow{DA} \times \mathbf{T}_A + \overrightarrow{DB} \times \mathbf{T}_B + \overrightarrow{DC} \times \mathbf{T}_C = \mathbf{0}:$$
$$(4T_B - 4T_C)\mathbf{i} + (-8T_A + 4T_B)\mathbf{k} = \mathbf{0} \tag{2}$$

Equating the coefficients of **i**, **j**, and **k** to zero in the above vector equations (1) and (2), we get a set of three simultaneous scalar equations (as expected), which yield

$$T_A = 200 \text{ lb} \qquad\qquad T_B = T_C = 400 \text{ lb} \quad \blacktriangleleft$$

Developmental Exercise

D5.36 Describe the situation as manifested in Example 5.14 in which the maximum number of statically determinate unknowns is equal to *three*.

EXAMPLE 5.15

A beam is supported by two eyebolts at A and C and is loaded as shown. Determine the reactions from the supports.

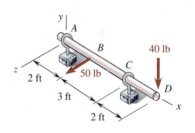

Solution. The free-body diagram of the beam may be drawn as shown. We note that the unknown forces acting on the free body are all parallel to the yz plane and all intersect the x axis. Therefore, $(\Sigma M_A)_x = 0$ and $\Sigma F_x = 0$ are trivial. We expect to have *four* independent equilibrium equations available for the solution of the *four* unknowns: A_y, A_z, C_y, and C_z. [Cf. case (2) in Sec. 5.12.] We have

$$\mathbf{A} = A_y\mathbf{j} + A_z\mathbf{k} \qquad \mathbf{B} = 50\mathbf{k} \qquad \mathbf{C} = C_y\mathbf{j} + C_z\mathbf{k} \qquad \mathbf{D} = -40\mathbf{j}$$
$$\overrightarrow{AB} = 2\mathbf{i} \qquad \overrightarrow{AC} = 5\mathbf{i} \qquad \overrightarrow{AD} = 7\mathbf{i}$$

For equilibrium, we write

$$\Sigma\mathbf{F} = \mathbf{A} + \mathbf{B} + \mathbf{C} + \mathbf{D} = \mathbf{0}:$$
$$(A_y + C_y - 40)\mathbf{j} + (A_z + C_z + 50)\mathbf{k} = \mathbf{0} \tag{1}$$
$$\Sigma\mathbf{M}_A = \overrightarrow{AB} \times \mathbf{B} + \overrightarrow{AC} \times \mathbf{C} + \overrightarrow{AD} \times \mathbf{D} = \mathbf{0}:$$
$$(-5C_z - 100)\mathbf{j} + (5C_y - 280)\mathbf{k} = \mathbf{0} \tag{2}$$

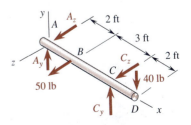

The above vector equations (1) and (2) together give a set of four simultaneous scalar equations (as expected), which yield

$$A_y = -16 \qquad A_z = -30 \qquad C_y = 56 \qquad C_z = -20$$

$$\mathbf{A} = -16\mathbf{j} - 30\mathbf{k} \text{ lb} \qquad \mathbf{C} = 56\mathbf{j} - 20\mathbf{k} \text{ lb} \blacktriangleleft$$

Developmental Exercise

D5.37   Describe the situation as manifested in Example 5.15 in which the maximum number of statically determinate unknowns is equal to *four*.

EXAMPLE 5.16

A ladder in a warehouse is supported by an unflanged wheel A resting against a smooth *shallow* track fixed to the wall and by two *shallow* flanged wheels B and C on a rail fixed to the floor. A man stands on the ladder and leans to one side to pick up a machine part from a high shelf. The combined mass of the ladder and the man holding the machine part is 96 kg, and the weight force acts as shown. Determine the reactions from the supports at A, B, and C.

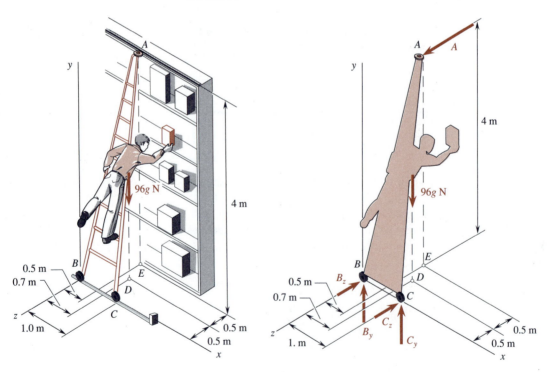

Solution. Since the ladder can move freely in the direction parallel to the x axis, there are no reaction components in the x direction. The free-body diagram of the ladder, the man, and the machine part may be drawn as shown. We note that the noncoplanar and nonconcurrent unknown reaction forces are all parallel to the yz plane but are not all parallel to a line and do not all intersect a common straight line. Therefore, only $\Sigma F_x = 0$ is trivial. We expect to have *five* independent equilibrium equations available for the solution of the *five* unknowns: A, B_y, B_z, C_y, and C_z. [Cf. case (3) in Sec. 5.12.] We have

$\mathbf{A} = A\mathbf{k}$ $\qquad \overrightarrow{BA} = 0.5\mathbf{i} + 4\mathbf{j} - \mathbf{k}$

$\mathbf{B} = B_y\mathbf{j} - B_z\mathbf{k}$ $\qquad \overrightarrow{BC} = \mathbf{i}$

$\mathbf{C} = C_y\mathbf{j} - C_z\mathbf{k}$ $\qquad \overrightarrow{BD} = 0.7\mathbf{i} - 0.5\mathbf{k}$ $\qquad \mathbf{W} = -96g\mathbf{j}$

For equilibrium, we write

$\Sigma\mathbf{F} = \mathbf{A} + \mathbf{B} + \mathbf{C} + \mathbf{W} = \mathbf{0}$:

$$(B_y + C_y - 96g)\mathbf{j} + (A - B_z - C_z)\mathbf{k} = \mathbf{0} \qquad (1)$$

$\Sigma\mathbf{M}_B = \overrightarrow{BA} \times \mathbf{A} + \overrightarrow{BC} \times \mathbf{C} + \overrightarrow{BD} \times \mathbf{W} = \mathbf{0}$:

$$(4A - 48g)\mathbf{i} + (-0.5A + C_z)\mathbf{j} + (C_y - 67.2g)\mathbf{k} = \mathbf{0} \qquad (2)$$

The above vector equations (1) and (2) together give a set of five simultaneous scalar equations (as expected), which yield

$A = 12g \qquad B_y = 28.8g \qquad B_z = 6g \qquad C_y = 67.2g \qquad C_z = 6g$

$$\mathbf{A} = 117.7\mathbf{k} \text{ N} \qquad \mathbf{B} = 283\mathbf{j} - 58.9\mathbf{k} \text{ N} \blacktriangleleft$$

$$\mathbf{C} = 659\mathbf{j} - 58.9\mathbf{k} \text{ N} \blacktriangleleft$$

Developmental Exercise

D5.38 Describe the situation as manifested in Example 5.16 in which the maximum number of statically determinate unknowns is equal to *five*.

EXAMPLE 5.17

The system shown carries a 110-lb crate at B and is supported by a ball-and-socket joint at A and two cables at C and D. Determine the tensions in cables CE and DF and the reaction from the support at A.

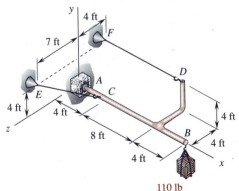

110 lb

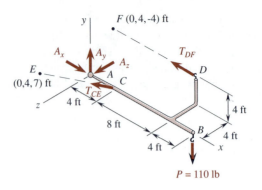

Solution.   The free-body diagram may be drawn as shown. We note that the noncoplanar and nonconcurrent unknown reaction forces either intersect or are parallel to the x axis but are not all parallel to a plane. Therefore, $(\Sigma M_A)_x = 0$ is trivial. We expect to have *five* independent equilibrium equations available for the solution of the *five* unknowns: T_{CE}, T_{DF}, A_x, A_y, and A_z. [Cf. case (4) in Sec. 5.12.] We have

$$\overrightarrow{CE} = -4\mathbf{i} + 4\mathbf{j} + 7\mathbf{k} \qquad \boldsymbol{\lambda}_{CE} = \frac{1}{9}(-4\mathbf{i} + 4\mathbf{j} + 7\mathbf{k})$$

$$\mathbf{T}_{CE} = \frac{T_{CE}}{9}(-4\mathbf{i} + 4\mathbf{j} + 7\mathbf{k}) \qquad \mathbf{T}_{DF} = -T_{DF}\mathbf{i}$$

$$\mathbf{A} = A_x\mathbf{i} + A_y\mathbf{j} + A_z\mathbf{k} \qquad \mathbf{P} = -110\mathbf{j}$$

$$\overrightarrow{AC} = 4\mathbf{i} \qquad \overrightarrow{AB} = 16\mathbf{i} \qquad \overrightarrow{AF} = 4\mathbf{j} - 4\mathbf{k}$$

For equilibrium, we write

$$\Sigma\mathbf{F} = \mathbf{T}_{CE} + \mathbf{T}_{DF} + \mathbf{A} + \mathbf{P} = 0:$$

$$\left(-\frac{4}{9}T_{CE} - T_{DF} + A_x\right)\mathbf{i} + \left(\frac{4}{9}T_{CE} + A_y - 110\right)\mathbf{j}$$

$$+ \left(\frac{7}{9}T_{CE} + A_z\right)\mathbf{k} = 0 \tag{1}$$

$$\Sigma\mathbf{M}_A = \overrightarrow{AC} \times \mathbf{T}_{CE} + \overrightarrow{AB} \times \mathbf{P} + \overrightarrow{AF} \times \mathbf{T}_{DF} = 0:$$

$$\left(-\frac{28}{9}T_{CE} + 4T_{DF}\right)\mathbf{j} + \left(\frac{16}{9}T_{CE} + 4T_{DF} - 1760\right)\mathbf{k} = 0 \tag{2}$$

The above vector equations (1) and (2) together give a set of five simultaneous scalar equations (as expected), which yield

$$T_{CE} = 360 \qquad T_{DF} = 280 \qquad A_x = 440 \qquad A_y = -50 \qquad A_z = -280$$

$$T_{CE} = 360 \text{ lb} \qquad T_{DF} = 280 \text{ lb} \qquad \mathbf{A} = 440\mathbf{i} - 50\mathbf{j} - 280\mathbf{k} \text{ lb} \quad \blacktriangleleft$$

Developmental Exercise

D5.39 Describe the situation as manifested in Example 5.17 in which the maximum number of statically determinate unknowns is equal to *five*.

EXAMPLE 5.18

Refer to the plate in Example 5.9. Determine the tension in the cable *DF* and the reaction from the support at *A*.

Solution. The free-body diagram of the plate is first drawn as shown. Following Sec. 5.12, we may replace each of the two unknown moment reaction components $\mathbf{M}_{Ay}$ and $\mathbf{M}_{Az}$ by a pair of forces which are at a small

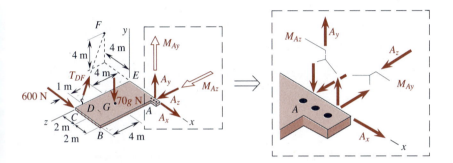

distance from each other and form a couple as shown. Thus, we see that the noncoplanar and nonconcurrent unknown reaction forces are not all parallel to a plane and their lines of action do not all intersect a common straight line. From case (5) in Sec. 5.12, we expect to have *six* independent scalar equilibrium equations for the solution of the *six* unknowns: T_{DF}, A_x, A_y, A_z, M_{Ay}, and M_{Az}. We have

$$\overrightarrow{DF} = -4\mathbf{i} + 4\mathbf{j} - 7\mathbf{k} \qquad \boldsymbol{\lambda}_{DF} = \frac{1}{9}(-4\mathbf{i} + 4\mathbf{j} - 7\mathbf{k})$$

$$\mathbf{T}_{DF} = \frac{T_{DF}}{9}(-4\mathbf{i} + 4\mathbf{j} - 7\mathbf{k}) \qquad \mathbf{A} = A_x\mathbf{i} + A_y\mathbf{j} + A_z\mathbf{k}$$

$$\mathbf{C} = 600\mathbf{i} \qquad \mathbf{G} = -70g\mathbf{j} \qquad \mathbf{M}_A = M_{Ay}\mathbf{j} + M_{Az}\mathbf{k}$$

$$\overrightarrow{AD} = -4\mathbf{i} + 7\mathbf{k} \qquad \overrightarrow{AB} = 8\mathbf{k} \qquad \overrightarrow{AG} = -2\mathbf{i} + 4\mathbf{k}$$

For equilibrium, we write

$$\Sigma\mathbf{F} = \mathbf{T}_{DF} + \mathbf{A} + \mathbf{C} + \mathbf{G} = \mathbf{0}:$$

$$\left(-\frac{4}{9}T_{DF} + A_x + 600\right)\mathbf{i} + \left(\frac{4}{9}T_{DF} + A_y - 70g\right)\mathbf{j}$$

$$+ \left(-\frac{7}{9}T_{DF} + A_z\right)\mathbf{k} = \mathbf{0} \tag{1}$$

$$\Sigma \mathbf{M}_A = \overrightarrow{AD} \times \mathbf{T}_{DF} + \mathbf{M}_A + \overrightarrow{AB} \times \mathbf{C} + \overrightarrow{AG} \times \mathbf{G} = \mathbf{0}:$$

$$\left(-\frac{28}{9} T_{DF} + 280g\right)\mathbf{i} + \left(-\frac{56}{9} T_{DF} + M_{Ay} + 4800\right)\mathbf{j}$$

$$+ \left(-\frac{16}{9} T_{DF} + M_{Az} + 140g\right)\mathbf{k} = \mathbf{0} \qquad (2)$$

The above vector equations (1) and (2) together give a set of six simultaneous scalar equations (as expected), which yield

$$T_{DF} = 90g \qquad M_{Ay} = 693.6 \qquad M_{Az} = 20g$$

$$A_x = -207.6 \qquad A_y = 30g \qquad A_z = 70g$$

$$T_{DF} = 883\ \text{N} \qquad \mathbf{M}_A = 694\mathbf{j} + 196.2\mathbf{k}\ \text{N}\cdot\text{m} \blacktriangleleft$$

$$\mathbf{A} = -208\mathbf{i} + 294\mathbf{j} + 687\mathbf{k}\ \text{N} \blacktriangleleft$$

Developmental Exercise

D5.40   Describe the situation as manifested in Example 5.18 in which the maximum number of statically determinate unknowns is equal to *six*.

EXAMPLE 5.19

The rigid body *ABCDEF* is supported by a cable *EG* and two ball-and-socket supports at *A* and *F*. The allowable tension in the cable is 6 kN. Determine the maximum mass *m* that can be supported at *D* as shown.

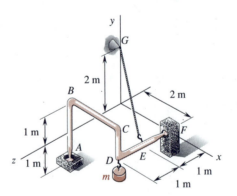

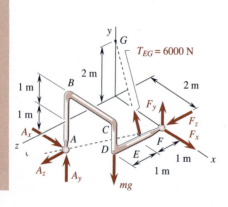

Solution.   When the maximum mass *m* is supported at *D*, the tension in the cable *EG* will be equal to 6000 N as indicated in the free-body diagram shown. We note that there are *seven* unknowns: A_x, A_y, A_z, F_x, F_y, F_z, and *m*; however, the maximum number of statically determinate unknowns for a rigid body in space is *six*. Therefore, the rigid body is statically indeterminate and we *may* be able to determine some, but not all, of the unknowns.

By inspection, we find that all unknowns except m can be prevented from entering an equilibrium equation when we equate to zero the sum of the moments of all forces about the axis AF. From the sketch, we have

$$\overrightarrow{EG} = -2\mathbf{i} + 2\mathbf{j} - \mathbf{k} \qquad \boldsymbol{\lambda}_{EG} = \frac{1}{3}(-2\mathbf{i} + 2\mathbf{j} - \mathbf{k}) \qquad T_{EG} = 6000$$

$$\mathbf{T}_{EG} = T_{EG}\boldsymbol{\lambda}_{EG} = 2000(-2\mathbf{i} + 2\mathbf{j} - \mathbf{k}) \qquad \mathbf{W} = -mg\mathbf{j} \qquad \overrightarrow{FE} = \mathbf{k}$$

$$\overrightarrow{FD} = 2\mathbf{k} \qquad \overrightarrow{AF} = 2\mathbf{i} + \mathbf{j} - 2\mathbf{k} \qquad \boldsymbol{\lambda}_{AF} = \frac{1}{3}(2\mathbf{i} + \mathbf{j} - 2\mathbf{k})$$

For equilibrium, we write

$$\Sigma M_{AF} = \boldsymbol{\lambda}_{AF} \cdot (\overrightarrow{FE} \times \mathbf{T}_{EG}) + \boldsymbol{\lambda}_{AF} \cdot (\overrightarrow{FD} \times \mathbf{W}) = -4000 + \frac{4}{3}mg = 0$$

$$m = 3000/g \qquad m = 306 \text{ kg} \blacktriangleleft$$

REMARK. Without using vector algebra to take moment about the axis AF, it would be rather difficult to solve the problem in this example.

Developmental Exercise

D5.41 If a rigid-body equilibrium problem requires only a partial solution, what should we do?

PROBLEMS

5.48 and 5.49* Is the bent bar ABC improperly or partially constrained for a vertical force $\mathbf{P}$ applied as shown?

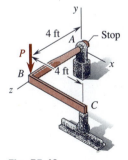

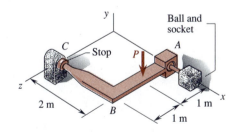

Fig. P5.48   Fig. P5.49*

5.50 Do Prob. 5.48 if the flanged wheel on rail at the end C is replaced with a ball-and-socket joint.

5.51* Do Prob. 5.49* if the wide bearing with stop at the end C is replaced with a wide bearing without stop.

Fig. P5.52   **Fig. P5.53***

5.52 and 5.53* A polygonal table has a total weight W and three vertical legs as shown. The table has equal sides and identical corners. Determine the maximum magnitude of a vertical force **P** which may act anywhere on the table without tipping it over.

5.54 It is known that the resultant weight force of the 200-kg ventilating duct shown passes through the point G. If $d = 3$ m, determine the tensions in the vertical wires supporting the duct at the points A, B, and C.

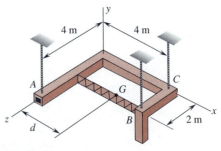

Fig. P5.54

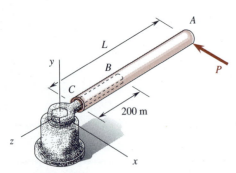

Fig. P5.56

5.55* Solve Prob. 5.54 if $d = 2$ m.

5.56 A 5-kg pipe is slipped over a wrench to loosen a stubborn bolt as shown. The friction between the pipe and the wrench handle is negligible. For $P = 400$ N and $L = 800$ mm, determine the reactions from the wrench handle to the pipe at the points B and C.

5.57 Solve Prob. 5.56 if $P = 300$ N and $L = 1$ m.

5.58 A uniform conic shell having a weight W and a base of radius r is suspended with three vertical wires of equal lengths L as shown. Determine the magnitude of the vertical moment **M** which is needed to rotate the shell through a small angle θ.

5.59* A 16-ft steel boom AB which weighs 100 lb is acted on by a force **P** of 340 lb as shown. Assuming the weight force acts through the midpoint of the boom, determine the tension in each cable and the reaction from the support at A.

5.60 A derrick carries a 12-Mg crate as shown. Determine the tensions in the cables BF and BG and the reaction from the support at A.

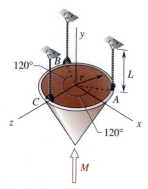

Fig. P5.58

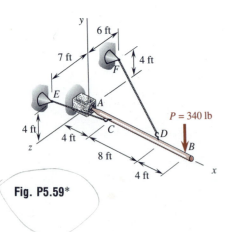

Fig. P5.59*

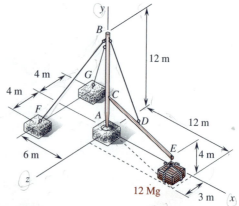

Fig. P5.60

5.61 Determine the reactions from the supporting rails at A, B, and C as shown.

5.62 through 5.67* *Using an appropriate moment equilibrium equation, deter-mine the tension in the cable supporting the rigid body shown.*

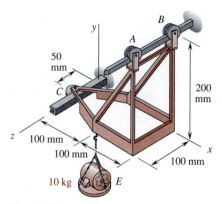

Fig. P5.61

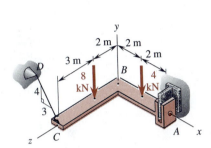

Fig. P5.62 and P5.68

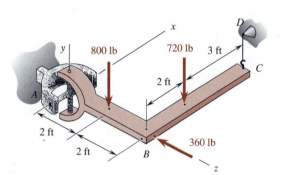

Fig. P5.63 and P5.69

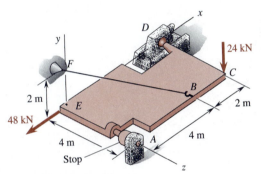

Fig. P5.64 and P5.70

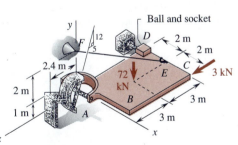

Fig. P5.65 and P5.71

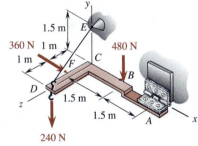

Fig. P5.66* and P5.72*

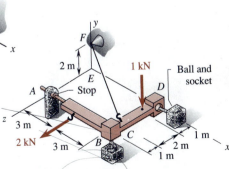

5.68 through 5.73* For the rigid body shown, determine the reaction from the support at A.

Fig. P5.67* and P5.73*

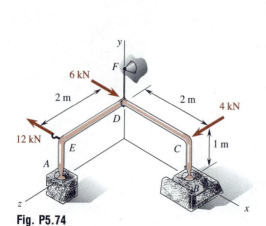

Fig. P5.74

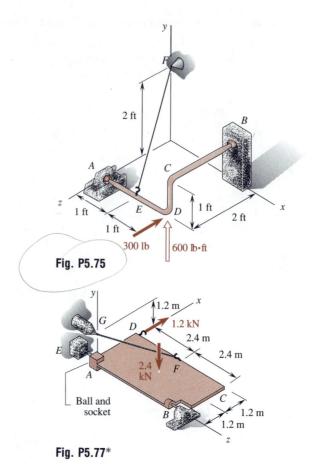

Fig. P5.75

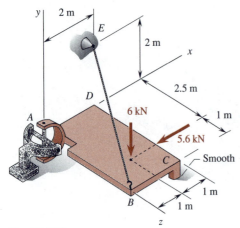

Fig. P5.76*

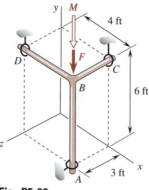

Fig. P5.77*

5.74 through 5.77* For the rigid body shown, determine the tension in the cable and the reactions from the supports at A and B.

5.78 Determine the reactions from the supports at A and B as shown if $\mathbf{M} = \mathbf{0}$.

5.79* Solve Prob. 5.78 if $\mathbf{M} = 450\mathbf{i} - 400\mathbf{k}$ N·m.

5.80 A rigid body in the shape of a triad is supported by three eyebolts as shown. Determine the reactions from the supports if $\mathbf{F} = -600\mathbf{j}$ lb and $\mathbf{M} = \mathbf{0}$.

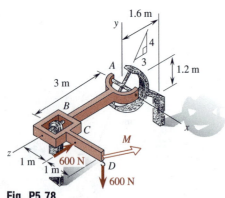

Fig. P5.78

Fig. P5.80

5.81* Solve Prob. 5.80 if $\mathbf{F} = -600\mathbf{j}$ lb and $\mathbf{M} = -120\mathbf{j}$ lb·ft.

5.82 Having disconnected a clogged drainpipe, a plumber introduced a pipe auger (or power snake) through the opening of the end at *A* as shown. The cutting head of the auger is connected by a heavy-duty flexible cable to an electric motor which uniformly rotates the cable the plumber forces into the pipe. The supports at *B*, *C*, and *D* may be taken as narrow bearings. The cleaning efforts at *A* are equivalent to a wrench of $\mathbf{F} = 60\mathbf{i}$ N and $\mathbf{M} = 200\mathbf{i}$ N·m. Determine the reactions from the supports at *B*, *C*, and *D* due to the cleaning operation.

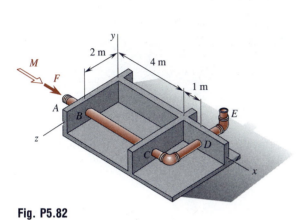

Fig. P5.82

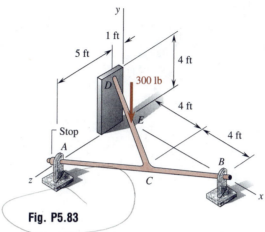

Fig. P5.83

5.83 A lever leans against a smooth vertical wall at *D* and is supported at *A* and *B* by narrow bearings as shown. A vertical force of 300 lb is applied at the midpoint *E* of the rod *CD*. Determine the reactions from the supports at *A*, *B*, and *D*.

5.84 A 30-kg rod *AB* is supported by a ball and socket at *A* and leans against the top of a vertical wall at *E* and the surface of a taller vertical wall at *B* as shown. Assume that friction forces between the rod and the walls are negligible and the weight force acts through the midpoint of the rod. Determine (a) the coordinates (x_E, y_E, z_E) of the point *E*, (b) the unit vector $\boldsymbol{\lambda}_E$ in the direction of the reaction from the support at *E*, (c) the reactions from the supports of the rod.

5.85 through 5.88* A rigid body is shown. (a) How many unknowns are there in the reactions from the supports? (b) How many independent equilibrium equations are there available for the solution of these unknowns? (c) Are there any statically indeterminate unknowns? (d) If any unknowns are statically determinate, calculate their values.

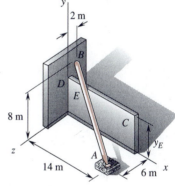

Fig. P5.84

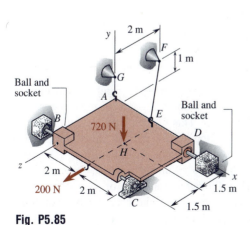

Fig. P5.85

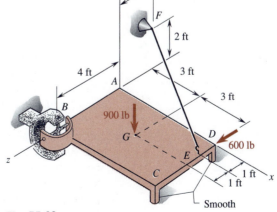

Fig. P5.86

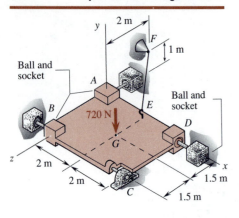

Fig. P5.87

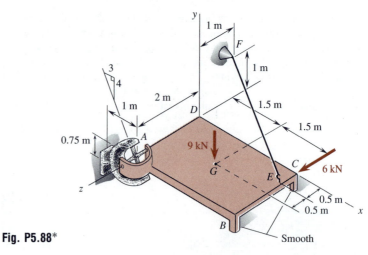

Fig. P5.88*

5.89 Three decorations are suspended by a system of wires as shown. Using the concept of equilibrium of a rigid body in space, determine the tensions in wires *AD*, *AE*, *BF*, *BG*, *CH*, and *CI*.

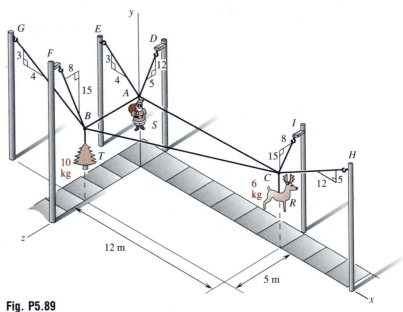

Fig. P5.89

5.14 Concluding Remarks

The fact surrounding the condition of equilibrium of a rigid body is that the *resultant force* and the *resultant moment,* about a point, of the force system acting on its free body are *equal to zero*. This fact provides us with the basis upon which the relations among the applied forces and the reactions from the supports and connections of the rigid body are established. The free-body diagram of a rigid body exhibits to us all the external forces that act together on the free body to bring about the condition or state of the rigid body as it existed in its space diagram. Thus, the establishment of those relations is efficaciously accomplished through the use of the free-body diagram of the rigid body.

The drawing of a good free-body diagram of a rigid body usually involves the exercise of good judgment on the way the supports or connections impose restrictions on the movements of the points of support of the body which is subjected to an applied force system. The recognition of the types of reactions, as discussed and illustrated in Secs. 5.5 and 5.11, is of fundamental importance in the drawing of correct free-body diagrams of rigid bodies in a plane and in space. Furthermore, it is well to *regard the drawing of appropriate free-body diagrams as an integral part of the effective study of the conditions of rigid bodies under the action of force systems whether they are in equilibrium or not.* Bodies not in equilibrium are studied in dynamics.

REVIEW PROBLEMS

5.90 and 5.91* For each rigid body shown, state (a) the unknowns in its free-body diagram; (b) the prospective equilibrium equations to be applied; (c) the prospective equilibrium equations, if any, which cannot be satisfied; (d) the one term (statically determinate, statically indeterminate, partially constrained, or improperly constrained) which best describes the body.

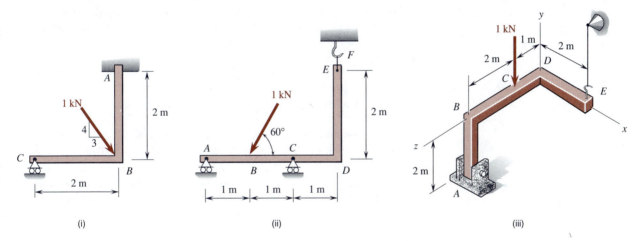

Fig. P5.90

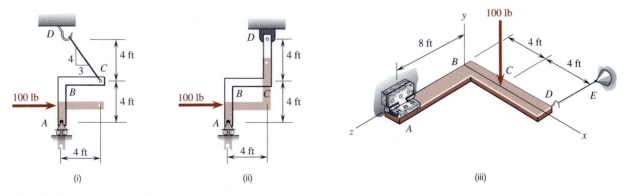

Fig. P5.91*

5.92* In each of the following rigid bodies, (a) identify (by the number of unknown components) the type of reaction from the support at A, (b) draw the free-body diagram of the body AB.

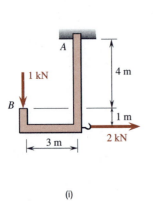

(i)

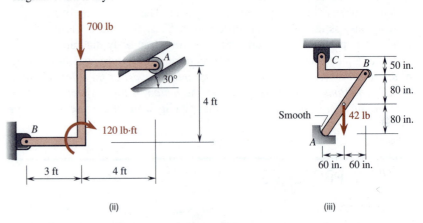

(ii)　　　　　　　　　　　　(iii)

Fig. P5.92* and P5.93

5.93 For each of the rigid bodies shown, determine the reaction from the support at A.

5.94* The resultant weight force $\mathbf{W}$ of the rod AB acts through the point G as shown. Verify that the transcendental equation governing the geometric parameters a, b, L, α and θ of the rod in equilibrium is

$$\left[\left(\frac{b}{a}\right)^2 - \sin^2(\theta + \alpha)\right]^{1/2}\left[\frac{L}{2a} + \sin(\theta + \alpha)\tan\theta - \cos(\theta + \alpha)\right]$$
$$+ \frac{1}{2}\tan\theta\sin 2(\theta + \alpha) + \sin^2(\theta + \alpha) - \left(\frac{b}{a}\right)^2 = 0$$

5.95* The resultant weight force $\mathbf{W}$ of the rod AB acts through the point G as shown. If $b = 2a$, $L = 3a$, $\alpha = 45°$, and the rod is in equilibrium, determine the angle θ. (*Hint.* $20° < \theta < 30°$.)[†]

5.96* Use the one term (statically determinate, statically indeterminate, partially constrained, or improperly constrained) that best describes the body shown if (a) the cable DE is removed, (b) the cable DF is removed, (c) both cables are removed, (d) nothing is removed.

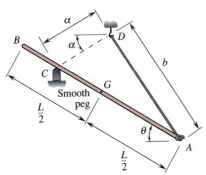

Fig. P5.94* and P5.95*

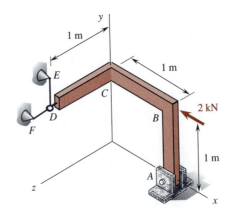

Fig. P5.96*

[†]Cf. App. E or Sec. B5.8 of App. B for the solution of a transcendental equation.

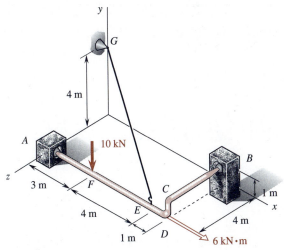

Fig. P5.97

5.97   For the bent rod shown, determine the tension in the cable *EG*.

5.98   Determine the reactions from the supports of the bent bar shown.

5.99   A plate *ABCD* is shown. (a) Determine whether the plate is statically inde-terminate. (b) If any unknowns are statically determinate, calculate their values.

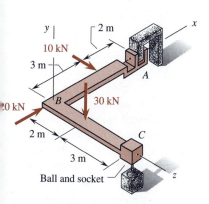

Fig. P5.98

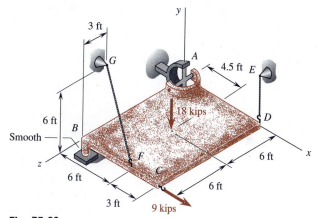

Fig. P5.99

Chapter 6

First Moments: Centroids and Centers of Gravity

KEY CONCEPTS

- Resultant of parallel forces, first moment of a quantity, principle of moments, geographic center, and population center.
- Centroids of lines and centers of gravity of slender members.
- Centroids of areas and centers of gravity of laminas.
- Distributed loads on beams.
- First and second theorems of Pappus-Guldinus.
- Centroids of volumes, centers of gravity of bodies, and centers of mass of bodies.
- Composite volumes and center of pressure.

This chapter deals with the first moments of various quantities and is focused on the *principle of moments*, which is synthesized from the concept of *equivalent systems of forces*. This principle is identified as the key to locating the resultant of parallel forces or any other force system, the centroids of geometric figures, and the centers of gravity as well as centers of mass of physical bodies. Common applications of this principle include those in determining (a) the resultant of distributed forces on a body; (b) the geographic center and the population center of a country; (c) the centroids of cross sections of beams and columns; and (d) the centers of gravity as well as the centers of mass of wires, plates, shells, machine parts, and various other physical bodies. Naturally, this chapter provides a prerequisite for the study of structures withstanding distributed loads or weight forces of given bodies.

6.1 Dimensional Homogeneity in Equations

The equation of a curve or surface is a mathematical statement of the general relationship to be satisfied by the values of the coordinates of any point on the curve or surface, as shown in Fig. 6.1. The equation of a curve or

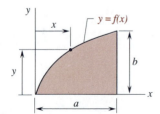

Fig. 6.1

176

surface is frequently used in engineering to describe the intensity of a distributed load, the shape of a figure or body, and other relations. However,
the resulting *dimensions* (and hence units) *of each term* in such an equation
must be the same. This requirement is usually met if the equation is mathematically homogeneous.[†] In any case, the constants including the implicit
1's (such as those in $y = x^2$) in an equation should be understood to carry
whatever dimensions necessary to ensure that *each term has the same resulting dimensions*.

Developmental Exercises

D6.1   A cantilever beam carries a load whose intensity w, in kN/m, varies continuously along its axis as shown. (a) What are the value and units of the constant
c? (b) What is the loading intensity w at $x = 1$ m?

D6.2   A paraboloid is shown. The equation of its curved surface is given by ay
$= x^2 + z^2$. What are the value and units of the constant a?

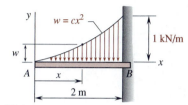

Fig. D6.1

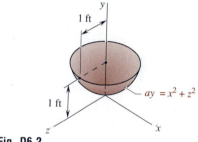

Fig. D6.2

6.2   Resultant of Parallel Forces:
Principle of Moments

We saw in Sec. 2.2 that, if the lines of action of two force vectors intersect
as shown in Fig. 2.8, the line of action of the resultant of these two forces
must pass through the point of intersection. However, the *resultant of a
system of parallel forces* cannot be located in that manner; it is located by
using Eqs. (5.4), which define equivalent systems of forces. As a general-
ization of the concept of the moment of a force, the *first moment*, or more
briefly the *moment*, *of a quantity* is defined as the product of an appropriate
distance and the quantity. This generalized definition of *moment* and the
concepts of *equivalence* and *equipollence* between two systems of forces in
Sec. 5.3 may here be synthesized into a general rule as follows: *The moment
of the "resultant" with respect to a point (or an axis, or a plane) is equal
to the sum of the moments of the "components" with respect to the same
point (or axis, or plane), where the "resultant" is equal to the sum of the
"components."* This *synthesized rule* is referred to as the *principle of moments*; i.e.,

> *Resultant* = Sum of the *Components*
>
> Moment of the *Resultant*   (6.1)
> = Sum of the Moments of the *Components*

where the moment may be taken with respect to any point, axis, or plane,
as appropriate. If the *components* are scalar quantities (e.g., lengths, areas,
volumes, or masses), the *resultant* is a total scalar quantity (e.g., total
length, total area, total volume, or total mass) equal to the algebraic sum of
the component scalars. If the *components* are vector quantities (e.g., forces),

---

[†]An equation is mathematically homogeneous if each term of the equation will contain t to the
same power as a factor when each variable in the equation is replaced by t times the variable,
where $t \neq 0$.

the *resultant* is a resultant vector quantity (e.g., resultant force) equal to the vector sum of the component vectors. When the components are *concurrent forces*, the principle of moments degenerates into *Varignon's theorem* presented in Secs. 4.3 and 4.11.

The principle of moments owes its origin to the concept of *equivalent systems of forces* and is primarily applied to *locate the position of the resultant relative to the components* in many situations. The centroids of lines, areas, and volumes, as well as the centers of mass and centers of gravity of bodies, are their respective "centers," which are defined and located according to the *principle of moments* in the sequel. Note that the *geographic center* of an area is simply the centroid of the area, whereas the *population center* of an area is a point determined according to the *principle of moments*, where the components are the entities of inhabitants and the resultant is the total number of inhabitants in the area.

EXAMPLE 6.1

If the resultant force **R** of the three vertical forces shown acts through the point $P(\bar{x}, 0, \bar{z})$, determine $\bar{x}$ and $\bar{z}$.

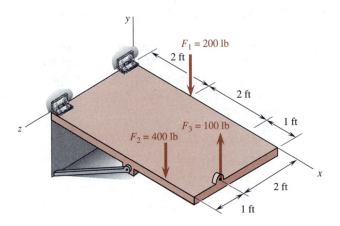

Solution.   The component forces are

$$\mathbf{F}_1 = -200\mathbf{j} \text{ lb} \qquad \mathbf{F}_2 = -400\mathbf{j} \text{ lb} \qquad \mathbf{F}_3 = 100\mathbf{j} \text{ lb}$$

The resultant force **R** is

$$\mathbf{R} = \mathbf{F}_1 + \mathbf{F}_2 + \mathbf{F}_3 \qquad \mathbf{R} = -500\mathbf{j} \text{ lb}$$

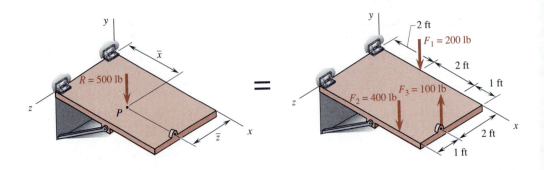

Applying the *principle of moments* and using the right-hand rule, we write

$$\Sigma M_z: \quad -\bar{x}(500) = -2(200) - 4(400) + 5(100) \qquad \bar{x} = 3 \text{ ft} \blacktriangleleft$$

$$\Sigma M_x: \quad \bar{z}(500) = 0(200) + 3(400) - 2(100) \qquad \bar{z} = 2 \text{ ft} \blacktriangleleft$$

Developmental Exercises

D6.3 Determine the distance $\bar{x}$ between the y axis and the line of action of the resultant force of the forces shown.

D6.4 The major towns in the county *PQRST*, as shown, are Oshkosh, Akin, and Butler, which have populations of 5000, 3000, and 2000, and coordinates of locations $O(0, 0)$ km, $A(-10, 20)$ km, and $B(10, -2)$ km, respectively. For this county, locate (a) its geographic center, (b) its approximate population center.

D6.5 (a) Define the *first moment of a quantity*. (b) Describe the *principle of moments*.

D6.6 Define the terms: (a) geographic center, (b) population center.

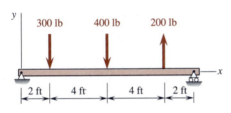

Fig. D6.3

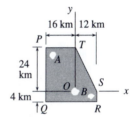

Fig. D6.4

LINES AND SLENDER MEMBERS

6.3 Centroids of Lines[†]

A line may be either a *straight line* or a *curved line*, which is often simply called a *curve*. The *centroid of a line* is the "center" of the line and is defined in accordance with the *principle of moments*. Let the length of the infinitesimal differential line element between the points $P(x, y, z)$ and P' $(x + dx, y + dy, z + dz)$ be dL as shown in Fig. 6.2. By the Pythagorean theorem, we can show that

$$dL = [(dx)^2 + (dy)^2 + (dz)^2]^{1/2}$$

The centroid $\bar{P}(\bar{x}_{el}, \bar{y}_{el}, \bar{z}_{el})$ of this line element may be taken as being at the midpoint of the line element between the points P and P', where the sub-

---

[†]A *centroid* indicates the "center" of a geometric figure, not of a physical body. Note that the suffix *-oid* usually means "resembling." Cf. Secs. 6.5, 6.9, 6.14, and 6.15.

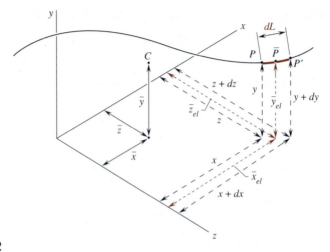

Fig. 6.2

script "*el*" stands for "element" and overbars are used with symbols which are associated with the centroid. Since the line element dL is an infinitesimal quantity, the coordinates of the centroid $\overline{P}$ may be written as

$$\overline{x}_{el} = x + \frac{dx}{2} \approx x \qquad \overline{y}_{el} = y + \frac{dy}{2} \approx y \qquad \overline{z}_{el} = z + \frac{dz}{2} \approx z$$

The *centroid* $C(\overline{x}, \overline{y}, \overline{z})$ of the line of total length L is, according to the *principle of moments*, defined by the following equations:

$$L = \int dL$$

$$\Sigma M_{yz}: \quad \overline{x}L = \int x\, dL$$

$$\Sigma M_{zx}: \quad \overline{y}L = \int y\, dL \qquad\qquad (6.2)$$

$$\Sigma M_{xy}: \quad \overline{z}L = \int z\, dL$$

The right-hand side integrals in the last three equations of Eqs. (6.2) give the so-called *first moments* of the line with respect to the yz, zx, and xy planes, respectively. Thus, the coordinates $(\overline{x}, \overline{y}, \overline{z})$ of the centroid C of the line can be obtained by dividing the total length L into the first moments of the line with respect to the coordinate planes. For a line lying in the xy plane, we have $\overline{z} = \overline{z}_{el} = z = 0$ and the last equation in Eqs. (6.2) becomes trivial. The first moments $\overline{x}L$ and $\overline{y}L$ may then be considered as being taken with respect to the y and x axes, respectively. For a line lying in a plane, dL is given by one of the following forms:

$$dL = \left[1 + \left(\frac{dy}{dx}\right)^2\right]^{1/2} dx \qquad\qquad (6.3)$$

$$dL = \left[\left(\frac{dx}{dy}\right)^2 + 1\right]^{1/2} dy \qquad\qquad (6.4)$$

$$dL = \left[r^2 + \left(\frac{dr}{d\theta}\right)^2\right]^{1/2} d\theta \qquad\qquad (6.5)$$

where (r, θ) are the polar coordinates.

The centroid C of a line is not necessarily located on the line. However, if the line is straight, the centroid is at the midpoint of the line. It is to be noted that the dimensions of the first moment of a line are *(length)²*. Furthermore, for the purpose of locating the centroid of a line, the *differential line element dL* and the *total length L* of a line segment are always made *positive*, even though the coordinates of the centroids of dL and L may be positive, negative, or zero.

EXAMPLE 6.2

Locate the centroid $C(\bar{x}, \bar{y})$ of the semicircular *arc* of radius R as shown.

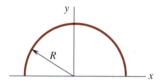

Solution. The semicircular arc is symmetrical about the y axis. Therefore, its centroid $C(\bar{x}, \bar{y})$ must lie on the y axis; i.e.,

$$\bar{x} = 0 \quad \blacktriangleleft$$

The length L of the semicircular arc is well-known; i.e.,

$$L = \pi R$$

For illustrative purposes, we will determine $\bar{y}$ in two different ways:[†]

(a) *Using rectangular coordinates*. The equation of the semicircular arc is

$$x^2 + y^2 = R^2 \quad \text{(same as that of a circle)}$$

By differentiation, we have

$$2x + 2y \frac{dy}{dx} = 0 \qquad \frac{dy}{dx} = -\frac{x}{y}$$

$$dL = \left[1 + \left(\frac{dy}{dx}\right)^2\right]^{1/2} dx = \frac{(y^2 + x^2)^{1/2}}{y} dx = \frac{R}{y} dx$$

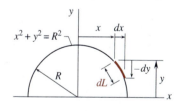

Applying the *principle of moments*, we write

$$\bar{y}L = \int y \, dL: \quad \bar{y}(\pi R) = \int_{-R}^{R} y\left(\frac{R}{y}\right) dx = 2R^2 \qquad \bar{y} = \frac{2R}{\pi} \quad \blacktriangleleft$$

[†]A third way to determine this $\bar{y}$ is given in Example 6.14.

(b) *Using polar coordinates.* In this case, we have

$$dL = R\,d\theta \qquad y = R\sin\theta$$

Applying the *principle of moments*, we write

$$\bar{y}L = \int y\,dL: \quad \bar{y}(\pi R) = R^2\int_0^\pi \sin\theta\,d\theta = 2R^2 \qquad \bar{y} = \frac{2R}{\pi} \blacktriangleleft$$

EXAMPLE 6.3

Let the point $C(\bar{x}, \bar{y})$ be the centroid and L be the length of the curve OA as shown. Making use of integral formulas (cf. Sec. B5.2 of App. B), determine the values of L, $\bar{x}$, and $\bar{y}$.

Solution. Differentiating the equation of the curve OA and using Eq. (6.3) as well as formula (t) in Sec. B5.2 of App. B, we write

$$\frac{dy}{dx} = 2x \qquad dL = (1 + 4x^2)^{1/2}\,dx$$

$$L = \int_0^1 (1 + 4x^2)^{1/2}\,dx = \frac{1}{4}\left[2\sqrt{5} + \ln(2 + \sqrt{5})\right] = 1.4789$$

$$L = 1.479 \text{ m} \blacktriangleleft$$

By the *principle of moments* and formulas (a) and (u) in Sec. B5.2 of App. B, we write

$$\bar{x}L = \int x\,dL: \quad \bar{x}(1.4789) = 2\int_0^1 x\left(x^2 + \frac{1}{4}\right)^{1/2}dx = \frac{1}{12}(5\sqrt{5} - 1)$$

$$\bar{y}L = \int y\,dL: \quad \bar{y}(1.4789) = 2\int_0^1 x^2\left(x^2 + \frac{1}{4}\right)^{1/2}dx$$

$$= \frac{1}{64}\left[18\sqrt{5} - \ln(2 + \sqrt{5})\right]$$

$$\bar{x} = 0.574 \text{ m} \qquad \bar{y} = 0.410 \text{ m} \blacktriangleleft$$

Developmental Exercise

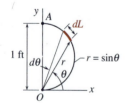

D6.7 Let the point $C(\bar{x}, \bar{y})$ be the centroid and L be the length of the curve OA as shown. Determine (a) dL in terms of $d\theta$, (b) L, (c) $\bar{x}L$, (d) $\bar{x}$, (e) $\bar{y}L$, (f) $\bar{y}$.

Fig. D6.7

6.4   Composite Lines

A *composite line* of total length L is one which is composed of line segments of lengths $L_1, L_2, \ldots, L_n$, where

$$L = L_1 + L_2 + \cdots + L_n \qquad (6.6)$$

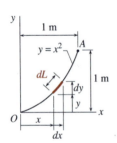

If the centroids of the line segments of lengths $L_1, L_2, \ldots, L_n$, as shown in Fig. 6.3, are located at the points $C_1(\bar{x}_1, \bar{y}_1, \bar{z}_1), C_2(\bar{x}_2, \bar{y}_2, \bar{z}_2), \ldots, C_n(\bar{x}_n, \bar{y}_n, \bar{z}_n)$, respectively, then the *centroid* $C(\bar{x}, \bar{y}, \bar{z})$ of the composite line is, according to the *principle of moments*, defined by the following equations:

$$\Sigma M_{yz}: \quad \bar{x}L = \bar{x}_1 L_1 + \bar{x}_2 L_2 + \cdots + \bar{x}_n L_n$$

$$\Sigma M_{zx}: \quad \bar{y}L = \bar{y}_1 L_1 + \bar{y}_2 L_2 + \cdots + \bar{y}_n L_n \qquad (6.7)$$

$$\Sigma M_{xy}: \quad \bar{z}L = \bar{z}_1 L_1 + \bar{z}_2 L_2 + \cdots + \bar{z}_n L_n$$

For a line lying in the xy plane, we have $\bar{z} = \bar{z}_1 = \bar{z}_2 = \cdots = \bar{z}_n = 0$ and the last equation in Eqs. (6.7) becomes trivial. Note that the centroid of a composite line, as well as the centroid of a curved line, is generally not located on the line. Moreover, a composite line may have an empty (or negative) line as a component. In this case, the length of the empty line is treated as negative. For use in determining the centroids of composite lines, the centroids of some circular arcs are shown in Fig. 6.4. If desired, a short computer program may be used to locate the centroid of a composite line as shown in Fig. 6.5.

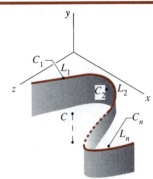

Fig. 6.3

Quarter-circular arc

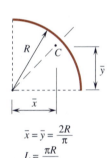

$$\bar{x} = \bar{y} = \frac{2R}{\pi}$$

$$L = \frac{\pi R}{2}$$

Semicircular arc

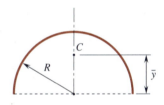

$$\bar{y} = \frac{2R}{\pi}$$

$$L = \pi R$$

Segment of circular arc

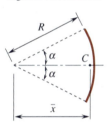

$$\bar{x} = \frac{R \sin \alpha}{\alpha}$$

$$L = 2\alpha R$$

Fig. 6.4

```
10 REM   A COMPOSITE LINE WITH N COMPONENT LINES: Enter N, L1, x1, y1, z1, L2,
15 REM   x2, y2, z2, L3, x3, y3, z3, L4, x4, y4, z4, ..., as data in Line 100.
20 REM   To output to the printer, change PRINT to LPRINT in the program.
25 READ N:DIM L(N),X(N),Y(N),Z(N):FOR I=1 TO N:READ L(I),X(I),Y(I),Z(I):L=L+L(I)
30 XL=XL+X(I)*L(I):YL=YL+Y(I)*L(I):ZL=ZL+Z(I)*L(I):NEXT I:X=XL/L:Y=YL/L:Z=ZL/L
35 PRINT:PRINT"The component lines and the coordinates of their centroids are"
40 FOR I = 1 TO N: PRINT "L ("; I; ") =";: PRINT USING " ##.######^^^^"; L(I),
45 PRINT SPC(8); "x, y, z = "; X(I); Y(I); Z(I): NEXT I: S$ = "      "
50 PRINT"The centroid of the composite line is located at":PRINT "x = ";X;S$;
55 PRINT "y = ";Y;: PRINT S$;"z = ";Z:        REM  (File Name: CENTL)
100 DATA 3,1.479,0.574,0.410,0,1,-0.4,0,0.3,3.141593,-1.43662,-1,0.6

The component lines and the coordinates of their centroids are
L ( 1 ) =   1.479000E+00        x, y, z =  .574   .41   0
L ( 2 ) =   1.000000E+00        x, y, z = -.4    0    .3
L ( 3 ) =   3.141593E+00        x, y, z = -1.43662 -1   .6
The centroid of the composite line is located at
x = -.723114        y = -.4510561       z =  .3887412
```

Fig. 6.5 (Cf. the solution in Example 6.4.)

EXAMPLE 6.4

Making use of the solutions in Examples 6.2 and 6.3, locate the centroid $C(\bar{x}, \bar{y}, \bar{z})$ of the composite line *PQRST* as shown.

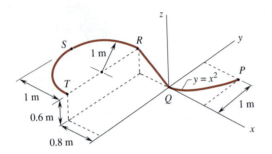 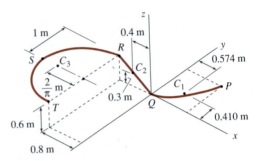

Solution. Let L_1, L_2, and L_3 be the lengths of the component lines *PQ*, *QR*, and *RST*. From the solutions in Examples 6.2 and 6.3 and for the coordinate system shown, we have $L_1 = 1.4789$ m, $L_2 = 1$ m, $L_3 = \pi$ m, and the centroids of these component lines are respectively located at $C_1(0.574, 0.410, 0)$ m, $C_2(-0.4, 0, 0.3)$ m, and $C_3(-0.8 - 2/\pi, -1, 0.6)$ m as shown. The total length is

$$L = L_1 + L_2 + L_3 \qquad L = 5.621 \text{ m}$$

Applying the *principle of moments* in the form of Eqs. (6.7), we write

$$\Sigma M_{yz}: \quad \bar{x}(5.621) = 0.574(1.479) + (-0.4)(1) + (-0.8 - 2/\pi)\pi$$

$$\Sigma M_{zx}: \quad \bar{y}(5.621) = 0.410(1.479) + 0(1) + (-1)(\pi)$$

$$\Sigma M_{xy}: \quad \bar{z}(5.621) = 0(1.479) + 0.3(1) + 0.6(\pi)$$

These three equations yield

$$\bar{x} = -0.723 \text{ m} \qquad \bar{y} = -0.451 \text{ m} \qquad \bar{z} = 0.389 \text{ m} \blacktriangleleft$$

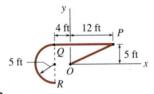

Fig. D6.8

Developmental Exercise

D6.8   Locate the centroid $C(\bar{x}, \bar{y})$ of the composite line *OPQR* as shown.

6.5   Centers of Gravity of Slender Members

The *center of gravity* of a body is the center of the gravitational attraction experienced by the body. In other words, the resultant weight force of the body acts at the center of gravity of the body. A slender member is a body where any linear measurement of its transverse cross section (e.g., diameter) is negligibly small compared with its length. Slender members may include wires, bars, rods, tubes, and pipes.

The *mass density* ρ_L of a slender member is expressed as mass per unit length, while its *weight density* γ_L is expressed as weight per unit length.

By Newton's second law, we note that

$$\gamma_L = \rho_L g \qquad (6.8)$$

where g is the gravitational acceleration at the location of the slender member. Unless otherwise stated, each slender member will be assumed to have a constant value of γ_L and be situated near the surface of the earth where $g = 9.81 \text{ m/s}^2$. Thus, the *centers of gravity* of slender members are taken to be coincident with the centroids of their center lines or axes, which are located as explained in Secs. 6.3 and 6.4. If the slender members were not uniform as assumed, they should be treated as bodies which are studied in Sec. 6.14.

EXAMPLE 6.5

A slender tube *ABCDE* is bent into the shape as shown. The tube has a mass density of 1 kg/m and is held in equilibrium by three vertical wires at *A*, *B*, and *D*. Determine the tension in each wire.

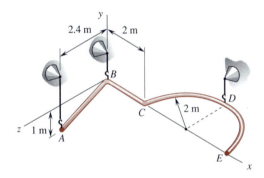
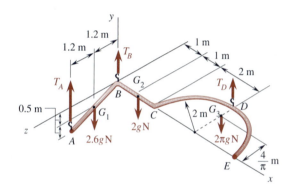

Solution.   The center of gravity of the slender tube may be taken as coincident with the centroid of its center line. By the *principle of moments*, the resultant weight force of the tube is equivalent to the sum of the weight forces of the segments *AB*, *BC*, and *CDE*, which are located at G_1, G_2, and G_3 as indicated. Since the tube has a mass density of 1 kg/m, the weight forces of these segments are 2.6g N, 2g N, and $2\pi g$ N, respectively, as indicated, where the distance between G_3 and the x axis is $2R/\pi = 4/\pi$ m (see Fig. 6.4). For equilibrium of the free body drawn, we write

$\Sigma M_z = 0: \quad 4T_D - 4(2\pi g) - 1(2g) = 0 \qquad\qquad T_D = 66.543$

$\Sigma M_x = 0: \quad 2T_D - \dfrac{4}{\pi}(2\pi g) + 1.2(2.6g) - 2.4T_A = 0 \qquad T_A = 35.506$

$\Sigma F_y = 0: \quad T_A + T_B + T_D - 2.6g - 2g - 2\pi g = 0 \qquad T_B = 4.715$

$$T_A \doteq 35.5 \text{ N} \qquad T_B = 4.72 \text{ N} \qquad T_D = 66.5 \text{ N} \blacktriangleleft$$

REMARK. The above results may similarly be obtained by replacing the component weight forces with the resultant weight force of the entire tube.

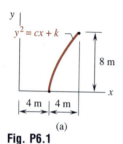

Fig. D6.10

Developmental Exercises

D6.9 Locate the center of gravity $G(\bar{x}, \bar{y}, \bar{z})$ of the slender tube in Example 6.5. Use it to check the answers obtained in that example.

D6.10 A bent rod $ABCDE$ weighing 10 lb/ft is suspended as shown. Determine the tensions in the wires at A, B, and C.

PROBLEMS

6.1 For the curves shown, what are the values and units of the constants c and k in each case?

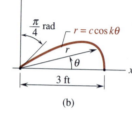

(a)   (b)

Fig. P6.1

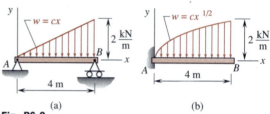

(a)   (b)

Fig. P6.2

6.2 For the beams shown, what is the loading intensity w at $x = 0.5$ m in each case?

6.3 through 6.5 Using the *principle of moments*, locate the resultant force $\mathbf{R}$ of the parallel forces acting as shown.

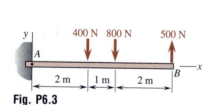

Fig. P6.3

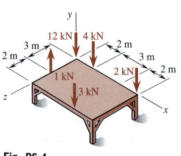

Fig. P6.4

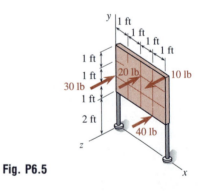

Fig. P6.5

6.6 through 6.8 Let the point $C(\bar{x}, \bar{y})$ be the centroid of the curved line shown. Determine by *integration* the value of $\bar{y}$.

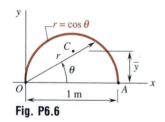

Fig. P6.6

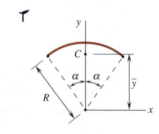

Fig. P6.7

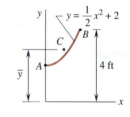

Fig. P6.8

6.9 through 6.17* Locate the centroid of the composite line shown.

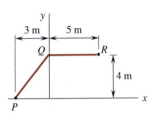

Fig. P6.9

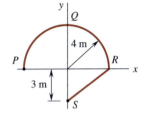

Fig. P6.10

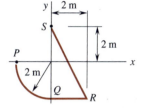

Fig. P6.11*

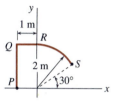

Fig. P6.12

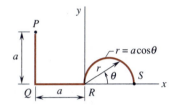

Fig. P6.13

Fig. P6.14*

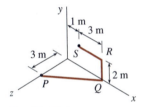

Fig. P6.15

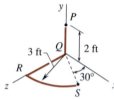

Fig. P6.16

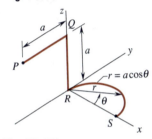

Fig. P6.17*

6.18 A bent rod ABC is suspended by a wire at B as shown. Determine the length L for which the portion AB of the bent rod is horizontal.

6.19 A bent rod ABC is suspended by a wire at B as shown. Determine the length L for which the portion BC of the bent rod is horizontal.

6.20* A wire is bent as shown and is suspended by a string at O. Determine the angle ϕ for which the plane of the wire is horizontal.

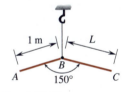

Fig. P6.18 and P6.19

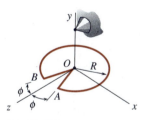

Fig. P6.20*

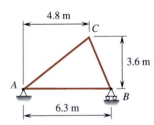

Fig. P6.21

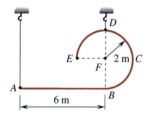

Fig. P6.22

6.21 through 6.23* The rigid body shown has a mass density of 1 kg/m. Determine the reactions from the supports.

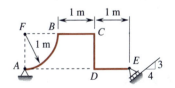

Fig. P6.23*

6.24 through 6.26* The rigid body shown has a weight density of 20 lb/ft. Determine the reactions from the supports.

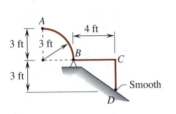

Fig. P6.24

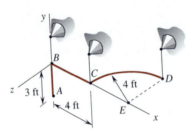

Fig. P6.25

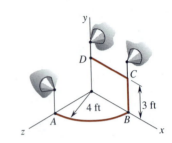

Fig. P6.26*

AREAS AND LAMINAS

6.6 Orders of Differential Area Elements

A differential area element dA may be of the first or second order. A *first-order differential area element* is an *infinitesimal areal strip* which is expressed in terms of a differential of a coordinate variable. It may be illustrated as shown in Fig. 6.6.

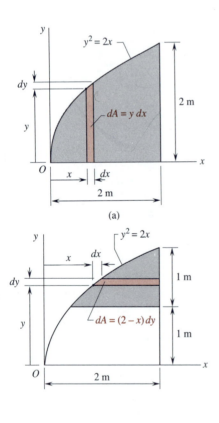

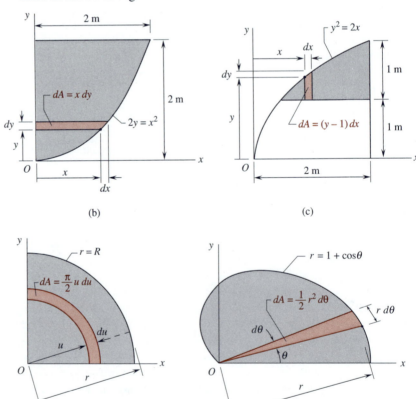

Fig. 6.6

A *second-order differential area element* is an *infinitesimal areal patch* which is expressed in terms of the product of two differentials of coordinate variables. Second-order differential area elements are generally of the shapes of a quadrilateral area and may be illustrated as shown in Fig. 6.7. In Fig. 6.7(a), note that the variables $(\xi, \eta)^\dagger$ denote the rectangular coordinates of the point E where the *second-order differential area element dA* is located, while the variables (x, y) denote the rectangular coordinates of *a point on the boundary curve* of the area A. In Fig. 6.7(b), note similarly that the variables $(u, \theta)^\dagger$ denote the polar coordinates of the point E where the *second-order differential area element dA* is located, while the variables (r, θ) denote the polar coordinates of *a point on the boundary curve* of the area A.

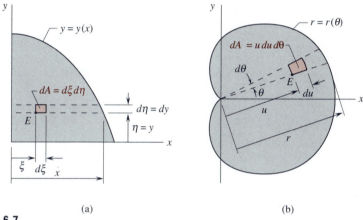

(a)   (b)

Fig. 6.7

Developmental Exercises

D6.11   Define the terms: (a) first-order differential area element, (b) second-order differential area element.

D6.12   Write, in terms of y and dy, the first-order differential area element dA parallel to the x axis for each of the areas shown.

D6.13   Write, in terms of x and dx, the first-order differential area element dA parallel to the y axis for each of the areas shown.

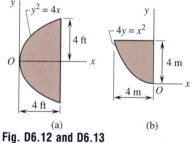

(a)   (b)

Fig. D6.12 and D6.13

6.7   Centroids of Areas

An *area* may be a plane area or a curved area. A curved area is generally called a *surface area*, or simply a *surface*. The *centroid of an area* is the "center" of the area and is defined in accordance with the principle of moments. Let the point $P(\bar{x}_{el}, \bar{y}_{el}, \bar{z}_{el})$ be the centroid of the differential area *element dA* of a given surface of total area A as shown in Fig. 6.8, where dA may be of the *second* or the *first* order. It is important to make sure that

---

†These variables can be used as the running (or dummy) variables for the integration (cf. Example 6.8).

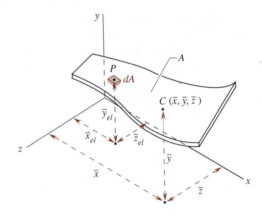

Fig. 6.8

$(\bar{x}_{el}, \bar{y}_{el}, \bar{z}_{el})$ are the *coordinates of the centroid* of dA if dA is of the *first order* (cf. Examples 6.6 and 6.8). The centroid $C(\bar{x}, \bar{y}, \bar{z})$ of the surface is, according to the *principle of moments*, defined by the following equations:

$$A = \int dA$$

$$\Sigma M_{yz}: \quad \bar{x}A = \int \bar{x}_{el}\, dA$$

$$\Sigma M_{zx}: \quad \bar{y}A = \int \bar{y}_{el}\, dA \qquad (6.9)$$

$$\Sigma M_{xy}: \quad \bar{z}A = \int \bar{z}_{el}\, dA$$

The integrals in the last three equations in Eqs. (6.9) give the *first moments* of the surface with respect to the yz, zx, and xy planes, respectively.[†] Thus, by dividing the total area A into the first moments of the surface, we can obtain the coordinates $(\bar{x}, \bar{y}, \bar{z})$ of the centroid C of the surface. For an area lying in the xy plane, we have $z = \bar{z} = \bar{z}_{el} = 0$ and the last equation in Eqs. (6.9) becomes trivial. It should be noted that, for the purpose of locating the centroid of an area with no holes, the *values* of dA and A are always made *positive*. However, the coordinates of the centroids of dA and A may be positive, negative, or zero.

EXAMPLE 6.6

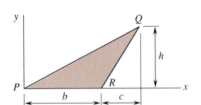

Determine the ordinate $\bar{y}$ of the centroid $C(\bar{x}, \bar{y})$ of the triangular area *PQR* as shown.

Solution. From geometry, we know that the area of the triangle is

$$A = \frac{1}{2}\, bh$$

By similar triangles in the sketch shown, we write

[†]*Second moments* of areas (i.e., moments of first moments) are presented in Chap. 7.

$$\frac{\xi}{b} = \frac{h - y}{h} \qquad \xi = b - \frac{b}{h}y$$

$$dA = \xi \, dy = \left(b - \frac{b}{h}y\right) dy \qquad \overline{y}_{el} = y + \frac{dy}{2} \approx y$$

Applying the *principle of moments*, we write

$$\overline{y}A = \int \overline{y}_{el} \, dA: \qquad \overline{y}\left(\frac{1}{2}bh\right) = \int_0^h y\left(b - \frac{b}{h}y\right) dy = \frac{1}{6}bh^2$$

$$\overline{y} = \frac{h}{3} \quad \blacktriangleleft^\dagger$$

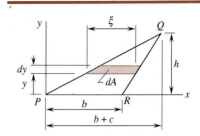

REMARK. The solution shows that, for any triangular area, the distance be-tween its centroid C and a base is equal to one third of the corresponding height. Notice that any side of a triangle may be regarded as a *base*, and the length of the altitude to that base is the *corresponding height*. Further-more, the centroid C must lie on any median drawn. The answer in this example is widely used as a *formula*.

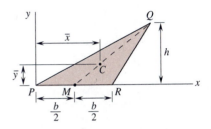

EXAMPLE 6.7

The vertices of an arbitrary triangle are located at $P(x_1, y_1)$, $Q(x_2, y_2)$, and $R(x_3, y_3)$. For the triangular area PQR, determine the formulas for (a) its area A, (b) the coordinates $\overline{x}$ and $\overline{y}$ of its centroid C.

Solution. Since $A = \frac{1}{2}|\overrightarrow{PQ} \times \overrightarrow{PR}|$ as shown (cf. Sec. 4.7), we write

$$\overrightarrow{PQ} = (x_2 - x_1)\mathbf{i} + (y_2 - y_1)\mathbf{j}$$

$$\overrightarrow{PR} = (x_3 - x_1)\mathbf{i} + (y_3 - y_1)\mathbf{j}$$

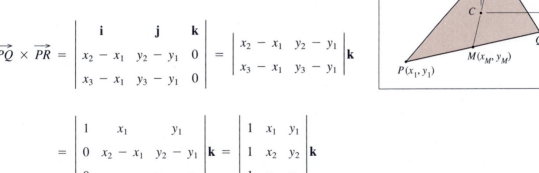

$$\overrightarrow{PQ} \times \overrightarrow{PR} = \begin{vmatrix} \mathbf{i} & \mathbf{j} & \mathbf{k} \\ x_2 - x_1 & y_2 - y_1 & 0 \\ x_3 - x_1 & y_3 - y_1 & 0 \end{vmatrix} = \begin{vmatrix} x_2 - x_1 & y_2 - y_1 \\ x_3 - x_1 & y_3 - y_1 \end{vmatrix} \mathbf{k}$$

$$= \begin{vmatrix} 1 & x_1 & y_1 \\ 0 & x_2 - x_1 & y_2 - y_1 \\ 0 & x_3 - x_1 & y_3 - y_1 \end{vmatrix} \mathbf{k} = \begin{vmatrix} 1 & x_1 & y_1 \\ 1 & x_2 & y_2 \\ 1 & x_3 & y_3 \end{vmatrix} \mathbf{k}$$

$$A = \frac{1}{2}|x_1(y_2 - y_3) + x_2(y_3 - y_1) + x_3(y_1 - y_2)| \quad \blacktriangleleft$$

---

†The centroid of a triangular *area* and the centroid of the triangular *line* coinciding with the boundaries of that triangular area are generally *different*.

The coordinates of the midpoint M between P and Q are given by

$$x_M = \frac{1}{2}(x_1 + x_2) \qquad y_M = \frac{1}{2}(y_1 + y_2)$$

By the solution in Example 6.6, we know that the centroid C lies on the median $\overline{MR}$, and $\overline{MC} = \frac{1}{3}\overline{MR}$. Thus, we write

$$\bar{x} = x_M + \frac{1}{3}(x_3 - x_M) = \frac{1}{2}(x_1 + x_2) + \frac{1}{3}\left[x_3 - \frac{1}{2}(x_1 + x_2)\right]$$

$$\bar{y} = y_M + \frac{1}{3}(y_3 - y_M) = \frac{1}{2}(y_1 + y_2) + \frac{1}{3}\left[y_3 - \frac{1}{2}(y_1 + y_2)\right]$$

$$\bar{x} = \frac{1}{3}(x_1 + x_2 + x_3) \qquad \bar{y} = \frac{1}{3}(y_1 + y_2 + y_3) \quad \blacktriangleleft$$

REMARK. The above formulas for A, $\bar{x}$, and $\bar{y}$ are useful in locating the centroid of *any* plane area using *triangular finite elements* (e.g., Probs. 6.40 through 6.42).

EXAMPLE 6.8

Locate the centroid $C(\bar{x}, \bar{y})$ of the semicircular *area* of radius R as shown.

Solution. By symmetry, the centroid $C(\bar{x}, \bar{y})$ must lie on the y axis; i.e.,

$$\bar{x} = 0 \quad \blacktriangleleft$$

The area A of the semicircle is well-known; i.e., $A = \frac{1}{2}\pi R^2$. For illustrative purposes, $\bar{y}$ will here be determined in four different ways:[†]

(a) *Using rectangular coordinates and single integration.* We first choose a first-order differential area element dA as shown. We see that

$$dA = y\,dx \qquad \bar{y}_{el} \approx \frac{1}{2}y$$

Applying the *principle of moments*, we write

$$\bar{y}A = \int \bar{y}_{el}\,dA:$$

$$\bar{y}\left(\frac{1}{2}\pi R^2\right) = \frac{1}{2}\int_{-R}^{R} y^2\,dx$$

$$= \frac{1}{2}\int_{-R}^{R}(R^2 - x^2)\,dx = \frac{2}{3}R^3$$

$$\bar{y} = \frac{4R}{3\pi} \quad \blacktriangleleft$$

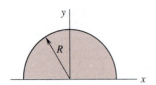

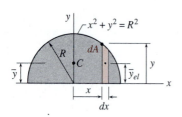

[†]A fifth way to determine the ordinate $\bar{y}$ is given in Example 6.15.

(b) *Using polar coordinates and single integration.* In polar coordinates, the equation of the curved boundary is $r = R$. The first-order differential area element dA as shown may be taken as a triangular area. Thus,

$$dA = \frac{1}{2}(R\,d\theta)R = \frac{1}{2}R^2\,d\theta$$

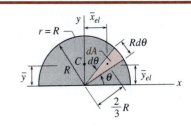

Making use of the solution in Example 6.6, we write

$$\bar{y}_{el} \approx \left(\frac{2}{3}R\right)(\sin\theta) = \frac{2}{3}R\sin\theta$$

Applying the *principle of moments*, we write

$$\bar{y}A = \int \bar{y}_{el}\,dA: \quad \bar{y}\left(\frac{1}{2}\pi R^2\right) = \frac{1}{3}R^3\int_0^\pi \sin\theta\,d\theta = \frac{2}{3}R^3$$

$$\boxed{\bar{y} = \frac{4R}{3\pi}} \blacktriangleleft$$

(c) *Using rectangular coordinates and double integration.* We here choose a second-order differential area element dA as shown. We see that

$$dA = d\xi\,d\eta = dx\,d\eta = d\eta\,dx \qquad \bar{y}_{el} = \eta + \frac{1}{2}d\eta \approx \eta$$

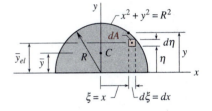

Applying the *principle of moments*, we write

$$\bar{y}A = \int \bar{y}_{el}\,dA: \quad \bar{y}\left(\frac{1}{2}\pi R^2\right) = \int_{x=-R}^{R}\int_{\eta=0}^{y}\eta\,d\eta\,dx = \int_{x=-R}^{R}\frac{1}{2}y^2\,dx$$

$$= \frac{1}{2}\int_{-R}^{R}(R^2 - x^2)\,dx = \frac{2}{3}R^3$$

$$\boxed{\bar{y} = \frac{4R}{3\pi}} \blacktriangleleft$$

(d) *Using polar coordinates and double integration.* The second-order differential area element dA as shown may be taken as a rectangle with sides du and $u\,d\theta$. Thus,

$$dA = du(u\,d\theta) = u\,du\,d\theta \qquad \bar{y}_{el} \approx u\sin\theta$$

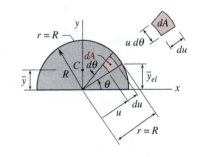

Applying the *principle of moments*, we write

$$\bar{y}A = \int \bar{y}_{el}\,dA: \quad \bar{y}\left(\frac{1}{2}\pi R^2\right) = \int_{\theta=0}^{\pi}\int_{u=0}^{R}u^2\sin\theta\,du\,d\theta$$

$$= \int_{\theta=0}^{\pi}\frac{1}{3}R^3\sin\theta\,d\theta = \frac{2}{3}R^3$$

$$\boxed{\bar{y} = \frac{4R}{3\pi}} \blacktriangleleft$$

REMARK. From the answers in Examples 6.2 and 6.8, it is to be noted that the centroid of a *semicircular arc* does not coincide with the centroid of a *semicircular area* with the same radius.

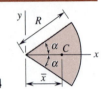

Fig. D6.14

Developmental Exercise

D6.14   The point C is the centroid of the area of the circular sector shown. Using integration in polar coordinates, (a) verify that $\bar{x} = (2R \sin\alpha)/(3\alpha)$, (b) check the value of $\bar{x}$ for $\alpha = \pi/2$.

EXAMPLE 6.9

Locate the centroid $C(\bar{x}, \bar{y}, \bar{z})$ of the curved surface of a hemispherical thin shell of radius R as shown.

Solution.   By symmetry, we know that the centroid C must be situated on the x axis. Thus, we have

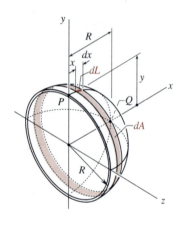

$$\bar{y} = \bar{z} = 0 \quad \blacktriangleleft$$

The surface of this shell can be generated by revolving the quarter circular arc PQ about the x axis one revolution as shown. The arc PQ is given by

$$x^2 + y^2 = R^2$$

The area element dA may be taken as the surface of a circular band obtained by revolving the element dL about the x axis one revolution. Thus, we write

$$2x + 2y\frac{dy}{dx} = 0 \qquad \frac{dy}{dx} = -\frac{x}{y} \qquad \bar{x}_{el} \approx x$$

$$dL = \left[1 + \left(\frac{dy}{dx}\right)^2\right]^{1/2} dx = \frac{1}{y}(y^2 + x^2)^{1/2} dx = \frac{R}{y} dx$$

$$dA = 2\pi y\, dL = 2\pi R\, dx \qquad A = \int dA = 2\pi R \int_0^R dx = 2\pi R^2$$

Applying the *principle of moments*, we write

$$\bar{x}A = \int \bar{x}_{el}\, dA: \quad \bar{x}(2\pi R^2) = \int x\, dA = 2\pi R \int_0^R x\, dx = \pi R^3$$

$$\bar{x} = \frac{R}{2} \quad \blacktriangleleft$$

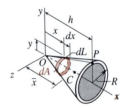

Fig. D6.15

Developmental Exercise

D6.15   The point C is the centroid of the surface area A of the right circular conic thin shell with an open base as shown. Determine (a) dA in terms of x and dx, (b) A, (c) $\bar{x}A$, (d) $\bar{x}$.

6.8   Composite Areas

A *composite area* of total surface area A is one which is composed of component surface areas $A_1, A_2, \ldots, A_n$, where

$$A = A_1 + A_2 + \cdots + A_n \tag{6.10}$$

If the centroids of the component surface areas $A_1, A_2, \ldots, A_n$, as shown in Fig. 6.9, are located at the points $C_1(\overline{x}_1, \overline{y}_1, \overline{z}_1)$, $C_2(\overline{x}_2, \overline{y}_2, \overline{z}_2)$, $\ldots$, $C_n(\overline{x}_n, \overline{y}_n, \overline{z}_n)$, respectively, then the centroid $C(\overline{x}, \overline{y}, \overline{z})$ of the composite surface area is, according to the *principle of moments*, defined by the following equations:

$$\Sigma M_{yz}: \quad \overline{x}A = \overline{x}_1 A_1 + \overline{x}_2 A_2 + \cdots + \overline{x}_n A_n$$

$$\Sigma M_{zx}: \quad \overline{y}A = \overline{y}_1 A_1 + \overline{y}_2 A_2 + \cdots + \overline{y}_n A_n \qquad (6.11)$$

$$\Sigma M_{xy}: \quad \overline{z}A = \overline{z}_1 A_1 + \overline{z}_2 A_2 + \cdots + \overline{z}_n A_n$$

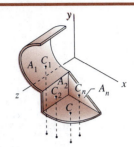

Fig. 6.9

For a composite area lying in the xy plane, we have $\overline{z} = \overline{z}_1 = \overline{z}_2 = \cdots = \overline{z}_n = 0$ and the last equation in Eqs. (6.11) becomes trivial. Furthermore, note that one or more of the component areas of a composite area may be negative. For use in determining the centroids of composite areas, the centroids of areas of some common shapes are shown in Fig. 6.10.

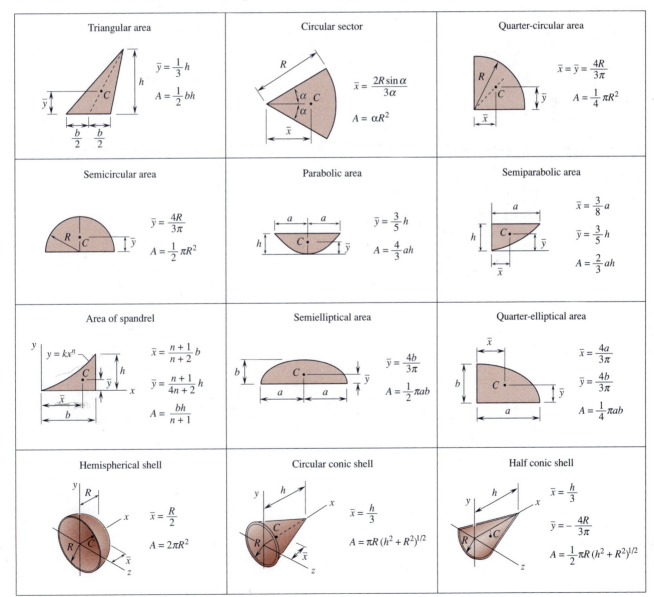

Fig. 6.10 Centroids of areas of common shapes.

If desired, a short computer program may be used to locate the centroid of a composite area as shown in Fig. 6.11, where the routine is the same as that shown in Fig. 6.5. Note in Line 100 of the program in Fig. 6.11 that the data are entered in the following order: N, A_1, $\bar{x}_1$, $\bar{y}_1$, $\bar{z}_1$, A_2, $\bar{x}_2$, $\bar{y}_2$, $\bar{z}_2$, A_3, $\bar{x}_3$, $\bar{y}_3$, $\bar{z}_3$, A_4, $\bar{x}_4$, $\bar{y}_4$, $\bar{z}_4$, A_5, $\bar{x}_5$, $\bar{y}_5$, and $\bar{z}_5$.

```
10 REM   A COMPOSITE AREA WITH N COMPONENT AREAS: Enter N, A1, x1, y1, z1, A2,
15 REM   x2, y2, z2, A3, x3, y3, z3, A4, x4, y4, z4, ..., as data in Line 100.
20 REM   To output to the printer, change PRINT to LPRINT in the program.
25 READ N:DIM A(N),X(N),Y(N),Z(N):FOR I=1 TO N:READ A(I),X(I),Y(I),Z(I):A=A+A(I)
30 XA=XA+X(I)*A(I):YA=YA+Y(I)*A(I):ZA=ZA+Z(I)*A(I):NEXT I:X=XA/A:Y=YA/A:Z=ZA/A
35 PRINT:PRINT"The component areas and the coordinates of their centroids are"
40 FOR I = 1 TO N: PRINT "A ("; I; ") =";: PRINT USING " ##.######^^^^"; A(I),
45 PRINT SPC(8); "x, y, z = "; X(I); Y(I); Z(I): NEXT I: S$ = "          "
50 PRINT"The centroid of the composite area is located at":PRINT "x = ";X;S$;
55 PRINT "y = ";Y;: PRINT S$;"z = ";Z:              REM  (File Name: CENTA)
100 DATA 5,9,2,1,0,36,3,0,3,-14.1372,3,0,4.72676,48,6,-4,3,-3.141593,6,-2,2
```

```
The component areas and the coordinates of their centroids are
A ( 1 ) =   9.000000E+00        x, y, z =  2  1  0
A ( 2 ) =   3.600000E+01        x, y, z =  3  0  3
A ( 3 ) =  -1.413720E+01        x, y, z =  3  0  4.72676
A ( 4 ) =   4.800000E+01        x, y, z =  6 -4  3
A ( 5 ) =  -3.141593E+00        x, y, z =  6 -2  2
The centroid of the composite area is located at
x =   4.658389        y = -2.333782        z =  2.362531
```

Fig. 6.11   (Cf. the solution in Example 6.11.)

EXAMPLE 6.10

Making use of the formulas in Fig. 6.10, locate the centroid $C(\bar{x}, \bar{y})$ of the area A between the semicircle and the semiellipse shown.

Solution. By symmetry, the centroid C must be on the x axis; i.e.,

$$\bar{y} = 0 \quad \blacktriangleleft$$

The area A may be treated as a composite area equal to a semielliptic

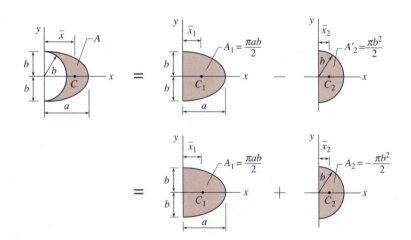

area A_1 minus a semicircular area A_2' or, equivalently, a semielliptic area A_1 plus a *negative* semicircular area A_2 as shown. Using the formulas in Fig. 6.10 and treating A_2 as negative, we write

$$\bar{x}_1 = \frac{4a}{3\pi} \qquad A_1 = \frac{\pi ab}{2} \qquad \bar{x}_2 = \frac{4b}{3\pi} \qquad A_2 = -\frac{\pi b^2}{2}$$

$$A = A_1 + A_2 = \frac{\pi ab}{2} - \frac{\pi b^2}{2} = \frac{\pi b}{2}(a - b)$$

Applying the *principle of moments*, we write

$$\bar{x}A = \bar{x}_1 A_1 + \bar{x}_2 A_2: \quad \bar{x}\left[\frac{\pi b}{2}(a - b)\right] = \frac{4a}{3\pi}\left(\frac{\pi ab}{2}\right) + \frac{4b}{3\pi}\left(-\frac{\pi b^2}{2}\right)$$

$$\bar{x} = \frac{4}{3\pi}(a + b) \quad \blacktriangleleft$$

REMARK. As $a \rightarrow b$, the above answer leads to $\bar{x} = 8b/(3\pi) = 0.849b$, instead of $2b/\pi = 0.637b$, which is the distance between the centroid of a semicircular arc of radius b and its center. This is to be expected because as $a \rightarrow b$ the above area approaches a thin strip of the shape of a *crescent*, instead of the shape of a uniform semicircular arc.

Developmental Exercise

D6.16   Making use of the formulas in Fig. 6.10, locate the centroid $C(\bar{x}, \bar{y})$ of the area as shown.

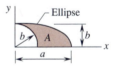

Fig. D6.16

6.9   Centers of Gravity of Laminas

We know that a *lamina* (such as a thin plate or a thin shell) is characterized by the feature that its thickness is negligibly small compared with its width or height. The *mass density* ρ_A of a lamina is expressed as mass per unit area, while the *weight density* γ_A of the lamina is expressed as weight per unit area. By Newton's second law, we note that

$$\gamma_A = \rho_A g \qquad (6.12)$$

where g is the gravitational acceleration.

Unless otherwise stated, each lamina will be assumed to have a constant value of γ_A and be situated near the surface of the earth where $g = 9.81$ m/s². Thus, the *centers of gravity of laminas* may be taken to be coincident with the centroids of the surface areas of the corresponding laminas. If the laminas were not uniform as assumed, they should be treated as bodies which are studied in Sec. 6.14.

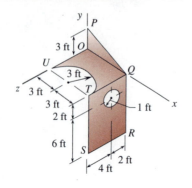

EXAMPLE 6.11

Locate the center of gravity of the thin sheet of aluminum *OPQRSTUO* which is cut and bent into the shape as shown.

Solution.   The center of gravity $G(\bar{x}, \bar{y}, \bar{z})$ of the thin sheet of aluminum *OPQRSTUO* is coincident with the centroid of the composite area *OPQRSTUO*. We may treat this composite area as the algebraic sum of five component areas as shown. The centroid of each component area is first determined and entered in the accompanying table, where A_3 and A_5 are negative because they are to be subtracted from the other areas. The coordinates of the centroid of the composite area may be determined from the *principle of moments* as follows:

Component	A, ft²	$\bar{x}$, ft	$\bar{y}$, ft	$\bar{z}$, ft	$\bar{x}A$, ft³	$\bar{y}A$, ft³	$\bar{z}A$, ft³
A_1	9	2	1	0	18	9	0
A_2	36	3	0	3	108	0	108
A_3	$-9\pi/2$	3	0	$6 - (4/\pi)$	-42.41	0	-66.82
A_4	48	.6	-4	3	288	-192	144
A_5	$-\pi$	6	-2	2	-18.85	6.28	-6.28
Sum	$\Sigma A = 75.72$	$\cdots$	$\cdots$	$\cdots$	$\Sigma \bar{x}A = 352.74$	$\Sigma \bar{y}A = -176.72$	$\Sigma \bar{z}A = 178.90$

$$\bar{x} = \frac{\Sigma \bar{x}A}{\Sigma A} = \frac{352.74}{75.72} \qquad \bar{y} = \frac{\Sigma \bar{y}A}{\Sigma A} = \frac{-176.72}{75.72} \qquad \bar{z} = \frac{\Sigma \bar{z}A}{\Sigma A} = \frac{178.90}{75.72}$$

$$\bar{x} = 4.66 \text{ ft} \qquad \bar{y} = -2.33 \text{ ft} \qquad \bar{z} = 2.36 \text{ ft} \blacktriangleleft^\dagger$$

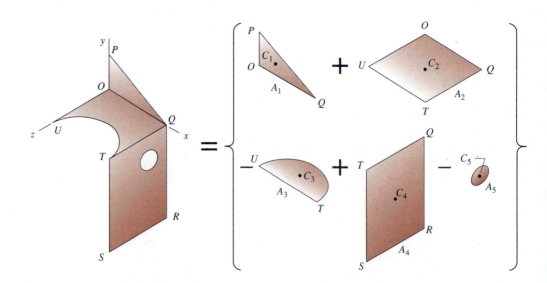

[†]Cf. the answers obtained via a computer program in Fig. 6.11.

EXAMPLE 6.12

A plate having a mass density of 10 kg/m² is cut into the shape *ABCDEFA* and is suspended as shown. Determine the tension in each cable.

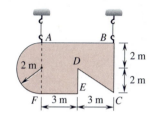

Solution. The center of gravity of the plate may be taken as coincident with the centroid of its surface area. By treating the plate as comprising a semicircular part, a rectangular part, and a *negative* triangular part, we write the areas and weight forces of these three parts as follows:

$$A_1 = \frac{1}{2}\pi(2)^2 = 2\pi \qquad A_2 = 6(4) = 24 \qquad A_3 = -\frac{1}{2}(3)(2) = -3$$

$$\mathbf{W}_1 = 20\pi g \text{ N} \downarrow \qquad \mathbf{W}_2 = 240g \text{ N} \downarrow \qquad \mathbf{W}_3 = 30g \text{ N} \uparrow$$

The direction of $\mathbf{W}_3$ is upward because it is contributed from the negative triangular part. Recalling the formulas specifying the locations of the centroids of semicircular and triangular areas, we draw the free-body diagram of the plate as shown. For equilibrium, we write

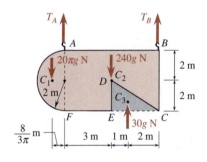

$$+ \circlearrowright \Sigma M_A = 0: \; 6T_B + \frac{8}{3\pi}(20\pi g) - 3(240g) + 4(30g) = 0$$

$$T_A = 1783 \text{ N} \blacktriangleleft$$

$$+ \circlearrowright \Sigma M_B = 0: \; -6T_A + (6 + \frac{8}{3\pi})(20\pi g) + 3(240g) - 2(30g) = 0$$

$$T_B = 894 \text{ N} \blacktriangleleft$$

Check $\quad + \uparrow \Sigma F_y = 0: \quad 1783 + 894 - 20\pi g - 240g + 30g \approx 0$

REMARK. For finding reactions from the supports, it is *not* necessary to locate the center of gravity of the *entire* composite body.

Developmental Exercise

D6.17 A thin plate *ABCDEA* having a mass density of 5 kg/m² is suspended as shown. Determine the tension in each wire.

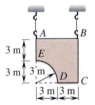

Fig. D6.17

PROBLEMS

6.27 through 6.35* Determine by integration the centroid of the area shown.

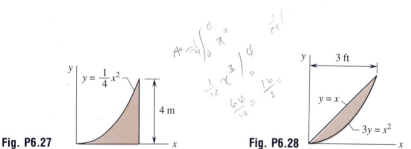

Fig. P6.27   Fig. P6.28   Fig. P6.29*

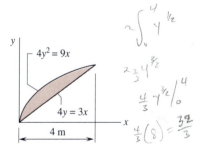

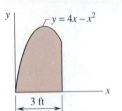

Fig. P6.30

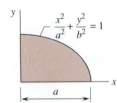

Fig. P6.31

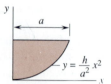

Fig. P6.32*

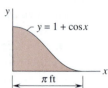

Fig. P6.33

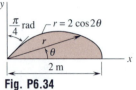

Fig. P6.34

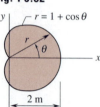

Fig. P6.35*

6.36 through 6.44* Locate the centroid of the area shown.

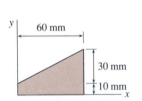

Fig. P6.36

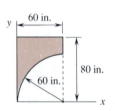

Fig. P6.37

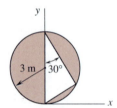

Fig. P6.38*

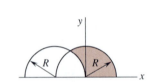

Fig. P6.39*

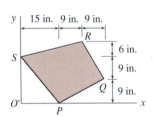

Fig. P6.40†

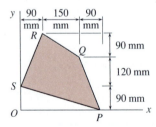

Fig. P6.41*†

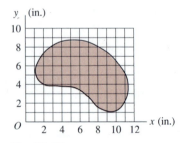

Fig. P6.42†

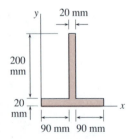

Fig. P6.43

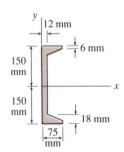

Fig. P6.44*

†*Suggestion*. Use *triangular finite elements* and apply the formulas in Example 6.7. (The result may be checked by using the program AREA described in App. F.)

6.45 through 6.47* The body shown is cut from a thin plate having a mass density of 5 kg/m². Determine the tension in each wire.

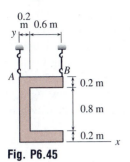

Fig. P6.45

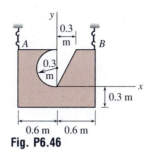

Fig. P6.46

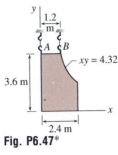

Fig. P6.47*

6.48 and 6.49 Determine by *integration* the centroid of the *surface* of revolution generated by revolving the line *AB* as shown about the *x* axis one revolution.

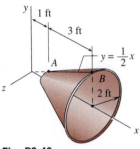

Fig. P6.48

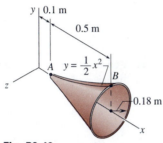

Fig. P6.49

6.50 through 6.53 Locate the center of gravity of the sheet metal form shown.

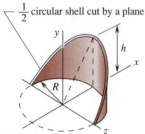

Fig. P6.50

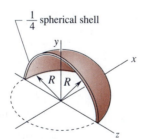

Fig. P6.51*

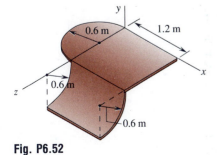

Fig. P6.52

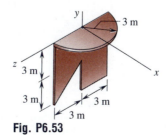

Fig. P6.53

6.10 Distributed Loads on Beams[†]

We know that a *load* is an applied force or moment carried by a body. A load whose intensity of action either remains constant or varies along the axis of a beam is called a *distributed load* on the beam. Generally, a distributed load on a beam is represented by a *loading diagram* indicating the load intensity $w = w(x)$ as a function of the position x on the beam as shown in Fig. 6.12, where the direction of the load is indicated by the arrow signs in the diagram. The load intensity is expressed as the load per unit length of the beam (e.g., N/m or lb/ft).

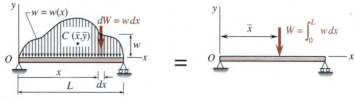

Fig. 6.12

As illustrated in Fig. 6.12, the *resultant weight W* supported by the beam is given by the *total area of the loading diagram*; i.e.,

$$W = \int_0^L w \, dx \tag{6.13}$$

The resultant weight force **W** *acts through the centroid C of the loading diagram*, where $\bar{x}$ is given by the *principle of moments*; i.e.,

$$\Sigma M_O: \quad \bar{x}W = \int_0^L xw \, dx \tag{6.14}$$

Clearly, centroids are useful in the study of distributed loads.

EXAMPLE 6.13

Sacks of cement are stacked on a plank which is supported at each end so that it acts like a beam. The distributed load exerted by the sacks of cement

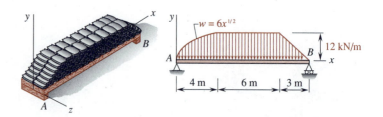

[†]Note that distributed loads do not act on beams only, but may also act on plates, shells, gates, and others. Cf. Example 6.17.

is essentially constant in the transverse direction of the plank and may be represented by the loading diagram shown. Determine the reactions from the supports at A and B.

Solution. By treating the area of the loading diagram as a composite area, we may replace the resultant force $\mathbf{W}$ at C with three concentrated forces $\mathbf{P}_1$, $\mathbf{P}_2$, and $\mathbf{P}_3$ at C_1, C_2, and C_3 as shown in the free-body diagram.

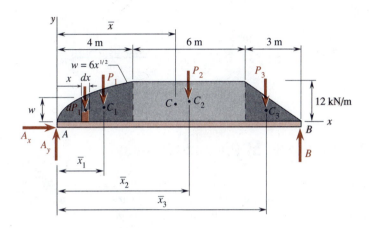

$$P_1 = \int dP_1 = \int_0^4 w \, dx = \int_0^4 6x^{1/2} \, dx = 32$$

$$P_2 = 12(6) = 72 \qquad P_3 = \frac{1}{2}(12)(3) = 18$$

Applying the *principle of moments*, we write

$$+\curvearrowright \Sigma M_A: \quad \bar{x}_1 P_1 = \int \bar{x}_{el} \, dP_1 = \int_0^4 xw \, dx = 6 \int_0^4 x^{1.5} \, dx = 76.8$$

$$\bar{x}_2 = 4 + \frac{6}{2} = 7 \qquad \bar{x}_3 = 4 + 6 + \frac{3}{3} = 11$$

For equilibrium of the beam, we write

$$+\curvearrowright \Sigma M_A = 0: \quad 13B - \bar{x}_1 P_1 - \bar{x}_2 P_2 - \bar{x}_3 P_3 = 0$$

$$\mathbf{B} = 59.9 \text{ kN} \uparrow \blacktriangleleft$$

$$\xrightarrow{+} \Sigma F_x = 0: \quad A_x = 0$$

$$\mathbf{A}_x = \mathbf{0} \blacktriangleleft$$

$$+\uparrow \Sigma F_y = 0: \quad A_y + B - P_1 - P_2 - P_3 = 0$$

$$\mathbf{A}_y = 62.1 \text{ kN} \uparrow \blacktriangleleft$$

REMARK. The abscissas $\bar{x}_1$ and $\bar{x}$ of $\mathbf{P}_1$ and $\mathbf{W}$ do not need to be explicitly found in this particular problem because our aim is to find the reactions from A and B, not the location of $\mathbf{W}$. Since the areas and centroids of rectangles and triangles are well-known, no integration is needed in finding P_2, P_3, $\bar{x}_2$, and $\bar{x}_3$.

Developmental Exercises

D6.18   Refer to Example 6.13. Determine the value of $\bar{x}$.

D6.19 and D6.20   A beam AB supports a distributed load as shown. Determine the reaction from the support at B.

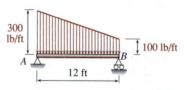

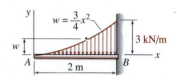

Fig. D6.19   **Fig. D6.20**

6.11   Theorems of Pappus-Guldinus[†]

A *surface of revolution* is a curved area which can be generated by revolving a line segment in a plane, called the *generating curve*, about a nonintersecting axis in that plane. This axis is called the *axis of revolution*. Let the line PQ of length L in the xy plane, as shown in Fig. 6.13, be the generating curve and the x axis be the axis of revolution. When the line PQ is revolved through an angle θ (in radians) about the x axis, the line element dL on the line PQ will generate an areal strip whose area dA is

$$dA = \theta y\, dL \tag{6.15}$$

By the *principle of moments*, the total area generated may be written as

$$A = \theta \int y\, dL \qquad A = \theta \bar{y} L \tag{6.16}$$

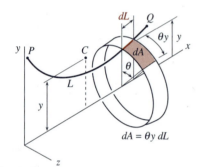

Fig. 6.13

where $\bar{y}$ is the ordinate of the centroid of the generating curve PQ, which *does not cross* the x axis. If the generating curve PQ is revolved one revolution, then $\theta = 2\pi$. Since the distance traveled by the centroid C of the generating curve is equal to $\theta\bar{y}$, we have the following *first theorem of Pappus-Guldinus*: *The area of any surface of revolution is equal to the length of the generating curve times the distance traveled by the centroid of the generating curve in generating the area.*[††]

A *body of revolution* is a body whose volume can be generated by revolving a plane area, called the *generating area*, about a nonintersecting axis in the plane of the area. This axis is called the *axis of revolution*. Let A be the total area of the generating area in the xy plane, as shown in Fig. 6.14, and the x axis be the axis of revolution. When the generating area is revolved through an angle θ (in radians) about the x axis, the area element

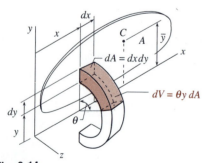

Fig. 6.14

[†]These two theorems are named after the geometer Pappus (born in Alexandria, Egypt; flourished A.D. 300–350 in the silver age of Greek mathematics) and the mathematician Habakkuk Guldinus (or Paul Guldin, born 1577 in Saint Gall, Switzerland; died 1643 in Graz, Austria). Cf. Sec. 6.15 for additional comments.

[††]The projection of the centroid of the generating curve on the axis of revolution is, in general, not coincident with the centroid of the surface of revolution, even when $\theta = 2\pi$.

dA on the generating area will generate a volumetric strip whose length is θy and whose volume is

$$dV = \theta y \, dA \qquad (6.17)$$

By the *principle of moments*, the total volume generated may be written as

$$V = \theta \int y \, dA \qquad \boxed{V = \theta \bar{y} A} \qquad (6.18)$$

where $\bar{y}$ is the ordinate of the centroid of the generating area, which *does not cross* the *x* axis. If the generating area is revolved one revolution, then $\theta = 2\pi$. Recognizing that $\theta\bar{y}$ is the distance traveled by the centroid *C* of the generating area, we have the following *second theorem of Pappus-Guldinus*: *The volume of any body of revolution is equal to the generating area times the distance traveled by the centroid of the generating area in generating the volume.*[†]

EXAMPLE 6.14

Using the *first theorem of Pappus-Guldinus*, determine the ordinate $\bar{y}$ of the centroid $C(0, \bar{y})$ of the semicircular *arc* of radius *R* as shown in Example 6.2.

Solution. We know that the surface area of a sphere of radius *R* is $4\pi R^2$ and the length of the semicircular arc of radius *R* as shown is πR. By considering the spherical area as a surface of revolution obtained by revolving the semicircular arc as the generating curve about the *x* axis through $\theta = 2\pi$ rad, we write, according to the *first theorem of Pappus-Guldinus*,

$$A = \theta\bar{y}L: \quad 4\pi R^2 = 2\pi(\bar{y})(\pi R) \qquad \bar{y} = \frac{2R}{\pi} \blacktriangleleft$$

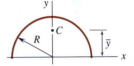

Developmental Exercises

D6.21   Describe the *first theorem of Pappus-Guldinus*.

D6.22   The point *C* as shown is the centroid of $\overline{OP}$, which is to be revolved one revolution about the *x* axis to generate a conic surface of area *A*. Determine (a) $\bar{y}$ in terms of *R*, (b) the distance traveled by *C*, (c) the area *A*.

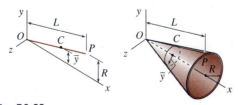

Fig. D6.22

[†]The projection of the centroid of the generating area on the axis of revolution is, in general, not coincident with the centroid of the volume of the body of revolution, even when $\theta = 2\pi$.

EXAMPLE 6.15

Using the *second theorem of Pappus-Guldinus*, determine the ordinate $\bar{y}$ of the centroid $C(0, \bar{y})$ of the semicircular area of radius R as shown in Example 6.8.

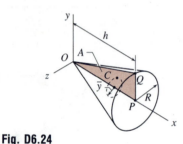

Solution.   We know that the volume of a sphere of radius R is $\frac{4}{3}\pi R^3$ and the area of a semicircular area of radius R as shown is $\frac{1}{2}\pi R^2$. By considering the spherical volume as the volume of a body of revolution obtained by revolving the semicircular area as the generating area about the x axis through $\theta = 2\pi$ rad, we write, according to the *second theorem of Pappus-Guldinus*,

$$V = \theta\bar{y}A: \quad \frac{4}{3}\pi R^3 = 2\pi(\bar{y})\left(\frac{1}{2}\pi R^2\right) \qquad \bar{y} = \frac{4R}{3\pi} \quad \blacktriangleleft$$

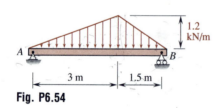

Fig. D6.24

Developmental Exercises

D6.23   Describe the *second theorem of Pappus-Guldinus*.

D6.24   The point C as shown is the centroid of the area A of the triangle OPQ, which is revolved one revolution about the x axis to generate a conic volume V. Determine (a) $\bar{y}$ in terms of R, (b) the distance traveled by C, (c) the volume V.

PROBLEMS

6.54 through 6.59*   Determine the reactions from the supports of the beam loaded as shown.

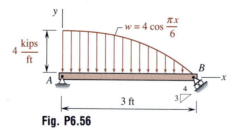

Fig. P6.54

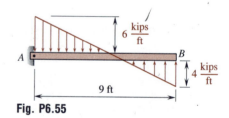

Fig. P6.55

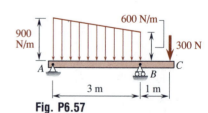

Fig. P6.56

Fig. P6.57

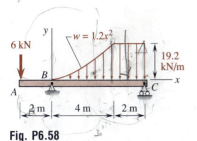

Fig. P6.58

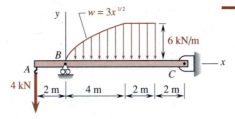

Fig. P6.59*

6.60 through 6.62* Making use of a theorem of Pappus-Guldinus, determine the area of the *entire* surface of the body of revolution which has the hatched area shown as its generating area.

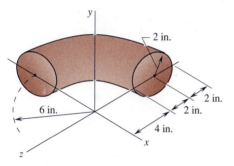

Fig. P6.60 and P6.63

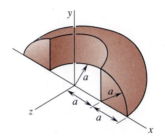

Fig. P6.61 and P6.64

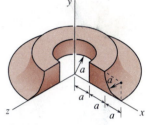

Fig. P6.62* and P6.65*

6.63 through 6.65* Making use of a theorem of Pappus-Guldinus, determine the volume of the body of revolution which has the hatched area shown as its generating area.

6.66 A machine part is of the same shape as that obtained by revolving the area shown about the y axis one revolution. It is made of an alloy whose mass density is 7 Mg/m³. Determine the total mass of the part.

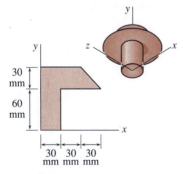

Flg. P6.66

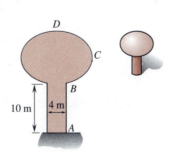

Fig. P6.67

6.67 The water storage tank shown is a shell of revolution with the composite line *ABCD* as the generating curve. The curved line segment *BCD* has a length of 12 m and its centroid is 3.5 m from the centerline of the tank. If the tank is to be sprayed with a coat of paint at a coverage of 10 m² per liter, how many liters of paint will be used?

VOLUMES AND BODIES

6.12 Centroids of Volumes

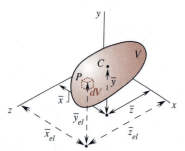

Fig. 6.15

We know that a *volume* is an amount of space and is measured in cubic lengths (e.g., m³). Let the point $P(\bar{x}_{el}, \bar{y}_{el}, \bar{z}_{el})$ be the centroid of the differential volume element dV of a given volume V as shown in Fig. 6.15, where dV is usually of the *third* or the *first* order. It is important to make sure that $(\bar{x}_{el}, \bar{y}_{el}, \bar{z}_{el})$ are the *coordinates of the centroid* of dV if dV is of the *first order* (cf. Example 6.16). The centroid $C(\bar{x}, \bar{y}, \bar{z})$ of the volume is, according to the *principle of moments*, defined by the following equations:

$$V = \int dV$$

$$\Sigma M_{yz}: \quad \bar{x}V = \int \bar{x}_{el}\, dV$$

(6.19)

$$\Sigma M_{zx}: \quad \bar{y}V = \int \bar{y}_{el}\, dV$$

$$\Sigma M_{xy}: \quad \bar{z}V = \int \bar{z}_{el}\, dV$$

where the last three integrals give the *first moments* of the volume about the coordinate planes. The *values* of dV and V are, for the purpose of locating the centroid of a volume, always made *positive*. However, the coordinates of the centroids of dV and V may be positive, negative, or zero.

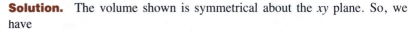

EXAMPLE 6.16

Locate the centroid $C(\bar{x}, \bar{y}, \bar{z})$ of the volume of the half right circular cone with a height h and a base of radius R as shown.

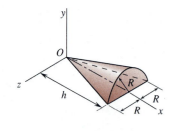

Solution. The volume shown is symmetrical about the xy plane. So, we have

$$\bar{z} = 0 \quad \blacktriangleleft$$

From the similar triangles *OEF* and *OIJ* in the sketch, we write

$$\frac{r}{R} = \frac{x}{h} \qquad r = \frac{R}{h}x$$

The volume element dV is in the form of a semicircular disk. Thus, we write

$$dV = \frac{1}{2}\pi r^2\, dx = \frac{\pi}{2}\left(\frac{R}{h}\right)^2 x^2\, dx$$

$$V = \int dV = \frac{\pi}{2}\left(\frac{R}{h}\right)^2 \int_0^h x^2\, dx = \frac{\pi R^2 h}{6}$$

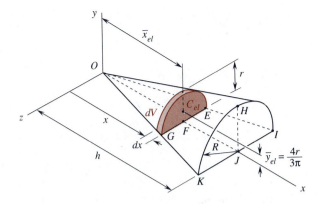

The coordinates of the centroid $C_{el}(\bar{x}_{el}, \bar{y}_{el}, \bar{z}_{el})$ of dV are

$$\bar{x}_{el} \approx x \qquad \bar{y}_{el} \approx \frac{4r}{3\pi} = \frac{4R}{3\pi h}x \quad (\text{Cf. Example 6.8.}) \qquad \bar{z}_{el} = 0$$

Applying the *principle of moments*, we write

$$\bar{x}V = \int \bar{x}_{el}\, dV: \quad \bar{x}\left(\frac{\pi R^2 h}{6}\right) = \frac{\pi}{2}\left(\frac{R}{h}\right)^2 \int_0^h x^3\, dx = \frac{\pi R^2 h^2}{8}$$

$$\bar{y}V = \int \bar{y}_{el}\, dV: \quad \bar{y}\left(\frac{\pi R^2 h}{6}\right) = \frac{2R^3}{3h^3} \int_0^h x^3\, dx = \frac{R^3 h}{6}$$

$$\bar{x} = \frac{3}{4}h \qquad \bar{y} = \frac{R}{\pi} \quad \blacktriangleleft$$

REMARK. The volume of the half right circular cone *OHIJK* may be obtained by revolving the right triangular area *OIJ* as the generating area about the *x* axis through π rad. The centroid of this generating area is at a point which is $h/3$ from its base *IJ*; however, the centroid of the half right circular cone is at a point which is $h/4$ from its base *HIJK*.

Developmental Exercises

D6.25 The point *C* as shown is the centroid of the right circular solid cone of volume *V*. Determine (a) dV in terms of *x* and dx, (b) *V*, (c) $\bar{x}V$, (d) $\bar{x}$.

D6.26 Refer to the conic shell in D6.15 and the solid cone in D6.25. Does the centroid of the curved surface area of a conic shell with an open base coincide with the centroid of the volume of a solid cone with identical height and base?

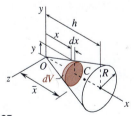

Fig. D6.25

6.13 Composite Volumes and Center of Pressure

A *composite volume* of total volume *V* is one which has component volumes $V_1, V_2, \ldots, V_n$, where

$$V = V_1 + V_2 + \cdots + V_n \qquad (6.20)$$

Let the centroids of the component volumes $V_1, V_2, \ldots, V_n$ be located at the points $C_1(\bar{x}_1, \bar{y}_1, \bar{z}_1)$, $C_2(\bar{x}_2, \bar{y}_2, \bar{z}_2)$, $\ldots$, $C_n(\bar{x}_n, \bar{y}_n, \bar{z}_n)$, respectively. Then, the centroid $C(\bar{x}, \bar{y}, \bar{z})$ of the composite volume is, according to the *principle of moments*, defined by the following equations:

$$\Sigma M_{yz}: \quad \bar{x}V = \bar{x}_1 V_1 + \bar{x}_2 V_2 + \cdots + \bar{x}_n V_n$$

$$\Sigma M_{zx}: \quad \bar{y}V = \bar{y}_1 V_1 + \bar{y}_2 V_2 + \cdots + \bar{y}_n V_n \qquad (6.21)$$

$$\Sigma M_{xy}: \quad \bar{z}V = \bar{z}_1 V_1 + \bar{z}_2 V_2 + \cdots + \bar{z}_n V_n$$

Similar to the case of a composite area, one or more of the component volumes of a composite volume may be negative in some situations. For use in determining the centroids of composite volumes, the centroids of volumes of some common shapes are listed as shown in Fig. 6.16.

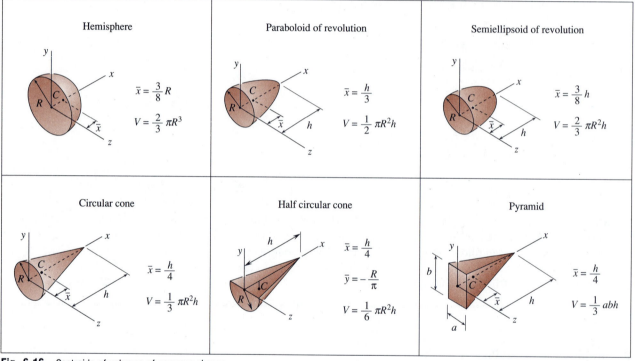

Fig. 6.16　Centroids of volumes of common shapes.

Fluids that are essentially incompressible are called *liquids*; otherwise, *gases*. The statics of liquids is generally referred to as *hydrostatics*,[†] which is usually treated in detail in *fluid mechanics* and *hydraulics*. An important property in hydrostatics is that the *pressure p* (in force per unit area) exerted by the liquid (e.g., water) on a submerged surface is *perpendicular* to the submerged surface as shown in Fig. 6.17 and is given by

$$p = \gamma h \qquad (6.22)^{††}$$

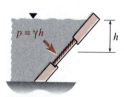

Fig. 6.17

[†]The history of the science of *hydrostatics* may trace back to the oft-quoted anecdotal bathtub discovery of the solution by Archimedes (287–212 B.C.) of Syracuse, Sicily, for King Hieron's problem concerning the purity of his gold crown.

[††]The pressure p is usually expressed in N/m^2 or kN/m^2 in the SI, and in lb/ft^2 or lb/in^2 in the U.S. customary system of units. Note that 1 N/m^2 = 1 pascal (Pa) and 1 lb/in^2 = 1 psi.

where γ is the specific weight (in force per unit volume) of the liquid and h is the vertical distance (or depth) measured from the free surface of the liquid to the point on the submerged surface under consideration. By Newton's second law, we note that

$$\gamma = \rho g \qquad (6.23)$$

where ρ is the mass density of the liquid and g is the gravitational acceleration. The *total volume of the pressure diagram* constructed according to this property gives the *magnitude of the resultant force of the pressure*. Furthermore, *the resultant force of the pressure acts through the centroid of volume of the pressure diagram*. In hydrostatics, the intersection of the line of action of the resultant force of the pressure and the submerged surface is known as the *center of pressure*.

EXAMPLE 6.17

A dam is holding back impounded water of mass density $\rho = 1\ \text{Mg/m}^3$. At the bottom of the dam as shown, there is a vertical gate which is 0.6 m high and 1 m wide. The gate is hinged along its top A and stopped along the bottom B. Determine (a) the magnitude of the reaction from the bottom B of the gate, (b) the distance $\bar{y}$ between the center of pressure and the bottom edge of the gate.

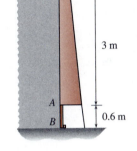

Solution. The specific weight of the water is $\gamma = \rho g = g\ \text{kN/m}^3$. Note that the depths at A and B are $h_A = 3$ m and $h_B = 3.6$ m, respectively. Thus, the pressures at A and B are

$$p_A = \gamma h_A = g(3) = 3g\ (\text{kN/m}^2) \qquad p_B = \gamma h_B = g(3.6) = 3.6g\ (\text{kN/m}^2)$$

The resultant force $\mathbf{R}$ exerted by the water on the gate is a force which acts through the centroid C of the volume of the pressure diagram shown and which has a magnitude equal to the volume of the pressure diagram.

By treating the volume of the pressure diagram as a composite volume, we may replace the resultant force $\mathbf{R}$ at C with two concentrated forces $\mathbf{F}_1$ and $\mathbf{F}_2$ at C_1 and C_2 as shown in the free-body diagrams, where

$$F_1 = p_A(0.6)(1) = 1.8g\ (\text{kN}) \qquad F_2 = \tfrac{1}{2}(p_B - p_A)(0.6)(1) = 0.18g\ (\text{kN})$$

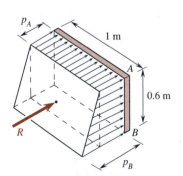

Pressure diagram on the gate

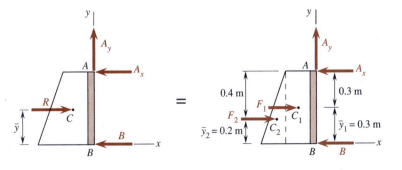

Free-body diagrams of the gate

For equilibrium of the gate, we write

$$+\,\text{↻}\ \Sigma M_A = 0: \qquad 0.3F_1 + 0.4F_2 - 0.6B = 0$$

$$B = 1.02g \qquad\qquad B = 10.01 \text{ kN} \ \blacktriangleleft$$

To determine $\bar{y}$, we apply the *principle of moments* as follows:

$$R = F_1 + F_2 = 1.8g + 0.18g = 1.98g$$

$$\bar{y}R = \bar{y}_1 F_1 + \bar{y}_2 F_2 = 0.3(1.8g) + 0.2(0.18g)$$

$$\bar{y} = \frac{0.576g}{1.98g} \qquad\qquad \bar{y} = 0.291 \text{ m} \ \blacktriangleleft$$

Developmental Exercises

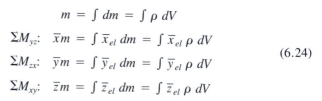

D6.27   Refer to Example 6.17. Determine the reaction component $\mathbf{A}_x$.

D6.28   Making use of the formulas in Fig. 6.16, locate the centroid of the composite volume consisting of a cone and a cylinder as shown.

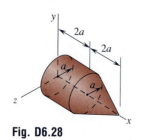

Fig. D6.28

6.14   Centers of Mass and Centers of Gravity of Bodies

A *body* can be viewed as a mass occupying a volume and having a shape. Let ρ (expressed as mass per unit volume) be the *mass density* of a body of total mass m as shown in Fig. 6.18. The *center of mass* $G(\bar{x}, \bar{y}, \bar{z})$ of the body is, according to the *principle of moments*, defined by the following equations:

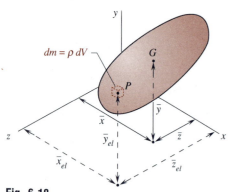

Fig. 6.18

$$m = \int dm = \int \rho\, dV$$

$$\Sigma M_{yz}: \quad \bar{x}m = \int \bar{x}_{el}\, dm = \int \bar{x}_{el}\, \rho\, dV$$

$$\Sigma M_{zx}: \quad \bar{y}m = \int \bar{y}_{el}\, dm = \int \bar{y}_{el}\, \rho\, dV \qquad (6.24)$$

$$\Sigma M_{xy}: \quad \bar{z}m = \int \bar{z}_{el}\, dm = \int \bar{z}_{el}\, \rho\, dV$$

If the mass density of the body is constant throughout, the *center of mass* of the body is clearly coincident with the centroid of the volume occupied by the body. Applying the *principle of moments*, we can determine the center of mass of a composite body in a similar manner as explained in Secs. 6.4, 6.8, and 6.13.

The weight force $\mathbf{W}$ of a body and hence the location of its center of gravity G is dependent on the gravitational acceleration vector $\mathbf{g}$. In situations where $\mathbf{g}$ varies significantly within the space occupied by a body of total weight W as shown in Fig. 6.19, the center of gravity $G(\bar{x}, \bar{y}, \bar{z})$ of the body is, according to the *principle of moments*, defined by the following equations:

$$W = \int dW = \int \rho g \, dV$$

$$\Sigma M_{yz}: \quad \bar{x}W = \int \bar{x}_{el} \, dW = \int \bar{x}_{el} \rho g \, dV$$

$$\Sigma M_{zx}: \quad \bar{y}W = \int \bar{y}_{el} \, dW = \int \bar{y}_{el} \rho g \, dV \qquad (6.25)$$

$$\Sigma M_{xy}: \quad \bar{z}W = \int \bar{z}_{el} \, dW = \int \bar{z}_{el} \rho g \, dV$$

If **g** has negligibly small variations throughout the space occupied by the body, the *center of gravity G* of the body is clearly *coincident* with the *center of mass* of the same body. In most engineering applications, the center of gravity and the center of mass may be taken as the same point without introducing appreciable errors. The determination of the *center of gravity of a composite body* can similarly be carried out according to the *principle of moments* as in Secs. 6.4, 6.8, and 6.13.

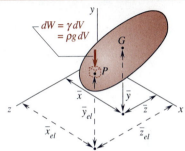

Fig. 6.19

EXAMPLE 6.18

A tetrahedral body has a mass density $\rho = 15z$ kg/m³ and is bounded by the inclined plane $x + 2y + 4z = 8$ and the three coordinate planes, where the unit of length is meter. Locate the center of mass of the body.

Solution. Since $\rho = 15z$ kg/m³, the center of mass of the body will not coincide with the centroid of its volume. We first let *dm* be situated at (x, y, z) such that

$$\bar{x}_{el} \approx x \qquad \bar{y}_{el} \approx y \qquad \bar{z}_{el} \approx z$$

$$dm = \rho \, dV = 15z \, dx \, dy \, dz$$

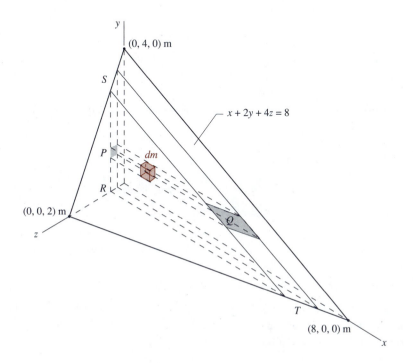

Then, we use the following procedure to obtain the total mass m:

1. Keeping y and z constant, integrate with respect to x from P (where $x = 0$) to Q (where $x = 8 - 2y - 4z$) to add the third-order mass elements and obtain the mass of the filament PQ.
2. Keeping z constant, integrate with respect to y from RT (where $y = 0$) to S (where $x = 0$, $2y + 4z = 8$ or $y = 4 - 2z$) to add the masses of the filaments and obtain the mass of the slab RST.
3. Integrate with respect to z from the back (where $z = 0$) to the front (where $z = 2$) of the tetrahedron to add the masses of the slabs and obtain the total mass m of the tetrahedral body. Thus, we write

$$m = \int_{z=0}^{2} \int_{y=0}^{4-2z} \int_{x=0}^{8-2y-4z} 15z \; dx \; dy \; dz$$

$$= \int_{z=0}^{2} \int_{y=0}^{4-2z} 15z(8 - 2y - 4z) \; dy \; dz$$

$$= \int_{z=0}^{2} 15z(16 - 16z + 4z^2) \; dz = 80 \quad \text{(kg)}$$

By the *principle of moments*, the mass center $G(\bar{x}, \bar{y}, \bar{z})$ of the tetrahedral body may similarly be determined by triple integration as follows:

$$\Sigma M_{yz}: \quad \bar{x}m = \int_{z=0}^{2} \int_{y=0}^{4-2z} \int_{x=0}^{8-2y-4z} 15zx \; dx \; dy \; dz$$

$$= 128 \quad \text{(kg} \cdot \text{m)} \qquad\qquad \bar{x} = 1.6 \text{ m} \blacktriangleleft$$

$$\Sigma M_{zx}: \quad \bar{y}m = \int_{z=0}^{2} \int_{y=0}^{4-2z} \int_{x=0}^{8-2y-4z} 15zy \; dx \; dy \; dz$$

$$= 64 \quad \text{(kg} \cdot \text{m)} \qquad\qquad \bar{y} = 0.8 \text{ m} \blacktriangleleft$$

$$\Sigma M_{xy}: \quad \bar{z}m = \int_{z=0}^{2} \int_{y=0}^{4-2z} \int_{x=0}^{8-2y-4z} 15z^2 \; dx \; dy \; dz$$

$$= 64 \quad \text{(kg} \cdot \text{m)} \qquad\qquad \bar{z} = 0.8 \text{ m} \blacktriangleleft$$

Developmental Exercise

D6.29 Work Example 6.18 if the mass density of the body is (a) $\rho = 3$ kg/m^3, (b) $\rho = y$ kg/m^3.

EXAMPLE 6.19

The body of a machine part is of the same shape as that obtained by revolving the area shown about the x axis. The machine part has a mass density $\rho = 7.7$ Mg/m^3. Determine the tension in each cable.

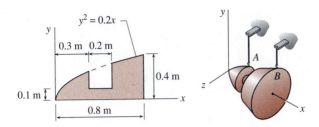

Solution. We first note that ρ is constant and $\mathbf{g}$ is essentially constant throughout the space occupied by the body. Thus, the center of mass, the center of gravity, and the centroid of volume of the body may be taken as coincident. Let V_1, V_2, and V_3 be the component volumes of the portions of the machine part lying in the ranges $0 \le x \le 0.3$ m, 0.3 m $\le x \le 0.5$ m, and 0.5 m $\le x \le 0.8$ m, respectively. From the sketch shown, we have

$$V_2 = \pi(0.1)^2(0.2) = 0.002\pi$$

$$dV = \pi y^2 dx = 0.2\pi x\, dx \qquad \bar{x}_{el} \approx x$$

$$V_1 = \int dV = 0.2\pi \int_0^{0.3} x\, dx = 0.009\pi$$

$$V_3 = \int dV = 0.2\pi \int_{0.5}^{0.8} x\, dx = 0.039\pi$$

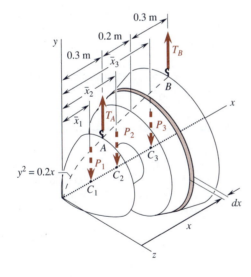

Applying the *principle of moments*, we write

$$\bar{x}_1 V_1 = \int \bar{x}_{el}\, dV:$$

$$\bar{x}_1(0.009\pi) = 0.2\pi \int_0^{0.3} x^2\, dx = 0.0018\pi \qquad \bar{x}_1 = 0.2 \text{ m}$$

$$\bar{x}_3 V_3 = \int \bar{x}_{el}\, dV:$$

$$\bar{x}_3 (0.039\pi) = 0.2\pi \int_{0.5}^{0.8} x^2\, dx = 0.0258\pi \qquad \bar{x}_3 = \frac{43}{65} \text{ m}$$

By inspection, we have

$$\bar{x}_2 = \left(0.3 + \frac{0.2}{2}\right) \text{ m} = 0.4 \text{ m}$$

Since $\rho = 7.7$ Mg/m^3, the three equivalent concentrated weights at C_1, C_2, and C_3 are

$$P_1 = \rho V_1 g = 0.0693\pi g \quad \text{(kN)} \qquad P_2 = \rho V_2 g = 0.0154\pi g \quad \text{(kN)}$$

$$P_3 = \rho V_3 g = 0.3003\pi g \quad \text{(kN)}$$

For equilibrium, we write

$$+ \circlearrowright \Sigma M_A = 0: \quad (0.3 - 0.2)(0.0693\pi g) - (0.4 - 0.3)(0.0154\pi g)$$

$$- \left(\frac{43}{65} - 0.3\right)(0.3003\pi g) + 0.5T_B = 0$$

$$+ \circlearrowright \Sigma M_B = 0: \quad (0.8 - 0.2)(0.0693\pi g) - 0.5T_A$$

$$+ (0.8 - 0.4)(0.0154\pi g) + \left(0.8 - \frac{43}{65}\right)(0.3003\pi g) = 0$$

$$T_A = 5.51 \text{ kN} \qquad T_B = 6.36 \text{ kN} \blacktriangleleft$$

Check $\qquad + \uparrow \Sigma F_y = T_A + T_B - P_1 - P_2 - P_3 \approx 0$

Developmental Exercise

D6.30 Locate the center of gravity of the machine part described in Example 6.19.

PROBLEMS

6.68 and 6.69 Determine by integration the centroid of volume of the body whose curved surface is defined by the equation as indicated. (*Hint.* Let dV be a thin slab perpendicular to the z axis.)

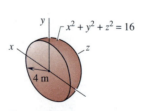

Fig. P6.68

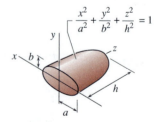

Fig. P6.69

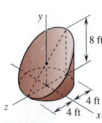

Fig. P6.70

6.70 A body is obtained by cutting a circular cylinder of radius 4 ft along an oblique plane as shown. Locate the centroid of volume of the body. (*Hint.* Let dV be a thin slab parallel to the xy plane.)

6.71 A body of revolution is of the shape of that obtained by revolving a quarter circular area about the y axis through 180° as shown. Locate the centroid of volume of the body. (*Hint.* Let dV be a half cylindrical shell parallel to the y axis.)

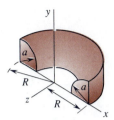

Fig. P6.71

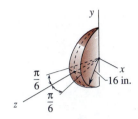

Fig. P6.72

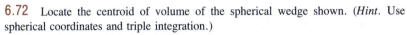

Fig. P6.73*

6.72 Locate the centroid of volume of the spherical wedge shown. (*Hint.* Use spherical coordinates and triple integration.)

6.73* Determine by integration the centroid of volume generated by revolving the area shown in the xy plane about the x axis one revolution.

6.74 The homogeneous body shown is composed of a solid circular cone of mass m_C and a solid hemispherical base of mass m_B. For $r = 80$ mm and $h = 480$ mm, determine the minimum ratio of m_B/m_C if the body is to return to an upright position after its longitudinal axis has been pushed to a horizontal position.

6.75 and 6.76 Locate the center of gravity of the homogeneous body shown.

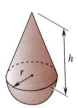

Fig. P6.74

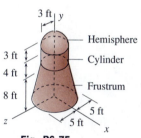

Fig. P6.75

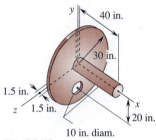

Fig. P6.76

NOTE. In Probs. 6.77 through 6.85*, take the specific weight and mass density for water to be $\gamma = 62.4$ lb/ft^3 and $\rho = 1$ Mg/m^3, and those for concrete to be $\gamma_c = 150$ lb/ft^3 and $\rho_c = 2.4$ Mg/m^3, respectively.

6.77 through 6.79* The cross section of a concrete dam is shown where a seal exists at A and no water pressure is present under the dam. Determine the maximum value of the depth d of water for which the dam will not overturn about B.

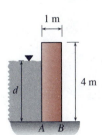

Fig. P6.77

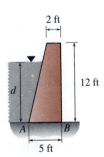

Fig. P6.78

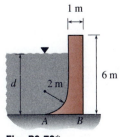

Fig. P6.79*

6.80 through 6.82 The gate *AB* for holding water has a uniform width *w* as indicated and is hinged at *A*. Determine the minimum compressive force to be exerted by the hydraulic cylinder *CD* to keep the gate shut.

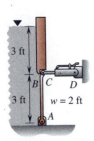

Fig. P6.80

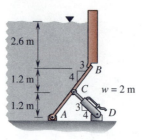

Fig. P6.81*

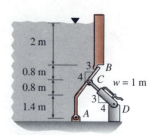

Fig. P6.82

6.83 The automatic valve *ABC* consists of a square plate pivoted at *B* as shown. Determine the depth *d* of water which will cause the valve to open.

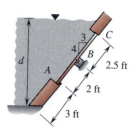

Fig. P6.83

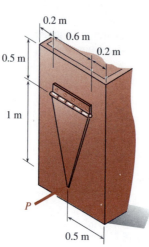

Fig. P6.84

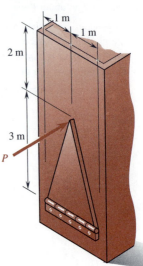

Fig. P6.85*

6.84 and 6.85* The tank shown is filled with water. Determine the magnitude of the horizontal force **P** required to keep the gate closed.

6.86 An industrial oil of mass density ρ is stored in a cylindrical tank to capacity as shown. Determine (a) the magnitude of the resultant force **P** of the oil pressure acting on one of the circular ends of the tank, (b) the distance between the center of pressure and the bottom of the tank. (Cf. Prob. 6.70.)

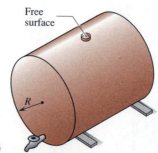

Fig. P6.86

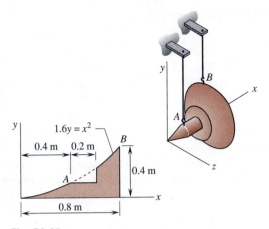

Fig. P6.87

6.87 and 6.88* A body of revolution is of the same shape as that obtained by revolving the area shown about the x axis one revolution. The body has a mass density of 5 Mg/m³ and is hung by two wires as shown. Determine the tension in each wire.

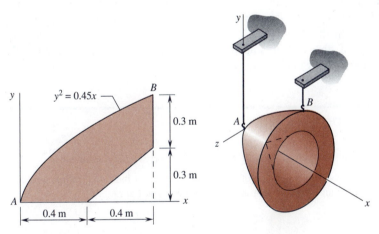

Fig. P6.88*

6.89 and 6.90 The body shown has a variable mass density ρ as indicated. Using triple integration, locate the center of mass of the body.

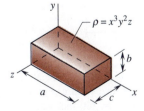

Fig. P6.89

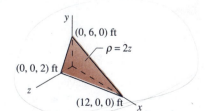

Fig. P6.90

6.15   Concluding Remarks

The focus throughout this chapter has been the *principle of moments*. It may roughly be stated as follows: *The resultant is equal to the sum of the components, and the moment of the resultant is equal to the sum of the moments of the components.* This principle is synthesized from the concept of *equivalent systems* of forces and degenerates into Varignon's theorem as a special case in which the components are concurrent forces. The resultants of any force systems (including distributed loads), the geographic center, and the population center are defined and located according to the *principle of moments*. Moreover, the centroids of lines, areas, and volumes as well as the centers of gravity of bodies are their respective "centers" which are defined and located according to the *principle of moments*.

The distances traveled by the centroids of generating curves and generating areas are related to the corresponding surface areas of revolution and volumes of revolution by the first and the second theorems of Pappus-Guldinus, respectively. Historical studies indicate that Pappus lived in Alexandria, Egypt around the end of the third century. The essence of those two theorems was first enunciated by him in Book VII of his work *Synagoge* (Συναγωγή) or *Collection*. The *Collection* contains his mathematical papers arranged in eight books of which Book I and the first 13 propositions (out of 26) of Book II have been lost. However, the said two theorems received little attention until Guldinus published Volume II of his four-volume work *Centrobaryca seu de centro gravitatis trium specierum quantitatis continuae* (in Vienna, Austria, 1635–1641) without reference to, and possibly without knowledge of, the work by Pappus. In some books, those two theorems are named after Pappus only.

REVIEW PROBLEMS

6.91   Locate the resultant force **R** of the parallel forces acting as shown.

6.92   The body shown is made of a thin rod weighing 4 lb/ft. Determine the reactions from the hinge support at A and the smooth support at D.

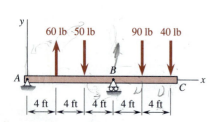

Fig. P6.91

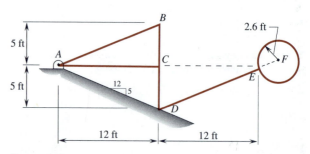

Fig. P6.92

6.93   A bent rod having a mass density of 1 kg/m and a sheet of metal having a mass density of 5 kg/m² are welded together and suspended from a rope in such a manner that the metal sheet is horizontal as shown. If $h = 100$ mm and $\bar{x} = 350$ mm, determine (a) L, (b) $\bar{z}$.

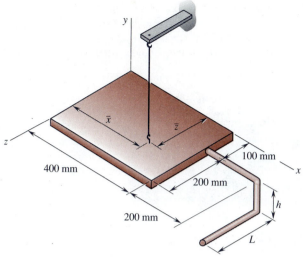

Fig. P6.93

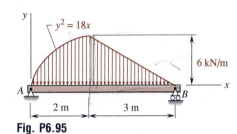

Fig. P6.95

6.94* Solve Prob. 6.93 if $h = 200$ mm and $\bar{x} = 360$ mm.

6.95 A beam is under the action of a distributed load as shown. Determine the reactions from the supports at A and B.

6.96* Locate the centroid of the area shown. (*Hint.* Cf. the footnote on p. 200.)

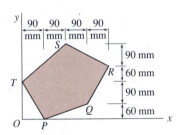

Fig. P6.96*

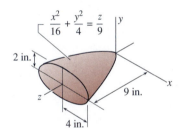

Fig. P6.97

6.97 Determine by integration the centroid of volume of the body whose curved surface is defined by the equation as indicated.

6.98* Locate the centroid of volume of the half torus shown.

6.99 Locate the centroid of volume of the conic wedge shown.

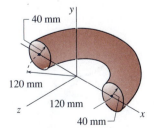

Fig. P6.98*

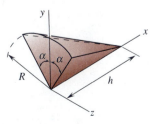

Fig. P6.99

6.100 A 4-m long tank is used to store a pool of oil with a specific weight of 9 kN/m³ and a pool of salt water with a specific weight of 10 kN/m³ by using a partition whose cross section is shown. The supporting cables are not stressed when the tank is empty. Neglecting the weight of the partition, determine the tension in each cable when $a = b = 1$ m.

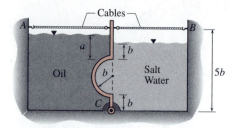

Fig. P6.100

6.101* Refer to Prob. 6.100. Determine the ratio of a/b for which the cables are unstressed.

6.102 A homogeneous body of revolution is of the same shape as that obtained by revolving the area shown about the x axis one revolution. Locate the center of gravity of the body.

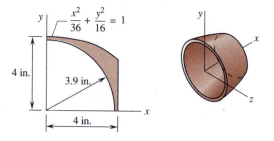

Fig. P6.102

Second Moments: Moments of Inertia[†]

Chapter 7

KEY CONCEPTS

- Terms related to a plane area with respect to certain axes: moment of inertia, polar moment of inertia, radius of gyration, and product of inertia.
- Parallel-axis theorems for area moments and products of inertia.
- Moments and products of inertia of composite areas.
- Rotation of axes for area moments and products of inertia.
- Mohr's circle for area moments and products of inertia.
- Moment of inertia and radius of gyration of a mass.
- Parallel-axis theorem for mass moments of inertia.

The study of many situations in engineering involves the use of quantities referred to as second moments, or moments of inertia, of areas and masses. For instance, moments of inertia of areas are used in expressing the stresses and strains experienced by beams in bending and shafts in torsion, the critical loads for the buckling of columns, and the locations of centers of pressure on nonhorizontal submerged surfaces. In kinetics, moments of inertia of masses are used in expressing the effective moments, the kinetic energies, and the momenta associated with bodies having rotational motion. One can see that moments of inertia of areas and masses are useful concepts which are employed in the study of mechanics of materials, fluid mechanics, and dynamics.

[†]Although it is logical for second moments to follow first moments in the sequence of study, this chapter may be covered later, such as after Chap. 9, without affecting continuity. If circumstances warrant, one may skip the mass moments of inertia in the last part of this chapter and take up that topic, as provided in the first part of Chap. 16, in *Dynamics*.

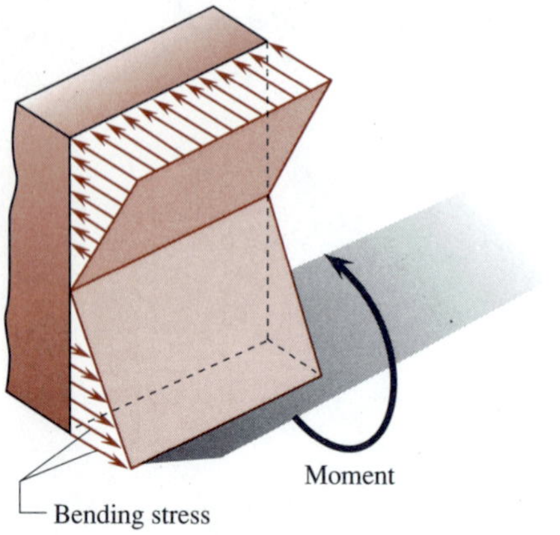

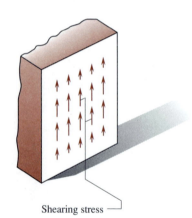

Moment

Bending stress

Fig. 7.1

AREA MOMENTS OF INERTIA[†]

7.1 Situations Involving Area Moments of Inertia

From time to time, engineers encounter situations in which the intensity of a distributed force (e.g., pressure or stress) over an area is proportional to the distance between the force and a certain axis. Consequently, the resultant moment of the forces distributed over the area about that axis involves an integral of the form $\int(\text{distance})^2\,d(\text{area})$. This integral is known as the *second moment*, or the *moment of inertia*, of the area about that axis.[††] An area has *no mass* and, hence, *no inertia*. However, because of mathematical similarity in the definitions (cf. Secs. 7.2 and 7.11), the term *moment of inertia* continues to be used by most engineers for both *areas* and *masses*. Moreover, note that an area moment of inertia is usually defined for a *plane area* with respect to an axis either in or perpendicular to the plane of the area.

In mechanics, there are many situations which involve the use of moments of inertia of an area in expressing certain physical quantities. We shall here mention some of those situations without dwelling on their details. In mechanics of materials, (a) the moment of inertia of the cross-sectional area of a beam about a centroidal axis is a quantity used in expressing the stresses and strains developed in the beam due to bending and shearing as shown in Figs. 7.1 and 7.2, (b) the polar moment of inertia of the cross-sectional area of a circular shaft about its center as the pole is a quantity used in expressing torsional shearing stresses and strains and the angle of twist in the shaft as shown in Figs. 7.3 and 7.4, and (c) the smallest moment of inertia of the

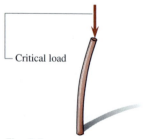

Shearing stress

Fig. 7.2

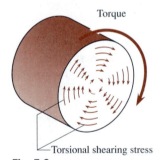

Torque

Torsional shearing stress

Fig. 7.3

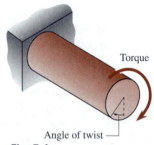

Torque

Angle of twist

Fig. 7.4

cross-sectional area of a column about a centroidal axis is a quantity used in expressing the critical load for buckling of a column as shown in Fig. 7.5. In fluid mechanics, the moment of inertia of the area of a nonhorizontal

Critical load

Fig. 7.5

[†]Also called *second moments of areas*. Cf. Sec. 7.1.

[††]Note that $\int(\text{distance})\,d(\text{area})$ gives the *first moment* of the area; $\int(\text{distance})^2\,d(\text{area})$ is, therefore, termed the *second moment* of the area.

submerged plane surface about an axis in the free surface of the liquid is a quantity used in locating the center of pressure on the submerged surface as shown in Fig. 7.6.

Developmental Exercise

D7.1 Name two situations in which area moments of inertia are used.

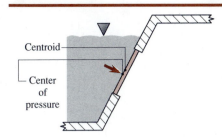

Fig. 7.6

7.2 Moments of Inertia of an Area

The *moment of inertia of an area* with respect to (or about) an axis is defined as the integral of the second-order differential area element times the square of the distance between that axis and the area element, as shown in Fig. 7.7. When it is feasible to subdivide a first-order differential area element of an area into an infinite number of second-order differential area

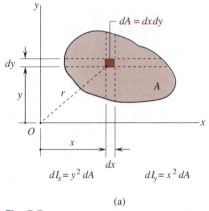

$$dI_x = y^2\, dA \qquad dI_y = x^2\, dA$$

(a)

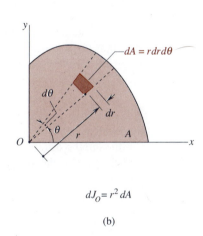

$$dJ_O = r^2\, dA$$

(b)

Fig. 7.7

elements which are *equidistant* from an axis, the *moment of inertia* of the area about that axis may be defined as the integral of the first-order differential area element times the square of that *equidistance*, as shown in Fig. 7.8. The moments of inertia of an area with respect to the x and y axes, which lie in the plane of the area, are customarily denoted by I_x and I_y, respectively.[†] Referring to Figs. 7.7 and 7.8, we write

---

[†]Sometimes, I_x and I_y are written with double subscripts as I_{xx} and I_{yy}, respectively, to signify that they are components of a second-order tensor (cf. the footnotes on p. 237).

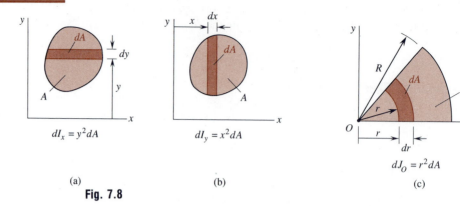

$$dI_x = y^2\,dA$$

(a)

$$dI_y = x^2\,dA$$

(b)

$$dJ_O = r^2\,dA$$

(c)

Fig. 7.8

$$I_x = \int y^2\,dA \tag{7.1}$$

$$I_y = \int x^2\,dA \tag{7.2}$$

The moments of inertia I_x and I_y defined in Eqs. (7.1) and (7.2) constitute a set of *rectangular moments of inertia* of the area.

The moment of inertia of an area about an axis perpendicular to the plane of the area is called the *polar moment of inertia* of the area about that axis. The polar moment of inertia of the area A about the z axis (or the pole at the origin O) is usually denoted by J_O. Referring to Figs. 7.7 and 7.8, we write

$$J_O = \int r^2\,dA \tag{7.3}$$

where r denotes a polar coordinate.[†] It is clear from Eqs. (7.1) through (7.3) that the dimensions of moments of inertia of an area are $(length)^4$.

Although the differential area element dA in Eqs. (7.1) through (7.3) may be of the *first* or the *second* order, as illustrated in Figs. 7.7 and 7.8, *it is important to make sure that dA is an areal strip parallel to the axis of interest if dA is of the first order for single integration.* Of course, double integration is required when dA is of the second order. The equation relating J_O, I_x, and I_y may be derived as follows:

$$J_O = \int r^2\,dA = \int (x^2 + y^2)\,dA = \int x^2\,dA + \int y^2\,dA = I_y + I_x$$

$$J_O = I_x + I_y \tag{7.4}$$

Note that the area element dA in Eqs. (7.1) through (7.3) must always be made positive regardless of the quadrant in which it is situated. Since the square of the distance is always positive, the rectangular as well as polar moments of inertia of areas must always be *positive*. Furthermore, the moment of inertia of an area about an axis which passes through its centroid is called a *centroidal moment of inertia* of the area, which is frequently marked with an overbar; e.g., $\bar{I}_x$, $\bar{I}_y$, and $\bar{J}_O$ signify that x and y are centroidal axes and O is the centroid of the given area.

[†]This is why J_O was given the name *polar moment of inertia*.

Developmental Exercises

D7.2   Define the moment of inertia of an area about an axis.

D7.3   What is a polar moment of inertia of an area?

EXAMPLE 7.1

Determine the centroidal moments of inertia $\bar{I}_x$ and $\bar{I}_y$ of a rectangular area with base b and height h parallel to the x and y centroidal axes, respectively.

Solution.   For illustrative purposes, $\bar{I}_x$ and $\bar{I}_y$ will here be determined in two different ways:

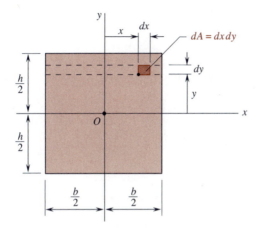

(a) *Using double integration.* We here choose a second-order differential area element dA as shown. By definition, we write

$$\bar{I}_x = \int y^2 \, dA = \int_{-h/2}^{h/2} \int_{-b/2}^{b/2} y^2 \, dx \, dy = \int_{-h/2}^{h/2} by^2 \, dy$$

$$\bar{I}_x = \frac{1}{12} bh^3 \quad \blacktriangleleft$$

$$\bar{I}_y = \int x^2 \, dA = \int_{-h/2}^{h/2} \int_{-b/2}^{b/2} x^2 \, dx \, dy = \int_{-h/2}^{h/2} \frac{1}{12} b^3 \, dy$$

$$\bar{I}_y = \frac{1}{12} b^3 h \quad \blacktriangleleft$$

(b) *Using single integration.* For determining $\bar{I}_x$, we choose dA to be a first-order areal strip parallel to the x axis as shown. By definition, we write

$$\bar{I}_x = \int y^2 \, dA = \int_{-h/2}^{h/2} y^2 \, (b \, dy) = b \int_{-h/2}^{h/2} y^2 \, dy$$

$$\bar{I}_x = \frac{1}{12} bh^3 \quad \blacktriangleleft$$

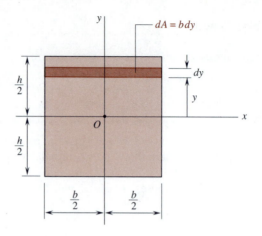

For determining $\bar{I}_y$, we choose dA to be a first-order areal strip parallel to the y axis as shown. By definition, we write

$$\bar{I}_y = \int x^2 \, dA = \int_{-b/2}^{b/2} x^2 \, (h \, dx) = h \int_{-b/2}^{b/2} x^2 \, dx$$

$$\bar{I}_y = \frac{1}{12} b^3 h \quad \blacktriangleleft$$

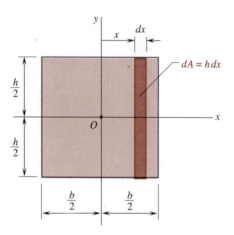

REMARK. The above results are well-known and widely used as *formulas*.

Developmental Exercise

D7.4   Refer to Example 7.1 and use Eq. (7.4). What is the centroidal polar moment of inertia $\bar{J}_O$ of a rectangle of base b and height h?

EXAMPLE 7.2

Determine the centroidal polar moment of inertia $\bar{J}_O$ of a circular area of radius R.

Solution.   For illustrative purposes, $\bar{J}_O$ will here be determined in two different ways:

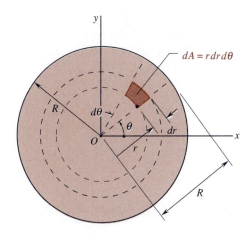

(a) *Using double integration*. We here choose a second-order dA in polar coordinates as shown. By definition, we write

$$\bar{J}_O = \int r^2 \, dA = \int_0^{2\pi} \int_0^R r^2 \, (r \, dr \, d\theta) = \int_0^{2\pi} \frac{1}{4} R^4 \, d\theta$$

$$\bar{J}_O = \frac{1}{2} \pi R^4 \quad \blacktriangleleft$$

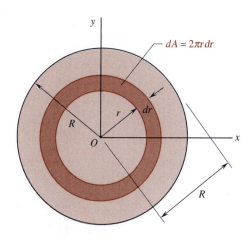

(b) *Using single integration*. We first choose a *ring-shaped* first-order differential area element dA as shown. By definition, we write

$$\bar{J}_O = \int r^2 \, dA = \int_0^R r^2 (2\pi r \, dr)$$

$$= 2\pi \int_0^R r^3 \, dr \qquad\qquad \boxed{\bar{J}_O = \frac{1}{2} \pi R^4} \blacktriangleleft$$

REMARK. If we choose to integrate with respect to θ first, we write

$$\bar{J}_O = \int r^2 \, dA = \int_0^R \int_0^{2\pi} r^3 \, d\theta \, dr = \int_0^R r^3 (2\pi) \, dr = \frac{1}{2} \pi R^4$$

Since the circular area is symmetrical about the x and y axes, we know that $\bar{I}_x = \bar{I}_y$. Thus, based on the above result for $\bar{J}_O$ and Eq. (7.4), we obtain for a circular area with its centroid at the origin that

$$\bar{I}_x = \bar{I}_y = \frac{1}{2} \bar{J}_O = \frac{1}{4} \pi R^4$$

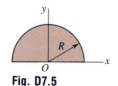

Fig. D7.5

Developmental Exercise

D7.5 Determine the polar moment of inertia J_O of the semicircular area shown.

EXAMPLE 7.3

The cross-sectional area of a slender rod is shown. Determine (a) the moments of inertia I_x and I_y of the area about the x and y axes, respectively; (b) the polar moment of inertia J_O of the area about the origin O.

Solution. To determine I_x, we choose a first-order dA which is *parallel* to the x axis as shown. We write

$$dA = x \, dy = \left(\frac{64}{5} y \right)^{1/2} dy$$

$$I_x = \int y^2 \, dA = \int_0^5 y^2 \left(\frac{64}{5} y \right)^{1/2} dy = \frac{2000}{7}$$

$$\boxed{I_x = 286 \text{ mm}^4} \blacktriangleleft$$

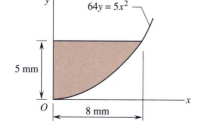

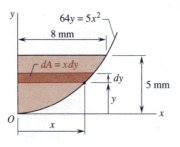

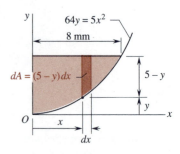

To determine I_y, we choose a first-order dA which is *parallel* to the y axis as shown. We write

$$dA = (5 - y)\,dx = \left[5 - \left(\frac{5}{64}x^2\right)\right]dx$$

$$I_y = \int x^2\,dA = \int_0^8 x^2\left[5 - \left(\frac{5}{64}x^2\right)\right]dx = \frac{1024}{3}$$

$$I_y = 341 \text{ mm}^4 \blacktriangleleft$$

By Eq. (7.4), we write

$$J_O = I_x + I_y = \frac{2000}{7} + \frac{1024}{3} = \frac{13168}{21}$$

$$J_O = 627 \text{ mm}^4 \blacktriangleleft$$

Developmental Exercises

D7.6 Determine the moments of inertia I_x and I_y of the triangular area shown.

D7.7 In Fig. 7.8(a), if $\bar{x}_{el}$ is the abscissa of the centroid of the first-order area element dA, is it correct to write $I_y = \int (\bar{x}_{el})^2\,dA$?

D7.8 In Fig. 7.8(b), if $\bar{y}_{el}$ is the ordinate of the centroid of the first-order area element dA, is it correct to write $I_x = \int (\bar{y}_{el})^2\,dA$?

D7.9 Refer to Fig. 6.6(a). Is $I_x = \int y^2\,(y\,dx)$?

D7.10 Refer to Fig. 6.6(b). Is $I_y = \int x^2\,(x\,dy)$?

D7.11 Refer to Fig. 6.6(f). Is $J_O = \int (2r/3)^2\,(\frac{1}{2}r^2\,d\theta)$?

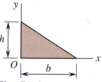

Fig. D7.6

7.3 Radius of Gyration of an Area

The *radius of gyration* of an area about an axis is a distance whose square multiplied by the area gives the moment of inertia of the area about that axis. For an area A in the xy plane as shown in Fig. 7.9, we write

$$I_x = k_x^2\,A \tag{7.5}$$

$$I_y = k_y^2\,A \tag{7.6}$$

$$J_O = k_O^2\,A \tag{7.7}$$

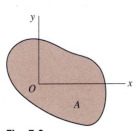

Fig. 7.9

where k_x, k_y, and k_O are the radii of gyration of A about the x axis, the y axis, and the pole O, respectively. Recalling from Eq. (7.4) that J_O is equal to the sum of I_x and I_y, we readily observe that

$$k_O^2 = k_x^2 + k_y^2 \tag{7.8}$$

For computing the moment of inertia about an axis, an area may be imagined to be a *thin areal strip*, with the same total area, whose elements are equidistant from that axis and the *equidistance* is equal to the radius of gyration of the area about that axis as shown in Fig. 7.10. The radius of gyration of an area about an axis is a measure of the spread of the area from that axis.[†]

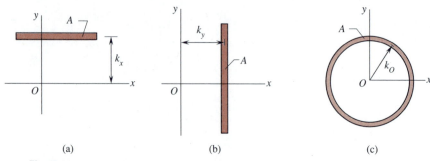

(a)   (b)   (c)

Fig. 7.10

EXAMPLE 7.4

Determine the radii of gyration k_x, k_y, and k_O of the cross-sectional area in Example 7.3.

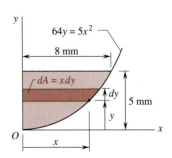

Solution. We first determine the area A of that cross section as follows:

$$dA = x \, dy = \left(\frac{64}{5}\right)^{1/2} y^{1/2} \, dy$$

$$A = \int dA = \left(\frac{64}{5}\right)^{1/2} \int_0^5 y^{1/2} \, dy \qquad A = \frac{80}{3} \text{ mm}^2$$

From the solution in Example 7.3, we write

$$I_x = \frac{2000}{7} \text{ mm}^4 \qquad I_y = \frac{1024}{3} \text{ mm}^4 \qquad J_O = \frac{13168}{21} \text{ mm}^4$$

$$\frac{2000}{7} = k_x^2 \left(\frac{80}{3}\right) \qquad \frac{1024}{3} = k_y^2 \left(\frac{80}{3}\right) \qquad \frac{13168}{21} = k_O^2 \left(\frac{80}{3}\right)$$

$$k_x = 3.27 \text{ mm} \qquad k_y = 3.58 \text{ mm} \qquad k_O = 4.85 \text{ mm} \quad \blacktriangleleft$$

Developmental Exercises

D7.12 Define the radius of gyration of an area about an axis.

D7.13 Determine the centroidal radii of gyration $\bar{k}_x$ and $\bar{k}_y$ of the rectangular area in Example 7.1.

D7.14 Determine the centroidal polar radius of gyration $\bar{k}_O$ of the circular area in Example 7.2.

[†]A centroidal radius of gyration of an area represents the *standard deviation* of the distances of the area elements from the centroidal axis.

PROBLEMS

7.1 through 7.6   Using *double integration*, determine the moment of inertia I_x of the area shown.

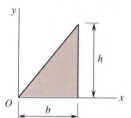

Fig. P7.1, P7.7*,
P7.13, and P7.19*

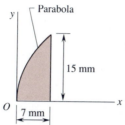

Fig. P7.2, P7.8,
P7.14, and P7.20*

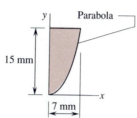

Fig. P7.3, P7.9,
P7.15*, and P7.21*

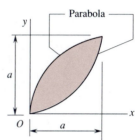

Fig. P7.4, P7.10,
P7.16, and P7.22*

7.7* through 7.12*   Using *double integration*, determine the moment of inertia I_y of the area shown.

7.13 through 7.18*   Using *double integration*, determine the polar moment of inertia J_O of the area shown.

7.19* through 7.24*   Using *single integration*, determine the moments of inertia I_x and I_y of the area shown.

$$y = x(2-x)$$

Fig. P7.5, P7.11, P7.17, and P7.23*

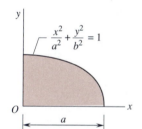

Fig. P7.6, P7.12*,
P7.18*, and P7.24*

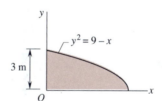

Fig. P7.25

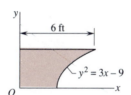

Fig. P7.26

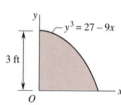

Fig. P7.27*

7.25 through 7.27*   Determine the moment of inertia I_x and the radius of gyration k_x of the area shown.

7.28 through 7.30*   Determine the moment of inertia I_y and the radius of gyration k_y of the area shown.

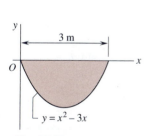

Fig. P7.28

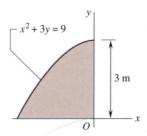

Fig. P7.29

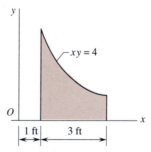

Fig. P7.30*

7.31 through 7.33* Determine the moment of inertia I_x and the radius of gyration k_x of the area shown.

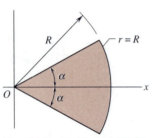

Fig. P7.31, P7.34, and P7.37

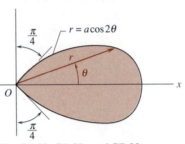

Fig. P7.32, P7.35, and P7.38

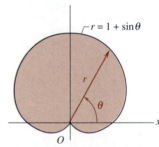

Fig. P7.33*, P7.36*, and P7.39*

7.34 through 7.36* Determine the moment of inertia I_y and the radius of gyration k_y of the area shown.

7.37 through 7.39* Determine the polar moment of inertia J_O and the polar radius of gyration k_O of the area shown.

7.4 Parallel-Axis Theorem: Area Moments of Inertia

Let the point $C(\bar{x}, \bar{y})$ be the centroid of an area A in the xy plane as shown in Fig. 7.11, where the centroidal axes x' and y' are parallel to the coordinate axes x and y, respectively. From this figure, we write

$$ y = y' + \bar{y} \qquad x = x' + \bar{x} $$

By definitions in Sec. 7.2 and the *principle of moments*, we write

$$ I_x = \int y^2 \, dA = \int (y' + \bar{y})^2 \, dA $$
$$ = \int y'^2 \, dA + 2\bar{y} \int y' \, dA + \bar{y}^2 \int dA $$
$$ = \bar{I}_{x'} + 2\bar{y}(\bar{y}'A) + \bar{y}^2 A $$
$$ I_y = \int x^2 \, dA = \int (x' + \bar{x})^2 \, dA $$
$$ = \int x'^2 \, dA + 2\bar{x} \int x' \, dA + \bar{x}^2 \int dA $$
$$ = \bar{I}_{y'} + 2\bar{x}(\bar{x}'A) + \bar{x}^2 A $$
$$ J_O = I_x + I_y = (\bar{I}_{x'} + \bar{I}_{y'}) + 2(\bar{x}\bar{x}' + \bar{y}\bar{y}')A + (\bar{x}^2 + \bar{y}^2)A $$

where $\bar{x}'$ and $\bar{y}'$ are the coordinates of the centroid of the area in the $Cx'y'$ coordinate system. Since the point C is the centroid of the area A, we have $\bar{x}' = \bar{y}' = 0$. We note from Fig. 7.11 that $\bar{J}_C = \bar{I}_{x'} + \bar{I}_{y'}$ and $\bar{r}^2 = \bar{x}^2 + \bar{y}^2$. Thus, the above equations reduce to

$$ \boxed{I_x = \bar{I}_{x'} + A\bar{y}^2} \qquad \boxed{I_y = \bar{I}_{y'} + A\bar{x}^2} \qquad \boxed{J_O = \bar{J}_C + A\bar{r}^2} \qquad (7.9) $$

These results are referred to as the *parallel-axis theorem for area moments of inertia*, which may be stated as follows: *The moment of inertia I of an*

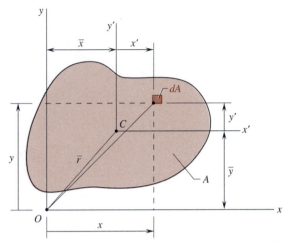

Fig. 7.11

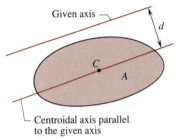

area A about a given axis is equal to the moment of inertia $\bar{I}$ of the area about a centroidal axis parallel to the given axis plus the product Ad^2 of the area A and the square of the distance d between the two axes, as shown in Fig. 7.12; i.e.,

$$I = \bar{I} + Ad^2 \qquad (7.10)$$

Fig. 7.12

If a, b, and c are three parallel axes and the axis c is a centroidal axis of the area A as shown in Fig. 7.13, then the moments of inertia I_a, I_b, and $\bar{I}_c$ of the area A have the following relations:

$$I_a = \bar{I}_c + Ad_{ca}^2 \qquad I_b = \bar{I}_c + Ad_{cb}^2$$

$$I_a = I_b + A(d_{ca}^2 - d_{cb}^2) \qquad (7.11)$$

where d_{ca} and d_{cb} are the distances from the centroidal axis c to the parallel axes a and b, respectively. Equations (7.10) and (7.11) are useful *transfer formulas* for moments of inertia.

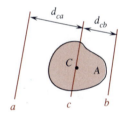

Fig. 7.13

EXAMPLE 7.5

The area considered in Example 7.3 is shown here for further study. Determine (a) I_x by choosing dA to be an areal strip perpendicular to the x axis, (b) I_y by choosing dA to be an areal strip perpendicular to the y axis. (Note in Example 7.3 that the areal strip dA in each case was chosen to be *parallel* to the axis about which the moment of inertia is computed.)

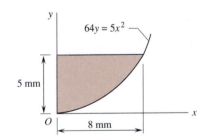

Solution. The areal strip dA perpendicular to the x axis as shown may be treated as a rectangular area of width $b = dx$ and height $h = (5 - y)$. The distance between the horizontal centroidal x'' axis of dA and the x axis is $d = \frac{1}{2}(5 + y)$. Thus, applying the formula $I_x = \frac{1}{12} bh^3$ obtained in Example 7.1 and the parallel-axis theorem as expressed in Eq. (7.10), we write

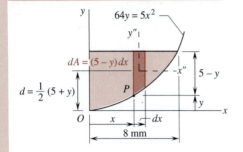

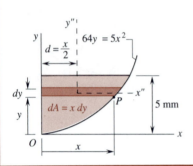

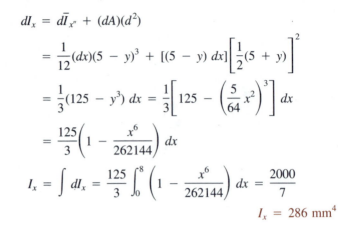

$$dI_x = d\bar{I}_{x''} + (dA)(d^2)$$

$$= \frac{1}{12}(dx)(5 - y)^3 + [(5 - y)\,dx]\left[\frac{1}{2}(5 + y)\right]^2$$

$$= \frac{1}{3}(125 - y^3)\,dx = \frac{1}{3}\left[125 - \left(\frac{5}{64}x^2\right)^3\right]dx$$

$$= \frac{125}{3}\left(1 - \frac{x^6}{262144}\right)dx$$

$$I_x = \int dI_x = \frac{125}{3}\int_0^8\left(1 - \frac{x^6}{262144}\right)dx = \frac{2000}{7}$$

$$I_x = 286 \text{ mm}^4 \blacktriangleleft$$

Next, we similarly let the areal strip dA be perpendicular to the y axis as shown. Treating dA as a rectangular area, we write

$$dI_y = d\bar{I}_{y''} + (dA)(d^2)$$

$$= \frac{1}{12}(dy)(x^3) + (x\,dy)\left(\frac{x}{2}\right)^2 = \frac{1}{3}x^3\,dy$$

$$= \frac{1}{3}\left(\frac{64}{5}y\right)^{3/2}dy = \frac{512}{15\sqrt{5}}y^{3/2}\,dy$$

$$I_y = \int dI_y = \frac{512}{15\sqrt{5}}\int_0^5 y^{3/2}\,dy = \frac{1024}{3}$$

$$I_y = 341 \text{ mm}^4 \blacktriangleleft$$

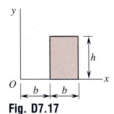

Fig. D7.17

Fig. D7.18

Developmental Exercises

D7.15 Describe the parallel-axis theorem for area moments of inertia.

D7.16 Using Eqs. (7.9), show that the coordinates $\bar{x}$ and $\bar{y}$ of the centroid of an area are generally less than its radii of gyration k_y and k_x, respectively (i.e., $\bar{y} < k_x$, $\bar{x} < k_y$).

D7.17 A rectangular area is shown. Using the results in Example 7.1 and the parallel-axis theorem, determine (a) I_x, (b) I_y, and (c) J_O of this rectangular area.

D7.18 For the area A with its centroid at C and the parallel axes a and b as shown, it is known that $I_a = 25$ ft^4. What is the value of I_b?

7.5 Product of Inertia of an Area

The product of inertia of an area is defined with respect to a specified pair of perpendicular axes which lie in the same plane as that of the area. The *product of inertia of an area A with respect to the x and y axes*, as shown in Fig. 7.14, is denoted by I_{xy} and is defined as the integral of the product

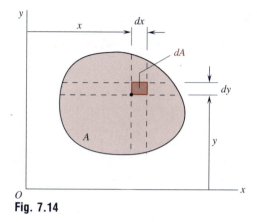

Fig. 7.14

of the *second-order* differential area element dA and the coordinates x and y defining the location of dA; i.e.,

$$I_{xy} = \int xy \, dA \qquad (7.12)^{\dagger}$$

Since dA in Eq. (7.12) is *not of the first order*, the basic definition for I_{xy} would call for *double integration*. However, I_{xy} may be determined with *single integration* via a theorem in Sec. 7.6. Note that dA is inherently positive, but x and y may be positive, negative, or zero. Consequently, the product of inertia of an area may be *positive*, *negative*; or *zero*. The dimensions of the product of inertia of an area are clearly $(length)^4$.

   When an area A is symmetrical with respect to one or both of the x and y axes, the product of inertia I_{xy} of A is zero. This is true because the contributions to the product of inertia from each pair of area elements dA and dA', as shown in Fig. 7.15, which are on the opposite sides of, and located at equal distances from, the (x or y) axis of symmetry cancel out. Thus, the integral for this product of inertia reduces to zero.[††]

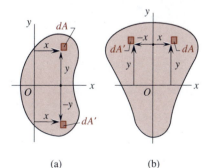

Fig. 7.15   (a)   (b)

[†]In tensor analysis, the product of inertia is mathematically defined as $I_{xy} = -\int xy \, dA$ so that the laws of transformation of second-order tensors may be followed by the moments and products of inertia.

[††]The product of inertia I_{xy} may be taken as a measure of *dissymmetry* or *imbalance* of the area with respect to the x and y axes.

EXAMPLE 7.6

Determine the product of inertia I_{xy} of the cross-sectional area in Example 7.3.

Solution. The area with a *second-order dA* is sketched as shown. By definition, we write

$$I_{xy} = \int xy \, dA = \int_{y=0}^{5} \int_{x=0}^{(64y/5)^{1/2}} xy \, dx \, dy$$

$$= \int_0^5 \left(\frac{1}{2} x^2 y \right) \Bigg|_{x=0}^{x=(64y/5)^{1/2}} dy$$

$$= \int_0^5 \frac{1}{2} \left(\frac{64}{5} \, y \right) y \, dy = \frac{800}{3} \qquad I_{xy} = 267 \text{ mm}^4 \quad \blacktriangleleft$$

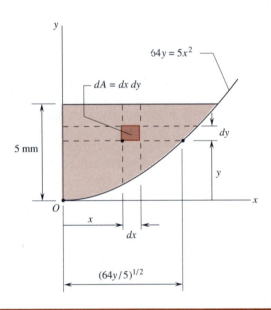

Developmental Exercises

D7.19   Define the product of inertia I_{xy} of an area A.

D7.20   Determine the product of inertia I_{xy} of each of the areas shown.

D7.21   Determine the product of inertia I_{xy} of the area shown.

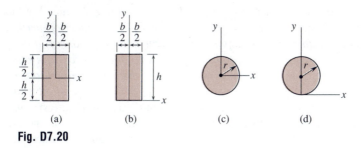

Fig. D7.20

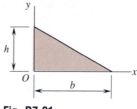

Fig. D7.21

7.6 Parallel-Axis Theorem: Area Products of Inertia

An area A in the xy plane with its centroid at the point $C(\bar{x}, \bar{y})$ is shown in Fig. 7.16. Noting the dA in this figure and Eq. (7.12), we write

$$I_{xy} = \int xy \, dA = \int (x' + \bar{x})(y' + \bar{y}) \, dA$$

$$= \int x'y' \, dA + \bar{y} \int x' \, dA + \bar{x} \int y' \, dA + \bar{x}\bar{y} \int dA$$

$$= \bar{I}_{x'y'} + \bar{y}(\bar{x}'A) + \bar{x}(\bar{y}'A) + \bar{x}\bar{y}A$$

$$= \bar{I}_{x'y'} + 0 + 0 + \bar{x}\bar{y}A$$

$$\boxed{I_{xy} = \bar{I}_{x'y'} + A\bar{x}\,\bar{y}} \qquad (7.13)$$

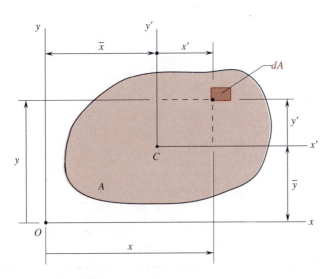

Fig. 7.16

This result is referred to as the *parallel-axis theorem* for area products of inertia. This theorem may be stated as follows: *The product of inertia I_{xy} of an area A with respect to any two perpendicular axes x and y in the plane of the area is equal to the product of inertia $\bar{I}_{x'y'}$ of the area with respect to two perpendicular centroidal axes x' and y', which are parallel to the x and y axes, respectively, plus the product $A\bar{x}\bar{y}$ of the area A and the two coordinates $\bar{x}$ and $\bar{y}$ of the centroid of the area in the xy coordinate system.* Note that $\bar{x}$ and $\bar{y}$ may be positive, negative, or zero depending on the location of the centroid C; however, the area A is always positive unless it is an empty area of a composite area.

EXAMPLE 7.7

Using *single integration*, determine the product of inertia I_{xy} of the cross-sectional area in Example 7.3. (Cf. Example 7.6.)

Solution. The area with a first-order dA is shown. The centroid of dA is at $(x/2, y)$, and the differential centroidal product of inertia $d\bar{I}_{x''y''}$ of dA is, by symmetry, equal to zero. Applying the *parallel-axis theorem* for area products of inertia, we write

$$dI_{xy} = d\bar{I}_{x''y''} + (dA)\left(\frac{x}{2}\right)(y) = 0 + \frac{1}{2}x^2 y\, dy = \frac{32}{5}y^2\, dy$$

$$I_{xy} = \int dI_{xy} = \frac{32}{5}\int_0^5 y^2\, dy = \frac{800}{3} \qquad I_{xy} = 267\text{ mm}^4 \blacktriangleleft$$

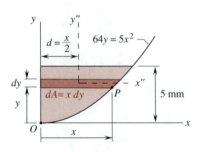

EXAMPLE 7.8

For the area shown, it is known from the solutions in Examples 7.3, 7.4, and 7.6 that $I_x = 2000/7$ mm^4, $I_y = 1024/3$ mm^4, $J_O = 13168/21$ mm^4, $A = 80/3$ mm^2, and $I_{xy} = 800/3$ mm^4. Determine for the area and axes shown the following quantities: (a) $\bar{I}_{x'}$, (b) $\bar{I}_{y'}$, (c) $\bar{J}_C$, (d) $\bar{I}_{x'y'}$, (e) $\bar{k}_{x'}$, (f) $\bar{k}_{y'}$, (g) $\bar{k}_C$.

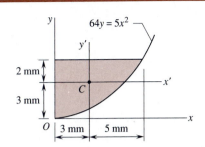

Solution. Applying the parallel-axis theorems, we write

$$\frac{2000}{7} = \bar{I}_{x'} + \frac{80}{3}(3)^2 \qquad \bar{I}_{x'} = \frac{320}{7} \qquad \bar{I}_{x'} = 45.7 \text{ mm}^4 \blacktriangleleft$$

$$\frac{1024}{3} = \bar{I}_{y'} + \frac{80}{3}(3)^2 \qquad \bar{I}_{y'} = \frac{304}{3} \qquad \bar{I}_{y'} = 101.3 \text{ mm}^4 \blacktriangleleft$$

$$\frac{13168}{21} = \bar{J}_C + \frac{80}{3}(3^2 + 3^2) \qquad \bar{J}_C = \frac{3088}{21} \qquad \bar{J}_C = 147.0 \text{ mm}^4 \blacktriangleleft$$

$$\frac{800}{3} = \bar{I}_{x'y'} + \frac{80}{3}(3)(3) \qquad \bar{I}_{x'y'} = \frac{80}{3} \qquad \bar{I}_{x'y'} = 26.7 \text{ mm}^4 \blacktriangleleft$$

Note that $\bar{J}_C = \bar{I}_{x'} + \bar{I}_{y'}$ is satisfied. The centroidal radii of gyration $\bar{k}_{x'}$, $\bar{k}_{y'}$, and $\bar{k}_C$ may be determined as follows:

$$\bar{I}_{x'} = \bar{k}_{x'}^2 A: \qquad \frac{320}{7} = \bar{k}_{x'}^2\left(\frac{80}{3}\right) \qquad \bar{k}_{x'} = 1.309 \text{ mm} \blacktriangleleft$$

$$\bar{I}_{y'} = \bar{k}_{y'}^2 A: \qquad \frac{304}{3} = \bar{k}_{y'}^2\left(\frac{80}{3}\right) \qquad \bar{k}_{y'} = 1.949 \text{ mm} \blacktriangleleft$$

$$\bar{J}_C = \bar{k}_C^2 A: \qquad \frac{3088}{21} = \bar{k}_C^2\left(\frac{80}{3}\right) \qquad \bar{k}_C = 2.35 \text{ mm} \blacktriangleleft$$

Developmental Exercises

D7.22 Describe the parallel-axis theorem for area products of inertia.

D7.23 For the area and the area element dA shown, determine (a) dA, (b) dI_{xy}, (c) I_{xy}.

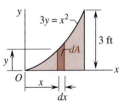

Fig. D7.23

7.7 Composite Areas

We know that a composite area has two or more component areas. The integrals representing the moments and the product of inertia of a composite area A are to be evaluated over its component areas $A_1, A_2, \ldots, A_n$. Thus, the moment of inertia I of the composite area A about a given axis is equal to the sum of the moments of inertia $I_1, I_2, \ldots, I_n$ of its component areas about the same axis; i.e.,

$$I = I_1 + I_2 + \cdots + I_n \tag{7.14}$$

Table 7.1 Properties of Some Rolled-Steel Structural Sections* (Refer to Fig. 7.17.)

Designation	Area A, in^2	Depth d, in.	Width b, in.	I_x in^4	I_y in^4	$\bar{x}$ in.	$\bar{y}$ in.
W 21 × 44	13.0	20.66	6.500	843	20.7		
W 16 × 31	9.12	15.88	5.525	375	12.4		
W 12 × 26	7.65	12.22	6.490	204	17.3		
W 10 × 30	8.84	10.47	5.810	170	16.7		
W 8 × 21	6.16	8.28	5.270	75.3	9.77		
S 18 × 70	20.6	18.00	6.251	926	24.1		
S 12 × 50	14.7	12.00	5.477	305	15.7		
S 12 × 35	10.3	12.00	5.078	229	9.87		
S 8 × 23	6.77	8.00	4.171	64.9	4.31		
S 7 × 20	5.88	7.00	3.860	42.4	3.17		
C 15 × 40	11.8	15.00	3.520	349	9.23	0.777	
C 15 × 33.9	9.96	15.00	3.400	315	8.13	0.787	
C 10 × 20	5.88	10.00	2.739	78.9	2.81	0.606	
C 9 × 15	4.41	9.00	2.485	51.0	1.93	0.586	
C 6 × 13	3.83	6.00	2.157	17.4	1.05	0.514	
		Weight lb/ft					
L 8 × 8 × 1	15.0	51.0		89.0	89.0	2.37	2.37
L 6 × 6 × 1	11.0	37.4		35.5	35.5	1.86	1.86
L 5 × 5 × $\frac{1}{2}$	4.75	16.2		11.3	11.3	1.43	1.43
L 8 × 6 × $\frac{3}{4}$	9.94	33.8		63.4	30.7	1.56	2.56
L 6 × 4 × $\frac{3}{4}$	6.94	23.6		24.5	8.68	1.08	2.08

*Courtesy of the American Institute of Steel Construction, Chicago, Ill.

If I is the product of inertia of the composite area A with respect to two given perpendicular axes, it may similarly be determined by adding algebraically the products of inertia I_1, I_2, . . . , I_n of its component areas with respect to those two axes; i.e., Eq. (7.14) holds true for both the moments and the product of inertia of a composite area.

Sometimes, it is desirable to let a composite area have *empty areas* (i.e., negative areas) as some of its component areas. In such a case, the values of the moments and products of inertia of the *empty areas* should change their signs before they are algebraically added to the values of those of the other component areas to obtain the moments and the product of inertia of the composite area.

The cross sections of many structural members are of the form of composite areas. The moment of inertia of the cross-sectional area of a structural member is an important property of the member. Some typical rolled-steel structural sections are shown in Fig. 7.17, and their properties for designing are shown in Table 7.1. Furthermore, the formulas for centroidal moments and products of inertia in Fig. 7.18 along with the parallel-axis theorems may be applied to determine the moments and products of inertia of composite areas in many problems.

242

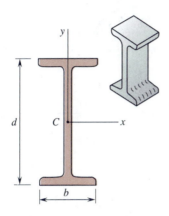

Wide-Flange Shapes (W Shapes)

A Wide-Flange Shape is designated by the letter
W followed first by its *nominal* depth in inches
and then its weight in lb/ft; e.g., W 16 x 31 designates
a Wide-Flange Shape which has a *nominal* depth of
16 in. and weighs 31 lb/ft. Its actual depth may be
greater or less than its nominal depth. (Cf. Table 7.1.)

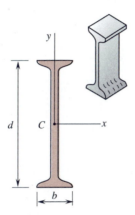

American Standard Shapes (S Shapes)

An American Standard Shape is designated by the
letter S followed first by its depth in inches
and then its weight in lb/ft; e.g., S 18 x 70
designates an American Standard Shape which is
18 in. deep and weighs 70 lb/ft.

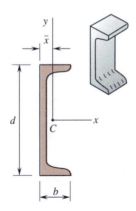

American Standard Channels (C Shapes)

An American Standard Channel is designated by the
letter C followed first by its depth in inches
and then its weight in lb/ft; e.g., C 15 x 33.9
designates an American Standard Channel which is
15 in. deep and weighs 33.9 lb/ft.

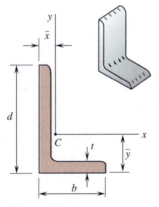

Angles (L Shapes)

An Angle is designated by the letter L followed
first by the lengths d and b of its legs and then
the thickness of its legs, all in inches; e.g.,
L 8 x 6 x $\frac{3}{4}$ designates an Angle which has one leg
equal to 8 in. long, the other leg equal to 6 in.
long, and the thickness of its legs equal to $\frac{3}{4}$ in.
Some angles may have equal legs.

Fig. 7.17   Some rolled-steel structural sections.

Parallelogram and Rectangular area

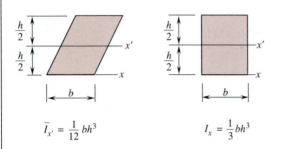

$$\bar{I}_{x'} = \frac{1}{12}bh^3 \qquad I_x = \frac{1}{3}bh^3$$

Elliptic area

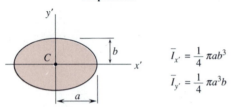

$$\bar{I}_{x'} = \frac{1}{4}\pi ab^3$$
$$\bar{I}_{y'} = \frac{1}{4}\pi a^3 b$$

Triangular area

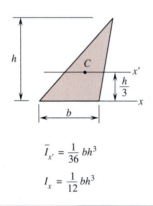

$$\bar{I}_{x'} = \frac{1}{36}bh^3$$
$$I_x = \frac{1}{12}bh^3$$

Semicircular area

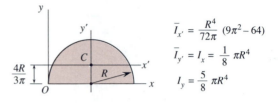

$$\bar{I}_{x'} = \frac{R^4}{72\pi}(9\pi^2 - 64)$$
$$\bar{I}_{y'} = I_x = \frac{1}{8}\pi R^4$$
$$I_y = \frac{5}{8}\pi R^4$$

Circular area

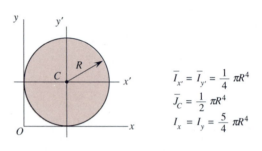

$$\bar{I}_{x'} = \bar{I}_{y'} = \frac{1}{4}\pi R^4$$
$$J_C = \frac{1}{2}\pi R^4$$
$$I_x = I_y = \frac{5}{4}\pi R^4$$

Right triangular areas

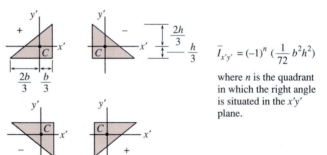

$$\bar{I}_{x'y'} = (-1)^n \left(\frac{1}{72}b^2h^2\right)$$

where n is the quadrant in which the right angle is situated in the $x'y'$ plane.

Quarter circular areas

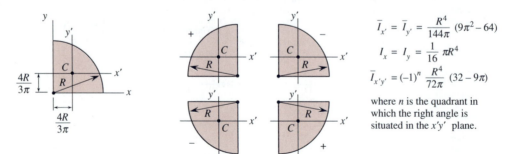

$$\bar{I}_{x'} = \bar{I}_{y'} = \frac{R^4}{144\pi}(9\pi^2 - 64)$$
$$I_x = I_y = \frac{1}{16}\pi R^4$$
$$\bar{I}_{x'y'} = (-1)^n \frac{R^4}{72\pi}(32 - 9\pi)$$

where n is the quadrant in which the right angle is situated in the $x'y'$ plane.

Fig. 7.18 Moments and products of inertia of areas of common shapes.

EXAMPLE 7.9

To increase the strength of the S7×20 rolled-steel beam, a rectangular plate is fastened to its upper flange as shown. Determine $\bar{I}_{x'}$ and $\bar{k}_{x'}$ of the composite section of the beam with respect to its centroidal axis x' as indicated.

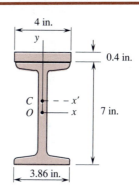

Solution. The area of the rectangular plate is $A_1 = 1.6 \text{ in}^2$. From Table 7.1, the area of S7×20 is $A_2 = 5.88 \text{ in}^2$. The total area is $A = A_1 + A_2 = 7.48 \text{ in}^2$. By inspection, $\bar{y}_1 = 3.7$ in. and $\bar{y}_2 = 0$. The ordinate $\bar{y}$ is determined from the *principle of moments*: $\bar{y}A = \bar{y}_1 A_1 + \bar{y}_2 A_2$. We write

$$\bar{y}(7.48) = 3.7(1.6) + 0(5.88) \qquad \bar{y} = 0.791 \text{ in.}$$

As shown in the figure, we have

$$d_1 = \bar{y}_1 - \bar{y} = 2.909 \text{ in.} \qquad d_2 = \bar{y} = 0.791 \text{ in.}$$

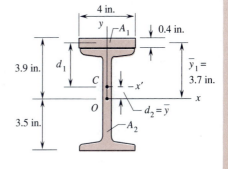

From Table 7.1, the centroidal moment of inertia of S7×20 is $(\bar{I}_x)_2 = 42.4 \text{ in}^4$. Applying the *parallel-axis theorem,* we write

$$(\bar{I}_{x'})_1 = \frac{1}{12} bh^3 + A_1 d_1^2 = \frac{1}{12}(4)(0.4)^3 + 1.6(2.909)^2 = 13.56$$

$$(\bar{I}_{x'})_2 = (\bar{I}_x)_2 + A_2 d_2^2 = 42.4 + 5.88(0.791)^2 = 46.08$$

For the composite section, we write

$$\bar{I}_{x'} = (\bar{I}_{x'})_1 + (\bar{I}_{x'})_2 \qquad \bar{I}_{x'} = 59.6 \text{ in}^4 \quad \blacktriangleleft$$

$$\bar{I}_{x'} = \bar{k}_x^2 A \qquad \bar{k}_{x'} = 2.82 \text{ in.} \quad \blacktriangleleft$$

Developmental Exercises

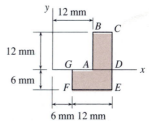

D7.24 Determine I_x and k_x of the composite area shown by treating the areas $ABCD$ and $GDEF$ as the component areas.

D7.25 Determine I_y and k_y of the composite area shown.

Fig. D7.24 and D7.25

EXAMPLE 7.10

A composite area is shown. Making use of the solutions in Examples 7.1, 7.3, 7.4, and 7.6, determine for the composite area and axes shown the following quantities: (a) I_x, (b) I_{xy}, (c) k_x.

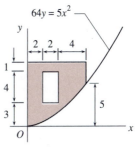

Dimensions in mm

Solution. We may treat this composite area as the algebraic sum of three component areas A_1, A_2, and A_3 as shown, where A_3 is an empty area. From the solutions in Examples 7.1, 7.3, 7.4, and 7.6, we write

$$A_1 = \frac{80}{3} \text{ mm}^2 \qquad (I_x)_1 = \frac{2000}{7} \text{ mm}^4 \qquad (I_{xy})_1 = \frac{800}{3} \text{ mm}^4$$

$$A_2 = 24 \text{ mm}^2 \qquad A_3 = 8 \text{ mm}^2$$

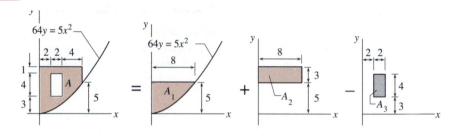

Dimensions in mm

$$(I_x)_2 = \frac{1}{12}(8)(3)^3 + A_2\left(5 + \frac{3}{2}\right)^2 \qquad (I_x)_2 = 1032 \text{ mm}^4$$

$$(I_{xy})_2 = 0 + A_2\left(\frac{8}{2}\right)\left(5 + \frac{3}{2}\right) \qquad (I_{xy})_2 = 624 \text{ mm}^4$$

$$(I_x)_3 = \frac{1}{12}(2)(4)^3 + A_3\left(3 + \frac{4}{2}\right)^2 \qquad (I_x)_3 = \frac{632}{3} \text{ mm}^4$$

$$(I_{xy})_3 = 0 + A_3\left(2 + \frac{2}{2}\right)\left(3 + \frac{4}{2}\right) \qquad (I_{xy})_3 = 120 \text{ mm}^4$$

Thus, for the composite area, we write

$$A = A_1 + A_2 - A_3 = 42.67 \text{ mm}^2$$

$$I_x = (I_x)_1 + (I_x)_2 - (I_x)_3 \qquad\qquad I_x = 1107 \text{ mm}^4 \blacktriangleleft$$

$$I_{xy} = (I_{xy})_1 + (I_{xy})_2 - (I_{xy})_3 \qquad\qquad I_{xy} = 771 \text{ mm}^4 \blacktriangleleft$$

$$I_x = k_x^2 A \qquad\qquad k_x = 5.09 \text{ mm} \blacktriangleleft$$

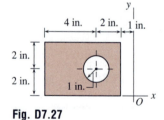

Fig. D7.27

Developmental Exercises

D7.26 Refer to Example 7.10. Determine (a) I_y, (b) k_y.

D7.27 Determine the values of I_x, I_y, and I_{xy} of the composite area shown.

EXAMPLE 7.11[†]

The vertices of an arbitrary triangle are located at $P(x_1, y_1)$, $Q(x_2, y_2)$, and $R(x_3, y_3)$. Determine I_x of the triangular area PQR.

Solution. Treating $\triangle PQR$ as a composite area as shown, we write

$$\triangle PQR = \square PSTU - \triangle PSQ - \triangle QTR - \triangle RUP$$

[†]The formulas for I_x, I_y, and I_{xy} in Example 7.11 and D7.28 and D7.29 are useful in determining the moments and the product of inertia of *any* plane area using *triangular finite elements* (e.g., Probs. 7.67 through 7.78*).

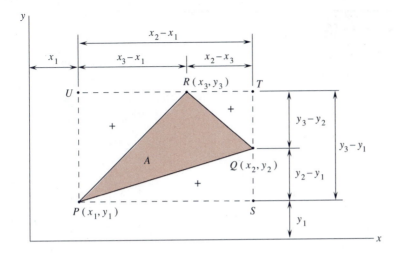

Applying appropriate formulas in Fig. 7.18 and the parallel-axis theorems, we write

$$I_x = \left[\frac{1}{12}(x_2 - x_1)(y_3 - y_1)^3 + (x_2 - x_1)(y_3 - y_1)\left(\frac{y_3 + y_1}{2}\right)^2\right]$$

$$- \left[\frac{1}{36}(x_2 - x_1)(y_2 - y_1)^3 + \frac{1}{2}(x_2 - x_1)(y_2 - y_1)\left(y_1 + \frac{y_2 - y_1}{3}\right)^2\right]$$

$$- \left[\frac{1}{36}(x_2 - x_3)(y_3 - y_2)^3 + \frac{1}{2}(x_2 - x_3)(y_3 - y_2)\left(y_3 - \frac{y_3 - y_2}{3}\right)^2\right]$$

$$- \left[\frac{1}{36}(x_3 - x_1)(y_3 - y_1)^3 + \frac{1}{2}(x_3 - x_1)(y_3 - y_1)\left(y_3 - \frac{y_3 - y_1}{3}\right)^2\right]$$

which can be simplified and expressed as

$$I_x = \frac{A}{12}[(y_1 + y_2)^2 + (y_2 + y_3)^2 + (y_3 + y_1)^2] \quad \blacktriangleleft$$

where A is the area of $\triangle PQR$ and is (cf. Example 6.7) given by

$$A = \frac{1}{2}|x_1(y_2 - y_3) + x_2(y_3 - y_1) + x_3(y_1 - y_2)|$$

Developmental Exercises

D7.28 Refer to Example 7.11. Show that

$$I_y = \frac{A}{12}[(x_1 + x_2)^2 + (x_2 + x_3)^2 + (x_3 + x_1)^2]$$

D7.29 Refer to Example 7.11. Show that

$$I_{xy} = \frac{A}{12}[(x_1 + x_2 + x_3)(y_1 + y_2 + y_3) + x_1y_1 + x_2y_2 + x_3y_3]$$

PROBLEMS

7.40 through 7.42* Determine the moment of inertia I_y of the area shown. (*Hint.* Choose dA to be an areal strip perpendicular to the y axis.)

7.43 through 7.45* Determine the moment of inertia I_x of the area shown. (*Hint.* Choose dA to be an areal strip perpendicular to the x axis.)

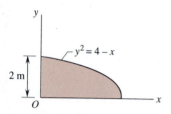

Fig. P7.40 and P7.46

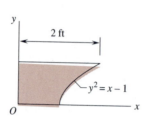

Fig. P7.41 and P7.47

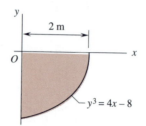

Fig. P7.42* and P7.48*

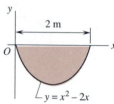

Fig. P7.43 and P7.49

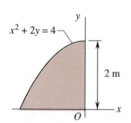

Fig. P7.44* and P7.50

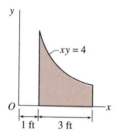

Fig. P7.45* and P7.51*

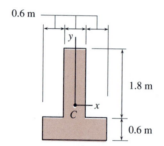

Fig. P7.52 and P7.58

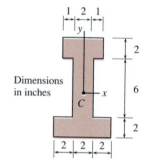

Fig. P7.53 and P7.59

7.46 through 7.51* Determine the product of inertia I_{xy} of the area shown. (*Hint.* One way is easier than the other in the integration involved.)

7.52 through 7.57* Determine the centroidal moment of inertia $\bar{I}_x$ and the centroidal radius of gyration $\bar{k}_x$ of the area shown.

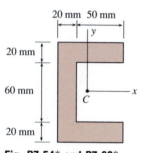

Fig. P7.54* and P7.60*

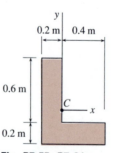

**Fig. P7.55, P7.61,
and P7.64**

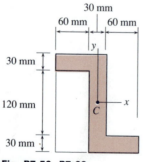

**Fig. P7.56, P7.62,
and P7.65**

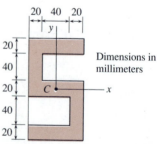

**Fig. P7.57*, P7.63*
and P7.66***

7.58 through 7.63* Determine the centroidal moment of inertia $\bar{I}_y$ and the centroidal radius of gyration $\bar{k}_y$ of the area shown.

7.64 through 7.66* Determine the centroidal product of inertia $\bar{I}_{xy}$ of the area shown.

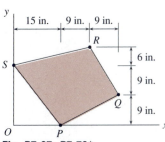

Fig. P7.67, P7.70*,
P7.73*, and P7.76

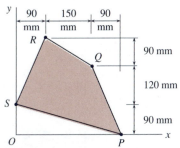

Fig. P7.68*, P7.71,
P7.74, and P7.77*

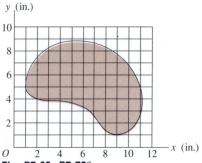

Fig. P7.69, P7.72*
P7.75, and P7.78*

7.67 through 7.69 Determine the moment of inertia I_x of the composite area shown.[†]

7.70* through 7.72* Determine the moment of inertia I_y of the composite area shown.[†]

7.73* through 7.75* Determine the product of inertia I_{xy} of the composite area shown.[†]

7.76 through 7.78* Determine the polar moment of inertia J_O of the composite area shown.[†]

7.79 through 7.81 Determine the centroidal moments of inertia $\bar{I}_x$ and $\bar{I}_y$ and polar moment of inertia $\bar{J}_C$ of the section shown.

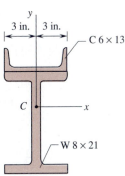

Fig. P7.79

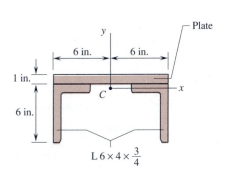

Fig. P7.80

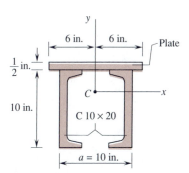

Fig. P7.81 and P7.82

7.82 For the section shown, determine the distance a for which $\bar{I}_x = \bar{I}_y$.

[†]*Suggestion.* Use *triangular finite elements* and apply the formulas in Example 7.11, D7.28, and D7.29. (The result may be checked by using the program AREA described in App. F.)

★7.8 Rotation of Axes

An area A in the xy plane is shown in Fig. 7.19, where the set of axes u and v is obtained by rotating the set of axes x and y about the origin O through an angle θ as indicated. We shall here derive the moments and the

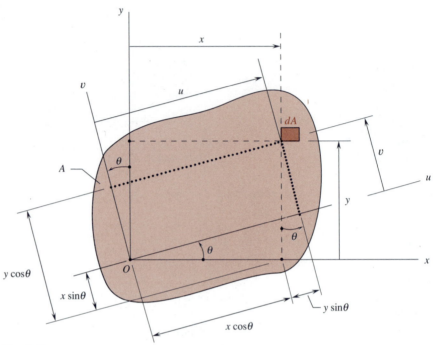

Fig. 7.19

product of inertia I_u, I_v, and I_{uv} in terms of I_x, I_y, I_{xy}, and θ. From Fig. 7.19, we have

$$u = x \cos\theta + y \sin\theta \qquad v = -x \sin\theta + y \cos\theta \qquad (7.15)$$

By the definitions of moments and products of inertia, we write

$$I_u = \int v^2 \, dA = \int (-x \sin\theta + y \cos\theta)^2 \, dA$$

$$I_u = I_x \cos^2\theta - 2I_{xy} \sin\theta \cos\theta + I_y \sin^2\theta \qquad (7.16)$$

$$I_v = \int u^2 \, dA = \int (x \cos\theta + y \sin\theta)^2 \, dA$$

$$I_v = I_y \cos^2\theta + 2I_{xy} \sin\theta \cos\theta + I_x \sin^2 \qquad (7.17)$$

$$I_{uv} = \int uv \, dA = \int (x \cos\theta + y \sin\theta)(-x \sin\theta + y \cos\theta) \, dA$$

$$I_{uv} = (I_x - I_y) \sin\theta \cos\theta + (\cos^2\theta - \sin^2\theta)I_{xy} \qquad (7.18)$$

Applying the trigonometric relations $2 \sin\theta \cos\theta = \sin 2\theta$, $\cos^2\theta - \sin^2\theta = \cos 2\theta$, $\cos^2\theta = \frac{1}{2}(1 + \cos 2\theta)$, and $\sin^2\theta = \frac{1}{2}(1 - \cos 2\theta)$ to Eqs. (7.16) through (7.18), we get the following relations:

$$I_u = \frac{1}{2}(I_x + I_y) + \frac{1}{2}(I_x - I_y)\cos 2\theta - I_{xy}\sin 2\theta$$

$$I_v = \frac{1}{2}(I_x + I_y) - \frac{1}{2}(I_x - I_y)\cos 2\theta + I_{xy}\sin 2\theta$$

(7.19)

$$I_{uv} = \frac{1}{2}(I_x - I_y)\sin 2\theta + I_{xy}\cos 2\theta$$

$$I_u + I_v = I_x + I_y$$

Developmental Exercises

D7.30   Is it true that the formula for I_v in Eqs. (7.19) may be obtained by replacing θ with $\theta + \pi/2$ in the formula for I_u?

D7.31   Refer to the area A in Fig. 7.19. Suppose that $I_x = 260$ mm^4, $I_y = 240$ mm^4, and $I_u = 200$ mm^4. What is the value of I_v?

★7.9   Representation by Mohr's Circle

The formulas in Eqs. (7.19) may graphically be represented by the so-called *Mohr's circle*,[†] after the German engineer Otto Mohr (or Christian Otto Mohr, 1835–1918). Mohr's circle for an area A in the xy plane, as shown in Fig. 7.20, is a circle which is drawn with the line segment $\overline{XY}$ as a diameter as shown in Fig. 7.21, where *the coordinates of X and Y are*

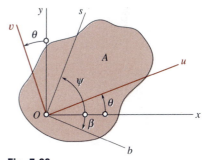

Fig. 7.20

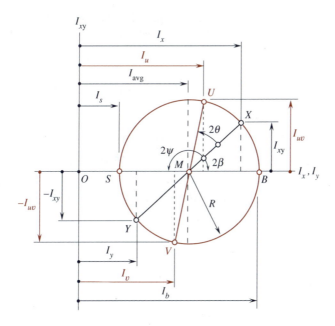

Fig. 7.21

[†]The graphical scheme of Mohr's circle is also used in studying stresses and strains in mechanics of materials, elasticity, etc.

(I_x, I_{xy}) and $(I_y, -I_{xy})$, *respectively.* The midpoint M of the diameter $\overline{XY}$ is on the horizontal axis and is the center of Mohr's circle. Thus, the points X and Y on Mohr's circle are related to the x and y axes on the area A, respectively. The points U and V in Fig. 7.21 are reached by moving the points X and Y counterclockwise along Mohr's circle until the angles XMU and YMV are both equal to 2θ. The points U and V on Mohr's circle are, therefore, related to the u and v axes on the area A, respectively. For the area A in Fig. 7.20, Mohr's circle in Fig. 7.21 indicates the following: *The abscissas of the points U and V are equal to I_u and I_v, respectively, and the ordinate of the point U (or the negative of the ordinate of the point V) is equal to I_{uv}.*

I. SIGN CONVENTION. On the area A and on Mohr's circle, the *positive directions* for corresponding angles are the *same*, although the *magnitudes* of the angles on Mohr's circle are *twice* as large as the corresponding angles on the area A. Note that the points B and S on Mohr's circle in Fig. 7.21 are related to the b and s axes on the area A in Fig. 7.20, respectively.[†] Furthermore, each set of coordinate axes in Fig. 7.20 and the z axis will form a right-handed system. We observe that (1) θ is always measured *counterclockwise* from the x axis to the u axis, (2) β is always measured *clockwise* from the x axis to the b axis, and (3) positive ψ is measured *counterclockwise* from the x axis to the s axis. In Figure 7.21, we note *correspondingly* that (1') 2θ is always measured *counterclockwise* from the radius $\overline{MX}$ to the radius $\overline{MU}$, (2') 2β is always measured *clockwise* from the radius $\overline{MX}$ to the radius $\overline{MB}$, and (3') positive 2ψ is measured *counterclockwise* from the radius $\overline{MX}$ to the radius $\overline{MS}$. From Fig. 7.21, we have

$$2\beta + 2\psi = 180° \tag{7.20}$$

Since 2θ and 2β are always positive, we see that 2ψ may be negative.[††]

II. PROOF. Referring to Fig. 7.21, we can verify the validity of representing the formulas in Eqs. (7.19) by Mohr's circle as follows:

$$I_{avg} = \tfrac{1}{2}(I_x + I_y) \tag{7.21}$$

$$R = [(I_x - I_{avg})^2 + I_{xy}^2]^{1/2} = \tfrac{1}{2}[(I_x - I_y)^2 + 4I_{xy}^2]^{1/2} \tag{7.22}$$

$$
\begin{aligned}
I_u &= I_{avg} + R\cos(2\theta + 2\beta) \\
&= I_{avg} + R(\cos 2\theta \cos 2\beta - \sin 2\theta \sin 2\beta) \\
&= I_{avg} + [(R\cos 2\beta)\cos 2\theta - (R\sin 2\beta)\sin 2\theta] \\
&= I_{avg} + [(I_x - I_{avg})\cos 2\theta - I_{xy}\sin 2\theta] \\
&= \tfrac{1}{2}(I_x + I_y) + \tfrac{1}{2}(I_x - I_y)\cos 2\theta - I_{xy}\sin 2\theta \qquad \text{Q.E.D.}[†††]
\end{aligned}
$$

[†]The names B and S are chosen to indicate that the moment of inertia I_b about the b axis is the *biggest*, while the moment of inertia I_s about the s axis is the *smallest*.

[††]Since 2β is always positive, we have $0 \leq 2\beta < 360°$. Thus, we find from Eq. (7.20) that $2\psi < 0$ if $2\beta > 180°$. When $2\psi < 0$, the actual rotation from $\overline{MX}$ to $\overline{MS}$ through -2ψ would be *clockwise* (cf. Example 7.12).

[†††]The abbreviation "Q.E.D." stands for the Latin words "quod erat demonstrandum," meaning "which was to be proved."

$$I_v = I_{avg} - R\cos(2\theta + 2\beta)$$
$$= I_{avg} - [(I_x - I_{avg})\cos2\theta - I_{xy}\sin2\theta]$$
$$= \tfrac{1}{2}(I_x + I_y) - \tfrac{1}{2}(I_x - I_y)\cos2\theta + I_{xy}\sin2\theta \qquad \text{Q.E.D.}$$

$$I_{uv} = R\sin(2\theta + 2\beta) = R(\sin2\theta\cos2\beta + \cos2\theta\sin2\beta)$$
$$= (R\cos2\beta)\sin2\theta + (R\sin2\beta)\cos2\theta$$
$$= (I_x - I_{avg})\sin2\theta + I_{xy}\cos2\theta$$
$$= \tfrac{1}{2}(I_x - I_y)\sin2\theta + I_{xy}\cos2\theta \qquad \text{Q.E.D.}$$

$$I_u + I_v = [\tfrac{1}{2}(I_x + I_y) + \tfrac{1}{2}(I_x - I_y)\cos2\theta - I_{xy}\sin2\theta]$$
$$+ \tfrac{1}{2}(I_x + I_y) - \tfrac{1}{2}(I_x - I_y)\cos2\theta + I_{xy}\sin2\theta]$$
$$= I_x + I_y \qquad \text{Q.E.D.}$$

EXAMPLE 7.12

For the Z section shown, it is known that $I_x = 592$ in^4, $I_y = 92$ in^4, and $I_{xy} = -150$ in^4. Using Mohr's circle, determine the moments and product of inertia I_u, I_v, and I_{uv} with respect to the axes indicated.

Solution. Using $X(I_x, I_{xy})$ and $Y(I_y, -I_{xy})$ as the end points of a diameter, we may draw Mohr's circle for the Z section about O as shown.[†] We see that

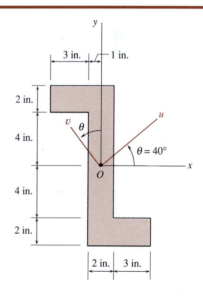

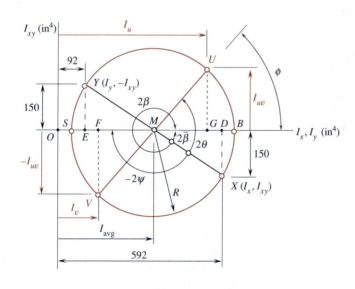

[†]The angles 2β and 2ψ discussed in Sec. 7.9 are also indicated on this circle for reference, where $2\beta = 360° - 2\bar{\beta} = 329.04°$, $2\beta + 2\psi = 180°$, and $2\psi = -149.04°$. These angles may be used as examples in studying Sec. 7.10, particularly Table 7.2.

$$I_{avg} = \tfrac{1}{2}(592 + 92) = 342$$

$$\overline{MD} = 592 - I_{avg} = 250 \qquad \overline{DX} = 150$$

$$R = (\overline{MD}^2 + \overline{DX}^2)^{1/2} = 291.5$$

$$\tan 2\bar{\beta} = \frac{\overline{DX}}{\overline{MD}} = \frac{150}{250} \qquad 2\bar{\beta} = 30.96°$$

$$\theta = 40° \qquad \phi = 2\theta - 2\bar{\beta} = 49.04°$$

Thus, we have

$$I_u = \overline{OG} = \overline{OM} + \overline{MG} = I_{avg} + R\cos\phi = 533.1$$

$$I_v = \overline{OF} = \overline{OM} - \overline{FM} = I_{avg} - R\cos\phi = 150.9$$

$$I_{uv} = \overline{GU} = R\sin\phi = 220.1$$

$$I_u = 533 \text{ in}^4 \qquad I_v = 150.9 \text{ in}^4 \qquad I_{uv} = 220 \text{ in}^4 \blacktriangleleft^\dagger$$

Developmental Exercises

D7.32 The points X and Y are the end points of a diameter of Mohr's circle for an area A and are related to the x and y axes on the area A. What are the coordinates of X and Y?

D7.33 Using the formulas in Eqs. (7.19), show that (a) $(I_u - I_{avg})^2 + I_{uv}^2 = R^2$, (b) $(I_v - I_{avg})^2 + I_{uv}^2 = R^2$, where I_{avg} and R are as defined in Eqs. (7.21) and (7.22).

D7.34 The equations established in D7.33 indicate that the locus of the points (I_u, I_{uv}) and $(I_v, -I_{uv})$ is a circle, called Mohr's circle, with its radius equal to R. What are the coordinates of the center of this circle?

★7.10 Principal Moments and Axes of Inertia

Referring to Fig. 7.21, we note that the maximum and minimum values of the moment of inertia I are given by the abscissas of the extreme right point B and the extreme left point S on Mohr's circle. They are

$$I_{max} = I_b = I_{avg} + R \qquad I_{min} = I_s = I_{avg} - R$$

$$I_{max} = \tfrac{1}{2}(I_x + I_y) + \left\{[\tfrac{1}{2}(I_x - I_y)]^2 + I_{xy}^2\right\}^{1/2} \tag{7.23}$$

$$I_{min} = \tfrac{1}{2}(I_x + I_y) - \left\{[\tfrac{1}{2}(I_x - I_y)]^2 + I_{xy}^2\right\}^{1/2} \tag{7.24}$$

These two extremum values of I are called the *principal moments of inertia* of the area A about the point O in Fig. 7.20.

---

†Cf. the answers obtained via a computer program in Fig. 7.22.

The axes b and s with respect to which the moments of inertia are maximum and minimum are called the *principal axes of inertia* of the area A about the point O. The directions of the principal axes are called the *principal directions*. Letting θ_b and θ_s be the *directional angles* of the principal axes b and s with respect to the x axis in Fig. 7.20, we write

$$\theta_b = 360° - \beta \qquad \theta_s = \psi$$

From Fig. 7.21, we see that

$$\tan 2\beta = \frac{I_{xy}}{I_x - I_{avg}} \qquad \tan 2\beta = \frac{2I_{xy}}{I_x - I_y} \qquad (7.25)$$

where the ranges of values of 2β are summarized in Table 7.2.

Table 7.2   Ranges of Values of 2β

$I_{xy} \geq 0$	$I_x \geq I_y$	$0 \leq 2\beta \leq 90°$
	$I_x < I_y$	$90° < 2\beta \leq 180°$
$I_{xy} < 0$	$I_x \leq I_y$	$180° < 2\beta \leq 270°$
	$I_x > I_y$	$270° < 2\beta < 360°$

Using Eqs. (7.19) and setting $\frac{d}{d\theta}(I_u) = 0$ or $I_{uv} = 0$, we readily find that

$$\tan 2\theta = -\frac{2I_{xy}}{I_x - I_y} \qquad (7.26)$$

From Eqs. (7.25) and (7.26), we note that $\tan 2\theta = -\tan 2\beta = \tan(-2\beta)$ or $\theta = -\beta$. This means that a principal axis (i.e., the b axis) is reached when the x axis in Fig. 7.20 is rotated *clockwise* about the origin O through an angle of β. Thus, Eqs. (7.20) and (7.25) conform to the result in Eq. (7.26) as derived with calculus. As the ordinates of the points B and S on Mohr's circle are zero, it is important to note that *the product of inertia is zero with respect to the principal axes*. If desired, a computer program may be used to compute I_{max}, I_{min}, β, and ψ, as well as I_u, I_v, and I_{uv} from given values of I_x, I_y, I_{xy}, and θ as shown in Fig. 7.22.

```
10 REM            AREA MOMENTS OF INERTIA            (File Name: AMOI)
15 REM  Enter Ix, Iy, Ixy, and Theta (in degrees) as data in Line 100
20 REM  To output to the printer, change PRINT to LPRINT in the program.
25 READ IX,IY,XY,TH: I1=(IX+IY)/2: I2=(IX-IY)/2: T2=TH*3.141593/90
30 I3=I2*COS(T2)-XY*SIN(T2):IU=I1+I3:IV=I1-I3:UV=I2*SIN(T2)+XY*COS(T2)
35 R = SQR(I2^2+XY^2): MX = I1 + R: MN = I1 - R: S$ = "    ": PRINT
40 PRINT "Ix =";IX;S$; "Iy =";IY;S$; "Ixy = ";XY;S$; "Theta = ";TH
45 PRINT "Iu =";IU;S$;"Iv =";IV;S$;"Iuv = ";UV:B=ATN(XY/I2)*90/3.141593
50 IF XY >= 0 AND IX >= IY THEN P = 90 - B: GOTO 70
55 IF XY >= 0 AND IX < IY THEN B = B + 90: P = 90 - B: GOTO 70
60 IF XY < 0 AND IX <= IY THEN B = B + 90: P = 90 - B: GOTO 70
65 IF XY < 0 AND IX > IY THEN B = B + 180: P = 90 - B
70 PRINT "Imax =";MX;S$; "Imin =";MN;S$; "Beta =";B;S$; "Psi = ";P
100 DATA 592,92,-150,40

Ix = 592      Iy = 92      Ixy = -150      Theta =   40
Iu = 533.1332      Iv = 150.8669      Iuv =   220.1548
Imax = 633.5476      Imin = 50.4524      Beta = 164.5181      Psi = -74.51813
```

Fig. 7.22   (Cf. the solution in Example 7.12.)

EXAMPLE 7.13

A slender aluminum rod with *round ends* has a cross-sectional area A as shown. It is known from mechanics of materials that the magnitude of the

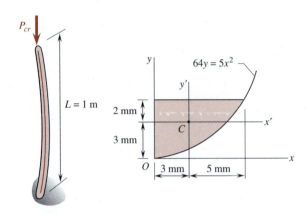

critical force $\mathbf{P}_{cr}$ for the buckling of the rod is given by $P_{cr} = (\pi^2 E \bar{I}_{min})/L^2$ where L is the length of the rod, E is the elastic modulus of the rod and is equal to 69 GPa, and $\bar{I}_{min}$ is the minimum moment of inertia of A about an axis through its centroid C as shown. This value of P_{cr} is known as Euler's buckling load. It is also known from the solution in Example 7.8 that $\bar{I}_{x'} = 45.71$ mm^4, $\bar{I}_{y'} = 101.33$ mm^4, and $\bar{I}_{x'y'} = 26.67$ mm^4. Determine (a) the value of P_{cr}, (b) the values of β and ψ specifying the directions of the principal axes of inertia b' and s' about the centroid C. Show the principal axes b' and s' on the area A in a sketch.

Solution. The data show that $L = 1$ m $= 10^3$ mm and $E = 69$ GPa $= 69 \times 10^9$ N/m$^2 = 69 \times 10^3$ N/mm^2. Thus, the determination of P_{cr} now requires a prior determination of $\bar{I}_{min}$. Using the points $(\bar{I}_{x'}, \bar{I}_{x'y'})$ and $(\bar{I}_{y'}, -\bar{I}_{x'y'})$ as the end points of the diameter $\overline{X'Y'}$, we may draw Mohr's

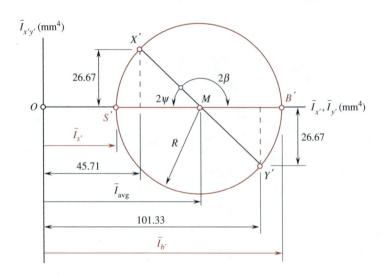

circle for the area A about C as shown. From this circle, we write

$$\bar{I}_{avg} = \frac{1}{2}(45.71 + 101.33) \qquad \bar{I}_{avg} = 73.52 \text{ mm}^4$$

$$R^2 = (\bar{I}_{avg} - 45.71)^2 + (26.67)^2 \qquad R = 38.53 \text{ mm}^4$$

$$\bar{I}_{min} = \bar{I}_{s'} = \bar{I}_{avg} - R \qquad \bar{I}_{min} = 34.99 \text{ mm}^4$$

Thus, we have

$$P_{cr} = \frac{\pi^2 (69 \times 10^3 \text{ N/mm}^2)(34.99 \text{ mm}^4)}{(10^3 \text{ mm})^2}$$

$$\tan 2\psi = \frac{26.67}{73.52 - 45.71}$$

$$2\beta = 180° - 2\psi$$

$$P_{cr} = 23.8 \text{ N} \qquad \psi = 21.9° \qquad \beta = 68.1° \blacktriangleleft$$

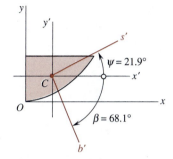

Any force greater than $\mathbf{P}_{cr}$ plus a small lateral disturbance will cause the rod to bow out (i.e., buckle) and bend about the s' axis which makes an angle of $\psi = 21.9°$ with the x' axis as shown in the sketch.

Developmental Exercises

D7.35 Refer to Example 7.12. Using Mohr's circle as drawn in the solution of that example, (a) determine for the Z section the principal moments of inertia I_b and I_s about the origin O, (b) show the principal axes b and s on the Z section in a sketch.

D7.36 Letting 45.71, 101.33, 26.67, and 0 be the data for Ix, Iy, Ixy, and θ, run the program in Fig. 7.22 to verify the values obtained for $\bar{I}_{min}$, ψ, and β in Example 7.13.

PROBLEMS

7.83* through 7.91* Using Eqs. (7.19), determine the centroidal moments of inertia $\bar{I}_u$ and $\bar{I}_v$ and the centroidal product of inertia $\bar{I}_{uv}$ for the area and axes shown.

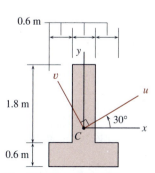

Fig. P7.83* and P7.92*

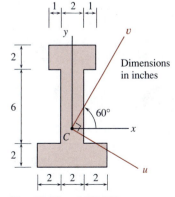

Fig. P7.84 and P7.93*

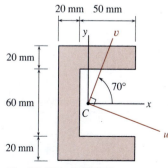

Fig. P7.85 and P7.94*

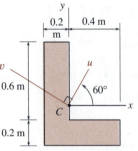

Fig. P7.86, P7.95*, and P7.101

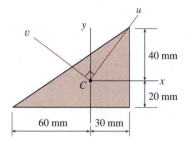

Fig. P7.87*, P7.96*, and P7.102*

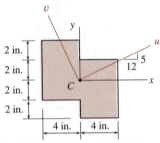

Fig. P7.88, P7.97*, and P7.103

7.92* through 7.100*   Using Mohr's circle, determine the centroidal moments of inertia $\bar{I}_u$ and $\bar{I}_v$ and the centroidal product of inertia $\bar{I}_{uv}$ for the area and axes shown.

7.101 through 7.106   For the area shown, determine (a) the centroidal principal moments of inertia $\bar{I}_b$ and $\bar{I}_s$, (b) the angles β and ψ locating the centroidal principal axes of inertia b and s on the area, respectively. Indicate the b and s axes on the area in a sketch.

7.107   For a given area in the xy plane, it is known that $I_x = 14 \times 10^6$ mm^4, $I_y = 8 \times 10^6$ mm^4, $I_{max} = 16 \times 10^6$ mm^4, and $I_{xy} > 0$. Using Mohr's circle, determine (a) I_{xy}, (b) I_{min}, (c) the angles β and ψ locating the principal axes of inertia b and s, respectively. Indicate the b and s axes on the xy plane.

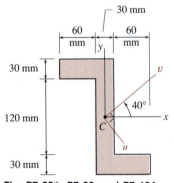

Fig. P7.89*, P7.98, and P7.104

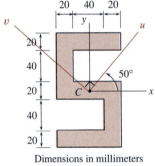

Dimensions in millimeters

Fig. P7.90*, P7.99, and P7.105*

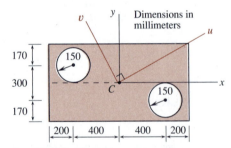

Fig. P7.91*, P7.100*, and P7.106

MASS MOMENTS OF INERTIA†

7.11   Moments of Inertia of a Mass

In addition to having a mass, a rigid body has a definite shape and size. Thus, under the action of nonconcurrent forces $\mathbf{F}_1$, $\mathbf{F}_2$, . . . , $\mathbf{F}_n$ as shown in Fig. 7.23, a rigid body will generally be accelerated to rotate as well as

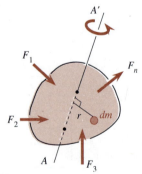

Fig. 7.23

†Cf. the footnote on p. 223.

translate. The tendency of a body to resist a translational acceleration is, of course, measured by the *mass* of the body. On the other hand, the tendency of a body to resist a rotational acceleration about an axis is measured by a quantity called the *moment of inertia* of the body about that axis. This quantity, encountered in kinetics of rigid bodies having rotational motion, is an integral of the form ∫ (distance)² d(mass) or

$$I = \int r^2 \, dm \qquad (7.27)$$

where *dm* is a differential mass element of the body, *r* is the shortest distance between *dm* and the axis of rotation, say *AA'*, and the integration extends over the entire body. The moment of inertia of a body defined in Eq. (7.27) is often referred to as the *mass moment of inertia*.

The differential mass element *dm* illustrated in Fig. 7.23 for Eq. (7.27) is of the *third order*. However, *dm* in Eq. (7.27) often takes the form of a *first-order* slender strip, thin ring, or thin shell if the particles in *dm* can be taken as equidistant from the axis of rotation, with *r* being the *equidistance* within the framework of first-order accuracy.

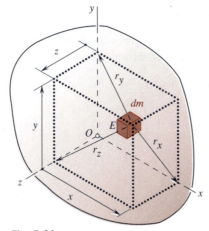

Fig. 7.24

The moments of inertia of a body about the *x*, *y*, and *z* axes are customarily denoted by I_x, I_y, and I_z, respectively.[†] As shown in Fig. 7.24, the rectangular coordinates of the differential mass element *dm* at the point *E* are (*x*, *y*, *z*), and the shortest distances between *dm* and the *x*, *y*, and *z* axes are r_x, r_y, and r_z, respectively. We write

$$\begin{aligned}
I_x &= \int r_x^2 \, dm = \int (y^2 + z^2) \, dm \\
I_y &= \int r_y^2 \, dm = \int (z^2 + x^2) \, dm \qquad (7.28) \\
I_z &= \int r_z^2 \, dm = \int (x^2 + y^2) \, dm
\end{aligned}$$

[†]Sometimes they are written with double subscripts as I_{xx}, I_{yy}, and I_{zz} to signify that they are components of a second-order tensor.

Note from Eqs. (7.27) and (7.28) that the moment of inertia of a body about any axis is always a *positive scalar quantity*, whose dimensions are *mass* × *(length)²*, because *dm* is always taken as positive regardless of the position in which it is situated. The primary units for the moment of inertia of a body are $kg \cdot m^2$ in the SI and $slug \cdot ft^2$ (or $lb \cdot ft \cdot s^2$) in the U.S. customary system. Unless otherwise stated, it will be assumed that each body considered is homogeneous and has a constant mass density.

Developmental Exercise

D7.37   Show that $1 \ slug \cdot ft^2 = 1 \ lb \cdot ft \cdot s^2 = 1.356 \ kg \cdot m^2$.

EXAMPLE 7.14

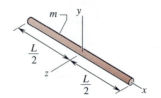

A slender rod of length L and mass m is shown. Determine the moments of inertia I_x, I_y, and I_z of the rod about the axes as indicated.

Solution. The mass density of the rod may be expressed as $\rho_L = m/L$. We note that the y and z coordinates of the mass element dm may be taken as zero throughout the rod as shown. Thus, we write

$$I_x = \int r_x^2 \, dm = \int (y^2 + z^2) \, dm = 0$$

$$I_y = \int r_y^2 \, dm = \int_{-L/2}^{L/2} x^2 \, \rho_L \, dx = \frac{1}{12} \rho_L L^3 = \frac{1}{12}\left(\frac{m}{L}\right)L^3$$

By symmetry, we note that $I_z = I_y$. Thus, we have

$$I_x = 0 \qquad I_y = I_z = \frac{1}{12} mL^2 \qquad \blacktriangleleft$$

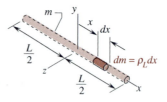

Developmental Exercise

D7.38   Define the moment of inertia of a body about an axis.

EXAMPLE 7.15

A thin rectangular plate of mass m lies in the zx plane as shown. Determine I_z of the plate.

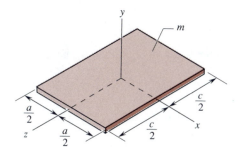

Solution. The mass density of the plate may be expressed as $\rho_A = m/(ca)$. Choosing dm to be a slender rod as shown, we write

$$I_z = \int r_z^2 \, dm = \int_{-a/2}^{a/2} x^2 \rho_A c \, dx = \frac{1}{12} \rho_A c a^3$$

$$I_z = \frac{1}{12} m a^2 \blacktriangleleft$$

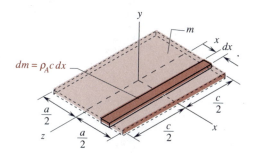

REMARK. For the thin plate in this example, we note that $y \approx 0$ throughout the plate. By Eq. (7.28), we write

$$I_x = \int z^2 \, dm \qquad I_y = \int (z^2 + x^2) \, dm \qquad I_z = \int x^2 \, dm$$

Thus, for such a plate, we have $I_y = I_z + I_x$.

Developmental Exercise

D7.39 Determine I_x of the plate in Example 7.15.

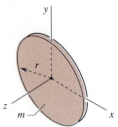

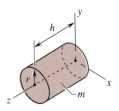

EXAMPLE 7.16

A thin circular disk of radius r and mass m lies in the xy plane as shown. Determine I_x, I_y, and I_z of the disk.

Solution. Since the thin circular disk lies in the xy plane with $z \approx 0$, we readily deduce from Eqs. (7.28) that

$$I_z = I_x + I_y$$

By symmetry, we note that $I_x = I_y$. Thus, we have

$$I_z = 2I_x \qquad I_x = I_y = \frac{1}{2} I_z$$

The mass density of the disk may be expressed as $\rho_A = m/(\pi r^2)$. Referring to the sketch, we write

$$I_z = \int r_z^2 \, dm = \int_0^r u^2 \rho_A (2\pi u) \, du = \frac{1}{2} \rho_A \pi r^4 = \frac{1}{2}\left(\frac{m}{\pi r^2}\right)\pi r^4$$

$$I_z = \frac{1}{2} mr^2 \qquad I_x = I_y = \frac{1}{4} mr^2 \quad \blacktriangleleft$$

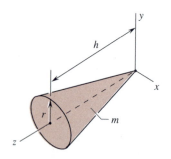

Fig. D7.40

Developmental Exercise

D7.40   A circular cylinder of total mass m and radius r is shown. Applying the results obtained in Example 7.16, determine I_z of the cylinder.

EXAMPLE 7.17

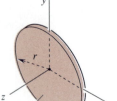

A circular cone having a mass m, a height h, and a base of radius r is shown. Using the solution for I_z in Example 7.16 as a formula, determine I_z of the cone.

Solution. Choosing dm to be a thin disk as shown in a different perspective for better clarity, we first determine the mass density ρ and dm in terms

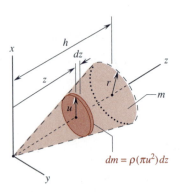

of z as follows:

$$u = \frac{r}{h} z \qquad dm = \rho(\pi u^2) \, dz = \frac{\rho \pi r^2}{h^2} z^2 \, dz$$

$$m = \int dm = \frac{\rho \pi r^2}{h^2} \int_0^h z^2 \, dz = \frac{\rho \pi r^2 h}{3} \qquad \rho = \frac{3m}{\pi r^2 h} \qquad dm = \frac{3m}{h^3} z^2 \, dz$$

Using the solution for I_z in Example 7.16 as a formula, we write

$$dI_z = \frac{1}{2} (dm) \, u^2 = \frac{3mr^2}{2h^5} z^4 \, dz$$

$$I_z = \int dI_z = \frac{3mr^2}{2h^5} \int_0^h z^4 \, dz = \frac{3mr^2}{2h^5}(\frac{1}{5} h^5) \qquad I_z = \frac{3}{10} mr^2 \quad \blacktriangleleft$$

Developmental Exercise

D7.41   For the circular cone in Example 7.17, is $I_z = I_x + I_y$?

EXAMPLE 7.18

A tetrahedron of mass density $\rho = 15$ slugs/ft^3 is bounded by the planes $x + 2y + 4z = 8$, $x = 0$, $y = 0$, and $z = 0$, where the lengths are in feet. Determine I_x of the tetrahedron.

Solution.   Let the differential mass element dm of the tetrahedron be chosen as sketched. We have

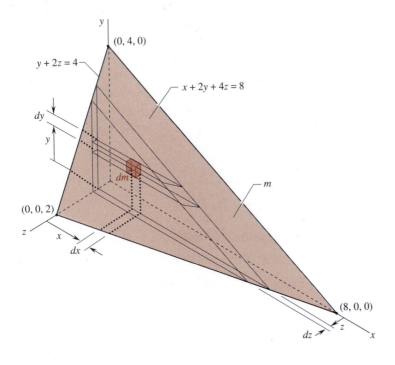

$$dm = \rho \, dV = \rho \, dx \, dy \, dz = 15 \, dx \, dy \, dz$$

$$I_x = \int (y^2 + z^2) \, dm$$

$$= 15 \int_{z=0}^{2} \int_{y=0}^{4-2z} \int_{x=0}^{8-2y-4z} (y^2 + z^2) \, dx \, dy \, dz$$

$$= 15 \int_{z=0}^{2} \int_{y=0}^{4-2z} (y^2 + z^2)(8 - 2y - 4z) \, dy \, dz$$

$$= 20 \int_{0}^{2} (5z^4 - 28z^3 + 60z^2 - 64z + 32) \, dz$$

$$I_x = 320 \text{ slug} \cdot \text{ft}^2 \quad \blacktriangleleft$$

Developmental Exercise

D7.42 Determine I_y of the tetrahedron in Example 7.18.

7.12 Radius of Gyration of a Mass

The *radius of gyration* of a body about an axis is a distance whose square multiplied by the mass of the body gives the moment of inertia of the body about that axis. For a body of mass m as shown in Fig. 7.25, we write

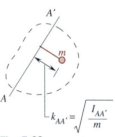

Fig. 7.25

$$I_x = k_x^2 m \qquad I_y = k_y^2 m \qquad I_z = k_z^2 m \qquad (7.29)$$

where k_x, k_y, and k_z are the radii of gyration of the body about the x, y, and z axes, respectively.

The radius of gyration $k_{AA'}$ of a body of mass m about an axis AA' is illustrated in Fig. 7.26, where $I_{AA'}$ is the moment of inertia of the body about the axis AA'. Note that the moment of inertia of a body about an axis will remain unchanged if the entire mass of the body is concentrated at a point which is away from that axis at a *distance* equal to the *radius of gyration* of the body about that axis. Clearly, the dimension of the radius of gyration is *length*; and it is measured in meters (m) in the SI and in feet (ft) in the U.S. customary system.

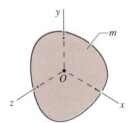

$$k_{AA'} = \sqrt{\frac{I_{AA'}}{m}}$$

Fig. 7.26

EXAMPLE 7.19

Refer to Example 7.14. Determine k_y of the rod.

Solution. From the solution in Example 7.14, we know that $I_y = \frac{1}{12} mL^2$. Thus, we write

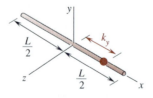

$$\frac{1}{12} mL^2 = k_y^2 m$$

$$k_y = \frac{\sqrt{3}}{6} L = 0.289L \quad \blacktriangleleft$$

EXAMPLE 7.20

Determine k_z of the disk in Example 7.16.

Solution. From the solution in Example 7.16, we know that $I_z = \frac{1}{2} mr^2$. Thus, we write

$$\frac{1}{2} mr^2 = k_z^2 m$$

$$k_z = \frac{\sqrt{2}}{2} r = 0.707r \quad \blacktriangleleft$$

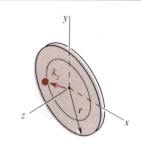

Developmental Exercises

D7.43 Define the radius of gyration of a body about an axis.

D7.44 Determine k_z of the cone in Example 7.17.

D7.45 The radius of gyration of a 50-kg flywheel about its axis of rotation A is $k_A = 0.4$ m. Determine I_A of the flywheel.

D7.46 Does it make sense for the radius of gyration of any wheel about its axle to be greater than its radius? (*Hint*. Cf. Example 7.20.)

PROBLEMS

NOTE. The following problems are to be solved *by integration*.

7.108* and 7.109* Determine I_x and I_y of the slender rod of mass m as shown.

7.110 Determine I_x, I_y, and I_z of the thin plate of mass m as shown.

7.111* through 7.116* Determine I_z and k_z of the body of mass m as shown.

Fig. P7.108*

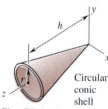

Fig. P7.109*

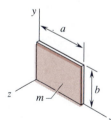

Fig. P7.110

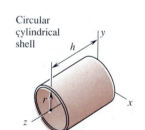

Circular cylindrical shell

Fig. P7.111*

Circular conic shell

Fig. P7.112*

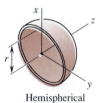

Hemispherical shell

Fig. P7.113*

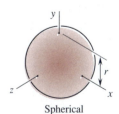

Spherical shell

Fig. P7.114*

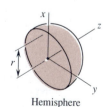

Hemisphere

Fig. P7.115*

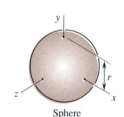

Sphere

Fig. P7.116*

7.117 A body of revolution and its generating area are shown. If the specific weight is $\gamma = 322$ lb/ft^3, determine I_x and k_x of the body.

7.118 An ellipsoid and its generating area are shown. If the mass density is $\rho = 6$ Mg/m^3, determine I_x and k_x of the ellipsoid.

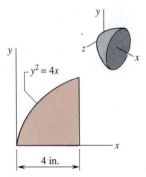

Fig. P7.117

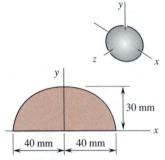

Fig. P7.118

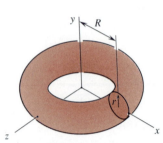

Fig. P7.119*

7.119* Determine I_y of the torus shown if its mass is m. (*Hint*. Choose dm to be a thin shell around the y axis and refer to Sec. B5.2 of App. B. if necessary.)

7.120 and 7.121* Determine I_x of the body shown if its mass is m.

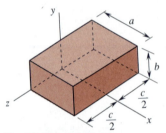

Fig. P7.120 and P7.122

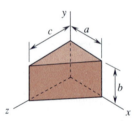

Fig. P7.121* and Fig. P7.123

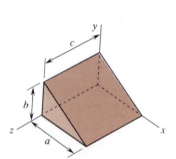

Fig. P7.124 and P7.125*

7.122 and 7.123 Determine I_y and k_y of the body shown if its mass is m.

7.124 The mass density of the prism shown is $\rho = 4$ Mg/m^3. If $a = 0.4$ m, $b = 0.3$ m, and $c = 0.6$ m, determine I_x and k_x of the prism.

7.125* The mass density of the prism shown is $\rho = 5$ Mg/m^3. If $a = 80$ mm, $b = 60$ mm, and $c = 120$ mm, determine I_y and k_y of the prism.

7.13 Parallel-Axis Theorem: Mass Moments of Inertia

Let the point $G(\bar{x}, \bar{y}, \bar{z})$ be the center of mass of a body of mass m as shown in Fig. 7.27, where the *central axes* x', y', and z' of the body are parallel to the fixed coordinate axes x, y, and z, respectively.[†] The coordinates of

---

[†]The word *central* indicates something associated with a center (e.g., center of mass or center of gravity), while the word *centroidal* indicates something associated with the *centroid* of a geometric figure (e.g., line, area, or volume). Thus, *central axes* of a body are axes passing through the center of mass of the body, while *centroidal axes* of a body are axes passing through the centroid of the volume of the body. Note that the *center of mass* of a body does not necessarily coincide with the *centroid* of the volume of the body. If these two points coincide in a body, then the *centroidal axes* are also the *central axes* of the body.

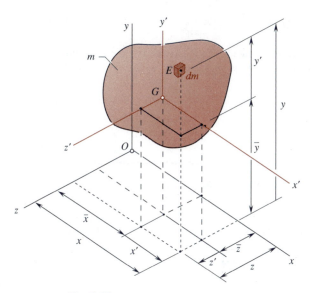

Fig. 7.27

the differential mass element dm at the point E are (x, y, z) in the $Oxyz$ fixed coordinate system but are (x', y', z') in the $Gx'y'z'$ central coordinate system as indicated. Thus, we write

$$x = \bar{x} + x' \qquad y = \bar{y} + y' \qquad z = \bar{z} + z' \qquad (7.30)$$

$$I_x = \int (y^2 + z^2)\, dm = \int [(\bar{y} + y')^2 + (\bar{z} + z')^2]\, dm$$

$$= \int (y'^2 + z'^2)\, dm + 2\bar{y} \int y'\, dm + 2\bar{z} \int z'\, dm + (\bar{y}^2 + \bar{z}^2) \int dm$$

$$= \bar{I}_{x'} + 2\bar{y}\bar{y}'m + 2\bar{z}\bar{z}'m + (\bar{y}^2 + \bar{z}^2)m$$

where $\bar{I}_{x'}$ represents the moment of inertia of the body about the central axis x', $\bar{y}'$ represents the distance between G and the $z'x'$ plane, and $\bar{z}'$ represents the distance between G and the $x'y'$ plane. Since G lies in the $z'x'$ and $x'y'$ planes, we see that $\bar{y}' = \bar{z}' = 0$. Thus, we write

$$I_x = \bar{I}_{x'} + m(\bar{y}^2 + \bar{z}^2) \qquad (7.31)$$

and similarly,

$$I_y = \bar{I}_{y'} + m(\bar{z}^2 + \bar{x}^2) \qquad I_z = \bar{I}_{z'} + m(\bar{x}^2 + \bar{y}^2) \qquad (7.32)$$

From Fig. 7.27, we see that the sum $\bar{y}^2 + \bar{z}^2$ in Eq. (7.31) represents the square of the distance between the x and x' axes. Similarly, the sums $\bar{z}^2 + \bar{x}^2$ and $\bar{x}^2 + \bar{y}^2$ represent the squares of the distances between the y and y' axes, and the z and z' axes, respectively.

The formulas in Eqs. (7.31) and (7.32) are called the *parallel-axis theorem* for mass moments of inertia, which may be stated as follows: *The moment of inertia I of a body of mass m about any given axis is equal to the moment of inertia $\bar{I}$ of the body about a central axis parallel to the given axis plus the product md^2, where d is the distance between those two axes,* as shown in Fig. 7.28; i.e.,

$$I = \bar{I} + md^2 \qquad (7.33)$$

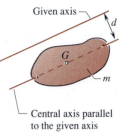

Fig. 7.28

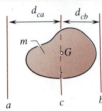

Fig. D7.47

Developmental Exercise

D7.47   If a, b, and c are three parallel axes and the axis c passes through the center of mass G of a body of mass m as shown, show that

$$I_a = I_b + m(d_{ca}^2 - d_{cb}^2)$$

EXAMPLE 7.21

Determine I_x of the circular cone in Example 7.17.

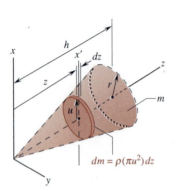

Solution.   Let dm be a thin disk and x' be a central axis of dm as shown, where the radius of the disk is $u = rz/h$ and its mass density is $\rho = 3m/(\pi r^2 h)$. Applying the *parallel-axis theorem* and the solution for I_x in Example 7.16 as a formula, we write

$$dI_x = d\bar{I}_{x'} + (dm)\, z^2 = \frac{1}{4}\,(dm)\, u^2 + (dm)\, z^2$$

$$= \left(\frac{1}{4}\, u^2 + z^2\right)(\rho \pi u^2\, dz) = \frac{\rho \pi r^2}{4h^4}\,(r^2 + 4h^2)z^4\, dz$$

$$I_x = \int dI_x = \frac{\rho \pi r^2}{4h^4}(r^2 + 4h^2)\int_0^h z^4\, dz = \frac{\rho \pi r^2 h}{20}(r^2 + 4h^2)$$

$$= \frac{3m}{\pi r^2 h}\,\frac{\pi r^2 h}{20}(r^2 + 4h^2)$$

$$I_x = \frac{3}{20}\, m(r^2 + 4h^2) \quad \blacktriangleleft$$

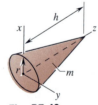

Fig. D7.49

Developmental Exercises

D7.48   Describe the parallel-axis theorem for mass moments of inertia.

D7.49   The circular cone shown has a mass m. Determine I_x of the cone.

7.14   Composite Bodies

A composite body has several component bodies. The moment of inertia I of a composite body about an axis is equal to the sum of the moments of inertia I_1, I_2, . . . , I_n of its component bodies about that axis; i.e.,

$$I = I_1 + I_2 + \cdots + I_n \tag{7.34}$$

For use in determining the moments of inertia of composite bodies, the moments of inertia of homogeneous bodies of some common shapes are given in Fig. 7.29, where the plates, disks, rings, and shells are understood to be thin, and m is the total mass of the body shown.

Slender rod

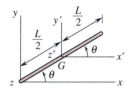

$$I_x = \frac{1}{3} mL^2 \sin^2\theta \qquad\qquad I_y = \frac{1}{3} mL^2 \cos^2\theta$$

$$I_z = \frac{1}{3} mL^2 \qquad\qquad\qquad \bar{I}_{x'} = \frac{1}{12} mL^2 \sin^2\theta$$

$$\bar{I}_{y'} = \frac{1}{12} mL^2 \cos^2\theta \qquad\qquad \bar{I}_{z'} = \frac{1}{12} mL^2$$

Circular ring

$$I_x = I_z = \frac{1}{2} mr^2$$

$$I_y = mr^2$$

Rectangular parallelepiped

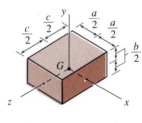

$$I_x = \frac{1}{12} m(b^2 + c^2)$$

$$I_y = \frac{1}{12} m(c^2 + a^2)$$

$$I_z = \frac{1}{12} m(a^2 + b^2)$$

Semicircular ring

$$\bar{z} = \frac{2r}{\pi} \qquad I_x = I_z = \frac{1}{2} mr^2 \qquad I_y = mr^2$$

$$\bar{I}_{x'} = \frac{1}{2\pi^2} mr^2(\pi^2 - 8)$$

$$\bar{I}_{y'} = \frac{1}{\pi^2} mr^2(\pi^2 - 4)$$

Rectangular plate

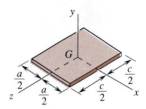

$$I_x = \frac{1}{12} mc^2 \qquad I_y = \frac{1}{12} m(c^2 + a^2) \qquad I_z = \frac{1}{12} ma^2$$

Circular disk

$$I_y = \frac{1}{2} mr^2$$

$$I_x = I_z = \frac{1}{4} mr^2$$

Circular cylinder

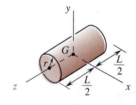

$$I_z = \frac{1}{2} mr^2$$

$$I_x = I_y = \frac{1}{12} m(3r^2 + L^2)$$

Semicircular disk

$$\bar{z} = \frac{4r}{3\pi}$$

$$I_y = \frac{1}{2} mr^2$$

$$I_x = I_z = \frac{1}{4} mr^2$$

$$\bar{I}_{x'} = \frac{1}{36\pi^2} mr^2(9\pi^2 - 64)$$

$$\bar{I}_{y'} = \frac{1}{18\pi^2} mr^2(9\pi^2 - 32)$$

Fig. 7.29 (a) Moments of inertia of bodies of common shapes.

Circular conic shell

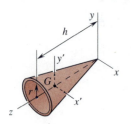

$$\bar{z} = \frac{2}{3}h$$

$$I_z = \frac{1}{2}mr^2$$

$$I_x = I_y = \frac{1}{4}m(r^2 + 2h^2)$$

$$\bar{I}_{x'} = \bar{I}_{y'} = \frac{1}{36}m(9r^2 + 2h^2)$$

Circular cone

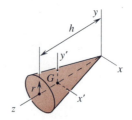

$$\bar{z} = \frac{3}{4}h$$

$$I_z = \frac{3}{10}mr^2$$

$$I_x = I_y = \frac{3}{20}m(r^2 + 4h^2)$$

$$\bar{I}_{x'} = \bar{I}_{y'} = \frac{3}{80}m(4r^2 + h^2)$$

Circular cylindrical shell

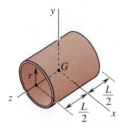

$$I_z = mr^2$$

$$I_x = I_y = \frac{1}{12}m(6r^2 + L^2)$$

Sphere

$$I_x = I_y = I_z = \frac{2}{5}mr^2$$

Triangular plate

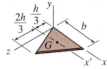

$$\bar{I}_{x'} = \frac{1}{18}mh^2$$

$$I_x = \frac{1}{6}mh^2$$

Hemispherical shell

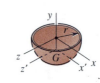

$$\bar{y} = -\frac{1}{2}r$$

$$I_x = I_y = I_z = \frac{2}{3}mr^2$$

$$\bar{I}_{x'} = \bar{I}_{z'} = \frac{5}{12}mr^2$$

Spherical shell

$$I_x = I_y = I_z = \frac{2}{3}mr^2$$

Hemisphere

$$\bar{y} = -\frac{3}{8}r$$

$$I_x = I_y = I_z = \frac{2}{5}mr^2$$

$$\bar{I}_{x'} = \bar{I}_{z'} = \frac{83}{320}mr^2$$

Half torus

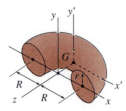

$$\bar{z} = -\frac{r^2 + 4R^2}{2\pi R}$$

$$I_x = I_z = \frac{1}{8}m(5r^2 + 4R^2)$$

$$I_y = \frac{1}{4}m(3r^2 + 4R^2)$$

Torus

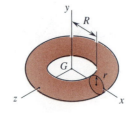

$$I_x = I_z = \frac{1}{8}m(5r^2 + 4R^2)$$

$$I_y = \frac{1}{4}m(3r^2 + 4R^2)$$

Fig. 7.29 (b) Moments of inertia of bodies of common shapes.

EXAMPLE 7.22

The mass density of the slender bent rod shown is $\rho_L = 3.75$ kg/m. Determine I_y and k_y of the rod.

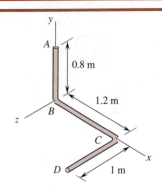

Solution. The rod may be treated as a composite body consisting of the segments AB, BC, and CD as the component bodies. Thus, we write

$$I_y = (I_y)_{AB} + (I_y)_{BC} + (I_y)_{CD}$$

Since the segment AB lies along the y axis, we have $(I_y)_{AB} = 0$. To find $(I_y)_{BC}$ and $(I_y)_{CD}$ of the segments BC and CD, we first establish their central axes y' and y'' which are parallel to the y axis. Then, we apply the *parallel-axis theorem* and the formulas for a slender rod to write

$$(I_y)_{BC} = (\bar{I}_{y'})_{BC} + m_{BC}\,\overline{BE}^2 = \frac{1}{12} m_{BC}\,\overline{BC}^2 + m_{BC}\,\overline{BE}^2$$

$$= \frac{1}{12}\left[3.75(1.2)\right](1.2)^2 + \left[3.75(1.2)\right](0.6)^2 = 2.16$$

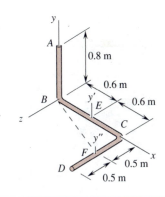

$$(I_y)_{CD} = (\bar{I}_{y''})_{CD} + m_{CD}\,\overline{BF}^2 = \frac{1}{12} m_{CD}\,\overline{CD}^2 + m_{CD}(\overline{BC}^2 + \overline{CF}^2)$$

$$= \frac{1}{12}\left[3.75(1)\right](1)^2 + \left[3.75(1)\right]\left[(1.2)^2 + (0.5)^2\right] = 6.65$$

$$I_y = 0 + 2.16 + 6.65 \qquad\qquad I_y = 8.81 \text{ kg} \cdot \text{m}^2 \blacktriangleleft$$

$$I_y = k_y^2 m: \quad 8.81 = k_y^2(3.75)(0.8 + 1.2 + 1)$$
$$k_y = 0.885 \text{ m} \blacktriangleleft$$

Developmental Exercises

D7.50 Determine I_x and k_x of the bent rod in Example 7.22.

D7.51 Determine I_z and k_z of the bent rod in Example 7.22.

EXAMPLE 7.23

A thin machine part of mass density $\rho_A = 4.83$ lbm/ft^2 is shown, where $a = 6$ in. Determine I_y of the machine part.

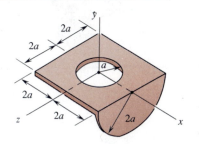

Solution. Let the machine part be a composite body consisting of a square plate of mass m_1, a semicircular plate of mass m_2, and an *empty* circular plate of mass m_3. Then, we have

$$m_1 = \rho_A(4a)^2 = 16\rho_A a^2$$

$$m_2 = \rho_A\left[\frac{1}{2}\pi(2a)^2\right] = 2\pi\rho_A a^2 \qquad m_3 = \rho_A(\pi a^2) = \pi\rho_A a^2$$

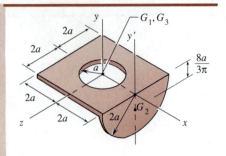

where $\rho_A = 0.15$ slug/ft^2 and $a = 0.5$ ft. Applying the *parallel-axis theorem* and appropriate formulas in Fig. 7.29, we write

$$I_y = (I_y)_1 + (I_y)_2 - (I_y)_3 = (I_y)_1 + \left[(\bar{I}_{y'})_2 + m_2 d_2^2\right] - (I_y)_3$$

$$= \frac{1}{12} m_1 \left[(4a)^2 + (4a)^2\right] + \left[\frac{1}{4} m_2(2a)^2 + m_2(2a)^2\right] - \frac{1}{2} m_3 a^2$$

$$= \frac{1}{6}(256 + 57\pi)\rho_A a^4 \qquad\qquad I_y = 0.680 \text{ slug} \cdot \text{ft}^2 \blacktriangleleft$$

Developmental Exercises

D7.52 Determine I_x and k_x of the machine part in Example 7.23.

D7.53 Determine I_z and k_z of the machine part in Example 7.23.

EXAMPLE 7.24

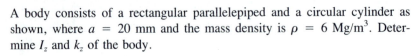

A body consists of a rectangular parallelepiped and a circular cylinder as shown, where $a = 20$ mm and the mass density is $\rho = 6$ Mg/m^3. Determine I_z and k_z of the body.

Solution. Let the component bodies be the rectangular parallelepiped of mass m_1 and the circular cylinder of mass m_2. Then, we have

$$m_1 = \rho(4a)(6a)(2a) = 48\rho a^3$$

$$m_2 = \rho(\pi a^2)(4a) = 4\pi\rho a^3$$

$$m = m_1 + m_2 = (48 + 4\pi)\rho a^3$$

where $\rho = 6000$ kg/m^3 and $a = 0.02$ m. Applying the *parallel-axis theorem* and appropriate formulas in Fig. 7.29, we write

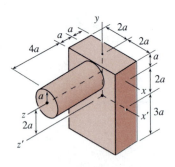

$$I_z = (I_z)_1 + (I_z)_2 = \left[(\bar{I}_{z'})_1 + m_1 d_1^2\right] + (I_z)_2$$

$$= \left\{\frac{1}{12} m_1 \left[(4a)^2 + (6a)^2\right] + m_1(2a)^2\right\} + \frac{1}{2} m_2 a^2$$

$$= (400 + 2\pi)\rho a^5 = (400 + 2\pi)(6000)(0.02)^5$$

$$I_z = 7.80 \times 10^{-3} \text{ kg} \cdot \text{m}^2 \blacktriangleleft$$

$$I_z = k_z^2 m: \quad (400 + 2\pi)\rho a^5 = k_z^2(48 + 4\pi)\rho a^3$$

$$k_z = 51.8 \text{ mm} \blacktriangleleft$$

Developmental Exercises

D7.54 Determine I_x and k_x of the body in Example 7.24.

D7.55 Determine I_y and k_y of the body in Example 7.24.

PROBLEMS

NOTE. Unless otherwise specified, the formulas given in Fig. 7.29 as well as the parallel-axis theorem may be applied in solving the following problems of mass moments of inertia.

7.126 through 7.128 Determine I_x and I_y of the body made of slender rods as shown.

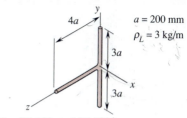

a = 200 mm
ρ_L = 3 kg/m

Fig. P7.126 and P7.129*

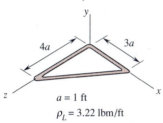

a = 1 ft
ρ_L = 3.22 lbm/ft

Fig. P7.127* and P7.130

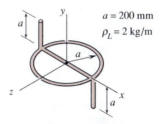

a = 200 mm
ρ_L = 2 kg/m

Fig. P7.128 and P7.131

7.129* through 7.131 Determine I_z and k_z of the body made of slender rods as shown.

7.132 through 7.134 Determine I_x and I_y of the sheet metal part shown.

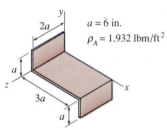

a = 6 in.
ρ_A = 1.932 lbm/ft^2

Fig. P7.132 and P7.135*

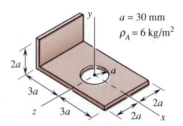

a = 30 mm
ρ_A = 6 kg/m^2

Fig. P7.133* and P7.136

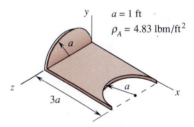

a = 1 ft
ρ_A = 4.83 lbm/ft^2

Fig. P7.134 and P7.137

7.135* through 7.137 Determine I_z and k_z of the sheet metal part shown.

7.138 through 7.140 Determine I_y and k_y of the thin shell shown.

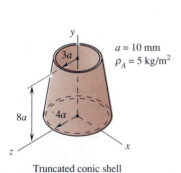

a = 10 mm
ρ_A = 5 kg/m^2

Truncated conic shell

Fig. P7.138 and P7.141*

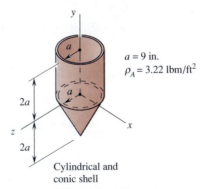

a = 9 in.
ρ_A = 3.22 lbm/ft^2

Cylindrical and
conic shell

Fig. P7.139* and P7.142

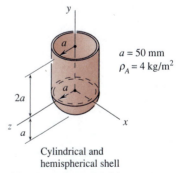

a = 50 mm
ρ_A = 4 kg/m^2

Cylindrical and
hemispherical shell

Fig. P7.140 and P7.143

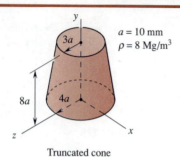

Truncated cone
Fig. P7.144 and P7.147*

$a = 10$ mm
$\rho = 8$ Mg/m^3

7.141* through **7.143**   Determine I_x and k_x of the thin shell shown.

7.144 through **7.146**   Determine I_y and k_y of the body of revolution shown.

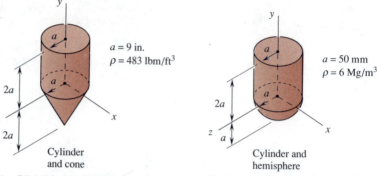

Cylinder
and cone
Fig. P7.145* and P7.148

$a = 9$ in.
$\rho = 483$ lbm/ft^3

Cylinder and
hemisphere
Fig. P7.146 and P7.149

$a = 50$ mm
$\rho = 6$ Mg/m^3

7.147* through **7.149**   Determine I_x and k_x of the body of revolution shown.

7.150   A horizontal hole of radius 1 in. is drilled through a body as shown. If $\rho = 14$ slugs/ft^3, determine I_z and k_z of the body.

7.151   A flywheel is of the same shape and size as that obtained by revolving the area shown about the x axis one revolution, where $r_1 = 250$ mm, $r_2 = 200$ mm, $r_3 = 50$ mm, $r_4 = 25$ mm, $b_1 = 150$ mm, and $b_2 = 25$ mm. If $\rho = 7$ Mg/m^3, determine I_x and k_x of the flywheel.

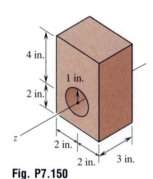

Fig. P7.150

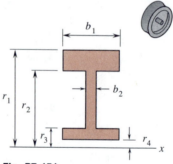

Fig. P7.151

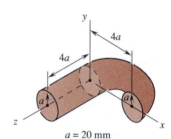

Fig. P7.152

7.152   A link consists of a cylinder and a half torus as shown. It is made of an alloy with $\rho = 6$ Mg/m^3. Determine I_y and I_z of the link.

7.153*   Using single integration, verify the formula $I_x = \frac{1}{12}m(3r^2 + L^2)$ for the circular cylinder in Fig. 7.29.

7.154*   Using single integration, verify the formula $I_x = \frac{1}{12}m(6r^2 + L^2)$ for the circular cylindrical shell in Fig. 7.29.

7.155*   Using single integration, verify the formula $I_x = \frac{1}{8}m(5r^2 + 4R^2)$ for the torus in Fig. 7.29. (*Hint.* Refer to the *hint* in Prob. 7.119*.)

7.15   Concluding Remarks

The *moment of the first moment* of a quantity is called the *second moment* of the quantity. The *moment of inertia of an area* about a given axis may be thought of as the integral sum of the second moments of its differential areas about the given axis. Similarly, the *moment of inertia of a mass* about a given axis may be thought of as the integral sum of the second moments of its differential masses about the given axis.

The moments of inertia of an area are primarily the mathematical properties of the area with respect to the given axes. The prime motivation to study the moments of inertia of areas comes from the fact that such quantities have many practical applications in studying certain situations in engineering, such as those mentioned in Sec. 7.1. The necessity to find the principal moments and principal axes of inertia of an area associated with a certain point (e.g., the centroid of the area) in solving some problems (e.g., the buckling of columns) leads naturally to the introduction and study of the products of inertia of areas. Additionally, one will find in fluid mechanics that both the moments and the product of inertia of an area are used in expressing the coordinates of the *center of pressure* if neither of the coordinate axes is an axis of symmetry of a nonhorizontal submerged plane surface.

As a graphical scheme, Mohr's circle is shown to provide a convenient means for studies involving the rotation of axes for moments and products of inertia. Furthermore, it is a good geometric aid in determining the principal moments and principal axes of inertia.

The second part of this chapter presents the moments of inertia of a mass, which is mathematically similar to the moments of inertia of an area presented in the first part of the chapter. This material is intended to provide students with the prerequisite background for the study of plane kinetics of rigid bodies in dynamics. However, at the instructor's discretion, this material may either be covered along with area moments of inertia or be skipped in a course in statics. For convenience of those who prefer to postpone covering mass moments of inertia until the study of dynamics, this material is repeated in the first part of Chap. 16, just before it is needed in the study of plane kinetics of rigid bodies. The concepts of products and principal moments of inertia of a mass are needed in the study of kinetics of rigid bodies in three dimensions. Therefore, they are timely covered in Chap. 19.

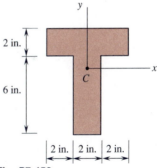

Fig. P7.156

REVIEW PROBLEMS

7.156   For the area and axes shown, determine the centroidal moments of inertia $\bar{I}_x$ and $\bar{I}_y$, and the centroidal radii of gyration $\bar{k}_x$ and $\bar{k}_y$.

7.157 and 7.158*   For the area and axes shown, determine the centroidal moments of inertia $\bar{I}_x$, $\bar{I}_y$, $\bar{I}_u$, and $\bar{I}_v$, and the centroidal products of inertia $\bar{I}_{xy}$ and $\bar{I}_{uv}$.

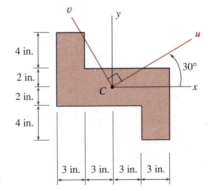

Fig. P7.157 and P7.159

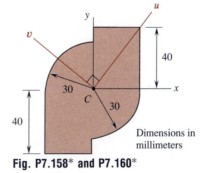

Fig. P7.158* and P7.160*

Dimensions in millimeters

7.159 and 7.160* For the area shown, determine (a) the centroidal principal moments of inertia $\bar{I}_b$ and $\bar{I}_s$, (b) the angles β and ψ locating the b and s axes.

7.161 For the area shown, determine the moments of inertia I_x and I_y, and the product of inertia I_{xy}. (*Hint.* Cf. the footnote to Prob. 7.67.)

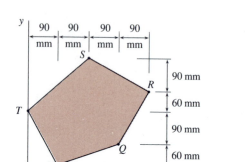

Fig. P7.161

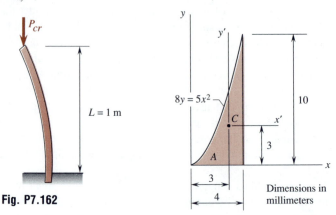

Fig. P7.162

7.162 A slender steel rod with one *fixed end* and one free end has a length of 1 m and a cross-sectional area A as shown. The magnitude of the critical force $\mathbf{P}_{cr}$ for the buckling of this cantilever rod is given by the formula

$$P_{cr} = \frac{\pi^2 E \bar{I}_{min}}{4L^2}$$

where L is the length of the rod, E is the elastic modulus of the rod and is equal to 200 GPa, and $\bar{I}_{min}$ is the minimum moment of inertia of A about an axis through its centroid C as shown. Determine the value of P_{cr}.

7.163* For a given area in the xy plane, it is known that $I_y = 30 \times 10^6$ mm^4, $I_{xy} = 15 \times 10^6$ mm^4, and $I_{min} = 5 \times 10^6$ mm^4. Using Mohr's circle, determine (a) I_x, (b) I_{max}, (c) the angles β and ψ locating the principal axes b and s, respectively. Indicate the b and s axes on the xy plane.

7.164* Using Eqs. (7.19), prove the invariance property that

$$I_u I_v - I_{uv}^2 = I_x I_y - I_{xy}^2$$

7.165* Using Eqs. (7.19), (7.23), and (7.24), prove the invariance property that

$$I_{max} I_{min} = I_x I_y - I_{xy}^2 = I_u I_v - I_{uv}^2$$

7.166 A slender rod weighing 1.61 lb/ft is used to form a composite body as shown. Determine the moment of inertia I_x and the radius of gyration k_x of the body.

7.167* Refer to Prob. 7.166. Determine I_y and k_y of the body.

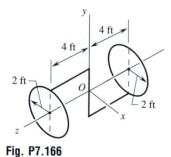

Fig. P7.166

7.168 A sheet of metal is cut and bent into a machine component as shown. If its mass density is $\rho_A = 16$ kg/m^2, determine I_x and k_x of the component.

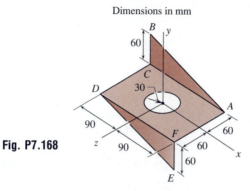

Dimensions in mm

Fig. P7.168

7.169 Refer to Prob. 7.168. Determine I_y and k_y of the component.

7.170* The mass density of the steel machine element is $\rho = 7.85$ Mg/m^3. Determine I_x and k_x of the element.

7.171 Refer to Prob. 7.170*. Determine I_y and k_y of the element.

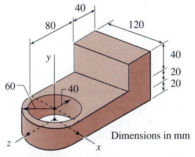

Dimensions in mm

Fig. P7.170*

Chapter 8

Structures

KEY CONCEPTS

- Truss, rigidly connected truss, pin-connected truss, idealized truss, simple truss, compound truss, complex truss, frame, machine, beam, and cable.
- Methods of joints and sections for analyzing plane trusses.
- Method of joints for analyzing space trusses.
- Equilibrium of frames and machines in a plane.
- Shear and bending-moment diagrams for beams.
- Parabolic and catenary cables.

A structure is a body, or an assemblage of component parts, which withstands applied loads. When the component parts are all straight two-force bodies, the assemblage is called an *idealized truss*. When one or more of the component parts are multiforce members, the assemblage is called a *frame* if all of the members are immovable, and a *machine* if some or all of its members are movable. The concepts developed previously in studying the equilibrium of particles and rigid bodies are adapted, in this chapter, to the study of the conditions of equilibrium of trusses, frames, machines, beams, and cables. In particular, useful methods are presented for the study of plane and space trusses, and a general procedure is given for solving equilibrium problems of frames and machines acted on by coplanar but nonconcurrent forces. The latter part of the chapter is devoted to the study of shears and bending moments in beams as well as to the analysis of cables. Since trusses, frames, machines, beams, and cables are basic structures, their analysis is fundamental and important.

TRUSSES

8.1 Trusses and Their Uses

A *joint* of a structure is a connection where two or more members of the structure are fastened together. A *truss* is a structure composed of a number of straight members connected (e.g., pinned, nailed, riveted, bolted, or welded) together at their ends by the joints of the structure where the loads are directly applied to the centers of the various joints, not to the portions of the members between the joints. Note that a joint may be a pin plus the end portions of the members being connected by the pin. The pin by itself is not a joint, although it is the center of the joint.

In modern engineering practice, wooden truss members are often nailed or bolted together at the joints and metal truss members are often riveted, bolted, or welded to gusset plates at the joints to achieve a greater degree of rigidity for the structure as illustrated in Fig. 8.1(a). A truss whose members cannot freely rotate at the joints is called a *rigidly connected truss*. In elementary studies, truss members are, however, taken as being pin-connected at the joints as illustrated in Fig. 8.1(b). A truss that uses a smooth pin at each joint to connect the truss members is called a *pin-connected truss*.

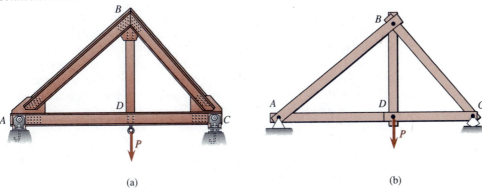

(a)   (b)

Fig. 8.1

Trusses are frequently used in roofs, bridges, stadiums, towers, cranes, etc. Some illustrations of trusses are shown in Fig. 8.2. Note in many drawings that a solid line is used to denote a member and a small circle is used to denote a joint. If a small circle is absent at the intersection of two solid lines, there is no connection of the members at that intersection.

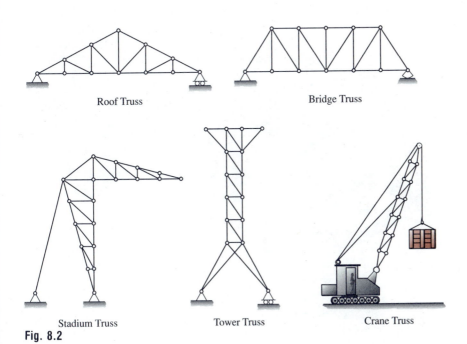

Roof Truss   Bridge Truss

Stadium Truss   Tower Truss   Crane Truss

Fig. 8.2

Developmental Exercises

D8.1   Define the terms: (a) joint, (b) truss, (c) rigidly connected truss, (d) pin-connected truss.

D8.2   Give three examples where trusses are used.

8.2   Usual Assumptions for Trusses

The *usual assumptions for trusses* in elementary studies are as follows:

1. Every member of a truss is straight and is connected at each end to a joint by a smooth circular pin on the longitudinal centroidal axis of the member. (No member is continuous through a joint, and deformations of members due to induced internal strains are taken as negligible.)
2. The weights of members of a truss are taken as being negligible compared with the applied loads. (If the weights of truss members were to be taken into account, they would be assumed to be applied to the joints, half of the weight of the member being applied to each of the two joints to which the member is connected.)
3. All loads are applied or transmitted to the joints only. (This assumption is usually met by providing a floor system to transmit loads between the joints. The manner in which floor beams and stringers are arranged to transmit loads to the joints of an actual bridge truss is illustrated in Fig. 8.3.)

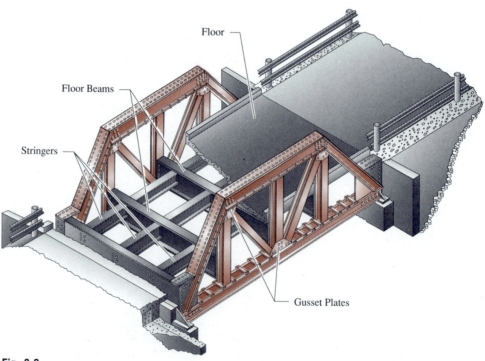

Fig. 8.3

A truss satisfying the above usual assumptions is called an *idealized truss*. Clearly, *members of an idealized truss are all two-force bodies*. It is to be pointed out that results of studies based on the idealized version of the rigidly connected truss are relatively easy to obtain and are useful for practical purposes. In this text, all trusses are to be taken as idealized trusses.

Developmental Exercises

D8.3   What are the usual assumptions for trusses?

D8.4   What is an idealized truss?

8.3   Rigid Truss, Plane Truss, and Simple Truss

A truss is said to have *collapsed* if its original configuration has undergone a great change due to loading. A *rigid truss* is a truss which does not collapse under the action of loads. The trusses shown in Fig. 8.4(a) and (b) are nonrigid. They may be made rigid by providing them with one more member and one more component of external constraint, respectively, as shown in Fig. 8.4(c) and (d).

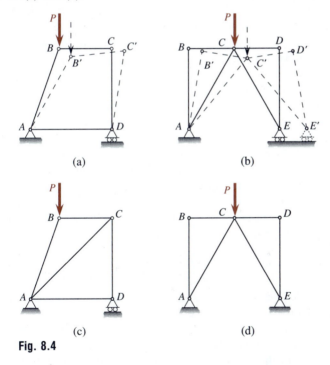

Fig. 8.4

The most basic two-dimensional rigid truss is a *triangular truss* which has three members and three joints as shown in Fig. 8.5(a). A *plane truss* is a two-dimensional truss; its members and the applied forces all lie in a plane. A *simple truss*[†] is a rigid plane truss which can be built from a triangular truss by repeating appropriately the addition of *two new members*

[†]The term *simple truss* denotes a plane truss. A *simple space truss* is defined in Sec. 8.10.

and *one new joint* at a time. This is sequentially illustrated in parts (a), (b), and (c) in Fig. 8.5.

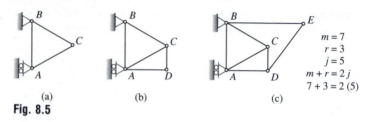

(a)   (b)   (c)

Fig. 8.5

$m = 7$
$r = 3$
$j = 5$
$m + r = 2j$
$7 + 3 = 2(5)$

Note that in building a simple truss the two new members are fastened to two separate existing joints, and the new joint connecting these two new members does not lie on the line containing those two existing joints. Suppose that m is the number of members and j is the number of joints in a simple plane truss. Then, the total number of new members and the total number of new joints, which are added to the original triangular truss, are $m - 3$ and $j - 3$, respectively. By the definition of a *simple truss*, we must, therefore, have $m - 3 = 2(j - 3)$, or

$$m + 3 = 2j \qquad (8.1)$$

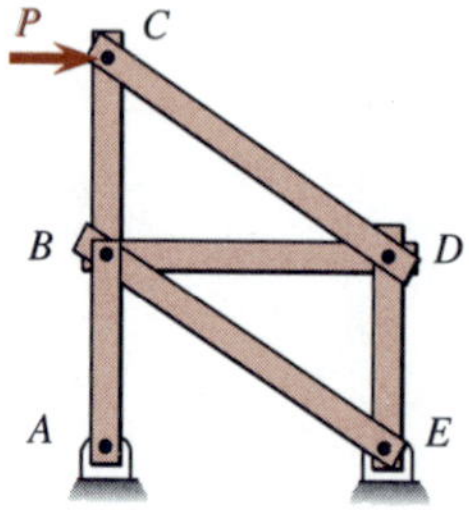

In associating Eq. (8.1) with a simple truss, it is necessary to define a *truss member* as a straight member which has *each of its ends connected to a joint* (which joins two or more members), rather than just a *hinge*. If one end of a member is connected to a joint but its other end is connected to a hinge, then that member is to be regarded as a *link* support, not as a truss member, of a simple truss. For example, the member *AB* in Fig. 8.6 needs to be regarded as a *link* support of the truss, not as a truss member, if the truss is to be classified as a simple truss; otherwise, Eq. (8.1) will not be satisfied.

Since truss members are two-force bodies, the forces in the free-body diagram of each joint must form a system of concurrent forces. The analysis of the equilibrium of *each joint* is, therefore, similar to the analysis of the equilibrium of a *particle*. For equilibrium of a truss lying in the *xy* plane, there are *two* scalar equations of equilibrium to be satisfied by the forces in the free-body diagram of *each* joint. They are

$$\Sigma F_x = 0 \qquad \Sigma F_y = 0 \qquad \begin{array}{c} (3.4)^\dagger \\ \text{(repeated)} \end{array}$$

This means that we may write $2j$ equilibrium equations from the j free-body diagrams of the j joints of a plane truss.

Suppose that a plane truss has j joints, m unknown forces in the m members, and r unknown components of reactions from the supports. The equilibrium of all of the j joints of the truss implies and ensures that the entire truss as a rigid body is also in equilibrium. We readily see that when the $(m + r)$ unknowns satisfy those $2j$ equilibrium equations of the joints, they will also automatically satisfy the equations of equilibrium of the entire truss

Fig. 8.6

[†]Note that the *x* and *y* axes may be chosen to be any pair of orthogonal axes having any desirable orientations. Furthermore, either of these two force equilibrium equations may be replaced with a moment equilibrium equation about any point other than the pin in the joint. (Cf. Example 8.1.)

or any portion of the truss. Thus, the *necessary condition* for a plane truss to be statically determinate is

$$m + r = 2j \tag{8.2}$$

This equation reduces to Eq. (8.1) when $r = 3$ (e.g., a truss with a hinge support and a roller support).

Developmental Exercises

D8.5   Define (a) a rigid truss, (b) a plane truss, (c) a simple truss.

D8.6   A simple truss has 15 members. How many joints does it have?

D8.7   A statically determinate plane truss has 5 joints and 4 components of reaction from its supports. How many members does it have?

8.4   Compound and Complex Trusses

A *compound truss* is a rigid plane truss which consists of two or more simple trusses directly connected together by linking joints, or linking members, or both, but cannot be built with the same procedure as that used in building a simple truss. Some illustrations of compound trusses are shown in Fig. 8.7, where component simple trusses are shaded and *LJ* denotes a linking joint and *LM* denotes a linking member.

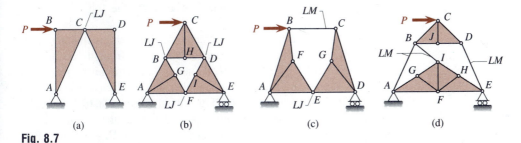

(a)   (b)   (c)   (d)

Fig. 8.7

A *complex truss* is a rigid plane truss which cannot be classified as a simple truss or a compound truss. Some complex trusses are illustrated in Fig. 8.8. Note that there is no joint at the middle of each truss shown in

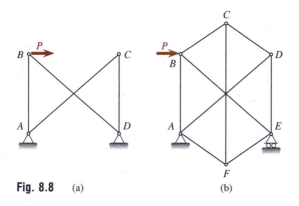

Fig. 8.8   (a)   (b)

this figure. The compound and complex trusses shown in Figs. 8.7 and 8.8 are statically determinate and can be shown to satisfy Eq. (8.2). We recall that the member *AB* in Fig. 8.6 needs to be regarded as a *link* support if the truss shown is to be classified as a *simple truss*. On the other hand, we can now regard the member *AB* as a truss member by classifying that truss as a *complex truss*, which satisfies Eq. (8.2) and is statically determinate.

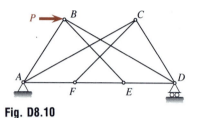

Fig. D8.10

Developmental Exercises

D8.8   Do the two trusses shown in Fig. 8.8 contain any simple trusses?

D8.9   Define (a) a compound truss, (b) a complex truss.

D8.10   The truss shown contains simple trusses. (a) Name those simple trusses. (b) Name the linking members connecting those simple trusses. (c) Is the truss shown a compound or a complex truss?

8.5   Forces in Truss Members

Since all loads are applied or transmitted to a truss at its joints only, any force induced in a truss member must come from the two pins at the two joints to which the member is connected. Thus, *each truss member is a two-force body*, as pointed out earlier. The *orientation* of the force in each truss member is the same as the axis of that member; only its *magnitude* and its *sense* need to be specified or determined. Therefore, the force in a truss member is also called an *axial force*.

If the axial force in a truss member tends to *stretch* the member or *pull* the end part of the member in an isolated joint away from its connecting pin, such as the force $\mathbf{F}_{AE}$ in Fig. 8.9, the axial force is a *tensile force* and the member is said to be in tension. A member that is in tension is called a *tensile member*. On the other hand, if the axial force in a truss member tends to *compress* the member or *push* the end part of the member in an isolated joint toward its connecting pin, such as the force $\mathbf{F}_{AB}$ or $\mathbf{F}_{AD}$ in Fig. 8.9, the axial force is a *compressive force* and the member is said to be in compression. A member that is in compression is called a *compressive member*.

In the analysis of trusses, each unknown axial force may initially be assumed to be either a tensile or a compressive force. If the value of the magnitude of the unknown axial force is algebraically found to be positive in the subsequent solution, the assumption is correct; if negative, we first drop the negative sign in the obtained value for the *magnitude* and then reverse the assumed *sense* of the axial force (i.e., change from tensile to compressive or vice versa) in the final assertion of the axial force.

The design criteria for tensile and compressive members of a truss are different. Therefore, each nonzero answer for the axial force in the solution of each truss problem should be labeled with a *T* when it is a tensile force and with a *C* when it is a compressive force. Note that the names of the joints to which a truss member is connected are used as the subscripts of the letter *F* to *label the magnitude of the axial force* in the truss member. The order of the subscripts is immaterial; e.g., $F_{AB} = F_{BA}$ = magnitude of axial force in the member which is connected to the joints *A* and *B*.

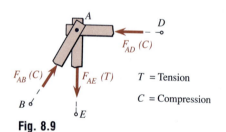

T = Tension

C = Compression

Fig. 8.9

Developmental Exercises

D8.11   Define the terms: (a) tensile force, (b) tensile member, (c) compressive force, (d) compressive member.

D8.12   How is the magnitude of the axial force in the member CD of a truss labeled in this chapter?

D8.13   Is $F_{GH} = F_{HG}$?

8.6   Zero-Force Members at T and V Joints

A *T joint* is a joint which connects three members where only two of them are collinear as shown in Fig. 8.10. *If no external load is applied to a T joint which is in equilibrium, the force in the noncollinear member must be zero.* This must be true; otherwise, the summation of the components of forces acting on the free body of the T joint in the direction perpendicular to the axis along the two collinear members cannot be zero. For example, if $F_{AD} \neq 0$ in Fig. 8.10, then $\Sigma F_y = -F_{AD}\sin\theta \neq 0$ because $\sin\theta \neq 0$. Since $\Sigma F_y = 0$ for equilibrium, we see that we must have $F_{AD} = 0$.

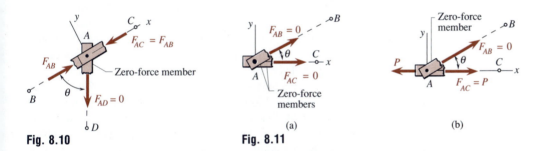

Fig. 8.10   Fig. 8.11

A *V joint* is a joint which connects two noncollinear members as shown in Fig. 8.11. *If no external load is applied to a V joint which is in equilibrium, the forces in those two members must be zero.* This is illustrated in Fig. 8.11(a) and must be true; otherwise, the summation of the components of forces acting on the free body of the V joint in the direction perpendicular to the axis of one of the two members, say ΣF_y, cannot be zero. Furthermore, when a V joint is acted on by a single force which is collinear with one of its two members, the force in the other member must be zero if equilibrium exists. This is illustrated in Fig. 8.11(b) and can be verified in a similar manner.

Developmental Exercises

D8.14   For equilibrium, verify that $F_{AC} = F_{AB}$ in Fig. 8.10 and $F_{AC} = P$ in Fig. 8.11(b).

D8.15   Which of the members in the truss loaded as shown are zero-force members?

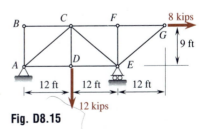

Fig. D8.15

8.7   Method of Joints for Plane Trusses

The equilibrium of each joint of a plane truss is similar to the equilibrium of a particle in a plane because the force system acting at each joint is either zero or coplanar and concurrent. The method used to determine the unknown axial forces or other parameters of a truss by considering the equilibrium of the individual joints, one at a time, is called the *method of joints*. This is one of the two basic methods for analyzing plane trusses; the other method is presented in Sec. 8.8.

According to Eq. (8.2), the *m* unknown axial forces and the *r* unknown components of reactions from the supports of a statically determinate plane truss with *j* joints may be determined by solving a set of 2*j* simultaneous equations of equilibrium written for those *j* joints. However, it is generally preferable to avoid solving a large set of simultaneous equations if possible. We see from Eqs. (3.4) that, for a plane truss, no more than two unknowns can be solved from the consideration of equilibrium of just one joint. Thus, whenever possible, we choose for each step of the solution an appropriate joint whose free-body diagram contains *no more than two unknowns*. This process is continued until all desired unknowns have been determined and checked. To facilitate such a process, we *sometimes* (but not always) need to determine beforehand the components of reactions from the supports of the truss by treating the entire truss as a rigid body in equilibrium.

We know that *when all joints of a truss are in equilibrium, the truss as a whole must be in equilibrium.* Therefore, redundant (or dependent) equations will be present in the set of equilibrium equations which are written for each joint of a truss *and* for the entire truss. However, the redundant equations may serve to check the solutions for the unknowns.

When the method of joints is applied to analyze a compound or complex truss, we may *sometimes* need to solve a set of more than two simultaneous equations which are coupled. For example, even if the *three* reaction components of the compound truss shown in Fig. 8.7(c) have been determined beforehand, the *method of joints* as applied to this truss will still lead to *eleven* unknown axial forces to be determined from the coupled equilibrium equations written for the various joints. A similar situation exists for the complex truss shown in Fig. 8.8(b). (Cf. Probs. 8.24 and 8.25*.)

EXAMPLE 8.1

Using the method of joints, determine the forces in members *AB* and *AE* of the simple cantilever truss shown.

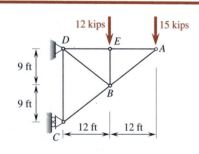

Solution. We do not need to find the reactions from the supports because the free-body diagram of the joint *A* as shown contains only two unknowns which are being sought. Note that the unknown forces are here arbitrarily assumed to be tensile forces. For equilibrium, we write

$$+\uparrow \Sigma F_y = 0: \quad -\frac{3}{5} F_{AB} - 15 = 0 \qquad F_{AB} = -25$$

$$\xrightarrow{+} \Sigma F_x = 0: \quad -F_{AE} - \frac{4}{5} F_{AB} = 0 \qquad F_{AE} = 20$$

Since F_{AE} is positive but F_{AB} is negative, the answers are

$$F_{AB} = 25 \text{ kips } C \qquad F_{AE} = 20 \text{ kips } T \quad \blacktriangleleft$$

where C denotes *compression* and T denotes *tension*.

REMARK. The value of F_{AE} may directly be obtained with a moment equilibrium equation as follows:

$$+ \, \circlearrowleft \, \Sigma M_B = 0: \quad 9F_{AE} - 12(15) = 0 \quad F_{AE} = 20 \text{ kips } T \quad \blacktriangleleft$$

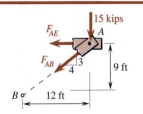

EXAMPLE 8.2

Using the *method of joints*, determine the force in each member of the simple truss shown.

Solution. We need to determine the reactions from the supports before we can choose for study an appropriate joint whose free-body diagram contains no more than two unknowns. From the free-body diagram of the entire truss in equilibrium as shown, we write

$$+ \, \circlearrowleft \, \Sigma M_A = 0: \quad 9\left(\frac{4}{5}\right)F - 6(120) = 0 \qquad F = 100$$

$$\xrightarrow{+} \Sigma F_x = 0: \quad A_x + \frac{3}{5}F = 0 \qquad\qquad A_x = -60$$

$$+\uparrow \Sigma F_y = 0: \quad A_y - 120 + \frac{4}{5}F = 0 \qquad A_y = 40$$

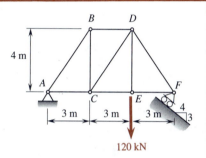

Thus, the reactions are $\mathbf{F} = 100 \text{ kN } \nearrow$, $\mathbf{A}_x = 60 \text{ kN } \leftarrow$, and $\mathbf{A}_y = 40 \text{ kN } \uparrow$.

The force in each member of the truss may now be determined by considering the equilibrium of the various joints as follows:

Joint A. Using the solutions above, we draw the free-body diagram of the joint A as shown. For equilibrium, we write

$$+\uparrow \Sigma F_y = 0: \quad \frac{4}{5}F_{AB} + 40 = 0 \qquad\qquad F_{AB} = -50$$

$$\xrightarrow{+} \Sigma F_x = 0: \quad \frac{3}{5}F_{AB} + F_{AC} - 60 = 0 \qquad F_{AC} = 90$$

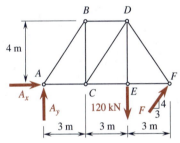

Noting the assumed senses of the unknown axial forces in the free-body diagram and the signs of the determined values of F_{AB} and F_{AC}, we write

$$F_{AB} = 50 \text{ kN } C \qquad F_{AC} = 90 \text{ kN } T \quad \blacktriangleleft$$

Joint B. Knowing that the force in the member AB is a 50-kN compressive force, we draw the free-body diagram of the joint B as shown, where the force in the member BD is assumed to be compressive. For equilibrium, we write

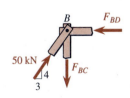

$$\xrightarrow{+} \Sigma F_x = 0: \quad \frac{3}{5}(50) - F_{BD} = 0 \qquad F_{BD} = 30$$

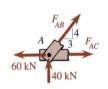

$$+\uparrow \Sigma F_y = 0: \quad \frac{4}{5}(50) - F_{BC} = 0 \qquad F_{BC} = 40$$

$$F_{BC} = 40 \text{ kN } T \qquad F_{BD} = 30 \text{ kN } C \quad \blacktriangleleft$$

Joint C. Knowing the forces in members *BC* and *AC*, we draw the free-body diagram of the joint *C* as shown. For equilibrium, we write

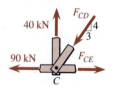

$$+\uparrow \Sigma F_y = 0: \quad 40 - \frac{4}{5}F_{CD} = 0 \qquad F_{CD} = 50$$

$$\xrightarrow{+} \Sigma F_x = 0: \quad F_{CE} - \frac{3}{5}F_{CD} - 90 = 0 \qquad F_{CE} = 120$$

$$F_{CD} = 50 \text{ kN } C \qquad F_{CE} = 120 \text{ kN } T \quad \blacktriangleleft$$

Joint D. Knowing the forces in members *BD* and *CD*, we draw the free-body diagram of the joint *D* as shown. For equilibrium, we write

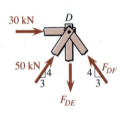

$$\xrightarrow{+} \Sigma F_x = 0: \quad 30 + \frac{3}{5}(50) - \frac{3}{5}F_{DF} = 0 \qquad F_{DF} = 100$$

$$+\uparrow \Sigma F_y = 0: \quad \frac{4}{5}(50) - F_{DE} + \frac{4}{5}F_{DF} = 0 \qquad F_{DE} = 120$$

$$F_{DE} = 120 \text{ kN } T \qquad F_{DF} = 100 \text{ kN } C \quad \blacktriangleleft$$

Joint E. The free-body diagram of this joint is drawn as shown. For equilibrium, we write

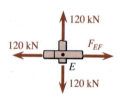

$$\xrightarrow{+} \Sigma F_x = 0: \quad F_{EF} - 120 = 0 \qquad F_{EF} = 120 \text{ kN } T \quad \blacktriangleleft$$

$$+\uparrow \Sigma F_y = 0: \quad 120 - 120 = 0$$

Joint F. The free-body diagram of this joint is drawn as shown. For equilibrium, we write

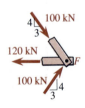

$$\xrightarrow{+} \Sigma F_x = 0: \quad \frac{3}{5}(100) + \frac{3}{5}(100) - 120 = 0$$

$$+\uparrow \Sigma F_y = 0: \quad \frac{4}{5}(100) - \frac{4}{5}(100) = 0$$

REMARK. We have written three equilibrium equations for the entire truss as a rigid body to determine the three unknown components of reactions (i.e., A_x, A_y, and F) in the beginning of the solution. This is why we have the last three extra equations which serve to check the results already obtained. If we considered only the equilibrium of the six joints but not that of the entire truss, we would have to solve a set of *twelve* simultaneous equations for the *nine* unknown axial forces and the *three* unknown components of reactions.

Developmental Exercises

D8.16 Using the *method of joints*, determine the force in each member of the simple truss shown.

D8.17 A complex truss is shown. Using the *method of joints*, express F_{CF} and F_{CD} in terms of F_{EF} with the assumption that the members CF, CD, and EF are in tension.

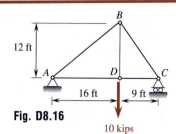

Fig. D8.16

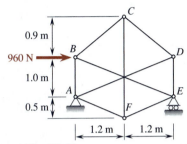

Fig. D8.17

PROBLEMS

8.1 through 8.12* Using the *method of joints*, determine the force in each member of the truss shown.

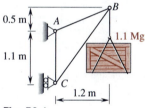

Fig. P8.1

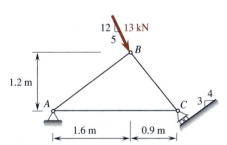

Fig. P8.2

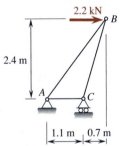

Fig. P8.3*

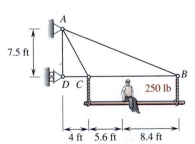

Fig. P8.4

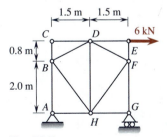

Fig. P8.5

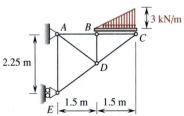

Fig. P8.6*

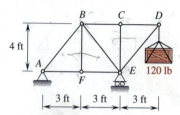

Fig. P8.7

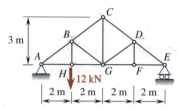

Fig. P8.8

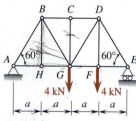

Fig. P8.9*

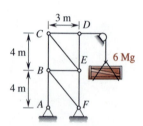

Fig. P8.10

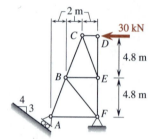

Fig. P8.11

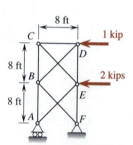

Fig. P8.12* and P8.13*

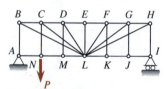

Fig. P8.14 and P8.18

8.13* through 8.17* Determine whether the truss shown is a simple truss, a compound truss, or a complex truss.

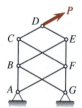

Fig. P8.16* and P8.20*

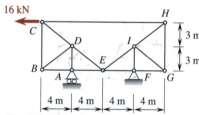

Fig. P8.17* and P8.21

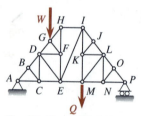

Fig. P8.15 and P8.19*

8.18 through 8.20* Determine the zero-force members in the truss shown.

8.21 Using the *method of joints*, determine the force in each member of the truss shown. (*Hint.* Determine the reactions first.)

8.22 and 8.23* Using the *method of joints*, determine the force in each member of the complex truss shown.

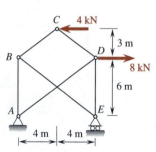

Fig. P8.22

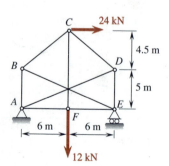

Fig. P8.23*

8.24 and 8.25* Using the *method of joints*, determine the force in each member of the compound truss shown.[†]

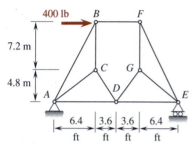

Fig. P8.24

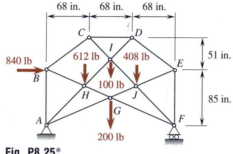

Fig. P8.25*

8.8 Method of Sections for Plane Trusses

The method used to determine the unknown axial forces or other parameters of a truss by considering the equilibrium of a portion of the truss as a rigid body which contains two or more joints is called the *method of sections*. This method often provides a shortcut in determining the forces in certain truss members. Furthermore, this method can be used in conjunction with the method of joints to eliminate the need of solving a large number of simultaneous equations in determining the forces in members of some compound and complex trusses alluded to in Sec. 8.7. In using the method of sections, we *sometimes* (but not always) need to determine beforehand some of the components of reactions from the supports of the truss. We shall describe the *moment* and *shear approaches* in this method.[††]

I. MOMENT APPROACH.   Oftentimes, we can isolate a portion of a plane truss by cutting along a section through certain members *whose axes all intersect at a common point*, except for the axes of the members whose axial forces are already known and for the axis of the member whose axial force is to be determined. If such is the case, we may proceed to consider the equilibrium of the free body of the isolated portion of the truss where pertinent components of reactions, if any, are determined beforehand from the free body of the entire truss. The desired axial force may then be determined by *summing to zero the moments of all forces acting on the free body of the isolated portion about the common point of intersection*.

[†]*Suggestion.* Do not determine beforehand the components of reactions from the supports. Treat them as undetermined unknowns so that a set of $2j$ equations in $2j$ unknowns will be obtained, where j is the number of joints. A computer may then be used to solve these equations (cf. App. D). Afterward, use the equilibrium equations for the entire truss to check the solutions for the reaction components.

[††]These two approaches simply pinpoint the shortest ways possible to solve for the unknowns in the subtruss as a rigid body.

EXAMPLE 8.3

Determine the forces in members *BD* and *CE* of the truss in Example 8.2.

Solution. We first determine **F** and then isolate the portion *DEF* of the truss to determine F_{BD} and F_{CE} by using the *moment approach* as follows:

Entire truss

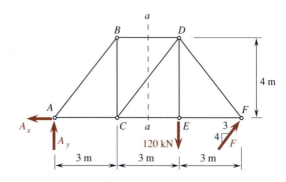

$$+ \circlearrowleft \; \Sigma M_A = 0: \quad 9\left(\frac{4}{5}\right)F - 6(120) = 0 \qquad \mathbf{F} = 100 \text{ kN} \nearrow$$

Portion DEF

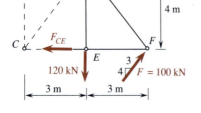

$$+ \circlearrowleft \; \Sigma M_C = 0: \quad 4F_{BD} - 3(120) + 6\left(\frac{4}{5}\right)(100) = 0$$

$$+ \circlearrowleft \; \Sigma M_D = 0: \quad -4F_{CE} + 4\left(\frac{3}{5}\right)(100) + 3\left(\frac{4}{5}\right)(100) = 0$$

These two equations yield $F_{BD} = -30$ and $F_{CE} = 120$. Thus, we write

$$F_{BD} = 30 \text{ kN } C \qquad F_{CE} = 120 \text{ kN } T \; \blacktriangleleft$$

REMARK. These answers check those obtained in Example 8.2. Note that the portion *ABC* could have been used in the solution if the reaction components $\mathbf{A}_x$ and $\mathbf{A}_y$ had been determined beforehand.

Developmental Exercise

D8.18 Determine F_{BD} in the truss in Example 8.1.

EXAMPLE 8.4

Determine the forces in members *AF*, *FN*, and *JK* of the compound truss shown.

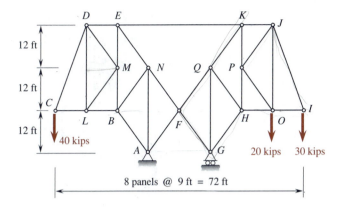

Solution. We first determine **G** and then isolate the portions *KQFGHOIJ* and *JPOI* of the truss to determine F_{AF}, F_{FN}, and F_{JK} by using the *moment approach* as follows:

Entire truss

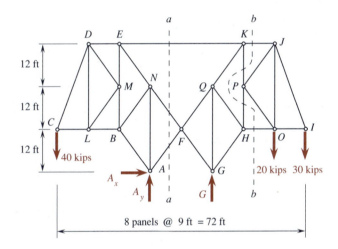

$$+\,\text{\Large ↺}\;\Sigma M_A = 0: \quad 18G + 27(40) - 36(20) - 45(30) = 0$$
$$G = 55 \qquad \mathbf{G} = 55 \text{ kips} \uparrow$$

Portion KQFGHOIJ

We transmit $\mathbf{F}_{AF}$ to R, as shown on next page, and then write

$$+\,\text{\Large ↺}\;\Sigma M_E = 0: \quad -48\!\left(\frac{3}{5}\right)\!F_{AF} + 27(55) - 45(20) - 54(30) = 0$$

$$F_{AF} = -35.9 \qquad F_{AF} = 35.9 \text{ kips } C \;\blacktriangleleft$$

Furthermore, we transmit $\mathbf{F}_{FN}$ to E and then write

$$+\,\text{\Large ↺}\;\Sigma M_K = 0: \quad -36\!\left(\frac{4}{5}\right)\!F_{FN} - 9(55) - 9(20) - 18(30) = 0$$

$$F_{FN} = -42.2 \qquad F_{FN} = 42.2 \text{ kips } C \;\blacktriangleleft$$

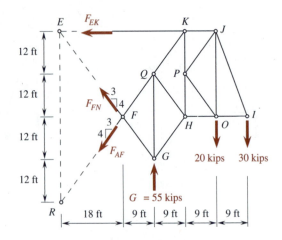

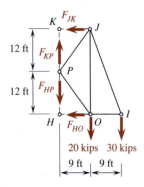

Portion JPOI

$$+\circlearrowleft \Sigma M_H = 0: \quad 24F_{JK} - 9(20) - 18(30) = 0$$

$$F_{JK} = 30 \qquad F_{JK} = 30 \text{ kips } T \blacktriangleleft$$

REMARK. The same equation for determining the value of F_{JK} could similarly be obtained if we had chosen to isolate the portion *JPHOI* of the truss for consideration and set $\Sigma M_H = 0$.

Developmental Exercise

D8.19   Determine F_{EK}, F_{HO}, and F_{DE} in the truss in Example 8.4.

II. SHEAR APPROACH.

At some other time, we can isolate a portion of a plane truss by cutting along a section through certain members *whose axes are all parallel to each other*, except for the members whose axial forces are already known and for the member whose axial force is to be determined. If such is the case, we may proceed to consider the equilibrium of the free body of the isolated portion of the truss where pertinent components of reactions, if any, are determined beforehand from the free body of the entire truss. The desired axial force may then be determined by *summing to zero the components of all forces acting on the free body of the isolated portion in the direction perpendicular to those parallel members*. (Note here that reaction components that are parallel to those parallel members do not need to be determined beforehand.)

EXAMPLE 8.5

Determine the force F_{CD} in the truss in Example 8.2.

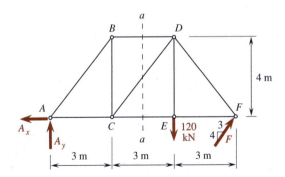

Solution. We first determine $\mathbf{A}_y$ from the entire truss and then isolate the portion *ABC* of the truss to determine F_{CD} by using the *shear approach* as follows:

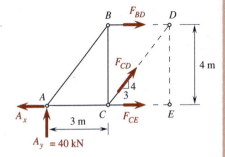

Entire truss

$$+ \circlearrowleft \Sigma M_F = 0: \quad -9A_y + 3(120) = 0$$

$$A_y = 40 \qquad \mathbf{A}_y = 40 \text{ kN} \uparrow$$

Portion ABC

$$+ \circlearrowleft \Sigma F_y = 0: \quad \frac{4}{5} F_{CD} + 40 = 0$$

$$F_{CD} = -50 \qquad F_{CD} = 50 \text{ kN } C \blacktriangleleft$$

REMARK. We do not need to find the value of $\mathbf{A}_x$ because $\mathbf{A}_x$ is parallel to the horizontal members *BD* and *CE*.

Developmental Exercises

D8.20 Determine F_{CN} in the truss shown.

D8.21 Which of the following best describes the value of F_{OK} in the truss shown? (a) 12 kips *T*, (b) 12 kips *C*, (c) 15 kips *T*, (d) 15 kips *C*, (e) statically indeterminate.

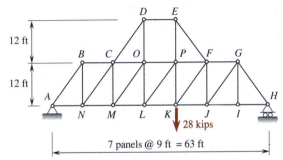

Fig. D8.20 and D8.21

EXAMPLE 8.6

Determine the force in the member *EJ* of the compound truss shown.

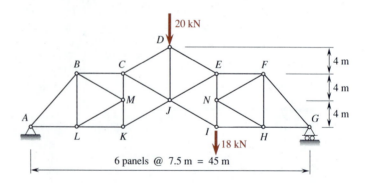

Solution. We first determine **G** and then isolate the portion *ENIHGF* of the truss to determine F_{EJ} by using the *shear approach* as follows:

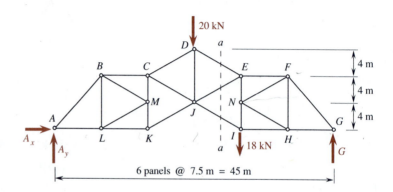

Entire truss

$$+\circlearrowright \ \Sigma M_A = 0: \quad 45G - 4(7.5)(18) - 3(7.5)(20) = 0$$

$$G = 22 \qquad \textbf{G} = 22 \text{ kN} \uparrow$$

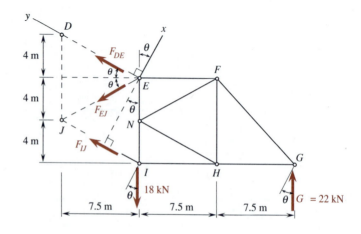

Portion ENIHGF

$$\tan\theta = \frac{4}{7.5} \qquad \theta = 28.07°$$

$$+\nearrow \Sigma F_x = 0: \quad -F_{EJ}\sin 2\theta - 18\cos\theta + 22\cos\theta = 0$$

$$F_{EJ} = 4.25 \qquad F_{EJ} = 4.25 \text{ kN } T \blacktriangleleft$$

Developmental Exercise

D8.22   Determine F_{CJ} in the truss in Example 8.6.

8.9 X Joints and Successive Method

An *X joint*, as shown in Fig. 8.12, is a joint which connects four members in two collinear pairs lying in two intersecting straight lines. *If there is no external force applied to an X joint which is in equilibrium, the tensile or compressive forces in members of a collinear pair of members must be equal*: however, the forces in this pair may not be equal to the forces in the other pair. This must be true; otherwise, the summation of the components of forces acting on the free body of the X joint in the direction perpendicular to either of the said intersecting straight lines cannot be zero. This property is useful in analyzing trusses containing X joints.

At times, the determination of the axial force in a specified member of a truss cannot be achieved by using either the method of joints or the method of sections only once. We may need to use the method of sections, or the method of joints, or both, to determine the axial forces in certain other nearby members before we can finally determine the specified axial force. The method which successively employs the method of sections, or the method of joints, or both, in determining a specified unknown axial force or parameter of a truss is called the *successive method*.

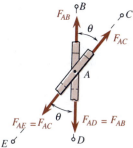

Fig. 8.12

EXAMPLE 8.7

Determine the force in the member *DE* of the compound truss shown.

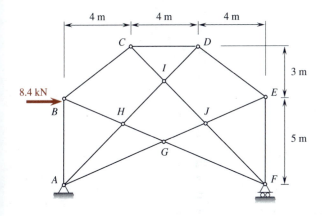

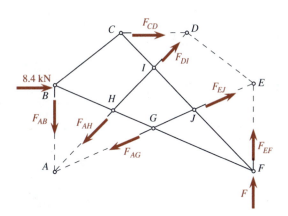

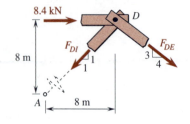

Solution. The value of F_{DE} can be determined by applying the *successive method* to the portion *BCIJFGH* and the joint *D*, which are isolated as shown on the preceding page and this page, respectively.

Portion BCIJFGH

Recalling the property associated with X joints, we see that $F_{AH} = F_{HI} = F_{DI}$ and $F_{AG} = F_{GJ} = F_{EJ}$. Since there is *no net effect* on a rigid body from any pair of collinear forces which are equal in magnitude and opposite in sense, we may *disregard* the unknown forces $\mathbf{F}_{AH}$, $\mathbf{F}_{DI}$, $\mathbf{F}_{AG}$, and $\mathbf{F}_{EJ}$ acting on the free body of the isolated portion shown.

$$\xrightarrow{+} \Sigma F_x = 0: \quad F_{CD} + 8.4 = 0 \qquad F_{CD} = -8.4 \qquad F_{CD} = 8.4 \text{ kN } C$$

Joint D

$$+ \circlearrowleft \ \Sigma M_A = 0: \quad -8\left(\frac{4}{5}\right)F_{DE} - 8\left(\frac{3}{5}\right)F_{DE} - 8(8.4) = 0$$

$$F_{DE} = -6 \qquad\qquad F_{DE} = 6 \text{ kN } C \ \blacktriangleleft^{\dagger}$$

Developmental Exercises

D8.23 Determine F_{BH} in the truss in Example 8.7.

D8.24 Determine F_{AG} in the truss in Example 8.7.

EXAMPLE 8.8

Determine the force F_{FN} in the truss in Example 8.6.

Solution. Having previously determined that $\mathbf{G} = 22$ kN $\uparrow$, we may apply the *successive method* to determine F_{EF} first and then F_{FN}.

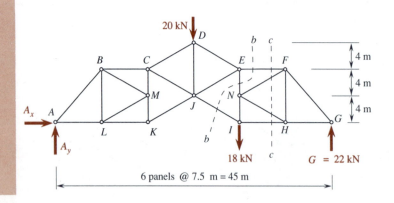

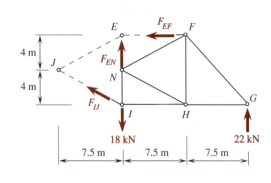

---

†Without using the property associated with X joints, as described in Sec. 8.9, much more effort would be needed to obtain the solution.

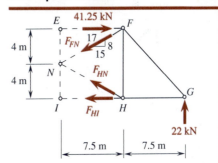

Portion FNIHG

$$+\circlearrowleft \ \Sigma M_I = 0: \quad 8F_{EF} + 15(22) = 0 \qquad F_{EF} = -41.25$$

$$F_{EF} = 41.25 \text{ kN } C$$

Portion FHG

$$+\circlearrowleft \ \Sigma M_H = 0: \quad 8\left(\frac{15}{17}\right)F_{FN} - 8(41.25) + 7.5(22) = 0$$

$$F_{FN} = 23.4 \qquad F_{FN} = 23.4 \text{ kN } T \ \blacktriangleleft$$

Developmental Exercises

D8.25 Determine F_{DJ} in the truss in Example 8.6.

D8.26 Determine F_{LM} in the truss in Example 8.6.

★8.10   Space Trusses

A *space truss* is a truss whose members do not all lie in a plane. The most basic rigid space truss is a *tetrahedron truss*, which has six members and four joints as shown in Fig. 8.13(a). In elementary studies of space trusses, the ends of truss members are assumed to be held by ball-and-socket joints, as illustrated in Fig. 8.13(b). Thus, each member of a space truss is treated as a two-force body.

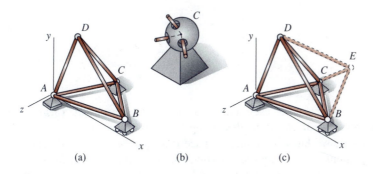

(a)   (b)   (c)

Fig. 8.13

A *simple space truss* is a rigid space truss which can be constructed from a tetrahedron truss by repeating appropriately the addition of *three new members* and *one new joint* at a time, as shown in Fig. 8.13(c). Note that the three new members are first fastened at their ends to three separate existing joints and then connected together at their other ends by the new joint. If m and j are the number of members and the number of joints in a simple space truss, respectively, we must have $m - 6 = 3(j - 4)$, or

$$m + 6 = 3j \tag{8.3}$$

Similar to Eq. (8.2), the *necessary condition* for a space truss to be statically determinate can be shown to be

$$m + r = 3j \tag{8.4}$$

where r is the number of unknown components of reaction from the supports. Space trusses considered in this chapter are statically determinate.

Developmental Exercises

D8.27 Define (a) a space truss, (b) a simple space truss.

D8.28 A simple space truss has 10 joints. How many truss members does it have?

★8.11 Method of Joints for Simple Space Trusses

The equilibrium of each joint of a space truss is similar to the equilibrium of a particle in space because the force system acting at each joint is either zero or concurrent and noncoplanar. In other words, the equilibrium of each joint of a space truss is described by Eq. (3.3) or its scalar version, Eqs. (3.5). Thus, we have

$$\Sigma \mathbf{F} = \mathbf{0} \tag{3.3}$$
(repeated)

$$\Sigma F_x = 0 \qquad \Sigma F_y = 0 \qquad \Sigma F_z = 0 \tag{3.5}$$
(repeated)

for the equilibrium of each joint of a space truss. From Eqs. (3.5), we infer that there are $3j$ equations describing the equilibrium of the j joints. From Eq. (8.3), we know in a simple space truss that $m + 6 = 3j$. Thus, those $3j$ equations may be used to determine the m unknown axial forces in the m members plus (up to) six unknown components of reactions from the supports of a simple space truss. This is the logic which forms the basis of the *method of joints for analyzing simple space trusses*.

Frequently, the effort required to solve the $3j$ equations may considerably be reduced by first determining the pertinent unknown reactions from the supports and then considering the equilibrium of the joints in such an order that each time no more than three new unknowns are involved. In this manner, we can avoid solving a large system of simultaneous equations.

EXAMPLE 8.9

A force $\mathbf{P} = 20\mathbf{i} - 60\mathbf{j} + 30\mathbf{k}$ kN is applied to the joint D of the space truss shown. Determine (a) the reactions from the supports, (b) the force in each member.

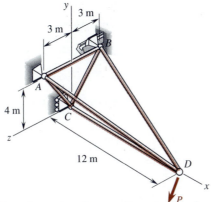

Solution. We first determine the reactions in the free-body diagram as shown.

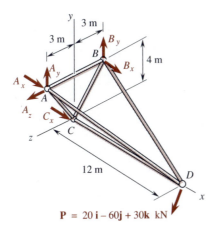

We write

$$\Sigma\mathbf{F} = 0: \quad (A_x + B_x + C_x + 20)\mathbf{i}$$
$$+ (A_y + B_y - 60)\mathbf{j} + (A_z + 30)\mathbf{k} = 0 \tag{1}$$

$$\Sigma\mathbf{M}_A = 0: \quad (-6\mathbf{k}) \times (B_x\mathbf{i} + B_y\mathbf{j}) + (-4\mathbf{j} - 3\mathbf{k}) \times (C_x\mathbf{i})$$
$$+ (12\mathbf{i} - 4\mathbf{j} - 3\mathbf{k}) \times (20\mathbf{i} - 60\mathbf{j} + 30\mathbf{k}) = 0$$

$$(6B_y - 300)\mathbf{i} - (6B_x + 3C_x + 420)\mathbf{j} + (4C_x - 640)\mathbf{k} = 0 \tag{2}$$

Equations (1) and (2) give a set of six scalar equations, which yield $A_x = -30$, $A_y = 10$, $A_z = -30$, $B_x = -150$, $B_y = 50$, and $C_x = 160$; i.e.,

$$\mathbf{A} = -30\mathbf{i} + 10\mathbf{j} - 30\mathbf{k} \text{ kN} \quad \blacktriangleleft$$

$$\mathbf{B} = -150\mathbf{i} + 50\mathbf{j} \text{ kN} \qquad \mathbf{C} = 160\mathbf{i} \text{ kN} \quad \blacktriangleleft$$

For use in the method of joints, we have

$$\vec{AB} = -6\mathbf{k} \qquad\qquad \lambda_{AB} = -\mathbf{k} = -\lambda_{BA}$$

$$\vec{AC} = -4\mathbf{j} - 3\mathbf{k} \qquad\qquad \lambda_{AC} = -\tfrac{1}{5}(4\mathbf{j} + 3\mathbf{k}) = -\lambda_{CA}$$

$$\vec{AD} = 12\mathbf{i} - 4\mathbf{j} - 3\mathbf{k} \qquad\qquad \lambda_{AD} = \tfrac{1}{13}(12\mathbf{i} - 4\mathbf{j} - 3\mathbf{k}) = -\lambda_{DA}$$

$$\vec{BC} = -4\mathbf{j} + 3\mathbf{k} \qquad\qquad \lambda_{BC} = \tfrac{1}{5}(-4\mathbf{j} + 3\mathbf{k}) = -\lambda_{CB}$$

$$\vec{BD} = 12\mathbf{i} - 4\mathbf{j} + 3\mathbf{k} \qquad\qquad \lambda_{BD} = \tfrac{1}{13}(12\mathbf{i} - 4\mathbf{j} + 3\mathbf{k}) = -\lambda_{DB}$$

$$\vec{CD} = 12\mathbf{i} \qquad\qquad \lambda_{CD} = \mathbf{i} = -\lambda_{DC}$$

The unknown axial forces may be determined as follows:

Joint D

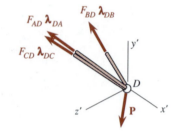

$$\Sigma \mathbf{F} = \mathbf{0}: \quad \left(-\frac{12}{13}F_{AD} - \frac{12}{13}F_{BD} - F_{CD} + 20\right)\mathbf{i}$$

$$+ \left(\frac{4}{13}F_{AD} + \frac{4}{13}F_{BD} - 60\right)\mathbf{j}$$

$$+ \left(\frac{3}{13}F_{AD} - \frac{3}{13}F_{BD} + 30\right)\mathbf{k} = \mathbf{0}$$

This vector equilibrium equation gives a set of three scalar equations, which yield $F_{AD} = 32.5$, $F_{BD} = 162.5$, and $F_{CD} = -160$. Thus, we have

$$F_{AD} = 32.5 \text{ kN } T \qquad F_{BD} = 162.5 \text{ kN } T \qquad F_{CD} = 160.0 \text{ kN } C \quad \blacktriangleleft$$

Joint C

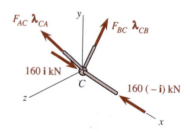

$$\Sigma \mathbf{F} = \mathbf{0}: \quad (160 - 160)\mathbf{i} + \left(\frac{4}{5}F_{AC} + \frac{4}{5}F_{BC}\right)\mathbf{j}$$

$$+ \left(\frac{3}{5}F_{AC} - \frac{3}{5}F_{BC}\right)\mathbf{k} = \mathbf{0}$$

This vector equilibrium equation gives *one trivial* and *two nontrivial* scalar equations, which yield

$$F_{AC} = F_{BC} = 0 \quad \blacktriangleleft$$

Joint B

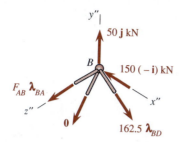

$$\Sigma \mathbf{F} = \mathbf{0}: \quad \left[\frac{12}{13}(162.5) - 150\right]\mathbf{i} + \left[-\frac{4}{13}(162.5) + 50\right]\mathbf{j}$$

$$+ \left[F_{AB} + \frac{3}{13}(162.5)\right]\mathbf{k} = \mathbf{0}$$

This vector equilibrium equation gives *two trivial* and *one nontrivial* scalar equations, which yield

$$F_{AB} = -37.5 \qquad F_{AB} = 37.5 \text{ kN } C \blacktriangleleft$$

Besides the above trivial scalar equations, the equilibrium of the joint *A* may be used to further check the obtained answers as follows:

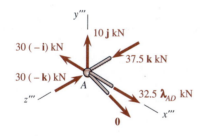

$$\Sigma \mathbf{F} = \left[\frac{12}{13}(32.5) - 30\right]\mathbf{i} + \left[-\frac{4}{13}(32.5) + 10\right]\mathbf{j}$$

$$+ \left[37.5 - \frac{3}{13}(32.5) - 30\right]\mathbf{k}$$

$$= 0\mathbf{i} + 0\mathbf{j} + 0\mathbf{k} = \mathbf{0}$$

REMARK. If we did not find the reactions from the supports first, we would need to solve simultaneous equations in determining the unknown axial forces in the truss.

Developmental Exercise

D8.29 Each member of the truss shown is 1 m long. (a) Determine the lengths of $\overline{AM}$, $\overline{OA}$, and $\overline{OD}$. (b) By symmetry, what should be the value of the y component of $\mathbf{F}_{DA}$ so that the joint D is in equilibrium? (c) Determine F_{AB}, F_{AC}, and F_{BC}.

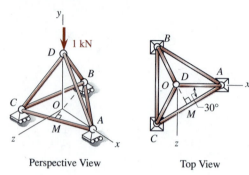

Perspective View   Top View

Fig. D8.29

PROBLEMS

8.26 Determine the forces in members BK and CJ of the truss shown.

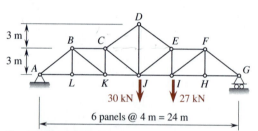

Fig. P8.26 and P8.27

8.27 Determine the forces in members EI and DJ of the truss shown.

8.28* Determine the forces in members AB and BF of the truss shown.

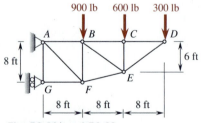

Fig. P8.28* and P8.29

8.29 Determine the forces in members AF and BE of the truss shown.

8.30 Determine the forces in members *BL* and *CL* of the truss shown.

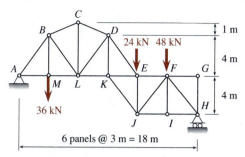

Fig. P8.30 and P8.31*

8.31* Determine the forces in members *DK* and *EK* of the truss shown.

8.32 Determine the forces in members *CO* and *EP* of the Baltimore bridge truss shown.

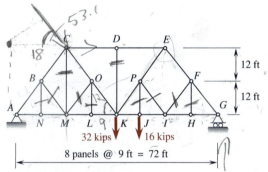

Fig. P8.32 and P8.33*

8.33* Determine the forces in members *CM* and *EI* of the truss shown.

8.34 Determine the forces in members *AB* and *FG* of the truss shown.

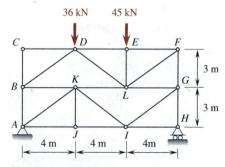

Fig. P8.34 and P8.35

8.35 Determine the forces in members *GI* and *DL* of the truss shown.

8.36* Determine the forces in members *DK* and *BJ* of the truss shown.

8.37 Determine the forces in members *FG* and *CJ* of the truss shown.

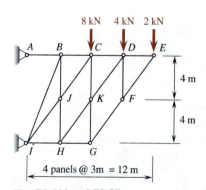

Fig. P8.36* and P8.37

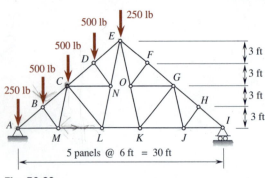

Fig. P8.38

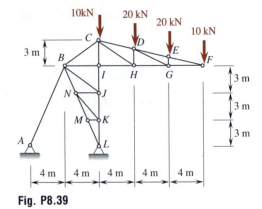

Fig. P8.39

 8.38   Determine the forces in members *EN*, *LM*, and *GK* of the Fink roof truss shown.

8.39   Determine the forces in members *BC*, *BI*, and *DH* of the truss shown.

8.40* through 8.43*   Determine the forces in members *AB*, *CD*, and *EF* of the truss shown.

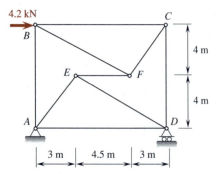

Fig. P8.40*

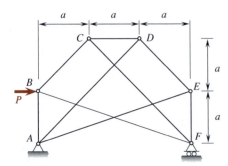

Fig. P8.41

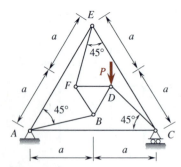

Fig. P8.42

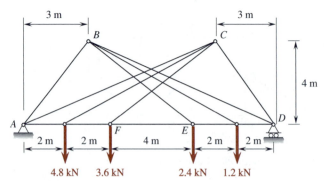

Fig. P8.43*

8.44   Determine the forces in members *BH* and *AI* of the truss when the crate is in equilibrium as shown.

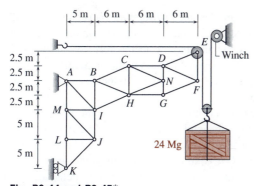

Fig. P8.44 and P8.45*

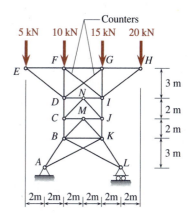

Fig. P8.46

8.45* Determine the forces in members *GH* and *CN* of the truss when the crate is in equilibrium as shown.

8.46 The members *FI* and *DG* of the tower truss shown are very slender and can act only in tension, and only one of them acts at a time; such members are known as *counters*. Determine the forces in the member *IJ* and the counter which is acting under the given loading. (*Hint.* First determine F_{IJ}, then isolate the portion *GIH* of the truss and consider $\Sigma F_y = 0$.)

8.47* and **8.48** Determine the zero-force members in the space truss shown.

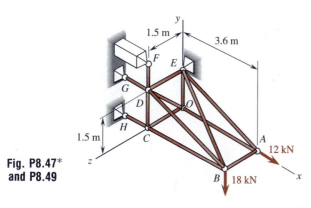

Fig. P8.47*
and P8.49

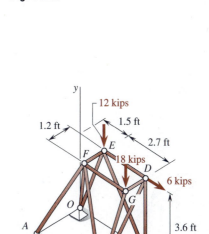

Fig. P8.48 and P8.50*

8.49 Determine the forces in members *BD*, *BE*, *BA*, and *BC* of the space truss shown.

8.50* Determine the forces in all nonzero-force members of the space truss shown.

FRAMES AND MACHINES

8.12   Structures Containing Multiforce Members

A *multiforce member* is a member or body which carries a force system and cannot be classified as a two-force body or as a particle. A *frame* is a structure which contains one or more multiforce members and is generally fully constrained. Frames are designed to support loads and are usually stationary. A *machine* is a structure which contains one or more movable multiforce members. Machines, which may be in the form of simple tools or complicated mechanisms, are designed to transmit or modify forces and motions. Machine problems in statics are usually associated with the transmission or modification of forces.

In studying frames and machines, we have the following *usual assumptions*: (a) the weights of members are negligible compared with the applied loads; (b) the connecting pin in each joint is circular and frictionless; (c) if a force acts at a joint, it is applied to the pin in the joint.

Members of a frame or machine are frequently detached for the analysis. For clarity in the sketch, we shall adopt a graphical convention that (1) *a pin or a pinhole keeping a pin is indicated by a solid circular dot*, (2) *a pinhole without a pin in it is indicated by a small circular curve*. By letting appropriate detached members keep certain pins in their pinholes, we often have the advantage of doing away with the need to consider separately the equilibrium of a pin connecting *three or more* members. However, the interacting forces between a pin and the walls of the pinholes in a joint can be studied by disassembling the joint, as illustrated in perspective views in Fig. 8.14 and in front views using the adopted graphical convention in Fig. 8.15, where Newton's third law is obeyed.

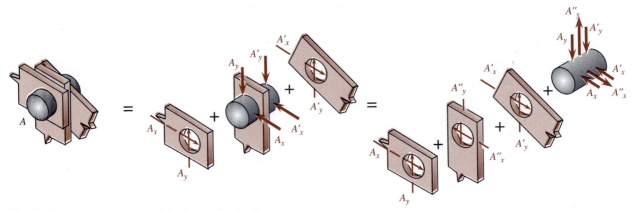

Fig. 8.14   Interacting forces in a joint (perspective views).

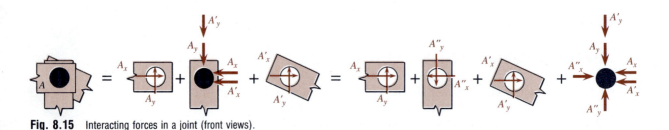

Fig. 8.15   Interacting forces in a joint (front views).

In modern structures, members of frames are seldom pin-connected. They are generally fastened at the joints with nails, bolts, rivets, or welding to achieve a greater degree of rigidity for the structure. However, results of studies based on the pin-connected version of the frame are relatively easy to obtain and are useful for practical purposes.

Developmental Exercises

D8.30 Define the terms: (a) multiforce member, (b) frame, (c) machine.

D8.31 What are the usual assumptions for frames and machines?

D8.32 As a graphical convention, what do we use to indicate (a) a pin in a pinhole, (b) a pinhole without a pin in it?

8.13   Force System Acting on a Detached Member

The force system in the free-body diagram of a detached member of a frame or machine generally includes the following:

1. the reactions to this detached member from its supports, if any;
2. the reactions to the walls of the pinholes of this detached member from the pins which this detached member does not keep;
3. the reactions from the walls of the pinholes of other members to the pins which this detached member keeps;
4. the external concentrated forces, if any, that are applied to the pins which this detached member keeps;
5. the external loads, if any, which are applied to this detached member.

Although the directions of unknown reaction components are initially assumed, it is of paramount importance to note that the free-body diagrams of the detached members of a frame or machine must be drawn in a *coordinated* manner consistent with *Newton's third law*. The letter designating the joint or support is generally used with appropriate subscripts (e.g., x or y) to label the unknown components of action and reaction at that joint or support. Furthermore, when a joint connects three or more multiforce members, superscripts (e.g., primes) in addition to the subscripts may be used along with the letter to name different unknown components of actions and reactions at that joint.

EXAMPLE 8.10

A frame is loaded as shown. Letting the member ABC keep the pin at C, draw the free-body diagram of each member separately.

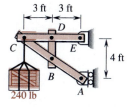

Solution.   According to Sec. 8.13 and Newton's third law, we draw the free-body diagram of each member as shown below. Note that (1) the member BD is a *two-force body* while the other members are multiforce members, (2) the *directions* of the unknown force components are assumed and indicated by the arrow signs consistent with *Newton's third law* (e.g., the action and reaction between the pinhole at B in the member ABC and

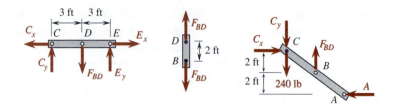

the pin at *B* in the member *BD* are equal in magnitude and opposite in direction), (3) the *magnitudes* of the unknown force components are labeled with appropriate letters and subscripts, and (4) the *pin* at *C* kept by the member *ABC* is acted on by the external concentrated force of 240 lb as well as the reaction components from the member *CDE*.

EXAMPLE 8.11

A frame is loaded as shown. Letting the member *BEF* keep the pin at *B*, draw the free-body diagram of each member separately.

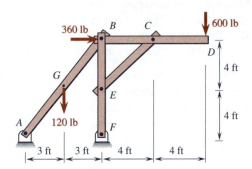

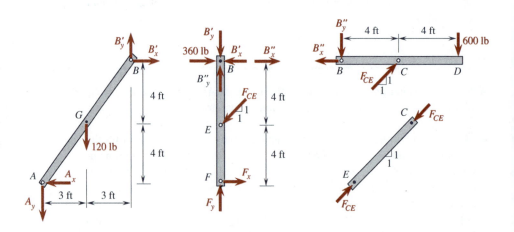

Solution.   According to Sec. 8.13 and Newton's third law, we draw the free-body diagram of each detached member as shown. Note here that (1) the member *CE* is a *two-force body* while the others are all multiforce members; (2) the *directions* of the unknown force components are assumed and *Newton's third law* is obeyed; (3) the *magnitudes* of the unknown force components are labeled with appropriate letters and subscripts or both subscripts and superscripts; (4) the *magnitudes* of the unknown force components exerted between the pin at *B* and the member *AB* are labeled as B_x' and B_y', while the magnitudes of those exerted between the pin at *B* and the member *BCD* are labeled as B_x'' and B_y''; and (5) the *pin* at *B* kept by the member *BEF* carries the 360-lb force as well as the reactions from the members *AB* and *BCD*.

Developmental Exercises

D8.33 and 8.34   A frame is shown. Draw an appropriate free-body diagram of the member *BCD* if (a) it keeps no pins, (b) it keeps all the pins it can keep.

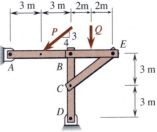

Fig. D8.33

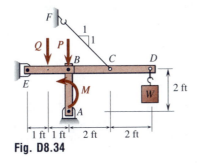

Fig. D8.34

8.14   Method of Solution for Frames and Machines in a Plane

In studying a frame or machine which is in *equilibrium in a plane*, it is helpful to keep in mind that the forces acting on it are coplanar forces and the *maximum number of independent scalar equilibrium equations* that can be written from its free-body diagram under consideration is:

- *three* if the unknown forces[†] are nonconcurrent and nonparallel;
- *two* if the unknown forces are concurrent but noncollinear;
- *two*, again, if the unknown forces are all parallel to each other and noncollinear but do not all form couples;
- *one* if the unknown forces all pair up to form couples;
- *one*, again, if there is only one unknown force or moment;
- *zero* if there is no unknown force or moment.

[†]If any of the three characteristics of a force is unknown, the force is an unknown force.

Naturally, the equilibrium of all of the members of a frame or machine implies and ensures the equilibrium of the entire frame or machine as a rigid body, and vice versa. The scalar equilibrium equations written from the free-body diagrams of all individual members *and* the entire frame or machine are, therefore, not independent. Some of them may, however, serve to check the solutions for the unknowns.

Although certain problems can best be solved by a special set of steps, the following procedure may serve as a guide in solving *equilibrium problems of frames and machines in a plane*:

1. *Identify two-force bodies*, if any, in the system.
2. *Draw appropriate free-body diagrams* which involve all of the unknowns to be determined.
3. *Count* the number of *unknowns* and the maximum number of independent scalar equilibrium *equations* that can be written from the free-body diagrams drawn.
4. If necessary, *draw additional free-body diagrams* of detached members *until* the number of unknowns is equal to the number of independent scalar equilibrium equations.
5. *Write* the set of *simultaneous* scalar equilibrium *equations* from the free-body diagrams drawn.
6. *Solve* the simultaneous equations for the *unknowns* to be determined.

Developmental Exercise

D8.35   A frame is in equilibrium in a plane. Under what condition(s) can we write from its free-body diagram the following maximum number of independent scalar equilibrium equations: (a) three, (b) only two, (c) only one, (d) zero?

EXAMPLE 8.12

A frame is loaded as shown. Determine the reactions from the pins at C, D, and E to the horizontal member *CDE*.

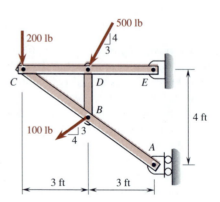

Solution. The free-body diagram of the member *CDE* with no pins contains six unknowns: C_x, C_y, D_x, D_y, E_x, and E_y. So, we need to draw additional free-body diagrams and seek six independent scalar equilibrium equations in these six unknowns, if possible. We find that it is possible to do so as follows:

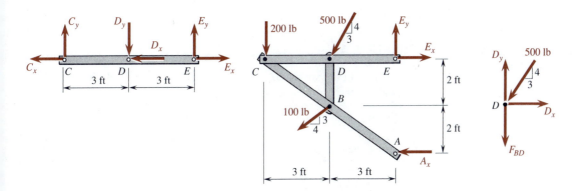

Entire frame

$$+\uparrow \Sigma F_y = 0: \quad E_y - \frac{4}{5}(500) - 200 - \frac{3}{5}(100) = 0$$

$$+ \circlearrowleft \Sigma M_A = 0: \quad -4E_x + 4\left(\frac{3}{5}\right)(500) + 3\left(\frac{4}{5}\right)(500) + 6(200)$$

$$+2\left(\frac{4}{5}\right)(100) + 3\left(\frac{3}{5}\right)(100) = 0$$

These two equations yield $E_y = 660$ and $E_x = 985$.

Pin at D

$$\xrightarrow{+} \Sigma F_x = 0: \quad D_x - \frac{3}{5}(500) = 0 \qquad D_x = 300$$

Member CDE

$$+ \circlearrowleft \Sigma M_D = 0: \quad 3E_y - 3C_y = 0 \qquad C_y = 660$$

$$\xrightarrow{+} \Sigma F_x = 0: \quad E_x - D_x - C_x = 0 \qquad C_x = 685$$

$$+\uparrow \Sigma F_y = 0: \quad E_y - D_y + C_y = 0 \qquad D_y = 1320$$

Thus, the reactions from the pins at *C*, *D*, and *E* to the member *CDE* are

$$\mathbf{C} = -685\mathbf{i} + 660\mathbf{j} \text{ lb} \qquad \mathbf{D} = -300\mathbf{j} - 1320\mathbf{j} \text{ lb} \blacktriangleleft$$

$$\mathbf{E} = 985\mathbf{i} + 660\mathbf{j} \text{ lb} \blacktriangleleft$$

REMARK. Note that the 500-lb concentrated force is applied to the pin at *D*. Furthermore, this pin receives a reaction from the two-force member *BD* as well as two reaction components from the multiforce member *CDE*, according to *Newton's third law*, as shown.

Developmental Exercises

D8.36   Determine the reactions from the pins to the member *ABC* of the frame in Example 8.10.

D8.37   Determine the reactions from the pins to the member *AGB* of the frame in Example 8.11.

EXAMPLE 8.13

If the tension in the spring *HI* of the vise grip shown is 17 N, determine the magnitude of the gripping forces at *G*.

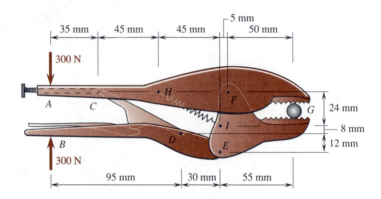

Solution.   Let the magnitude of the gripping forces be *G*. To find *G*, we need to draw free-body diagrams of some detached members. Note that the member *CD* is a *two-force body* because the release lever is not exerting any force on it. The free-body diagram of the upper jaw as shown contains four unknowns: G, F_x, F_y, and F_{CD}. Thus, we need to draw at least one more free-body diagram and seek four independent scalar equilibrium equations in these four unknowns, if possible. We find that it is possible to do so as follows:

Upper jaw

$$+\,\backsim\,\Sigma M_C = 0:\quad 145G - 95F_y - 45\left(\frac{24}{51}\right)(17) + 35(300) = 0 \qquad (1)$$

$$\overset{+}{\rightarrow}\,\Sigma F_x = 0:\quad F_x - \frac{60}{68}F_{CD} + \frac{45}{51}(17) = 0 \qquad (2)$$

$$+\uparrow\,\Sigma F_y = 0:\quad G - F_y + \frac{32}{68}F_{CD} - \frac{24}{51}(17) - 300 = 0 \qquad (3)$$

Lower jaw

$$+\,\backsim\,\Sigma M_E = 0:\quad -55G + 44F_x + 5F_y + 20\left(\frac{45}{51}\right)(17) = 0 \qquad (4)$$

Solving these four simultaneous equations, we obtain

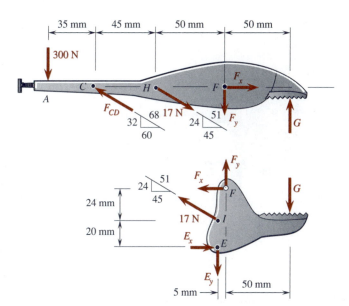

$$G = 8712 \text{ N} \qquad F_x = 9360 \text{ N} \qquad F_y = 13404 \text{ N} \qquad F_{CD} = 10625 \text{ N}$$

$$G = 8.71 \text{ kN} \blacktriangleleft$$

Thus, we see that the two 300-N applied forces have been magnified about 29 times when they are transmitted to the rod at the jaws of the vise grip.

EXAMPLE 8.14

The loader holds a 2.4-kip load which acts through the point G as shown. The machine is symmetrical about a vertical plane passing through the boom $ABCDE$ and has two sets of linkages and hydraulic cylinders. Determine the compressive forces in the cylinders KL and BM.

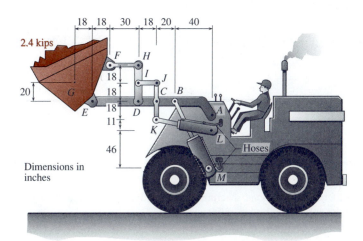

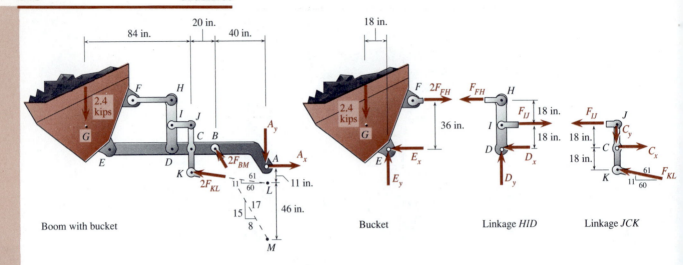

Boom with bucket   Bucket   Linkage *HID*   Linkage *JCK*

Solution. We first draw the free-body diagrams needed to solve for the unknowns F_{KL} and F_{BM}. Note that we indicate $2F_{BM}$, $2F_{KL}$, and $2F_{FH}$ in the first two free-body diagrams because there are two sets of linkages and hydraulic cylinders to hold the load. We may solve this problem in the following sequence:

Bucket

$$+ \ \circlearrowleft \ \Sigma M_E = 0: \quad 18(2.4) - 36(2F_{FH}) = 0 \qquad F_{FH} = 0.6 \text{ kips}$$

Linkage HID

$$+ \ \circlearrowleft \ \Sigma M_D = 0: \quad 36(F_{FH}) - 18F_{IJ} = 0 \qquad\qquad F_{IJ} = 1.2 \text{ kips}$$

Linkage JCK

$$+ \ \circlearrowleft \ \Sigma M_C = 0: \quad 18F_{IJ} - 18\left(\frac{60}{61}\right)F_{KL} = 0 \qquad F_{KL} = 1.22 \text{ kips} \ \blacktriangleleft$$

Boom with bucket

We first transmit $2\mathbf{F}_{KL}$ to the point L and $2\mathbf{F}_{BM}$ to the point M; then, we apply Varignon's theorem to write

$$+ \ \circlearrowleft \ \Sigma M_A = 0: \quad 144(2.4) - 11\left(\frac{60}{61}\right)(2F_{KL}) - 57\left(\frac{8}{17}\right)(2F_{BM}) = 0$$

$$F_{BM} = 5.95 \text{ kips} \ \blacktriangleleft$$

Developmental Exercise

D8.38 Determine the total shearing force (i.e., the resultant reaction) supported by the pin at *A* of the loader in Example 8.14.

PROBLEMS

8.51 through 8.54* Letting the member *ABC* of the frame shown keep all the pins it can keep, draw the free-body diagram of each member as detached.

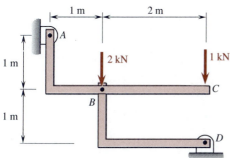

Fig. P8.51 and P8.55

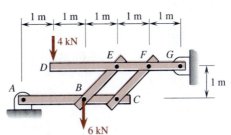

Fig. P8.52* and P8.56*

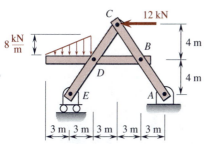

Fig. P8.53* and P8.57

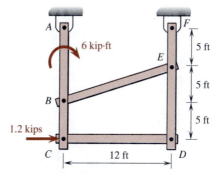

Fig. P8.54* and P8.58

8.55 through 8.58 Determine the reactions *from* the pins *to* the member *ABC* of the frame shown.

8.59 Two bent pipes are shown. Neglecting the friction between them in the portion *CD*, determine the reactions from the supports at *A* and *F*.

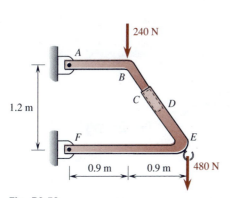

Fig. P8.59

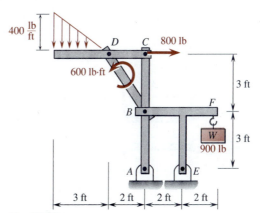

Fig. P8.60

8.60 Determine the reactions from the pins to the member *ABC* as shown.

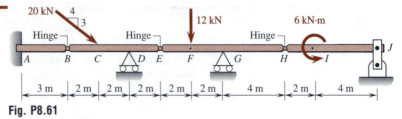

Fig. P8.61

8.61 and 8.62* Determine the reactions from the supports at A and G of the hinged beam shown.[†]

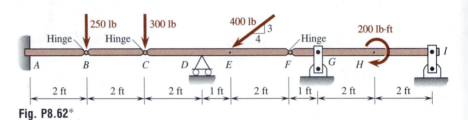

Fig. P8.62*

8.63* and 8.64 Neglecting friction, determine the reactions from the supports at A, B, and C if the weight of each pipe shown is 100 lb.

Fig. P8.63*   Fig. P8.64

8.65 and 8.66* Determine the magnitude of the forces exerted on the rod at A as shown.

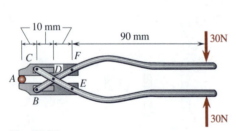

Fig. P8.65

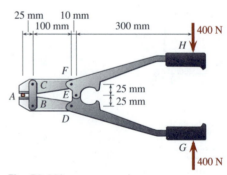

Fig. P8.66*

[†]A beam of this type is also known as a *Gerber beam (Gerberbalken)*, after the German engineer Heinrich Gerber (1832–1912).

8.67   Determine the forces exerted at C and D on the member BCD of the tongs shown.

8.68   Two shafts are connected by a universal joint at D and supported by narrow bearings without stops at B and C and by a narrow bearing with stop at E as shown. When the arm of the crosspiece attached to the shaft AD is vertical and the system is in equilibrium, determine (a) the magnitude of the torque $\mathbf{M}_F$, (b) the reactions from the three bearings.

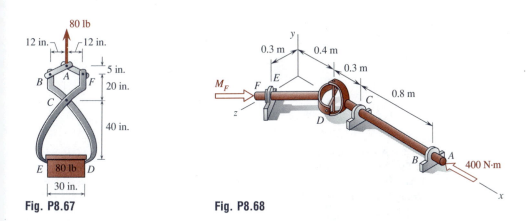

Fig. P8.67   **Fig. P8.68**

8.69*   Solve Prob. 8.68 for the case in which the crosspiece attached to the shaft DF is vertical.

8.70   Determine the magnitude of the force $\mathbf{F}$ required to release the self-locking brace of a shelf which supports a 60-lb crate as shown.

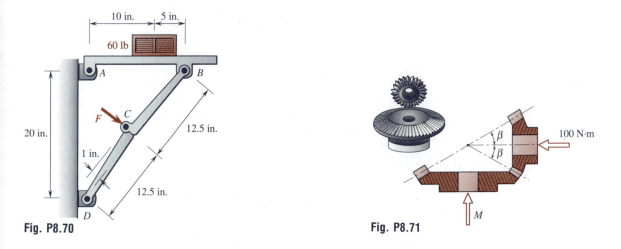

Fig. P8.70   **Fig. P8.71**

8.71   The pair of meshing bevel gears shown is in equilibrium. If the pitch angle $\beta = 30°$, determine the magnitude of the torque $\mathbf{M}$.

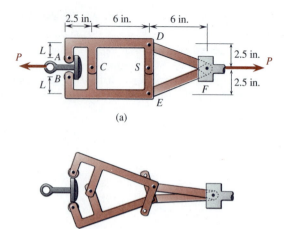

(a)

(b)

Fig. P8.72

8.72   A safety device to prevent overloading is shown in (a). The breaking strength of the shear pin at S is 240 lb. When $P = P_{max}$, the shear pin at S will break and thus release the load as shown in (b). If $L = 1.5$ in., determine the value of P_{max}. (*Hint.* Note symmetry in the system.)

8.73*   Solve Prob. 8.72 if $L = 1.25$ in.

8.74   If the gear train shown is in equilibrium, determine the tension T.

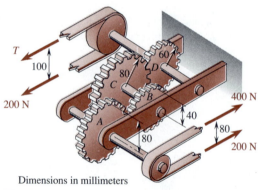

Dimensions in millimeters

Fig. P8.74

8.75   A 99-N force **P** is exerted on the handle of the crusher as shown. For $a/b = 40/9$, determine the magnitude of the crushing force **Q**.

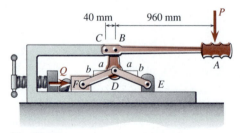

Fig. P8.75

8.76* Solve Prob. 8.75 for $a/b = 60/11$.

8.77 A bar *DE* is to be cut by the cutter shown. Determine the magnitude of the cutting force in terms of the applied force **P**.

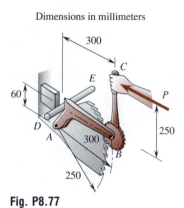

Dimensions in millimeters

Fig. P8.77

8.78 Half of the weight *W* of an automobile hood is held in equilibrium with the hinge counterbalance arrangement as shown. Find the force in the spring *CH*.

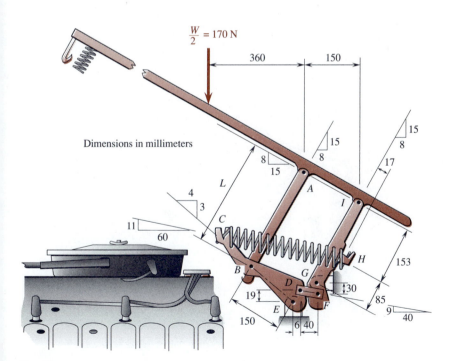

Dimensions in millimeters

Fig. P8.78

8.79 Determine the forces in the hydraulic cylinders *DE* and *CF* of the dump truck in the position shown.

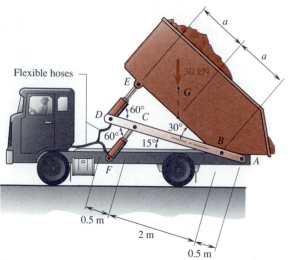

Fig. P8.79

8.80 Three hydraulic cylinders are used to control the action of the backhoe bucket. For a 1170-N load acting as shown, determine (a) the force developed in each cylinder, (b) the reactions from the pins at *A* and *D* to the member *ABCD*.

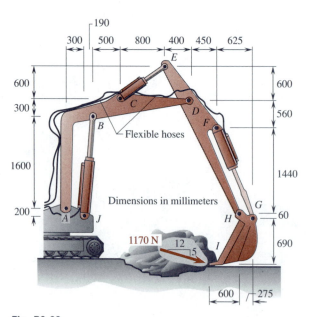

Fig. P8.80

BEAMS AND CABLES

★8.15   Internal Forces and Moments in Multiforce Members

When loads are applied to a structure, internal forces and moments are developed in the cross sections of its multiforce members which act to withstand the effects of the applied loads. We shall here study such forces and moments in the cross sections of multiforce members.

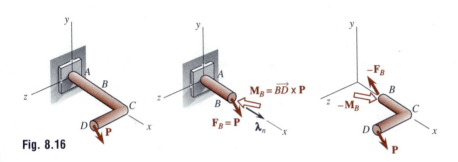

Fig. 8.16

In Fig. 8.16(a), a cantilever bent rod $ABCD$ carrying a force $\mathbf{P}$ at its free end D is shown. Using the procedure of *resolving a given force into a force-moment system at another point*, as presented in Sec. 5.1, we find that the force $\mathbf{F}_B$ and the moment $\mathbf{M}_B$ acting in the cross section at B on the end of the portion AB of the bent rod are

$$\mathbf{F}_B = \mathbf{P} \qquad \mathbf{M}_B = \overrightarrow{BD} \times \mathbf{P} \qquad (8.5)$$

which are shown in Fig. 8.16(b).[†] By Newton's third law, the force and the moment acting in the cross section at B on the end of the portion BCD must be $-\mathbf{F}_B$ and $-\mathbf{M}_B$, respectively, and are shown in Fig. 8.16(c). Thus, the force and the moment acting in a cross section of a member may, equivalently, be determined by *considering the equilibrium* of a portion of the member as divided by the cross section.[††]

Unless otherwise stated, all cross sections are to be understood as right cross sections. Let the *unit vector* perpendicular to the cross section at B and pointing outward at the end of the chosen portion AB be $\boldsymbol{\lambda}_n$ as shown in

---

[†]If there are other forces as well as moments acting on the portion BCD of the bent rod, each of the forces may similarly be resolved into a force at B and a moment. The vector sum of the forces brought to B by the resolutions will give the value of $\mathbf{F}_B$. Similarly, the vector sum of the moments acting on the portion BCD and the moments brought forth by the resolutions will give the value of $\mathbf{M}_B$.

[††]If the member is not in equilibrium, the forces and moments in its cross sections should be computed from the kinetics of the member.

Fig. 8.16(b). By resolving the force $\mathbf{F}_B$ as well as the moment $\mathbf{M}_B$ in the cross section at B into two components, one perpendicular and the other parallel to the cross section, we write

$$\mathbf{F}_B = \mathbf{F}_\perp + \mathbf{F}_\| \qquad \mathbf{M}_B = \mathbf{M}_\perp + \mathbf{M}_\| \qquad (8.6)$$

where

$$
\begin{aligned}
\mathbf{F}_\perp &= (\boldsymbol{\lambda}_n \cdot \mathbf{F}_B)\boldsymbol{\lambda}_n & \mathbf{F}_\| &= \mathbf{F}_B - \mathbf{F}_\perp \\
\mathbf{M}_\perp &= (\boldsymbol{\lambda}_n \cdot \mathbf{M}_B)\boldsymbol{\lambda}_n & \mathbf{M}_\| &= \mathbf{M}_B - \mathbf{M}_\perp
\end{aligned}
\qquad (8.7)
$$

The perpendicular component $\mathbf{F}_\perp$ is called the *normal force* or the *axial force*, while the parallel component $\mathbf{F}_\|$ is called the *shear* or the *shearing force* in the cross section. On the other hand, the perpendicular component $\mathbf{M}_\perp$ is called the *torque*, while the parallel component $\mathbf{M}_\|$ is called the *bending moment* in the cross section. (Cf. Sec. 4.18.)

EXAMPLE 8.15

In the cross section at B on the end of the portion AB of the cantilever bent rod shown, determine (a) the internal force $\mathbf{F}_B$, (b) the normal force $\mathbf{F}_\perp$, (c) the shear $\mathbf{F}_\|$, (d) the internal moment $\mathbf{M}_B$, (e) the torque $\mathbf{M}_\perp$, (f) the bending moment $\mathbf{M}_\|$.

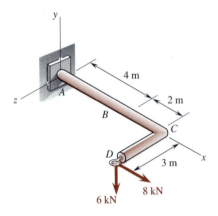

Solution. Applying Eqs. (8.5) and (8.7), we write

$$\mathbf{P} = 8\mathbf{i} - 6\mathbf{j} \text{ kN} \qquad \boldsymbol{\lambda}_n = \mathbf{i}$$

$$\mathbf{F}_B = \mathbf{P} \qquad\qquad \mathbf{F}_B = 8\mathbf{i} - 6\mathbf{j} \text{ kN} \blacktriangleleft$$

$$\mathbf{F}_\perp = (\boldsymbol{\lambda}_n \cdot \mathbf{F}_B)\boldsymbol{\lambda}_n \qquad\qquad \mathbf{F}_\perp = 8\mathbf{i} \text{ kN} \blacktriangleleft$$

$$\mathbf{F}_\| = \mathbf{F}_B - \mathbf{F}_\perp \qquad\qquad \mathbf{F}_\| = -6\mathbf{j} \text{ kN} \blacktriangleleft$$

$$\overrightarrow{BD} = 2\mathbf{i} + 3\mathbf{k} \text{ m} \qquad \mathbf{M}_B = \overrightarrow{BD} \times \mathbf{P}$$

$$\mathbf{M}_B = 18\mathbf{i} + 24\mathbf{j} - 12\mathbf{k} \text{ kN·m} \blacktriangleleft$$

$$\mathbf{M}_\perp = (\boldsymbol{\lambda}_n \cdot \mathbf{M}_B)\boldsymbol{\lambda}_n \qquad\qquad \mathbf{M}_\perp = 18\mathbf{i} \text{ kN·m} \blacktriangleleft$$

$$\mathbf{M}_\| = \mathbf{M}_B - \mathbf{M}_\perp \qquad \mathbf{M}_\| = 24\mathbf{j} - 12\mathbf{k} \text{ kN·m} \blacktriangleleft$$

Developmental Exercise

D8.39   A short section of semicircular shell having a mass of 10 kg rests on a smooth horizontal surface as shown. In the cross section at B on the end of the portion AB of the shell, determine (a) the outward unit vector $\boldsymbol{\lambda}_n$ which is perpendicular to the cross section, (b) the internal force $\mathbf{F}_B$, (c) the normal force $\mathbf{F}_\perp$, (d) the shear $\mathbf{F}_\parallel$, (e) the internal moment $\mathbf{M}_B$, (f) the torque $\mathbf{M}_\perp$, (g) the bending moment $\mathbf{M}_\parallel$.

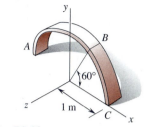

Fig. D8.39

★8.16   Shears and Bending Moments in Beams

A *beam* is a structural member designed to support mainly transverse loads. In general, beams are *long*, *straight*, *prismatic*, and *horizontal*. Since transverse loads are perpendicular to the axes of beams, they will induce shears and bending moments in the cross sections of beams.

A *load* may be a force or a moment, which may be distributed or concentrated on a beam. The basic types of *loads carried by beams*, as illustrated in Fig. 8.17, are usually expressed in (1) N/m or lb/ft for *distributed forces*, (2) N or lb for *concentrated forces*, (3) N·m/m or lb·ft/ft for *distributed moments*, and (4) N·m or lb·ft for *concentrated moments*. Note that the symbols w and $\tilde{m}$ in Fig. 8.17 denote the intensities of the distributed force and the distributed moment, respectively. Furthermore, the distributed moment $\tilde{m}$ and the concentrated moment T may graphically be represented as shown in parts (b) and (c) of Fig. 8.17.

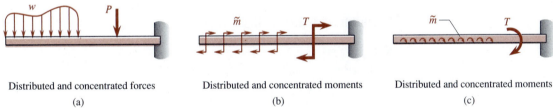

Distributed and concentrated forces	Distributed and concentrated moments	Distributed and concentrated moments
(a)	(b)	(c)

Fig. 8.17

For studying shears and bending moments in the cross sections of beams which are horizontal and under the action of force systems in the vertical plane, we adopt the following *sign convention*:

1. *The shear in a cross section of a beam is positive when it acts upward on the left end of the right portion* (or downward on the right end of the left portion) *of the beam as divided by the cross section.*
2. *The bending moment in a cross section of a beam is positive when it tends to cause compression at the top* (or tension at the bottom) *of the beam at that cross section.*

Positive shear and positive bending moment	Effect of positive shear	Effect of positive bending moment
(a)	(b)	(c)

Fig. 8.18

The positive shear and positive bending moment in a cross section are illustrated in Fig. 8.18(a), where Newton's third law is observed. Note that the effects of the positive shear and positive bending moment on a beam as illustrated in Fig. 8.18(b) and (c) are somewhat exaggerated. If the shear or bending moment is not zero and cannot be classified as positive, it is negative.

According to the above sign convention and Sec. 8.15, two *general rules* for computing shears and bending moments may be stated as follows:

1. The *shear in a cross section* of a beam is equal to the algebraic sum of all forces which lie *on a chosen side* of the cross section of the free body of the beam. If the *left side* is chosen, *upward forces are positive* while downward forces are negative. If the *right side* is chosen, the reverse is true.
2. The *bending moment in a cross section* of a beam is equal to the algebraic sum of moments, about the centroid of the cross section, of all forces and couples which lie *on a chosen side* of the cross section of the free body of the beam. The moment is positive if its action tends to cause *compression at the top* of the beam at that cross section.

Developmental Exercises

D8.40   Are beams usually vertical, horizontal, or inclined?

D8.41   Name the four basic types of loads carried by beams.

EXAMPLE 8.16

Determine the shears and the bending moments in the cross sections at E and F of the beam shown.

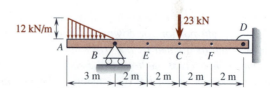

Solution. The reactions from the supports are determined by considering the equilibrium of the free body of the entire beam; they are

$$\mathbf{B} = 34 \text{ kN} \uparrow \qquad \mathbf{D} = 7 \text{ kN} \uparrow$$

$$\Sigma M_B = -8R_D + 4(23) - 2(18) = 6$$

$$\Sigma M_D = 8R_B - 18(10) - 4(23) = 0$$

$$R_B = \frac{272}{8} = 34 \text{ kN}$$

Suppose that we choose the *left side* of E to compute the shear V_E and the bending moment M_E. Noting the replacement of the distributed load with its resultant force of 18 kN at G as shown, we write

$$V_E = -18 + 34 \qquad\qquad V_E = +16 \text{ kN} \blacktriangleleft$$

$$M_E = -4(18) + 2(34) \qquad M_E = -4 \text{ kN·m} \blacktriangleleft$$

To find the shear V_F and the moment M_F in the cross section at F, we note that it is simpler to work with the *right side* of F. Thus, we write

$$V_F = -7 \qquad\qquad V_F = -7 \text{ kN} \blacktriangleleft$$

$$M_F = 2(7) \qquad\qquad M_F = +14 \text{ kN·m} \blacktriangleleft$$

Note that these same results may be obtained by considering the *equilibrium* of a portion of the member.

Developmental Exercises

D8.42 Refer to Example 8.16. Compute V_E and M_E by working with the *right* side of E.

D8.43 Refer to Example 8.16. Compute V_F and M_F by working with the *left* side of F.

EXAMPLE 8.17

A beam with a bracket is loaded as shown. Determine the equations for the shear and bending moment in the cross sections of the portions AB and BC.

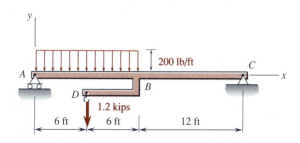

Solution. The 1.2-kip force at D is resolved into a 1.2-kip force and a moment of 7.2 kip·ft at B as shown. The reactions from the supports as shown are determined by considering the equilibrium of the free body of the entire beam.

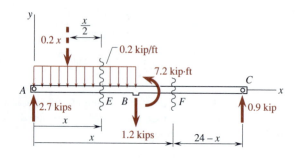

Portion AB. Let an arbitrary cross section in this portion be located at E which is at a distance x from A. Summing the forces as well as the moments about E of the forces which lie on the *left side* of E, we write

$$V_{AB} = 2.7 - 0.2x \qquad\qquad V_{AB} = 2.7 - 0.2x \text{ kips} \blacktriangleleft$$

$$M_{AB} = x(2.7) - \frac{x}{2}(0.2x) \qquad M_{AB} = 2.7x - 0.1x^2 \text{ kip·ft} \blacktriangleleft$$

Portion BC. Let an arbitrary cross section in this portion be located at F which is at a distance x from A. Noting that it is simpler to work with the *right side* of F, we write

$$V_{BC} = -0.9 \qquad\qquad V_{BC} = -0.9 \text{ kips} \blacktriangleleft$$

$$M_{BC} = (24 - x)(0.9) \qquad M_{BC} = 21.6 - 0.9x \text{ kip·ft} \blacktriangleleft$$

Developmental Exercises

D8.44 Describe the general rule for computing the shear in a cross section of a beam.

D8.45 Describe the general rule for computing the bending moment in a cross section of a beam.

D8.46 Is it true that an upward force on a beam always has a positive contribution to the bending moment in any cross section of the beam?

*8.17 Shear and Bending-Moment Diagrams

A graph showing the shear V in the cross section of a beam as a function of the position coordinate x of the cross section is called a *shear diagram*. Similarly, a graph showing the bending moment M in the cross section of a beam as a function of the position coordinate x of the cross section is called a *bending-moment diagram*. The curves in these diagrams are called the *shear* and *bending-moment curves*, respectively.

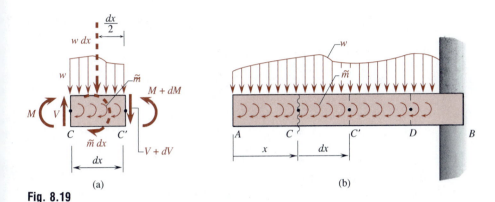

Fig. 8.19

Note that a concentrated force of magnitude P on a beam may be considered as a distributed force of intensity $w = P/\Delta x$ over a very short length Δx of the beam. Similarly, a concentrated moment of magnitude T on a beam may be considered as a distributed moment of intensity $\tilde{m} = T/\Delta x$ over a very short length Δx of the beam. Thus, let us consider the free body shown in Fig. 8.19(a) for the differential beam element CC' of the beam in Fig. 8.19(b), where the beam is subjected to a *downward* distributed force of intensity w and a *clockwise* distributed moment of intensity $\tilde{m}$. Since the beam element as well as the beam is in equilibrium, we refer to Fig. 8.19(a) and write

$$+\uparrow \Sigma F_y = 0: \quad V - w\,dx - (V + dV) = 0$$

$$\frac{dV}{dx} = -w \tag{8.8}$$

$$+\,\circlearrowleft\, \Sigma M_{C'} = 0: \quad -M - (dx)V + \frac{dx}{2}(w\,dx) - \tilde{m}\,dx + (M + dM) = 0$$

$$dM = V\,dx + \tilde{m}\,dx - \frac{w}{2}(dx)^2 \approx V\,dx + \tilde{m}\,dx$$

$$\frac{dM}{dx} = V + \tilde{m} \tag{8.9}^{\dagger}$$

Let $x = x_C$, $V = V_C$, $M = M_C$ at the point C and $x = x_D$, $V = V_D$, $M = M_D$ at the point D of the beam. Integrating Eqs. (8.8) and (8.9) from

---

[†]This equation reduces to $dM/dx = V$ if there is no distributed moment acting on the beam. However, without this equation, the slopes of the lines in the bending-moment diagram in Example 8.21 would be inexplicable. Further discussions on distributed moments and couple stresses may be found in texts on continuum mechanics.

the point C to the point D of the beam in Fig. 8.19(b), we obtain

$$V_D - V_C = -\int_{x_C}^{x_D} w \, dx \qquad (8.10)$$

$$M_D - M_C = \int_{x_C}^{x_D} V \, dx + \int_{x_C}^{x_D} \tilde{m} \, dx \qquad (8.11)$$

From Eqs. (8.8) and (8.9), we can show that

$$\frac{d^2V}{dx^2} = -\frac{dw}{dx} \qquad (8.12)$$

$$\frac{d^2M}{dx^2} = -w + \frac{d\tilde{m}}{dx} \qquad (8.13)$$

Notice in Eqs. (8.8) through (8.13) that w and $\tilde{m}$ are *positive* when the distributed force and the distributed moment act in the *downward* and *clockwise* directions, respectively; otherwise, they are *negative*. From these six equations, we state the following six *relations*, which are useful in plotting shear and bending-moment curves:

1. The *slope of the shear curve* at any point is equal to negative of the intensity of the distributed force on the beam at that point. Thus, the shear curve has a vertical slope at the point where a concentrated force is applied.
2. The *slope of the bending-moment curve* at any point is equal to the sum of the shear in the cross section and the intensity of the distributed moment on the beam at that point. Furthermore, the bending-moment curve has a vertical slope at the point where a concentrated moment is applied.
3. The *change in shear* from C to D of the beam is equal to negative of the sum of all forces on the beam between C and D.
4. The *change in bending moment* from C to D of the beam is equal to the "area" bounded by the shear curve and the x axis between C and D plus the sum of all moments on the beam between C and D.
5. The *concavity of the shear curve* in a certain portion is upward if $dw/dx < 0$ in that portion of the beam; and downward, if $dw/dx > 0$. Of course, the portion of the shear curve is a straight line if $dw/dx = 0$ in that portion.
6. The *concavity of the bending-moment curve* in a certain portion is upward if $-w + d\tilde{m}/dx > 0$ in that portion of the beam; and downward, if $-w + d\tilde{m}/dx < 0$. Of course, the portion of the bending-moment curve is a straight line if $-w + d\tilde{m}/dx = 0$ in that portion.

From the six relations, we infer that the shear and bending-moment diagrams have the following special *features*: (a) a shear curve has vertical jumps *at and only at* the points where concentrated forces are applied to the beam, (b) *each vertical jump in the shear curve* at a point is equal to the concentrated force at that point, (c) a bending-moment curve has vertical jumps *at and only at* the points where concentrated moments are applied to the beam, (d) *each vertical jump in the bending-moment curve* at a point is equal to the concentrated moment at that point, (e) *each point of inflection* (if any) *in the shear curve* is located at the point where the value of dw/dx

on the beam changes its sign, and (f) *each point of inflection* (if any) *in the bending-moment curve* is located at the point where the value of $-w + d\tilde{m}/dx$ on the beam changes its sign.

Developmental Exercise

D8.47 In proceeding from the left to the right along a shear curve of a beam, does the vertical jump in the curve at a certain point go in the same direction and have the same magnitude as the concentrated force there?

EXAMPLE 8.18

Draw the shear and bending-moment diagrams for the beam shown.

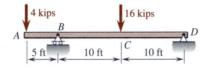

Solution. The reactions are first determined. The shear and bending-moment diagrams are then drawn according to Secs. 8.16 and 8.17. In particular, note that the vertical jumps in the shear curve are equal to the corresponding concentrated forces.

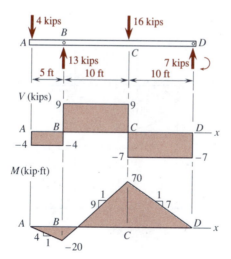

Developmental Exercise

D8.48 Describe the six *relations* which are useful in drawing shear and bending-moment diagrams.

EXAMPLE 8.19

Draw the shear and bending-moment diagrams for the beam in Example 8.16.

Solution. We first draw the free-body diagram of the beam as shown. The shear and bending-moment diagrams are then drawn according to Secs. 8.16 and 8.17. Note that, instead of computing the *area* bounded by the shear curve and the *x* axis in the range $0 \le x \le 3$ m, we may compute the value of the bending moment at B directly from the free body of the beam. Thus, replacing the distributed force with its resultant force at G, we write

$$M_B = -2(18) \qquad M_B = -36 \text{ kN·m}$$

Furthermore, note that (1) $dV/dx = 0$ at $x = 3$ m, (2) $dM/dx = 0$ at $x = 0$, (3) the shear curve between A and B is concave upward, and (4) the bending-moment curve between A and B is concave downward.

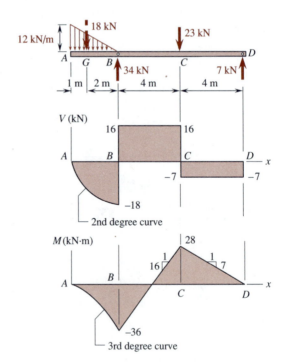

Developmental Exercise

EXAMPLE 8.20

Draw the shear and bending-moment diagrams for the beam in Example
8.17.

Solution.   The 1.2-kip force at D is replaced with a 1.2-kip force and a
moment of 7.2 kip·ft at B as shown. After drawing the free-body diagram
of the beam as shown, the shear and bending-moment diagrams are drawn
according to Secs. 8.16 and 8.17. In particular, note that the vertical jumps
in the shear and bending-moment curves at $x = 12$ ft are equal to the concentrated force and concentrated moment on the beam at B, respectively.
Furthermore, note that the bending-moment curve between A and B is concave downward.

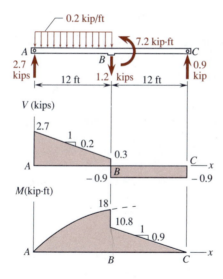

Developmental Exercises

EXAMPLE 8.21

A beam is subjected to a uniformly distributed moment of intensity 10 kN·m/m as shown. Draw the shear and bending-moment diagrams for the beam.

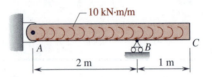

Solution. We first draw the free-body diagram as shown. The shear and bending-moment diagrams are then drawn according to Secs. 8.16 and 8.17. Note that the beam experiences a constant shear of -15 kN in the portion BC. In particular, note from Eq. (8.9) that the slopes of the bending-moment curve are

$$\left(\frac{dM}{dx}\right)_{AB} = (V + \tilde{m})_{AB} = -15 + 10 = -5 \text{ (kN·m/m)}$$

$$\left(\frac{dM}{dx}\right)_{BC} = (V + \tilde{m})_{BC} = 0 + 10 = 10 \text{ (kN·m/m)}$$

Furthermore, note from Eq. (8.11) that the changes in bending moment between key points are

$$M_B - M_A = \int_{x_A}^{x_B} V \, dx + \int_{x_A}^{x_B} \tilde{m} \, dx$$

$$= -15(2) + 10(2) = -10 \text{ (kN·m)}$$

$$M_C - M_B = \int_{x_C}^{x_B} V \, dx + \int_{x_C}^{x_B} \tilde{m} \, dx$$

$$= 0 + 10(1) = 10 \text{ (kN·m)}$$

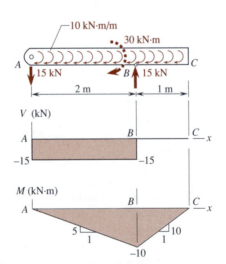

Developmental Exercise

D8.52   Work Example 8.21 if the beam is cantilevered at A and the roller support at B is removed.

PROBLEMS

8.81   In the cross section at A of the cantilever bent rod shown, determine (a) the normal force $\mathbf{F}_\perp$, (b) the shear $\mathbf{F}_\parallel$, (c) the torque $\mathbf{M}_\perp$, (d) the bending moment $\mathbf{M}_\parallel$.

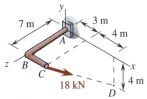

Fig. P8.81

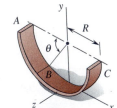

Fig. P8.82*

8.82*   A short section of semicircular shell having a weight W and a radius R is shown. In the cross section at B on the end of the portion BC of the shell, determine, in terms of W, R, and θ, (a) the normal force $\mathbf{F}_\perp$, (b) the shear $\mathbf{F}_\parallel$, (c) the torque $\mathbf{M}_\perp$, (d) the bending moment $\mathbf{M}_\parallel$.

8.83*   Refer to Prob. 8.82*. Determine (a) the value of θ for which the normal force in the cross section at B is maximum, (b) the magnitude of the maximum normal force.

8.84* through 8.98*   Draw the shear and bending-moment diagrams for the beam shown.

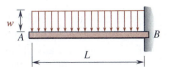

Fig. P8.84*

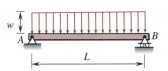

Fig. P8.85

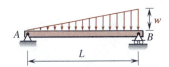

Fig. P8.86*

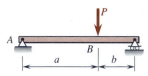

Fig. P8.87*

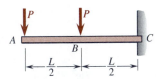

Fig. P8.88

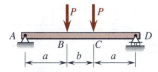

Fig. P8.89*

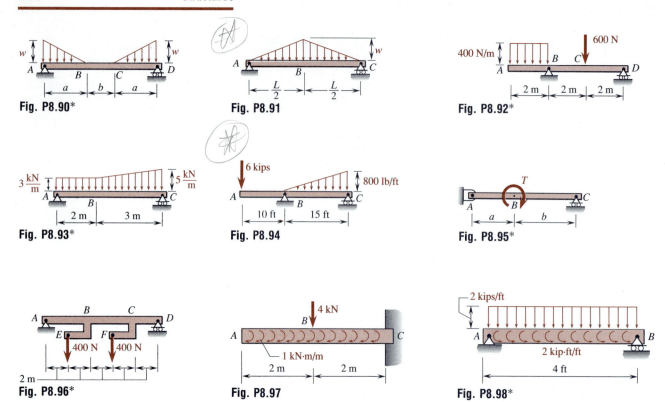

Fig. P8.90*   **Fig. P8.91**   **Fig. P8.92***

Fig. P8.93*   **Fig. P8.94**   **Fig. P8.95***

Fig. P8.96*   **Fig. P8.97**   **Fig. P8.98***

8.99 Determine the equations for the shear and bending moment in the beam shown.

8.100* The shear diagram for a beam, which has no moments applied between the ends, is shown. Draw for the beam (a) the free-body diagram, (b) the bending-moment diagram.

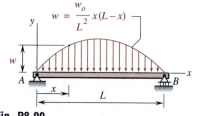

Fig. P8.99

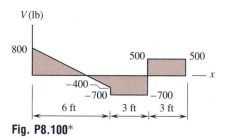

Fig. P8.100*

*8.18 Cables and the Governing Differential Equation

A *cable* is a cordlike element which is assumed to have great tensile strength and negligible resistance to bending. In other words, a cable is usually assumed to be a flexible structural member and the *force developed* anywhere in the cable is always a *tensile force tangent to the centerline of the cable there*. In elementary study, the elongation of the cable under loading is assumed to be negligible; i.e., all cables are assumed to be *inextensible* as well as *flexible*.

Cables are commonly used in suspension bridges, transmission lines, telephone lines, guy wires, aerial tramways, cranes, derricks, and many other applications. The applied load, tension, span, sag, and length are important parameters in the design of cables.

A cable may carry *vertical concentrated loads* as shown in Fig. 8.20(a), or it may carry a *vertical distributed load* as shown in Fig. 8.20(b). In some cases, the weight of the cable is taken as negligible compared with the applied loads. In other cases, the weight of the cable may be an appreciable load or the sole load and cannot be neglected. For instance, the weight of the cable in Fig. 8.20(c) is the sole load carried by the cable.

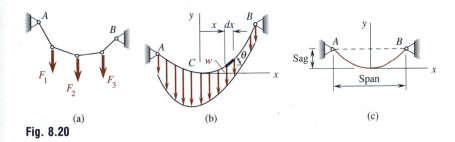

(a)   (b)   (c)

Fig. 8.20

When a cable carries only concentrated loads, such as that illustrated in Fig. 8.20(a), the weight of the cable is usually taken as negligible. In such a case, any portion of the cable between successive loads is considered as a two-force body, and the force of tension developed in the cable is directed along the cable. Thus, the method of solution outlined in Sec. 3.10 is applicable to the solution of problems involving cables carrying only concentrated loads. Furthermore, the concept of rigid-body equilibrium is also applicable as illustrated in Example 5.10.

The equilibrium of a cable carrying a *vertical distributed load w* may be studied by considering the free body of a typical differential element of the cable as shown in Fig. 8.21, where T is the tension in the cable and w is the intensity of the vertical distributed load *measured in force per unit of*

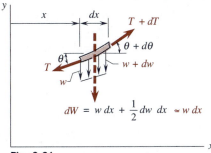

Fig. 8.21

horizontal length. Note here that *the positive direction of w is downward.* For equilibrium of the element as shown, we write

$$\xrightarrow{+} \Sigma F_x = 0: \quad (T + dT)\cos(\theta + d\theta) - T\cos\theta = 0$$

$$+\uparrow \Sigma F_y = 0: \quad (T + dT)\sin(\theta + d\theta) - T\sin\theta - w\,dx = 0$$

Expanding the sine and cosine of the sum of two angles, recognizing that $\cos d\theta \approx 1$ and $\sin d\theta \approx d\theta$, and neglecting higher-order terms in these two equations, we obtain

$$dT\cos\theta - T\sin\theta\,d\theta = 0$$

$$dT\sin\theta + T\cos\theta\,d\theta - w\,dx = 0$$

which may be written as

$$d(T\cos\theta) = 0 \tag{8.14}$$

$$d(T\sin\theta) - w\,dx = 0 \tag{8.15}$$

Equation (8.14) indicates the property that

$$T\cos\theta = T_o \tag{8.16}$$

where T_o is a constant equal to the tension in the cable at the point where θ and hence the slope of the cable are zero. This means that the horizontal component of the tension in the cable is a constant T_o, which is clear because *the cable carries only vertical loads.* Since $-1 \leq \cos\theta \leq 1$, it is obvious that the *minimum tension* in the cable is T_o. Noting that $\tan\theta = dy/dx$, we may obtain from Eqs. (8.15) and (8.16) that

$$d\left(T_o \frac{dy}{dx}\right) - w\,dx = 0$$

$$\boxed{\frac{d^2y}{dx^2} = \frac{w}{T_o}} \tag{8.17}$$

Equation (8.17) is the *governing differential equation* for the cable carrying a vertical distributed load. The conditions at the fixed ends of the cable are called the *boundary conditions* of the cable. The shape of the cable is defined by the functional relation $y = f(x)$ which satisfies Eq. (8.17) and the boundary conditions of the cable.

EXAMPLE 8.22

The cable *AE* loaded as shown is in equilibrium. Determine (a) the elevations y_B and y_D, (b) the tension T_{AB} in the portion *AB*.

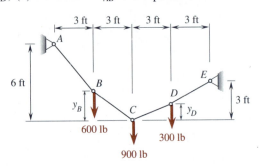

Solution. We first note that there are six unknowns involved in the equilibrium of the cable. They are the two unknown elevations y_B and y_D and the four unknown tensions T_{AB}, T_{BC}, T_{CD}, and T_{DE}. The equilibrium of each of the three loading points B, C, and D yields two independent equilibrium equations. Thus, there are six independent equations for the solution of the six unknowns. However, this problem may more directly be solved by applying the concept of the equilibrium of rigid bodies as follows:

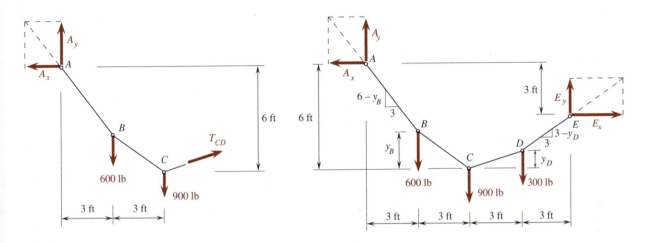

Portion ABC

$$+\circlearrowleft \ \Sigma M_C = 0: \quad 6A_x - 6A_y + 3(600) = 0$$

Entire cable

$$+\circlearrowleft \ \Sigma M_E = 0: \quad 3A_x - 12A_y + 9(600) + 6(900) + 3(300) = 0$$

$$\overset{+}{\rightarrow} \ \Sigma F_x = 0: \quad -A_x + E_x = 0$$

$$+\uparrow \ \Sigma F_y = 0: \quad A_y + E_y - 600 - 900 - 300 = 0$$

Solving these four equations, we obtain

$$A_x = 900 \text{ lb} \qquad A_y = 1200 \text{ lb} \qquad E_x = 900 \text{ lb} \qquad E_y = 600 \text{ lb}$$

Since portions AB and DE are two-force bodies, the resultant reaction from A must be equal to $-\mathbf{T}_{AB}$, and the resultant reaction from E must be equal to $-\mathbf{T}_{ED}$. Noting the slopes of the portions AB and ED as indicated in the free-body diagram of the entire cable, we write

$$\frac{6-y_B}{3} = \frac{A_y}{A_x} = \frac{1200}{900} \qquad y_B = 2 \text{ ft} \ \blacktriangleleft$$

$$\frac{3-y_D}{3} = \frac{E_y}{E_x} = \frac{600}{900} \qquad y_D = 1 \text{ ft} \ \blacktriangleleft$$

$$T_{AB}^2 = A_x^2 + A_y^2 = (900)^2 + (1200)^2 \qquad T_{AB} = 1500 \text{ lb} \ \blacktriangleleft$$

Developmental Exercises

D8.53 What are the usual assumptions for cables?

D8.54 Refer to Example 8.22. Determine the tension T_{DE} in the portion DE of the cable.

D8.55 Refer to Example 8.22. Determine the tensions T_{BC} and T_{CD} in the portions BC and CD of the cable, respectively.

D8.56 Refer to the cable in Fig. 8.20(b). Let T_A, T_B, and T_C be the tensions in the cable at A, B, and C, respectively. Using Eq. (8.16) as a guide, determine which of the following is true: (a) $T_A > T_B > T_C$, (b) $T_B > T_C > T_A$, (c) $T_C > T_A > T_B$, (d) $T_B > T_A > T_C$, (e) $T_A = T_B = T_C$.

D8.57 In Eq. (8.17), is the positive direction of the distributed load w the same as that of the ordinate y?

D8.58 In Eq. (8.17), is w measured along the cable or along the horizontal?

*8.19 Parabolic Cable

Let us here consider a cable AB carrying a *vertical load distributed with a constant intensity w along the horizontal* as shown in Fig. 8.22(a). The weight of the cable is not uniformly distributed along the horizontal but is assumed to be negligibly small compared with the load carried. Such a cable closely simulates the cables of a suspension bridge where the weight of the roadway is the dominant load and may be assumed to be uniformly distributed along the horizontal.

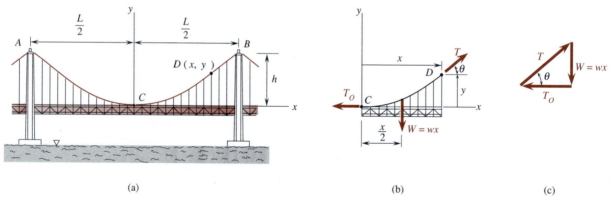

(a)   (b)   (c)

Fig. 8.22

We shall show that the aforementioned cable AB hangs in a *parabolic arc*. As shown in Fig. 8.22(a), the cable has a span L and sag h, and the origin of coordinates is at the *lowest point* C of the span where the slope dy/dx of the cable is zero. Knowing that w and T_o are not functions of x, we may integrate Eq. (8.17) to yield

$$\frac{dy}{dx} = \frac{w}{T_o}x + C_1$$

$$y = \frac{w}{2T_o}x^2 + C_1 x + C_2$$

where C_1 and C_2 are constants of integration. Since $dy/dx = y = 0$ at $x = 0$, we find that $C_1 = C_2 = 0$. Thus, we have

$$y = \frac{w}{2T_o}x^2 \qquad (8.18)$$

which we see is a *vertical parabola*. Note that this equation may also be obtained by considering the equilibrium of the portion CD, shown in Fig. 8.22(b), as follows:

$$+\circlearrowright \ \Sigma M_D = 0: \quad \frac{x}{2}(wx) - yT_o = 0$$

Solving for y from this equation, we obtain Eq. (8.18). The boundary conditions of the cable at A and B require that $y = h$ when $x = \pm L/2$. Substitution of these values of y and x into Eq. (8.18) yields

$$T_o = \frac{wL^2}{8h} \qquad y = \frac{4h}{L^2}x^2 \qquad (8.19)$$

The equilibrium of the portion CD shown in Fig. 8.22(b) requires that the three force vectors there form a right triangle as shown in Fig. 8.22(c). Thus, the tension T is given by

$$T^2 = T_o^2 + (wx)^2 = \left(\frac{wL^2}{8h}\right)^2 + (wx)^2$$

$$T = \frac{w}{8h}(64h^2x^2 + L^4)^{1/2} \qquad (8.20)$$

The tension T is maximum when $x = L/2$. Hence, the maximum tension is

$$T_{\max} = \frac{wL}{8h}(16h^2 + L^2)^{1/2} \qquad (8.21)$$

The length S of the complete cable AB may be studied from the relation

$$ds = [(dx)^2 + (dy)^2]^{1/2} = \left[1 + \left(\frac{dy}{dx}\right)^2\right]^{1/2} dx$$

We find from Eqs. (8.19) that $dy/dx = 8hx/L^2$. Thus, we write

$$S = 2\int_0^{L/2} ds = 2\int_0^{L/2} \left(1 + \frac{64h^2x^2}{L^4}\right)^{1/2} dx$$

$$= 2\int_0^{L/2} \left(1 + \frac{32h^2x^2}{L^4} - \frac{512h^4x^4}{L^8} + \cdots\right) dx$$

$$S = L\left[1 + \frac{8}{3}\left(\frac{h}{L}\right)^2 - \frac{32}{5}\left(\frac{h}{L}\right)^4 + \cdots\right] \qquad (8.22)$$

The series in Eq. (8.22) can be shown to converge for values of $h/L \leq \frac{1}{4}$. In most cases, h is much smaller than $L/4$, and the first three terms of the series will give a sufficiently good approximation. If the supports of a parabolic cable have *different elevations*, the relations expressed in the preceding equations may be applied to the part of the cable on one side of the *lowest point* of the cable. The other side is studied separately.

We know that a cable hanging under its own weight is not loaded uniformly along the horizontal. However, the shape of such a cable may be approximated by a parabola when the cable is sufficiently taut. Cables hanging under their own weight are studied in Sec. 8.20.

EXAMPLE 8.23

The mass density of the pipe shown is 10 kg/m. Determine (a) the minimum and maximum tensions in the cable, (b) the length S_{AC} of the segment AC of the cable.

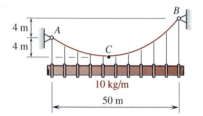

Solution. Let the lowest point C of the cable be chosen as the origin of the coordinates as shown. The intensity of the load is $w = 10g$ N/m $= 98.1$ N/m measured along the horizontal, and the shape of the cable is a *parabola* defined by Eq. (8.18); i.e.,

$$y = \frac{98.1}{2T_o} x^2$$

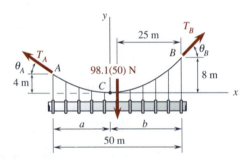

where T_o is the minimum tension in the cable at C. Imposing the boundary conditions of the cable at A and B on this equation and noting the relation between a and b, we write

$$4 = \frac{98.1}{2T_o}(-a)^2 \qquad 8 = \frac{98.1}{2T_o}(b)^2 \qquad a + b = 50$$

These three equations yield

$$a = 20.711 \text{ m} \qquad b = 29.289 \text{ m} \qquad T_o = 5.26 \times 10^3 \text{ N}$$

$$T_{\min} = 5.26 \text{ kN} \blacktriangleleft$$

We note from Eq. (8.16) that T is maximum when θ is as close to 90° as possible. Since θ is closest to 90° at B, we write

$$\tan \theta_B = \left.\frac{dy}{dx}\right|_{x=b} = \frac{98.1}{T_o} b = \frac{98.1(29.289)}{5.26 \times 10^3} \qquad \theta_B = 28.6°$$

$$T_{\max} = T_B = T_o \sec \theta_B = 5.99 \times 10^3 \text{ N}$$

$$T_{\max} = 5.99 \text{ kN} \blacktriangleleft$$

Note that this answer may alternatively be obtained by substituting $L = 2b = 58.58$ m, $h = 8$ m, and $w = 98.1$ N/m into Eq. (8.21).

The length S_{AC} of the segment AC of the cable may be computed by treating it as the left half of a parabolic cable of span $2a = 41.42$ m. Thus, substituting $S = 2S_{AC}$, $L = 41.42$ m, and $h = 4$ m into Eq. (8.22), we write

$$2S_{AC} = 41.42\left[1 + \frac{8}{3}\left(\frac{4}{41.42}\right)^2 - \frac{32}{5}\left(\frac{4}{41.42}\right)^4 + \cdots\right]$$

$$S_{AC} = 21.2 \text{ m} \blacktriangleleft$$

Developmental Exercises

D8.59   In using Eq. (8.18) to define the shape of a parabolic cable, how is the origin of coordinates chosen?

D8.60   Refer to Example 8.23. Determine the tension T_A in the cable at the support A.

D8.61   Refer to Example 8.23. Determine (a) the length S_{CB} of the segment CB of the cable, (b) the total length S_{AB} of the cable.

*8.20   Catenary Cable

Let us now consider a cable AB carrying a vertical load distributed with a constant intensity γ_L along the cable itself as shown in Fig. 8.23(a). The load intensity γ_L is *measured in force per unit of length along the cable*. A cable hanging under its own weight is loaded in this way. We shall show that such a cable hangs in the shape of a *hyperbolic cosine curve—a catenary*.

Choosing the lowest point C of the cable as the origin of coordinates, we draw the free-body diagram of an arbitrary segment CD of the cable as shown in Fig. 8.23(b), where T_o is still the minimum tension in the cable at

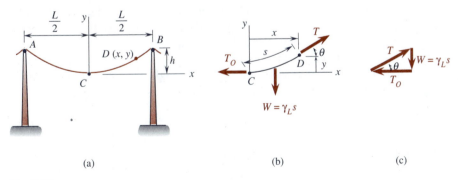

(a)   (b)   (c)

Fig. 8.23

C, s is the arc length of the cable measured from C to D, and the present load intensity parameter γ_L is related to the load intensity parameter w in Eq. (8.17) as follows:

$$\gamma_L \, ds = w \, dx \qquad w = \gamma_L \frac{ds}{dx}$$

$$w = \gamma_L \left[1 + \left(\frac{dy}{dx} \right)^2 \right]^{1/2} \tag{8.23}$$

For simplicity in the subsequent studies, we introduce the constant

$$c = \frac{T_o}{\gamma_L} \tag{8.24}$$

Substituting Eq. (8.23) into Eq. (8.17) and using Eq. (8.24), we write

$$\frac{d^2y}{dx^2} = \frac{1}{c} \left[1 + \left(\frac{dy}{dx} \right)^2 \right]^{1/2}$$

Substitution of $u = dy/dx$ into this equation yields

$$\frac{du}{dx} = \frac{1}{c}(1 + u^2)^{1/2} \qquad (1 + u^2)^{-1/2} \, du = \frac{1}{c} \, dx$$

Integrating from the point C (where $x = 0$ and $dy/dx = u = 0$) to the point D of the cable in Fig. 8.23(a), we write

$$\int_0^u (1 + u^2)^{-1/2} \, du = \frac{1}{c} \int_0^x dx$$

$$\ln[u + (1 + u^2)^{1/2}] = \frac{x}{c} \qquad e^{x/c} = u + (1 + u^2)^{1/2}$$

Solving for u, we find

$$u = \frac{dy}{dx} = \frac{1}{2}(e^{x/c} - e^{-x/c}) = \sinh \frac{x}{c}$$

This slope of the cable may be integrated from $C(0, 0)$ to $D(x, y)$ of the cable in Fig. 8.23(a) to yield

$$y = c \left(\cosh \frac{x}{c} - 1 \right) \tag{8.25}$$

Thus, the shape of the cable AB hanging under its own weight is defined by Eq. (8.25), which is a vertical *hyperbolic cosine curve* shifted a distance c downward and is commonly called a *catenary*.

The equilibrium of the portion CD shown in Fig. 8.23(b) requires that the three force vectors there form a right triangle as shown in Fig. 8.23(c). From this figure and Eq. (8.24), we write

$$\frac{dy}{dx} = \tan \theta = \frac{\gamma_L s}{T_o} = \frac{s}{c} \qquad s = c \frac{dy}{dx} \quad .$$

which together with Eq. (8.25) yields

$$s = c \sinh \frac{x}{c} \tag{8.26}$$

The tension T in the cable may be obtained from Fig. 8.23(c), Eq. (8.24), and Eq. (8.26) as follows:

$$T^2 = T_o^2 + (\gamma_L s)^2 = T_o^2 \left[1 + \left(\frac{s}{c} \right)^2 \right] = T_o^2 \left(1 + \sinh^2 \frac{x}{c} \right)$$

$$= T_o^2 \cosh^2 \frac{x}{c}$$

$$T = T_o \cosh \frac{x}{c} \qquad (8.27)$$

From Eqs. (8.24), (8.25), and (8.27), we also obtain that

$$T = T_o + \gamma_L y \qquad (8.28)$$

If the supports of a catenary cable have *different elevations*, the relations expressed in the preceding equations may be applied to the part of the cable on one side of the *lowest point* of the cable. The other side is studied separately. However, the analysis of such a catenary cable often leads to simultaneous transcendental equations whose solution would require a numerical method and a computer to be practical.

EXAMPLE 8.24

The transmission wire AB as shown has a mass density of 1 kg/m measured along its own length. Determine (a) the minimum and maximum tensions in the wire, (b) the total length of the wire.

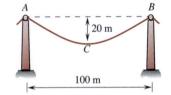

Solution. Let the lowest point C of the wire be chosen as the origin of the coordinates as shown. The intensity of the load is $\gamma_L = 1g$ N/m = 9.81 N/m measured along the wire. Since the slope and the coordinates of the wire at C are all zero, the shape of the wire is a *catenary* defined by

$$y = c \left(\cosh \frac{x}{c} - 1 \right)$$

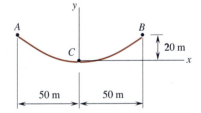

Imposing the boundary conditions at B (or A) on this equation, we write

$$20 = c \left(\cosh \frac{50}{c} - 1 \right)$$

This transcendental equation may be solved by trial and error, a numerical method (cf. Sec. B5.8 of App. B), or with a *digital root finder* to yield[†]

$$c = 65.59$$

[†]The solution of that transcendental equation with a *digital root finder* is illustrated in App. E.

From Eq. (8.24), we write

$$T_o = c\gamma_L = 65.59(9.81) = 643.4$$

$$T_{\min} = T_o \qquad\qquad T_{\min} = 643 \text{ N} \blacktriangleleft$$

Since the maximum tension occurs at B (or A), we apply Eq. (8.27) to obtain

$$T_{\max} = T_B = T_o \cosh \frac{x_B}{c} = 643.4 \cosh \frac{50}{65.59} = 839.6$$

$$T_{\max} = 840 \text{ N} \blacktriangleleft$$

Note that this answer may also be obtained from Eq. (8.28). Applying Eq. (8.26), we find the total length S_{AB} of the wire as follows:

$$S_{AB} = 2s_B = 2c \sinh \frac{x_B}{c} = 2(65.59) \sinh \frac{50}{65.59} = 109.97$$

$$S_{AB} = 110.0 \text{ m} \blacktriangleleft$$

Developmental Exercises

D8.62 Refer to Example 8.24. If the sag h is reduced to 10 m, determine (a) the constant c as defined in Eq. (8.24), (b) the maximum tension in the wire.

D8.63 Refer to Example 8.24. Determine S_{AB} using Eq. (8.22) as an approximation.

PROBLEMS

8.101* The cable AE is loaded as shown. If $h_C = 25$ ft, determine (a) the horizontal component of the reaction from the support at E, (b) the maximum tension in the cable.

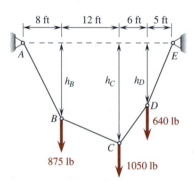

Fig. P8.101* and P8.102

8.102 If the cable shown has a maximum tension of 1560 lb, determine the values of h_B, h_C, and h_D.

8.103 If $a = 5.76$ m in the cable shown, determine (a) the load P, (b) the maximum tension in the cable.

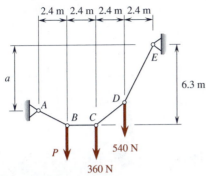

2.4 m 2.4 m 2.4 m 2.4 m

a

E

6.3 m

A

B C

D

P

360 N

540 N

Fig. P8.103 and P8.104

8.104 If $P = 200$ N in the cable shown, determine the distance a.

8.105* The main span of a suspension bridge consists of a uniform roadway suspended from two cables. The load carried by *each* cable is 15 Mg/m along the horizontal. If the span is 1300 m and the sag is 150 m, determine the minimum and maximum tensions in each cable.

8.106 Refer to Prob. 8.105*. Determine the length of each cable used in the main span of the suspension bridge.

8.107* Applying Formula (b) in Sec. B5.6 of App. B, show that Eq. (8.18) for the parabolic cable can be obtained from Eq. (8.25) for the catenary cable by taking the first two terms in the series expansion of cosh x/c if the sag is very small and the cable is sufficiently taut.

8.108 and 8.109 A cable supports a uniformly distributed load with respect to the horizontal as shown. Knowing that the slope of the cable at C is zero, determine the minimum and maximum tensions in the cable.

8.110 and 8.111 A cable supports a uniformly distributed load along the horizontal as shown. Knowing that the slope of the cable at C is zero, determine the total length of the cable.

8.112* A cable supports a distributed load along the horizontal as shown. The intensity of the load at B is w_o per unit length and decreases uniformly to zero at C. The slope of the cable at C is zero. Determine (a) the equation of the curve assumed by the cable, (b) the maximum tension in the cable.

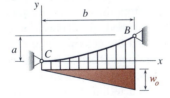

y

b

B

a

C

x

w_o

Fig. P8.112*

8.113 A 10-m cord hangs under its own weight as shown. If the span is $L = 2$ m, determine the sag h. [*Hint.* Use Eqs. (8.25) and (8.26). Start with $c \approx 0.3$ as a trial solution.][†]

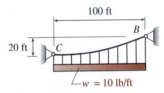

100 ft

B

20 ft C

$w = 10$ lb/ft

Fig. P8.108 and P8.110

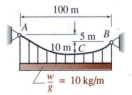

100 m

A

5 m B

10 m C

$\dfrac{w}{g} = 10$ kg/m

Fig. P8.109 and P8.111

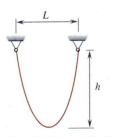

L

h

Fig. P8.113 and P8.114*

[†]Cf. App. E or Sec. B5.8 of App. B for the solution of a transcendental equation.

8.114* A 10-m cord hangs under its own weight as shown. If the sag is $h = 4$ m, determine the span L.

8.115 Cables are hung under their own weight between transmission towers as shown. If each cable has a weight density of 10 lb/ft and the resultant force **R** exerted by the cables at B is vertical, determine the sag h of the cable BC.

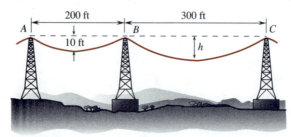

Fig. P8.115

8.116 Refer to Prob. 8.115. Determine (a) the angle which the cable just to the left of B makes with the horizontal, (b) the angle which the cable just to the right of B makes with the horizontal, (c) the magnitude of **R**.

8.117* Refer to Prob. 8.115. Determine the total length of each of the cables AB and BC.

8.118 Solve Prob. 8.115 by assuming $w \approx \gamma_L$ in Eq. (8.23) and using the relations for the parabolic cable as an approximation.

8.119 Solve Prob 8.116 by assuming $w \approx \gamma_L$ in Eq. (8.23) and using the relations for the parabolic cable as an approximation.

8.120* Solve Prob. 8.117* by assuming $w \approx \gamma_L$ in Eq. (8.23) and using the relations for the parabolic cable as an approximation.

8.21 Concluding Remarks

This chapter covers the study of the conditions of equilibrium of trusses, frames, machines, beams, and cables. Such a study entails applications of all the concepts of equilibrium of particles and rigid bodies in a plane and in space, which were treated in Chaps. 1 through 5.

Specifically, the method of joints for analyzing plane trusses and simple space trusses is established from the *concept of equilibrium of particles in a plane and in space*. The method of sections for analyzing the forces in members of plane trusses and the determination of reactions from the supports of trusses, frames, machines, beams, and cables under the action of co-planar force systems are based on the *concept of equilibrium of rigid bodies in a plane*. Furthermore, the determination of reactions from the supports of space trusses is based on the *concept of equilibrium of rigid bodies in space*.

In retrospect, it is clear that the study of the conditions of equilibrium of frames and machines in a plane is a direct extension of the analysis of simple rigid-body equilibrium problems covered in Chap. 5. Such a study is more advanced in nature. However, it is greatly facilitated by the drawing of pertinent free-body diagrams with strict observance of Newton's third law.

Although the shears and bending moments in beams can be determined from the concept of equilibrium, we have emphasized the use of the concept of resolving a given force into a force-moment system at another point to determine those quantities. The two general rules in Sec. 8.16 and the six relations listed in Sec. 8.17 are intended to offer helpful guides in drawing the shear and bending-moment diagrams of beams.

Cables carrying concentrated loads have, in fact, been studied to some extent in Chap. 3 as well as in Example 5.8. They are only briefly discussed in this chapter. The preceding three sections are mainly devoted to the study of cables carrying distributed loads. It is true that many of the equations derived for parabolic and catenary cables are frequently used as formulas if problems involving such cables are to be efficiently solved. Furthermore, the analysis of catenary cables often leads to transcendental equations, which offer students the opportunity of finding roots by trial and error, a numerical method, or a *digital root finder*.[†]

REVIEW PROBLEMS

8.121* Refer to the compound truss in Prob. 8.21. Using the method of sections, determine the forces in members *CH*, *DE*, and *AE*.

8.122* Refer to the compound truss in Prob. 8.24. Using the method of sections, determine the forces in members *BF*, *CD*, and *AD*.

8.123 Determine the forces in members *GH*, *EI*, *BK*, *KJ*, and *CL* of the truss shown.

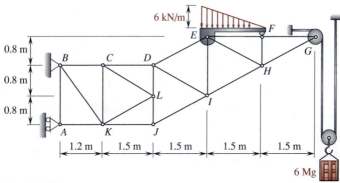

Fig. P8.123

8.124 Determine the reactions from the pins to the member *BEG* of the frame shown.

8.125 A pair of compound-lever snips is used to cut a wire at *M* as shown. For the gripping forces of 200 N applied at *A* and *B*, determine the magnitude of the cutting forces exerted by the blades on the wire at *M*.

8.126* Draw the shear and bending-moment diagrams for the beam shown.

[†]Cf. App. E.

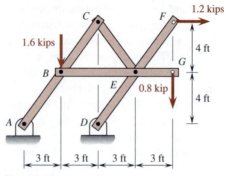

Fig. P8.124

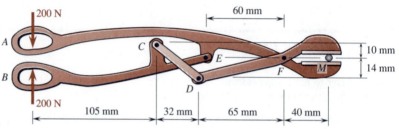

Fig. P8.125

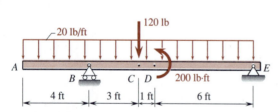

Fig. P8.126*

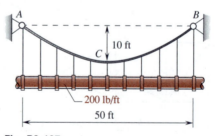

Fig. P8.127

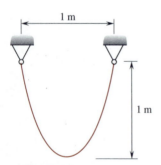

Fig. P8.128

8.127   A cable supports a section of pipeline which weighs 200 lb/ft as shown. Determine (a) the maximum tension in the cable, (b) the total length of the cable.

8.128   A rope hanging under its own weight has a span of 1 m and a sag of 1 m as shown. Determine the total length of the rope.

Friction

Chapter 9

The study of the conditions of physical systems often requires that the effects of friction be taken into account if more accurate results are to be obtained. This chapter is concerned with the conditions of equilibrium of systems involving dry friction between contacting surfaces. In other words, friction problems considered in this chapter are essentially equilibrium problems which may partially be governed by the laws of dry friction. Thus, it is important to be able to appropriately introduce the equations representing the laws of dry friction when the circumstances warrant. Although the effects of dry friction are sometimes undesirable, there are numerous situations in which such effects are not only desirable but also necessary. By thinking of the friction needed to walk and run, the braking needed to stop, the transmission of power through belt drives, and the like, we readily see that friction is ever-present in our daily activities and in many engineering systems.

FRICTION BETWEEN RIGID BODIES

9.1 State of Sliding Surfaces

Friction is a phenomenon in which the surfaces of contact of two bodies resist relative sliding, or the tendency toward relative sliding. We assumed in preceding chapters that surfaces in contact were either *frictionless* or *rough*. This assumption was an idealization. Most surfaces of contact have some asperities which interlock to produce resistance to relative mo-

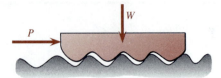

Fig. 9.1

tion between the surfaces. Such asperities may be magnified as illustrated in Fig. 9.1.

The state of two sliding surfaces is said to be in *full film lubrication* when sufficient lubricant is present to completely separate the asperities of the two surfaces. It is said to be in *solid film lubrication* if a solid lubricant, such as graphite, is used. It is said to be in *boundary lubrication* when a significant amount of fluid lubricant is present, but opposing asperities penetrate the fluid film and come in contact. When no significant amount of lubricant covers the interface, the state of two sliding surfaces is said to be in *dry friction* or *Coulomb friction*, after the French engineer and physicist Charles Augustin Coulomb (1736–1806). In this chapter, we shall be concerned with systems involving dry friction.

Developmental Exercise

D9.1 Define the terms: (a) friction, (b) full film lubrication, (c) solid film lubrication, (d) boundary lubrication, (e) dry friction.

9.2 Friction Force

A *friction force* is the force component which is tangent to the contact surfaces of two bodies and *opposes* the sliding motion, or the tendency toward sliding motion, of one contact surface relative to the other. Like other forces, friction forces always occur in pairs, one on each of the contact surfaces. The *sense of the friction force* acting on one contact surface is always opposite to the sense or tendency of its relative sliding motion with respect to the other contact surface. When sliding between the contact surfaces occurs, the friction force is called a *kinetic friction force*, which will be denoted by $\mathbf{F}_k$; otherwise, it is called a *static friction force*, which will be denoted by $\mathbf{F}$.

As an illustration, let us consider a block of weight W which is initially at rest on a rough surface as shown in Fig. 9.2(a). Upon application of the horizontal force $\mathbf{P}$ as shown, the block will either move or tend to move to the right. Thus, the friction force $\mathbf{F}$ acting on the block, as shown in Fig. 9.2(b), is directed to the left to oppose the motion, or tendency toward motion, of the block to the right. By Newton's third law, we note that the friction force $\mathbf{F}$ and the normal force $\mathbf{N}$ acting on the block are related to the friction force $\mathbf{F}'$ and the normal force $\mathbf{N}'$ acting on the support as follows: $F' = F$, $\mathbf{F}' = -\mathbf{F}$; $N' = N$, $\mathbf{N}' = -\mathbf{N}$.

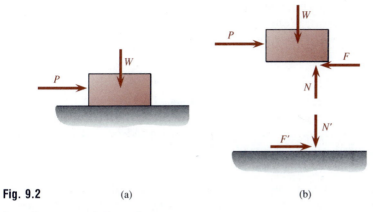

Fig. 9.2   (a)   (b)

Developmental Exercises

D9.2   Define the terms: (a) friction force, (b) kinetic friction force, (c) static friction force.

D9.3†   The wedge A having a weight W_A is driven by the horizontal force **P** to raise the body B having a weight W_B as shown. All contact surfaces are rough. Draw the free-body diagrams of (a) the wedge A, (b) the body B, (c) the wedge A and the body B together. (*Hint.* The senses of the forces must be correctly indicated.)

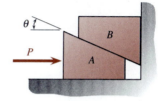

Fig. D9.3

9.3   Laws of Dry Friction

The elementary *laws of dry friction* are applicable to friction surfaces with no significant amount of lubricant. They are, to a first-order approximation, applicable to *boundary lubrication* and *solid film lubrication*. We may state these laws as follows:[††]

1. For bodies not in relative motion, the magnitude of the friction force **F** reaches a maximum value F_m *when slipping impends*; i.e.,

$$F \leq F_m \qquad (9.1)$$

$$F_m = \mu_s N \qquad (9.2)^{†††}$$

[†]Continued in D9.16.

[††]These laws are attributed to Charles Augustin Coulomb, whose friction memoir *Théorie des Machines Simples* was declared in the spring of 1781 by the Academy of Sciences in Paris, France, to be the winner of the double prize for the solution of "problems of friction of sliding and rolling surfaces, the resistance to bending in cords, and the application of these solutions to simple machines used in the navy." One of Coulomb's two-term formulas for friction between sliding surfaces is, in his notation, written as $F = A + (P/\mu)$, where F is the maximum friction force, P is the normal force, μ is the *inverse* of the coefficient of friction, and A is a constant force accounting for adhesive or cohesive effects. However, we shall neglect the adhesive or cohesive effects in our study. For a detailed account, see C. S. Gillmor, *Coulomb and the Evolution of Physics and Engineering in Eighteenth-Century France* (Princeton, N.J.: Princeton University Press, 1971).

[†††]In some mechanics literature, Eqs. (9.2) and (9.3) have been written merely as $F = \mu N$, where F represents either F_m or F_k and μ represents either μ_s or μ_k as appropriate. Note that the F in $F = \mu N$ is *not* the same as the F we use here to denote the magnitude of the general friction force **F** where slipping does not necessarily occur or impend. [Cf. Eqs. (9.1) and (9.2).]

where N is the magnitude of the normal force $\mathbf{N}$ acting on the surface of contact, and μ_s is called the *coefficient of static friction*.

2. For bodies in relative motion, the magnitude of the kinetic friction force $\mathbf{F}_k$ is given by

$$F_k = \mu_k N \qquad (9.3)$$

where μ_k is called the *coefficient of kinetic friction*.

3. The values of μ_s and μ_k are dependent on the materials involved as well as the nature of the surfaces in contact, but independent of the sliding speed and size of the area of contact.

4. In general, we have $\mu_s > \mu_k$.

According to these laws, a horizontal force $\mathbf{P}$ applied to a block of weight W resting on a horizontal dry surface, as shown in Fig. 9.2(a), will cause a friction force $\mathbf{F}$ to be developed on the block as shown in Fig. 9.3(a). The relation between F and P is given by the graph shown in Fig. 9.3(b) where the same scale is used for both F and P axes.

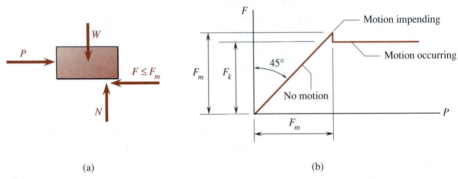

(a)   (b)

Fig. 9.3

Note in Fig. 9.3 that the maximum value of the friction force which can be developed is designated by F_m; i.e., $F_m = F_{\max}$. As P is increased from 0 to F_m, we see that $F = P \le F_m$ and the block, which is originally at rest, remains at rest. At $F = P = F_m$, slipping of the block impends. When $P > F_m$, we see that $F = F_k < F_m$ and the block is being accelerated to the right by the net unbalanced force of magnitude $(P - F_k)$. However, if P is then decreased to F_k, the block will move with a constant velocity. The approximate ranges of values of μ_s for several dry surfaces are given in Table 9.1.

Table 9.1 Approximate Ranges of Values of Coefficients of Static Friction

Dry Surfaces	μ_s	Dry Surfaces	μ_s
Metal on metal	0.2–1.2	Wood on leather	0.2–0.5
Metal on wood	0.2–0.6	Stone on stone	0.4–0.7
Metal on stone	0.3–0.7	Rubber on asphalt	0.7–1.1
Metal on leather	0.3–0.6	Rubber on concrete	0.6–0.9
Wood on wood	0.2–0.6	Rubber on ice	0.05–0.2

Note. The values of μ_s are mostly less than 1, but they can exceed 1 as indicated. For metal on metal, the wide variation in the value of μ_s depends on the metals involved (e.g., copper on copper, mild steel on aluminum, hard steel on lead).

Developmental Exercises

D9.4   Define the terms: (a) coefficient of static friction, (b) coefficient of kinetic friction.

D9.5   Is it correct to say that the friction force is *always* equal to the product of the coefficient of friction and the normal force acting on the surface of contact?

D9.6   A man pushes an 80-lb crate with a horizontal force **P** as shown. He finds that the crate is on the verge of moving when $P = 32$ lb. When the crate is moving, he finds that he needs to keep P at 28 lb if the crate is to move at a constant velocity. Determine the values of (a) μ_s, (b) μ_k, (c) **F** (the friction force) acting on the crate at rest if $P = 25$ lb, (d) **F** acting on the crate which is moving to the left but is slowing down when a 25-lb force **P** as shown is acting on the crate.

D9.7   Give a brief summary of the laws of dry friction.

Fig. D9.6

9.4   Angles of Friction

The *angle of static friction*, denoted by ϕ_s, is the angle between the normal force **N** and the reaction **R**, where **R** is the resultant force of **N** and the maximum static friction force $\mathbf{F}_m$ developed on the surface of contact. Similarly the *angle of kinetic friction*, denoted by ϕ_k, is the angle between the normal force **N'** and the reaction **R'**, where **R'** is the resultant force of **N'** and the kinetic friction force $\mathbf{F}_k$ developed on the surface of contact. For two given contact surfaces, we generally have

$$\phi_s > \phi_k \tag{9.4}$$

The angles ϕ_s and ϕ_k are shown in Fig. 9.4, where **R** is vertical, but **R'** is not. We write

$$\tan\phi_s = \frac{F_m}{N} = \frac{\mu_s N}{N} \qquad \tan\phi_k = \frac{F_k}{N'} = \frac{\mu_k N'}{N'}$$

$$\boxed{\tan\phi_s = \mu_s} \qquad \boxed{\tan\phi_k = \mu_k} \tag{9.5}$$

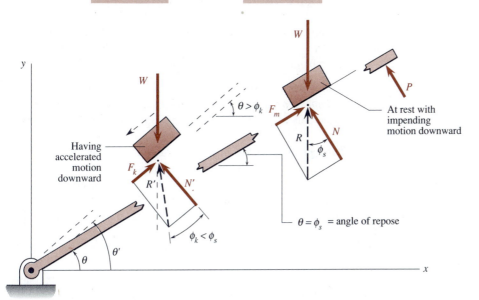

Fig. 9.4

In Fig. 9.4, if the block is originally at rest and the angle θ between the inclined plank and the horizontal is slowly increased from relatively small values to ϕ_s, slipping of the block down the inclined plank will impend at $\theta = \phi_s$. The angle of inclination of the plank corresponding to impending motion is sometimes called the *angle of repose*. Clearly, the angle of repose is equal to ϕ_s. If the block is originally at rest, it will remain at rest for all $\theta < \phi_s$; but it will have motion down the inclined plank if $\theta > \phi_s$.

Developmental Exercises

D9.8 Define the terms: (a) angle of static friction, (b) angle of kinetic friction, (c) angle of repose.

D9.9 The vase on the table starts to slide when the table a young boy lifts makes an angle of 20° with the horizontal as shown. Between the vase and the table, what are the values of (a) ϕ_s, (b) μ_s?

D9.10 The brick shown moves with a constant velocity if $\theta = 30°$, but it will accelerate down the plank when θ exceeds 30°. What are the values of ϕ_k and μ_k between the brick and the plank?

Fig. D9.9

Fig. D9.10

9.5 Types of Friction Problems

Friction problems considered in statics are mainly *equilibrium problems involving dry friction*. For developing systematic solutions, we may classify most friction problems into the following three types:

- *Type I. Relative motion (slipping, tipping, or rolling) in the system is not assured to impend or occur.* In problems of this type, it is usually necessary to ascertain whether the relative motion in the system will impend or occur.
- *Type II. The type and location of relative motion which impends or occurs in the system are known.* In problems of this type, we know that (1) friction forces have magnitudes equal to F_m or F_k, as appropriate, on surfaces where slipping impends or occurs, and (2) the resultant reaction from the support acts through a corner or along an edge of the body if tipping or rolling impends or occurs at that support.
- *Type III. Relative motion in the system is known to impend or occur, but the type and location of relative motion which impends or occurs are not specified.* In problems of this type, the various possible ways for relative motion to start should be analyzed and checked.

Developmental Exercise

D9.11   Describe briefly each of the three types of friction problems.

9.6   Method of Solution for Type I Problems

The following suggested steps are generally helpful in *analyzing and solving type I friction problems* described in Sec. 9.5:

1. *Assume* that the system is at *rest*.
2. *Solve the problem as an equilibrium problem* without using laws of dry friction. (The general procedure for solving equilibrium problems in previous chapters may be followed.)
3. If appropriate, *compute the maximum value F_m of the friction force* which may be developed on each contact surface.
4. *Check the initial assumption* in step (1) by appropriately comparing the magnitude F of the friction force, obtained under that assumption, with its maximum value F_m obtained in step (3). If $F \leq F_m$, the initial assumption is correct, and the problem is solved. If $F > F_m$, equilibrium does not exist and the body must have kinetic friction; therefore, the correct value of F is F_k.

EXAMPLE 9.1

A 100-lb block is initially at rest on the horizontal support before a horizontal force **P** is applied as shown. Knowing that $\mu_s = 0.30$ and $\mu_k = 0.25$ between the surfaces of contact, determine the magnitude F of the friction force acting on the block if the magnitude of **P** is (a) 20 lb, (b) 27 lb, (c) 30 lb, (d) 40 lb, (e) 45 lb.

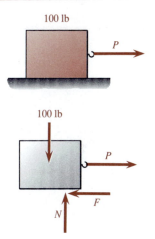

Solution.   From the statement of the problem, relative motion of the block is not assured to impend or occur. Thus, this is a *type I friction problem*. Since the block cannot move in the vertical direction, we may draw the free-body diagram of the block as shown and write

$$+\uparrow \Sigma F_y = 0: \quad N - 100 = 0 \qquad N = 100 \text{ lb}$$

Although the block may be moved to the right by **P**, we know that $F \leq P$. By the laws of dry friction, we write

$$F_m = \mu_s N = 0.30(100) \qquad F_m = 30 \text{ lb}$$

$$F_k = \mu_k N = 0.25(100) \qquad F_k = 25 \text{ lb}$$

(a) In this case, $P = 20 \text{ lb} < F_m$. Thus, the block is in equilibrium. We write

$$\xrightarrow{+} \Sigma F_x = 0: \quad P - F = 0 \qquad F = 20 \text{ lb} \blacktriangleleft$$

(b) In this case, $P = 27 \text{ lb} < F_m$. Thus, the block is in equilibrium even though $P > F_k$. This is so because the block is initially at rest. We write

$$\xrightarrow{+} \Sigma F_x = 0: \quad P - F = 0 \qquad F = 27 \text{ lb} \blacktriangleleft$$

(c) In this case, $P = 30$ lb $= F_m$. Thus, slipping impends but the block is still in equilibrium. We write

$$\xrightarrow{+} \Sigma F_x = 0: \quad P - F = 0 \qquad\qquad F = 30 \text{ lb} \blacktriangleleft$$

(d) In this case, $P = 40$ lb $> F_m$. Thus, slipping occurs and the block has kinetic friction. We write

$$F = F_k \qquad\qquad F = 25 \text{ lb} \blacktriangleleft$$

(e) In this case, $P = 45$ lb $> F_m$. Thus, slipping occurs and the block has kinetic friction. We write

$$F = F_k \qquad\qquad F = 25 \text{ lb} \blacktriangleleft$$

Developmental Exercise

D9.12 A 26-lb force acts on a 100-lb weight which is initially at rest as shown. If $\mu_s = 0.22$, determine (a) N, (b) F, (c) F_m, (d) whether the weight will move.

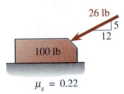

26 lb
5
12
100 lb

$\mu_s = 0.22$

Fig. D9.12

EXAMPLE 9.2

A 10-lb horizontal force **P** is applied to a homogeneous body of weight W which is initially at rest as shown. If $W = 35$ lb and $\mu_s = 0.3$ between the body and the support, determine whether or not motion of the body will impend or occur.

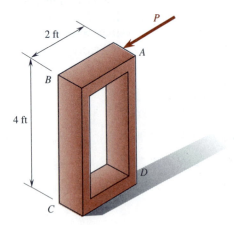

Solution. We first note that, under the action of **P**, the body could slide, tip, or simply remain at rest. Thus, this is a *type I friction problem*. We may analyze and solve this problem as follows:

(a) *Force* **P** *required to cause sliding of the body to impend.* From the free-body diagram shown below, where we set $F = 0.3 N$, we write

$$+\uparrow \Sigma F_y = 0: \quad N - 35 = 0$$

$$\xrightarrow{+} \Sigma F_x = 0: \quad 0.3N - P = 0$$

$$N = 35 \text{ lb} \qquad P = 10.5 \text{ lb} > 10 \text{ lb}$$

Therefore, sliding does not impend or occur.

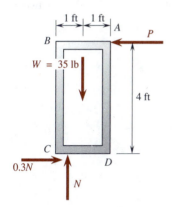
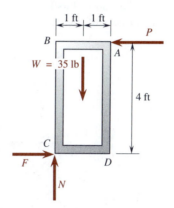

(b) *Force* **P** *required to cause tipping of the body to impend.* In this case, the resultant reaction from the floor must act through the corner C as shown above. We write

$$+\curvearrowleft \Sigma M_C = 0: \quad 4P - 1(35) = 0$$

$$P = 8.75 \text{ lb} < 10 \text{ lb}$$

Conclusion: Motion of the body will occur by tipping. ◄

Developmental Exercise

D9.13   Work Example 9.2 if (a) $W = 41$ lb and $\mu_s = 0.24$, (b) $W = 41$ lb and $\mu_s = 0.25$, (c) $W = 39$ lb and $\mu_s = 0.26$.

9.7   Method of Solution for Type II Problems

The following suggested steps are generally helpful in *analyzing and solving type II friction problems* described in Sec. 9.5:

1. *Write the equations governing the impending slipping or kinetic friction* between all contact surfaces where impending slipping or kinetic friction is known to exist.
2. *Solve the problem as an equilibrium problem* by incorporating appropriately the friction equations in step (1) in the solution. (The general pro-

cedure for solving equilibrium problems in previous chapters may be followed.)

3. *Decide,* if required, *the range of a parameter* by checking the obtained values with the equations governing the impending slipping or kinetic friction.

EXAMPLE 9.3

Determine the range of the magnitude of the horizontal force **P** which will keep the 200-lb block in equilibrium as shown, where $\mu_s = 0.25$ between the surfaces of contact.

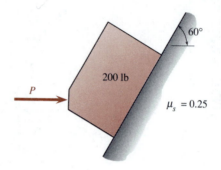

Solution.   We first find that the angle of static friction is

$$\phi_s = \tan^{-1}\mu_s = \tan^{-1} 0.25 = 14.04° < 60°$$

This reveals that the block will slide down the incline without the application of the force **P**. Clearly, slipping of the block downward or upward on the incline will impend when the magnitude of **P** reaches its lower bound or upper bound, respectively. Thus, this is a *type II friction problem*.

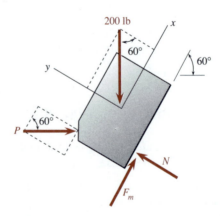

Case (a). Impending motion down the incline. The free-body diagram in this case is drawn as shown. We write

Friction law: $F_m = 0.25N$

$+\nearrow \Sigma F_x = 0$: $P \cos 60° + F_m - 200 \sin 60° = 0$

$+\nwarrow \Sigma F_y = 0$: $-P \sin 60° + N - 200 \cos 60° = 0$

Solution of these three equations yields

$$P = 206.8 \text{ lb} \qquad N = 279.1 \text{ lb} \qquad F_m = 69.8 \text{ lb}$$

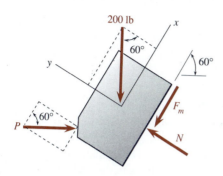

Case (b). Impending motion up the incline. The free-body diagram in this case is drawn as shown. We write

Friction law: $F_m = 0.25N$

$+\nearrow \Sigma F_x = 0$: $P \cos 60° - F_m - 200 \sin 60° = 0$

$+\nwarrow \Sigma F_y = 0$: $-P \sin 60° + N - 200 \cos 60° = 0$

Solution of these three equations yields

$$P = 699.2 \text{ lb} \qquad N = 705.5 \text{ lb} \qquad F_m = 176.4 \text{ lb}$$

From the results in the above two cases, we conclude that the solution to the problem is

$$207 \text{ lb} \leq P \leq 699 \text{ lb} \quad \blacktriangleleft$$

Developmental Exercise

D9.14   Refer to the block in Example 9.3. Determine the magnitude of **P** for which the friction force between the block and the incline is zero.

EXAMPLE 9.4

A 400-kg machine is mounted on an 80-kg horizontal beam as shown. In order to raise the left end of the beam somewhat, a wedge is driven under the base at A by a horizontal force **P**. Knowing that $\mu_s = 0.3$ between all surfaces of contact, determine (a) the maximum value of θ of the wedge for which the base at B will not move, (b) the magnitude of **P** required to drive the wedge when θ is maximum.

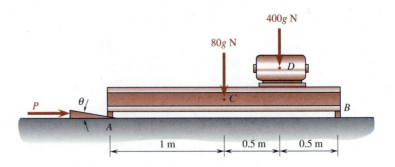

Solution. When the wedge has the desired maximum value of θ and is under the action of the required force **P**, the base at A will have an *impending slipping* on the wedge while the base at B will have an *impending slipping* on the horizontal support. Thus, this is a *type II friction problem*. The free-body diagrams of the wedge and the body AB at impending slipping are drawn as shown. Note that the friction law $F_m = \mu_s N$ has been employed

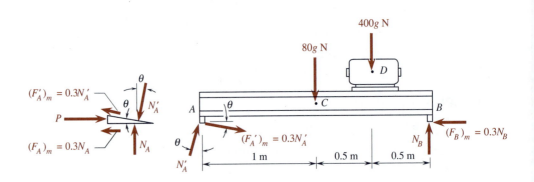

in writing friction forces in the free-body diagrams. In the two free-body diagrams, there are five unknowns: N_A, N_A', N_B, P, and θ. They may be determined as follows:

Wedge

$$\xrightarrow{+}\ \Sigma F_x = 0:\quad P - 0.3N_A - N_A' \sin\theta - 0.3N_A' \cos\theta = 0 \tag{1}$$

$$+\uparrow\ \Sigma F_y = 0:\quad N_A - N_A' \cos\theta + 0.3N_A' \sin\theta = 0 \tag{2}$$

Body AB

$$\xrightarrow{+}\ \Sigma F_x = 0:\quad N_A' \sin\theta + 0.3N_A' \cos\theta - 0.3N_B = 0 \tag{3}$$

$$+\uparrow\ \Sigma F_y = 0:\quad N_A' \cos\theta - 0.3N_A' \sin\theta + N_B - 80g - 400g = 0 \tag{4}$$

$$+\ \Sigma M_A = 0:\quad 2N_B - 1(80g) - 1.5(400g) = 0 \tag{5}$$

Solving these five simultaneous equations, we obtain

$$N_A = 1373 \text{ N} \qquad N_B = 3335 \text{ N} \qquad \theta = 19.38°$$
$$N'_A = 1628 \text{ N} \qquad P = 1413 \text{ N}$$
$$\theta_{max} = 19.38° \qquad P = 1.413 \text{ kN} \blacktriangleleft$$

Developmental Exercises

D9.15 Work Example 9.4 if $\mu_s = 0.2$.

D9.16 Refer to the system in D9.3. Determine the magnitude of **P** required to raise the body B if $\theta = 30°$, $W_A = W_B = 100$ lb, and $\mu_s = 0.35$ between all surfaces of contact.

D9.17 A force **P** pulls on a 10-kg block which rests on a horizontal support as shown. For $\mu_s = 0.22$, determine the minimum value of P for which the block will start to move.

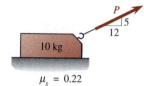

Fig. D9.17

PROBLEMS

9.1 through 9.3* The body shown is homogeneous and is tilted on the rough surface by the horizontal force **P**. The angle α is equal to 30° when sliding impends. Determine the value of μ_s between the surfaces of contact.

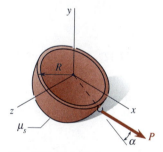

Hemispherical shell

Fig. P9.1 and P9.4

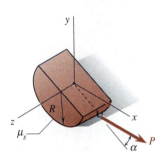

Semicircular cylinder

Fig. P9.2 and P9.5*

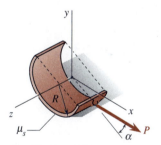

Semicircular pipe

Fig. P9.3* and P9.6

9.4 through 9.6 The body shown is homogeneous and is tilted on the rough surface by the horizontal force **P**. If $\mu_s = 0.5$ between the surfaces of contact, determine the angle α when sliding impends.

9.7 through 9.12* A 58-N horizontal force **P** is applied to a 10-kg homogeneous body which is initially at rest as shown. If $\mu_s = 0.6$ between the body and the support, determine whether or not motion of the body will impend or occur.

9.13 through 9.18 An adequate horizontal force **P** is applied to move the homogeneous body which is initially at rest as shown. Determine the maximum allowable value of μ_s for which tipping will not occur.

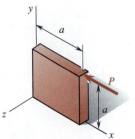

Fig. P9.7, P9.13 and P9.44*

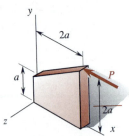

Fig. P9.8, P9.14*, and P9.45

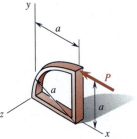

Fig. P9.9*, P9.15 and P9.46

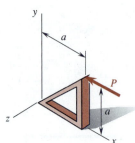

Fig. P9.10, P9.16 and P9.47*

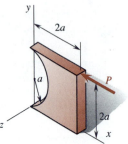

Fig. P9.11, P9.17* and P9.48

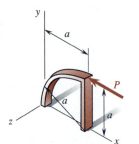

Fig. P9.12*, P9.18 and P9.49

9.19*   A section of pipe weighing 40 lb rests against a curb of height $h = 0.2$ ft. It is known that $\mu_s = 0.45$ between the pipe and the curb. Can an adequate horizontal force **P** roll the pipe over the curb as shown?

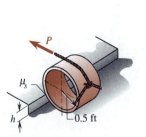

Fig. P9.19* and P9.20

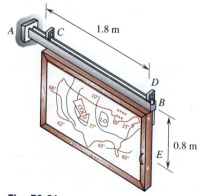

Fig. P9.21

9.20 A section of pipe weighing 60 lb rests against a curb of height $h = 0.1$ ft. It is desired to apply a horizontal force **P** to roll the pipe over the curb as shown. Determine the required minimum values of (a) μ_s between the pipe and the curb, (b) the magnitude of **P**.

9.21 The values of μ_s between the track AB and the 8-kg board shown are 0.2 at C and 0.25 at D. If the board is to be moved to the right, determine the horizontal force **P** which must be applied to the handle at E.

9.22* Solve Prob. 9.21 if the board is to be moved to the left.

9.23 Two crates on the ramp are released from rest as shown. It is known that $\mu_s = 0.4$ and $\mu_k = 0.35$ between the ramp and the 60-kg crate A, and $\mu_s = 0.3$ and $\mu_k = 0.25$ between the ramp and the 40-kg crate B. For $\theta = 20°$, determine (a) whether the crates will move, (b) the friction force acting on each crate.

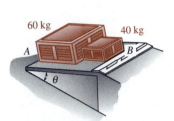

Fig. P9.23

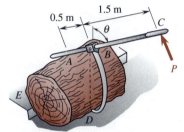

Fig. P9.25

9.24* Solve Prob. 9.23 if $\theta = 19°$.

9.25 A cant hook is used to lift a log as shown. It is known that $\mu_s = 0.4$ between the log and the device at A. Determine the minimum value of θ for which the log may be lifted by an adequate force **P** as shown.

9.26 The weights of the blocks A and B, as shown, are 200 lb and 52 lb, respectively. The system is initially at rest and $\mu_s = 0.4$ between all surfaces of contact. Determine the magnitude of the horizontal force **P** that will cause motion of the block A to impend.

9.27 A 20-lb body A rests on a wedge B of negligible weight and against a smooth vertical wall as shown. If $\mu_s = 0.25$ between A and B as well as between B and the horizontal support, determine whether the wedge B is self-locking in the position shown when (a) $\theta = 25°$, (b) $\theta = 30°$, (c) $\theta = 35°$.

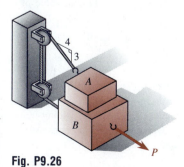

Fig. P9.26

Fig. P9.27

9.28   It is known that the 20.8-lb body A and the 21.7-lb body B are initially at rest as shown and $\mu_s = 0.1$ between all surfaces of contact. Determine the magnitude of **P** that will cause the body A to have an impending motion down the incline.

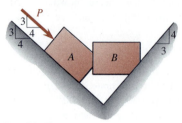

Fig. P9.28

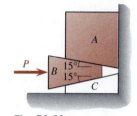

Fig. P9.29

9.29   If the bodies A and B as shown have masses $m_A = 100$ kg and $m_B \approx 0$ and $\mu_s = 0.25$ between all surfaces of contact, determine the range of the magnitudes of **P** if equilibrium exists.

9.30*   The body A shown weighs 60 lb and $\mu_s = 0.25$ between all surfaces of contact. Neglecting the weight of the wedge B, determine the minimum magnitude of **P** which will cause an upward motion of the body A to impend.

9.31   During a manufacturing process, a piece of plywood with a thickness of h is to be further compressed to a final thickness of 20 mm by steel rollers driven as shown. It is known that $\mu_s = 0.15$ between the plywood and either of the rollers. Determine the maximum value of h for which the plywood will be pulled into the rollers by the friction forces.

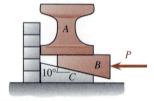

Fig. P9.30*

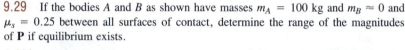
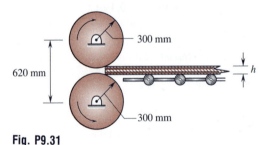

Fig. P9.31

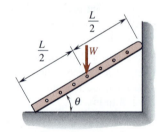

Fig. P9.32 and P9.33*

9.32   A ladder of weight W and length L is placed against a vertical wall as shown, where $\mu_s = 0.5$ between all surfaces of contact. What is the minimum value of θ for which the ladder will not slide down?

9.33*   A ladder is shown, where $\mu_s = 0.4$ between all surfaces of contact. Determine whether the ladder will be in equilibrium for $\theta = 45°$.

9.34   A 10-kg movable chalkboard $ABCD$ is normally supported by two 5-kg counterweights E and F by means of cords and pulleys as shown. One day, the left cord is broken and the chalkboard is found to remain in the position shown due to friction. Assume that the chalkboard is slightly smaller than the fixed guiding frame and it will bind only at the points A and C. For equal values of μ_s between the surfaces of contact at A and C, determine the smallest possible value of μ_s.

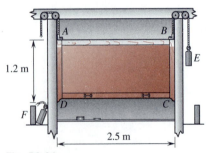

Fig. P9.34

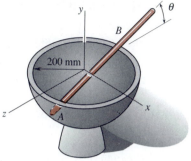

Fig. P9.35

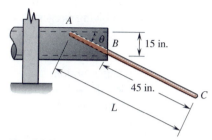

Fig. P9.37

9.35 A 600-mm uniform slender rod is placed in a hemispherical bowl which has an interior radius of 200 mm as shown. If the achievable maximum value of the angle θ is 30°, determine μ_s between the rod and the bowl.

9.36* Refer to the system in Prob. 9.35. Determine the value of μ_s between the rod and the bowl if the achievable minimum value of θ is 10°.

9.37 A uniform slender rod ABC of length L is held by a horizontal pipe which has an interior diameter of 15 in. as shown. Determine the minimum value of μ_s existing between the rod and the pipe if $L = 75$ in.

9.38* Refer to the system in Prob. 9.37. Determine the minimum value of μ_s existing between the rod and the pipe if $L = 84$ in.

9.39 Two uniform rods, each of weight W, are held by frictionless pins at A and B as shown. For $\mu_s = 0.4$ between the surfaces of contact at C, determine whether the rods will remain in equilibrium if $\theta = 50°$.

9.40* Refer to the system in Prob. 9.39. Determine the required minimum value of μ_s in terms of the angle θ if the system is to remain at rest.

9.41 For $\mu_s = 0.25$ between the body A of mass m_A and the incline as shown, determine the range of values of m_A for which the 10-kg body B will remain in equilibrium.

9.42 It is known that the mass of the crate A is 10 kg, the mass of the cylinder B is 6 kg, and $\mu_s = 0.55$ between the crate and the horizontal support as shown. For $0 \le \theta \le 90°$, determine the range of values of θ for which the crate will not move.

9.43 If the mass of the collar A is $m_A = 16$ kg and $\mu_s = 0.25$ between the collars and the rods, determine the range of values of the mass m_B of the collar B for which the collars as shown will not move.

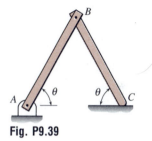

Fig. P9.39

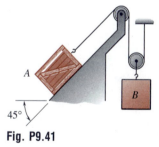

Fig. P9.41

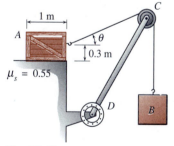

Fig. P9.42

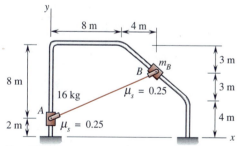

Fig. P9.43

9.8   Method of Solution for Type III Problems

The following suggested steps are generally helpful in *analyzing and solving type III friction problems* described in Sec. 9.5:

1. *List all of the possible ways* in which the system can have the stated relative motion.
2. *Find the solution for each of the possible ways* listed in step (1). Each of these possible ways now corresponds to a type II friction problem or a simple equilibrium problem.
3. *Select the correct solution* which satisfies the situation or condition stated in the problem.

EXAMPLE 9.5

A horizontal force **P** is applied to a 100-lb homogeneous quarter circular disk which is initially at rest as shown. If $\mu_s = 0.6$ between the disk and the support, determine the minimum magnitude of **P** required to move the disk.

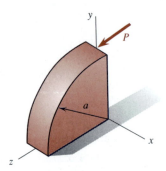

Solution.   When the magnitude of **P** is large enough, the disk will be caused to have impending slipping or tipping. However, it is not yet known whether slipping or tipping impends first. Thus, this is a *type III friction problem*.

Case (a). Assume that slipping impends first. From the free-body diagram shown on the preceding page, we write

$$+\uparrow \Sigma F_y = 0: \quad N - 100 = 0 \qquad N = 100 \text{ lb}$$

$$+\swarrow \Sigma F_z = 0: \quad P - 0.6N = 0 \qquad P = 60 \text{ lb}$$

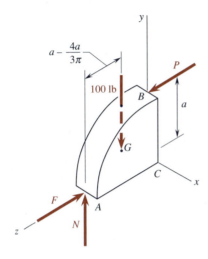

Case (b). Assume that tipping impends first. In this case, the components of reaction from the support pass through the corner at A as shown in the free-body diagram above. Thus, we write

$$+ \circlearrowright \Sigma M_A = 0: \quad aP - [a - 4a/(3\pi)](100) = 0$$

$$P = 57.6 \text{ lb}$$

Decision. The smaller of the two values of P obtained in the above two cases is the minimum magnitude of **P** required. Thus, we have

$$P_{\min} = 57.6 \text{ lb} \quad \blacktriangleleft$$

Developmental Exercise

D9.18 A horizontal force **P** acts on a 120-lb cabinet which rests on a floor as shown. It is known that $\mu_s = 0.21$. (a) What is the magnitude of **P** if slipping impends? (b) Where will the resultant floor reaction act if tipping impends? (c) What is the magnitude of **P** if tipping impends? (d) What is the smallest magnitude of **P** which will cause the cabinet to move?

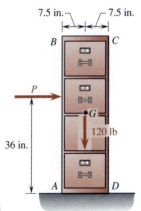

Fig. D9.18

EXAMPLE 9.6

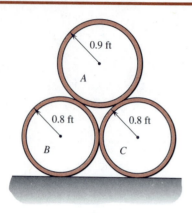

The cylindrical pipe A is to be stacked on top of the cylindrical pipes B and C as shown. The pipes have weights $W_A = 270$ lb, $W_B = W_C = 240$ lb, respectively. Assume that μ_s is the same between all contact surfaces. Determine the smallest value of μ_s for which the pipe A will not fall down.

Solution.   We first recognize that the motion or collapse of the pipe A downward simultaneously with the motion of the pipes B and C outward will impend when μ_s equals the smallest value. However, the type and location of such motion (slipping or rolling) are not yet known. Thus, this is a *type III friction problem*.

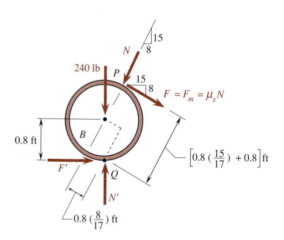

Case (a). Assume that slipping of the pipe A on the pipes B and C impends, while the pipes B and C tend to move outward by rolling (and possibly slipping) on the horizontal support. The free-body diagram of the pipe B in this case is drawn as shown above. Note that slipping is now assumed to impend at the point P. (Slipping does not necessarily impend at the point Q.) The friction law, $F_m = \mu_s N$, is indicated at P in the free-body diagram. Thus, we write

$$+\circlearrowleft \ \Sigma M_Q = 0: \quad 0.8(\tfrac{8}{17})(N) - [0.8(\tfrac{15}{17}) + 0.8](\mu_s N) = 0$$

$$\mu_s = \tfrac{1}{4} = 0.25$$

Case (b). Assume that slipping of the pipes B and C on the horizontal support impends, while the pipe A tends to move downward by either rolling or slipping on the pipes B and C. In this case, let us first find the normal component of the reaction from the support to the pipe B. From the free-body diagram of the system shown on the next page, we write

$$+\circlearrowleft \ \Sigma M_R = 0: \quad -1.6(N') + 1.6(240) + 0.8(270) = 0$$

$$N' = 375 \text{ lb}$$

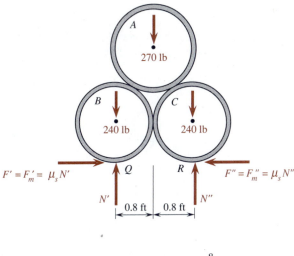

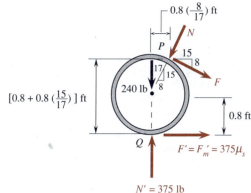

Having determined N', we may draw the free-body diagram of the pipe B in this case as shown above. The friction law, $F'_m = \mu_s N' = 375\mu_s$, has been indicated in this diagram for the assumed impending slipping at the point Q. (Slipping does not necessarily impend at the point P.) We write

$+ \circlearrowleft \Sigma M_P = 0$: $-0.8(\frac{8}{17})(375 - 240)$

$+ [0.8 + 0.8(\frac{15}{17})](375\mu_s) = 0$

$\mu_s = 0.09 < 0.25$

Decision. The larger of the two values of μ_s obtained in the above two cases is the smallest μ_s required. Thus, the answer to this problem is

$$\mu_s = 0.25 \quad \blacktriangleleft$$

REMARK. If this problem asked for the minimum value of μ_s between the pipes and the horizontal support as well as the minimum value of μ_s between the pipes themselves so that the pipe A could be stacked, then this problem would become a type II friction problem. On the other hand, if the values of μ_s between all surfaces of contact were given and the problem asked whether the pipes as stacked would stay together, then this problem would become a type I friction problem.

Developmental Exercise

D9.19 If μ_s is the same between all surfaces of contact, determine its minimum value for which the three identical pipes may be placed as shown.

Fig. D9.19

PROBLEMS

9.44* through **9.49** (*The figures for these problems are shown and specified along with those for Probs. 9.7 through 9.12* on p. 364.*) The body shown is homogeneous and weighs 18 lb. It is initially at rest when the horizontal force **P** acts on it. If $\mu_s = 0.6$ between the body and the support, determine the minimum magnitude of **P** required to cause the body to have impending motion.

9.50* If $\mu_s = 0.4$ between the 14-kg uniform plank ABC and either of the two joists shown, determine the minimum magnitude of the horizontal force **P** required to move the plank.

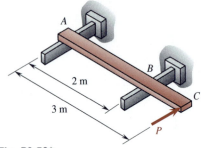

Fig. P9.50*

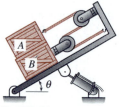

Fig. P9.51

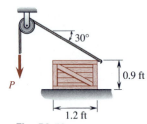

Fig. P9.52

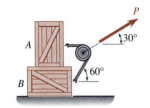

Fig. P9.53

9.51 Bodies A and B are of the same weight W. If $\mu_s = 0.3$ between A and B and between B and the adjustable frame shown, determine the value of θ at which relative motion between A and B impends.

9.52 If $\mu_s = 0.5$ between the 40-lb crate and the floor shown, determine the minimum magnitude of **P** required to cause an impending motion of the crate.

9.53 The value of μ_s is 0.35 between the 10-kg crate A and the 20-kg crate B, and is 0.2 between the crate B and the floor as shown. Determine the minimum value of the magnitude of **P** required to cause impending motion of the system.

9.54 A 20-kg concrete block rests on an incline as shown. If $\mu_s = 0.2$, determine the maximum magnitude of the horizontal force **P** for which the block remains in equilibrium.

9.55* Solve Prob. 9.54 if $\mu_s = 0.3$.

9.56 It is desired to stack the 20-kg cylinder A on top of the 45-kg cylinders B and C as shown. Assume that μ_s is the same between all surfaces of contact. Determine the minimum value of μ_s for which the cylinders may remain in equilibrium as shown.

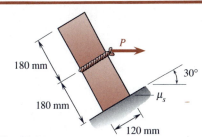

Fig. P9.54

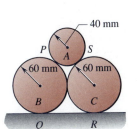

Fig. P9.56

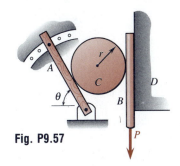

Fig. P9.57

9.57 A cylinder C having a small weight W_C and a vertical bar B having negligible weight are placed between an adjustable guide A and a vertical wall D as shown. Assume that $\mu_s = 0.5$ between all surfaces of contact. Determine the minimum value of the angle θ if the system is to be self-locking for any magnitude of the vertical force **P** applied to the bar B.

9.58* Refer to the system in Prob. 9.57. Determine the required minimum value of μ_s in terms of the angle θ if the system is to be self-locking for any magnitude of **P**.

9.59* Two links are used to connect the 20-kg block A and the 10-kg block B as shown. If $\mu_s = 0.25$ between all contact surfaces, determine the maximum magnitude of the vertical force **P** which will not cause motion of the system to impend.

9.60 A section of pipe having a mass of 26 kg is held on an incline as shown. It is known that $\mu_s = 0.3$ between all contact surfaces. If $\theta = 40°$, determine the minimum magnitude of **P** required to prevent downward motion from occurring.

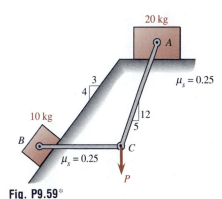

Fig. P9.59*

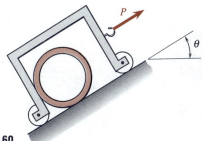

Fig. P9.60

9.61 Solve Prob. 9.60 if $\theta = 50°$.

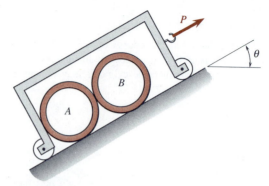

Fig. P9.62

9.62   Two identical sections of pipes A and B, each weighing 72 lb, are held on an incline as shown. If $\mu_s = 0.2$ between all contact surfaces and $\theta = 30°$, determine the minimum magnitude of **P** required to prevent downward motion from occurring.

9.63*   Solve Prob. 9.62 if $\theta = 20°$.

★9.9   Dry Friction in Square-Threaded Screws

A *screw* is a machine which consists of a head and a cylindrical or tapered shank that has at least one helical thread built on its surface. Screws are used for fastenings and for transmitting power or motion. A *square-threaded screw* is a screw whose shank is cylindrical and whose thread has a square or rectangular cross section. Square-threaded screws are generally more efficient than screws of other types in transmitting power or motion.

A dry or partially lubricated square-threaded jack is shown in Fig. 9.5(a), where an axial load **W** is to be raised by applying a torque **M** about the axis

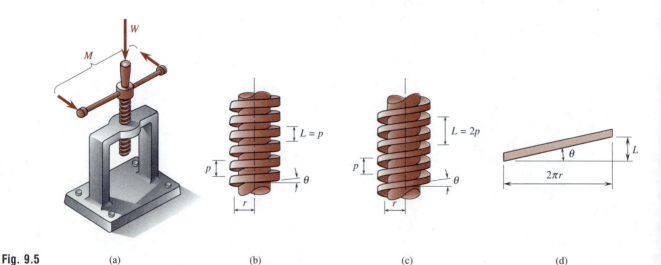

Fig. 9.5   (a)   (b)   (c)   (d)

of the screw. The *lead*, denoted by L, is the distance through which the
screw advances in one turn; the *pitch*, denoted by p, is the distance measured between two consecutive threads. These are illustrated in Fig. 9.5(b)
for a *single-threaded screw* and in Fig. 9.5(c) for a *double-threaded screw*.
For an *n-threaded screw*, we have

$$L = np \qquad (9.6)$$

If one turn of the thread is unwrapped from the screw and stretched as
shown in Fig. 9.5(d), we see that the *lead angle* θ is given by

$$\tan \theta = \frac{L}{2\pi r} \qquad (9.7)$$

where r is the mean radius of the thread.

The action of a screw is closely related to the friction developed in the
threads. When the applied torque $\mathbf{M}$ in Fig. 9.5(a) is sufficiently large so
that raising of the axial load $\mathbf{W}$ impends, the reaction $\Delta\mathbf{R}$ from the thread
of the jack frame to a small representative portion of the thread of the screw
may be depicted as shown in Fig. 9.6, where ϕ_s is the angle of static friction. Similar reactions exist on all portions of the screw thread where contact
with the thread of the jack frame occurs. Let the component of $\Delta\mathbf{R}$ perpendicular to the axis of the screw be $\Delta\mathbf{R}_\perp$. Then, the magnitude of $\Delta\mathbf{R}_\perp$ is
$\Delta R_\perp = \Delta R \sin(\theta + \phi_s)$. The torque about the axis of the screw produced
by $\Delta\mathbf{R}$ is $r(\Delta R_\perp) = [r \sin(\theta + \phi_s)]\Delta R$. The total torque produced by all
reactions is $\Sigma(r \sin(\theta + \phi_s)\Delta R] = [r \sin(\theta + \phi_s)]\Sigma(\Delta R)$. Therefore, equilibrium of moments in the axial direction of the screw in Fig. 9.6 yields

$$M = [r \sin(\theta + \phi_s)]\, \Sigma(\Delta R)$$

On the other hand, equilibrium of forces in the axial direction of the screw
in Fig. 9.6 requires that

$$W = \Sigma[\Delta R \cos(\theta + \phi_s)] = [\cos(\theta + \phi_s)]\Sigma(\Delta R)$$

Dividing M by W and rearranging, we obtain the magnitude of the torque required to *raise* the axial load $\mathbf{W}$ on the jack (i.e., to *advance* the screw) as

$$M = Wr \tan(\theta + \phi_s) \qquad (9.8)$$

The action of the entire screw may be simulated by using the unwrapped
thread of the screw on an incline as a model as shown in Fig. 9.7. The

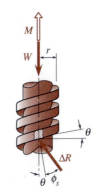

Fig. 9.6

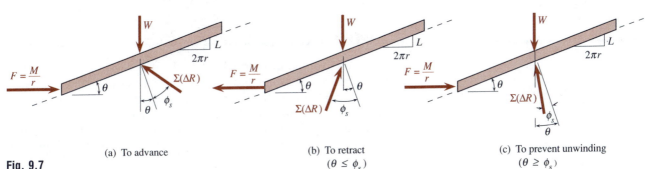

(a) To advance

(b) To retract
$(\theta \leq \phi_s)$

(c) To prevent unwinding
$(\theta \geq \phi_s)$

Fig. 9.7

equivalent force required to move this unwrapped thread on the fixed incline is $F = M/r$. By considering the equilibrium of the unwrapped thread in Fig. 9.7(a), we can also obtain Eq. (9.8). When the torque is removed, the screw will remain in place and be self-locking if $\theta \leq \phi_s$.

If $\theta \leq \phi_s$, the torque **M** required to *lower* the axial load **W** on the jack (i.e., to *retract* the screw) can readily be obtained as

$$M = Wr \tan(\phi_s - \theta) \qquad (9.9)$$

by considering the equilibrium of the free body in Fig. 9.7(b). If $\theta > \phi_s$, the screw will unwind by itself, and we see from Fig.9.7(c) that the torque required to *prevent unwinding* is given by

$$M = Wr \tan(\theta - \phi_s) \qquad (9.10)$$

The important relations which are used in studying situations involving the impending motion of square-threaded screws are summarized in Fig. 9.8, where **W** and **M** form a wrench.

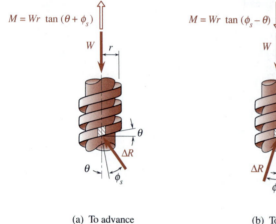

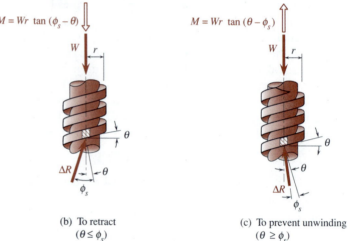

(a) To advance　　　　　　(b) To retract　　　　(c) To prevent unwinding
　　　　　　　　　　　　　　$(\theta \leq \phi_s)$　　　　　　$(\theta \geq \phi_s)$

Fig. 9.8

Developmental Exercises

D9.20　Define the *lead*, the *pitch*, and the *lead angle* of a square-threaded screw with n threads.

D9.21　For a square-threaded screw with n threads, show that Eq. (9.8) may equivalently be written as $M = Wr(np + 2\pi\mu_s r)/(2\pi r - \mu_s np)$.

EXAMPLE 9.7

The scissors jack shown supports a load **P** from a car. The screw has a double square thread with a mean diameter of 10 mm and a pitch of 2 mm. The static coefficient of friction between threads is 0.2. The jack is symmetrical about a vertical plane containing the axis of the screw and has two sets of linkages. If a torque $M = 10$ N·m is required to raise **P** as shown, determine (a) the magnitude of **P**, (b) the torque M' required to lower **P**.

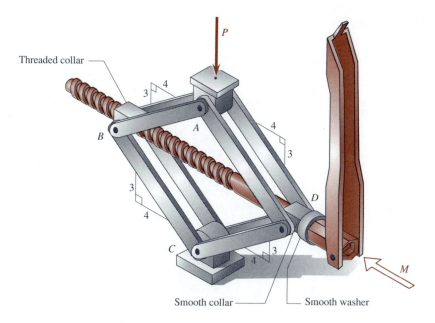

Threaded collar

P

4

3

A

B

4

3

3

4

D

C

3

4

M

Smooth collar — Smooth washer

Solution. We note that the pairs of linkages *AB*, *AD*, *BC*, and *CD* are all two-force members. Although the application of the torque **M** will cause the front and back members in the pairs of linkages to develop different axial forces, we shall combine the axial forces in each pair into a single force and denote the combined axial forces in those pairs of linkages by $\mathbf{F}_{AB}$, $\mathbf{F}_{AD}$, $\mathbf{F}_{BC}$, and $\mathbf{F}_{CD}$, respectively.

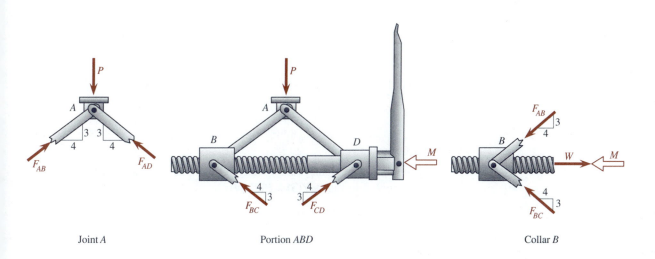

Joint *A*   Portion *ABD*   Collar *B*

Joint A

$$\xrightarrow{+}\ \Sigma F_x = 0: \quad \frac{4}{5}F_{AB} - \frac{4}{5}F_{AD} = 0 \qquad F_{AB} = F_{AD}$$

$$+\uparrow \Sigma F_y = 0: \quad \frac{3}{5}F_{AB} + \frac{3}{5}F_{AD} - P = 0 \qquad F_{AB} = F_{AD} = \frac{5}{6}P$$

Portion ABD

$$\xrightarrow{+}\ \Sigma F_x = 0: \quad -\frac{4}{5}F_{BC} + \frac{4}{5}F_{CD} = 0 \qquad F_{BC} = F_{CD}$$

$$+\uparrow \Sigma F_y = 0: \quad \frac{3}{5}F_{BC} + \frac{3}{5}F_{CD} - P = 0 \qquad F_{BC} = F_{CD} = \frac{5}{6}P$$

Collar B

$$\xrightarrow{+}\ \Sigma F_x = 0: \quad W - \frac{4}{5}F_{AB} - \frac{4}{5}F_{BC} = 0 \qquad W = \frac{4}{3}P$$

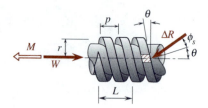

In the free-body diagram of the collar B, we may move the force $\mathbf{W}$ and torque $\mathbf{M}$ to the left end of the screw as shown, where $r = 5$ mm, $p = 2$ mm, $L = 2p = 4$ mm, and

$$\tan \theta = \frac{L}{2\pi r} = \frac{4}{2\pi(5)} \qquad \theta = 7.256°$$

$$\tan \phi_s = \mu_s = 0.2 \qquad \phi_s = 11.310°$$

$$M = 10 \text{ N·m} \qquad W = \tfrac{4}{3}P$$

From the relation summarized in Fig. 9.8(a), we write

$$10 = \tfrac{4}{3}P(\tfrac{5}{1000}) \tan (7.256° + 11.310°)$$

$$P = 4.466 \times 10^3 \qquad\qquad P = 4.47 \text{ kN} \blacktriangleleft$$

Since $\phi_s > \theta$, the screw will not unwind by itself. Thus, the torque M' required to lower the load $\mathbf{P}$ may be computed by applying the relation summarized in Fig. 9.8(b) as follows:

$$M' = \tfrac{4}{3}(4.466 \times 10^3)(\tfrac{5}{1000}) \tan (11.310° - 7.256°)$$

$$M' = 2.11 \text{ N·m} \blacktriangleleft$$

Developmental Exercise

D9.22   Work Example 9.7 for the case in which the screw has a single thread.

*9.10 Axle Friction in Journal Bearings

A *journal bearing* is a bearing that provides lateral support to a rotating axle or shaft in contrast to axial or thrust support. The friction between an axle and its journal bearings is referred to as *axle friction*.

Let a dry or partially lubricated journal bearing be shown in Fig. 9.9 where the clearance between the axle and bearing is greatly exaggerated. As the axle begins to turn in the direction shown, it rolls up the inner surface of the bearing until slippage occurs. The point of contact remains more or less at a fixed position A as rotation of the axle continues. To keep the axle rotating at a constant speed, we find that a torque $\mathbf{M}$ must be applied as shown. The torque $\mathbf{M}$ and the lateral load $\mathbf{W}$ on the axle at the bearing cause a reaction $\mathbf{R}$ at A. For equilibrium, $\mathbf{W}$ and $\mathbf{R}$ must be *equal in magnitude but not collinear*. Note that $\mathbf{R}$ is tangent to a small circle of radius r_f. This circle is called the *circle of friction* of the axle and bearing and is generally independent of $\mathbf{W}$. The angle between $\mathbf{R}$ and its normal component $\mathbf{N}$ is the angle of kinetic friction ϕ_k. For the axle rotating at a constant speed, we write

$$+\uparrow \Sigma F_y = 0: \quad R - W = 0 \qquad R = W$$

$$+\circlearrowright \Sigma M_A = 0: \quad r_f W - M = 0 \qquad M = r_f W \qquad (9.11)$$

Noting that $\mu_k = \tan\phi_k$ and $\sin\phi_k \approx \tan\phi_k$ for small values of ϕ_k, we write

$$r_f = r \sin\phi_k \approx r \tan\phi_k \qquad r_f \approx r\mu_k \qquad (9.12)$$

In situations involving impending slipping in the journal bearing, we replace μ_k with μ_s; i.e.,

$$r_f \approx r\mu_s \qquad (9.13)$$

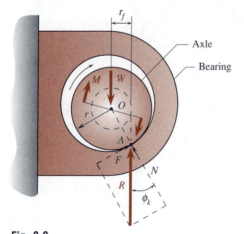

Fig. 9.9

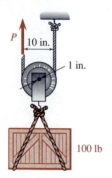

EXAMPLE 9.8

A 100-lb crate is suspended from a pulley of diameter 10 in. as shown. For $\mu_s = 0.2$ between the pulley and the axle of diameter 1 in., determine the minimum force **P** required to maintain equilibrium of the system.

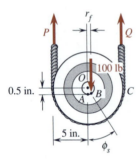

Solution. When the magnitude of **P** has the minimum value required to maintain equilibrium of the system, the pulley tends to rotate counterclockwise about the axle or, equivalently, the axle tends to rotate clockwise relative to the pulley and to climb from A to B. Thus, the free-body diagram of the pulley may be drawn as shown, where Q denotes the cable tension on the right pulley. As this problem involves impending slipping, rather than kinetic friction, we write

$$r_f \approx r\mu_s = (0.5 \text{ in.})(0.2) = 0.1 \text{ in.}$$

$$+\circlearrowleft \ \Sigma M_C = 0: \quad -10P + (5 - 0.1)(100) = 0$$

$$\mathbf{P} = 49 \text{ lb} \uparrow \quad \blacktriangleleft$$

Developmental Exercises

D9.23 Refer to Example 9.8. Determine the maximum force **P** consistent with equilibrium.

D9.24 The kinetic coefficient of friction between a 1-in.-diameter axle and its bearing is 0.15. The load on the axle at the bearing is 300 lb. Determine (a) the torque required to keep the axle rotating at a constant speed, (b) the radius of the friction circle.

★9.11 Disk Friction in Thrust Bearings

A *thrust bearing* is a bearing that gives axial or thrust support to a rotating shaft or disk. Examples of thrust bearings include pivot bearings, collar bearings, disk brakes, and disk clutches. The friction between a disk or a shaft and its thrust bearing is referred to as *disk friction*.

Let us consider a hollow shaft rotating against a pivot bearing as shown in Fig. 9.10. The hollow shaft of inside radius R_i and outside radius R_o is acted on by a *wrench* consisting of an axial force **P** and a torque **M** which keeps the shaft rotating at a constant speed. Assuming that the normal pres-

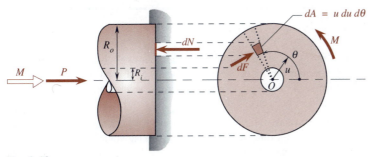

Fig. 9.10

sure p between the contact surfaces is uniform, we find that

$$p = \frac{P}{\pi(R_o^2 - R_i^2)} \qquad (9.14)$$

The normal force $d\mathbf{N}$ and the friction force $d\mathbf{F}$ acting on the differential area dA have magnitudes $p\,dA$ and $\mu_k p\,dA$, respectively, where μ_k is the coefficient of kinetic friction. The moment of $d\mathbf{F}$ about the axis of the shaft is $u\,dF = \mu_k p u\,dA = \mu_k p u(u\,du\,d\theta) = \mu_k p u^2\,du\,d\theta$, and the total moment is

$$M = \mu_k p \int_{\theta=0}^{2\pi} \int_{u=R_i}^{R_o} u^2\,du\,d\theta = \frac{2}{3}\pi\mu_k p(R_o^3 - R_i^3)$$

$$M = \frac{2\mu_k P(R_o^3 - R_i^3)}{3(R_o^2 - R_i^2)} \qquad (9.15)$$

If the shaft is solid and has a radius R, we let $R_i = 0$ and $R_o = R$ in Eq. (9.15) to obtain

$$M = \tfrac{2}{3}\mu_k PR \qquad (9.16)$$

Thus, the largest torque which can be transmitted by a disk clutch without causing slippage is

$$M = \tfrac{2}{3}\mu_s PR \qquad (9.17)$$

Developmental Exercise

D9.25 When a disk clutch passes its wearing-in period, the normal pressure p may be assumed to be inversely proportional to the radial distance u from its center; i.e., $p = c/u$ where c is a constant. For the disk clutch shown, show that (a) $c = P/(2\pi R)$, (b) the largest torque which may be transmitted after the wearing-in period is given by $M = \tfrac{1}{2}\mu_s PR$.

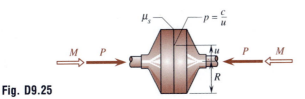

Fig. D9.25

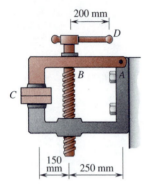

Fig. P9.64

200 mm

D

B   *A*

C

150 mm   250 mm

Fig. P9.65

PROBLEMS

9.64   Two plates are fastened with a bolt and nut as shown. Assume that the bolt has a single square thread of mean diameter 0.5 in. and a lead of 0.1 in. For $\mu_s = 0.35$ between threads, determine the torque which must be applied to the bolt and nut to induce a bolt tension of 1 kip.

9.65   The screw of the vise shown has a double square thread of mean diameter 20 mm and a pitch of 4 mm. A clamping force of magnitude F is induced at C when a horizontal force of magnitude Q is applied normal to the handle at D to tighten the vise. For $\mu_s = 0.3$ between threads and neglecting effects of friction at B, determine the ratio of F/Q.

9.66*   Solve Prob. 9.65 for $\mu_s = 0.2$.

9.67   The worm gear shown consists of a worm AB (i.e., a threaded segment of shaft) and a gear wheel C which is rigidly attached to a drum D. The worm AB has a single square thread with a mean radius of 30 mm and a pitch of 10 mm. For $\mu_s = 0.2$ between the thread of AB and the teeth of C, determine the torque M which must be applied to the shaft EF in order to raise the 200-kg crate shown.

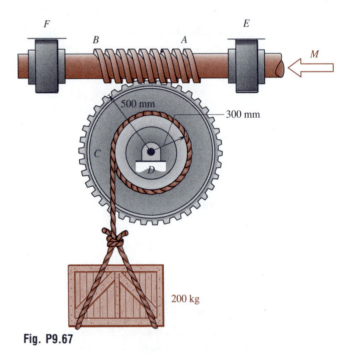

Fig. P9.67

9.68*   In Prob. 9.67, determine the torque which must be applied to the shaft EF in order to lower the crate.

9.69   The jack shown supports a load $\mathbf{P}$ of 6 kN from a car. The screw has a single square thread with a mean radius of 5 mm and a pitch of 2 mm. The jack is symmetrical about a vertical plane containing the axis of the screw and has two sets of linkages. For $\mu_s = 0.25$ between threads, determine the torque M required (a) to raise the load $\mathbf{P}$, (b) to lower the load $\mathbf{P}$.

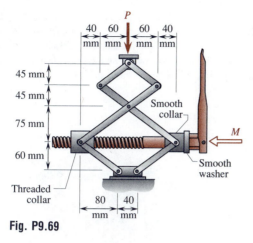

Fig. P9.69

9.70 A bushing of outside radius 1.5 in. fits loosely on a horizontal shaft of radius 1 in. as shown. No slippage occurs between the rope and the bushing when the force **P** starts to raise the 100-lb crate. For $\mu_s = 0.1$ between the shaft and the bushing, determine the magnitude of **P** if (a) $\theta = 0$, (b) $\theta = 90°$.

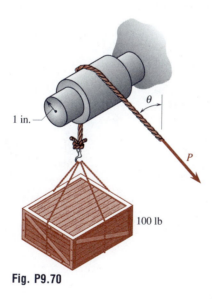

Fig. P9.70

9.71 A 100-kg crate is to be raised as shown. Each of the 100-mm-diameter pulleys rotates on a 20-mm-diameter axle. For $\mu_s = 0.15$ between the pulley and axle, determine the magnitude of **P** required to raise the crate.

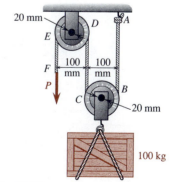

Fig. P9.71

9.72* A cart having four steel wheels of 30-in. diameter is to be moved on rails as shown. Each steel wheel supports a 4-in.-diameter axle and carries an axle load of 2 kips. For $\mu_k = 0.02$ between the wheel and axle, determine the magnitude of the horizontal force **P** required to move the cart at a constant velocity.

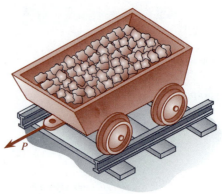

Fig. P9.72*

9.73 A collared shaft and its cross section in a thrust bearing are shown. If a wrench consisting of a 1500-lb thrust **P** and a 100-lb·ft torque **M** will cause rotation of the shaft to impend, determine the value of μ_s between the collar and the bearing.

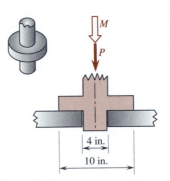

Fig. P9.73

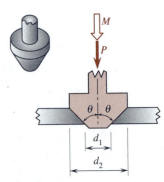

Fig. P9.74*

9.74* A shaft and its cross section in a conical pivot bearing are shown. The coefficient of kinetic friction is μ_k and the bearing pressure is uniform. If a wrench consisting of a thrust **P** and a torque **M** will keep the shaft rotating at a constant speed, show that

$$M = \frac{(d_1^2 + d_1 d_2 + d_2^2)\mu_k P}{3(d_1 + d_2)\sin\theta}$$

9.75 A shaft and its cross section in a spherical thrust bearing are shown, where the coefficient of static friction is μ_s. The bearing pressure p may be assumed to be equal to $K \sin\theta$. The shaft carries a wrench consisting of **P** and **M**. Determine the expressions of (a) K, (b) the magnitude of **M** required to start rotating the shaft.

9.76 A 20-kg electric floor polisher is shown. Assume that the normal pressure between the floor and the ring-shaped area of the attached polishing disk is uniform and $\mu_k = 0.2$. Determine the magnitude Q of the forces in the couple required to prevent rotation of the handle during polishing.

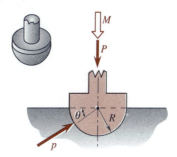

Fig. P9.75

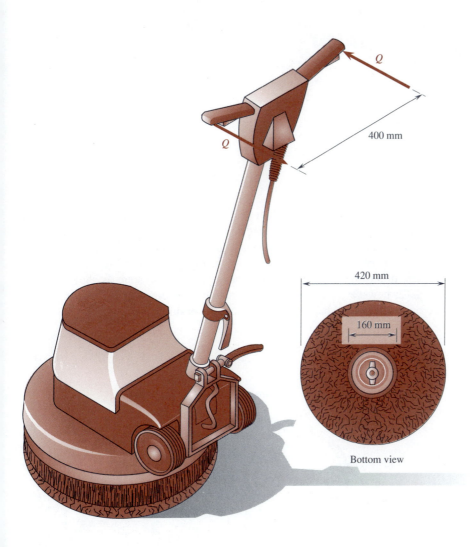

400 mm

420 mm

160 mm

Bottom view

Fig. P9.76

BELT FRICTION

9.12 Flat Belts

Friction forces developed between a belt and a drum have various engineering applications (e.g., belt drives and brakes). The relationship between the two tensile forces T_1 and T_2 in the two parts of a flat belt (or rope), which

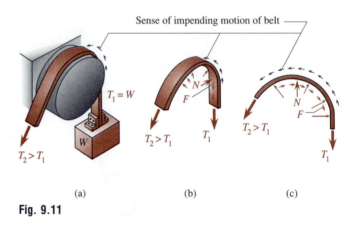

Fig. 9.11

are not in contact with the drum as shown in Fig. 9.11(a), can be established when *slipping* of the belt around the drum *impends* (and the conditions of equilibrium still hold true). A perspective view and a front view of the free-body diagram of the belt are shown in Fig. 9.11(b) and (c), respectively. The drum may have either constant or varying curvature as shown in Fig. 9.12; however, the coefficient of friction is assumed to be the same at all points of contact between the belt and the drum.

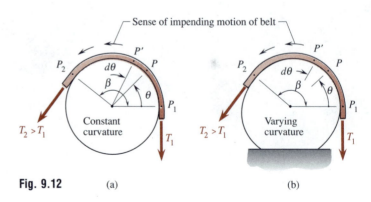

Fig. 9.12   (a)   (b)

The free-body diagram of the infinitesimal belt element PP' in Fig. 9.12 is shown in Fig. 9.13. For equilibrium of this element, we write

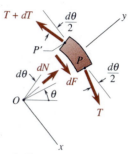

$$+\searrow \Sigma F_x = 0: \quad T\cos\frac{d\theta}{2} + dF - (T + dT)\cos\frac{d\theta}{2} = 0 \quad (9.18)^\dagger$$

$$+\nearrow \Sigma F_y = 0: \quad dN - T\sin\frac{d\theta}{2} - (T + dT)\sin\frac{d\theta}{2} = 0 \quad (9.19)$$

$$\text{Friction law:} \quad dF = \mu_s\, dN \quad (9.20)$$

Fig. 9.13

Recognizing that $\cos(d\theta/2) \approx 1$, $\sin(d\theta/2) \approx d\theta/2$, and neglecting second-order infinitesimals in comparison with first-order infinitesimals, we obtain, from the above equations,

$$\frac{dT}{T} = \mu_s\, d\theta \quad (9.21)$$

By integrating both sides of Eq. (9.21) from P_1 to P_2 of the flat belt, as shown in Fig. 9.12, we write

$$\int_{T_1}^{T_2} \frac{dT}{T} = \mu_s \int_0^\beta d\theta \qquad \ln\frac{T_2}{T_1} = \mu_s \beta \quad (9.22)$$

Thus, the equation governing the impending slipping of a flat belt may be written as

$$T_2 = T_1\, e^{\mu_s \beta} \quad (9.23)$$

where e (equal to $2.718\ldots$) is the base of natural logarithms, μ_s is the coefficient of static friction between the belt and the drum, β is the *angle of contact* (i.e., the angle between the radii of curvature of the drum, which are drawn at the two points P_1 and P_2 where the belt leaves the drum), and T_1 and T_2 are, respectively, the smaller tensile force and the larger tensile force in the two parts of the belt not in contact with the drum.

In using the above formula for flat-belt friction, it is to be noted that the angle of contact β must be in radians and the belt must have impending relative motion with respect to the drum. If a belt or rope is wrapped n times around a post, β is equal to $2n\pi$. Furthermore, this belt friction formula is *valid for any shape of surface of the drum* provided that the coefficient of friction is the same at all points of contact. In kinetic friction, if the mass of the belt is negligible, we may simply replace μ_s with μ_k and write

$$T_2 = T_1\, e^{\mu_k \beta} \quad (9.24)$$

Developmental Exercise

D9.26 Write the equation governing the impending slipping of a flat belt around a drum and define the symbols used in the equation.

---

†This equation or its equivalent can also be obtained from Fig. 9.13 by applying $\Sigma M_O = 0$.

EXAMPLE 9.9

If $\mu_s = 0.25$ between all contact surfaces and the body A shown has a mass of 80 kg, determine the maximum mass of the body B for which the system will remain in equilibrium.

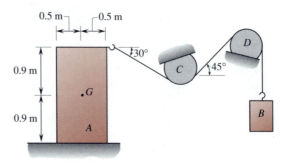

Solution. When the body B has the said maximum mass, its weight will cause the body A to have either *impending slipping* or *impending tipping*. Thus, this is a *type III friction problem*. We may solve this problem by first determining the tension T_{AC} between A and C that will cause A to have impending motion (slipping or tipping).

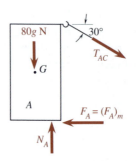

Case (a). Assume that slipping of the body A impends. (See above.)

Friction law: $(F_A)_m = 0.25 N_A$

$$\xrightarrow{+} \Sigma F_x = 0: \quad T_{AC} \cos 30° - (F_A)_m = 0$$

$$+\uparrow \Sigma F_y = 0: \quad N_A - T_{AC} \sin 30° - 80g = 0$$

Solution of these three equations yields

$$T_{AC} = 264.8 \text{ N} \qquad N_A = 917.2 \text{ N} \qquad (F_A)_m = 229.3 \text{ N}$$

Case (b). Assume that tipping of the body A impends. In this case, the normal reaction component is N_P and the friction force is F_P. They act through the edge P of the body A as shown. We have

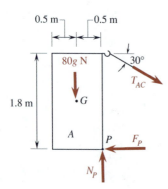

$$+ \, \circlearrowright \Sigma M_P = 0: \quad 0.5(80g) - 1.8(T_{AC} \cos 30°) = 0$$

$$T_{AC} = 251.7 \text{ N} < 264.8 \text{ N}$$

Decision. The value of T_{AC} at which the motion of the body A impends is the smaller value of T_{AC} obtained in the above two cases. This means that the impending motion is in the form of *impending tipping*; so, we take

$$T_{AC} = 251.7 \text{ N}$$

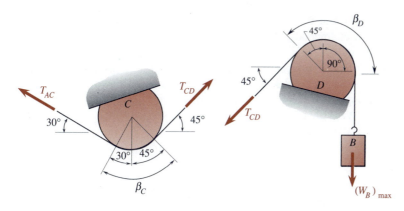

As indicated, the angles of contact of the cord around the drums are

$$\beta_C = 30° + 45° = 75° = \frac{5\pi}{12} \text{ rad}$$

$$\beta_D = 45° + 90° = 135° = \frac{3\pi}{4} \text{ rad}$$

By repeated applications of Eq. (9.23), we write

$$(W_B)_{\text{max}} = T_{CD} \, e^{\mu_s \beta_D} = (T_{AC} \, e^{\mu_s \beta_C}) \, e^{\mu_s \beta_D}$$

$$= 251.7 \, e^{0.25(5\pi/12)} \, e^{0.25(3\pi/4)}$$

$$= 629.3 = (m_B)_{\text{max}} \, (9.81)$$

$$(m_B)_{\text{max}} = 64.1 \text{ kg} \blacktriangleleft$$

Developmental Exercises

D9.27   A boy and a girl pull on the ends of a rope which is wrapped around a post as shown. When the girl pulls with a 5-lb force on one end of the rope, the boy finds that he needs to exert a 50-lb force on the other end of the rope if he wants to pull her toward him. Determine (a) the angle of contact β, (b) the value of μ_s between the rope and the post.

D9.28   It is known that $\mu_s = 0.25$ between the elliptic drum and the rope which carries a 10-kg block on one side and is held by a force $\mathbf{P}$ on the other side as shown. Determine (a) the angle of contact β, (b) the minimum and the maximum values of P if equilibrium exists.

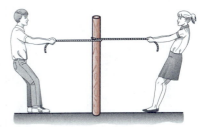

Fig. D9.27

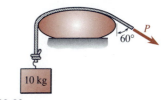

Fig. D9.28

9.13   V Belts

In automobile engines and other machines, the belts used for belt drives are frequently V-shaped. Let α be the angle between the two friction sides of the V belt as shown in Fig. 9.14(a). The V belt is in contact with the sides of the groove of the pulley; it does not touch the bottom of the groove. The condition of impending slipping of the V belt in the groove of the pulley will be studied.

The perspective view of the free-body diagram of an infinitesimal V-belt element subtending an angle $d\theta$ in Fig. 9.14(a) is shown in Fig. 9.14(b). The orthographic projections of this free-body diagram are shown in Fig. 9.14(c). We write

$$\xrightarrow{+} \Sigma F_x = 0: \quad T \cos \frac{d\theta}{2} + 2\, dF - (T + dT) \cos \frac{d\theta}{2} = 0 \qquad (9.25)$$

$$+\uparrow \Sigma F_y = 0: \quad 2(dN) \sin \frac{\alpha}{2} - T \sin \frac{d\theta}{2} - (T + dT) \sin \frac{d\theta}{2} = 0 \quad (9.26)$$

Friction law: $dF = \mu_s\, dN$ $\qquad (9.27)$

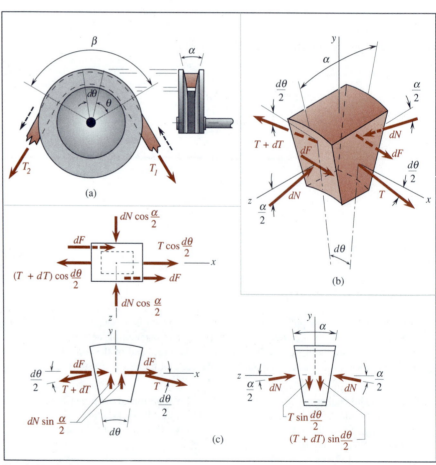

Fig. 9.14

From Eqs. (9.25) through (9.27), we can similarly obtain the equation governing the impending slipping of a V belt as

$$T_2 = T_1 \, e^{\mu_s \, \beta / \sin \, (\alpha/2)} \tag{9.28}$$

where T_1, T_2, μ_s, and β have the same meaning as those used in Sec. 9.12. If kinetic friction occurs, μ_s is replaced by μ_k and we get

$$T_2 = T_1 \, e^{\mu_k \, \beta / \sin \, (\alpha/2)} \tag{9.29}$$

It is readily seen from Eqs. (9.23), (9.24), (9.28), (9.29), and Fig. 9.14 that when $\alpha \to \pi$ rad, the formulas governing the V-belt friction do reduce to the formulas governing the flat-belt friction.

Developmental Exercise

D9.29   Write the equation governing the impending slipping of a V belt around a pulley and define the symbols used in the equation.

EXAMPLE 9.10

A V belt with its cross section shown is to deliver a torque of 100 N·m to a lathe pulley at A from a motor pulley at B rotating counterclockwise at a constant speed. If $R = 400$ mm, $r = 40$ mm, and $\mu_s = 0.3$ between the belt and the pulleys, determine (a) the minimum tensile strength required of the V belt, (b) the driving torque of the motor.

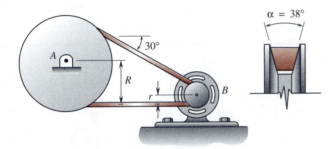

Solution.   In solving this problem, we need to investigate the condition under which the *difference* between the magnitudes of the tensile forces in the portions of the V belt not in contact with the pulleys is maximum. When such a difference in maximum, impending slipping exists between the V belt and the motor pulley, rather than the lathe pulley, because $\beta_B < \beta_A$ as shown. Thus, the present problem is a *type II friction problem*.

Note that the reaction torque $\mathbf{M}_A$ of 100 N·m from the shaft of the lathe pulley and the driving torque $\mathbf{M}_B$ of the motor have been indicated in the free-body diagrams of the pulleys on the next page. The tensile forces T_2 and T_1 are related to each other by the condition of impending slipping of the V belt around the motor pulley at B. From the given data, we have

$$\mu_s = 0.3 \qquad \alpha = 38° \qquad \beta_B = 180° - 30° = 150° = \frac{5\pi}{6} \text{ rad}$$

$$\text{V-belt friction:} \quad T_2 = T_1 \, e^{0.3(5\pi/6)/\sin \, (38°/2)} \tag{1}$$

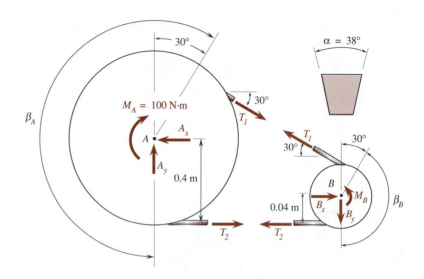

Since the lathe pulley and the motor pulley have no angular acceleration, they are in rotational equilibrium. Thus, we write

Lathe pulley, $+\circlearrowleft \Sigma M_A = 0$: $0.4T_2 - 0.4T_1 - 100 = 0$ (2)

Motor pulley, $+\circlearrowleft \Sigma M_B = 0$: $M_B + 0.04T_1 - 0.04T_2 = 0$ (3)

Solution of the above three simultaneous equations yields

$$T_1 = 24.60 \qquad T_2 = 274.60 \qquad M_B = 10$$

Minimum tensile strength required = 275 N ◀
Driving torque of the motor = 10 N·m ◀

Developmental Exercise

D9.30 In a car engine as shown, the crankshaft pulley C, which rotates clockwise, delivers power to pulleys A and F. When the car is accelerated, a screeching noise is heard due to slipping of the V belt around the pulley C. Determine the ratio T_{CF}/T_{CA} of belt tensions in portions CF and CA when (a) the screeching impends, (b) the screeching noise is being heard.

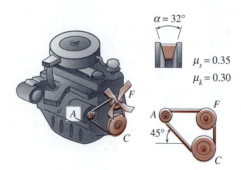

Fig. D9.30

PROBLEMS

9.77 A rope is fastened to a crate of mass m and passed over a drum as shown. If $m = 100$ kg and $\mu_s = 0.1$ between all surfaces of contact, what is the smallest magnitude of the vertical force $\mathbf{Q}$ which will move the crate?

9.78* A man fastens a rope to a crate of weight W and passes it over a drum as shown. The force $\mathbf{Q}$ exerted by the man on the rope may reach a maximum value of 150 lb. If $\mu_s = 0.1$ between all surfaces of contact, determine the maximum weight of the crate which can be moved by the man.

9.79 A ship pulls on a hawser with a force of 10 kips as shown. For $\mu_s = 0.25$ between the hawser and the bollard, determine the number of full turns a dock worker has to wrap the hawser around the bollard if he is to hold the ship by exerting on his side of the hawser a force $\mathbf{P}$ of magnitude not greater than 20 lb.[†]

9.80 If $\mu_s = 0.3$ between the rope and the pipes shown, determine the minimum horizontal force $\mathbf{P}$ required to hold the 10-kg mass in equilibrium.

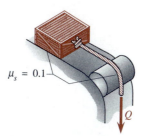

Fig. P9.77and P9.78*

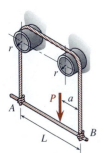

Fig. P9.79

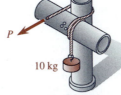

Fig. P9.80 and P9.81*

9.81* If $\mu_s = 0.3$ between the rope and the pipes shown, determine the minimum horizontal force $\mathbf{P}$ required to pull the 10-kg mass upward from rest.

9.82 through 9.84* A force $\mathbf{P}$ is applied to a horizontal bar AB which has a negligible weight and is supported by a rope passing around cantilever pipes as shown. For $a = L/4$, determine the minimum value of μ_s between the rope and the pipes for which the bar will remain horizontal.

Fig. P9.82 and P9.85

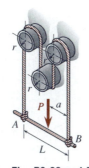

Fig. P9.83 and P9.86*

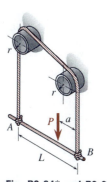

Fig. P9.84* and P9.87

9.85 through 9.87 A force $\mathbf{P}$ is applied to a horizontal bar AB which has a negligible weight and is supported by a rope passing around cantilever pipes as shown. For $\mu_s = 0.2$ between the rope and the pipes, determine the minimum ratio of a/L for which the bar will remain horizontal.

[†]Note that a *bollard* is a thick post mounted on a wharf and cannot be rotated, while a *capstan* is a type of windlass which can be rotated manually or by motor.

9.88 through 9.90* A vertical force **P** is applied to a rope which passes around cantilever pipes and is connected to a 100-kg homogeneous body *B* as shown. The values of μ_s between the rope and the pipes and between the body and the horizontal support are 0.2 and 0.27, respectively. Determine the minimum magnitude of **P** required to cause impending motion.

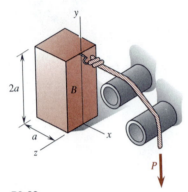

Fig. P9.88

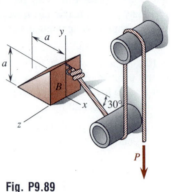

Fig. P9.89

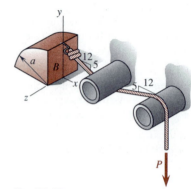

Fig. P9.90*

9.91 The two drums *C* and *D* are rigidly fastened together and are free to rotate about a frictionless bearing at *O* as shown. The body *A* weighs 100 lb. The value of μ_s is 0.1 between the belt and the small drum *D* and 0.2 between the belt and the large drum *C*. Determine the minimum magnitude of the force **P** which will lift the body *A*.

9.92* Solve Prob. 9.91 if the value of μ_s is 0.15 between the belt and the small drum *D* and 0.3 between the belt and the large drum *C*.

9.93 A constant torque of 100 N·m is applied to the axle of a flywheel to keep it rotating at a constant speed. The coefficients of friction between the brake band and the drum of the flywheel as shown are $\mu_s = 0.3$ and $\mu_k = 0.25$. Determine the magnitude of the force **P** acting at the hand bar if the direction of the applied torque and the rotation of the flywheel are clockwise.

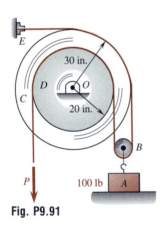

Fig. P9.91

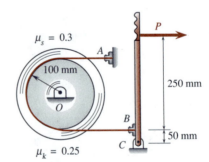

Fig. P9.93

9.94 Solve Prob. 9.93 if the direction of the applied torque and the rotation of the flywheel are counterclockwise.

9.95 A 20-lb force is applied to the lever at *A* and $\mu_k = 0.5$ between the flat belt and the drum which is attached to a flywheel as shown. A torque **M** is applied to the axle of the flywheel. Determine the torque **M** required to keep the flywheel rotating counterclockwise at a constant speed.

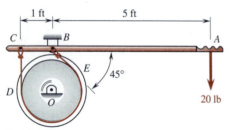

Fig. P9.95

9.96* Refer to the system in Prob. 9.95. Determine the torque **M** required to keep the flywheel rotating clockwise at a constant speed.

9.97 A 100-lb weight C and a body D of weight W are connected by a flat belt which passes around a shaft A and a fixed cylindrical drum B as shown. For $\mu_s = 0.3$ and $\mu_k = 0.2$ between all surfaces of contact, determine the maximum weight W which can be raised if the shaft A is rotated from rest.

9.98 A 10-kg body A and a body B of mass m_B are connected by a rope which passes around a smooth pulley C and a fixed drum D as shown. If $\mu_s = 0.2$ between all surfaces of contact and $m_B = 17$ kg, determine the minimum magnitude of the force **P** which will cause impending motion between the rope and the drum D.

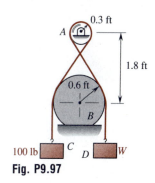

Fig. P9.97

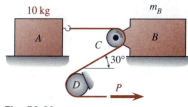

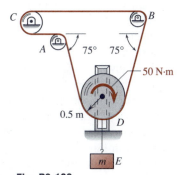

Fig. P9.98

9.99* Solve Prob. 9.98 for $m_B = 18$ kg.

9.100 A flat belt is used to deliver a torque from the pulley D to a load at the pulley C. It passes around two smooth idler pulleys A and B as shown. The value of μ_s between the belt and the pulley D is 0.3. A block E of mass m is attached to the axle of the pulley D by a cable to keep the pulley D from slipping on the belt. What is the minimum mass m of the block E which will prevent slippage from occurring when a 50 N·m torque is applied to the pulley D?

Fig. P9.100

9.101* Refer to the system in Prob. 9.100. Determine the minimum value of μ_s between the belt and the pulley C if the belt is not to slip around the pulley C.

9.102 The tape of a recorder is driven by the pulley A with a torque of 0.3 N·m as shown. The maximum tension which is allowed to be developed in the tape is 40 N. Determine the minimum value of μ_s required between the tape and the pulley A.

9.103* Solve Prob. 9.102 if the maximum tension which is allowed to be developed in the tape is 30 N.

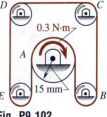

Fig. P9.102

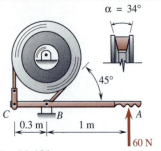

α = 34°

45°

C
B
A
0.3 m   1 m
60 N

Fig. P9.104

9.104   A V belt passes around a grooved wheel and is connected to form a brake as shown. For $\mu_k = 0.3$ between the belt and the wheel, determine the resisting torque of the brake if the wheel is turning clockwise.

9.105* Solve Prob. 9.104 if the wheel is turning counterclockwise.

9.106   A V belt of negligible mass is used to deliver a torque from the motor at B to a load at the pulley A as shown. If the maximum allowable belt tension is 900 lb and $\mu_s = 0.3$, determine the largest torque which can be delivered to the pulley A.

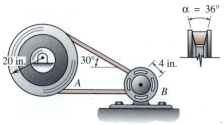

α = 36°

20 in.   30°   4 in.
A
B

Fig. P9.106

9.107* Solve Prob. 9.106 if the maximum allowable belt tension is 700 lb and $\mu_s = 0.2$.

9.14   Concluding Remarks

A dry friction force is simply the tangential component of the reaction between two nonlubricated or partially lubricated surfaces of contact of two bodies. This tangential component of the reaction is found, through experiments, to be proportional to the normal component of the reaction *when and only when* slipping between the two contact surfaces impends or is occurring. The constant of proportionality is called the *coefficient of friction*.

Many equilibrium problems involving dry friction are really statically indeterminate problems because they cannot be solved by using only the equations of equilibrium. These statically indeterminate problems may be solved if the equations representing the laws of dry friction are appropriately introduced. There are many other classes of statically indeterminate equilibrium problems which are not treated in this text. They may be found in texts on mechanics of materials, advanced structural analysis, elasticity, etc.

The classification of friction problems into certain types is very helpful in developing methods for solving them. Such a classification is similar to the classification of problems in other subjects (e.g., differential equations) into certain types for the study of their solutions. Thus, a careful review of Secs. 9.5 through 9.8 is recommended.

The resistance to rolling of a wheel arising from the deformation between the wheel and its supporting surface is referred to as the *rolling resistance*. This resistance is not due to tangential friction forces and therefore is a phenomenon entirely different from that of dry friction. As illustrated in Fig. 9.15 where θ is usually small, the force **P** necessary to initiate and maintain

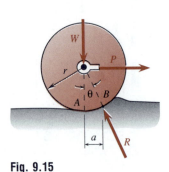

W
P
r
θ B
A
a   R

Fig. 9.15

rolling at constant velocity can be found by writing

$$+\circlearrowleft \Sigma M_B = 0: \quad aW - (r\cos\theta)P \approx aW - rP = 0$$

$$\mu_r = \frac{a}{r} \qquad P = \mu_r W \qquad (9.30)$$

where the moment arm of **P** is taken as r and **W** is the load on the wheel. The ratio $\mu_r = a/r$ is referred to as the *coefficient of rolling resistance*, which may vary with the speed of travel, the radius of the wheel, the elastic and plastic properties of the wheel and its support, and the roughness of the surface of support. Some tests indicate that the distance a varies only slightly with the radius r. Thus, the distance a has also been referred to as the coefficient of rolling resistance. Note that a has the dimension of length, while μ_r is dimensionless.

REVIEW PROBLEMS

9.108 If $\mu_s = 0.5$ between the 26-lb homogeneous body and the support as shown, determine the minimum magnitude of the horizontal force **P** required to cause the body to have impending motion.

9.109 Two horizontal forces **F** and $-$**F** are applied to a 20-kg horizontal cylinder A which rests on a 10-kg vertical cylinder B as shown. The value of μ_s between the cylinder A and its surfaces of contact is 0.35, and that between the cylinder B and the horizontal support is 0.25. Determine the smallest magnitude of **F** for which the cylinder A will start to move.

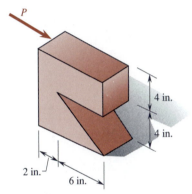

Fig. P9.108

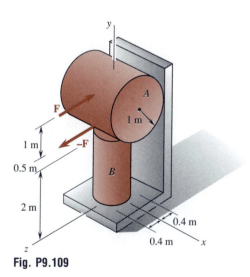

Fig. P9.109

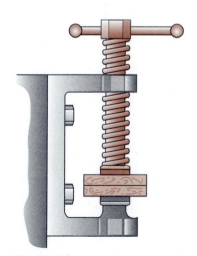

Fig. P9.110

9.110 The screw of the clamp shown has a single square thread, a mean diameter of 12 mm, and a pitch of 3 mm. If the clamp is tightened with a torque of 1 N·m and $\mu_s = 0.25$ between the threads, determine (a) the magnitude of the force pressing the two pieces of wood together, (b) the torque required to loosen the clamp if self-locking exists.

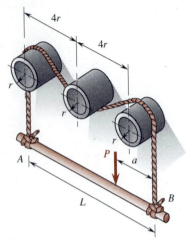

Fig. P9.111* and P9.112*

9.111* If the weight of the bar AB is negligible and $a = L/4$, determine the minimum value of μ_s between the rope and the pipes for which the bar under the action of the force **P** will remain horizontal as shown.

9.112* If the weight of the bar AB is negligible and $\mu_s = 0.2$ between the rope and the pipes, determine the minimum ratio of a/L for which the bar under the action of the force **P** will remain horizontal as shown.

9.113 A strap wrench is used to grip a pipe as shown. If $a = 50$ mm and $\mu_s = 0.25$ between the strap and the pipe, determine the maximum value of b for which the wrench will be self-locking.[†]

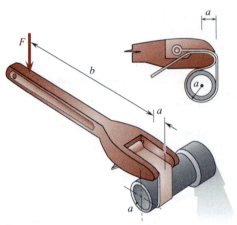

Fig. P9.113 and P9.114*

9.114* Show that the strap wrench shown will be self-locking if the value of μ_s between the strap and the pipe exceeds that given by the equation

$$3\pi\mu_s + 2 \ln \mu_s = 0$$

Determine that minimum value of μ_s.[††]

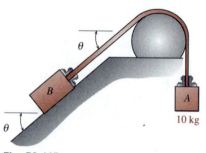

Fig. P9.115

[†]Note that the strap is not to slip on the pipe, however large the force **F** may be. At impending slipping, we have $T_2 > T_1 = \mu_s N_w$ where N_w is the normal force exerted by the wrench on the strap at the top of the pipe.

[††]Cf. App. E or Sec. B5.8 of App. B for the solution of a transcendental equation.

9.115 The mass of body A is 10 kg and $\mu_s = 0.5$ between all surfaces of contact. For $\theta = 35°$ as shown, determine the range of values of the mass of body B for which the system will remain in equilibrium.

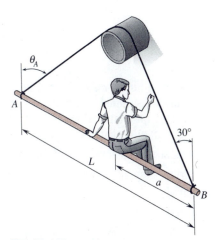

Fig. P9.116

9.116 A man sits on a horizontal bar of negligible weight as shown. For $a = L/3$, determine the angle θ_A and the minimum value of μ_s between the rope and the pipe for which the bar will remain horizontal.

9.117 The car engine shown is at rest. Determine the ratio T_{AF}/T_{FC} of the tensions in the portions AF and FC of the V belt when a mechanic manually rotates the fan clockwise and the belt slips around the fan pulley F.

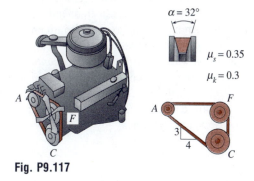

Fig. P9.117

Chapter 10

Virtual Work

KEY CONCEPTS

- Displacement, work of a force, work of a couple, compatible virtual displacement, constrained virtual displacement, virtual work, potential energy, gravitational potential energy, elastic potential energy, and applied potential energy.
- Computation of virtual displacements using differential calculus.
- Computation of virtual displacements using displacement center.
- Principle of virtual work.
- Principle of potential energy.
- Stability of equilibrium of conservative systems.

In mechanics, *work* is done by a force or a moment to (or on) a body during a displacement of the body; and no work can be done to a body which remains stationary. Having carefully defined work, we proceed to derive the *principle of virtual work*, which is then applied to solve equilibrium problems of *frames* as well as *machines*. Here, by giving a free body in equilibrium an appropriate compatible virtual displacement and setting the corresponding total virtual work to zero, we can frequently determine a specific unknown quantity in a seemingly complex equilibrium problem without having to solve a system of simultaneous equations. This is an outstanding feature of the principle of virtual work. Later, a *quantitative definition* for the potential energy of a body is given, and the *principle of potential energy* is deduced. In the last part, the potential energy and its expansion into a *Taylor's series* are employed to establish the criteria for gauging the *stability of equilibrium* of conservative systems.

PRINCIPLE OF VIRTUAL WORK

10.1 Displacements

The *displacement* of a body is the change of position of the body.[†] When the body is in translation or being treated as a particle, the displacement of the body is simply a linear displacement. The *linear displacement* of a par-

---

[†]Unless otherwise stated, the displacement of a body is to be understood as being measured from a fixed position in an inertial reference frame. Virtual displacements are discussed in Sec. 10.4.

ticle moving from the point A along *any* path (e.g., the path C) to the point B is a vector $\mathbf{q} = \overrightarrow{AB}$ which is directed from the point A to the point B and has a magnitude equal to the shortest distance between the points A and B as shown in Fig. 10.1. In terms of the position vectors $\mathbf{r}_A$ and $\mathbf{r}_B$ of A and B, we write

$$\mathbf{q} = \mathbf{r}_B - \mathbf{r}_A \qquad (10.1)$$

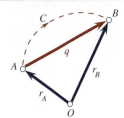

Fig. 10.1

The *angular displacement* of a body is the change of orientation of the body; it is produced by the rotation of the body. The *direction of the angular displacement* $\Delta\theta$ of a rigid body, as shown in Fig. 10.2, is defined to be parallel to the extended thumb of the right hand when the other fingers of the right hand are curled in the direction of rotation of the body. Note that finite angular displacements of a rigid body are *not* vector quantities because they do not follow the commutative law as well as the parallelogram law for the addition of vectors.[†] However, *infinitesimal angular displacements* $\overrightarrow{d\theta}_1$ and $\overrightarrow{d\theta}_2$ (as well as their time rates of change) of a rigid body are vectors because they do obey the laws for the addition of vectors.[††] We have

$$\overrightarrow{d\theta}_1 + \overrightarrow{d\theta}_2 = \overrightarrow{d\theta}_2 + \overrightarrow{d\theta}_1 \qquad (10.2)$$

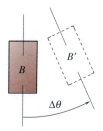

Fig. 10.2

The *total distance traveled* by a body is the total accumulated length of path traversed by the body. Similarly, the *total angle turned through* by a body is the total accumulated angle rotated through by the body.

Developmental Exercises

D10.1 Define the terms: (a) displacement, (b) linear displacement, (c) angular displacement, (d) total distance traveled, (e) total angle turned through.

D10.2 A particle travels along the x axis from the point A first to the point B, then back to the point C, as shown. (a) What is the displacement of the particle? (b) What is the total distance traveled by the particle?

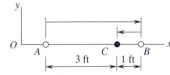

Fig. D10.2

D10.3 A rod is pinned to a support at A. The rod rotates from the position AB first to the position AC, then back to the position AD, as shown. (a) What is the angular displacement of the rod? (b) What is the total angle turned through by the rod?

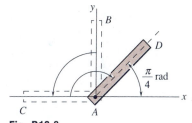

Fig. D10.3

10.2 Work of a Force[†††]

The *work of a force* $\mathbf{F}$ *acting on a particle during a differential displacement* $d\mathbf{r}$ *of the particle* from the point A to a neighboring point A', as shown in Fig. 10.3, is defined as an infinitesimal scalar quantity dU equal to the dot product of $\mathbf{F}$ and $d\mathbf{r}$; i.e.,

$$dU = \mathbf{F} \cdot d\mathbf{r} \qquad (10.3)$$

[†]Cf. D1.3.

[††]Cf. App. C for the proof that they are vectors.

[†††]Note that *work* is done *by* a force *to* (or *on*) a body. It is the *force* (not the *body*) which does work; a body *receives* work from the force acting on it during its displacement from one position to another. If the action of the force on a particle and the displacement of the particle do not occur simultaneously, no work is done by the force to the particle. The *work of a force* acting on a body is the *work done by a force* to a body.

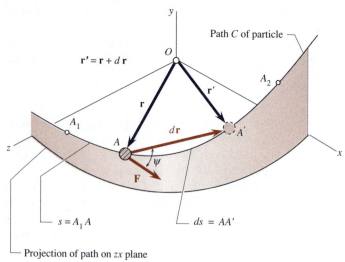

Fig. 10.3

Alternatively, we may write $dU = F \cos\psi \, ds$, where $ds = |d\mathbf{r}|$ and ψ is the angle between $\mathbf{F}$ and $d\mathbf{r}$. Anyway, the *work $U_{1\to2}$ of a force $\mathbf{F}$ acting on a particle during a finite displacement of the particle* from the point A_1 along a path C to the point A_2, as shown in Fig. 10.3, is

$$U_{1\to2} = \int_C \mathbf{F} \cdot d\mathbf{r} \qquad (10.4)$$

where the label C at the integral sign signifies that the work $U_{1\to2}$ done by $\mathbf{F}$ to the particle is generally dependent on the path C along which the particle travels from A_1 to A_2. If $U_{1\to2}$ is dependent on the coordinates of A_1 and A_2, but not on those of C, then $\mathbf{F}$ is referred to as a *conservative force* (which is discussed in Sec. 10.9) and we may write

$$U_{1\to2} = \int_{A_1}^{A_2} \mathbf{F} \cdot d\mathbf{r} \qquad (10.5)$$

For example, if $\mathbf{F}$ is a friction force, which is a nonconservative force, Eq. (10.4) will generally yield a different work for each different path C. By definition, the work of a force is a scalar and has the dimensions of *force × length*. The primary units for work are joules (1 joule $= 1$ J $= 1$ N·m) in the SI and foot-pounds (ft·lb) in the U.S. customary system.[†]

The *work of a force $\mathbf{F}$ acting on a rigid body* is equal to the sum of works done by $\mathbf{F}$ to the particles of the rigid body. When work is done to a body by a force, *the particle (or particles) of the body*, on which the force acts, *must have a nonzero component of displacement in the direction of the force during the action of the force*; otherwise, no work is done. Note that

---

[†]The joule (J) is the primary unit of *work* and *energy* in the SI. Note that *moment* is not a form of energy. Although 1 J $= 1$ N·m, the primary units of *moment* of a force are N·m, *not* J, in the SI. In the U.S. customary system, the primary units are usually lb·ft for *moment*, but ft·lb for *work*. [However, some do not strictly follow the conventions for lb·ft and ft·lb, which are in reverse order to those shown in Eqs. (4.18) and (10.4).]

the work done by any force to a stationary rigid body by sliding on its surface is zero. (Cf. D10.6 through D10.8.)

The work as defined above may be referred to as a *kinetic work*, which is shown in dynamics to be directly related to the *change of kinetic energy* of the body under consideration. Since velocity is the time rate of increase of displacement (or the time rate of change of position), it should be noted that *the work of a force acting on a rigid body is zero if the particle of the rigid body, on which the force acts, has, during the instant of action of the force, a zero velocity.*

If the force **F** is *constant* in magnitude and direction, the work $U_{1 \to 2}$ of **F** acting on a particle during its motion from the point A_1 along a path C to the point A_2, as shown in Fig. 10.4, is

$$U_{1 \to 2} = \int_C \mathbf{F} \cdot d\mathbf{r} = \mathbf{F} \cdot \int_C d\mathbf{r} = \mathbf{F} \cdot \mathbf{q}$$

$$U_{1 \to 2} = Fq \cos \psi \qquad (10.6)$$

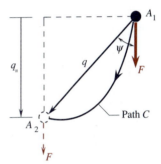

Fig. 10.4

where **q** is the displacement of the particle from A_1 to A_2 and ψ is the angle between the positive directions of **F** and **q**. We readily observe that this work $U_{1 \to 2}$ is positive if ψ is acute and negative if ψ is obtuse. There are three cases which are of special interest: (1) if **F** and **q** have the same direction, then $U_{1 \to 2} = Fq$; (2) if **F** and **q** have opposite directions, then $U_{1 \to 2} = -Fq$; (3) if **F** is perpendicular to **q**, then $U_{1 \to 2} = 0$. Using $q_{\parallel}$ to denote the component of the displacement of the particle parallel to the constant force **F**, we may write Eq. (10.6) as

$$U_{1 \to 2} = Fq_{\parallel} \qquad (10.7)$$

where

$$q_{\parallel} = q \cos \psi \qquad (10.8)$$

Note that the value of $q_{\parallel}$ is negative when $90° < \psi \le 180°$.

Furthermore, *the work of a force is equal to the sum of the works of its component forces, and conversely.* This property is analogous to *Varignon's theorem* and can be proved by applying the distributive property of dot products of vectors to the definition of the work of a force and the relation between the force and its component forces.[†]

EXAMPLE 10.1

If $\mu_k = 0.2$ between the 100-lb body and the horizontal support as shown, determine the works U_P, U_W, U_N, and U_F done to the body by the 50-lb

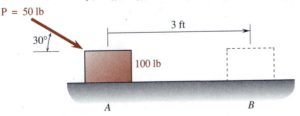

[†]Some refer to this property as the *theorem of work*.

pushing force **P**, the 100-lb weight force **W** of the body, the normal reaction **N**, and the friction force **F**, respectively, during the 3-ft displacement of the body from A to B.[†]

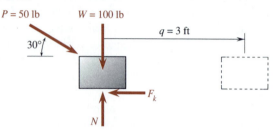

$P = 50$ lb $W = 100$ lb

$q = 3$ ft

$30°$

F_k

N

Solution. Since the body can move only in the horizontal direction, it is in equilibrium in the vertical direction. We write

$$+\uparrow \Sigma F_y = 0: \quad N - 100 - 50 \sin 30° = 0 \qquad N = 125 \text{ lb}$$

$$\text{Friction law:} \quad F_k = \mu_k N \qquad\qquad\qquad F_k = 25 \text{ lb}$$

The angles ψ_P, ψ_W, ψ_N, and ψ_F between the displacement **q** of the body and the forces **P**, **W**, **N**, and **F**$_k$, respectively, are

$$\psi_P = 30° \qquad \psi_W = 90° \qquad \psi_N = 90° \qquad \psi_F = 180°$$

Noting that the forces considered are constant and $U_{1\to 2} = Fq \cos\psi$, we write

$$U_P = Pq \cos\psi_P = 50(3)(\sqrt{3}/2) \qquad U_W = Wq \cos\psi_W = 100(3)(0)$$

$$U_N = Nq \cos\psi_N = 125(3)(0) \qquad U_F = F_k q \cos\psi_F = 25(3)(-1)$$

$$U_P = 129.9 \text{ ft·lb} \qquad U_W = U_N = 0 \qquad U_F = -75 \text{ ft·lb} \quad \blacktriangleleft$$

Alternative solution for U_P. Resolving the 50-lb force **P** into two rectangular components $\mathbf{P}_x = 50 \cos 30°$ lb → and $\mathbf{P}_y = 50 \sin 30°$ lb ↓, we write

$$U_P = P_x q \cos 0° + P_y q \cos 90° = (50 \cos 30°)(3)(1) + 0$$

$$U_P = 129.9 \text{ ft·lb} \quad \blacktriangleleft$$

[†]Cf. the third footnote on p. 401.

Developmental Exercises

D10.4   Define the work of a force acting on a (a) particle, (b) rigid body.

D10.5   A 100-N force **F** acts on a particle P during a 2-m displacement **q** of the particle as shown. Compute the work of **F**.

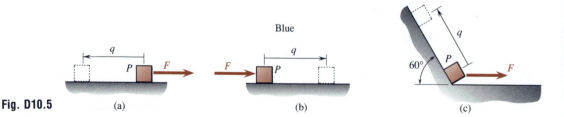

Blue

q q q

P F F P $60°$ P F

Fig. D10.5   (a)   (b)   (c)

EXAMPLE 10.2

The tension in the cord which holds a spring of modulus $k = 300$ N/m in compression as shown is 60 N. A package slides from the position A down the incline and displaces the right end of the spring from B to C by an amount $\delta = 0.6$ m before it comes to a stop. Determine the work U_s of the spring force $\mathbf{F}_s$ acting on the package during its displacement from A to C.

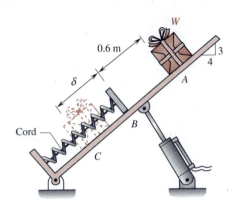

Solution. We first draw the free-body diagram of the package as shown. Note that the spring force $\mathbf{F}_s$ will act on the package only when the package comes to the region between B and C. The other forces are ever present during the displacement.

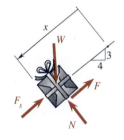

Initially, the tensile force in the cord and the compressive force in the spring must have the same magnitude; i.e., $F_{s1} = 60$ N. For a spring modulus $k = 300$ N/m, the magnitude of the initial deflection x_1 of the spring is given by

$$x_1 = F_{s1}/k = 60/300 \qquad x_1 = 0.2 \text{ m}$$

The final deflection x_2 of the spring is given by

$$x_2 = x_1 + 0.6 \qquad x_2 = 0.8 \text{ m}$$

The magnitude of the spring force $\mathbf{F}_s$ in terms of the deflection x of the spring is $F_s = kx = 300x$. The differential displacement $d\mathbf{r}$ of the package and $\mathbf{F}_s$ acting on the package are opposite in sense. Thus, the angle ψ_s between $d\mathbf{r}$ and $\mathbf{F}_s$ is 180°. Noting that $|d\mathbf{r}| = dx$, we write

$$U_s = \int_C \mathbf{F}_s \cdot d\mathbf{r} = -\int_{x_1}^{x_2} F_s \, dx = -\int_{0.2}^{0.8} 300x \, dx$$

$$U_s = -90 \text{ J} \blacktriangleleft$$

REMARK. By the concept of "area" in calculus, the value of U_s may also be obtained from the "area" of the trapezoid in the spring-force diagram as shown. Using the formula for the area of a trapezoid (cf. Sec. B2.3 of App. B), we write

$$U_s = -\tfrac{1}{2}(60 + 240)(0.6)$$

$$U_s = -90 \text{ J} \blacktriangleleft$$

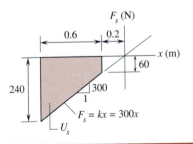

Developmental Exercises

D10.6 A line is drawn with a piece of chalk on a fixed chalkboard. Is the work *greater than*, *equal to*, or *less than* zero when it is done by the friction force to (a) the chalkboard, (b) the chalk?

D10.7 A 13-kg wheel *rolls without slipping* down an incline through a 2-m displacement as shown. The free-body diagram of the wheel and the normal and friction forces acting on the incline are also shown. If $\mu_s = 0.6$ and $\mu_k = 0.5$, determine during that displacement the work done by (a) **W** to the wheel, (b) **N** to the wheel, (c) **F** to the wheel, (d) **N′** to the incline, (e) **F′** to the incline.

D10.8 Is the work of a constant force **F** during the displacement **q** *of its point of application* on a stationary rigid body equal to $\mathbf{F} \cdot \mathbf{q}$?

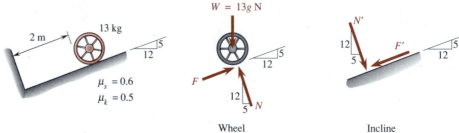

Fig. D10.7

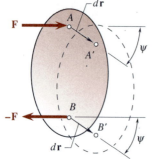

Fig. 10.5

10.3 Work of a Couple[†]

If a rigid body is in a translational motion, its orientation must remain unchanged, and the displacements of all particles of the rigid body are equal. Thus, *the work of a couple acting on a rigid body in translation is always equal to zero.* This is obviously true because, as shown in Fig. 10.5, the total work of the two forces **F** and −**F** in the couple acting on the rigid body through the common displacement d**r** is always zero; i.e.,

$$dU = \mathbf{F} \cdot d\mathbf{r} + (-\mathbf{F}) \cdot d\mathbf{r} = 0 \tag{10.9}$$

Now, consider the case in which a rigid body is acted on by a couple of moment **M** and, at the same time, undergoes an infinitesimal angular displacement $\overrightarrow{d\theta}$ which is parallel to **M**. Let the couple be represented by the forces **F** and −**F** which are tangent to the arcs of rotation at the points A and B as shown in Fig. 10.6. By Eq. (10.7), the work of the couple acting on the rigid body during the angular displacement $d\theta$ is

$$dU = |-\mathbf{F}|(-R\,d\theta) + |\mathbf{F}|(R + r)\,d\theta = Fr\,d\theta$$

Since the product Fr is equal to the magnitude of the moment **M** of the couple, we write

$$dU = M\,d\theta = \mathbf{M} \cdot \overrightarrow{d\theta} \tag{10.10}$$

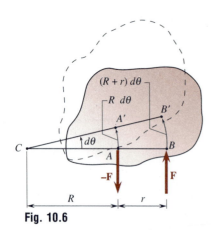

Fig. 10.6

[†]The work of a couple is the same as the work of the *moment* of the couple. (Cf. the footnote on p. 90.)

Note that $dU = 0$ if $\mathbf{M} \perp \vec{d\theta}$. Thus, *the work $U_{1 \to 2}$ of a moment* $\mathbf{M}$, *or a couple with moment* $\mathbf{M}$, acting on a rigid body during its rotation from the angular position θ_1 to the angular position θ_2 is

$$U_{1 \to 2} = \int_{\theta_1}^{\theta_2} \mathbf{M} \cdot \vec{d\theta} \qquad (10.11)$$

If the moment $\mathbf{M}$ is *constant*, the work $U_{1 \to 2}$ of $\mathbf{M}$ acting on a rigid body during its finite angular displacement $\Delta\theta$, which is expressed in radians and is in the same direction as that of $\mathbf{M}$, is

$$U_{1 \to 2} = M \, \Delta\theta \qquad (10.12)$$

Clearly, $U_{1 \to 2}$ is negative if M and $\Delta\theta$ have opposite senses. Note that the work of a moment, or a couple, is a scalar whose dimensions are also *force × length* because angles are taken as dimensionless.

Developmental Exercise

D10.9 Define the work of a couple acting on a rigid body.

EXAMPLE 10.3

Determine the total work dU of the force system acting on the bent bar shown if it undergoes an infinitesimal angular displacement $\vec{d\theta} = d\theta \circlearrowleft$ about the point B.

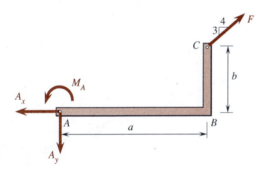
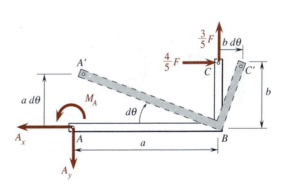

Solution. The angular displacement of the bent bar is drawn as shown. By resolving $\mathbf{F}$ at C into two components and applying Eqs. (10.7) and (10.12), we write

$$dU = A_x(0) + A_y(-a\,d\theta) + M(-d\theta) + \frac{4}{5}F(b\,d\theta) + \frac{3}{5}F(0)$$

$$dU = \left(\frac{4}{5}Fb - A_y a - M\right) d\theta \quad \blacktriangleleft$$

Developmental Exercises

D10.10   Work Example 10.3 if the bent bar undergoes an infinitesimal linear displacement $\overrightarrow{dx} = dx \rightarrow$.

D10.11   Work Example 10.3 if the bent bar undergoes an infinitesimal linear displacement $\overrightarrow{dy} = dy \uparrow$.

PROBLEMS

10.1   A 10-lb force **P** is applied to a 40-lb crate to move it through a 30-ft displacement **q** on a smooth horizontal surface as shown. Determine the work of (a) **P**, (b) all external forces.

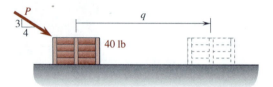

Fig. P10.1

10.2   Solve Prob. 10.1 if $P = 20$ lb, $q = 18$ ft, and the surfaces of contact are rough with $\mu_k = 0.2$.

10.3 through 10.6   The crate undergoes a displacement **q** on a ramp as shown. With the coefficient of friction between the crate and the ramp as indicated, determine the total work of all external forces acting on the crate during this displacement.

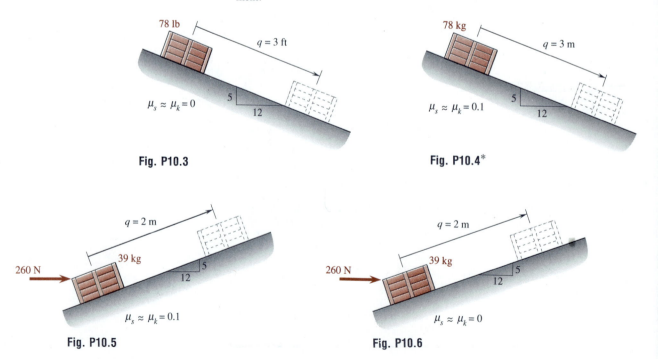

Fig. P10.3

Fig. P10.4*

Fig. P10.5

Fig. P10.6

10.7   A cord is used to tie a box to a wall with a compressed spring separating (but not connecting) them as shown. The spring has a free length $L = 1.5$ m and a modulus $k = 20$ N/m. Neglecting friction, determine the work of the spring force acting on the box after the cord is severed.

10.8* Solve Prob. 10.7 if $k = 15$ N/m.

10.9 and 10.10   The spring BC has a free length L and a modulus k as shown. As the bar is released to rotate from its horizontal position AB to its vertical position AB', the total work done to the bar is known to be zero. Determine the mass of the bar.

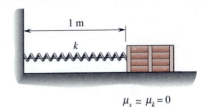

$\mu_s \approx \mu_k = 0$

Fig. P10.7

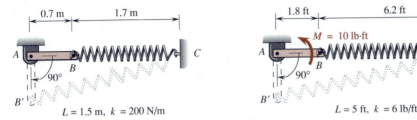

$L = 1.5$ m, $k = 200$ N/m

Fig. P10.9

$L = 5$ ft, $k = 6$ lb/ft

Fig. P10.10

10.11* Solve Prob. 10.9 if the total work done to the bar is known to be 9 J.

10.12   Determine the total work dU of the forces and moments acting on the frame shown if it undergoes an infinitesimal linear displacement of $\overrightarrow{dy} = dy$ ft $\uparrow$.

10.13* Solve Prob. 10.12 if the frame undergoes an infinitesimal linear displacement $\overrightarrow{dx} = dx$ ft $\rightarrow$.

10.14* Solve Prob. 10.12 if the frame rotates about the point A through a differential angle $\overrightarrow{d\theta} = d\theta \uparrow$.

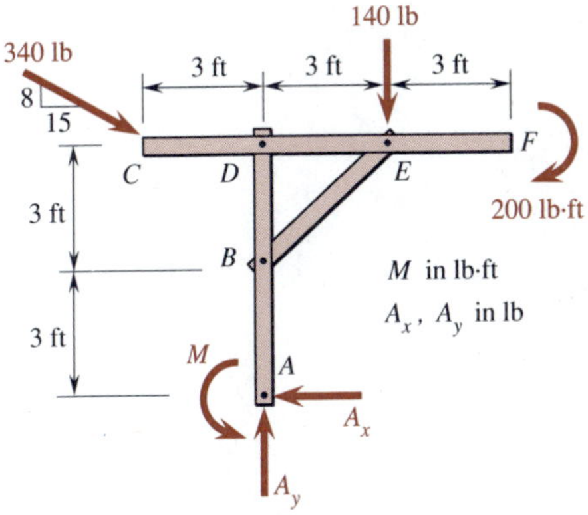

M in lb·ft
A_x, A_y in lb

Fig. P10.12

10.4   Virtual Displacements

An *actual displacement* of a body, as the name implies, gives the actual change of position of the body. A *virtual displacement* of a body is a given or imaginary differential displacement, which is possible but does not necessarily take place under actual motion, of the body. There are *linear* and *angular virtual displacements*; they are vector quantities.[†] The symbol $\delta \mathbf{r}$ is used to denote the virtual displacement of a particle at the position $\mathbf{r}$; it distinguishes the virtual displacement from the actual differential displacement $d\mathbf{r}$ of the particle. For a particle located at the point (x, y), the components of its virtual displacement are usually denoted by $\overrightarrow{\delta x}$ and $\overrightarrow{\delta y}$. Similarly, the angular virtual displacement of a line or body, whose angular position coordinate is θ, is denoted by $\overrightarrow{\delta \theta}$.

A *compatible virtual displacement* of a body is a set of imaginary *first-order* differential displacements, which conforms to the integrity (i.e., no breakage or rupture) of its free body within the framework of first-order accuracy, where the body may be a particle, a rigid body, or a set of connected particles or rigid bodies.[††] Note that a compatible virtual displacement

---

[†]Cf. App. C.
[††]Second- and higher-order virtual displacements are considered in studying the *stability of equilibrium* later in this chapter.

of a body does not necessarily conform to the constraints at the supports in the space diagram. On the other hand, a *constrained virtual displacement* of a body is a compatible virtual displacement of the body which conforms to the constraints at the supports in the space diagram, as illustrated in Fig. 10.7.

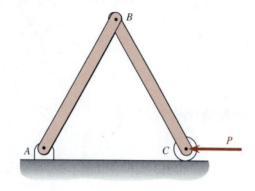

(a) A linkage

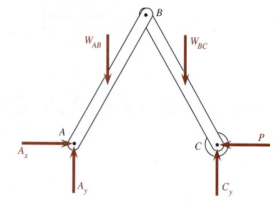

(b) Free body

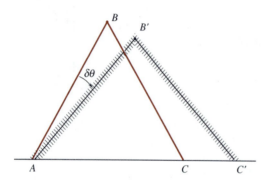

(c) This shows a *constrained*, as well as compatible, *virtual displacement* of the linkage.

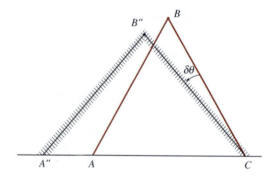

(d) This shows a *compatible*, but not a constrained, *virtual displacement* of the linkage.

Fig. 10.7

EXAMPLE 10.4

A beam AB is given an angular virtual displacement $\vec{\delta\theta}$ about its end A so that its end B moves to the point B' as shown. The line segment $\overline{AB'B''}$ is the hypotenuse of the right triangle ABB''. In terms of a Maclaurin's series (cf. Sec. B5.5 of App. B), we have

$$\sec(\delta\theta) = 1 + \frac{1}{2!}(\delta\theta)^2 + \text{(higher-order terms of } \delta\theta)$$

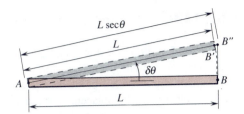

Determine (a) the approximate length of the line segment $\overline{B'B''}$ in terms of L and $\delta\theta$, (b) whether it is permissible for the length and position of the displaced beam to be represented by the hypotenuse $\overline{AB''}$ in considering questions of equilibrium, (c) the virtual displacement $\boldsymbol{\delta}_B$ of the end B.

Solution. Since $\delta\theta$ is infinitesimal, we write

$$\sec(\delta\theta) \approx 1 + \frac{1}{2!}(\delta\theta)^2 = 1 + \frac{1}{2}(\delta\theta)^2$$

Thus, the approximate length of $\overline{B'B''}$ is

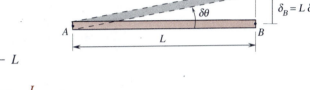

$$\overline{B'B''} = L \sec(\delta\theta) - L \approx L\left[1 + \frac{1}{2}(\delta\theta)^2\right] - L$$

$$\overline{B'B''} \approx \frac{L}{2}(\delta\theta)^2 \quad \blacktriangleleft$$

This shows that $\overline{B'B''}$ is a second-order quantity in $\delta\theta$. Therefore, in considering questions of equilibrium, we may neglect $\overline{B'B''}$ and take $\overline{AB'}$ as being equal to $\overline{AB''}$. In other words, *it is permissible for the length and the position of the displaced beam to be represented by the hypotenuse $\overline{AB''}$*. The arc BB' and the line segment $\overline{BB''}$ are taken to be equal because $\delta\theta$ is infinitesimal. Thus, the virtual displacement of the end B is

$$\delta_B = \overline{BB''} = L\,\delta\theta \qquad \boldsymbol{\delta}_B = L\,\delta\theta \uparrow \quad \blacktriangleleft$$

Developmental Exercises

D10.12 Define the terms: (a) virtual displacement, (b) compatible virtual displacement, (c) constrained virtual displacement.

D10.13 A hinge B connects two beams AB and BC as shown. An angular virtual displacement $\overrightarrow{\delta\theta}$ is given to the beam AB about its end A, and the end C of the beam BC is to remain on the x axis. The exact configuration of the displaced beams is indicated by the line $AB'C'$ as shown. In terms of a Maclaurin's series, we have

$$\cos(\delta\theta) = 1 - \frac{1}{2!}(\delta\theta)^2 + \frac{1}{4!}(\delta\theta)^4 - \cdots$$

For considerations of equilibrium, determine (a) the approximate length of the line segment $\overline{CC'}$ in terms of L and $\delta\theta$, (b) whether it is permissible for the lengths and the positions of the displaced beams to be represented by the lines $\overline{AB''}$ and $\overline{B''C}$ as shown, (c) the virtual displacement $\boldsymbol{\delta}_B$ of the hinge B.

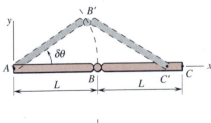

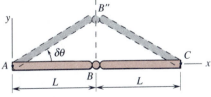

Fig. D10.13

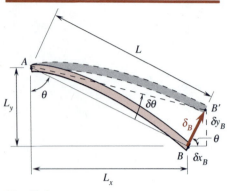

Fig. 10.8

10.5 Components of a Linear Virtual Displacement

From time to time, we need to compute the rectangular components of a linear virtual displacement of a point on a rigid body which is given an angular virtual displacement about a fixed point. Let $\overrightarrow{\delta x_B}$ and $\overrightarrow{\delta y_B}$ be the x and y components of the linear virtual displacement $\boldsymbol{\delta}_B$ of the point B on the rigid body AB which is given an angular virtual displacement $\overrightarrow{\delta\theta}$ about the fixed point A as shown in Fig. 10.8. From this figure, we write

$$\delta x_B = \delta_B \cos\theta = (L\,\delta\theta)\cos\theta = (L\cos\theta)\,\delta\theta = L_y\,\delta\theta$$

$$\delta y_B = \delta_B \sin\theta = (L\,\delta\theta)\sin\theta = (L\sin\theta)\,\delta\theta = L_x\,\delta\theta$$

$$\boxed{\delta x_B = L_y\,\delta\theta} \qquad \boxed{\delta y_B = L_x\,\delta\theta} \qquad (10.13)$$

The formulas in Eqs. (10.13) show that *the x (or y) component of the linear virtual displacement of the point B on a rigid body which is given an angular virtual displacement $\overrightarrow{\delta\theta}$ about a fixed point A is equal to the projection of the line segment $\overline{AB}$ in the y (or x) direction times the infinitesimal angle $\delta\theta$.* Alternatively, we may use the cross product to write

$$\boxed{\boldsymbol{\delta}_B = \overrightarrow{\delta\theta} \times \overrightarrow{AB}} \qquad (10.14)$$

where $\boldsymbol{\delta}_B = \delta x_B \mathbf{i} + \delta y_B \mathbf{j}$, $\overrightarrow{\delta\theta} = \delta\theta\mathbf{k}$, and $\overrightarrow{AB} = L_x\mathbf{i} - L_y\mathbf{j}$.

EXAMPLE 10.5

It is desired to give the member *ABCDE* of the free body shown a clockwise angular virtual displacement $\overrightarrow{\delta\theta}$ about the pinhole at A with the pinhole at F allowed to move only along the x axis. Determine in terms of $\delta\theta$ the linear virtual displacements $\overrightarrow{\delta x_F}$, $\overrightarrow{\delta y_C}$, $\overrightarrow{\delta x_D}$, and $\overrightarrow{\delta y_E}$ of the referred points on the free body.

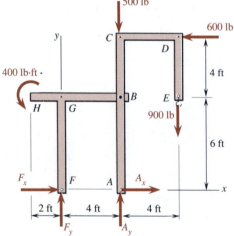

Solution. The desired virtual displacements of the free body are sketched to first-order accuracy as shown. Using Eqs. (10.13) and noting that $\delta x_F = \delta x_B$, we write

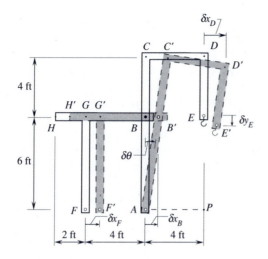

$$\delta x_B = (\overline{AB})_y\ \delta\theta = \overline{AB}\ \delta\theta = 6\ \delta\theta = \delta x_F \qquad \overrightarrow{\delta x_F} = 6\ \delta\theta\ \text{ft} \rightarrow \ \blacktriangleleft$$

$$\delta y_C = (\overline{AC})_x\ \delta\theta = \overline{AA}\ \delta\theta = 0\ (\delta\theta) \qquad\qquad \overrightarrow{\delta y_C} = \mathbf{0}\ \blacktriangleleft$$

$$\delta x_D = (\overline{AD})_y\ \delta\theta = \overline{AC}\ \delta\theta = 10\ \delta\theta \qquad \overrightarrow{\delta x_D} = 10\ \delta\theta\ \text{ft} \rightarrow \ \blacktriangleleft$$

$$\delta y_E = (\overline{AE})_x\ \delta\theta = \overline{AP}\ \delta\theta = 4\ \delta\theta \qquad \overrightarrow{\delta y_E} = 4\ \delta\theta\ \text{ft} \downarrow \ \blacktriangleleft$$

REMARK. The preceding virtual displacements may also be obtained by using the cross product. For instance, we write

$$\boldsymbol{\delta}_D = \overrightarrow{\delta\theta} \times \overrightarrow{AD} = (-\delta\theta\mathbf{k}) \times (4\mathbf{i} + 10\mathbf{j})$$
$$= -4\ \delta\theta\mathbf{j} + 10\ \delta\theta\mathbf{i} = \delta x_D\mathbf{i} + \delta y_D\mathbf{j}$$

Thus, $\delta x_D = 10\ \delta\theta$ and $\delta y_D = -4\ \delta\theta$.

Developmental Exercises

D10.14 Let the entire body in Example 10.5 be rigidly rotated about F through $\delta\theta\ \circlearrowright$. Determine in terms of $\delta\theta$ the linear virtual displacements $\overrightarrow{\delta y_C}$, $\overrightarrow{\delta x_D}$, $\overrightarrow{\delta y_E}$, $\overrightarrow{\delta x_A}$, and $\overrightarrow{\delta y_A}$ of the referred points on the free body.

D10.15 Using the cross product, determine $\overrightarrow{\delta x_E}$ and $\overrightarrow{\delta y_E}$ in Example 10.5.

10.6 Differential Calculus and Displacement Center[†]

The relations among the virtual displacements of certain particles or members of a system need to be determined in many applications. They can frequently be found by using *differential calculus* or the *displacement center*, or both, which are explained in the following paragraphs.

[†]For a detailed treatment, see Jong, I. C., and C. W. Crook, "Introducing the Concept of Displacement Center in Statics," *Engineering Education,* ASEE, vol. 80, no. 4, May/June 1990, pp. 477-479.

For a body or system allowed to have only a single degree of freedom, the use of *differential calculus* may be implemented as follows: (a) study the manner in which the virtual displacements of the system are to be specified; (b) select a governing position variable whose differential is the key parameter in specifying the virtual displacements; (c) write the expressions or equations which relate the relevant position variables of the particles or members to the governing position variable selected in step (b); (d) take the differentials of the expressions or equations established in step (c); (e) solve for the differentials of the desired relevant position variables in terms of the differential of the governing position variable. For illustrations, see Examples 10.6 and 10.11.

The *displacement center* of a body (or a member of a system) is the center of rotation of the body when it undergoes an angular virtual displacement and the points on it incur linear virtual displacements. Such a center is located at the intersection of two straight lines which are perpendicular to two linear virtual displacement vectors of two different points of that body which pass through those two points, respectively. For example, if a body AB undergoes an angular virtual displacement $\vec{\delta\theta}$ so that the points A and B have linear virtual displacements $\delta\mathbf{r}_A$ and $\delta\mathbf{r}_B$, respectively, as shown in Fig. 10.9, the displacement center D of the body is located at the intersection of the two straight lines AD and BD which are perpendicular to $\delta\mathbf{r}_A$ and $\delta\mathbf{r}_B$, respectively.

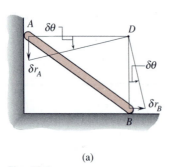

(a)

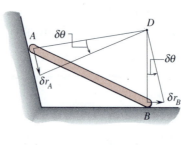

(b)

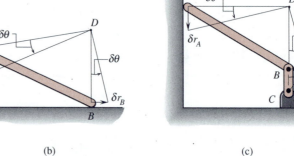

(c)

Fig. 10.9

Note in Fig. 10.9 that the angular virtual displacements of the lines AD, BD, and any other lines drawn from the body to the point D must all be equal to $\delta\theta$. If a system consists of two or more members, *each member will generally have a different displacement center, and each displacement center has zero virtual displacement*. In Fig. 10.9(c), note that the displacement center of the member BC is at C.

EXAMPLE 10.6[†]

It is desired to give the member AB of the free body shown a counterclockwise angular virtual displacement $\vec{\delta\theta}$ about the pinhole at A with the pinhole at C moving only along the x axis. Determine in terms of $\delta\theta$ (a) the angular

[†]Continued in D10.24.

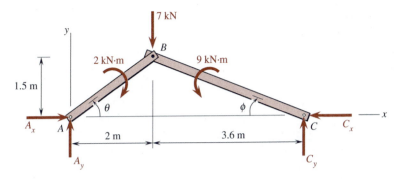

virtual displacement $\overrightarrow{\delta\phi}$ of the member BC, (b) the linear virtual displacement $\overrightarrow{\delta x_C}$ of the pinhole at C.

Solution. For illustrative purposes, let the problem in this example be solved in the following two ways:

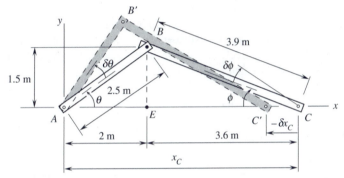

(a) *Using differential calculus.* The two angular position variables θ and ϕ may be related as follows:

$$\overline{BE} = 2.5 \sin\theta = 3.9 \sin\phi$$

Taking the differentials, we write

$$2.5 \, (\cos\theta) \, \delta\theta = 3.9 \, (\cos\phi)\delta\phi$$

Since $\cos\theta = \dfrac{2}{2.5}$ and $\cos\phi = \dfrac{3.6}{3.9}$, we obtain

$$\delta\phi = \frac{2.5 \cos\theta}{3.9 \cos\phi} \, \delta\theta = \frac{5}{9} \, \delta\theta \qquad\qquad \overrightarrow{\delta\phi} = \frac{5}{9} \, \delta\theta \,\circlearrowright \; \blacktriangleleft$$

The variable x_C defining the position of the pinhole at C is given by $x_C = 2.5 \cos\theta + 3.9 \cos\phi$. Taking the differentials, we write

$$\delta x_C = -2.5 \sin\theta \, \delta\theta - 3.9 \sin\phi \, \delta\phi$$

Since $\sin\theta = \dfrac{1.5}{2.5}$, $\sin\phi = \dfrac{1.5}{3.9}$, and $\delta\phi = \dfrac{5}{9} \, \delta\theta$, we have

$$\delta x_C = -\frac{7}{3} \, \delta\theta \qquad\qquad \overrightarrow{\delta x_C} = \frac{7}{3} \, \delta\theta \; \text{m} \leftarrow \; \blacktriangleleft$$

Note that the negative sign in the value of δx_C indicates that the direction of $\overrightarrow{\delta x_C}$ is opposite to the increasing direction of x_C.

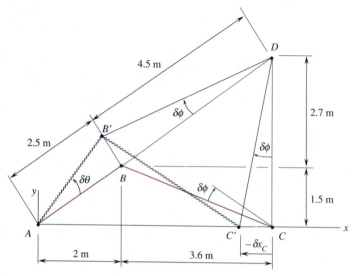

(b) *Using displacement center*. Based on the specified virtual displacements of the system, we see that the displacement centers of members AB and BC are at A and D, respectively, as shown in the sketch for the centerlines of the members. We note that $\overline{BB'} = \overline{AB}\,\delta\theta = \overline{BD}\,\delta\phi$. Thus, we write

$$2.5\,\delta\theta = 4.5\,\delta\phi \qquad \delta\phi = \frac{2.5}{4.5}\,\delta\theta \qquad \vec{\delta\phi} = \frac{5}{9}\,\delta\theta \;\circlearrowleft \quad \blacktriangleleft$$

$$\delta x_C = -\overline{CD}\,\delta\phi = -4.2\left(\frac{5}{9}\,\delta\theta\right) \qquad \vec{\delta x}_C = \frac{7}{3}\,\delta\theta \;\text{m} \leftarrow \quad \blacktriangleleft$$

Developmental Exercise

D10.16[†]   It is desired to give the member ABC of the free body shown a counterclockwise angular virtual displacement $\vec{\delta\theta}$ about the pinhole at A with the pinhole at E moving only along the x axis. (a) Relate the variables θ and ϕ which define the angular positions of members ABC and BDE. (b) Determine in terms of $\delta\theta$ (1) the angular virtual displacement $\vec{\delta\phi}$, (2) the linear virtual displacements $\vec{\delta x}_E$ and $\vec{\delta y}_C$. (c) Using displacement center, verify the answers obtained in (b).

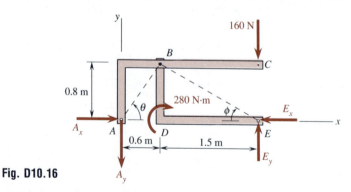

Fig. D10.16

[†]Continued in D10.25.

10.7 Principle of Virtual Work[†]

The work of a force or a moment acting on a body during a virtual displacement of the body is called a *virtual work*. Let δU be the *total virtual work* of the forces $\mathbf{F}_1, \mathbf{F}_2, \ldots, \mathbf{F}_n$ acting on a particle during a virtual displacement $\delta \mathbf{r}$ of the particle from the point A to the point A' as shown in Fig. 10.10. Then, we write

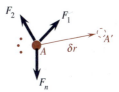

Fig. 10.10

$$\delta U = \mathbf{F}_1 \cdot \delta \mathbf{r} + \mathbf{F}_2 \cdot \delta \mathbf{r} + \cdots + \mathbf{F}_n \cdot \delta \mathbf{r}$$

$$= (\mathbf{F}_1 + \mathbf{F}_2 + \cdots + \mathbf{F}_n) \cdot \delta \mathbf{r}$$

$$= \mathbf{R} \cdot \delta \mathbf{r}$$

where $\mathbf{R}$ is the resultant force. If equilibrium exists, we have $\mathbf{R} = \mathbf{0}$; therefore, $\delta U = \mathbf{0} \cdot \delta \mathbf{r} = 0$ is a necessary condition. At the same time, $\delta U = 0$ for any $\delta \mathbf{r}$ is a sufficient condition for equilibrium because we must have $\mathbf{R} = \mathbf{0}$ if $\delta U = \mathbf{R} \cdot \delta \mathbf{r} = 0$ for any $\delta \mathbf{r}$. Thus, *if a particle is in equilibrium, the total virtual work of all forces acting on the particle is zero for any virtual displacement of the particle, and conversely.*

By Newton's third law, all internal forces acting on the particles of a rigid body or a system of pin-connected rigid bodies must exist in pairs, each of which consists of two collinear forces that are equal in magnitude and opposite in direction; and the collinear forces in each pair remain at a fixed distance from each other within a rigid body. Since pin connections are taken as frictionless, we see that *the total virtual work of all internal forces must be zero during any compatible virtual displacement of a rigid body or a system of pin-connected rigid bodies.*

Furthermore, all particles forming a body must be in equilibrium when the body is in equilibrium. Thus, the total virtual work of all external and internal forces acting on all particles of the body in equilibrium must be zero during any compatible virtual displacement of the body. Since the total virtual work of all *internal forces* is zero, the total virtual work of all *external forces* acting on that body must also be zero; i.e.,

$$\delta U = 0 \qquad\qquad (10.15)$$

The properties stated above lead to the *principle of virtual work*[††] in statics, which may be stated as follows: *If a body is in equilibrium, the total virtual work of the external force system acting on its free body during any compatible virtual displacement of its free body is equal to zero, and conversely*, where the body may be a particle, a set of connected particles, a rigid body, a frame, or a machine.

[†]Recent historical studies show that on February 26, 1715, the Swiss mathematician Johann (Jean) Bernoulli (1667–1748) communicated to Pierre Varignon (1654–1722) the principle of virtual velocities in analytical form for the first time. That was the forerunner of the principle of virtual work today. The approach to mechanics based on the principle of virtual work was formally treated by Joseph Louis Lagrange (1736–1813) in his *Mécanique Analytique* published in 1788.

[††]Cf. the *energy criterion* in Sec. 10.15.

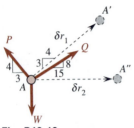

P δr_1 Q
4   4
4   3   8
3   15
A δr_2
A'
A''
W

Fig. D10.19

Developmental Exercises

D10.17   What is a virtual work?

D10.18   Describe the *principle of virtual work* in statics.

D10.19   A particle at A is in equilibrium as shown, where $W = 280$ N. (a) Write the analytical expressions of $\mathbf{P}$ and $\mathbf{Q}$. (b) If the particle is given a virtual displacement $\delta\mathbf{r} = (\delta r_1)(\frac{4}{5}\mathbf{i} + \frac{3}{5}\mathbf{j})$, which is perpendicular to $\mathbf{P}$, compute the total virtual work δU of $\mathbf{P}$, $\mathbf{Q}$, and $\mathbf{W}$. (c) Solve for Q by setting $\delta U = 0$. (d) Having solved for Q, determine P by letting $\delta\mathbf{r} = \delta r_2\mathbf{i}$ and setting $\delta U = 0$.

D10.20   The free body of a bent rod and its virtual displacement are shown. (a) Determine the virtual displacements $\overrightarrow{\delta x_C}$, $\overrightarrow{\delta y_C}$, and $\overrightarrow{\delta\phi}$ in terms of $\delta\theta$. (b) Determine $\mathbf{M}_A$ if the bent rod is in equilibrium.

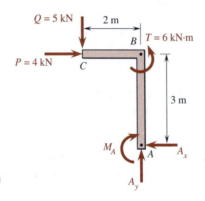

$Q = 5$ kN   2 m
B $T = 6$ kN·m
$P = 4$ kN C
3 m
M_A A A_x
A_y

Fig. D10.20

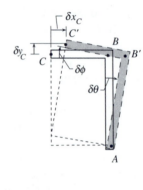

δx_C
C'
δy_C B
C B'
$\delta\phi$
$\delta\theta$
A

10.8   Applications to Equilibrium Problems

The principle of virtual work in statics frequently enables us to solve for unknown quantities in seemingly complex equilibrium problems of frames and machines without having to solve simultaneous equations. The following suggested steps are generally helpful in applying the *principle of virtual work* to the solution of many equilibrium problems:

1. *Draw* the appropriate *free-body diagram* of the system.
2. *Give*, if possible, the free body or its centerlines *a compatible virtual displacement* in such a manner that only one unknown parameter, besides the given loading, will be involved in doing the virtual work. (If more than one unknown parameter is involved, the need to solve simultaneous equations will arise.)
3. *Mark and compute* the pertinent components of the *virtual displacements* at the various locations where forces and moments are acting. (This step may involve considerations of geometry, trigonometry, and differential calculus or displacement center.)
4. *Write* the expression of the *total virtual work* in terms of the unknown parameter and the virtual displacement.
5. *Set the total virtual work to zero* and solve for the unknown parameter in the resulting equation.
6. *Repeat steps* (2) *through* (5) until all required unknown parameters have been determined.

EXAMPLE 10.7

A frame is loaded as shown. Determine the reactions from the supports at A
and C.

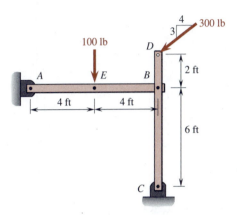

Solution. The free-body diagram of the entire system is drawn as shown,
where the original 300-lb force at D is replaced by its rectangular compo-
nents. The unknown reaction components are $\mathbf{A}_x$, $\mathbf{A}_y$, $\mathbf{C}_x$, and $\mathbf{C}_y$.

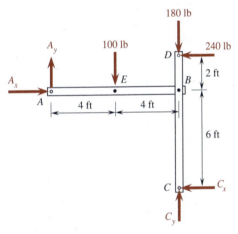

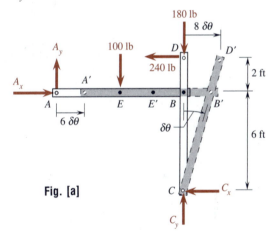

Fig. [a]

(a) To determine $\mathbf{A}_x$, let the member CD be given a clockwise angular virtual displacement $\overrightarrow{\delta\theta}$ about the pinhole at C and the member AB remain
horizontal as shown in Fig. [a].[†] From the sketch, we write

$$\overline{AA'} = \overline{BB'} = 6\,\delta\theta \qquad \overline{DD'} = 8\,\delta\theta$$

Applying the *principle of virtual work*, we write

$$\delta U = A_x(6\,\delta\theta) + 240(-8\,\delta\theta) = 0$$

$$A_x = 320 \qquad\qquad \mathbf{A}_x = 320\text{ lb} \rightarrow \blacktriangleleft$$

(b) To determine $\mathbf{A}_y$, let the member AB be given a counterclockwise angular virtual displacement $\overrightarrow{\delta\theta}$ about the pin at B and the member DBC
remain unmoved as shown in Fig. [b].[†] From the sketch and by the

---

[†]This is just a *compatible* (not constrained) *virtual displacement*.

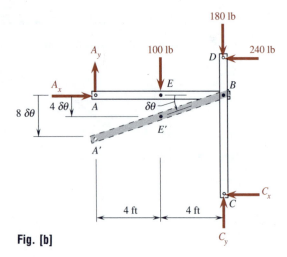

Fig. [b]

principle of virtual work, we write

$$\delta U = A_y(-8\ \delta\theta) + 100(4\ \delta\theta) = 0$$

$$A_y = 50 \qquad\qquad \mathbf{A_y} = 50\ \text{lb} \uparrow \ \blacktriangleleft$$

(c) To determine $\mathbf{C}_x$, let the member *DBC* be given a clockwise angular virtual displacement $\overrightarrow{\delta\theta}$ about the pin at *B* and the member *AB* remain unmoved as shown in Fig. [c] (see footnote on p. 419). From the sketch and by the *principle of virtual work*, we write

$$\delta U = C_x(6\ \delta\theta) + 240(-2\ \delta\theta) = 0$$

$$C_x = 80 \qquad\qquad \mathbf{C_x} = 80\ \text{lb} \leftarrow \ \blacktriangleleft$$

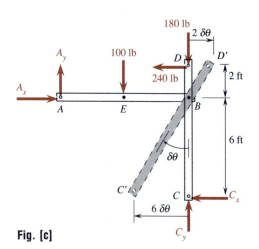

Fig. [c]

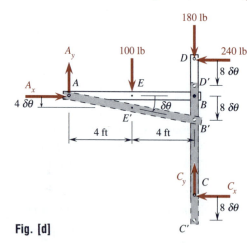

Fig. [d]

(d) To determine $\mathbf{C}_y$, let the member *AB* be given a clockwise angular virtual displacement $\overrightarrow{\delta\theta}$ about the end at *A* and the member *DBC* remain vertical as shown in Fig. [d] (see footnote on p. 419). From the sketch and by the *principle of virtual work*, we write

$$\delta U = C_y(-8\ \delta\theta) + 180(8\ \delta\theta) = 100(4\ \delta\theta) = 0$$

$$C_y = 230 \qquad\qquad \mathbf{C_y} = 230\ \text{lb} \uparrow \ \blacktriangleleft$$

REMARK. The above solutions for A_x, A_y, C_x, and C_y did not involve the solution of simultaneous equations; i.e., these unknowns have been found independently of each other. Moreover, the same results would be obtained if the directions of the virtual displacements were reversed.

Developmental Exercise

D10.21   If the free body shown in Example 10.5 is in equilibrium, determine the unknown force $\mathbf{F}_x$ acting on that free body.

EXAMPLE 10.8

A frame is shown. Determine the reaction from the support at C.

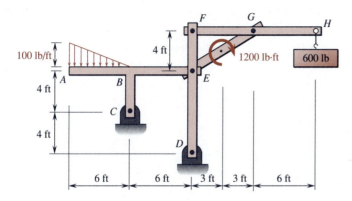

Solution.   We first draw the free-body diagram of the entire system as shown. Note that the distributed load over the portion AB of the frame has been replaced in the free-body diagram by its resultant force which is equal to the area of the loading diagram and acts through its centroid. The unknown reaction components are $\mathbf{C}_x$, $\mathbf{C}_y$, $\mathbf{D}_x$, and $\mathbf{D}_y$. We need to determine $\mathbf{C}_x$ and $\mathbf{C}_y$.

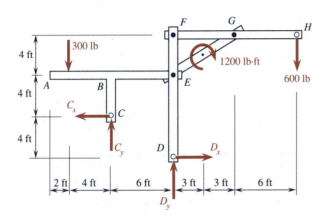

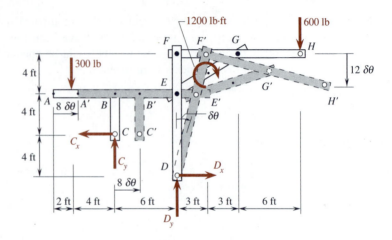

(a) To determine C_x, let the portion *DEFGH* of the frame be given a clockwise angular virtual displacement $\overrightarrow{\delta\theta}$ about the pinhole at *D* and the orientation of the portion *ABCE* of the frame be unchanged.[†] Note that *ABCE* of the frame has only a linear virtual displacement to the right equal to the length of $\overline{EE'}$. From the sketch and by the *principle of virtual work*, we write

$$\delta U = C_x(-8\ \delta\theta) + 1200(\delta\theta) + 600(12\ \delta\theta) = 0$$

$$C_x = 1050 \qquad\qquad\qquad \mathbf{C_x = 1050\ lb} \leftarrow \blacktriangleleft$$

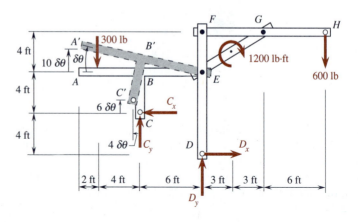

(b) To determine C_y, let the portion *ABCE* be given a clockwise angular virtual displacement $\overrightarrow{\delta\theta}$ about the pin *E* and the other portion be unmoved.[†] This is satisfactory because C_x is already known. From the sketch and by the *principle of virtual work*, we write

$$\delta U = 300(-10\ \delta\theta) + C_x(4\ \delta\theta) + C_y(6\ \delta\theta) = 0$$

$$C_y = 500 - \tfrac{2}{3}C_x = 500 - \tfrac{2}{3}(1050) = -200$$

$$\mathbf{C_y = 200\ lb} \downarrow \blacktriangleleft$$

[†]This is just a *compatible* (not constrained) *virtual displacement*.

Developmental Exercise

D10.22 Refer to the frame in Example 10.8. Determine D_x and D_y.

EXAMPLE 10.9

Determine the reactions from the supports at A and D of the hinged beam shown.[†]

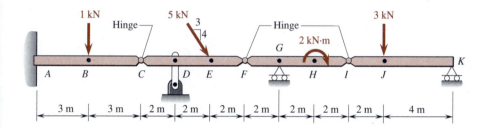

Solution. The free-body diagram of the entire system is drawn as shown. Note that the 5-kN force in the space diagram has been replaced by its rectangular components at E in the free-body diagram. The unknown reaction components are M_A, A_x, A_y, D, G, and K.

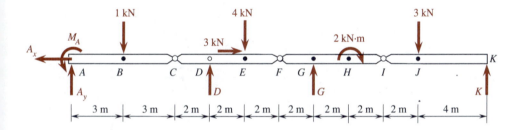

To use the *principle of virtual work* to determine M_A, A_x, A_y, and D, we choose four appropriate *compatible* (not constrained) *virtual displacements* of the centerline of the hinged beam as shown. From the virtual displacement shown in Fig. (a), we write

$$\delta U = M_A\,(\delta\theta) + 1(-3\,\delta\theta) + 4(6\,\delta\theta)$$
$$+\ 2(-6\,\delta\theta) + 3(-16\,\delta\theta) = 0$$

$$M_A = 39 \qquad\qquad \mathbf{M_A = 39\ kN\cdot m} \;\curvearrowright\; \blacktriangleleft$$

From the virtual displacement shown in Fig. (b) on the next page, we write

$$\delta U = A_x(-\delta x) + 3(\delta x) = 0$$

$$A_x = 3 \qquad\qquad \mathbf{A_x = 3\ kN} \leftarrow \;\blacktriangleleft$$

[†]Cf. the footnote on p. 318.

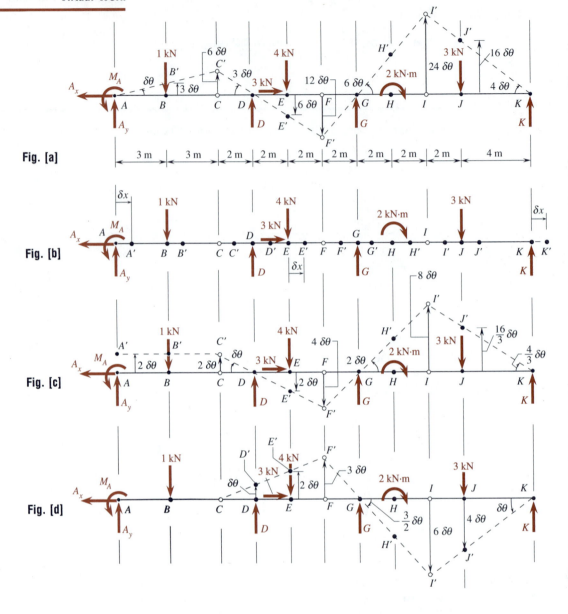

Fig. [a]

Fig. [b]

Fig. [c]

Fig. [d]

From the virtual displacement shown in Fig. [c], we write

$$\delta U = A_y(2\ \delta\theta) + 1(-2\ \delta\theta) + 4(2\ \delta\theta)$$
$$+ 2(-2\ \delta\theta) + 3(-\tfrac{16}{3}\ \delta\theta) = 0$$

$$A_y = 7 \qquad\qquad \mathbf{A}_y = 7\ \text{kN} \uparrow \quad \blacktriangleleft$$

From the virtual displacement shown in Fig. [d], we write

$$\delta U = D(\delta\theta) + 4(-2\ \delta\theta) + 2(\tfrac{3}{2}\ \delta\theta) + 3(4\ \delta\theta) = 0$$

$$D = -7 \qquad\qquad \mathbf{D} = 7\ \text{kN} \downarrow \quad \blacktriangleleft$$

Developmental Exercise

D10.23   Determine the reactions from the supports at G and K of the beam in Example 10.9.

EXAMPLE 10.10

Determine the moment **M** for which the system shown is in equilibrium.

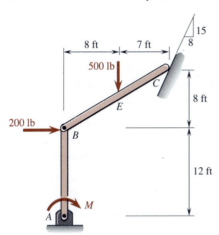

Solution.   The free-body diagram and a *constrained virtual displacement* of the centerlines of the members of the system are sketched as shown, where the *displacement centers* of the members AB and BC are at A and D, respectively.[†] Note that the angular virtual displacements for AB and BC are

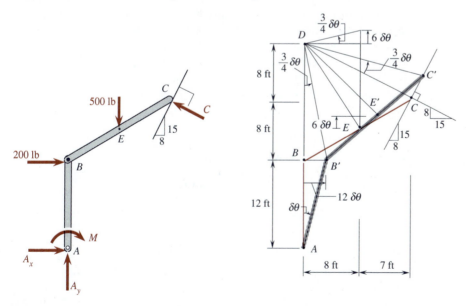

[†]Note that a *constrained virtual displacement* is also a *compatible virtual displacement*. (Cf. Sec. 10.4.)

$\delta\theta \downdownarrows$ and $\frac{3}{4}\delta\theta \circlearrowright$, respectively. From the sketches and by the *principle of virtual work*, we write

$$\delta U = M(\delta\theta) + 200(12\ \delta\theta) + 500(-6\ \delta\theta)$$
$$= (M - 600)\ \delta\theta = 0$$
$$M = 600 \qquad \mathbf{M} = 600\ \text{lb·ft} \ \circlearrowright \ \blacktriangleleft$$

Developmental Exercises

D10.24 If the free body shown in Example 10.6 is in equilibrium, determine the unknown force $\mathbf{C}_x$ acting on that free body.

D10.25 If the free body shown in D10.16 is in equilibrium, determine the unknown force $\mathbf{E}_x$ acting on that free body.

EXAMPLE 10.11

Collars A and B, each weighing 20 lb, are connected by a thin wire AB and may slide freely on the smooth bent rod as shown. Determine the magnitude of $\mathbf{F}$ consistent with equilibrium.

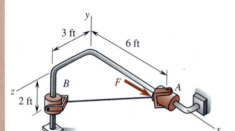

Solution. Since the collar A slides along the x axis and the collar B slides along the vertical portion of the rod, the coordinates of A and B may be written as $A(x_A, 0, 0)$ ft and $B(0, y_B, 3)$ ft, respectively. In the position shown, we have

$$x_A = 6\ \text{ft} \qquad y_B = -2\ \text{ft}$$

The length of the wire AB can readily be found to be 7 ft. The free-body diagram of the collars and the wire is sketched as shown. Note that $\mathbf{B}_{zx}$ is the reaction from the bent rod to the collar B and is parallel to the zx plane. In order to prevent $\mathbf{B}_{zx}$, $\mathbf{A}_y$, and $\mathbf{A}_z$ from entering the expression of δU, we give the collar A a virtual displacement $\overrightarrow{\delta x_A}$ and the collar B a corresponding virtual displacement $\overrightarrow{\delta y_B}$ as shown. Using the Pythagorean theorem, we have

$$x_A^2 + y_B^2 + 3^2 = 7^2$$

Taking the differentials, we write

$$2x_A\ \delta x_A + 2y_B\ \delta y_B = 0$$
$$\delta y_B = -\frac{x_A}{y_B}\ \delta x_A = -\left(\frac{6}{-2}\right)\delta x_A = 3\delta x_A$$

From the sketch and by the *principle of virtual work*, we write

$$\delta U = F(\delta x_A) + 20(-\delta y_B) = [F - 20(3)]\ \delta x_A = 0$$
$$F = 60\ \text{lb} \ \blacktriangleleft$$

Developmental Exercise

D10.26   Using the *principle of virtual work*, do Example 3.11. (*Hint.* Let the coordinates of A and B be $A(0, y_A, 12)$ ft and $B(x_B, y_B, 0)$ ft, where $y_B = 12 - \frac{3}{4}x_B$.)

PROBLEMS

NOTE. The following problems are to be solved using the *principle of virtual work*.

10.15 and 10.16   Determine the reaction from the support at A as shown.

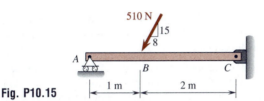

Fig. P10.15

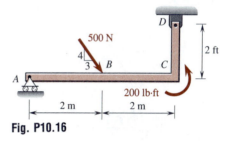

Fig. P10.16

10.17* A 24-kg ladder, with its center of gravity at its midpoint G, is shown. Neglecting friction forces at the supports, determine the magnitude of the horizontal force **P** necessary to keep the ladder from sliding.

10.18   A 96-lb ladder with its center of gravity at its midpoint G is kept from sliding by a 14-lb horizontal force at its bottom as shown. Neglecting friction forces at the supports, determine the distance x.

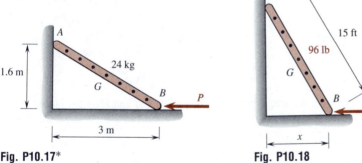

Fig. P10.17*

Fig. P10.18

10.19 through 10.23*   Determine the magnitude of the force **P** for which the linkage as shown is in equilibrium.

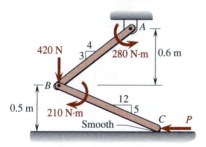

Fig. P10.19

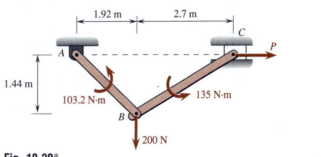

Fig. 10.20*

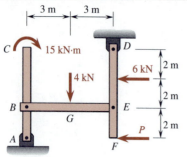

Fig. P10.21

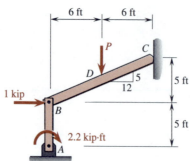

Fig. P10.22

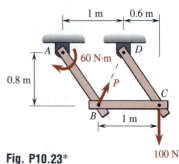

Fig. P10.23*

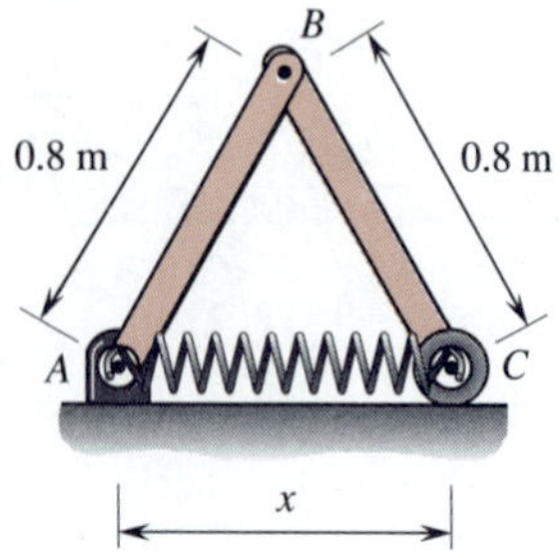

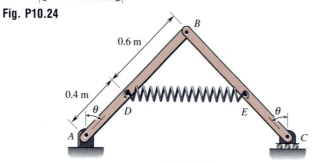

Fig. P10.24

10.24 Two 20-kg uniform bars AB and BC are connected and supported as shown. The spring AC has a free length of 0.4 m and a modulus of 400 N/m. Determine the values of x consistent with equilibrium.[†]

10.25 and 10.26* Two 10-kg uniform bars AB and BC are connected and supported as shown. The spring DE can develop either tension or compression and has a free length of 0.6 m as well as a modulus of 900 N/m. Determine the values of θ consistent with equilibrium.[†]

Fig. P10.25

Fig. P10.26*

10.27 Determine the reactions from the supports at A and H of the hinged beam shown.

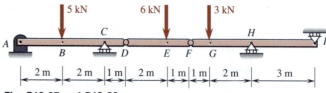

Fig. P10.27 and P10.28*

10.28* Determine the reactions from the supports at C and I of the hinged beam shown.

10.29 Determine the reactions from the supports at A and G of the hinged beam shown.

[†]The solution of this problem leads to a quartic equation. (Cf. App. E or Sec. B5.8 of App. B for the solution of a nonlinear equation.)

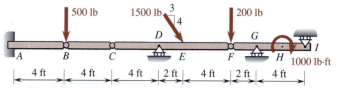

Fig. P10.29 and P10.30*

10.30* Determine the reactions from the supports at D and I of the hinged beam shown.

10.31 Determine the reactions from the supports at A and B of the frame shown.

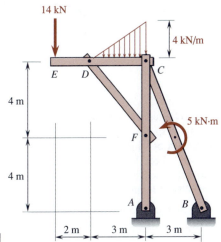

Fig. P10.31

10.32 Determine the magnitude of the force exerted by the cutter on the bolt at E as shown.

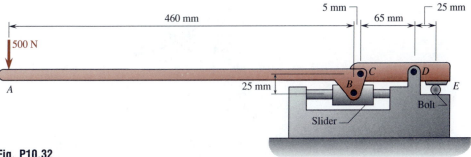

Fig. P10.32

10.33 A body of weight W is suspended at B of the system as shown. The spring has a modulus k of 72 lb/ft and is neither stretched nor compressed when AB and BC are horizontal. Determine the value of W if the system is in equilibrium when $\theta = 30°$.

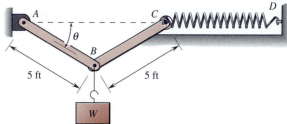

Fig. P10.33

10.34* Refer to Prob. 10.33. Determine the angle θ if $W = 400$ lb when the system is in equilibrium.[†]

10.35 and 10.36* Determine the magnitude of the force **P** required to hold the structure as shown in equilibrium.

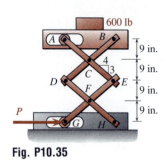

Fig. P10.35

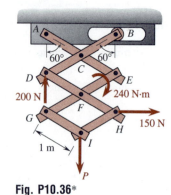

Fig. P10.36*

10.37 A hydraulic cylinder FG is used to adjust the height of a mobile platform which supports a crate of 1 Mg as shown. The point C is the midpoint of the supporting members AE and BD, each of which is 2 m long. For $\theta = 60°$, determine the magnitude of the compressive force $\mathbf{F}_{GF}$ in the hydraulic cylinder. (*Hint.* The virtual work of $\mathbf{F}_{GF}$ may be calculated by first computing δx_F and δy_F and then applying the *theorem of work* alluded to on p. 403.)

10.38 Collars A and B, each weighing 35 lb, are connected by a thin wire AB and may slide freely on the smooth bent rod. Determine the magnitude of **F** consistent with equilibrium as shown.

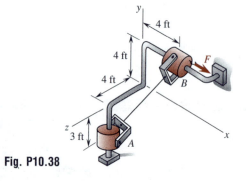

Fig. P10.38

Fig. P10.37

POTENTIAL ENERGY AND STABILITY

*10.9   Conservative System and Potential Energy

An *analytic force field* is a region in space where a body experiences a determinable single-valued force or force system at every position in the region. Since the kinetic friction force developed on a body at a given position may have different values in magnitude and direction, a kinetic friction force field is not an analytic force field.

[†]The footnote on p. 428 applies.

A *conservative force field* is an analytic force field in which the work of
the force in the field acting on a body traveling in the field is dependent on
the initial and final positions of the body, not on the path traveled by the
body. The force of a conservative force field is called a *conservative force*.
In the rest of this chapter, we shall deal with topics related to the equilibrium as well as the stability of equilibrium of structures subjected to the
following conservative forces: gravitational forces, elastic spring forces, and
applied constant loads.

Any force which acts on a body but does no work to the body during a
displacement of the body is called a *nonworking force*. For instance, when
a body moves on a surface, the normal force exerted by the surface on the
body does no work to the body and is a nonworking force. A *conservative
system* is a system where the body or assemblage of bodies is subjected to
only conservative forces and nonworking forces, if any.

For practical use, the *potential energy of a body* in a conservative force
field relative to an appropriate reference datum is defined as *the amount of
work which the body will receive from the conservative force if the body
travels from its present position to the reference datum.*[†] Thus, the potential
energy has the same dimensions and units as the work of a force. The potential energy of a body situated at the reference datum is, of course, equal
to zero.

Developmental Exercises

D10.27 Define (a) analytic force field, (b) conservative force field, (c) conservative force, (d) nonworking force, (e) conservative system.

D10.28 Define the potential energy of a body.

★10.10   Gravitational Potential Energy

We know that the force exerted on a body in a *gravitational force field* is
the weight force of the body. As a body travels from one position to a
nearby position in the vicinity of the surface of the earth, the weight force
of the body is essentially a downward vertical force of constant magnitude.
Thus, the *gravitational potential energy V_g of a body* with a weight W near
the surface of the earth at the position A of altitude y above the reference
datum, as shown in Fig. 10.11, is equal to the work which the body will
receive from the constant weight force $\mathbf{W}$ of the body if the body travels

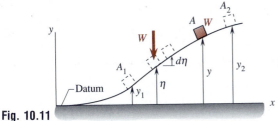

Fig. 10.11

[†]This is a *quantitative definition. Qualitatively*, the potential energy of a body is the energy
possessed by the body by virtue of its position relative to the reference datum in the conservative force field.

from its present position at $\eta = y$ to the reference datum at $\eta = 0$. We write

$$V_g = \int_y^0 W(-d\eta) \qquad \boxed{V_g = Wy} \qquad (10.16)$$

where V_g is positive only if the datum chosen is situated between the body and the center of the earth. The work $U_{1\to2}$ of the weight force $\mathbf{W} = W \downarrow$ acting on the body traveling from A_1 to A_2 is found to be related to the *change in gravitational potential energy*, ΔV_g, as follows:

$$U_{1\to2} = \int_{y_1}^{y_2} W(-d\eta) = -(Wy_2 - Wy_1) = -\left[(V_g)_2 - (V_g)_1 \right]$$

$$U_{1\to2}' = -\Delta V_g \qquad (10.17)$$

EXAMPLE 10.12

A body of weight W is at an altitude y above the surface of the earth whose radius is R. Choosing the surface of the earth as the reference datum, show that Wy is a good approximation for the potential energy V_g of the body if $y \ll R$.

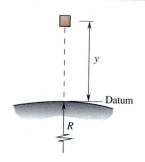

Solution. The weight W of a body is usually measured on the surface of the earth. Letting O be the center of the earth as shown, m be the mass of the body, and M be the mass of the earth, we may apply Newton's law of gravitation and the definition of V_g to write

$$F = \frac{GMm}{r^2} \qquad W = F|_{r=R} = \frac{GMm}{R^2}$$

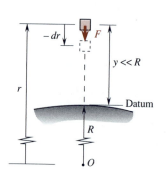

$$V_g = \int_{R+y}^{R} F(-dr) = -GMm \int_{R+y}^{R} r^{-2}\, dr$$

$$= GMm\left(\frac{1}{R} - \frac{1}{R+y} \right)$$

$$= \frac{GMmy}{R(R+y)}$$

Since $y \ll R$, we have $R + y \approx R$ and

$$V_g \approx \frac{GMm}{R^2} y = Wy \qquad \text{Q.E.D.} \blacktriangleleft$$

Developmental Exercises

D10.29   Define the gravitational potential energy of a body near the surface of the earth.

D10.30   When will the gravitational potential energy be positive?

⋆10.11   Elastic Potential Energy

The force exerted by an elastic spring on a body attached to the spring is proportional to the deformation in length of the spring (cf. Sec. 3.5). A body subjected to the force of an elastic spring is said to be in an *elastic force field*. The *reference datum in an elastic force field* is generally chosen to be the position of the body where the spring attached to the body is neutral (i.e., neither stretched nor compressed). For a body at A connected by a spring with a modulus k and an elongation x as shown in Fig. 10.12, the *elastic potential energy* V_e *of the body* is equal to the work which the

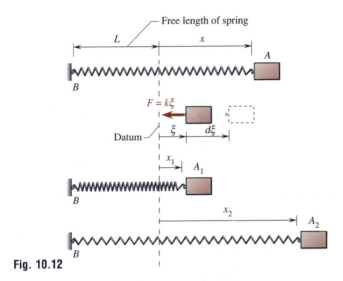

Fig. 10.12

body will receive from the spring force $\mathbf{F} = k\xi \leftarrow$ if the body travels from its present position at $\xi = x$ to the reference datum at $\xi = 0$. Thus, we have

$$V_e = \int_x^0 k\xi(-d\xi) \qquad \boxed{V_e = \frac{1}{2}kx^2} \qquad (10.18)$$

Similarly, we find that the work $U_{1\to2}$ of the spring force $\mathbf{F}$ acting on the body traveling from A_1 to A_2 is related to the *change in elastic potential energy*, ΔV_e, as follows:

$$U_{1\to2} = \int_{x_1}^{x_2} k\xi(-d\xi) = -\left(\frac{1}{2}kx_2^2 - \frac{1}{2}kx_1^2\right) = -[(V_e)_2 - (V_e)_1]$$

$$U_{1\to2} = -\Delta V_e \qquad (10.19)$$

In the case where a torsional spring of modulus K is used to supply a restoring moment $M = K\theta$ to a body which undergoes an angular rotation θ from the neutral position of the spring as shown in Fig. 10.13, the *elastic potential energy* V_e *of the body* can similarly be shown to be

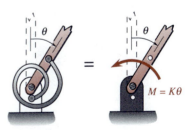

Fig. 10.13

$$V_e = \int_\theta^0 K\theta(-d\theta) \qquad \boxed{V_e = \frac{1}{2}K\theta^2} \qquad (10.20)$$

where the angle θ is measured in radians and the modulus K is measured in moment per radian (e.g., N·m/rad or lb·ft/rad). Of course, Eq. (10.19) is also true in the case of a torsional spring.

EXAMPLE 10.13

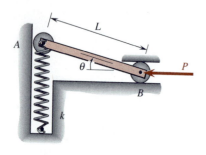

A bar of negligible weight is attached on one end to a spring of modulus k and acted on the other end by a constant force **P** as shown. The spring is neutral when $\theta = 0$. For the position shown, determine the elastic potential energy V_e of the bar in terms of θ.

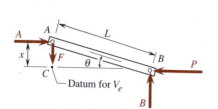

Solution. From the sketch shown, we see that the spring has an elongation of

$$x = \overline{CA} = L \sin\theta$$

Thus, we write

$$V_e = \frac{1}{2} kx^2 = \frac{1}{2} k(L \sin\theta)^2$$

$$V_e = \frac{1}{2} kL^2 \sin^2\theta \qquad \blacktriangleleft$$

Developmental Exercises

D10.31   Define the elastic potential energy of a body.

D10.32   How is the reference datum for the elastic potential energy of a body generally chosen?

★10.12   Applied Potential Energy

Besides gravitational and elastic force fields, there are other conservative force fields which exist to exert conservative forces on various bodies. A body deflected by an *applied conservative force* is said to possess an applied potential energy because the body will receive a definite amount of work from the applied conservative force if it returns to its undeflected state. The *applied potential energy V_a of a body* in a deflected state under the action of an applied conservative force is equal to the work which the body will receive from the applied conservative force if the body travels from its present deflected position to its undeflected position of equilibrium, which is the reference datum. However, we shall here focus our attention on applied conservative forces which are constant in magnitude and direction. Note that an *applied constant force* is a constant force which is suddenly or instantaneously applied to a body, and its magnitude as well as direction remains the same thereafter.

Let us now consider a body AB carrying an applied constant force **P** as shown in Fig. 10.14(a). The *applied potential energy V_a of the body* situated in the deflected position $A'B$ under the action of the applied constant force

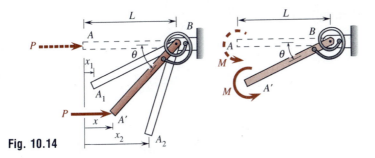

Fig. 10.14

P is equal to the work which the body will receive from **P** if the body returns from its present deflected position $A'B$ to its undeflected neutral position AB of equilibrium, which is the reference datum; i.e.,

$$V_a = \int_x^0 P \, dx \qquad \boxed{V_a = -Px} \qquad (10.21)$$

The work $U_{1\to2}$ of the applied constant force **P** on the member moving from the position where $x = x_1$ to the position where $x = x_2$ is related to the *change in the applied potential energy*, ΔV_a, as follows:

$$U_{1\to2} = \int_{x_1}^{x_2} P \, dx = -(-Px_2 + Px_1) = -[(V_a)_2 - (V_a)_1]$$

$$\boxed{U_{1\to2} = -\Delta V_a} \qquad (10.22)$$

In the case where an applied constant moment **M** acts on a body AB as shown in Fig. 10.14(b), the *applied potential energy V_a of the body* can similarly be shown to be

$$\boxed{V_a = -M\theta} \qquad (10.23)$$

Of course, Eq. (10.22) is also true in the case of an applied constant moment **M**.

EXAMPLE 10.14

Refer to the bar in Example 10.13. Determine in terms of θ the applied potential energy V_a of the bar under the action of the applied constant force **P**.

Solution.   From the sketch shown, we see that when the bar returns from its present deflected position AB to its original undeflected neutral position CD of equilibrium, the right end of the bar moves from the point B to the point D. The distance from B to D is

$$x = \overline{BD} = \overline{CD} - \overline{CB} = L - L\cos\theta$$

Thus, we write

$$V_a = -Px = -P(L - L\cos\theta)$$

$$V_a = -PL(1 - \cos\theta) \quad \blacktriangleleft$$

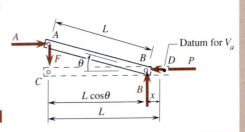

Developmental Exercises

D10.33   Is the applied potential energy defined for a body under the action of nonconservative forces?

D10.34   Define the applied potential energy of a body.

★10.13   Principle of Potential Energy

The *potential energy V* of a body in a conservative system is the sum of the gravitational, elastic, and applied potential energies of the body; i.e.,

$$V = V_g + V_e + V_a \tag{10.24}$$

By its definition, we note that the body in a conservative system receives work only from the conservative forces when it undergoes a displacement. From Eqs. (10.16) through (10.24), we can say that any differential work dU done by the conservative forces to the body, which has a differential displacement, is equal to the negative of the corresponding differential increment dV in the potential energy of the body; i.e.,

$$dU = -dV \tag{10.25}$$

In the case of a virtual displacement, Eq. (10.25) becomes

$$\delta U = -\delta V \tag{10.26}$$

For a single-degree-of-freedom system, whose position can adequately be defined by a single variable θ, we write

$$V = V(\theta) \qquad \delta V = \frac{dV}{d\theta} \delta\theta$$

Since $\delta\theta \neq 0$, the principle of virtual work for the equilibrium of the system leads to $\delta U = -\delta V = 0$ and

$$\frac{dV}{d\theta} = 0 \tag{10.27}$$

Thus, *the derivative of the potential energy of a conservative system in equilibrium with respect to its position variable is zero.*[†] This property, a *corollary* of the principle of virtual work, is referred to as the *principle of potential energy. If the position of a system depends on n independent variables $\theta_1, \theta_2, \ldots, \theta_n$, the system is said to possess n degrees of freedom.* These independent variables are called the *generalized coordinates* of the system. For equilibrium of an *n*-degree-of-freedom conservative system, we have $V = V(\theta_1, \theta_2, \ldots, \theta_n)$ and

$$\delta U = -\delta V = -\left(\frac{\partial V}{\partial \theta_1} \delta\theta_1 + \frac{\partial V}{\partial \theta_2} \delta\theta_2 + \cdots + \frac{\partial V}{\partial \theta_n} \delta\theta_n \right) = 0$$

[†]This means that the potential energy has an *extremum value* when the conservative system is in equilibrium.

Since the virtual displacements $\delta\theta_1$, $\delta\theta_2$, . . . , $\delta\theta_n$ are independent of one another, it is necessary that the partial derivatives of V with respect to each independent variable be equal to zero for equilibrium; i.e.,

$$\frac{\partial V}{\partial\theta_1} = 0 \qquad \frac{\partial V}{\partial\theta_2} = 0 \quad \cdots \quad \frac{\partial V}{\partial\theta_n} = 0 \qquad (10.28)$$

These equations represent the *principle of potential energy* for a conservative system with n degrees of freedom.

EXAMPLE 10.15

The spring shown has a modulus k and a free length L, and each of the two bars has a weight W, where $W < 4kb$. Determine the values of θ for which equilibrium exists. (Cf. Example 10.18 for stability analysis.)

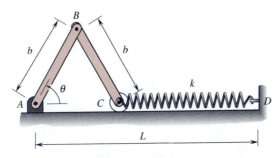

Solution.   The system shown is a conservative system. The conservative forces are the resultant weight forces at the midpoints of the bars and the spring force at the point C as shown. Others are nonworking forces. Since the free length of the spring is L, the contraction of the spring is x as shown. Choosing the line AC and the point A as the reference datums for V_g and V_e of the system, respectively, we write

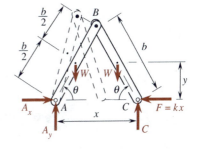

$$y = \frac{b}{2}\sin\theta \qquad V_g = 2Wy = Wb\sin\theta$$

$$x = 2b\cos\theta \qquad V_e = \frac{1}{2}kx^2 = 2kb^2\cos^2\theta$$

$$V = V_g + V_e = Wb\sin\theta + 2kb^2\cos^2\theta$$

By the *principle of potential energy*, we write

$$\frac{dV}{d\theta} = b\cos\theta(W - 4kb\sin\theta) = 0$$

Thus, the positions of equilibrium are given by

$$\cos\theta = 0 \qquad\qquad \theta = \frac{\pi}{2} \;\blacktriangleleft$$

$$W - 4kb\sin\theta = 0 \qquad \theta = \sin^{-1}\frac{W}{4kb} \;\blacktriangleleft$$

REMARK. In the first position, the points A and C coincide. Since $W < 4kb$, the second position does exist.

Developmental Exercise

D10.35 A uniform bent rod ABC has a total weight $3W$ and is hung by a wire at B as shown. (a) Choosing the horizontal line passing through B as the reference datum, express the potential energy V of the bent rod as a function of θ. (b) Determine the value of θ for which the bent rod may be in equilibrium.

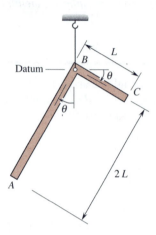

Fig. D10.35

EXAMPLE 10.16

Refer to the system in Example 10.13. Determine the values of θ for which the bar may be in equilibrium if $\mathbf{P} = \frac{1}{2}\, kL$.

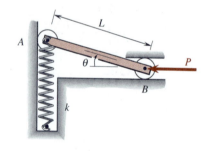

Solution. The system shown is a conservative system. The elastic and applied potential energies of the bar, as established in Examples 10.13 and 10.14, are

$$V_e = \frac{1}{2}\, kL^2 \sin^2\theta \qquad V_a = -PL(1 - \cos\theta)$$

Since $P = \frac{1}{2}\, kL$, we write the potential energy of the bar as follows:

$$V = V_e + V_a = \frac{1}{2}\, kL^2 \sin^2\theta - PL(1 - \cos\theta)$$

$$= \frac{1}{2}\, kL^2(\sin^2\theta + \cos\theta - 1)$$

For equilibrium, we apply the *principle of potential energy* to write

$$\frac{dV}{d\theta} = kL^2 \sin\theta\left(\cos\theta - \frac{1}{2}\right) = 0$$

$$\sin\theta = 0 \qquad\qquad \theta = 0 \ \blacktriangleleft$$

$$\cos\theta - \frac{1}{2} = 0 \qquad\qquad \theta = 60° \ \blacktriangleleft$$

Developmental Exercise

D10.36 Describe the *principle of potential energy*.

★10.14 Stability of Equilibrium[†]

Three types of equilibrium have been discussed in Sec. 5.6. They are *stable equilibrium*, *unstable equilibrium*, and *neutral equilibrium*. We may illustrate them again as shown in Fig. 10.15, where each of the three pin-supported uniform bars has a weight W and a length L. The corresponding potential energies of these bodies are plotted in Fig. 10.16.

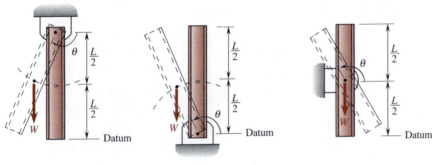

(a) Stable equilibrium
at $\theta = \pi/2$

(b) Unstable equilibrium
at $\theta = \pi/2$

(c) Neutral equilibrium
at $\theta = \pi/2$

Fig. 10.15

We observe from Figs. 10.15 and 10.16 that (a) the potential energy of the bar at its position of *stable* equilibrium is *minimum*, (b) the potential energy of the bar at its position of *unstable* equilibrium is *maximum*, and (c) the potential energy of the bar at its position of *neutral* equilibrium is *constant*. These observed results are quite general for conservative systems because *the conservative force in the force field always tends to do positive work to the system and thus to decrease the potential energy of the system* [cf. Eq. (10.25)].

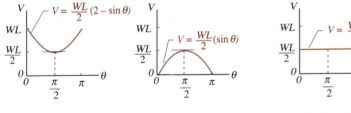

(a) Stable equilibrium
at $\theta = \pi/2$

(b) Unstable equilibrium
at $\theta = \pi/2$

(c) Neutral equilibrium
at $\theta = \pi/2$

Fig. 10.16

[†]This section deals only with the stability of equilibrium of *rigid bodies* subjected to conservative force fields where gravitational, elastic, and applied potential energies of the rigid body are definable. The loss of stability of *deformable bodies* through *buckling* or *flutter* is usually treated in mechanics of materials, structural stability, and theory of vibrations.

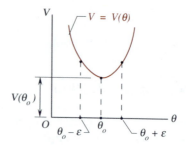

Fig. 10.17

For a single-degree-of-freedom conservative system with potential energy V and generalized coordinate θ, we may employ the conditions under which a function is minimum, maximum, or constant in calculus to summarize the conditions for the equilibrium of the system as follows:

$$\frac{dV}{d\theta} = 0 \quad \text{and} \quad \frac{d^2V}{d\theta^2} > 0: \quad \text{stable equilibrium}$$

$$\frac{dV}{d\theta} = 0 \quad \text{and} \quad \frac{d^2V}{d\theta^2} < 0: \quad \text{unstable equilibrium} \tag{10.29}$$

$$\frac{dV}{d\theta} = \frac{d^2V}{d\theta^2} = \cdots = 0: \quad \text{neutral equilibrium}$$

If we expand the potential energy function $V(\theta)$ into a Taylor's series (cf. Sec. B5.5 of App. B) around an equilibrium position at $\theta = \theta_o$ and transpose the first term, we have

$$V(\theta) - V(\theta_o) = (\theta - \theta_o) V'(\theta_o) + \frac{1}{2!}(\theta - \theta_o)^2 V''(\theta_o)$$

$$+ \frac{1}{3!}(\theta - \theta_o)^3 V'''(\theta_o) + \frac{1}{4!}(\theta - \theta_o)^4 V^{iv}(\theta_o)$$

$$+ \cdots + \frac{1}{n!}(\theta - \theta_o)^n V^{(n)}(\theta_o) \tag{10.30}$$

where the superscripts of V indicate derivatives of V with respect to θ. We see from Eq. (10.30) and Fig. 10.17 that the potential energy is a minimum at $\theta = \theta_o$ if $V(\theta) - V(\theta_o)$ is positive for all values of θ different from θ_o in the interval $\theta_o - \varepsilon \leq \theta \leq \theta_o + \varepsilon$, where ε is a positive number as small as we please so that the series converges rapidly. If $V'(\theta_o) = V''(\theta_o) = 0$ and $V'''(\theta_o) \neq 0$, then $V(\theta) - V(\theta_o)$ as well as $(\theta - \theta_o)^3$ changes its sign as θ increases through θ_o; and $V(\theta_o)$ is neither a minimum nor a maximum. If $V'(\theta_o) = V''(\theta_o) = V'''(\theta_o) = 0$ and $V^{iv}(\theta_o) \neq 0$, then the sign of $V(\theta) - V(\theta_o)$ is dependent on that of $V^{iv}(\theta_o)$. Therefore, *if the first and the second derivatives of V with respect to θ are zero while the first nonzero higher-order derivative of V with respect to θ is of even order and has a positive value, then the value of V at $\theta = \theta_o$ is a minimum and the equilibrium is still stable; otherwise, the equilibrium is unstable.*

For a two-degree-of-freedom conservative system with generalized coordinates θ_1 and θ_2, the potential energy $V(\theta_1, \theta_2)$ is minimum and hence the equilibrium is stable if the following are simultaneously satisfied:[†]

[†]Cf. Sec. 10.15 and the second footnote on p. 447.

$$\frac{\partial V}{\partial \theta_1} = \frac{\partial V}{\partial \theta_2} = 0$$

$$\frac{\partial^2 V}{\partial \theta_1^2} \frac{\partial^2 V}{\partial \theta_2^2} - \left(\frac{\partial^2 V}{\partial \theta_1 \partial \theta_2}\right)^2 > 0 \qquad (10.31)$$

$$\frac{\partial^2 V}{\partial \theta_1^2} > 0 \qquad \text{or} \qquad \frac{\partial^2 V}{\partial \theta_2^2} > 0$$

These criteria correspond to the special case of $n = 2$ in the criteria stated later in Sec. 10.15. Furthermore, note that

$$\frac{\partial^2 V}{\partial \theta_1^2} \quad \text{and} \quad \frac{\partial^2 V}{\partial \theta_2^2}$$

must be of the same sign whenever

$$\frac{\partial^2 V}{\partial \theta_1^2} \frac{\partial^2 V}{\partial \theta_2^2} - \left(\frac{\partial^2 V}{\partial \theta_1 \partial \theta_2}\right)^2 > 0$$

is satisfied.

EXAMPLE 10.17

A thin hemispherical shell of mean radius R and weight W is placed in equilibrium on the top of a fixed semicylindrical drum of radius R as shown. Assuming that slipping does not occur between the shell and the drum, determine the stability of equilibrium of the shell in the position shown.

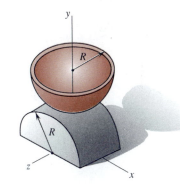

Solution.   Since the shell does not slip on the drum, the arcs BA and BA' are equal and the slightly displaced position of the shell can be defined by the generalized coordinate θ as shown. From Fig. 6.10, we know that the center of gravity G of the shell is at a distance $R/2$ from its center C as shown. Choosing the fixed support as the reference datum, we may write the potential energy V of the shell as

$$V = Wy = W\left(2R\cos\theta - \frac{1}{2}R\cos 2\theta\right)$$

By differentiation, we write

$$\frac{dV}{d\theta} = WR(-2\sin\theta + \sin 2\theta) \qquad \frac{d^2V}{d\theta^2} = WR(-2\cos\theta + 2\cos 2\theta)$$

At $\theta = 0$, which is the position of equilibrium, we find that

$$\frac{dV}{d\theta} = \frac{d^2V}{d\theta^2} = 0$$

Thus, we continue to investigate the values of higher-order derivatives at $\theta = 0$:

$$\left.\frac{d^3V}{d\theta^3}\right|_{\theta=0} = WR(2\sin\theta - 4\sin 2\theta)\Big|_{\theta=0} = 0$$

$$\left.\frac{d^4V}{d\theta^4}\right|_{\theta=0} = WR(2\cos\theta - 8\cos 2\theta)\Big|_{\theta=0} = -6WR < 0$$

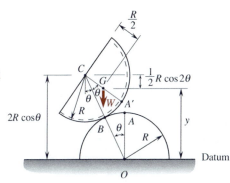

Since the first nonzero higher-order derivative is of even order but has a *negative* value, we conclude that V is maximum.

The equilibrium of the hemispherical shell at $\theta = 0$ is unstable. ◀

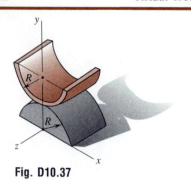

Fig. D10.37

Developmental Exercise

D10.37   A thin semicylindrical shell of mean radius R and weight W is placed in equilibrium on the top of a fixed semicylindrical drum of radius R as shown. Assuming that slipping does not occur between the shell and the drum, determine the stability of equilibrium of the shell in the position shown.

EXAMPLE 10.18

Refer to the system in Example 10.15, where it is found that the system with $W < 4kb$ may be in equilibrium in the positions $\theta = \pi/2$ and $\theta = \sin^{-1} W/(4kb)$. Assume that the bars cannot rotate as a single rigid body about A when A and C coincide at $\theta = \pi/2$. Determine the stability of the system in these two positions.

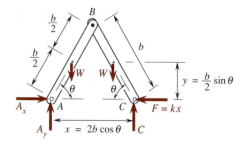

Solution.   The potential energy of the bars, as established in Example 10.15, is

$$V = Wb \sin\theta + 2kb^2 \cos^2\theta$$

By differentiation, we write

$$\frac{dV}{d\theta} = Wb \cos\theta - 4kb^2 \cos\theta \sin\theta$$

$$\frac{d^2V}{d\theta^2} = -Wb \sin\theta - 4kb^2(- \sin^2\theta + \cos^2\theta)$$

Noting that $\cos^2\theta = 1 - \sin^2\theta$, we write

$$\frac{d^2V}{d\theta^2} = 8kb^2 \sin^2\theta - Wb \sin\theta - 4kb^2$$

(a) *Stability in the position $\theta = \pi/2$.* We have $\dfrac{dV}{d\theta} = 0$. Since $\sin\theta = 1$ and $W < 4kb$, we write

$$\frac{d^2V}{d\theta^2} = 8kb^2 - Wb - 4kb^2 = 4kb^2 - Wb = b(4kb - W) > 0$$

The equilibrium of the system at $\theta = \dfrac{\pi}{2}$ is stable.   ◀

(b) *Stability in the position* $\theta = \sin^{-1} \dfrac{W}{4kb}$. We have $\dfrac{dV}{d\theta} = 0$. Since $\sin\theta = \dfrac{W}{4kb}$ and $W < 4kb$, we write

$$\frac{d^2V}{d\theta^2} = 8kb^2\left(\frac{W}{4kb}\right)^2 - Wb\left(\frac{W}{4kb}\right) - 4kb^2 = \frac{W^2}{4k} - 4kb^2$$

$$= \frac{1}{4k}(W + 4kb)(W - 4kb) < 0$$

The equilibrium of the system at $\theta = \sin^{-1}\dfrac{W}{4kb}$ is unstable. ◀

Developmental Exercise

D10.38 Refer to the system in Example 10.15. Determine the stability of equilibrium of the system in the position $\theta = \pi/2$ if $W = 4kb$.

EXAMPLE 10.19

Refer to the system in Example 10.13. Determine the range of the magnitudes of **P** for which the equilibrium of the system is stable in the position $\theta = 0$.

Solution. The potential energy of the bar, as established in Example 10.16, is

$$V = \frac{1}{2} kL^2 \sin^2\theta - PL(1 - \cos\theta)$$

By differentiation, we write

$$\frac{dV}{d\theta} = kL^2 \sin\theta \cos\theta - PL \sin\theta$$

$$\frac{d^2V}{d\theta^2} = kL^2(\cos^2\theta - \sin^2\theta) - PL \cos\theta$$

We note that $\dfrac{dV}{d\theta} = 0$ at $\theta = 0$. For *stable* equilibrium in the position $\theta = 0$, we write

$$\left.\frac{d^2V}{d\theta^2}\right|_{\theta = 0} = kL^2 - PL > 0 \qquad P < kL \quad ◀$$

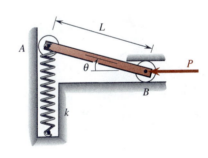

Developmental Exercise

D10.39 Refer to the system studied in Examples 10.13 and 10.19. Note that $d^2V/d\theta^2 = 0$ if $P = kL$. Determine the stability of equilibrium of the system in the position $\theta = 0$ if $P = kL$.

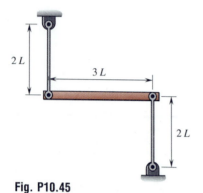

Fig. P10.45

PROBLEMS

10.39* through 10.44* Using the *principle of potential energy*, solve the following problems:

10.39*	Prob. 10.24.	**10.40***	Prob. 10.25
10.41*	Prob. 10.26*.	**10.42***	Prob. 10.33.
10.43*	Prob. 10.34*.	**10.44***	Prob. 10.35.

10.45 through 10.47* Determine the stability of equilibrium of the uniform bar *AB* of weight *W* supported by two vertical light links as shown.

Fig. P10.46   Fig. P10.47*

10.48 through 10.51* The homogeneous body *A* is placed in equilibrium on the top of the body *B* as shown. Assuming that the radii *a* and *b* are equal and slipping does not occur between these two bodies, determine the stability of equilibrium of the body *A* in the position shown.

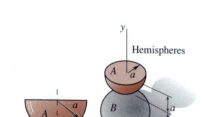

Hemispheres

Fig. P10.48 and P10.52*

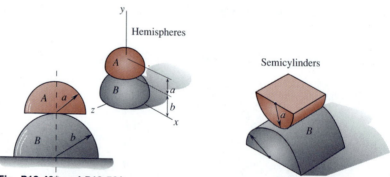

Hemispheres

Fig. P10.49* and P10.53*

Semicylinders

Fig. P10.50 and P10.54

Semicylinders

Fig. P10.51* and P10.55*

10.52* through 10.55* The homogeneous body *A* is placed in equilibrium on the top of the body *B* as shown. Assuming that slipping does not occur between these two bodies, determine the range of values of the ratio *a/b* of the radii for which this equilibrium of the body *A* is stable.

10.56 A homogeneous composite body is composed of a hemisphere of radius *b* an a cylinder of height *a* as shown. Determine the range of values of the ratio *a/b* for which the equilibrium of the body is stable in its upright position.

10.57 A container is composed of a hemispherical shell of radius *b* and a cylindrical shell of height *a*, both made from the same sheet metal, as shown. Determine the range of values of the ratio *a/b* for which the equilibrium of the body is stable in its upright position.

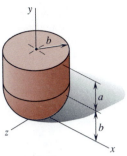

Fig. P10.56

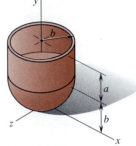

Fig. P10.57

10.58   A torsional spring with modulus $K = 100$ N·m/rad is installed at the base of an inverted pendulum which carries a mass $m = 10$ kg at a height h as shown. It is known that the pendulum in its upright position as shown is in equilibrium. Determine the range of values of h for which this equilibrium is stable.

10.59*   A spring with modulus k is used to stabilize a bent bar which is under the action of a horizontal force **P** as shown. It is known that the bent bar is in equilibrium in the position shown. Determine the range of magnitudes of **P** for which this equilibrium is stable.

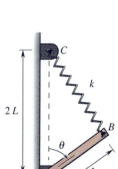

Fig. P10.58

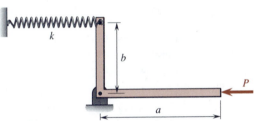

Fig. P10.59*

10.60   Determine the range of values of the modulus k of the spring for which the equilibrium of the platform AB of mass m as shown is stable.

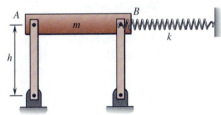

Fig. P10.60

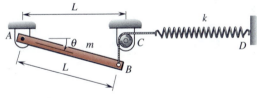

Fig. P10.61

10.61   A uniform bar of length L and mass m is supported as shown. The spring with modulus k is neutral when $\theta = 0$. Determine the angle θ for equilibrium and prove that this equilibrium is stable.

10.62   A uniform bar of length L and weight W is supported as shown. The spring with modulus k is neutral when $\theta = 0$. Determine all possible positions of equilibrium of the bar and their stability if $W = 2kL$.

10.63*   Solve Prob. 10.62 if $W = 3kL$.

10.64   Solve Prob. 10.62 if $W = \frac{8}{3}kL$.

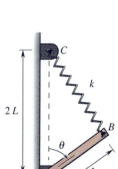

Fig. P10.62

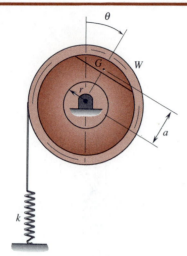

Fig. P10.65

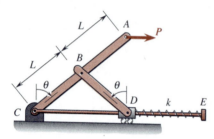

Fig. P10.69

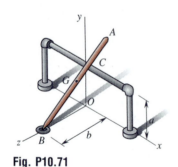

Fig. P10.71

10.65 An eccentric weight W is carried at G by a wheel which is held by a spring with modulus k wrapping around its drum of radius r as shown. The spring is neutral when $\theta = 0$. For $r = 6$ in., $a = 10$ in., $k = 2$ lb/in., and $W = 10$ lb, determine the positions of equilibrium and their stability.

10.66 A body of mass m is hung at the end A of the mechanism shown. The rod CE can freely slide through the collar pivoted at D. The spring with modulus k is compressed by the stopper at E and the collar at D and is neutral when $\theta = 0$. It is known that $k = 98.1$ N/m and $L = 800$ mm. Determine the value of m for which the mechanism will be in *neutral equilibrium* for any value of θ.

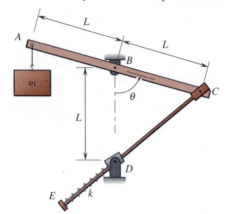

Fig. P10.66

10.67* Refer to the system in Prob. 10.66. Determine the range of values of m for which the equilibrium of the mechanism is stable in the position $\theta = 0$.

10.68 Refer to the system in Prob. 10.66. Determine the range of values of m for which the equilibrium of the mechanism is stable in the position $\theta = 180°$.

10.69 A horizontal force $\mathbf{P}$ is applied to the mechanism shown. The rod CE can freely slide through the collar D which rests on rollers. The spring with modulus k is compressed by the stopper at E and the collar D and is neutral when $\theta = 0$. Determine the positions of equilibrium of the mechanism and their stability if $P = kL$.

10.70* Refer to the system in Prob. 10.69. Determine the range of magnitudes of $\mathbf{P}$ for which the equilibrium of the mechanism is stable in the position $\theta = 90°$.

10.71 A uniform slender rod AB of length L and mass m is supported by a ball and socket at B and rests on a smooth horizontal member of a frame as shown. Determine the stability of equilibrium of the rod AB in the position shown.

10.15 Concluding Remarks

The concept of *work* is both important and subtle in mechanics. It is the *force* (not the *body*) which does work. A body *receives* work from a force only when the particle (of the body) acted on by the force has a displacement during the action of the force. In other words, if a force *slides* on a *stationary* body, no work is done by the force to the stationary body. This is discussed in Sec. 10.2 and emphasized in Example 10.1 and D10.6 through D10.8, which involve a variety of situations.

As an extension, we may here employ the concept of virtual work in this chapter to alternatively define the concept of *equivalent systems of forces*, presented in Sec. 5.3, as follows: *If each of any two systems of forces will do the same amount of total virtual work to a body for the same given compatible virtual displacement of the body, then those two systems of forces are equivalent; otherwise, they are either equipollent or nonequivalent.* This alternative definition is referred to as the *energy criterion* for equivalent systems of forces.[†] We see that, when the compatible virtual displacement is a linear virtual displacement of the entire body, the energy criterion will yield the condition that the resultant force of each of the two equivalent systems of forces must be the same (i.e., $\mathbf{R} = \mathbf{R}'$). We also see that, when the compatible virtual displacement is an angular virtual displacement of the entire body about a given point, the energy criterion will yield the condition that the resultant moment of each of the two equivalent systems of forces about a given point must be the same [i.e., $\mathbf{M}_P^R = (\mathbf{M}_P^R)'$]. Thus, the conditions of equivalent systems of forces as defined in Eqs. (5.4) are fully covered by the *energy criterion*. Moreover, the energy criterion will clearly lead to the *principle of virtual work*, as presented in Sec. 10.7, when *the force system* in the free-body diagram of a body is *equivalent to a zero wrench*.

The *potential energy* of a body is a vital concept in mechanics. The quantitative definition for the potential energy of a body as given in Sec. 10.9 is a working definition devised to enable one to quickly acquire a firm grasp of that concept. Note that the potential energy is not defined for a body in a nonconservative force field. The *principle of potential energy* for conservative systems is a direct consequence of the *principle of virtual work*; the former is a corollary of the latter.

The criteria for the stability of equilibrium of a conservative system are expressed in terms of the potential energy of the system. Such criteria are discussed and summarized in Sec. 10.14 for single- and two-degree-of-freedom conservative systems. For an *n*-degree-of-freedom conservative system with generalized coordinates $\theta_1, \theta_2, \ldots, \theta_n$, it can be shown that the potential energy $V(\theta_1, \theta_2, \ldots, \theta_n)$ is minimum and hence the equilibrium is stable if the following criteria, which consist of *n* equations and *n* inequalities, are simultaneously satisfied:

$$\frac{\partial V}{\partial \theta_1} = \frac{\partial V}{\partial \theta_2} = \cdots = \frac{\partial V}{\partial \theta_n} = 0 \tag{10.32}$$

$$\begin{vmatrix} \dfrac{\partial^2 V}{\partial \theta_1^2} & \dfrac{\partial^2 V}{\partial \theta_1 \partial \theta_2} & \cdots & \dfrac{\partial^2 V}{\partial \theta_1 \partial \theta_i} \\[2mm] \dfrac{\partial^2 V}{\partial \theta_2 \partial \theta_1} & \dfrac{\partial^2 V}{\partial \theta_2^2} & \cdots & \dfrac{\partial^2 V}{\partial \theta_2 \partial \theta_i} \\[2mm] \cdots & \cdots & \cdots & \cdots \\[2mm] \dfrac{\partial^2 V}{\partial \theta_i \partial \theta_1} & \dfrac{\partial^2 V}{\partial \theta_i \partial \theta_2} & \cdots & \dfrac{\partial^2 V}{\partial \theta_i^2} \end{vmatrix} > 0 \tag{10.33}$$

where the left-hand side of the inequality sign is a determinant of the *i*th order and $i = 1, 2, \ldots, n$ to yield the *n* inequalities.[††] It is obvious that

---

[†] In this alternative definition, the body may be a particle, a rigid body, or a set of connected particles or rigid bodies.

[††] For a detailed treatment, see G. J. Simitses, *An Introduction to the Elastic Stability of Structures* (Englewood Cliffs, N.J.: Prentice-Hall, 1976).

the stability criteria become increasingly complex for systems with three or more degrees of freedom. The stability of equilibrium is a measure of the quality of equilibrium; hence, the stability analysis of equilibrium represents an in-depth analysis of equilibrium. Thus, this is logically the last chapter in the study of statics.

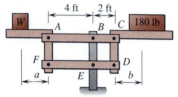

Fig. P10.72

REVIEW PROBLEMS

10.72 Determine the weight W which balances the 180-lb load as shown.[†] Does the value of W depend on the values of a and b?

10.73 and 10.74 Determine the reactions from the support at A of the structure shown.

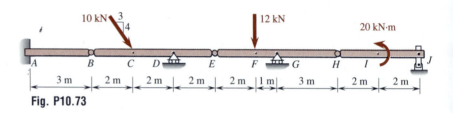

Fig. P10.73

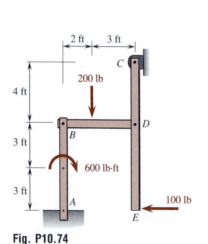

Fig. P10.74

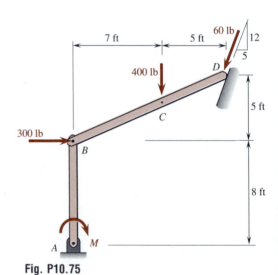

Fig. P10.75

[†]This type of balance is also known as a *Roberval balance*, after the French mathematician Gilles Personne De Roberval (1602–1675).

10.75   Determine the moment **M** for which the linkage as shown is in equilibrium.

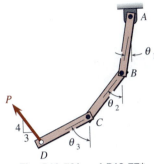

Fig. P10.76* and P10.77*

10.76*   Uniform rods, each having a weight W and a length L, are connected as shown. For equilibrium, determine the angle θ_2.

10.77*   Uniform rods, each having a weight W and a length L, are connected as shown. For equilibrium, determine the angle θ_1.

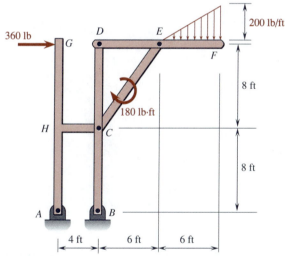

Fig. P10.78*

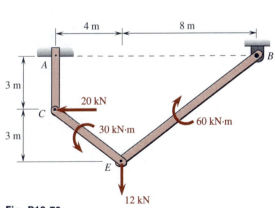

Fig. P10.79

10.78* and 10.79   Determine the reactions from the supports at A and B of the frame shown.

10.80   A bar of negligible weight is supported and loaded as shown. The free length of the spring with modulus k is L. Determine the range of magnitudes of the constant force **P** for which the equilibrium of the bar is stable in the position shown.

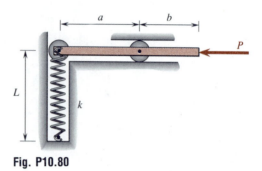

Fig. P10.80

10.81   A cylinder of radius r and weight Q is suspended from the ceiling to *stabilize* the equilibrium of a uniform slender rod of length L and weight W which is hinged at its base and leans against a vertical wall as shown. Determine (a) the value of the derivative $d\theta/d\phi$ as evaluated in the position $\theta = \phi = 0$, (b) the range of values of Q required.

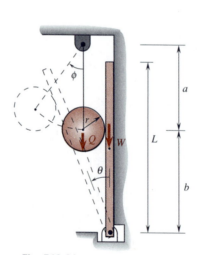

Fig. P10.81

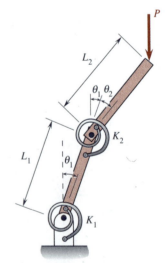

Fig. P10.82

10.82   A double pendulum lies in a smooth *horizontal* plane as shown. The two bars are hinged with torsional springs at the support and the joint. The torsional springs have moduli K_1 and K_2 as indicated. The system is in equilibrium when $\theta_1 = \theta_2 = 0$. For $L_1 = 2L$, $L_2 = L$, and $K_1 = K_2 = K$, determine the range of magnitudes of the horizontal force **P** for which the equilibrium position is stable.

10.83*   Solve Prob. 10.82 if $L_1 = L_2 = L$, $K_1 = 2K$, and $K_2 = K$.

10.84*   Using the *principle of virtual work*, solve Prob. 3.55.

10.85*   Using the *principle of virtual work*, solve Prob. 3.56.

Kinematics of Particles

Chapter 11

KEY CONCEPTS

- Rectilinear motion, curvilinear motion, absolute motion, relative motion, kinematic quantity, velocity, speed, acceleration, deceleration, jerk, and initial conditions.
- Rectilinear motion of a particle whose acceleration is a function of (1) time, (2) position, (3) velocity, or (4) a constant.
- Dependent rectilinear motions of particles.
- Graphical solution of rectilinear-motion problems.
- Descriptions of the curvilinear motion of a particle using (1) rectangular components, (2) tangential and normal components, (3) radial and transverse components, or (4) cylindrical components.

Dynamics is a subject dealing with conditions of bodies under the action of unbalanced force systems. The phase of dynamics which deals with the study of the motion of bodies without considering the cause of motion is called *kinematics*. The phase of dynamics which involves the study of the relation between the motion of a body and the force system causing the motion is called *kinetics*. Nevertheless, kinetics usually contains some kinematics. A kinematic quantity is a measurable quantity in kinematics, such as time, displacement, velocity, acceleration, and jerk. The concepts and methods of computation of various kinematic quantities of particles, which move along straight lines or along curved paths, are presented in this chapter. These concepts and methods form a useful background which will facilitate the study of kinetics in the subsequent chapters.

11.1 Types of Motion of a Particle[†]

A particle is said to be in *motion* if it does not remain in a fixed position during any finite interval of time. The motion of a particle may generally be classified according to its path. A *rectilinear motion* is a motion along a straight-line path. A *curvilinear motion* is a motion along a curved path. When a particle moves in a plane, it is said to have a *plane motion*; otherwise, its motion is a *motion in three dimensions*.

[†]A *particle* is an *idealized body* whose size and shape can be disregarded without introducing appreciable errors in the description and prediction of its state of rest or motion (cf. Sec. 1.6).

A motion may also be classified according to the fixity of the reference frame.[†] An *absolute motion* is a motion in a fixed reference frame, while a *relative motion* is a motion in a moving reference frame. In kinematics, any reference frame may be designated as *fixed*; all others not rigidly attached to it are regarded as *moving*.

Developmental Exercise

D11.1[††] Define the terms: (a) rectilinear motion, (b) curvilinear motion, (c) absolute motion, (d) relative motion.

RECTILINEAR MOTION

11.2 Kinematic Quantities

A *kinematic quantity* is a measurable quantity in kinematics, which can be defined without involving force or mass. Examples of kinematic quantities are position, distance, displacement, total distance traveled, time, speed, velocity, acceleration, deceleration, and jerk. We note that some of the kinematic quantities (e.g., displacement, velocity, and acceleration) do possess magnitudes and directions and are vector quantities (cf. Sec. 1.2).

In rectilinear motion of a particle, the *orientations* of the directions of those vector kinematic quantities must, however, *lie along the given straight-line path*. In such a motion, only the *magnitude* and the *sense* of each vector kinematic quantity are the nontrivial information to be described or predicted. Because of this special feature, the kinematics of particles in rectilinear motion can be studied using essentially *scalar analysis* rather than vector analysis.

Developmental Exercises

D11.2   What is a kinematic quantity? Give three examples.

D11.3   Why can the kinematics of particles in rectilinear motion be studied using essentially scalar analysis?

11.3 Position, Displacement, and Velocity

The *position* of a particle is the location of the particle described by using a coordinate system. To facilitate the study of the rectilinear motion of a particle, we usually lay the *x* axis along the straight-line path of the particle

---

[†]A *reference frame* is a frame (or body) of reference in which one or more coordinate systems may be embedded. For conciseness, the phrase "in a reference frame" is, from time to time, used to mean "with respect to a reference frame." An *inertial reference frame* is a frame of reference in which Newton's laws of motion are valid or highly accurate (cf. Sec. 3.1).

[††]The developmental exercises (D's) may be *utilized* or *skipped* as explained in the Preface. Whenever a developmental exercise asks for the definition of a term or description of a law or principle, students are encouraged to answer in their *own words*, which reflect their understanding of the concept involved. Answers to selected developmental exercises are given at the end of the text.

and choose a convenient point O on the path as the fixed origin, as shown in Fig. 11.1. At the time t, the particle is situated at the point P whose position is given by the *coordinate* x, where $x = x(t)$. Note that the magnitude of x indicates the *distance* between the origin O and the particle at P. The value of x is positive if P is reached from O by moving along the axis in the positive direction; otherwise, it is negative. Since the coordinate x completely defines the position of the particle in rectilinear motion, it is referred to as the *position* (or the *position coordinate*) of the particle. The position x has the dimension *length* and is usually measured in *meters* (m) in the SI and in *feet* (ft) in the U.S. customary system. The time t is usually measured in *seconds* (s), both in the SI and in the U.S. customary system.

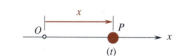

Fig. 11.1

The *displacement* of a particle is the change of position of the particle. Suppose that, over the time interval Δt, the particle in rectilinear motion travels from the point P to the point P' as shown in Fig. 11.2. In terms of the positions x and x' of the points P and P', the *displacement*, denoted by Δx, of the particle from P to P' is given by

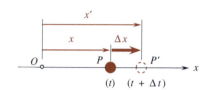

Fig. 11.2

$$\Delta x = x' - x \qquad (11.1)$$

The displacement Δx is positive if $\overrightarrow{PP'}$ points in the positive direction of the x axis; otherwise, Δx is negative. The displacement of a particle has both magnitude and direction and is a vector. The *total distance traveled* x_T by a particle is the total length of path traversed by the particle; it is always a positive scalar quantity. Clearly, Δx and x_T have the same dimension and units as those of x.

The *velocity* of a particle is defined as the time rate of change of position (or the time rate of displacement) of the particle. The *average velocity* v_{avg} of a particle having a displacement Δx over a time interval Δt is given by

$$v_{\text{avg}} = \frac{\Delta x}{\Delta t} \qquad (11.2)$$

As Δt becomes smaller and approaches zero in the limit, v_{avg} becomes the *instantaneous velocity* v of the particle; i.e.,

$$v = \lim_{\Delta t \to 0} \frac{\Delta x}{\Delta t} \qquad v = \frac{dx}{dt} = \dot{x} \qquad (11.3)^{\dagger}$$

The instantaneous velocity is what we usually call the *velocity*.

A positive value of the velocity v indicates that the value of the position x is increasing algebraically and the particle is moving in the positive direction of the x axis as shown in Fig. 11.3. On the other hand, a negative value of v indicates that x is decreasing algebraically and the particle is moving in the negative direction as shown in Fig. 11.4. The velocity of a particle has both magnitude and direction and is a vector. The magnitude of the velocity is known as the *speed*. If the *total distance traveled* by a particle over the time interval Δt is x_T, the *average speed* $|v|_{\text{avg}}$ of the particle is defined as

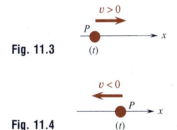

Fig. 11.3

Fig. 11.4

$$|v|_{\text{avg}} = \frac{x_T}{\Delta t} \qquad (11.4)$$

---

†A quantity with n dots over it represents the nth derivative of the quantity with respect to the time variable t.

The velocity and the speed have the dimensions *length/time*. Their primary units are m/s in the SI and ft/s in the U.S. customary system.

EXAMPLE 11.1

The position x of a particle in rectilinear motion is related to the time t by the equation $x = t^3 - 6t^2 - 36t + 116$, where x is measured in meters and t in seconds. Over the time interval $0 \le t \le 8$ s, determine Δx, x_T, v_{avg}, $|v|_{avg}$, and the velocity v_7 at $t = 7$ s.

Solution. We compute the displacement Δx of the particle by subtracting its initial position x_0 at $t = 0$ from its final position x_8 at $t = 8$ s as follows:

$$x_0 = x|_{t=0} = 0 - 0 - 0 + 116 = 116$$

$$x_8 = x|_{t=8} = (8)^3 - 6(8)^2 - 36(8) + 116 = -44$$

$$\Delta x = x_8 - x_0 \qquad\qquad \Delta x = -160 \text{ m} \quad \blacktriangleleft$$

The negative sign in the value of Δx indicates that the displacement points in the negative direction during the interval as shown.[†]

To find the total distance traveled x_T, we need to first find out whether the particle has changed its direction of motion over the interval $0 \le t \le 8$ s. The velocity v of the particle is given by

$$v = \dot{x} = 3t^2 - 12t - 36 = 3(t - 6)(t + 2)$$

From the sketch, we note that $v < 0$ (the particle moves in the negative direction) for $0 \le t < 6$ s, $v = 0$ (the particle stops) at $t = 6$ s, and $v > 0$ (the particle moves in the positive direction) for 6 s $< t \le 8$ s. From $t = 0$ to $t = 6$ s, the distance traveled is given by

$$\Delta x_{0 \to 6} = x_6 - x_0 = x|_{t=6} - x|_{t=0} = -100 - 116$$

$$\Delta x_{0 \to 6} = -216 \text{ m} = 216 \text{ m in the negative direction}$$

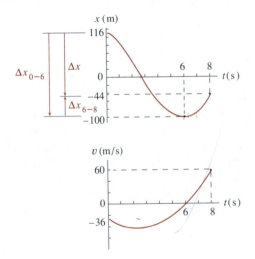

[†]Note that the *digital root finder* in App. E may also be utilized to plot the approximate x-t and v-t curves as explained in Sec. 11.16.

From $t = 6$ s to $t = 8$ s, the distance traveled is given by

$$\Delta x_{6 \to 8} = x_8 - x_6 = x|_{t=8} - x|_{t=6} = -44 - (-100)$$

$$\Delta x_{6 \to 8} = 56 \text{ m in the positive direction}$$

Therefore, the total distance traveled from $t = 0$ to $t = 8$ s is

$$x_T = |\Delta x_{0 \to 6}| + |\Delta x_{6 \to 8}| = 216 + 56 \qquad x_T = 272 \text{ m} \blacktriangleleft$$

From $t = 0$ to $t = 8$ s, we have $\Delta t = 8$ s. To compute v_{avg}, we write

$$v_{\text{avg}} = \frac{\Delta x}{\Delta t} = \frac{-160}{8} \qquad v_{\text{avg}} = -20 \text{ m/s} \blacktriangleleft$$

The negative sign in the value of v_{avg} indicates that the average velocity points in the negative direction.

To find the average speed of the particle from $t = 0$ to $t = 8$ s, we write

$$|v|_{\text{avg}} = \frac{x_T}{\Delta t} = \frac{272}{8} \qquad |v|_{\text{avg}} = 34 \text{ m/s} \blacktriangleleft$$

It should be noted that, in general, $|v_{\text{avg}}| \leq |v|_{\text{avg}}$. For the velocity of the particle at $t = 7$ s, we write

$$v_7 = \dot{x}|_{t=7} = 3(7)^2 - 12(7) - 36 \qquad v_7 = 27 \text{ m/s} \blacktriangleleft$$

Developmental Exercises

D11.4 Define for a particle the terms: (a) position, (b) displacement, (c) total distance traveled, (d) velocity, (e) average velocity, (f) speed, (g) average speed.

D11.5 The rectilinear motion of a particle is given by the relation $x = -t^3 + 3t^2 + 24t - 50$, where x is measured in feet and t in seconds. Determine (a) the velocity v_6 of the particle at $t = 6$ s, (b) the time τ at which the velocity will be zero, (c) the position x_τ at $t = \tau$, (d) the total distance traveled x_T from $t = 0$ to $t = 6$ s.

D11.6 A particle moves from the origin O first to the point P and then back to the point Q as shown during a 5-s time interval. Over this time interval, determine Δx, x_T, v_{avg}, and $|v|_{\text{avg}}$.

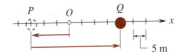

Fig. D11.6

11.4 Acceleration and Jerk

The *acceleration* of a particle is defined as the time rate of change of velocity of the particle. Suppose that, over the time interval Δt, a particle moves from the point P to the point P' while its velocity changes from v to $v + \Delta v$ as shown in Fig. 11.5. The *average acceleration* a_{avg} of the particle over Δt is given by

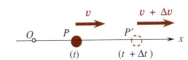

Fig. 11.5

$$a_{\text{avg}} = \frac{\Delta v}{\Delta t} \qquad (11.5)$$

As Δt becomes smaller and approaches zero in the limit, a_{avg} becomes the *instantaneous acceleration* a of the particle; i.e.,

$$a = \lim_{\Delta t \to 0} \frac{\Delta v}{\Delta t} \qquad \boxed{a = \frac{dv}{dt} = \dot{v}} \qquad (11.6)$$

The instantaneous acceleration is what we usually call the *acceleration*. Substituting Eq. (11.3) into Eq. (11.6), we write

$$a = \frac{d}{dt}\left(\frac{dx}{dt}\right) \qquad \boxed{a = \frac{d^2x}{dt^2} = \ddot{x}} \qquad (11.7)$$

$$a = \frac{dv}{dt} = \frac{dv}{dx}\frac{dx}{dt} = \frac{dv}{dx}v \qquad \boxed{a = v\frac{dv}{dx}} \qquad (11.8)^{\dagger}$$

Note from calculus that when $a = \dot{v} > 0$, the value of v must be increasing algebraically; and it means physically that the particle is either moving faster and faster in the positive direction or moving slower and slower in the negative direction. Furthermore, when $a = \dot{v} < 0$, the value of v must be decreasing algebraically; and it means physically that the particle is either moving slower and slower in the positive direction or moving faster and faster in the negative direction. The term *deceleration* is an alias of acceleration when the speed is decreasing.[††] The acceleration of a particle has both magnitude and direction and is a vector. Moreover, the acceleration has the dimensions *length/time²*, and its primary units are m/s² in the SI and ft/s² in the U.S. customary system.

The time rate of change of acceleration is referred to as the *jerk*. However, no standard names are given to the second- and higher-order time derivatives of acceleration. The jerk of a vehicle is often used as a measure of discomfort to be experienced by the passengers.

EXAMPLE 11.2

Refer to Example 11.1. Determine (a) the accelerations a_0 and a_7 at $t = 0$ and $t = 7$ s, respectively; (b) a_{avg} over the time interval $0 \le t \le 7$ s.

Solution.   The position of the particle in Example 11.1 is

$$x = t^3 - 6t^2 - 36t + 116$$

Thus, we write

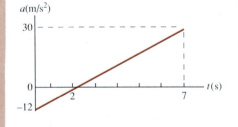

$$v = \dot{x} = 3t^2 - 12t - 36 \qquad a = \dot{v} = 6t - 12$$

$$a_0 = a|_{t=0} = -12 \qquad\qquad a_0 = -12 \text{ m/s}^2 \blacktriangleleft$$

$$a_7 = a|_{t=7} = 30 \qquad\qquad a_7 = 30 \text{ m/s}^2 \blacktriangleleft$$

To find a_{avg}, we first find Δv and then divide it by Δt as follows:

[†]It can happen that $dv/dx \to \infty$ as $v \to 0$ (e.g., the instant a body is dropped from rest). Thus, it is *false* that $a = 0$ whenever $v = 0$.

[††]Note that *deceleration* of a body is not equal to the negative of the acceleration of the body, just, in a way, as "debar" is not the opposite of "bar," and "inflammable" is not the opposite of "flammable." The deceleration of a body is simply the acceleration of the body whose velocity is opposite to the acceleration at the time considered.

$$v_0 = v|_{t=0} = -36 \qquad v_7 = v|_{t=7} = 27$$

$$\Delta v = v_7 - v_0 = 63 \qquad \Delta t = 7 - 0 = 7$$

$$a_{\text{avg}} = \frac{\Delta v}{\Delta t} = \frac{63}{7} \qquad\qquad a_{\text{avg}} = 9 \text{ m/s}^2 \blacktriangleleft$$

As a check, note that $a_{\text{avg}} = (a_0 + a_7)/2 = 9$ m/s^2.

Developmental Exercise

D11.7   Refer to D11.5. Determine (a) a_0 and a_6 at $t = 0$ and $t = 6$ s, respectively; (b) a_{avg} over the time interval $0 \le t \le 6$ s.

EXAMPLE 11.3

The velocity v of a recoiling piston in a cylinder filled with oil, as shown, is given by the relation $v = v_0 - bx$, where x is the position of the piston and v_0 and b are positive constants. Determine the acceleration a of the piston at $x = v_0/4b$.

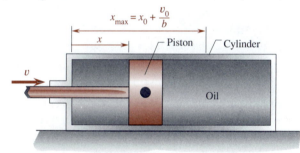

Solution.   Since v is given as a function of x, we write

$$a = v\frac{dv}{dx} = (v_0 - bx)\frac{d}{dx}(v_0 - bx) = -b(v_0 - bx)$$

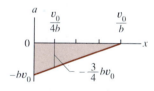

At $x = v_0/(4b)$, we obtain

$$a = -\frac{3}{4}bv_0 \blacktriangleleft$$

Developmental Exercise

D11.8   Define the *acceleration*, *deceleration*, and *jerk* of a particle.

11.5   Determination of Rectilinear Motion

Problems of the rectilinear motion of a particle may be classified according to the type of acceleration of the particle. In each problem, the conditions that $x = x_0$ and $v = v_0$ at the time $t = 0$ are called the *initial conditions*.

An acceleration may be a function of time, position, velocity, or simply a constant; it may also be a function of several variables.

(1) *Acceleration is a given function of time, $a = f(t)$.*
For this class of problems, we first use Eq. (11.6) to write

$$dv = a\,dt = f(t)\,dt \qquad \int_{v_0}^{v} dv = \int_{0}^{t} f(t)\,dt$$

$$v = v_0 + \int_{0}^{t} f(t)\,dt \qquad (11.9)$$

which expresses v as a function of t. Then, we use Eq. (11.3) to write

$$dx = v\,dt \qquad \int_{x_0}^{x} dx = \int_{0}^{t} v\,dt$$

$$x = x_0 + \int_{0}^{t} v\,dt \qquad (11.10)$$

Developmental Exercises

D11.9 The acceleration of a particle moving along a straight line is $a = A\cos\omega t$, where A and ω are constants. The initial conditions are $v = v_0$ and $x = x_0$. Express the velocity v and the position x as functions of the time t.

D11.10 Do D11.9 if $a = A\sin\omega t$ and $v = x_0 = 0$.

(2) *Acceleration is a given function of position, $a = f(x)$.*
For this class of problems, we first use Eq. (11.8) to write

$$v\,dv = a\,dx = f(x)\,dx \qquad \int_{v_0}^{v} v\,dv = \int_{x_0}^{x} f(x)\,dx$$

$$v^2 = v_0^2 + 2\int_{x_0}^{x} f(x)\,dx \qquad (11.11)$$

which yields v in terms of x. Then, we use Eq. (11.3) to write

$$dt = \frac{dx}{v} \qquad \int_{0}^{t} dt = \int_{x_0}^{x} \frac{dx}{v}$$

$$t = \int_{x_0}^{x} \frac{dx}{v} \qquad (11.12)$$

which provides the relation between x and t.

EXAMPLE 11.4

The acceleration of an oscillating weight W, as shown, is $a = -\omega^2 x$, where ω is a constant. If the initial conditions are $x_0 \neq 0$ and $v_0 = 0$, express the position x as a function of the time t.

Solution. The oscillation of the weight W is a rectilinear motion whose acceleration a is given as a function of the position x. Noting that $a = f(x) = -\omega^2 x$, $x_0 \neq 0$, and $v_0 = 0$, we utilize the derived results in Eqs. (11.11) and (11.12) to write

$$v^2 = 0 + 2\int_{x_0}^{x} (-\omega^2 x)\, dx = \omega^2 (x_0^2 - x^2)$$

$$t = \int_{x_0}^{x} \frac{dx}{v} = \frac{1}{\pm\omega} \int_{x_0}^{x} \frac{dx}{(x_0^2 - x^2)^{1/2}}$$

The last integral can be evaluated by letting $x = x_0 \sin\theta$ and changing the corresponding limits of integration. We write

$$\pm\omega t = \int_{x_0}^{x} \frac{dx}{(x_0^2 - x^2)^{1/2}} = \int_{\pi/2}^{\theta} \frac{x_0 \cos\theta\, d\theta}{x_0 \cos\theta} = \theta - \frac{\pi}{2} = \sin^{-1}\frac{x}{x_0} - \frac{\pi}{2}$$

$$\frac{x}{x_0} = \sin\left(\pm\omega t + \frac{\pi}{2}\right) = \cos(\pm\omega t) = \cos\omega t$$

$$x = x_0 \cos \omega t \quad ◄$$

This result shows that the motion of the weight W is sinusoidal. This type of motion is known as a *simple harmonic motion*.

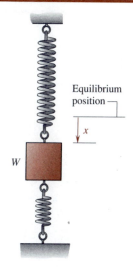

Equilibrium position

x

W

Developmental Exercises

D11.11 Refer to Example 11.4. Show that the solution $x = x_0 \cos \omega t$ satisfies the initial conditions and yields the acceleration as given.

D11.12 Work Example 11.4 if $x_0 = 0$ and $v_0 \neq 0$.

(3) *Acceleration is a given function of velocity, $a = f(v)$.*
For this class of problems, we first use Eq. (11.6) to write

$$dt = \frac{dv}{a} = \frac{dv}{f(v)} \qquad \int_0^t dt = \int_{v_0}^{v} \frac{dv}{f(v)}$$

$$t = \int_{v_0}^{v} \frac{dv}{f(v)} \tag{11.13}$$

which expresses t as a function of v. The position x can be related to the velocity v by integrating Eq. (11.8) as follows:

$$dx = \frac{v\, dv}{a} = \frac{v\, dv}{f(v)} \qquad \int_{x_0}^{x} dx = \int_{v_0}^{v} \frac{v\, dv}{f(v)}$$

$$x = x_0 + \int_{v_0}^{v} \frac{v\, dv}{f(v)} \tag{11.14}$$

EXAMPLE 11.5

The acceleration a of a recoiling piston in a cylinder filled with oil is given by $a = -bv$, where b is a positive constant and v is the velocity of the piston as shown. If $x_0 \neq 0$ and $v_0 \neq 0$, express and plot v and x as functions of the time t.

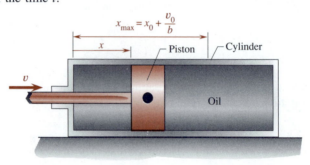

Solution. Noting that $a = f(v) = -bv$, $x_0 \neq 0$, and $v_0 \neq 0$, we utilize the derived results in Eqs. (11.13) and (11.14) to write

$$t = \int_{v_0}^{v} \frac{dv}{-bv} = -\frac{1}{b} \ln \frac{v}{v_0}$$

$$\frac{v}{v_0} = e^{-bt} \qquad\qquad v = v_0\, e^{-bt} \quad \blacktriangleleft$$

$$x = x_0 + \int_{v_0}^{v} \frac{v\, dv}{-bv} = x_0 - \frac{1}{b}(v - v_0) = x_0 - \frac{v_0}{b}(e^{-bt} - 1)$$

$$x = x_0 + \frac{v_0}{b}(1 - e^{-bt}) \quad \blacktriangleleft$$

Developmental Exercise

D11.13   The acceleration a of a body immersed in a fluid is given by $a = -bv^2$, where v is the velocity of the body and b is a positive constant. If $x_0 = 0$ and $v_0 \neq 0$, express v and x as functions of t.

(4) *Acceleration is a constant, a = a_c.*

For this class of problems, we use Eqs. (11.6) and (11.3) to write

$$dv = a\, dt = a_c\, dt \qquad \int_{v_0}^{v} dv = a_c \int_{0}^{t} dt$$

$$v = v_0 + a_c t \qquad (11.15)$$

$$dx = v\, dt \qquad \int_{x_0}^{x} dx = \int_{0}^{t} (v_0 + a_c t)\, dt$$

$$x = x_0 + v_0 t + \tfrac{1}{2} a_c t^2 \qquad (11.16)$$

Furthermore, by substituting a_c for $f(x)$ in Eq. (11.11), we obtain

$$v^2 = v_0^2 + 2a_c(x - x_0) \qquad (11.17)$$

Equations (11.15) through (11.17) are utilized as formulas when $a = a_c$.

EXAMPLE 11.6

A stone is thrown upward with $v_0 = 40$ ft/s at A off the side of a cliff as shown. Knowing that the stone is subjected to a constant downward acceleration of 32.2 ft/s^2 due to gravity, determine (a) the velocity v and elevation x of the stone as functions of the time t, (b) the maximum height h reached by the stone, (c) the velocity v_D of the stone just before it hits the bottom of the cliff, and (d) the corresponding time t_D.

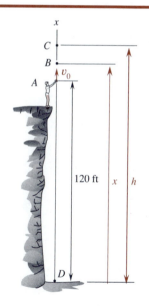

Solution. We note that the motion of the stone is rectilinear with a constant acceleration $a_c = -32.2$ ft/s^2. By using the point D as the origin of the x axis as given, we have (at $t = 0$) the initial conditions

$$x_0 = 120 \text{ ft} \qquad v_0 = 40 \text{ ft/s}$$

Utilizing Eqs. (11.15) and (11.16), we write

$$v = 40 - 32.2t \qquad x = 120 + 40t - 16.1t^2 \quad \blacktriangleleft$$

where v, x, and t are measured in ft/s, ft, and s, respectively.

As the stone reaches its highest point C, its velocity v becomes zero and its elevation x is equal to h. Utilizing Eq. (11.17), we write

$$0^2 = (40)^2 + 2(-32.2)(h - 120) \qquad h = 144.8 \text{ ft} \quad \blacktriangleleft$$

Just before the stone hits the bottom of the cliff, the elevation x becomes zero and the velocity v is denoted by v_D. Thus, we write

$$v_D^2 = (40)^2 + 2(-32.2)(0 - 120)$$

$$v_D = \pm 96.58 \qquad v_D = -96.6 \text{ ft/s} \quad \blacktriangleleft$$

where we take only the negative value of v_D because the direction of v_D must be downward. To find the corresponding time t_D, we let $v = v_D = -96.58$ ft/s in Eq. (11.15) and write

$$-96.58 = 40 + (-32.2)t_D \qquad t_D = 4.24 \text{ s} \quad \blacktriangleleft$$

REMARK. The value of t_D may also be found by letting $x = x_D = 0$ in Eq. (11.16). This yields $t_D = 4.24$ or -1.757. The negative value of t_D is an extraneous root to be discarded; therefore, the same result of $t_D = 4.24$ s is obtained.

Developmental Exercise

D11.14   Refer to Example 11.6. Determine (a) v_A of the stone as it falls back to the position at A, (b) the corresponding time t_A.

(5) *Acceleration is a given function of time, position, and velocity; $a = f(t, x, v)$.*

   For this class of problems, the procedure for obtaining the solutions generally depends on the explicit form of the function $f(t, x, v)$ and usually involves the solution of a differential equation as given by $a = f(t, x, v)$. Because of the potential involvement of advanced mathematical skills in obtaining the solutions, a detailed study of this class of problems is beyond the level of this text. However, a sample case in this class is described in Prob. 11.28, which may be solved by those who can solve first-order differential equations.

PROBLEMS†

11.1 and 11.2* The position x of a particle moving along a straight line is given as shown. Determine the time, position, and acceleration when the velocity is zero.

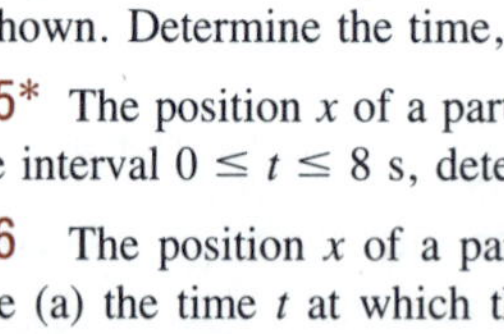

(x in m, t in s)

$$x = t^3 - 9t^2 + 24t - 10$$

O P

Fig. P11.1 and P11.3

(x in mm, t in s)

$$x = t^3 - 12t^2 + 45t + 5$$

O P

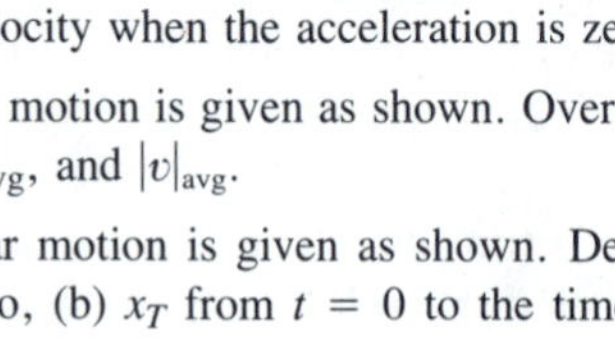

Fig. P11.2* and P11.4

11.3 and 11.4   The position x of a particle moving along a straight line is given as shown. Determine the time, position, and velocity when the acceleration is zero.

11.5*   The position x of a particle in rectilinear motion is given as shown. Over the time interval $0 \le t \le 8$ s, determine Δx, x_T, v_{avg}, and $|v|_{\text{avg}}$.

11.6   The position x of a particle in rectilinear motion is given as shown. Determine (a) the time t at which the velocity is zero, (b) x_T from $t = 0$ to the time at which the acceleration is zero.

11.7   The acceleration of a particle in rectilinear motion is directly proportional to the square of the time t. At $t = 0$, its velocity is -32 m/s. At $t = 2$ s, the particle comes to the origin O and its velocity becomes zero. Express the position x of the particle as a function of t.

(x in in., t in s)

$$x = 4t^3 - 54t^2 + 216t + 17$$

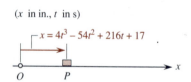

O P

Fig. P11.5* and P11.6

---

†Answers are provided at the end of the text for all problems except those marked with an asterisk.

11.8 A particle is given an acceleration $a = 24t - 72$ along a straight line, where a is measured in in./s^2 and t in seconds. If $v_0 = 0$ and $x_0 = 75$ in., determine (a) the time t_1 when the velocity v is again zero, (b) x_T from $t = 0$ to $t = 8$ s.

11.9* A particle is subjected to a sinusoidal acceleration $a = A \sin \omega t$ along a straight line, where A and ω are constants. If $v_0 = x_0 = 0$, determine (a) the position x as a function of the time t, (b) the maximum speed of the particle.

11.10* The position of an oscillating particle is given by $x = X \sin(\omega t + \phi)$, where X, ω, and ϕ are constants and represent the amplitude, the circular frequency, and the phase angle, respectively. Determine the values of X and ϕ in terms of x_0, v_0, and ω.

11.11 The acceleration of an oscillating particle is given by $a = -\omega^2 x$, where ω is a constant. If $x_0 = 3$ m, $v_0 = 0$, and $v = -12$ m/s when $x = 0$, determine (a) the value of ω, (b) the time elapsed during which the particle returns from $x = 3$ m to $x = 0$.

11.12 A particle is given an acceleration $a = 24 - 6x^2$, where a is measured in m/s^2 and x in m. If $x_0 = v_0 = 0$, determine (a) the velocity when $x = 3$ m, (b) the first position where the velocity is again zero, (c) the first position where the velocity is maximum, (d) the first maximum velocity.

11.13* Solve Prob. 11.12 if $a = 36 - 24x + 3x^2$.

11.14 The acceleration of a particle moving in the vertical direction under the effect of gravity is $a = -gR^2/r^2$, where r is the distance from the center of the earth to the particle, $R = 6.37$ Mm which is the radius of the earth, and $g = 9.81$ m/s^2 which is the gravitational acceleration near the surface of the earth. Based on this formula where air resistance is neglected, determine the height h reached by a bullet fired vertically upward from the surface of the earth with the following muzzle velocities: (a) 500 m/s, (b) 5 km/s, (c) 10 km/s.

11.15* A particle is released from rest at a height h above the surface of the earth. Using the formula for the acceleration $a = -gR^2/r^2$ as defined in Prob. 11.14, (a) determine the speed of the particle as it strikes the earth, (b) deduce the approximate value of this speed if $h \ll R$ (i.e., $R + h \approx R$), (c) evaluate the time required for the particle to fall to the surface of the earth if $h = R$.

11.16 The *escape velocity* v_{esc} of a particle is the minimum velocity with which a particle projected vertically will not return to the earth; i.e., the velocity of an escaping particle will approach zero when its height approaches infinity as shown. Using the formula $a = -gR^2/r^2$ as defined in Prob. 11.14, determine the value of v_{esc} of a particle projected from the surface of the earth in (a) km/s, (b) mi/s.

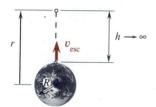

Fig. P11.16

Fig. P11.17

11.17 The hydroplane shown has a touchdown speed of 90 mi/h upon contact with the water. This speed is to be reduced to 15 mi/h in a distance of 1500 ft. The deceleration of the hydroplane on the water is observed to be proportional to the square of its velocity; i.e., $a = -kv^2$, where k is a constant related to the size and shape of the landing gear vanes. Determine (a) the value of k if a is measured in ft/s^2 and v in ft/s, (b) the time elapsed in slowing down from 90 mi/h to 15 mi/h.

11.18* Due to gravity and air resistance, the downward acceleration of an object is found to be given by $a = 9.81 - 0.0003v^2$, where a is measured in m/s^2 and v in m/s. If the object is released from rest and keeps falling downward as shown, determine its velocity after it has fallen 300 m.

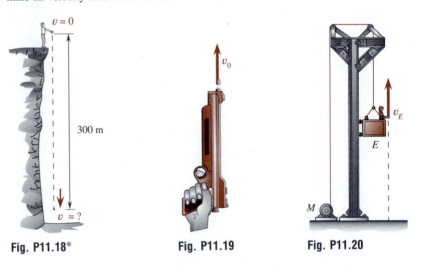

Fig. P11.18*   **Fig. P11.19**   **Fig. P11.20**

11.19 A pellet is fired from an air pistol with a muzzle velocity $v_0 = 350$ ft/s as shown. The muzzle of the pistol is 8 ft above the ground at the time the trigger is pulled. Neglecting air resistance and taking $g = 32.2$ ft/s^2, determine (a) the maximum altitude reached by the pellet, (b) the time elapsed since firing for the pellet to return to the ground.

11.20 An open elevator is moving up at a constant speed $v_E = 15$ ft/s as shown. A ball is released from the elevator and reaches the ground in 3 s. Determine (a) the height at which the ball was released, (b) the speed with which the ball strikes the ground.

11.21 A boy gently pushes a stone off the top of a cliff down to a lake as shown. The time elapsed from the instant the stone is pushed off to the instant the boy hears the splashing sound of the stone is 5 s. Taking the speed of sound in the air to be 332 m/s and $g = 9.81$ m/s^2, determine the height h of the cliff above the water surface.

Fig. P11.21

11.22* The motorcyclist shown is traveling at 72 km/h when a traffic light 300 m ahead of him turns red. The red light is timed to stay for 20 s in his direction. If he wishes to decelerate uniformly so that he reaches the traffic light just as it turns green again, determine (a) the deceleration of the motorcycle, (b) the speed of the motorcycle as it passes the traffic light.

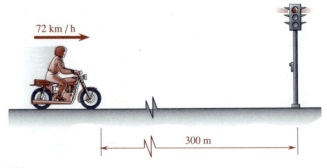

Fig. P11.22*

11.23* The acceleration of a particle moving in a viscous fluid as shown is given by $a = -2v$, where a is measured in ft/s^2 and v in ft/s. If $x_0 = 2$ ft and $v_0 = 10$ ft/s, determine (a) the displacement of the particle before it comes to rest, (b) the time required for its velocity to be reduced to 1 ft/s, (c) the time required for its velocity to become zero.

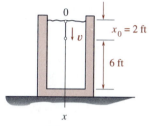

Fig. P11.23*   **Fig. P11.24**

11.24 The touchdown speed of 180 km/h of an airplane upon contact with the runway as shown is to be reduced to 18 km/h at a constant rate of 4 m/s^2 by operating thrust reversers as well as brakes of the airplane. Determine (a) the time elapsed from touchdown to the time at which the speed is reduced to 18 km/h, (b) the distance on the runway covered by the airplane during that time interval.

11.25* The car shown travels 800 m in 40 s while being accelerated at a constant rate of 0.25 m/s^2. For this interval of travel, determine (a) its initial velocity, (b) its final velocity.

11.26 The car shown is traveling at 90 km/h and is to be stopped over a distance of 125 m through a constant deceleration. Determine (a) the rate of its deceleration, (b) the time required for the car to stop.

Fig. P11.25* and P11.26

11.27 The single-stage rocket shown is launched vertically from the ground, and its thrust is designed to give the rocket a constant upward acceleration of 15 ft/s^2 for 30 s before the fuel is exhausted. Neglecting air resistance, determine the maximum altitude reached by the rocket.

11.28 If the decaying thrust and air resistance are taken into account, the upward acceleration of a rocket during its atmospheric flight is given by $a = ke^{-ct} - bv - g$, where k, c, and b are constants, and the gravitational acceleration g is taken as essentially constant. The rocket is launched from the ground with initial conditions $x_0 = v_0 = 0$. Determine the velocity v and the position x of the rocket as functions of the time t. (*Hint.* Since $a = dv/dt$, we have $dv/dt + bv = ke^{-ct} - g$. Solve this differential equation for v, then integrate to obtain x.)

Fig. P11.27

11.6 Relative Rectilinear Motion

The *relative motion* of any two particles is the *difference of the motions* of those two particles in a given reference frame. The *position of the particle A relative to the particle B*, as shown in Fig. 11.6, is denoted by $x_{A/B}$. We write

$$x_{A/B} = x_A - x_B \quad \text{or} \quad \boxed{x_A = x_B + x_{A/B}} \quad (11.18)$$

The time rate of change of $x_{A/B}$ is called the *velocity of A relative to B* and is denoted by $v_{A/B}$. Differentiating Eq. (11.18), we write

$$\boxed{v_A = v_B + v_{A/B}} \quad (11.19)$$

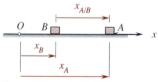

Fig. 11.6

The time rate of change of $v_{A/B}$ is called the *acceleration of A relative to B* and is denoted by $a_{A/B}$. Differentiating Eq. (11.19), we write[†]

$$a_A = a_B + a_{A/B} \qquad (11.20)$$

Note that the same coordinate system and the same time (or clock) have been used in describing the motion of each particle in the system. Thus, when $x_{A/B} > 0$, it means that A is positioned on the positive side of B; and when $v_{A/B} < 0$, it means that A is moving to the negative side of B.

Developmental Exercise

11.15   What does it mean when (a) $x_{A/B} < 0$, (b) $v_{C/D} > 0$, (c) $a_{D/C} < 0$?

EXAMPLE 11.7

An open elevator is being lowered at a constant rate of 2 m/s. A man in this elevator throws a handball upward with a velocity of 18 m/s relative to the elevator when he is 15 m above the ground. Determine (a) the time t and the position x_E at which the man will be able to catch the ball, (b) the velocity $v_{B/E}$ of the ball relative to the elevator when the ball is caught.

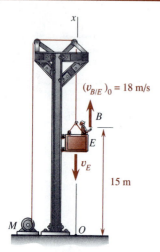

$(v_{B/E})_0 = 18$ m/s

Solution.   Noting that the positive direction is upward, we may write the initial conditions for the ball B and elevator E as follows:

$$(x_B)_0 = 15 \qquad (x_E)_0 = 15 \qquad (v_E)_0 = -2 \qquad (v_{B/E})_0 = 18$$

$$(v_{B/E})_0 = (v_B)_0 - (v_E)_0 \qquad 18 = (v_B)_0 - (-2) \qquad (v_B)_0 = 16$$

The acceleration of the elevator is $a_E = 0$, but the acceleration of the ball is $a_B = -g = -9.81$ m/s². Thus, we write

$$v_E = -2 \qquad\qquad x_E = 15 - 2t$$

$$v_B = 16 - 9.81t \qquad x_B = 15 + 16t - \frac{9.81}{2}t^2$$

At the time the ball is caught, we have $x_E = x_B$. Thus, we set

$$15 - 2t = 15 + 16t - \frac{9.81}{2}t^2$$

which yields two roots: $t = 0$ and $t = 3.67$ s. Since $t = 0$ is a trivial solution, the time the ball is caught must be

$$t = 3.67 \text{ s} \blacktriangleleft$$

At this time, we have

$$x_E = 15 - 2(3.67) \qquad\qquad x_E = 7.66 \text{ m} \blacktriangleleft$$

$$v_B = 16 - 9.81(3.67) = -20$$

$$v_{B/E} = v_B - v_E = -20 - (-2) \qquad v_{B/E} = -18 \text{ m/s} \blacktriangleleft$$

[†]Note in Eqs. (11.18) through (11.20) that the subscript A in the left-hand member is formally equal to the product of the subscripts B and A/B in the right-hand member; i.e., $A = B(A/B)$. This is a useful way of remembering the equations of relative motion.

11.7 Dependent Rectilinear Motions

Through the use of pulleys and cordlike elements (e.g., cords, strings, ropes, wires, cables, belts, or chains), we may have a system of connected particles which execute *dependent rectilinear motions*, as illustrated in Fig. 11.7. We recall that cordlike elements are generally assumed to be perfectly flexible, inextensible, and of negligible mass. Thus, a very important condition to be satisfied here is that the *length of each cordlike element*, as expressed in terms of the position coordinates of the particles, *must be constant*. Such a condition is called the *constraint condition* in the dependent rectilinear motions of particles.

In Fig. 11.7, we note that the length of the cord *CDEFGB* must be constant. Since the lengths of the portions *DE* and *FG* remain constant, it follows that the sum of the lengths of the segments $\overline{CD}$, $\overline{EF}$, and $\overline{GB}$ is constant. Observing that each of the lengths of $\overline{CD}$ and $\overline{EF}$ differs from x_A only by a constant, and that, similarly, the length of $\overline{GB}$ differs from x_B only by a constant, we write

$$2x_A + x_B = \text{constant}$$

Such an equation is a *constraint equation on the positions* of the particles in the system. Because of this constraint equation, only one of the two position coordinates may be chosen arbitrarily. The system in Fig. 11.7 has, therefore, only *one degree of freedom*. By differentiating the above constraint equation on the positions, we obtain the *constraint equations on the velocities and accelerations* of the particles in that system as follows:

$$2\frac{dx_A}{dt} + \frac{dx_B}{dt} = 0 \quad \text{or} \quad 2v_A + v_B = 0$$

$$2\frac{dv_A}{dt} + \frac{dv_B}{dt} = 0 \quad \text{or} \quad 2a_A + a_B = 0$$

where the positive directions of the dependent variables are downward because their directions of increase are downward as shown.

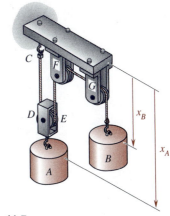

Fig. 11.7

EXAMPLE 11.8

The three cylinders *A*, *B*, and *C* are connected by cords and pulleys. At the instant shown, the cylinders *A* and *C* have $v_A = 4$ m/s ↓, $v_C = 6$ m/s ↑, $a_A = 3$ m/s² ↑, and $a_C = 8$ m/s² ↓. Determine v_B and a_B of the cylinder *B* at this instant.

Solution. The position coordinates for the cylinders *A*, *B*, *C*, and the pulley *D* are established as shown on the next page. Since the length of the cord *EFGHID* is constant, we write

$$2x_A + x_D = k_1 \tag{1}$$

where k_1 is a constant. Letting the height from the bottom support to the top support be h, which is a constant, we see that $\overline{JK} = h - x_D$. Since the length of the cord *JKLMNOPC* is constant, it follows that the sum of the lengths of the segments, $\overline{JK}$, $\overline{LM}$, $\overline{NO}$, and $\overline{PC}$ must be constant. We write

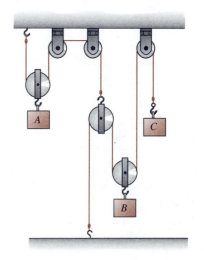

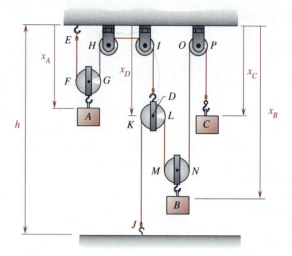

$$(h - x_D) + (x_B - x_D) + x_B + x_C = \bar{k}_2$$

$$2x_B + x_C - 2x_D = k_2 \qquad (2)$$

where $\bar{k}_2$ and k_2 are constants. Multiplying Eq. (1) by 2 and then adding it to Eq. (2), we can eliminate x_D and obtain

$$4x_A + 2x_B + x_C = 2k_1 + k_2$$

Differentiating with respect to time, we write

$$4v_A + 2v_B + v_C = 0 \qquad 4a_A + 2a_B + a_C = 0$$

Since the increasing directions of the dependent variables as shown are downward, their positive directions are all downward. Thus, we have $v_A = 4$ m/s, $v_C = -6$ m/s, $a_A = -3$ m/s^2, and $a_C = 8$ m/s^2. Substituting these values into the above equations, we obtain $v_B = -5$ m/s and $a_B = 2$ m/s^2.

$$\mathbf{v}_B = 5 \text{ m/s} \uparrow \qquad \mathbf{a}_B = 2 \text{ m/s}^2 \downarrow \blacktriangleleft$$

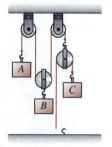

Fig. D11.17

Developmental Exercises

D11.16 How many degrees of freedom does the system in Example 11.8 have?

D11.17 At the instant shown, the cylinders A and C move with $\mathbf{v}_A = 5$ m/s $\downarrow$ and $\mathbf{v}_C = 2$ m/s $\uparrow$. Determine $\mathbf{v}_B$ of the cylinder B at this instant.

PROBLEMS

11.29 through 11.31* At the instant shown, the velocities and accelerations of the cylinders A and C are $\mathbf{v}_A = 2$ ft/s ↑, $\mathbf{a}_A = 1$ ft/s² ↓, $\mathbf{v}_C = 7$ ft/s ↓, and $\mathbf{a}_C = 3$ ft/s² ↑, respectively. Determine $\mathbf{v}_B$ and $\mathbf{a}_B$ of the cylinder B at this instant.

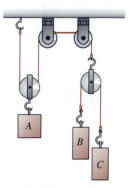

Fig. P11.29

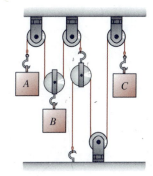

Fig. P11.30

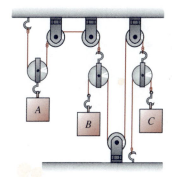

Fig. P11.31*

11.32 The block A as shown moves with $\mathbf{v}_A = 1.2$ m/s ←. Determine (a) $\mathbf{v}_B$ of the block B, (b) $\mathbf{v}_C$ and $\mathbf{v}_D$ of the points C and D on the cable, (c) $\mathbf{v}_{A/B}$.

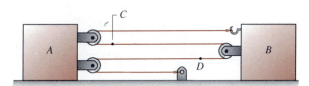

Fig. P11.32 and P11.33*

11.33* The block A as shown starts from rest and moves to the left with a constant acceleration. After 5 s, it is found that $\mathbf{v}_{A/B} = 300$ mm/s →. Determine $\mathbf{a}_A$ and $\mathbf{a}_B$ at this instant, and Δx_B during the first 4 s.

11.34 The motor at M as shown winds the cable at a constant rate of 700 mm/s. If the block B moves with $\mathbf{v}_B = 100$ mm/s ←, determine $\mathbf{v}_A$.

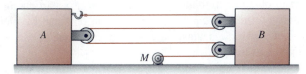

Fig. P11.34

11.35 The motor at M as shown starts from rest and winds the cable with a constant acceleration of 2 m/s² for 4 s. Determine (a) Δx_E of the elevator E over the time interval $0 \le t \le 4$ s, (b) $\mathbf{v}_{E/C}$ of E relative to the counterweight C at $t = 4$ s.

11.36 A crate D is hoisted by a man who holds one end of the rope and walks to the right with a constant speed of 0.5 m/s as shown. The length of the rope ABC is 14 m. Determine $\mathbf{v}_D$ and $\mathbf{a}_D$ of the crate when its bottom passes the corner at E.

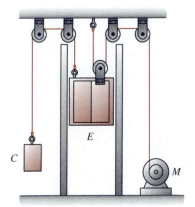

Fig. P11.35

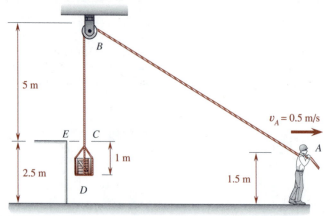

Fig. P11.36

11.37* A truck travels with a constant speed of 54 km/h on a straight road as shown. A car traveling at 90 km/h is 30 m behind the truck when the car's brakes are suddenly applied. If the road condition allows the car to decelerate only at the constant rate of 1.25 m/s², determine (a) whether a rear-end collision could occur, (b) if so, the velocity of the car relative to the truck just before the collision.

Fig. P11.37*

★11.8   Graphical Solution[†]

When the acceleration or velocity of a particle in rectilinear motion is given by different linear functions over several time intervals, graphical representations of the motion can be very helpful. A *position curve* is a graph obtained by plotting the position *x* of a particle against the time *t*. A *velocity curve* is a graph obtained by plotting the velocity *v* of a particle against the time *t*. Similarly, an *acceleration curve* is a graph obtained by plotting the acceleration *a* of a particle against the time *t*. These three curves are called the *motion curves* of the particle.

By calculus, we deduce the following six *useful relations* for the graphical solution of rectilinear-motion problems:

(1) The *slope of the position curve* at any time is equal to the ordinate of the velocity curve at that time. This is true because we have

$$\frac{dx}{dt} = v \qquad (11.21)$$

———————————————

[†]A five-point star indicates an optional section. Cf. the Preface.

(2) The *slope of the velocity curve* at any time is equal to the ordinate of the acceleration curve at that time. This is true because we have

$$\frac{dv}{dt} = a \tag{11.22}$$

The relations described in (1) and (2) are illustrated in Fig. 11.8.

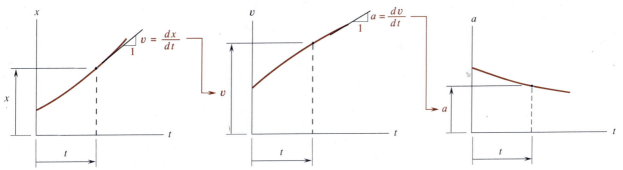

Fig. 11.8

(3) The *change in velocity* from the time t_1 to the t_2 is equal to the algebraic sum of the "areas" bounded by the acceleration curve and the time axis from t_1 to t_2.[†] This is true because, by integrating Eq. (11.22), we have

$$v_2 - v_1 = \int_{t_1}^{t_2} a \, dt \tag{11.23}$$

(4) The *change in position* from the time t_1 to the time t_2 is equal to the algebraic sum of the "areas" bounded by the velocity curve and the time axis from t_1 to t_2.[††] This is true because, by integrating Eq. (11.21), we have

$$x_2 - x_1 = \int_{t_1}^{t_2} v \, dt \tag{11.24}$$

The relations described in (3) and (4) are illustrated in Fig. 11.9.

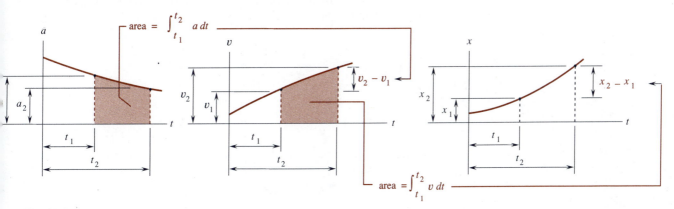

Fig. 11.9

[†]The dimensions of such "areas" are *length/time*.

[††]The dimension of such "areas" is *length*.

(5) The *concavity of the position curve* is upward if the corresponding acceleration is positive; it is downward if the latter is negative. This is true because, from Eqs. (11.21) and (11.22), we have

$$\frac{d^2x}{dt^2} = a \qquad (11.25)$$

Furthermore, the position curve will have a *point of inflection* at the instant the acceleration changes its sign.

(6) The *total distance traveled* from the time t_1 to the time t_2 is equal to the sum of the absolute values of the "areas" bounded by the velocity curve and the time axis from t_1 to t_2. This is true because the total distance traveled x_T is the total length of the path traversed. In terms of an integral, we write

$$x_T = \int_{t_1}^{t_2} |v|\, dt \qquad (11.26)$$

In terms of the "areas" $A_1, A_2, \ldots, A_n$ bounded by the velocity curve and the time axis, as shown in Fig. 11.10, we write

$$x_T = |A_1| + |A_2| + \cdots + |A_n| \qquad (11.27)$$

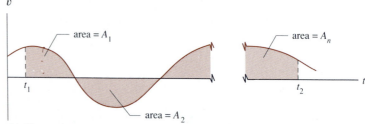

Fig. 11.10 (The dimension of "areas" in this figure is *length*.)

EXAMPLE 11.9

A particle moves along a straight line with the velocity shown. If $x_0 = 20$ ft, (a) draw the acceleration and position curves for $0 \leq t \leq 14$ s, (b) determine Δx and x_T over the interval $0 \leq t \leq 12$ s.

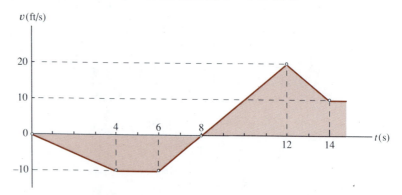

Solution. Using the first five relations listed in Sec. 11.8, we can readily draw the acceleration and position curves as shown. The given velocity curve is here repeated and lined up in the time scale with these two curves for easy reference. In particular, note in the position curve that (1) the slope

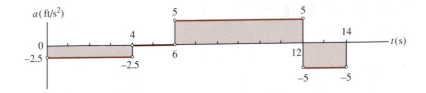

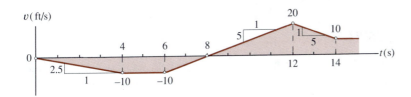

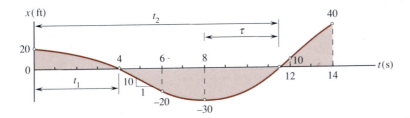

is zero at $t = 0$ and $t = 8$ s, (2) the concavity is downward during $0 < t < 4$ s and 12 s $< t < 14$ s, (3) the concavity is upward during 6 s $< t < 12$ s, (4) a point of inflection exists at $t = 12$ s, and (5) the curve becomes a straight line during 4 s $\leq t \leq 6$ s, in which the acceleration is zero.

Having drawn the position curve, we see that the positions of the particle at $t = 0$ and $t = 12$ s are $x_0 = 20$ ft and $x_{12} = 10$ ft. Thus, the displacement over the time interval $0 \leq t \leq 12$ s is

$$\Delta x = x_{12} - x_0 \qquad \Delta x = -10 \text{ ft} \blacktriangleleft$$

Alternatively, we can obtain Δx by summing algebraically the "areas" in the v-t diagram from $t = 0$ to $t = 12$ s; i.e.,

$$\Delta x = -\tfrac{1}{2}(4)(10) - 2(10) - \tfrac{1}{2}(2)(10) + \tfrac{1}{2}(4)(20) \text{ ft} = -10 \text{ ft} \quad \text{Q.E.D.}$$

From the position curve drawn, we see that the particle moves from $x = 20$ ft at $t = 0$ to $x = -30$ ft at $t = 8$ s, then back to $x = 10$ ft at $t = 12$ s. Thus, the total distance traveled over that time interval is

$$x_T = |x_8 - x_0| + |x_{12} - x_8| = |-30 - 20| + |10 - (-30)|$$

$$x_T = 90 \text{ ft} \blacktriangleleft$$

Of course, we can alternatively obtain x_T by summing the absolute values of the "areas" in the v-t diagram from $t = 0$ to $t = 12$ s; i.e.,

$$x_T = \left|-\tfrac{1}{2}(4)(10)\right| + |-2(10)| + \left|-\tfrac{1}{2}(2)(10)\right| + \left|\tfrac{1}{2}(4)(20)\right|$$

$$x_T = 90 \text{ ft} \qquad \text{Q.E.D.}$$

Developmental Exercises

D11.18   Refer to Example 11.9. Over the time interval $0 \leq t \leq 13$ s, determine Δx and x_T.

D11.19   Refer to Example 11.9. By inspecting the position curve drawn, we see that the particle passes through the origin first at $t = t_1 = 4$ s and later at $t = t_2 = 8 + \tau$ s, as shown. Applying the fourth relation in Sec. 11.8, determine the values of τ and t_2.

EXAMPLE 11.10

A motorcycle is 15 m behind a car when both are traveling at a constant speed of 72 km/h on a straight road as shown. Seeing an obstacle looming ahead, the driver of the car suddenly applies the brakes, causing the car to decelerate at the constant rate of 3 m/s^2. Two seconds after seeing the car's brake lights activated, the motorcyclist applies the brakes to produce a constant deceleration of a_m and is able to barely avoid a collision. Determine the value of a_m.

Solution.   Let the time be measured from the instant the car's brake lights are activated. Furthermore, let τ be the duration (in seconds) from the instant the brakes of the motorcycle are applied to the instant the motorcycle almost hits the car. Noting that 72 km/h = 20 m/s, we draw the velocity curves for the car and the motorcycle as solid and dashed lines, respectively, as shown.

We know that the motorcycle is 15 m behind the car before the brake lights are activated. The fact that the motorcycle and the car have barely avoided a collision implies that the distance which the motorcycle has traveled more than the car during the braking maneuver must be equal to 15 m.

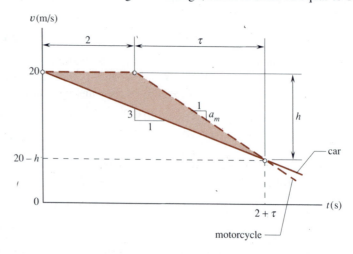

Thus, applying the sixth relation in Sec. 11.8 to the car and motorcycle, we see that the shaded triangular "area" shown in the velocity diagram must be equal to 15 m. From the geometry of this shaded "area" and the slope triangles as indicated, we write

$$\tfrac{1}{2}(2)h = 15 \qquad h = 3(2 + \tau) \qquad h = \tau a_m$$

These equations yield $h = 15$, $\tau = 3$, $a_m = 5$. We write

$$a_m = 5 \text{ m/s}^2 \quad \blacktriangleleft$$

Developmental Exercises

D11.20 Refer to Example 11.10. During the braking maneuver, determine the total distance traveled by (a) the car, (b) the motorcycle.

D11.21 Describe the six useful relations for the graphical solution of rectilinear motion problems.

PROBLEMS

11.38 through 11.40* The acceleration curve of a particle in rectilinear motion is shown. If $x_0 = -2$ m and $v_0 = -4$ m/s, (a) draw the velocity and position curves of the particle for $0 \leq t \leq 12$ s, (b) determine the value of t for which the particle passes through the origin.

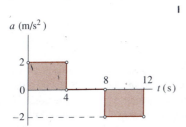

Fig. P11.38

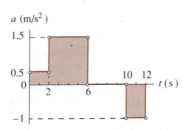

Fig. P11.39*

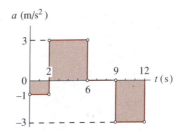

Fig. P11.40*

11.41 and 11.42* The velocity curve of a particle in rectilinear motion is shown. If $x_0 = 16$ ft, draw the acceleration and position curves of the particle for $0 \leq t \leq 15$ s and determine (a) Δx and x_T of the particle over the interval $0 \leq t \leq 12$ s, (b) the two values t_1 and t_2 of t for which the particle passes through the origin.

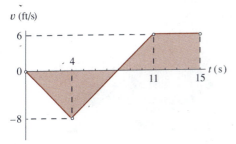

Fig. P11.41

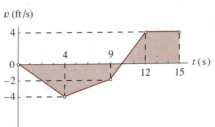

Fig. P11.42*

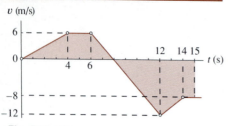

Fig. P11.43*

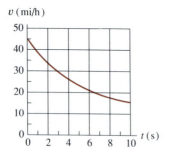

Fig. P11.44

11.43* A particle moves in a straight line with the velocity as shown. If $x_0 = -12$ m, draw its position curve for $0 \le t \le 15$ s and determine (a) Δx and x_T of the particle over the interval $0 \le t \le 11$ s, (b) the two values t_1 and t_2 of t for which the particle passes through the origin.

11.44 A motorist is traveling on a straight highway which is being repaired. He reduces his speed from 45 mi/h, when he passes the first flagger, to 16 mi/h 10 s later, when he passes the second flagger. His speed during this interval is recorded as shown. Determine (a) his deceleration at $t = 6$ s, (b) the distance between the first and the second flaggers.

11.45 A motorist traveling at 90 km/h notices that the traffic signal 450 m ahead of him turns red. He knows that it is timed to stay red for 20 s. If he desires to pass the signal at 90 km/h just as it turns green again, determine (a) the smallest common rate of deceleration and acceleration he should have, (b) the minimum speed reached in this maneuver.

11.46 By mistake, cars A and B are traveling toward each other in a single-lane portion of a road with speeds as shown. When $L = 210$ m, both drivers realize the situation and apply their brakes at the same time. If the decelerations of A and B are 1.5 m/s² and 1.2 m/s², respectively, (a) verify that a head-on collision will occur, (b) the velocity of A relative to B just before collision, (c) the distance traveled by each car while decelerating.

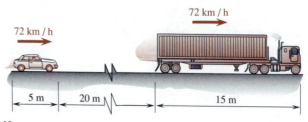

Fig. P11.46 and P11.47

11.47 By mistake, cars A and B are traveling toward each other in a single-lane portion of a road with speeds as shown. When $L = 210$ m, both drivers realize the situation and apply their brakes at the same time. If their decelerations are constant and they stop simultaneously just in time to barely avoid a head-on collision, determine (a) the deceleration of each car, (b) the distance traveled by each car while decelerating.

11.48 A car and a truck travel at a constant speed of 72 km/h on a road; the car is 20 m behind the truck as shown. The driver of the car wishes to pass the truck and place his car 20 m in front of the truck, and then resume the speed of 72 km/h. The maximum acceleration and deceleration of the car are 2 m/s² and 2.5 m/s², respectively. Determine the shortest time in which the passing operation can be completed if the car does not exceed a speed of 90 km/h at any time.

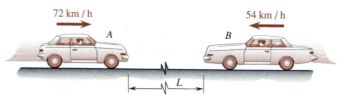

Fig. P11.48

11.49 The time rate of change of acceleration is referred to as the *jerk*,[†] which is often regarded as a cause of passenger discomfort. If the jerk of an elevator is limited to ± 2 ft/s^3, determine the shortest time in which an elevator starting from rest can rise (or descend) 13.5 ft and stop.

CURVILINEAR MOTION

11.9   Position Vector, Velocity, and Acceleration

The curvilinear motion of a particle is the motion of the particle along a curved path. The position P occupied by the particle at a given time t is defined by the *position vector* $\mathbf{r}$, which is a vector drawn from the origin O of a fixed coordinate system to the point P as shown in Fig. 11.11. At a later time $t + \Delta t$, the particle occupies the position P' which is defined by the position vector $\mathbf{r}'$. The vector $\Delta \mathbf{r}$, directed from P to P', represents the *change in the position vector* over the time interval Δt; i.e., from Fig. 11.11 and the vector addition (cf. the *triangle rule* in Sec. 2.2), we have

$$\mathbf{r}' = \mathbf{r} + \Delta \mathbf{r} \qquad \boxed{\Delta \mathbf{r} = \mathbf{r}' - \mathbf{r}} \qquad (11.28)$$

This vector $\Delta \mathbf{r}$ is called the *displacement* of the particle over the time interval Δt. Note that $\Delta \mathbf{r}$ accounts for the changes in both the *direction* and the *magnitude* of the position vector over Δt.

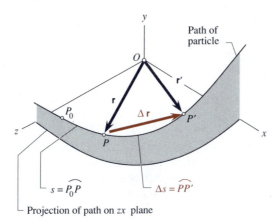

Fig. 11.11

Similar to that in rectilinear motion, the *velocity* of a particle in curvilinear motion in a reference frame is defined as the time rate of change of position (or the time rate of displacement) of the particle in that reference frame. As shown in Fig. 11.12(a), the *average velocity* $\mathbf{v}_{avg}$ of a particle having a displacement $\Delta \mathbf{r}$ over a time interval Δt is given by

$$\mathbf{v}_{avg} = \frac{\Delta \mathbf{r}}{\Delta t} \qquad (11.29)$$

[†]Cf. Sec. 11.4.

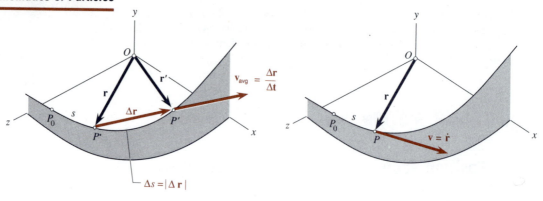

Fig. 11.12   (a)   (b)

Since $\Delta\mathbf{r}$ is a vector and Δt a scalar, the quotient $\Delta\mathbf{r}/\Delta t$ (and hence $\mathbf{v}_{avg}$) is a vector directed from P through P' as shown. As Δt becomes smaller and approaches zero in that limit, $\mathbf{v}_{avg}$ becomes the *instantaneous velocity* $\mathbf{v}$ of the particle as shown in Fig. 11.12(b); i.e.,

$$\mathbf{v} = \lim_{\Delta t \to 0} \frac{\Delta\mathbf{r}}{\Delta t} \qquad \boxed{\mathbf{v} = \frac{d\mathbf{r}}{dt} = \dot{\mathbf{r}}} \qquad (11.30)$$

This instantaneous velocity is what we usually call the *velocity*.

Since P' gets closer to P as Δt becomes smaller, the velocity $\mathbf{v}$ obtained at the limit of $\Delta t \to 0$ must therefore be *tangent to the path* of the particle at P. The magnitude v of the velocity $\mathbf{v}$ is called the *speed* of the particle. Noting that the magnitude of $\Delta\mathbf{r}$ approaches the length Δs of the arc PP' as Δt approaches zero, we write

$$v = \lim_{\Delta t \to 0} \frac{|\Delta\mathbf{r}|}{\Delta t} = \lim_{\Delta t \to 0} \frac{\Delta s}{\Delta t} \qquad \boxed{v = \frac{ds}{dt} = \dot{s}} \qquad (11.31)$$

Similar to that in rectilinear motion, the *acceleration* of a particle in curvilinear motion in a reference frame is defined as the time rate of change of velocity of the particle in that reference frame. Suppose that, over the time interval Δt, a particle moves from the point P to the point P' while its velocity changes from $\mathbf{v}$ to $\mathbf{v}'$ as shown in Fig. 11.13(a). By drawing the

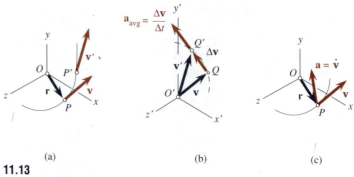

(a)   (b)   (c)

Fig. 11.13

vectors $\mathbf{v}$ and $\mathbf{v}'$ from a common point O' as shown in Fig. 11.13(b), we see that the vector $\Delta\mathbf{v}$, which is drawn from the tip Q of $\mathbf{v}$ to the tip Q' of $\mathbf{v}'$, represents the *change in the velocity* over Δt; i.e.,

$$\Delta\mathbf{v} = \mathbf{v}' - \mathbf{v} \qquad (11.32)$$

Note that the vector $\Delta\mathbf{v}$ accounts for the changes in both the *direction* and the *magnitude* of the velocity over Δt. The *average acceleration* $\mathbf{a}_{avg}$ of the particle over Δt is given by

$$\mathbf{a}_{avg} = \frac{\Delta\mathbf{v}}{\Delta t} \qquad (11.33)$$

which is a vector directed from Q through Q' as shown. As Δt becomes smaller and approaches zero in the limit, $\mathbf{a}_{avg}$ becomes the *instantaneous acceleration* $\mathbf{a}$ of the particle as shown in Fig. 11.13(c); i.e.,

$$\mathbf{a} = \lim_{\Delta t \to 0} \frac{\Delta\mathbf{v}}{\Delta t} \qquad \boxed{\mathbf{a} = \frac{d\mathbf{v}}{dt} = \dot{\mathbf{v}}} \qquad (11.34)$$

This instantaneous acceleration is what we usually call the *acceleration*.

For our study, the *path* of a particle may be viewed as a curve described by the tip of its position vector, which is drawn from a fixed origin O, as illustrated in Fig. 11.14(a). When the velocity vectors of a particle at its various positions are drawn to scale from a fixed origin O', as illustrated in Fig. 11.14(b), the curve described by the tip of the velocity vector is called the *hodograph* of the motion. We readily see from Figs. 11.13 and 11.14 that the acceleration of a particle is a vector tangent to the hodograph of the motion, *not* the path of the particle; the exception is in the case of rectilinear motion.

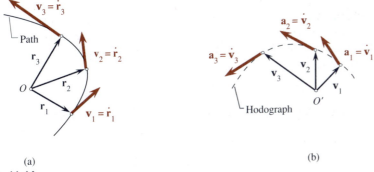

(a)   (b)

Fig. 11.14

The only definable motion of a particle, whose size and shape are to be disregarded, is translation along a path. For two particles A and B in space, as shown in Fig. 11.15, the motion of B in a nonrotating reference frame attached to A (e.g., $Ax'y'z'$)[†] is called the *motion of B relative to A*. The *position vector of B relative to A* is denoted by $\mathbf{r}_{B/A}$. The position vectors $\mathbf{r}_A$ and $\mathbf{r}_B$ of A and B are related to $\mathbf{r}_{B/A}$ in the fixed reference frame $Oxyz$ as follows:[†]

$$\mathbf{r}_B = \mathbf{r}_A + \mathbf{r}_{B/A} \qquad (11.35)$$

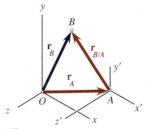

Fig. 11.15

[†]For convenience, a reference frame is often named after the coordinate system which is the only one embedded in it. (Cf. Sec. 11.1.)

In terms of the *velocity* and *acceleration of B relative to A,* we write

$$\dot{\mathbf{r}}_B = \dot{\mathbf{r}}_A + \dot{\mathbf{r}}_{B/A} \qquad \text{or} \qquad \boxed{\mathbf{v}_B = \mathbf{v}_A + \mathbf{v}_{B/A}} \qquad (11.36)$$

$$\dot{\mathbf{v}}_B = \dot{\mathbf{v}}_a + \dot{\mathbf{v}}_{B/A} \qquad \text{or} \qquad \boxed{\mathbf{a}_B = \mathbf{a}_A + \mathbf{a}_{B/A}} \qquad (11.37)$$

More precisely, the vectors $\mathbf{r}_{B/A}$, $\mathbf{v}_{B/A}$, and $\mathbf{a}_{B/A}$ denote, respectively, the position vector, the velocity, and the acceleration of the particle B in a nonrotating reference frame (e.g., $Ax'y'z'$) attached to the particle A.

Developmental Exercises

D11.22 Define for a particle in curvilinear motion the following terms: (a) position vector, (b) displacement, (c) average velocity, (d) velocity, (e) average acceleration, (f) acceleration, (g) hodograph.

D11.23 Define the vectors (a) $\mathbf{r}_{B/A}$, (b) $\mathbf{v}_{B/A}$, (c) $\mathbf{a}_{B/A}$.

EXAMPLE 11.11

Instruments in an airplane show that the plane is flying north with an airspeed of 250 km/h (relative to the air). Meanwhile, the radar on the ground indicates that the plane is moving at a speed of 220 km/h in a direction 10° west of north. Determine the wind velocity.

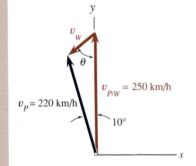

Solution. We know from the given data that the velocity of the plane relative to the wind is $\mathbf{v}_{P/W} = 250$ km/h ↑ and that the velocity of the plane relative to the ground is $\mathbf{v}_P = 200$ km/h ⬉ 80°. Denoting the wind velocity by $\mathbf{v}_W$, we have

$$\mathbf{v}_{P/W} + \mathbf{v}_W = \mathbf{v}_P$$

as illustrated. Applying the laws of cosines and sines, we write

$$v_W^2 = (220)^2 + (250)^2 - 2(220)(250)\cos 10°$$

$$v_W = 50.71 \qquad \frac{220}{\sin\theta} = \frac{v_W}{\sin 10°} \qquad \theta = 48.9°$$

$$\mathbf{v}_W = 50.7 \text{ km/h} \ \text{⬋} \ 41.1° \blacktriangleleft$$

Alternative Solution. Using analytical expressions, we write

$$\mathbf{v}_{P/W} = 250\mathbf{j} \qquad \mathbf{v}_P = 220(-\sin 10° \ \mathbf{i} + \cos 10° \ \mathbf{j})$$

$$\mathbf{v}_W = \mathbf{v}_P - \mathbf{v}_{P/W} = -38.20\mathbf{i} - 33.34\mathbf{j}$$

$$\mathbf{v}_W = 50.7 \text{ km/h} \ \text{⬋} \ 41.1° \blacktriangleleft$$

Developmental Exercises

D11.24   Name the type of motion of a particle if the hodograph of its motion is (a) a point, (b) a straight line, (c) a circle with radius equal to v.

D11.25   The velocity and acceleration vectors of a car at five separate instants are shown. Which instants indicate that the car is turning a corner?

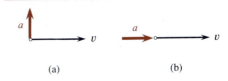

(a)   (b)

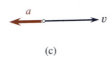

(c)

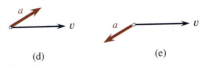

(d)   (e)

Fig. D11.25

11.10   Derivatives of Vector Functions[†]

If to each value of t (such as time) there corresponds a vector $\mathbf{F}$, we say that $\mathbf{F}$ is a *vector function* of t; i.e., $\mathbf{F} = \mathbf{F}(t)$. We note from Eq. (11.30) that the velocity $\mathbf{v}$ of a particle in curvilinear motion is simply equal to the time derivative of its position $\mathbf{r}$, which is a vector function of the time t. Similarly, we note from Eq. (11.34) that the acceleration $\mathbf{a}$ of a particle is simply equal to the time derivative of its velocity $\mathbf{v}$. Thus, it is quite concise when we use vector calculus, in addition to vector algebra, to derive such equations and formulas. In this section, we shall go over a few rules governing the differentiation of sums and products of vector functions.

For a *scalar function* $f = f(t)$, we recall that its derivative is

$$\frac{df}{dt} = \lim_{\Delta t \to 0} \frac{\Delta f}{\Delta t} = \lim_{\Delta t \to 0} \frac{f(t + \Delta t) - f(t)}{\Delta t} \tag{11.38}$$

For a *vector function* $\mathbf{F} = \mathbf{F}(t)$, we define in a given reference frame that[††]

$$\frac{d\mathbf{F}}{dt} = \lim_{\Delta t \to 0} \frac{\Delta \mathbf{F}}{\Delta t} = \lim_{\Delta t \to 0} \frac{\mathbf{F}(t + \Delta t) - \mathbf{F}(t)}{\Delta t} \tag{11.39}$$

For the *sum of two vector functions* $\mathbf{A}(t)$ *and* $\mathbf{B}(t)$, we similarly define

$$\frac{d(\mathbf{A} + \mathbf{B})}{dt} = \lim_{\Delta t \to 0} \frac{\Delta(\mathbf{A} + \mathbf{B})}{\Delta t}$$

$$= \lim_{\Delta t \to 0} \frac{\mathbf{A}(t + \Delta t) + \mathbf{B}(t + \Delta t) - [\mathbf{A}(t) + \mathbf{B}(t)]}{\Delta t}$$

$$= \lim_{\Delta t \to 0} \frac{\mathbf{A}(t + \Delta t) - \mathbf{A}(t)}{\Delta t} + \lim_{\Delta t \to 0} \frac{\mathbf{B}(t + \Delta t) - \mathbf{B}(t)}{\Delta t}$$

$$= \lim_{\Delta t \to 0} \frac{\Delta \mathbf{A}}{\Delta t} + \lim_{\Delta t \to 0} \frac{\Delta \mathbf{B}}{\Delta t}$$

$$\frac{d(\mathbf{A} + \mathbf{B})}{dt} = \frac{d\mathbf{A}}{dt} + \frac{d\mathbf{B}}{dt} \tag{11.40}$$

[†]Those who are familiar with the derivatives of vector functions may proceed directly to Sec. 11.11.

[††]The time rate of change of a scalar function has the same value in all reference frames. However, the time rate of change of a vector function may have different values in different references frames (cf. Chap. 15).

For the *product of f(t) and* **A**(t), we write

$$\frac{d(f\mathbf{A})}{dt} = \lim_{\Delta t \to 0} \frac{\Delta(f\mathbf{A})}{\Delta t} = \lim_{\Delta t \to 0} \frac{(f + \Delta f)(\mathbf{A} + \Delta\mathbf{A}) - f\mathbf{A}}{\Delta t}$$

$$= \lim_{\Delta t \to 0} \left(\frac{\Delta f}{\Delta t} \mathbf{A} \right) + \lim_{\Delta t \to 0} \left(f \frac{\Delta\mathbf{A}}{\Delta t} \right)$$

$$\frac{d(f\mathbf{A})}{dt} = \frac{df}{dt} \mathbf{A} + f \frac{d\mathbf{A}}{dt} \tag{11.41}$$

The derivatives of the *dot product* and of the *cross product* of two vector functions **A**(t) and **B**(t) may be obtained in a similar manner; i.e.,

$$\frac{d(\mathbf{A} \cdot \mathbf{B})}{dt} = \frac{d\mathbf{A}}{dt} \cdot \mathbf{B} + \mathbf{A} \cdot \frac{d\mathbf{B}}{dt} \tag{11.42}$$

$$\frac{d(\mathbf{A} \times \mathbf{B})}{dt} = \frac{d\mathbf{A}}{dt} \times \mathbf{B} + \mathbf{A} \times \frac{d\mathbf{B}}{dt} \tag{11.43}$$

Note that the order of the factors in Eq. (11.43) must be maintained because the cross product is not commutative.

The unit vectors **i**, **j**, and **k** associated with the x, y, and z axes of a fixed rectangular coordinate system are constant vectors having a magnitude of 1 and fixed directions. Their derivatives are therefore zero. However, note that the derivative of a vector having a constant magnitude but *varying direction* is generally *nonzero*. To find $d\mathbf{A}/dt$, we first express the vector **A** in terms of its rectangular components and then apply Eq. (11.41) as follows:

$$\mathbf{A} = A_x\mathbf{i} + A_y\mathbf{j} + A_z\mathbf{k}$$

$$\frac{d\mathbf{A}}{dt} = \frac{dA_x}{dt}\mathbf{i} + \frac{dA_y}{dt}\mathbf{j} + \frac{dA_z}{dt}\mathbf{k} \tag{11.44}$$

It should be pointed out that all equations in this section are valid whether the independent variable t is a time variable or just a parameter.

Developmental Exercises

D11.26 If $\mathbf{A} = t^2\mathbf{i} + 2\mathbf{k}$ and $\mathbf{B} = t^3\mathbf{j} - 3t\mathbf{k}$, evaluate at $t = 1$ (a) $\frac{d}{dt}(\mathbf{A} + \mathbf{B})$, (b) $\frac{d}{dt}(\mathbf{A} \cdot \mathbf{B})$, (c) $\frac{d}{dt}(\mathbf{B} \times \mathbf{A})$.

D11.27 If $\frac{d}{dt}(\mathbf{A} \times \mathbf{B}) = -4t^3\mathbf{i} - 2t\mathbf{k}$ and $\mathbf{B} \times \frac{d\mathbf{A}}{dt} = 2t^3\mathbf{i} + 2t\mathbf{k}$, what is the value of $\frac{d\mathbf{B}}{dt} \times \mathbf{A}$?

11.11 Rectangular Components

The curvilinear motion of particles may be described by using one of several different coordinate systems embedded in a reference frame. The choice of a particular coordinate system is usually dependent on the geometry of the path, the way the acceleration is generated, and other given data. The rectangular coordinate system is particularly useful when the rectangular components of acceleration are independently generated or determined. The resulting curvilinear motion is, then, obtained by vector additions of the x, y, and z rectangular components of the acceleration, velocity, and position vectors.

Recall that the unit vectors **i**, **j**, and **k** along the fixed xyz rectangular coordinate axes are constant vectors. For a particle moving along a space curve as illustrated in Fig. 11.16, we may, therefore, represent the position vector **r**, the velocity **v**, and the acceleration **a** of the particle as follows:

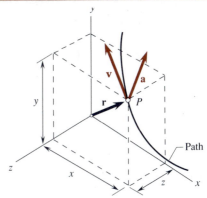

Fig. 11.16

$$\mathbf{r} = x\mathbf{i} + y\mathbf{j} + z\mathbf{k}$$

$$\mathbf{v} = \dot{\mathbf{r}} = \dot{x}\mathbf{i} + \dot{y}\mathbf{j} + \dot{z}\mathbf{k} \qquad (11.45)$$

$$\mathbf{a} = \dot{\mathbf{v}} = \ddot{\mathbf{r}} = \ddot{x}\mathbf{i} + \ddot{y}\mathbf{j} + \ddot{z}\mathbf{k}$$

Thus, the scalar components of **v** and **a** are

$$v_x = \dot{x} \qquad v_y = \dot{y} \qquad v_z = \dot{z}$$

$$(11.46)$$

$$a_x = \dot{v}_x = \ddot{x} \qquad a_y = \dot{v}_y = \ddot{y} \qquad a_z = \dot{v}_z = \ddot{z}$$

The magnitudes of **v** and **a** are

$$v = (v_x^2 + v_y^2 + v_z^2)^{1/2} \qquad a = (a_x^2 + a_y^2 + a_z^2)^{1/2} \qquad (11.47)$$

If the coordinates x, y, and z of a particle are independently generated as functions of the time t, then for any value of t we may combine them and their time derivatives to obtain **r**, **v**, and **a** as given in Eqs. (11.45). On the other hand, if the acceleration components a_x, a_y, and a_z are given as functions of the time t, we may, by integration and imposition of the given initial conditions, first obtain v_x, v_y, and v_z and then x, y, and z. The curvilinear motion described by using a fixed rectangular coordinate system is, in fact, a combination of three simultaneous rectilinear motions in the x, y, and z directions. Thus, methods developed for studying the rectilinear motion of particles may be applied to each of them separately.

If a particle undergoes curvilinear motion in the xy plane, we have

$$z = \dot{z} = \ddot{z} = 0 \qquad v_z = a_z = 0 \qquad (11.48)$$

and the relations in Eqs. (11.45) through (11.47) may correspondingly be simplified. The curvilinear motion of particles in a plane is further discussed in the next three sections.

EXAMPLE 11.12

The plane curvilinear motion of a particle is defined by the parametric equations $x = 50 - 2t^2$ and $y = 30t - 5t^2$, where x and y are the coordinates measured in meters and t is the parameter measured in seconds. Plot the path traveled by the particle for the interval $0 \leq t \leq 5$ s and determine its velocity and acceleration at $t = 5$ s.

Solution. By differentiation with respect to t, we have

$$v_x = \dot{x} = -4t \qquad a_x = \ddot{x} = -4$$

$$v_y = \dot{y} = 10(3 - t) \qquad a_y = \ddot{y} = -10$$

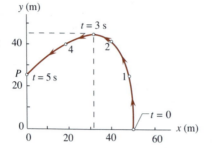

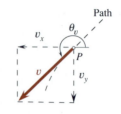

Since $v_y = 0$ at $t = 3$ s, the curved path has a peak at this instant. By calculating corresponding values of x and y for several values of t, we plot the path as shown. The velocity $\mathbf{v}$ and the acceleration $\mathbf{a}$ of the particle at P, where $t = 5$ s, are sketched as shown. For $t = 5$ s, we find that

$$v_x = -20 \qquad v_y = -20 \qquad\qquad v = 28.3 \text{ m/s} \blacktriangleleft$$

$$\tan \theta_v = \frac{-20}{-20} \qquad\qquad\qquad \theta_v = 225° \blacktriangleleft$$

$$a_x = -4 \qquad a_y = -10 \qquad\qquad a = 10.77 \text{ m/s}^2 \blacktriangleleft$$

$$\tan \theta_a = \frac{-10}{-4} \qquad\qquad\qquad \theta_a = 248.2° \blacktriangleleft$$

Developmental Exercise

D11.28   Refer to Example 11.12. Determine the velocity and acceleration of the particle at $t = 2$ s.

EXAMPLE 11.13

The motions of the two pegs A and B are constrained in such a way that they remain both in the vertical slot of the member C and in the elliptic path at all times as shown. If the member C moves with a constant velocity of 2 ft/s →, determine $\mathbf{v}_{B/A}$ and $\mathbf{a}_{B/A}$ of B relative to A when $x = 3$ ft.

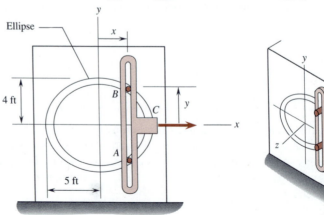

Solution. Since the path of B must be elliptic, we have

$$\frac{x^2}{5^2} + \frac{y^2}{4^2} = 1 \tag{1}$$

$$\frac{2}{25} x\dot{x} + \frac{1}{8} y\dot{y} = 0 \tag{2}$$

$$\frac{2}{25} (\dot{x}^2 + x\ddot{x}) + \frac{1}{8} (\dot{y}^2 + y\ddot{y}) = 0 \tag{3}$$

The constraints require that the x component of the motion of B be the same as that of the member C which has a velocity of 2 ft/s → and a zero acceleration. Consequently, we have $\dot{x} = 2$ ft/s, $\ddot{x} = 0$. From Eq. (1), we find that $y = 3.2$ ft when $x = 3$ ft. Substituting these values into Eqs. (2) and (3), we get

$$\dot{y} = -1.2 \text{ ft/s} \qquad \ddot{y} = -1.25 \text{ ft/s}^2$$

By vector addition and symmetry, we write

$$\mathbf{v}_B = 2\mathbf{i} - 1.2\mathbf{j} \qquad \mathbf{v}_A = 2\mathbf{i} + 1.2\mathbf{j}$$

$$\mathbf{v}_{B/A} = \mathbf{v}_B - \mathbf{v}_A = -2.4\mathbf{j} \qquad v_{B/A} = 2.4 \text{ ft/s} \downarrow \ \blacktriangleleft$$

$$\mathbf{a}_B = -1.25\mathbf{j} \qquad \mathbf{a}_A = 1.25\mathbf{j}$$

$$\mathbf{a}_{B/A} = \mathbf{a}_B - \mathbf{a}_A = -2.5\mathbf{j} \qquad a_{B/A} = 2.5 \text{ ft/s}^2 \downarrow \ \blacktriangleleft$$

Developmental Exercise

D11.29 Refer to Example 11.13. If $v_C = 2$ ft/s → and $a_C = 0.5$ ft/s^2 → when $x = 3$ ft, determine $\mathbf{a}_{B/A}$ at this instant.

11.12 Free Flight of a Projectile

A *projectile* is a body projected (e.g., thrown, cast, or impelled) through the air or space. The *powered flight* of a projectile is the motion of a body by virtue of the action of a thrust in addition to the inertia of the body and the actions of the gravitational force and air resistance, if any. The *free flight* of a projectile is the motion of a body by virtue of the inertia of the body and the actions of the gravitational force and air resistance, if any. A free flight is a *flight without a thrust* or an *unpowered flight*. The free-body diagrams of space vehicle in powered flight and in free flight are illustrated in Fig. 11.17, where $\mathbf{W}$ is the gravitational force, $\mathbf{F}_T$ is the thrust, and $\mathbf{F}_A$ is the friction force from the air.

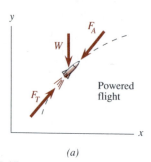
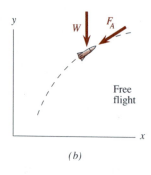

Powered flight

Free flight

(a)   *(b)*

Fig. 11.17

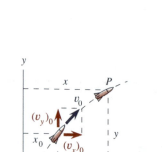

Fig. 11.18

The usual *assumptions* for the elementary study of the *free flight of a projectile* are as follows:

1. The projectile travels in a vertical plane near the surface of the earth; and the surface of the earth over which the projectile flies may be taken as flat and horizontal.
2. The friction between the air and the projectile is negligible.
3. The acceleration of the projectile is generated only by gravity; this acceleration points vertically downward and has a magnitude $g = 9.81$ m/s^2 or 32.2 ft/s^2.

Thus, the free flight of a projectile may readily be studied by using an xy rectangular coordinate system fixed to the surface of the earth and noting that the scalar components of its acceleration are

$$a_x = 0 \qquad a_y = -g \qquad (11.49)$$

Let the initial position of the projectile be at (x_0, y_0) and the scalar components of the initial velocity $\mathbf{v}_0$ be $(v_x)_0$ and $(v_y)_0$, as shown in Fig. 11.18. With these conditions and Eqs. (11.49), we can readily show that

$$v_x = \dot{x} = (v_x)_0 \qquad v_y = \dot{y} = (v_y)_0 - gt \qquad (11.50)$$

$$x = x_0 + (v_x)_0 t \qquad y = y_0 + (v_y)_0 t - \tfrac{1}{2} g t^2 \qquad (11.51)$$

$$v_y^2 = (v_y)_0^2 - 2g(y - y_0) \qquad (11.52)$$

We recognize that Eqs. (11.51) form the parametric equations of a parabola; i.e., the trajectory is *parabolic*. Of course, the trajectory would not be precisely parabolic if air resistance were taken into account.

Suppose that a projectile is launched at A with a muzzle velocity $\mathbf{v}_0$ as shown in Fig. 11.19, where the initial conditions are $x_0 = y_0 = 0$, $(v_x)_0 = v_0 \cos\theta$, and $(v_y)_0 = v_0 \sin\theta$. Substituting these conditions into Eqs. (11.51), we obtain

$$x = (v_0 \cos\theta)t \qquad y = (v_0 \sin\theta)t - \tfrac{1}{2} g t^2$$

For the projectile to hit a target at B, we have $y = h$ when $x = L$; i.e.,

$$L = (v_0 \cos\theta)t \qquad h = (v_0 \sin\theta)t - \tfrac{1}{2} g t^2$$

where t is the flying time of the projectile from A to B. Eliminating t between these two equations and simplifying, we get

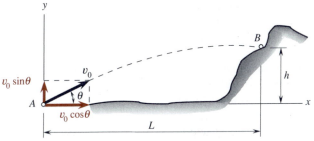

Fig. 11.19

$$\tan^2\theta - \frac{2v_0^2}{gL}\tan\theta + \left(1 + \frac{2v_0^2 h}{gL^2}\right) = 0 \qquad (11.53)$$

We infer from Eq. (11.53) that, for appropriate values of v_0, L, and h, it is possible for the projectile to hit the target at B in Fig. 11.19 in two different ways corresponding to two different values of the firing angle θ. A computer program to find such different values of θ as well as the corresponding flying times is shown in Fig. 11.20.

```
10 REM FIRING ANGLES & FLYING TIMES OF A PROJECTILE   (File Name: PROJECT)
15 REM          Refer to Eq. (11.53) and enter V0 (in meters/second),
20 REM          L (in meters), and h (in meters) as data in Line 100
25 REM   To output to the printer, change PRINT to LPRINT in the program.
30 READ V0,L,H: B=-2*V0^2/(9.81#*L): C=1-B*H/L: D=B^2-4*C: IF D<0 THEN 60
35 E1=(-B+SQR(D))/2:E2=(-B-SQR(D))/2:A1=ATN(E1):A2=ATN(E2):F=18C/3.141593
40 P1=A1*F: P2=A2*F: T1=L/(V0*COS(A1)): T2=L/(V0*COS(A2)): S$="          "
45 PRINT: PRINT "V0 ="; V0; "m/s"; S$; "L ="; L; "m"; S$; "h ="; H; "m"
50 PRINT "Firing angles (in degrees) ="; P1; "and"; P2
55 PRINT "Corresponding flying times (in seconds) ="; T1; "and"; T2: END
60 PRINT: PRINT "The projectile cannot reach the target.": END
100 DATA 300,4000,1000
```

```
V0 = 300 m/s        L = 4000 m        h = 1000 m
Firing angles (in degrees) = 76.15708 and 27.87916
Corresponding flying times (in seconds) = 55.72724 and 15.08405
```

Fig. 11.20 (Cf. the solution in Example 11.14 and the answer to D11.31.)

EXAMPLE 11.14

A projectile is launched at A with a muzzle velocity $\mathbf{v}_0$ as shown. If $v_0 = 300$ m/s, $L = 4$ km, and $h = 1$ km, determine the firing angle θ which will permit the projectile to hit the target at B.

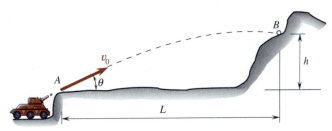

Solution. Substituting $v_0 = 300$ m/s, $L = 4000$ m, $h = 1000$ m, and $g = 9.81$ m/s^2 into Eq. (11.53), we obtain

$$\tan^2\theta - 4.587 \tan\theta + 2.147 = 0$$

which yields $\tan\theta = 4.058$ or 0.529. Thus, we find two solutions: $\theta = 76.16°$ and $\theta = 27.88°$.

$$\theta_1 = 76.2° \qquad \theta_2 = 27.9° \qquad \blacktriangleleft$$

This means that the target will be hit if either of these two firing angles is used as shown.

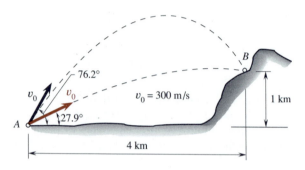

Developmental Exercises

D11.30 What are the usual assumptions for the elementary study of the free flight of a projectile?

D11.31 Refer to Example 11.14. Determine the flying time of the projectile from A to B if the firing angle θ is (a) 76.16°, (b) 27.88°.

D11.32 Refer to Example 11.14. Determine the time t at which the projectile attains the maximum height in its trajectory if the firing angle θ is (a) 76.16°, (b) 27.88°.

D11.33 Using the program in Fig. 11.20, "try" to find θ in Example 11.14 for the same data except that (a) $v_0 = 225$ m/s, (b) $v_0 = 220$ m/s.

PROBLEMS

11.50* The plane curvilinear motion of a particle is defined by $x = 16t - 2t^2$ and $y = 36 - t^2$, where x and y are measured in feet and t in seconds. Plot the *path* traveled by the particle for the interval $0 \le t \le 6$ s and determine its velocity and acceleration at $t = 6$ s.

11.51 A particle moves in the xy plane. The x component of its initial velocity is $(v_x)_0 = 18$ ft/s, the x component of its acceleration is $a_x = 6t$, and its y coordinate is $y = 4t^3 - 8t$, where t is measured in seconds, y in feet, and a_x in ft/s^2. Determine the velocity and acceleration of the particle when $t = 2$ s.

11.52 The position vector of a particle is given by $\mathbf{r} = (2 - t^3)\mathbf{i} + (3t - 1)\mathbf{j}$, where t is measured in seconds and r in meters. Determine the velocity and acceleration of the particle at $t = 2$ s.

11.53* The position vector **r** of a particle is given by $\mathbf{r} = (A \cos\omega t)\mathbf{i} + (B \sin\omega t)\mathbf{j}$, where t is the time variable, and A, B, and ω are positive constants. It is known that $A > B$. Show that (a) the path of the particle is an ellipse with A and B as the major and minor axes, (b) the acceleration is directed toward the origin, (c) the magnitude of the acceleration is proportional to r, not r^2.

11.54* The motion of a particle is defined by the position vector $\mathbf{r} = 2t\mathbf{i} + (4t^2 + 9t^4)\mathbf{j} + 3t^2\mathbf{k}$, where r is measured in feet and t in seconds. Show that the path of the particle lies on the paraboloid as shown. Determine the magnitudes of the velocity and acceleration when $t = 1$ s.

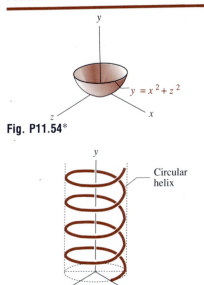

Fig. P11.54*

11.55 The motion of a particle is defined by the position vector $\mathbf{r} = 40 \cos t\mathbf{i} + 9t\mathbf{j} - 40 \sin t\mathbf{k}$, where r is measured in meters and t in seconds, and the path is a circular helix as shown. Determine the magnitudes of the velocity and acceleration of the particle.

11.56 The motions of two pins A and B are constrained in such a way that they remain in the vertical slot of the member C and in the circular path at all times as shown. If $\mathbf{v}_{B/A} = 40$ mm/s $\downarrow$ when $x = 100$ mm, determine $\mathbf{v}_C$ of the member C.

Fig. P11.55

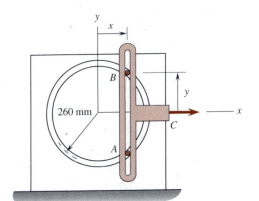

Fig. P11.56

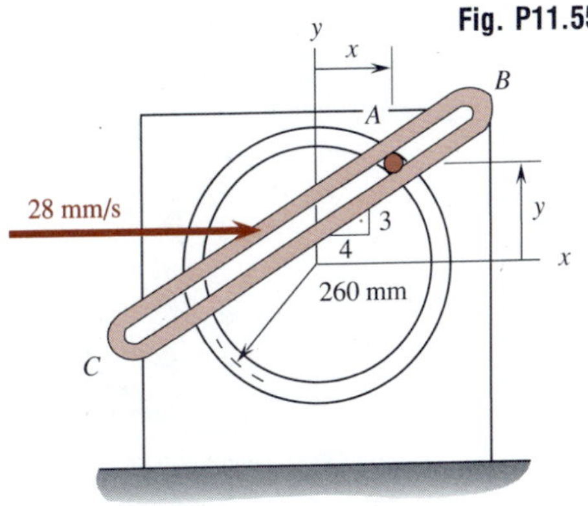

Fig. P11.58*

11.57 Refer to Prob. 11.56. If $\mathbf{v}_{B/A} = 40$ mm/s $\downarrow$ and $\mathbf{a}_{B/A} = 48$ mm/s^2 $\downarrow$ when $x = 100$ mm, determine $\mathbf{a}_C$ of the member C.

11.58* The motion of the pin A is constrained in such a way that it remains in the inclined slot of the member BC as well as in the circular path at all times as shown. It is known that the member BC maintains its slope of 3/4 and translates with a constant velocity of 28 mm/s $\rightarrow$. At $t = 0$, the y intercept of the inclined slot is 40 mm. (a) Show that the equation of the straight line along the inclined slot at any time t is given by $4y = 3(x - 28t) + 160$. (b) Determine $\mathbf{v}_A$ of the pin A when $x = 100$ mm.

11.59 Refer to Prob. 11.58*. Determine $\mathbf{a}_A$ of the pin A when $x = 100$ mm.

11.60 Water is discharged at A from a pressure tank with a horizontal velocity v_0 as shown. If $L = 4$ m, determine the range of values of v_0 for which the water will enter the drainage opening BC.

11.61 Water is discharged at A from a pressure tank with a velocity $v_0 = 5$ m/s $\rightarrow$ as shown. Determine the range of values of L for which the water will enter the drainage opening BC.

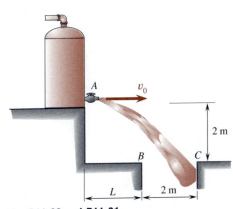

Fig. P11.60 and P11.61

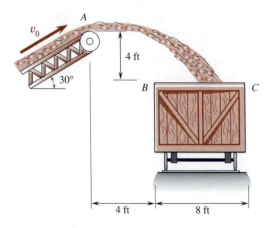

Fig. P11.62

11.62   A freight car is being loaded with coal as shown. Determine the range of values of the constant speed v_0 of the conveyor belt for which the coal will enter the opening BC of the freight car.

11.63   A ball is kicked from A with an initial velocity $\mathbf{v}_0$ which makes an angle θ with the horizontal as shown. Determine (a) the range L, traveled by the ball from A to B, in terms of v_0, θ, and g; (b) the value of θ for which L is maximum; (c) the maximum value of L in terms of v_0 and g; (d) the maximum height, reached in traveling over the maximum range L, in terms of v_0 and g.

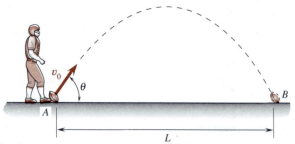

Fig. P11.63

11.64*   A ball is kicked from A with an initial velocity $\mathbf{v}_0$ inside a horizontal tunnel of height h as shown. If $v_0^2 > 4gh$, determine in terms of v_0, h, and g, (a) the angle θ for which the range L traveled by the ball from A to B is maximum, (b) the maximum value of L.

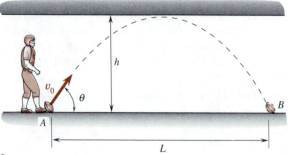

Fig. P11.64*

11.65   Solve Prob. 11.64* if $v_0^2 = 4gh$.

11.66   Solve Prob. 11.64* if $v_0 = 12$ m/s and $h = 5$ m.

11.67   Solve Prob. 11.64* if $v_0 = 15$ m/s and $h = 5$ m.

11.68*   A bale of hay is to be dropped from an airplane A so that it lands at B on a snow-covered ground of a ranch as shown. If the airplane is flying horizontally at an altitude $h = 400$ ft with a speed $v_A = 120$ mi/h, determine the angle θ between the horizontal and the line of sight AB when the bale of hay is dropped.

11.69   Solve Prob. 11.68* if $h = 100$ m and $v_A = 180$ km/h.

11.70   An airplane A is flying horizontally at an altitude of 2 km and with a constant speed of 900 km/h on a path which passes directly over an antiaircraft gun at B. The gun fires a shell with a muzzle velocity of 550 m/s and hits the airplane at C. Knowing that the firing angle of the gun is 60°, determine the angle θ between the horizontal and the line of sight BA when the gun is fired.

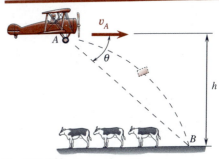

Fig. P11.68*

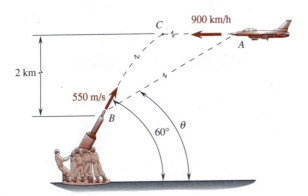

Fig. P11.70

11.71*   Refer to Prob. 11.70. Determine the velocity and acceleration of the shell relative to the airplane at the time of impact at C.

11.72   A rifleman fires a bullet with a muzzle velocity of 1800 ft/s as shown. If the barrel of the rifle is aimed at B and the bullet strikes at C, determine the distance δ between B and C.

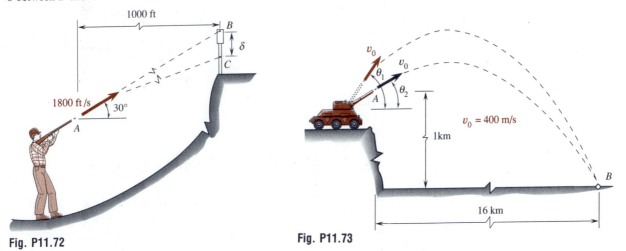

Fig. P11.72   **Fig. P11.73**

11.73   A gun fires a shell with a muzzle velocity of 400 m/s as shown. Determine (a) the two values of the firing angle θ with which the shell will hit the target at B, (b) the time of flight of the shell from A to B corresponding to each value of θ.

11.74   An airplane has a cruising airspeed of 130 mi/h. It leaves an airport to fly to another which is 186 mi due north of the first. A wind of 50 mi/h is blowing steadily from the east. Determine the shortest time which the trip will take.

11.75*   A 2-km wide river flows at a speed of 6 km/h as shown. A motorboat is used to go from A to B. The boat can travel at a maximum speed of 16 km/h relative to the water. Determine (a) the minimum time which the trip will take, (b) the angle θ at which the boat should be oriented.

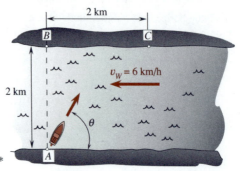

Fig. P11.75*

11.76   Solve Prob. 11.75* if the motorboat is used to go from A to C.

11.77   Cars B and C are traveling at constant speeds along straight highways as shown. The car C passes over the point A on the bridge at the time $t = 0$, and the car B passes under the point A at the time $t = 2$ s. Determine (a) the velocity of the car B relative to the car C, (b) the change in position of the car B relative to the car C from $t = 2$ s to $t = 4$ s, (c) the distance between the two cars when $t = 5$ s.

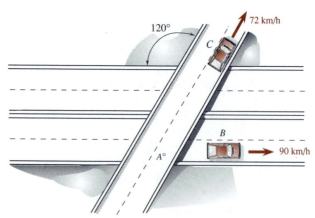

Fig. P11.77

11.13   Tangential and Normal Components

We saw in Sec. 11.9 that the velocity of a particle is a vector tangent to the path of the particle but that the acceleration vector is, in general, not tangent to the path. To facilitate the study of curvilinear motion, the acceleration vector is sometimes resolved into components directed, respectively, along the *tangent* and the *normal* to the path of the particle. We shall denote the unit vector directed along the tangent by $\mathbf{e}_t$ and the unit vector directed along the normal toward the center of curvature of the path by $\mathbf{e}_n$.

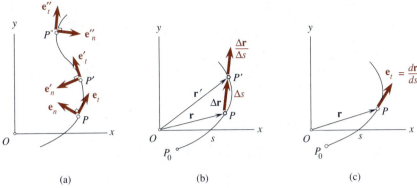

Fig. 11.21

The *tangential unit vector* $\mathbf{e}_t$ and the *normal unit vector* $\mathbf{e}_n$ are considered to be attached to the particle and to move along the path with the particle as seen in Fig. 11.21(a), where the particle moves from P to P' to P''. Unlike the unit vectors $\mathbf{i}$ and $\mathbf{j}$ in a fixed rectangular coordinate system, the tangential and normal unit vectors $\mathbf{e}_t$ and $\mathbf{e}_n$ will generally change their directions as the particle moves from one point to another along the curved path. Thus, $\mathbf{e}_t$ and $\mathbf{e}_n$ are *unit vectors* but *not constant vectors.*[†] From parts (b) and (c) of Fig. 11.21, we see that $\mathbf{e}_t$ may be obtained from the derivative of the position vector $\mathbf{r}$ with respect to the path variable s. We write

$$\Delta\mathbf{r} = \mathbf{r}' - \mathbf{r} \qquad \lim_{\Delta s \to 0} |\Delta\mathbf{r}| = \Delta s$$

$$\mathbf{e}_t = \lim_{\Delta s \to 0} \frac{\Delta\mathbf{r}}{\Delta s} = \frac{d\mathbf{r}}{ds} \qquad \boxed{\frac{d\mathbf{r}}{ds} = \mathbf{e}_t} \qquad (11.54)$$

$$|\mathbf{e}_t| = \lim_{\Delta s \to 0} \frac{|\Delta\mathbf{r}|}{|\Delta s|} = \lim_{\Delta s \to 0} \frac{|\Delta s|}{|\Delta s|} = 1$$

Thus, $\mathbf{e}_t$ as given by Eq. (11.54) is indeed a unit vector and is tangent to the path of the particle at the point P.

Let ρ be the *radius of curvature* of the curved path at the point P as shown in Fig. 11.22(a). In this figure, we note that (1) the tangential unit vectors at P and its neighboring point P' are denoted by $\mathbf{e}_t$ and $\mathbf{e}_t'$, respec-

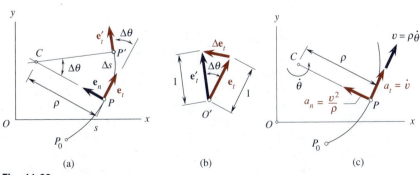

Fig. 11.22

[†]Recall that a vector is changed whenever its *magnitude*, or its *direction*, or *both* are changed.

tively, (2) the normal unit vector $\mathbf{e}_n$ at the point P is directed toward the center of curvature C, and (3) the angle between the radii of curvature at P and P' is denoted by $\Delta\theta$. The angle between $\mathbf{e}_t$ and $\mathbf{e}_t'$ is, of course, also $\Delta\theta$. By drawing $\mathbf{e}_t$ and $\mathbf{e}_t'$ from the point O' as shown in Fig. 11.22(b), we see that the change in the tangential unit vector $\mathbf{e}_t$ as the particle moves from P to P' is a vector $\Delta\mathbf{e}_t$ as shown. From parts (a) and (b) of Fig. 11.22, we write

$$\Delta s = PP' = \rho\,\Delta\theta \qquad \Delta\mathbf{e}_t = \mathbf{e}_t' - \mathbf{e}_t \approx \Delta\theta\,\mathbf{e}_n$$

$$\frac{d\theta}{ds} = \lim_{\Delta s \to 0} \frac{\Delta\theta}{\Delta s} = \lim_{\Delta s \to 0} \frac{\Delta\theta}{\rho\,\Delta\theta} = \frac{1}{\rho}$$

$$\frac{d\mathbf{e}_t}{d\theta} = \lim_{\Delta\theta \to 0} \frac{\Delta\mathbf{e}_t}{\Delta\theta} = \lim_{\Delta\theta \to 0} \frac{\Delta\theta\,\mathbf{e}_n}{\Delta\theta} = \mathbf{e}_n$$

$$\boxed{\frac{d\theta}{ds} = \frac{1}{\rho}} \qquad \boxed{\frac{d\mathbf{e}_t}{d\theta} = \mathbf{e}_n} \tag{11.55}$$

Note that $1/\rho$ represents the curvature κ of the path at the point P.

Using the preceding equations, we derive the velocity $\mathbf{v}$ and the acceleration $\mathbf{a}$ of a particle in curvilinear motion as follows:

$$v = \frac{ds}{dt} = \lim_{\Delta t \to 0} \frac{\Delta s}{\Delta t} = \lim_{\Delta t \to 0} \frac{\rho\,\Delta\theta}{\Delta t} = \rho\frac{d\theta}{dt} = \rho\dot{\theta}$$

$$\mathbf{v} = \frac{d\mathbf{r}}{dt} = \frac{d\mathbf{r}}{ds}\frac{ds}{dt} = \mathbf{e}_t v = v\mathbf{e}_t$$

$$\mathbf{a} = \frac{d\mathbf{v}}{dt} = \frac{d}{dt}(v\mathbf{e}_t) = \frac{dv}{dt}\mathbf{e}_t + v\frac{d\mathbf{e}_t}{dt}$$

$$= \frac{dv}{dt}\mathbf{e}_t + v\frac{d\mathbf{e}_t}{d\theta}\frac{d\theta}{ds}\frac{ds}{dt}$$

$$= \dot{v}\mathbf{e}_t + v(\mathbf{e}_n)\left(\frac{1}{\rho}\right)(v) = \dot{v}\mathbf{e}_t + \frac{v^2}{\rho}\mathbf{e}_n$$

$$= \mathbf{a}_t + \mathbf{a}_n = a_t\mathbf{e}_t + a_n\mathbf{e}_n$$

$$v = \frac{ds}{dt} = \rho\dot{\theta} \qquad a_t = \dot{v} \qquad a_n = \frac{v^2}{\rho} = \rho\dot{\theta}^2 \tag{11.56}$$

$$\boxed{\mathbf{v} = v\mathbf{e}_t} \qquad \boxed{\mathbf{a} = \dot{v}\mathbf{e}_t + \frac{v^2}{\rho}\mathbf{e}_n} \tag{11.57}$$

where a_t and a_n are, respectively, the scalar components of the acceleration in the tangential and normal directions of the path of the moving particle as shown in Fig. 11.22(c). Note in this figure that (1) $\mathbf{v}$ is always *tangent to the path* as the particle moves along the path, (2) the sense of $\mathbf{a}_t$ is the same as that of $\mathbf{v}$ if $\dot{v}$ is positive, (3) $\mathbf{a}_n$ is always directed from P toward the center of curvature C, and (4) $\dot{\theta}$ is geometrically the angular speed of the radius of curvature $\rho = \overline{CP}$. If the path is a circular arc, the radius of curvature ρ becomes a constant equal to the radius of the circular arc, and $\dot{\theta}$ is equal to the angular speed of the radius of curvature joining the center C and the particle at P.

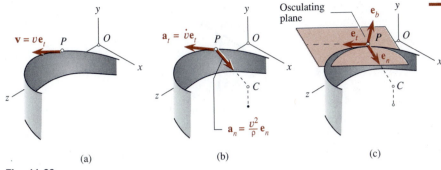

Fig. 11.23

The relations given in Eqs. (11.56) and (11.57) still hold in the case of a particle moving along a space curve as shown in parts (a) and (b) of Fig. 11.23, where e_t is directed along the tangent to the curve at the point P, and e_n is perpendicular to e_t and points toward the center of curvature C of the curve at P. The plane containing e_t and e_n is called the *osculating plane* at P as shown in Fig. 11.23(c), where e_b is called the *binormal unit vector* at P and is given by

$$e_b = e_t \times e_n \qquad (11.58)^\dagger$$

Thus, e_t, e_n, and e_b form a right-handed trihedral. In distinction from e_b, the vector e_n is sometimes called the *principal normal unit vector*. It is obvious from Eqs. (11.56) that both the acceleration and the velocity vectors lie entirely in the osculating plane at P, and the acceleration is always directed to the *concave side* of the curved path.

Developmental Exercises

D11.34   Are the unit vectors e_t and e_n *constant* vectors?

D11.35   Define the terms: (a) osculating plane, (b) binormal unit vector, (c) principal normal unit vector.

D11.36   The rod AB rotates clockwise inside a fixed pipe as shown. The direction of the velocity of the end B relative to the end A is (a) ↑, (b) ↓, (c) ↙, (d) ↖, (e) →, (f) ←.

Fig. D11.36

EXAMPLE 11.15

A nozzle discharges a stream of water with an initial velocity of 20 m/s as shown. Determine the radius of curvature ρ_A of the stream as it leaves the nozzle at A.

Solution.   Immediately after leaving the nozzle, the stream is subjected to the downward gravitational acceleration $a = g = 9.81$ m/s^2. Thus, the nor-

---

†In differential geometry, the magnitude of the derivative de_b/ds is called the *torsion* τ of the space curve, and $1/\tau$ is called the *radius of torsion* σ.

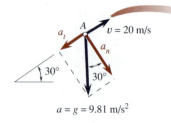

$$a = g = 9.81 \text{ m/s}^2$$

mal component of acceleration of the stream as it leaves the nozzle at A is

$$a_n = 9.81 \cos 30° = 4.905 \sqrt{3}$$

Since $v = 20$ m/s at A, we write

$$a_n = \frac{v^2}{\rho} \qquad 4.905 \sqrt{3} = \frac{(20)^2}{\rho_A} \qquad \rho_A = 47.1 \text{ m} \quad \blacktriangleleft$$

Developmental Exercises

D11.37 Refer to Example 11.15. Determine the radius of curvature ρ_B of the stream at the point B where the height of the stream is maximum.

D11.38 A particle moves with a *constant speed* of 3 m/s along a circular path of radius 2 m. What is the magnitude of its acceleration?

11.14 Radial and Transverse Components

In some problems of plane motion of particles, it is desirable to use polar coordinates to describe the motion. The position of the particle at P is defined by the polar coordinates r and θ as shown in parts (a) and (b) of Fig. 11.24, where r is a length and θ is an angle in radians. The unit vectors in the *radial and transverse* directions are denoted by $\mathbf{e}_r$ and $\mathbf{e}_\theta$, respectively, as shown.

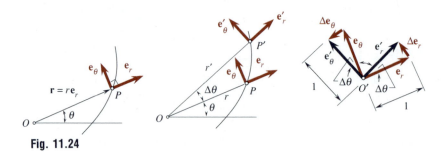

Fig. 11.24

Note that the *radial unit vector* $\mathbf{e}_r$ is directed outward along $\overrightarrow{OP}$, *and the transverse unit vector* $\mathbf{e}_\theta$ is obtained by rotating $\mathbf{e}_r$ through 90° counterclockwise. As the particle moves from the point P to its neighboring point P', the vectors $\mathbf{e}_r$ and $\mathbf{e}_\theta$ change to $\mathbf{e}'_r$ and $\mathbf{e}'_\theta$ by amounts $\Delta\mathbf{e}_r$ and $\Delta\mathbf{e}_\theta$, respectively, as shown in parts (b) and (c) of Fig. 11.24. A derivation similar to the one employed in establishing Eqs. (11.55) will yield the relations

$$\frac{d\mathbf{e}_r}{d\theta} = \mathbf{e}_\theta \qquad \frac{d\mathbf{e}_\theta}{d\theta} = -\mathbf{e}_r \qquad (11.59)$$

Expressing the position vector $\mathbf{r}$ of the particle at P as

$$\mathbf{r} = r\mathbf{e}_r \qquad (11.60)$$

and using Eqs. (11.59) and (11.60), we derive the velocity and acceleration of the particle as follows:

$$\mathbf{v} = \frac{d\mathbf{r}}{dt} = \frac{dr}{dt}\mathbf{e}_r + r\frac{d\mathbf{e}_r}{dt} = \frac{dr}{dt}\mathbf{e}_r + r\frac{d\mathbf{e}_r}{d\theta}\frac{d\theta}{dt} = \dot{r}\mathbf{e}_r + r\mathbf{e}_\theta\dot{\theta}$$

$$\mathbf{v} = \dot{r}\mathbf{e}_r + r\dot{\theta}\mathbf{e}_\theta \qquad (11.61)$$

$$\mathbf{a} = \frac{d\mathbf{v}}{dt} = \ddot{r}\mathbf{e}_r + \dot{r}\frac{d\mathbf{e}_r}{d\theta}\frac{d\theta}{dt} + \dot{r}\dot{\theta}\mathbf{e}_\theta + r\ddot{\theta}\mathbf{e}_\theta + r\dot{\theta}\frac{d\mathbf{e}_\theta}{d\theta}\frac{d\theta}{dt}$$

$$= \ddot{r}\mathbf{e}_r + \dot{r}\mathbf{e}_\theta\dot{\theta} + \dot{r}\dot{\theta}\mathbf{e}_\theta + r\ddot{\theta}\mathbf{e}_\theta + r\dot{\theta}(-\mathbf{e}_r)\dot{\theta}$$

$$\mathbf{a} = (\ddot{r} - r\dot{\theta}^2)\mathbf{e}_r + (r\ddot{\theta} + 2\dot{r}\dot{\theta})\mathbf{e}_\theta \qquad (11.62)$$

If we write

$$\mathbf{v} = \mathbf{v}_r + \mathbf{v}_\theta = v_r\mathbf{e}_r + v_\theta\mathbf{e}_\theta \qquad \mathbf{a} = \mathbf{a}_r + \mathbf{a}_\theta = a_r\mathbf{e}_r + a_\theta\mathbf{e}_\theta$$

we see that the r and θ components of the velocity and acceleration are

$$v_r = \dot{r} \qquad v_\theta = r\dot{\theta} \qquad (11.63)$$

$$a_r = \ddot{r} - r\dot{\theta}^2 \qquad a_\theta = r\ddot{\theta} + 2\dot{r}\dot{\theta} \qquad (11.64)$$

In the case of a particle moving along a circle with its center at the origin O, we have $r = $ constant, $\dot{r} = \ddot{r} = 0$, and the formulas in Eqs. (11.61) and (11.62) degenerate to

$$\mathbf{v} = r\dot{\theta}\mathbf{e}_\theta \qquad \mathbf{a} = -r\dot{\theta}^2\mathbf{e}_r + r\ddot{\theta}\mathbf{e}_\theta \qquad (11.65)$$

EXAMPLE 11.16

As the gear A rolls on the fixed gear B, the point P on the gear A describes a cardioid defined by the equations $r = 1 + \cos\pi t$ and $\theta = \pi t$, where r is measured in feet, θ in radians, and t in seconds. When $t = 0$, P is situated at P_0 as shown. Determine the velocity and acceleration of P when $t = \frac{1}{4}$ s.

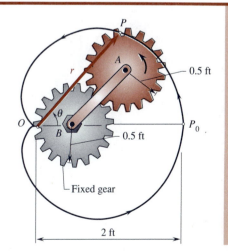

Solution.   When $t = \frac{1}{4}$ s, we have

$r = 1 + \cos\pi t = 1 + 1/\sqrt{2}$ $\theta = \pi t = \pi/4$

$\dot{r} = -\pi\sin\pi t = -\pi/\sqrt{2}$ $\dot{\theta} = \pi$

$\ddot{r} = -\pi^2\cos\pi t = -\pi^2/\sqrt{2}$ $\ddot{\theta} = 0$

$v_r = \dot{r} = -\pi/\sqrt{2} = -2.22$ $v_\theta = r\dot{\theta} = (1 + 1/\sqrt{2})\pi = 5.36$

$v = (v_r^2 + v_\theta^2)^{1/2} = 5.80$ $\beta = \tan^{-1}(v_\theta/|v_r|) = 67.5°$

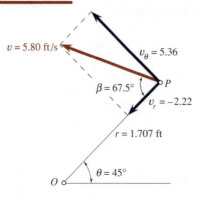

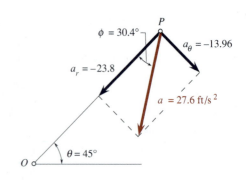

$$a_r = \ddot{r} - r\dot{\theta}^2 = -23.8 \qquad a_\theta = r\ddot{\theta} + 2\dot{r}\dot{\theta} = -13.96$$

$$a = (a_r^2 + a_\theta^2)^{1/2} = 27.6 \qquad \phi = \tan^{-1}(|a_\theta|/|a_r|) = 30.4°$$

Besides those shown in the sketch, we may write

$$\mathbf{v} = 5.80 \text{ ft/s } \measuredangle \; 22.5° \qquad \mathbf{a} = 27.6 \text{ ft/s}^2 \; \measuredangle \; 75.4° \quad \blacktriangleleft$$

Developmental Exercises

D11.39   In polar coordinates, is it true that (a) $a_r = \dot{v}_r$, (b) $a_\theta = \dot{v}_\theta$?

D11.40   Refer to Example 11.16. Determine the velocity and acceleration of P when $t = \frac{1}{2}$ s.

11.15   Cylindrical Components

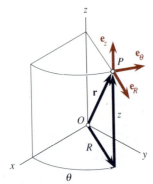

Fig. 11.25

In some cases, the motion of a particle along a space curve is best described by using a cylindrical coordinate system. In such cases, the position of a particle P in space is defined by its cylindrical coordinates R, θ, and z, as shown in Fig. 11.25, where $\mathbf{e}_R$, $\mathbf{e}_\theta$, and $\mathbf{e}_z$ are the *radial*, *transverse*, and *axial unit vectors* attached to P, respectively. Observe that $\mathbf{e}_R$ and $\mathbf{e}_\theta$ are unit vectors parallel to the xy plane; they are not constant vectors because their directions may change as P moves from one point to another. However, the axial unit vector $\mathbf{e}_z$, which is identical with the cartesian unit vector $\mathbf{k}$ in the fixed xyz coordinate system, is a constant vector because it is constant in direction and magnitude as P moves.

Expressing the position vector $\mathbf{r}$ of the particle P in terms of its radial and axial components, we write

$$\mathbf{r} = R\mathbf{e}_R + z\mathbf{e}_z \tag{11.66}$$

A procedure similar to the one used in deriving Eqs. (11.61) and (11.62) will lead to the formulas for the velocity $\mathbf{v}$ and acceleration $\mathbf{a}$ as

$$\mathbf{v} = \dot{R}\mathbf{e}_R + R\dot{\theta}\mathbf{e}_\theta + \dot{z}\mathbf{e}_z \tag{11.67}$$

$$\mathbf{a} = (\ddot{R} - R\dot{\theta}^2)\mathbf{e}_R + (R\ddot{\theta} + 2\dot{R}\dot{\theta})\mathbf{e}_\theta + \ddot{z}\mathbf{e}_z \tag{11.68}$$

where the coefficients of $\mathbf{e}_R$, $\mathbf{e}_\theta$, and $\mathbf{e}_z$ are respectively the *radial*, transverse, and *axial components* of motion of the particle.

EXAMPLE 11.17

The motion of a particle along a spiral path on the surface of a right circular frustrum is defined by the cylindrical coordinates $R = 4 + \frac{1}{4}t^2$, $\theta = 2\pi t$, $z = t^2$, where R and z are measured in meters, θ in radians, and t in seconds. Determine the magnitudes of the velocity and acceleration of the particle when $t = 2$ s.

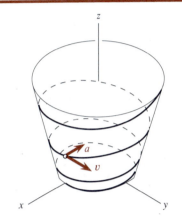

Solution. The velocity $\mathbf{v}$ and the acceleration $\mathbf{a}$ are given by Eqs. (11.67) and (11.68). When $t = 2$ s, we write

$$R = 4 + \tfrac{1}{4}t^2 = 5 \qquad \theta = 2\pi t = 4\pi \qquad z = t^2 = 4$$

$$\dot{R} = \tfrac{1}{2}t = 1 \qquad \dot{\theta} = 2\pi \qquad \dot{z} = 2t = 4$$

$$\ddot{R} = \tfrac{1}{2} \qquad \ddot{\theta} = 0 \qquad \ddot{z} = 2$$

$$\mathbf{v} = \dot{R}\mathbf{e}_R + R\dot{\theta}\mathbf{e}_\theta + \dot{z}\mathbf{e}_z = \mathbf{e}_R + 10\pi\mathbf{e}_\theta + 4\mathbf{e}_z$$

$$v^2 = (1)^2 + (10\pi)^2 + (4)^2 \qquad v = 31.7 \text{ m/s} \blacktriangleleft$$

$$\mathbf{a} = (\ddot{R} - R\dot{\theta}^2)\mathbf{e}_R + (R\ddot{\theta} + 2\dot{R}\dot{\theta})\mathbf{e}_\theta + \ddot{z}\mathbf{e}_z$$

$$= (\tfrac{1}{2} - 20\pi^2)\mathbf{e}_R + 4\pi\mathbf{e}_\theta + 2\mathbf{e}_z$$

$$a^2 = (\tfrac{1}{2} - 20\pi^2)^2 + (4\pi)^2 + (2)^2 \qquad a = 197.3 \text{ m/s}^2 \blacktriangleleft$$

Developmental Exercises

D11.41 Refer to Example 11.17. At the instant $t = 2$ s, is it true that $\mathbf{e}_R = \mathbf{i}$, $\mathbf{e}_\theta = \mathbf{j}$, and $\mathbf{e}_z = \mathbf{k}$?

D11.42 Refer to Example 11.17 and the answer to D11.41. For the instant $t = 2$ s, determine (a) $\mathbf{v} \times \mathbf{a}$, (b) the binormal unit vector $\mathbf{e}_b$, (c) the angle between $\mathbf{e}_b$ and the z axis, (d) the angle between the osculating plane and the xy plane.

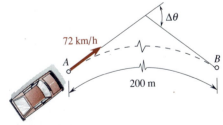

Fig. P11.78*

PROBLEMS

11.78* A ball is thrown upward at the position A from a car traveling at a constant velocity as shown. Determine the radius of curvature of the path of the ball when the ball (a) leaves the position A, (b) passes through its highest point B.

11.79 A car travels at a constant speed of 72 km/h along the curve AB of constant curvature as shown. An accelerometer mounted in the car indicates that the magnitude of acceleration of the car is 2.5 m/s². Determine the change of direction $\Delta\theta$ of the car from A to B.

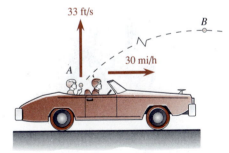

Fig. P11.79

Fig. P11.80

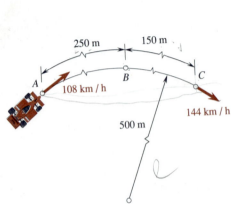

Fig. P11.82

11.80   A car travels with a steady speed of v over a vertical curve whose radius of curvature is 100 m as shown. If the magnitude of acceleration of the car is not to exceed $0.5g$, determine the maximum value of v permitted.

11.81*   Upon entering a curve of 200-m radius at a speed of 72 km/h, a motorist applies his brakes to decrease his speed at a constant rate of 2 m/s². Determine the magnitude of the total acceleration of the car when its speed is 54 km/h.

11.82   A racing car travels along the curve ABC of radius 500 m as shown. The speed of the car is increased at a constant rate from 108 km/h at A to 144 km/h at C. Determine the magnitude of the total acceleration of the car when it passes through the point B.

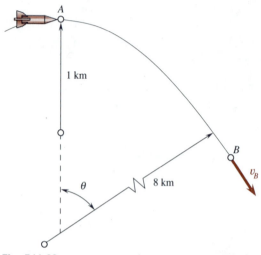

Fig. P11.83

11.83   A projectile travels along the trajectory AB near the surface of the earth as shown. The radius of curvature at the highest point A of the trajectory is 1 km, and that at the point B is 8 km. Determine the speed v_B and the angle θ between the radii as indicated.

11.84   A spacecraft will coast indefinitely in a circular "parking" orbit around the earth if it travels with $a_n = g(R/r)^2$, where $g = 32.2$ ft/s², R = radius of the earth = 3.96×10^3 mi, and r = distance between the spacecraft and the center of the earth. The altitude h of a space shuttle in one of its flights in such an orbit is 250 mi as shown. Determine (a) the speed v of the shuttle in the orbit, (b) the time τ required for the shuttle to coast in the orbit one revolution.

11.85*   The time required for the moon to revolve around the earth once is 27.32 d. Using the information given in Prob. 11.84 and assuming the orbit of the moon around the earth to be a circle,[†] estimate the mean distance between the center of the earth and the center of the moon.

11.86   The position vector of a particle is given by $\mathbf{r} = 12t^2\mathbf{i} + 3t^3\mathbf{j}$, where $\mathbf{r}$ is measured in meters and t in seconds. Determine the radius of curvature of the path for the position at $t = 2$ s.

11.87   A skier travels down a slope as shown, where x and y are measured in feet. If his normal component of acceleration at the position P is equal to $0.5g$, determine his speed at this position.

Fig. P11.84

[†]The orbit of the moon around the earth is close to an ellipse having an eccentricity of 0.055.

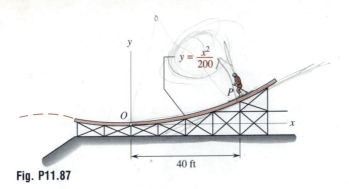

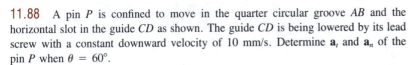

Fig. P11.87

11.88 A pin P is confined to move in the quarter circular groove AB and the horizontal slot in the guide CD as shown. The guide CD is being lowered by its lead screw with a constant downward velocity of 10 mm/s. Determine $\mathbf{a}_t$ and $\mathbf{a}_n$ of the pin P when $\theta = 60°$.

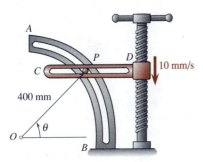

Fig. P11.88

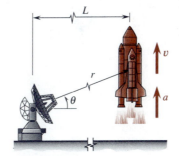

Fig. P11.89

11.89 A space shuttle is launched vertically by firing its main engine and two booster rockets and is tracked by the radar as shown. When $\theta = \pi/3$ rad, the measurements give $r = 10$ km, $\ddot{r} = 20$ m/s^2, and $\dot{\theta} = 0.02$ rad/s. Determine the velocity and acceleration of the shuttle at this position.

11.90 Determine the velocity of the shuttle in Prob. 11.89 in terms of L, θ, and $\dot{\theta}$.

11.91* Determine the acceleration of the shuttle in Prob. 11.89 in terms of L, θ, $\dot{\theta}$, and $\ddot{\theta}$.

11.92 through 11.94* A particle moves along a curved path defined by the polar coordinates r and θ as shown, where $\theta = (\pi/6)t$, r is measured in meters, θ in radians, and t in seconds. Determine the velocity and acceleration of the particle when it passes through the point P.

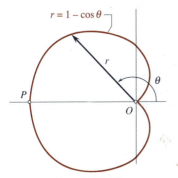

Fig. P11.92 and P11.95

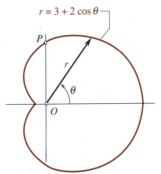

Fig. P11.93 and P11.96

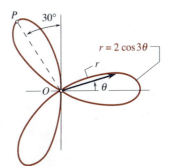

Fig. P11.94* and P11.97*

11.95 through 11.97* A particle moves along a curved path defined by the polar coordinates r and θ as shown, where $\theta = (\pi/12)t^2$, r is measured in feet, θ in radians, and t in seconds. Determine the magnitudes of the velocity and acceleration of the particle when $t = 2$ s.

11.98 The rod BP rotates clockwise at a constant rate of $\pi/4$ rad/s. The pin at P is free to slide along the slot of the arm OA as shown. Determine the values of $\dot{r}$, $\ddot{r}$, $\dot{\theta}$, and $\ddot{\theta}$ at the instant when $\phi = 90°$.

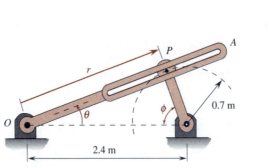

Fig. P11.98

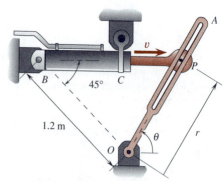

Fig. P11.99

11.99 The hydraulic cylinder gives the pin at P a constant velocity $\mathbf{v} = 60$ mm/s $\rightarrow$ as shown. The pin is free to slide along the slot in the arm OA. Determine the values of $\dot{r}$, $\ddot{r}$, $\dot{\theta}$, and $\ddot{\theta}$ at the instant when $\theta = 60°$.

11.100* The motion of a particle P along a circular helix as shown is defined by the cylindrical coordinates $R = 4$, $\theta = \pi t$, $z = t$, where R and z are measured in meters, θ in radians, and t in seconds. Determine the velocity and acceleration of the particle.

11.101 Refer to Prob. 11.100*. Determine the angle between the osculating plane and the xy plane.

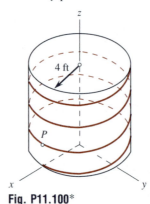

Fig. P11.100*

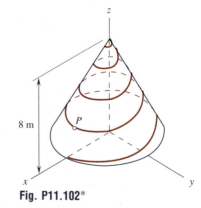

Fig. P11.102*

11.102* The path of the particle P as shown is defined by the cylindrical coordinates $R = 4 - \frac{1}{2}t$, $\theta = \pi t$, and $z = t$, where R and z are measured in meters, θ in radians, and t in seconds. Determine the magnitudes of the velocity and acceleration of the particle when $t = 4$ s.

11.103 The motion of the end P of the 20-m boom AP of the crane as shown is defined by the cylindrical coordinates $R = 20 \sin(\pi/30) t$, $\theta = (\pi/15) t$, and $z = 1 + 20 \cos(\pi/30)t$, where R and z are measured in meters, θ in radians, and t in seconds. Determine the magnitudes of the velocity and acceleration of the end P when $\theta = \pi/3$ rad.

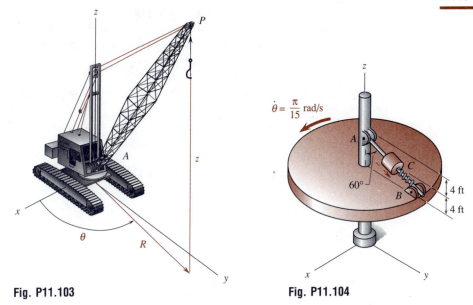

Fig. P11.103   **Fig. P11.104**

11.104   A collar C oscillates along a rod AB as shown. The turntable rotates with a constant speed $\pi/15$ rad/s about the z axis. At the instant shown, the collar C is moving downward with a velocity of 2 ft/s and an acceleration of 3 ft/s² relative to the rod AB. Determine the magnitudes of the velocity and acceleration of the collar relative to the ground at this instant. (*Hint*. Use cylindrical coordinates.)

11.16   Concluding Remarks

The concepts of position, displacement, total distance traveled, time, velocity, speed, acceleration, deceleration, and jerk are very fundamental. They represent the basic kinematic quantities studied in this chapter. It is pointed out that when the acceleration or velocity of a rectilinear motion is given by different linear functions of time over several intervals, the use of graphical solution (described in Sec. 11.8) may be advantageous. In such a case, the six useful relations for the graphical solution given in that section should be very helpful in achieving effective solutions.

The use of different systems of components to describe the curvilinear motion of particles is presented in the latter half of this chapter. The free flight of a projectile as well as certain constrained motions can conveniently be described by using rectangular components. The use of tangential and normal components is often desirable when velocity, acceleration, and radius of curvature are to be related. The use of the system of radial and transverse components and the system of cylindrical components is largely influenced by the geometry of the path of the particle. We shall see in Chap. 12 that the free flight of a spacecraft is best described using radial and transverse components.

Nowadays, programmable calculators and microcomputers are often utilized in analyzing and solving problems. In addition to finding roots of any nonlinear equations, the *digital root finder* in App. E may be utilized to plot the approximate shapes of some plane curves. For example, we may use the 40-column version of that root finder (named DIGIT40 and stored on the

TUTORIAL diskette as described in App. F) to plot the motion curves in Example 11.1. The *x-t* curve, given by $x = t^3 - 6t^2 - 36t + 116$, may be plotted as shown in Fig. 11.26 by writing Line 1000 of the program as

$$1000 \; Y = X\char`\^3 - 6*X\char`\^2 - 36*X + 116$$

The *v-t* curve, given by $v = 3t^2 - 12t - 36$, may be plotted as shown in Fig. 11.27 by writing Line 1000 of the program as

$$1000 \; Y = 3*X\char`\^2 - 12*X - 36$$

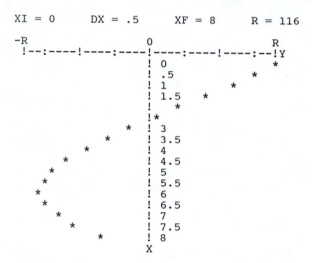

Fig. 11.26   1000 Y = X^3 - 6*X^2 - 36*X + 116

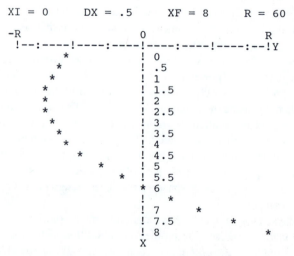

Fig. 11.27   1000 Y = 3*X^2 - 12*X - 36

When the program is run, the computer will ask for the values of XI, DX, and XF, which are, respectively, the initial, the incremental, and the final values of the independent variable X of the function Y. Thereafter, the computer takes care of the rest. Note that the Y and X axes in Figs. 11.26 and 11.27[†] should be understood as the x and t, and the v and t, axes, respectively, in the context of the present application.

REVIEW PROBLEMS

11.105 The acceleration of an oscillator is given by $a = -\omega^2 x$, where ω is a constant. If $v = 6$ m/s when $x = 0$ and $v = 0$ when $x = 2$ m, determine (a) the value of ω, (b) the time elapsed during which the oscillator moves from $x = 2$ m back to $x = 0$.

Fig. P11.105

11.106 The velocities of the cylinders A and C as shown are $\mathbf{v}_A = 1$ m/s ↑ and $\mathbf{v}_C = 6$ m/s ↓, respectively. Determine $\mathbf{v}_B$ of the cylinder B.

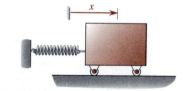

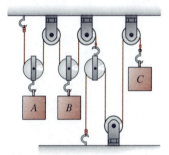

Fig. P11.106

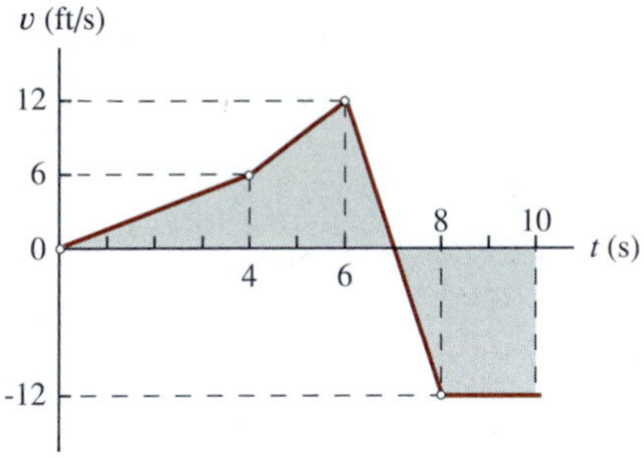

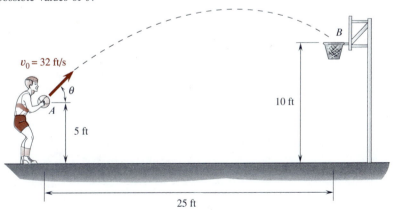

Fig. P11.107

11.107 A particle moves in a straight line with the velocity as shown. For $x_0 = -21$ ft and over the time interval $0 \leq t \leq 10$ s, determine Δx and x_T, as well as the two values t_1 and t_2 of t for which the particle passes through the origin.

11.108 A basketball is thrown with an initial velocity $\mathbf{v}_0$ and enters the basket at B as shown. Determine the possible values of θ.

Fig. P11.108

[†]It is true that one may use other plotters to plot these curves. However, most plotters are not intended to find roots to many digits of accuracy as the *digital root finder* in App. E.

11.109   An earth satellite has a velocity $\mathbf{v}_B$ as it passes through the end B of the semiminor axis of its orbit as shown. Knowing that the acceleration of the satellite is $\mathbf{a} = \mathbf{a}_r = -g(R/r)^2\mathbf{e}_r$, where $g = 9.81$ m/s^2, determine the radius of curvature of the orbit at B.

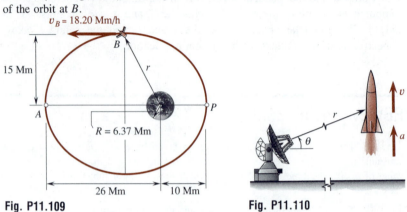

Fig. P11.109   **Fig. P11.110**

11.110   A rocket moves vertically as shown. When $\theta = 30°$, the radar measurements give $r = 8$ km, $\dot{\theta} = 0.025$ rad/s, and $\ddot{\theta} = 0.005$ rad/s^2. Determine the velocity and acceleration of the rocket.

Kinetics of Particles: Force and Acceleration

Chapter 12

KEY CONCEPTS

- Effective force, effective-force diagram, vector-diagram equation, dynamic equilibrium, inertia force, central-force motion, areal speed, eccentricity of orbit, and period of orbit.
- Method of force and acceleration for particles.
- Forces and accelerations in rectilinear motion of particles.
- Forces and accelerations in curvilinear motion of particles using rectangular components, tangential and normal components, radial and transverse components, or cylindrical components.
- Gravitational central-force motion of spacecraft and satellites.

Newton's second law and his law of gravitation are two major fundamental laws in kinetics. They and the fundamental laws used in statics form the physical foundation of kinetics. Early in this chapter, Newton's second law is applied to show that the effective force on a particle is the resultant force on the particle. The effective-force diagram together with the free-body diagram is then employed to formulate the *method of force and acceleration* for particles. Later, Newton's second law and his law of gravitation are jointly applied to study the central-force motion of spacecraft and satellites. We note that present-day space technology involves the applications of space shuttles, geosynchronous satellites, surveying satellites, space stations, etc. Thus, it is easy to see that the applications of central-force motion have significant impacts on our modern living and technology.

NEWTON'S SECOND LAW[†]

12.1 Method of Force and Acceleration

A particle will accelerate if the forces acting on it are not balanced. To be more specific, let $\Sigma\mathbf{F}$ denote the *resultant force* of a system of forces $\mathbf{F}_1$, $\mathbf{F}_2$, . . . , $\mathbf{F}_n$ acting on a particle of *mass m*. This particle will move with

---

[†]Newton's work, entitled *Philosophiæ Naturalis Principia Mathematica*, was written in Latin and published in 1687; it was revised in 1713 and 1726. For a modern translation, see Florian Cajori, tran., *Sir Isaac Newton's Mathematical Principles of Natural Philosophy and His System of the World* (Berkeley: University of California Press, 1934).

an *acceleration* **a** if $\Sigma\mathbf{F} \neq \mathbf{0}$. In an inertial reference frame, we may apply *Newton's second law* to the particle to write

$$\Sigma\mathbf{F} = m\mathbf{a} \qquad (12.1)$$

which is called the *equation of motion* and is illustrated in Fig. 12.1.

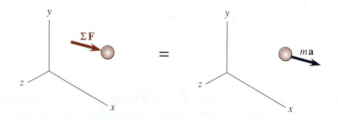

Fig. 12.1   Newton's second law

Note that the *resultant force* of a system of forces acting on a particle is *not* necessarily an actual force applied to the particle. Rather, it is merely an *equivalent force* representing the sum of the forces in the system; a force equal to the resultant force may replace the system of applied forces without changing the net effect on the particle. Thus, it is important to recognize that the *effective force* on a particle is the *resultant force* on the particle. The resultant force $\Sigma\mathbf{F}$ and the vector $m\mathbf{a}$ are, by Eq. (12.1), equivalent to each other. For this reason, the vector $m\mathbf{a}$ is also referred to as the *effective force* on (or of) the particle. Moreover, because of Eq. (12.1), the terms *effective force* and *resultant force* are sometimes interchangeably used. Nevertheless, because of the special *abstract* nature of the quantities $\Sigma\mathbf{F}$ and $m\mathbf{a}$, there are people who refrain from using the word "force" in naming them and refer to them simply as the *resultant* and the $m\mathbf{a}$ *vector*, respectively.

The use of the vector $m\mathbf{a}$ as the expression for the *effective force* or the *resultant force* is especially desirable in cases where the acceleration of the particle is known or to be determined. This may be depicted in Fig. 12.2, where the separate diagram showing the applied forces acting on the particle is, as before, called the *free-body diagram*, and the separate diagram showing the effective force $m\mathbf{a}$, or its components, on the particle is referred to as the *effective-force diagram* of the particle.[†] Notice here that the equality of any two diagrams containing vectors, such as those in Figs. 12.1 and 12.2, constitutes a *vector-diagram equation*.

Resolving each force $\mathbf{F}_i$ ($i = 1, 2, \dots, n$) and the acceleration **a** in Eq. (12.1) into rectangular components, we write

$$\Sigma(F_x\mathbf{i} + F_y\mathbf{j} + F_z\mathbf{k}) = m(a_x\mathbf{i} + a_y\mathbf{j} + a_z\mathbf{k})$$

$$\Sigma F_x = ma_x \qquad \Sigma F_y = ma_y \qquad \Sigma F_z = ma_z \qquad (12.2)$$

We refer to Eqs. (12.2) as a set of *scalar equations of motion* of the particle.

[†]Besides the free-body diagram defined in statics, the effective-force diagram mentioned here and the impulse and momentum diagrams to be defined in subsequent chapters are very helpful geometrical aids in the study of plane kinetics. An *effective-force diagram* is also called a *resultant-force diagram*, an *equivalent-force diagram*, a *second free-body diagram*, or a *kinetic diagram*. In some texts, such a diagram is left unnamed.

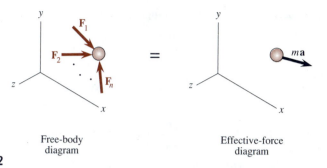

Free-body
diagram

Effective-force
diagram

Fig. 12.2

The method of solution which directly applies Newton's second law to particles in the form of equations of motion, or a vector-diagram equation consisting of a *free-body diagram* and an *effective-force diagram*, is called the *method of force and acceleration* for particles. This method is useful whenever forces and accelerations are to be directly related.

Developmental Exercises

D12.1 Define for a particle the terms: (a) effective force, (b) effective-force diagram, (c) vector-diagram equation.

D12.2 Describe the *method of force and acceleration* for particles.

12.2 Consistent Systems of Kinetic Units

A *kinetic quantity* is a measurable quantity in kinetics. *Kinetic units* are units for kinetic quantities. In writing Newton's second law in the form of Eq. (12.1), we cannot arbitrarily choose the units of length, mass, time, and force. If the units are arbitrarily chosen, the magnitude of the resultant force $\Sigma\mathbf{F}$ required to move a particle of mass m with an acceleration $\mathbf{a}$ will *not* be numerically equal to the product ma; it will only be proportional to this product. However, the set of units for length, mass, time, and force, which satisfy Eq. (12.1), is said to form a *consistent system of kinetic units*. Two consistent systems of kinetic units are in use by American engineers during the transition period; they are the SI and the U.S. customary system. Since both systems have been discussed in detail in Chap. 1, we shall describe them only briefly in this section.

The SI *base units* used in elementary mechanics are those for length, mass, and time, and are called, respectively, the *meter* (m), *kilogram* (kg), and *second* (s). The unit *newton* (N) for force is a *derived unit. One newton* is defined as the net force required to give a body with a mass of 1 kg an acceleration of 1 m/s². By Eq. (12.1), we write

$$1 \text{ N} = 1 \text{ kg} \cdot \text{m/s}^2 \tag{12.3}$$

In the U.S. customary system of units, the *base units* used in elementary mechanics are those for length, force, and time, and are called, respectively, the *foot* (ft), *pound* (lb), and *second* (s). The *second* here is the same as that in the SI. The foot is defined as

$$1 \text{ ft} = 0.3048 \text{ m} \tag{12.4}$$

One pound-mass (1 lbm) is legally defined as

$$1 \text{ lbm} = 0.45359237 \text{ kg} \tag{12.5}$$

One pound of force is equal to the weight of a body with one pound-mass at a location where the measured gravitational acceleration is equal to the standard gravitational acceleration, which is 9.80665 m/s^2. Thus, we write

$$1 \text{ lb} = 1 \text{ lbm} \cdot \frac{9.80665}{0.3048} \text{ ft/s}^2 \approx 32.2 \text{ lbm} \cdot \text{ft/s}^2 \tag{12.6}$$

The amount of mass which will accelerate at the rate of 1 ft/s^2 when it is acted on by a net force of 1 lb is defined as *one slug*. We write

$$1 \text{ lb} = 1 \text{ slug} \cdot \text{ft/s}^2 \tag{12.7}$$

$$1 \text{ slug} = \frac{9.80665}{0.3048} \text{ lbm} = \frac{9.80665}{0.3048} (0.45359237) \text{ kg} \tag{12.8}$$

For practical purposes, we write

$$1 \text{ slug} \approx 32.2 \text{ lbm} \approx 14.59 \text{ kg} \tag{12.9}$$

Note that *slug* is a *derived unit* and the abbreviations for *pound-mass* and *pound* are lbm and lb, respectively. The two consistent systems of kinetic units discussed above are summarized in Table 12.1.

Table 12.1 Consistent Systems of Kinetic Units*

Name of System	Length	Mass	Time	Force
SI	meter (m)	kilogram (kg)	second (s)	newton (N)
U.S. Customary System	foot (ft)	slug	second (s)	pound (lb)

*The SI is termed an *absolute* system since mass is taken as an absolute or base quantity. The U.S. customary system is termed a *gravitational* system since force *as measured from gravity* is taken as a base quantity.

For the relation between weight and mass of a body, we use Eq. (12.1) to write

$$W = mg \tag{12.10}$$

where W is the weight of a body with a mass m subjected to the gravitational acceleration g. Near the surface of the earth, we take

$$g = 9.81 \text{ m/s}^2 = 32.2 \text{ ft/s}^2 \tag{12.11}$$

Developmental Exercises

D12.3 Name the units which form a consistent system of kinetic units in (a) the SI, (b) the U.S. customary system.

D12.4 How much does a body on the surface of the earth weigh, in the primary or base unit of the corresponding system of units, if its mass is (a) 10 kg, (b) 10 Mg, (c) 10 mg, (d) 10 g, (e) 10 slugs, (f) 10 lbm?

12.3   Dynamic Equilibrium: Inertia Force

Whether balanced or not, the system of forces acting on a particle is always equivalent to the effective force $m\mathbf{a}$, which is the product of the mass m and the acceleration $\mathbf{a}$ of the particle as illustrated in Fig. 12.2. The negative of the vector $m\mathbf{a}$ is called the *inertia vector* of a particle and is denoted by $(m\mathbf{a})_{\text{rev}}$, where the subscript indicates that it points in the reversed direction of $m\mathbf{a}$; i.e.,

$$(m\mathbf{a})_{\text{rev}} = -m\mathbf{a} \qquad (12.12)$$

If the inertia vector $(m\mathbf{a})_{\text{rev}}$ is added to the free-body diagram, we obtain *a system of balanced vectors* as shown in Fig. 12.3. This is true because, by Eqs. (12.1) and (12.12), the resultant of the vectors in such a system is

$$\Sigma\mathbf{F} + (m\mathbf{a})_{\text{rev}} = \mathbf{0} \qquad (12.13)^{\dagger}$$

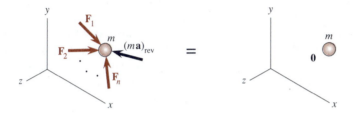

Fig. 12.3   Dynamic equilibrium

A particle perceived as being under the action of the applied forces *and* the inertia vector is said to be in a state of *dynamic equilibrium*, and the equations of equilibrium in statics may similarly be applied. In application, an inertia vector is often called an *inertia force*, as in advanced dynamics and vibration analysis.

Note that the *inertia force* of a body B simply represents (by Newton's third law) the *reaction* of the body B to the action of the resultant force on B.[††] Since such a reaction of B is an action exerted *by* (not *on*) B, the inertia force of B is *not* a force to be included in the free-body diagram of B. If it is included there, the resultant of the system of vectors on B becomes zero, even though the acceleration of B is nonzero. Therefore, a number of people, who stress the fundamental concept of *force* (as defined in Newton's second law) for beginners of dynamics, avoid calling the *inertia vector* an *inertia force*. Furthermore, many people today prefer to think in terms of dynamics (rather than in terms of statics) when they analyze dynamics problems.

[†]Note that d'Alembert's principle (to be covered in Sec. 16.5) was originally written for the motion of *a system of linked masses*, where the sum of interaction forces vanishes. Quotation of Eq. (12.13), which is written for a *single particle*, as d'Alembert's principle has been criticized. For translation of parts of d'Alembert's *Traité de Dynamique*, see R. M. Rosenberg, ''d'Alembert and Others on d'Alembert's Principle,'' *Engineering Education*, April 1968, pp. 959–960.

[††]By Newton's third law, *each action is always matched by a reaction*; i.e., forces always occur in pairs of action and reaction. In this regard, the *inertia force* is clearly a *reactive mechanical force*, rather than a *fictitious* (or *mythical*) force.

Anyhow, one should note that an effective force or a resultant force is simply an *equivalent force*. But, an inertia force is a *reactive force*, which is mainly used in formulating the so-called *dynamic equilibrium* of a body. Although the vectors acting on a body in dynamic equilibrium are balanced, the acceleration of that body is not necessarily zero.

Developmental Exercise

D12.5   Define for a particle the terms: (a) inertia vector, (b) dynamic equilibrium, (c) inertia force.

12.4   Rectilinear Motion of a Particle

The *method of force and acceleration* referred to in Sec. 12.1 may now be used to solve kinetic problems of particles moving along straight paths, where forces and accelerations need to be related. If we choose the x axis to be the straight path of motion of a particle of mass m moving with an acceleration $\mathbf{a}$, as illustrated in Fig. 12.4, we have $a_x = a$, $a_y = a_z = 0$, and Eqs. (12.2) become

$$\Sigma F_x = ma \qquad \Sigma F_y = 0 \qquad \Sigma F_z = 0 \qquad (12.14)$$

Note that the *free-body diagram* of the particle in Fig. 12.4(a) accounts for all the applied forces acting on the particle, while the *effective-force diagram* in Fig. 12.4(b) indicates the effective force (or its components) on the particle. If the sense of any unknown is not known, it may first be assumed and then verified at the end of the solution.[†]

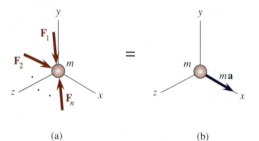

Fig. 12.4   (a)   (b)

In dependent rectilinear motions of particles as discussed in Sec. 11.7, the *constraint equation on the accelerations* of the particles is usually obtained by taking the second derivative of the constraint equation on the positions of the particles with respect to the time. The positive direction of each unknown acceleration in the constraint equation is assumed to point in the positive direction of the corresponding position coordinate of the parti-

---

[†]If the value of an unknown quantity is found to be negative in the solution, the magnitude of this unknown quantity is simply equal to the absolute value of the solved value, but the true sense of this unknown quantity is opposite to that initially assumed. However, this practice may cease to be valid when the unknown quantity is a *friction force* at a location where slipping *impends* or *occurs*.

cle. For consistency, an *unknown effective force* in dependent rectilinear motions *must* also be assumed to point in the positive direction of the corresponding position coordinate if the above-mentioned *constraint equation* is used.[†] Furthermore, the *positive direction* chosen in the summation of components of vectors in a vector-diagram equation must be the *same* for each side of the equal sign. With these precautions, we can avoid most of the errors and difficulties commonly encountered.

EXAMPLE 12.1

A truck is traveling at a constant speed of 72 km/h as shown. Determine the required minimum coefficient of static friction $(\mu_s)_{min}$ between the crate and the flatbed trailer if the truck is to be brought to a stop through a constant deceleration over a distance of 100 m without causing the crate to shift on the flat bed.

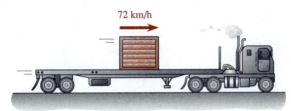

72 km/h

Solution. Since the crate is not to shift on the flatbed, the crate and the truck must have the same motion at all times. The initial speed is $v_0 = 72$ km/h $= 20$ m/s. The magnitude a of the constant deceleration may be determined from $v^2 = v_0^2 + 2a_c(x - x_0)$, where $v = 0$, $v_0 = 20$ m/s, $a_c = -a$, and $x - x_0 = 100$ m. Thus, we write

$$0^2 = (20)^2 + 2(-a)(100) \qquad a = 2$$

The acceleration of the crate is, therefore, $\mathbf{a} = 2$ m/s² ←.

Assume that the mass of the crate is m. Under the stated condition, slipping of the crate on the flatbed impends. We first equate the free-body diagram to the effective-force diagram as shown. By *equating the sums of the y components of the vectors* (denoted by ΣV_y) in the vector-diagram equation, we write

$$+\uparrow \Sigma V_y:{}^{\dagger\dagger} \qquad N - mg = 0 \qquad (1)$$

Similarly, *equating the sums of the x components of the vectors* (denoted by ΣV_x) in the vector-diagram equation, we write

$$\xleftarrow{+} \Sigma V_x:{}^{\dagger\dagger} \qquad (\mu_s)_{min}N = m(2) \qquad (2)$$

Equations (1) and (2) yield

$$(\mu_s)_{min} = \frac{2}{g} = \frac{2}{9.81} \qquad (\mu_s)_{min} = 0.204 \blacktriangleleft$$

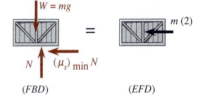

$W = mg$

$N \quad (\mu_s)_{min} N$

(FBD)

m (2)

(EFD)

[†] This is illustrated in Example 12.3.

[††] Henceforth, the letter V following the summation sign Σ stands for the vectors being considered.

Developmental Exercise

D12.6   Work Example 12.1 if the constant speed is 54 km/h.

EXAMPLE 12.2

The blocks A and B are released to move from rest as shown. The coefficient of kinetic friction is 0.2 between the cord and the drum C as well as between the block A and the horizontal support. Determine the acceleration of the block B.

Solution.   Since the cord is considered to be inextensible, the accelerations $\mathbf{a}_A$ and $\mathbf{a}_B$ of the blocks A and B must have the same magnitude. Letting $a_A = a_B = a$, we first draw the following pertinent diagrams:

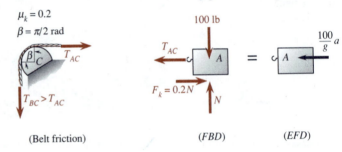

Then, we continue the solution as follows:

Block B.

$$+\downarrow \Sigma V_y: \qquad 50 - T_{BC} = \frac{50}{g}a \qquad (1)$$

Belt friction around the drum C, $T_2 = T_1 e^{\mu_k \beta}$.

$$T_{BC} = T_{AC} e^{0.2(\pi/2)} \qquad (2)$$

Block A.

$$+\uparrow \Sigma V_y: \qquad N - 100 = 0 \qquad (3)$$

$$\xleftarrow{+} \Sigma V_x: \qquad T_{AC} - 0.2N = \frac{100}{g}a \qquad (4)$$

With $g = 32.2$ (ft/s²), the above four simultaneous equations yield

$$N = 100 \qquad T_{AC} = 32.1 \qquad T_{BC} = 43.9 \qquad a = 3.90 = a_B$$

$$\mathbf{a}_B = 3.90 \text{ ft/s}^2 \downarrow \blacktriangleleft$$

Developmental Exercise

D12.7   Work Example 12.2 if friction between the block A and its support is negligible.

EXAMPLE 12.3

The blocks A and B are released to move from rest as shown. Neglecting effects of friction, determine the acceleration $\mathbf{a}_A$ of the block A and the tension F in the cord BCD.

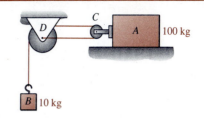

Solution. We first note that the motions of the blocks as shown are subject to the constraint equation on the position coordinates

$$2x_A + x_B = \text{constant}$$

which, through differentiations, yields the *constraint equation on the accelerations*

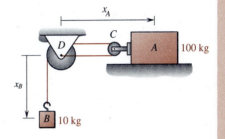

$$2a_A + a_B = 0 \qquad (1)$$

Note in Eq. (1) that the positive sense of a_A is to the right, while that of a_B is downward. For Eq. (1) to hold true, a_A and a_B must have opposite signs. We should let them "run the algebraic course." But we expect a_A to have a negative value in the final solution because the actual acceleration of the block A must be directed to the left. Thus, we continue the solution as follows:

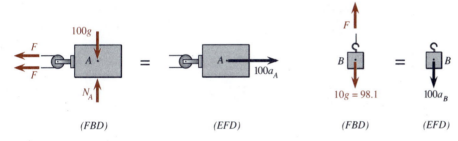

(FBD)   (EFD)   (FBD)   (EFD)

Block A.

$$\overset{+}{\rightarrow} \Sigma V_x: \qquad -F - F = 100 a_A \qquad (2)$$

Block B.

$$+\downarrow \Sigma V_y: \qquad 98.1 - F = 10 a_B \qquad (3)$$

Equations (1), (2), and (3) yield

$$a_A = -1.401 \qquad a_B = 2.80 \qquad F = 70.1$$

Since the solved value of a_A is negative (as expected), the true sense of $\mathbf{a}_A$ is to the left. Therefore, we write

$$\mathbf{a}_A = 1.401 \text{ m/s}^2 \leftarrow \qquad F = 70.1 \text{ N} \quad \blacktriangleleft$$

REMARK. The positive sense of a_A in Eq. (1) is inherently the same as the positive sense of x_A, which is to the right. Therefore, the sense of the unknown effective force $100 a_A$ must be assumed to point to the right in the effective-force diagram for the sake of algebraic consistency with Eq. (1). Furthermore, notice that $F = 70.1$ N and the weight $W_B = 10g = 98.1$ N > 70.1 N. This makes sense because the block B is not in static equilibrium, but accelerating downward.

Developmental Exercise

D12.8   Refer to Example 12.3. Determine the mass of the block B if the mass of the block A remains the same but $\mathbf{a}_A = 2 \text{ m/s}^2 \leftarrow$.

PROBLEMS

12.1*   The package A is weighed on a lever scale while the package B is weighed on a spring scale as shown. The scales are fastened to the bar CD which is held horizontally by the force $\mathbf{P}$. When the bar CD is at rest, each scale indicates a reading of 2.5 lb. Determine the reading of each scale when the bar CD is accelerated upward at the rate of 10 ft/s².

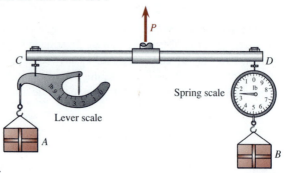

Spring scale

Lever scale

Fig. P12.1*

12.2   Refer to Prob. 12.1*. If the spring scale indicates a reading of 2 lb, determine (a) the acceleration of the bar CD, (b) the reading indicated by the lever scale.

12.3   The device shown is used as an accelerometer. It consists of a 200-mg plunger AB which compresses the spring when the housing of the device is given an acceleration to the right. Assuming that effects of friction are negligible, determine the required spring modulus k which fits the calibration of 1 mm of deflection of the plunger for every 1 m/s² of acceleration.

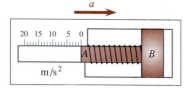

Fig. P12.3

12.4   Determine the acceleration of the cylinder A in each of the three cases illustrated.

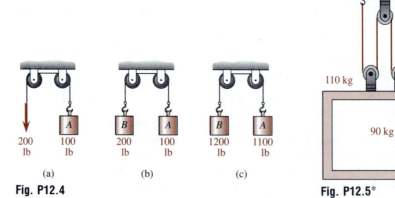

(a)	(b)	(c)

200 lb | 100 lb | 200 lb | 100 lb | 1200 lb | 1100 lb

(a)   (b)   (c)

Fig. P12.4

110 kg

90 kg

Fig. P12.5*

12.5*   A 90-kg workman inside a 110-kg cage pulls the hoisting cable with a force of 500 N as shown. Determine the acceleration of the cage.

12.6 A 20-lb block *B* is suspended from a 5-ft cord attached to a 50-lb cart *A* which can move freely down the incline. Determine for the instant the system is released from rest (a) the acceleration of the cart, (b) the acceleration of the block, (c) the tension in the cord.

12.7* A car climbs the hill *AB* at a constant speed as shown. If the driver does not change the throttle setting or shift gears, determine the magnitude of the acceleration of the car immediately after entering (a) the level section *BC*, (b) the downhill section *CD*.

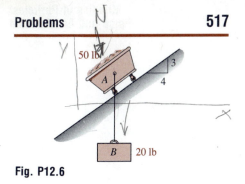

Fig. P12.6

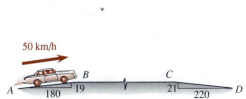

Fig. P12.7*

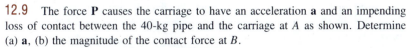

Fig. P12.8*

12.8* If the frame has an acceleration **a** and the tensions in the ropes *CA* and *CB* are such that $T_{CA} = 4T_{CB}$, determine (a) **a**, (b) T_{CB}.

12.9 The force **P** causes the carriage to have an acceleration **a** and an impending loss of contact between the 40-kg pipe and the carriage at *A* as shown. Determine (a) **a**, (b) the magnitude of the contact force at *B*.

12.10 A truck is traveling at a constant speed of 54 km/h down a hill as shown. If $\mu_s = 0.5$ between the crate and the flatbed trailer, determine the shortest distance in which the truck can be brought to a stop without causing the crate to shift on the flatbed.

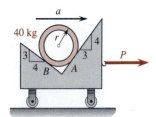

Fig. P12.9

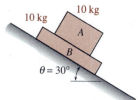

Fig. P12.10

12.11* and 12.12 If μ_k equals 0.2 between the blocks *A* and *B*, and 0.3 between the block *B* and the fixed support as shown, where the blocks are moving, determine the magnitude of acceleration of each block.

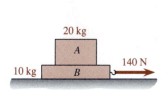

Fig. P12.11*

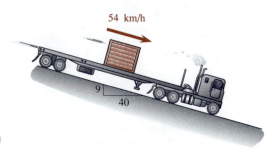

Fig. P12.12

12.13 Solve Prob. 12.12 if $\theta = 20°$.

12.14 The downward motion of a 200-lb carriage in its smooth vertical guides is controlled by a cable as shown. If $\mu_k = 0.15$ between the cable and the pegs, determine the magnitude of the force **P** which allows (a) a constant velocity downward, (b) a constant acceleration of 3 ft/s² ↓ .

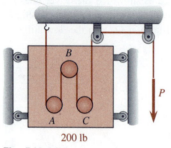

Fig. P12.14

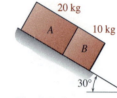

Fig. P12.15

12.15 The blocks A and B are sliding down an incline as shown. If μ_k equals 0.2 between the block A and the incline, and 0.3 between the block B and the incline, determine (a) the acceleration of the blocks, (b) the force exerted by the block A on the block B.

12.16* The truck shown suddenly accelerates forward for a moment, and slipping occurs between the crates and the flatbed. If μ_k equals 0.2 between the crate A and the flatbed, and 0.3 between the crate B and the flatbed, determine for the moment in which slipping occurs (a) the minimum acceleration of the truck, (b) the force exerted by A on B.

Fig. P12.16*

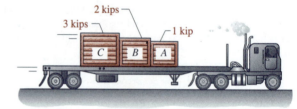

Fig. P12.17

12.17 The truck shown is traveling forward at a constant speed when the brakes are applied causing the crates to slide on the flatbed. If the values of μ_k between the flatbed and the crates A, B, and C are 0.3, 0.25, and 0.2, respectively, determine for the period of braking (a) the minimum magnitude of the deceleration of the truck, (b) the force exerted by A on B, (c) the force exerted by C on B.

12.18 and 12.19* Neglecting effects of friction in the system shown, determine (a) the acceleration of each block, (b) the tension in the cable.

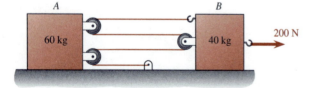

Fig. P12.18 and P12.20*

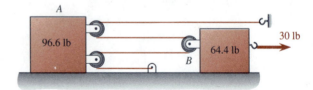

Fig. P12.19* and P12.21

12.20* and 12.21 The blocks shown are moving to the right. If $\mu_k = 0.1$ between the support and the blocks, determine (a) the acceleration of each block, (b) the tension in the cable.

12.22 through 12.24 In the system shown, the cylinder B moves with $\mathbf{a}_B = 4$ ft/s^2 $\downarrow$. If the cylinder A weighs 161 lb, determine (a) the weight of the cylinder B, (b) the tension in each cable.

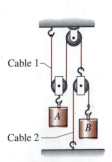

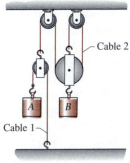

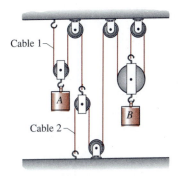

Fig. P12.22 and P12.25   **Fig. P12.23* and P12.26**   **Fig. P12.24 and P12.27***

12.25 through 12.27* In the system shown, the cylinder A moves with $\mathbf{a}_A = 0.6$ m/s^2 $\uparrow$. If the mass of the cylinder B is 60 kg, determine (a) the mass of the cylinder A, (b) the tension in each cable.

12.28 and 12.29 The system shown is released from rest. Determine the acceleration of each cylinder and the tension in each cable.[†]

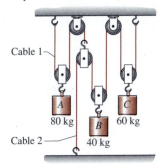

Fig. P12.28

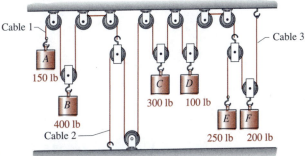

Fig. P12.29

12.30* A river boat of mass m is anchored as shown where the water flows with a constant velocity $\mathbf{v}$. The horizontal component of the force exerted on the boat by the anchor chain is $\mathbf{P}$. Suppose that the friction force between the boat and the water is proportional to the velocity of the boat relative to the water. Determine the time required for the boat to attain a speed of $v/3$ if the anchor chain suddenly breaks. (*Hint.* Use the frame fixed to the flowing water as an inertial reference frame.)

Fig. P12.30*

12.31* A steel ball of mass m is released from rest at the surface O of a tank of viscous fluid as shown. During sinking, the ball experiences an upward viscous friction force of magnitude cv, where c is a constant and v is the speed of the ball. Determine the position coordinate x of the ball in terms of m, c, the gravitational acceleration g, and the time t.

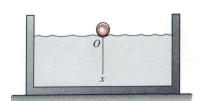

Fig. P12.31*

[†]Cf. App. D for computer solution of resulting simultaneous equations.

12.5 Curvilinear Motion of a Particle

We saw in Chap. 11 that the curvilinear motion of a particle may be described by using one of the several sets of components, such as rectangular components, tangential and normal components, radial and transverse components, or cylindrical components. The choice of a particular set of components usually depends on the given conditions in the problem and is one of the basic decisions to be made at the beginning of the solution of a curvilinear-motion problem. Once the choice is made, the *method of force and acceleration* may then be applied to establish the equations necessary to yield the desired solution.

Resolving the forces acting on the particle and the acceleration of the particle into appropriate sets of components, as illustrated in Figs. 12.5 through 12.8, we may write the scalar versions of Eq. (12.1) for the motion of the particle with mass m as follows:

(1) Using *rectangular components* (cf. Fig. 12.5 and Sec. 11.11), we have

$$\Sigma F_x = ma_x \qquad \Sigma F_y = ma_y \qquad \Sigma F_z = ma_z \qquad (12.15)$$

where $a_x = \ddot{x}$, $a_y = \ddot{y}$, and $a_z = \ddot{z}$.

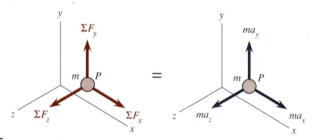

Fig. 12.5

(2) Using *tangential and normal components* (cf. Fig. 12.6 and Sec. 11.13), we have

$$\Sigma F_t = ma_t \qquad \Sigma F_n = ma_n \qquad (12.16)$$

where $a_t = \dot{v}$ and $a_n = v^2/\rho$.

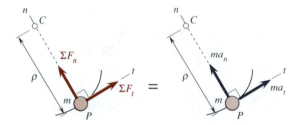

Fig. 12.6

(3) Using *radial and transverse components* (cf. Fig. 12.7 and Sec. 11.14), we have

$$\Sigma F_r = ma_r \qquad \Sigma F_\theta = ma_\theta \qquad (12.17)$$

where $a_r = \ddot{r} - r\dot{\theta}^2$ and $a_\theta = r\ddot{\theta} + 2\dot{r}\dot{\theta}$.

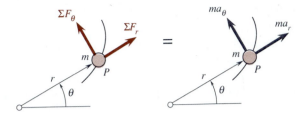

Fig. 12.7

(4) Using *cylindrical components* (cf. Fig. 12.8 and Sec. 11.15), we have

$$\Sigma F_R = ma_R \qquad \Sigma F_\theta = ma_\theta \qquad \Sigma F_z = ma_z \qquad (12.18)$$

where $a_R = \ddot{R} - R\dot{\theta}^2$, $a_\theta = R\ddot{\theta} + 2\dot{R}\dot{\theta}$, and $a_z = \ddot{z}$.

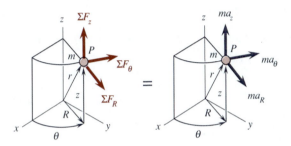

Fig. 12.8

Equations (12.15) through (12.18) are sets of *scalar equations of motion* of a particle in the chosen coordinate systems.

EXAMPLE 12.4

The resultant external force acting on a 2-kg particle in space is $\mathbf{F} = 12t\mathbf{i} - 24t^2\mathbf{j} - 40t^3\mathbf{k}$ N, where t is the time measured in seconds. The particle is at rest at the origin when $t = 0$. Determine the acceleration component $\mathbf{a}_y$, the velocity component $\mathbf{v}_y$, and the coordinate y of the particle when $t = 4$ s.

Solution. The y component of the resultant external force is $F_y = -24t^2$ N, and the mass of the particle is $m = 2$ kg. Referring to the diagrams shown, we write

$$+\uparrow \Sigma V_y: \qquad -24t^2 = 2a_y \qquad a_y = -12t^2$$

The initial conditions are $v_y = y = 0$ at $t = 0$. Thus, we write

$$\int_0^{v_y} dv_y = \int_0^t a_y \, dt \qquad v_y = \int_0^t (-12t^2) \, dt = -4t^3$$

$$\int_0^y dy = \int_0^t v_y \, dt \qquad y = \int_0^t (-4t^3) \, dt = -t^4$$

At $t = 4$ s, we obtain

$$a_y = -12t^2 = -12(4)^2 \qquad \mathbf{a_y} = -192\mathbf{j} \text{ m/s}^2 \blacktriangleleft$$

$$v_y = -4t^3 = -4(4)^3 \qquad \mathbf{v_y} = -256\mathbf{j} \text{ m/s} \blacktriangleleft$$

$$y = -t^4 = -(4)^4 \qquad y = -256 \text{ m} \blacktriangleleft$$

Developmental Exercises

D12.9 Refer to Example 12.4. Determine for the particle at $t = 4$ s (a) $\mathbf{a}_x$, (b) $\mathbf{v}_x$, (c) x.

D12.10 Refer to Example 12.4. Determine for the particle at $t = 4$ s (a) $\mathbf{a}_z$, (b) $\mathbf{v}_z$, (c) z.

EXAMPLE 12.5

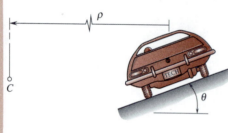

A car traveling at the *rated speed* of a road should have no lateral friction force developed between its tires and the road surface. If a highway curve has a radius of curvature $\rho = 200$ m, a rated speed $v = 90$ km/h, and a banking angle θ as shown, determine the value of θ.

Solution. The rated speed is $v = 90$ km/h $= 25$ m/s. The normal component of acceleration of the car is $a_n = v^2/\rho = (25)^2/200$ m/s^2. Thus, the

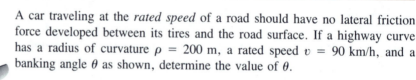

free-body diagram of the car is equated to its effective-force diagram as shown. We write

$$+\uparrow \Sigma V_y: \qquad N \cos\theta - W = 0$$

$$\xleftarrow{+} \Sigma V_n: \qquad N \sin\theta = \frac{W}{g} \frac{(25)^2}{200}$$

These two equations yield

$$\tan\theta = \frac{(25)^2}{200g} = \frac{(25)^2}{200(9.81)} \qquad \theta = 17.67° \blacktriangleleft$$

Developmental Exercise

D12.11 Determine the rated speed of a highway curve of radius $\rho = 100$ m and a banking angle $\theta = 20°$.

EXAMPLE 12.6

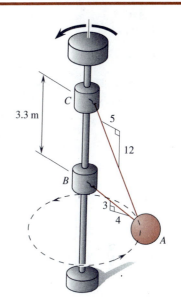

A sphere at A is made to revolve in a horizontal circle at a constant speed v by two wires AB and AC, which are fastened to a vertical shaft as shown. Determine the minimum value of the speed v for which both wires are taut.

Solution. Let the center of the horizontal circle be at the point D as shown. The radius r of the circle can be determined as follows:

$$\frac{12}{5}r - \frac{3}{4}r = 3.3$$

Thus, $r = 2$ m. Since the sphere revolves in the horizontal circle with a constant speed, its acceleration components are $a_t = 0$, $a_n = v^2/r = v^2/2$.

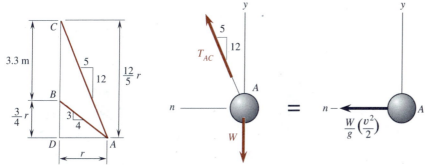

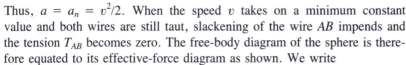

Thus, $a = a_n = v^2/2$. When the speed v takes on a minimum constant value and both wires are still taut, slackening of the wire AB impends and the tension T_{AB} becomes zero. The free-body diagram of the sphere is therefore equated to its effective-force diagram as shown. We write

$$+\uparrow \Sigma V_y: \qquad \frac{12}{13}T_{AC} - W = 0$$

$$\xleftarrow{+} \Sigma V_n: \qquad \frac{5}{13}T_{AC} = \frac{W}{g}\left(\frac{v^2}{2}\right)$$

These two equations yield

$$v^2 = \frac{5g}{6} = \frac{5(9.81)}{6} \qquad v = 2.86 \text{ m/s} \blacktriangleleft$$

Developmental Exercise

D12.12 Refer to Example 12.6. Determine the maximum value of the speed v for which both wires are taut.

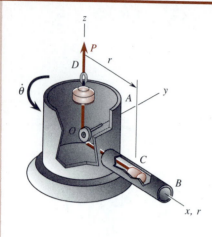

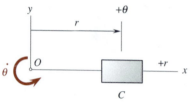

EXAMPLE 12.7

The vertical tube A and the horizontal tube B are connected as shown. The tubes are rotated about the z axis at a constant rate of $\dot{\theta} = 2.5$ rad/s. The radial position r of the 2-kg plug C inside the tube B is controlled by the rope COD which passes around the pulley at O mounted inside the tube A. At the instant shown, $r = 4$ m and the end D is pulled by $\mathbf{P}$ with v_D = 3 m/s ↑ and $a_D = 5$ m/s² ↓. For this instant, determine the tension in the rope and the components of the force $\mathbf{F}$ exerted by the tube B on the plug C if effects of friction are negligible.

Solution. The acceleration of the plug C, which moves in the horizontal plane, can conveniently be determined by using radial and transverse components. At the instant shown, we have from the given data that $r = 4$ m, $\dot{r} = -3$ m/s, $\ddot{r} = 5$ m/s², $\dot{\theta} = 2.5$ rad/s, and $\ddot{\theta} = 0$. Thus,

$$a_r = \ddot{r} - r\dot{\theta}^2 = -20 \text{ m/s}^2 \qquad a_\theta = r\ddot{\theta} + 2\dot{r}\dot{\theta} = -15 \text{ m/s}^2$$

The negative signs in the values of a_r and a_θ indicate that the actual senses of a_r and a_θ are opposite to the positive senses of r and θ, respectively.

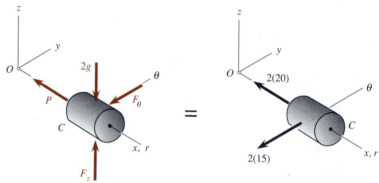

Noting that the tension in the rope is P and the mass of the plug C is $m = 2$ kg, we equate the free-body diagram of the plug C to its effective-force diagram as shown. We write

$$+\searrow \Sigma V_r: \qquad -P = -2(20) \qquad\qquad P = 40 \text{ N} \blacktriangleleft$$

$$+\nearrow \Sigma V_\theta: \qquad -F_\theta = -2(15) \qquad\qquad \mathbf{F_\theta} = -30\mathbf{e_\theta} \text{ N} \blacktriangleleft$$

$$+\uparrow \Sigma V_z: \qquad F_z - 2g = 0 \qquad\qquad \mathbf{F_z} = 19.62\mathbf{k} \text{ N} \blacktriangleleft$$

Developmental Exercise

D12.13 Refer to Example 12.7. (a) Is the z component of the acceleration of the plug C equal to 9.81 m/s² ↓ or **0**? (b) What is the z component of the resultant force acting on the plug C?

EXAMPLE 12.8

The path of a 2-kg particle in space is known to be a helix defined by the cylindrical coordinates: $R = 40$, $\theta = (\pi/4)t$, and $z = (9\pi/4)t$, where R

and z are measured in meters, θ is the angle in radians, and t is the time in seconds. Determine the resultant external force $\mathbf{F}$ acting on the particle.

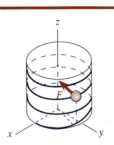

Solution.   From the given cylindrical coordinates of the particle, we have

$$R = 40 \qquad \dot{R} = 0 \qquad \ddot{R} = 0$$

$$\theta = \frac{\pi}{4}\,t \qquad \dot{\theta} = \frac{\pi}{4} \qquad \ddot{\theta} = 0$$

$$z = \frac{9\pi}{4}\,t \qquad \dot{z} = \frac{9\pi}{4} \qquad \ddot{z} = 0$$

The acceleration $\mathbf{a}$ of the particle is

$$\mathbf{a} = (\ddot{R} - R\dot{\theta}^2)\mathbf{e}_R + (R\ddot{\theta} + 2\dot{R}\dot{\theta})\mathbf{e}_\theta + \ddot{z}\mathbf{e}_z$$

$$= -\frac{5\pi^2}{2}\,\mathbf{e}_R$$

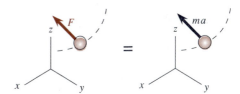

The mass of the particle is $m = 2$ kg. By Newton's second law, we write

$$\mathbf{F} = m\mathbf{a} = 2\left(-\frac{5\pi^2}{2}\,\mathbf{e}_R\right) \qquad \mathbf{F} = -49.3\mathbf{e}_R \text{ N} \blacktriangleleft$$

Developmental Exercise

D12.14   Refer to Example 12.8. Determine (a) the expression of the tangential unit vector $\mathbf{e}_t$ in terms of the transverse unit vector $\mathbf{e}_\theta$ and the axial unit vector $\mathbf{e}_z$, (b) the angle between the velocity vector $\mathbf{v}$ and the xy plane, (c) the radius of curvature ρ of the helical path.

12.6   Systems of Particles

The forces acting on the particle P_i of a system of n particles are illustrated in Fig. 12.9(a), where $\mathbf{F}_i$ is the resultant of the external forces exerted on P_i, and $\mathbf{f}_{i1}, \ldots, \mathbf{f}_{in}$ are the internal forces exerted on P_i by other particles

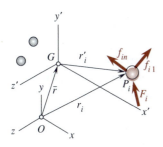

(a)   (b)

Fig. 12.9

$P_1, \ldots, P_n$ of the system (where $\mathbf{f}_{ii}$ has no meaning and is set equal to zero). According to Newton's second law, the *resultant* of the external and internal forces acting on the particle P_i is equal to the *effective force* $m_i\mathbf{a}_i$ on that particle, which is illustrated in Fig. 12.9(b). Thus, we write

$$\mathbf{F}_i + \sum_{j=1}^{n} \mathbf{f}_{ij} = m_i\mathbf{a}_i \tag{12.19}$$

Noting that the force vectors in Fig. 12.9(a) are equivalent to the effective force in Fig. 12.9(b), we may take moments about the mass center G of the system of particles and also write

$$\mathbf{r}'_i \times \mathbf{F}_i + \sum_{j=1}^{n} (\mathbf{r}'_i \times \mathbf{f}_{ij}) = \mathbf{r}'_i \times m_i\mathbf{a}_i \tag{12.20}$$

where $\mathbf{r}'_i$ is the position vector of the particle P_i in the nonrotating reference frame $Gx'y'z'$. Since the equivalence relation established in Fig. 12.9 holds true for any particle, it follows that the system of all the external and internal forces acting on the various particles is *equivalent* to the system of the effective forces on the particles, as illustrated in Fig. 12.10.

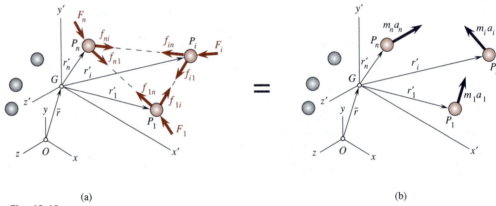

(a)   (b)

Fig. 12.10

By Newton's third law, the internal forces $\mathbf{f}_{ij}$ must occur in collinear pairs which are equal in magnitude and opposite in direction. Thus, adding all the internal forces of the system and summing their moments about the mass center G, respectively, we write

$$\sum_{i=1}^{n}\sum_{j=1}^{n} \mathbf{f}_{ij} = \mathbf{0} \qquad \sum_{i=1}^{n}\sum_{j=1}^{n} (\mathbf{r}'_i \times \mathbf{f}_{ij}) = \mathbf{0} \tag{12.21}$$

We see from Eqs. (12.21) that the sum of the internal forces themselves and the sum of their moments about a given point, such as G, are zero.

Letting $i = 1, 2, \ldots, n$ in Eq. (12.19), we obtain a set of n equations. Adding these n equations member by member, using the first of Eqs. (12.21), and denoting the sum of the external forces by $\Sigma\mathbf{F}$, we get

$$\Sigma\mathbf{F} = \sum_{i=1}^{n} m_i\mathbf{a}_i \tag{12.22}$$

Proceeding similarly with Eq. (12.20) and using the second of Eqs. (12.21), we readily find that the sum of moments about G of the external forces is

$$\Sigma \mathbf{M}_G = \sum_{i=1}^{n} (\mathbf{r}_i' \times m_i \mathbf{a}_i) \qquad (12.23)$$

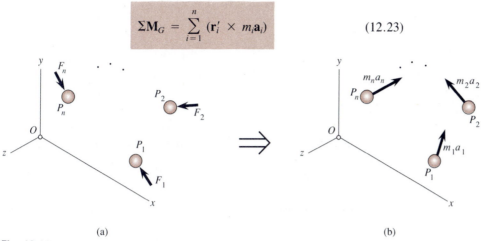

(a)   (b)

Fig. 12.11

Furthermore, it can similarly be shown that the *system of the external forces* $\mathbf{F}_i$ and the *system of effective forces* $m_i \mathbf{a}_i$, as illustrated in Fig. 12.11(a) and (b), have the same resultant force and the same resultant moment about *any* fixed point. In particular, we can readily show that the sum of moments about the fixed origin O of the external forces is

$$\Sigma \mathbf{M}_O = \sum_{i=1}^{n} (\mathbf{r}_i \times m_i \mathbf{a}_i) \qquad (12.24)$$

However, we observe here that the external force $\mathbf{F}_i$ itself acting on a given particle P_i and the effective force $m_i \mathbf{a}_i$ on the same particle P_i are generally *not* equal. Therefore, with internal forces excluded, the two systems of vectors in parts (a) and (b) of Fig. 12.11 do *not* have the same physical effect on the individual particles; these two systems are *equipollent*, not equivalent (cf. Sec. 5.3).

Note in Fig. 12.11 that the sign $\Rightarrow$ denotes *equipollence* between the system of vectors on its left side and the system of vectors on its right side. Although two equipollent systems of forces acting on a rigid body are also *equivalent*, two equipollent systems of forces acting on a set of independent particles are not necessarily equivalent.

Developmental Exercise

D12.15 Is the system of *external forces* acting on a system of particles *equivalent* to the system of *effective forces* on the particles in the system?

12.7 Principle of Motion of the Mass Center

For a system of n particles with masses $m_1, m_2, \ldots, m_n$ and position vectors $\mathbf{r}_1, \mathbf{r}_2, \ldots, \mathbf{r}_n$ as illustrated in Fig. 12.12(a), we may determine

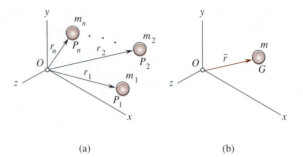

Fig. 12.12   (a)   (b)

the total mass m of the system and the position vector $\bar{\mathbf{r}}$ of the center of mass G of the system, as illustrated in Fig. 12.12(b), by applying the *principle of moments* as follows:

$$m = \sum_{i=1}^{n} m_i \qquad m\bar{\mathbf{r}} = \sum_{i=1}^{n} m_i \mathbf{r}_i \qquad (12.25)$$

By taking the time derivatives of the second of Eqs. (12.25), we have

$$m\dot{\bar{\mathbf{r}}} = \sum_{i=1}^{n} m_i \dot{\mathbf{r}}_i \qquad m\ddot{\bar{\mathbf{r}}} = \sum_{i=1}^{n} m_i \ddot{\mathbf{r}}_i$$

Denoting the velocities, $\dot{\bar{\mathbf{r}}}$, $\dot{\mathbf{r}}_i$, and accelerations, $\ddot{\bar{\mathbf{r}}}$, $\ddot{\mathbf{r}}_i$, by $\bar{\mathbf{v}}$, $\mathbf{v}_i$, $\bar{\mathbf{a}}$, and $\mathbf{a}$, ($i = 1, 2, \ldots, n$), respectively, we write

$$m\bar{\mathbf{v}} = \sum_{i=1}^{n} m_i \mathbf{v}_i \qquad m\bar{\mathbf{a}} = \sum_{i=1}^{n} m_i \mathbf{a}_i \qquad (12.26)$$

where $\bar{\mathbf{v}}$ and $\bar{\mathbf{a}}$ are the velocity and acceleration of the mass center G, respectively. Thus, from Eq. (12.22) and the second of Eqs. (12.26), we have

$$\boxed{\Sigma \mathbf{F} = m\bar{\mathbf{a}}} \qquad (12.27)$$

Equation (12.27) expresses an important property that *the resultant force of all the external forces acting on a system of particles is equal to the total mass of the system times the acceleration of the mass center of the system.* This property, a corollary to Newton's second and third laws, is called the *principle of motion of the mass center* of a system of particles.[†] In other words, the mass center of a system of particles is a point which moves as if the entire mass of the system and all the external forces were concentrated at that point. Since a rigid body may be considered as a system of particles, the principle of motion of the mass center is, of course, applicable to a rigid body.

By Eqs. (12.22) and (12.27), we know that the resultant of the system of external forces in Fig. 12.11(a) and the resultant of the system of effec-

---

[†]Under the action of internal forces only (i.e., $\mathbf{F}_i = \mathbf{0}$), this principle degenerates into *corollary* IV to Newton's laws of motion (cf. pp. 19–20 of Newton's *Principia* cited in the footnote on p. 507). In some mechanics literature, this principle, or its equivalent, is also called the *generalized Newton's second law* of motion for a mass system, or *Euler's first law*, after Leonhard Euler (1707–1783). For the latter, see C. Truesdell, *Essays in the History of Mechanics* (New York: Springer-Verlag, 1968).

tive forces in Fig. 12.11(b) are both identically equal to the *resultant effective force* $m\bar{\mathbf{a}}$, in magnitude, direction, and line of action, as illustrated in Fig. 12.13. However, the derivation of Eq. (12.27) does not involve the moments of the external forces or the effective forces which are generally nonconcurrent. Thus, the resultant of the external forces, as well as the resultant of the effective forces, *does not*, in general, pass through the mass center G of the system of particles on which they act.

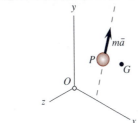

Fig. 12.13

EXAMPLE 12.9

A system consists of three particles P_1, P_2, and P_3. At the instant shown, the particles are acted on by the external forces $\mathbf{F}_1 = 10\mathbf{i} - 15\mathbf{j}$ lb, $\mathbf{F}_2 = -13\mathbf{j} + 20\mathbf{k}$ lb, $\mathbf{F}_3 = 18\mathbf{i} - 12\mathbf{j} + 5\mathbf{k}$ lb, and have velocities $\mathbf{v}_1 = 5\mathbf{i} + 8\mathbf{j}$ ft/s, $\mathbf{v}_2 = 3\mathbf{i} - 6\mathbf{k}$ ft/s, $\mathbf{v}_3 = 2\mathbf{j} - 2\mathbf{k}$ ft/s, respectively. For the mass center of the system at the instant shown, determine the analytical expressions of (a) its position vector $\bar{\mathbf{r}}$, (b) its velocity $\bar{\mathbf{v}}$, (c) its acceleration $\bar{\mathbf{a}}$.

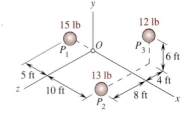

Solution. Applying the *principle of moments*, in the form of Eqs. (12.25), we write

$$m = \frac{15}{g} + \frac{13}{g} + \frac{12}{g} = \frac{40}{g}$$

$$\frac{40}{g}\bar{\mathbf{r}} = \frac{15}{g}(-5\mathbf{i}) + \frac{13}{g}(10\mathbf{i} + 8\mathbf{k}) + \frac{12}{g}(10\mathbf{i} + 6\mathbf{j} - 4\mathbf{k})$$

$$\bar{\mathbf{r}} = 1.875\mathbf{i} + 1.8\mathbf{j} + 1.4\mathbf{k} \text{ ft} \quad \blacktriangleleft$$

Applying the *principle of moments*, as in the first of Eqs. (12.26), and the *principle of motion of the mass center*, as in Eq. (12.27), we write

$$\frac{40}{g}\bar{\mathbf{v}} = \frac{15}{g}(5\mathbf{i} + 8\mathbf{j}) + \frac{13}{g}(3\mathbf{i} - 6\mathbf{k}) + \frac{12}{g}(3\mathbf{j} - 2\mathbf{k})$$

$$\bar{\mathbf{v}} = 2.85\mathbf{i} + 3.9\mathbf{j} - 2.55\mathbf{k} \text{ ft/s} \quad \blacktriangleleft$$

$$(10\mathbf{i} - 15\mathbf{j}) + (-13\mathbf{j} + 20\mathbf{k}) + (18\mathbf{i} - 12\mathbf{j} + 5\mathbf{k}) = \frac{40}{g}\bar{\mathbf{a}}$$

$$\bar{\mathbf{a}} = (0.7\mathbf{i} - \mathbf{j} + 0.625\mathbf{k})g$$

$$\bar{\mathbf{a}} = 22.5\mathbf{i} - 32.2\mathbf{j} + 20.1\mathbf{k} \text{ ft/s}^2 \quad \blacktriangleleft$$

Developmental Exercises

D12.16 Refer to Example 12.9. For the instant shown, determine (a) the moment $\mathbf{M}_G$ of the resultant effective force $m\mathbf{a}$ about the mass center G of the system, (b) whether $m\mathbf{a}$ passes through G.

D12.17 Describe the *principle of motion of the mass center*.

PROBLEMS

12.32 and 12.33* The resultant external force acting on the particle P in space is $\mathbf{F} = 240t^2\mathbf{i} - 400t^3\mathbf{j} + 300t^4\mathbf{k}$ N, where t is the time in seconds. The particle is at rest at the position shown when $t = 0$. Determine the position coordinates (x, y, z) of the particle when $t = 3$ s.

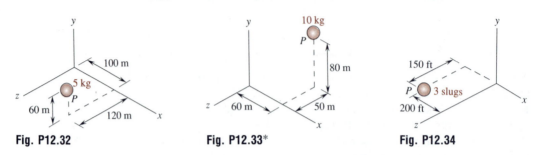

Fig. P12.32   **Fig. P12.33***   **Fig. P12.34**

12.34 The resultant external force acting on the particle P in space is $\mathbf{F} = 180t\mathbf{i} + 60t^3\mathbf{j} - 90t^4\mathbf{k}$ lb. When $t = 0$, the particle is situated at the position shown and has an initial velocity $\mathbf{v}_0 = 200\mathbf{j} + 500\mathbf{k}$ ft/s. Determine the position vector $\mathbf{r}$ of the particle when $t = 4$ s.

12.35* A 2-kg particle moves along the curve $x = 4\sin\pi t$, $y = 4\cos\pi t$, and $z = 0$, where x and y are measured in meters and t is the time in seconds. Determine the magnitude of the resultant force $\mathbf{R}$ acting on the particle and show that $\mathbf{R}$ is always directed toward the origin O.

12.36 A pendulum composed of a thin wire and a bob of mass 5 kg swings in a vertical plane. If the bob has a speed of 6 m/s in the position shown, determine the tension in the wire.

12.37 A small bob of weight 10 lb is revolving in a horizontal circle at a constant speed as shown. Determine (a) the tension in the wire AB, (b) the speed of the bob.

Fig. P12.36

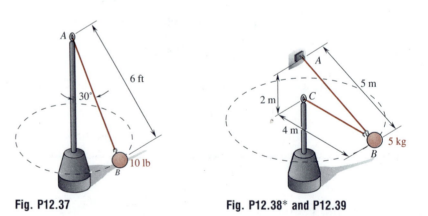

Fig. P12.37   **Fig. P12.38* and P12.39**

12.38* A small bob supported by two wires is revolving in a horizontal circle at a constant speed as shown. Determine the range of values of the speed of the bob for which no wire becomes slack.

12.39 A small bob supported by two wires is revolving in a horizontal circle at a constant speed of 9 m/s as shown. Knowing that the mass of the bob is 5 kg, determine the tension in each wire.

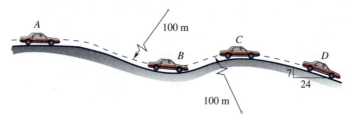

Fig. P12.40

12.40 Four cars are all traveling at a constant speed of 72 km/h along a road as shown. If $\mu_k = 0.5$ between all tires and the road and the brakes of each car are suddenly applied and skidding occurs at the respective positions shown, determine the ratios k_A, k_B, k_C, and k_D of the magnitudes of the tangential decelerations of the cars at A, B, C, and D to the magnitude g of the gravitational acceleration, respectively.

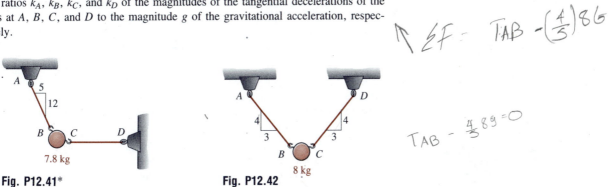

Fig. P12.41*   **Fig. P12.42**

12.41* **and 12.42** A small sphere is suspended by two wires as shown. Determine (a) the tension in the wire AB before the wire CD is cut, (b) the tension in the wire AB and the acceleration of the sphere immediately after the wire CD is cut.

12.43 A jet airplane for the training of astronauts flies at high altitude with a constant speed of 600 mi/h, where the gravitational acceleration is 32.1 ft/s². To simulate the zero-contact-force (i.e., apparent weightlessness) condition in the cabin of the airplane for a brief period of time, the pilot maneuvers to fly along a vertical curve as shown. Determine for this brief period (a) the radius of curvature of the path of flight, (b) the rate of change of the orientation of the airplane in degrees per second.

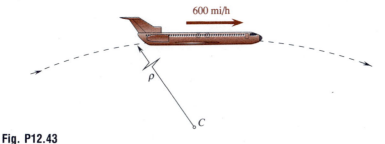

Fig. P12.43

12.44 A small plane flies along a vertical circular path of 150-m radius as shown. As the plane passes through the bottom point B, the pilot experiences an apparent weight equal to three times his actual weight. Determine the speed v_B of the plane as it passes through B. (*Hint.* The pilot's apparent weight is equal to the reaction from his seat.)

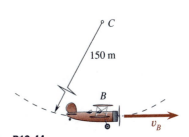

Fig. P12.44

12.45* Determine the minimum value of μ_s between the chip P and the vertical wall which will keep the chip from sliding down when the cylindrical vessel is rotating at 1 rev/s.

12.46 A slider S can move freely along a circular ring which is rotating at a constant rate $\dot{\theta}_0$ about the vertical axis OA as shown. If the angle β between the vertical axis and OS is 60°, determine the value of $\dot{\theta}_0$ in revolutions per second.

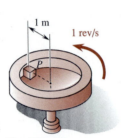

Fig. P12.45*

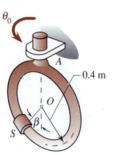

Fig. P12.46

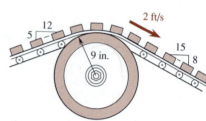

Fig. P12.47

12.47 A conveyor belt moves a series of small objects over an idler of radius 9 in. as shown. If the belt moves at a constant speed of 2 ft/s and no objects slip with respect to the belt, determine the minimum value of μ_s between the objects and the belt.

12.48* A racetrack turn of radius 150 m is banked at an angle as shown. If $\mu_s = 0.8$ between the tires and the pavement, determine the range of values of the speed v at which the race car will not skid sideways when rounding the turn.

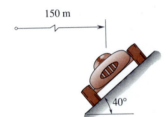

Fig. P12.48*

12.49 A car rounds a flat unbanked turn of radius 100 m at 72 km/h. Determine the minimum value of μ_s between the tires and the pavement if no sideway skidding occurs.

12.50 A rocket moving in a vertical plane is being propelled by a thrust $\mathbf{T}$ of 10 kips and is experiencing an air resistance $\mathbf{F}$ of 1.9 kips as shown. At this instant, the rocket has a mass m of 150 slugs and a velocity $\mathbf{v}$ of 2 mi/s. If the gravitational acceleration at this altitude is $\mathbf{g} = 28.6 \text{ ft/s}^2 \downarrow$, determine the radius of curvature ρ of the path and the magnitude a_t of the tangential component of acceleration of the rocket at the instant shown. [*Hint.* $m(d\mathbf{v}/dt) = \mathbf{F} + m\mathbf{g} + \mathbf{T}$.]

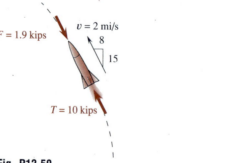

Fig. P12.50

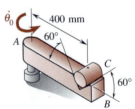

Fig. P12.51*

12.51* The horizontal arm AB is rotating with a constant angular speed $\dot{\theta}_0$ as shown. If the cylinder C is to remain in the V-shaped groove, determine the maximum allowable value of $\dot{\theta}_0$ in revolutions per second.

12.52 A 5-kg collar C mounted on a horizontal arm AB is connected by a spring to the end A of the arm as shown. The collar is oscillating along the arm while the arm is rotating with a constant angular speed $\dot{\theta}_0 = \frac{1}{2}$ rev/s. At this instant, $r =$ 1 m, $\dot{r} = 0.8$ m/s, and $\ddot{r} = -1.2$ m/s^2. If $\mu_k = 0.35$ between the collar and the rod, determine the friction force $\mathbf{F}_f$ and the spring force $\mathbf{F}_s$ acting on the collar at the instant shown.

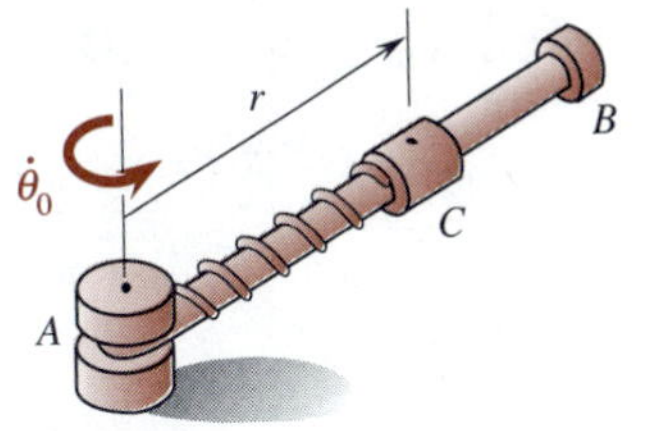

Fig. P12.52

12.53* Solve Prob. 12.52 if $\dot{\theta}_0 = 1$ rev/s.

12.54 A collar C of mass $m_C = 4$ kg, which is mounted on the frictionless arm AB, is pulled inward by a cord that passes through an eyebolt at the end A as shown. At the instant shown, the arm AB is rotating in a horizontal plane with $\dot{\theta} =$ 2 rad/s and $\ddot{\theta} = 3$ rad/s^2, while the end D of the cord is pulled upward with $\dot{y} =$ 1 m/s and $\ddot{y} = 2.5$ m/s^2. If the tension in the cord is $T = 34$ N at this instant, determine (a) the radial distance r, (b) the magnitude of the horizontal force $\mathbf{Q}$ acting on C by the arm AB.

12.55* Solve Prob. 12.54 if $m_C = 16.1$ lbm, $\dot{\theta} = 2.5$ rad/s, $\ddot{\theta} = 2$ rad/s^2, $\dot{y} =$ 3 ft/s, $\ddot{y} = 5$ ft/s^2, and $T = 15$ lb.

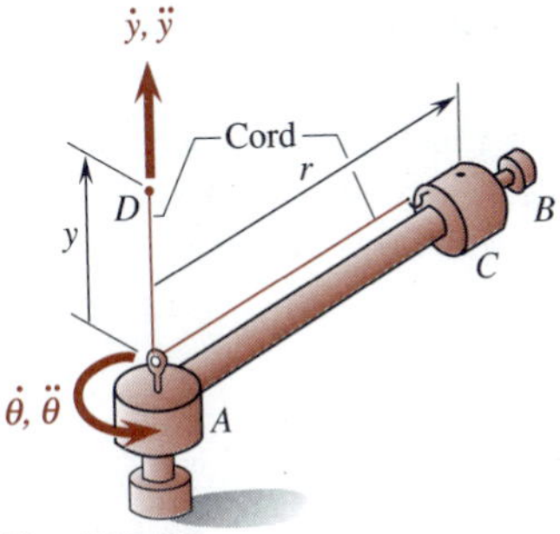

Fig. P12.54

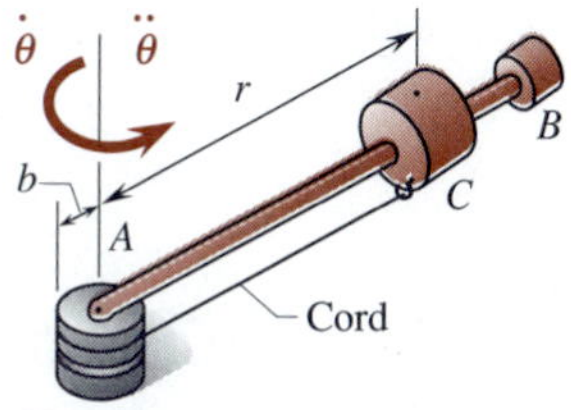

Fig. P12.56

12.56 A collar C of weight $W_C = 8.05$ lb, which is mounted on the frictionless arm AB, is pulled inward by a cord that wraps around a fixed drum of radius $b = 3$ in. as the arm AB rotates in a horizontal plane as shown, where $\dot{\theta} =$ 2 rad/s and $\ddot{\theta} = 4$ rad/s^2. If $r = 3$ ft at this instant, determine (a) the tension T in the cord, (b) the magnitude of the horizontal force $\mathbf{Q}$ acting on C by the arm AB. (*Hint.* The motion of the collar C relative to the arm AB is given by $\dot{r} = -b\dot{\theta}$ and $\ddot{r} = -b\ddot{\theta}$.)

12.57* Solve Prob. 12.56 if $W_C = 19.62$ N, $b = 80$ mm, $\dot{\theta} = 2.5$ rad/s, $\ddot{\theta} =$ 5 rad/s^2, and $r = 1.6$ m.

12.58 A collar C of mass $m_C = 5$ kg is mounted on the spring BC and oscillates along the frictionless rod AB as shown. The rod AB is connected to a horizontal turntable and a vertical shaft OD. The turntable rotates with a constant speed $\dot{\theta}_0 = \pi$ rad/s about the axis of the shaft. At the instant shown, $b = 0.8$ m and the collar C is moving downward with a velocity $\mathbf{v}_{C/AB} = 2$ m/s ↘ and an acceleration $\mathbf{a}_{C/AB} = 3$ m/s^2 ↘ relative to the rod AB. Determine for this instant (a) the resultant force $\mathbf{F}$ acting on the collar C, (b) the tensile or compressive force $\mathbf{F}_s$ developed in the spring. (*Hint.* Use cylindrical coordinates and note for this instant shown that

$$\mathbf{e}_R = \mathbf{j} \qquad \mathbf{e}_\theta = -\mathbf{i} \qquad \mathbf{e}_z = \mathbf{k}$$

The magnitude of $\mathbf{F}_s$ may be determined by considering $\mathbf{F} \cdot \boldsymbol{\lambda}_{BA}$, where $\boldsymbol{\lambda}_{BA}$ is the unit vector in the direction of $\overrightarrow{BA}$.)

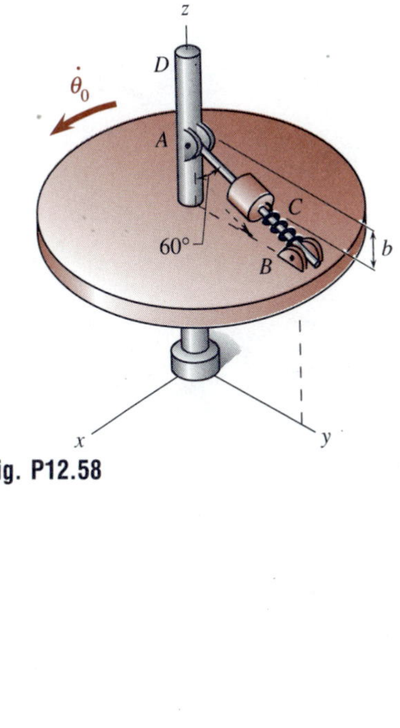

Fig. P12.58

12.59 Solve Prob. 12.58 if $m_C = 48.3$ lbm, $\dot{\theta}_0 = 2\pi$ rad/s, $\mathbf{v}_{C/AB} = 3$ ft/s ↘, $\mathbf{a}_{C/AB} = 4$ ft/s^2 ↘, and $b = 2$ ft.

12.60* Solve Prob. 12.58 if friction exists between the collar C and the rod AB, where $\mu_k = 0.3$.

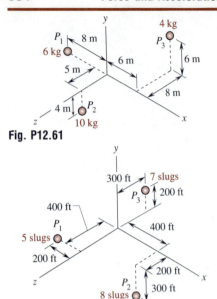

Fig. P12.61

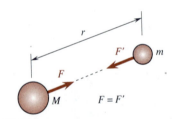

Fig. P12.64*

12.61 A system in space consists of three particles P_1, P_2, and P_3. At the instant shown, the particles are acted on by the external forces $\mathbf{F}_1 = -60\mathbf{j} + 90\mathbf{k}$ N, $\mathbf{F}_2 = -50\mathbf{i} - 100\mathbf{j} - 40\mathbf{k}$ N, $\mathbf{F}_3 = 30\mathbf{i} - 40\mathbf{j}$ N, and have velocities $\mathbf{v}_1 = 4\mathbf{i} - 6\mathbf{j}$ m/s, $\mathbf{v}_2 = 5\mathbf{j} - 3\mathbf{k}$ m/s, $\mathbf{v}_3 = 2\mathbf{i} - 7\mathbf{k}$ m/s, respectively. For the instant shown, determine for the mass center of the system (a) its position vector $\bar{\mathbf{r}}$, (b) its velocity $\bar{\mathbf{v}}$, (c) its acceleration $\bar{\mathbf{a}}$.

12.62* Refer to Prob. 12.61. For the instant shown, determine the analytical expressions of (a) the resultant effective force $m\bar{\mathbf{a}}$ of the system of particles, (b) the moment $\mathbf{M}_G$ of $m\bar{\mathbf{a}}$ about the mass center G of the system.

12.63 Refer to Prob. 12.61. For the instant shown, determine the accelerations $\mathbf{a}_{1/G}$, $\mathbf{a}_{2/G}$, and $\mathbf{a}_{3/G}$ of the particles P_1, P_2, and P_3 relative to the mass center G, respectively.

12.64* A system in space consists of three particles P_1, P_2, and P_3. At the instant shown, the particles are acted on by the external forces $\mathbf{F}_1 = 150\mathbf{i} - 40\mathbf{j}$ lb, $\mathbf{F}_2 = 132\mathbf{i} - 160\mathbf{j} + 240\mathbf{k}$ lb, $\mathbf{F}_3 = 280\mathbf{k}$ lb, respectively. For the mass center of the system at the instant shown, determine the analytical expressions of (a) its position vector $\bar{\mathbf{r}}$, (b) its acceleration $\bar{\mathbf{a}}$.

12.65 Refer to Prob. 12.64*. For the instant shown, determine the accelerations $\mathbf{a}_{1/G}$, $\mathbf{a}_{2/G}$, and $\mathbf{a}_{3/G}$ of the particles P_1, P_2, and P_3 relative to the mass center G, respectively.

CENTRAL-FORCE MOTION

12.8 Gravitational Force

A *gravitational force* is a naturally occurring force of mutual attraction between any two masses in the universe. Such a force is described in *Newton's law of gravitation*, which states that two particles are mutually attracted by a pair of opposite forces that are directed along the line joining the particles and have the same magnitude which is directly proportional to the product of their masses and inversely proportional to the square of the distance between them.[†] For two particles with masses M and m separated by a distance r, as shown in Fig. 12.14, the magnitude of the attractive force may be written as

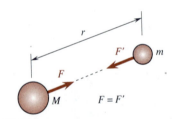

Fig. 12.14

$$F = G\frac{Mm}{r^2} \qquad (12.28)$$

where G is the *constant of gravitation*. Experiments show that

$$G = 66.726 \times 10^{-12} \frac{\text{m}^3}{\text{kg} \cdot \text{s}^2} \qquad (12.29)^{[††]}$$

[†]The statement of this law is a synthesis from *Book III*, *The System of the World* in Newton's *Principia* as cited in the footnote on p. 507. In certain astronomical studies, this law is replaced by the law of gravitation as covered in the *general* theory of relativity (1915) of Albert Einstein (1879–1955). (Cf. Sec. 1.14.)

[††]The accepted value of G as of 1982 is $G = (66.726 \pm 0.005) \times 10^{-12}$ m³/(kg·s²).

Developmental Exercise

D12.18 Using Eqs. (12.10), (12.11), (12.28), and (12.29), *estimate* the mass of the earth whose mean radius is 6.37 megameters.

12.9 Motion Under a Central Force

A *central force* is a force in a central force field, which is a region in space where a body is always acted on by a force directed toward or away from a common fixed point, which is called the *center of force*. A *central-force motion* is the motion of a particle under a central force, such as that shown in Fig. 12.15(a). A *gravitational central-force motion* is a central-force motion where the central force is a gravitational force. The motion of the planets around the sun and the motion of spacecraft and satellites around the earth can be studied as gravitational central-force motions. An *elastic central-force motion* is one where the central force is an elastic spring force. An example of elastic central-force motion is found in the motion of a particle attached to a stretched elastic cord on a frictionless horizontal surface, where the other end of the cord is connected to a fixed point.

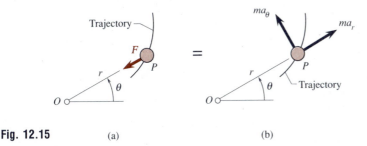

Fig. 12.15   (a)   (b)

For studying a gravitational or elastic central-force motion of a particle P with mass m, we may select the center of force as the origin O and use radial and transverse components to describe the motion as shown in Fig. 12.15. In this case, the transverse component $\mathbf{F}_\theta$ of the central force $\mathbf{F}$ is always zero. Referring to Fig. 12.15, we write

$$+\nwarrow \Sigma V_\theta: \qquad 0 = ma_\theta \qquad a_\theta = r\ddot{\theta} + 2\dot{r}\dot{\theta} = \frac{1}{r}\frac{d}{dt}(r^2\dot{\theta}) = 0$$

$$r^2\dot{\theta} = h \tag{12.30}$$

where $m \neq 0$, $r \neq 0$, and h is a constant. Thus, for any given central-force motion, $\dot{\theta}$ must decrease when r is increased, and vice versa. However, h has a different constant value for each different trajectory.[†] Since $v_\theta = r\dot{\theta}$, we may write the equivalent version of Eq. (12.30) as

$$rv_\theta = h \tag{12.31}$$

which is an important equation in the study of central-force motion.

[†]The trajectory of a spacecraft in free flight (i.e., coasting) becomes a different one whenever a noncentral force (e.g., thrust of a rocket) momentarily acts on the "coasting" spacecraft.

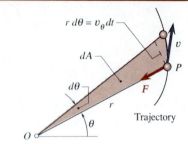

Fig. 12.16

Either of Eqs. (12.30) or (12.31) lends itself to an interesting geometric interpretation. As shown in Fig. 12.16, the radius vector **r** of a particle P moving under a central force **F** sweeps an infinitesimal area $dA = \frac{1}{2}r^2\, d\theta = \frac{1}{2}rv_\theta\, dt$ as it rotates about the center of force O through an angle $d\theta$ during the time interval dt. Thus, the time rate at which area is swept by **r** is

$$\dot{A} = \frac{dA}{dt} = \frac{1}{2}r^2\frac{d\theta}{dt} = \frac{1}{2}r^2\dot{\theta} = \frac{1}{2}rv_\theta$$

By Eq. (12.31), we find that

$$\dot{A} = \frac{1}{2}h \tag{12.32}$$

The quantity $\dot{A}$ is called the *areal speed* of the particle, which is a constant as the particle moves along a trajectory in central-force motion. Therefore, *the radius vector of a particle in central-force motion sweeps equal areas in equal times*. This geometric property may be referred to as the *conservation of areal speed* in central-force motion.[†]

Using the cross product of two vectors and Eqs. (12.31) and (12.32), we write

$$h = |\mathbf{r} \times \mathbf{v}| \qquad \dot{A} = \frac{1}{2}|\mathbf{r} \times \mathbf{v}| \tag{12.33}$$

$$\dot{\mathbf{A}} = \frac{1}{2}\mathbf{r} \times \mathbf{v} \tag{12.34}$$

where $\dot{\mathbf{A}}$ is called the *areal velocity* of the particle. We find from Eqs. (12.33) and (12.34) that *when a particle is in central-force motion, its areal velocity, as well as its areal speed, is constant.*

The *usual assumptions* made in elementary studies of the central-force motion of a *spacecraft* (which is here used to denote any *space vehicle, space shuttle, satellite,* or *space station*) passing by or revolving around a *planet* (such as the earth) are as follows:

1. The mass m of the spacecraft compared with the mass M of the planet is negligibly small; i.e., $m \ll M$.
2. The altitude of the spacecraft is high enough so that the air drag is negligible.
3. The spacecraft is not too far from the planet so that the dominant external force acting on the spacecraft is exerted by the planet; i.e., other external forces acting on it are negligible.
4. The duration of flight of the spacecraft around the planet is relatively short compared with the period of time it takes the planet to revolve around the sun once.
5. The acceleration of the center of the planet with respect to the sun is very small,[††] and a nonrotating reference frame attached to the center of the planet may serve as an inertial reference frame (cf. Sec. 3.1).

[†]This is *Kepler's second law* in Sec. 12.12. Cf. Eq. (14.23) in Sec. 14.6 regarding the interpretation of h as the *angular momentum per unit mass* of the particle about the center of force O and the *conservation of angular momentum* of a particle in central-force motion.

[††]The average magnitude of the acceleration of the center of the earth relative to the sun is only about 6 mm/s².

In the elementary study of the *central-force motion of a spacecraft around the moon*, the moon is treated as if it were a planet and the above usual assumptions are applied accordingly. Of course, the motion of the planets around the sun may similarly be studied.

Developmental Exercises

D12.19 Refer to Eqs. (12.30) and (12.31). (a) Are these two equations true for *any* type of central-force motion? (b) Does the value of h remain the same for *different* trajectories?

D12.20 Describe the *usual assumptions* made in elementary studies of the central-force motion of a spacecraft coasting near a planet.

EXAMPLE 12.10

A spacecraft in free flight around the earth has a velocity $\mathbf{v}_P$ parallel to the surface of the earth as it descends to the lowest point at P as shown. The radius of the earth is $R = 3.96 \times 10^3$ mi. Determine the speed v_Q of the spacecraft as it ascends to the highest point at Q.

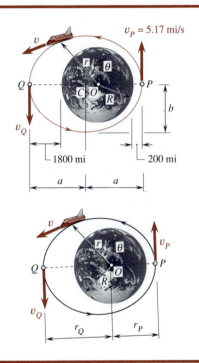

Solution. Since the spacecraft is in free flight, it is under the action of only the gravitational force. Thus, the spacecraft is having a *gravitational central-force motion*. The velocity of the spacecraft is always tangent to its orbit. From the sketch shown, we note that, at the extremities P and Q, the velocity $\mathbf{v}$ of the spacecraft is perpendicular to the radius vector $\mathbf{r}$. Thus, we have

$$(v_\theta)_P = v_P = 5.17 \text{ mi/s} \qquad (v_\theta)_Q = v_Q$$

Referring to the sketch, we write

$$h_P = h_Q: \qquad r_P(v_\theta)_P = r_Q(v_\theta)_Q \qquad r_P v_P = r_Q v_Q$$

$$v_Q = \frac{r_P v_P}{r_Q} = \frac{(3.96 \times 10^3 + 200)(5.17)}{3.96 \times 10^3 + 1800}$$

$$v_Q = 3.73 \text{ mi/s} \blacktriangleleft$$

Developmental Exercises

D12.21 Define the terms: (a) central force, (b) center of force, (c) central-force motion, (d) gravitational central-force motion, (e) elastic central-force motion, (f) areal speed, (g) areal velocity.

D12.22 Refer to Example 12.10. Determine (a) the areal speed $\dot{A}$ of the spacecraft, (b) the semimajor axis a of the elliptic orbit, (c) the distance c between the center C of the elliptic orbit and the center of force O which is a focus of the ellipse, (d) the semiminor axis b of the elliptic orbit, (e) the area A of the plane enclosed by the elliptic orbit, (f) the time τ required for the spacecraft to revolve around the earth once. (*Hint.* $a^2 - b^2 = c^2$, $A = \pi ab$, and $\tau = A/\dot{A}$.)

12.10   Governing Differential Equation

The differential equation defining the family of trajectories of a particle, which is set into central-force motion with various possible initial conditions, is called the *governing differential equation* for the central-force motion of the particle. The actual trajectory of the particle is a branch of the family of trajectories, which corresponds to the solution of the governing differential equation satisfying the actual initial conditions of the particle. We shall here derive such a differential equation valid for any type of central-force motion.

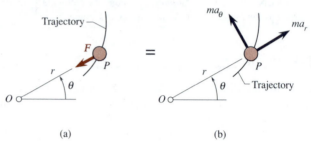

(a)   (b)

Fig. 12.15 (repeated)

The free-body diagram of a particle P with mass m moving under a central force $\mathbf{F}$ is equated to its effective-force diagram as shown in Fig. 12.15, where $\mathbf{F}$ is directed toward the center of force O. Referring to this figure, we write

$$+\nearrow \ \Sigma V_r: \qquad -F = ma_r \qquad (12.35)$$

$$+\nwarrow \ \Sigma V_\theta: \qquad 0 = ma_\theta \qquad (12.36)$$

where $a_r = \ddot{r} - r\dot\theta^2$ and $a_\theta = r\ddot\theta + 2\dot r\dot\theta$. We note that Eqs. (12.35) and (12.36) represent a set of *simultaneous* nonlinear ordinary differential equations defining the family of trajectories of a particle in central-force motion. The independent variable in these equations is the time variable t. But, as we saw in Sec. 12.9, Eq. (12.36) leads to Eq. (12.30); i.e.,

$$r^2\dot\theta = h \qquad (12.30)$$
$$\text{(repeated)}$$

where h is a constant. We may write Eq. (12.35) explicitly as

$$m(\ddot{r} - r\dot\theta^2) = -F \qquad (12.37)$$

To solve Eq. (12.37), we first introduce a variable u as

$$u = r^{-1} \qquad (r \neq 0) \qquad (12.38)$$

From Eqs. (12.30) and (12.38), we write

$$\dot\theta = u^2 h \qquad r = u^{-1} \qquad \frac{dr}{d\theta} = \frac{d}{d\theta}(u^{-1}) = -u^{-2}\frac{du}{d\theta}$$

$$\dot{r} = \frac{dr}{d\theta}\dot{\theta} = \left(-u^{-2}\frac{du}{d\theta}\right)(u^2 h) = -h\frac{du}{d\theta}$$

$$\ddot{r} = \frac{d\dot{r}}{d\theta}\dot{\theta} = \left(-h\frac{d^2u}{d\theta^2}\right)(u^2 h) = -h^2 u^2 \frac{d^2u}{d\theta^2}$$

Substitution of the above expressions into Eq. (12.37) leads to

$$\frac{d^2u}{d\theta^2} + u = \frac{F}{mh^2 u^2} \qquad (12.39)$$

This is the *governing differential equation* for any type of central-force motion, where $u = 1/r$ and the constant $h = r^2\dot{\theta} = rv_\theta$. In deriving Eq. (12.39), the central force **F** was assumed to be an *attractive force* directed toward the center of force O. If **F** is a repulsive force directed away from O, we replace F with $-F$ in Eq. (12.39). If F is a known function of r and hence u, Eq. (12.39) will take the form of a nonhomogeneous second-order ordinary differential equation in u and θ.

Developmental Exercise

D12.23 Is Eq. (12.39) the governing differential equation for either the elastic or the gravitational central-force motion?

12.11 Trajectories of Spacecraft

A spacecraft is in a *powered flight* when its launching rockets or its on-board rockets are producing thrusts to itself.[†] On the other hand, a spacecraft is in a *free flight* when the only external force acting on it is the gravitational attractive force. Thus, a spacecraft is in a *gravitational central-force motion* when it is in a free flight.

Let us consider a spacecraft S with mass m moving under the gravitational central force **F** exerted by the earth of mass M as shown in Fig. 12.17. By Newton's law of gravitation, the magnitude of **F** is given by

$$F = G\frac{Mm}{r^2} = GMmu^2$$

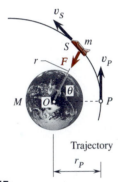

Fig. 12.17

where G is the constant of gravitation and $u = 1/r$. Substituting this expression of F into Eq. (12.39), we obtain

$$\frac{d^2u}{d\theta^2} + u = \frac{GM}{h^2} \qquad (12.40)^{[††]}$$

which is a nonhomogeneous linear second-order ordinary differential equation with constant coefficients.

[†]The launching rockets of a spacecraft may include on-board and booster rockets, and the on-board rockets may include main rockets and auxiliary rockets.

[††]Those who are familiar with the general solution of such a differential equation may skip the ensuing explanation and go directly to Eq. (12.44).

The general solution of Eq. (12.40) is given by the sum of a complementary solution u_c and a particular solution u_p; i.e., $u = u_c + u_p$. The particular solution u_p is any special solution which we can find for Eq. (12.40). Since the derivative of a constant is zero, we readily find that the simplest particular solution is

$$u_p = \frac{GM}{h^2} \tag{12.41}$$

The complementary solution u_c is given by the general solution of the corresponding homogeneous equation; i.e.,

$$\frac{d^2 u_c}{d\theta^2} + u_c = 0 \tag{12.42}$$

Since Eq. (12.42) is a *second-order* homogeneous differential equation, its general solution (which defines a family of curves) should contain *two* arbitrary constants of integration. By writing Eq. (12.42) as $u_c = -d^2 u_c/d\theta^2$, we forthwith infer that any function u_c of θ, which is equal to the negative of the second derivative of the function u_c with respect to θ, is a solution of Eq. (12.42). Noting that the second derivatives of the functions $\cos\theta$ and $\sin\theta$ with respect to θ are equal to $-\cos\theta$ and $-\sin\theta$, respectively, we readily conclude that the general solution of Eq. (12.42) and hence the complementary solution of Eq. (12.40) may be written as

$$u_c = C \cos\theta + D \sin\theta \tag{12.43}$$

where C and D are the two arbitrary constants of integration. Thus, we combine Eqs. (12.41) and (12.43) to obtain the general solution as

$$u = \frac{1}{r} = C \cos\theta + D \sin\theta + \frac{GM}{h^2} \tag{12.44}$$

Using the polar coordinate system established in Fig. 12.17, we note that the radial distance r of the trajectory at P is minimum, and the velocity $\mathbf{v}_P$ of the spacecraft at P is perpendicular to the radius vector $\mathbf{r}_P$. Thus, the *conditions at P* are

$$r = r_P \qquad \dot{r} = 0 \qquad \theta = 0 \qquad \dot{\theta} = \dot{\theta}_P = \frac{v_P}{r_P} \neq 0$$

Differentiating Eq. (12.44) with respect to the time t, we write

$$-r^{-2}\dot{r} = -C\dot{\theta} \sin\theta + D\dot{\theta} \cos\theta \tag{12.45}$$

Imposing the aforementioned *conditions at P* on Eq. (12.45), we get $0 = 0 + D\dot{\theta}_P$. Since $\dot{\theta}_P \neq 0$, we must have $D = 0$. Thus, Eq. (12.44) reduces to

$$\frac{1}{r} = C \cos\theta + \frac{GM}{h^2} \tag{12.46}$$

$$\frac{1}{r} = \frac{GM}{h^2}(1 + \varepsilon \cos\theta) \tag{12.47}^\dagger$$

[dagger]Note that the $+$ sign in Eq. (12.47) becomes $-$ if the polar coordinate system is chosen in such a way that r has a maximum value when $\theta = 0$.

where

$$\varepsilon = \frac{Ch^2}{GM} \tag{12.48}$$

Recall that Eq. (12.47) represents a *conic section* in polar coordinates, and ε is its *eccentricity*.[†] Therefore, *the trajectory of a spacecraft in free flight is a conic section* (e.g., a circle, ellipse, parabola, or hyperbola). The various shapes of the conic sections are illustrated in Fig. 12.18. Note in Figs. 12.17 and 12.18 that the origin O, which is the center of force, is a *focus* of the conic sections, and the polar axis is an axis of symmetry of the conic sections. Recalling that $-1 \le \cos\theta \le 1$ and referring to Eq. (12.47), we can reason as follows:

1. In the case of $\varepsilon > 1$, r becomes infinite when $\cos\theta = -1/\varepsilon$. Thus, the conic section is a *hyperbola* if $\varepsilon > 1$.
2. In the case of $\varepsilon = 1$, r becomes infinite when $\theta = 180°$. Thus, the conic section is a *parabola* if $\varepsilon = 1$.
3. In the case of $0 < \varepsilon < 1$, r remains finite for any value of θ. Thus, the conic section is an *ellipse* if $0 < \varepsilon < 1$.
4. In the case of $\varepsilon = 0$, r has a constant value of $h^2/(GM)$ for any value of θ. Thus, the conic section is a *circle* if $\varepsilon = 0$.

Since the radius vector of an *orbiting* spacecraft must never become infinite, *the trajectory of a spacecraft in orbit must be either an ellipse or a circle.* Such an unopen trajectory is called an *orbit*.

Suppose that a spacecraft S is *orbiting along an elliptic trajectory* around the earth as shown in Fig. 12.19, where the origin O and the point O' are the foci, the point C is the centroid of the ellipse, the point P on the orbit closest to the earth is called the *perigee*, the point Q on the orbit farthest away from the earth is called the *apogee*, the semimajor axis is $\overline{CP} = a$, the semiminor axis is $\overline{CM} = b$, the distance between the centroid C and either focus is c, and $r_P + r_Q = 2a$. Thus, we write

$$a = \frac{1}{2}(r_P + r_Q) \tag{12.49}$$

$$c = \overline{CO} = \overline{CP} - \overline{OP} = a - r_P = \frac{1}{2}(r_P + r_Q) - r_P$$

$$c = \frac{1}{2}(r_Q - r_P) \tag{12.50}$$

It can be shown that the sum of the distances from each of the foci to any point of the ellipse is constant. By symmetry, we note that $\overline{OM} = \overline{O'M}$ and $\overline{QO'} = \overline{OP}$. Thus, from Fig. 12.19 and Eqs. (12.49) and (12.50), we write

$$\overline{OM} + \overline{O'M} = \overline{OP} + \overline{O'P}$$

$$2\,\overline{OM} = \overline{QO'} + \overline{O'P} = 2a \qquad \overline{OM} = a$$

$$b^2 = \overline{CM}^2 = \overline{OM}^2 - \overline{CO}^2 = a^2 - c^2 = r_P r_Q$$

$$b = (r_P r_Q)^{1/2} \tag{12.51}$$

[†]Cf. Eq. (12.52) and Secs. B4.2 through B4.6 of App. B.

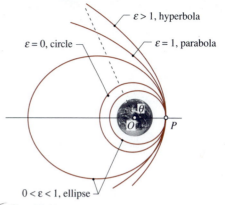

Fig. 12.18

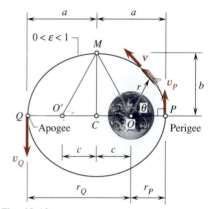

Fig. 12.19

Since the eccentricity of an ellipse is $\varepsilon = c/a$ and a circle is a special case of an ellipse, the *eccentricity of orbit* is defined to be

$$\varepsilon = (r_Q - r_P)/(r_Q + r_P) \tag{12.52}$$

The elliptic area enclosed by the orbit is $A = \pi ab$, or

$$A = \frac{\pi}{2}(r_P + r_Q)(r_P r_Q)^{1/2} \tag{12.53}$$

The time required for a spacecraft or satellite to travel along a complete orbit is called the *period of orbit* and is denoted by τ. The areal speed of an orbiting spacecraft is $\dot{A} = h/2$. Thus, we write

$$\tau = A/\dot{A}$$

$$\tau = \frac{\pi(r_P + r_Q)(r_P r_Q)^{1/2}}{h} \tag{12.54}$$

When the spacecraft passes through the *perigee* P, we have $r = r_P$ and $\theta = 0$. Imposing these conditions on Eq. (12.47), we write

$$\frac{1}{r_P} = \frac{GM}{h^2}(1 + \varepsilon) \tag{12.55}$$

Moreover, when the spacecraft passes through the *apogee* Q, we have $r = r_Q$ and $\theta = 180°$. Imposing these conditions on Eq. (12.47), we write

$$\frac{1}{r_Q} = \frac{GM}{h^2}(1 - \varepsilon) \tag{12.56}$$

Adding Eqs. (12.55) and (12.56), we get

$$\frac{1}{r_P} + \frac{1}{r_Q} = \frac{2GM}{h^2} \tag{12.57}[†]$$

Since $h = rv_\theta$, the constant h in Eq. (12.57) can be written as

$$h = r_P v_P = r_Q v_Q \tag{12.58}$$

The weight W of any body with mass m on the surface of the earth is equal to the magnitude of the attractive force $\mathbf{F}$ exerted on the body by the earth which has a mass M and a radius R. Thus, we apply Newton's second law and Newton's law of gravitation to write

$$W = mg = F = \frac{GM}{R^2} m$$

$$GM = gR^2 \tag{12.59}$$

where $g = 9.81$ m/s^2 or 32.2 ft/s^2 and $R = 6.37$ Mm or 3.96×10^3 mi.[††]

[†]This equation is very useful in subsequent studies.

[††]These values of g and R cannot directly be used to evaluate GM in Eq. (12.59) if M is not the mass of the earth. (Cf. Probs. 12.78 through 12.84.)

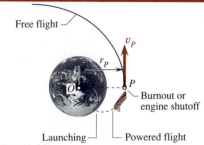

Fig. 12.20

The constant h and the eccentricity ε in Eq. (12.47) are the parameters which characterize the trajectory of a spacecraft. These parameters may be determined from the initial conditions of its free flight. In most actual launches, the velocity $\mathbf{v}_P$ of the spacecraft at the end of its powered flight is programmed to be perpendicular to its radius vector $\mathbf{r}_P$ as shown in Fig. 12.20.[†]

If the spacecraft follows a parabolic trajectory, the conditions at P are $r = r_P$, $\theta = 0$, $v_P = v_{\text{par}}$, $h = r_P v_P = r_P v_{\text{par}}$, and $\varepsilon = 1$. Imposing these conditions on Eq. (12.47) and using Eq. (12.59), we write

$$\frac{1}{r_P} = \frac{GM}{r_P^2 v_{\text{par}}^2}(1 + 1) \qquad v_{\text{par}} = (2GM/r_P)^{1/2}$$

$$v_{\text{par}} = (2gR^2/r_P)^{1/2} \qquad\qquad (12.60)$$

We can easily check that the trajectory becomes hyperbolic if $v_P > v_{\text{par}}$, and elliptic or circular if $v_P < v_{\text{par}}$. Since the value of v_{par} obtained for the parabolic trajectory is the smallest value of v_P for which the spacecraft will not return to its starting point, the velocity with magnitude equal to v_{par} is called the *escape velocity* $\mathbf{v}_{\text{esc}}$ at $r = r_P$; i.e.,

$$v_{\text{esc}} = (2gR^2/r_P)^{1/2} \qquad\qquad (12.61)$$

If the spacecraft follows a circular trajectory, the conditions at P are $r = r_P$, $\theta = 0$, $v_P = v_{\text{cir}}$, $h = r_P v_P = r_P v_{\text{cir}}$, and $\varepsilon = 0$. Imposing these conditions on Eq. (12.47) and using Eq. (12.59), we write

$$\frac{1}{r_P} = \frac{GM}{r_P^2 v_{\text{cir}}^2} \qquad v_{\text{cir}} = (GM/r_P)^{1/2}$$

$$v_{\text{cir}} = (gR^2/r_P)^{1/2} \qquad\qquad (12.62)$$

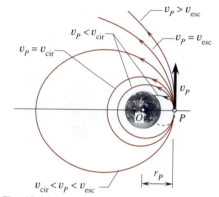

Fig. 12.21

The various trajectories corresponding to different values of the velocity $\mathbf{v}_P$ at the vertex P are illustrated in Fig. 12.21. If $v_P \ll v_{\text{cir}}$, the spacecraft may, of course, reenter the earth's atmosphere. If so, the air drag becomes significant, and the assumption (2) in the latter part of Sec. 12.9 will no longer be valid. The computer program in Fig. 12.22 on the next page is written for an orbiting spacecraft; it may be used to compute the period τ, the eccentricity ε, the maximum speed v_P at the perigee P, and the minimum speed v_Q at the apogee Q of the spacecraft when the altitudes of P and Q in its orbit around the earth are known.

Developmental Exercises

D12.24   Name the conic section if ε is (a) 2, (b) 1, (c) 0.5, (d) 0.

D12.25   Define the terms: (a) orbit, (b) eccentricity of orbit, (c) period of orbit, (d) escape velocity.

D12.26   Derive Eq. (12.61) from Eq. (12.57) by using Eq. (12.59) and letting $r_Q \to \infty$ as well as $h = r_P v_{\text{esc}}$.

D12.27   Derive Eq. (12.62) from Eq. (12.57). (*Hint.* Cf. D12.26.)

[†]Oblique launchings of spacecraft are studied in Chaps. 13 and 14.

```
10 REM          COMPUTING PERIOD, ECCENTRICITY, AND EXTREMUM SPEEDS
15 REM    OF A SPACECRAFT FROM THE ALTITUDES OF ITS PERIGEE AND APOGEE
20 REM     Refer to Eqs. (12.52), (12.54), (12.57), and (12.59).  Let
25 REM     "ap" and "aq" be the altitudes of the perigee and apogee of
30 REM     the spacecraft orbiting around the earth.  Enter "ap" & "aq"
35 REM     (in miles) as data in Line 100.  To output to the printer,
40 REM     change PRINT to LPRINT in the program.      (File name: SPACFT)
45 READ AP,AQ: R=3960: RP=AP+R: RQ=AQ+R: E=(RQ-RP)/(RQ+RP): G=32.2/5280
50 GM=G*R^2: H=SQR(2*GM*RP*RQ/(RP+RQ)): T=3.141593*(RP+RQ)*SQR(RP*RQ)/H
55 VP=H/RP: VQ=H/RQ: M$="miles": S$="        ": PRINT: PRINT "ap =";AP;M$;
60 PRINT S$;"aq =";AQ;M$: PRINT "Period =";T/3600;"hours";S$;
65 PRINT "Eccentricity =";E: J$="miles/second   or ": K$="miles/hour"
70 PRINT "Vp =";VP;J$;VP*3600;K$: PRINT "Vq =";VQ;J$;VQ*3600;K$
100 DATA 250,500

ap = 250 miles        aq = 500 miles
Period = 1.610849 hours      Eccentricity = 2.883507E-02
Vp = 4.834349 miles/second  or   17403.66 miles/hour
Vq = 4.563366 miles/second  or   16428.12 miles/hour
```

Fig. 12.22 (Cf. the solution in Example 12.11.)

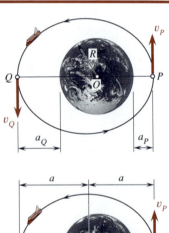

EXAMPLE 12.11

The altitudes of the perigee P and apogee Q in the orbit of a space shuttle are $a_P = 250$ mi and $a_Q = 500$ mi as shown. For such an orbit, determine (a) the period τ, (b) the eccentricity ε, (c) the maximum speed v_P, (d) the minimum speed v_Q.

Solution.　Since the radius of the earth is $R = 3.96 \times 10^3$ mi, $a_P = 250$ mi, $a_Q = 500$ mi, $g = 32.2/5280$ mi/s^2, and $GM = gR^2$, we write

$$r_P = R + a_P \qquad r_P = 4.21 \times 10^3 \text{ mi}$$

$$r_Q = R + a_Q \qquad r_Q = 4.46 \times 10^3 \text{ mi}$$

$$\frac{1}{r_P} + \frac{1}{r_Q} = \frac{2GM}{h^2} = \frac{2gR^2}{h^2} \qquad h = 20.35 \times 10^3 \text{ mi}^2/\text{s}$$

$$\tau = \frac{\pi(r_P + r_Q)(r_P r_Q)^{1/2}}{h} = 5.799 \times 10^3 \text{ s}$$

$$\tau = 1.611 \text{ hours} \blacktriangleleft$$

$$\varepsilon = (r_Q - r_P)/(r_Q + r_P) \qquad h = r_P v_P = r_Q v_Q$$

$$\varepsilon = 0.0288 \qquad v_P = 4.83 \text{ mi/s} \qquad v_Q = 4.56 \text{ mi/s} \blacktriangleleft$$

Developmental Exercise

D12.28　Refer to Example 12.11. Determine for the orbit (a) the semimajor axis a, (b) the semiminor axis b, (c) the elliptic area A, (d) the areal speed $\dot{A}$, (e) the period τ as computed from $A/\dot{A}$.

EXAMPLE 12.12

A spacecraft is describing a circular orbit of radius 8 Mm around the earth when its on-board rockets are fired momentarily at P to increase its speed from $(v_P)_{cir}$ to $(v_P)_{ell}$ as shown, where $R = 6.37$ Mm. This maneuver sends the spacecraft along an elliptic trajectory to the apogee at Q. Upon reaching Q, its on-board rockets are fired momentarily again to increase its speed from $(v_Q)_{ell}$ to $(v_Q)_{cir}$ to insert itself into a circular orbit of radius 20 Mm as shown. Determine the values of $(v_P)_{cir}$, $(v_P)_{ell}$, $(v_Q)_{ell}$, $(v_Q)_{cir}$, and the time t_{PQ} required for the spacecraft to travel from P to Q.

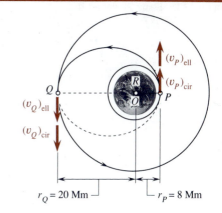

Solution. From the sketch shown, we have $r_P = 8$ Mm $= 8 \times 10^3$ km and $r_Q = 20$ Mm $= 20 \times 10^3$ km. Thus, Eqs. (12.57) and (12.59) are best for the solution. Since the radius of the earth is $R = 6.37$ Mm $= 6.37 \times 10^3$ km and $g = 9.81$ m/s^2 $= 9.81 \times 10^{-3}$ km/s^2, we write

$$\frac{1}{r_P} + \frac{1}{r_P} = \frac{2gR^2}{r_P^2(v_P)_{cir}^2} \qquad (v_P)_{cir} = 7.05 \text{ km/s} \quad \blacktriangleleft$$

At the perigee P and apogee Q of the elliptic trajectory, we write

$$\frac{1}{r_P} + \frac{1}{r_Q} = \frac{2gR^2}{r_P^2(v_P)_{ell}^2} \qquad (v_P)_{ell} = 8.43 \text{ km/s} \quad \blacktriangleleft$$

$$h = r_P(v_P)_{ell} = r_Q(v_Q)_{ell} \qquad (v_Q)_{ell} = 3.37 \text{ km/s} \quad \blacktriangleleft$$

$$\frac{1}{r_Q} + \frac{1}{r_Q} = \frac{2gR^2}{r_Q^2(v_Q)_{cir}^2} \qquad (v_Q)_{cir} = 4.46 \text{ km/s} \quad \blacktriangleleft$$

The time t_{PQ} required to travel from P to Q is equal to one half of the period of orbit along the entire ellipse. Thus, we write

$$t_{PQ} = \frac{\pi(r_P + r_Q)(r_P r_Q)^{1/2}}{2r_P(v_P)_{ell}} = 8.25 \times 10^3 \text{ s}$$

$$t_{PQ} = 2.29 \text{ hours} \quad \blacktriangleleft$$

Developmental Exercise

D12.29   Refer to Example 12.12. Determine the period of orbit of the spacecraft along the circle of radius (a) 8 Mm, (b) 20 Mm.

★12.12   Kepler's Laws of Planetary Motion

The theory developed in preceding sections for central-force motion of space-craft around the earth is also applicable to the study of the motion of planets around the sun. The properties deducible from Eqs. (12.32) and (12.47), with M representing the mass of the sun, were first discovered by the German astronomer Johann (or Johannes) Kepler (1571–1630) from astronomical observations of the motions of planets. Kepler's discovery served not only as background but also as proving ground for Newton (1642–1727) in

Table 12.2   Some Planetary Data

Planet	Semimajor Axis of Orbit (AU)[†]	Eccentricity of Orbit	Period of Orbit around the Sun	Period of Rotation about own Axis	Mass Relative to the Earth
Mercury	0.3871	0.205629	87.97 d	58.65 d	0.0552
Venus	0.7233	0.006787	224.70 d	−243.0 d[††]	0.8144
Earth	1	0.016721	365.26 d	23 h 56 min 4 s	1[†††]
Mars	1.5237	0.093379	686.98 d	24 h 37 min 23 s	0.1077
Jupiter	5.203	0.0485	11.86 y	9 h 50.5 min (at equator)	317.7
Saturn	9.539	0.0556	29.46 y	0.446 d	95.15
Uranus	19.18	0.0472	84.01 y	−0.717 d[††]	14.4
Neptune	30.06	0.0086	164.8 y	0.67 d	17.2
Pluto	39.44	0.25	247.7 y	6.39 d	0.003

[†]One AU (astronomical unit of distance) is defined to be the length of the semimajor axis of the Earth's orbit around the Sun; i.e.,

$$1 \text{ AU} = 149.5 \text{ Gm} = 149.5 \times 10^6 \text{ km}$$
$$= 92.9 \times 10^6 \text{ mi}$$

[††]The negative sign indicates retrograde motion.

[†††]The mass of the Earth itself is 5.976×10^{24} kg. (The Moon of the Earth has a mass of 7.35×10^{22} kg. Its orbit around the Earth has semimajor axis = 384.398 Mm, eccentricity = 0.055, and period = 27.32 d.)

formulating his law of gravitation as well as laws of motion. For reference, some planetary data are given in Table 12.2, and the mean distances (i.e., the semimajor axes of orbits) of the planets from the sun are illustrated in Fig. 12.23.

Kepler's three *laws of planetary motion* may be stated as follows:

1. *First law.* The orbit of each planet is an ellipse with the sun at one of its foci.
2. *Second law.* The radius vector drawn from the sun to a planet sweeps equal areas in equal times.
3. *Third law.* The squares of the periods of orbits of the planets are proportional to the cubes of the semimajor axes of their orbits.

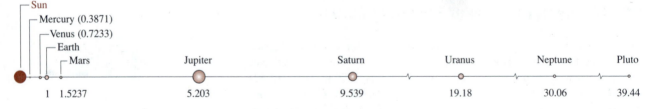

Mean distances of the planets from the sun (in AU)

Fig. 12.23

Note that Kepler's first law states the particular case of $0 < \varepsilon < 1$ in the result established in Eq. (12.47). His second law is equivalent to the statement that the areal speed of each planet is constant, which is the result established in Eq. (12.32). Kepler's third law may be proved by using Eqs. (12.49), (12.51), (12.54), and (12.57).

Developmental Exercises

D12.30 Do the following two steps to prove Kepler's third law of planetary motion: (a) From Eqs. (12.49), (12.51), and (12.57), show that $h^2 = GMb^2/a$. (b) Using this expression for h^2 and Eq. (12.54), show that $\tau^2 = [4\pi^2/(GM)]a^3$, hence $\tau_i^2/\tau_j^2 = a_i^3/a_j^3$ where the subscripts refer to the ith and the jth planets.

D12.31 Using the data given in Table 12.2, determine the shortest distance r_P and the longest distance r_Q existing between the centers of Mercury and the Sun in miles.

D12.32 Using the data given in Table 12.2, verify Kepler's third law of planetary motion for the planets Mercury and Earth.

PROBLEMS

NOTE. The motion of any spacecraft mentioned in the following problems is measured in a nonrotating reference frame attached to the center of force in the attracting body, such as the earth, and the usual assumptions stated in Sec. 12.9 are implied. Whenever needed, use $R = 6.37$ Mm or 3.96×10^3 mi for the radius of the earth.

12.66 and 12.67*[†] The minimum and maximum altitudes of the orbit of an earth satellite S are indicated as shown. Determine (a) the eccentricity ε of the orbit, (b) the maximum speed v_{max} of the satellite.

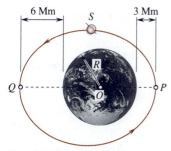

Fig. P12.66 and P12.68*

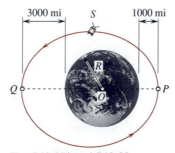

Fig. P12.67* and P12.69

12.68* and 12.69 The minimum and maximum altitudes of the orbit of an earth satellite S are indicated as shown. Determine for the satellite (a) its minimum speed, (b) its areal speed, (c) the elliptic area enclosed by its orbit, (d) its period of orbit.

12.70 A *Landsat S* describes a circular orbit around the earth as shown. If its period of orbit is 103 min, determine (a) its altitude a_0 in miles, (b) its speed v_{cir} in miles per second.

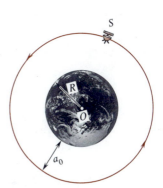

Fig. P12.70 and P12.71

[†]Cf. the computer program in Fig. 12.22.

12.71   Determine the altitude a_0 of a geosynchronous satellite in both SI and U.S. customary units.[†]

12.72*   Determine the speed of a geosynchronous satellite in both SI and U.S. customary units.[†]

12.73   It is known that the radius of the planet Mars is 2.11×10^3 mi. Making use of the data in Table 12.2, determine the altitude of a synchronous satellite of the planet Mars.

12.74*   Making use of the data in Table 12.2, determine the speed of a synchronous satellite of the planet Mars.

12.75*   A spacecraft is describing a circular orbit of 12-Mm radius around the earth when its rockets are fired momentarily to increase the speed of the spacecraft by 1 km/s. The spacecraft is thus inserted into an elliptic orbit as shown. For the spacecraft in this elliptic orbit, determine (a) its maximum altitude, (b) its period of orbit.

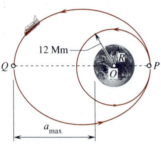
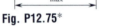

Fig. P12.75*

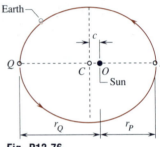

Fig. P12.76

12.76   The earth revolves around the sun in an elliptic orbit as shown. Making use of the data in Table 12.2, determine (a) the distance c between the centroid C of the elliptic orbit and the center O of the sun, (b) the minimum and maximum radial distances r_P and r_Q, (c) the shortest distance existing between the surface of the earth and the surface of the sun if the radius of the sun is 696 Mm.

12.77   Refer to Prob. 12.76. Determine (a) the areal speed $\dot{A}$ of the earth in its orbit around the sun, (b) the ratio of the mass of the sun to the mass of the earth.

12.78   A spacecraft approaches the planet Venus along a parabolic trajectory SQ as shown. As it reaches the vertex Q of the trajectory SQ, its retrorockets are fired momentarily to decrease its speed and insert itself into an elliptic orbit as indicated. The ratio of the mass of Venus to the mass of the earth is known to be 0.8144. Determine the speed of the spacecraft (a) as it approaches Q, (b) after the firing of retrorockets, (c) as it approaches P.

12.79*   A spacecraft approaches the planet Mars along a hyperbolic trajectory SQ of eccentricity $\varepsilon = 3$ as shown. As it reaches the vertex Q of the trajectory SQ, its retrorockets are fired momentarily to decrease its speed and insert itself into an elliptic orbit as indicated. The ratio of the mass of Mars to the mass of the earth is known to be 0.1077. Determine for the spacecraft (a) its speed as it approaches Q, (b) its speed after the firing of retrorockets, (c) its period along the elliptic orbit.

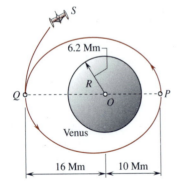

Fig. P12.78

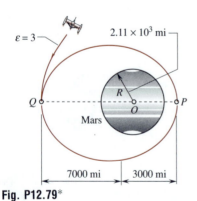

Fig. P12.79*

[†]A *geosynchronous satellite* is a synchronous satellite which travels from west to east around the center of the *earth* in a circular orbit directly above the equator with a period of orbit equal to one sidereal day (i.e., 23 h 56 min 4 s) and thus appears to be stationary relative to the earth. Communication and weather satellites are examples of geosynchronous satellites.

12.80 In returning from a moon-exploration mission, the lunar excursion module (LEM) is first propelled by its rocket engines from the moon's surface along a curved path to the point P and then travels in free flight along an elliptic path to rendezvous at Q with the command module which is traveling in a circular "parking" orbit around the moon as shown. The velocities of the LEM at the points P and Q are perpendicular to the radius vectors. The mass of the moon is 0.01230 times the mass of the earth. Determine (a) the speed of the LEM as it begins its free flight at P, (b) the speed of the LEM as it reaches Q, (c) the relative speed with which the command module approaches the LEM at Q.

12.81 Refer to Prob. 12.80. Determine (a) the period of orbit of the command module, (b) the time it takes the LEM to travel from P to Q.

12.82 After the rendezvous of the command module and the LEM at the point Q as described in Prob. 12.80, the astronauts in the LEM move into the command module and turn the spacecraft around so that the LEM faces the rear. When the spacecraft reaches the point S as shown, the LEM is separated from the command module and is cast adrift by reducing its speed from $(v_S)_{cir}$ to $(v_S)_{ell}$. Consequently, the LEM crashes on the moon's surface at the point K. Using the mass and radius of the moon given in Prob. 12.80, determine the value of $(v_S)_{ell}$ if K is located at $\theta = 60°$. [*Hint.* The point S is the apogee of the crash trajectory, and the $+$ sign in Eq. (12.47) becomes $-$.]

12.83* Solve Prob. 12.82 if K is located at $\theta = 120°$.

12.84 Refer to Prob. 12.82. If $(v_S)_{ell} = 1.52$ km/s, determine the value of θ at which the LEM will crash on the moon's surface.

12.85* A spacecraft is in free flight along an elliptic orbit of eccentricity ε around the earth as shown. Show that the speeds v_P and v_Q of the spacecraft at the perigee P and the apogee Q may be written as

$$v_P = R[(g/a)(1 + \varepsilon)/(1 - \varepsilon)]^{1/2} = R[gr_Q/(ar_P)]^{1/2}$$
$$v_Q = R[(g/a)(1 - \varepsilon)/(1 + \varepsilon)]^{1/2} = R[gr_P/(ar_Q)]^{1/2}$$

12.86 A spacecraft launched for interplanetary exploration begins its free flight along a conic section of eccentricity ε from the vertex P as shown. For any position S of the spacecraft on its free flight, express in terms of ε, r_P, g, the radius R of the earth, and the angular position coordinate θ for the position S, (a) the areal speed $\dot{A}$ of the spacecraft, (b) the speed of the spacecraft, (c) the angle ψ between **v** and the radius vector **r**. [*Hint.* One may find $\dot{\theta}$ from Eq. (12.30) and $\dot{r}$ by differentiating Eq. (12.47) with respect to the time t.]

12.87 A spacecraft begins its free flight along a parabolic trajectory (i.e., $\varepsilon = 1$) from the vertex P, a distance r_P from the center O of the earth as shown. For any position S of the spacecraft on its trajectory, express the time t elapsed since its flight from P in terms of r_P, g, the radius R of the earth, and the angular position coordinate θ for the position S. [*Hint.* Solve for t using the relation

$$\frac{d\theta}{dt} = \dot{\theta} = \frac{h}{r^2} = \frac{G^2M^2}{h^3}(1 + \cos\theta)^2 = \frac{4G^2M^2}{h^3}\cos^4(\theta/2)$$

and formula (1) in Sec. B5.2 of App. B.]

12.88* Refer to Prob. 12.86. The spacecraft crosses a point on the orbit of geosynchronous satellites when $r = \overline{OS} = 42.1$ Mm. For $\varepsilon = 1$ and $r_P = 6.86$ Mm, determine (a) the speed v_P of the spacecraft at P, (b) the angular position coordinate θ and the speed v of the spacecraft as it crosses the orbit of geosynchronous satellites.

12.89 Refer to Prob. 12.87. For $r = \overline{OS} = 42.1$ Mm = radius of the orbit of geosynchronous satellites and $r_P = 6.86$ Mm, determine the time it takes the spacecraft to fly from P along the parabolic trajectory to S.

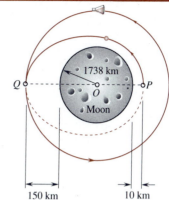

Fig. P12.80

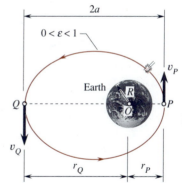

Fig. P12.82

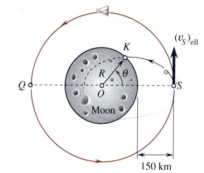

Fig. P12.85*

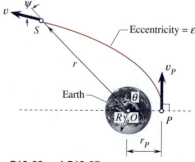

Fig. P12.86 and P12.87

12.13   Concluding Remarks

Newton's second law of motion is a keystone in kinetics. This law is used to formulate the *method of force and acceleration* at the beginning of the chapter. The crux of the *method of force and acceleration* is that the *system of force vectors* in the free-body diagram of a body (which may be a particle or a system of particles) is either equivalent or equipollent to the *system of effective-force vectors* in the effective-force diagram of the body. Those two systems are equivalent if the body is a single particle; otherwise, they are equipollent.

In any case, the said *system of force vectors* and *system of effective-force vectors* must have the *same resultant force* and the *same resultant moment* about any given point. Thus, it is *necessary* that (a) the sum of components of forces in the free-body diagram in any desirable direction is equal to the sum of components of effective forces in the effective-force diagram in the same direction, (b) the sum of moments of the forces in the free-body diagram about any desirable point (or axis) is equal to the sum of moments of the effective forces in the effective-force diagram about the same point (or axis).

For a partial solution of a problem, judiciously repeated applications of one or both of the above two necessary conditions will usually yield equations from which the unknowns may more directly be solved. Note that the resultant effective force vector does not necessarily pass through the mass center of the system of particles. Furthermore, the drawing of the effective-force diagram as well as the free-body diagram is an *enlightening step* in the application of the method of force and acceleration.

The study of gravitational central-force motion of spacecraft is presented in the latter half of the chapter. Such a study is accomplished by joining Newton's law of gravitation with Newton's second law of motion. It should be noted that Eqs. (12.30), (12.31), (12.32), (12.47), (12.52), (12.54), (12.57), and (12.59) are particularly helpful in such a study. However, we should be aware of the *usual assumptions* stated in the latter part of Sec. 12.9 when we use these equations.

REVIEW PROBLEMS

12.90   In the system shown, it is observed that the acceleration of the cylinder A is 1 m/s² ↓ . If the mass of the cylinder B is 30 kg, determine (a) the mass of the cylinder A, (b) the tension in cable 1 as indicated.

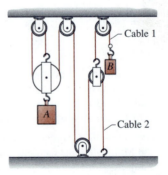

Fig. P12.90

12.91   A train consists of an 80-Mg locomotive A, a 45-Mg car B, and a 35-Mg car C as shown. If the train is accelerating to the right at 0.4 m/s², determine (a) the magnitude of the tractive friction force between the locomotive A and the tracks, (b) the force in the coupling between A and B, (c) the force in the coupling between B and C.

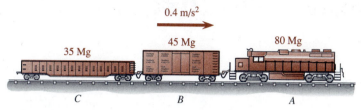

0.4 m/s²

35 Mg   45 Mg   80 Mg

C   B   A

Fig. P12.91

12.92   A highway curve of radius 100 m is banked at an angle as shown. If $\mu_s = 0.4$ between the tires and the pavement, determine the maximum speed at which the car will not skid sideways when rounding the curve.

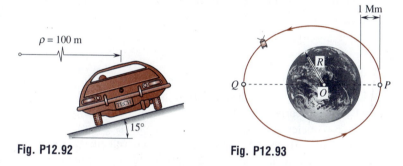

$\rho = 100$ m

15°

Fig. P12.92

1 Mm

R

Q   O   P

Fig. P12.93

12.93   A spacecraft in free flight around the earth as shown has a maximum speed of 8 km/s. Determine for the spacecraft (a) its maximum altitude, (b) its minimum speed, (c) the eccentricity of its orbit, (d) its areal speed, (e) its period of orbit.

12.94   A spacecraft approaches the moon along a parabolic trajectory SP as shown. As it reaches the vertex P of the parabola, its retrorockets are fired momentarily to decrease its speed and insert itself into a circular orbit as indicated. The mass of the moon is 0.01230 times the mass of the earth. Determine for the spacecraft (a) its speed as it approaches P, (b) its speed after the firing of retrorockets, (c) its period of orbit.

12.95*   Using the data given in Table 12.2, verify Kepler's third law of planetary motion for the planets (a) Venus and Earth, (b) Mars and Earth.

12.96   Using the data given in Table 12.2, determine the maximum speed of the center of the earth with respect to a nonrotating reference frame attached to the center of the sun.

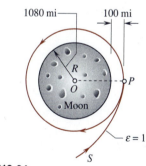

1080 mi   100 mi

R

O   P

Moon

$\varepsilon = 1$

S

Fig. P12.94

Chapter 13

Kinetics of Particles: Work and Energy

KEY CONCEPTS

- Work of a force, power, efficiency, kinetic energy of a particle, conservative force field, conservative system, potential energy, total mechanical energy, and constrained virtual displacement.
- Principle of work and energy for particles.
- Principle of conservation of energy for particles.
- Principle of virtual work in kinetics for studying forces and accelerations in one-degree-of-freedom systems of particles.

Work and energy are scalar quantities with specific meaning in mechanics. Through the concept of *power*, the subtlety in defining the work of a force is discussed in detail early on in the chapter. Then, the *principle of work and energy* for particles is presented. This principle is useful in solving kinetic problems of particles where displacement and velocity (among others) are to be directly related without initially involving acceleration and time. Later, the concept of potential energy is quantitatively defined. The *principle of conservation of energy* is derived as a special case of the principle of work and energy as it is applied to conservative systems. Furthermore, the *principle of virtual work in kinetics* is shown to be very useful in studying the motion of systems of connected particles with one degree of freedom, where *force* and *acceleration* are to be directly related. It is presented as a complement to the method of force and acceleration studied in Chap. 12.

WORK AND KINETIC ENERGY[†]

13.1 Work of a Force

The *work of a force* **F** *acting on a particle during a differential displacement* d**r** *of the particle* from the point A to a neighboring point A', as shown in

---

[†] The concept of work was presented in Chap. 10. For convenience, we repeat here portions of that material relevant to the current study.

Fig. 13.1, is defined as an infinitesimal scalar quantity dU equal to the dot product of $\mathbf{F}$ and $d\mathbf{r}$; i.e.,

$$dU = \mathbf{F} \cdot d\mathbf{r} \qquad (13.1)^\dagger$$

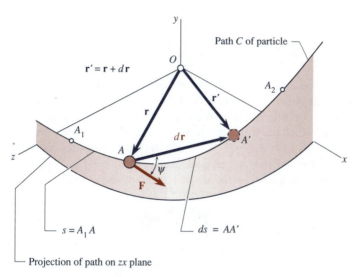

Fig. 13.1

Alternatively, we may write $dU = F\cos\psi\,ds$, where $ds = |d\mathbf{r}|$ and ψ is the angle between $\mathbf{F}$ and $d\mathbf{r}$. Anyway, the *work $U_{1\to2}$ of a force $\mathbf{F}$ acting on a particle during a finite displacement of the particle* from the point A_1 along a path C to the point A_2, as shown in Fig. 13.1, is

$$U_{1\to2} = \int_C \mathbf{F} \cdot d\mathbf{r} \qquad (13.2)$$

where the label C at the integral sign signifies that the work $U_{1\to2}$ done by $\mathbf{F}$ to the particle is generally dependent on the path along which the particle travels from A_1 to A_2. (Cf. D13.4 and D13.5.) If $U_{1\to2}$ is dependent on the coordinates of A_1 and A_2 but not on those of C, then $\mathbf{F}$ is referred to as a *conservative force* and we may write

$$U_{1\to2} = \int_{A_1}^{A_2} \mathbf{F} \cdot d\mathbf{r} \qquad (13.3)$$

The concept of conservative force is further discussed in Sec. 13.8.

If $\mathbf{F}$ is a friction force, which is a nonconservative force, Eq. (13.2) will generally yield a different work for each different path C. By definition, the work of a force is a scalar and has the dimensions of *force* $\times$ *length*. The

---

†In kinetics, *work* is done *by* a force *to* (or *on*) a body which has a *mass*. It is the *force* (not the *body*) which does work; a body *receives* work from the force acting on it during its displacement from one position to another. If the action of the force on a particle and the displacement of the particle do not occur simultaneously, no work is done by the force to the particle. The *work of a force* acting on a body is the *work done by a force* to a body. Moreover, *no work* can be done to a body which is *stationary* or has *no mass*.

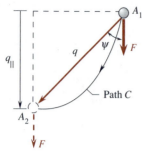

Fig. 13.2

primary units for work are joules (1 joule = 1 J = 1 N·m) in the SI and foot-pounds (ft·lb) in the U.S. customary system.[†]

If the force **F** is *constant* in magnitude and direction, the work $U_{1 \to 2}$ of **F** acting on a particle during its motion from the point A_1 along a path C to the point A_2, as shown in Fig. 13.2, is

$$U_{1 \to 2} = \int_C \mathbf{F} \cdot d\mathbf{r} = \mathbf{F} \cdot \int_C d\mathbf{r} = \mathbf{F} \cdot \mathbf{q}$$

$$U_{1 \to 2} = Fq \cos\psi \qquad (13.4)$$

where **q** is the displacement of the particle from A_1 to A_2 and ψ is the angle between the positive directions of **F** and **q**. We readily observe that this work $U_{1 \to 2}$ is positive if ψ is acute and negative if ψ is obtuse. There are three cases which are of special interest: (1) if **F** and **q** have the same direction, then $U_{1 \to 2} = Fq$; (2) if **F** and **q** have opposite directions, then $U_{1 \to 2} = -Fq$; (3) if **F** is perpendicular to **q**, then $U_{1 \to 2} = 0$. Using $\mathbf{q}_\parallel$ to denote the component of the displacement of the particle parallel to the constant force **F**, we may write Eq. (13.4) as

$$U_{1 \to 2} = Fq_\parallel \qquad (13.5)$$

where

$$q_\parallel = q \cos\psi \qquad (13.6)$$

Note that the value of $q_\parallel$ is negative when $90° < \psi \leq 180°$.

Furthermore, *the work of a force is equal to the sum of the works of its component forces, and conversely*. This property is analogous to *Varignon's theorem* and can be proved by applying the distributive property of dot products of vectors to the definition of the work of a force and the relation between the force and its component forces.[††]

EXAMPLE 13.1

If $\mu_k = 0.2$ between the 100-lb body and the horizontal support as shown, determine the works U_P, U_W, U_N, and U_F done to the body by the 50-lb pushing force **P**, the 100-lb weight force **W** of the body, the normal reaction **N**, and the friction force **F**, respectively, during the 3-ft displacement of the body from A to B.[†††]

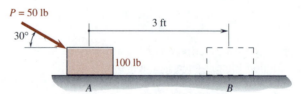

[†]The joule (J) is the primary unit of *work* and *energy* in the SI. Note that *moment* is not a form of energy. Although 1 J = 1 N·m, the primary units of *moment* of a force are N·m, *not* J, in the SI. In the U.S. customary system, the primary units are usually lb·ft for *moment*, but ft·lb for *work*. [However, some do not strictly follow the conventions for lb·ft and ft·lb, which are in reverse order to those shown in Eqs. (4.18) and (13.2).]

[††]Some refer to this property as the *theorem of work*.

[†††]Cf. the footnote on p. 553.

Solution. Since the body can move only in the horizontal direction, it is in equilibrium in the vertical direction. We write

$$+\uparrow \Sigma F_y = 0: \quad N - 100 - 50 \sin 30° = 0 \quad N = 125 \text{ lb}$$

$$\text{Friction law:} \quad F_k = \mu_k N \quad\quad\quad F_k = 25 \text{ lb}$$

The angles ψ_P, ψ_W, ψ_N, and ψ_F between the displacement **q** of the body and the forces **P**, **W**, **N**, and $\mathbf{F}_k$, respectively, are

$$\psi_P = 30° \quad\quad \psi_W = 90° \quad\quad \psi_N = 90° \quad\quad \psi_F = 180°$$

Noting that the forces considered are constant and $U_{1\to2} = Fq\cos\psi$, we write

$$U_P = Pq\cos\psi_P = 50(3)(\sqrt{3}/2)$$

$$U_W = Wq\cos\psi_W = 100(3)(0)$$

$$U_N = Nq\cos\psi_N = 125(3)(0)$$

$$U_F = F_k q\cos\psi_F = 25(3)(-1)$$

$$U_P = 129.9 \text{ ft·lb} \quad\quad U_W = U_N = 0 \quad\quad U_F = -75 \text{ ft·lb} \blacktriangleleft$$

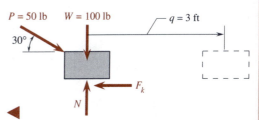

Alternative Solution for U_P. Resolving the 50-lb force **P** into two rectangular components $\mathbf{P}_x = 5\cos30°$ lb $\rightarrow$ and $\mathbf{P}_y = 50\sin30°$ lb $\downarrow$, we write

$$U_P = P_x q \cos 0° + P_y q \cos 90° = (50\cos 30°)(3)(1) + 0$$

$$U_P = 129.9 \text{ ft·lb} \blacktriangleleft$$

Developmental Exercises

D13.1 Define the work of a force acting on a particle.

D13.2 A 100-N force **F** acts on a particle P during a 2-m displacement **q** of the particle as shown. Compute the work of **F**.

D13.3 A 260-lb crate moves 10 ft down an incline as shown. Resolving the weight force **W** into two components, one parallel and the other perpendicular to the incline, compute the work of **W**.

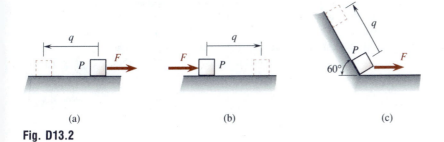

Fig. D13.2

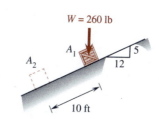

Fig. D13.3

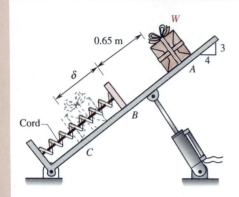

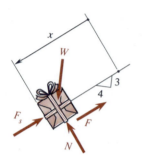

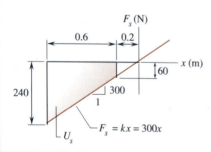

EXAMPLE 13.2

The tension in the cord which holds a spring of modulus $k = 300$ N/m in compression as shown is 60 N. A package slides from the position A down the incline and displaces the right end of the spring from B to C by an amount $\delta = 0.6$ m before it comes to a stop. Determine the work U_s of the spring force $\mathbf{F}_s$ acting on the package during its displacement from A to C.

Solution. We first draw the free-body diagram of the package as shown. Note that the spring force $\mathbf{F}_s$ will act on the package only when the package comes to the region between B and C. The other forces are ever present during the displacement.

Initially, the tensile force in the cord and the compressive force in the spring must have the same magnitude; i.e., $F_{s1} = 60$ N. For a spring modulus $k = 300$ N/m, the magnitude of the initial deflection x_1 of the spring is given by

$$x_1 = F_{s1}/k = 60/300 \qquad x_1 = 0.2 \text{ m}$$

The final deflection x_2 of the spring is given by

$$x_2 = x_1 + 0.6 \qquad x_2 = 0.8 \text{ m}$$

The magnitude of the spring force $\mathbf{F}_s$ in terms of the deflection x of the spring is $F_s = kx = 300x$. The differential displacement $d\mathbf{r}$ of the package and $\mathbf{F}_s$ acting on the package are opposite in sense. Thus, the angle ψ_s between $d\mathbf{r}$ and $\mathbf{F}_s$ is 180°. Noting that $|d\mathbf{r}| = dx$, we write

$$U_s = \int_C \mathbf{F}_s \cdot d\mathbf{r} = -\int_{x_1}^{x_2} F_s \, dx = -\int_{0.2}^{0.8} 300x \, dx$$

$$U_s = -90 \text{ J} \blacktriangleleft$$

REMARK. By the concept of "area" in calculus, the value of U_s may also be obtained from the "area" of the trapezoid in the spring-force diagram as shown. Using the formula for the area of a trapezoid (cf. Sec. B2.3 of App. B), we write

$$U_s = -\frac{1}{2}(60 + 240)(0.6)$$

$$U_s = -90 \text{ J} \blacktriangleleft$$

Developmental Exercises

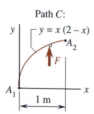

Fig. D13.4   Fig. D13.5

D13.4 A vertical force $\mathbf{F} = 6x\mathbf{j}$ N acts on a particle during its motion from A_1 to A_2 along the path C defined by $y = x$ as shown. For computing the work $U_{1 \to 2}$ of $\mathbf{F}$ acting on the particle, we note that $\mathbf{r} = x\mathbf{i} + y\mathbf{j}$, $d\mathbf{r} = dx\mathbf{i} + dy\mathbf{j}$. Determine (a) dy in terms of dx along C, (b) $\mathbf{F} \cdot d\mathbf{r}$ in terms of x and dx, (c) the work $U_{1 \to 2}$ using Eq. (13.2) where the integral is evaluated from $x = 0$ at A_1 to $x = 1$ m at A_2.

D13.5 Do D13.4 if the path C is defined by $y = x(2 - x)$ as shown. Is the force $\mathbf{F}$ under consideration a *conservative force*?

13.2   Power and the Subtlety in Defining Work

Power is defined as the time rate of doing work. If ΔU is the work of a force **F** during the time interval Δt, the average power during this interval is $P_{avg} = \Delta U / \Delta t$. As $\Delta t \to 0$, we obtain the power P at any instant as

$$P = \frac{dU}{dt} \tag{13.7}$$

From Eqs. (13.1) and (13.7), we have $P = dU/dt = \mathbf{F} \cdot d\mathbf{r}/dt$. Since $d\mathbf{r}/dt$ is equal to the *velocity* **v** *of the particle* acted on by **F**, we write

$$P = \mathbf{F} \cdot \mathbf{v} \tag{13.8}$$

Such a power is called a *mechanical power*.[†]
   As defined above, power is clearly a scalar quantity. In SI, the primary unit of power is the *watt* (W). Note that

$$1 \text{ W} = 1 \text{ J/s} = 1 \text{ N·m/s} \tag{13.9}$$

In the U.S. customary system, power is expressed in ft·lb/s or in *horse-power* (hp). Note that

$$1 \text{ hp} = 550 \text{ ft·lb/s} \tag{13.10}$$

$$1 \text{ hp} = 745.7 \text{ W} = 0.7457 \text{ kW} \tag{13.11}$$

The *efficiency* η of a machine is defined as the ratio of the power output of the machine to the power input to the machine; i.e.,

$$\eta = \frac{P_{output}}{P_{input}} \tag{13.12}$$

If the work is done at a constant rate, η may be defined as the ratio of the output work of the machine to the input work to the machine; i.e.,

$$\eta = \frac{U_{output}}{U_{input}} \tag{13.13}$$

Efficiency is always less than 1 because every machine operates with some loss of energy due to the negative work done by kinetic friction forces. This negative work is converted into heat energy, which is subsequently dissipated to the surroundings.
   Since $dU = P\,dt$, we may use Eq. (13.8) to write the work done $U_{1\to2}$ during the time interval from t_1 to t_2 as follows:

$$U_{1\to2} = \int_{t_1}^{t_2} \mathbf{F} \cdot \mathbf{v} \, dt \tag{13.14}$$

where **v** *is the velocity of the particle on which the force* **F** *acts*. Note that **v** here is, in general, *not* the same as the velocity of the point of application of **F** if **F** does *not* act continually on the same particle. The distinction here is a *subtlety* as well as a *necessity* in the definition of work. (Cf. D13.8.)

[†]Note that *electric power* = *voltage · current*.

This was emphasized and reemphasized by mechanics educators.[†] Since $U_{1\to2} = 0$ when $\mathbf{v} = \mathbf{0}$, we see that *no work is done to any body which is stationary during the time the force acts on it.*

For example, let us consider a rigid wheel having a weight W and *rolling without slipping* down a stationary rigid incline, as shown in Fig. 13.3(a). The reaction from the incline to the wheel is denoted by $\mathbf{R}$ as shown in Fig. 13.3(b). We see that the path of each particle on the rim of the wheel is a cycloid with a cusp at the point P where the particle of the wheel comes in contact with the incline, whereas the path of the point of application of $\mathbf{R}$ is a straight line along the incline, as shown in Fig. 13.3(c). The *velocity $\mathbf{v}$ of the particle* of the wheel on which $\mathbf{R}$ acts is, at the instant of action by $\mathbf{R}$, equal to *zero*. This is true because $\mathbf{v}$ *reverses its direction* and hence becomes zero at the cusp at P in Fig. 13.3(c).[††] Consequently, according to Eq. (13.14), *the reaction $\mathbf{R}$ does no work to the rolling wheel*, despite the fact that the point P of application of the reaction $\mathbf{R}$ moves down the incline with the wheel. On the other hand, the *reaction $\mathbf{R}$* in Fig. 13.3 *will do negative work to the wheel if the wheel slides down the incline.* In this case, the particle of the wheel at P on which $\mathbf{R}$ acts has, at the instant of action by $\mathbf{R}$, a velocity $\mathbf{v}$ pointing down the incline such that $\mathbf{R}\cdot\mathbf{v} = Rv\cos\psi < 0$, where $90° < \psi < 180°$ as shown in Fig. 13.3(d).

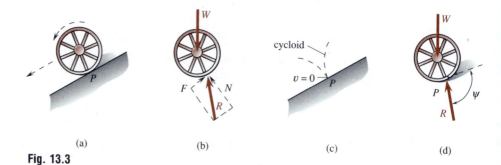

(a)   (b)   (c)   (d)

Fig. 13.3

As another example, let us consider a spinning grinder being slowed down with a stationary rod which is pressed against the grinder at B as shown in Fig. 13.4(a). In such a system, *the kinetic friction force $\mathbf{F}_B$ exerted by the rod at B on the grinder*, as shown in Fig. 13.4(b), *does negative work to the grinder* because the particle of the grinder in contact with the rod at B has a velocity $\mathbf{v}$ opposite to $\mathbf{F}_B$ which acts on it. However, *the kinetic friction force $\mathbf{F}_B'$ exerted by the grinder on the rod at B* (where $\mathbf{F}_B' = -\mathbf{F}_B$ and $F_B' = F_B$), as shown in Fig. 13.4(c), *does no work to the rod* because the rod has no velocity. Meanwhile, *the normal forces $\mathbf{N}_B$ and $\mathbf{N}_B'$*

[†]Osgood, W. F., *Mechanics*, Macmillian, New York, 1937, pp. 261–263; and Langhaar, H. L., "The Role of Rigor in the Teaching of Mechanics," *Mechanics Monograph M-4*, ASEE, 1980, pp. 1–10.

[††]Note that the particle of the wheel at P *does not slide* on the incline and has, at that instant, the same velocity as the incline which is stationary. Furthermore, notice that once the particle of the rolling wheel at P moves upward, the force $\mathbf{R}$ no longer acts on it. Thus, no work is done because the action of the force and the displacement of the particle do not occur simultaneously.

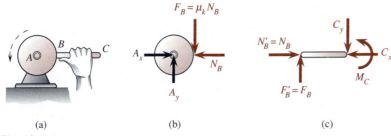

$F_B = \mu_k N_B$

$N'_B = N_B$

$F'_B = F_B$

(a)   (b)   (c)

Fig. 13.4

as shown (where $N_B = -N'_B$ and $N_B = N'_B$) *do no work to the grinder and
the rod*, respectively, because the particles under the action of the normal
forces have no velocity components parallel to the normal forces.

Developmental Exercises

D13.6   Define (a) power, (b) efficiency.

D13.7   If a constant force **F** does not act continually on the same particle, is the
work of **F** equal to $\mathbf{F} \cdot \mathbf{q}$ where **q** is the displacement of **F**?

D13.8   An instructor draws a line with a piece of chalk on a chalkboard which
is fixed to a wall of the classroom. Determine whether each of the following
quantities is *positive*, *zero*, or *negative*: (a) the work done to the chalkboard by
the force exerted by the chalk, (b) the work done to the chalk by the reaction
from the chalkboard, (c) the work done to the chalk by the force exerted by the
instructor.

EXAMPLE 13.3

An electric motor at A is used to hoist a 200-kg crate C. At the instant
shown, the crate C has a velocity of 0.4 m/s ↑ and an acceleration of 0.2
m/s² ↑ , and the wattmeter at B shows a power input of 850 W. For this
instant, determine (a) the velocity of the point D indicated on the cable, (b)
the power output of the motor, (c) the efficiency η of the motor.

Solution.   At the instant shown, the portion of the cable from D to E must
have a constant length. Thus, we write

$$2x_C + x_D = \text{constant} \qquad 2v_C + v_D = 0$$

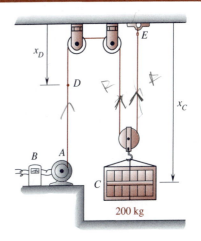

F F

$=$

200g N   200 (0.2)

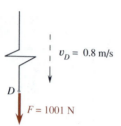

$v_D = 0.8$ m/s

D

$F = 1001$ N

Since the positive directions are downward, we have

$$v_C = -0.4 \text{ m/s} \qquad v_D = -2v_C = 0.8 \text{ m/s}$$

$$\mathbf{v}_D = 0.8 \text{ m/s} \downarrow \quad \blacktriangleleft$$

Denoting the tension in the cable by F and referring to the free-body and effective-force diagrams drawn, we write

$$+\uparrow \Sigma V_y:^\dagger \qquad F + F - 200g = 200(0.2) \qquad F = 1001 \text{ N}$$

Thus, at the instant shown, the motor exerts a force of $\mathbf{F} = 1001 \text{ N} \downarrow$ on the point D of the cable and moves it with a velocity of $\mathbf{v}_D = 0.8$ m/s $\downarrow$ as shown. We write

$$P_{\text{output}} = \mathbf{F} \cdot \mathbf{v} = (-1001\mathbf{j}) \cdot (-0.8\mathbf{j}) = 800.8$$

$$P_{\text{output}} = 801 \text{ W} \quad \blacktriangleleft$$

$$\eta = \frac{P_{\text{output}}}{P_{\text{input}}} = \frac{800.8}{850} \qquad \eta = 0.942 \quad \blacktriangleleft$$

Developmental Exercise

D13.9 Refer to Example 13.3. Suppose that all the data remain the same except that the acceleration of the crate C is 0.2 m/s^2 $\downarrow$ and the power input to the motor is 800 W. For the instant shown, determine (a) the power output of the motor, (b) the efficiency η of the motor.

13.3 Work of a Gravitational Force

To study the work of a gravitational force acting on a body whose motion involves large changes in altitude in the gravitational force field, let us employ the description using radial and transverse components as presented in Sec. 11.14. We consider a particle of mass m which occupies the position A at a certain instant and is moving under a gravitational force $\mathbf{F}$ exerted at a distance r by a body of mass M at a fixed position O as shown in Fig. 13.5(a), where $m \ll M$. We shall compute the work $U_{1\to2}$ of $\mathbf{F}$ acting on the particle during its motion from the position A_1 (where $r = r_1$) to the position A_2 (where $r = r_2$).

Note that the radial and transverse unit vectors are $\mathbf{e}_r$ and $\mathbf{e}_\theta$, respectively. From the first of Eqs. (11.59) and Fig. 13.5, we write

$$d\mathbf{e}_r = d\theta \mathbf{e}_\theta \qquad \mathbf{r} = r\mathbf{e}_r \qquad d\mathbf{r} = dr\mathbf{e}_r + r(d\theta \mathbf{e}_\theta)$$

$$d\mathbf{r} = dr\mathbf{e}_r + r d\theta \mathbf{e}_\theta \tag{13.15}$$

Applying Newton's law of gravitation, we write

$$\mathbf{F} = -GMmr^{-2}\mathbf{e}_r \tag{13.16}$$

---
†Cf. the second footnote on p. 513.

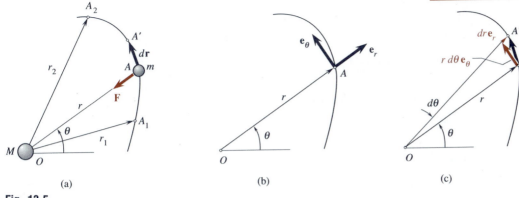

Fig. 13.5

Thus, the work $U_{1\to2}$ of $\mathbf{F}$ acting on the particle during its motion from A_1 to A_2 may be computed as follows:

$$dU = \mathbf{F} \cdot d\mathbf{r} \qquad (13.17)$$

$$U_{1\to2} = \int_C \mathbf{F} \cdot d\mathbf{r} = -GMm \int_{r_1}^{r_2} r^{-2}\, dr$$

$$U_{1\to2} = GMm\left(\frac{1}{r_2} - \frac{1}{r_1}\right) \qquad (13.18)$$

which shows that the work of a gravitational force acting on a body moving from A_1 to A_2 does not depend on the path connecting A_1 and A_2. Therefore, a gravitational force is a *conservative force*.

Developmental Exercises

D13.10   Refer to Fig. 13.5. (a) Is the position vector $\mathbf{r} = r\mathbf{e}_r$? (b) Is the magnitude $|\mathbf{r}| = r$? (c) Is $|d\mathbf{r}| = dr$? (d) Is $|d\mathbf{r}| = d|\mathbf{r}|$?

D13.11   Determine the work of the gravitational force acting on each 1 kg of mass of a geosynchronous satellite during its flight from the surface of the earth to its orbit, which is 35.8 Mm above the equator.

13.4   Work of a Spring Force

The force exerted by an elastic spring on a body attached to the spring is proportional to the deformation in length of the spring.[†] For a body in the position A connected to a fixed point at B by a spring with a modulus k and an elongation x as shown in Fig. 13.6(a), the spring force acting on the body is $\mathbf{F} = kx \leftarrow$; i.e.,

$$F = kx \qquad (13.19)^{\dagger\dagger}$$

[†]Unless otherwise noted, all springs will be assumed to be linear.

[††]Under dynamic conditions, the error introduced by using the relation $F = kx$ is small if the mass of the spring is small compared with other moving masses in the system.

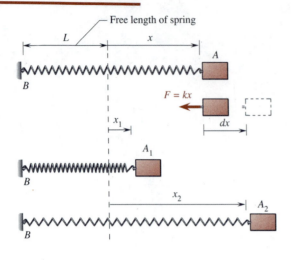

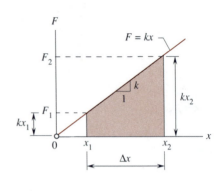

(a)   (b)

Fig. 13.6

The spring modulus k is usually expressed in N/m or kN/m if SI units are used and in lb/ft or lb/in. if U.S. customary units are used. However, care should be taken to express k and x in a consistent system of kinetic units (cf. Table 12.1).

The work $U_{1 \to 2}$ of the spring force $\mathbf{F}$ acting on the attached body during a finite displacement of the body from the position A_1 (where $x = x_1$) to the position A_2 (where $x = x_2$) is obtained as follows:

$$dU = F(-dx) = -kx\,dx \qquad U_{1 \to 2} = -k \int_{x_1}^{x_2} x\,dx$$

$$U_{1 \to 2} = -\tfrac{1}{2} k(x_2^2 - x_1^2) \tag{13.20}$$

Furthermore, we may obtain $U_{1 \to 2}$ by computing the area of the trapezoid shown in the *spring-force diagram* in Fig. 13.6(b), where $F_1 = kx_1$, $F_2 = kx_2$, and $\Delta x = x_2 - x_1$. Since $U_{1 \to 2}$ is negative when $|x_2| > |x_1|$, we may use the formula for the area of a trapezoid to write

$$U_{1 \to 2} = -\tfrac{1}{2}(F_1 + F_2)\,\Delta x \tag{13.21}$$

It is important to note that the work of the spring force acting on a body is dependent on the changes in length (i.e., x_1 and x_2) of the spring, not on the initial and final orientations of the spring.

EXAMPLE 13.4

A collar P is mounted on the horizontal bar AB and is attached to an elastic cord POD as shown. The elastic cord has a free length $L = 10$ m and a spring modulus $k = 400$ N/m. Using the basic definition of work as given in Eq. (13.2), determine the work $U_{1 \to 2}$ of the spring force $\mathbf{F}$ of the elastic cord acting on the collar during its displacement from A to B.

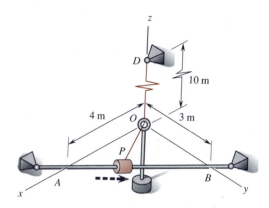

Solution. The path C of the collar P in moving from A to B is given by $y = -\frac{3}{4}x + 3$. The position vector of P is

$$\mathbf{r} = x\mathbf{i} + y\mathbf{j} = x\mathbf{i} + (-\tfrac{3}{4}x + 3)\mathbf{j}$$

Since $k = 400$ N/m, we write

$$\mathbf{F} = -kr\mathbf{e}_r = -k\mathbf{r} = -400x\mathbf{i} + 400(\tfrac{3}{4}x - 3)\mathbf{j}$$

$$d\mathbf{r} = dx\mathbf{i} - \tfrac{3}{4}dx\mathbf{j} = (\mathbf{i} - \tfrac{3}{4}\mathbf{j})\,dx$$

$$\mathbf{F} \cdot d\mathbf{r} = (-625x + 900)\,dx$$

$$U_{1\rightarrow2} = \int_C \mathbf{F} \cdot d\mathbf{r} = \int_4^0 (-625x + 900)\,dx = 1400$$

$$U_{1\rightarrow2} = 1400 \text{ J} \blacktriangleleft$$

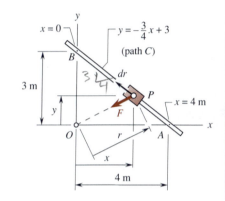

EXAMPLE 13.5

Using a *spring-force diagram*, work Example 13.4.

Solution. Since the free length of the elastic cord is $L = 10$ m and $\overline{OD} = 10$ m, we readily infer that the net elongation of the elastic cord is $r = \overline{OP}$. Therefore, the spring force is $F = kr = 400r$. At the position A,

$-400xi + 30(x)j - 1200j$

$-400r$

bening

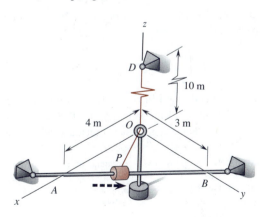

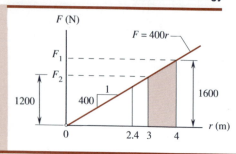

$r = r_1 = \overline{OA} = 4$ m; and at the position B, $r = r_2 = \overline{OB} = 3$ m. In moving from A to B, r first decreases from $r_1 = 4$ m to $r_{min} = \frac{3}{5} r_1 = 2.4$ m and then increases to $r_2 = 3$ m. Since $r_2 < r_1$, the total work $U_{1\to2}$ of the spring force acting on the collar during its displacement from A to B must be positive. Thus, by computing the area of the shaded trapezoid in the spring-force diagram shown, we write

$$U_{1\to2} = \tfrac{1}{2}(1200 + 1600)(1) \qquad U_{1\to2} = 1400 \text{ J} \blacktriangleleft$$

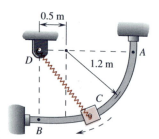

Fig. D13.12

Developmental Exercise

D13.12 A spring CD of free length 1 m and modulus 1 kN/m is attached to a sliding collar C of mass 10 kg as shown. Determine the works U_W and U_s of the weight force **W** of the collar and the spring force $\mathbf{F}_s$, respectively, acting on the collar during its motion from the position A to the position B.

13.5   Kinetic Energy of a Particle

The *kinetic energy*, denoted by T, of a particle with a mass m and a speed v is defined to be $\frac{1}{2}mv^2$, which is equal to *the total work done to the particle in bringing it from its state of rest to its state of motion with the speed v.* We write

$$\boxed{T = \tfrac{1}{2}mv^2} \tag{13.22}^{\dagger}$$

For verification, let the resultant force acting on the particle of mass m be **F** as shown in Fig. 13.7. Employing the description using tangential and normal components as presented in Sec. 11.13, we write

$$dU = \mathbf{F} \cdot d\mathbf{r} = (F_t\mathbf{e}_t + F_n\mathbf{e}_n) \cdot (ds\, \mathbf{e}_t) = F_t\, ds$$

$$= ma_t\, ds = m\frac{dv}{dt}\, ds = m\frac{dv}{ds}\frac{ds}{dt}\, ds = m\frac{ds}{dt}\, dv$$

$$\boxed{dU = mv\, dv} \tag{13.23}$$

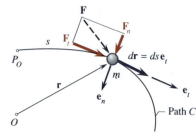

Fig. 13.7

The total work done to the particle in bringing it from a state of rest to a state having a speed v is, therefore,

$$U_{0\to v} = \int_C dU = m \int_0^v v\, dv = \tfrac{1}{2}\, mv^2 \qquad \text{Q.E.D.}$$

The kinetic energy T as defined in Eq. (13.22) is a *scalar quantity* which is independent of the direction of the velocity and is always *positive* or *zero*. Since a reference frame translating with a constant velocity in an inertial

---

†It may be of interest to note that the relativistic formula for the kinetic energy of a particle with a speed v is $T = m_0c^2[(1 - v^2/c^2)^{-1/2} - 1]$ where c is the speed of light and m_0 is the mass of the particle at rest. By binomial expansion, it can be shown that $T \approx \frac{1}{2} m_0 v^2$ if $v \ll c$. This was referred to in Sec. 1.14.

reference frame is also an inertial reference frame, the kinetic energy of a body is clearly a *relative quantity* whose value is dependent on the inertial reference frame chosen for the description. However, *a common inertial reference frame should be used in solving a kinetic problem.*[†]

It can readily be verified that the kinetic energy has the same dimensions and units as the work of a force. This means that the primary units for the kinetic energy are joules (J) in the SI and foot-pounds (ft·lb) in the U.S. customary system.

Developmental Exercises

D13.13 Define the kinetic energy of a particle.

D13.14 Show that the dimensions of kinetic energy are *force* × *length*.

D13.15 Using Eq. (13.23), determine the total work which must be done to a particle with a mass m and a speed v to bring it to a state of rest.

D13.16 Which of the following are *false*? (a) The kinetic energy points in the same direction as the velocity. (b) Both work and kinetic energy are measured in the same units. (c) Both work and kinetic energy can be positive, zero, or negative. (d) The work is always positive or zero. (e) The kinetic energy is always positive or zero.

13.6 Principle of Work and Energy

A differential work dU done by the resultant force $\mathbf{F}$ to a particle with a mass m and a speed v will result in a differential change dv in the speed of the particle. Such a relation was given by Eq. (13.23). By integrating this differential equation from the position A_1 (where $v = v_1$) to the position A_2 (where $v = v_2$) as shown in Fig. 13.8, we write

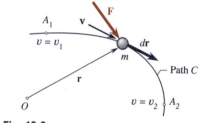

Fig. 13.8

$$U_{1\to2} = \int_C dU = m \int_{v_1}^{v_2} v \, dv$$

$$U_{1\to2} = \tfrac{1}{2} m v_2^2 - \tfrac{1}{2} m v_1^2 \tag{13.24}$$

where $U_{1\to2}$ is the work of $\mathbf{F}$ acting on the particle during its motion from A_1 to A_2. Since the kinetic energies of the particle at A_1 and A_2 are $T_1 = \tfrac{1}{2} m v_1^2$ and $T_2 = \tfrac{1}{2} m v_2^2$, respectively, we may write Eq. (13.24) as

$$U_{1\to2} = T_2 - T_1 \tag{13.25}$$

Equation (13.25) expresses the *principle of work and energy* for a particle. Since the work of the resultant force is equal to the sum of the works of its component forces (cf. Sec. 13.1), this principle may be stated as follows: *The total work of all forces acting on a particle during an interval of its motion is equal to the change in kinetic energy of the particle during*

[†]Any reference frame which is either fixed or practically translating with a constant velocity relative to the sun is, for most engineering purposes, an inertial reference frame. (Cf. Sec. 3.1.)

that interval. Rearranging the terms in Eq. (13.25), we write

$$T_1 + U_{1\rightarrow 2} = T_2 \tag{13.26}$$

Thus, the *principle of work and energy* for a particle may alternatively be stated as follows: *The kinetic energy of a particle at A_1 plus the total work of all forces acting on the particle during its motion from A_1 to A_2 is equal to the kinetic energy of the particle at A_2.* Since Newton's second law of motion is a basis upon which this principle is derived, this principle should obviously be applied *only* in an inertial reference frame. For example, the *speed* used to compute the kinetic energy and the *displacement* used to compute the work should be measured in a common inertial reference frame.

We note that the principle of work and energy directly relates *displacement* and *velocity* (among others) without explicitly involving acceleration and time. Furthermore, we refer to the procedure which uses the principle of work and energy to analyze and solve appropriate kinetic problems as the *method of work and energy.*

EXAMPLE 13.6

A 100-lb crate is given an initial speed $v_1 = 4$ ft/s down a chute from the position A_1 as shown. If $\mu_k = 0.3$ between the crate and the chute, determine the speed v_2 of the crate when it reaches the position A_2.

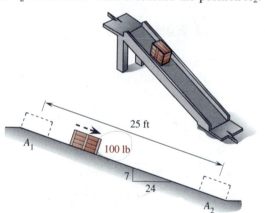

Solution. We note that the crate does not move perpendicularly to the chute; i.e., the crate is in equilibrium in the y direction. Noting in the diagram that $\cos\theta = 24/25$, we write

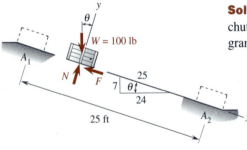

$$+\nearrow \Sigma F_y = 0: \qquad N - 100\cos\theta = 0 \qquad N = 96 \text{ lb}$$

$$\text{Friction law:} \qquad F_k = \mu_k N = 0.3N \qquad F_k = 28.8 \text{ lb}$$

$$\text{Kinetic energy at } A_1: \qquad T_1 = \tfrac{1}{2} mv_1^2 = \frac{1}{2}\left(\frac{100}{32.2}\right)(4)^2$$

$$\text{Kinetic energy at } A_2: \qquad T_2 = \tfrac{1}{2} mv_2^2 = \frac{1}{2}\left(\frac{100}{32.2}\right)v_2^2$$

Total work $(U = Fq\cos\psi)$ of **W**, **N**, and **F**:

$$U_{1\rightarrow 2} = 100(25)\left(\frac{7}{25}\right) + 0 + 28.8(25)(-1)$$

Applying the *principle of work and energy*, $T_1 + U_{1 \to 2} = T_2$, we write

$$\frac{1}{2}\left(\frac{100}{32.2}\right)(4)^2 + \left[100(25)\left(\frac{7}{25}\right) + 0 + 28.8(25)(-1)\right]$$

$$= \frac{1}{2}\left(\frac{100}{32.2}\right)v_2^2$$

$$v_2 = 1.766 \text{ ft/s} \quad \blacktriangleleft$$

Developmental Exercise

D13.17   Describe the *principle of work and energy* for a particle.

EXAMPLE 13.7

A spacecraft is in free flight around the earth as shown, where $a_1 = 3000$ km and $a_2 = 6000$ km. When the spacecraft passes through the position A_1, its speed is $v_1 = 6.27$ km/s. Determine the speed v_2 of the spacecraft when it passes through the position A_2.

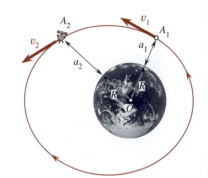

Solution.   The only force acting on the spacecraft in free flight is the gravitational force from the earth whose radius is $R = 6370$ km. We have

$$r_1 = \overline{OA_1} = R + a_1 = 6370 + 3000 \qquad r_1 = 9370 \text{ km}$$

$$r_2 = \overline{OA_2} = R + a_2 = 6370 + 6000 \qquad r_2 = 12370 \text{ km}$$

Let the mass of the spacecraft be m. The work of the gravitational force acting on the spacecraft during its motion from A_1 to A_2 is given by Eq. (13.18), where $GM = gR^2$. Applying the *principle of work and energy*, $T_1 + U_{1 \to 2} = T_2$, we write

$$\tfrac{1}{2} m v_1^2 + \left[GMm\left(\frac{1}{r_2} - \frac{1}{r_1}\right) \right] = \tfrac{1}{2} m v_2^2$$

$$\tfrac{1}{2} v_1^2 + gR^2\left(\frac{1}{r_2} - \frac{1}{r_1}\right) = \tfrac{1}{2} v_2^2$$

$$\tfrac{1}{2}(6.27)^2 + \frac{9.81}{1000}(6370)^2\left(\frac{1}{12370} - \frac{1}{9370}\right) = \tfrac{1}{2} v_2^2$$

$$v_2 = 4.33 \text{ km/s} \quad \blacktriangleleft$$

Developmental Exercise

D13.18   If the minimum speed of the spacecraft in Example 13.7 is 3.99 km/s, determine the maximum altitude of its orbit.

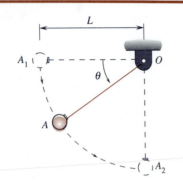

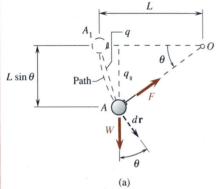

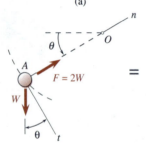

(a)

EXAMPLE 13.8

If the pendulum of weight W is released from rest from a horizontal position OA_1 and swings in a vertical plane to OA_2 as shown, determine (a) the speed v of the bob in terms of g, L, and θ in the range $0 \le \theta \le 90°$, (b) the angle θ at which the tension F in the cord is equal to $2W$, (c) the tension F_2 in the cord when the bob reaches the position A_2.

Solution. As the pendulum swings down, the tensile force $\mathbf{F}$ exerted by the cord on the bob, as shown in Fig. (a), is always perpendicular to the differential displacement $d\mathbf{r}$ (and hence the path) of the bob. Thus, $\mathbf{F}$ does no work to the bob. The work is done only by the weight force $\mathbf{W}$. Applying the *principle of work and energy*, $T_{A_1} + U_{A_1 \to A} = T_A$, we write

$$0 + [0 + W(L \sin\theta)] = \frac{1}{2}\left(\frac{W}{g}\right)v^2$$

$$v^2 = 2gL\sin\theta \qquad v = (2gL\sin\theta)^{1/2} \quad \blacktriangleleft$$

The normal component of acceleration of the bob at A is

$$a_n = \frac{v^2}{\rho} = \frac{2gL\sin\theta}{L} = 2g\sin\theta$$

When $\theta = 90°$, we have $a_n = 2g$. These two values of a_n are used in drawing the effective-force diagram as shown in Figs. (b) and (c).

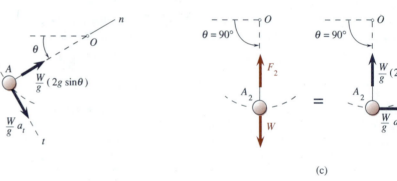

(b)   (c)

To find the angle θ at which the tension F in the cord is equal to $2W$, we refer to the vector-diagram equation in Fig. (b) and write

$$+\nearrow \ \Sigma V_n: \qquad 2W - W\sin\theta = \frac{W}{g}(2g\sin\theta)$$

$$\sin\theta = \frac{2}{3}$$

$$\theta = 41.8° \quad \blacktriangleleft$$

To find the tension F_2 in the cord when the bob reaches the position A_2, we refer to the vector-diagram equation in Fig. (c) and write

$$+\uparrow \ \Sigma V_y: \qquad F_2 - W = \frac{W}{g}(2g) \qquad F_2 = 3W \quad \blacktriangleleft$$

Developmental Exercise

D13.19   Refer to Example 13.8. Determine the angle θ at which the tension F in the cord is equal to $2.5W$.

13.7   System of Particles

If a system consists of several particles, we may consider each particle separately and apply the principle of work and energy to each particle. By adding side by side all equations of work and energy, in the form of Eq. (13.26), for all particles in the system, we obtain the *principle of work and energy* for a system of particles as follows:

$$T_1 + U_{1\to2} = T_2 \qquad (13.27)$$

where T_1 represents the arithmetic sum of the *initial* kinetic energies of all the particles, T_2 represents the arithmetic sum of the *final* kinetic energies of all the particles, $U_{1\to2}$ represents the algebraic sum of the works of all the forces acting on the various particles *during the interval*, and the forces include both *internal* and *external* forces to the system (cf. Sec. 3.3). We know that internal forces occur in pairs of forces equal in magnitude and opposite in direction. If the particles in the system are connected by inextensible cordlike elements or links, such pairs of internal forces in them must be tangent to their centerlines, which do not elongate nor shrink. Consequently, the total work of the internal forces to such a system is zero, and $U_{1\to2}$ is equal to the total work of the external forces only.

The kinetic energy T of a system of n particles may be written as

$$T = \frac{1}{2}\sum_{i=1}^{n} m_i v_i^2 \qquad (13.28)$$

where m_i is the mass and v_i is the speed, in the inertial reference frame $Oxyz$, of the particle P_i as shown in Fig. 13.9.[†] Note that

$$\mathbf{v}_i = \bar{\mathbf{v}} + \mathbf{v}_i' \qquad (13.29)$$

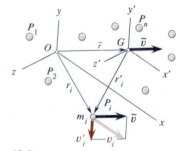

Fig. 13.9

where $\bar{\mathbf{v}}$ is the velocity of the mass center G of the system in $Oxyz$, and $\mathbf{v}_i'$ is the velocity of P_i in the reference frame $Gx'y'z'$, which translates with G and keeps the same orientation in $Oxyz$. Since $v_i^2 = \mathbf{v}_i \cdot \mathbf{v}_i$, we may use Eqs. (13.28) and (13.29) to write

$$T = \frac{1}{2}\sum_{i=1}^{n}[m_i(\bar{\mathbf{v}} + \mathbf{v}_i') \cdot (\bar{\mathbf{v}} + \mathbf{v}_i')]$$

$$= \frac{1}{2}\left(\sum_{i=1}^{n} m_i\right)\bar{v}^2 + \bar{\mathbf{v}} \cdot \left(\sum_{i=1}^{n} m_i\mathbf{v}_i'\right) + \frac{1}{2}\sum_{i=1}^{n} m_i v_i'^2$$

We recognize that the first sum is equal to the total mass m of the system. By the *principle of moments*, we see that the second sum is equal to $m\bar{\mathbf{v}}'$, where $\bar{\mathbf{v}}'$ represents the velocity of G in $Gx'y'z'$, which is clearly zero.

[†]Cf. the footnote on p. 479.

Thus, we have

$$T = \tfrac{1}{2} m\bar{v}^2 + \frac{1}{2} \sum_{i=1}^{n} m_i v_i'^2 \qquad (13.30)$$

Equation (13.30) establishes the property that the kinetic energy of a system of particles (including the case of a rigid body) may be perceived to be composed of two parts: (a) *the kinetic energy of the system moving with G in the inertial reference frame Oxyz*, and (b) *the kinetic energy of the system moving relative to G in the translating reference frame Gx'y'z'.*

EXAMPLE 13.9

The blocks A and B are released to move from rest as shown. If $\mu_k = 0.3$ between the block A and the support, determine the velocity $\mathbf{v}_A$ of the block A when the block B strikes the ground.

Solution. The free-body diagram of the entire system and the expected motions are first drawn and indicated as shown below. Similar to that in Example 12.3, the constraint condition is

$$2x_A + x_B = \text{constant}$$

from which we have

$$2 \, \Delta x_A + \Delta x_B = 0 \qquad 2v_A + v_B = 0$$

Thus, when the block B has a displacement of $\Delta x_B = 3$ m downward, the displacement of the block A is $\Delta x_A = -\tfrac{1}{2} \Delta x_B = -1.5$ m. The negative sign indicates that the block A displaces to the left as shown. For the block B, we have $v_B = -2v_A$. The block A as shown is in equilibrium in the vertical direction. Hence, we have

$$N_A = W_A = 100g = 981 \text{ N} \qquad F_A = \mu_k N_A = 294.3 \text{ N}$$

$$W_B = 20g = 196.2 \text{ N}$$

During the interval of motion, the forces $\mathbf{W}_A$, $\mathbf{N}_A$, $\mathbf{D}_x$, and $\mathbf{D}_y$ do no work; only the forces $\mathbf{F}_A$ and $\mathbf{W}_B$ do work. We note that the system is initially at

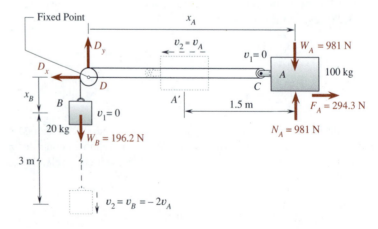

rest. Applying the *principle of work and energy*, $T_1 + U_{1 \to 2} = T_2$, we write

$$0 + [294.3(-1.5) + 196.2(3)]$$
$$= \tfrac{1}{2}(100)v_A^2 + \tfrac{1}{2}(20)(-2v_A)^2$$

$$v_A = 1.279 \qquad \mathbf{v}_A = 1.279 \text{ m/s} \leftarrow \blacktriangleleft$$

Developmental Exercise

D13.20 Refer to Example 13.9. Determine the displacement of the block B when the speed of the block A is increased from 0 to 1 m/s.

EXAMPLE 13.10

Two particles A and B are mounted on a cross, which in turn is mounted on a cart as shown. The cross rotates at a constant angular speed $\dot\theta_0 = 4$ rad/s relative to the cart, while the cart moves along the x axis at a constant velocity $\mathbf{v}_0 = 3$ ft/s ↘. If each particle weighs 8.05 lb, determine the total kinetic energy T of the *particles A and B* in the reference frame $Oxyz$ fixed to the ground.

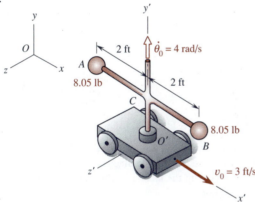

Solution. By symmetry, we know that the mass center of the particles A and B is located at the point C which has a velocity $\mathbf{v} = 3$ ft/s ↘ in $Oxyz$. The speed of each particle relative to the mass center C is

$$v' = r\dot\theta_0 = 2(4) \qquad v' = 8 \text{ ft/s}$$

The total mass of A and B is

$$m = m_A + m_B = \frac{8.05 + 8.05}{32.2} \text{ slug} = \frac{16.1}{32.2} \text{ slug}$$

Thus, the total kinetic energies of A and B in $Oxyz$ is [cf. Eq. (13.30)]

$$T = \frac{1}{2}\left(\frac{16.1}{32.2}\right)(3)^2 + \frac{1}{2}\left[\frac{8.05}{32.2}(8)^2 + \frac{8.05}{32.2}(8)^2\right]$$

$$T = 18.25 \text{ ft·lb} \blacktriangleleft$$

Developmental Exercises

D13.21 Refer to Example 13.10. (a) Determine the total kinetic energy T' of the particles A and B in the reference frame $O'x'y'z'$ which is fixed to the cart. (b) Is $O'x'y'z'$ an inertial reference frame?

D13.22 Refer to Example 13.10. If the cart weighs 50 lb and the weights of the cross and the wheels are negligible, what is the total kinetic energy of the system in $Oxyz$?

PROBLEMS

13.1* Using the *principle of work and energy*, solve Prob. 12.10.

13.2 An astronaut drops a 5-kg rock from a height of 2 m above the surface of the moon, where the gravitational acceleration is 1.618 m/s². Later, this same rock is brought back to the earth and dropped from the same height of 2 m above the ground. Determine the kinetic energy T and the speed v of the rock immediately before it strikes the surface of (a) the moon, (b) the earth.

13.3* A 10-kg package is given an initial speed of 1 m/s in the position A down a slope to the position B where its speed becomes 5 m/s as shown. Determine the work U_f of the friction force acting on the package during its motion from A to B.

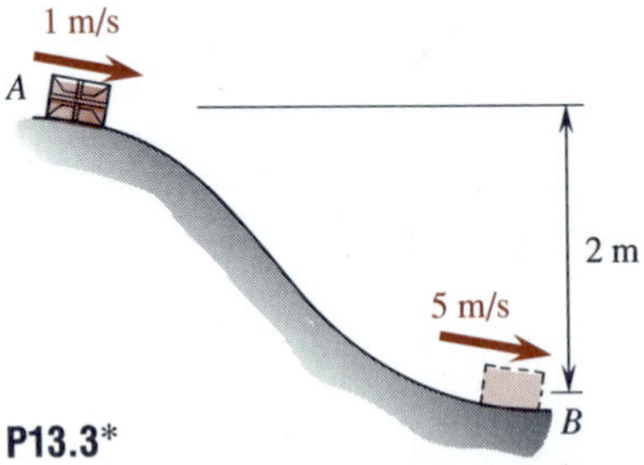

$T_A + W_{A \to B} = T_B$

Fig. P13.3*

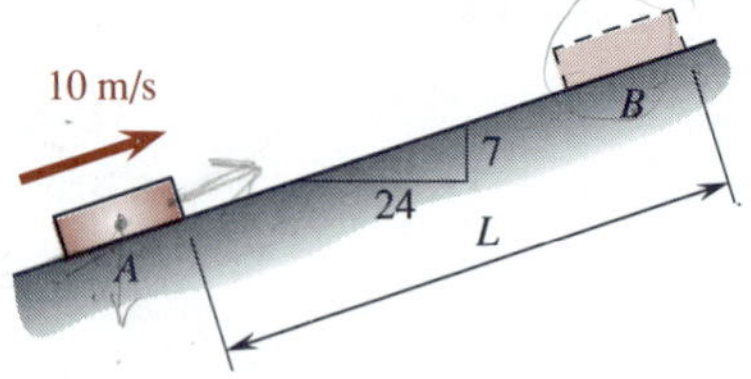

Fig. P13.4

13.4 A disk is projected with an initial velocity of 10 m/s up an incline as shown. If $\mu_k = 0.2$ between the disk and the incline, determine (a) the maximum distance L which the disk will travel up the incline, (b) the speed of the disk as it slides back through its original position.

13.5 A 20-lb collar C is moved from rest from the position A to the position B by a cable CDE which passes around a pulley at D and is pulled downward by a constant force of 50 lb at E as shown. Neglecting effects of friction, determine the speed of the collar as it passes through B.

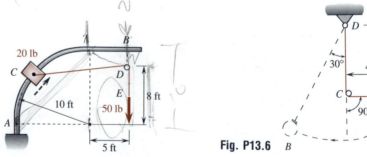

Fig. P13.5

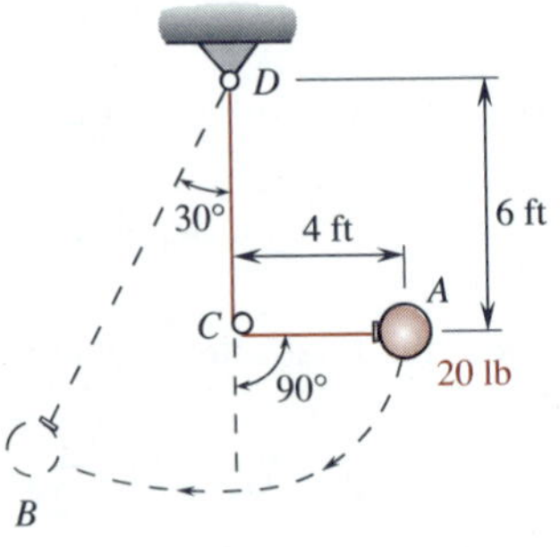

Fig. P13.6

13.6 The cord of a pendulum is deflected by a peg at C when it is released from the position A as shown. As it swings to the position B, determine (a) the speed of the bob, (b) the tension in the cord.

13.7* and 13.8 Determine the speed of the block A after it is moved 2 m from rest by the horizontal force **P** as shown.

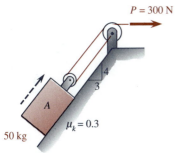

$P = 300$ N

4
3

A

$\mu_k = 0.3$

50 kg

Fig. P13.7*

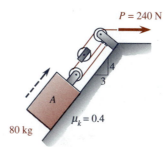

$P = 240$ N

4
3

A

$\mu_k = 0.4$

80 kg

Fig. P13.8

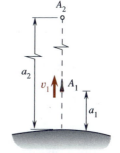

A_2

a_2

v_1 A_1

a_1

Fig. P13.9* and P13.10

13.9* A projectile in the position A_1 has an altitude $a_1 = 80$ km and a velocity $\mathbf{v}_1 = 8$ km/s ↑ as shown. Neglecting air resistance, determine the highest altitude a_2 reached by the projectile.

13.10 At engine burnout, a rocket has an altitude $a_1 = 75$ mi and an upward speed v_1 as shown. If the rocket is to reach a geosynchronous satellite at an altitude $a_2 = 22.2 \times 10^3$ mi and air resistance above the altitude a_1 is negligible, determine the value of v_1.

13.11 and 13.12* A satellite describes an elliptic orbit around the earth as shown. In the position S, it has an altitude a_S and a speed v_S as indicated. Determine the speed v_P of the satellite when it passes through the perigee P.

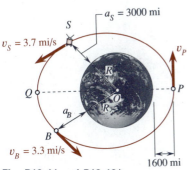

$a_S = 3000$ mi

S

$v_S = 3.7$ mi/s

v_P

R

Q O P

a_B R

B

$v_B = 3.3$ mi/s

1600 mi

Fig. P13.11 and P13.13*

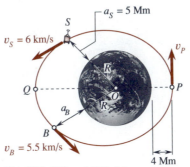

$a_S = 5$ Mm

S

$v_S = 6$ km/s

v_P

R

Q O P

a_B R

B

$v_B = 5.5$ km/s

4 Mm

Fig. P13.12* and P13.14

13.13* and 13.14 A satellite describes an elliptic orbit around the earth as shown. In the position S, it has an altitude a_S and a speed v_S as indicated. When it passes through the point B, its speed is v_B as indicated. Determine the altitude a_B of the point B.

13.15 Refer to Prob. 13.11. Determine the altitude a_Q of the apogee Q.

13.16 Refer to Prob. 13.12. Determine the altitude a_Q of the apogee Q.

13.17 and 13.18* A spring of free length 2.5 m and modulus 200 N/m is attached to a sliding collar C of mass 20 kg as shown. The collar is released from rest at A to slide downward. Neglecting effects of friction, determine the speed of the collar as it passes through B.

13.19 A spring of free length 10 m and modulus 120 N/m is attached to a sliding collar C of mass 20 kg as shown. The collar is released from rest at A. Neglecting effects of friction, determine the speed of the collar as it passes through B.

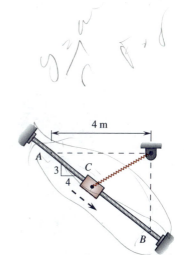

4 m

A

3

C

4

B

Fig. P13.17

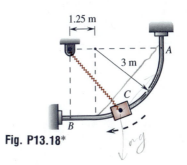

1.25 m

A

3 m

C

B

Fig. P13.18*

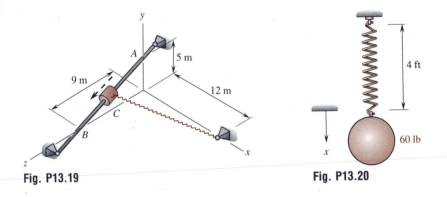

Fig. P13.19

Fig. P13.20

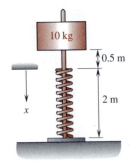

Fig. P13.21

13.20   A 60-lb sphere connected to a spring, which has a free length of 4 ft and a modulus of 120 lb/ft, is to be released from rest in the position $x = 0$ as shown. Determine (a) the maximum elongation of the spring, (b) the maximum speed of the sphere. (*Hint.* When the speed is maximum, what is the resultant force acting on the sphere?)

13.21   A 10-kg cylindrical collar is released from rest in the position shown and drops onto the spring which has a free length of 2 m and a modulus of 300 N/m. Determine (a) the maximum speed of the collar, (b) the maximum contraction of the spring.

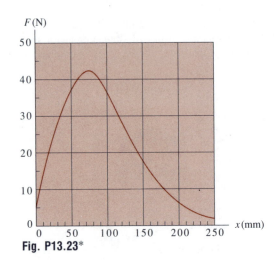

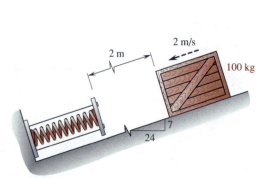

Fig. P13.22*

Fig. P13.23*

13.22*   A spring is precompressed 0.2 m by a light board and restraining cables as shown. The modulus of the spring is 2 kN/m, and $\mu_k = 0.25$ between the 100-kg crate and the incline. If the speed of the crate is 2 m/s when it is 2 m from the spring, determine the maximum additional deflection of the spring.

13.23*   The length of the barrel of an air pistol is 250 mm. Upon firing, the magnitude of the net force F acting on the pellet varies with the position x of the pellet in the barrel as shown. If the diameter of the bore is 4.5 mm and the mass of the pellet is 500 mg, determine (a) the maximum pressure in the barrel, (b) the muzzle speed of the pellet. (*Hint.* Compute the area by dividing it into a series of trapezoids, or by using the program AREA in the Supplementary Software.)

13.24   Solve Prob. 13.20 if the spring is a hard spring where the magnitude of the spring force F developed due to an elongation x is given as shown.

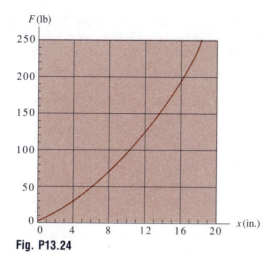

Fig. P13.24

13.25 and 13.26* The blocks are released from rest in the position shown. Determine the speed of the block A when the block B is stopped by impact.

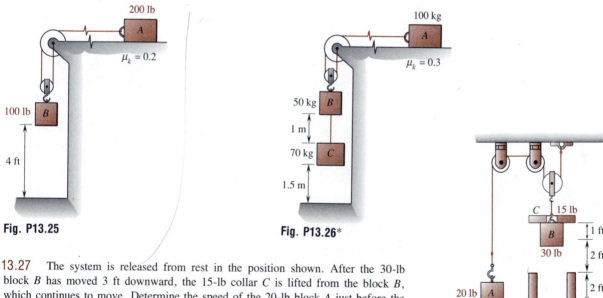

Fig. P13.25   **Fig. P13.26***

Fig. P13.27

13.27 The system is released from rest in the position shown. After the 30-lb block B has moved 3 ft downward, the 15-lb collar C is lifted from the block B, which continues to move. Determine the speed of the 20-lb block A just before the block B strikes the bottom of the pit.

13.28 and 13.29 The system is at rest when the 400-N force is applied to the block B as shown. Neglecting effects of friction, determine the speed of the block B when the block A has moved 1.5 m.

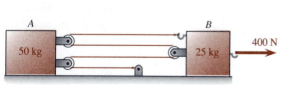

Fig. P13.28 and P13.30*

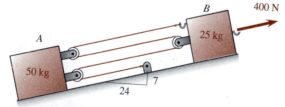

Fig. P13.29 and P13.31

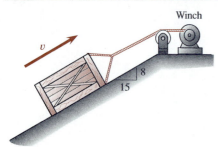

Fig. P13.32* and P13.33

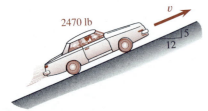

Fig. P13.35 and P13.36

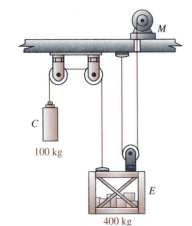

Fig. P13.37* and P13.38

13.30* and **13.31** The system is at rest when the 400-N force is applied to the block B as shown. If $\mu_k = 0.25$ between the blocks and the support, determine the speed of the block B when the block A has moved 1.5 m.

13.32* A power winch hoists a 340-kg crate up an incline at a constant speed $v = 0.5$ m/s as shown. If the power input to the winch is 1.5 kW and $\mu_k = 0.4$ between the crate and the incline, determine (a) the power output of the winch, (b) the efficiency of the winch.

13.33 A power winch hoists a 340-kg crate up an incline at a constant speed $v = 1$ m/s as shown. The power output of the winch is known to be 2 kW. Determine μ_k between the crate and the incline.

13.34 Refer to Prob. 13.33. If the power output of the winch is suddenly increased from 2 kW to 3 kW, determine the acceleration which the crate may momentarily experience.

13.35 A 2470-lb car with four-wheel drive climbs an incline as shown. The output power of the car is known to be 100 hp and the wheels do not slip on the incline. Neglecting effects of drag resistance, determine the constant speed v attained by the car.

13.36 A 2470-lb car with four-wheel drive climbs an incline as shown. As it moves forward, it experiences a drag resistance from the wind $F_D = 0.007v^2$ lb, where v is the speed of the car measured in ft/s. The output power of the car is known to be 100 hp, and the wheels do not slip on the incline. Determine the constant speed v attained by the car.[†]

13.37* The elevator E with its freight has a total mass of 400 kg and is hoisted by its 100-kg counterweight C and the motor at M as shown. If the motor has an efficiency of 0.8, determine the power that must be supplied to the motor when the elevator (a) is moving upward at a constant speed of 2 m/s, (b) has an instantaneous velocity of 2 m/s ↑ and an acceleration of 1 m/s² ↑.

13.38 The elevator E with its freight has a total mass of 400 kg and is hoisted by its 100-kg counterweight C and the motor at M as shown. If the output power of the motor is 6 kW at the instant the velocity of the elevator is 2 m/s ↑, determine the acceleration of the elevator.

CONSERVATION OF ENERGY AND SPECIAL TOPIC[††]

13.8 Conservative System and Potential Energy

An *analytic force field* is a region in space where a body experiences a determinable single-valued force or force system at every position in the region. Since the kinetic friction force developed on a body at a given position may have different values in magnitude and direction, a kinetic friction force field is not an analytic force field. A *conservative force field* is an analytic force field in which the work of the force in the field acting on a body traveling in the field is dependent on the initial and final positions of

---

[†]Cf. App. E for computer solution of resulting cubic equation.

[††]The concepts of potential energy and virtual work were presented in Chap. 10. For convenience, we repeat here portions of that material relevant to the current study.

the body, not on the path traveled by the body. The force in a conservative force field is called a *conservative force*.

It can be shown that the necessary and sufficient condition for the force $\mathbf{F} = \mathbf{F}(x,\ y,\ z) = F_x\mathbf{i} + F_y\mathbf{j} + F_z\mathbf{k}$ to be conservative is $\nabla \times \mathbf{F} = \mathbf{0}$, where the symbol ∇ is a vector operator called "del." Note that

$$\nabla = \mathbf{i}\,\frac{\partial}{\partial x} + \mathbf{j}\,\frac{\partial}{\partial y} + \mathbf{k}\,\frac{\partial}{\partial z}$$

and the cross product $\nabla \times \mathbf{F}$ is called the *curl* of $\mathbf{F}$. Thus, it can be shown that $\mathbf{F}$ is conservative *if and only if*[†]

$$\frac{\partial F_z}{\partial y} = \frac{\partial F_y}{\partial z} \qquad \frac{\partial F_x}{\partial z} = \frac{\partial F_z}{\partial x} \qquad \frac{\partial F_y}{\partial x} = \frac{\partial F_x}{\partial y} \qquad (13.31)$$

Moreover, one can show that $\mathbf{F} = \nabla\phi$, where $\phi = \phi\ (x,\ y,\ z)$, if $\mathbf{F}$ is conservative. For example, $\mathbf{F} = x\mathbf{i}$ is conservative, but $\mathbf{F} = x\mathbf{j}$ is not. In the rest of this chapter, we shall deal with the following conservative forces: gravitational forces, elastic spring forces, and applied constant loads.

Any force which acts on a body but does no work to the body during a displacement of the body is called a *nonworking force*. For instance, when a body moves on a surface, the normal force exerted by the surface on the body does not work to the body and is a nonworking force. A *conservative system* is a system where the body or assemblage of bodies is subjected to only conservative forces and nonworking forces, if any.

For practical use, the *potential energy of a body* in a conservative force field relative to an appropriate reference datum is defined as *the amount of work which the body will receive from the conservative force if the body travels from its present position to the reference datum.*[††] Thus, the potential energy as well as the kinetic energy has the same dimensions and units as the work of a force. The potential energy of a body situated at the reference datum is, of course, equal to zero.

Developmental Exercises

D13.23 Define (a) analytic force field, (b) conservative force field, (c) conservative force, (d) nonworking force, (e) conservative system, (f) curl of $\mathbf{F}$.

D13.24 Define the potential energy of a body.

13.9 Gravitational Potential Energy

We know that the force exerted on a body in a *gravitational force field* is the weight force of the body. As a body travels from one position to a nearby position in the vicinity of the surface of the earth, the weight force of the body is essentially a downward vertical force of constant magnitude.

[†]Spiegel, M. R., *Vector Analysis and an Introduction to Tensor Analysis*, Schaum's Outline Series, McGraw-Hill, New York, 1959, pp. 89–91.

[††]This is a *quantitative definition*. Qualitatively, the potential energy of a body is the energy possessed by the body by virtue of its position relative to the reference datum in the conservative force field. A potential energy is expressible as a function of spatial coordinates only.

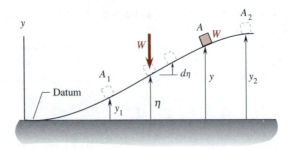

Fig. 13.10

Thus, the *gravitational potential energy V_g of a body* with a weight W near the surface of the earth at the position A of altitude y above the reference datum, as shown in Fig. 13.10, is equal to the work which the body will receive from the constant weight force **W** of the body if the body travels from its present position at $\eta = y$ to the reference datum at $\eta = 0$. We write

$$V_g = \int_y^0 W\,(-d\eta) \qquad \boxed{V_g = Wy} \qquad (13.32)$$

where V_g is positive only if the datum chosen is situated between the body and the center of the earth. The work $U_{1\to2}$ of the weight force $\mathbf{W} = W \downarrow$ acting on the body traveling from A_1 to A_2 is found to be related to the *change in gravitational potential energy*, ΔV_g, as follows:

$$U_{1\to2} = \int_{y_1}^{y_2} W\,(-d\eta) = -(Wy_2 - Wy_1) = -[(V_g)_2 - (V_g)_1]$$

$$U_{1\to2} = -\Delta V_g \qquad (13.33)$$

In the case of the gravitational central-force motion of a particle (e.g., a spacecraft), we should take into account the variation of the gravitational force with the radial distance r drawn from the center of force O to the particle at the position A as shown in Fig. 13.5. The gravitational potential energy of such a particle is defined relative to *a position at infinity* (i.e., $r = \infty$) *as the reference datum*. Thus, the *potential energy V_g of a particle in gravitational central-force motion* at a radial distance r from the center

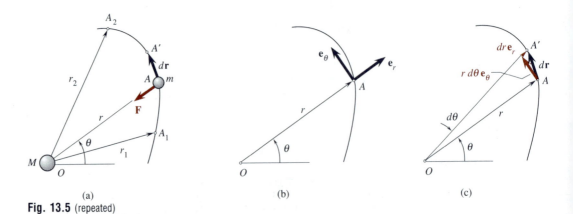

(a) (b) (c)

Fig. 13.5 (repeated)

of force O is obtained by computing the work which the particle will receive from the gravitational force $\mathbf{F}$ if the particle travels from its present position to the reference datum at $r = \infty$. Using Eqs. (13.15) through (13.17), we write

$$V_g = \int_r^{\infty} (-GMmr^{-2})\, dr \qquad \boxed{V_g = -\frac{GMm}{r}} \qquad (13.34)$$

Note that V_g in Eq. (13.34) is always negative because the work of the gravitational central force $\mathbf{F}$ acting on the particle during its travel from its present position to the reference datum at $r = \infty$ is always negative. The work $U_{1\to2}$ of such a force $\mathbf{F}$ acting on the particle traveling from A_1 (where $r = r_1$) to A_2 (where $r = r_2$) is found to be related to the change in V_g as follows:

$$\Delta V_g = (V_g)_2 - (V_g)_1 = -\frac{GMm}{r_2} + \frac{GMm}{r_1}$$

$$U_{1\to2} = \int_{r_1}^{r_2} (-GMmr^{-2})\, dr = -\left(-\frac{GMm}{r_2} + \frac{GMm}{r_1}\right)$$

$$U_{1\to2} = -\Delta V_g \qquad (13.35)$$

Note that Eqs. (13.33) and (13.35) are of the same form.

EXAMPLE 13.11

A body of weight W is at an altitude y above the surface of the earth whose radius is R. Choosing the surface of the earth as the reference datum, show that Wy is truly a good approximation for the potential energy V_g of the body if $y \ll R$.

Solution. The weight W of the body is usually measured on the surface of the earth. Let O be the center of the earth, as shown, m be the mass of the body, and M be the mass of the earth. We may apply Newton's law of gravitation and the definition of V_g to write

$$F = \frac{GMm}{r^2} \qquad W = F|_{r=R} = \frac{GMm}{R^2}$$

$$V_g = \int_{R+y}^{R} F(-dr) = -GMm \int_{R+y}^{R} r^{-2}\, dr$$

$$= GMm\left(\frac{1}{R} - \frac{1}{R+y}\right) = \frac{GMmy}{R(R+y)}$$

Since $y \ll R$, we have $R + y \approx R$ and

$$V_g \approx \frac{GMm}{R^2} y = Wy \qquad \text{Q.E.D.} \blacktriangleleft$$

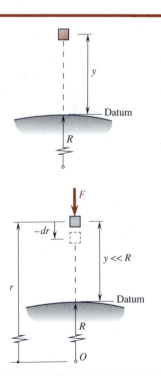

D13.25   Define the *gravitational potential energy* of a body.

D13.26   Where is the *reference datum* for the gravitational potential energy of a spacecraft revolving around the earth?

D13.27   Define the *potential energy* of a particle in gravitational central-force motion.

13.10   Elastic Potential Energy

The force exerted by an elastic spring on a body attached to the spring is proportional to the deformation in length of the spring (cf. Sec. 13.4). A body subjected to the forces of an elastic spring is said to be in an *elastic force field*. The *reference datum in an elastic force field* is generally chosen to be the position of the body where the spring attached to the body is neutral. For a body at A connected to B by a spring with a modulus k and an elongation x as shown in Fig. 13.11, the *elastic potential energy V_e of the body* is equal to the work which the body will receive from the spring force $\mathbf{F} = k\xi \leftarrow$ if the body travels from its present position at $\xi = x$ to the reference datum at $\xi = 0$. Thus, we have

$$V_e = \int_x^0 k\xi \, (-d\xi) \qquad \boxed{V_e = \tfrac{1}{2} kx^2} \qquad (13.36)^\dagger$$

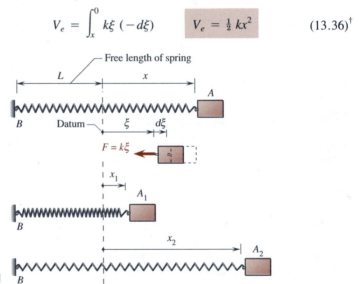

Fig. 13.11

Similarly, we find that the work $U_{1\to2}$ of the spring force $\mathbf{F}$ acting on the body traveling from A_1 to A_2 is related to the *change in elastic potential energy*, ΔV_e, as follows:

$$U_{1\to2} = \int_{x_1}^{x_2} k\xi \, (-d\xi) = -(\tfrac{1}{2} kx_2^2 - \tfrac{1}{2} kx_1^2)$$

$$= -[(V_e)_2 - (V_e)_1]$$

$$U_{1\to2} = -\Delta V_e \qquad (13.37)$$

---

† If a body is attached to a *nonlinear spring*, the elastic potential energy of the body may be obtained by integrating the work which the body will receive from the nonlinear spring force if the body travels from its present position to the reference datum. (Cf. Probs. 13.62 and 13.63.)

In the case where a torsional spring of modulus K is used to supply a restoring moment of magnitude $M = K\theta$ to a member AB which carries a body (a lumped mass as a particle) at the end A and undergoes an angular rotation θ from the neutral position of the spring as shown in Fig. 13.12, the *elastic potential energy* V_e *of the member* can similarly be shown to be

$$V_e = \int_\theta^0 K\theta\,(-d\theta) \qquad \boxed{V_e = \tfrac{1}{2} K\theta^2} \qquad (13.38)$$

where the angle θ is measured in radians and the modulus K is measured in moment per radian (e.g., N·m/rad or lb·ft/rad). Of course, Eq. (13.37) is also true in the case of a torsional spring.

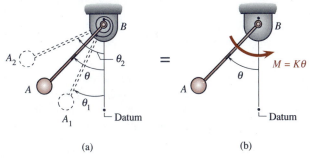

(a)   (b)

Fig. 13.12

Developmental Exercises

D13.28   How is the *reference datum* for the elastic potential energy of a body in an elastic force field generally chosen?

D13.29   Define the *elastic potential energy* of a body.

13.11   Applied Potential Energy

Besides gravitational and elastic force fields, there are other conservative force fields which exist to exert conservative forces on various bodies. A body deflected by an *applied conservative force* $\mathbf{F}$ is said to possess an applied potential energy because the body will receive a definite amount of work from the applied conservative force if it returns to its undeflected state. In this case, we must have $\nabla \times \mathbf{F} = \mathbf{0}$.[†] The *applied potential energy* V_a *of a body* in a deflected state under the action of an applied conservative force is equal to the work which the body will receive from the applied conservative force if the body travels from its present deflected position to its undeflected position of equilibrium, which is the reference datum. However, we shall here focus our attention on applied conservative forces which are constant in magnitude and direction. Note that an *applied constant force* is a constant force which is suddenly or instantaneously applied to a body, and its magnitude as well as direction remains the same thereafter.

Let us now consider a body (a lumped mass as a particle) which is carried by a member AB at its end A and is acted on by an applied constant force $\mathbf{P}$

[†]Cf. Eqs. (13.31).

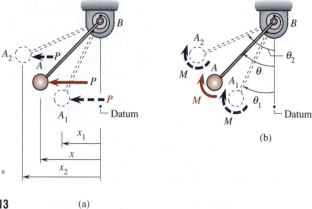

Fig. 13.13   (a)

as shown in Fig. 13.13(a). The *applied potential energy* V_a *of the body* in its deflected position under the action of the applied constant force **P** is equal to the work which the body will receive from **P** if it travels from its present deflected position to its *undeflected neutral equilibrium position*, which is the *reference datum*; i.e.,

$$V_a = \int_x^0 P\,dx \qquad \boxed{V_a = -Px} \qquad (13.39)$$

The work $U_{1\to2}$ of the applied constant force **P** acting on the body during its motion from the position where $x = x_1$ to the position where $x = x_2$ is related to the *change in the applied potential energy*, ΔV_a, as follows:

$$U_{1\to2} = \int_{x_1}^{x_2} P\,dx = -(-Px_2 + Px_1) = -[(V_a)_2 - (V_a)_1]$$

$$U_{1\to2} = -\Delta V_a \qquad (13.40)$$

In the case where an applied constant moment **M** acts on a body which is carried by a member *AB* at its end *A* as shown in Fig. 13.13(b), the *applied potential energy* V_a *of the body* can similarly be shown to be

$$\boxed{V_a = -M\theta} \qquad (13.41)$$

Of course, Eq. (13.40) is also true in the case of an applied constant moment **M**.

EXAMPLE 13.12[†]

A constant horizontal force **P** is suddenly applied to a particle of weight W which is attached to a slender rod *AB* of length L and a spring *BC* of free length L and modulus k as shown. When the particle is deflected from the position B to the position B' by **P**, determine the gravitational potential

---

[†]To be continued in Example 13.15.

energy V_g, the elastic potential energy V_e, and the applied potential energy V_a of the particle as a function of θ.

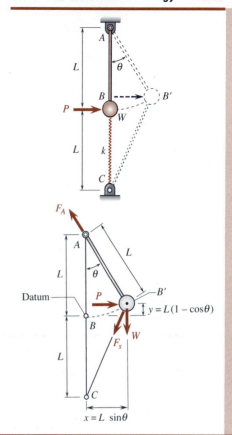

Solution. The system is originally in neutral equilibrium in its vertical configuration ABC. When the constant horizontal force $\mathbf{P}$ is suddenly applied to the particle to deflect it from the point B to the point B', the particle is subjected to an applied constant force field in addition to the gravitational and elastic force fields, and the system is conservative. Thus, we choose the position B as the reference datum for V_g, V_e, and V_a. Referring to the sketch shown, we write

$$V_g = Wy \qquad V_g = WL(1 - \cos\theta) \blacktriangleleft$$

The elongation of the spring is

$$x_s = \overline{CB'} - \overline{CB} = [(2L)^2 + L^2 - 2(2L)L \cos\theta]^{1/2} - L$$

$$= L[(5 - 4\cos\theta)^{1/2} - 1]$$

Thus, we write

$$V_e = \tfrac{1}{2} kx_s^2$$

$$V_e = \tfrac{1}{2} kL^2[(5 - 4\cos\theta)^{1/2} - 1]^2 \blacktriangleleft$$

The force $\mathbf{F}_A$ as shown is directed along the line segment $\overline{B'A}$ because the rod is taken as a massless extension of the particle. The force $\mathbf{F}_A$ does no work and has no contribution to V_a. Thus, we write

$$V_a = -Px \qquad V_a = -PL \sin\theta \blacktriangleleft$$

Developmental Exercises

D13.30   Is an applied constant load *gradually* or *suddenly* applied to a body?

D13.31   Define the *applied potential energy* of a body.

13.12   Conservation of Energy

The *potential energy* V of the body in a conservative system is the sum of the gravitational, elastic, and applied potential energies of the body; i.e.,

$$V = V_g + V_e + V_a \tag{13.42}$$

By definition, we note that the body in a conservative system receives work only from the conservative forces when it undergoes a displacement. From Eqs. (13.32) through (13.42), we readily see that *any work $U_{1\to2}$ of the conservative forces* acting on the body, which moves from the position 1 to the position 2 in the conservative force field, *is equal to the negative of the change in the potential energy of the body*; i.e.,

$$U_{1\to2} = -(V_2 - V_1) \tag{13.43}$$

This means that, for a body in a conservative system, the principle of work and energy may be expressed in a modified form. Substituting Eq. (13.43) into Eq. (13.26) and rearranging, we obtain

$$T_1 + V_1 = T_2 + V_2 \qquad (13.44)$$

The sum of the kinetic energy T and the potential energy V of a body is called the *total mechanical energy* of the body. The modified form of the principle of work and energy for a conservative system, as expressed in Eq. (13.44), is called the *principle of conservation of energy*, which can be stated as follows: *The total mechanical energy of the body in a conservative system remains constant during its motion from one position to another*. For a conservative system consisting of several particles, the energies in Eq. (13.44) are to be understood to represent the sums of the energies of all particles in the system. In particular, V_1 and V_2 should include the potential energies associated with the internal as well as external forces acting on the particles of the system.

EXAMPLE 13.13

Using the *principle of conservation of energy*, work Example 13.7.

Solution. From the given data in Example 13.7, we have $v_1 = 6.27$ km/s, $r_1 = 9370$ km, and $r_2 = 12370$ km. We recognize that the system is conservative. Let the mass of the spacecraft be m. Applying the *principle of conservation of energy* and recalling that $GM = gR^2$, we write

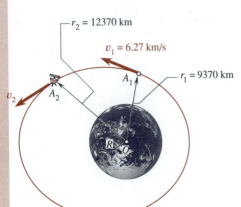

$$T_1 + V_1 = T_2 + V_2: \qquad \tfrac{1}{2}mv_1^2 - \frac{GMm}{r_1} = \tfrac{1}{2}mv_2^2 - \frac{GMm}{r_2}$$

$$\tfrac{1}{2}v_1^2 - \frac{gR^2}{r_1} = \tfrac{1}{2}v_2^2 - \frac{gR^2}{r_2}$$

$$\tfrac{1}{2}(6.27)^2 - \frac{(9.81/1000)(6370)^2}{9370}$$

$$= \tfrac{1}{2}v_2^2 - \frac{(9.81/1000)(6370)^2}{12370}$$

$$v_2 = 4.33 \text{ km/s} \blacktriangleleft$$

Developmental Exercises

D13.32   Define the *total mechanical energy* of a body.

D13.33   Is the *principle of conservation of energy* applicable to a system involving kinetic friction forces?

EXAMPLE 13.14

A 4-kg collar is released from rest in the position A_1 and is slid up a smooth rod by a 50-N horizontal force as shown. The spring attached to the collar has a free length $L = 0.9$ m and a modulus $k = 40$ N/m. Determine the speed v_2 of the collar as it passes through the position A_2.

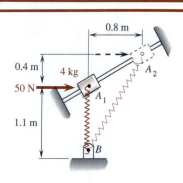

Solution.   Since the rod is smooth, the system is conservative. Let us choose the position A_1 as the reference datum for both V_g and V_a. Noting that the elongations of the spring at A_1 and A_2 are $x_1 = (1.1 - 0.9)$ m $= 0.2$ m and $x_2 = (1.7 - 0.9) = 0.8$ m, respectively, we may compute the potential energies V_1 and V_2 of the collar at A_1 and A_2, respectively, as follows:

$$(V_g)_1 = (V_a)_1 = 0$$

$$(V_e)_1 = \tfrac{1}{2}kx_1^2 = \tfrac{1}{2}(40)(0.2)^2 \qquad (V_e)_1 = 0.8 \text{ J}$$

$$(V_g)_2 = Wy = 4g(0.4) \qquad (V_g)_2 = 15.696 \text{ J}$$

$$(V_a)_2 = -Px = -50(0.8) \qquad (V_a)_2 = -40 \text{ J}$$

$$(V_e)_2 = \tfrac{1}{2}kx_2^2 = \tfrac{1}{2}(40)(0.8)^2 \qquad (V_e)_2 = 12.8 \text{ J}$$

$$V_1 = (V_g)_1 + (V_e)_1 + (V_a)_1 = 0.8 \text{ J}$$

$$V_2 = (V_g)_2 + (V_e)_2 + (V_a)_2 = -11.504 \text{ J}$$

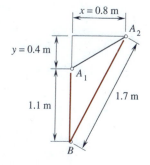

The kinetic energies of the collar at A_1 and A_2 are

$$T_1 = 0 \qquad T_2 = \tfrac{1}{2}mv_2^2 = \tfrac{1}{2}(4)v_2^2 = 2v_2^2$$

Applying the *principle of conservation of energy*, we write

$$T_1 + V_1 = T_2 + V_2: \qquad 0 + 0.8 = 2v_2^2 - 11.504$$

$$v_2 = 2.48 \text{ m/s} \qquad \blacktriangleleft$$

Developmental Exercise

D13.34   Describe the *principle of conservation of energy*.

EXAMPLE 13.15

Refer to Example 13.12. It is known that the particle comes to a stop when $\theta = 30°$. For $W = 40$ lb, $L = 4$ ft, and $P = 30$ lb, determine the modulus k of the spring if (a) **P** is suddenly applied as stated, (b) **P** is gradually applied (i.e., the magnitude of **P** is gradually increased from 0 to its final value of 30 lb).

Solution.   We first consider the case where **P** is suddenly applied. Such a system is clearly conservative. Applying the *principle of conservation of energy* to the particle which moves from B to B' (where the velocity, not

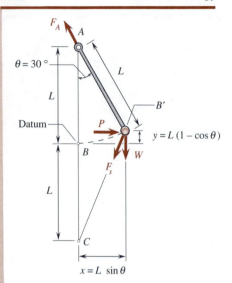

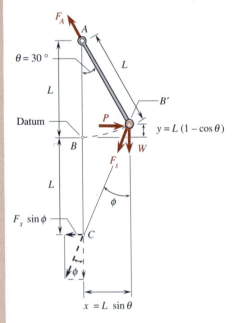

acceleration, is zero), we write

$$T_B + V_B = T_{B'} + V_{B'}$$

Since $T_B = V_B = T_{B'} = 0$ as inferred from the problem, we must have $V_{B'} = 0$. Using the results in Example 13.12, we write

$$V_{B'} = V_g + V_e + V_a$$

$$= WL(1 - \cos\theta) + \tfrac{1}{2}kL^2[(5 - 4\cos\theta)^{1/2} - 1]^2 - PL\sin\theta$$

$$= 40(4)(1 - \cos 30°) + \tfrac{1}{2}k(4)^2[(5 - 4\cos 30°)^{1/2} - 1]^2$$

$$- 30(4)\sin 30°$$

$$= 0$$

$$k = 84.17 \qquad\qquad k = 84.2 \text{ lb/ft} \blacktriangleleft$$

In the case where **P** is gradually applied, **P** is *no longer* an applied constant force; the magnitude of **P** is gradually increased from 0 to its final value of 30 lb at which *static equilibrium* of the particle exists. In the free-body diagram as shown, we have

$$\tan\phi = \frac{L\sin\theta}{L + L(1 - \cos\theta)} = \frac{\sin 30°}{2 - \cos 30°} \qquad \phi = 23.79°$$

From the solution in Example 13.12, we see that the elongation of the spring in the position CB' is

$$x_s = \overline{CB'} - \overline{CB} = L[(5 - 4\cos 30°)^{1/2} - 1]$$

$$= 4[(5 - 4\cos 30°)^{1/2} - 1]$$

Thus, we have

$$x_s = 0.9573 \text{ ft} \qquad F_s = kx_s = 0.9573k$$

For studying equilibrium of the particle at B', we transmit $\mathbf{F}_s$ to the point C and write

$$+\circlearrowleft \Sigma M_A = 0: \qquad (L\cos\theta)P - 2L(F_s\sin\phi)$$

$$- (L\sin\theta)W = 0$$

$$(4\cos 30°)(30) - 2(4)(0.9573k)(\sin 23.79°)$$

$$- 4(\sin 30°)(40) = 0$$

$$k = 7.744 \qquad\qquad k = 7.74 \text{ lb/ft} \blacktriangleleft$$

REMARK. The results obtained in this example show a great difference between the values of the modulus k of the spring required to limit the deflection of the structure (i.e., $\theta \le 30°$) when the 30-lb force **P** is allowed to be first suddenly (i.e., dynamically) applied and, the next time, to be gradually (i.e., statically) applied.

Developmental Exercise

D13.35   A 20-lb collar is released from rest in the position A_1 and is slid up a smooth rod by a 30-lb horizontal force as shown. The spring attached to the collar has a free length $L = 2.5$ ft and a modulus $k = 40$ lb/ft. Determine the speed v_2 of the collar as it passes through the position A_2.

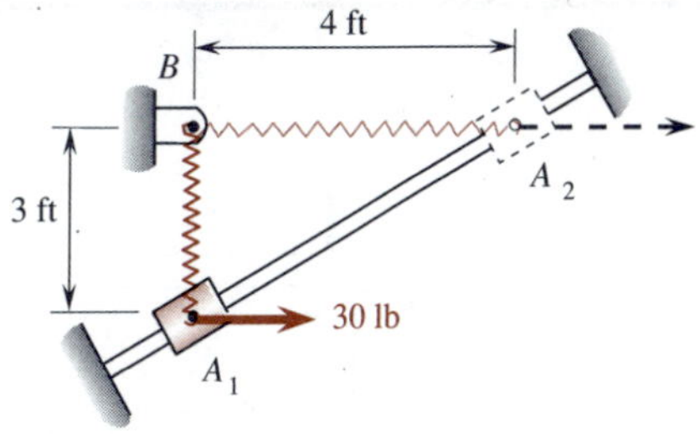

Fig. D13.35

*13.13   Virtual Work in Kinetics: Force and Acceleration[†]

A *compatible virtual displacement* of a body is a set of imaginary *first-order* differential displacements, which conforms to the integrity (i.e., no breakage or rupture) of its free body within the framework of first-order accuracy, where the body may be a particle, a rigid body, or a set of connected particles or rigid bodies. Note that a compatible virtual displacement of a body does not necessarily conform to the constraints at the supports in the space diagram. However, a *constrained virtual displacement* of a body is a compatible virtual displacement of the body which conforms to the constraints at the supports in the space diagram. Note that a constrained virtual displacement of a body may differ from the *actual differential displacement* of the body in *sense*; i.e., the latter is a special case of the former. (Cf. Sec. 10.4.)

We know that when two systems of forces acting on two identical sets of connected particles are equipollent, they must have *equal resultant forces* and *equal resultant moments* about any corresponding given point. Moreover, we can always give a constrained virtual displacement to a one-degree-of-freedom system consistent with its infinitesimal displacement which will actually take place in the next infinitesimal period of time. Thus, it follows logically from the equipollence described in Sec. 12.6 that, *during a constrained virtual displacement of a set of connected particles with one degree of freedom, the total virtual work δU of all the external and internal forces acting on the various particles in the set must be equal to the total virtual work δU_{eff} of all the effective forces on the various particles in the set*; i.e.,

$$\delta U = \delta U_{\text{eff}} \tag{13.45}$$

If the set of particles under consideration is connected by *inextensible cords* or *links*, the total virtual work of the internal forces is zero, and δU in Eq. (13.45) reduces to the *total virtual work of external forces only*. The property expressed in Eq. (13.45) is here referred to as the *principle of virtual work in kinetics* for particles. Such a principle complements the traditional *method of force and acceleration* in Chap. 12.

Since the system of forces acting on a body in equilibrium is *equivalent* to a zero force and a zero moment (i.e., a zero wrench), the virtual displacement used in the principle of virtual work in statics for studying problems of equilibrium can be *any chosen compatible virtual displacement*. Naturally, any constrained virtual displacement is always a compatible virtual displacement, but the converse is not true.

[†]Cf. Secs. 10.4 and 10.7.

EXAMPLE 13.16

Using the *principle of virtual work in kinetics*, work Example 12.3.

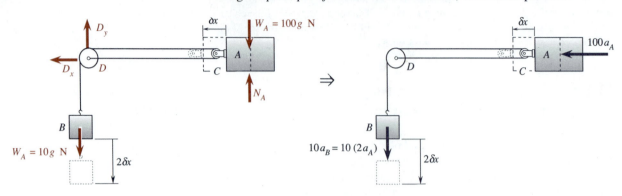

Solution. Let the free-body diagram and the effective-force diagram of the entire system be drawn first, where the pulley at D is assumed to have negligible mass. From the constraints at the supports and the force system in the free-body diagram, we note that the system has only one degree of freedom and the block A will accelerate to the left with a magnitude a_A while the block B will accelerate downward with a magnitude $a_B = 2a_A$. Likewise, the *constrained virtual displacement* consists of $\delta x \leftarrow$ for the block A and $2\,\delta x \downarrow$ for the block B. These are all indicated in the diagrams as shown. Applying the *principle of virtual work in kinetics* and noting that $\mathbf{W}_A$, $\mathbf{N}_A$, $\mathbf{D}_x$, and $\mathbf{D}_y$ do no work during the *constrained virtual displacement*, we write

$$\delta U: \qquad 10g(2\,\delta x) = 100a_A(\delta x) + 10(2a_A)(2\,\delta x) \qquad (1)$$

$$a_A = \frac{g}{7} = \frac{9.81}{7} \qquad\qquad \mathbf{a}_A = 1.401 \text{ m/s}^2 \leftarrow \quad \blacktriangleleft$$

To find the tension F in the cord BCD, we give the block B a constrained virtual displacement $\delta x_B \downarrow$ as shown below and apply the *principle of virtual work in kinetics* to the block B to write

$$\delta U: \qquad 10g(\delta x_B) + F(-\delta x_B) = 10(2a_A)(\delta x_B) \qquad (2)$$

$$F = 10g - 20a_A = 70.1 \qquad F = 70.1 \text{ N} \quad \blacktriangleleft$$

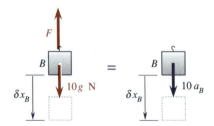

REMARK. Equation (1) for finding a_A *cannot* directly be written from the method of force and acceleration, while Eq. (2) for finding F is equivalent to equating the sums of the y components of the vectors in the method of force and acceleration.

Developmental Exercises

D13.36 Define a *constrained virtual displacement* of a system.

D13.37 Describe the *principle of virtual work in kinetics* for particles.

D13.38 Refer to Example 13.9. Assume that $\mu_k = 0.1$ between the block A and the horizontal support. Applying the *principle of virtual work in kinetics*, determine the acceleration of the block A.

PROBLEMS

13.39* through 13.50* Using the *principle of conservation of energy*, solve the following problems:

13.39* Prob. 13.2.	13.40* Prob. 13.5.
13.41* Prob. 13.9*.	13.42* Prob. 13.10.
13.43* Prob. 13.11.	13.44* Prob. 13.12.
13.45* Prob. 13.13*.	13.46* Prob. 13.14*.
13.47* Prob. 13.17.	13.48* Prob. 13.18*.
13.49* Prob. 13.19.	13.50* Prob. 13.21.

13.51 Neglecting air resistance, determine (a) the magnitude v_{esc} of the escape velocity of a projectile fired from the surface of the earth at the North Pole, as shown, in terms of the gravitational acceleration g on the surface of the earth and the radius R, (b) whether v_{esc} is dependent on θ.

13.52* A spacecraft of mass m is in free flight around the earth as shown. Show that the speeds v_P and v_Q of the spacecraft at the perigee P and the apogee Q, respectively, are

$$v_P = R\{2gr_Q/[r_P(r_P + r_Q)]\}^{1/2}$$

$$v_Q = R\{2gr_P/[r_Q(r_P + r_Q)]\}^{1/2}$$

and that the total mechanical energy is $E = -gmR^2/(r_P + r_Q)$. (*Hint.* The reference datum for V_g is at $r = \infty$.)

Fig. P13.51

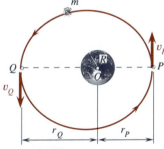

Fig. P13.52*

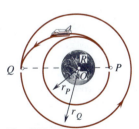

Fig. P13.53

13.53 A spacecraft of mass m is initially in free flight along a circular orbit of radius r_P around the earth. In order to transfer the spacecraft to a larger circular orbit of radius r_Q by placing it on a transitional semielliptic path PQ as shown, amounts of energy ΔE_P and ΔE_Q must momentarily be imparted to the spacecraft at P and Q, respectively. Knowing that the total mechanical energy of the spacecraft traveling along the semielliptic path PQ is $E = -gmR^2/(r_P + r_Q)$, determine the values of (a) ΔE_P, (b) ΔE_Q, (c) the sum $\Delta E'$ of ΔE_P and ΔE_Q. (Cf. Prob. 13.52*.)

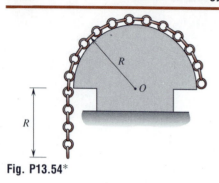

Fig. P13.54*

13.54* The chain is released from rest in the position shown. Neglecting effects of friction, determine the speed of the chain in terms of g and R as its last link leaves the left edge of the semicircular support. (*Hint.* Choose the horizontal line through O as the reference datum for V_g.)

13.55 Solve Prob. 13.54* for (a) $R = 2$ m, (b) $R = 2$ ft.

13.56* A spring of free length 1.2 m is attached to a collar of mass 20 kg. The collar is released from rest in the position A_1 to slide down a smooth vertical rod as shown. If the lowest position reached by the collar is A_2 where $b = 2$ m, determine the modulus k of the spring.

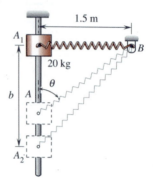

Fig. P13.56* and P13.57

13.57 A spring of free length 1.2 m and modulus 1 kN/m is attached to a collar of mass 20 kg. The collar is released from rest in the position A_1 to slide down a smooth vertical rod as shown. If the lowest position reached by the collar is A_2, determine the distance b. (*Hint.* It leads to a cubic equation in b.)[†]

13.58 Refer to Prob. 13.57. If the collar attains a maximum speed v_{max} as it passes the position A, determine (a) the distance $\overline{A_1A}$, (b) v_{max}. (*Hint.* It leads to a quartic equation in $\sin\theta$.)[†]

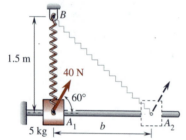

Fig. P13.59 and P13.60

13.59 A 5-kg collar is released from rest in the position A_1 and is slid to the right on a smooth rod by a constant force of 40 N as shown. The spring attached to the collar has a modulus of 50 N/m, and the farthest position reached by the collar is A_2 where $b = 2$ m. Determine the free length L of the spring.

13.60 A 5-kg collar is released from rest in the position A_1 and is slid to the right on a smooth rod by a constant force of 40 N as shown. The spring attached to the collar has a free length of 1.3 m and a modulus of 50 N/m. Determine the speed of the collar as it passes through the position A_2 where $b = 2$ m.

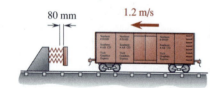

Fig. P13.61*

13.61* A bumper composed of three springs is to stop a 40-Mg box car which is traveling at a speed of 1.2 m/s as shown. One of the springs is 80 mm shorter than the other two. The modulus of the shorter spring is 1.5 MN/m, while the modulus of each of the two longer springs is 1 MN/m. Determine the maximum deflection of the bumper.

13.62 A bumper made of a hard spring is to stop a 30-ton gondola car which is traveling at a speed of 4 ft/s as shown. Such a spring develops a restoring force $F = 100x + 12x^3$ kips when it is compressed by an amount of x ft. Determine the maximum deflection of the bumper.

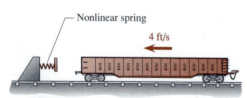

Fig. P13.62

13.63 Solve Prob. 13.62 if the spring is a soft spring which develops a restoring force $F = 100x - 12x^3$ kips when it is compressed by an amount of x ft.

[†]Cf. App. E for computer solution of resulting nonlinear equation.

13.64 A 4-kg slider is connected to *two* identical springs as shown. Each of the two springs has a modulus of 5.1 kN/m, and a force of 240 N is required to hold the slider in equilibrium in the position A_1 as shown. The 240-N force is later suddenly removed to release the slider so that it travels to the right in the smooth horizontal slot. Determine the speed of the slider as it passes through the position A_2.

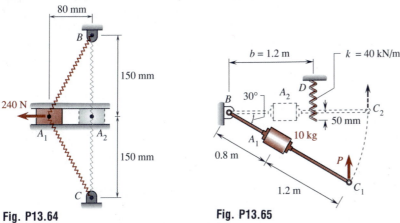

Fig. P13.64   **Fig. P13.65**

13.65 A light rod with a fixed 10-kg lumped mass is rotated from rest in the inclined position BA_1C_1 about the hinge at B by a vertical constant force **P** as shown. The vertical spring at D has a modulus of 40 kN/m. Having compressed the spring 50 mm, the rod is brought to a stop in the horizontal position BA_2C_2. Determine (a) the magnitude of **P**, (b) whether P is dependent on b.

13.66* A 10-kg plunger is released from rest in the position shown and is stopped by the nest of two springs. The outer spring has a modulus of 4 kN/m and the inner one a modulus of 3 kN/m. If the inner spring is adjusted to give $b = 100$ mm before the impact, determine the maximum deflection δ of the outer spring.

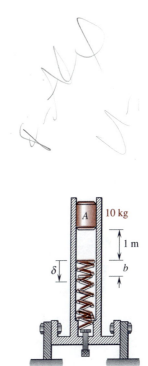

Fig. P13.66*

13.67 The 40-lb sphere as shown is released from rest when $\theta = 0$. Knowing that the spring has a free length of 30 in. and a modulus of 6 lb/in., determine the speed of the sphere when $\theta = 90°$.

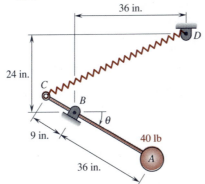

Fig. P13.67

13.68* through 13.77* Using the *principle of virtual work in kinetics*, solve the following problems:

13.68* Prob. 12.18.	13.69* Prob. 12.19*.
13.70* Prob. 12.20*.	13.71* Prob. 12.21.
13.72* Prob. 12.22.	13.73* Prob. 12.23*.
13.74* Prob. 12.24.	13.75* Prob. 12.25.
13.76* Prob. 12.26.	13.77* Prob. 12.27*.

13.14   Concluding Remarks

This chapter covers three principles related to work and energy for studying the motion of particles. They are (1) the *principle of work and energy*, which is derived directly from Newton's second law of motion; (2) the *principle of conservation of energy*, which is a special case of the principle of work and energy as it is applied to conservative systems; and (3) the *principle of virtual work in kinetics*, which is an extension of that in statics. In any case, an *inertial reference frame* should properly be used in describing the various quantities.

Note that a spring is usually idealized as having negligible mass. Having negligible mass, a spring is *not a body* in the usual sense. Rather, a spring is a device which provides an elastic force field for the body attached to it. It is the spring force, rather than the spring, that does the work to the body which is attached to the spring. By the principle of work and energy, *any net amount of work done by forces to a body must be exactly equal to the amount of change of the kinetic energy of the body*. Therefore, *no work can be done to a stationary body or a body with no mass*. If we say that work is done to a spring, we would have to take the mass of the spring into account; otherwise, the requirement that the net amount of work done to the spring be equal to the amount of change of kinetic energy of the spring cannot be satisfied.

Furthermore, note that the potential energy of a body is defined *only* for a body in a conservative force field. The gravitational, elastic, and applied potential energies of a body are *numerically* equal to the amounts of work which the body will receive from the forces in the respective conservative force fields *if* the body travels from its present position to the respective reference datums.

In mechanics, force is exerted by one body on another body; however, it is the *force* (not the *body*) which does *work* to a body having a *mass* and a nonzero component of *displacement* parallel to the force while the force is acting on it. Therefore, phrases such as "the work done by a body," "the work done on a (massless) spring," and "the capacity of a body to do work" are considered as colloquial. Work, kinetic energy, and potential energy are subtle as well as vital concepts in kinetics. A clear understanding of these concepts is essential to the proper applications of principles related to work and energy.

REVIEW PROBLEMS

13.78   A brick is released from rest at A_1 to slide down a chute as shown. The brick leaves the chute at A_2 with a horizontal velocity and strikes the ground at A_3. Determine the coefficient of kinetic friction μ_k between the brick and the chute.

13.79   A 120-lb cylinder C is released from rest when the spring with a modulus of 40 lb/ft is unstressed as shown. Determine (a) the maximum deflection of the cylinder, (b) the maximum speed attained by the cylinder.

13.80*   A satellite is in free flight around the earth as shown. In the position S, it has an altitude and a speed as indicated. Determine the altitude a_Q of its apogee.

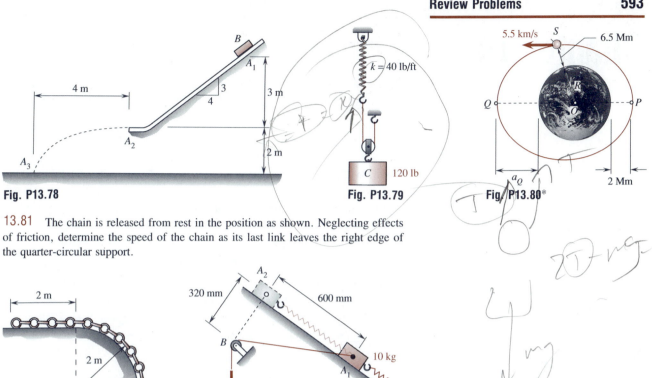

Fig. P13.78

Fig. P13.79

Fig. P13.80*

13.81 The chain is released from rest in the position as shown. Neglecting effects of friction, determine the speed of the chain as its last link leaves the right edge of the quarter-circular support.

Fig. P13.81

Fig. P13.82*

13.82* A 10-kg block is released from rest in the position A_1 to be moved up a smooth incline by the constant force **P** of 500 N as shown. The attached spring has a modulus of 400 N/m and is stretched 120 mm when the block is in the position A_1. Determine the speed of the block as it passes through the position A_2.

13.83* Using the *principle of virtual work in kinetics*, solve Prob. 12.90.

13.84 A step-type escalator as shown is designed to transport 120 persons per minute at a constant speed $v = 1.5$ ft/s. The overall efficiency of the escalator is estimated to be 70%, and a 200% *overload* is to be allowed. Assuming that the average weight per person is 160 lb, determine the required capacity of the motor of the escalator.

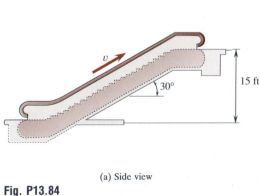

(a) Side view

Fig. P13.84

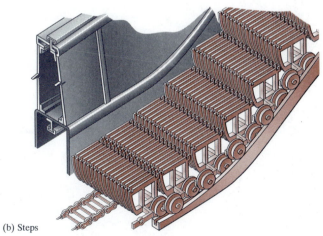

(b) Steps

Chapter 14

Kinetics of Particles: Impulse and Momentum

KEY CONCEPTS

- Impulse, momentum, linear impulse, linear momentum, angular impulse, angular momentum, impulsive force, direct central impact, oblique central impact, and coefficient of restitution.
- Principle of impulse and momentum for particles.
- Conservation of angular momentum in central-force motion.
- Direct and oblique central impacts between particles.
- Principle of generalized virtual work for studying impulses and momenta in one-degree-of-freedom systems of particles.
- Kinetics of variable systems of particles.

Impulse and momentum are vector quantities with specific meaning in mechanics. They are first defined and then used in formulating the *principle of impulse and momentum*, which is an effective tool for solving kinetic problems where *time* and *velocity* (among others) are to be directly related without initially involving acceleration and displacement. Furthermore, such a principle can be applied to provide useful equations in analyzing problems of central-force motion of particles and problems of impact. Later, the *principle of generalized virtual work* for particles is propounded as a complement to the principle of impulse and momentum for systems of connected particles with one degree of freedom. The last part of this chapter covers the motions of variable systems (e.g., turbines, jet engines, and rockets). The study of variable systems is a step into the study of advanced engineering systems.

IMPULSE AND MOMENTUM

14.1   Linear Impulse and Linear Momentum

The *linear impulse of a force* $\mathbf{F}$ acting on a particle during the time interval $t_1 \le t \le t_2$ is denoted by $\mathbf{Imp}_{1 \to 2}$ and is defined as the integral of $\mathbf{F}$ times the differential time dt evaluated over that time interval; i.e.,

$$\mathbf{Imp}_{1\rightarrow 2} = \int_{t_1}^{t_2} \mathbf{F}\, dt \qquad (14.1)$$

If the force $\mathbf{F}$ acting on the particle remains *constant* during the interval $\Delta t = t_2 - t_1$, we readily obtain

$$\mathbf{Imp}_{1\rightarrow 2} = \mathbf{F}\, \Delta t \qquad (14.2)$$

The *linear impulse* defined above is often simply referred to as the *impulse*. Physically, the *linear impulse of a force* on a particle during a time interval is a measure of the *cumulative effectiveness or tendency of the force to change the state of motion* of the particle.[†]

Clearly, the dimensions of the linear impulse are *force* × *time*, and its primary units are newton-seconds (N·s) in the SI and pound-seconds (lb·s) in the U.S. customary system. If $\mathbf{F}$ does not have a constant direction, the direction of the linear impulse $\mathbf{Imp}_{1\rightarrow 2}$ is dependent on the final value of the integral in Eq. (14.1). If $\mathbf{F}$ is constant, the direction of $\mathbf{Imp}_{1\rightarrow 2}$ is identical with the direction of $\mathbf{F}$.

The *linear momentum* of a particle is denoted by $\mathbf{L}$ and is defined as the product of its mass m and its velocity $\mathbf{v}$; i.e.,

$$\mathbf{L} = m\mathbf{v} \qquad (14.3)$$

The linear momentum defined above is a vector quantity possessed by the particle and is tangent to the path of the particle as shown in Fig. 14.1. Physically, the *linear momentum* of a particle indicates the *strength as well as the direction of motion* of the particle as observed from the inertial reference frame. Note that the dimensions of linear momentum are *mass* × *velocity*. Using Newton's second law, we write

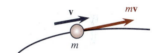

Fig. 14.1

$$\text{Mass} \times \text{Velocity} = \frac{\text{Force} \times (\text{Time})^2}{\text{Length}} \times \frac{\text{Length}}{\text{Time}} = \text{Force} \times \text{Time}$$

This shows that the linear momentum and linear impulse have equivalent dimensions, and they may have the same units. The *linear momentum* defined above is often simply referred to as the *momentum*.

By Newton's second law, the resultant force $\Sigma\mathbf{F}$ acting on a particle of mass m is given by $\Sigma\mathbf{F} = m\mathbf{a}$, where $\mathbf{a}$ is the acceleration of the particle. Noting that $\mathbf{a} = d\mathbf{v}/dt$ and recognizing m as being constant, we may express Newton's second law in terms of the linear momentum of the particle as

$$\Sigma\mathbf{F} = \frac{d}{dt}(m\mathbf{v}) \qquad (14.4)$$

Substituting Eq. (14.3) into Eq. (14.4) and using an overdot to denote the time derivative, we may write *Newton's second law* as

$$\Sigma\mathbf{F} = \dot{\mathbf{L}} \qquad (14.5)$$

[†]Note that *impulse* is exerted *by* a force *on* a body. It is the *force* (not the *body*) which exerts the impulse; a body *receives* impulse from the force acting on it during a time interval whether or not the body has any velocity or displacement during that time interval. Besides, *impulse of a force* and *impulse exerted by a force* are interchangeably used.

Thus, the resultant force acting on a particle is equal to the time rate of change of the linear momentum of the particle. Newton's second law expressed in the form of Eq. (14.4) or (14.5) is *valid* in *newtonian mechanics* for a particle with a constant mass m, but is *required* in *relativistic mechanics* for a particle whose mass m is a function of its speed v.[†] However, Eq. (14.4) or (14.5) should not, in general, be used to solve problems involving the motion of a variable system gaining or losing particles (and hence masses), such as rockets.[††]

We can readily see from Eq. (14.4) that $m\mathbf{v}$ is a constant vector when $\Sigma\mathbf{F} = \mathbf{0}$. In other words, the linear momentum of a particle remains constant in magnitude and direction if the resultant force acting on the particle is zero. This is an alternative statement of *Newton's first law* of motion.

Developmental Exercises

D14.1 Define the terms: (a) linear impulse, (b) linear momentum.

D14.2 The force $\mathbf{F} = 2t\mathbf{i} - 2\mathbf{j} - 3t^2\mathbf{k}$ N acts on a particle during the time interval $0 \le t \le 3$ s, where t is the time measured in seconds. Determine the linear impulse of $\mathbf{F}$ during that interval.

D14.3 Determine the linear momentum of a particle if it has (a) a weight of 20 lb and a velocity of $20\mathbf{i} - 25\mathbf{j} + 30\mathbf{k}$ ft/s near the surface of the earth, (b) a mass of 30 mg and a velocity of $1500\,\mathbf{j} - 800\mathbf{k}$ m/s.

14.2 Principle of Impulse and Momentum

The motion of a particle of mass m and velocity $\mathbf{v}$ under the action of the resultant force $\Sigma\mathbf{F}$ of a system of forces $\mathbf{F}_1, \mathbf{F}_2, \ldots, \mathbf{F}_n$ is defined by Eq. (14.4). Multiplying both sides of this equation by the differential time dt and integrating from a time t_1 to a time t_2, we write

$$\Sigma\mathbf{F}\,dt = d(m\mathbf{v}) \qquad \int_{t_1}^{t_2} \Sigma\mathbf{F}\,dt = m\mathbf{v}_2 - m\mathbf{v}_1$$

where $\mathbf{v}_1$ and $\mathbf{v}_2$ are the velocities of the particle at the times t_1 and t_2, respectively. Thus, we have

$$m\mathbf{v}_1 + \int_{t_1}^{t_2} \Sigma\mathbf{F}\,dt = m\mathbf{v}_2 \qquad (14.6)$$

Using the symbols defined in Eqs. (14.1) and (14.3), we may write Eq. (14.6) as

$$\mathbf{L}_1 + \Sigma\mathbf{Imp}_{1\rightarrow2} = \mathbf{L}_2 \qquad (14.7)$$

[†]In relativistic mechanics, the mass m of a particle moving with a speed v is given by $m = m_0/(1 - v^2/c^2)^{1/2}$, where c is the speed of light and m_0 is the mass of the particle at rest in an inertial reference frame. The mass m_0 is called the *rest mass*, while the mass m is called the *relativistic mass* of the particle, which is a function of v.

[††]Cf. Eqs. (14.43) and (14.44).

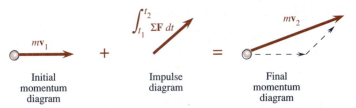

Fig. 14.2

The property expressed in Eq. (14.6) or (14.7) and illustrated in Fig. 14.2 is called the *principle of impulse and momentum* for a particle, which may be stated as follows: *The initial momentum of a particle plus the impulses received by the particle during a time interval is equivalent to the final momentum of the particle.* In other words, the net impulse received by a particle during a time interval is equivalent to the change in momentum of the particle during that time interval.

Unlike work and kinetic energy, impulse and momentum are vector quantities. In obtaining an analytic solution, we may replace Eq. (14.6) with its equivalent component equations

$$(mv_x)_1 + \int_{t_1}^{t_2} \Sigma F_x \, dt = (mv_x)_2$$

$$(mv_y)_1 + \int_{t_1}^{t_2} \Sigma F_y \, dt = (mv_y)_2$$

$$(mv_z)_1 + \int_{t_1}^{t_2} \Sigma F_z \, dt = (mv_z)_2$$

Geometrically, the integrals in the above equations are equal to the shaded areas indicated in the force-time diagrams in Fig. 14.3.

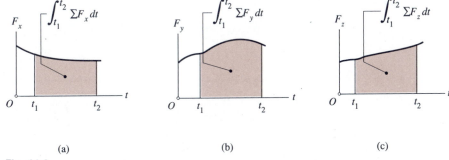

Fig. 14.3

We note that the principle of impulse and momentum directly relates *time* and *velocity* (among others) without initially involving acceleration and displacement. Furthermore, we refer to the procedure which uses the principle of impulse and momentum to analyze and solve a proper type of problem as the *method of impulse and momentum*. In using such a method, it is recommended that appropriate *momentum and impulse diagrams*, such as those shown in Fig. 14.2, *be drawn in formulating the solution*.

36 km/h

8
15

EXAMPLE 14.1

A man on a light-weight racing bicycle is coasting at a speed of 36 km/h down an incline as shown when he applies the brakes. He reduces his speed to 18 km/h in 10 s, and the combined mass of him and the bicycle is known to be 85 kg. Neglecting the rotational effects of the wheels, determine the average magnitude F of the total braking force exerted by the incline on the tires.

Solution. We note that 36 km/h = 10 m/s and 18 km/h = 5 m/s. Since the rotational effects of the wheels are to be neglected, we may treat the man and the bicycle as a *particle*. Applying the *principle of impulse and momentum* to the man and the bicycle during the 10-s interval, we have

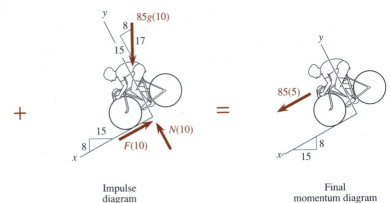

| Initial | Impulse | Final |
| momentum diagram | diagram | momentum diagram |

$$+\swarrow \ \Sigma V_x:^{\dagger} \quad 85(10) + [\tfrac{8}{17}(85g)(10) - F(10)] = 85(5)$$

$$F = 435 \text{ N} \ \blacktriangleleft$$

Developmental Exercise

D14.4   Describe the *principle of impulse and momentum* for a particle.

EXAMPLE 14.2

A force of magnitude $F = 38t$ lb. is applied to a 100-lb crate as shown, where t is measured in seconds. The crate is at rest when $t = 0$, and the direction of **F** is always parallel to the incline. If $\mu_s = 0.5$ and $\mu_k = 0.4$ between the crate and the incline, determine that the time t_2 at which the speed of the crate is 16.1 ft/s.

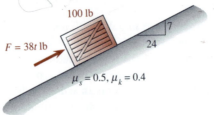

100 lb

$F = 38t$ lb

7
24

$\mu_s = 0.5, \mu_k = 0.4$

---
†Cf. the second footnote on p. 513.

Solution. Assuming that slipping of the crate impends at $t = t_1$ and observing that the free-body diagram as shown is in equilibrium, we write

$$+\nwarrow \Sigma F_y = 0: \quad N - \frac{24}{25}(100) = 0 \quad N = 96 \text{ lb}$$

$$+\nearrow \Sigma F_x = 0: \quad 38t_1 - \frac{7}{25}(100) - 0.5N = 0 \quad t_1 = 2 \text{ s}^\dagger$$

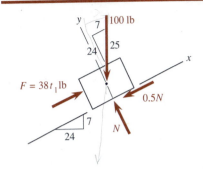

Since the crate has no motion perpendicular to the incline, we have $N = 96$ lb, and $F_k = \mu_k N = 0.4(96)$ lb $= 38.4$ lb. Applying the *principle of impulse and momentum* to the crate during $2 \text{ s} < t \leq t_2$, we have

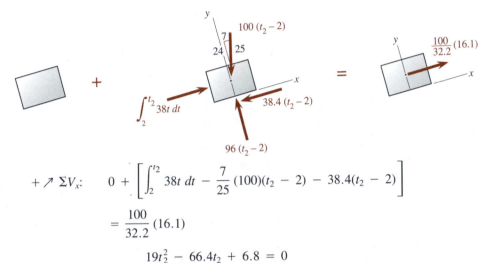

$$+\nearrow \Sigma V_x: \quad 0 + \left[\int_2^{t_2} 38t \, dt - \frac{7}{25}(100)(t_2 - 2) - 38.4(t_2 - 2) \right]$$

$$= \frac{100}{32.2}(16.1)$$

$$19t_2^2 - 66.4t_2 + 6.8 = 0$$

which yields two roots: $t_2 = 3.39$ s and $t_2 = 0.1056$ s. Since $t_2 > 2$ s, we discard the second root and write

$$t_2 = 3.39 \text{ s} \quad \blacktriangleleft$$

Developmental Exercise

D14.5 A package slides down an incline with a speed of 1.5 m/s at the time $t = 0$ as shown. If $\mu_k = 0.35$ between the package and the incline, determine the time t_2 at which the package comes to a stop.

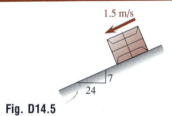

Fig. D14.5

14.3 Angular Impulse and Angular Momentum

To study the motion of a particle about a point, it is useful to know the moments of the forces acting on the particle about that point. The *angular impulse* of a force $\mathbf{F}$ acting on a particle P about a fixed point A, as shown in Fig. 14.4(a), during the time interval $t_1 \leq t \leq t_2$ is denoted by $(\mathbf{Aimp}_A)_{1 \to 2}$ and is defined as the integral of the moment of $\mathbf{F}$ about A times

---

†The crate starts to move at $t = t_1 = 2$ s. During the interval $0 \leq t \leq 2$ s, the resultant force acting on the crate is zero.

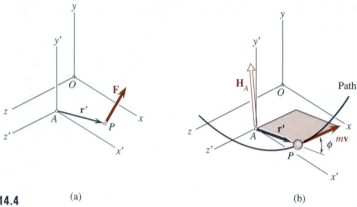

Fig. 14.4   (a)   (b)

the differential time dt evaluated over that interval. We write

$$(\mathbf{Aimp}_A)_{1\to2} = \int_{t_1}^{t_2} \mathbf{r}' \times \mathbf{F}\ dt \qquad (14.8)$$

where $\mathbf{r}'$ is the position vector of any convenient point, say P, on the line of action of $\mathbf{F}$ in the reference frame $Ax'y'z'$ as shown. Denoting the moment of $\mathbf{F}$ about A by $\mathbf{M}_A$, we may write Eq. (14.8) as

$$(\mathbf{Aimp}_A)_{1\to2} = \int_{t_1}^{t_2} \mathbf{M}_A\ dt \qquad (14.9)$$

If the force $\mathbf{F}$, and hence its moment $\mathbf{M}_A$, remains *constant* during the interval $\Delta t = t_2 - t_1$, we write

$$\mathbf{M}_A = \mathbf{r}' \times \mathbf{F} = d_s F\ \boldsymbol{\lambda}$$

$$(\mathbf{Aimp}_A)_{1\to2} = \mathbf{M}_A\ \Delta t \qquad (14.10)$$

where d_s is the shortest distance between A and the line of action of $\mathbf{F}$, and $\boldsymbol{\lambda}$ is a unit vector pointing in the direction perpendicular to the plane containing A and $\mathbf{F}$. An angular impulse is a vector quantity; its dimensions are *force* $\times$ *length* $\times$ *time*, and its primary units are newton-meter-seconds (N·m·s) in the SI and pound-foot-seconds (lb·ft·s) in the U.S. customary system.

The *angular momentum* of a particle about any point A is denoted by $\mathbf{H}_A$ and is defined as the *moment of the linear momentum* $m\mathbf{v}$ of the particle about the point A, as shown in Fig. 14.4(b);[†] i.e.,

$$\mathbf{H}_A = \mathbf{r}' \times m\mathbf{v} \qquad (14.11)$$

where m and $\mathbf{v}$ are the mass and velocity of the particle P, and $\mathbf{r}'$ is the position vector of P in the reference frame $Ax'y'z'$.

It is obvious that the dimensions of angular momentum are *length* $\times$ *mass* $\times$ *velocity*. Using Newton's second law, we write

---

[†]Note that a *hollow arrow* is used, for easy distinction, to represent an *angular momentum*, an *angular impulse*, and a *moment* in this text.

$$\text{Length} \times \text{Mass} \times \text{Velocity}$$

$$= \text{Length} \times \frac{\text{Force} \times (\text{Time})^2}{\text{Length}} \times \frac{\text{Length}}{\text{Time}}$$

$$= \text{Force} \times \text{Length} \times \text{Time}$$

Thus, the angular momentum and angular impulse have equivalent dimensions, and they may have the same units. Because of its definition, the *angular momentum* is also called the *moment of momentum*.

EXAMPLE 14.3

A system consists of three particles A, B, and C. At the instant considered, they are located as shown and have velocities $\mathbf{v}_A = 5\mathbf{i} - 8\mathbf{k}$ m/s, $\mathbf{v}_B = 4\mathbf{i} + v_y\mathbf{j} + v_z\mathbf{k}$ m/s, and $\mathbf{v}_C = 6\mathbf{i} + 3\mathbf{j} - 2\mathbf{k}$ m/s, respectively. If the masses of these particles are $m_A = 3$ kg, $m_B = 1$ kg, and $m_C = 2$ kg, respectively, and the resultant angular momentum $\mathbf{H}_O$ of the system about the origin O is parallel to the z axis, determine the value of $\mathbf{H}_O$.

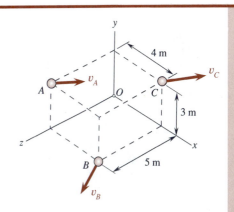

Solution. Since the resultant angular momentum $\mathbf{H}_O$ of the system is equal to the vector sum of the angular momenta of the individual particles in the system, we write

$$\mathbf{H}_O = \mathbf{r}_A \times m_A\mathbf{v}_A + \mathbf{r}_B \times m_B\mathbf{v}_B + \mathbf{r}_C \times m_C\mathbf{v}_C$$

$$= (3\mathbf{j} + 5\mathbf{k}) \times (3)(5\mathbf{i} - 8\mathbf{k})$$
$$+ (4\mathbf{i} + 5\mathbf{k}) \times (1)(4\mathbf{i} + v_y\mathbf{j} + v_z\mathbf{k})$$
$$+ (4\mathbf{i} + 3\mathbf{j}) \times (2)(6\mathbf{i} + 3\mathbf{j} - 2\mathbf{k})$$

$$= -(5v_y + 84)\mathbf{i} + (111 - 4v_z)\mathbf{j} + (4v_y - 57)\mathbf{k}$$

Since $\mathbf{H}_O$ is parallel to the z axis, we set

$$-(5v_y + 84) = 0 \qquad 111 - 4v_z = 0$$

These two equations yield $v_y = -16.8$ and $v_z = 27.75$. Thus, we have

$$\mathbf{H}_O = (4v_y - 57)\mathbf{k} = [4(-16.8) - 57]\mathbf{k}$$

$$\mathbf{H}_O = -124.2\mathbf{k} \text{ kg·m}^2/\text{s} \blacktriangleleft$$

Developmental Exercises

D14.6 Define the angular impulse about a point as exerted by (a) a force, (b) a moment.

D14.7 Define the angular momentum of a particle about a point.

D14.8 Refer to Example 14.3. Determine the value of $\mathbf{H}_O$ if it is known to be parallel to the x axis at the instant considered.

14.4   System of Particles

The principle of impulse and momentum for a particle, as expressed in Eq. (14.6) and illustrated in Fig. 14.2, may be applied to each individual particle in a system of n particles as illustrated in Fig. 14.5. Note in this figure that $\mathbf{F}_i$ is the resultant force acting on the particle P_i at the time t, and $(m_i \mathbf{v}_i)_1$ and $(m_i \mathbf{v}_i)_2$ are the momenta of P_i at the times t_1 and t_2, respectively.

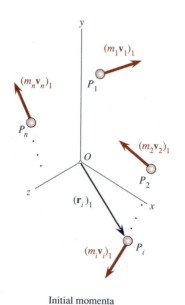

Initial momenta

Fig. 14.5   (a)

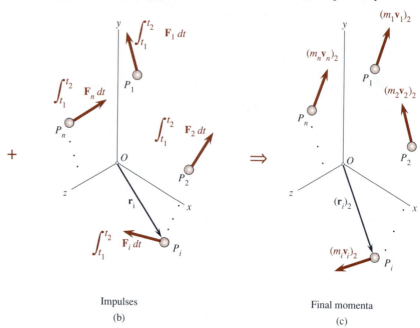

Impulses

(b)

Final momenta

(c)

By superposition of the principle of impulse and momentum as applied to the individual particles in a system, we see that

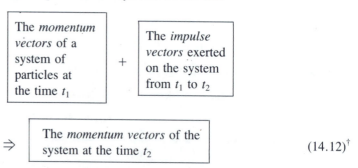

The *momentum vectors* of a system of particles at the time t_1	$+$	The *impulse vectors* exerted on the system from t_1 to t_2

$\Rightarrow$ | The *momentum vectors* of the system at the time t_2 | $(14.12)^\dagger$

Since internal forces in a system of particles occur in collinear pairs of forces equal in magnitude and opposite in sense, their sum and their resultant moment about any point are zero at all times. This means that only the impulses of the external forces need to be considered in applying Eq. (14.12). Thus, Eq. (14.12) may be simplified and written as

Syst Momenta₁ + Syst Ext Imp₁₋₂ ⇒ Syst Momenta₂   (14.13)

---

†Recall that the double arrow sign $\Rightarrow$ indicates *equipollence*.

If no external force acts on the system, Eq. (14.13) reduces to

$$\textbf{Syst Momenta}_1 \quad \Rightarrow \quad \textbf{Syst Momenta}_2 \qquad (14.14)$$

which expresses the *conservation of momentum*.

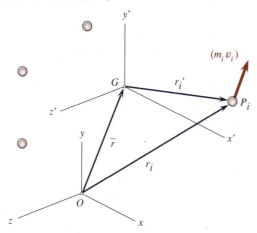

Fig. 14.6

Referring to Fig. 14.6, we define the linear momentum $\textbf{L}$ and the angular momentum $\textbf{H}_G$ about the mass center G of the system of particles in the fixed reference frame $Oxyz$ as

$$\textbf{L} = \sum_{i=1}^{n} m_i \textbf{v}_i \qquad \textbf{H}_G = \sum_{i=1}^{n} (\textbf{r}_i' \times m_i \textbf{v}_i) \qquad (14.15)$$

We note that the position vectors $\textbf{r}_i$ and $\bar{\textbf{r}}$ and the velocities $\textbf{v}_i$ and $\bar{\textbf{v}}$ of the particle P_i and the mass center G are related to each other by

$$\textbf{r}_i = \bar{\textbf{r}} + \textbf{r}_i' \qquad \textbf{v}_i = \bar{\textbf{v}} + \textbf{v}_i' \qquad (14.16)$$

Applying Eqs. (12.22) and (14.15), we find that the sum of the external forces acting on the system of particles is

$$\Sigma \textbf{F} = \sum_{i=1}^{n} m_i \textbf{a}_i = \sum_{i=1}^{n} \frac{d}{dt} (m_i \textbf{v}_i) = \frac{d}{dt} \sum_{i=1}^{n} m_i \textbf{v}_i = \frac{d\textbf{L}}{dt}$$

$$\boxed{\Sigma \textbf{F} = \dot{\textbf{L}}} \qquad (14.17)$$

Applying Eqs. (14.15), (14.16), and (12.23), as well as the *principle of moments*, we see that the time derivative of the angular momentum $\textbf{H}_G$ about the mass center G of a system of particles of total mass m taken in $Oxyz$ is

$$\dot{\textbf{H}}_G = \frac{d}{dt} \sum_{i=1}^{n} (\textbf{r}_i' \times m_i \textbf{v}_i) = \sum_{i=1}^{n} (\textbf{v}_i' \times m_i \textbf{v}_i + \textbf{r}_i' \times m_i \textbf{a}_i)$$

$$= \sum_{i=1}^{n} [(\textbf{v}_i - \bar{\textbf{v}}) \times m_i \textbf{v}_i] + \sum_{i=1}^{n} (\textbf{r}_i' \times m_i \textbf{a}_i)$$

$$= \sum_{i=1}^{n} (\textbf{v}_i \times m_i \textbf{v}_i) - \bar{\textbf{v}} \times \sum_{i=1}^{n} m_i \textbf{v}_i + \sum_{i=1}^{n} (\textbf{r}_i' \times m_i \textbf{a}_i)$$

$$= \textbf{0} - \bar{\textbf{v}} \times m\bar{\textbf{v}} + \Sigma \textbf{M}_G = 0 - 0 + \Sigma \textbf{M}_G$$

Therefore, the sum of moments about G of the external forces is

$$\Sigma \mathbf{M}_G = \dot{\mathbf{H}}_G \tag{14.18}$$

Defining the angular momentum $\mathbf{H}_O$ about the fixed origin O for the system in Fig. 14.6 as

$$\mathbf{H}_O = \sum_{i=1}^{n} (\mathbf{r}_i \times m_i \mathbf{v}_i) \tag{14.19}$$

and applying Eq. (12.24), we find that the time derivative of $\mathbf{H}_O$ in $Oxyz$ is

$$\dot{\mathbf{H}}_O = \frac{d}{dt} \sum_{i=1}^{n} (\mathbf{r}_i \times m_i \mathbf{v}_i) = \sum_{i=1}^{n} (\mathbf{v}_i \times m_i \mathbf{v}_i + \mathbf{r}_i \times m_i \mathbf{a}_i)$$

$$= \sum_{i=1}^{n} (\mathbf{0} + \mathbf{r}_i \times m_i \mathbf{a}_i) = \sum_{i=1}^{n} (\mathbf{r}_i \times m_i \mathbf{a}_i) = \Sigma \mathbf{M}_O$$

Thus, the sum of moments about a fixed point O of the external forces is

$$\Sigma \mathbf{M}_O = \dot{\mathbf{H}}_O \tag{14.20}$$

Equations (14.17), (14.18), and (14.20) are fundamental equations of motion expressed in terms of the time derivatives of momenta of a system.

EXAMPLE 14.4

An 80-Mg locomotive A coasting at 0.65 m/s as shown strikes, and is coupled with, a 50-Mg box car B which is at rest with its brakes released. If the coupling is completed in 0.5 s, determine (a) the speed v_2 of the locomotive and the box car after the coupling, (b) the magnitude F of the average force exerted on the box car during the coupling.

Solution. We note that 80 Mg = 80×10^3 kg and 50 Mg = 50×10^3 kg. For simplicity, let us denote the weight forces of the locomotive and the box car by $\mathbf{W}_A$ and $\mathbf{W}_B$, and the resultant reactions from the tracks to the locomotive and the box car by $\mathbf{R}_A$ and $\mathbf{R}_B$, respectively. Applying the *principle of impulse and momentum* to the locomotive and the box car as a system during the 0.5-s interval of coupling, we have

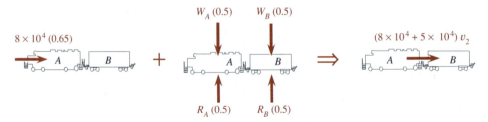

Notice that the coupling forces between A and B are *internal forces* and have no net contribution to the impulse received by A and B as a system. We write

$$\overset{+}{\rightarrow} \Sigma V_x: \quad (80 \times 10^3)(0.65) + 0 = (80 \times 10^3 + 50 \times 10^3)v_2$$

$$v_2 = 0.4 \text{ m/s} \blacktriangleleft$$

By applying the same principle to the box car alone during the 0.5-s interval of coupling, we have

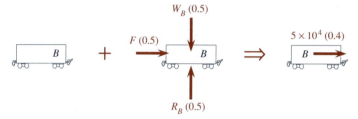

Notice that the coupling force $\mathbf{F}$ is an *external force* to the box car alone and contributes impulse to the box car. We write

$$\overset{+}{\rightarrow} \Sigma V_x: \quad 0 + F(0.5) = (50 \times 10^3)(0.4)$$

$$F = 40 \times 10^3 \text{ N} \qquad F = 40 \text{ kN} \blacktriangleleft$$

Developmental Exercise

D14.9 Refer to Example 14.4. Using the result that $v_2 = 0.4$ m/s, determine the value of F by considering the locomotive alone.

EXAMPLE 14.5

The blocks A and B as shown are released to move from rest at the time $t = 0$. If $\mu_k = 0.3$ between the block A and the support, determine the speed v_A of the block A when $t = 4$ s.

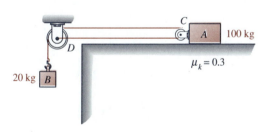

Solution. From the constraints of the system, we readily see that the velocities of the blocks A and B are $\mathbf{v}_A = v_A \leftarrow$ and $\mathbf{v}_B = 2v_A \downarrow$, respectively. Since the block A has no motion perpendicular to the support, the normal reaction to the block A is $\mathbf{N}_A = 100g$ N $\uparrow$, and the kinetic friction force acting on the block A is $\mathbf{F}_k = 0.3(100g)$ N $\rightarrow$. Denoting the tension in the cord BCD by F and applying the *principle of impulse and momentum*

to the blocks A and B, separately, during the interval $0 \leq t \leq 4$ s, we have

Block A

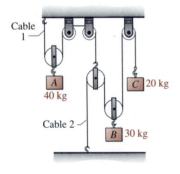

$$\overset{+}{\underset{\leftarrow}{}} \Sigma V_x: \qquad 0 + [F(4) + F(4) - 0.3(100g)(4)] = 100v_A \qquad (1)$$

Block B

$$+\downarrow \Sigma V_y: \qquad 0 + [20g(4) - F(4)] = 20(2v_A) \qquad (2)$$

Solving Eqs. (1) and (2), we obtain

$$F = 174.4 \text{ N} \qquad\qquad v_A = 2.18 \text{ m/s} \quad \blacktriangleleft$$

Developmental Exercise

D14.10 Work Example 14.5 if the velocity of the block B is 1 m/s $\downarrow$ at $t = 0$.

EXAMPLE 14.6

The cylinders A, B, and C as shown are released from rest at the time $t = 0$. Determine the tension in each cable and the velocity of each cylinder at the time $t = 3$ s.

Solution. As derived in the solution in Example 11.8, we have the following constraint equation on the velocities:

$$4v_A + 2v_B + v_C = 0 \qquad (1)$$

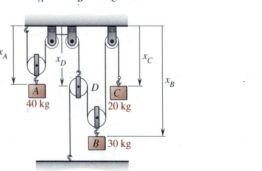

Notice that the positive directions of the unknown velocities are downward and all pulleys are assumed to have negligible mass.

Let the tensions in cables 1 and 2 be F_1 and F_2, respectively. Since the pulley D is taken to have negligible mass, it has a negligible effective force as shown. We write

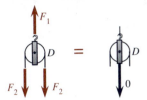

$$+\uparrow \Sigma V_y: \qquad F_1 - F_2 - F_2 = 0 \qquad (2)$$

Applying the *principle of impulse and momentum* to each individual cylinder during the interval $0 \leq t \leq 3$ s, we have

Cylinder A

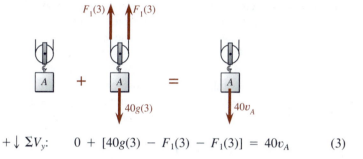

$$+\downarrow \Sigma V_y: \qquad 0 + [40g(3) - F_1(3) - F_1(3)] = 40v_A \qquad (3)$$

Cylinder B

$$+\downarrow \Sigma V_y: \qquad 0 + [30g(3) - F_2(3) - F_2(3)] = 30v_B \qquad (4)$$

Cylinder C

$$+\downarrow \Sigma V_y: \qquad 0 + [20g(3) - F_2(3)] = 20v_C \qquad (5)$$

Solution of the preceding five equations in five unknowns yields

$$F_1 = 235.44 \qquad F_2 = 117.72 \qquad v_A = -5.886$$

$$v_B = 5.886 \qquad v_C = 11.772$$

$$\mathbf{F_1 = 235\ N} \qquad \mathbf{F_2 = 117.7\ N} \qquad \mathbf{v_A = 5.89\ m/s \uparrow} \blacktriangleleft$$

$$\mathbf{v_B = 5.89\ m/s \downarrow} \qquad \mathbf{v_C = 11.77\ m/s \downarrow} \blacktriangleleft$$

Developmental Exercise

D14.11 Work Example 14.6 if the initial velocities of the cylinders A and B are $v_A = 1$ m/s $\downarrow$ and $v_B = 2$ m/s $\downarrow$, respectively, when the cylinders are released at the time $t = 0$.

EXAMPLE 14.7[†]

A projectile of mass $m = 100$ kg is traveling in space. At the time $t = 0$, it is observed to pass through the origin O of an inertial reference frame with a constant velocity $\mathbf{v}_0 = 756\mathbf{i}$ m/s. Following an explosion at $t = t_1$, where $0 < t_1 < 1.3$ s, the projectile separates into three parts, A, B, and C, of masses $m_A = 40$ kg, $m_B = 24$ kg, and $m_C = 36$ kg, respectively. At $t = 1.3$ s, it is observed that the position vectors of B and C are $\mathbf{r}_B = 1218\mathbf{i} - 420\mathbf{j} - 840\mathbf{k}$ m and $\mathbf{r}_C = 698\mathbf{i} - 320\mathbf{j} + 560\mathbf{k}$ m, and that the y component of the velocity $\mathbf{v}_A$ of A is $(v_A)_y = 675\mathbf{j}$ m/s. Neglecting the effect of gravity, determine the position vector $\mathbf{r}_A$ of A and the velocities $\mathbf{v}_A$, $\mathbf{v}_B$, and $\mathbf{v}_C$ of A, B, and C, respectively, at $t = 1.3$ s.

Solution. We observe that there is no external force acting on the system.

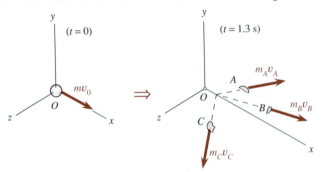

By the *principle of motion of the mass center* of a system of particles as expressed in Eq. (12.27), we know that the mass center G of the system must move with the constant velocity $\mathbf{v}_0 = 756\mathbf{i}$ m/s along the x axis. At $t = 1.3$ s, the position vector of G must, therefore, be $\bar{\mathbf{r}} = 982.8\mathbf{i}$ m. Applying the *principle of moments*, we write

$$m\bar{\mathbf{r}} = m_A\mathbf{r}_A + m_B\mathbf{r}_B + m_C\mathbf{r}_C$$

$$100(982.8\mathbf{i}) = 40\mathbf{r}_A + 24(1218\mathbf{i} - 420\mathbf{j} - 840\mathbf{k})$$

$$+ 36(698\mathbf{i} - 320\mathbf{j} + 560\mathbf{k})$$

$$\mathbf{r}_A = 1098\mathbf{i} + 540\mathbf{j} \text{ m} \blacktriangleleft$$

Since there is no external force, *both the linear momentum and the angular momentum of the system are conserved*; i.e., the initial momentum $m\mathbf{v}_0$ is equipollent to the system of final momenta as shown. Equating the

---

[†]The complete solution of this example is somewhat lengthy. *Skip* this example and Probs. 14.39 and 14.40* if circumstances warrant. However, the technical intricacy here is pedagogically valuable. Note that the solution involves the direct applications of (a) the principle of motion of the mass center of a system of particles, (b) the principle of moments, (c) the conservation of linear momentum, (d) the conservation of angular momentum, and (e) Newton's first law.

sums of the linear momentum vectors, we write

$$\mathbf{L}_1 = \mathbf{L}_2: \qquad m\mathbf{v}_0 = m_A\mathbf{v}_A + m_B\mathbf{v}_B + m_C\mathbf{v}_C$$

$$100(756\mathbf{i}) = 40[(v_A)_x\mathbf{i} + 675\mathbf{j} + (v_A)_z\mathbf{k}]$$

$$+ 24[(v_B)_x\mathbf{i} + (v_B)_y\mathbf{j} + (v_B)_z\mathbf{k}]$$

$$+ 36[(v_C)_x\mathbf{i} + (v_C)_y\mathbf{j} + (v_C)_z\mathbf{k}]$$

which yields, after simplification, three scalar equations:

$$10(v_A)_x + 6(v_B)_x + 9(v_C)_x = 18900 \tag{1}$$

$$2(v_B)_y + 3(v_C)_y = -2250 \tag{2}$$

$$10(v_A)_z + 6(v_B)_z + 9(v_C)_z = 0 \tag{3}$$

Equating the sums of the angular momentum vectors about O, we write

$$(\mathbf{H}_O)_1 = (\mathbf{H}_O)_2:$$

$$\bar{\mathbf{r}} \times m\mathbf{v}_0 = \mathbf{r}_A \times m_A\mathbf{v}_A + \mathbf{r}_B \times m_B\mathbf{v}_B + \mathbf{r}_C \times m_C\mathbf{v}_C$$

$$\mathbf{0} = 40\begin{vmatrix} \mathbf{i} & \mathbf{j} & \mathbf{k} \\ 1098 & 540 & 0 \\ (v_A)_x & 675 & (v_A)_z \end{vmatrix} + 24\begin{vmatrix} \mathbf{i} & \mathbf{j} & \mathbf{k} \\ 1218 & -420 & -840 \\ (v_B)_x & (v_B)_y & (v_B)_z \end{vmatrix}$$

$$+ 36\begin{vmatrix} \mathbf{i} & \mathbf{j} & \mathbf{k} \\ 698 & -320 & 560 \\ (v_C)_x & (v_C)_y & (v_C)_z \end{vmatrix}$$

which, after expansion and simplification, yields

$$15(v_A)_z + 14(v_B)_y - 7(v_B)_z - 14(v_C)_y - 8(v_C)_z = 0 \tag{4}$$

$$610(v_A)_z + 280(v_B)_x + 406(v_B)_z - 280(v_C)_x$$
$$+ 349(v_C)_z = 0 \tag{5}$$

$$300(v_A)_x - 140(v_B)_x - 406(v_B)_y - 160(v_C)_x$$
$$- 349(v_C)_y = 411750 \tag{6}$$

The above scalar equations (1) through (6) contain eight unknowns. For a complete solution, we need at least two additional equations.

Since no force acts on the parts A, B, and C, immediately after the explosion at $t = t_1$, they must, by *Newton's first law*, move away with *constant* velocities from the point of explosion on the x axis. At $t = 1.3$ s, the y and z components of the position vectors of A, B, and C must, therefore, be equal to the y and z components of the velocities of A, B, and C multiplied by the elapsed time $(1.3 - t_1)$ s since the explosion, respectively. From the given data and the solution for $\mathbf{r}_A$, we have $y_A = 540$ m, $z_A = 0$, $y_B = -420$ m, $z_B = -840$ m, $y_C = -320$ m, and $z_C = 560$ m. Since $(v_A)_y = 675$ m/s is given and the elapsed time since the explosion is common to all parts, we write

$$1.3 - t_1 = \frac{540}{675} = \frac{0}{(v_A)_z} = \frac{-420}{(v_B)_y} = \frac{-840}{(v_B)_z}$$

$$= \frac{-320}{(v_C)_y} = \frac{560}{(v_C)_z} \tag{7}$$

Solving Eqs. (7), we obtain

$$(v_A)_z = 0 \qquad (v_B)_y = -525 \qquad (v_B)_z = -1050$$

$$(v_C)_y = -400 \qquad (v_C)_z = 700$$

which satisfy Eqs. (2), (3), and (4) and render Eqs. (5) and (6) to

$$(v_B)_x - (v_C)_x = 650 \tag{5'}$$

$$15(v_A)_x - 7(v_B)_x - 8(v_C)_x = 2950 \tag{6'}$$

Solving Eqs. (1), (5'), and (6') simultaneously, we obtain

$$(v_A)_x = 900 \qquad (v_B)_x = 1050 \qquad (v_C)_x = 400$$

Thus, the velocities of parts A, B, and C at $t = 1.3$ s are

$$\mathbf{v}_A = 900\mathbf{i} + 675\mathbf{j} \text{ m/s} \qquad \mathbf{v}_B = 1050\mathbf{i} - 525\mathbf{j} - 1050\mathbf{k} \text{ m/s} \blacktriangleleft$$

$$\mathbf{v}_C = 400\mathbf{i} - 400\mathbf{j} + 700\mathbf{k} \text{ m/s} \blacktriangleleft$$

Developmental Exercise

D14.12 Refer to Example 14.7. Based on the solution, determine the time t_1 and the abscissa x_1 at which the explosion occurred.

14.5 Impulsive Motion

An *impulsive force* is a relatively large force which acts suddenly on a body for a relatively short time interval and causes a significant change in the momentum of the body. Thus, forces due to collision are impulsive forces. The motion of a body caused by impulsive forces is called an *impulsive motion*. *Nonimpulsive forces* include weight forces and those which do not cause a significant change in the momentum of the body during a relatively short time interval.

For instance, when a tennis ball of mass m coming with a velocity of $\mathbf{v}_1$ is hit, the contact between the ball and the racket lasts for only a very short time interval Δt. However, the average value of the contact force $\mathbf{F}$ exerted by the racket on the ball is relatively large, and the impulse $\mathbf{F}\,\Delta t$ is large enough to cause a change in momentum from $m\mathbf{v}_1$ to $m\mathbf{v}_2$ as illustrated in Fig. 14.7, where $W = mg =$ weight of the ball. Here, the contact force $\mathbf{F}$ is an impulsive force, while the weight force $\mathbf{W}$ is a *nonimpulsive force*. Note that *a nonimpulsive force in juxtaposition with an impulsive force is negligible*. Thus, we set $W\,\Delta t \approx 0$ in Fig. 14.7.

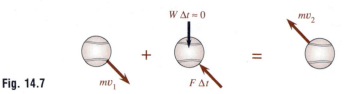

Fig. 14.7

If $\Sigma\mathbf{F}$ is the vector sum of the average impulsive forces acting on a particle of mass m during the time interval Δt, Eq. (14.6) becomes

$$m\mathbf{v}_1 + \Sigma\mathbf{F}\,\Delta t = m\mathbf{v}_2 \qquad (14.21)$$

In general, an *impulsive force* is a force having a sudden surge in magnitude during a relatively short time interval Δt, whereas a *nonimpulsive force* is small compared with an impulsive force and remains essentially constant during the time interval Δt.

EXAMPLE 14.8

A 2.59-g rifle bullet B is fired with a velocity of 300 m/s into a 500-g sphere A which is suspended by an inextensible wire and is at rest as shown. Knowing that the bullet becomes embedded in the sphere in 1 ms, determine (a) the velocity $\mathbf{v}_2$ of the sphere immediately after the bullet becomes embedded, (b) the average impulsive tension developed in the wire.

Solution. Since the wire is taken to be inextensible, the velocity $\mathbf{v}_2$ of the sphere must be perpendicular to the wire and point to the left. We note that the mass of the bullet is $m_B = 2.59 \text{ g} = 2.59 \times 10^{-3}$ kg, the mass of the sphere is $m_A = 500 \text{ g} = 0.5$ kg, and the time interval is $\Delta t = 1$ ms $= 10^{-3}$ s. Letting the average impulsive tension in the wire be F and applying the *principle of linear impulse and momentum* to the bullet and the sphere as a system during the time interval $0 \le t \le 10^{-3}$ s, we have

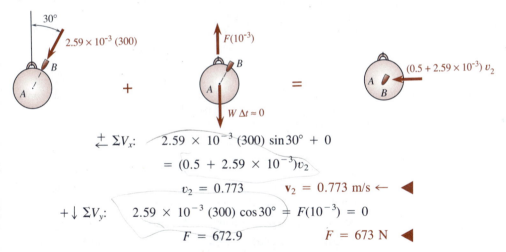

$$\xleftarrow{+}\ \Sigma V_x: \quad 2.59 \times 10^{-3}\,(300)\sin 30° + 0$$
$$= (0.5 + 2.59 \times 10^{-3})v_2$$
$$v_2 = 0.773 \qquad \mathbf{v}_2 = 0.773 \text{ m/s} \leftarrow \ \blacktriangleleft$$
$$+\!\downarrow \Sigma V_y: \quad 2.59 \times 10^{-3}\,(300)\cos 30° - F(10^{-3}) = 0$$
$$F = 672.9 \qquad F = 673 \text{ N} \ \blacktriangleleft$$

Developmental Exercises

D14.13 Define the terms: (a) impulsive force, (b) impulsive motion.

D14.14 A 5-oz baseball is traveling with a velocity as shown in (a) just before it is hit by a bat. After having been hit, it travels with a velocity as shown in (b). If the impact lasts for 20 ms, determine the average impulsive force exerted on the baseball.

Fig. D14.14

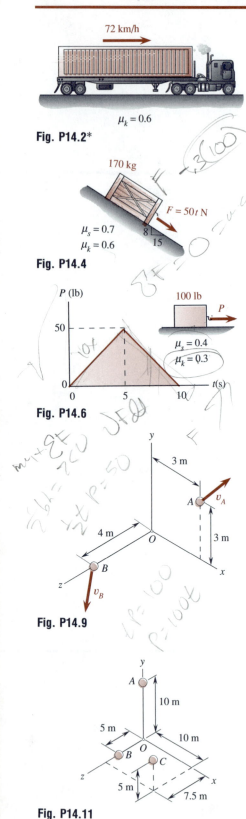

72 km/h

$\mu_k = 0.6$

Fig. P14.2*

170 kg

$F = 50t$ N

$\mu_s = 0.7$
$\mu_k = 0.6$

8
15

Fig. P14.4

P (lb)

100 lb

50

P

$\mu_s = 0.4$
$\mu_k = 0.3$

0
0   5   10 t(s)

Fig. P14.6

y

3 m

A v_A

4 m O 3 m

B x

z v_B

Fig. P14.9

y

A

10 m

5 m

O 10 m

B C

z

5 m x

7.5 m

Fig. P14.11

PROBLEMS

14.1   A 2400-lb car increases its speed uniformly on a straight road from 0 to 45 mi/h in 8 s. Neglecting the rotational effects of the wheels, determine the resultant friction force acting on the tires of the car.

14.2*   A truck is traveling at 72 km/h as shown when the brakes are fully applied, causing all wheels to skid. Determine the time required for the truck to stop if $\mu_k = 0.6$ between the tires and the road surface.

14.3   Refer to Problem. 14.2*. Assume that the brakes are fully applied for only 2 s and then released. Determine the speed v_2 of the truck when the brakes are released.

14.4   A force of magnitude $F = 50t$ N is applied to a 170-kg crate as shown, where t is measured in seconds. The crate is at rest when $t = 0$, and the direction of **F** is always parallel to the incline. It is known that $\mu_s = 0.7$ and $\mu_k = 0.6$ between the crate and the incline. Determine the time t_2 at which the speed of the crate is 4 m/s.

14.5*   Refer to Prob. 14.4. Determine the speed v_2 of the crate when $t = 12$ s.

14.6   The magnitude of a force **P** varies as shown. The force **P** is applied to a 100-lb block at the time $t = 0$. It is known that the block is at rest when $t = 0$, and that $\mu_s = 0.4$ and $\mu_k = 0.3$ between the block and the support. Determine (a) the time t_1 at which the block starts to move, (b) the time t_2 at which the block attains its maximum speed, (c) the maximum speed attained by the block, (d) the time t_3 at which the block stops again.

14.7   The linear momentum of a 2-kg particle is given by $\mathbf{L} = (28 - 3t^2)\mathbf{i} + (16 + 6t)\mathbf{j} - (8 + 4t)\mathbf{k}$ N·s, where t is measured in seconds. When $t = 2$ s, determine (a) the resultant force **F** acting on the particle, (b) the magnitude of the acceleration of the particle, (c) the speed of the particle.

14.8*   The resultant force acting on a 16.1-lb particle is given by $\mathbf{F} = (8 + 2t)\mathbf{i} + (4 + t^2)\mathbf{j} + (10 - 2t)\mathbf{k}$ lb, where t is measured in seconds. If the initial velocity of the particle is $\mathbf{v}_1 = 24\mathbf{i} - 6\mathbf{j} + 48\mathbf{k}$ ft/s at $t = 0$, determine (a) the time t_2 at which the velocity of the particle is parallel to the xy plane, (b) the velocity $\mathbf{v}_2$ of the particle at $t = t_2$.

14.9   A system consists of a 2-kg particle A and a 3-kg particle B. At the instant shown, the velocities of A and B are $\mathbf{v}_A = 2\mathbf{i} + v_y\mathbf{j} + v_z\mathbf{k}$ m/s and $\mathbf{v}_B = 3\mathbf{i} - 4\mathbf{j} + 5\mathbf{k}$ m/s. For this instant, determine (a) the numerical values of v_y and v_z for which the resultant angular momentum $\mathbf{H}_O$ of the system about O is parallel to the y axis, (b) the corresponding value of $\mathbf{H}_O$.

14.10*   Solve Prob. 14.9 if $\mathbf{H}_O$ is parallel to the x axis.

14.11   A system consists of three particles A, B, and C, which have masses $m_A = 5$ kg, $m_B = 3$ kg, $m_C = 2$ kg, and velocities $\mathbf{v}_A = 3\mathbf{i} + 5\mathbf{j} - 10\mathbf{k}$ m/s, $\mathbf{v}_B = 4\mathbf{i} - 7\mathbf{j} + 8\mathbf{k}$ m/s, $\mathbf{v}_C = 6\mathbf{i} + 2\mathbf{j} - 7\mathbf{k}$ m/s, respectively, at the instant shown. For this instant, determine the resultant angular momentum $\mathbf{H}_G$ of the system about its mass center G.

14.12   A 10-kg object is traveling with a velocity of 40 m/s ↓ when it explodes into a 6-kg fragment A and a 4-kg fragment B as shown. Knowing that immediately after the explosion the fragments travel with velocities as indicated, determine the angles θ_A and θ_B.

14.13   A boy tosses a 2-kg sandbag with a velocity of 5 m/s ↘ onto an 18-kg wagon, which is at rest, as shown. Knowing that the wagon can move freely, determine the speed of the wagon after the sandbag has slid to a relative stop on the wagon.

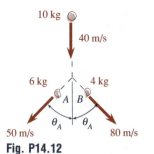

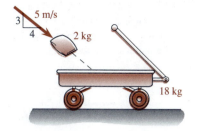

Fig. P14.12

Fig. P14.13

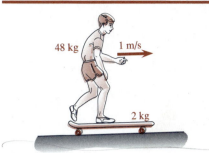

Fig. P14.14*

14.14* A 48-kg boy steps with a horizontal velocity of 1 m/s onto his 2-kg skateboard which is at rest as shown. Determine the speed of the skateboard when the boy rides steadfastly on it.

14.15 A 5-lb skateboard and a 95-lb boy, who stands on it, are initially at rest. At the instant shown, the boy jumps off the skateboard with a velocity $v_{B/S}$ of 10 ft/s relative to the skateboard. If the jump is completed in 0.5 s, determine (a) the speed v_S of the skateboard just after the jump, (b) the magnitude of the average horizontal force exerted on the skateboard by the boy, (c) the magnitude of the average total force exerted on the ground by the wheels of the skateboard during the jump.

14.16 Two railroad cars are traveling as shown. Knowing that coupling occurs and is completed in 0.4 s, determine (a) the speed of the coupled cars, (b) the magnitude of the average force exerted by the car A on the car B during the coupling, (c) the percentage of the total loss of kinetic energy of the cars during the coupling.

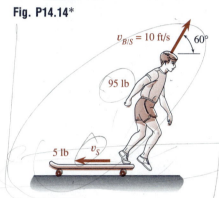

Fig. P14.15

$mv_1 + \int F \, dt = mv_2$

$151.1 \, lb$

Fig. P14.16

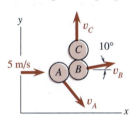

Fig. P14.17

14.17 The pucks B and C are at rest side by side on a horizontal plane when the puck A strikes the puck B with a velocity of 5 m/s $\rightarrow$ as shown. After the impact, the pucks are observed to move with velocities as indicated, where $v_A = 1.5\mathbf{i} - 2\mathbf{j}$ m/s. If the pucks have equal masses, determine the magnitudes of v_B and v_C.

14.18* A 30-g bullet is fired with a speed v_0 into a 10-kg block which is at rest as shown. If $\theta = 30°$ when the block with the embedded bullet swings to its highest position, determine (a) the speed v_0, (b) the percentage of loss of the kinetic energy during the impact.

14.19 A 1-oz bullet is fired with a speed v_0 into a 15-lb block which is at rest as shown. The block with the embedded bullet is observed to move 2 ft to the right before coming to a stop. If $\mu_k = 0.3$ between the block and the support, determine the speed v_0 of the bullet.

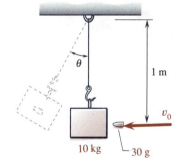

Fig. P14.18*

$95\cos 60\,(10) = \dfrac{5}{32.2}^2 \, x$

$= x \, 32.2$

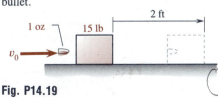

Fig. P14.19

14.20 A 10-g bullet is fired with a speed $v_1 = 260$ m/s at a 2-kg block as shown. The bullet pierces through the block and causes the block to slide a distance of 0.2 m. If $\mu_k = 0.3$ between the block and the support, determine the speed v_2 with which the bullet leaves the block.

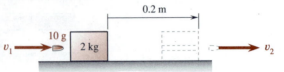

Fig. P14.20

14.21* A 2-kg particle, which is moving on a smooth horizontal surface with a constant velocity of 8 m/s ←, is struck by an impulsive force **F** that acts to the right on the particle for 10 ms. If **F** varies with the time t as shown, determine the resulting velocity of the particle.[†]

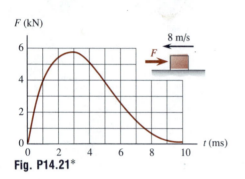

Fig. P14.21*

14.22* The block A of mass m_A and the block B of mass m_B are held to compress a spring as shown. The spring is not attached to the blocks, and effects of friction between the support and the blocks are negligible. When the blocks are suddenly released, show that $v_A/v_B = T_A/T_B = m_B/m_A$ where v_A and v_B are the speeds after separation, and T_A and T_B are the kinetic energies of the blocks A and B, respectively.

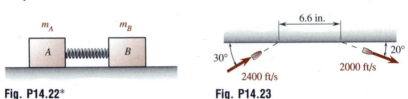

Fig. P14.22*   **Fig. P14.23**

14.23 A 1-oz bullet strikes a hard surface with a velocity of 2400 ft/s ∡ 30° and incurs a 6.6-in. scratch on the surface before ricocheting with a velocity of 2000 ft/s ⦨ 20° as shown. Assuming an average speed of 2200 ft/s during contact, determine the average impulsive force exerted by the hard surface on the bullet.

14.24 A 60-g tennis ball is traveling with a velocity of 10 m/s ∡ 30° just before it is hit by a racket. After having been hit, it travels with a velocity of 20 m/s ⦨ 30° as shown. If the contact between the ball and the racket lasts for 50 ms, determine the average impulsive force exerted by the racket on the ball.

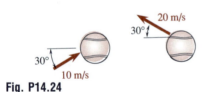

Fig. P14.24

14.25 and 14.26* The system shown is released from rest at the time $t = 0$. Neglecting friction, determine the speed of the block A when $t = 3$ s.

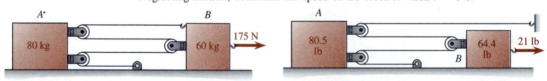

Fig. P14.25 and P14.27*   **Fig. P14.26* and P14.28**

14.27* A system of blocks is shown. At the time $t = 0$, the velocity of the block B is 2 m/s →. If $\mu_k = 0.1$ between the support and the blocks, determine the speed of the block A when $t = 3$ s.

14.28 A system of blocks is shown. At the time $t = 0$, the velocity of the block B is 5 ft/s →. If $\mu_k = 0.1$ between the support and the blocks, determine the speed of the block B when $t = 3$ s.

[†]Cf. the *hint* in Prob. 13.23*.

14.29 through 14.31 A system of cylinders is shown. The initial velocity of the cylinder B is 2 ft/s ↓ when the system is released at the time $t = 0$. If the cylinder A has a weight of 64.4 lb and a velocity of 3 ft/s ↓ when $t = 4$ s, determine the weight of the cylinder B.

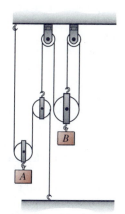

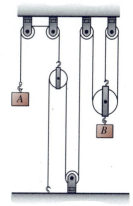

Fig. P14.29 and P14.32   **Fig. P14.30* and P14.33**   **Fig. P14.31 and P14.34***

14.32 through 14.34* A system of cylinders is shown. The initial velocity of the cylinder A is 3 m/s ↑ when the system is released at the time $t = 0$. If the cylinder B has a mass of 40 kg and a velocity of 2 m/s ↑ when $t = 4$ s, determine the mass of the cylinder A.

14.35 and 14.36 The system shown is released from rest at the time $t = 0$. Determine the tension in each cable and the velocity of each cylinder when $t = 2$ s.[†]

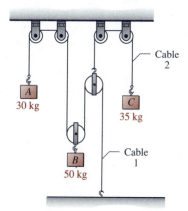

Fig. P14.35

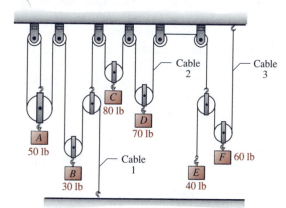

Fig. P14.36

14.37 A 100-lb object is falling with a velocity of 100 ft/s ↓ when it explodes into three fragments, A, B, and C, of weights $W_A = 38$ lb, $W_B = 27$ lb, and $W_C = 35$ lb, respectively. Immediately after the explosion, the velocities of the fragments are directed as shown. Determine the speed of each fragment immediately after the explosion.

14.38* A 50-kg shell is traveling with a velocity $\mathbf{v} = 60\mathbf{i} + 40\mathbf{j} + 120\mathbf{k}$ m/s when it explodes into three fragments, A, B, and C, of masses $m_A = 25$ kg, $m_B = 15$ kg, and $m_C = 10$ kg, respectively. Immediately after the explosion, the

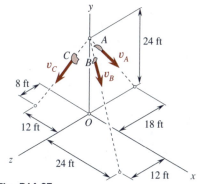

Fig. P14.37

[†]Cf. App. D for computer solution of resulting simultaneous equations.

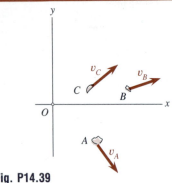

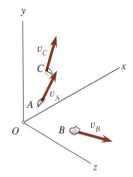

Fig. P14.39

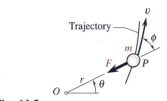

Fig. P14.40*

velocities of A and B are observed to be $\mathbf{v}_A = -30\mathbf{i} + 240\mathbf{j} + 120\mathbf{k}$ m/s and $\mathbf{v}_B = 180\mathbf{i} - 90\mathbf{j} - 60\mathbf{k}$ m/s. Determine the velocity $\mathbf{v}_C$ of C immediately after the explosion.

14.39 A 55-kg object traveling in space is observed at the time $t = 0$ to pass through the origin O of an inertial reference frame Oxy with a constant velocity $\mathbf{v}_0$ along the x axis. Following an explosion at $t = t_1$, where $0 < t_1 < 4.5$ s, the object separates into three fragments, A, B, and C, of masses $m_A = 17$ kg, $m_B = 8$ kg, and $m_C = 30$ kg, respectively, as shown. At $t = 4.5$ s, it is observed that the position vectors of A and B are $\mathbf{r}_A = 135\mathbf{i} - 102\mathbf{j}$ m and $\mathbf{r}_B = 225\mathbf{i} + 48\mathbf{j}$ m, and the velocities of A, B, and C are $\mathbf{v}_A = 30\mathbf{i} + (v_A)_y\mathbf{j}$ m/s, $\mathbf{v}_B = 60\mathbf{i} + (v_B)_y\mathbf{j}$ m/s, and $\mathbf{v}_C = 22\mathbf{i} + (v_C)_y\mathbf{j}$ m/s. Neglecting the effect of gravity, determine (a) the velocity $\mathbf{v}_0$, (b) the position vector $\mathbf{r}_C$ of C at $t = 4$ s, (c) the time t_1, (d) the numerical values of $(v_A)_y$, $(v_B)_y$, and $(v_C)_y$.

14.40* A 100-kg object traveling in space is observed at the time $t = 0$ to pass through the origin O of an inertial reference frame $Oxyz$ with a constant velocity $\mathbf{v}_0 = 624\mathbf{i}$ m/s. Following an explosion at $t = t_1$, where $0 < t_1 < 1.75$ s, the object separates into three fragments, A, B, and C, of masses $m_A = 39$ kg, $m_B = 36$ kg, and $m_C = 25$ kg, respectively, as shown. At $t = 1.75$ s, it is observed that the position vectors of A and B are $\mathbf{r}_A = 568\mathbf{i} + 300\mathbf{j} + 75\mathbf{k}$ m and $\mathbf{r}_B = 1118\mathbf{i} - 325\mathbf{j} + 650\mathbf{k}$ m, and that the z component of the velocity $\mathbf{v}_C$ of C is $(v_C)_z = -1053\mathbf{k}$ m/s. Neglecting the effect of gravity, determine the velocities $\mathbf{v}_A$ and $\mathbf{v}_B$ of A and B, respectively, after the explosion.

14.6 Central-Force Motion

We saw in Sec. 12.8 that, when a particle P of mass m moves under a central force $\mathbf{F}$, the line of action of $\mathbf{F}$ always passes through the center of force O as shown in Fig. 14.8. Consequently, the moment, and hence the angular impulse, of $\mathbf{F}$ about O is always zero during the motion. From this fact and Eq. (14.20), we see that, *when a particle moves under a central force, the angular momentum of the particle about the center of force must remain constant during the motion.* The magnitude of this angular momentum about the center of force O is

$$H_O = |\mathbf{r} \times m\mathbf{v}| = mrv \sin\phi = \text{constant} \qquad (14.22)$$

where ϕ is the angle between the radius vector $\mathbf{r}$ and the velocity $\mathbf{v}$ of the particle P. Since $v_\theta = v \sin\phi = r\dot\theta$ and m is a constant, we deduce from Eqs. (14.22) that

$$rv \sin\phi = rv_\theta = r^2\dot\theta = h \qquad (14.23)$$

where h is the same constant appearing in Eqs. (12.30) through (12.33). We saw in Eq. (12.32) that *the constant h is equal to twice the areal speed of the particle P.* On the other hand, we see in Eqs. (14.23) that *the constant h represents the angular momentum per unit mass of a particle moving under a central force about the center of force.* This interpretation provides us with an additional insight into the physical significance of the constant h.

If a particle of mass m moves under a *conservative central force* (e.g., a gravitational central force or an elastic central force) from the position A_1 to the position A_2 as illustrated in Fig. 14.9, we note that (a) the *angular momentum* of the particle about the center of force O is *conserved*, and (b) the *total mechanical energy* of the particle is *conserved*. In this case, we

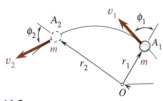

Fig. 14.8

Fig. 14.9

write

$$h_1 = h_2: \quad r_1 v_1 \sin\phi_1 = r_2 v_2 \sin\phi_2 \quad\quad (14.24)$$

$$\text{Energy:} \quad T_1 + V_1 = T_2 + V_2 \quad\quad (13.44)$$
(repeated)

In gravitational central-force motion, Eq. (13.44) may be written as

$$\frac{1}{2} m v_1^2 - \frac{GMm}{r_1} = \frac{1}{2} m v_2^2 - \frac{GMm}{r_2}$$

$$\frac{1}{2} v_1^2 - \frac{GM}{r_1} = \frac{1}{2} v_2^2 - \frac{GM}{r_2} \quad\quad (14.25)$$

where M is the mass of the attracting body at the center of force O, $m \ll M$, and G is the constant of gravitation. In the case of an elastic central force, Eq. (13.44) may take the form

$$\tfrac{1}{2} m v_1^2 + \tfrac{1}{2} k x_1^2 = \tfrac{1}{2} m v_2^2 + \tfrac{1}{2} k x_2^2 \quad\quad (14.26)$$

where k is the spring modulus, and x_1 and x_2 are the changes in length of the spring from its natural length at A_1 and A_2, respectively.

EXAMPLE 14.9

A spacecraft starts its free flight from the position A_1 where $v_1 = 5$ mi/s, $a_1 = 1000$ mi, and $\phi_1 = 75°$ as shown. Determine the maximum and minimum altitudes of the orbit of the spacecraft.

Solution.   We recognize that the spacecraft moves under a gravitational central force, and the system is conservative. The center of force is located at O as shown. Let an extremum altitude be in the position A_2, where $\mathbf{v}_2 \perp \mathbf{r}_2$. Recalling that the radius of the earth is $R = 3960$ mi, we write

$$r_1 = R + a_1 = 3960 + 1000 \quad\quad r_1 = 4960 \text{ mi}$$

$$h_1 = h_2: \quad r_1 v_1 \sin\phi_1 = r_2 v_2 \sin\phi_2$$

$$4960(5) \sin 75° = r_2 v_2 \sin 90° \quad\quad (1)$$

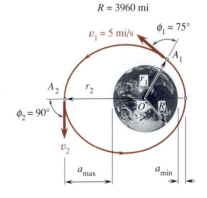

$$\text{Energy:} \quad \frac{1}{2} v_1^2 - \frac{GM}{r_1} = \frac{1}{2} v_2^2 - \frac{GM}{r_2}$$

$$\frac{1}{2} v_1^2 - \frac{gR^2}{r_1} = \frac{1}{2} v_2^2 - \frac{gR^2}{r_2}$$

$$\frac{1}{2} (5)^2 - \frac{32.2(3960)^2}{5280(4960)} = \frac{1}{2} v_2^2 - \frac{32.2(3960)^2}{5280(r_2)} \quad\quad (2)$$

Elimination of v_2 between Eqs. (1) and (2) leads to

$$6.781 r_2^2 - 9.563 \times 10^4 r_2 + 2.869 \times 10^8 = 0$$

which yields the two roots $r' = 9.77 \times 10^3$ mi and $r'' = 4.33 \times 10^3$ mi. These are the maximum and minimum values of r. Thus, we write

$$a_{\max} = r' - R \quad\quad a_{\max} = 5.81 \times 10^3 \text{ mi} \quad \blacktriangleleft$$

$$a_{\min} = r'' - R \quad\quad a_{\min} = 0.37 \times 10^3 \text{ mi} \quad \blacktriangleleft$$

Developmental Exercises

D14.15 Refer to Example 14.9. Determine the angular momentum per unit mass of the spacecraft about the center of the earth.

D14.16 Refer to Example 14.9. Determine the maximum and minimum speeds of the spacecraft in its orbit.

EXAMPLE 14.10

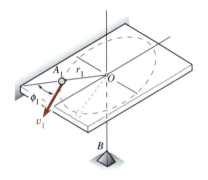

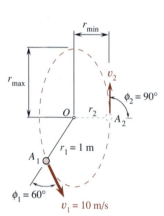

A 1-kg sphere attached to an elastic cord of modulus $k = 100$ N/m moves along a closed path on the horizontal plate as shown. The cord is undeformed when the sphere is situated over the small hole at O. In the position shown, $r_1 = 1$ m, $v_1 = 10$ m/s, and $\phi_1 = 60°$. Knowing that friction is negligible, determine the maximum and minimum distances r_{max} and r_{min} of the sphere from O.

Solution. We recognize that the sphere moves under an elastic central force, and the system is conservative. The center of force is located at O. Let an extremum distance of the sphere from O occur in the position A_2, where $\mathbf{v}_2 \perp \mathbf{r}_2$. We write

$$h_1 = h_2: \qquad r_1 v_1 \sin\phi_1 = r_2 v_2 \sin\phi_2$$

$$1(10) \sin 60° = r_2 v_2 \sin 90° \tag{1}$$

$$\text{Energy:} \qquad \tfrac{1}{2}mv_1^2 + \tfrac{1}{2}kx_1^2 = \tfrac{1}{2}mv_2^2 + \tfrac{1}{2}kx_2^2$$

$$\tfrac{1}{2}(1)(100)^2 + \tfrac{1}{2}(100)(1)^2 = \tfrac{1}{2}(1)v_2^2 + \tfrac{1}{2}(100)r_2^2 \tag{2}$$

Elimination of v_2 between Eqs. (1) and (2) leads to

$$50r_2^4 - 100r_2^2 + 37.5 = 0$$

which yields $r_2 = \pm 1.225$ m and ± 0.707 m. Thus, we write

$$r_{max} = 1.225 \text{ m} \qquad r_{min} = 0.707 \text{ m} \blacktriangleleft$$

REMARK. It can be shown that the path described by the sphere in an ellipse. However, the center of force is located at the geometric center O, rather than one of the foci, of the ellipse. (Cf. Example 14.9.)

Developmental Exercises

D14.17 Refer to Example 14.10. Determine the angular momentum of the sphere about the point O.

D14.18 Refer to Example 14.10. Determine the maximum and minimum speeds of the sphere moving along its path.

PROBLEMS

14.41* A spacecraft describes an elliptic orbit around the earth as shown. In the position S, its velocity $\mathbf{v}_S$ and the vertical make an angle $\phi_S = 70°$, its altitude is $a_S = 5000$ mi, and its speed is $v_S = 3.2$ mi/s. Determine the maximum and the minimum altitudes of its orbit.

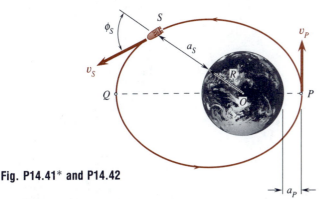

Fig. P14.41* and P14.42

14.42 A spacecraft describes an elliptic orbit around the earth as shown. As it passes through the perigee P, its altitude is $a_P = 3$ Mm and its speed is $v_P = 7.3$ km/s. Determine (a) the angle ϕ_S between its velocity $\mathbf{v}_S$ and the vertical as it passes through the position S where $a_S = 7.5$ Mm, (b) its speed v_S.

14.43 A spacecraft starts its free flight around the earth in the position S at an altitude $a_S = 1500$ mi and with a velocity $\mathbf{v}_S$ of magnitude 4 mi/s as shown. Determine (a) the minimum altitude a_{min} of its orbit if the angle ϕ_S between $\mathbf{v}_S$ and the vertical is 85°, (b) the range of values of ϕ_S if a_{min} is to be no less than 300 mi.

14.44 A spacecraft starts its free flight around the earth in the position S at an altitude $a_S = 1500$ mi and with a velocity v_S of magnitude 4 mi/s as shown. If the minimum altitude a_{min} of its orbit is 150 mi, determine (a) the angle ϕ_S between $\mathbf{v}_S$ and the vertical, (b) the eccentricity of orbit, (c) the period of orbit.

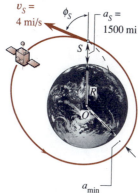

Fig. P14.43 and P14.44

14.45* A *Landsat L* is describing a circular orbit of altitude 900 km around the earth as shown. At an appropriate time, a space shuttle is launched to rendezvous with the *Landsat*. If the shuttle starts its free flight in the position S at an altitude of 200 km and with a velocity $\mathbf{v}_S$ of magnitude 5 km/s, determine the angle ϕ_S with which the trajectory of the shuttle will be tangent at A to the orbit of the *Landsat*.

14.46 A *Landsat L* is describing a circular orbit of altitude 900 km around the earth as shown. At an appropriate time, a space shuttle is launched to rendezvous with the *Landsat*. If the shuttle starts its free flight in the position S at an altitude of 200 km and with a velocity $\mathbf{v}_S$ which makes an angle $\phi_S = 65°$ with the vertical, determine the magnitude of $\mathbf{v}_S$ with which the trajectory of the shuttle will be tangent at A to the orbit of the *Landsat*.

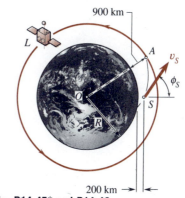

Fig. P14.45* and P14.46

14.47 An elastic cord having a free length $L = 1.2$ m and a modulus $k = 100$ N/m is attached to a fixed point at O and a 0.4-kg disk which is given an initial velocity $\mathbf{v}_1 = 6\mathbf{j}$ m/s at P_1 to move on a horizontal smooth surface in the xy plane as shown. Determine (a) the speed v_2 of the disk when the cord becomes slack at P_2, (b) the shortest distance d_s that the disk will come to O.

14.48* Refer to Prob. 14.47. Determine the magnitude of $\mathbf{v}_1$ as shown if the shortest distance that the disk will come to O is $d_s = 1$ m.

14.49 Refer to Prob. 14.47. Determine the minimum magnitude of $\mathbf{v}_1$ as shown if the elastic cord is to remain taut at all times.

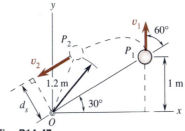

Fig. P14.47

14.50 An elastic cord having a free length $L = 25$ in. and a modulus $k = 1$ lb/in. is attached to a fixed point at O and a 1-lb disk which is given an initial velocity $\mathbf{v}_1 = v_1\mathbf{j}$ at P_1 to move on a horizontal smooth surface in the xy plane as shown. If the maximum distance between O and the trajectory of the disk is $\overline{OP_2} = 30$ in. as indicated, determine (a) the value of v_1, (b) the radius of curvature of the trajectory at P_2.

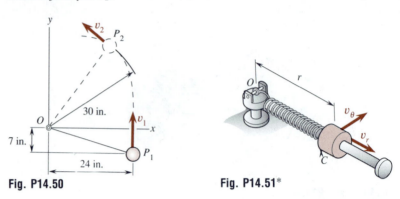

Fig. P14.50   **Fig. P14.51***

14.51* A spring having a free length $L = 0.5$ m and a modulus $k = 400$ N/m is mounted on a horizontal cantilever rod as shown. One end of the spring is attached to the fixed point O, and its other end is attached to a 2-kg collar C. The rod can rotate freely about the vertical axis through O. The system is set in motion with $r = 0.3$ m, $v_r = 0$, and $v_\theta = 6$ m/s. Neglecting effects of friction and the mass of the rod, determine the velocity components of the collar when $r = 0.8$ m.

14.52 Refer to Prob. 14.51*. Determine (a) the maximum distance r_{max} between the point O and the collar during the ensuing motion, (b) the speed of the collar when $r = r_{max}$.[†]

IMPACT AND SPECIAL TOPIC

14.7 Impact

An *impact* is a forceful collision or contact between two bodies over a very short period of time. Several types of impacts are illustrated in Figs. 14.10 through 14.13.

The *line of impact* is the straight line normal to the contacting surfaces during the impact. The lines of impact in Figs. 14.10 through 14.13 are indicated by dashed lines. A *direct impact* is an impact where the velocities of the points of contact of the two colliding bodies are both directed along the line of impact. An *oblique impact* is an impact where the velocity of at least one of the points of contact is not directed along the line of impact; it is an impact where one body strikes another body with a glancing blow. A *central impact* is an impact where the mass centers of the two colliding bodies are both located on the line of impact. An *eccentric impact* is an impact where at least one of the mass centers of the two colliding bodies is not located on the line of impact. Direct and oblique *central impacts* involve motions of *particles* and are considered in the next two sections, whereas

---

[†]Cf. App. E for computer solution of resulting quartic equation.

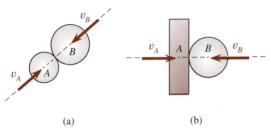

(a)   (b)

Direct central impact

Fig. 14.10

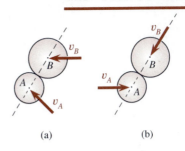

(a)   (b)

Oblique central impact

Fig. 14.11

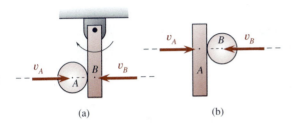

(a)   (b)

Direct eccentric impact

Fig. 14.12

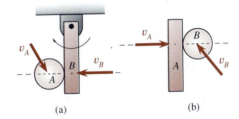

(a)   (b)

Oblique eccentric impact

Fig. 14.13

direct and oblique *eccentric impacts* involve motions of *rigid bodies* and are considered in Chap. 18.

When the two colliding particles are considered together as a system, we note that there is no external impulsive force acting on the system during the impact. Therefore, *the total linear momentum of any two particles must remain the same immediately before and after the impact*; and we may write Eq. (14.14) for two colliding particles A and B as

$$m_A\mathbf{v}_A + m_B\mathbf{v}_B = m_A\mathbf{v}_A' + m_B\mathbf{v}_B' \qquad (14.27)$$

where m_A and m_B are their masses, $\mathbf{v}_A$ and $\mathbf{v}_B$ are their velocities just before the impact, and $\mathbf{v}_A'$ and $\mathbf{v}_B'$ are their velocities just after the impact. If the two particles A and B have a *direct central impact*, they must move along the same straight line; and we have

$$m_A v_A + m_B v_B = m_A v_A' + m_B v_B' \qquad (14.28)$$

Developmental Exercise

D14.19 Define the terms: (a) impact, (b) line of impact, (c) direct impact, (d) oblique impact, (e) central impact, (f) eccentric impact.

14.8 Direct Central Impact

In general, the process of *direct central impact* of two particles involves a period of *deformation* and a period of *restitution*, as illustrated in Figs. 14.14 and 14.15. The impulses exerted by the force of impact on either particle during these two periods are generally unequal.

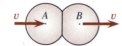

(a) Motion before
impact

(b) Motion at maximum
deformation

(a) Motion after
impact

Fig. 14.14

(a) Momenta and impulse during
period of deformation

(b) Momenta and impulse during
period of restitution

Fig. 14.15

Letting the impulsive force acting on A by B during the period of deformation be $\mathbf{D}$ and the common velocity at the end of deformation be $\mathbf{v}$, as shown in Fig. 14.15(a), we write

$$\xrightarrow{+} \Sigma V_x: \qquad m_A v_A - \int D \, dt = m_A v \qquad (14.29)$$

where the integral extends over the period of deformation. Likewise, letting the impulsive force acting on A by B during the period of restitution be $\mathbf{R}$, as shown in Fig. 14.15(b), we write

$$\xrightarrow{+} \Sigma V_x: \qquad m_A v - \int R \, dt = m_A v'_A \qquad (14.30)$$

where the integral extends over the period of restitution. Generally, *the impulse $\int R \, dt$ during restitution* is smaller than *the impulse $\int D \, dt$ during deformation.* The ratio of these two impulses is called the *coefficient of restitution* and is denoted by e.[†] We write

$$e = \frac{\int R \, dt}{\int D \, dt} \qquad (14.31)$$

Solving for the integrals in Eqs. (14.29) and (14.30) and substituting them into Eq. (14.31), we write

$$v - v'_A = e(v_A - v) \qquad (14.32)$$

(a) Momenta and impulse during
period of deformation

(b) Momenta and impulse during
period of restitution

Fig. 14.16

The momenta and impulses during the periods of deformation and restitution of the particle B are depicted in Fig. 14.16, from which we similarly obtain the relation

$$v'_B - v = e(v - v_B) \qquad (14.33)$$

[†]The coefficient of restitution e is usually determined from experiments. It is employed to facilitate the elementary study of the phenomenon of impact. The value of e is highly dependent on the two materials involved; moreover, it is also dependent on the relative velocity and the geometry of the two colliding bodies. Experiments show that e may approach 1 if the relative velocity just before the impact approaches zero.

Adding Eqs. (14.32) and (14.33) side by side, we get

$$v_B' - v_A' = e(v_A - v_B) \qquad (14.34)$$

A problem of direct central impact is usually solved by applying Eqs. (14.28) and (14.34) simultaneously, where the positive sense chosen for each of the quantities v_A, v_B, v_A', and v_B' must, of course, be the same.

A *perfectly plastic impact* is an impact where $e = 0$; i.e., the two particles cling together and have a common velocity after the impact. On the other hand, a *perfectly elastic impact* is an impact where $e = 1$; i.e., the magnitude of the relative velocity just before the impact is equal to that just after the impact. In a perfectly elastic impact, it can be shown that the total kinetic energy of the two particles, as well as their total momentum, is conserved. In impacts where $e < 1$, only their total momentum is conserved; their total kinetic energy will suffer a loss.

EXAMPLE 14.11

A 20-lb block A is released from rest in the position shown and hits the 40-lb sphere B which is at rest. If $e = 0.8$ between A and B, determine the velocities of A and B just after the impact.

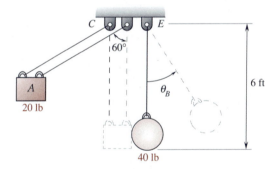

Solution. Letting v_A be the speed of the block A just before it hits the sphere B and applying the *principle of work and energy* to the block A before the impact, we write

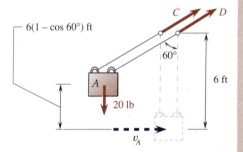

$$T_1 + U_{1\to2} = T_2: \qquad 0 + 20[6(1 - \cos 60°)] = \frac{1}{2}\left(\frac{20}{32.2}\right)v_A^2$$

$$v_A = 13.90 \qquad \mathbf{v}_A = 13.90 \text{ ft/s} \rightarrow$$

The impact which ensues is a *direct central impact*. The speed of the sphere B just before the impact is $v_B = 0$. Let us choose the positive sense of each velocity to be to the right and assume the velocities of A and B after the impact to be $\mathbf{v}_A' = v_A' \rightarrow$ and $\mathbf{v}_B' = v_B' \rightarrow$. Applying the *principle of impulse and momentum* to A and B as a system during the impact, we have

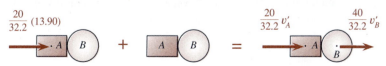

Notice that the net impulse exerted on A and B as a system is zero. We write

$$\xrightarrow{+}\ \Sigma V_x: \qquad \frac{20}{32.2}\,(13.90) + 0 = \frac{20}{32.2}\,v'_A + \frac{40}{32.2}\,v'_B \tag{1}$$

Since it is given that $e = 0.8$, we apply Eq. (14.34) to write

$$v'_B - v'_A = 0.8(13.90 - 0) \tag{2}$$

Solving Eqs. (1) and (2), we obtain $v'_A = -2.78$ and $v'_B = 8.34$. Recalling that we have chosen the positive sense to be to the right, we write

$$\mathbf{v}'_A = 2.78 \text{ ft/s} \leftarrow \qquad \mathbf{v}'_B = 8.34 \text{ ft/s} \rightarrow \quad \blacktriangleleft$$

Developmental Exercises

D14.20 Define the terms: (a) coefficient of restitution, (b) perfectly plastic impact, (c) perfectly elastic impact.

D14.21 Refer to Example 14.11. Determine the angle θ_B when B reaches its highest position.

D14.22 Refer to Example 14.11. If the velocity of the block A just after the impact is zero, what is the value of e?

14.9 Oblique Central Impact

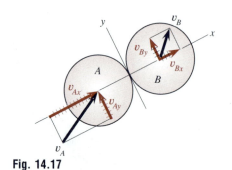

Fig. 14.17

When two particles A and B have an oblique central impact, their velocities $\mathbf{v}_A$ and $\mathbf{v}_B$ just before the impact are not both directed along the line of impact. Let us choose the x axis to be along the line of impact of the particles A and B as illustrated in Fig. 14.17. Then, the y axis, which lies in the plane of motion, is along the common tangent to the surfaces in contact.

The elementary study of the oblique impact between any two bodies is based on the *usual assumption* that the surfaces of contact are perfectly smooth so that *impulses due to friction between surfaces of contact are negligible during the impact*. The only impulsive forces acting on them are the forces of impact between them, which obey Newton's third law.

For analysis, let the particles A and B in Fig. 14.17 have, respectively, masses m_A and m_B, velocities $\mathbf{v}_A$ and $\mathbf{v}_B$ before the impact, and velocities $\mathbf{v}'_A$ and $\mathbf{v}'_B$ after the impact, where

$$\mathbf{v}_A = v_{Ax}\mathbf{i} + v_{Ay}\mathbf{j} \qquad \mathbf{v}_B = v_{Bx}\mathbf{i} + v_{By}\mathbf{j}$$
$$\mathbf{v}'_A = v'_{Ax}\mathbf{i} + v'_{Ay}\mathbf{j} \qquad \mathbf{v}'_B = v'_{Bx}\mathbf{i} + v'_{By}\mathbf{J}$$

During the *oblique central impact* of A and B, we recognize that:

(1) The impulse exerted on either particle in the y direction is zero; hence the momentum of either particle in the y direction is conserved. As the mass of either particle is constant, we have

$$v_{Ay} = v'_{Ay} \qquad v_{By} = v'_{By} \tag{14.35}$$

(2) The total impulse exerted in the x direction on both particles as a system is, by Newton's third law, equal to zero; hence the total momentum in the x direction is conserved during the impact. We write

$$m_A v_{Ax} + m_B v_{Bx} = m_A v'_{Ax} + m_B v'_{Bx} \qquad (14.36)$$

(3) The components of velocities in the x direction, immediately before and after the impact, are related by the equation

$$v'_{Bx} - v'_{Ax} = e(v_{Ax} - v_{Bx}) \qquad (14.37)^{\dagger}$$

The unknown quantities in a problem of oblique central impact are usually determined by solving Eqs. (14.35) through (14.37) simultaneously.

Developmental Exercise

D14.23   What is the usual assumption on which the elementary study of the oblique impact between any two bodies is based?

EXAMPLE 14.12

The velocities of the disks A and B just before their impact are shown. If $e = 0.8$, determine the velocities $\mathbf{v}'_A$ and $\mathbf{v}'_B$ of the disks A and B just after the impact.

Solution.   The impact between A and B is an oblique central impact. Choosing the x axis to be along the line of impact, we may indicate the components of velocities of A and B just before impact as shown; i.e., $v_{Ax} = 3\sqrt{3}$ m/s, $v_{Ay} = 3$ m/s, $v_{Bx} = -2$ m/s, and $v_{By} = -2\sqrt{3}$ m/s. The masses of A and B are $m_A = 4$ kg and $m_B = 5$ kg. Applying Eqs. (14.35) through (14.37), we write

$$3 = v'_{Ay} \qquad -2\sqrt{3} = v'_{By}$$

$$4(3\sqrt{3}) + 5(-2) = 4v'_{Ax} + 5v'_{Bx}$$

$$v'_{Bx} - v'_{Ax} = 0.8[3\sqrt{3} - (-2)]$$

From these four equations, we obtain

$$v'_{Ax} = -2 \qquad v'_{Ay} = 3 \qquad v'_{Bx} = 3.757 \qquad v'_{By} = -3.464$$

Adding vectorially the velocity components of each disk, we obtain

$$\mathbf{v}'_A = 3.61 \text{ m/s} \ \measuredangle \ 56.3° \ \blacktriangleleft$$

$$\mathbf{v}'_B = 5.11 \text{ m/s} \ \measuredangle \ 42.7° \ \blacktriangleleft$$

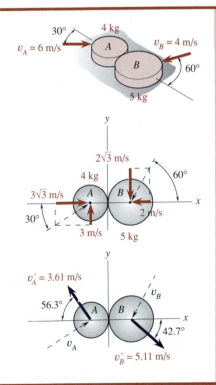

Developmental Exercise

D14.24   Refer to Example 14.12. Determine the percentage of the loss in kinetic energy of the disks due to the impact.

---

†Cf. Eq. (14.34).

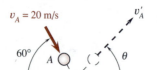

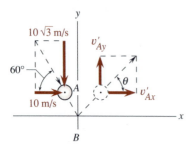

EXAMPLE 14.13

A steel ball A strikes an embedded steel plate B with a velocity $\mathbf{v}_A$ of 20 m/s and rebounds from it with a velocity $\mathbf{v}'_A$ as shown. If $e = 0.7$ between the ball and the plate, determine the velocity $\mathbf{v}'_A$.

Solution. We first resolve the velocities before and after the impact into components as shown. Since the surfaces of contact are to be assumed as being smooth, the ball experiences no impulse in the x direction. Its momentum and hence velocity in the x direction are conserved. We write

$$v'_{Ax} = v_{Ax} = 20\cos 60° \qquad v'_{Ax} = 10 \text{ m/s} \rightarrow$$

As the steel plate is embedded in the support, the mass of the plate (and the earth) is practically infinite. The conservation of momentum in the y direction will lead to the sensible conclusion that the velocity of the plate is always essentially zero. Therefore, we have

$$v'_{By} = v_{By} = 0 \qquad v_{Ay} = -20\sin 60° = -10\sqrt{3}$$

The components of velocities in the y direction are, by Eq. (14.34), related to e by $v'_{By} - v'_{Ay} = e(v_{Ay} - v_{By})$. Thus, we write

$$0 - v'_{Ay} = 0.7(-10\sqrt{3} - 0) \qquad v'_{Ay} = 7\sqrt{3} \text{ m/s} \uparrow$$

$$\mathbf{v}'_A = \mathbf{v}'_{Ax} + \mathbf{v}'_{Ay} = 10\mathbf{i} + 7\sqrt{3}\mathbf{j} \text{ m/s}$$

$$\mathbf{v}'_A = 15.72 \text{ m/s} \measuredangle 50.5° \blacktriangleleft$$

Developmental Exercise

D14.25 Refer to Example 14.13. If the ball rebounds from the plate with a velocity $\mathbf{v}'_A = 10\sqrt{2}$ m/s $\measuredangle$ 45°, determine the coefficient of restitution e between the ball and the plate.

*14.10 Generalized Virtual Work: Impulse and Momentum

In Sec. 13.13, we made the first application of the concept of *virtual work* to the study of certain kinetic problems. As an extension, we here define the *generalized virtual work* $\delta U'$ of a *non-force vector* $\mathbf{L}$ (e.g., a linear impulse or linear momentum vector) acting on or associated with a particle during a virtual displacement $\delta\mathbf{r}$ of the particle from the position A to the position A', as shown in Fig. 14.18, as the dot product of $\mathbf{L}$ and $\delta\mathbf{r}$; i.e.,

$$\delta U' = \mathbf{L} \cdot \delta\mathbf{r} = L\cos\psi\,\delta s \tag{14.38}$$

where $\delta s = |\delta\mathbf{r}|$ and ψ is the angle between $\mathbf{L}$ and $\delta\mathbf{r}$.

By equivalence and equipollence in the *principle of impulse and momentum* in Secs. 14.2 and 14.4 and the nature of *constrained virtual displacement* for a one-degree-of-freedom system in Sec. 13.13, it follows logically

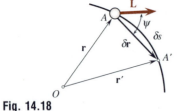

Fig. 14.18

that, *during a constrained virtual displacement of a one-degree-of-freedom system of particles, the generalized virtual work of all the momentum vectors in the initial momentum diagram of the system plus the generalized virtual work of all the impulse vectors in the impulse diagram of the system must be equal to the generalized virtual work of all the momentum vectors in the final momentum diagram of the system.* This property is referred to as the *principle of generalized virtual work* for particles, which complements the traditional *principle of impulse and momentum* for particles.

EXAMPLE 14.14

Using the *principle of generalized virtual work*, do Example 14.5.

Solution. Let the vector-diagram equation for the momenta and impulses during the interval $0 \le t \le 4$ s for the entire system be drawn first.

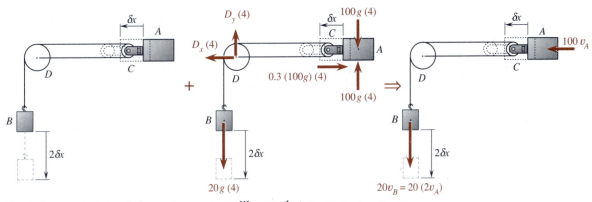

From the constraints of the system, we readily see that $\mathbf{v}_A = v_A \leftarrow$, $\mathbf{v}_B = 2v_A \downarrow$, and the constrained virtual displacements of A and B are $\delta x \leftarrow$ and $2\,\delta x \downarrow$, respectively. Applying the *principle of generalized virtual work*, we write

$$\delta U': \quad 0 + [0.3(100g)(4)(-\delta x) + 20g(4)(2\,\delta x)]$$

$$= 100v_A(\delta x) + 20(2v_A)(2\,\delta x)$$

$$v_A = \tfrac{2}{9}g \qquad\qquad v_A = 2.18 \text{ m/s} \blacktriangleleft$$

REMARK. The contrasting feature between the solutions in Examples 14.5 and 14.14 is that the application of the *principle of generalized virtual work* enables us to *uncouple the unknown* v_A without mathematically involving the tension F in the cord BCD.

Developmental Exercises

D14.26   Describe the *principle of generalized virtual work* for particles.

D14.27   The velocity of the cylinder A shown is $\mathbf{v}_A = 2$ m/s $\downarrow$ when the system is released at the time $t = 0$. Using the *principle of generalized virtual work*, determine the value of $\mathbf{v}_A$ when $t = 2$ s.

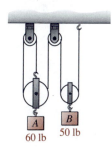

Fig. D14.27   60 lb   50 lb

PROBLEMS

14.53 A ball is thrown vertically downward with an initial speed of v_0 at a height of 2 m above a floor and rebounds from the floor to the same height of 2 m before falling downward again. If $e = 0.7$ between the ball and the floor, determine the value of v_0.

14.54* and 14.55 If $e = 0.8$ between the sliders A and B as shown, determine (a) the velocity of each slider after the impact, (b) the percentage of loss in kinetic energy during the impact.

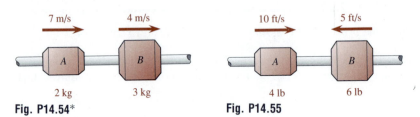

Fig. P14.54*   **Fig. P14.55**

14.56 A sphere A is released from rest in the position shown and strikes the block B which is at rest. If $e = 0.75$ between A and B and $\mu_k = 0.5$ between B and the support, determine (a) the velocity of A just after the impact, (b) the displacement of B after the impact.

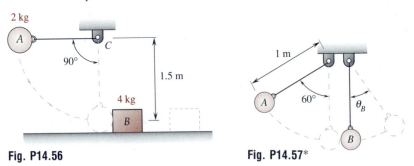

Fig. P14.56   **Fig. P14.57***

14.57* The sphere A is released from rest in the position shown and strikes an identical sphere B which is at rest. If $\theta_B = 40°$ when B reaches its highest position after the impact, determine (a) the value of e between A and B, (b) the velocity of A just after the impact.

14.58 The sphere A is released from rest in the position shown and strikes an identical sphere B which is at rest. If the wire suspending B becomes horizontal when B reaches its highest position after the impact, determine (a) the value of e between A and B, (b) the velocity of A just after the impact.

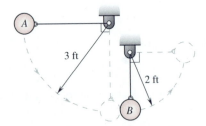

Fig. P14.58

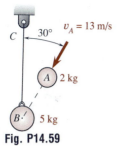

Fig. P14.59

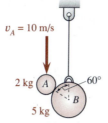

Fig. P14.60

14.59 and 14.60 A sphere is traveling with a velocity v_A when it strikes a sphere B which is at rest as shown. If $e = 0.65$ between A and B, determine the velocity of each sphere just after the impact.[†]

14.61 The block B, which can move freely in the horizontal direction, is at rest when it is struck by a ball A traveling with a velocity of 6 m/s ↓ as shown. If $e = 0.6$ between A and B, determine the velocities of A and B just after the impact.

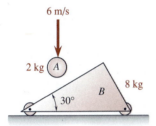

Fig. P14.61

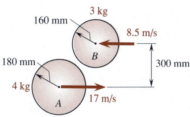

Fig. P14.62*

14.62* If $e = 0.7$ between the disks A and B as shown, determine the velocities of A and B just after the impact.

14.63 A steel ball A is dropped from rest in the position shown. The ball strikes the embedded steel plate B, rebounds from B, and barely clears the wall C as shown. If $e = 0.8$ between A and B, determine the distance x.

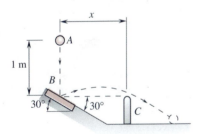

Fig. P14.63

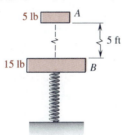

Fig. P14.64*

14.64* The disk A is dropped from rest in the position shown. It falls onto the disk B which is resting on a spring of modulus $k = 120$ lb/ft. If $e = 0$ between A and B, determine the maximum deflection of the disk B.

14.65 The ram A of a pile driver is dropped from rest in the position shown. It falls onto the pile B and drives it 100 mm into the ground. Determine the magnitude of the average resistance of the ground to the penetration by the pile if $e = 0$ between A and B.

[†]Cf. Example 14.8.

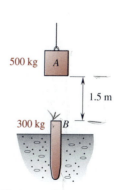

Fig. P14.65

14.66* through 14.75*   Using the *principle of generalized virtual work*, solve the following problems:

14.66* Prob. 14.25.	14.67* Prob. 14.26*.
14.68* Prob. 14.27*.	14.69* Prob. 14.28.
14.70* Prob. 14.29.	14.71* Prob. 14.30*.
14.72* Prob. 14.31.	14.73* Prob. 14.32.
14.74* Prob. 14.33.	14.75* Prob. 14.34*.

VARIABLE SYSTEMS[†]

★14.11   Variable Systems of Particles

We observe that each of the systems of particles heretofore studied contains the same particles and has the same mass during its motion. However, the operations of many engineering systems (e.g., vanes, pipes, nozzles, turbines, conveyors, fans, propellers, jet engines, chains, and rockets) do involve motions of *systems which are continuously acquiring or ejecting particles, or doing both at the same time*. Such systems are called *variable systems of particles*. A variable system of particles whose total mass remains constant is called a *system with steady mass flow*. On the other hand, a variable system of particles whose total mass is increasing or decreasing is called a *system with variable mass*.

    Each variable system of particles constantly involves the *flow of mass*. The flow of mass, in turn, involves the change of momentum of the particles as well as the exertion of impulse on the particles in the flow. Accordingly, the *principle of impulse and momentum*, among others, should be very useful in the analysis of the motion of variable systems of particles. However, there is a catch: the principles in kinetics heretofore established are directly applicable to only *constant systems* which neither gain nor lose particles. Therefore, we employ a strategy which seeks to analyze the motion of an *auxiliary constant system* isolated from the variable system during a short time interval Δt of its motion. The specific procedure for the analysis is explained and illustrated in Secs. 14.12 and 14.13 for two major types of variable systems.

Developmental Exercises

D14.28   What is a variable system of particles? Give three examples.

D14.29   How does a system with steady mass flow differ from a system with variable mass?

[†]The topics related to *variable systems* in Secs. 14.11 through 14.13 may be studied at any later time, or omitted if circumstances warrant, without affecting continuity to the rest of this text.

*14.12   Systems with Steady Mass Flow

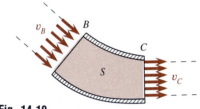

A *steady mass flow* is a condition of flow of mass where the time rate at which mass enters a given volume is equal to the time rate at which mass leaves the same volume. In other words, when there exists a steady mass flow through a volume, *the total mass in the volume remains unchanged, although the particles occupying the volume keep changing*. The volume in question may be the space in any fixed or moving rigid container, the space between blades, or a certain space associated with a vane. Let us consider *a steady mass flow* through the rigid container *BC* as shown in Fig. 14.19. The continuity of flow requires that the *time rates of mass flow dm/dt* through cross sections *B* and *C* be equal; i.e.,

Fig. 14.19

$$\frac{dm}{dt} = \rho_B A_B v_B = \rho_C A_C v_C \qquad (14.39)$$

where ρ_B and ρ_C are the mass densities of the stream of flow at *B* and *C*, A_B and A_C are the (right) cross-sectional areas at *B* and *C*, and v_B and v_C are the speeds of the stream at *B* and *C*, respectively.

In order to examine the resultant force $\Sigma\mathbf{F}$ exerted on the particles in contact with the container, we consider the particles situated in the container and denote them by *S* as shown in Fig. 14.19. We see that *S* is *a variable system of particles* because it continuously gains and loses particles (at the same rate). Thus, the principles in kinetics established so far are *not* directly applicable to *S*. We need to define and isolate *an auxiliary constant system of particles* for analysis.

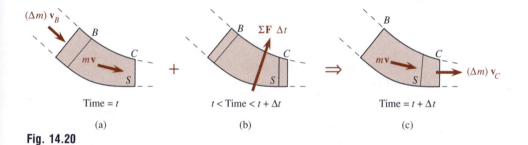

(a)	(b)	(c)
Time = t	t < Time < t + Δt	Time = t + Δt

Fig. 14.20

As illustrated in Fig. 14.20(a), we consider an auxiliary system of particles, which is *a system consisting of two parts*: (1) the particles which are situated in the container *BC* at the time *t* and have a total mass *m*, and (2) the particles which will enter the container *BC* during the next short time interval Δt and have a total mass Δm. Since the flow through the container is a *steady mass flow*, the total amount of mass of the particles situated in the container at the time $t + \Delta t$ must still be equal to *m*, and the total amount of mass of the particles which have left the container from the time *t* to the time $t + \Delta t$ must be equal to Δm. At the time $t + \Delta t$, the auxiliary system has a configuration as illustrated in Fig. 14.20(c). Clearly, the same particles are involved in the auxiliary system in its motion during the interval Δt. The principle of impulse and momentum may, therefore, be applied to the foregoing *auxiliary constant system of particles* during the interval Δt, as shown in Fig. 14.20, where $\mathbf{v}_B$ and $\mathbf{v}_C$ are the velocities of the stream of flow at B and C, respectively, the resultant momentum of the particles situated in the container is represented by $m\mathbf{v}$ and is the same in (a) as in (c), and the resultant impulse $\Sigma \mathbf{F} \, \Delta t$ in (b) does not necessarily pass through the center of mass of the system.

Note that, unless otherwise stated, fluid friction in the system is assumed to be negligible, and the impulse of weight forces are assumed to be also negligible. As indicated in Fig. 14.20, we write

$$(\Delta m)\mathbf{v}_B + m\mathbf{v} + \Sigma \mathbf{F} \, \Delta t = m\mathbf{v} + (\Delta m)\mathbf{v}_C$$

$$(\Delta m)\mathbf{v}_B + \Sigma \mathbf{F} \, \Delta t = (\Delta m)\mathbf{v}_C \qquad (14.40)$$

Referring to Fig. 14.20 and Eq. (14.40), we conclude that the system of vectors consisting of the momentum $(\Delta m)\mathbf{v}_B$ of the particles entering S during the time interval Δt and the impulses of the forces exerted on S during Δt is *equipollent* to the momentum $(\Delta m)\mathbf{v}_C$ of the particles leaving S during the same time interval Δt.

Dividing both sides of Eq. (14.40) by Δt and letting $\Delta t \to 0$, we find that the resultant force acting on S is

$$\Sigma \mathbf{F} = \frac{dm}{dt}(\mathbf{v}_C - \mathbf{v}_B) \qquad (14.41)$$

where dm/dt is given by Eqs. (14.39) for systems with steady mass flow. Note that *a consistent system of kinetic units* must be used in Eqs. (14.40) and (14.41), and $\mathbf{v}_B$ and $\mathbf{v}_C$ must be measured in *an inertial reference frame*, which may be fixed or translating with a constant velocity. Furthermore, we may express

$$\frac{dm}{dt} = \rho Q = \rho A v \qquad (14.42)$$

where ρ is the mass density of the stream of flow, Q is the time rate of volumetric flow of the stream (in volume per unit time), A is the cross-sectional area of the stream, and v is the speed of the stream.

We can see that the principle of impulse and momentum is very useful in determining the force developed when a stream of water or other fluid is deflected by a blade or a vane. *The mass of fluid striking the blade during the time interval Δt depends on the cross-sectional area and velocity of the stream, the mass density of the fluid, the velocity of the blade, and whether a single blade or a series of blades mounted on the periphery of a wheel is involved.*

For example, if the stream strikes a fixed blade as shown in Fig. 14.21(a), the mass of fluid striking the blade during the interval Δt is

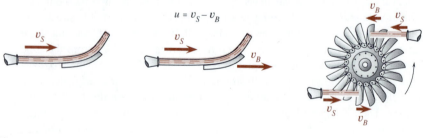

Fig. 14.21   (a)   (b)   (c)

$\Delta m = \rho A v_S \, \Delta t$, where v_S is the speed of the stream. If *a single blade* is moving such that the stream strikes the blade with a speed u relative to the blade as shown in Fig. 14.21(b), then $\Delta m = \rho A u \, \Delta t$. When streams of fluid strike *a series of blades* mounted on the periphery of a wheel as shown in Fig. 14.21(c), all the fluid in each stream strikes the blades during the interval Δt, and $\Delta m = \rho A v_S \, \Delta t$ from each stream.

The vector diagram equation for the momenta and impulses during the interval Δt for the *auxiliary constant system of particles* is a very helpful tool for gaining geometric and physical perception into the problem situation of variable systems. We shall use it in the examples.

EXAMPLE 14.15

A nozzle discharges a stream of water of cross-sectional area $A = 100 \text{ mm}^2$ with a speed $v = 50$ m/s, and the stream is deflected by a fixed vane as shown. The mass density of water is $\rho = 1 \text{ Mg/m}^3$. Determine the resultant force **F** exerted on the stream by the fixed vane.

Solution. The system in the problem is a system with steady mass flow. The mass density of water is $\rho = 1 \text{ Mg/m}^3 = 10^3 \text{ kg/m}^3$, the cross-sectional area of the stream is $A = 100 \text{ mm}^2 = 10^{-4} \text{ m}^2$, and the speed of the stream is $v = 50$ m/s. Thus, the time rate of mass flow in the stream is

$$\frac{dm}{dt} = \rho A v = 10^3 (10^{-4})(50) \qquad \frac{dm}{dt} = 5 \text{ kg/s}$$

Let the *auxiliary constant system of particles* be a system consisting of (1) the particles of the stream which is in contact with the vane at the time t and has a total mass m, and (2) the particles of the stream which will be in contact with the vane during the next short time interval Δt and has a total mass Δm. Applying the *principle of impulse and momentum* to such a system during the interval Δt, we have

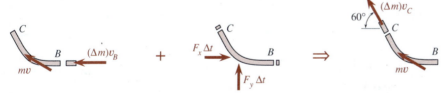

Noting that $v_B = v_C = v = 50$ m/s and canceling the momentum $m\mathbf{v}$ which appears on both sides, we write

$$\xrightarrow{+} \Sigma V_x: \qquad -(\Delta m)v_B + F_x \, \Delta t = -(\Delta m)v_C \cos 60°$$

$$F_x = \frac{dm}{dt} v (1 - \cos 60°) = 5(50)(1 - \cos 60°) = 125$$

$$+\uparrow \Sigma V_y: \qquad F_y \, \Delta t = (\Delta m)v_C \sin 60°$$

$$F_y = \frac{dm}{dt} v \sin 60° = 5(50) \sin 60° = 216.5$$

$$\mathbf{F} = F_x\mathbf{i} + F_y\mathbf{j} \qquad \mathbf{F} = 250 \text{ N} \angle 60° \quad \blacktriangleleft$$

Developmental Exercises

D14.30 In a system with steady mass flow, how do you define and isolate an *auxiliary constant system of particles* for analysis?

D14.31 Work Example 14.15 if the vane moves with a constant velocity of 10 m/s ←. (*Hint.* Use the vane as the inertial reference frame.)

EXAMPLE 14.16[†]

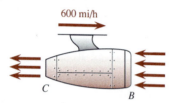

A jet airplane is cruising at a speed of 600 mi/h in level flight. Each of its jet engines scoops in air at the rate of 180 lb/s and ejects it with a speed of 2000 ft/s relative to the airplane. Neglecting effect of fuel consumption, determine for each jet engine (a) the magnitude of the propulsive force generated, (b) the total horsepower generated relative to the ground.

Solution. With effect of fuel consumption neglected, the system may be treated as one with steady mass flow. Furthermore, we may use the *jet engine* as an inertial reference frame because it moves at a constant velocity. Thus, the velocities of the air at B and C are

$$\mathbf{v}_B = -600\mathbf{i} \text{ mi/h} = -880\mathbf{i} \text{ ft/s} \qquad \mathbf{v}_C = -2000\mathbf{i} \text{ ft/s}$$

The time rate of mass flow through the jet engine is

$$\frac{dm}{dt} = \frac{180}{32.2} \text{ slugs/s}$$

Let the *auxiliary constant system of particles* be a system consisting of (1) the air particles in the jet engine at the time t which have a total mass m, and (2) the air particles which will enter the jet engine during the next short time interval Δt and have a total mass Δm. Applying the *principle of impulse and momentum* to such a system during the interval Δt, we have

Notice that the force $\mathbf{F}$ acting on the air is equal in magnitude but opposite in direction to the propulsive force acting on the engine. We write

$$\xleftarrow{+} \Sigma V_x: \qquad mv + (\Delta m)v_B + F\,\Delta t = (\Delta m)v_C + mv$$

$$F = \frac{dm}{dt}(v_C - v_B) = \frac{180}{32.2}(2000 - 880)$$

$$F = 6260.9 \text{ lb} \qquad\qquad F = 6.26 \text{ kips} \blacktriangleleft$$

Relative to the ground, the total horsepower P generated consists of two parts: (1) the horsepower P_1 to propel the airplane, and (2) the horsepower P_2 to eject the air particles out of the engine. The velocities of the airplane and the ejected air are $\mathbf{v}_A = 880\mathbf{i}$ ft/s and $\mathbf{v}_E = (880 - 2000)\mathbf{i}$ ft/s = $-1120\mathbf{i}$ ft/s. Recalling that 1 hp = 550 ft·lb/s, we write

---

[†]Effect of fuel consumption is considered in Example 14.20.

$$P_1 = \frac{Fv_A}{550} = \frac{6260.9(880)}{550} = 10017$$

$$P_2 = \frac{\dfrac{1}{2}\dfrac{dm}{dt}v_E^2}{550} = \frac{\dfrac{1}{2}\left(\dfrac{180}{32.2}\right)(1120)^2}{550} = 6375$$

$$P = P_1 + P_2 = 16392 \qquad P = 16.39 \times 10^3 \text{ hp} \blacktriangleleft$$

Developmental Exercise

D14.32 Refer to Example 14.16. What is the total horsepower generated by each jet engine relative to the jet airplane?

EXAMPLE 14.17

A nozzle discharges water at the rate $Q = 2$ gal/s with a speed $v = 60$ ft/s, and the stream is deflected by the vane BC as shown. The combined weight force of the vane and the water it supports is $\mathbf{W} = 100$ lb $\downarrow$ acting through the point G. It is known that the specific weight of water is $\gamma = 62.4$ lb/ft^3 and 1 ft^3 = 7.48 gal. Determine the reaction from the support at C.

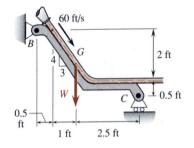

Solution. The mass density ρ and the time rate of volumetric flow Q of the stream of water are

$$\rho = \frac{\gamma}{g} = \frac{62.4}{32.2} \text{ slugs/ft}^3 \qquad Q = \frac{2}{7.48} \text{ ft}^3/\text{s}$$

Thus, we write

$$\frac{dm}{dt} = \rho Q = 0.5182 \text{ slug/s} \qquad v_B = v_C = v = 60 \text{ ft/s}$$

Let the *auxiliary constant system of particles* be a system consisting of (1) the stream of water in contact with the vane at the time t which has a total mass m, (2) the particles of the stream which will be in contact with the vane during the next short time interval Δt and have a total mass Δm, and (3) the vane BC, where the combined weight of the vane and the water it supports is $W = 100$ lb. Applying the *principle of impulse and momentum* to such a system during the interval Δt, we have

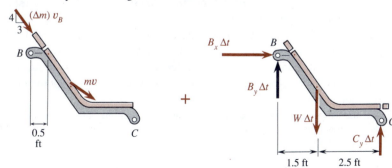

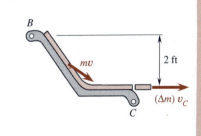

Canceling the momentum $m\mathbf{v}$ which appears on both sides, we write

$$+\circlearrowleft\ \Sigma M_B: \qquad -0.5\left[\frac{4}{5}(\Delta m)v_B\right] - 1.5(W\ \Delta t) + 4(C_y\ \Delta t)$$

$$= 2[(\Delta m)v_C]$$

$$-0.5\left[\frac{4}{5}\left(\frac{dm}{dt}\right)v\right] - 1.5W + 4C_y = 2\left(\frac{dm}{dt}\right)v$$

$$C_y = 56.15 \qquad\qquad \mathbf{C} = 56.2\ \text{lb}\ \uparrow\ \blacktriangleleft$$

Developmental Exercise

D14.33 Refer to Example 14.17. Determine the components of reaction from the support at B.

★14.13 Systems with Variable Mass: Motion of Rockets

A *system with variable mass* is a variable system of particles where the system is gaining mass by continuously absorbing particles or is losing mass by continuously ejecting particles. In other words, the particles, as well as the total mass, in a system with variable mass do not remain constant. In order for the principles in kinetics established earlier to be applicable, we again need to define and isolate an *auxiliary constant system of particles* for analysis.

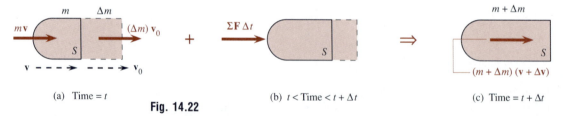

Fig. 14.22

(a) Time $= t$ (b) $t <$ Time $< t + \Delta t$ (c) Time $= t + \Delta t$

As illustrated in Fig. 14.22(a), we consider an auxiliary system of particles, which is *a system consisting of two parts:* (1) the particles which are contained in the system S at the time t and have a total mass m moving in an inertial reference frame with a velocity $\mathbf{v}$, and (2) the particles which will be absorbed by S during the next short time interval Δt and have a total mass Δm moving in the inertial reference frame with a velocity $\mathbf{v}_0$. During the interval Δt, the resultant $\Sigma\mathbf{F}$ of the external forces acting on the auxiliary constant system (as a whole) will have imparted to the system an impulse of $\Sigma\mathbf{F}\ \Delta t$ as shown in Fig. 14.22(b).[†] At the time $t + \Delta t$, the auxiliary constant system has a velocity $\mathbf{v} + \Delta\mathbf{v}$ and a momentum

---

[†]Note that the forces of interaction between S and the particles being absorbed are internal forces to the *auxiliary constant system.* Clearly, $\Sigma\mathbf{F}$ does *not* include the *internal force* exerted on S by the particle being absorbed.

$(m + \Delta m)(\mathbf{v} + \Delta \mathbf{v})$ as shown in Fig. 14.22(c). Since the same particles are involved, the *principle of impulse and momentum* may, therefore, be applied to the *auxiliary constant system of particles* during the interval Δt, as depicted in Fig. 14.22. We write

$$\overset{+}{\underset{\rightarrow}{}} \Sigma V_x: \qquad m\mathbf{v} + (\Delta m)\mathbf{v}_0 + \Sigma \mathbf{F}\,\Delta t = (m + \Delta m)(\mathbf{v} + \Delta \mathbf{v})$$

$$\Sigma \mathbf{F}\,\Delta t = m\,\Delta \mathbf{v} + \Delta m(\mathbf{v} - \mathbf{v}_0) + (\Delta m)\,\Delta \mathbf{v}$$

Neglecting the second-order term, dividing through by Δt, and letting $\Delta t \rightarrow 0$, we obtain the equation of motion

$$\Sigma \mathbf{F} = m\frac{d\mathbf{v}}{dt} + \frac{dm}{dt}\mathbf{u} \qquad (14.43)^\dagger$$

where

$$\mathbf{u} = \mathbf{v} - \mathbf{v}_0 \qquad (14.44)^\dagger$$

Here, dm/dt represents the *time rate of increase of mass* in S by absorbing particles, and $\mathbf{u}$ is *the velocity of S relative to the particles which are being absorbed by S*. Note that the positive senses of $\Sigma \mathbf{F}$, $\mathbf{v}$, and $\mathbf{u}$ are all the same (e.g., to the right) when the system is in rectilinear motion. By writing Eq. (14.43) as

$$\Sigma \mathbf{F} - \frac{dm}{dt}\mathbf{u} = m\frac{d\mathbf{v}}{dt} \qquad (14.45)$$

we see that the action exerted on S by the particles being absorbed is equivalent to *a thrust of* $-(dm/dt)\mathbf{u}$ *to slow down the motion of S*.

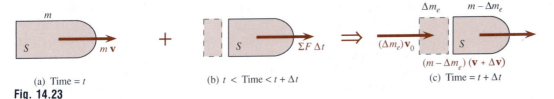

(a) Time $= t$ (b) $t <$ Time $< t + \Delta t$ (c) Time $= t + \Delta t$

Fig. 14.23

On the other hand, the system S of total mass m, which moves in an inertial reference frame with a velocity $\mathbf{v}$ and is acted on by the resultant external force $\Sigma \mathbf{F}$, may be ejecting mass at the rate of dm_e/dt, where the velocity of the ejected particles in the inertial reference frame is $\mathbf{v}_0$. In this case, we may likewise apply the *principle of impulse and momentum* to the *auxiliary constant system of particles* as illustrated in Fig. 14.23, from which we similarly obtain the relation

$$\Sigma \mathbf{F} = m\frac{d\mathbf{v}}{dt} - \frac{dm_e}{dt}\mathbf{u} \qquad (14.46)$$

where $\mathbf{u}$ is the same as that defined in Eq. (14.44). Note in Eq. (14.46) that dm_e/dt represents the *time rate of mass being ejected* from S, and $\mathbf{u}$ is the velocity of S relative to the particles which have just been ejected from S.

---

†From Eqs. (14.43) and (14.44), it is clear that Newton's second law in the form of Eq. (14.4) is applicable to a system with variable mass *only when* $\mathbf{u} = \mathbf{v}$ or $\mathbf{v}_0 = \mathbf{0}$ in the inertial reference frame.

The positive senses of $\Sigma \mathbf{F}$, $\mathbf{v}$, and $\mathbf{u}$ are still all the same (e.g., to the right) when the system is in rectilinear motion. By writing Eq. (14.46) as

$$\Sigma \mathbf{F} + \frac{dm_e}{dt} \mathbf{u} = m \frac{d\mathbf{v}}{dt} \tag{14.47}$$

we readily see that the action exerted on S by the particles being ejected is equivalent to a *thrust of* $(dm_e/dt)\mathbf{u}$ *to speed up the motion of S*. A *rocket* furnishes an example of such a system S. The burning of jet fuel and ejecting it into the exhaust of a *jet airplane* furnishes another example of a system S ejecting mass, although the fuel consumption is usually a small percentage of the air consumption in jet engines.

A comparison of Eqs. (14.43) with (14.46) readily reveals that

$$\frac{dm_e}{dt} \doteq -\frac{dm}{dt} \tag{14.48}$$

By considering the *time rate of mass being ejected* as the negative of the time rate of change of mass, we readily see that Eq. (14.46) is a special case of Eq. (14.43).

EXAMPLE 14.18

A chain of mass ρ_L per unit length and total length L lies in a pile on the floor. Starting from $y = 0$, one end of the chain is lifted vertically with a constant velocity $\mathbf{v}_0$ by a variable force $\mathbf{P}$. Determine the magnitude of $\mathbf{P}$ as a function of y.

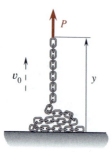

Solution. For an open-link chain as shown, each link on the floor acquires its velocity abruptly through the impact with its upper link which lifts it off the floor. Thus, the lifting of such a chain involves an energy loss. We may choose the *auxiliary constant system of particles* to be a system consisting of (1) the part of the chain already off the floor which has a total mass $\rho_L y$ and total weight $\rho_L y g$, and (2) the part of the chain which will be set in motion during the next short time interval Δt and has a total mass $\rho_L v_0 \Delta t$. Applying the *principle of impulse and momentum* to such a system during the interval Δt, we have

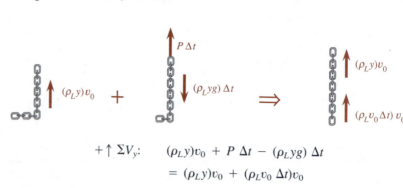

$$+\uparrow \Sigma V_y: \qquad (\rho_L y)v_0 + P \Delta t - (\rho_L y g) \Delta t$$
$$= (\rho_L y)v_0 + (\rho_L v_0 \Delta t)v_0$$

$$P = \rho_L(gy + v_0^2) \blacktriangleleft$$

REMARK. If the open-link chain here were replaced with an ideal rope of the same length L and the same mass ρ_L per unit length as shown, the elements

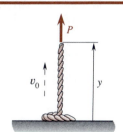

of the rope would acquire their upward velocity in a gradual and continuous manner. No impact would occur and there would be no energy loss. Applying the *principle of conservation of energy* in Sec. 13.12 to this entire ideal rope and choosing the floor to be the reference datum for the potential energies V_g and V_a, we write

$$T + V = T + V_g + V_a = \text{constant}$$

$$\frac{d}{dy}(T + V_g + V_a) = 0$$

where

$$T = \text{kinetic energy} = \tfrac{1}{2}\, \rho_L y v_0^2$$

$$V_g = \text{gravitational potential energy} = \rho_L y g\left(\frac{y}{2}\right)$$

$$V_a = \text{applied potential energy} = \int_y^0 P \, dy$$

Thus, we have[†]

$$\frac{d}{dy}\left(\tfrac{1}{2}\, \rho_L y v_0^2 + \tfrac{1}{2}\, \rho_L g y^2 + \int_y^0 P \, dy\right) = 0$$

$$\tfrac{1}{2}\, \rho_L v_0^2 + \rho_L g y - P = 0$$

$$P = \rho_L(g y + \tfrac{1}{2}\, v_0^2) \quad \blacktriangleleft$$

which is smaller than the result obtained for the open-link chain by an amount of $(\rho_L v_0^2)/2$.

Developmental Exercise

D14.34   Refer to Example 14.18. Determine (a) the work done by **P** to the chain from $y = 0$ to $y = L$, (b) the kinetic energy of the chain when $y = L$, (c) the potential energy of the chain when $y = L$ with the floor being used as a reference datum, (d) the total loss of energy due to successions of impacts between links during the lifting of the chain, (e) the magnitude of the reaction from the floor when $0 < y < L$.

EXAMPLE 14.19

An experimental rocket has a gross mass $m_0 = 1$ Mg, including a total fuel $m_f = 900$ kg. The rocket is fired vertically up from the north pole at the time $t = 0$. The fuel is consumed at a constant rate $q = 15$ kg/s and ejected

---

[†]To differentiate the integral, we apply Leibnitz's rule and write

$$\frac{d}{dy}\int_y^0 P \, dy = \frac{d}{dy}\int_y^0 P(y) \, dy = \frac{d}{dy}\int_y^0 P(t) \, dt$$

$$= 0 + 0 - P(y)\frac{dy}{dy} = -P(y) = -P$$

(Cf. Sec. B5.9 of App. B for Leibnitz's rule.)

with a speed of 2 km/s relative to the rocket. Neglecting the air resistance and the variation of gravity with altitude, determine the velocity of the rocket when (a) half of the fuel is consumed, (b) the total fuel is consumed.

Solution. From the given data, we have $m_0 = 1$ Mg $= 1000$ kg, $dm_e/dt = q = 15$ kg/s, and $\mathbf{u} = 2$ km/s $\uparrow = 2000$ m/s $\uparrow$. At the time t, the total mass of the rocket is $m = m_0 - qt$, the resultant external force is $\Sigma\mathbf{F} = mg \downarrow = (m_0 - qt)g \downarrow$, the velocity of the rocket is $\mathbf{v} = v \uparrow$, and the velocity of the ejected fuel is $\mathbf{v}_0 = \mathbf{v} - \mathbf{u} = (v - u) \uparrow$. The mass of the ejected fuel during the next short time interval Δt is $\Delta m_e = q \Delta t$. Applying the *principle of impulse and momentum* during the interval Δt, we have

$$mv = (m_0 - qt)v \qquad W \Delta t = (m_0 - qt)g \Delta t$$

$$(m - \Delta m_e)(v + \Delta v)$$
$$= (m_0 - qt - q \Delta t)(v + \Delta v)$$

$$\Delta m_e v_0 = q(\Delta t)(v - u)$$

$+\uparrow \Sigma V_y:$ $\qquad (m_0 - qt)v - (m_0 - qt)g \Delta t$

$$= (m_0 - qt - q \Delta t)(v + \Delta v) + q(\Delta t)(v - u)$$

Dividing through by Δt and letting $\Delta t \to 0$, we write

$$-(m_0 - qt)g = (m_0 - qt)\frac{dv}{dt} - qu \qquad (1)$$

Separating variables and integrating, we write

$$\int_0^t \left(\frac{qu}{m_0 - qt} - g\right) dt = \int_0^v dv$$

$$v = u \ln \left(\frac{m_0}{m_0 - qt}\right) - gt$$

When half of the fuel is consumed, we have $t = t_1$ and $v = v_1$. Thus, $t_1 = (m_f/2)/q = (900/2)/15$ s $= 30$ s and

$$v_1 = 2000 \ln \left[\frac{1000}{1000 - 15(30)}\right] - 9.81(30)$$

$$\mathbf{v}_1 = 901 \text{ m/s} \uparrow \quad \blacktriangleleft$$

When the total fuel is consumed, $t = t_2$ and $v = v_2$. We have $t_2 = m_f/q = 900/15$ s $= 60$ s and

$$v_2 = 2000 \ln \left[\frac{1000}{1000 - 15(60)}\right] - 9.81(60)$$

$$\mathbf{v}_2 = 4.02 \text{ km/s} \uparrow \quad \blacktriangleleft$$

Developmental Exercise

D14.35   Refer to Example 14.19. Show that Eq. (1) in the solution can be obtained by applying Eq. (14.43) to the rocket as described.

EXAMPLE 14.20

Refer to Example 14.16. Assume that each jet engine consumes fuel at the rate of 1.2 lb/s and ejects it into the exhaust with the same speed of 2000 ft/s relative to the airplane. Taking this fuel consumption into account, determine the magnitude F' of the propulsive force developed by each jet engine.

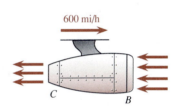

600 mi/h

Solution.   At any instant, the operations of each jet engine may be considered to consist of two parts: (1) operation with steady mass flow, and (2) operation with variable mass. From the solution in Example 14.16, we know that the operation of each jet engine with steady mass flow of air yields a propulsive force of magnitude $F = 6260.9$ lb. The operation of each jet engine with variable mass involves the loss of mass by consuming fuel at the rate of 1.2 lbm/s and ejecting it into the exhaust with the speed of 2000 ft/s relative to the airplane. By Eq. (14.46) or (14.47), we know that this operation will generate an additional propulsive force of magnitude

$$\Delta F = \frac{dm_e}{dt} u = \frac{1.2}{32.2} \text{(slug/s)(2000 ft/s)} = 74.5 \text{ lb}$$

Thus, with fuel consumption taken into account, we write

$$F' = F + \Delta F = 6335.4 \text{ lb} \qquad F' = 6.34 \text{ kips} \ \blacktriangleleft$$

Note that $\Delta F/F \approx 0.012$ or about 1.2% in the present case.

Developmental Exercise

D14.36   What is the thrust on a rocket which consumes fuel at the rate of 15 kg/s and is ejecting it at a speed of 2 km/s relative to itself?

PROBLEMS

NOTE: When any of the following problems involves a stream of water, use $\rho = 1$ Mg/m^3 for its mass density in SI units and $\gamma = 62.4$ lb/ft^3 for its specific weight (or weight density) in U.S. customary units. Unless otherwise stated, the motion of any rocket in a problem is measured in a nonrotating reference frame attached to the center of the earth, and the variation of gravity with altitude as well as air resistance during the flight of the rocket is assumed to be negligible.

14.76 through 14.78   A stream of water of cross-sectional area $A = 1.96$ in^2 and velocity $\mathbf{v} = 100$ ft/s $\rightarrow$ is split into two equal streams by a vane which is held at a fixed position by a force $\mathbf{R}$ as shown. Determine the magnitude of $\mathbf{R}$.

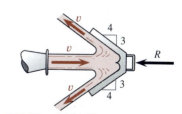

Fig. P14.76 and P14.79

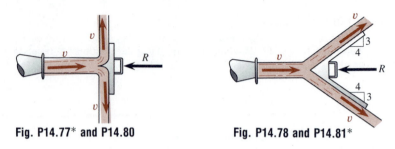

Fig. P14.77* and P14.80   **Fig. P14.78 and P14.81***

14.79 through 14.81* A stream of water of cross-sectional area $A = 400$ mm^2 and velocity $\mathbf{v} = 50$ m/s $\rightarrow$ is split into two equal streams by a vane which is held by a force $\mathbf{R}$ and moves with a constant velocity of 10 m/s $\rightarrow$. Determine the magnitude of $\mathbf{R}$.

14.82* The nozzle shown discharges a stream of water at the rate of $Q = 4$ L/s with a velocity $\mathbf{v} = 30$ m/s $\rightarrow$. The stream is split into two branches by a smooth vane which is held at a fixed position by a force $\mathbf{R}$ as shown. Determine (a) the magnitude of $\mathbf{R}$, (b) the rates of flow Q_1 and Q_2 of the two branch streams. (*Hint*. The force $\mathbf{R}$ is perpendicular to the vane.)

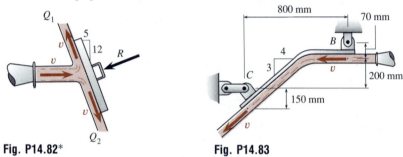

Fig. P14.82*   **Fig. P14.83**

14.83 A nozzle discharges a stream of water at the rate of $Q = 0.9$ m^3/min with a velocity $\mathbf{v} = 18$ m/s $\leftarrow$, and the stream is deflected by the vane BC as shown. Determine the reactions from the supports at B and C.

14.84 Sand falls onto a chute at the rate of 100 kg/s with a velocity $\mathbf{v}_D = 4.5$ m/s $\downarrow$ and leaves the chute with a velocity $\mathbf{v}_C = 3$ m/s $\leftarrow$ as shown. The combined weight force of the chute and the sand in it is $W = 2$ kN $\downarrow$ acting through the point G. Determine the reactions from the supports at B and C.

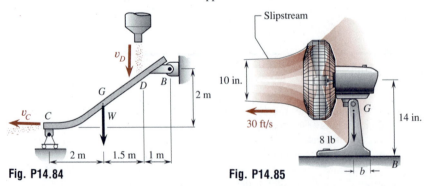

Fig. P14.84   **Fig. P14.85**

14.85 The slipstream of an 8-lb electric fan has a diameter of 10 in. and a velocity of 30 ft/s relative to the fan as shown. Assuming that air weighs 0.076 lb/ft^3, determine the minimum value of the distance b for which tipping of the fan will not occur. (*Hint*. The velocity of the air entering the fan may be assumed to be negligibly small.)

14.86* The helicopter shown has a mass of 4.5 Mg and produces a slipstream of 10-m diameter when it is hovering in midair. Assuming that the mass density of the air is 1.22 kg/m^3, determine the speed of the air in the slipstream. (Cf. the *hint* in Prob. 14.85.)

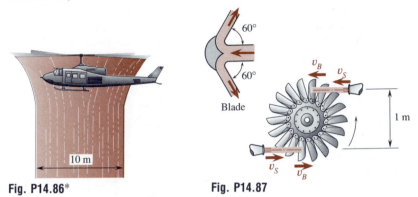

Fig. P14.86*   **Fig. P14.87**

14.87 Each of the two nozzles shown discharges a stream of water at the rate $Q = 1$ m^3/s with a speed $v_S = 30$ m/s, and the stream is deflected by a series of blades mounted on the periphery of a turbine wheel as shown. If the wheel rotates at 240 rev/min, determine (a) the input power from the streams of water, (b) the output power of the turbine, (c) the mechanical efficiency.

14.88 A jet aircraft is equipped with movable vanes as reverse thrusters to provide reverse thrust for deceleration during landing.[†] At the instant shown, the speed of the aircraft is 225 km/h, each of its engines scoops in air at the rate of 50 kg/s and discharges it with a velocity of 600 m/s relative to the reverse thrusters as shown. Determine the reverse thrust generated by each engine at this instant.

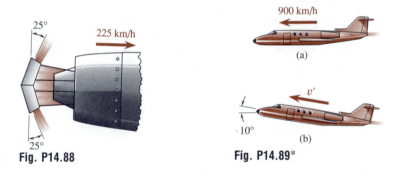

Fig. P14.88   **Fig. P14.89***

14.89* A 5-Mg jet aircraft cruising at a constant speed of 900 km/h in a level flight, as shown in Fig. (a), scoops in air at the rate of 100 kg/s and discharges it with a velocity of 800 m/s relative to the aircraft. (i) Determine the total drag due to air friction in such a flight. (ii) Assuming that the drag is proportional to the square of the speed, determine the constant speed v' with which the aircraft may climb at an angle of 10° as shown in Fig. (b) without additional power.

14.90 A jet aircraft is cruising at a constant speed of 900 km/h in a level flight as shown. Each of its three engines of equal capacities discharges air with a velocity of 600 m/s relative to the aircraft. Assuming that the drag is proportional to the square of the speed, determine the cruising speed v' of the aircraft if the engine at its tail is shut off so that only the two engines under its wings are in operation.

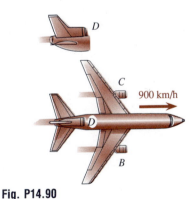

Fig. P14.90

[†]Reverse thrusters are also deployed to provide reverse thrust for the backward motion of the jet aircraft at the terminal of an airport.

14.91　Solve Prob. 14.89* if it is to be taken into account that the jet fuel is consumed at the rate of 1 kg/s and ejected in the exhaust with the same velocity of 800 m/s relative to the aircraft.

14.92　A chain of mass ρ_L per unit length and total length L is released from rest when $y = L$, where y is the distance between the floor and the top end of the chain as shown. Determine as a function of y the magnitude of the reaction from the floor as the chain is falling. (*Hint.* The portion of the chain which is still off the floor is in a state of free fall.)

Fig. P14.92

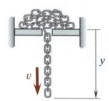

Fig. P14.93

14.93　The pile of chain shown has a mass ρ_L per unit length and a total length L. One end of the chain falls through a small hole in its support when the chain is released from rest at the time $t = 0$. Neglecting effects of friction, determine (a) the speed of the falling chain as a function of the length y, (b) the length y as a function of the time t, (c) the total energy loss as the last link leaves the hole.

14.94　The chain shown has a mass ρ_L per unit length and a total length L. One end of the chain is attached to a hook at H, and the other end is released from rest when $y = 0$. Determine as a function of y the magnitude of the reaction from the hook at H as the chain is falling. (*Hint.* The falling portion of the chain is in a state of free fall through a distance of $2y$ and with a speed v, while the time rate of mass added to the stationary portion of the chain is only $\frac{1}{2}\rho_L v$.)

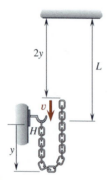

Fig. P14.94

14.95*　A space shuttle is launched vertically by firing its main engines and two booster rockets as shown. The main engines consist of three identical rocket engines, each of which burns the hydrogen-oxygen liquid propellant at the rate of 1200 lb/s and ejects it with a velocity of 12.5×10^3 ft/s relative to the engines. Determine the total thrust provided by the main engines.

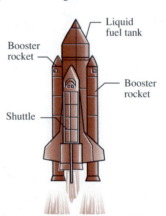

Fig. P14.95* and P14.96

14.96　A space shuttle is launched vertically by firing its main engines and two booster rockets as shown. Each booster rocket is used to provide 2×10^6 lb of thrust by burning solid propellant and ejecting it with a velocity of 9000 ft/s relative to the rocket. Determine the rate of fuel consumption by each booster rocket in tons per second.

14.97   The rocket engine of a lunar landing module is operated in such a way that it hovers motionlessly moments before its touchdown on the surface of the moon as shown. The fuel burned in the rocket is ejected with a velocity of 2500 m/s. Knowing that the gravitational acceleration on the moon is 1.618 m/s², determine the rate of fuel consumption when the total mass of the module is 18 Mg.

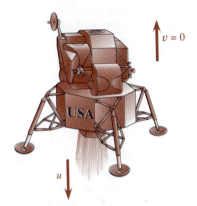

Fig. P14.97

14.98*   A rocket of gross mass 2 Mg, including 1.8 Mg of fuel, is fired vertically from the ground at the time $t = 0$. If fuel is consumed at the rate of 25 kg/s and ejected with a velocity of 2500 m/s relative to the rocket, determine the acceleration and velocity of the rocket when (a) $t = 36$ s, (b) $t = 72$ s.

14.99   Suppose that the rocket in Prob. 14.98* is redesigned as a two-stage rocket consisting of rockets A and B as shown. Each of the rockets A and B has a gross mass of 1 Mg, including 0.9 Mg of fuel. The two-stage rocket is likewise launched vertically from the ground. When the rocket A ejects its last fuel particle, its shell is released and the rocket B is ignited. Knowing that fuel is likewise consumed at the rate of 25 kg/s and ejected with a velocity of 2500 m/s relative to the rocket, determine the speed of the rocket B when (a) the rocket A is released, (b) the last fuel particle in B is ejected.

Fig. P14.99

14.14   Concluding Remarks

The *principle of impulse and momentum* for particles is the major principle in this chapter. It has been shown to be very suitable for attacking (1) *kinetic problems where time and velocity* (among others) *are to be directly related*, (2) *problems of central-force motion*, and (3) *problems of impact*. The *drawing* of appropriate *vector-diagram equations* for the momenta and impulses in the system during the time interval considered is recommended because it provides a useful geometric aid in the solutions of such problems.

The *principle of generalized virtual work* for particles propounded in Sec. 14.10 is an extension of the principle of virtual work. This principle often allows us to obtain direct solutions of some kinetic quantities in problems of impulse and momentum related to systems of particles with one degree of freedom. The advantage of such a principle lies in the fact that many of the unknown impulse vectors can be excluded when we write the generalized virtual work equation for the momenta and impulses in the system during a constrained virtual displacement. Besides, the *conservation of energy* may be applied to provide an additional equation in the solution of a problem when the system is conservative.

Variable systems are rather common in modern technology. The motion of a variable system of particles is usually analyzed by considering the motion of *an auxiliary constant system* isolated from the variable system during a short time interval Δt. The *principle of impulse and momentum* is then applied to solve the problem.

With this chapter, we have now completed the study of major basic physical laws and principles used in elementary kinetics of particles. A thorough comprehension of these laws and principles as applied to systems of particles will provide a helpful background for the subsequent study of the kinetics of rigid bodies.

REVIEW PROBLEMS

14.100 The magnitude of a force **P** varies as shown. This force is applied to a 5-kg block at the time $t = 0$ when the block is at rest. If $\mu_s = 0.6$ and $\mu_k = 0.5$ between the block and the support, determine (a) the maximum speed attained by the block, (b) the time at which the block stops again.

14.101 A 20-g bullet is fired with a velocity $\mathbf{v}_0$ at the 2-kg block B and the 4-kg block C which are resting on a smooth surface as shown. The bullet pierces B and lodges in C. If the bullet causes B and C to start moving with velocities of 3 m/s and 2 m/s, respectively, determine the magnitude of $\mathbf{v}_0$.

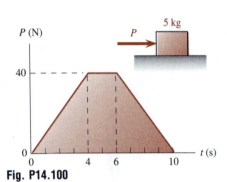

Fig. P14.100

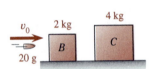

Fig. P14.101

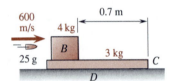

Fig. P14.102

14.102 A 25-g bullet is fired with a velocity of 600 m/s at the 4-kg block B which is resting on a 3-kg plate C as shown. The bullet lodges in B at the time $t = 0$ and causes B to slide on C and C to slide on the horizontal support D. If $\mu_k = 0.6$ between B and C and friction between C and D is negligible, determine (a) the time t_1 at which B ceases to slide on C, (b) the displacement of C during the interval $0 \le t \le t_1$, (c) the displacement of B relative to C during the same interval.

14.103 Solve Prob. 14.102 if $\mu_k = 0.1$ between C and D is taken into account.

14.104* The velocity of the cylinder A shown is $v_A = 2$ m/s ↓ when the system is released at the time $t = 0$. Determine the tension in each cable and the velocity of each cylinder when $t = 5$ s.

14.105 Refer to Prob. 14.104*. Using the *principle of generalized virtual work* for particles, determine the time t_1 at which the velocities of the cylinders become zero.

14.106 A 22-lb body slides with negligible friction on a horizontal xy plane. The body is observed to pass through the origin O with a velocity $\mathbf{v}_0 = 15\mathbf{i}$ ft/s at the time $t = 0$. Following an explosion at $t = t_1$, where $0 < t_1 < 5$ s, the body separates into three fragments, A, B, and C, as shown, of weights $W_A = 9$ lb, $W_B = 8$ lb, and $W_C = 5$ lb, respectively. At $t = 5$ s, it is observed that the position vectors of B and C are $\mathbf{r}_B = 48\mathbf{i} + 36\mathbf{j}$ ft and $\mathbf{r}_C = 75\mathbf{i} - 90\mathbf{j}$ ft, and their velocities are $\mathbf{v}_B = 6\mathbf{i} + (v_B)_y\mathbf{j}$ ft/s and $\mathbf{v}_C = 15\mathbf{i} + (v_C)_y\mathbf{j}$ ft/s. Determine the position vector $\mathbf{r}_A$ and the velocity $\mathbf{v}_A$ of A at $t = 5$ s.

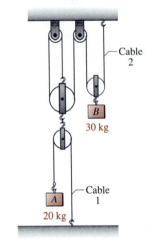

Fig. P14.104*

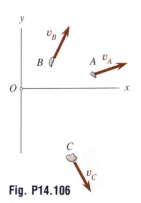

Fig. P14.106

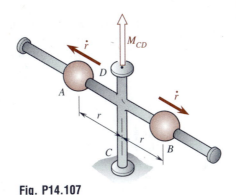

Fig. P14.107

14.107 Two particles A and B, each having a mass of 4 kg, are moving out radially at a constant rate of $\dot{r} = 0.5$ m/s as shown. At the same time, the system is rotating freely about the vertical rod CD with an angular speed $\dot{\theta}$. At the time $t = 0$, it is known that $r = 1$ m, $\dot{\theta} = 3$ rad/s, and a torque $M_{CD} = 3t^2 + 4t$ N·m in the same sense as $\dot{\theta}$ is applied to the vertical rod CD, where t is measured in seconds. Neglecting the masses of the rods, determine the value of $\dot{\theta}$ when $r = 2$ m.

14.108 A spacecraft starts its free flight around the earth in the position S at an altitude $a_S = 5000$ mi and with a velocity $\mathbf{v}_S$ as shown. If the magnitude of $\mathbf{v}_S$ is equal to the speed v_{cir} of a satellite describing a circular orbit passing through S, determine the maximum and minimum altitudes of the spacecraft in its elliptic orbit.

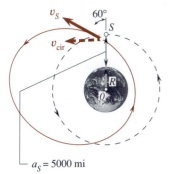

Fig. P14.108

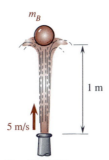

Fig. P14.109

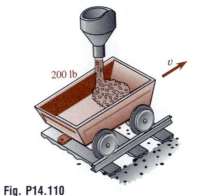

Fig. P14.110

14.109 A vertical jet of water is issued with a velocity of 5 m/s from a nozzle of 20-mm inside diameter and supports a ball at a height of 1 m as shown. Knowing that the mass density of water is 1 Mg/m³, determine the mass m_B of the ball. (*Hint.* The diameter of the jet near the ball is larger than that near the nozzle. However, the time rate of mass flow dm/dt just below the ball is equal to that at the nozzle.)

14.110 A 200-lb cart moving freely on horizontal tracks is loaded by dropping gravel vertically into it from a stationary chute, as shown, at a constant rate of 80 lb/s. The speed of the cart is 2 ft/s when the loading begins at the time $t = 0$. Determine (a) the time t_1 at which the speed of the cart is reduced to 1 ft/s, (b) the acceleration of the cart at $t = t_1$.

14.111 An inverted pendulum formed by attaching a sphere of weight W to a light rod can rotate freely about the hinge O as shown. The system is released from the vertical position OA_1 to fall to the horizontal position OA_2 where impact with the support occurs. If the highest position to which it rebounds is OA_3, determine the coefficient of restitution between the sphere and the support.

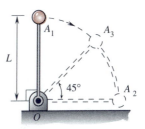

Fig. P14.111

14.112* A chain of mass ρ_L per unit length and total length L is held at one end by a force $\mathbf{P}$ as shown. Starting with $y = L$, the chain is lowered with a constant velocity of magnitude v_0. Determine as a function of y (a) the magnitude of $\mathbf{P}$, (b) the magnitude of the reaction from the floor as the chain is lowered.

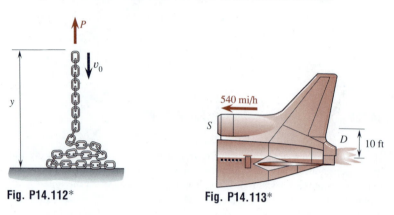

Fig. P14.112*   Fig. P14.113*

14.113* The jet engine at the tail of an airplane cruising at a speed of 540 mi/h scoops in air at S at the rate of 150 lb/s and discharges it at D with a velocity of 2500 ft/s relative to the engine. Determine the magnitude and line of action of the propulsive thrust generated by the engine.

Plane Kinematics of Rigid Bodies

Chapter 15

KEY CONCEPTS

- Translation, rotation, general plane motion, angular velocity, angular acceleration, Chasles' theorem, linkage equation, velocity center, space centrode, body centrode, acceleration center, and Coriolis acceleration.
- Methods of solution for kinematic quantities of rigid bodies in plane motion using one or more of the following: (1) relations between linear and angular motions, (2) equations of relative motion, (3) linkage equation for velocities, (4) velocity center, (5) linkage equation for accelerations, (6) acceleration center, and (7) parametric method.
- Plane kinematics using rotating reference frames.

The plane kinematics of rigid bodies is a study of the motion of rigid bodies as taking place in a single plane without considering the cause of motion. Traditional methods of analysis using nonrotating reference frames are covered first. Moreover, the technique using acceleration center is introduced. This technique is particularly useful in analyzing systems of rigid bodies which have nonzero angular accelerations but zero angular velocities at the instant under consideration. Later, the method of analysis using rotating reference frames is presented. This method is useful in many situations, especially in the analysis of accelerations of a system where two members rotate as well as slide at a guided connection. The Coriolis acceleration and other components of acceleration involved in such a system are interpreted and illustrated.

15.1   Types of Plane Motion of a Rigid Body

A rigid body is said to be in *motion* if the position of any particle of the body or the orientation of any line drawn on the body is changed during a finite time interval. A body is at *rest* if and only if its velocity and acceleration are both equal to zero. A rigid body is said to be in *plane motion* if each particle of the body remains at a constant distance from a fixed reference plane during the motion. Thus, the plane motion of a rigid body may be analyzed as taking place in a single plane.[†]

In general, the plane motion of a rigid body may be classified as one of the following three types:

(1) *Translation.* The motion of a rigid body is said to be a *translation* if the orientation of any line drawn on the body remains unchanged during the motion. In translation, the paths of the particles forming the body must be either parallel straight lines or congruent curves. If all paths are parallel straight lines, as in Fig. 15.1(a), the motion is called a *rectilinear translation*; if all paths are congruent curves, as in Fig. 15.1(b), the motion is called a *curvilinear translation*. In either case, the motion of the body is completely specified by the motion of any single point in the body.

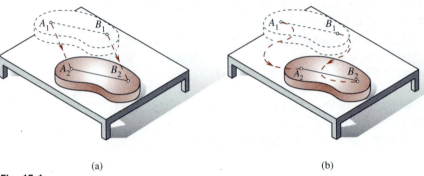

(a)   (b)

Fig. 15.1

(2) *Rotation.* The motion of a rigid body is said to be a *rotation* if all particles of the body move along circles centered on a fixed axis, which is perpendicular to the plane of motion and is called the *axis of rotation*. A rotation is an angular motion. If the axis of rotation passes through the rigid body, as in Fig. 15.2(a), the particles located on the axis will not move. If the axis of rotation is located outside the rigid body, as in Fig. 15.2(b), all particles of the rigid body will move.

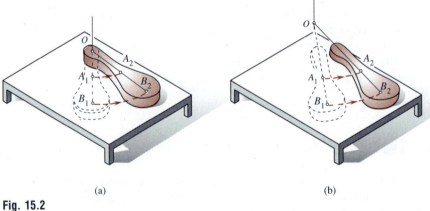

(a)   (b)

Fig. 15.2

[†]The kinematics of a rigid body requiring three spatial coordinates for its description is covered in Chap. 19.

(3) *General plane motion.* Any plane motion which is neither a translation nor a rotation is referred to as a *general plane motion*. When a rigid body is in general plane motion, the body may be viewed as undergoing a combined motion of a translation parallel to a fixed plane and a rotation about an axis perpendicular to the fixed plane, as in Fig. 15.3. Many problems of plane motion may be studied using nonrotating reference frames, but there are others whose study would require the use of rotating reference frames.

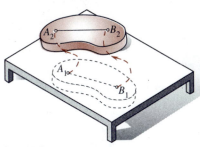

Fig. 15.3

Developmental Exercises

D15.1 Define the following terms related to a rigid body: (a) translation, (b) rotation, (c) general plane motion.

D15.2 Identify the type of plane motion for each of the following:

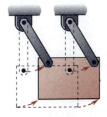
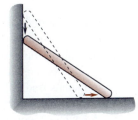

| Compound pendulum | Parallel-link swinging plate | Sliding rod | Sliding block |
| (a) | (b) | (c) | (d) |

Fig. D15.2

USE OF NONROTATING REFERENCE FRAMES

15.2 Translation

Let *A* and *B* be any two of the particles of a rigid body in either rectilinear or curvilinear translation parallel to the *xy* plane as shown in Fig. 15.4. Denoting respectively the position vectors of *A* and *B* by $\mathbf{r}_A$ and $\mathbf{r}_B$, and the position vector of *B* relative to *A* by $\mathbf{r}_{B/A}$, as shown, we write

$$\mathbf{r}_B = \mathbf{r}_A + \mathbf{r}_{B/A} \tag{15.1}$$

Since *A* and *B* belong to the same rigid body, the direction and magnitude of the vector $\mathbf{r}_{B/A}$ must remain unchanged during the translation of the rigid body. Thus, $\mathbf{r}_{B/A}$ is a constant vector and the increments of $\mathbf{r}_B$ and $\mathbf{r}_A$ in Eq. (15.1) are related by

$$\Delta\mathbf{r}_B = \Delta\mathbf{r}_A \tag{15.2}$$

which expresses the property that all points of a rigid body *in translation* during any time interval have the *same displacement*. By differentiating Eq. (15.1) with respect to the time *t*, we obtain the velocities

$$\mathbf{v}_B = \mathbf{v}_A \tag{15.3}$$

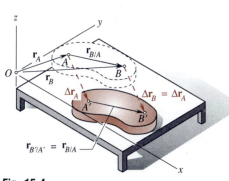

Fig. 15.4

Differentiating Eq. (15.3) with respect to t, we obtain the accelerations

$$\mathbf{a}_B = \mathbf{a}_A \qquad (15.4)$$

Equations (15.3) and (15.4) indicate that, *when a rigid body is in translation during a certain time interval, all points of the body must have the same velocity and the same acceleration at any instant during that time interval.*

Developmental Exercise

D15.3   The rod BC of the linkage rotates about the pin at C, and the ends A and B of the rod AB have the same velocity but different accelerations at the instant shown. Is the rod AB in translation at this instant?

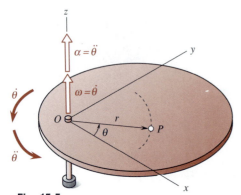

Fig. D15.3

15.3   Rotation

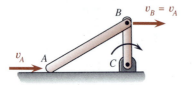

Let us consider a thin rigid slab which rotates about the z axis as shown in Fig. 15.5, where the reference frame $Oxyz$ is fixed in space. Note that the angular position coordinate θ for the line segment $\overline{OP}$ on the slab serves to specify the angular position of the slab. The value of θ must be measured in radians (rad) when a circular arc length s subtended by the angle θ and a radius r is given by the relation $s = r\theta$. Otherwise, θ may also be measured in degrees (°) or revolutions (rev). We note that

$$1 \text{ rev} = 2\pi \text{ rad} = 360° \qquad (15.5)$$

Fig. 15.5

The *angular velocity* of a rigid body is a vector denoted by $\boldsymbol{\omega}$ and is defined as the time rate of change of the angular position coordinate θ of the rigid body. The direction of $\boldsymbol{\omega}$ is given by the extended thumb of the right hand when the other fingers of the right hand are curled in the direction of rotation of the body. Referring to Fig. 15.5, we write

$$\boldsymbol{\omega} = \omega \mathbf{k} \qquad \omega = \frac{d\theta}{dt} = \dot{\theta} \qquad (15.6)^\dagger$$

where the scalar quantity ω is called the *angular speed* of the body.

The *angular acceleration* of a rigid body is a vector denoted by $\boldsymbol{\alpha}$ and is defined as the time rate of change of the angular velocity $\boldsymbol{\omega}$ of the body. Since $\mathbf{k}$ is a constant unit vector, we write

$$\boldsymbol{\alpha} = \frac{d\boldsymbol{\omega}}{dt} = \frac{d}{dt}(\omega \mathbf{k}) = \frac{d\omega}{dt}\mathbf{k}$$

$$\boldsymbol{\alpha} = \alpha \mathbf{k} \qquad \alpha = \frac{d\omega}{dt} = \dot{\omega} \qquad (15.7)$$

Applying the chain rule of differentiation, we write

$$\alpha = \frac{d\omega}{dt} = \frac{d\omega}{d\theta}\frac{d\theta}{dt} = \frac{d\omega}{d\theta}\omega \qquad \alpha = \omega\frac{d\omega}{d\theta} \qquad (15.8)$$

---

†Cf. the angular velocities defined by Eqs. (15.34) and (15.40).

Usually, the angular velocity ω is measured in rad/s or in rpm, while the angular acceleration α is measured in rad/s^2. In drawings, the angular velocity and angular acceleration vectors are represented either by *curved arrows* in the plane of rotation or by *hollow arrows* perpendicular to the plane of rotation, as shown in Fig. 15.5.

The orientations of ω and α of a body rotating about a fixed axis are always parallel to the axis of rotation. In this case, only the *magnitudes* and *senses* of ω and α are the nontrivial information to be described or predicted. Consequently, the rotation of a body about a fixed axis can be studied using essentially *scalar analysis*.

The scalar equations in Eqs. (15.6) through (15.8) should be recognized as mathematically analogous to Eqs. (11.3), (11.6), and (11.8), which define the rectilinear motion of particles. Thus, all relations described for the rectilinear motion of particles are applicable to the case of rotation of a rigid body about a fixed axis. In application, we may simply replace the linear kinematic quantities x, v, and a in Chap. 11 with their respective counterparts θ, ω, and α in rotation.

If the angular acceleration is a constant, $\alpha = \alpha_c$, the scalar equations in Eqs. (15.6) through (15.8) may readily be integrated to yield

$$\omega = \omega_0 + \alpha_c t \tag{15.9}$$

$$\theta = \theta_0 + \omega_0 t + \tfrac{1}{2}\alpha_c t^2 \tag{15.10}$$

$$\omega^2 = \omega_0^2 + 2\alpha_c(\theta - \theta_0) \tag{15.11}$$

where t is the time variable, and ω_0 and θ_0 are the angular velocity and the angular position coordinate at $t = 0$.

EXAMPLE 15.1

A grinding wheel is rotating at the rate of 1500 rpm ↻ when the power is cut off at the time $t = 0$. Because of friction, the wheel decelerates at a constant rate α_c and comes to rest in 500 rev. Determine α_c and the time t at which the wheel comes to rest.

Solution. The angular speed of the grinding wheel at $t = 0$ is

$$\omega_0 = 1500 \, \frac{\text{rev}}{\text{min}} \cdot \frac{2\pi \text{ rad}}{1 \text{ rev}} \cdot \frac{1 \text{ min}}{60 \text{ s}} = 50\pi \text{ rad/s}$$

When it comes to rest, its final speed is $\omega = 0$. Its angular displacement $\Delta\theta$ from the time it starts to slow down until the time it comes to rest is

$$\Delta\theta = \theta - \theta_0 = 500 \text{ rev} \cdot \frac{2\pi \text{ rad}}{1 \text{ rev}} = 1000\pi \text{ rad}$$

Since the wheel slows down at a constant rate, we write

$$\omega^2 = \omega_0^2 + 2\alpha_c(\theta - \theta_0): \qquad 0^2 = (50\pi)^2 + 2\alpha_c(1000\pi)$$

$$\alpha_c = -1.25\pi \qquad \alpha_c = 3.93 \text{ rad/s}^2 \text{ ↻} \blacktriangleleft$$

To find the time t at which the wheel comes to rest, we write

$$\omega = \omega_0 + \alpha_c t: \qquad 0 = 50\pi + (-1.25\pi)t$$

$$t = 40 \text{ s} \blacktriangleleft$$

15.4 Linear and Angular Motions

A thin rigid slab which lies in the xy plane and rotates about the z axis is shown in Fig. 15.6, where the reference frame $Oxyz$ is fixed in space. In this case, all points of the slab move along concentric circles with centers all located at the origin O. As the slab in Fig. 15.6 rotates about the z axis, the point P of the slab will describe a circular path of radius $\overline{OP}$ with center at O. We here use the notations $r = \overline{OP} = \rho$ for the radius of curvature of the path of P, $\omega = \dot{\theta}$ for the angular speed of $\overline{OP}$, and $\alpha = \dot{\omega} = \ddot{\theta}$ for the time rate of change of the angular speed. Of course, the angular velocity and angular acceleration of $\overline{OP}$ are $\boldsymbol{\omega} = \omega\mathbf{k}$ and $\boldsymbol{\alpha} = \alpha\mathbf{k}$, respectively, as indicated in parts (a) and (b) of Fig. 15.6.

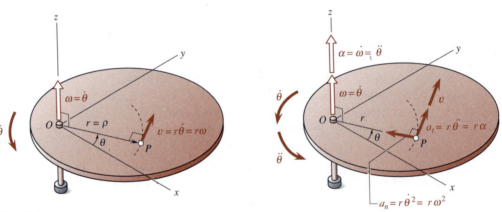

Fig. 15.6

The relations between the linear motion of the point P and the angular motion of the line segment $\overline{OP}$, which exemplifies the angular motion of the rigid slab, can be established from the relations expressed in Eqs. (11.56). Substituting the aforementioned notations into Eqs. (11.56), we write

$$v = r\omega \qquad a_t = r\alpha \qquad a_n = r\omega^2 \qquad (15.12)$$

where v, a_t, and a_n are, respectively, the speed, the tangential component of acceleration, and the normal component of acceleration of the point P. The quantities v, a_t, and a_n have been illustrated in Fig. 15.6(a) and (b). Note that *the resultant acceleration* $\mathbf{a}$, which is the vector sum of $\mathbf{a}_t$ and $\mathbf{a}_n$, *is always directed toward the concave side of the path of* P.

The velocity $\mathbf{v}$ and the acceleration components $\mathbf{a}_t$ and $\mathbf{a}_n$ of P may be related to its position vector $\mathbf{r}$, the angular velocity $\boldsymbol{\omega}$ of $\mathbf{r}$, and the angular acceleration $\boldsymbol{\alpha}$ of $\mathbf{r}$ through the use of vector analysis. By the definition of the cross product of two vectors, we see that $\mathbf{v}$ is obtained by crossing $\boldsymbol{\omega}$ into $\mathbf{r}$ as shown in Fig. 15.7(a). This cross product gives precisely the correct magnitude and direction of $\mathbf{v}$. Thus, we write

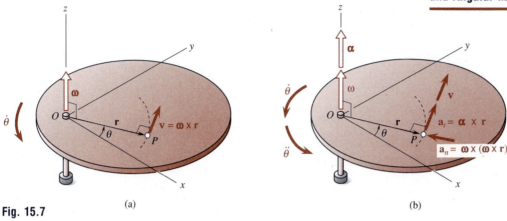

Fig. 15.7

$$\mathbf{v} = \boldsymbol{\omega} \times \mathbf{r} \qquad\qquad (15.13)$$

Note that $\mathbf{r} \times \boldsymbol{\omega} = -\mathbf{v}$. Since the acceleration $\mathbf{a}$ of P is, by definition, the time rate of change of $\mathbf{v}$, we write

$$\mathbf{a} = \dot{\mathbf{v}} = \frac{d}{dt}(\boldsymbol{\omega} \times \mathbf{r}) = \dot{\boldsymbol{\omega}} \times \mathbf{r} + \boldsymbol{\omega} \times \dot{\mathbf{r}}$$

$$= \boldsymbol{\alpha} \times \mathbf{r} + \boldsymbol{\omega} \times \mathbf{v}$$

$$= \boldsymbol{\alpha} \times \mathbf{r} + \boldsymbol{\omega} \times (\boldsymbol{\omega} \times \mathbf{r}) = \mathbf{a}_t + \mathbf{a}_n$$

where $\boldsymbol{\alpha} = \dot{\boldsymbol{\omega}}$, the angular acceleration of $\mathbf{r}$. Thus, we have

$$\mathbf{a}_t = \boldsymbol{\alpha} \times \mathbf{r} \qquad \mathbf{a}_n = \boldsymbol{\omega} \times (\boldsymbol{\omega} \times \mathbf{r}) \qquad (15.14)$$

which are shown in Fig. 15.7(b). Note that Eqs. (15.13) and (15.14) are the vector versions of Eqs. (15.12).

In plane motion of a rigid body, the directions of $\boldsymbol{\omega}$ and $\boldsymbol{\alpha}$ are parallel to each other because both of them are perpendicular to the plane of motion. In three-dimensional motion of a rigid body, the directions of $\boldsymbol{\omega}$ and $\boldsymbol{\alpha}$ are generally not parallel to each other.

EXAMPLE 15.2

A load L is raised from rest through a distance of 9 m by a cable wrapped around the drum of a hoisting system as shown. Determine the angular displacements of the gears A and B in raising the load.

Solution. From the sketch shown on next page, we see that the distance s traveled by the load L must be equal to the length of the arc CC' which corresponds to the part of the cable wound around the drum during the raising of the load. Thus, the angular displacement $\Delta\theta_B$ of the gear B in raising the load L through a distance $s = 9$ m is

$$\Delta\theta_B = \frac{s}{\overline{HC}} = \frac{9}{0.36} \qquad \Delta\theta_B = 25 \text{ rad } \circlearrowright \blacktriangleleft$$

The length of the arc DD' subtended by $\Delta\theta_B$, on the gear B, is s' $= \overline{HD} \, \Delta\theta_B$, while the length of the arc EE' subtended by the corresponding

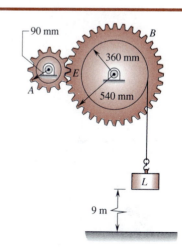

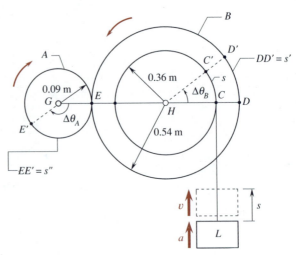

angular displacement $\Delta\theta_A$ of the gear A is $s'' = \overline{GE}\,\Delta\theta_A$. Since the lengths s' and s'' of the arcs rolled out by the gears A and B must be equal, we write

$$\overline{HD}\,\Delta\theta_B = \overline{GE}\,\Delta\theta_A \qquad 0.54(25) = 0.09\,\Delta\theta_A$$

$$\Delta\theta_A = 150 \text{ rad} \;\blacktriangleleft$$

Developmental Exercise

D15.5 Work Example 15.2 if the radius of the gear A is equal to 180 mm.

EXAMPLE 15.3

Refer to Example 15.2. Suppose that the gear A of the hoist motor accelerates from rest at a constant rate to a speed of 150 rpm in 5 s and then remains at this speed. Determine the total time required to raise the load L from rest through the distance of 9 m.

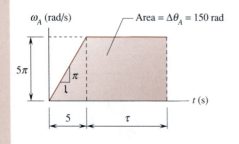

Solution. Since $\omega_A = 0$ at the time $t = 0$ and $\omega_A = 150$ rpm $= 5\pi$ rad/s at $t = 5$ s, we draw the angular velocity curve for the gear A as shown. From the solution in Example 15.2, we know that the angular displacement of the gear A in raising the load through 9 m is $\Delta\theta_A = 150$ rad. By analogy to that in Sec. 11.8, we may equate the 150-rad angular displacement to the shaded trapezoidal area as shown. We write

$$150 = \tfrac{1}{2}[(5 + \tau) + \tau](5\pi) \qquad \tau = 7.05$$

Thus, the total time required is

$$t = 5 + \tau = 5 + 7.05 \qquad t = 12.05 \text{ s} \;\blacktriangleleft$$

Developmental Exercise

D15.6 Refer to Example 15.3. Determine the velocity of the load L when $t = 5$ s.

EXAMPLE 15.4

Refer to Example 15.2, where $\omega_A = 6\omega_B$ and $\alpha_A = 6\alpha_B$. Suppose that, at a certain instant, the gears A and B have $\omega_A = 2\pi$ rad/s $\circlearrowleft$, $\alpha_A = 6\pi$ rad/s^2 $\circlearrowleft$, and $\omega_B = \pi/3$ rad/s $\circlearrowright$, $\alpha_B = \pi$ rad/s^2 $\circlearrowright$, respectively. The point of contact between the two gears is E. For this instant, determine the acceleration $\mathbf{a}_E'$ of the particle at E on the gear A and the acceleration $\mathbf{a}_E''$ of the particle at E on the gear B.

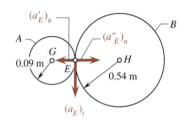

Solution. Since no slipping occurs at E between the two gears, the tangential component $(\mathbf{a}_E)_t$ of the acceleration of the particle at E on either gear must be the same and may be computed from the motion of either gear. From the given data and the sketch as shown, we write

$$(a_E)_t = r_A \alpha_A = 0.09(6\pi) = 0.54\pi$$

The normal components $(a_E')_n$ and $(a_E'')_n$ of the accelerations of the particles at E on the gears A and B, respectively, are different. We write

$$(a_E')_n = r_A \omega_A^2 = 0.09(2\pi)^2 = 0.36\pi^2$$

$$(a_E'')_n = r_B \omega_B^2 = 0.54(\pi/3)^2 = 0.06\pi^2$$

Adding vectorially the normal and tangential components, we write

$$\mathbf{a}_E' = -(a_E')_n \mathbf{i} - (a_E)_t \mathbf{j} = -0.36\pi^2 \mathbf{i} - 0.54\pi \mathbf{j}$$

$$\mathbf{a}_E'' = (a_E'')_n \mathbf{i} - (a_E)_t \mathbf{j} = 0.06\pi^2 \mathbf{i} - 0.54\pi \mathbf{j}$$

$$\mathbf{a}_E' = 3.94 \text{ m/s}^2 \ \nearrow \ 25.5° \qquad \mathbf{a}_E'' = 1.797 \text{ m/s}^2 \ \nwarrow \ 70.8° \ \blacktriangleleft$$

We see here that $\mathbf{a}_E'$ and $\mathbf{a}_E''$ are considerably different.

Developmental Exercise

D15.7 A system consisting of a gear and a rack is shown. If the rack moves with $\mathbf{v} = 16$ in./s $\rightarrow$ and $\mathbf{a} = 120$ in./s^2 $\rightarrow$ at the given instant, determine ω, α, and $\mathbf{a}_P$ of the point P of the gear.

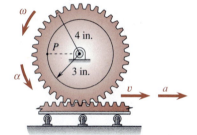

Fig. D15.7

PROBLEMS

15.1 A flywheel is attached to the shaft of a motor which has a rated speed of 2400 rpm. When the motor is turned on, it accelerates at a constant rate from rest and reaches its rated speed after executing 80 rev; and when the motor is turned off, it decelerates at a constant rate and comes to rest after executing 900 rev. Determine the time required for the flywheel (a) to reach its rated speed, (b) to come to rest.

15.2 The angular position coordinate of a swinging pendulum as shown is given by $\theta = (\pi/3) \cos 2\pi t$, where θ is measured in radians and t in seconds. Determine the magnitudes of the maximum angular velocity and the maximum angular acceleration of the pendulum.

Fig. P15.2

15.3* At the instant shown, the gear A rotates with $\omega_A = 2$ rad/s $\circlearrowright$ and the gear B rotates with $\alpha_B = 6$ rad/s^2 $\circlearrowleft$. Determine $\mathbf{v}_D$, $\mathbf{v}_E$, $\mathbf{a}_D$, and $\mathbf{a}_E$ of the points D and E at this instant.

15.4 The gear A of the system shown executes 200 rev over a time interval Δt, while its angular velocity is being increased from 300 rpm $\circlearrowright$ to 900 rpm $\circlearrowright$ at a constant rate α_c. Determine α_c and Δt.

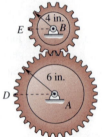

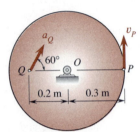

Fig. P15.3* and P15.4   **Fig. P15.5 and P15.6***

15.5 A circular disk rotates about the point O as shown. At the instant shown, the point P moves with $v_P = 0.9$ m/s $\uparrow$ and $\mathbf{a}_Q$ of the point Q of the disk is directed as indicated. Determine α of the disk and the magnitude of $\mathbf{a}_Q$ at this instant.

15.6* At the instant shown, $\mathbf{a}_Q$ of the point Q of the rotating disk is directed as indicated and has a magnitude of 10 m/s^2. Determine the speed v_P of the point P and α of the disk at this instant.

15.7 A small object P rests on a horizontal disk as shown. At the time $t = 0$, the disk accelerates at a constant rate $\alpha_c = \frac{1}{2}\mathbf{k}$ rad/s^2 from rest until $t = 5$ s, then it rotates freely without any angular acceleration. It is known that P will start to slide on the disk when its resultant acceleration exceeds 3 m/s^2. Determine the maximum radial distance r for which sliding of P on the disk will not occur.

15.8 The gears A, B, and C, of radii $r_A = 500$ mm, $r_B = 200$ mm, and $r_C = 150$ mm, respectively, all rotate about fixed axes parallel to the x axis as shown. At a certain instant, the gear A rotates with $\omega_A = 6\mathbf{i}$ rad/s and the gear C rotates with $\alpha_C = 10\mathbf{i}$ rad/s^2. Determine ω_B and α_B of the gear B at this instant.

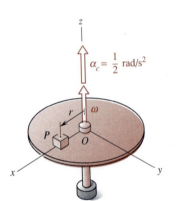

Fig. P15.7

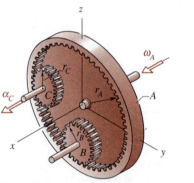

Fig. P15.8

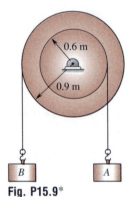

Fig. P15.9*

15.9* A 0.6-m radius drum carrying the load A is rigidly attached to a 0.9-m radius pulley carrying the load B as shown. At the time $t = 0$, the load B moves with a velocity of 2 m/s $\downarrow$ and a constant acceleration of 3 m/s^2 $\downarrow$. Over the time interval $0 \leq t \leq 2$ s, determine (a) the number of revolutions executed by the pulley, (b) the displacement of the load A.

15.10　The relation between ω and θ of a cam rotating about a fixed axis is plotted as shown. Estimate the magnitudes of $\boldsymbol{\alpha}$ of the cam when (a) $\theta = 2$ rad, (b) $\theta = 6$ rad.[†]

15.11*　Refer to Prob. 15.10. Estimate the magnitudes of $\boldsymbol{\alpha}$ of the cam when (a) $\theta = 4$ rad, (b) $\theta = 8$ rad.[†]

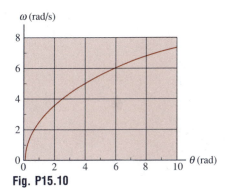

Fig. P15.10

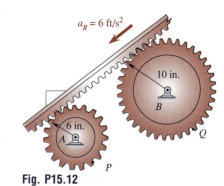

Fig. P15.12

15.12　The rack R moves with $\mathbf{a}_R = 6$ ft/s² ↙ on the gears A and B. At the instant shown, the point P of the gear A has an acceleration of magnitude 10 ft/s². Determine the magnitude of the acceleration of the point Q of the gear B at this instant.

15.13*　A roll of carpet material is unrolled by pulling its end downward with a constant speed v_0 as shown, where h denotes the thickness of the carpet and r denotes the radius of the carpet roll at the instant under consideration. Determine the angular acceleration of the carpet roll in terms of the parameters v_0, h, and r. [*Hint.* Note that $v_0 = r\omega$, $\alpha = \dot{\omega}$, and $h(v_0\,dt) = 2\pi r(-dr)$ from the change in volume.]

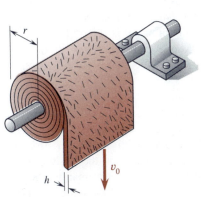

Fig. P15.13*

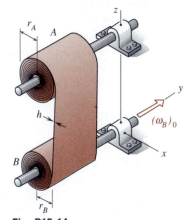

Fig. P15.14

15.14　The roll of paper at A is being transferred to that at B by turning the paper roll at B with a constant angular speed $(\omega_B)_0$ as shown. The radii of the paper rolls at A and B are r_A and r_B, respectively, and the thickness of the paper is h. Determine α_A of the paper roll at A at the instant shown in terms of the parameters r_A, r_B, $(\omega_B)_0$, and h. [*Hint.* The time rates of change $\dot{r}_A$ and $\dot{r}_B$ are related to each other by the fact that, during the time interval dt, the volume of paper per unit width unrolled from the roll at A is equal to that taken up by the roll at B; i.e., $2\pi r_A(-dr_A) = 2\pi r_B(dr_B) = hr_B(\omega_B)_0\,dt$.]

[†]Estimate by applying Eq. (15.8).

15.15 A circular cylinder is rotating about the fixed axis AB with a constant angular speed $\omega_0 = 13$ rad/s as shown, where AB passes through the origin O and the unit vector along AB is $\boldsymbol{\lambda}_{AB} = \frac{1}{13}(-5\mathbf{i} + 12\mathbf{k})$. At the instant considered, the coordinates of the point P of the cylinder are (3, 2, 6) m. Determine $\mathbf{v}_P$ and $\mathbf{a}_P$ of P at this instant. [*Hint*. Apply Eqs. (15.13) and (15.14).]

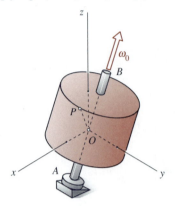

Fig. P15.15

15.16 Refer to Prob. 15.15. At the instant considered, determine (a) the shortest distance d_s between the point P and the axis AB, (b) v, a_t, a_n, and a of the point P. [*Hint*. Use Eqs. (15.12).]

15.5 General Plane Motion: Chasles' Theorem

The most general motion of a rigid body is equivalent to a translation of the body with some point of the body (or its massless extension) *plus a rotation of the body about an axis passing through that point, and the order of such translation and rotation is immaterial.* This property is referred to as *Chasles' theorem*, after the French geometer Michel Chasles (1793–1880).[†] This theorem will here be applied to the study of the general plane motion of rigid bodies.

Let a body executing a general plane motion as shown in Fig. 15.8 be considered. Over a certain time interval, two given points P and Q of the body will have moved, respectively, from P_1 to P_2 and from Q_1 to Q_2. This motion is equivalent to a translation of the body with the point P from P_1 to P_2 plus a rotation of the body about the point P from the position P_2Q_1' to the position P_2Q_2; of course, it is also equivalent to a rotation about P first plus a translation with P later as shown. The actual motion of the body is a combination of translation and rotation simultaneously.

The motion of the body in Fig. 15.8 may also be represented in Fig. 15.9. In the latter figure, the general plane motion of the body is shown to be equivalent to a translation of the body with the point Q from Q_1 to Q_2 plus a rotation of the body about the point Q from the position Q_2P_1' to the position Q_2P_2; of course, it is also equivalent to a rotation about Q first plus

---

[†]We accept Chasles' theorem as a more-or-less-obvious axiom. This theorem is valid in studying both the kinematics and the kinetics of rigid bodies. When this theorem is used in studying the plane kinetics of rigid bodies, the point of reference mentioned in the theorem is usually the mass center G of the rigid body, although the theorem remains valid when the point of reference is not the mass center. An illustration for the latter case is provided in D16.23. For further treatment, see Beggs, J. S., *Advanced Mechanism*, Macmillan, New York, 1966.

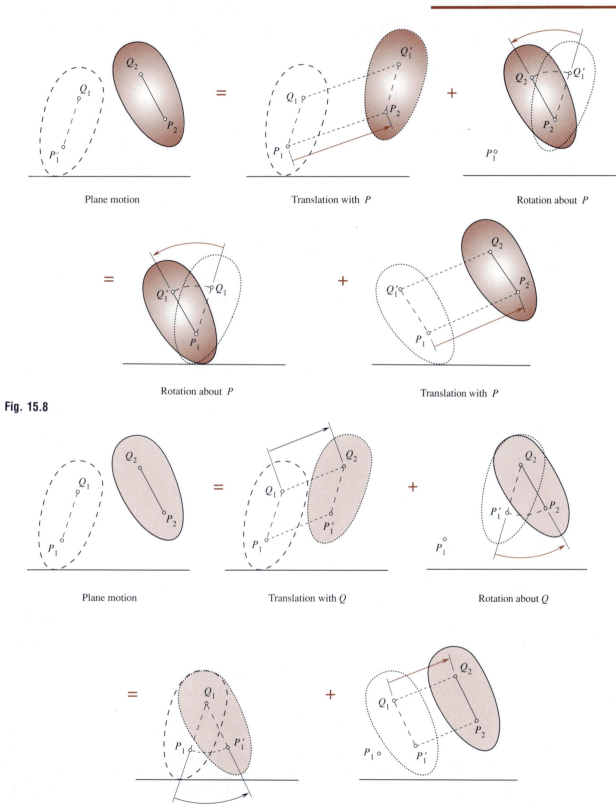

Plane motion   Translation with P   Rotation about P

Rotation about P   Translation with P

Fig. 15.8

Plane motion   Translation with Q   Rotation about Q

Rotation about Q   Translation with Q

Fig. 15.9

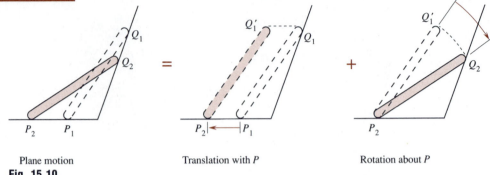

Plane motion   Translation with P Rotation about P

Fig. 15.10

a translation with Q later as shown. Note in Figs. 15.8 and 15.9 that the amounts of translation with different points may be unequal; the amounts of rotation about those points are always equal.

As additional examples, a sliding rod is shown in Fig. 15.10 and a rolling cylinder is shown in Fig. 15.11. Explanations for these two figures are similar to those given in the preceding paragraphs.

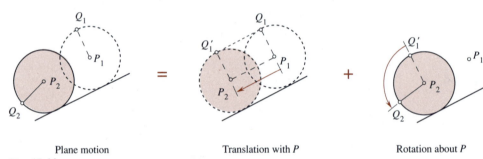

Plane motion   Translation with P Rotation about P

Fig. 15.11

Developmental Exercises

D15.8 Describe Chasles' theorem.

D15.9 Show that the motion of the sliding rod in Fig. 15.10 is also equivalent to a rotation about the end Q plus a translation with Q.

D15.10 Show that the motion of the cylinder in Fig. 15.11 is also equivalent to a translation with the point Q plus a rotation about Q.

15.6 Velocities in Relative Motion

Suppose that the points A and B are two different points of a rigid slab in plane motion with an angular velocity $\boldsymbol{\omega} = \omega\mathbf{k}$, as shown in Fig. 15.12. By Chasles' theorem, we may view the plane motion of the rigid slab as a rotation of the slab about A plus a translation of the slab with A. Such a viewpoint is depicted in the graphical equation shown. Thus, we may express the velocity $\mathbf{v}_B$ of B as the velocity $\mathbf{v}_{B/A}$ of B relative to A plus the velocity $\mathbf{v}_A$ of A; i.e.,

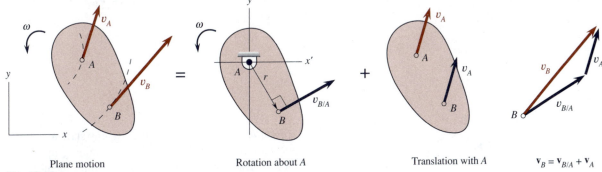

Plane motion   Rotation about A Translation with A $\mathbf{v}_B = \mathbf{v}_{B/A} + \mathbf{v}_A$

Fig. 15.12

$$\mathbf{v}_B = \mathbf{v}_{B/A} + \mathbf{v}_A \qquad\qquad (15.15)^{\dagger}$$

which is equivalent to Eq. (11.36). Furthermore, letting $r = r_{B/A} = \overline{AB}$ in Fig. 15.12 and using the cross product of two vectors, we see that

$$v_{B/A} = r\omega \qquad\qquad \mathbf{v}_{B/A} = \boldsymbol{\omega} \times \mathbf{r} \qquad\qquad (15.16)$$

where $\mathbf{r} = \mathbf{r}_{B/A} = \overrightarrow{AB}$. Note that (1) *the velocity* $\mathbf{v}_{B/A}$ *of B relative to A is due solely to the rotational part of the motion of the slab* (i.e., the value of $\mathbf{v}_{B/A}$ vanishes whenever $\boldsymbol{\omega} = \mathbf{0}$), (2) *the direction of* $\mathbf{v}_{B/A}$ *is perpendicular to* $\overrightarrow{AB}$, and (3) *the vectors* $\boldsymbol{\omega}$, $\overrightarrow{AB}$, *and* $\mathbf{v}_{B/A}$ (in this order) *form a right-handed trihedral.* As a matter of fact, we may view $\mathbf{v}_{B/A}$ as the velocity of the tip B of the radius vector $\overrightarrow{AB}$ which rotates with the angular velocity $\boldsymbol{\omega}$ about A as a "fixed point."

For a system consisting of a single member or several pin-connected members, the equation linking the motions of the key points (e.g., the connecting pins) of the system is referred to as a *linkage equation* for the motion of the system. Such an equation ensures the compatibility of motion of the linkage with its constraint conditions. For the linkage illustrated in Fig. 15.13, we note that the pins at A and E are fixed; therefore, their velocities

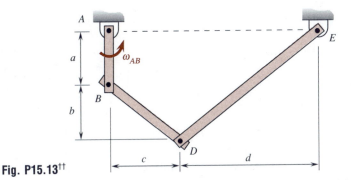

Fig. P15.13††

---

†The order of terms in Eq. (15.15) is preferable to that in Eq. (11.36) in establishing the *linkage equation* for velocities of a system, such as Eq. (15.17).

††In *mechanism*, the linkage shown is called a *four-bar linkage*, where AE is the *frame* or *ground link*, AB is the *driver* or *crank*, BD is the *coupler link*, and DE is the *follower* or *output link*.

$\mathbf{v}_A$ and $\mathbf{v}_E$ are zero. Taking these constraint conditions $\mathbf{v}_A = \mathbf{v}_E = \mathbf{0}$ into account and applying Eq. (15.15), we write

$$\mathbf{v}_B = \mathbf{v}_{B/A} + \mathbf{v}_A = \mathbf{v}_{B/A}$$

$$\mathbf{v}_B = \mathbf{v}_{B/D} + \mathbf{v}_D = \mathbf{v}_{B/D} + (\mathbf{v}_{D/E} + \mathbf{v}_E) = \mathbf{v}_{B/D} + \mathbf{v}_{D/E}$$

Thus, we have

$$\mathbf{v}_{B/A} = \mathbf{v}_{B/D} + \mathbf{v}_{D/E} \qquad (15.17)$$

which is the *linkage equation* for velocities of the system shown in Fig. 15.13. Such an equation may be used as a basis for determining the linear velocities of certain points and the angular velocities of certain members in a system.

EXAMPLE 15.5

The end B of the rod AB is pulled with $\mathbf{v}_B = 0.9$ m/s $\rightarrow$. For the position shown, determine $\boldsymbol{\omega}$ of the rod and $\mathbf{v}_A$ of the end A of the rod.

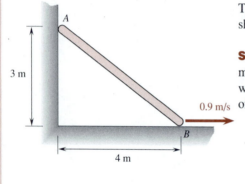

Solution. From the constraints of the rod AB, we see that $\mathbf{v}_A$ of its end A must be directed downward and $\boldsymbol{\omega}$ of the rod must be counterclockwise when $\mathbf{v}_B = 0.9$ m/s $\rightarrow$ as given. Thus, the *linkage equation* for velocities of the rod may be written as

$$\mathbf{v}_B = \mathbf{v}_{B/A} + \mathbf{v}_A$$

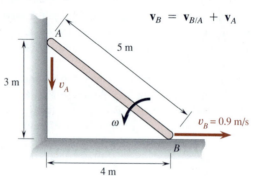

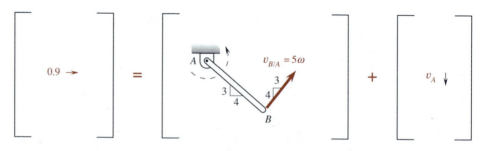

This vector equation contains two unknowns: ω and v_A, which may be solved from the two ensuing scalar equations as follows:

$$\xrightarrow{+} \Sigma V_x: \qquad 0.9 = \tfrac{3}{5}(5\omega) + 0 \qquad \omega = 0.3$$

$$+\uparrow \Sigma V_y: \qquad 0 = \tfrac{4}{5}(5\omega) - v_A \qquad v_A = 1.2$$

$$\omega = 0.3 \text{ rad/s } \circlearrowleft \qquad \mathbf{v}_A = 1.2 \text{ m/s } \downarrow \quad \blacktriangleleft$$

Developmental Exercise

D15.11   What is a *linkage equation* for the motion of a system consisting of a single member or several pin-connected members?

EXAMPLE 15.6

Refer to the system in Fig. 15.13. If $a = b = 3$ ft, $c = 4$ ft, $d = 8$ ft, and $\omega_{AB} = 10$ rad/s $\circlearrowright$ in the position shown, determine (a) ω_{BD} and ω_{DE} of the members BD and DE, (b) $\mathbf{v}_D$ of the pin at D.

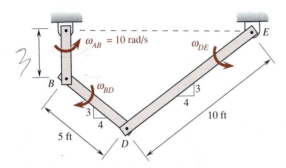

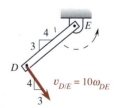

Solution.   We first indicate ω_{AB} as shown. The directions of ω_{BD} and ω_{DE} are perpendicular to the plane of motion, and their senses may initially be assumed as indicated. Since the *linkage equation* for velocities of the system has already been derived as Eq. (15.17), we here use that equation and write

$$\mathbf{v}_{B/A} = \mathbf{v}_{B/D} + \mathbf{v}_{D/E}$$

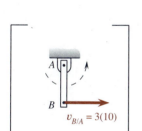

$$+\uparrow \Sigma V_y: \qquad 0 = \tfrac{4}{5}(5\omega_{BD}) - \tfrac{4}{5}(10\omega_{DE})$$

$$\xrightarrow{+} \Sigma V_x: \qquad 3(10) = \tfrac{3}{5}(5\omega_{BD}) + \tfrac{3}{5}(10\omega_{DE})$$

which yield $\omega_{BD} = 5$ and $\omega_{DE} = 2.5$. Thus, we write[†]

$$\omega_{BD} = 5 \text{ rad/s } \circlearrowright \qquad \omega_{DE} = 2.5 \text{ rad/s } \circlearrowright \quad \blacktriangleleft$$

To find $\mathbf{v}_D$ of the pin at D, we write, from the sketch as shown,

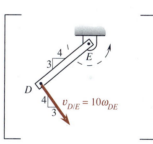

$$\mathbf{v}_D = \mathbf{v}_{D/E} + \mathbf{v}_E = \mathbf{v}_{D/E} = 10(2.5) \; \diagdown\, \tan^{-1}(4/3)$$

$$\mathbf{v}_D = 25 \text{ ft/s } \diagdown\, 53.1° \quad \blacktriangleleft$$

[†]If either of ω_{BD} and ω_{DE} were found to be negative, its initially assumed sense would have to be reversed in reporting the answers.

Developmental Exercise

D15.12 Suppose that A and B are two different points of a rigid slab in plane motion. (a) What is the angular velocity of the rigid slab if $\mathbf{v}_{B/A} = \mathbf{0}$? (b) Define $\mathbf{v}_{B/A}$ with a cross product.

EXAMPLE 15.7

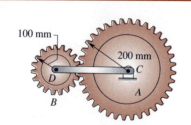

The gear A of the system shown rotates with $\boldsymbol{\omega}_A = 180$ rpm $\circlearrowright$, and the connecting arm CD rotates with $\boldsymbol{\omega}_{CD} = 60$ rpm $\circlearrowleft$. Determine $\boldsymbol{\omega}_B$ of the gear B.

Solution. We note that

$$\omega_A = 180 \text{ rpm } \circlearrowright = 6\pi \text{ rad/s } \circlearrowright$$
$$\omega_{CD} = 60 \text{ rpm } \circlearrowleft = 2\pi \text{ rad/s } \circlearrowleft$$

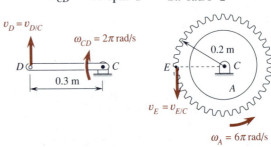

Seeing that the point C is fixed, we write

$$\mathbf{v}_D = \mathbf{v}_{D/C} + \mathbf{v}_C = \mathbf{v}_{D/C} = 0.3(2\pi) \uparrow = 0.6\pi \uparrow$$
$$\mathbf{v}_E = \mathbf{v}_{E/C} + \mathbf{v}_C = \mathbf{v}_{E/C} = 0.2(6\pi) \downarrow = 1.2\pi \downarrow$$

Since E is the point of contact between the gears A and B, the velocity of the point E on these two gears must be the same. From the velocities of D and E on the gear B as shown, we readily see that $\boldsymbol{\omega}_B$ of the gear B must be clockwise as indicated. We write

$$\mathbf{v}_D = \mathbf{v}_{D/E} + \mathbf{v}_E$$

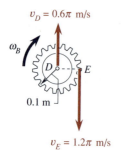

$$+\uparrow \Sigma V_y: \qquad 0.6\pi = 0.1\omega_B - 1.2\pi$$
$$\omega_B = 18\pi \text{ rad/s} = 540 \text{ rpm} \qquad \boldsymbol{\omega}_B = 540 \text{ rpm } \circlearrowleft \quad \blacktriangleleft$$

Developmental Exercise

D15.13 Refer to Example 15.7. If $\boldsymbol{\omega}_{CD} = 60$ rpm $\circlearrowleft$ and the gear B is in curvilinear translation, determine $\boldsymbol{\omega}_A$.

PROBLEMS

15.17 The end B of the rod AB slides with $v_B = 2.8$ m/s →. For the position shown, determine ω of the rod and v_A of the end A.

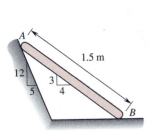

Fig. P15.17

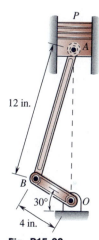

Fig. P15.18*

15.18* A gear rolls on a stationary rack as shown. If the center A of the gear moves with $v_A = 1.5$ m/s →, determine the velocities of the points B, C, D, and E of the gear.

15.19 A cable is wrapped around a drum which is mounted on a gear that rolls on a stationary rack as shown. The end D of the cable is pulled with $v_D = 0.9$ m/s →. For the position shown, determine v_A of the center A and v_B of the point B of the gear.

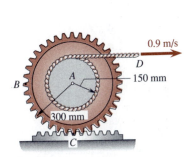

Fig. P15.19

Fig. P15.20

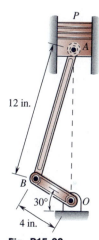

Fig. P15.22

15.20 A triangular plate moves in the xy plane. For the position shown, the corners P and Q move with $v_P = 4.2\mathbf{i} + 0.9\mathbf{j}$ m/s and $v_Q = 1.2\mathbf{i} + (v_Q)_y\mathbf{j}$ m/s, respectively. Determine ω of the plate and $(v_Q)_y$ for this position.

15.21* Refer to Prob. 15.20. For the position shown, (a) determine v_R of the corner R, (b) locate the point $C(x, y)$ of the plate where $v_C = \mathbf{0}$.

15.22 The crank OB of an internal-combustion engine rotates with $\omega_{OB} = 1200$ rpm ↻. For the position shown, determine ω_{AB} of the connecting rod AB and v_P of the piston P.

15.23 The planetary gear train shown consists of a sun gear S which is fixed, an annular gear A which is driven with $\omega_A = 1800$ rpm ↻, three identical planet gears P, Q, and R, and a carrier C which is connected to the centers of the planet gears. For $a = 60$ mm and $b = 40$ mm, determine ω_P of the planet gear P and ω_C of the carrier C.

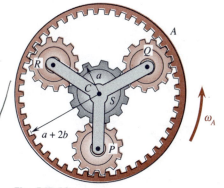

Fig. P15.23

15.24* Refer to the planetary gear train in Prob. 15.23. If $a/b = 2$, determine the ratio of ω_C of the carrier C to ω_A of the annular gear A.

15.25 The end A of the rack AB moves with a constant velocity of 8.75 in./s $\rightarrow$. For the position shown, determine ω_{AB} of AB and ω_O of the gear.

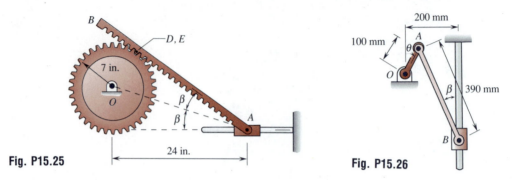

Fig. P15.25

Fig. P15.26

15.26 The crank OA of a slider-crank mechanism rotates with a constant angular velocity of 3.6 rad/s $\circlearrowleft$. Determine ω_{AB} of the connecting rod AB and v_B of the slider B when (a) $\theta = 0$, (b) $\theta = 30°$.

15.27* Refer to Prob. 15.26. Determine ω_{AB} and v_B when (a) $\theta = 90°$, (b) $\theta = 180°$.

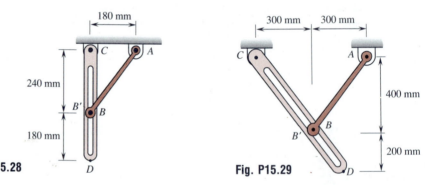

Fig. P15.28

Fig. P15.29

15.28 and 15.29 In the position shown, the crank AB rotates with $\omega_{AB} = 10$ rad/s $\circlearrowleft$. Suppose that B' is a point of the member CD coinciding with the pin at B. (a) Is the component of v_B of the pin at B in the direction perpendicular to the slot equal to $v_{B'}$? (b) Determine v_D of the point D.

15.30* and 15.31 In the position shown, the link AB rotates with $\omega_{AB} = 6$ rad/s $\circlearrowleft$. Determine ω_{BD} and ω_{DE} of the links BD and DE.

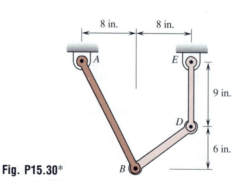

Fig. P15.30*

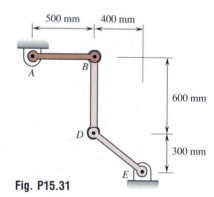

Fig. P15.31

15.32 In the position shown, the end A of the bar AD moves with $v_A = 1134$ mm/s →. Determine v_B of the end B of the bar BD and ω_{CD} of the bar CD.

15.33* The bar CD of the mechanism shown rotates with $\omega_{CD} = 4.5$ rad/s ↺. Determine ω_{AD} and ω_{BD} of the bars AD and BD.

15.34 The rod AB slides in the sleeve S of the member DE. In the position shown, the link OB rotates with $\omega_{OB} = 3$ rad/s ↻. Determine v_E of the point E. (*Hint.* $\omega_{AB} = \omega_{DE}$.)

15.35* In the position shown, the member DE containing a sleeve S rotates with $\omega_{DE} = 4$ rad/s ↻. Determine ω_{OB} of the link OB.

15.36 In the position shown, the arm ABC of the mechanism rotates with $\omega_{ABC} = 3$ rad/s ↺. Determine v_P of the point P of the smaller gear.

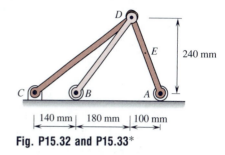

140 mm | 180 mm | 100 mm

Fig. P15.32 and P15.33*

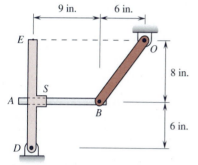

Fig. P15.34 and P15.35*

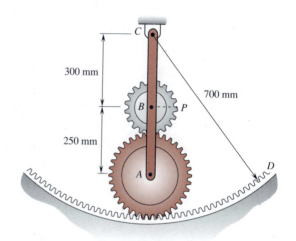

Fig. P15.36

15.7 Velocity Center

An *instantaneous center of zero velocity* of a rigid slab in plane motion is a point of the slab (or its massless extension), denoted by C as shown in Fig. 15.14, which lies in the plane of motion and has a *zero velocity* at the instant considered. The axis which passes through such a center and is perpendic-

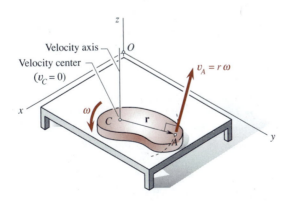

Fig. 15.14

ular to the plane of motion of the slab is called the *instantaneous axis of zero velocity* of the slab at the instant considered. For brevity, the term the *instantaneous center* (or axis) *of zero velocity* of a body will simply be referred to as the *velocity center* (or axis) of the body.[†] Naturally, the velocity of any point *A* of a rigid slab in plane motion is simply equal to the velocity of the point *A* relative to the velocity center *C* of the slab; i.e.,

$$\mathbf{v}_A = \mathbf{v}_{A/C} \qquad (15.18)$$

Furthermore, rods, bars, links, cranks, disks, plates, gears, wheels, and bodies of other shapes may be regarded as special cases of the rigid slab in the present study.

In general, each rigid slab in a mechanism has its own velocity center, which is generally different from that of another slab. If two or more points qualify as velocity centers of a slab, the slab must have no velocity at the instant considered. However, the velocity center of a rigid slab does *not* necessarily have zero acceleration.

Theorem of velocity center.[††] *If the velocities $\mathbf{v}_A$ and $\mathbf{v}_B$ of any two points A and B of a rigid slab in plane motion are not parallel to each other, then the velocity center C of the slab at the instant considered is located at the intersection of $\overrightarrow{AC}$ and $\overrightarrow{BC}$, where $\overrightarrow{AC} \perp \mathbf{v}_A$ and $\overrightarrow{BC} \perp \mathbf{v}_B$.* (See Fig. 15.15.)

Proof. To prove that the velocity $\mathbf{v}_C$ of the point *C* is zero when *C* is located as described, we first apply Chasles' theorem to write

$$\mathbf{v}_A = \mathbf{v}_{A/C} + \mathbf{v}_C \qquad \mathbf{v}_B = \mathbf{v}_{B/C} + \mathbf{v}_C \qquad (15.19)$$

As $\overrightarrow{AC} \perp \mathbf{v}_A$ and $\overrightarrow{BC} \perp \mathbf{v}_B$, we infer from Eqs. (15.16) that

$$\mathbf{v}_{A/C} \parallel \mathbf{v}_A \qquad \mathbf{v}_{B/C} \parallel \mathbf{v}_B \qquad (15.20)$$

For the conditions in (15.20) and Eqs. (15.19) to hold true simultaneously, we must require that

$$\mathbf{v}_C \parallel \mathbf{v}_A \qquad \mathbf{v}_C \parallel \mathbf{v}_B \qquad (15.21)$$

Since $\mathbf{v}_A$ and $\mathbf{v}_B$ are, as stated, *not* parallel to each other and the conditions in (15.21) must be true, we logically conclude that $\mathbf{v}_C = \mathbf{0}$.   Q.E.D.

Corollary I. (Refer to the theorem.) *The speeds of A and B are $v_A = \overline{CA}\omega$ and $v_B = \overline{CB}\omega$, where ω is the angular speed of the slab at the instant considered.*

Corollary II. (Refer to the theorem.) *If $\mathbf{v}_A$ is not perpendicular to the line AB, but $\mathbf{v}_A \parallel \mathbf{v}_B$, then the velocity center C is located at infinity and the angular velocity of the slab is zero.*

Corollary III. (Refer to the theorem.) *If $\mathbf{v}_A \parallel \mathbf{v}_B$ and $\mathbf{v}_A \perp \overline{AB}$, then the velocity center C is located at the intersection of the line AB and the line joining the extremities of the vectors $\mathbf{v}_A$ and $\mathbf{v}_B$ which are drawn to scale from A and B, respectively,* as shown in Fig. 15.16.

Fig. 15.15

[†]Cf. the *acceleration center* and the *acceleration axis* in Sec. 15.9.

[††]In the text, the symbols ⊥ and ∥ denote "is perpendicular to" and "is parallel to," respectively.

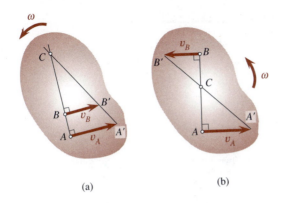

Fig. 15.16   (a)   (b)

As a slab undergoes a general plane motion, the position of the velocity center *C* will generally change from time to time. The curve representing the locus of the velocity center in space is called the *space centrode*, and the curve representing the locus of the velocity center on the slab is called the *body centrode*. These two centrodes are illustrated in Fig. 15.17 for a general slab and in Fig. 15.18 for a wheel rolling without slipping on a rigid surface, where the *circumference* of the wheel is the body centrode and the *track* of the wheel is the space centrode. Note that these two centrodes are always tangent to each other at the velocity center *C*, and that, as the motion proceeds, *the body centrode appears to roll on the space centrode*.

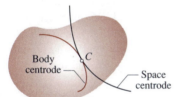

Fig. 15.17

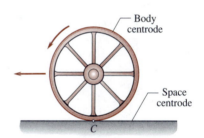

Fig. 15.18

EXAMPLE 15.8

Using *velocity center*, work Example 15.5.

Solution.   From the constraint condition of the rod *AB*, we note that $\mathbf{v}_A$ must be directed downward when $\mathbf{v}_B$ is directed to the right as given. Thus, the velocity center *C* of the rod is the point of intersection of the lines *AC* and *BC* as shown, where $\overrightarrow{AC} \perp \mathbf{v}_A$ and $\overrightarrow{BC} \perp \mathbf{v}_B$. For computing velocities at the instant considered, the rod may be assumed to rotate about *C* as a fixed point as shown. We write

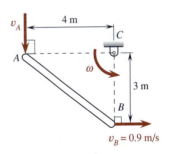

$$v_B = 0.9 = \overline{CB}\omega = 3\omega \qquad \omega = 0.3 \text{ rad/s} \circlearrowright \blacktriangleleft$$

$$v_A = \overline{CA}\omega = 4(0.3) \qquad \mathbf{v}_A = 1.2 \text{ m/s} \downarrow \blacktriangleleft$$

Developmental Exercises

D15.14   Define the terms: (a) velocity center, (b) space centrode, (c) body centrode.

D15.15   Describe the *theorem of velocity center* and its three corollaries.

EXAMPLE 15.9

Using *velocity center*, work Example 15.6.

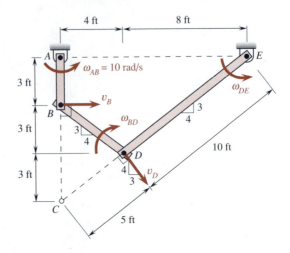

Solution. The system in Fig. 15.13 together with the given data in Example 15.6 may be drawn and indicated as shown above. Since the pins at A and E are fixed, the points A and E are the *velocity centers* of the links AB and DE, respectively. Next, we note that $\mathbf{v}_B \perp \overrightarrow{BA}$ and $\mathbf{v}_D \perp \overrightarrow{DE}$. Thus, by the theorem of velocity center, the *velocity center* of the link BD is the point of intersection C of the lines AB and DE extended as shown. For computing velocities at the instant considered, the links AB, BD, and DE may be assumed to rotate about their velocity centers A, C, and E, respectively, as fixed points as shown in Figs. (a), (b), and (c).

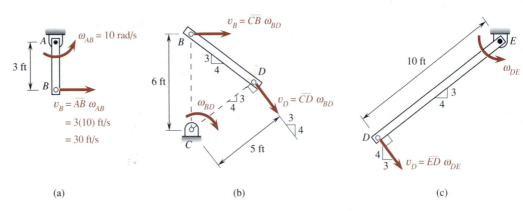

(a)   (b)   (c)

From Figs. (a) and (b), we write

$$v_B = 30 = \overline{CB}\, \omega_{BD} = 6\omega_{BD}$$

$$\omega_{BD} = 5 \text{ rads/s} \,\circlearrowright \; \blacktriangleleft$$

From Figs. (b) and (c), we write

$$v_D = \overline{CD}\, \omega_{BD} = 5(5) = \overline{ED}\, \omega_{DE} = 10\omega_{DE}$$

$$\mathbf{v}_D = 25 \text{ ft/s} \,\measuredangle\, 53.1° \qquad \omega_{DE} = 2.5 \text{ rad/s} \,\circlearrowright \; \blacktriangleleft$$

Developmental Exercise

EXAMPLE 15.10

A drum of radius 120 mm is mounted on a wheel of radius 180 mm as shown, where a rope is wound around the drum. The end E of the rope is pulled with a constant velocity $\mathbf{v}_E = 0.6$ m/s →, and the wheel rolls without slipping. Using *velocity center*, determine (a) $\mathbf{v}_D$ of the point D, (b) the rate at which the rope is being wound or unwound.

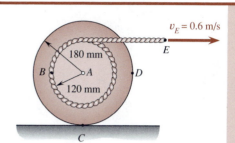

Solution. As the wheel rolls without slipping, the point C of the wheel in contact with the support has zero velocity; therefore, the point C is the *velocity center* of the body consisting of the wheel and the drum. Since the rope may be taken as inextensible, the point F on the top of the drum and the end E of the rope should have the same velocity; i.e.,

$$\mathbf{v}_F = \mathbf{v}_E = 0.6 \text{ m/s} \rightarrow$$

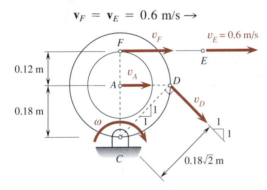

For computing velocities at the instant considered, the wheel and the drum may be assumed to rotate with $\boldsymbol{\omega}$ about the velocity center C as a fixed point as shown. We write

$$v_F = \overline{CF}\omega = 0.3\omega = 0.6 \qquad \omega = 2 \text{ rad/s } \circlearrowleft$$
$$v_D = \overline{CD}\omega = 0.18\sqrt{2}(2)$$
$$\mathbf{v}_D = 0.509 \text{ m/s } \searrow 45° \blacktriangleleft$$
$$v_A = \overline{CA}\omega = 0.18(2) \qquad \mathbf{v}_A = 0.36 \text{ m/s} \rightarrow$$

Since $v_E > v_A$, we see that the rope is being *unwound* at the rate of

$$v_{E/A} = v_E - v_A = 0.6 - 0.36$$
$$v_{E/A} = 0.24 \text{ m/s } \blacktriangleleft$$

Developmental Exercises

PROBLEMS

15.37* through 15.50* Using *velocity center*, solve the following problems:

15.37* Prob. 15.17.	15.38* Prob. 15.18*.
15.39* Prob. 15.19.	15.40* Prob. 15.22.
15.41* Prob. 15.25.	15.42* Prob. 15.26.
15.43* Prob. 15.27*.	15.44* Prob. 15.30*.
15.45* Prob. 15.31.	15.46* Prob. 15.32.
15.47* Prob. 15.33*.	15.48* Prob. 15.34.
15.49* Prob. 15.35*.	15.50* Prob. 15.36.

15.51 A driving wheel of a car is shown. The car travels at 15 mi/h → while the wheel spins with $\omega = 300$ rpm ↻ on a slippery road. Determine (a) the position of the velocity center C of the wheel, (b) $\mathbf{v}_B$ of the point B of the wheel in contact with the road.

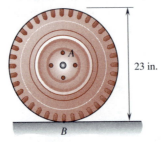

Fig. P15.51

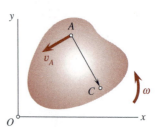

Fig. P15.52*

15.52* A rigid slab is in plane motion with an angular velocity $\boldsymbol{\omega}$ as shown. Knowing the vector identity $\mathbf{P} \times (\mathbf{Q} \times \mathbf{R}) = \mathbf{Q}(\mathbf{P} \cdot \mathbf{R}) - \mathbf{R}(\mathbf{P} \cdot \mathbf{Q})$, show that $\overrightarrow{AC} = \boldsymbol{\omega} \times \mathbf{v}_A/\omega^2$, where C is the velocity center and A is a particle of the slab.

15.53 At the instant shown, the quadrilateral plate in plane motion has an angular velocity of 3 rad/s ↻ and its corner A has a velocity $\mathbf{v}_A$ of 0.45 m/s as indicated. Determine (a) the distance $\overline{CA}$ between the velocity center C and the corner A, (b) $\mathbf{v}_B$ of the corner B, (c) $\mathbf{v}_D$ of the corner D.

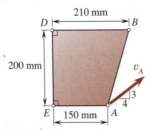

Fig. P15.53

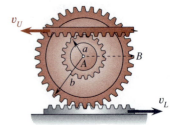

Fig. P15.54

15.54 A double gear is engaged with an upper rack and a lower rack as shown. The upper rack is pulled with $\mathbf{v}_U = 260$ mm/s ← while the lower rack is pulled with $\mathbf{v}_L = 100$ mm/s →. If $a = 60$ mm and $b = 120$ mm, determine (a) the distance between the velocity center C and the center A of the gear, (b) $\boldsymbol{\omega}$ of the gear, (c) $\mathbf{v}_B$ of the point B.

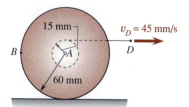

Fig. P15.55

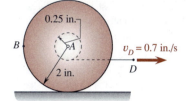

Fig. P15.56*

15.55 and 15.56* The end D of the string wound around the core of a yoyo is pulled with $\mathbf{v}_D$ as shown. If the yoyo rolls without slipping, determine $\boldsymbol{\omega}$ of the yoyo and $\mathbf{v}_B$ of the point B.

15.57 The end A of the slender rod AB leaning against a fixed semicylindrical surface as shown is pulled with $\mathbf{v}_A = 0.6$ m/s $\rightarrow$. If $x = 2.25$ m, determine $\boldsymbol{\omega}_{AB}$ of AB and $\mathbf{v}_B$ of the end B.

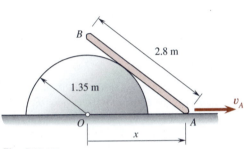

Fig. P15.57

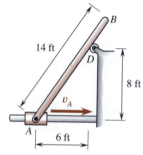

Fig. P15.58

15.58 At the instant shown, the collar A slides with $\mathbf{v}_A = 1.5$ ft/s $\rightarrow$. Determine (a) $\boldsymbol{\omega}_{AB}$ of the rod AB, (b) $\mathbf{v}_D$ of the point D in contact with the roller support, (c) $\mathbf{v}_B$ of the end B.

15.59* The rod AB is mounted on sliders as shown, where the slider at A is moved with $\mathbf{v}_A = 0.24$ m/s $\rightarrow$. If $x = 0.3$ m, determine $\boldsymbol{\omega}_{AB}$ of the rod and $\mathbf{v}_B$ of the end B.

15.60 Solve Prob. 15.59* if $x = 0.4$ m.

15.61 At the instant shown, the crank OA rotates with $\boldsymbol{\omega}_{OA} = 5$ rad/s $\circlearrowleft$. Determine $\mathbf{v}_B$ of collar B and $\mathbf{v}_M$ of the midpoint M of the link AB.

15.62* At the instant shown, the collar B moves with $\mathbf{v}_B = 7$ in./s $\downarrow$. Determine $\boldsymbol{\omega}_{OA}$ of the crank OA and $\mathbf{v}_M$ of the midpoint M of the link AB.

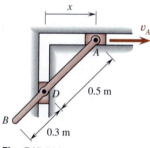

Fig. P15.59*

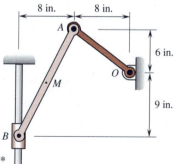

Fig. P15.61 and P15.62*

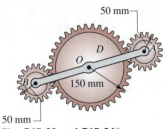

Fig. P15.63 and P15.64*

50 mm

150 mm

50 mm

15.63 The gear D shown rotates about O with $\boldsymbol{\omega}_D$ = 4 rad/s $\circlearrowright$, while the arm AOB rotates about O with $\boldsymbol{\omega}_{AOB}$ = 2 rad/s $\circlearrowleft$. Determine $\boldsymbol{\omega}_A$ of the smaller gear at A and $\boldsymbol{\omega}_B$ of the smaller gear at B.

15.64* The gear D shown rotates about O with a constant angular velocity $\boldsymbol{\omega}_D$ = 6 rad/s $\circlearrowleft$. Determine the angular velocity $\boldsymbol{\omega}_{AOB}$ of the arm AOB for which the smaller gears at A and B will have curvilinear translation.

15.65 At the instant shown, the crank AB rotates with $\boldsymbol{\omega}_{AB}$ = 2.4 rad/s $\circlearrowright$. Determine $\boldsymbol{\omega}_{DE}$ of the crank DE and $\mathbf{v}_F$ of the point F of the link BD.

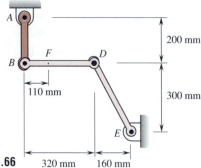

Fig. P15.65 and P15.66

200 mm

300 mm

110 mm

320 mm   160 mm

15.66 At the instant shown, the pin at D moves with a speed of 408 mm/s. Determine (a) the angular speed of the crank AB, (b) the speed of the point F of the link BD.

15.8   Accelerations in Relative Motion

Suppose that the points A and B are two different points of a rigid slab in plane motion with an angular velocity $\boldsymbol{\omega} = \omega\mathbf{k}$ and an angular acceleration $\boldsymbol{\alpha} = \alpha\mathbf{k}$, as shown in Fig. 15.19. By Chasles' theorem, we may view the plane motion of the rigid slab as a rotation of the slab about A plus a translation of the slab with A. Such a viewpoint is depicted in the graphical equation shown. Thus, we may express the acceleration $\mathbf{a}_B$ of B in the ref-

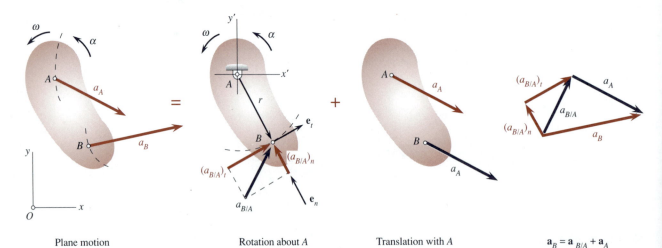

Plane motion

Rotation about A

Translation with A

$\mathbf{a}_B = \mathbf{a}_{B/A} + \mathbf{a}_A$

Fig. 15.19

erence frame Oxy as the acceleration $\mathbf{a}_{B/A}$ of B relative to A plus the acceleration $\mathbf{a}_A$ of A; i.e.,

$$\mathbf{a}_B = \mathbf{a}_{B/A} + \mathbf{a}_A \qquad (15.22)^\dagger$$

Recall that, more precisely, $\mathbf{a}_{B/A}$ denotes the acceleration of B in a nonrotating reference frame (e.g., $Ax'y'$) which translates with A.

Clearly, $\mathbf{a}_{B/A}$ is due solely to the rotational part of the motion of the slab. As a matter of fact, we may view $\mathbf{a}_{B/A}$ as the acceleration of the tip B of the radius vector $\overrightarrow{AB}$ which rotates with the angular velocity $\boldsymbol{\omega}$ and the angular acceleration $\boldsymbol{\alpha}$ about A as a "fixed point." Thus, $\mathbf{a}_{B/A}$ *is equivalent to the acceleration of B in circular motion about A as a fixed center, and is composed of a tangential component* $(\mathbf{a}_{B/A})_t$ *and a normal component* $(\mathbf{a}_{B/A})_n$; i.e.,

$$\mathbf{a}_{B/A} = (\mathbf{a}_{B/A})_t + (\mathbf{a}_{B/A})_n \qquad (15.23)$$

$$(\mathbf{a}_{B/A})_t = r\alpha\mathbf{e}_t \qquad (\mathbf{a}_{B/A})_n = r\omega^2\mathbf{e}_n \qquad (15.24)$$

where $r = \overline{AB}$, and $\mathbf{e}_t$ and $\mathbf{e}_n$ are the tangential and normal unit vectors pertaining to the circular arc at B, respectively, as indicated in Fig. 15.19. Since $\boldsymbol{\omega} = \omega\mathbf{k}$ and $\boldsymbol{\alpha} = \alpha\mathbf{k}$, we may use the cross product of vectors to write

$$(\mathbf{a}_{B/A})_t = \boldsymbol{\alpha} \times \mathbf{r} \qquad (\mathbf{a}_{B/A})_t = \alpha\mathbf{k} \times \mathbf{r} \qquad (15.25)$$

$$(\mathbf{a}_{B/A})_n = \boldsymbol{\omega} \times (\boldsymbol{\omega} \times \mathbf{r}) \qquad (\mathbf{a}_{B/A})_n = -\omega^2\mathbf{r} \qquad (15.26)$$

where $\mathbf{r} = \overrightarrow{AB}$. Note that (1) *the magnitude of* $(\mathbf{a}_{B/A})_t$ *is* $r\alpha$, (2) *the magnitude of* $(\mathbf{a}_{B/A})_n$ *is* $r\omega^2$, (3) *the direction of* $(\mathbf{a}_{B/A})_t$ *is perpendicular to* $\overrightarrow{AB}$, (4) *the vectors* $\boldsymbol{\alpha}$, $\overrightarrow{AB}$, *and* $(\mathbf{a}_{B/A})_t$ *(in this order) form a right-handed trihedral,* and (5) *the normal component* $(\mathbf{a}_{B/A})_n$ *is always directed from B toward A regardless of the senses of* $\boldsymbol{\omega}$ *and* $\boldsymbol{\alpha}$.

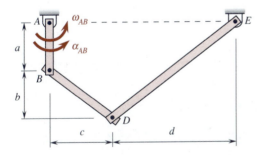

Fig. 15.20

Suppose that the link AB in Fig. 15.13 rotates with $\boldsymbol{\omega}_{AB}$ and $\boldsymbol{\alpha}_{AB}$ as shown in Fig. 15.20, where the pins at A and E are fixed. Taking the constraint conditions $\mathbf{a}_A = \mathbf{a}_E = \mathbf{0}$ into account and applying Eq. (15.22), we write

$$\mathbf{a}_B = \mathbf{a}_{B/A} + \mathbf{a}_A = \mathbf{a}_{B/A}$$

$$\mathbf{a}_B = \mathbf{a}_{B/D} + \mathbf{a}_D = \mathbf{a}_{B/D} + (\mathbf{a}_{D/E} + \mathbf{a}_E) = \mathbf{a}_{B/D} + \mathbf{a}_{D/E}$$

---

†This equation is equivalent to Eq. (11.37). However, the order of terms in Eq. (15.22) is preferable to that in Eq. (11.37) in establishing the *linkage equation* for accelerations of a system, such as Eq. (15.27).

Thus, we have

$$\mathbf{a}_{B/A} = \mathbf{a}_{B/D} + \mathbf{a}_{D/E} \qquad (15.27)^\dagger$$

which is the *linkage equation* for accelerations of the system shown in Fig. 15.20. Such an equation may be used as a basis for determining the linear and angular accelerations in a system.

EXAMPLE 15.11

The end B of the rod AB is pulled with a constant velocity $\mathbf{v}_B = 0.9$ m/s →. For the position shown, determine $\boldsymbol{\alpha}$ of the rod and $\mathbf{a}_A$ of the end A of the rod.

Solution. We first need to know $\boldsymbol{\omega}$ of the rod. From the solution in Example 15.5 or 15.8, we have $\boldsymbol{\omega} = 0.3$ rad/s $\circlearrowright$, which is indicated in the sketch, where $\mathbf{a}_B = \mathbf{0}$. The constraint condition of the rod prompts us to initially assume that $\mathbf{a}_A = a_A \downarrow$ and $\boldsymbol{\alpha} = \alpha \circlearrowright$ as indicated. Thus, we write

$$\mathbf{a}_B = \mathbf{a}_{B/A} + \mathbf{a}_A$$

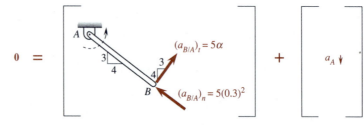

From this vector equation, we write

$$\xrightarrow{+} \Sigma V_x: \qquad 0 = \tfrac{3}{5}(5\alpha) - \tfrac{4}{5}(5)(0.3)^2 + 0 \qquad \alpha = 0.12$$

$$+\uparrow \Sigma V_y: \qquad 0 = \tfrac{4}{5}(5\alpha) + \tfrac{3}{5}(5)(0.3)^2 - a_A \qquad a_A = 0.75$$

$$\boldsymbol{\alpha} = 0.12 \text{ rad/s}^2 \circlearrowright \qquad \mathbf{a}_A = 0.75 \text{ m/s}^2 \downarrow \quad \blacktriangleleft$$

Developmental Exercise

D15.19 Determine $\boldsymbol{\alpha}$ and $\mathbf{a}_A$ in Example 15.11 if $\mathbf{a}_B = 0.6$ m/s² → and $\mathbf{v}_B = 0.9$ m/s →.

EXAMPLE 15.12

Refer to the system in Fig. 15.20. If $a = b = 3$ ft, $c = 4$ ft, $d = 8$ ft, and the link AB rotates with $\boldsymbol{\alpha}_{AB} = 12$ rad/s² $\circlearrowright$ as well as $\boldsymbol{\omega}_{AB} = 10$ rad/s $\circlearrowright$ in the position shown, determine (a) $\boldsymbol{\alpha}_{BD}$ and $\boldsymbol{\alpha}_{DE}$ of the links BD and DE, (b) $\mathbf{a}_D$ of the pin at D.

---

†Cf. Eq. (15.17).

Solution. We first need to know ω_{BD} and ω_{DE} in the position shown. From the solution in Example 15.6 or 15.9, we have $\omega_{BD} = 5$ rad/s $\circlearrowright$ and $\omega_{DE} = 2.5$ rad/s $\circlearrowleft$, which are indicated in the sketch.

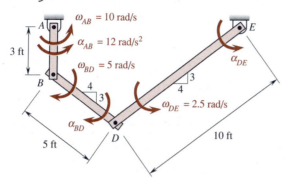

The directions of $\boldsymbol{\alpha}_{BD}$ and $\boldsymbol{\alpha}_{DE}$ of the links BD and DE may initially be assumed as indicated. Since the *linkage equation* for accelerations of the system has already been derived as Eq. (15.27), we write

$$\mathbf{a}_{B/A} = \mathbf{a}_{B/D} + \mathbf{a}_{D/E}$$

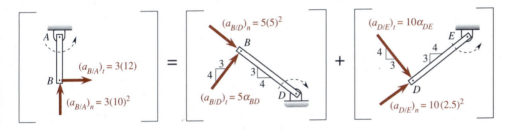

From this vector equation, we write

$$\xrightarrow{+} \Sigma V_x: \quad 3(12) = \tfrac{3}{5}(5\alpha_{BD}) + \tfrac{4}{5}(5)(5)^2$$
$$+ \tfrac{3}{5}(10\alpha_{DE}) + \tfrac{4}{5}(10)(2.5)^2$$

$$+\uparrow \Sigma V_y: \quad 3(10)^2 = \tfrac{4}{5}(5\alpha_{BD}) - \tfrac{3}{5}(5)(5)^2$$
$$- \tfrac{4}{5}(10\alpha_{DE}) + \tfrac{3}{5}(10)(2.5)^2$$

which yield $\alpha_{BD} = 23.188$ and $\alpha_{DE} = -30.594$. Since α_{DE} is negative, the true sense of $\boldsymbol{\alpha}_{DE}$ is opposite to that assumed. Thus, we write

$$\alpha_{BD} = 23.2 \text{ rad/s}^2 \circlearrowright \qquad \alpha_{DE} = 30.6 \text{ rad/s}^2 \circlearrowright \quad \blacktriangleleft$$

To find the acceleration of the pin at D, we write

$$\mathbf{a}_D = \mathbf{a}_{D/E} + \mathbf{a}_E = \mathbf{a}_{D/E} + \mathbf{0} = \mathbf{a}_{D/E}$$

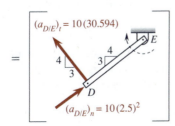

$$\mathbf{a}_D = [-\tfrac{3}{5}(10)(30.594) + \tfrac{4}{5}(10)(2.5)^2]\mathbf{i}$$
$$+ [\tfrac{4}{5}(10)(30.594) + \tfrac{3}{5}(10)(2.5)^2]\mathbf{j}$$
$$= -133.6\mathbf{i} + 282.3\mathbf{j}$$

$$\mathbf{a}_D = 312 \text{ ft/s}^2 \; \measuredangle \; 64.7° \blacktriangleleft$$

Developmental Exercises

D15.20 Refer to Example 15.12. Determine $\mathbf{a}_B$.

D15.21 Refer to Example 15.12. Determine the acceleration of the midpoint M of the bar DE.

EXAMPLE 15.13

Refer to Example 15.10. Suppose that the end E of the rope wound around the drum is pulled with $\mathbf{a}_E = 1.5$ m/s² → as well as $\mathbf{v}_E = 0.6$ m/s → at the instant shown, and the wheel rolls without slipping. For this instant, determine the accelerations of points A, C, and F of the wheel.

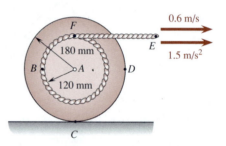

Solution. The corresponding angular velocity of the wheel has been determined in the solution of Example 15.10 as $\omega = 2$ rad/s ↻ which is indicated in the sketch. Furthermore, we may assume that the unknown angular acceleration α of the wheel to be clockwise as indicated. From the constraints on the wheel, we recognize the following conditions:

(a) The center A of the wheel must move in the horizontal direction; i.e.,

$$\mathbf{a}_A = a_A \rightarrow \tag{1}$$

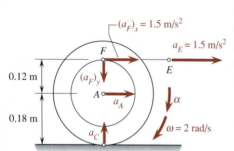

(b) As the wheel rolls without slipping, the point C of the wheel in contact with the support must have no tangential component of acceleration. However, C may have a normal component of acceleration; i.e.,

$$\mathbf{a}_C = a_C \uparrow \tag{2}$$

(c) The tangential component of the acceleration of the point F of the wheel should be equal to $\mathbf{a}_E$; i.e., $(a_F)_x = a_E = 1.5$ m/s² →. Assuming the normal component of $\mathbf{a}_F$ to be downward, we write

$$\mathbf{a}_F = [1.5 \rightarrow] + [(a_F)_y \downarrow] \tag{3}$$

Since $\mathbf{a}_F = \mathbf{a}_{F/A} + \mathbf{a}_A$, we write

$$[1.5 \rightarrow] + [(a_F)_y \downarrow] \;=\; \begin{bmatrix} \\ \\ \end{bmatrix} + [a_A \rightarrow]$$

$$+\downarrow \Sigma V_y: \qquad (a_F)_y = 0.12(2)^2 \qquad\qquad (4)$$

$$\overset{+}{\rightarrow} \Sigma V_x: \qquad 1.5 = 0.12\alpha + a_A \qquad\qquad (5)$$

Moreover, $\mathbf{a}_F = \mathbf{a}_{F/C} + \mathbf{a}_C$. Thus, we write

$$[1.5 \rightarrow] + [(a_F)_y \downarrow] \;=\; \begin{bmatrix} \\ \\ \end{bmatrix} + [a_C \uparrow]$$

$$\overset{+}{\rightarrow} \Sigma V_x: \qquad 1.5 = 0.3\alpha \qquad\qquad\qquad (6)$$

$$+\downarrow \Sigma V_y: \qquad (a_F)_y = 0.3(2)^2 - a_C \qquad\qquad (7)$$

Equations (4) through (7) yield $(a_F)_y = 0.48$, $\alpha = 5$, $a_C = 0.72$, and $a_A = 0.9$. By Eqs. (1) through (3), we write

$$\mathbf{a}_A = 0.9 \text{ m/s}^2 \rightarrow \qquad \mathbf{a}_C = 0.72 \text{ m/s}^2 \uparrow \quad \blacktriangleleft$$

$$\mathbf{a}_F = 1.5\mathbf{i} - 0.48\mathbf{j} \qquad \mathbf{a}_F = 1.575 \text{ m/s}^2 \; \diagdown \; 17.74° \quad \blacktriangleleft$$

REMARK. As $\mathbf{a}_C \neq 0$ and the support is stationary, we see that when a body rolls without slipping on another body, the points of contact have *equal* velocities and *equal* tangential components of acceleration, but *unequal* normal components of acceleration.

Developmental Exercises

D15.22   Refer to Example 15.13. Determine $\mathbf{a}_B$.

D15.23   Refer to Example 15.13. Determine $\mathbf{a}_D$.

PROBLEMS†

15.67   Refer to Prob. 15.17. In the position shown, the end B of the rod AB slides with $v_B = 2.8$ m/s $\rightarrow$ and $\mathbf{a}_B = 0.9$ m/s$^2 \rightarrow$. For this position, determine α of the rod and $\mathbf{a}_A$ of the end A.

---

†In this problem set, we include systems which are formed by taking accelerations into account in a number of previous systems.

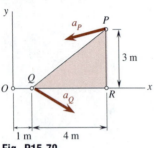

Fig. P15.70

15.68 Refer to Prob. 15.18*. In the position shown, the center A of the gear moves with $v_A = 1.5$ m/s → and $a_A = 1.2$ m/s² →. For this position, determine the accelerations of the points B, C, and E of the gear.

15.69* Refer to Prob. 15.19. In the position shown, the end D of the cable is pulled with $v_D = 0.9$ m/s → and $a_D = 0.63$ m/s² →. For this position, determine the accelerations of the points A, B, and C of the gear.

15.70 A triangular plate moves in the xy plane. In the position shown, the corners P and Q have accelerations $a_P = -14\mathbf{i} - 3\mathbf{j}$ m/s² and $a_Q = 17\mathbf{i} - 11\mathbf{j}$ m/s², respectively. For this position, determine (a) ω and α of the plate, (b) a_R of the corner R.

15.71 Refer to Prob. 15.70. For the position shown, determine the coordinates of the point $Z(x, y)$ of the plate whose acceleration $a_Z = 0$.

15.72* Refer to Prob. 15.22. In the position shown, the crank OB rotates with $\omega_{OB} = 60$ rpm ↺ and $\alpha_{OB} = 10$ rad/s² ↺. For this position, determine α_{AB} of the connecting rod AB and a_P of the position P.

15.73 Refer to Prob. 15.23. At the instant shown, the annular gear A is driven with $\omega_A = 120$ rpm ↻ and $\alpha_A = 20$ rad/s² ↻. For $a = 60$ mm and $b = 40$ mm, determine α_P and α_C of the planet gear P and the carrier C, respectively, and a_T of the particle T which belongs to the planet gear P and is in contact with the annular gear A at the instant shown.

15.74 Refer to Prob. 15.25. In the position shown, the end A of the rack AB moves with $v_A = 8.75$ in./s → and $a_A = 15$ in./s² →. For this position, determine α_{AB} of the rack AB and α_O of the gear O.

15.75 In the position shown, the crank OA of a slider-crank mechanism rotates with $\omega_{OA} = 6$ rad/s ↺ and $\alpha_{OA} = 5$ rad/s² ↺. For this position, determine α_{AB} of the connecting rod AB and a_B of the slider B.

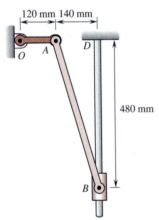

Fig. P15.75

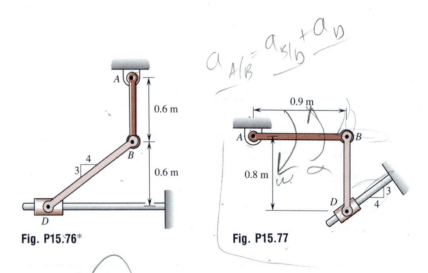

Fig. P15.76*

Fig. P15.77

15.76* and **15.77** In the position shown, the crank AB of a slider-crank mechanism rotates with $\omega_{AB} = 4$ rad/s ↺ and $\alpha_{AB} = 10$ rad/s² ↻. For this position, determine α_{BD} of the connecting rod BD and a_D of the slider D.

15.78 and **15.79*** In the position shown, the link AB rotates with $\omega_{AB} = 1.2$ rad/s ↺ and $\alpha_{AB} = 3$ rad/s² ↻. For this position, determine α_{BD} of the link BD and a_D of the pin at D.

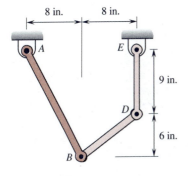

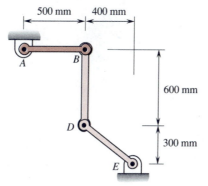

Fig. P15.78 and P15.80   Fig. P15.79* and P15.81*

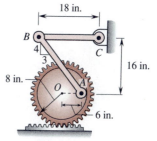

Fig. P15.84* and P15.85

15.80 and 15.81* In the position shown, the link AB is accelerated from rest (i.e., $\omega_{AB} = 0$) with $\alpha_{AB} = 3$ rad/s² $\circlearrowleft$. For this position, determine α_{BD} of the link BD and $\mathbf{a}_D$ of the pin at D.

15.82 Refer to Prob. 15.32. The end A of the bar AD moves with a constant velocity $\mathbf{v}_A = 1134$ mm/s →. For the position shown, determine (a) $\mathbf{a}_B$ of the end B of the bar BD, (b) the angular acceleration of each bar.

15.83 Refer to Prob. 15.32. In the position shown, the end A of the bar AD is accelerated from rest (i.e., $\mathbf{v}_A = 0$) with $\mathbf{a}_A = 2.268$ m/s² →. For this position, determine (a) $\mathbf{a}_B$ and $\mathbf{a}_E$ of the points B and the midpoint E of AD, (b) the angular acceleration of each bar.

15.84* In the position shown, the center O of the gear moves with $\mathbf{v}_O = 19.2$ in./s ← and $\mathbf{a}_O = 24$ in./s² → over the stationary rack. For this position, determine α_{AB} and α_{BC} of the links AB and BC.

15.85 In the position shown, the center O of the gear is accelerated from rest (i.e., $\mathbf{v}_O \neq 0$) with $\mathbf{a}_O = 48$ in./s² ← over the stationary rack. For this position, determine (a) α_{AB} and α_{BC} of the links AB and BC, (b) $\mathbf{a}_M$ of the midpoint M of AB.

15.86 Refer to Prob. 15.36. In the position shown, the arm ABC of the mechanism rotates with $\omega_{ABC} = 3$ rad/s $\circlearrowleft$ and $\alpha_{ABC} = 6$ rad/s² $\circlearrowleft$. For this position, determine (a) α_A of the gear at A, (b) α_B of the gear at B, (c) $\mathbf{a}_P$ of the point P.

15.9   Acceleration Center

An *instantaneous center of zero acceleration* of a rigid slab in plane motion is a point of the slab (or its massless extension), denoted by Z as shown in Fig. 15.21, which lies in the plane of motion and has a *zero acceleration* at the instant considered. The axis which passes through such a center and is perpendicular to the plane of motion of the slab is called the *instantaneous axis of zero acceleration* of the slab at the instant considered. For brevity, the term the *instantaneous center (or axis) of zero acceleration* of a body will simply be referred to as the *acceleration center (or axis)* of the body. Naturally, the acceleration of any point A of a rigid slab in plane motion is simply equal to the acceleration of the point A relative to the acceleration center Z of the slab; i.e.,

$$\mathbf{a}_A = \mathbf{a}_{A/Z} \tag{15.28}$$

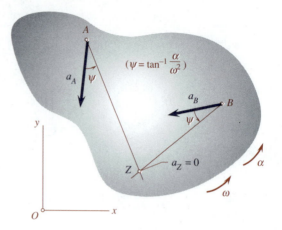

Fig. 15.21

Theorem of acceleration center.[†] *If the accelerations* $\mathbf{a}_A$ *and* $\mathbf{a}_B$ *of any two points A and B of a rigid slab, which is in plane motion with an angular velocity* $\boldsymbol{\omega}$ *and an angular acceleration* $\boldsymbol{\alpha}$*, are not parallel to each other, then the acceleration center Z of the slab at the instant considered is located at the intersection of* $\overrightarrow{AZ}$ *and* $\overrightarrow{BZ}$*, which make an angle* ψ *with* $\mathbf{a}_A$ *and* $\mathbf{a}_B$*, respectively, where*

$$\psi = \tan^{-1}\left(\frac{\alpha}{\omega^2}\right) \tag{15.29}$$

and the set of vectors $\mathbf{a}_A$*,* $\overrightarrow{AZ}$*, and* $\boldsymbol{\alpha}$*, as well as the set of vectors* $\mathbf{a}_B$*,* $\overrightarrow{BZ}$*, and* $\boldsymbol{\alpha}$*, forms a right-handed trihedral.* (See Fig. 15.21.)

Proof. To prove that the acceleration $\mathbf{a}_Z$ of the point Z is zero when Z is located as described, we first apply Chasles' theorem to write

$$\mathbf{a}_A = \mathbf{a}_{A/Z} + \mathbf{a}_Z \qquad \mathbf{a}_B = \mathbf{a}_{B/Z} + \mathbf{a}_Z \tag{15.30}$$

Then, as shown in Fig. 15.22, we let ψ' be the angle between $\mathbf{a}_{A/Z}$ and $\overline{AZ}$.

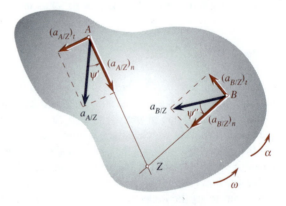

Fig. 15.22

[†]This theorem and its corollary, which follows, were presented in the authors' paper, "Teaching the Concepts of Acceleration Center and Virtual Work in Dynamics," *Engineering Education*, vol. 79, no. 3, April 1989, pp. 441–443. Furthermore, note that this theorem is mainly intended to provide a general basis for the corollary. We shall emphasize the application of the corollary.

The magnitude of the tangential and normal components of $\mathbf{a}_{A/Z}$ are

$$(a_{A/Z})_t = \overline{AZ}\alpha \qquad (a_{A/Z})_n = \overline{AZ}\omega^2$$

Thus, we have

$$\tan\psi' = \frac{(a_{A/Z})_t}{(a_{A/Z})_n} = \frac{\overline{AZ}\alpha}{\overline{AZ}\omega^2} \qquad \psi' = \tan^{-1}\left(\frac{\alpha}{\omega^2}\right)$$

Letting ψ'' be the angle between $\mathbf{a}_{B/Z}$ and $\overline{BZ}$ as shown in Fig. 15.22, we can, in a similar manner, show that $\psi'' = \tan^{-1}(\alpha/\omega^2)$. From the angle ψ as given in Eq. (15.29), we see that $\psi = \psi' = \psi''$. Therefore,

$$\mathbf{a}_{A/Z} \parallel \mathbf{a}_A \qquad \mathbf{a}_{B/Z} \parallel \mathbf{a}_B \qquad (15.31)$$

For the conditions in (15.31) and Eqs. (15.30) to hold true simultaneously, we must require that

$$\mathbf{a}_Z \parallel \mathbf{a}_A \qquad \mathbf{a}_Z \parallel \mathbf{a}_B \qquad (15.32)$$

Since $\mathbf{a}_A$ and $\mathbf{a}_B$ are, as stated, *not* parallel to each other and the conditions in (15.32) must be true, we logically conclude that $\mathbf{a}_Z = \mathbf{0}$.   Q.E.D.

Corollary. (Refer to the theorem.) *If a rigid slab at an instant is accelerated to undergo a plane motion from zero angular velocity, then $\omega = 0$, $\alpha \neq 0$, $\psi = 90°$, $\overrightarrow{AZ} \perp \mathbf{a}_A$, and $\overrightarrow{BZ} \perp \mathbf{a}_B$, where $\mathbf{a}_A$ and $\mathbf{a}_B$ are accelerations of the points A and B of the slab, and Z is the acceleration center of the slab at that instant. (See Fig. 15.23.) Furthermore, for any point P of the slab at that instant, we have*

$$a_P = \overline{ZP}\alpha \qquad \mathbf{a}_P = \boldsymbol{\alpha} \times \overrightarrow{ZP} \qquad (15.33)^\dagger$$

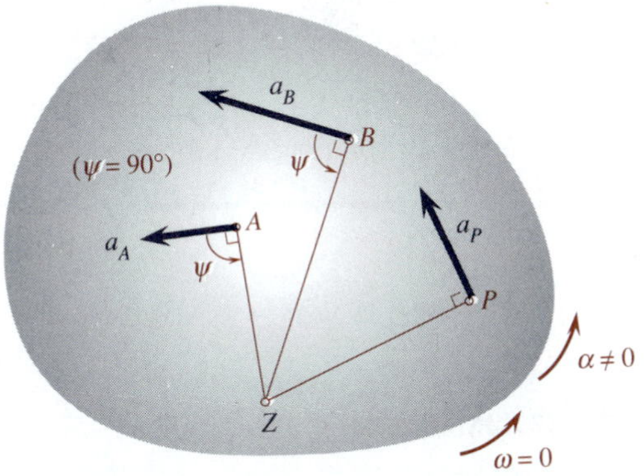

Fig. 15.23

Note that, at the instant considered, (1) any acceleration center Z must have a zero acceleration, (2) any point of a rigid slab (or its massless extension) having a zero acceleration is an acceleration center of the slab, (3) both the angular acceleration and angular velocity of a rigid slab are zero if two or more points of the slab have zero accelerations, (4) the velocity of an acceleration center is *not* necessarily zero. In fact, each rigid slab in a mechanism has its own acceleration center, which is generally different from that of another slab.

[†]This *corollary* resembles the theorem of velocity center in Sec. 15.7 and is *very useful* in cases where a rigid slab in plane motion has $\boldsymbol{\omega} = \mathbf{0}$ and $\boldsymbol{\alpha} \neq \mathbf{0}$ at the instant considered. Cf. Examples 15.14 through 15.16.

EXAMPLE 15.14

The rod AB is at rest when its end A is given $\mathbf{a}_A = 1.2$ m/s^2 → in the position as shown. For this position, determine (a) the acceleration center Z of the rod, (b) $\boldsymbol{\alpha}$ of the rod, (c) $\mathbf{a}_B$ and $\mathbf{a}_P$ of the end B and the midpoint P of the rod, respectively.

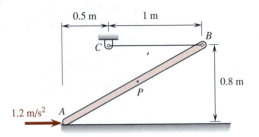

Solution. Since the rod is accelerated from rest, we see that it has $\boldsymbol{\omega} = \mathbf{0}$, $\boldsymbol{\alpha} \neq \mathbf{0}$, and $\psi = 90°$ in the position shown. From the constraint condition of the rod, we know that $\mathbf{a}_B = a_B \uparrow$. Thus, we draw the lines AZ and BZ, where $\overrightarrow{AZ} \perp \mathbf{a}_A$ and $\overrightarrow{BZ} \perp \mathbf{a}_B$, to locate the *acceleration center* at Z as shown. To be consistent with the direction of $\mathbf{a}_A$, the direction of $\boldsymbol{\alpha}$ must be counterclockwise as indicated. We write

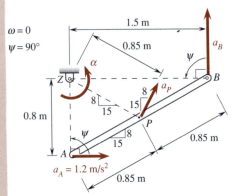

$$a_A = \overline{ZA}\alpha \qquad 1.2 = 0.8\alpha \qquad \alpha = 1.5 \text{ rad/s}^2 \circlearrowleft \blacktriangleleft$$

$$\mathbf{a}_B = \boldsymbol{\alpha} \times \overrightarrow{ZB} = 1.5\mathbf{k} \times (1.5\mathbf{i}) = 2.25\mathbf{j}$$

$$\mathbf{a}_P = \boldsymbol{\alpha} \times \overrightarrow{ZP} = 1.5\mathbf{k} \times (0.75\mathbf{i} - 0.4\mathbf{j}) = 0.6\mathbf{i} + 1.125\mathbf{j}$$

$$\mathbf{a}_B = 2.25 \text{ m/s}^2 \uparrow \qquad \mathbf{a}_P = 1.275 \text{ m/s}^2 \angle 61.9° \blacktriangleleft$$

REMARK. If the angular velocity of the rod is not zero in the position shown, then $\psi \neq 90°$ and the foregoing solution would be invalid.

Developmental Exercises

D15.24 Refer to the given data and the solution in Example 15.14. Verify that $\mathbf{a}_B = \mathbf{a}_{B/A} + \mathbf{a}_A$.

D15.25 Describe the *theorem of acceleration center* and its corollary.

EXAMPLE 15.15

The pin-connected bars are at rest when the bar OA is given $\boldsymbol{\alpha}_{OA} = 2$ rad/s^2 $\circlearrowleft$ in the position shown. For this position, determine (a) the acceleration center Z of the bar AB, (b) $\mathbf{a}_P$ of the midpoint P of the bar AB.

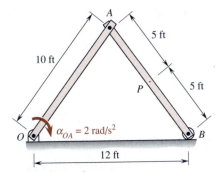

Solution. Clearly, the point O is the acceleration center of the bar OA because $\mathbf{a}_O = \mathbf{0}$. As the system is accelerated from rest, we have $\boldsymbol{\omega}_{OA} = \boldsymbol{\omega}_{AB} = \mathbf{0}$. Thus, we write

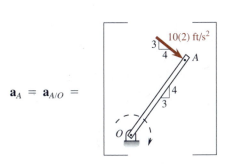

$$\mathbf{a}_A = \mathbf{a}_{A/O} =$$

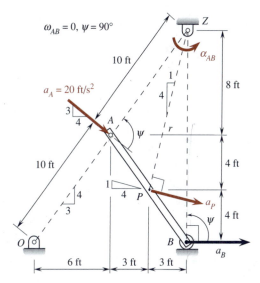

$$= 20 \text{ ft/s}^2 \; \text{⦨} \; 36.9°$$

The constraint condition of the bar AB requires that $\mathbf{a}_B = a_B \rightarrow$. Therefore, we locate the *acceleration center* of the bar AB at Z as shown, where $\overrightarrow{AZ} \perp \mathbf{a}_A$ and $\overrightarrow{BZ} \perp \mathbf{a}_B$; i.e.,

Z is 16 ft directly above B. ◀

From the sketch, we write

$$\overrightarrow{ZP} = -3\mathbf{i} - 12\mathbf{j} \text{ ft}$$

$$a_A = 20 = \overline{ZA}\alpha_{AB} = 10\alpha_{AB} \qquad \alpha_{AB} = 2$$

$$\mathbf{a}_P = \boldsymbol{\alpha}_{AB} \times \overrightarrow{ZP} = 2\mathbf{k} \times (-3\mathbf{i} - 12\mathbf{j}) = 2(-3\mathbf{j} + 12\mathbf{i})$$

$$\mathbf{a}_P = 24\mathbf{i} - 6\mathbf{j}$$

$$\mathbf{a}_P = 24.7 \text{ ft/s}^2 \; \text{⦨} \; 14.04°$$ ◀

REMARK. If the angular velocities of the bars are not zero in the position shown, then $\psi \neq 90°$ and the foregoing solution would be invalid.

Developmental Exercises

D15.26 Refer to Example 15.15. Determine $\mathbf{a}_B$ of the end point B.

D15.27 Is the velocity of an acceleration center equal to zero?

EXAMPLE 15.16

Using *acceleration center*, work Example 15.11.

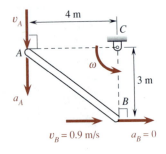

Solution. Since the end B of the rod AB is pulled with a constant velocity $v_B = 0.9$ m/s →, we have $\mathbf{a}_B = \mathbf{0}$ as indicated. The constraint condition of the rod requires that $\mathbf{v}_A = v_A \downarrow$ and $\mathbf{a}_A = a_A \downarrow$ as shown, where the point C is the velocity center of the rod. We have

$$v_B = 0.9 = \overline{CB}\omega = 3\omega \qquad \omega = 0.3 \text{ rad/s } \circlearrowleft$$

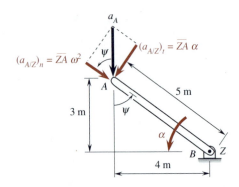

Since $\mathbf{a}_B = \mathbf{0}$, the *acceleration center* Z of the rod is, therefore, coincident with the point B. Referring to the second sketch, we write

$$\tan\psi = \frac{\alpha}{\omega^2} \qquad \frac{4}{3} = \frac{\alpha}{(0.3)^2}$$

$$a_A \cos\psi = \overline{ZA}\omega^2 \qquad a_A\left(\frac{3}{5}\right) = 5(0.3)^2$$

$$\alpha = 0.12 \text{ rad/s}^2 \circlearrowleft \qquad \mathbf{a}_A = 0.75 \text{ m/s}^2 \downarrow \blacktriangleleft$$

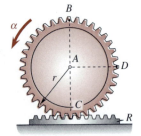

Fig. D15.28

Developmental Exercise

D15.28 A gear of radius $r = 120$ mm is accelerated from rest with $\alpha = 2$ rad/s² $\circlearrowleft$ as shown, where the rack R is fixed to the support. For this position, determine (a) the acceleration center Z of the gear, (b) $\mathbf{a}_B$ and $\mathbf{a}_D$.

15.10   Parametric Method

We know that a one-degree-of-freedom system is a system whose position and motion can be defined by a single independent parameter or by a set of n dependent parameters subject to $n - 1$ constraint equations. For a one-degree-of-freedom mechanism moving in the xy plane, the x and y coordinates of all key points of the mechanism can be expressed, directly or indirectly, in terms of a single independent parameter; and their velocities and accelerations can be obtained by taking time derivatives of their coordinates. Of course, the angular velocity and angular acceleration of any line in the mechanism are given by the first and the second time derivatives of the angular position coordinate of that line, respectively. The procedure employing appropriate parameters to define and analyze the motion of a system is referred to as the *parametric method*.

EXAMPLE 15.17[†]

The end B of the rod AB moves with a constant velocity $\mathbf{v}_B = 0.9$ m/s $\rightarrow$. Using the *parametric method*, determine for the position shown (a) ω and α of the rod, (b) $\mathbf{v}_A$ and $\mathbf{a}_A$ of the end A.

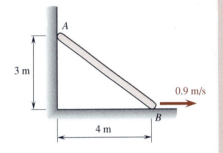

Solution.   The system under consideration is a one-degree-of-freedom system whose motion can be completely defined by the parameter θ as indicated. Let y_A and x_B be the position coordinates of the ends A and B of the rod as shown. In the position shown, we have $y_A = 3$ m, $x_B = 4$ m, $\theta = \tan^{-1}(3/4)$, $v_B = \dot{x}_B = 0.9$ m/s, and $a_B = \ddot{x}_B = 0$, where the positive (i.e., increasing) direction of θ is clockwise. We write

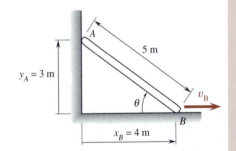

$$x_B = 5\cos\theta \qquad v_B = \dot{x}_B = -5(\sin\theta)\dot{\theta} = 0.9$$

$$\dot{\theta} = -0.3 \qquad \omega = \dot{\theta} \qquad \omega = 0.3 \text{ rad/s } \circlearrowleft \blacktriangleleft$$

$$a_B = \ddot{x}_B = -5[(\cos\theta)\dot{\theta}^2 + (\sin\theta)\ddot{\theta}]$$

$$= -5[\tfrac{4}{5}(-0.3)^2 + \tfrac{3}{5}\ddot{\theta}] = 0$$

$$\ddot{\theta} = -0.12 \qquad \alpha = \ddot{\theta} \qquad \alpha = 0.12 \text{ rad/s}^2 \circlearrowleft \blacktriangleleft$$

$$y_A = 5\sin\theta \qquad v_A = \dot{y}_A = 5(\cos\theta)\dot{\theta} = -1.2$$

$$a_A = \ddot{y}_A = 5[-(\sin\theta)\dot{\theta}^2 + (\cos\theta)\ddot{\theta}] = -0.75$$

Since v_A and a_A are found to be negative, the actual directions of $\mathbf{v}_A$ and $\mathbf{a}_A$ must point in the negative sense of y_A. Thus, we write

$$\mathbf{v}_A = 1.2 \text{ m/s } \downarrow \qquad \mathbf{a}_A = 0.75 \text{ m/s}^2 \downarrow \blacktriangleleft$$

Developmental Exercises

D15.29   Work Example 15.17 if the end B has $\mathbf{a}_B = 0.6$ m/s$^2 \rightarrow$ as well as $\mathbf{v}_B = 0.9$ m/s $\rightarrow$.

D15.30   The end B of the rod AB is pulled with a constant velocity $\mathbf{v}_B = 5.76$ ft/s $\rightarrow$ as shown. (a) Express x_B of the end B in terms of the angle θ. (b) Determine ω and α of the rod when $y_A = 12$ ft.

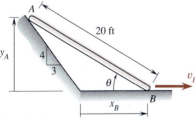

Fig. D15.30

[†]Cf. Examples 15.11 and 15.16.

EXAMPLE 15.18

A linkage moving in a plane is shown, where the crank AB rotates with a constant angular velocity $\omega_{AB} = 10$ rad/s $\circlearrowright$. Determine the angular velocities ω_{BD} and ω_{DE} of the coupler link BD and the output link DE when AB, BD, and DE all become collinear with the ground link AE.

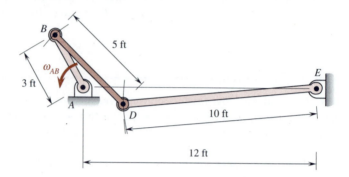

Solution. Moments before AB, BD, and DE all become collinear with the ground link AE, the configuration of AB, BD, and DE may be specified by the position coordinates θ_1, θ_2, and θ_3 as parameters as indicated in the sketch, where the positive directions of θ_1 and θ_2 are clockwise, but the positive direction of θ_3 is counterclockwise. The horizontal and vertical projections of AB, BD, and DE in this one-degree-of-freedom system require that θ_1, θ_2, and θ_3 be subject to the following two constraint equations:

$$-3\cos\theta_1 + 5\cos\theta_2 + 10\cos\theta_3 = 12 \tag{1}$$

$$3\sin\theta_1 + 10\sin\theta_3 = 5\sin\theta_2 \tag{2}$$

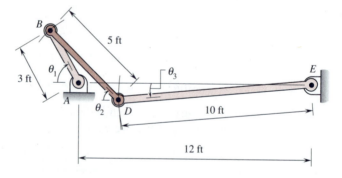

Since the values of θ_1, θ_2, and θ_3 under consideration are very small, we may use series expansion (cf. Sec. B3.1 of App. B) to approximate Eqs. (1) and (2) as follows:

$$-3(1 - \tfrac{1}{2}\theta_1^2) + 5(1 - \tfrac{1}{2}\theta_2^2) + 10(1 - \tfrac{1}{2}\theta_3^2) = 12 \tag{1'}$$

$$3\theta_1 + 10\theta_3 = 5\theta_2 \tag{2'}$$

where third-degree and higher terms in θ_1, θ_2 and θ_3 are neglected. Equations (1') and (2') yield

$$\theta_2 = \tfrac{1}{5}(1 \pm 2\sqrt{2})\theta_1 \qquad \theta_3 = \tfrac{1}{5}(-1 \pm \sqrt{2})\theta_1 \tag{3}$$

Differentiating with respect to the time, we write

$$\dot{\theta}_2 = \tfrac{1}{5}(1 \pm 2\sqrt{2})\dot{\theta}_1 \qquad \dot{\theta}_3 = \tfrac{1}{5}(-1 \pm \sqrt{2})\dot{\theta}_1 \qquad (4)$$

Note that $\dot{\theta}_1 = -10$ rad/s because $\omega_{AB} = 10$ rad/s ↻ . Thus, we have $\dot{\theta}_2 = -7.66$ rad/s or 3.66 rad/s and $\dot{\theta}_3 = -0.828$ rad/s or 4.83 rad/s; i.e., we obtain two sets of values:

$$\omega_{BD} = 7.66 \text{ rad/s } ↻ \qquad \text{and} \qquad \omega_{DE} = 0.828 \text{ rad/s } ↺ \quad ◀$$

or

$$\omega_{BD} = 3.66 \text{ rad/s } ↺ \qquad \text{and} \qquad \omega_{DE} = 4.83 \text{ rad/s } ↻ \quad ◀$$

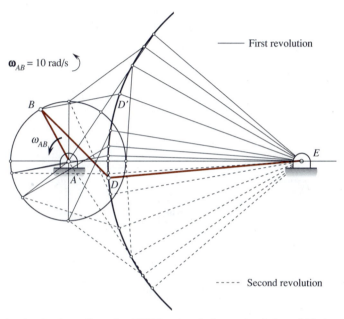

——— First revolution

$\omega_{AB} = 10$ rad/s ↻

B

ω_{AB}

D'

A

D

E

----- Second revolution

A plot showing that the configuration *ABDE* is repeated after two revolutions of *AB* about *A*

REMARK. Let us assume that *no sudden reversal* of rotation exists at the time the links become collinear. Then the *first set* of values is the solution for the case in which the configuration of *ABDE* (just before the links become collinear) is as depicted in the problem. The *second set* of values corresponds to the solution for the case in which the crank *AB* has completed one more revolution. Thus, each configuration of the entire linkage is repeated in every two revolutions of the crank *AB* as illustrated in the plot. Furthermore, it is well to point out that the problem in this example cannot be solved using the methods described in Secs. 15.6 and 15.7 because they yield only *one* nontrivial equation to relate the *two* unknowns: ω_{BD} and ω_{DE}.

Developmental Exercise

D15.31 Refer to Example 15.18. What are the values of α_{BD} and α_{DE} when *AB*, *BD*, and *DE* are collinear with *AE*?

PROBLEMS

15.87 through 15.89 At the instant shown, the end A of the rod AB is accelerated from rest with $\mathbf{a}_A$ as indicated. Using *acceleration center*, determine for this instant (a) $\boldsymbol{\alpha}$ of the rod, (b) $\mathbf{a}_B$ and $\mathbf{a}_D$ of the points B and D.

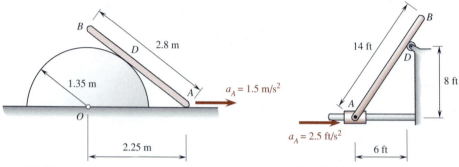

Fig. P15.87   **Fig. P15.88**

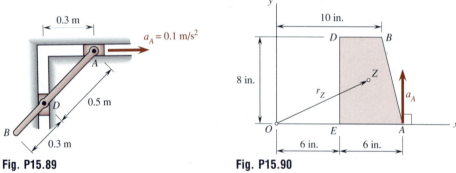

Fig. P15.89   **Fig. P15.90**

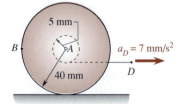

Fig. P15.91

15.90 The plate $ABDE$ moves in the xy plane. In the position shown, the plate rotates with $\boldsymbol{\omega} = 2$ rad/s ↻, $\boldsymbol{\alpha} = 3$ rad/s² ↻, and the corner A moves with $\mathbf{a}_A = 25$ in./s² ↑. For this position, determine the position vector $\mathbf{r}_Z$ of the acceleration center Z of the plate.

15.91 In the position shown, the double gear is accelerated from rest by pulling the upper rack with $\mathbf{a}_U = 132$ mm/s² ← and the lower rack with $\mathbf{a}_L = 84$ mm/s² →. If $a = 90$ mm and $b = 180$ mm, determine for this position (a) the distance $\overline{ZA}$ between the acceleration center Z and the center A of the gear, (b) $\boldsymbol{\alpha}$ of the gear, (c) $\mathbf{a}_B$ of the point B.

15.92* In the position shown, the yoyo is accelerated from rest by pulling the end D of the string, which is wound around its core, with $\mathbf{a}_D = 7$ mm/s² →. If the yoyo rolls without slipping, determine for this position $\boldsymbol{\alpha}$ of the yoyo and $\mathbf{a}_B$ of the point B.

Fig. P15.92*

15.93* through 15.96* Using *acceleration center*, solve the following problems:

15.93* Prob. 15.80.   15.94* Prob. 15.81*.
15.95* Prob. 15.83.   15.96* Prob. 15.85.

15.97 In the position shown, the slider is accelerated from rest with $\mathbf{a}_D$ = 210 mm/s² ←. Using *acceleration center*, determine for this position $\boldsymbol{\alpha}_{AB}$ of the crank AB and $\mathbf{a}_E$ of the point E.

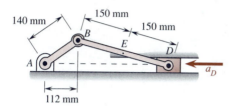

Fig. P15.97

15.98 The linkage is released from rest in the position shown, and the link OB acquires instantaneously $\alpha_{OB} = \alpha_{OB}$ rad/s² ↺ due to gravity. Using *acceleration center*, determine for the instant of release $\mathbf{a}_D$ and $\mathbf{a}_E$ of the points D and E in terms of α_{OB} as a parameter.

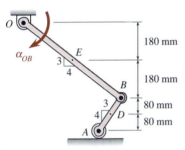

Fig. P15.98

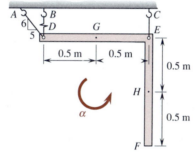

Fig. P15.99

15.99 The bent rod DEF is suspended by three wires AD, BD, and CE. At the instant shown, the wire BD is suddenly cut and the bent rod instantaneously accelerates with $\alpha = \alpha$ rad/s² ↻ due to gravity. Using *acceleration center*, determine for this instant $\mathbf{a}_G$ and $\mathbf{a}_H$ of the points G and H in terms of α as a parameter.

15.100* through 15.102 The end A of the rod AB is moved to the right with a speed v_A as shown. Using the *parametric method*, determine $\boldsymbol{\omega}_{AB}$ of the rod in terms of v_A, b, and x.

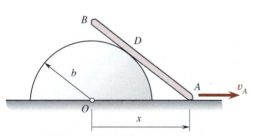

Fig. P15.100* and P15.103

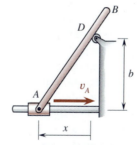

Fig. P15.101 and P15.104*

Fig. P15.102 and P15.105

15.103 through 15.105 The end A of the rod AB is moved to the right with a constant speed v_A as shown. Using the *parametric method*, derive an expression for α_{AB} of the rod in terms of v_A, b, and x.

15.106* The crank AB of the linkage shown rotates with a constant angular velocity $\omega_{AB} = 5$ rad/s $\circlearrowleft$. Using the *parametric method*, determine ω_{BD} and ω_{DE} of the coupler link BD and the output link DE when AB, BD, and DE all become collinear with the ground link AE.[†]

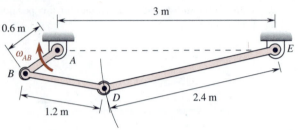

Fig. P15.106*

15.107 A linkage is shown, where the slider D moves along a smooth circular groove of radius 6 ft as indicated. It is known that the crank AB rotates with a constant angular velocity $\omega_{AB} = 2$ rad/s $\circlearrowright$. Using the *parametric method*, determine ω_{BD} of the link BD and $\mathbf{v}_D$ of the slider D when AB and BD become collinear with the line AO. (*Hint.* Can the line joining O and D be taken as a link?)[†]

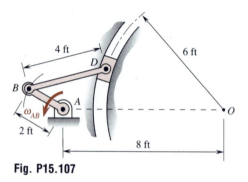

Fig. P15.107

USE OF ROTATING REFERENCE FRAMES

15.11 Time Derivatives of Rotating Unit Vectors

We recall that, in kinematics, any reference frame may be designated as *fixed*; all others not rigidly attached to it are regarded as *moving*.[††] A *moving reference frame* may be translating, rotating, or doing both, in a *fixed reference frame*. It is true that the time rate of change of a vector in a fixed reference frame is the *same* as that in a translating reference frame. However, the time rate of change of a vector in a rotating reference frame is generally *different* from that in a fixed or translating reference frame.

Let a rigid slab S move in the XY plane of the *fixed reference frame OXYZ* as shown in Fig. 15.24, where $\mathbf{I}$, $\mathbf{J}$, and $\mathbf{K}$ are the unit vectors directed along the axes of $OXYZ$. As a fundamental step, we first consider the time

---

[†]Cf. Example 15.18.

[††]A reference frame designated as fixed in *kinetics* is usually an *inertial reference frame* in which Newton's laws are directly applicable.

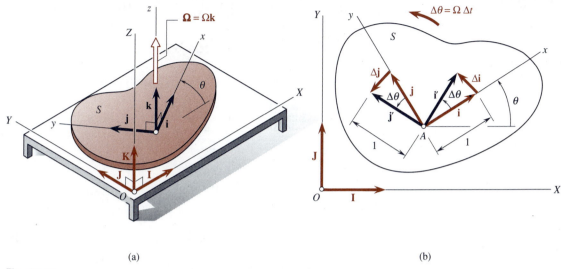

(a)   (b)

Fig. 15.24

rates of change of the unit vectors **i**, **j**, and **k** directed along the axes of a rectangular coordinate system *Axyz* which is embedded in the moving rigid slab *S* to form a *moving reference frame Axyz.*[†] For the study of plane kinematics where **K** = **k**, let the angular velocity of *Axyz* be

$$\boldsymbol{\Omega} = \Omega\mathbf{K} = \Omega\mathbf{k} \qquad (15.34)^{††}$$

and the velocity of the point *A* be $\mathbf{v}_A$ in *OXYZ*. Note that **k** is constant in *OXYZ*. The time rate of change of **k** in *OXYZ* is, therefore, zero; i.e.,

$$\left(\frac{d\mathbf{k}}{dt}\right)_{OXYZ} = (\dot{\mathbf{k}})_{OXYZ} = \dot{\mathbf{k}} = \mathbf{0} \qquad (15.35)$$

NOTE: Since **Ω** = Ω**k** is being used, the *Z* and *z* axes are included in the names of the reference frames. For subsequent studies, *a quantity, or the derivative of a quantity, without the name of a specific reference frame as subscripts is to be understood as one measured, or taken, in the fixed reference frame*, such as *OXYZ*.

The magnitudes of the unit vectors **i** and **j** are always equal to 1, a constant; however, their directions may change during the motion of *Axyz* in the *XY* plane as shown in Fig. 15.24(a). It is readily seen that the changes of **i** and **j** are due *only* to the *rotation* of *Axyz*. Thus, the changes Δ**i** and Δ**j** of **i** and **j** during the time interval Δ*t* are due to the incremental rotation Δθ of *Axyz* as shown in Fig. 15.24(b). From this figure and Eq. (15.34), we derive the time rates of change of **i** and **j** as follows:

$$\lim_{\Delta t \to 0} \frac{\Delta\theta}{\Delta t} = \dot{\theta} = \Omega \qquad \mathbf{j} = \mathbf{k} \times \mathbf{i} \qquad \mathbf{i} = -\mathbf{k} \times \mathbf{j}$$

$$\Delta\mathbf{i} = \mathbf{i}' - \mathbf{i} \approx \Delta\theta\mathbf{j}$$

[†]Cf. the footnote on p. 479.

[††]Cf. the angular velocity defined by Eq. (15.40).

$$\dot{\mathbf{i}} = \lim_{\Delta t \to 0} \frac{\Delta \mathbf{i}}{\Delta t} = \lim_{\Delta t \to 0} \frac{\Delta \theta \mathbf{j}}{\Delta t} = \Omega \mathbf{j} = \Omega \mathbf{k} \times \mathbf{i}$$

$$\boxed{\dot{\mathbf{i}} = \mathbf{\Omega} \times \mathbf{i}} \qquad (15.36)$$

$$\Delta \mathbf{j} = \mathbf{j}' - \mathbf{j} \approx -\Delta \theta \mathbf{i}$$

$$\dot{\mathbf{j}} = \lim_{\Delta t \to 0} \frac{\Delta \mathbf{j}}{\Delta t} = \lim_{\Delta t \to 0} \frac{-\Delta \theta \mathbf{i}}{\Delta t} = -\Omega \mathbf{i} = \Omega \mathbf{k} \times \mathbf{j}$$

$$\boxed{\dot{\mathbf{j}} = \mathbf{\Omega} \times \mathbf{j}} \qquad (15.37)$$

Due to the fact that $\mathbf{\Omega} \times \mathbf{k} = \Omega \mathbf{k} \times \mathbf{k} = \mathbf{0}$, we may also write Eqs. (15.35) as

$$\boxed{\dot{\mathbf{k}} = \mathbf{\Omega} \times \mathbf{k} = \mathbf{0}} \qquad (15.38)$$

Since $\mathbf{i}$, $\mathbf{j}$, and $\mathbf{k}$ are all constant in $Axyz$, we have

$$(\dot{\mathbf{i}})_{Axyz} = (\dot{\mathbf{j}})_{Axyz} = (\dot{\mathbf{k}})_{Axyz} = \mathbf{0} \qquad (15.39)$$

If $Axyz$ were to have a general motion in $OXYZ$, its angular velocity $\mathbf{\Omega}$ in $OXYZ$ would be given by

$$\mathbf{\Omega} = (\dot{\mathbf{j}} \cdot \mathbf{k})\mathbf{i} + (\dot{\mathbf{k}} \cdot \mathbf{i})\mathbf{j} + (\dot{\mathbf{i}} \cdot \mathbf{j})\mathbf{k} \qquad (15.40)[†]$$

which agrees with Eq. (15.34) as a special case for the plane kinematics of rigid bodies. However, we shall not be concerned with the derivation of Eq. (15.40) until Sec. 19.2.

Developmental Exercise

D15.32 In the notations adopted for subsequent studies, what does each of the following represent? (a) $OXYZ$, (b) $Axyz$, (c) $\mathbf{\Omega}$, (d) $(\dot{\mathbf{i}})_{OXYZ}$, (e) $(\dot{\mathbf{i}})_{Axyz}$, (f) $\dot{\mathbf{i}}$.

15.12 Time Derivatives of a Vector in Two Reference Frames

We first note that the *time derivative of a scalar function* ϕ *is* $\dot{\phi}$, which is still a scalar quantity whose value is the same in any reference frame. In other words, the time derivative of a scalar in a *moving reference frame Axyz* (which is rotating, translating, or doing both) is the same as that in a *fixed reference frame OXYZ*; i.e.,

$$\dot{\phi} = (\dot{\phi})_{OXYZ} = (\dot{\phi})_{Axyz} \qquad (15.41)$$

As an example of the property expressed in Eq. (15.41), let ϕ denote the

---

[†]This general definition was set forth by the dynamicist T. R. Kane in his book *Dynamics*, Holt, Rinehart and Winston, New York, 1968, p. 21. The formulation of this definition may be found in the development leading to Eq. (19.30).

temperature at a point of a region. We readily see that the time rate of change (i.e., the time derivative) of the temperature ϕ has (and must have) the same value at a given time in all moving and fixed reference frames which are used to describe it.

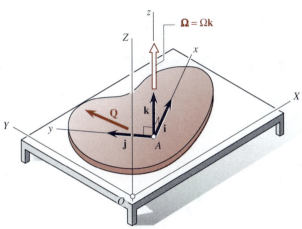

Fig. 15.25

Next, let us consider a vector $\mathbf{Q}$ in the xy plane of $Axyz$ which moves in the XY plane of $OXYZ$ as shown in Fig. 15.25, where $Axyz$ has an angular velocity $\mathbf{\Omega} = \Omega\mathbf{k}$ and the point A has a velocity $\mathbf{v}_A$ in $OXYZ$. Resolving $\mathbf{Q}$ into components parallel to the x and y axes, we write

$$\mathbf{Q} = Q_x\mathbf{i} + Q_y\mathbf{j} \tag{15.42}$$

In following the adopted notation, we denote by $\dot{\mathbf{Q}}$ the time derivative of Q in $OXYZ$, and by $(\dot{\mathbf{Q}})_{Axyz}$ the time derivative of $\mathbf{Q}$ in $Axyz$. Taking the time derivative first in $Axyz$, then in $OXYZ$, on both sides of Eq. (15.42) and applying Eqs. (15.36), (15.37), (15.39), and (15.41), we write

$$(\dot{\mathbf{Q}})_{Axyz} = (\dot{Q}_x)_{Axyz}\mathbf{i} + (\dot{Q}_y)_{Axyz}\mathbf{j}$$

$$= (\dot{Q}_x)_{OXYZ}\mathbf{i} + (\dot{Q}_y)_{OXYZ}\mathbf{j}$$

$$(\dot{\mathbf{Q}})_{Axyz} = \dot{Q}_x\mathbf{i} + \dot{Q}_y\mathbf{j} \tag{15.43}$$

$$\dot{\mathbf{Q}} = \dot{Q}_x\mathbf{i} + Q_x\mathbf{i} + \dot{Q}_y\mathbf{j} + Q_y\mathbf{j}$$

$$= \dot{Q}_x\mathbf{i} + Q_x(\mathbf{\Omega} \times \mathbf{i}) + \dot{Q}_y\mathbf{j} + Q_y(\mathbf{\Omega} \times \mathbf{j})$$

$$= (\dot{Q}_x\mathbf{i} + \dot{Q}_y\mathbf{j}) + \mathbf{\Omega} \times (Q_x\mathbf{i} + Q_y\mathbf{j})$$

$$\dot{\mathbf{Q}} = (\dot{\mathbf{Q}})_{Axyz} + \mathbf{\Omega} \times \mathbf{Q} \tag{15.44}$$

Thus, we see that the time derivative of the vector $\mathbf{Q}$ taken in the *fixed reference frame OXYZ* is equal to the sum of the following two parts: (a) the time derivative of $\mathbf{Q}$ taken in the *moving reference frame Axyz*, and (b) the cross product $\mathbf{\Omega} \times \mathbf{Q}$, where $\mathbf{\Omega}$ is the *angular velocity* of $Axyz$ in $OXYZ$. If $\mathbf{\Omega} = \mathbf{0}$, then $Axyz$ is nonrotating and Eq. (15.44) degenerates to $\dot{\mathbf{Q}} = (\dot{\mathbf{Q}})_{Axyz}$, which is a property we used in the first part of this chapter.

Developmental Exercises

D15.33 Notations are a means for effective communication. In the notations adopted, which of the following are always true?
(a) $\dot{Q}_x = (\dot{Q}_x)_{OXYZ}$, (b) $\dot{Q}_x = (\dot{Q}_x)_{Axyz}$, (c) $\dot{\mathbf{Q}} = (\dot{\mathbf{Q}})_{OXYZ}$, (d) $\dot{\mathbf{Q}} = (\dot{\mathbf{Q}})_{Axyz}$.

D15.34 Suppose that $\mathbf{Q}$ is a constant vector in $Axyz$, whose angular velocity in $OXYZ$ is $\boldsymbol{\Omega}$. Express the time derivative of $\mathbf{Q}$ taken in $OXYZ$.

15.13 Velocities in Different Reference Frames

Let us consider the motion of a particle B that moves along a path (e.g., a curved groove) C on a rigid slab S which is itself translating and rotating in the XY plane of the *fixed reference frame OXYZ* as shown in Fig. 15.26. To facilitate the study, we embed a coordinate system $Axyz$ in the slab S to form a *moving reference frame Axyz* as shown. The angular velocity of $Axyz$ in $OXYZ$ is $\boldsymbol{\Omega} = \Omega\mathbf{K} = \Omega\mathbf{k}$, and the velocity of the point A in $OXYZ$ is $\mathbf{v}_A$. In $OXYZ$, the positions of A and B are specified by the position vectors $\mathbf{r}_A$ and $\mathbf{r}_B$, while the position of B relative to A is specified by the position vector $\mathbf{r}_{B/A}$.

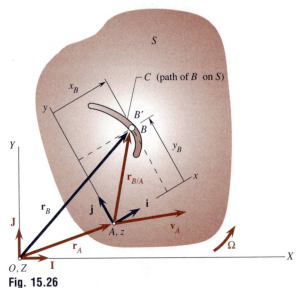

Fig. 15.26

It is well to recall in studies involving rotating reference frames that *a quantity, or the derivative of a quantity, without the name of a specific reference frame as subscripts is to be understood as one measured, or taken, in the fixed reference frame*, such as $OXYZ$. Besides, note that

$$\mathbf{v}_A = \dot{\mathbf{r}}_A \qquad \mathbf{v}_B = \dot{\mathbf{r}}_B \qquad \mathbf{v}_{B/A} = \dot{\mathbf{r}}_{B/A} \tag{15.45}$$

$$\mathbf{r}_{B/Axyz} = (\mathbf{r}_{B/A})_{Axyz} \qquad \mathbf{v}_{B/Axyz} = (\dot{\mathbf{r}}_{B/A})_{Axyz} \tag{15.46}$$

Suppose that B' is a point *of the slab* S, and B' *coincides* with the moving particle B at the instant considered, as shown in Fig. 15.26. It is obvious that the velocity of B' in $Axyz$, which is embedded in S, is zero; i.e.,

$$\mathbf{v}_{B'/Axyz} = (\dot{\mathbf{r}}_{B'/A})_{Axyz} = \mathbf{0} \tag{15.47}$$

The position vector of B' in $OXYZ$ is

$$\mathbf{r}_{B'} = \overrightarrow{OB'} = \mathbf{r}_A + \mathbf{r}_{B'/A} \qquad (15.48)$$

Thus, we apply Eqs. (15.44) through (15.48) to write

$$\mathbf{v}_{B'} = \dot{\mathbf{r}}_{B'} = \dot{\mathbf{r}}_A + \dot{\mathbf{r}}_{B'/A}$$

$$= \mathbf{v}_A + [(\dot{\mathbf{r}}_{B'/A})_{Axyz} + \boldsymbol{\Omega} \times \mathbf{r}_{B'/A}]$$

$$= \mathbf{v}_A + (\mathbf{0} + \boldsymbol{\Omega} \times \mathbf{r}_{B'/A})$$

$$\mathbf{v}_{B'} = \boldsymbol{\Omega} \times \mathbf{r}_{B'/A} + \mathbf{v}_A \qquad (15.49)$$

Since $\mathbf{r}_{B'/A} = \mathbf{r}_{B/A}$ *at the instant considered*, we find that the velocity of B' is given by

$$\mathbf{v}_{B'} = \boldsymbol{\Omega} \times \mathbf{r}_{B/A} + \mathbf{v}_A \qquad (15.50)$$

The position vector of the particle B in $OXYZ$, as shown in Fig. 15.26, is

$$\mathbf{r}_B = \mathbf{r}_A + \mathbf{r}_{B/A} \qquad (15.51)$$

Applying Eqs. (15.44) through (15.46), (15.50), and (15.51), we derive the velocity of B in $OXYZ$ as follows:

$$\mathbf{v}_B = \dot{\mathbf{r}}_B = \dot{\mathbf{r}}_A + \dot{\mathbf{r}}_{B/A}$$

$$= \mathbf{v}_A + [(\dot{\mathbf{r}}_{B/A})_{Axyz} + \boldsymbol{\Omega} \times \mathbf{r}_{B/A}]$$

$$= \mathbf{v}_{B/Axyz} + (\boldsymbol{\Omega} \times \mathbf{r}_{B/A} + \mathbf{v}_A)$$

$$\mathbf{v}_B = \mathbf{v}_{B/Axyz} + \mathbf{v}_{B'} \qquad (15.52)^{\dagger}$$

$$\mathbf{v}_B = \mathbf{v}_{B/Axyz} + \boldsymbol{\Omega} \times \mathbf{r}_{B/A} + \mathbf{v}_A \qquad (15.53)$$

We note that the velocity of B relative to A in $OXYZ$ is given by

$$\mathbf{v}_{B/A} = \dot{\mathbf{r}}_{B/A} = (\dot{\mathbf{r}}_{B/A})_{Axyz} + \boldsymbol{\Omega} \times \mathbf{r}_{B/A}$$

$$\mathbf{v}_{B/A} = \mathbf{v}_{B/Axyz} + \boldsymbol{\Omega} \times \mathbf{r}_{B/A} \qquad (15.54)$$

Substitution of Eq. (15.54) into Eq. (15.53) yields

$$\mathbf{v}_B = \mathbf{v}_{B/A} + \mathbf{v}_A \qquad (15.55)$$

In recapitulation of the symbols used, note that

$\mathbf{r}_{B/A} = \overrightarrow{AB}$ = position vector of B relative to A

$\mathbf{v}_A, \mathbf{v}_B$ = velocities of A and B in the *fixed* reference frame $OXYZ$

$\mathbf{v}_{B'}$ = velocity of the *coinciding point B'* in $OXYZ$

$\mathbf{v}_{B/A}$ = velocity of B relative to A in $OXYZ$

$\mathbf{v}_{B/Axyz}$ = velocity of B in the *moving* reference frame $Axyz$

$\boldsymbol{\Omega}$ = angular velocity of $Axyz$ in $OXYZ$

For motion in the XY plane, the various vectors in Eqs. (15.53) through (15.55) are illustrated in Fig. 15.27. In this figure, note that (1) the path C

---

†Note that the relation expressed in Eq. (15.52) is very *succinct*; it will provide a helpful background for understanding why the *Coriolis acceleration* in Eq. (15.64) is called a *complementary acceleration*.

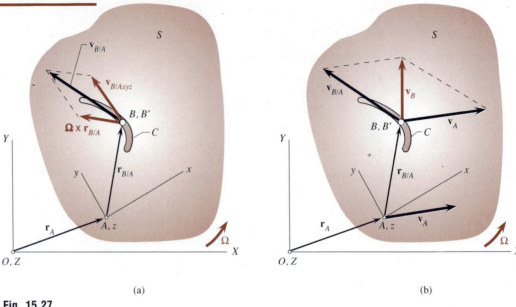

(a)   (b)

Fig. 15.27

is the curve traveled by the particle B in the moving reference frame $Axyz$, (2) the path C is generally different from the curve traveled by B in the fixed reference frame $OXYZ$, (3) $\mathbf{v}_{B/Axyz}$ is tangent to the path C, (4) $\mathbf{v}_{B/Axyz}$ as well as $\mathbf{v}_{B/A}$ is generally not perpendicular to the position vector $\mathbf{r}_{B/A}$, (5) $\boldsymbol{\Omega} \times \mathbf{r}_{B/A}$ is a vector perpendicular to $\mathbf{r}_{B/A}$, and (6) $\mathbf{v}_{B/A}$ is equal to $\mathbf{v}_{B/Axyz}$ only when $\boldsymbol{\Omega} = \mathbf{0}$. In summarizing velocities in different reference frames, we refer to Fig. 15.27 and write

$$\mathbf{v}_B = \underbrace{\mathbf{v}_{B/Axyz} + \boldsymbol{\Omega} \times \mathbf{r}_{B/A}} + \mathbf{v}_A \qquad (15.53)$$
$$\text{(repeated)}$$

$$\mathbf{v}_B = \underbrace{\mathbf{v}_{B/A} \qquad\qquad\qquad + \mathbf{v}_A} \qquad (15.55)$$
$$\text{(repeated)}$$

$$\mathbf{v}_B = \mathbf{v}_{B/Axyz} + \qquad\quad \mathbf{v}_{B'} \qquad (15.52)$$
$$\text{(repeated)}$$

EXAMPLE 15.19

In the position shown, the crank BC rotates with a constant angular velocity of 4.25 rad/s $\circlearrowright$, and its pin at B is confined to move in the slot of the link

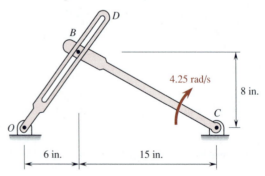

OD. Determine for this position $\boldsymbol{\omega}_{OD}$ of the link OD and the velocity $\mathbf{v}_{B/OD}$ of the pin at B relative to the link OD.

Solution. To solve this problem, we let $OXYZ$ *be fixed to the ground* and $Oxyz$ *be embedded in the link* OD, as shown. Since $\boldsymbol{\omega}_{BC} = 4.25$ rad/s $\boldsymbol{\Omega} = -4.25\mathbf{K}$ rad/s and $\mathbf{v}_C = \mathbf{0}$, we write

$$\mathbf{v}_B = \mathbf{v}_{B/C} = \boldsymbol{\omega}_{BC} \times \mathbf{r}_{B/C} = -4.25\mathbf{K} \times (-15\mathbf{I} + 8\mathbf{J})$$

$$\mathbf{v}_B = 34\mathbf{I} + 63.75\mathbf{J} \tag{1}$$

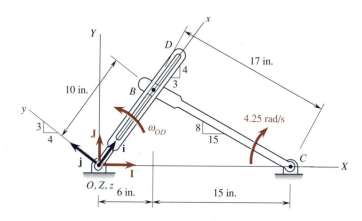

The angular velocity of $Oxyz$ is $\boldsymbol{\Omega} = \omega_{OD}\mathbf{k}$. Applying Eq. (15.53), we write

$$\mathbf{v}_B = \mathbf{v}_{B/Oxyz} + \boldsymbol{\Omega} \times \mathbf{r}_{B/O} + \mathbf{v}_O$$
$$= v_{B/OD}\mathbf{i} + \omega_{OD}\mathbf{k} \times (10\mathbf{i}) + \mathbf{0}$$
$$= v_{B/OD}\mathbf{i} + 10\omega_{OD}\mathbf{j}$$
$$= v_{B/OD}(\tfrac{1}{5})(3\mathbf{I} + 4\mathbf{J}) + 10\omega_{OD}(\tfrac{1}{5})(-4\mathbf{I} + 3\mathbf{J})$$
$$\mathbf{v}_B = (\tfrac{3}{5}v_{B/OD} - 8\omega_{OD})\mathbf{I} + (\tfrac{4}{5}v_{B/OD} + 6\omega_{OD})\mathbf{J} \tag{2}$$

From Eqs. (1) and (2), we obtain two ensuing scalar equations, which yield $v_{B/OD} = 71.4$ and $\omega_{OD} = 1.105$. Thus, we have

$$\boldsymbol{\omega}_{OD} = 1.105 \text{ rad/s} \circlearrowleft \qquad \mathbf{v}_{B/OD} = 71.4 \text{ in./s} \quad 53.1° \blacktriangleleft$$

REMARK. If B' is a point of the link OD coinciding with the pin at B, then, without using a rotating reference frame, we can solve for ω_{OD} by equating the velocity of B' to the component of the velocity of B in the direction perpendicular to the slot. On the other hand, it can be shown from the solution in this example that

$$\mathbf{v}_B = 71.4\mathbf{i} + 11.05\mathbf{j} \text{ in./s}$$
$$\mathbf{v}_{B'} = 11.05\mathbf{j} \text{ in./s}$$
$$\mathbf{v}_{B/B'} = \mathbf{v}_B - \mathbf{v}_{B'} = 71.4\mathbf{i} \text{ in./s}$$

Thus, we see that $\mathbf{v}_{B/OD} = \mathbf{v}_{B/B'}$. (Cf. the *Remark* in Example 15.20.)

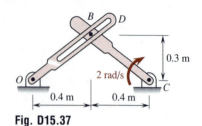

Fig. D15.37

Developmental Exercises

D15.35 Derive Eq. (2) in Example 15.19 by replacing $Oxyz$ with $B'xyz$ where B' is a point of the link OD coinciding with the pin at B.

D15.36 Both Eqs. (15.52) and (15.55) are useful formulas. Describe these two formulas in words.

D15.37 In the position shown, the crank BC rotates with a constant angular velocity $\omega_{BC} = 2$ rad/s $\curvearrowleft$. Making use of a rotating reference frame, determine ω_{OD} and $\mathbf{v}_{B/OD}$.

15.14 Accelerations in Different Reference Frames

Refer to Fig. 15.28 where a particle B moves along a path C on a rigid slab S which is itself translating and rotating in the XY plane of the *fixed reference frame* $OXYZ$. The coordinate system $Axyz$ is embedded in the slab S to form a *moving reference frame* $Axyz$, which has an angular acceleration $\dot{\boldsymbol{\Omega}} = \dot{\Omega}\mathbf{k}$ as well as an angular velocity $\boldsymbol{\Omega} = \Omega\mathbf{k}$ at the instant considered. The velocity and acceleration of the origin A of $Axyz$ are $\mathbf{v}_A$ and $\mathbf{a}_A$ in $OXYZ$, respectively.

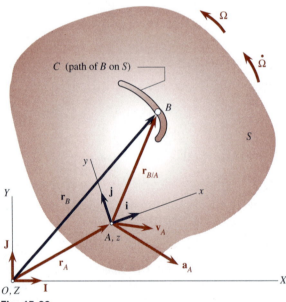

Fig. 15.28

In the ensuing study of acceleration in different reference frames, note that

$$\mathbf{a}_A = \dot{\mathbf{v}}_A \qquad \mathbf{a}_B = \dot{\mathbf{v}}_B \qquad \mathbf{a}_{B/A} = \dot{\mathbf{v}}_{B/A} \tag{15.56}$$

$$\mathbf{a}_{B/Axyz} = (\dot{\mathbf{v}}_{B/Axyz})_{Axyz} \tag{15.57}$$

Taking the time derivative in $OXYZ$ on both sides of Eq. (15.53) and applying Eqs. (15.44), (15.46), (15.56), and (15.57), we write

$$\dot{\mathbf{v}}_B = [(\dot{\mathbf{v}}_{B/Axyz})_{Axyz} + \mathbf{\Omega} \times \mathbf{v}_{B/Axyz}] + \dot{\mathbf{\Omega}} \times \mathbf{r}_{B/A}$$
$$+ \mathbf{\Omega} \times [(\dot{\mathbf{r}}_{B/A})_{Axyz} + \mathbf{\Omega} \times \mathbf{r}_{B/A}] + \dot{\mathbf{v}}_A$$

$$\mathbf{a}_B = \mathbf{a}_{B/Axyz} + \mathbf{\Omega} \times \mathbf{v}_{B/Axyz} + \dot{\mathbf{\Omega}} \times \mathbf{r}_{B/A}$$
$$+ \mathbf{\Omega} \times \mathbf{v}_{B/Axyz} + \mathbf{\Omega} \times (\mathbf{\Omega} \times \mathbf{r}_{B/A}) + \mathbf{a}_A$$

$$\mathbf{a}_B = \mathbf{a}_{B/Axyz} + \dot{\mathbf{\Omega}} \times \mathbf{r}_{B/A} + \mathbf{\Omega} \times (\mathbf{\Omega} \times \mathbf{r}_{B/A})$$
$$+ 2\mathbf{\Omega} \times \mathbf{v}_{B/Axyz} + \mathbf{a}_A \qquad (15.58)$$

Performing similar operations on Eqs. (15.54) and (15.55), we obtain

$$\mathbf{a}_{B/A} = \mathbf{a}_{B/Axyz} + \dot{\mathbf{\Omega}} \times \mathbf{r}_{B/A} + \mathbf{\Omega} \times (\mathbf{\Omega} \times \mathbf{r}_{B/A})$$
$$+ 2\mathbf{\Omega} \times \mathbf{v}_{B/Axyz} \qquad (15.59)$$

$$\mathbf{a}_B = \mathbf{a}_{B/A} + \mathbf{a}_A \qquad (15.60)$$

Note that $\mathbf{v}_{B/Axyz}$, $\mathbf{\Omega}$, and $\mathbf{r}_{B/A}$ are the same as those defined earlier, and

$\mathbf{a}_A$, $\mathbf{a}_B$ = accelerations of A and B in the *fixed* reference frame $OXYZ$

$\mathbf{a}_{B/A}$ = acceleration of B relative to A in $OXYZ$

$\mathbf{a}_{B/Axyz}$ = acceleration of B in the *moving* reference frame $Axyz$

$\dot{\mathbf{\Omega}}$ = angular acceleration of $Axyz$ in $OXYZ$

For motion in the XY plane, the various terms on the right side of Eq. (15.58) are illustrated in Fig. 15.29. Note that the vector sum of the four rust vectors at B gives the acceleration vector $\mathbf{a}_{B/A}$ of B relative to A in $OXYZ$ as defined in Eq. (15.59). Suppose that B' is a point *of* the slab S, and B' *coincides* with the moving particle B at the instant considered, as shown in Fig. 15.29. The acceleration of the *coinciding point* B' in $OXYZ$ may be derived by applying Eqs. (15.44), (15.47), (15.49), [not (15.50)], and

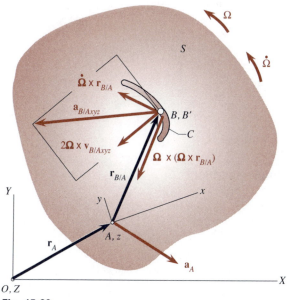

Fig. 15.29

(15.56) as follows:

$$\mathbf{a}_{B'} = \dot{\mathbf{v}}_{B'} = \dot{\mathbf{\Omega}} \times \mathbf{r}_{B'/A} + \mathbf{\Omega} \times \dot{\mathbf{r}}_{B'/A} + \dot{\mathbf{v}}_A$$

$$= \dot{\mathbf{\Omega}} \times \mathbf{r}_{B'/A} + \mathbf{\Omega} \times [(\dot{\mathbf{r}}_{B'/A})_{Axyz} + \mathbf{\Omega} \times \mathbf{r}_{B'/A}] + \mathbf{a}_A$$

$$= \dot{\mathbf{\Omega}} \times \mathbf{r}_{B'/A} + \mathbf{\Omega} \times [\mathbf{0} + \mathbf{\Omega} \times \mathbf{r}_{B'/A}] + \mathbf{a}_A$$

Since $\mathbf{r}_{B'/A} = \mathbf{r}_{B/A}$ *at the instant considered*, we find that the acceleration of B' in *OXYZ* is given by

$$\mathbf{a}_{B'} = \dot{\mathbf{\Omega}} \times \mathbf{r}_{B/A} + \mathbf{\Omega} \times (\mathbf{\Omega} \times \mathbf{r}_{B/A}) + \mathbf{a}_A \qquad (15.61)$$

It can readily be verified that the terms $\mathbf{\Omega} \times \mathbf{r}_{B/A}$ and $\mathbf{\Omega} \times (\mathbf{\Omega} \times \mathbf{r}_{B/A})$ in Eq. (15.61) represent, respectively, the *tangential* and the *normal components* of the *acceleration* $\mathbf{a}_{B'/A}$ of the coinciding point B' relative to A in *OXYZ*. We write

$$\mathbf{a}_{B'/A} = \dot{\mathbf{\Omega}} \times \mathbf{r}_{B/A} + \mathbf{\Omega} \times (\mathbf{\Omega} \times \mathbf{r}_{B/A}) \qquad (15.62)^{\dagger}$$

Substitution of Eq. (15.62) into Eq. (15.61) yields

$$\mathbf{a}_{B'} = \mathbf{a}_{B'/A} + \mathbf{a}_A \qquad (15.63)$$

Furthermore, substituting Eq. (15.61) into Eq. (15.58), we obtain

$$\mathbf{a}_B = \mathbf{a}_{B/Axyz} + \mathbf{a}_{B'} + 2\mathbf{\Omega} \times \mathbf{v}_{B/Axyz} \qquad (15.64)$$

Unlike Eq. (15.52), we see here that $\mathbf{a}_B \neq \mathbf{a}_{B/Axyz} + \mathbf{a}_{B'}$. The needed additional term $2\mathbf{\Omega} \times \mathbf{v}_{B/Axyz}$ is called the *complementary acceleration*[††] or, more often, the *Coriolis acceleration*, after the French mathematician and engineer Gaspard Gustave de Coriolis (1792–1843). In recapitulation, we refer to Fig. 15.29 and write

$$\mathbf{a}_B = \underbrace{\mathbf{a}_{B/Axyz} + \dot{\mathbf{\Omega}} \times \mathbf{r}_{B/A} + \mathbf{\Omega} \times (\mathbf{\Omega} \times \mathbf{r}_{B/A}) + 2\mathbf{\Omega} \times \mathbf{v}_{B/Axyz}} + \mathbf{a}_A \qquad \begin{matrix}(15.58)\\ \text{(repeated)}\end{matrix}$$

$$\mathbf{a}_B = \qquad\qquad\qquad\qquad \mathbf{a}_{B/A} \qquad\qquad\qquad\qquad + \mathbf{a}_A \qquad \begin{matrix}(15.60)\\ \text{(repeated)}\end{matrix}$$

$$\mathbf{a}_B = \mathbf{a}_{B/Axyz} + \underbrace{\dot{\mathbf{\Omega}} \times \mathbf{r}_{B/A} + \mathbf{\Omega} \times (\mathbf{\Omega} \times \mathbf{r}_{B/A})} + \mathbf{a}_A + 2\mathbf{\Omega} \times \mathbf{v}_{B/Axyz}$$

$$\mathbf{a}_B = \mathbf{a}_{B/Axyz} + \underbrace{\qquad\qquad \mathbf{a}_{B'/A} \qquad\qquad} + \mathbf{a}_A + 2\mathbf{\Omega} \times \mathbf{v}_{B/Axyz}$$

$$\mathbf{a}_B = \mathbf{a}_{B/Axyz} + \qquad\qquad \mathbf{a}_{B'} \qquad\qquad + 2\mathbf{\Omega} \times \mathbf{v}_{B/Axyz} \qquad \begin{matrix}(15.64)\\ \text{(repeated)}\end{matrix}$$

Developmental Exercises

D15.38 Suppose that B' is a point *of* the moving reference frame *Axyz*, and B' *coincides* with the moving particle B. (a) Is $\mathbf{v}_B = \mathbf{v}_{B/Axyz} + \mathbf{v}_{B'}$? (b) Is $\mathbf{a}_B = \mathbf{a}_{B/Axyz} + \mathbf{a}_{B'}$?

D15.39 Derive Eq. (15.62) by applying Eq. (15.44) and taking the second time derivative of $\mathbf{r}_{B'/A}$ in the fixed reference frame *OXYZ*.

[†]Cf. D15.39.

[††]Cf. Eqs. (15.22) and (15.52).

15.15 Interpretations for Coriolis Acceleration

Let us now focus our attention on some possible interpretations for the Coriolis acceleration $2\boldsymbol{\Omega} \times \mathbf{v}_{B/Axyz}$ as manifested in Eq. (15.64). We recall that the coordinate system $Axyz$ is embedded in the slab S to form a *moving reference frame Axyz*, whose angular velocity in the *fixed reference frame OXYZ* is $\boldsymbol{\Omega}$. Furthermore, we suppose that B' is a point *of* the slab S, and B' *coincides* with the moving particle B at the instant considered, as shown in Fig. 15.29. The *general interpretation* for the Coriolis acceleration $2\boldsymbol{\Omega} \times \mathbf{v}_{B/Axyz}$ is that *it represents the difference between the acceleration* $\mathbf{a}_{B/B'}$ *of B relative to B' in the fixed reference frame OXYZ and the acceleration* $\mathbf{a}_{B/Axyz}$ *of B in the moving reference frame Axyz*; i.e.,

$$2\boldsymbol{\Omega} \times \mathbf{v}_{B/Axyz} = \mathbf{a}_{B/B'} - \mathbf{a}_{B/Axyz} \qquad (15.65)$$

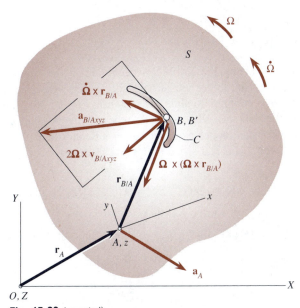

Fig. 15.29 (repeated)

To verify the truth of the above general interpretation, we first note from relative motion that

$$\mathbf{a}_{B/B'} = \mathbf{a}_B - \mathbf{a}_{B'} \qquad (15.66)$$

Rearranging Eq. (15.64) and employing Eq. (15.66), we write

$$2\boldsymbol{\Omega} \times \mathbf{v}_{B/Axyz} = (\mathbf{a}_B - \mathbf{a}_{B'}) - \mathbf{a}_{B/Axyz}$$
$$= \mathbf{a}_{B/B'} - \mathbf{a}_{B/Axyz}$$

which confirms the truth of the general interpretation.

In the case in which the particle B has a zero acceleration in the moving reference frame $Axyz$, the Coriolis acceleration $2\boldsymbol{\Omega} \times \mathbf{v}_{B/Axyz}$ lends itself to a *special interpretation* that *it simply represents the acceleration* $\mathbf{a}_{B/B'}$ *of B relative to B' in the fixed reference frame OXYZ*; i.e.,

$$\mathbf{a}_{B/Axyz} = \mathbf{0} \qquad 2\boldsymbol{\Omega} \times \mathbf{v}_{B/Axyz} = \mathbf{a}_{B/B'} \qquad (15.67)$$

The validity of the preceding special interpretation is obvious because it simply states a special case of Eq. (15.65).

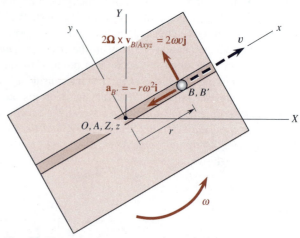

Fig. 15.30

To aid in visualizing the Coriolis acceleration in the simplest possible case in which it exists, let us consider a rotating slab with a slot in which a small particle B is confined to slide, as shown in Fig. 15.30, where $Axyz$ is embedded in the slab and $OXYZ$ is fixed to the ground. Assume that the slab rotates with a constant angular velocity $\boldsymbol{\omega} = \omega\mathbf{K}$ and the particle B moves along the slot with a constant speed v in $Axyz$. Then, referring to Fig. 15.30, we have

$$\boldsymbol{\Omega} = \omega\mathbf{K} = \omega\mathbf{k} \qquad \dot{\boldsymbol{\Omega}} = \mathbf{0} \qquad \mathbf{r}_{B/A} = r\mathbf{i}$$
$$\mathbf{v}_{B/Axyz} = v\mathbf{i} \qquad \mathbf{a}_{B/Axyz} = \mathbf{0} \qquad \mathbf{a}_A = \mathbf{0} \qquad (15.68)$$

Substituting Eqs. (15.68) into Eq. (15.58) and simplifying, we obtain

$$\mathbf{a}_B = -r\omega^2\mathbf{i} + 2\omega v\mathbf{j} \qquad (15.69)$$

Recall that B' is a point *of* the slab, and B' *coincides* with B at the instant considered, as shown in Fig. 15.30. By Eqs. (15.61) and (15.68), we obtain

$$\mathbf{a}_{B'} = -r\omega^2\mathbf{i} \qquad (15.70)$$

$$2\boldsymbol{\Omega} \times \mathbf{v}_{B/Axyz} = 2\omega v\mathbf{j} \qquad (15.71)$$

Substituting Eqs. (15.69) and (15.70) into Eq. (15.66), we get

$$\mathbf{a}_{B/B'} = 2\omega v\mathbf{j} \qquad (15.72)$$

Equations (15.71) and (15.72) clearly demonstrate that *the Coriolis acceleration* $2\boldsymbol{\Omega} \times \mathbf{v}_{B/Axyz}$ *simply represents the acceleration* $\mathbf{a}_{B/B'}$ *of B relative to B' in the fixed reference frame $OXYZ$ when the acceleration of B in the moving reference frame $Axyz$ is* $\mathbf{a}_{B/Axyz} = \mathbf{0}$.

Developmental Exercise

D15.40   Give general and special interpretations for the *Coriolis acceleration* of a particle.

EXAMPLE 15.20

Refer to Example 15.19. Making use of a rotating reference frame, determine, for the position shown, $\boldsymbol{\alpha}_{OD}$ of the link OD and $\mathbf{a}_{B/OD}$ of the pin at B relative to the link OD.

Solution. As we did in Example 15.19, we first let *OXYZ be fixed to the ground and Oxyz be embedded in the link OD*, as shown, where we have $\mathbf{k} = \mathbf{K}$. From Example 15.19, we know that the crank BC rotates with a constant angular velocity $\boldsymbol{\omega}_{BC} = -4.25\mathbf{k}$ rad/s; therefore, $\boldsymbol{\alpha}_{BC} = \mathbf{0}$. Besides, we know that, in the position shown,

$$\boldsymbol{\Omega} = \omega_{OD}\mathbf{k} = 1.105\mathbf{k} \text{ rad/s}$$

$$\mathbf{v}_{B/Oxyz} = v_{B/OD}\mathbf{i} = 71.4\mathbf{i} \text{ in./s}$$

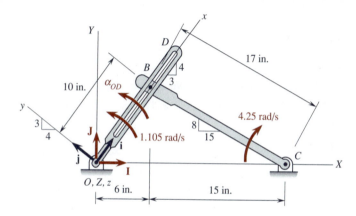

Since $\mathbf{a}_C = \mathbf{0}$ and $\boldsymbol{\alpha}_{BC} = \mathbf{0}$, we apply Eqs. (15.25) and (15.26) to write

$$\mathbf{a}_B = \mathbf{a}_{B/C} = (\mathbf{a}_{B/C})_t + (\mathbf{a}_{B/C})_n$$

$$= -\omega_{BC}^2\mathbf{r}_{B/C} = -(4.25)^2(-15\mathbf{I} + 8\mathbf{J})$$

$$\mathbf{a}_B = 270.94\mathbf{I} - 144.5\mathbf{J} \tag{1}$$

Next, we write

$$\dot{\boldsymbol{\Omega}} = \alpha_{OD}\mathbf{k} \qquad \mathbf{a}_{B/Oxyz} = \mathbf{a}_{B/OD} = a_{B/OD}\mathbf{i}$$

Since $\mathbf{a}_O = \mathbf{0}$ and $\mathbf{r}_{B/O} = 10\mathbf{i}$, we apply Eq. (15.58) to write

$$\mathbf{a}_B = \mathbf{a}_{B/Oxyz} + \dot{\boldsymbol{\Omega}} \times \mathbf{r}_{B/O} + \boldsymbol{\Omega} \times (\boldsymbol{\Omega} \times \mathbf{r}_{B/O})$$

$$\quad + 2\boldsymbol{\Omega} \times \mathbf{v}_{B/Oxyz} + \mathbf{a}_O$$

$$= a_{B/OD}\mathbf{i} + \alpha_{OD}\mathbf{k} \times 10\mathbf{i} + 1.105\mathbf{k} \times (1.105\mathbf{k} \times 10\mathbf{i})$$

$$\quad + 2(1.105\mathbf{k}) \times 71.4\mathbf{i} + \mathbf{0}$$

$$\mathbf{a}_B = (a_{B/OD} - 12.21)\mathbf{i} + (10\alpha_{OD} + 157.79)\mathbf{j} \tag{2}$$

Substituting $\mathbf{i} = \frac{1}{5}(3\mathbf{I} + 4\mathbf{J})$ and $\mathbf{j} = \frac{1}{5}(-4\mathbf{I} + 3\mathbf{J})$ into Eq. (2), we get

$$\mathbf{a}_B = (0.6a_{B/OD} - 8\alpha_{OD} - 133.56)\mathbf{I}$$

$$\quad + (0.8a_{B/OD} + 6\alpha_{OD} + 84.91)\mathbf{J} \tag{2'}$$

From Eqs. (1) and (2'), we obtain two ensuing scalar equations, which yield $a_{B/OD} = 59.17$ and $\alpha_{OD} = -46.12$. Thus, we have

$$\alpha_{OD} = 46.1 \text{ rad/s}^2 \ \circlearrowright \qquad\qquad \mathbf{a}_{B/OD} = 59.2 \text{ in./s}^2 \ \measuredangle \ 53.1° \ \blacktriangleleft$$

REMARK. Let B' be a point of the link OD coinciding with the pin at B. Then, from the solution in this example, it can be shown that

$$\mathbf{a}_B = 47.0\mathbf{i} - 303.4\mathbf{j} \text{ in./s}^2$$

$$\mathbf{a}_{B'} = -12.2\mathbf{i} - 461.2\mathbf{j} \text{ in./s}^2$$

$$\mathbf{a}_{B/B'} = \mathbf{a}_B - \mathbf{a}_{B'} = 59.2\mathbf{i} + 157.8\mathbf{j} \text{ in./s}^2$$

Clearly, $\mathbf{a}_{B/OD} \neq \mathbf{a}_{B/B'}$. However, by projection, we see that

$$a_{B/OD} = \mathbf{i} \cdot \mathbf{a}_{B/B'}$$

Developmental Exercises

D15.41 Derive Eq. (2) in Example 15.20 by replacing $Oxyz$ with $B'xyz$ where B' is a point of the link OD coinciding with the pin at B.

D15.42 Refer to D15.37. Determine α_{OD} and $\mathbf{a}_{B/OD}$.

EXAMPLE 15.21

The car B is rounding a circular curve while the car A is moving along a straight path with speeds and time rates of change of speed as indicated. For the instant shown, determine the velocity and acceleration of the car A in the coordinate system $Bxyz$ embedded in the car B.

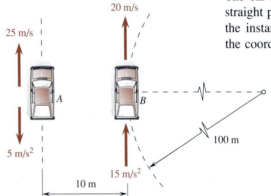

Solution. Since the motion of the car A relative to the car B is to be determined, we let $BXYZ$ *be fixed to the ground and* $Bxyz$ *be embedded in the car* B as shown. Applying Eqs. (15.53) and (15.58) to the motion of the car A relative to the car B, we write

$$\mathbf{v}_A = \mathbf{v}_{A/Bxyz} + \mathbf{\Omega} \times \mathbf{r}_{A/B} + \mathbf{v}_B \tag{1}$$

$$\mathbf{a}_A = \mathbf{a}_{A/Bxyz} + \dot{\mathbf{\Omega}} \times \mathbf{r}_{A/B} + \mathbf{\Omega} \times (\mathbf{\Omega} \times \mathbf{r}_{A/B})$$
$$+ 2\mathbf{\Omega} \times \mathbf{v}_{A/Bxyz} + \mathbf{a}_B \tag{2}$$

where $\mathbf{v}_{A/Bxyz}$ and $\mathbf{a}_{A/Bxyz}$ are the velocity and acceleration of the car A in $Bxyz$, and $\mathbf{\Omega}$ and $\dot{\mathbf{\Omega}}$ are the angular velocity and angular acceleration of $Bxyz$ (and hence the car B) in $BXYZ$, respectively. From the given data, we write

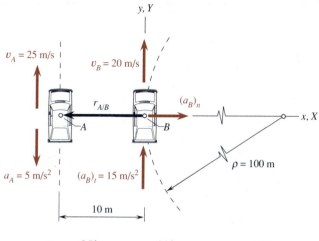

$$\mathbf{v}_A = 25\mathbf{j} \qquad \mathbf{v}_B = 20\mathbf{j} \qquad \mathbf{r}_{A/B} = -10\mathbf{i}$$

$$\mathbf{\Omega} = -\frac{v_B}{\rho}\mathbf{k} = -\frac{20}{100}\mathbf{k} = -0.2\mathbf{k}$$

Substituting these quantities into Eq. (1), we write

$$25\mathbf{j} = \mathbf{v}_{A/Bxyz} + (-0.2\mathbf{k}) \times (-10\mathbf{i}) + 20\mathbf{j}$$

$$\mathbf{v}_{A/Bxyz} = 3\mathbf{j} \text{ m/s} \quad \blacktriangleleft$$

Furthermore, from the given data, we write

$$(\mathbf{a}_B)_t = 15\mathbf{j} \qquad (\mathbf{a}_B)_n = \frac{v_B^2}{\rho}\mathbf{i} = \frac{(20)^2}{100}\mathbf{i} = 4\mathbf{i}$$

$$\mathbf{a}_B = (\mathbf{a}_B)_t + (\mathbf{a}_B)_n = 15\mathbf{j} + 4\mathbf{i} = 4\mathbf{i} + 15\mathbf{j}$$

$$\mathbf{a}_A = -5\mathbf{j}$$

$$\dot{\mathbf{\Omega}} = -\frac{(a_B)_t}{\rho}\mathbf{k} = -\frac{15}{100}\mathbf{k} = -0.15\mathbf{k}$$

Substituting these quantities and others into Eq. (2), we write

$$-5\mathbf{j} = \mathbf{a}_{A/Bxyz} + (-0.15\mathbf{k}) \times (-10\mathbf{i})$$

$$+ (-0.2\mathbf{k}) \times [-0.2\mathbf{k} \times (-10\mathbf{i})]$$

$$+ 2(-0.2\mathbf{k}) \times (3\mathbf{j}) + (4\mathbf{i} + 15\mathbf{j})$$

$$\mathbf{a}_{A/Bxyz} = -5.6\mathbf{i} - 21.5\mathbf{j} \text{ m/s}^2 \quad \blacktriangleleft$$

Developmental Exercises

D15.43   Refer to Example 15.21. For the instant shown, determine the velocity and acceleration of the car A in a nonrotating coordinate system translating with the car B.

D15.44   Refer to Example 15.21. For the instant shown, determine the velocity and acceleration of the car B in a coordinate system embedded in the car A.

PROBLEMS

15.108 through 15.110* In the position shown, the crank BC rotates with ω_{BC} as indicated, and its pin at B slides in the slot of the member AD. Determine for this position ω_{AD} of AD and $\mathbf{v}_{B/AD}$ of the pin at B relative to AD.

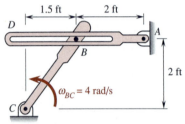

Fig. P15.108 and P15.111

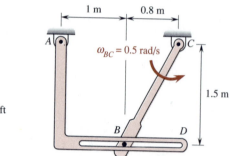

Fig. P15.109 and P15.112*

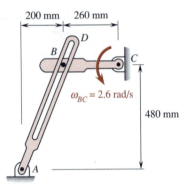

Fig. P15.110* and P15.113

15.111 through 15.113 In the position shown, the crank BC rotates with $\alpha_{BC} = 2$ rad/s² ↄ as well as ω_{BC} as indicated. For this position, determine α_{AD} of the member AD and $\mathbf{a}_{B/AD}$ of the pin at B relative to AD.

15.114* In this position shown, the rod BD moves downward through the fixed collar at C with $\mathbf{v}_{BD}$ as indicated, and its pin at B slides in the slot of the member AE. Determine for this position ω_{AE} of AE and $\mathbf{v}_{B/AE}$ of the pin at B relative to AE.

15.115 In the position shown, the rod BD moves with $\mathbf{a}_{BD} = 25$ mm/s² ↑ as well as $\mathbf{v}_{BD}$ as indicated. For this position, determine α_{AE} of the member AE and $\mathbf{a}_{B/AE}$ of the pin at B relative to AE.

15.116 The dumper pivoted at C is operated by the hydraulic cylinder AB. In the position shown, the piston rod is being extended with a velocity of 120 mm/s and an acceleration of 23.1 mm/s² relative to the cylinder. For this position, determine ω_D and α_D of the container D.

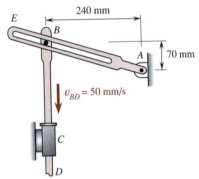

Fig. P15.114* and P15.115

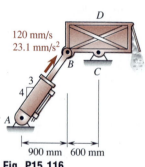

Fig. P15.116

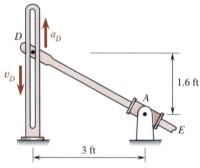

Fig. P15.117

15.117 The rod DE slides freely through the pivoted collar at A. In the position shown, the pin at its end D moves with $\mathbf{v}_D = 2.89$ ft/s ↓ and $\mathbf{a}_D = 5.78$ ft/s² ↑. For this position, determine ω and α of the rod DE.

15.118 In the position shown, the crank OD rotates with a constant angular velocity $\omega_{OD} = 1.69$ rad/s ↄ. For this position, determine (a) α_{DE} of the rod DE which slides freely through the pivoted collar at A, (b) $\mathbf{a}_B$ of the point B of the rod DE coinciding with A.

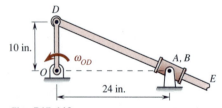

Fig. P15.118

15.119* The boats A and B are rounding circular curves of radius 50 m with equal constant speeds of 10 m/s on a lake. For the instant shown, determine the velocity and acceleration of the boat B in a reference frame embedded in the boat A.

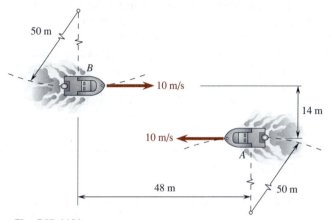

50 m

B

10 m/s

14 m

10 m/s

A

48 m

50 m

Fig. P15.119*

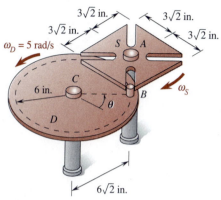

$3\sqrt{2}$ in.

$3\sqrt{2}$ in.

$3\sqrt{2}$ in.

$3\sqrt{2}$ in.

$\omega_D = 5$ rad/s

S A

C

6 in.

B ω_S

θ

D

$6\sqrt{2}$ in.

Fig. P15.120

15.120   The Geneva mechanism shown consists of a star wheel S, which rotates about A, and a driving wheel D, which rotates about C. The pin B attached to D engages one of the four radial slots of S in the position $\theta = 45°$ and turns S through 90° before disengaging itself from the slot in the position $\theta = 135°$. Note that the angular velocity of S is zero as B enters and leaves each slot of S. Thus, S produces only intermittent angular motion while D executes a constant angular velocity $\omega_D = 5$ rad/s $\circlearrowright$ as indicated. At the instant when $\theta = 60°$, determine (a) ω_S of the star wheel S, (b) the velocity of the pin B in a coordinate system embedded in S.

15.121*   Refer to Prob. 15.120. At the instant when $\theta = 60°$, determine (a) α_S of the star wheel S, (b) the acceleration of the pin B in a coordinate system embedded in S.

15.16   Concluding Remarks

It has been pointed out that kinematics deals with the study of motion of bodies without considering the cause of motion, while kinetics relates the motion of bodies to the force system causing the motion; however, kinetics generally contains some kinematics. Thus, this chapter provides a useful background for the study of kinetics of rigid bodies in the subsequent chapters.

Early on in the chapter, we note that the kinematics of a rigid body in rotation about a fixed axis is mathematically similar to that of the rectilinear motion of a particle. The analytical and graphical techniques presented in Chap. 11 are therefore applicable to the solutions of problems of rigid bodies rotating about fixed axes.

The major concepts and techniques for studying the plane motion of rigid bodies using nonrotating reference frames include the following: relations between linear and angular motions, equations of relative motion, linkage equations for velocities and accelerations, velocity and acceleration centers, and the parametric method. The technique using acceleration center is particularly useful in analyzing accelerations in a system at the instant the system is *accelerated from zero angular velocity*.

The latter part of this chapter is devoted to the study of plane kinematics involving rotating reference frames. In a system where rotations of two members as well as sliding between them occur at a guided connection, the technique using rotating reference frames often brings about an effective analysis of velocities and accelerations in the system. Furthermore, this technique is shown to be useful in the kinematic analysis of particles where the particles and the reference frame are moving independently of one another or along curved paths. Although the use of rotating reference frames is challenging and mind-bending at times, its development does set the stage for the study of motions of bodies in three dimensions in Chap. 19.

REVIEW PROBLEMS

15.122   The load L is pulled up the incline from rest through a distance of 6 ft by a cable wrapped around the drum D which is attached to the gear B of a hoisting system as shown. If the gear A drives the gear B from rest with a constant angular acceleration α_A from the time $t = 0$ to $t = 4$ s, determine (a) α_A, (b) the speed of the load L when $t = 4$ s.

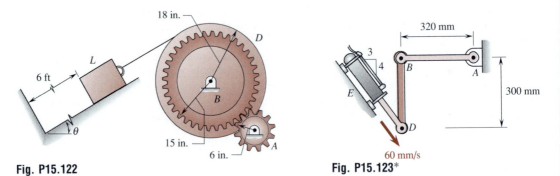

Fig. P15.122   Fig. P15.123*

15.123*   In the position shown, the piston rod of the hydraulic cylinder at E is being extended at a constant rate of 60 mm/s. For this position, determine ω_{AB} and ω_{BD} of the links AB and BD.

15.124   Refer to Prob. 15.123*. For the position shown, determine α_{AB} and α_{BD} of the links AB and BD.

15.125   The rod BD slides freely through the fixed collar at E. In the position shown, the wheel at O is rotated with a constant angular velocity of 3 rad/s ↻. For this position, determine $\mathbf{v}_B$ and $\mathbf{a}_B$ of the point B.

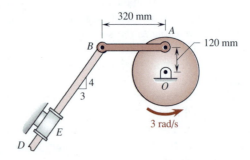

Fig. P15.125

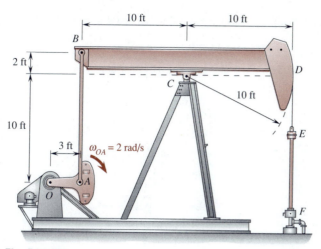

Fig. P15.126

15.126 The crank *OA* of the oil pumping unit rotates at a constant rate of 2 rad/s $\circlearrowright$. The pump rod *EF* is connected by cables *DE* to the sector *D* which is welded to the walking beam *BCD*. Note that the rod hanger at *E* as well as the pump rod always moves vertically during the pumping. For the position shown, determine $\mathbf{a}_E$ of the hanger at *E* and $\boldsymbol{\alpha}_{AB}$ of the connecting rod *AB*.

15.127* In the position shown, the link *AB* rotates with a constant angular velocity of 3.2 rad/s $\circlearrowright$. For this position, determine $\boldsymbol{\alpha}_{BD}$ and $\boldsymbol{\alpha}_{DE}$ of the links *BD* and *DE*.

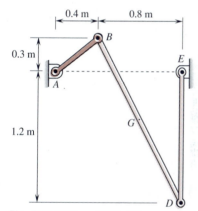

Fig. P15.127* and **P15.128**

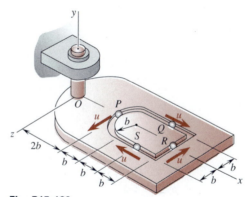

Fig. P15.129

15.128 In the position shown, the link *AB* is accelerated from rest with $\boldsymbol{\alpha}_{AB} = 4$ rad/s² $\circlearrowright$. Determine for this position (a) $\boldsymbol{\alpha}_{BD}$ and $\boldsymbol{\alpha}_{DE}$ of the links *BD* and *DE*, (b) $\mathbf{a}_G$ of the midpoint *G* of the link *BD*.

15.129 Each of the four particles *P*, *Q*, *R*, and *S* moves with a constant speed *u* relative to the slot of the rigid slab, which rotates about *O* with a constant counterclockwise angular velocity ω in the position shown. For this position and the directions of motion as indicated, determine the acceleration of each particle.

15.130* Solve Prob. 15.129 if the rigid slab rotates about *O* with a constant clockwise angular velocity ω in the position shown.

15.131 The crank *OD* rotates with a constant angular velocity $\omega_{OD} = 2.5$ rad/s $\circlearrowright$. When $\theta = 0$, determine $\boldsymbol{\alpha}_{DE}$ of the rod *DE*, which slides freely through the pivoted collar at *A*.

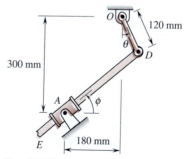

Fig. P15.131

15.132* Using the *parametric method* with θ and ϕ as dependent parameters, solve Prob. 15.131.

15.133 If the motion of the rod *OA* is as indicated, determine for the instant shown $\boldsymbol{\alpha}_{CD}$ of the rod *CD* and $\mathbf{a}_{A/CD}$ of the collar at *A* relative to the rod *CD*.

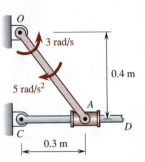

Fig. P15.133

Plane Kinetics of Rigid Bodies: Force and Acceleration

Chapter 16

KEY CONCEPTS

- Moment of inertia of a mass about an axis.
- Radius of gyration of a mass about an axis.
- Parallel-axis theorem for moments of inertia of a mass.
- Effective force-moment system on a rigid body in plane motion.
- D'Alembert's principle and the method of force and acceleration for rigid bodies in plane motion.
- Constrained general plane motion of rigid bodies.

Unlike a particle which may only be accelerated to translate by the concurrent forces acting on it, a rigid body may be accelerated to both translate and rotate by the nonconcurrent force system acting on it. We know that the innate property of a body to resist a translational acceleration is due to the *translational inertia* of the body, which is measured by its *mass*. Similarly, the innate property of a rigid body to resist a rotational acceleration is due to the *rotational inertia* of the rigid body, which is measured by its *moment of inertia*. Thus, mass moments of inertia are important characteristics of rigid bodies in kinetics; they are covered in the first part of the chapter. Later, we utilize Chasles' theorem, among other concepts, to establish the effective force-moment system acting on a rigid body in plane motion. The development leads to the property known as d'Alembert's principle, which underlies the *method of force and acceleration* for rigid bodies. Additional methods for studying plane kinetics of rigid bodies are presented in the next two chapters.

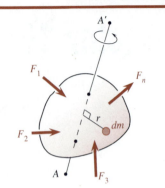

Fig. 16.1

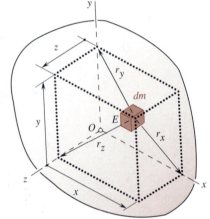

Fig. 16.2

MASS MOMENTS OF INERTIA[†]

16.1   Moments of Inertia of a Mass

In addition to having a mass, a rigid body has a definite shape and size. Thus, under the action of nonconcurrent forces $\mathbf{F}_1, \mathbf{F}_2, \ldots, \mathbf{F}_n$ as shown in Fig. 16.1, a rigid body will generally be accelerated to rotate as well as translate. The tendency of a body to resist a translational acceleration is, of course, measured by the *mass* of the body. On the other hand, the tendency of a body to resist a rotational acceleration about an axis is measured by a quantity called the *moment of inertia* of the body about that axis. This quantity, to be encountered in Sec. 16.5 and interpreted in Sec. 16.8, is an integral of the form $I = \int (\text{distance})^2 \, d(\text{mass})$ or

$$I = \int r^2 \, dm \tag{16.1}$$

where dm is a differential mass element of the body, r is the shortest distance between dm and the axis of rotation, say AA', and the integration extends over the entire body. The moment of inertia of a body defined in Eq. (16.1) is often referred to as the *mass moment of inertia*.

The differential mass element dm illustrated in Fig. 16.1 for Eq. (16.1) is of the *third order*. However, dm in Eq. (16.1) often takes the form of a *first-order* slender strip, thin ring, or thin shell if the particles in dm can be taken as equidistant from the axis of rotation, with r being the *equidistance* within the framework of first-order accuracy.

The moments of inertia of a body about the x, y, and z axes are customarily denoted by I_x, I_y, and I_z (sometimes, I_{xx}, I_{yy}, and I_{zz}), respectively. As shown in Fig. 16.2, the rectangular coordinates of the differential mass element dm at the point E are (x, y, z) and the shortest distances between dm and the x, y, and z axes are r_x, r_y, and r_z, respectively. We write

$$
\begin{aligned}
I_x &= \int r_x^2 \, dm = \int (y^2 + z^2) \, dm \\
I_y &= \int r_y^2 \, dm = \int (z^2 + x^2) \, dm \\
I_z &= \int r_z^2 \, dm = \int (x^2 + y^2) \, dm
\end{aligned}
\tag{16.2}
$$

Note from Eqs. (16.1) and (16.2) that the moment of inertia of a body about any axis is always a *positive scalar quantity*, whose dimensions are *mass* × *(length)*^{2}, because dm is always taken as positive regardless of the position in which it is situated. The primary units for the moment of inertia of a body are kg·m^2 in the SI and slug·ft^2 (or lb·ft·s^2) in the U.S. customary system. Unless otherwise stated, it will be assumed that each body considered is homogeneous and has a constant mass density.

Developmental Exercise

D16.1   Show that 1 slug·ft^2 = 1 lb·ft·s^2 = 1.356 kg·m^2.

[†]Those who have included the *mass moments of inertia* (cf. Secs. 7.11 through 7.14) in the study of *Statics* may proceed directly to Sec. 16.5. The *mass products of inertia* are needed and presented in Chap. 19.

EXAMPLE 16.1

A slender rod of length L and mass m is shown. Determine the moments of inertia I_x, I_y, and I_z of the rod about the axes as indicated.

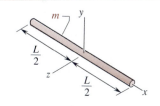

Solution. The mass density of the rod may be expressed as $\rho_L = m/L$. We note that the y and z coordinates of the mass element dm may be taken as zero throughout the rod as shown. Thus, we write

$$I_x = \int r_x^2 \, dm = \int (y^2 + z^2) \, dm = 0$$

$$I_y = \int r_y^2 \, dm = \int_{-L/2}^{L/2} x^2 \rho_L \, dx = \frac{1}{12}\rho_L L^3 = \frac{1}{12}\left(\frac{m}{L}\right)L^3$$

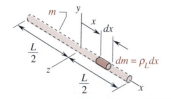

By symmetry, we note that $I_z = I_y$. Thus, we have

$$I_x = 0 \qquad I_y = I_z = \frac{1}{12}mL^2 \quad \blacktriangleleft$$

EXAMPLE 16.2

A thin rectangular plate of mass m lies in the zx plane as shown. Determine I_z of the plate.

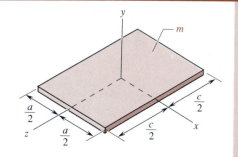

Solution. The mass density of the plate may be expressed as $\rho_A = m/(ca)$. Choosing dm to be a slender rod as shown, we write

$$I_z = \int r_z^2 \, dm = \int_{-a/2}^{a/2} x^2 \rho_A c \, dx = \frac{1}{12}\rho_A c a^3$$

$$I_z = \frac{1}{12}ma^2 \quad \blacktriangleleft$$

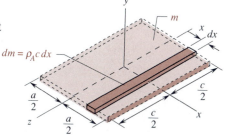

REMARK. For the thin plate in this example, we note that $y \approx 0$ throughout the plate. By Eq. (16.2), we write

$$I_x = \int z^2 \, dm$$

$$I_y = \int (z^2 + x^2) \, dm$$

$$I_z = \int x^2 \, dm$$

Thus, for such a plate, we have $I_y = I_z + I_x$.

Developmental Exercises

D16.2 Define the moment of inertia of a body about an axis.

D16.3 Determine I_x of the plate in Example 16.2.

EXAMPLE 16.3

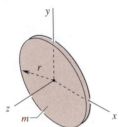

A thin circular disk of radius r and mass m lies in the xy plane as shown. Determine I_x, I_y, and I_z of the disk.

Solution. Since the thin circular disk lies in the xy plane with $z \approx 0$, we readily deduce from Eqs. (16.2) that

$$I_z = I_x + I_y$$

By symmetry, we note that $I_x = I_y$. Thus, we have

$$I_z = 2I_x \qquad I_x = I_y = \tfrac{1}{2}I_z$$

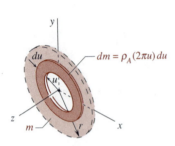

The mass density of the disk may be expressed as $\rho_A = m/(\pi r^2)$. Referring to the sketch, we write

$$I_z = \int r_z^2 \, dm = \int_0^r u^2 \rho_A (2\pi u) \, du$$

$$= \frac{1}{2} \rho_A \pi r^4 = \frac{1}{2}\left(\frac{m}{\pi r^2}\right)\pi r^4$$

$$I_z = \frac{1}{2}mr^2 \qquad I_x = I_y = \frac{1}{4}mr^2 \;\blacktriangleleft$$

Developmental Exercise

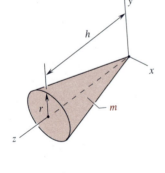

Fig. D16.4

D16.4 A circular cylinder of total mass m and radius r is shown. Applying the results obtained in Example 16.3, determine I_z of the cylinder.

EXAMPLE 16.4

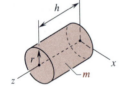

A circular cone having a mass m, a height h, and a base of radius r is shown. Using the solution for I_z in Example 16.3 as a formula, determine I_z of the cone.

Solution. Choosing dm to be a thin disk as shown in a different perspective for better clarity, we first determine the mass density ρ and dm, in terms

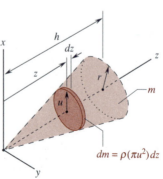

of z, as follows:

$$u = \frac{r}{h} z \qquad dm = \rho(\pi u^2)\, dz = \frac{\rho \pi r^2}{h^2} z^2\, dz$$

$$m = \int dm = \frac{\rho \pi r^2}{h^2} \int_0^h z^2\, dz = \frac{\rho \pi r^2 h}{3}$$

$$\rho = \frac{3m}{\pi r^2 h} \qquad dm = \frac{3m}{h^3} z^2\, dz$$

Using the solution for I_z in Example 16.3 as a formula, we write

$$dI_z = \frac{1}{2}\, (dm)\, u^2 = \frac{3mr^2}{2h^5} z^4\, dz$$

$$I_z = \int dI_z = \frac{3mr^2}{2h^5} \int_0^h z^4\, dz = \frac{3mr^2}{2h^5}\left(\frac{1}{5} h^5\right)$$

$$I_z = \frac{3}{10} mr^2 \quad \blacktriangleleft$$

Developmental Exercise

D16.5 For the circular cone in Example 16.4, is $I_z = I_x + I_y$?

EXAMPLE 16.5

A tetrahedron of mass density $\rho = 15$ slugs/ft^2 is bounded by the planes $x + 2y + 4z = 8$, $x = 0$, $y = 0$, and $z = 0$, where the lengths are in feet. Determine I_x of the tetrahedron.

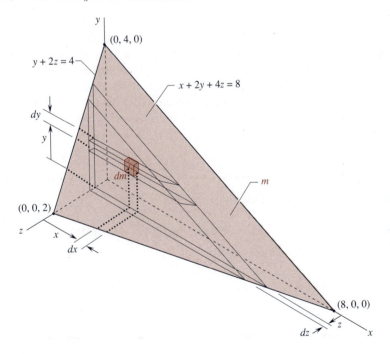

Solution. Let the differential mass element dm of the tetrahedron be chosen as sketched. We have $dm = \rho\, dV = \rho\, dx\, dy\, dz = 15\, dx\, dy\, dz$, and

$$I_x = \int (y^2 + z^2)\, dm$$

$$= 15 \int_{z=0}^{2} \int_{y=0}^{4-2z} \int_{x=0}^{8-2y-4z} (y^2 + z^2)\, dx\, dy\, dz$$

$$= 15 \int_{z=0}^{2} \int_{y=0}^{4-2z} (y^2 + z^2)(8 - 2y - 4z)\, dy\, dz$$

$$= 20 \int_{0}^{2} (5z^4 - 28z^3 + 60z^2 - 64z + 32)\, dz$$

$$I_x = 320 \text{ slug·ft}^2 \quad \blacktriangleleft$$

Developmental Exercise

D16.6 Determine I_y of the tetrahedron in Example 16.5.

16.2 Radius of Gyration of a Mass

The *radius of gyration* of a body about an axis is a distance whose square multiplied by the mass of the body gives the moment of inertia of the body about that axis. For a body of mass m as shown in Fig. 16.3, we write

$$I_x = k_x^2 m \qquad I_y = k_y^2 m \qquad I_z = k_z^2 m \qquad \text{(16.3)}$$

where k_x, k_y, and k_z are the radii of gyration of the body about the x, y, and z axes, respectively.

The radius of gyration $k_{AA'}$ of a body of mass m about an axis AA' is illustrated in Fig. 16.4, where $I_{AA'}$ is the moment of inertia of the body about the axis AA'. Note that the moment of inertia of a body about an axis will remain unchanged if the entire mass of the body is concentrated at a point which is away from that axis at a *distance* equal to the *radius of gyration* of the body about that axis. Clearly, the dimension of the radius of gyration is *length*, and it is measured in meters (m) in the SI and in feet (ft) in the U.S. customary system.

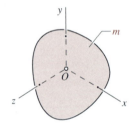

Fig. 16.3

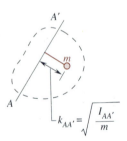

Fig. 16.4

EXAMPLE 16.6

Refer to Example 16.1. Determine k_y of the rod.

Solution. From the solution in Example 16.1, we know that $I_y = \frac{1}{12} mL^2$. Thus, we write

$$\tfrac{1}{12} mL^2 = k_y^2 m$$

$$k_y = \frac{\sqrt{3}}{6} L = 0.289L \quad \blacktriangleleft$$

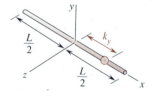

EXAMPLE 16.7

Determine k_z of the disk in Example 16.3.

Solution. From the solution in Example 16.3, we know that $I_z = \frac{1}{2} mr^2$. Thus, we write

$$\frac{1}{2} mr^2 = k_z^2 m$$

$$k_z = \frac{\sqrt{2}}{2} r = 0.707r \quad \blacktriangleleft$$

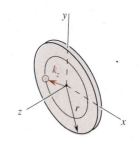

Developmental Exercises

D16.7 Define the radius of gyration of a body about an axis.

D16.8 Determine k_z of the cone in Example 16.4.

D16.9 The radius of gyration of a 50-kg flywheel about its axis of rotation A is $k_A = 0.4$ m. Determine I_A of the flywheel.

D16.10 Does it make sense for the radius of gyration of any wheel about its axle to be greater than its radius? (*Hint.* Cf. Example 16.7.)

PROBLEMS

NOTE. The following problems are to be solved by *integration*.

16.1* and 16.2* Determine I_x and I_y of the slender rod of mass m as shown.

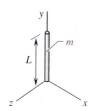

Fig. P16.1*

Fig. P16.2*

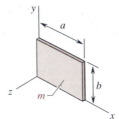

Fig. P16.3

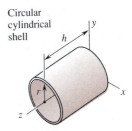

Circular cylindrical shell

Fig. P16.4*

16.3 Determine I_x, I_y, and I_z of the thin plate of mass m as shown.

16.4* through 16.9* Determine I_z and k_z of the body of mass m as shown.

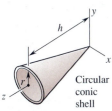

Circular conic shell

Fig. P16.5*

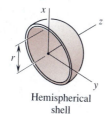

Hemispherical shell

Fig. P16.6*

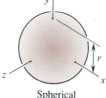

Spherical shell

Fig. P16.7*

Hemisphere

Fig. P16.8*

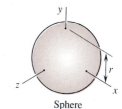

Sphere

Fig. P16.9*

16.10 A body of revolution and its generating area are shown. If the specific weight is $\gamma = 322$ lb/ft^3, determine I_x and k_x of the body.

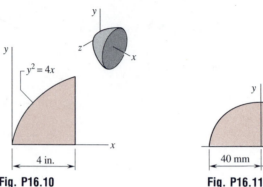

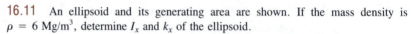

Fig. P16.10 **Fig. P16.11**

16.11 An ellipsoid and its generating area are shown. If the mass density is $\rho = 6$ Mg/m^3, determine I_x and k_x of the ellipsoid.

16.12* Determine I_y of the torus shown if its mass is m. (*Hint*. Choose dm to be a thin shell around the y axis. Cf. Sec. B5.2 of App. B.)

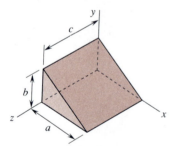

Fig. P16.12*

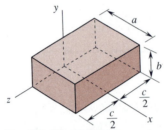

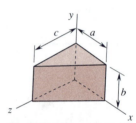

Fig. P16.13 and P16.15 **Fig. P16.14* and P16.16**

16.13 and 16.14* Determine I_x of the body shown if its mass is m.

16.15 and 16.16 Determine I_y and k_y of the body shown if its mass is m.

16.17 The mass density of the prism shown is $\rho = 4$ Mg/m^3. If $a = 0.4$ m, $b = 0.3$ m, and $c = 0.6$ m, determine I_x and k_x of the prism.

16.18* The mass density of the prism shown is $\rho = 5$ Mg/m^3. If $a = 80$ mm, $b = 6$ mm, and $c = 120$ mm, determine I_y and k_y of the prism.

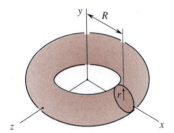

Fig. P16.17 and P16.18*

16.3 Parallel-Axis Theorem: Mass Moments of Inertia

Let the point $G(\bar{x}, \bar{y}, \bar{z})$ be the center of mass of a body of mass m as shown in Fig. 16.5 where the *central axes* x', y', and z' of the body are parallel to the fixed coordinate axes x, y, and z, respectively.[†] The coordinates of the

---

[†]The word *central* indicates something associated with a center (e.g., center of mass or center of gravity), while the word *centroidal* indicates something associated with the *centroid* of a geometric figure (e.g., line, area, or volume). Thus, *central axes* of a body are axes passing through the center of mass of the body, while *centroidal axes* of a body are axes passing through the centroid of the volume of the body. Note that the *center of mass* of a body does not necessarily coincide with the *centroid* of the volume of the body. If these two points coincide in a body, then the *centroidal axes* are also the *central axes* of the body.

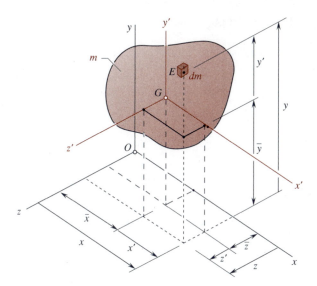

Fig. 16.5

differential mass element dm at the point E are (x, y, z) in the $Oxyz$ fixed coordinate system but are (x', y', z') in the $Gx'y'z'$ central coordinate system as indicated. Thus, we write

$$x = \bar{x} + x' \qquad y = \bar{y} + y' \qquad z = \bar{z} + z' \tag{16.4}$$

$$I_x = \int (y^2 + z^2)\, dm = \int [(\bar{y} + y')^2 + (\bar{z} + z')^2]\, dm$$

$$= \int (y'^2 + z'^2)\, dm + 2\bar{y} \int y'\, dm + 2\bar{z} \int z'\, dm + (\bar{y}^2 + \bar{z}^2) \int dm$$

$$= \bar{I}_{x'} + 2\bar{y}\,\bar{y}'m + 2\bar{z}\,\bar{z}'m + (\bar{y}^2 + \bar{z}^2)m$$

where $\bar{I}_{x'}$, represents the moment of inertia of the body about the central axis x', $\bar{y}'$ represents the distance between G and the $z'x'$ plane, and $\bar{z}'$ represents the distance between G and the $x'y'$ plane. Since G lies in the $z'x'$ and $x'y'$ planes, we see that $\bar{y}' = \bar{z}' = 0$. Thus, we write

$$I_x = \bar{I}_{x'} + m(\bar{y}^2 + \bar{z}^2) \tag{16.5}$$

and similarly,

$$I_y = \bar{I}_{y'} + m(\bar{z}^2 + \bar{x}^2) \quad I_z \qquad \bar{I}_{z'} + m(\bar{x}^2 + \bar{y}^2) \tag{16.6}$$

From Fig. 16.5, we see that the sum $\bar{y}^2 + \bar{z}^2$ in Eq. (16.5) represents the square of the distance between the x and x' axes. Similarly, the sums $\bar{z}^2 + \bar{x}^2$ and $\bar{x}^2 + \bar{y}^2$ represent the squares of the distances between the y and y' axes, and the z and z' axes, respectively.

The formulas in Eqs. (16.5) and (16.6) are called the *parallel-axis theorem* for mass moments of inertia, which may be stated as follows: *The moment of inertia I of a body of mass m about any given axis is equal to the moment of inertia $\bar{I}$ of the body about a central axis parallel to the given axis plus the product md^2, where d is the distance between those two axes,* as shown in Fig. 16.6; i.e.,

$$I = \bar{I} + md^2 \tag{16.7}$$

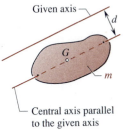

Given axis

d

G

m

Central axis parallel
to the given axis

Fig. 16.6

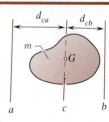

Fig. D16.11

D16.11   If a, b, and c are three parallel axes and the axis c passes through the center of mass G of a body of mass m as shown, show that

$$I_a = I_b + m(d_{ca}^2 - d_{cb}^2)$$

EXAMPLE 16.8

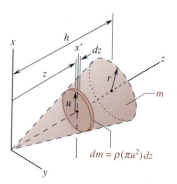

Determine I_x of the circular cone in Example 16.4.

Solution.   Let dm be a thin disk and x' be a central axis of dm as shown, where the radius of the disk is $u = (r/h) z$ and its mass density is $\rho = 3m/(\pi^2 h)$. Applying the *parallel-axis theorem* and the solution for I_x in Example 16.3 as a formula, we write

$$dI_x = d\bar{I}_{x'} + (dm) z^2 = \tfrac{1}{4} (dm) u^2 + (dm) z^2$$

$$= (\tfrac{1}{4} u^2 + z^2)(\rho \pi u^2 \, dz) = \frac{\rho \pi r^2}{4h^4}(r^2 + 4h^2)z^4 \, dz$$

$$I_x = \int dI_x = \frac{\rho \pi r^2}{4h^4}(r^2 + 4h^2) \int_0^h z^4 \, dz$$

$$= \frac{\rho \pi r^2 h}{20}(r^2 + 4h^2) = \frac{3m}{\pi r^2 h} \frac{\pi r^2 h}{20}(r^2 + 4h^2)$$

$$I_x = \frac{3}{20} m(r^2 + 4h^2) \quad \blacktriangleleft$$

Developmental Exercises

D16.12   Describe the *parallel-axis theorem* for mass moments of inertia.

D16.13   The circular cone shown has a mass m. Determine I_x of the cone.

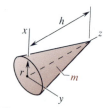

Fig. D16.13

16.4   Composite Bodies

A composite body has several component bodies. The moment of inertia I of a composite body about an axis is equal to the sum of the moments of inertia I_1, I_2, . . . , I_n of its component bodies about that axis; i.e.,

$$I = I_1 + I_2 + \cdots + I_n \tag{16.8}$$

For use in determining the moments of inertia of composite bodies, the moments of inertia of homogeneous bodies of some common shapes are given in Fig. 16.7, where the plates, disks, rings, and shells are understood to be thin, and m is the total mass of the body shown.

Slender rod

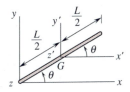

$$I_x = \frac{1}{3}\, mL^2 \sin^2\theta \qquad\qquad I_y = \frac{1}{3}\, mL^2 \cos^2\theta$$

$$I_z = \frac{1}{3}\, mL^2 \qquad\qquad \bar{I}_{x'} = \frac{1}{12}\, mL^2 \sin^2\theta$$

$$\bar{I}_{y'} = \frac{1}{12}\, mL^2 \cos^2\theta \qquad\qquad \bar{I}_{z'} = \frac{1}{12}\, mL^2$$

Circular ring

$$I_x = I_z = \frac{1}{2}\, mr^2$$

$$I_y = mr^2$$

Rectangular parallelepiped

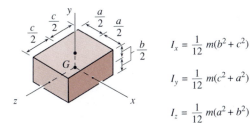

$$I_x = \frac{1}{12}\, m(b^2 + c^2)$$

$$I_y = \frac{1}{12}\, m(c^2 + a^2)$$

$$I_z = \frac{1}{12}\, m(a^2 + b^2)$$

Semicircular ring

$$\bar{z} = \frac{2r}{\pi} \qquad I_x = I_z = \frac{1}{2}\, mr^2 \qquad I_y = mr^2$$

$$\bar{I}_{x'} = \frac{1}{2\pi^2}\, mr^2(\pi^2 - 8)$$

$$\bar{I}_{y'} = \frac{1}{\pi^2}\, mr^2(\pi^2 - 4)$$

Rectangular plate

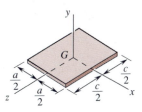

$$I_x = \frac{1}{12}\, mc^2 \qquad I_y = \frac{1}{12}\, m(c^2 + a^2) \qquad I_z = \frac{1}{12}\, ma^2$$

Circular disk

$$I_y = \frac{1}{2}\, mr^2$$

$$I_x = I_z = \frac{1}{4}\, mr^2$$

Circular cylinder

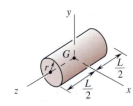

$$I_z = \frac{1}{2}\, mr^2$$

$$I_x = I_y = \frac{1}{12}\, m(3r^2 + L^2)$$

Semicircular disk

$$\bar{z} = \frac{4r}{3\pi}$$

$$I_y = \frac{1}{2}\, mr^2$$

$$I_x = I_z = \frac{1}{4}\, mr^2$$

$$\bar{I}_{x'} = \frac{1}{36\pi^2}\, mr^2(9\pi^2 - 64)$$

$$\bar{I}_{y'} = \frac{1}{18\pi^2}\, mr^2(9\pi^2 - 32)$$

Fig. 16.7(a)   Moments of inertia of bodies of common shapes.

Circular conic shell

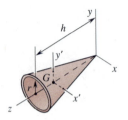

$$\bar{z} = \frac{2}{3} h$$

$$I_z = \frac{1}{2} mr^2$$

$$I_x = I_y = \frac{1}{4} m(r^2 + 2h^2)$$

$$\bar{I}_{x'} = \bar{I}_{y'} = \frac{1}{36} m(9r^2 + 2h^2)$$

Circular cone

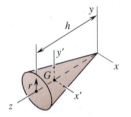

$$\bar{z} = \frac{3}{4} h$$

$$I_z = \frac{3}{10} mr^2$$

$$I_x = I_y = \frac{3}{20} m(r^2 + 4h^2)$$

$$\bar{I}_{x'} = \bar{I}_{y'} = \frac{3}{80} m(4r^2 + h^2)$$

Circular cylindrical shell

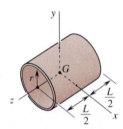

$$I_z = mr^2$$

$$I_x = I_y = \frac{1}{12} m(6r^2 + L^2)$$

Sphere

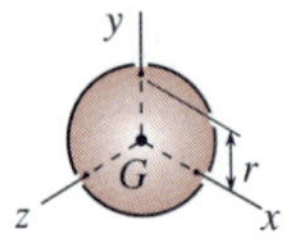

$$I_x = I_y = I_z = \frac{2}{5} mr^2$$

Hemispherical shell

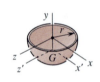

$$\bar{y} = -\frac{1}{2} r$$

$$I_x = I_y = I_z = \frac{2}{3} mr^2$$

$$\bar{I}_{x'} = \bar{I}_{z'} = \frac{5}{12} mr^2$$

Triangular plate

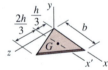

$$\bar{I}_{x'} = \frac{1}{18} mh^2$$

$$I_x = \frac{1}{6} mh^2$$

Hemisphere

$$\bar{y} = -\frac{3}{8} r$$

$$I_x = I_y = I_z = \frac{2}{5} mr^2$$

$$\bar{I}_{x'} = \bar{I}_{z'} = \frac{83}{320} mr^2$$

Spherical shell

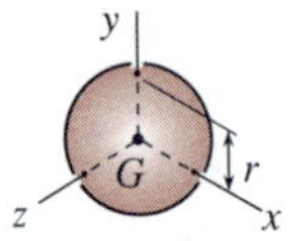

$$I_x = I_y = I_z = \frac{2}{3} mr^2$$

Half torus

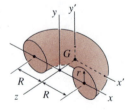

$$\bar{z} = -\frac{r^2 + 4R^2}{2\pi R}$$

$$I_x = I_z = \frac{1}{8} m(5r^2 + 4R^2)$$

$$I_y = \frac{1}{4} m(3r^2 + 4R^2)$$

Torus

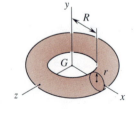

$$I_x = I_z = \frac{1}{8} m(5r^2 + 4R^2)$$

$$I_y = \frac{1}{4} m(3r^2 + 4R^2)$$

Fig. 16.7(b)   Moments of inertia of bodies of common shapes.

EXAMPLE 16.9

The mass density of the slender bent rod shown is $\rho_L = 3.75$ kg/m. Determine I_y and k_y of the rod.

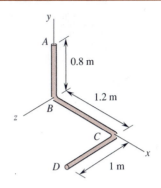

Solution. The rod may be treated as a composite body consisting of the segments AB, BC, and CD as the component bodies. Thus, we write

$$I_y = (I_y)_{AB} + (I_y)_{BC} + (I_y)_{CD}$$

Since the segment AB lies along the y axis, we have $(I_y)_{AB} = 0$. To find $(I_y)_{BC}$ and $(I_y)_{CD}$ of the segments BC and CD, we first establish their central axes y' and y'' which are parallel to the y axis. Then, we apply the *parallel-axis theorem* and the formulas for a slender rod to write

$$(I_y)_{BC} = (\bar{I}_{y'})_{BC} + m_{BC}\,\overline{BE}^2 = \tfrac{1}{12}\,m_{BC}\,\overline{BC}^2 + m_{BC}\,\overline{BE}^2$$

$$= \tfrac{1}{12}[3.75(1.2)](1.2)^2 + [3.75(1.2)](0.6)^2$$

$$= 2.16$$

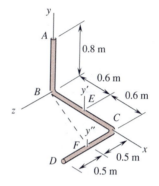

$$(I_y)_{CD} = (\bar{I}_{y''})_{CD} + m_{CD}\,\overline{BF}^2$$

$$= \tfrac{1}{12}\,m_{CD}\,\overline{CD}^2 + m_{CD}(\overline{BC}^2 + \overline{CF}^2)$$

$$= \tfrac{1}{12}[3.75(1)](1)^2 + [3.75(1)][(1.2)^2 + (0.5)^2]$$

$$= 6.65$$

$$I_y = 0 + 2.16 + 6.65 \qquad\qquad I_y = 8.81 \text{ kg·m}^2 \blacktriangleleft$$

$$I_y = k_y^2 m: \qquad 8.81 = k_y^2(3.75)(0.8 + 1.2 + 1)$$

$$k_y = 0.885 \text{ m} \blacktriangleleft$$

Developmental Exercises

D16.14 Deterime I_x and k_x of the bent rod in Example 16.9.

D16.15 Determine I_z and k_z of the bent rod in Example 16.9.

EXAMPLE 16.10

A thin machine part of mass density $\rho_A = 4.83$ lbm/ft^2 is shown, where $a = 6$ in. Determine I_y of the machine part.

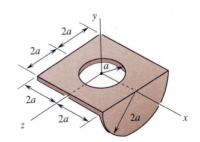

Solution. Let the machine part be a composite body consisting of a square plate of mass m_1, a semicircular plate of mass m_2, and an *empty* circular plate of mass m_3. Then, we have

$$m_1 = \rho_A(4a)^2 = 16\rho_A a^2$$

$$m_2 = \rho_A[\tfrac{1}{2}\,\pi(2a)^2] = 2\pi\rho_A a^2$$

$$m_3 = \rho_A(\pi a^2) = \pi\rho_A a^2$$

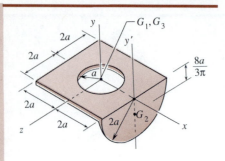

where $\rho_A = 0.15$ slug/ft^2 and $a = 0.5$ ft. Applying the *parallel-axis theorem* and appropriate formulas in Fig. 16.7, we write

$$I_y = (I_y)_1 + (I_y)_2 - (I_y)_3$$
$$= (I_y)_1 + [(\bar{I}_{y'})_2 + m_2 d_2^2] - (I_y)_3$$
$$= \tfrac{1}{12} m_1[(4a)^2 + (4a)^2]$$
$$+ [\tfrac{1}{4} m_2(2a)^2 + m_2(2a)^2] - \tfrac{1}{2} m_3 a^2$$
$$= \tfrac{1}{6}(256 + 57\pi)\rho_A a^4$$

$$I_y = 0.680 \text{ slug·ft}^2 \quad \blacktriangleleft$$

Developmental Exercises

D16.16   Determine I_x and k_x of the machine part in Example 16.10.

D16.17   Determine I_z and k_z of the machine part in Example 16.10.

EXAMPLE 16.11

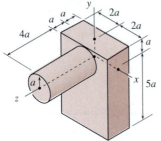

A body consists of a rectangular parallelepiped and a circular cylinder as shown, where $a = 20$ mm and the mass density is $\rho = 6$ Mg/m^3. Determine I_z and k_z of the body.

Solution.   Let the component bodies be the rectangular parallelepiped of mass m_1 and the circular cylinder of mass m_2. Then, we have

$$m_1 = \rho(4a)(6a)(2a) = 48\rho a^3$$
$$m_2 = \rho(\pi a^2)(4a) = 4\pi \rho a^3$$
$$m = m_1 + m_2 = (48 + 4\pi)\rho a^3$$

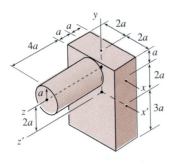

where $\rho = 6000$ kg/m^3 and $a = 0.02$ m. Applying the *parallel-axis theorem* and appropriate formulas in Fig. 16.7, we write

$$I_z = (I_z)_1 + (I_z)_2 = [(\bar{I}_{z'})_1 + m_1 d_1^2] + (I_z)_2$$
$$= \{\tfrac{1}{12} m_1[(4a)^2 + (6a)^2] + m_1(2a)^2\} + \tfrac{1}{2} m_2 a^2$$
$$= (400 + 2\pi)\rho a^5 = (400 + 2\pi)(6000)(0.02)^4$$

$$I_z = 7.80 \times 10^{-3} \text{ kg·m}^2 \quad \blacktriangleleft$$

$$I_z = k_z^2 m: \qquad (400 + 2\pi)\rho a^5 = k_z^2(48 + 4\pi)\rho a^3$$

$$k_z = 51.8 \text{ mm} \quad \blacktriangleleft$$

Developmental Exercises

D16.18   Determine I_x and k_x of the body in Example 16.11.

D16.19   Determine I_y and k_y of the body in Example 16.11.

PROBLEMS

NOTE: Unless otherwise specified, the formulas given in Fig. 16.7 as well as the parallel-axis theorem may be applied in solving the following problems of mass moments of inertia.

16.19 through 16.21   Determine I_x and I_y of the body made of slender rods as shown.

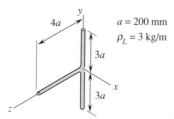

$a = 200$ mm
$\rho_L = 3$ kg/m

Fig. P16.19 and P16.22*

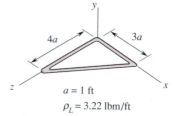

$a = 1$ ft
$\rho_L = 3.22$ lbm/ft

Fig. P16.20* and P16.23

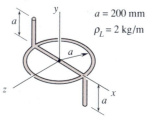

$a = 200$ mm
$\rho_L = 2$ kg/m

Fig. P16.21 and P16.24

16.22* through 16.24   Determine I_z and k_z of the body made of slender rods as shown.

16.25 through 16.27   Determine I_x and I_y of the sheet metal part shown.

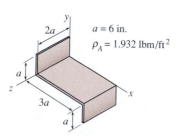

$a = 6$ in.
$\rho_A = 1.932$ lbm/ft^2

Fig. P16.25 and P16.28*

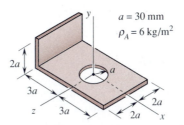

$a = 30$ mm
$\rho_A = 6$ kg/m^2

Fig. P16.26* and P16.29

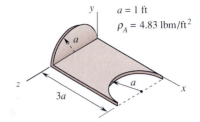

$a = 1$ ft
$\rho_A = 4.83$ lbm/ft^2

Fig. P16.27 and P16.30

16.28* through 16.30   Determine I_z and k_z of the sheet metal part shown.

16.31 through 16.33   Determine I_y and k_y of the thin shell shown.

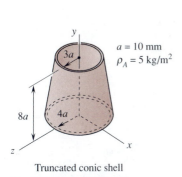

$a = 10$ mm
$\rho_A = 5$ kg/m^2

Truncated conic shell
Fig. P16.31 and P16.34*

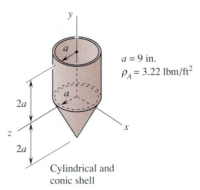

$a = 9$ in.
$\rho_A = 3.22$ lbm/ft^2

Cylindrical and
conic shell
Fig. P16.32* and P16.35

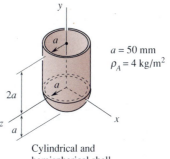

$a = 50$ mm
$\rho_A = 4$ kg/m^2

Cylindrical and
hemispherical shell
Fig. P16.33 and P16.36

16.34* through 16.36   Determine I_x and k_x of the thin shell shown.

16.37 through 16.39   Determine I_y and k_y of the body of revolution shown.

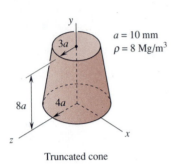

$a = 10$ mm
$\rho = 8$ Mg/m^3

Truncated cone

Fig. P16.37 and P16.40*

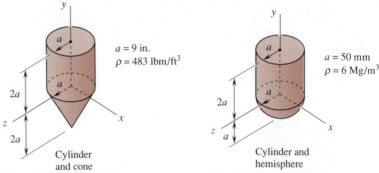

$a = 9$ in.
$\rho = 483$ lbm/ft^3

Cylinder
and cone

Fig. P16.38* and P16.41

$a = 50$ mm
$\rho = 6$ Mg/m^3

Cylinder and
hemisphere

Fig. P16.39 and P16.42

16.40* through 16.42   Determine I_x and k_x of the body of revolution shown.

16.43   A horizontal hole of radius 1 in. is drilled through a body as shown. If $\rho = 14$ slugs/ft^3, determine I_z and k_z of the body.

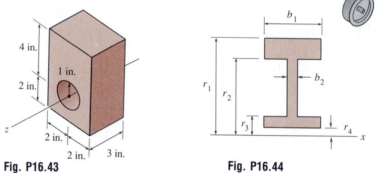

Fig. P16.43

Fig. P16.44

16.44   A flywheel is of the same shape and size as that obtained by revolving the area shown about the x axis one revolution, where $r_1 = 250$ mm, $r_2 = 200$ mm, $r_3 = 50$ mm, $r_4 = 25$ mm, $b_1 = 150$ mm, and $b_2 = 25$ mm. If $\rho = 7$ Mg/m^3, determine I_x and k_x of the flywheel.

16.45   A link consists of a cylinder and a half-torus as shown. It is made of an alloy with $\rho = 6$ Mg/m^3. Determine I_y and I_z of the link.

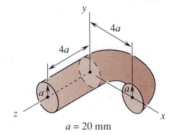

Fig. P16.45 $a = 20$ mm

16.46*   Using single integration, verify the formula $I_x = \frac{1}{12}m(3r^2 + L^2)$ for the circular cylinder in Fig. 16.7.

16.47*   Using single integration, verify the formula $I_x = \frac{1}{12}m(6r^2 + L^2)$ for the circular cylindrical shell in Fig. 16.7.

16.48*   Using single integration, verify the formula $I_x = \frac{1}{8}m(5r^2 + 4R^2)$ for the torus in Fig. 16.7). (*Hint*. Refer to the *hint* in Prob. 16.12*.)

FORCE AND ACCELERATION

16.5 Effective Force-Moment System on a Rigid Body

A rigid body may be viewed as a continuum composed of a large number of particles which remain at fixed distances from each other. Thus, a rigid body is a special case of a system of particles. We saw in Sec. 12.6 that *the system of external forces acting on a system of particles is equipollent to the system of effective forces on the particles*. This relation may be applied to the plane motion of a rigid body of mass m under the action of the external forces $\mathbf{F}_1, \mathbf{F}_2, \ldots , \mathbf{F}_n$ in an inertial reference frame Oxy, as shown in Fig. 16.8, where G is the mass center of the rigid body, $(\Delta m_i)\mathbf{a}_i$ is the effective force on the ith particle P_i having a mass Δm_i and an acceleration $\mathbf{a}_i$, $\bar{\mathbf{a}}$ is the acceleration of G, and $m\bar{\mathbf{a}}$ is the resultant effective force, which does not necessarily pass through G.

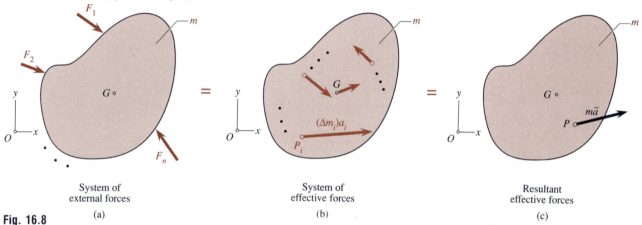

System of external forces	System of effective forces	Resultant effective forces
(a)	(b)	(c)

Fig. 16.8

Note in Fig. 16.8 that the *equal* sign $=$ (not the *equipollent* sign $\Rightarrow$) for *equivalence* is used because the particles of a rigid body are not free particles; they are interconnected at fixed distances. Applying *Chasles' theorem* in Sec. 15.5, we may treat the *plane motion* of a rigid body as being equal to the *translation* of the rigid body *with G* plus the *rotation* of the rigid body *about G*. In other words, the *system of effective forces* on a rigid body of mass m as illustrated in Fig. 16.8(b) may be treated as being equal to the vector sum of the following two systems: (1) the *translational system of effective forces* corresponding to the translation of the rigid body with G whose acceleration is $\bar{\mathbf{a}}$, and (2) the *rotational system of effective forces* corresponding to the rotation of the rigid body about G with an angular velocity $\boldsymbol{\omega}$ and an angular acceleration $\boldsymbol{\alpha}$. Such a relation is depicted in Fig. 16.9, where $Gx'y'$ is a nonrotating reference frame attached to G, and r_i' is the radial distance from G to the ith particle P_i whose mass is Δm_i and whose acceleration $\mathbf{a}_i$ is given by

$$\mathbf{a}_i = \bar{\mathbf{a}} + (r_i'\alpha\mathbf{e}_t + r_i'\omega^2\mathbf{e}_n) \qquad (16.9)$$

The vectors $\mathbf{e}_t$ and $\mathbf{e}_n$ in Eq. (16.9) and Fig. 16.9(c) are, respectively, the tangential and the normal unit vectors associated with the path of the ith particle P_i rotating about G.

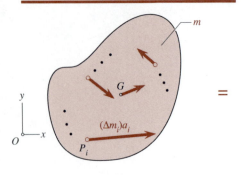
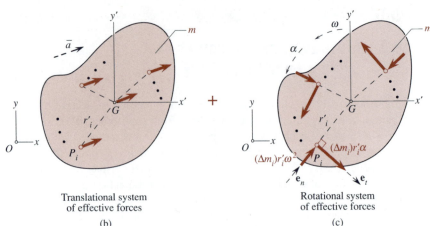

System of
effective forces

(a)

Translational system
of effective forces

(b)

Rotational system
of effective forces

(c)

Fig. 16.9

Letting $\mathbf{R}_{\text{eff,trans}}$ be the *resultant effective force* of the *translational* system of effective forces in Fig. 16.9(b), we write

$$\mathbf{R}_{\text{eff,trans}} = \sum_{i=1}^{n} (\Delta m_i)\bar{\mathbf{a}} = \bar{\mathbf{a}} \sum_{i=1}^{n} \Delta m_i = \bar{\mathbf{a}}m$$

$$\mathbf{R}_{\text{eff,trans}} = m\bar{\mathbf{a}} \tag{16.10}$$

Letting $(\mathbf{M}_G^R)_{\text{eff,trans}}$ be the *resultant effective moment* about G of the *translational* system of effective forces in Fig. 16.9(b), we write

$$(\mathbf{M}_G^R)_{\text{eff,trans}} = \sum_{i=1}^{n} \mathbf{r}_i' \times (\Delta m_i)\bar{\mathbf{a}}$$

$$= -\bar{\mathbf{a}} \times \sum_{i=1}^{n} \mathbf{r}_i' (\Delta m_i)$$

$$= -\bar{\mathbf{a}} \times \bar{\mathbf{r}}'m$$

where, by the *principle of moments* $\bar{\mathbf{r}}'$ is the position vector of G in $Gx'y'$. Since G is the origin, we see that $\bar{\mathbf{r}}' = \mathbf{0}$. Thus, we have

$$(\mathbf{M}_G^R)_{\text{eff,trans}} = \mathbf{0} \tag{16.11}$$

We find from Eqs. (16.10) and (16.11) that *the translational system of effective forces* in Fig. 16.9(b) *is equivalent to the resultant effective force* $\mathbf{R}_{\text{eff,trans}}$ *acting through* G as illustrated in Fig. 16.10.

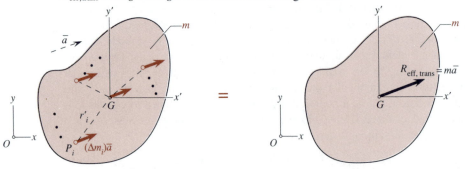

Translational system
of effective forces

Resultant of the
translational system

Fig. 16.10

Now, letting $\mathbf{R}_{\text{eff,rot}}$ be the *resultant effective force* of the *rotational* system of effective forces in Fig. 16.9(c) and applying Eq. (16.9) as well as the *principle of moments*, we write

$$\mathbf{R}_{\text{eff,rot}} = \sum_{i=1}^{n} [(\Delta m_i)r_i'\alpha\mathbf{e}_t + (\Delta m_i)r_i'\omega^2\mathbf{e}_n]$$

$$= \sum_{i=1}^{n} (\Delta m_i)(r_i'\alpha\mathbf{e}_t + r_i'\omega^2\mathbf{e}_n)$$

$$= \sum_{i=1}^{n} (\Delta m_i)(\mathbf{a}_i - \bar{\mathbf{a}})$$

$$= \sum_{i=1}^{n} \mathbf{a}_i(\Delta m_i) - \bar{\mathbf{a}} \sum_{i=1}^{n} (\Delta m_i)$$

$$= \bar{\mathbf{a}}m - \bar{\mathbf{a}}m$$

$$\mathbf{R}_{\text{eff,rot}} = \mathbf{0} \qquad (16.12)$$

Letting $(\mathbf{M}_G^R)_{\text{eff,rot}}$ be the *resultant effective moment* about G of the *rotational* system of effective forces in Fig. 16.9(c) and noting that $\mathbf{e}_n \times \mathbf{e}_n = \mathbf{0}$ and $\mathbf{e}_n \times \mathbf{e}_t = -\mathbf{k}$, we write

$$(\mathbf{M}_G^R)_{\text{eff,rot}} = \sum_{i=1}^{n} (-r_i'\mathbf{e}_n) \times [(\Delta m_i)r_i'\alpha\mathbf{e}_t + (\Delta m_i)r_i'\omega^2\mathbf{e}_n]$$

$$= \alpha(-\mathbf{e}_n \times \mathbf{e}_t) \sum_{i=1}^{n} r_i'^2(\Delta m_i) + \mathbf{0}$$

$$= (\alpha\mathbf{k}) \int r'^2 \, dm = \alpha\bar{I}$$

$$(\mathbf{M}_G^R)_{\text{eff,rot}} = \bar{I}\alpha \qquad (16.13)$$

where we let $n \to \infty$ in the last summation to obtain the *integral* which equals the moment of inertia $\bar{I}$ of the body about G. We find from Eqs. (16.12) and (16.13) that *the rotational system of effective forces* in Fig. 16.9(c) *is equivalent to the resultant effective moment* $(\mathbf{M}_G^R)_{\text{eff,rot}}$ *acting about* G as illustrated in Fig. 16.11.

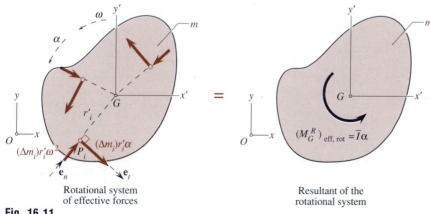

Rotational system
of effective forces

Resultant of the
rotational system

Fig. 16.11

By *Chasles' theorem* and Eqs. (16.10) and (16.12), we conclude that the *resultant effective force* acts through G and is given by

$$\mathbf{R}_{\text{eff}} = \mathbf{R}_{\text{eff,trans}} + \mathbf{R}_{\text{eff,rot}} = m\bar{\mathbf{a}} + \mathbf{0}$$

$$\mathbf{R}_{\text{eff}} = m\bar{\mathbf{a}} \tag{16.14}$$

Furthermore, by *Chasles' theorem* and Eqs. (16.11) and (16.13), we conclude that the coexisting *resultant effective moment* acting about G is given by

$$(\mathbf{M}_G^R)_{\text{eff}} = (\mathbf{M}_G^R)_{\text{eff, trans}} + (\mathbf{M}_G^R)_{\text{eff,rot}} = \mathbf{0} + \bar{I}\boldsymbol{\alpha}$$

$$(\mathbf{M}_G^R)_{\text{eff}} = \bar{I}\boldsymbol{\alpha} \tag{16.15}$$

The vectors $m\bar{\mathbf{a}}$ and $\bar{I}\boldsymbol{\alpha}$ in Eqs. (16.14) and (16.15) are respectively called the *effective force* and the *effective moment* at the mass center G of the rigid body. A diagram showing the vectors $m\bar{\mathbf{a}}$ and $\bar{I}\boldsymbol{\alpha}$ at G, or their equivalent at a different point, is referred to as the *effective-force diagram* of the rigid body, where the effective moment is dependent on the point chosen (cf. D16.20).

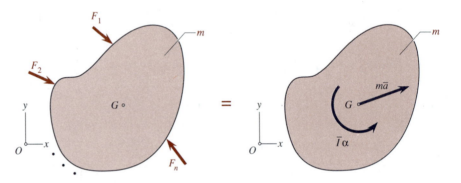

Fig. 16.12   Free–body diagram   Effective–force diagram

In general, we see that *the system of external forces* $\mathbf{F}_1, \mathbf{F}_2, \ldots, \mathbf{F}_n$ *acting on a rigid body in plane motion is equivalent to the effective force-moment system consisting of* $m\bar{\mathbf{a}}$ *acting through the mass center* G *and* $\bar{I}\boldsymbol{\alpha}$ *acting about* G *of the body*. This important property is illustrated in Fig. 16.12 and is a modern version of *d'Alembert's principle*, after the French mathematician Jean le Rond d'Alembert (1717–1783), even though d'Alembert's original statement was written in different terms for the motion of a *system of linked masses*.[†] However, d'Alembert's principle was not originally intended for the case of a *single particle*.[††]

[†]The original statement of d'Alembert's principle is seldom directly quoted; it requires interpretation for modern usage. D'Alembert essentially found that *the sum of the interaction forces in a system of connected bodies vanishes*. The modern version of his principle is really the consequence of that finding. The principle is contained in d'Alembert's *Traité de Dynamique*, which was first published in 1743 and later revised in 1758. A reprint of its second edition with an elaborate commentary and a bibliographical note by Thomas L. Hankins was published in 1968 by Johnson Reprint Corporation, New York. Despite some criticisms of d'Alembert's attempts to "geometrize" the world in his *Traité de Dynamique*, his work remains an important milestone in the history of mechanics.

[††]Cf. the first footnote on p. 511.

Developmental Exercises

D16.20   A rigid body of mass m in plane motion has an angular velocity $\boldsymbol{\omega}$, an angular acceleration $\boldsymbol{\alpha}$, and a moment of inertia $\bar{I}$ about its central axis perpendicular to the plane of motion. If the acceleration of its mass center G is $\bar{\mathbf{a}}$, give the proper effective force-moment system at G for the rigid body.

D16.21   Describe *d'Alembert's principle*.

16.6   Center of Percussion

Using the concept of equivalence in Sec. 5.1, we can move the effective force $m\bar{\mathbf{a}}$ at G in Fig. 16.12 to a certain point P plus the moment of $m\bar{\mathbf{a}}$ at G about P. If the moment of $m\bar{\mathbf{a}}$ at G about P is equal to $-\bar{I}\boldsymbol{\alpha}$, then we have reduced the effective force-moment system at G in Fig. 16.12 to an *equivalent single effective force $m\bar{\mathbf{a}}$ acting through the point P*, as shown in Fig. 16.13, where

$$\overrightarrow{PG} \times m\bar{\mathbf{a}} = -\bar{I}\boldsymbol{\alpha} \qquad (16.16)$$

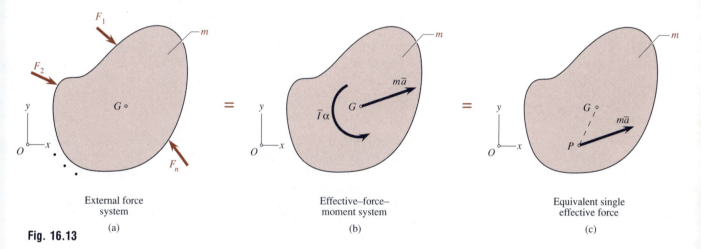

External force system	Effective–force–moment system	Equivalent single effective force
(a)	(b)	(c)

Fig. 16.13

In the case of a rigid body rotating in a plane about a fixed axis O, the point of intersection P of the *equivalent single effective force $m\bar{\mathbf{a}}$* and the line containing the fixed point O and the mass center G of the body is called the *center of percussion* of the body. The steps which may be used to locate the center of percussion are illustrated in Fig. 16.14, where $\boldsymbol{\omega}$ is the angular velocity, $\bar{k}$ is the central radius of gyration, and the distance from the mass center G to the *center of percussion P* is

$$d = \frac{\bar{k}^2}{\bar{r}} \qquad (16.17)$$

The *center of percussion P* is a unique point of the body (or its massless extension) through which the *equivalent single effective force $m\bar{\mathbf{a}}$* must pass. If a body pivoted at a fixed point O is suddenly struck by a relatively large force $\mathbf{F}$ through its *center of percussion P*, the free-body diagram and the

Center of Percussion

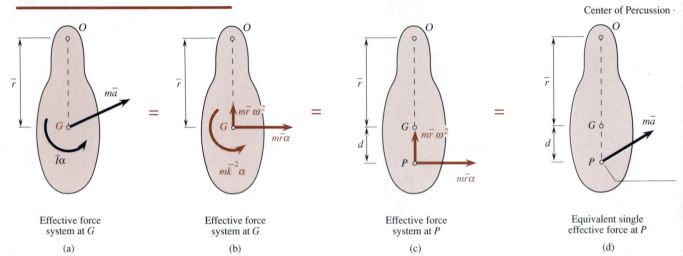

Effective force system at G		Effective force system at G		Effective force system at P		Equivalent single effective force at P
(a)		(b)		(c)		(d)

Fig. 16.14

effective-force diagram of the body may be drawn as shown in Fig. 16.15. Referring to this figure, we write

$$+\, \circlearrowleft \; \Sigma M_P: \qquad -(\bar{r} + d)O_t = 0$$

Since $\bar{r} + d \neq 0$, we see that $O_t = 0$; i.e.,

$$\mathbf{O}_t = \mathbf{0} \qquad\qquad (16.18)$$

This means that *any force striking a rigid body through its center of percussion will not induce any tangential component of reaction from its supporting pivot.* As an application, a baseball batter may avoid the sting in the grip by appropriately hitting the ball at the *center of percussion* (or the "sweet spot") of the bat. Furthermore, the concept of center of percussion is useful in designing impact devices and testing machines.

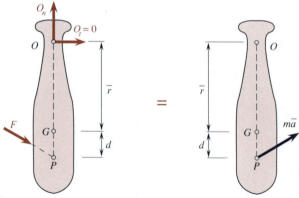

Fig. 16.15

Developmental Exercises

D16.22 If the central radius of gyration of a baseball bat AB is 7 in., determine the distance d between its *center of percussion P* and its *mass center G* as shown.

D16.23 A slender rod AB of mass m is at rest when the force $\mathbf{F}$ is applied to its lower end B as shown in the space diagram. Applying *Chasles' theorem* to the study of the rod with its hinged end A as the point of reference, we may draw the free-body diagram (FBD) and a series of effective-force diagrams (EFDs) of the rod having an angular acceleration $\boldsymbol{\alpha}$ as shown. Verify that the systems of vectors in EFD-1 through EFD-4 are equivalent to each other and have values as indicated.

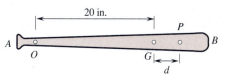

Fig. D16.22

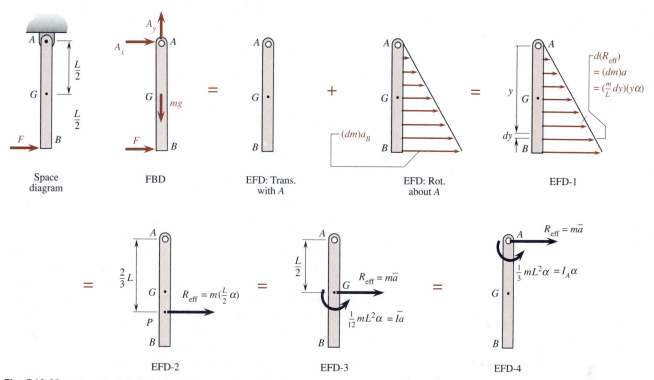

Fig. D16.23 (The point P in EFD-2 is the *center of percussion* of the rod.)

16.7 Method of Force and Acceleration

In plane kinetics of rigid bodies, an *effective-force diagram* may show *one* of the following: (a) the effective force-moment system acting at the mass center of a rigid body, (b) the effective force-moment systems acting at the various mass centers of the component parts of a composite rigid body, and (c) the effective force-moment systems acting at the various mass centers at the component parts of a set of connected rigid bodies. The *method of force and acceleration* for rigid bodies is a method of solution utilizing the *equivalence* or *equipollence* of the system of vectors in the free-body diagram of a body *to* the system of vectors in the corresponding effective-force diagram, where the body may be a simple rigid body, a composite rigid body, or a set of connected rigid bodies. Notice that the underlying principle in this method is *d'Alembert's principle*, which is illustrated in Fig. 16.12.

The equivalence depicted in Fig. 16.12 on p. 734 can be implemented algebraically by *equating* the sums of appropriate rectangular components of the vectors, as well as equating the sums of the moments of the vectors about a desirable point, as follows:

$$\Sigma V_x: \qquad \Sigma F_x = m\bar{a}_x \qquad (16.19)$$

$$\Sigma V_y: \qquad \Sigma F_y = m\bar{a}_y \qquad (16.20)$$

$$\Sigma M_G: \qquad \Sigma M_G = \bar{I}\alpha \qquad (16.21)$$

where the x and y axes can be any convenient pair of orthogonal axes in the plane of motion of the rigid body. If the moments are computed about a given point P, Eq. (16.21) is replaced by

$$\Sigma M_P: \qquad \Sigma M_P = \Sigma M_{P,\text{eff}} \qquad (16.22)$$

where the subscript "eff" stands for "effective." The point P in Eq. (16.22) is usually a propitious point, such as the point of intersection of two unknown vectors on the body. Using Eq. (16.22) and making an intelligent choice of P as the moment center, we can often reduce the computational effort involved in solving a kinetic problem. Thus, Eq. (16.22) is a powerful equation in rigid-body kinetics.

The *method of force and acceleration* for rigid bodies is best used in solving rigid-body problems where forces (including moments) and accelerations are to be directly related without initially involving displacement, velocity, and time. Furthermore, it should be noted that at most *three* independent scalar equations can be written from a vector-diagram equation for a rigid body in plane motion. For a set of connected rigid bodies, separate vector-diagram equations for the component parts of the set may sometimes be needed. In general, *the number of determinate unknowns contained in a given number of vector-diagram equations cannot exceed the total number of independent scalar equations which can be written* from (1) equating the sums of the appropriate rectangular components, as well as equating the sums of the moments of the vectors about a point, in each vector-diagram equation; (2) the laws of friction; (3) the rigid-body kinematics; and (4) the principle of virtual work in kinetics, which will be presented in the next chapter.

For a much clearer view of the situation in the problem under consideration and a much better command in setting up equations for the efficient solution of unknowns, it is recommended that both the *free-body diagram* and the *effective-force diagram* of the rigid body be drawn and equated to each other to form a *vector-diagram equation* as illustrated in Fig. 16.12. Such a geometric approach places the problem situation in good perspective. It generally makes the learning of kinetics less abstract than the approach which applies formulas directly without the aid of any diagram.

EXAMPLE 16.12

A 150-N horizontal force **P** is applied to a 30-kg file cabinet which is initially at rest on a floor as shown. If $\mu_k = 0.2$ and slipping occurs between the cabinet and the floor, determine (a) the acceleration $\bar{a}$ of its mass center G, (b) the *minimum* value of h for which the cabinet will not tip while it is being moved to the right.

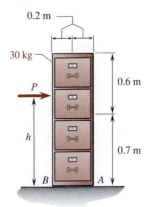

Solution. As the value of h is decreased to h_{min}, tipping of the cabinet about its corner at B will impend and the reaction from the floor will pass through B. Thus, we equate the *free-body diagram* to the *effective-force*

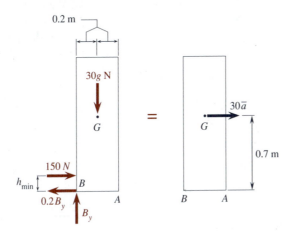

diagram for such a case as shown, where $\bar{I}\alpha = 0$ because rotation does not actually occur. We write

$$+\uparrow \Sigma V_y: \qquad B_y - 30g = 0$$

$$\xrightarrow{+} \Sigma V_x: \qquad 150 - 0.2B_y = 30\bar{a}$$

$$+\curvearrowright \Sigma M_B: \qquad h_{min}(150) + 0.2(30g) = 0.7(30\bar{a})$$

These three equations yield $B_y = 30g$, $\bar{a} = 3.308$ m/s², and $h_{min} = 0.0329$ m. Thus, we have

$$\bar{a} = 3.04 \text{ m/s}^2 \rightarrow \qquad h_{min} = 32.9 \text{ mm} \quad \blacktriangleleft$$

Developmental Exercise

D16.24   Refer to Example 16.12. Determine the *maximum* value of h for which the cabinet will not tip while it is being moved to the right.

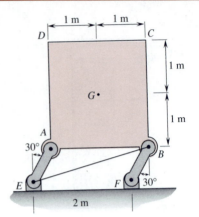

EXAMPLE 16.13

The 100-kg plate $ABCD$ with its mass center at G is held in the position shown by the wire BE and the links AE and BF whose masses are negligible. Determine the force in each link immediately after the wire BE is cut.

Solution. After the wire BE is cut, the plate will have a curvilinear translation, and the acceleration $\bar{\mathbf{a}}$ of G is equal to the acceleration $\mathbf{a}_A$ of A (or $\mathbf{a}_B$ of B) at any given instant. However, *immediately* after the wire BE is cut, the velocities of the plate and the links are zero; the acceleration $\bar{\mathbf{a}}$ being equal to $\mathbf{a}_A$ (or $\mathbf{a}_B$) is thus perpendicular to the link AE (or BF). Since the masses of the links AE and BF are negligible, the effective forces on these links are also negligible and the forces developed in them are taken as directed along their axes. Noting that the weight of the plate is $W = mg = 100(9.81)$ N $= 981$ N and letting $\bar{\mathbf{a}}$ be equal to $\bar{a} \; \searrow \; 30°$, we equate

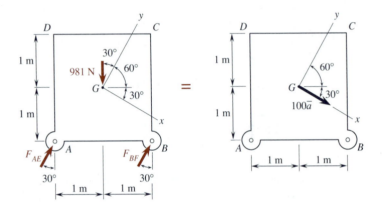

the *free-body diagram* of the plate (just after the wire is cut) to its *effective-force diagram* as shown, where the effective moment $\bar{I}\alpha$ is zero because the plate is in translation. We write

$$+\searrow \Sigma V_x: \qquad 981 \sin 30° = 100\bar{a} \qquad \bar{a} = 4.905 \text{ m/s}^2$$

$$+\curvearrowleft \Sigma M_A: \qquad 1(981) - 2(F_{BF} \cos 30°)$$

$$= 1(100\bar{a} \cos 30°) + 1(100\bar{a} \sin 30°)$$

$$+\nearrow \Sigma V_y: \qquad F_{AE} + F_{BF} - 981 \cos 30° = 0$$

$$F_{BF} = 179.5 \text{ N } C \qquad F_{AE} = 670 \text{ N } C \quad \blacktriangleleft$$

Developmental Exercise

D16.25 Refer to Example 16.13. Suppose that the mass of the plate is $m > 0$. Immediately after the wire BE is cut, (a) does the acceleration of the plate depend on the value of m? (b) do the forces in the links depend on the value of m?

EXAMPLE 16.14

The slender bent bar *ABC* supported as shown has a mass density $\rho_L = 10$ kg/m. Determine the angular acceleration of the bent bar immediately after the wire *BD* is cut.

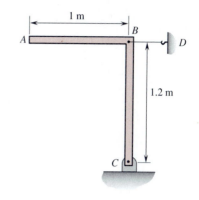

Solution. The bent bar *ABC* may be treated as a composite rigid body with *AB* and *BC* as its component parts. Since $\rho_L = 10$ kg/m, the masses of *AB* and *BC* are $m_{AB} = 10$ kg and $m_{BC} = 12$ kg, and their weights are $W_{AB} = 10g$ N and $W_{BC} = 12g$ N, respectively. *Immediately* after the wire *BD* is cut, the bent bar will lose balance and accelerate from rest with a counterclockwise angular acceleration **α** about the hinge at *C*, which is also its *acceleration center*. Thus, we equate the *free-body diagram* of the bent bar to its *effective-force diagram* as shown.

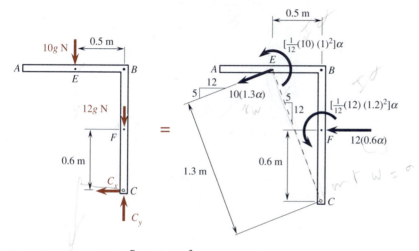

Note that the formula $\bar{I} = \frac{1}{12} mL^2$ for the central moment of inertia of a slender rod of mass *m* and length *L* is used in indicating the effective moment $\bar{I}\alpha$ about the center of mass of each of the component parts *AB* and *BC*. We write

$$+ \circlearrowleft \ \Sigma M_C: \qquad 0.5(10g) = 1.3[10(1.3\alpha)]$$
$$+ \ [\tfrac{1}{12}(10)(1)^2]\alpha$$
$$+ \ 0.6[12(0.6\alpha)]$$
$$+ \ [\tfrac{1}{12}(12)(1.2)^2]\alpha$$

$$5(9.81) = \frac{70.48}{3}\alpha \qquad \alpha = 2.09 \text{ rad/s}^2 \ \circlearrowleft \ \blacktriangleleft$$

Developmental Exercise

D16.26   Refer to Example 16.14. Determine the reaction from the support at *C* immediately after the wire *BD* is cut.

EXAMPLE 16.15

It is known that the cylinder B moves with $\mathbf{a}_B = 0.8$ m/s² ↓ when the system is released from rest as shown. Neglecting friction and the mass of the shaft CD, determine the moment of the inertia, $\bar{I}$ of the flywheel and its drum about the axis CD.

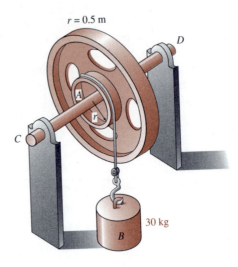

Solution. The corresponding angular acceleration α of the flywheel and its drum may be obtained as follows:

$$\alpha = a_B/r = 0.80/0.5 \qquad \alpha = 1.6 \text{ rad/s}^2 \ \text{⟳}$$

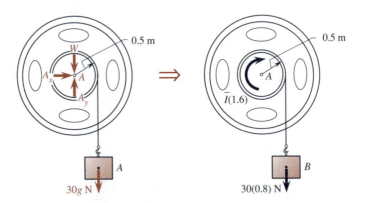

Denoting the weight of the flywheel and its drum by W, we equate the *free-body diagram* of the entire system to its *effective-force diagram* as shown, where $\bar{I}$ is the moment of inertia, and the *equipollence* sign ⟹ is used because the system is composed of connected rigid bodies. We write

$$+\text{⟳} \ \Sigma M_A: \qquad 0.5(30g) = 0.5(30)(0.8) + \bar{I}(1.6)$$

$$\bar{I} = 84.5 \qquad\qquad \bar{I} = 84.5 \text{ kg·m}^2 \ ◀$$

Developmental Exercise

D16.27 Refer to Example 16.15. Determine the tension in the cord when the system is released from rest.

16.8 Interpretations for Mass Moments of Inertia

Referring to Fig. 16.12 and substituting $\alpha = 1$ rad/s^2 into Eq. (16.21),

$$\Sigma M_G = \bar{I} \qquad (16.23)$$

The equation allows this interpretation: *the moment of inertia of a rigid body about its own central axis is numerically equal to the moment required to accelerate the body to rotate at the rate of* 1 rad/s^2 *about that axis.*

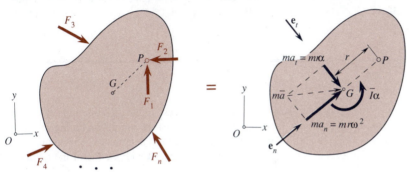
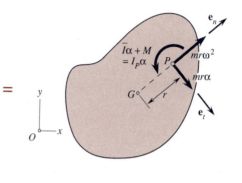

Fig. 16.16   (a)   (b)   (c)

If a rigid body of mass m rotates, at the instant considered, with an angular velocity ω and an angular acceleration α about a *fixed axis* passing through the point P, we may equate its *free-body diagram* to its *effective-force diagram* as illustrated in Fig. 16.16. Note in Fig. 16.16(b) that the effective force $m\bar{a}$ is, for convenience, replaced by its tangential component $ma_t = mr\alpha\mathbf{e}_t$ and its normal component $ma_n = mr\omega^2\mathbf{e}_n$. These two components as well as the effective moment $\bar{I}\alpha$ may, without losing equivalence, be moved from G to P if a moment $\mathbf{M}$ equal to the sum of moments about P due to ma_t and ma_n at G is added as shown in Fig. 16.16(c). Since ma_n contributes no moment about P and $\mathbf{e}_n \times \mathbf{e}_t = -\mathbf{k}$, we have

$$\mathbf{M} = \overrightarrow{PG} \times ma_t = -r\mathbf{e}_n \times mr\alpha\mathbf{e}_t = mr^2\alpha\mathbf{k}$$

Noting that the moment of inertia of the body about the fixed axis at P is $I_P = \bar{I} + mr^2$, we may write the effective moment of the equivalent effective force-moment system at P as

$$\Sigma\mathbf{M}_{P,\text{eff}} = \bar{I}\alpha\mathbf{k} + \mathbf{M} = (\bar{I} + mr^2)\alpha\mathbf{k} = I_P\alpha\mathbf{k}$$

$$\Sigma M_{P,\text{eff}} = I_P\alpha \qquad (16.24)$$

which is indicated in Fig. 16.16(c). Referring to this figure and Eqs. (16.22) and (16.24), we write

$$\Sigma M_P: \qquad \boxed{\Sigma M_P = I_P\alpha} \qquad (16.25)$$

Substituting $\alpha = 1$ rad/s^2 into Eq. (16.25), we have

$$\Sigma M_P = I_P \qquad (16.26)$$

which may be interpreted as follows: *the moment of inertia of a rigid body about any axis fixed in space is numerically equal to the moment required to accelerate the body to rotate at the rate of* 1 rad/s^2 *about that axis.* Since the acceleration axis (cf. Sec. 15.9) has zero acceleration, a similar interpretation for the moment of inertia of a body about its acceleration axis can be shown to be valid.

Developmental Exercise

D16.28 Refer to the relations established in Fig. 16.16, where the rigid body rotates about a fixed axis passing through the point P. Show that Chasles' theorem can be applied to yield the same results as shown in Fig. 16.16(c) when the point of reference mentioned in the theorem is the fixed point P, instead of the mass center G, of the body. [*Hint.* Cf. D16.23 and note in Sec. 16.5 that $\mathbf{R}_{\text{eff,trans}} = (\mathbf{M}_P^R)_{\text{eff,trans}} = \mathbf{0}$, $\mathbf{R}_{\text{eff,rot}} = m\overline{\mathbf{a}}$, and $(\mathbf{M}_P^R)_{\text{eff,rot}} = I_P\alpha$ in the present case.]

PROBLEMS

16.49   A plank AB of mass m is placed in a truck so that its lower end rests against a crate on the bed while its upper part rests against the top of the railing as shown. If the truck moves to the right with an acceleration $\mathbf{a}$, determine the maximum magnitude of $\mathbf{a}$ for which the plank will remain in contact with the top of the railing.

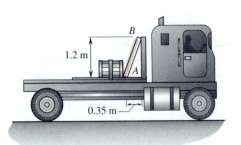

Fig. P16.49

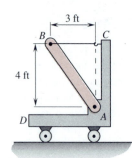

Fig. 16.50

16.50   The slender bar AB of weight W is supported as shown. If the cart CD is moved to the left with an acceleration a, determine the maximum magnitude of $\mathbf{a}$ for which the wire BC will remain taut.

16.51*   Refer to Prob. 16.50. If $W = 12.88$ lb and $\mathbf{a} = 10$ ft/s$^2 \rightarrow$, determine the tension in the wire BC and the reaction to the bar AB from the hinge at A.

16.52*   A crate of mass m is placed on the flat bed of a truck as shown, where the point G is the mass center of the crate. If the truck moves from rest with a forward acceleration $\mathbf{a}$ and $\mu_s = 0.35$ between the crate and the flat bed, determine the maximum magnitude of $\mathbf{a}$ for which the crate will remain unmoved relative to the truck.

16.53   Solve Prob. 16.52* if $\mu_s = 0.4$.

16.54   The forward motion of the 5000-lb forklift truck carrying a 2500-lb crate is being decelerated as shown. If $\mu_s = 0.5$ between the crate and the forklift, deter-

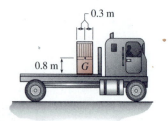

Fig. 16.52*

mine the maximum deceleration the truck can have without causing the crate to move
relative to the forklift.

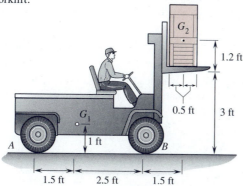

Fig. P16.54   1.5 ft   2.5 ft   1.5 ft

16.55* Refer to Prob. 16.54. Neglecting the rotational effects of the wheels, determine the vertical component of the reaction from the front wheels.

16.56 Two uniform rods AB and BC, each of weight 20 lb, are welded together
to form a 40-lb bent bar which is held by the wire CE and the links BE and CD as
shown. Neglecting the masses of the links, determine $\mathbf{a}_B$ of the pin at B and the
force in each link immediately after the wire CE is cut.

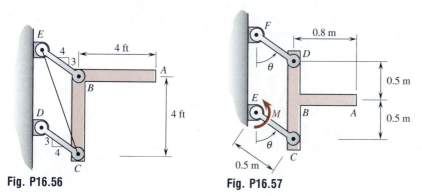

Fig. P16.56   **Fig. P16.57**

16.57 A 16-kg rod AB and a 20-kg rod CD are welded together to form a T bar
as shown. The T bar is raised from rest at $\theta = 0$ by means of the parallel links CE
and DF where a constant torque $\mathbf{M} = 360$ N·m $\circlearrowleft$ is applied to the link CE. Neglecting the masses of the parallel links, determine (a) α_{DF} of the link DF as a
function of θ, (b) the force in the link DF when $\theta = 120°$. [*Hint.* The tangential
component of the reaction from the pin at C to the T bar is $\mathbf{C}_t = (360/0.5)\mathbf{e}_t$ N.]

16.58 It is known that $\mu_s = 0.7$ between the road and the tires of the car as
shown. Determine the shortest time in which the car can travel from rest through a
distance of 100 m on a level road, assuming (a) front-wheel drive, (b) rear-wheel
drive, (c) four-wheel drive.

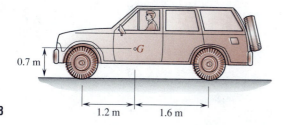

Fig. P16.58   1.2 m   1.6 m

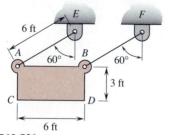

6 ft

E

F

A 60° B 60°

3 ft

C D

6 ft

Fig. P16.59*

16.59* The uniform rectangular plate $ABCD$ of weight W is suspended by two parallel wires and swings in a vertical plane. In the position shown, the tension in the wire AE is known to be equal to $0.1W$. Determine the angular speed of the wire AE in this position.

16.60 Solve Prob. 16.59* if the tension in the wire AE is known to be equal to $0.05W$.

16.61 The 1.4-Mg coal car with load is being towed by the counterweight E as shown. If the mass of the counterweight is 100 kg, determine the corresponding reactions from the supports to the small wheels at A and B.

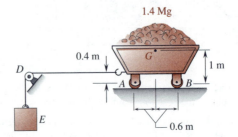

1.4 Mg

0.4 m

G

1 m

D

A B

E

0.6 m

Fig. P16.61

16.62* Solve Prob. 16.61 if the mass of the counterweight is 200 kg.

16.63 Each of the uniform bent bars AB has a weight W and is supported as shown. Determine the percentage of change in magnitude of the reaction from the support at C when the wire at A is suddenly cut in each case.

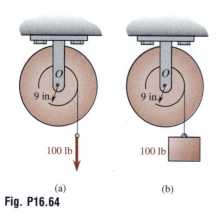

9 in.

O

100 lb

(a)

9 in.

O

100 lb

(b)

Fig. P16.64

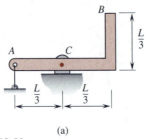

B

$\frac{L}{3}$

A C

$\frac{L}{3}$ $\frac{L}{3}$

(a)

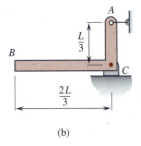

A

$\frac{L}{3}$

B

C

$\frac{2L}{3}$

(b)

Fig. P16.63

16.64 Each of the double pulleys shown weighs 322 lb and has a central radius of gyration of 15 in. about O. Determine the angular acceleration in each case.

16.65 Each of the bodies shown is uniform and has a total weight W. Determine the angular acceleration of each body and the reaction from its support at A when the body is released from rest in the position shown.

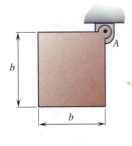

b

b

(a) Square plate

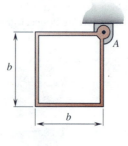

A

b

b

(b) Square wire

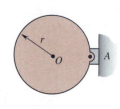

r

O

A

(c) Circular plate

Fig. P16.65

16.66 The uniform bent bar *BCD* weighs 15 lb and is connected by a cord *AB* to a 6-lb smooth collar *A* as shown. Neglecting effects of friction, determine (a) the minimum magnitude of the constant force **F** required to cause the cord *AB* to become collinear with the segment *BC*, (b) the corresponding tension in the cord.

16.67* Solve Prob. 16.66 if $\mu_k = 0.2$ between the bent bar and the support at *C*.

16.68 The trailer shown weighs 1500 lb and is attached at *H* to a rear-bumper hitch as shown. If the car pulls the trailer with an acceleration of 10 ft/s² ← and the lightweight wheels have negligible rotational effects, determine the force exerted on the trailer by the hitch at *H*.

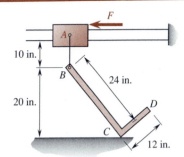

Fig. P16.66

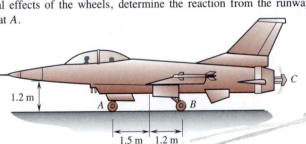

Fig. P16.68

16.69* Solve Prob. 16.68 if the brakes of the car are applied, causing the trailer to decelerate at 10 ft/s² →.

16.70 The gears *A* and *B* have masses of 10 kg and 30 kg and central radii of gyration of 80 mm and 150 mm, respectively. Determine the torque **M** acting on the gear *A* if the angular acceleration of the gear *B* is 2 rad/s² ↻.

16.71* Determine the reaction from the support at *A* when the bent rod weighing 0.5 lb/ft is released from rest in the position shown.

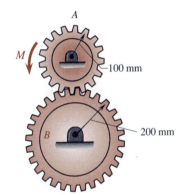

Fig. P16.70

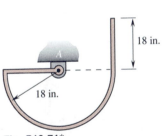

Fig. P16.71*

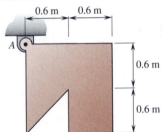

Fig. P16.72

16.72 Determine the reaction from the support at *A* when the plate with a mass density of 10 kg/m² is released from rest in the position shown.

16.73 A 5-Mg jet fighter aircraft decelerates at the rate of 3 m/s² by operating its thrust reverser at *C* along the runway without applying the mechanical brakes in its wheels at *A* and *B* as shown. Neglecting the aerodynamic forces on the aircraft and the rotational effects of the wheels, determine the reaction from the runway to the nose wheel at *A*.

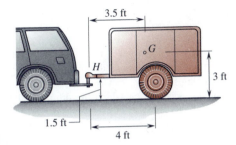

Fig. P16.73

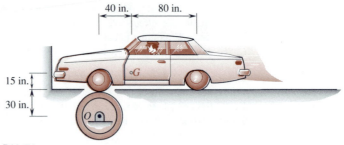

Fig. P16.74

16.74   A front-wheel drive car with a gross weight of 3600 lb is being tested with a dynamometer as shown. If the freely turning drum of the dynamometer has a moment of inertia $I_O = 1200$ slug·ft^2 and $\mu_s = 0.9$ between the tire and the drum, determine the maximum angular acceleration which the drum can have during the testing.

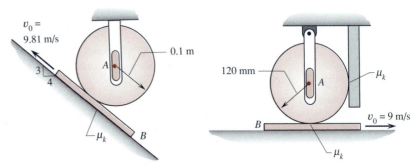

Fig. P16.75   **Fig. P16.76***

16.75 and 16.76*   At the time $t = 0$, the uniform cylinder A of weight W is lowered onto the plate B which moves with a constant velocity $\mathbf{v}_0$ as shown. If $\mu_k = 0.5$ between the surfaces of contact, determine (a) the angular acceleration of the cylinder immediately after contact, (b) the time t at which the cylinder starts rolling without slipping on the plate B.

16.77*   The moment of inertia of the 100-mm drum with flywheel as shown is 10 kg·m^2 about O, and $\mu_k = 0.4$ between the drum and the brake band. If the initial angular velocity is 600 rpm $\circlearrowleft$ and the system is to stop in 10 revolutions, determine the force **P** which must be applied to the point D of the hand bar.†

16.78   Solve Prob. 16.77* if the initial angular velocity is 600 rpm $\circlearrowright$.

16.79   The grooved wheel weighs 128.8 lb and has a radius of gyration of 9 in. about O as shown. It is known that $\mu_k = 0.4$ between the groove and the V belt as delineated, and the wheel rotates with $\omega = 600$ rpm $\circlearrowright$ when the 10-lb force **P** is applied to the point A of the lever. Determine the number of revolutions executed by the wheel during the braking.†

16.80*   Solve Prob. 16.79 if $\omega = 600$ rpm $\circlearrowleft$.

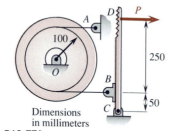

Dimensions
in millimeters

Fig. P16.77*

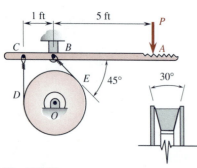

Fig. P16.79

---

†Recall the belt friction formulas in Secs. 9.12 and 9.13 that $T_2 = T_1 e^{\mu_k \beta}$ for a flat belt, and $T_2 = T_1 e^{\mu_k \beta/\sin(\alpha/2)}$ for a V belt.

16.81   The central moments of inertia of the gears A and B as shown are 10 kg·m^2 and 20 kg·m^2, respectively. If the mass of the block C is $m_C = 50$ kg and the system is released from rest, determine $\boldsymbol{\alpha}_A$ of the gear A.

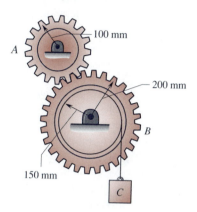

Fig. P16.81

16.82*   Solve Prob. 16.81 if $m_C = 100$ kg.

16.83   The 2-kg slender plank AB rests on a frictionless horizontal surface. If the end A at the instant shown has zero acceleration when a 10-N horizontal force acts on it at C as shown, determine (a) the length r, (b) $\mathbf{a}_B$ of the end B.

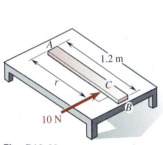

Fig. P16.83

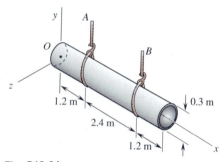

Fig. P16.84

16.84   A 300-kg thin-walled pipe is raised from *rest* by the cables of tall overhead cranes as shown. At the instant of raising, the cable A moves with $4\mathbf{j}$ m/s^2 and the pipe rotates with $2\mathbf{k}$ rad/s^2. Determine the corresponding tension in each cable.

16.85*   Refer to Prob. 16.84. Suppose that, at the instant of raising, the tensions in the cables A and B are 2 kN and 3 kN, respectively. Determine for that instant (a) the angular acceleration of the pipe, (b) the acceleration of each cable.

16.86*   A 300-kg hemispherical shell is raised from *rest* by the cables of tall overhead cranes as shown. At the instant of raising, the accelerations of the cables A and B are $3\mathbf{j}$ m/s^2 and $1\mathbf{j}$ m/s^2, respectively. Determine the corresponding tension in each cable.

16.87   Refer to Prob. 16.86*. Suppose that, at the instant of raising, the tensions in the cables A and B are 1.8 kN and 1.4 kN, respectively. Determine for that instant (a) the angular acceleration of the shell, (b) the acceleration of each cable.

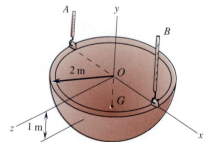

Fig. P16.86*

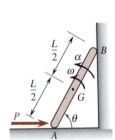

Fig. 16.17

16.9 Constrained General Plane Motion

A rigid body is said to be in a *constrained motion* if there exist definite
relations between the components of the acceleration **ā** of the mass center
G of the rigid body and its angular acceleration **α** as well as its angular
velocity **ω**. The rotation of a rigid body about a fixed axis is a special case
of constrained plane motion, which was covered in Secs. 16.7 and 16.8.
The constrained motion of rigid bodies which are in *general plane motion*
is studied in this section.

   An important step in the solution of a problem involving constrained
general plane motion is the *kinematic analysis* of the body in the problem.
For example, let us consider a slender rod *AB* of length *L* and mass *m* whose
extremities *A* and *B* are constrained to move on the horizontal surface and
the vertical wall as shown in Fig. 16.17. The rod is pushed by a force **P**
applied at *A* and has an angular velocity **ω** and an angular acceleration **α** in
the position defined by the angle θ. If effects of friction are negligible, we
may draw the vector-diagram equation containing the *free-body diagram* and
the *effective-force diagram* of the rod as shown in Fig. 16.18. Performing
the *kinetmatic analysis* (cf. Secs. 15.8 through 15.10), we can express $\bar{a}_x$
and $\bar{a}_y$ in terms of α and ω for a given position defined by θ. Substituting
the expressions for $\bar{a}_x$ and $\bar{a}_y$ into Fig. 16.18(b), we can reduce the total
number of unknowns in the vector-diagram equation to a minimum. If the
minimum number of unknowns in such an equation is three or less, the
unknowns may then be determined by using the *method of force and acceleration* for rigid bodies described in Sec. 16.7.

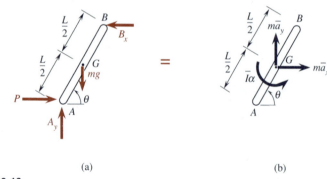

(a)   (b)

Fig. 16.18

   As another example, let us consider the rolling motion of a disk (or
wheel) on a rigid surface due to a horizontal force **P** applied as shown in
Fig. 16.19. The vector-diagram equation containing the *free-body diagram*
and *effective-force diagram* of the disk is shown in Fig. 16.20, where the

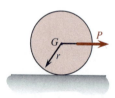

Fig. 16.19

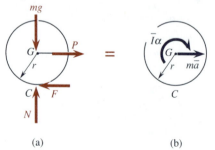

Fig. 16.20   (b)

symbols have the usual meaning. The relation between the friction force **F** and the normal force **N**, as well as that between the acceleration $\bar{a}$ of G and the angular acceleration α of the disk, may be summarized as follows:

1. $F \leq \mu_s N$ and $\bar{a} = r\alpha$ if it rolls without slipping.
2. $F = \mu_s N$ and $\bar{a} = r\alpha$ if it rolls with impending slipping.
3. $F = \mu_k N$ and $\bar{a} \neq r\alpha$ if it rolls and slips.

If the mass center G of the disk does not coincide with its geometric center O, then $\bar{a}$ in the above relations is replaced by the magnitude a_O of the acceleration $\mathbf{a}_O$ of O and $\bar{a}$ may be computed from the relation $\bar{\mathbf{a}} = \mathbf{a}_{G/O} + \mathbf{a}_O$.

EXAMPLE 16.16

The 20-kg rod AB is released from rest in the position shown. Neglecting effects of friction as well as the masses of the small wheels at A and B, determine the angular acceleration of the rod and the reaction from its support at A immediately after release.

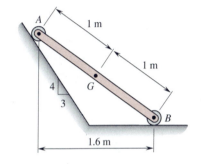

Solution.   Immediately after release, the rod has zero angular velocity, and its *acceleration center Z* (cf. Sec. 15.9) is located as shown. Assuming the angular acceleration α to be counterclockwise, we see that the acceleration $\bar{\mathbf{a}}$ of the mass center G is perpendicular to $\overline{ZG}$ and points to the lower right. The magnitude of $\bar{\mathbf{a}}$ is $\bar{a} = \overline{ZG}\alpha = 0.2\sqrt{97}\alpha$. Thus, we draw the vector-diagram equation containing the *free-body diagram* and the *effective-force diagram* of the rod as shown, where the unknowns are A, B, and α.

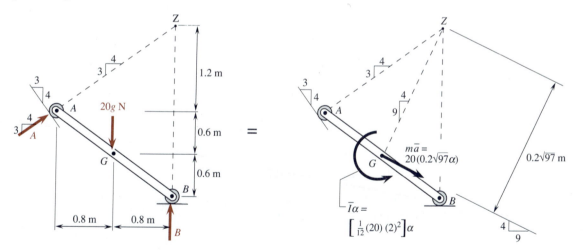

To determine α and **A**, we write

$$+ \text{\Large$\circlearrowleft$} \ \Sigma M_Z: \quad 0.8(20g) = [\tfrac{1}{12}(20)(2)^2]\alpha + 0.2\sqrt{97}\,[20(0.2\sqrt{97}\,\alpha)]$$

$$\alpha = 1.863 \qquad \alpha = 1.863 \text{ rad/s}^2 \ \text{\Large$\circlearrowleft$} \ \blacktriangleleft$$

$$\xrightarrow{+} \Sigma V_x: \quad \tfrac{4}{5}A = \frac{9}{\sqrt{97}}\left[20(0.2\sqrt{97}\,\alpha)\right]$$

$$A = 83.8 \text{ N} \ \angle \ 36.9° \ \blacktriangleleft$$

Developmental Exercise

D16.29 Refer to Example 16.16. Determine the reaction from the support at B immediately after release.

EXAMPLE 16.17

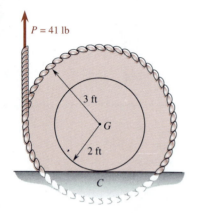

$P = 41$ lb

3 ft

G

2 ft

C

A cord is wrapped around the outer drum of a 96.6-lb wheel and pulled vertically with a constant force of 41 lb as shown. If the wheel has a central radius of gyration of 2.5 ft and rolls without slipping on the horizontal track, determine the angular acceleration of the wheel.

Solution. Since it rolls without slipping, we have $\bar{a} = r\alpha = 2\alpha$. Thus, we equate the *free-body diagram* of the wheel to its *effective-force diagram*

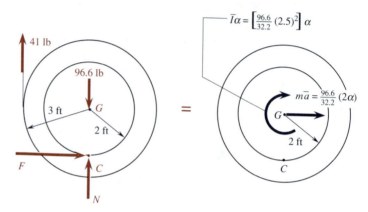

as shown, where the unknowns are N, F, and α. To determine α, we write

$$+\circlearrowleft \ \Sigma M_C: \qquad 3(41) = \left[\frac{96.6}{32.2}(2.5)^2\right]\alpha + 2\left[\frac{96.6}{32.2}(2\alpha)\right]$$

$$\alpha = 4 \qquad\qquad \alpha = 4 \text{ rad/s}^2 \ \circlearrowleft \ \blacktriangleleft$$

Developmental Exercise

D16.30 Refer to Example 16.17. Determine (a) the reaction components $\mathbf{F}$ and $\mathbf{N}$, (b) the minimum value of μ_s between the wheel and the track.

EXAMPLE 16.18[†]

The two slender bars OA and AB, each of length $L = 5$ ft and weight $W = 10$ lb, are released from rest in the position shown. Determine α_{OA} of the bar OA immediately after release.

[†]The problem in this example is solved by using the *principle of virtual work in kinetics* in Example 17.8.

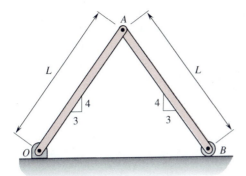

Solution. Immediately after release, the bars *OA* and *AB* have zero angular velocities and their *acceleration centers* are located at *O* and *Z*, respectively, as shown. By symmetry, we know that $\alpha_{OA} = -\alpha_{AB}$; so, we may assume that $\alpha_{OA} = \alpha \circlearrowleft$ and $\alpha_{AB} = \alpha \circlearrowright$.

In the vector-diagram equation consisting of the *free-body diagram* and the *effective-force diagram* for the entire system as shown, we note that there are four unknowns (O_x, O_y, B_y, and α) but only three independent

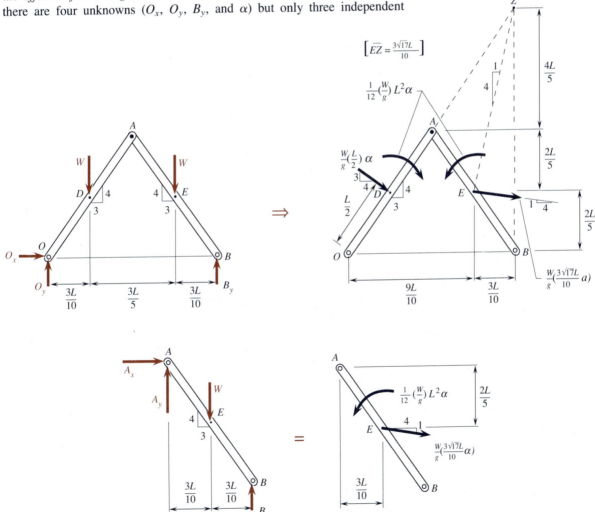

scalar equations to be written. Thus, we resort to drawing an additional vector-diagram equation for the bar AB as shown, which incurs two new unknowns (A_x and A_y) but provides three additional independent scalar equations. The six unknowns may therefore be solved from the six equations. However, to focus our effort on solving for α, we proceed as follows:

Entire system

$+\circlearrowleft \ \Sigma M_O$:

$$\frac{6L}{5} B_y - \frac{3L}{10} W - \frac{9L}{10} W$$

$$= -\frac{1}{12}\left(\frac{W}{g}\right)L^2\alpha + \frac{1}{12}\left(\frac{W}{g}\right)L^2\alpha - \frac{L}{2}\left[\frac{W}{g}\left(\frac{L}{2}\alpha\right)\right]$$

$$- \frac{2L}{5}\left\{\frac{4}{\sqrt{17}}\left[\frac{W}{g}\left(\frac{3\sqrt{17}L}{10}\alpha\right)\right]\right\} - \frac{9L}{10}\left\{\frac{1}{\sqrt{17}}\left[\frac{W}{g}\left(\frac{3\sqrt{17}L}{10}\alpha\right)\right]\right\}$$

Bar AB

$+\circlearrowleft \ \Sigma M_A$:

$$\frac{3L}{5} B_y - \frac{3L}{10} W$$

$$= \frac{1}{12}\left(\frac{W}{g}\right)L^2\alpha + \frac{2L}{5}\left\{\frac{4}{\sqrt{17}}\left[\frac{W}{g}\left(\frac{3\sqrt{17}L}{10}\alpha\right)\right]\right\}$$

$$- \frac{3L}{10}\left[\frac{1}{\sqrt{17}}\left(\frac{3\sqrt{17}L}{10}\alpha\right)\right]$$

Since $W = 10$ lb and $L = 5$ ft, the above two equations yield

$$B_y = \frac{217W}{292} = 7.43 \text{ lb} \qquad \alpha = \frac{45g}{146L} = 1.985 \text{ rad/s}^2$$

$$\alpha_{OA} = 1.985 \text{ rad/s}^2 \ \circlearrowleft \ \blacktriangleleft$$

Developmental Exercise

D16.31   Refer to Example 16.18. Determine the unknowns O_x and O_y.

EXAMPLE 16.19[†]

A 40-kg block C is supported by a 20-kg cylinder A and a 30-kg cylinder B as shown, where the cylinders have the same radius $r = 100$ mm. If the cylinders roll without slipping, determine the acceleration of the block C when a 50-N force $\mathbf{P}$ is applied to it, which is initially at rest.

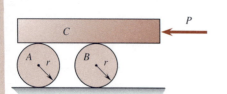

[†]The problem in this example is solved by using the *principle of virtual work in kinetics* in Example 17.9.

Solution. Since the cylinders roll without sliding and are of the same radius $r = 0.1$ m, they must have the same angular acceleration α which is related to the acceleration $\mathbf{a}_C$ of the block C by $a_C = 2r\alpha = 2(0.1)\alpha$ or $\alpha = 5a_C$. Thus, the vector-diagram equation consisting of the *free-body diagram* and the *effective-force diagram* for the entire system as well as that for each cylinder may be drawn as shown. In the three vector-diagram equations shown, there are nine unknowns: a_C, F_A, N_A, F_B, N_B, D_x, D_y, E_x, and E_y. However, our objective is mainly to find $\mathbf{a}_C$. Thus, we proceed:

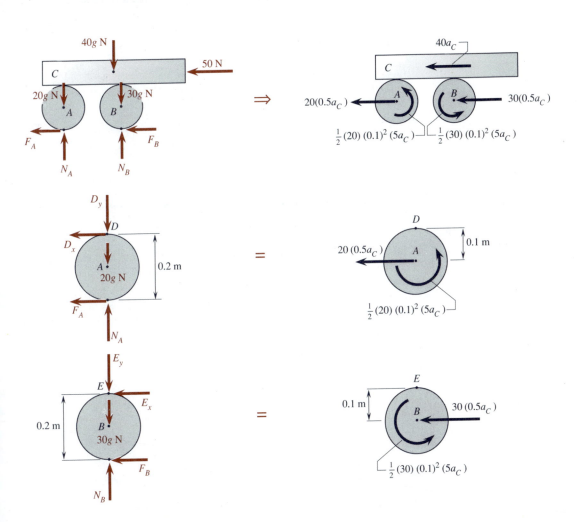

Entire system

$$\xleftarrow{+} \Sigma V_x: \qquad 50 + F_A + F_B = 40a_C + 20(0.5a_C) + 30(0.5a_C)$$

Cylinder A

$$+\,\backslash\!\!\!\!\text{)}\ \Sigma M_D: \qquad 0.2F_A = 0.1[20(0.5a_C)] - \tfrac{1}{2}(20)(0.1)^2(5a_C)$$

Cylinder B

$$+\,\backslash\!\!\!\!\text{)}\ \Sigma M_E: \qquad 0.2F_B = 0.1[30(0.5a_C)] - \tfrac{1}{2}(30)(0.1)^2(5a_C)$$

The above three equations yield

$$a_C = 0.851 \text{ m/s}^2 \qquad F_A = 2.13 \text{ N} \qquad F_B = 3.19 \text{ N}$$

$$\mathbf{a}_C = 0.851 \text{ m/s}^2 \leftarrow \blacktriangleleft$$

REMARK. Since the values of F_A and F_B are found to be positive, the directions of the friction forces $\mathbf{F}_A$ and $\mathbf{F}_B$ as indicated in the free-body diagrams are correct.

Developmental Exercises

D16.32 Refer to Example 16.19. Without knowing dimensions other than the radii of the cylinders, which of the nine unknowns in the vector-diagram equations could not be determined?

D16.33 Work Example 16.19 if the masses of the cylinders A and B are $m_A = m_B = 20$ kg.

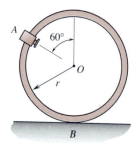

Fig. P16.88

PROBLEMS

16.88 A small collar of weight W is securely attached at A to a hoop of weight W and radius r. The system is released from rest in the position shown and no slipping occurs. Determine the reaction from the support at B immediately after release.

16.89* and 16.90 The rod AB has a weight W and is released from rest in the position shown. Neglecting effects of friction, determine the angular acceleration of the rod immediately after release.

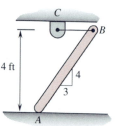

Fig. P16.89* and P16.91

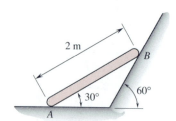

Fig. P16.90 and P16.92*

16.91 and 16.92* The rod AB has a weight W and is released from rest in the position shown. Knowing that $\mu_s = 0.1$ between all surfaces of contact, determine the normal component of reaction from the support at A immediately after release.

16.93 An aircraft lands on a level runway and covers it with a constant speed of 180 mph. The landing wheels are not turning before they touch the runway. If the length of the skid mark of the tires on the runway is 200 ft, and the normal force between the runway and the landing wheel of radius r is $8W$ where W is the weight of the wheel whose central radius of gyration is $0.7r$, determine the kinetic coefficient of friction between the tires and the runway.

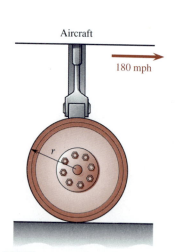

Fig. P16.93

16.94* A 50-kg wheel of radius 180 mm and central radius of gyration 120 mm is released from rest on the incline as shown. Knowing that the wheel rolls without slipping, determine (a) the acceleration of the mass center G, (b) the minimum value of μ_s compatible with this motion.

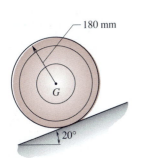

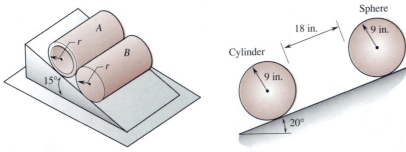

Fig. P16.94*   Fig. P16.95

16.95 A 100-kg pipe rolls without slipping in a circular support of radius 2.2 m. In the position shown, the pipe has an angular speed of 10 rad/s. Determine the reaction from the support.

16.96* The pipe A and the cylinder B are in contact when they are released from rest as shown. Knowing that $r = 120$ mm and both A and B roll without slipping, determine (a) the acceleration of the center of B relative to that of A, (b) the clear distance between them after 2 s.

Fig. P16.96*   Fig. P16.97

16.97 The bodies as shown are released from rest at the time $t = 0$. If they roll without slipping, determine the time at which the clearance between them becomes zero.

16.98 A pipe has a cord wrapped around it at the midsection and is released from rest in the position shown. Determine the time required for the pipe to descend 2 m to the ground.

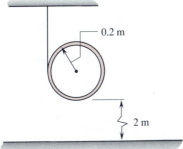

Fig. P16.98

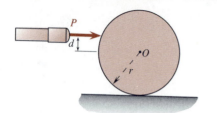

Fig. P16.99

16.99 A cue hits a billiard ball with a horizontal force **P** at a height which is a distance d above its center O as shown. If the ball is to roll without slipping regardless of the value of μ_s between the ball and the table, determine the value of d in terms of the radius r of the ball.

16.100 A hoop is spinning at 10 rad/s when it is set on a horizontal floor as shown. If $\mu_s = 0.1$ between the hoop and the floor, determine the time during which the hoop will both roll and slip on the floor.

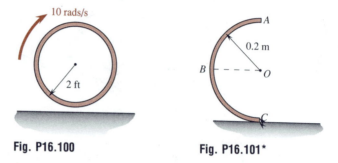

Fig. P16.100   **Fig. P16.101***

16.101* A half-section of pipe of mass 10 kg is released from rest in the position shown. If the section rolls without slipping, determine the acceleration of the point B and the reaction from the support at C immediately after release.

16.102 and 16.103* A cord is wrapped around the hub of the spool and pulled with a constant force **P** as shown. The spool weighs 36 lb and has a central radius of gyration of 4 in. about G. If $P = 13$ lb, $r = 3$ in., $R = 6$ in., and the spool rolls without slipping, determine (a) the acceleration of G, (b) the minimum value of μ_s between the spool and the floor.

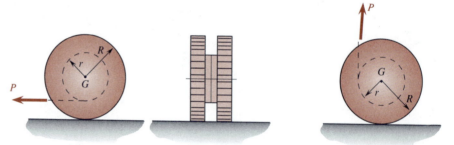

Fig. P16.102 and P16.104*   **Fig. P16.103* and P16.105**

16.104* and 16.105 A cord is wrapped around the hub of the spool and pulled with a constant force **P** as shown. The spool has a mass of 10 kg and a radius of gyration of 150 mm about its mass center G. If $P = 40$ N, $r = 100$ mm, $R = 200$ mm, and $\mu_s \approx \mu_k \approx 0.25$ between the spool and the floor, determine (a) whether the spool slips, (b) the acceleration of G.

16.106 The door on the passenger's side of the car is unintentionally left slightly open when the driver backs up the car from rest with a constant backward acceleration of $\mathbf{a} = 5$ ft/s² →. The door has a weight $W = 80$ lb, and its center of gravity is located at a distance $\bar{r} = 20$ in. from the hinge axis. If the central radius of gyration of the door is $\bar{k} = 12$ in., determine the angular velocity ω of the door immediately before it becomes wide open (i.e., $\theta = 90°$). [*Hint.* Cf. Eq. (15.8).]

16.107* Refer to Prob. 16.106. Determine the angular velocity ω of the door as it passes through the position $\theta = 45°$.

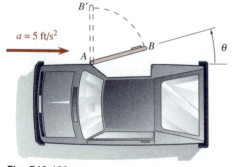

Fig. P16.106

16.108 A 10-kg disk has a radius of gyration of 120 mm about its mass center G. The disk rolls without slipping and has an angular velocity $\boldsymbol{\omega} = 5$ rad/s $\curvearrowright$ in the position shown, where $b = 60$ mm and $r = 200$ mm. Determine the corresponding angular acceleration of the disk and the reaction from the support at C.

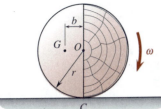

Fig. P16.108

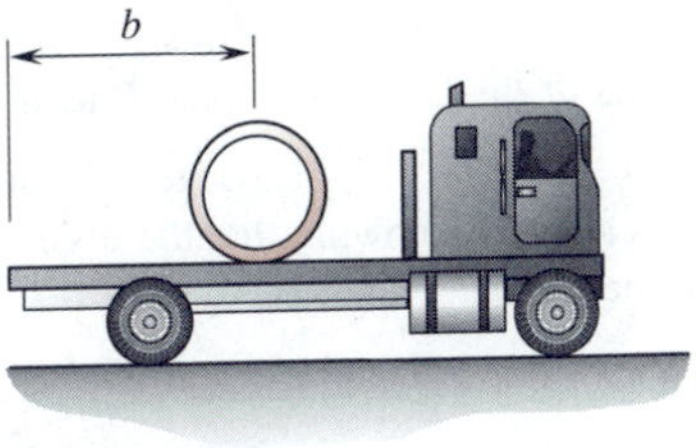

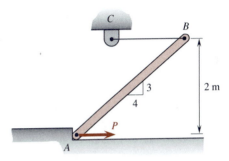

Fig. P16.109*

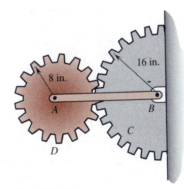

16.109* A pipe rests on the horizontal bed of a truck as shown. If the truck moves forward with a constant acceleration **a** and the pipe rolls without slipping, determine (a) the acceleration of the mass center of the pipe, (b) the distance which the truck will move before the pipe rolls off the truck.

16.110 Solve Prob. 16.109* for the case in which the pipe is replaced (i) by a cylinder, (ii) by a sphere.

16.111 The 20-kg rod AB is initially at rest. Neglecting effects of friction, determine the tension in the cable BC at the instant a 200-N force **P** is applied as shown.

16.112* Solve Prob. 16.111 if $\mu_s = 0.2$ between the rod and the horizontal surface.

16.113 The motion of the 10-kg slender rod ABC is guided by two rollers of negligible mass. A force **P** is applied to hold the rod in equilibrium as shown. Determine the angular acceleration of the rod and the reactions from its supports immediately after the force **P** is removed.

Fig. P16.111

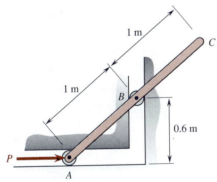

Fig. P16.113 and P16.114*

Fig. P16.115

16.114* In the position shown, the force **P** causes the 10-kg slender rod to move from rest with an angular acceleration of 5 rad/s² $\curvearrowright$. where the masses of the rollers at A and B are negligible. Determine the corresponding magnitude of **P**.

16.115 The connecting rod AB weighs 10 lb and the gear C is stationary. The gear D weighs 20 lb and has a central radius of gyration of 6 in. Determine $\mathbf{a}_A$ of the point A immediately after the system is released from rest in the position shown.

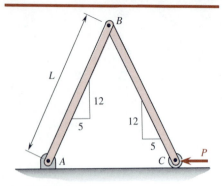

Fig. P16.116* and P16.117

16.116* Each of the two slender bars AB and BC weighs 20 lb and the roller at C has a negligible mass. A force $\mathbf{P}$ is applied at C to hold the system in equilibrium as shown where $L = 2.6$ ft. Determine α_{BC} of the bar BC and the reaction from the support at C immediately after the force $\mathbf{P}$ is removed.

16.117 Each of the two slender bars AB and BC has a mass of 10 kg and the roller at C has a negligible mass. In the position shown, the point C moves with a *constant* velocity of 1.2 m/s ←. If $L = 1.3$ m, determine the corresponding magnitude of the horizontal force $\mathbf{P}$ acting at C.

16.118 The 50-kg carriage C is supported by the 20-kg disk A and the 15-kg disk B as shown. Knowing that the disks roll without slipping, determine the acceleration of the carriage when the system is released from rest.

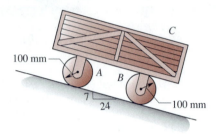

Fig. P16.118

16.119* Solve Prob. 16.118 if both disks A and B have the same mass of 20 kg.

16.120 and 16.121* The system shown consists of a section of pipe A weighing 15 lb and a cart B weighing 10 lb. If the system is released from rest and no slipping occurs between the pipe and the cart, determine (a) the angular acceleration of the pipe, (b) the acceleration of the cart, immediately after release.

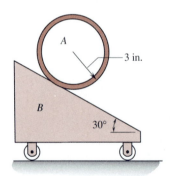

Fig. P16.120

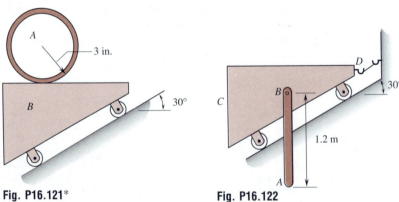

Fig. P16.121*   **Fig. P16.122**

16.122 A 10-kg rod AB is hinged to a 15-kg cart C as shown. If the cord at D is cut, determine (a) the angular acceleration of the rod, (b) the acceleration of the cart, immediately after cutting.

16.10 Concluding Remarks

Similar to the moment of inertia of an area, the *moment of inertia of a body* about a given axis may be thought of as the integral sum of the products of the differential masses of the body and the squares of their individual distances from the given axis. The moment of inertia of a body is also referred to as the *mass moment of inertia*. The definitions and the parallel-axis theorem related to moments of inertia of a body, as presented in the first part of this chapter, serve as a timely prerequisite to the study of the plane kinetics of rigid bodies.

Unlike the moment of inertia of an area (which does not lend itself to a simple interpretation), the moment of inertia of a body about its own central axis or an axis fixed in space can be considered as being numerically equal to *the moment required to accelerate the body to rotate at the rate of* 1 rad/s^2 *about that axis.* Physically, the *moment of inertia* of a body about a central or fixed axis is *a measure of the rotational inertia* of the body about that axis, while the *mass* of a body is *a measure of the translational inertia* of the body.

The second part of this chapter presents the *method of force and acceleration* for rigid bodies in plane motion, which is a direct extension of that presented in Chap. 12 for particles. The central concept in this method is that *the system of forces* (including moments, if any) *in the free-body diagram of a body is either equivalent or equipollent to the effective force-moment system(s),* at the mass center of the body or mass centers of its component parts, *in the corresponding effective-force diagram of the body.* Those two systems of vectors are equivalent if the body is a single rigid body; they are equipollent if the body is composed of a set of connected rigid bodies. In any case, those two systems have the *same resultant force* and the *same resultant moment* about any given point; and these conditions generally yield up to three independent scalar equations, such as Eqs. (16.19), (16.20), and (16.21) or (16.22), which are then used to solve for the unknowns in the problem considered.

Note that the drawing of the vector-diagram equation consisting of the *free-body diagram* and the *effective-force diagram* is an effective approach in learning plane kinetics of rigid bodies using the method of force and acceleration. Without drawing such a vector-diagram equation, we would have to give up the freedom of choice of moment center accorded by Eq. (16.22), which is a more general and powerful equation than Eq. (16.21). In fact, a dexterity in the application of Eq. (16.22) is a benchmark of excellence in one's understanding of the *method of force and acceleration* for rigid bodies in plane motion.

REVIEW PROBLEMS

16.123 A slender rod weighing 1.61 lb/ft is used to form a composite body as shown. Determine the moment of inertia I_x and the radius of gyration k_x of the body.

16.124* Refer to Prob. 16.123. Determine I_y and k_y of the body.

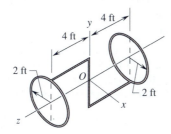

Fig. P16.123

16.125   A sheet of metal is cut and bent into a machine component as shown. If its mass density is $\rho_A = 16$ kg/m^2, determine I_x and k_x of the component.

16.126   Refer to Prob. 16.125. Determine I_y and k_y of the component.

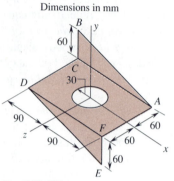

Fig. P16.125

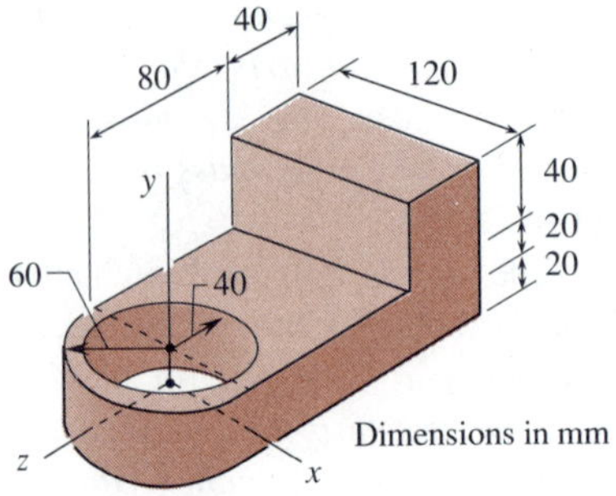

16.127*   The mass density of the steel machine element is $\rho = 7.85$ Mg/m^3. Determine I_x and k_x of the element.

16.128   Refer to Prob. 16.127*. Determine I_y and k_y of the element.

16.129   Two uniform rods AB and BC, weighing 1 lb/ft, are welded together to form a single rigid body which is held by three wires as shown. Determine the angular acceleration of the rigid body and the tension in the wires AD and BF immediately after the wire BE is cut.

Fig. P16.127*

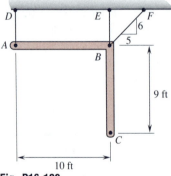

Fig. P16.129

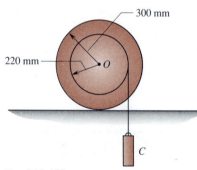

Fig. P16.130

16.130   A 10-kg cylinder C is suspended by a cord which is wrapped around the drum of a wheel as shown. The wheel with the drum has a total mass of 15 kg and a central radius of gyration $\bar{k}_O = 200$ mm. If the system is released from rest and the wheel rolls without slipping, determine the acceleration of the cylinder.

16.131*   The system shown is obtained by attaching a rigid guide AOB of negligible mass to a hinge at the point O of the system described in Prob. 16.130. Neglecting effects of friction between the cylinder C and the rigid guide and knowing that the wheel will roll without slipping, determine the acceleration of the cylinder when the system is released.

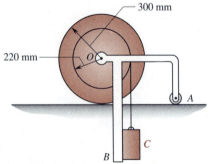

Fig. P16.131*

16.132* A 50-kg wooden block C is supported by a 20-kg cylinder A and a 15-kg cylinder B as shown. If no slipping occurs, determine the acceleration of the block when the system is released from rest.

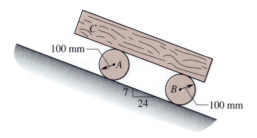

Fig. P16.132*

16.133 Solve Prob. 16.132* if both cylinders A and B have the same mass of 20 kg.

16.134 Each of the two bars OA and AB weighs 13 lb. The bars are released from rest in the position shown. Determine the angular acceleration of each bar and the reactions from the supports at O and B immediately after release.

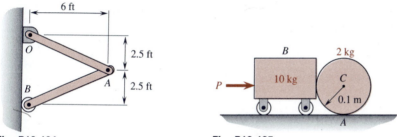

Fig. P16.134   **Fig. P16.135**

16.135 A 10-kg block B rests on rollers of negligible mass and is in contact with a 2-kg cylinder C as shown. It is known that $\mu_k \approx \mu_s = 0.5$ between B and C as well as between C and its support A. Determine the range of values of the magnitude of the horizontal force **P** which will cause the cylinder to roll without slipping.

16.136 Refer to Prob. 16.135. Determine the magnitude of the horizontal force **P** for which the cylinder will roll with an angular acceleration of 20 rad/s^2 ↺ on the support without sliding.

16.137 The 50-lb carriage D is supported by a 20-lb cylinder A, a 25-lb cylinder B, and a 30-lb cylinder C as shown, where the cylinders have the same radius of 4 in. If the cylinders roll without slipping when the 10-lb force acts on it, determine $\mathbf{a}_D$ of the carriage D.

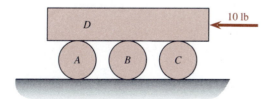

Fig. P16.137

16.138   The rods AB and BC weigh 10 lb and 20 lb, respectively. In the position shown, the end A of AB is acted on by a horizontal force $\mathbf{P}$ and moves with $\mathbf{v}_A = 5$ ft/s $\rightarrow$ and $\mathbf{a}_A = 10$ ft/s^2 $\rightarrow$. Neglecting effects of friction, determine the magnitude of $\mathbf{P}$ in this position.

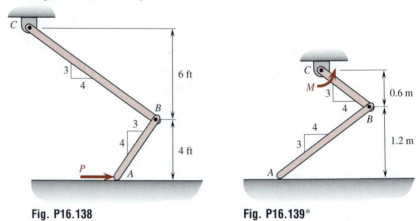

Fig. P16.138   **Fig. P16.139***

16.139*   The rods AB and BC have masses of 4 kg and 2 kg, respectively. In the position shown, a moment $\mathbf{M}$ acts on the rod BC and the end A of the rod AB moves with $\mathbf{v}_A = 4.8$ m/s $\rightarrow$ and $\mathbf{a}_A = 2$ m/s^2 $\rightarrow$. Neglecting effects of friction, determine the magnitude of $\mathbf{M}$ in this position.

Plane Kinetics of Rigid Bodies: Work and Energy

Chapter 17

KEY CONCEPTS

- Work of a force sliding on a rigid body, work of a couple acting on a rigid body, kinetic energy of a rigid body in plane motion, total mechanical energy of a rigid body, and power input to a rigid body by a force or by a moment.
- Principle of work and energy for rigid bodies.
- Principle of conservation of energy for rigid bodies.
- Principle of virtual work in kinetics for studying forces and accelerations in one-degree-of-freedom systems of rigid bodies.

We saw in Chap. 13 that the principle of work and energy and the principle of conservation of energy were useful in studies involving velocity changes of particles resulting from the cumulative effect of forces which move them through specific distances. By those principles, it is possible to determine the change in velocity by analyzing the work done and the change in kinetic energy for a general system, or by analyzing just the initial and final total mechanical energies for a conservative system. In these analyses, the acceleration and the time need not be explicitly considered. Such advantages still hold true when we extend those principles to rigid bodies in this chapter. Moreover, the principle of virtual work in kinetics for particles in Sec. 13.13 is extended to rigid bodies in Sec. 17.7. It is shown in Examples 17.8 and 17.9 that the virtual work principle can effectively *complement* the method of force and acceleration for systems of rigid bodies with one degree of freedom.

WORK AND KINETIC ENERGY

17.1 Work of a Force Acting on a Rigid Body

*The work of a force **F** acting on a rigid body* is defined to be the sum of the works done by **F** to the various particles of the rigid body. Because of similarities and close relations, several of the concepts and equations in Chap. 13 may be applied to compute the work of a force acting on a rigid body. For example, the work $U_{1\to2}$ of a force **F** acting on a particle of a rigid body during the displacement of the particle from the point A_1 along a path C to the point A_2 may be computed from the equation

$$U_{1\to2} = \int_C \mathbf{F} \cdot d\mathbf{r} \tag{17.1}$$

where $d\mathbf{r}$ is the differential displacement *of the particle*[†] (moving with the rigid body) in an inertial reference frame. If **F** is a conservative force, the value of $U_{1\to2}$ is dependent on the coordinates of A_1 and A_2, not on those of the path C. In this case, we write

$$U_{1\to2} = \int_{A_1}^{A_2} \mathbf{F} \cdot d\mathbf{r} \tag{17.2}$$

If the force **F** is a *constant force* as shown in Fig. 17.1, we write

$$U_{1\to2} = \int_C \mathbf{F} \cdot d\mathbf{r} = \mathbf{F} \cdot \int_C d\mathbf{r} = \mathbf{F} \cdot \mathbf{q} = Fq \cos\psi$$

$$\boxed{U_{1\to2} = Fq_{\parallel}} \tag{17.3}$$

where **q** is the displacement of the particle acted on by **F** from A_1 to A_2, $q_{\parallel} = q \cos\psi$ is the scalar component of **q** parallel to **F**, and ψ is the angle between the positive directions of **F** and **q** as indicated in Fig. 17.1. Since $d\mathbf{r} = (d\mathbf{r}/dt)dt = \mathbf{v}\,dt$, we may write Eq. (17.1) as follows:

$$U_{1\to2} = \int_{t_1}^{t_2} \mathbf{F} \cdot \mathbf{v}\,dt \tag{17.4}$$

where **v** is the velocity *of the particle* on which **F** acts, t is the time variable, and t_1 and t_2 are beginning and ending times of the interval during which **F** acts on the particle. Clearly, *no work* is done by a force to a rigid body which remains stationary during the action of the force.

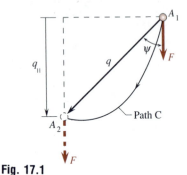

Fig. 17.1

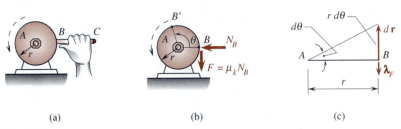

(a)   (b)   (c)

Fig. 17.2

    At times, a force may act on *more than one particle* of a rigid body during a certain time interval. This is illustrated in Fig. 17.2, where a stationary rod BC is pressed against a spinning grinder of radius r, which is mounted at A. When the grinder rotates through an angle θ, the friction

---

[†]The *pitfall* of defining $d\mathbf{r}$ in Eq. (17.1) as the differential displacement *of the point of application of the force* **F** is underscored in the last paragraph of this section, where **F** does not continually act on the same particle of a rigid body. Furthermore, note that the dimensions and units of the work of a force have been discussed in Sec. 13.1.

force **F** will have acted on *a series of particles* lying along the arc *BB'* on the rim of the grinder. Assume that the normal force $\mathbf{N}_B$ and the friction force **F** are *constant* forces and the unit vector in the direction of **F** is $\boldsymbol{\lambda}_F$. Applying the definition for the work of a force acting on a rigid body and referring to Fig. 17.2(c), we may compute *the work* $U_{1\to2}$ *of* **F** *acting on the grinder during its rotation through* θ as follows:

$$U_{1\to2} = \int_C \mathbf{F} \cdot d\mathbf{r} = \int_C F\,\boldsymbol{\lambda}_F \cdot d\mathbf{r}$$

$$= \int_C F(-r\,d\theta) = -Fr\int_0^\theta d\theta$$

$$U_{1\to2} = -Fr\theta \tag{17.5}$$

The negative sign in Eq. (17.5) is due to the fact that $\boldsymbol{\lambda}_F$ and $d\mathbf{r}$ are opposite in direction. The work of the normal force $\mathbf{N}_B$ acting on the grinder is clearly zero because $\mathbf{N}_B$ is always perpendicular to $d\mathbf{r}$. On the other hand, the friction force **F'** as well as the normal force $\mathbf{N}_B'$ exerted by the grinder on the stationary rod does no work to the rod because the rod remains stationary. (**F'** and $\mathbf{N}_B'$ are not shown in Fig. 17.2.) For that matter, note that *no work is done by any force during the interval in which the force acts at a stationary point or a velocity center of any body.*

Furthermore, it is to be pointed out that $d\mathbf{r}$ in Eq. (17.1) should *not* be taken as the differential displacement *of the point of application of* **F** which acts on a series of particles of a rigid body. For instance, the friction force **F** exerted by a writing chalk on a stationary chalkboard does *no* work to the chalkboard. If $d\mathbf{r}$ in Eq. (17.1) were taken as the differential displacement of the point of application of **F**, one would *erroneously* assert that "a positive work was done by the friction force **F** to the stationary chalkboard."

Developmental Exercises

D17.1   Define the work of a force acting on a rigid body.

D17.2   A rigid wheel *rolls without slipping* down a fixed incline. (a) Does the friction force at the point of contact do any work to the wheel? (b) Does the friction force do any work to the fixed incline? [*Hint*. Cf. Eq. (17.4).]

D17.3   Do D17.2 if the rigid wheel *slips* on the fixed incline.

17.2   Work of a Couple Acting on a Rigid Body

The forces acting on a rigid body in *plane motion* are generally nonconcurrent coplanar forces where some of them may pair up to form couples. Each couple may be represented by a pair of noncollinear parallel forces such as **F** and −**F** shown in Fig. 17.3. Since the displacements of all particles of a rigid body in translation must be equal, *the work of a couple acting on a rigid body in translation is always zero.* This is obviously true because the total work done by **F** and −**F** to the rigid body moving through the common displacement $d\mathbf{r}$ is always zero; i.e.,

$$dU = \mathbf{F} \cdot d\mathbf{r} + (-\mathbf{F}) \cdot d\mathbf{r} = 0 \tag{17.6}$$

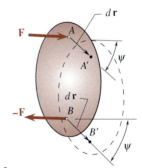

Fig. 17.3

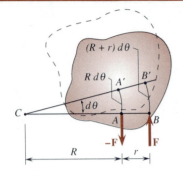

Fig. 17.4

Now, consider the case in which a rigid body is acted on by a couple of moment **M** and, at the same time, undergoes an infinitesimal angular displacement $\overrightarrow{d\theta}$ which is parallel to **M**. Let the couple be represented by the forces **F** and $-$**F** which are tangent to the arcs of rotation at the points A and B as shown in Fig. 17.4. By Eq. (17.3), the work of the couple acting on the rigid body during the angular displacement $\overrightarrow{d\theta}$ is

$$dU = |-\mathbf{F}|\,(-R\,d\theta) + |\mathbf{F}|\,(R + r)\,d\theta = Fr\,d\theta$$

Since the product Fr is equal to the magnitude of the moment **M**, we write

$$dU = M\,d\theta = \mathbf{M} \cdot \overrightarrow{d\theta} \tag{17.7}$$

Thus, *the work* $U_{1\rightarrow2}$ *of a couple of moment* **M** *acting on a rigid body rotating from the angular position* θ_1 *to the angular position* θ_2 *is*

$$U_{1\rightarrow2} = \int_{\theta_1}^{\theta_2} \mathbf{M} \cdot \overrightarrow{d\theta} \tag{17.8}$$

If the moment **M** is *constant*, the work $U_{1\rightarrow2}$ of **M** acting on a rigid body during its finite angular displacement $\Delta\theta$, which is expressed in radians and is in the same direction as that of **M**, is

$$U_{1\rightarrow2} = M\,\Delta\theta \tag{17.9}$$

Naturally, $U_{1\rightarrow2}$ is negative if M and $\Delta\theta$ have opposite senses. The work of a couple is a scalar whose dimensions are *force* $\times$ *length* and whose primary units are *joules* (1 joule $=$ 1 J $=$ 1 N·m) in the SI and *foot-pounds* (ft·lb) in the U.S. customary system.

Since $\overrightarrow{d\theta} = (d\theta/dt)\,dt = \boldsymbol{\omega}\,dt$, we may write Eq. (17.8) as follows:

$$U_{1\rightarrow2} = \int_{t_1}^{t_2} \mathbf{M} \cdot \boldsymbol{\omega}\,dt \tag{17.10}$$

where $\boldsymbol{\omega}$ is the angular velocity of the rigid body on which the couple acts, t is the time variable, and t_1 and t_2 are beginning and ending times of the interval during which the couple acts on the rigid body. Clearly, *no* work is done by a couple to the rigid body which remains stationary or in translation (where $\boldsymbol{\omega} = \mathbf{0}$) during the action of the couple.

Developmental Exercises

D17.4 Define the work of a couple acting on a rigid body.

D17.5 A couple of moment $\mathbf{M} = 10$ N·m $\circlearrowleft$ acts on a rigid body during its rotation through an angle $\Delta\theta$. Determine the work of the couple if (a) $\Delta\theta = 1.5$ rad $\circlearrowleft$, (b) $\Delta\theta = 1.5$ rad $\circlearrowright$, (c) $\Delta\theta = 30°$ $\circlearrowright$.

17.3 Kinetic Energy of a Rigid Body in Plane Motion

For the present study, let Oxy be an inertial reference frame and $\mathbf{v}_i$ be the velocity, in Oxy, of the ith particle P_i of mass Δm_i in a system of n particles which form a rigid body of mass m and moves in the xy plane, as shown in

Fig. 17.5. We write

$$\mathbf{v}_i = \bar{\mathbf{v}} + \mathbf{v}_i' \tag{17.11}$$

where $\bar{\mathbf{v}}$ is the velocity of the mass center G of the system in Oxy, and $\mathbf{v}_i'$ is the velocity of P_i in the nonrotating reference frame $Gx'y'$ which translates with G. By Eq. (13.30), we may write the kinetic energy T of the system of particles forming the rigid body in plane motion as

$$T = \frac{1}{2}\,m\bar{v}^2 + \frac{1}{2}\sum_{i=1}^{n}(\Delta m_i)v_i'^2 \tag{17.12}$$

Note that the speed of the particle in $Gx'y'$ is $v_i' = r_i'\omega$, where r_i' is the distance between P_i and G, and ω is the angular speed of the rigid body in Oxy at the instant considered. We have

$$\sum_{i=1}^{n}(\Delta m_i)v_i'^2 = \sum_{i=1}^{n}(\Delta m_i)(r_i'\omega)^2 = \omega^2\sum_{i=1}^{n}r_i'^2\,(\Delta m_i)$$

$$= \omega^2\int r'^2\,dm = \omega^2\bar{I}$$

where $\bar{I}$ is the central moment of inertia of the body, taken about the perpendicular axis through G. Therefore, Eq. (17.12) may be written as

$$T = \tfrac{1}{2}\,m\bar{v}^2 + \tfrac{1}{2}\,\bar{I}\omega^2 \tag{17.13}$$

where the term $\tfrac{1}{2}\,m\bar{v}^2$ represents the *translational kinetic energy* of the rigid body translating with G, and the term $\tfrac{1}{2}\bar{I}\omega^2$ represents the *rotational kinetic energy* of the rigid body rotating about G.

For computing velocities and hence the kinetic energy at a given instant, a rigid body in *general plane motion* may conveniently be taken as rotating with an angular velocity $\boldsymbol{\omega}$ about its *velocity axis* (i.e., instantaneous axis of zero velocity) as a "fixed axis" at that instant. By definition, the velocity axis is perpendicular to the plane of motion and passes through the *velocity center C* of the rigid body as shown in Fig. 17.6,[†] where the speed v_i of the ith particle P_i a distance r_i from C is given by $v_i = r_i\omega$. Denoting the mass of the ith particle P_i as Δm_i and applying the definition of moment of inertia, we write the kinetic energy of the rigid body as

$$T = \frac{1}{2}\sum_{i=1}^{n}(\Delta m_i)(r_i\omega)^2 = \frac{1}{2}\,\omega^2\sum_{i=1}^{n}r_i^2\,(\Delta m_i)$$

$$= \frac{1}{2}\,\omega^2\int r^2\,dm = \frac{1}{2}\,\omega^2 I_C$$

$$T = \tfrac{1}{2}\,I_C\omega^2 \tag{17.14}$$

where I_C is the moment of inertia of the rigid body about its velocity axis at C at the instant considered. Of course, *Eq. (17.14) is still valid if C happens to be a fixed point about which the rigid body rotates.* Note that

$$I_C = \bar{I} + md^2 \tag{17.15}[††]$$

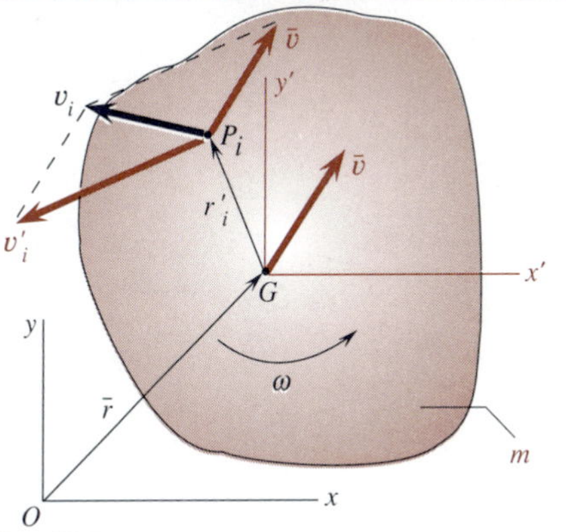

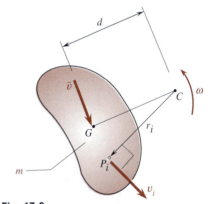

Fig. 17.5

Fig. 17.6

[†]Cf. Sec. 15.7.
[††]Cf. Sec. 16.3.

where d is the distance between G and $\overline{C}$ in Fig. 17.6. Substituting Eq. (17.15) into Eq. (17.14) and noting that $\overline{v} = \omega d$, we readily obtain Eq. (17.13). Clearly, Eq. (17.14) is equivalent to, but simpler in form than, Eq. (17.13).

Developmental Exercises

D17.6 Applying *Chasles' theorem* in Sec. 15.5 to the general plane motion of a rigid body with its *mass center G* as the point of reference, derive the formula given by Eq. (17.13).

D17.7 Suppose that the rigid body in Fig. 17.6 is nearly in translation; i.e., $\omega \approx 0$ and $d \approx \infty$. In this case, show that Eq. (17.14) yields $T \approx \frac{1}{2} m\overline{v}^2$.

17.4 Principle of Work and Energy

In Secs. 13.6 and 13.7, the principle of work and energy was first developed for a particle and then for a system of particles. Recalling that a rigid body may be regarded as a system of a large number of particles which are at fixed distances from each other, we see that Eq. (13.27) for a system of particles is also applicable to a rigid body. Thus, the *principle of work and energy* for a rigid body is also of the form

$$T_1 + U_{1\to2} = T_2 \qquad (17.16)$$

where $U_{1\to2}$ denotes the work done by all forces acting on the various particles of the rigid body during its displacement from the position 1 to the position 2, and T_1 and T_2 denote the total kinetic energies of the particles of the rigid body in the positions 1 and 2, respectively. Since the total work done by the internal forces holding together the various particles of a rigid body is zero, the term $U_{1\to2}$ in Eq. (17.16) may be taken as representing *the work done by the external forces* acting on the rigid body during the displacement considered. The kinetic energy terms T_1 and T_2 in Eq. (17.16) are to be computed for the rigid body in the system using Eq. (17.14) or (17.13).

When a problem involves a *system of rigid bodies*, each rigid body in the system may be considered separately, and Eq. (17.16) may be applied to each individual rigid body. However, *the principle of work and energy in the form of Eq. (17.16) is also applicable to the entire system of rigid bodies*, where T_1 and T_2 represent, respectively, the arithmetic sums of the kinetic energies of the rigid bodies forming the system in the positions 1 and 2, and $U_{1\to2}$ represents the work done by all the forces acting on the various rigid bodies, whether these forces are *internal* or *external* to the system as a whole. If the rigid bodies in a system are pin-connected, in mesh with one another as gears, or connected by cordlike elements, then the total work done by the internal forces is zero, and $U_{1\to2}$ reduces to the total work done by the *external forces* acting on the system.

The *principle of work and energy* for rigid bodies is particularly well adapted to the solution of kinetic problems of rigid bodies where *displacement* and *velocity* (among others) are to be directly related without initially involving acceleration and time.[†] An effective visual aid in computing the work done $U_{1 \to 2}$ in applying this principle is, of course, the appropriate *free-body diagrams*. Note that the procedure which uses the principle of work and energy to analyze and solve appropriate kinetic problems is often referred to as the *method of work and energy*.

EXAMPLE 17.1

The system shown is released from rest, where the 150-kg flywheel with a drum has a central radius of gyration $\bar{k}_O = 0.3$ m and the spring with a modulus $k = 1$ kN/m is unstretched. Determine the angular velocity of the flywheel when the end A of the cord is displaced by the amount $q = 0.6$ m $\downarrow$ due to the constant force $\mathbf{P} = 100$ N $\downarrow$.

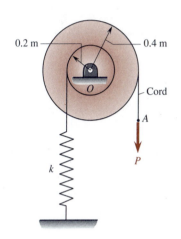

Solution. Since the problem directly relates velocity to displacement, it may be solved by applying the *principle of work and energy*. In the initial position, the kinetic energy is $T_1 = 0$. Let $\boldsymbol{\omega}$ be the angular velocity of the flywheel when the end A of the cord is displaced 0.6 m $\downarrow$. In this final position, the kinetic energy is $T_2 = \frac{1}{2} \bar{I} \omega^2 = \frac{1}{2} m \bar{k}^2 \omega^2$. During the motion, only the force $\mathbf{P}$ and the spring force $\mathbf{F}_s$ will do work because the point O remains unmoved. When the end A of the cord is displaced 0.6 m $\downarrow$, the corresponding angular displacement of the flywheel is

$$\Delta\theta = \frac{0.6}{0.4} \qquad \Delta\theta = 1.5 \text{ rad } \circlearrowleft$$

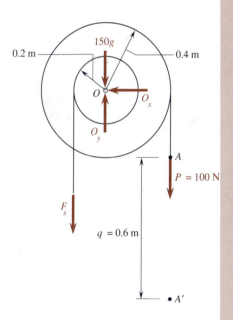

and the spring is stretched by the amount

$$x = r\,\Delta\theta = 0.2(1.5) \qquad x = 0.3 \text{ m}$$

Thus, the work done is $U_{1 \to 2} = Pq - \frac{1}{2}kx^2$. Applying $T_1 + U_{1 \to 2} = T_2$ to the flywheel and noting that $m = 150$ kg, $\bar{k}_O = 0.3$ m, $P = 100$ N, $q = 0.6$ m, $k = 1000$ N/m, and $x = 0.3$ m, we write

$$0 + [100(0.6) - \tfrac{1}{2}(1000)(0.3)^2] = \tfrac{1}{2}(150)(0.3)^2 \omega^2$$

$$\omega = 1.4907 \qquad \boldsymbol{\omega} = 1.491 \text{ rad/s } \circlearrowleft \quad \blacktriangleleft$$

[†]Note that the solutions of some problems may require the use of more than one principle.

Developmental Exercises

D17.8 Describe the *principle of work and energy* for (a) a rigid body, (b) a system of rigid bodies.

D17.9 Refer to Example 17.1. Determine (a) the maximum displacement of the end A of the cord due to the constant force $\mathbf{P} = 100$ N $\downarrow$, (b) the angular acceleration of the flywheel when the displacement of the end A of the cord is maximum. (*Hint*. A constant force is suddenly or instantaneously applied. The system will exhibit oscillatory rotation about O after it is released from rest.)

EXAMPLE 17.2

The mass center G of a 48.3-lb eccentric wheel of radius $r = 24$ in. is located at a distance $e = 7$ in. from its geometric center A as shown. As the wheel rolls without sliding, its angular velocity is observed to vary from $\omega_1 = 2$ rad/s $\circlearrowleft$ in the position 1 to $\omega_2 = 3.5$ rad/s $\circlearrowleft$ in the position 2. Determine the central radius of gyration $\bar{k}$ of the wheel.

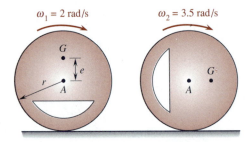

Solution. The central radius of gyration $\bar{k}$ is a parameter contained in the expression of the kinetic energy of the wheel. Thus, the *principle of work and energy* may be applied to solve for $\bar{k}$. In the position 1, the velocity center is at C_1 and

$$T_1 = \frac{1}{2} I_{C_1}\omega_1^2 = \frac{1}{2}(\bar{I} + md_1^2)\omega_1^2 = \frac{1}{2}(m\bar{k}^2 + md_1^2)\omega_1^2$$

$$= \frac{1}{2}m(\bar{k}^2 + d_1^2)\omega_1^2 = \frac{1}{2}\left(\frac{48.3}{32.2}\right)\left[\bar{k}^2 + \left(\frac{31}{12}\right)^2\right](2)^2$$

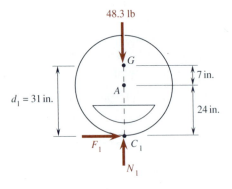

Position 1

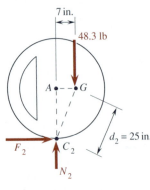

Position 2

In the position 2, the velocity center is at C_2 and

$$T_2 = \frac{1}{2} I_{C_2}\omega_2^2 = \frac{1}{2}(\bar{I} + md_2^2)\omega_2^2 = \frac{1}{2}(m\bar{k}^2 + md_2^2)\omega_2^2$$

$$T_2 = \frac{1}{2}m(\bar{k}^2 + d_2^2)\omega_2^2 = \frac{1}{2}\left(\frac{48.3}{32.2}\right)\left[\bar{k}^2 + \left(\frac{25}{12}\right)^2\right](3.5)^2$$

Recalling that *no work is done by any force which acts at a velocity center of any body*, we see that, during the motion, the friction and normal forces do no work to the wheel because they always act on the wheel at its velocity center. Thus, only the weight force does work. We have

$$U_{1\to2} = We = 48.3\left(\frac{7}{12}\right)$$

Substituting the above expressions into $T_1 + U_{1\to2} = T_2$, we write

$$\frac{1}{2}\left(\frac{48.3}{32.2}\right)\left[\bar{k}^2 + \left(\frac{31}{12}\right)^2\right](2)^2 + 48.3\left(\frac{7}{12}\right)$$

$$= \frac{1}{2}\left(\frac{48.3}{32.2}\right)\left[\bar{k}^2 + \left(\frac{25}{12}\right)^2\right](3.5)^2$$

$$\bar{k} = 1.15956 \text{ ft} \qquad\qquad \bar{k} = 13.91 \text{ in.} \blacktriangleleft$$

Developmental Exercises

D17.10 Work Example 17.2 if $\omega_2 = 3.4$ rad/s . Note the change in $\bar{k}$.

D17.11 Refer to Example 17.2. Show that the same values for the kinetic energies T_1 and T_2 can be obtained by using Eq. (17.13).

EXAMPLE 17.3

A 10-kg block B is suspended by a cord which is wrapped around the drum of a 100-kg flywheel A as shown. The flywheel is known to have a central radius of gyration $\bar{k}_O = 0.5$ m and the system is released from rest. Neglecting effects of friction, determine $\mathbf{v}_B$ of the block B after it has moved 2 m ↓.

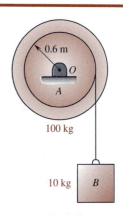

Solution. This problem directly relates displacement and velocity, but not acceleration and time. Thus, the *principle of work and energy* may be applied to obtain the solution. Let the *entire system* be considered and its *free-body diagram* as well as the positions 1 and 2 be sketched as shown. Since the system is released from rest, we have $T_1 = 0$. During the motion, the forces at O do no work; only the weight force of the block B does work. For the entire system, we write

$$U_{1\to2} = W_B q_B = 10(9.81)(2) \qquad U_{1\to2} = 196.2 \text{ J}$$

In the position 2, the block B translates with $\mathbf{v}_B = v_B$ ↓, while the flywheel rotates about the fixed axis at O with $\omega_A = (v_B/0.6)$ ↺. Therefore, for the

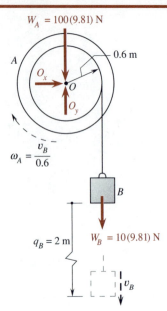

$W_A = 100(9.81)$ N

A

0.6 m

O_x

O

O_y

$\omega_A = \dfrac{v_B}{0.6}$

B

$q_B = 2$ m

$W_B = 10(9.81)$ N

v_B

entire system in this position, we write

$$T_2 = (T_2)_A + (T_2)_B = \tfrac{1}{2}(I_O)_A\,\omega_A^2 + \tfrac{1}{2}m_B v_B^2$$

$$= \tfrac{1}{2}\left[100(0.5)^2\right]\left(\frac{v_B}{0.6}\right)^2 + \tfrac{1}{2}(10)v_B^2 = 39.72v_B^2$$

Substituting the above expressions into $T_1 + U_{1\to 2} = T_2$, we write

$$0 + 196.2 = 39.72v_B^2 \qquad v_B = 2.22 \text{ m/s}$$

$$\mathbf{v}_B = 2.22 \text{ m/s} \downarrow \quad \blacktriangleleft$$

Alternative Solution.   This problem may also be solved by considering the *block B* and the *flywheel A* separately. Denoting the tension in the cord by F, we may draw their *free-body diagrams* and the positions 1 and 2 as

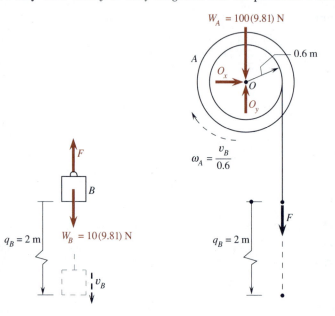

$W_A = 100(9.81)$ N

A

0.6 m

O_x

O

O_y

$\omega_A = \dfrac{v_B}{0.6}$

F

$q_B = 2$ m

F

B

$q_B = 2$ m

$W_B = 10(9.81)$ N

v_B

Block B   Flywheel A

shown. Applying $T_1 + U_{1\to 2} = T_2$ to the *block B*, we write

$$0 + [10(9.81)(2) + F(-2)] = \tfrac{1}{2}(10)v_B^2 \qquad (1)$$

Applying $T_1 + U_{1\to 2} = T_2$ to the *flywheel A*, we write

$$0 + F(2) = \frac{1}{2}\left[100(0.5)^2\right]\left(\frac{v_B}{0.6}\right)^2 \qquad (2)$$

Solving Eqs. (1) and (2), we obtain $v_B = 2.22$ m/s and $F = 85.8$ N. Thus,

$$\mathbf{v}_B = 2.22 \text{ m/s} \downarrow \quad \blacktriangleleft$$

Developmental Exercise

D17.12   Work Example 17.3 if $\bar{k}_O = 0.4$ m. Note the change in $\mathbf{v}_B$.

EXAMPLE 17.4

A 10-kg block B is suspended by a cord which passes around a frictionless pulley of negligible mass at D and is wrapped around the drum of a 100-kg wheel A as shown. The wheel with the drum has a central radius of gyration $\bar{k}_O = 0.5$ m and the system is released from rest. Knowing that the wheel rolls without slipping, determine $\mathbf{v}_B$ of the block B after it has moved 2 m $\downarrow$.

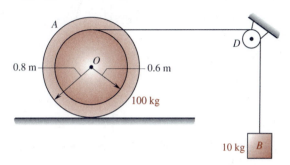

Solution. Clearly, the *principle of work and energy* may be applied to obtain the solution. Let the *entire system* be considered and its free-body diagram as well as the positions 1 and 2 be sketched as shown. Since the system is released from rest, we have $T_1 = 0$. During the motion, the *velocity center* of the wheel is located at the point of contact between the wheel and the horizontal support, the displacement of the mass center of the wheel is perpendicular to the weight force of the wheel, and the center of the pulley never moves. We recall that *no work is done by any force which acts at a stationary point or a velocity center of any body*. Thus, the weight force of the block B is the only force which does work. For the entire system, we write

$$U_{1 \to 2} = W_B q_B = 10(9.81)(2) \qquad U_{1 \to 2} = 196.2 \text{ J}$$

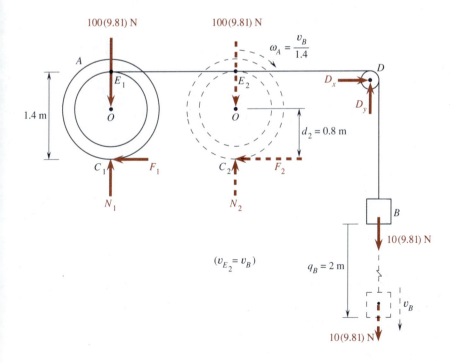

In the position 2, the block B translates with $\mathbf{v}_B = v_B \downarrow$, while the wheel rotates with $\boldsymbol{\omega}_A = (v_B/1.4)\,\curvearrowleft$ about its velocity center at C_2. For the entire system, we write

$$T_2 = (T_2)_A + (T_2)_B = \tfrac{1}{2}(I_{C_2})_A \omega_A^2 + \tfrac{1}{2} m_B v_B^2$$

$$= \tfrac{1}{2} m_A(\bar{k}_O^2 + d_2^2)\omega_A^2 + \tfrac{1}{2} m_B v_B^2$$

$$= \tfrac{1}{2}(100)\left[(0.5)^2 + (0.8)^2\right]\left(\frac{v_B}{1.4}\right)^2 + \tfrac{1}{2}(10)v_B^2$$

$$= 27.7 v_B^2$$

Substituting the above expressions into $T_1 + U_{1\to 2} = T_2$, we write

$$0 + 196.2 = 27.7 v_B^2 \qquad v_B = 2.66$$

$$\mathbf{v}_B = 2.66 \text{ m/s} \downarrow \quad \blacktriangleleft$$

Developmental Exercises

D17.13 Refer to Example 17.4. What is the displacement of the center of the wheel A when the block B has moved 2 m $\downarrow$?

D17.14 Refer to Example 17.4. Determine the tension in the cord by considering the block B and the wheel A separately.

EXAMPLE 17.5

A 15-lb gear B of central radius of gyration $\bar{k} = 4.5$ in. is in mesh with a stationary gear D as shown. The system is at rest when the 20-lb·ft torque is applied to the 10-lb bar OA. Determine $\boldsymbol{\omega}_{OA}$ of the bar OA after it is rotated to the horizontal position.

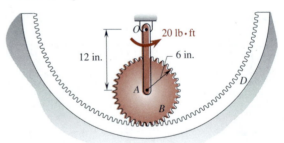

Solution.   We may apply the *principle of work and energy* to obtain the solution. Let the *entire system* be considered and its free-body diagram as well as the positions 1 and 2 be sketched as shown. Since the system starts from rest, we have $T_1 = 0$. During the motion, the *velocity center* of the bar OA is at O, while the *velocity center* of the gear B is at the point of contact between the gear B and the stationary gear D. Thus, we have[†]

[†]Recall that no work is done by any force acting at a velocity center.

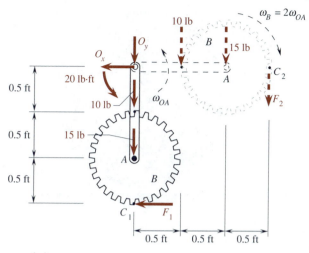

$$U_{1\to 2} = 20\left(\frac{\pi}{2}\right) + 10(-0.5) + 15(-1) \qquad U_{1\to 2} = 11.416 \text{ ft·lb}$$

In the position 2, the bar OA rotates with $\omega_{OA} = \omega_{OA} \curvearrowright$ and the speed of the point A is $v_A = \overline{OA}\,\omega_{OA} = 1\omega_{OA} = \overline{C_2A}\,\omega_B = 0.5\omega_B$. In other words, the gear B rotates with $\omega_B = 2\omega_{OA} \curvearrowleft$ about its velocity center C_2. We write

$$T_2 = (T_2)_{OA} + (T_2)_B = \frac{1}{2}(I_O)_{OA}\omega_{OA}^2 + \frac{1}{2}(I_{C_2})_B\omega_B^2$$

$$= \frac{1}{2}\left[\frac{1}{12}\left(\frac{10}{32.2}\right)(1)^2 + \frac{10}{32.2}(0.5)^2\right]\omega_{OA}^2$$

$$+ \frac{1}{2}\left(\frac{15}{32.2}\right)\left[\left(\frac{4.5}{12}\right)^2 + (0.5)^2\right](2\omega_{OA})^2$$

$$= 0.4157\omega_{OA}^2$$

Substituting the above expressions into $T_1 + U_{1\to 2} = T_2$, we write

$$0 + 11.416 = 0.4157\omega_{OA}^2 \qquad \omega_{OA} = 5.24$$

$$\omega_{OA} = 5.24 \text{ rad/s } \curvearrowright \quad \blacktriangleleft$$

Developmental Exercise

D17.15 Refer to Example 17.5. Determine ω_{OA} of the bar OA after it is rotated from the vertical position through 45°.

PROBLEMS

17.1 A bolt is tightened by applying a couple as shown. The wheel has a mass of 20 kg and a central radius of gyration of 280 mm. If the wheel is rotated by the couple through one-quarter revolution and friction is negligible, determine the angular velocity of the wheel.

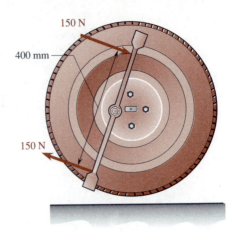

Fig. P17.1

17.2 and 17.3* The system shown is released from rest, where the 300-lb flywheel has a central radius of gyration $\bar{k}_O = 1.6$ ft. Determine the angular velocity of the flywheel after it has rotated 1 revolution.

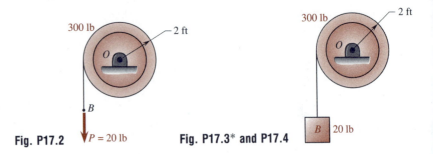

Fig. P17.2   **Fig. P17.3* and P17.4**

17.4   A hoisting engine has a flywheel, as shown, which weighs 300 lb and has a central radius of gyration $\bar{k}_O = 1.5$ ft. If the power is cut off when the flywheel is rotating at 10 rad/s $\circlearrowright$, determine the number of revolutions that the flywheel executes before stopping occurs.

17.5   A 30-kg gear of central radius of gyration $\bar{k}_O = 300$ mm is rolled up a rack from rest by a 200-N force **P** applied to a cord which is wrapped around the drum of the gear as shown, where $r_1 = 200$ mm and $r_2 = 400$ mm. Determine the speed of its center O after it has traveled 2 m up the rack.

Fig. P17.5 and P17.6*

17.6*   A 100-lb gear of central radius of gyration $\bar{k}_O = 0.75$ ft is rolled up a rack from rest by a constant force **P** applied to a cord which is wrapped around the drum of the gear as shown, where $r_1 = 0.5$ ft and $r_2 = 1$ ft. It is known that the angular velocity of the gear is 10 rad/s $\downarrow$ after its center O has traveled 4 ft up the rack. Determine the magnitude of **P**.

17.7*   A 60-g yo-yo of central radius of gyration $\bar{k}_O = 18$ mm is released from rest as shown. Assume that the string is wound around the central peg such that the mean radius at which it unravels is $r = 6$ mm. Determine the distance it must descend in order to attain an angular speed of 600 rpm.

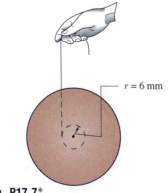

Fig. P17.7*

17.8   The driver and the jeep have a total mass of 750 kg, excluding the four wheels. Each wheel, with an all-terrain tire mounted on it, has a mass of 30 kg and a central radius of gyration of 0.24 m. The driver sets the jeep in neutral gear and lets it coast from rest down the slope for 50 m before applying the brakes. Neglecting the effects of friction and assuming that the wheels roll without slipping, determine the maximum speed attained by the jeep in km/h.

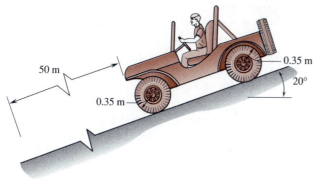

50 m

0.35 m

0.35 m

20°

Fig. P17.8

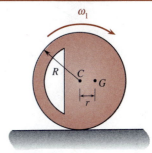

ω_1

R C G

r

Fig. P17.9

17.9 The mass center of a 16.1-lb wheel of radius R = 15 in. is at G as shown. The central radius of gyration of the wheel is $\bar{k}$ = 9 in. and r = 8 in. If the wheel rolls without slipping and has an angular velocity of ω_1 = 5 rad/s $\circlearrowright$ in the position shown, determine its angular velocity ω_2 in the next position where G is directly *below* its center C.

17.10* Refer to Prob. 17.9. Determine the angular velocity ω_2 of the wheel in the next position where G is directly *above* its center C.

17.11 A collar C of mass m_C is securely attached to a slender rod AB of mass m_{AB} as shown. The rod with the collar is released from rest to rotate downward from θ = 0 to θ = 180°. If m_{AB}/m_C = 4, determine the ratio r/L for which the angular velocity of the rod is maximum when θ = 180°.

17.12* A collar C of weight W_C is securely attached to a slender rod AB of weight W_{AB} as shown. The rod with the collar is released from rest to rotate downward from θ = 0 to θ = 180°. If W_{AB}/W_C = 4, r = 3 ft, and L = 4 ft, determine the angular velocity of the rod when θ = 180°.

17.13 Three identical rectangular plate glasses, each of mass 24 kg, are mounted on a vertical axle to form a three-door revolving door, which is initially at rest. A man passes through a door by pushing with a 50-N force **F** which is constantly directed perpendicular to the plane of the door as shown. If the angular velocity of the door is increased to 1.5 rad/s after the door is rotated through 90°, determine the resisting moment due to friction in the door.

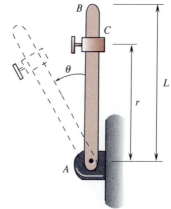

B

C

θ

L

r

A

Fig. P17.11 and P17.12*

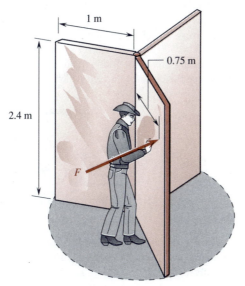

1 m

0.75 m

2.4 m

F

Fig. P17.13

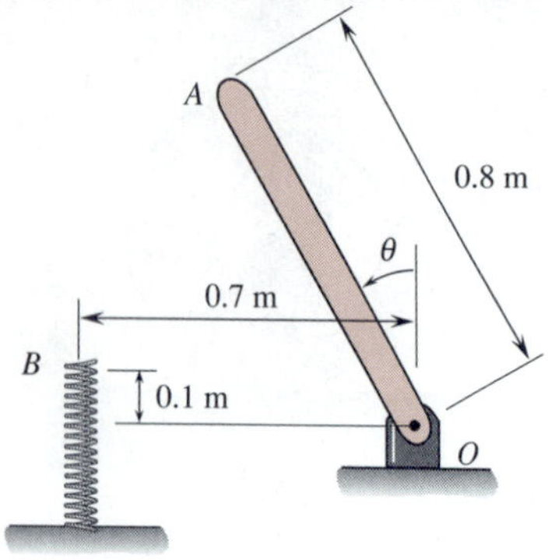

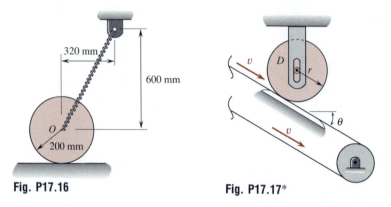

Fig. P17.14 and P17.15*

17.14   The 10-kg rod shown is given an angular velocity of 6 rad/s ↺ in the vertical position where $\theta = 0$. It hits the spring at B and is stopped after compressing it by 0.1 m. Determine the modulus of the spring.

17.15*   The 10-kg rod shown is given an angular velocity of 6 rad/s ↺ in the vertical position where $\theta = 0$. The spring at B has a modulus $k = 100$ kN/m. Determine the angular velocity of the rod as it passes through the position where $\theta = 90°$.

17.16   The disk shown has a mass of 20 kg and is attached to a spring which has a free length of 500 mm and a modulus of 600 N/m. If the disk is released from rest and rolls without slipping, determine the maximum speed attained by its center O.

Fig. P17.16

Fig. P17.17*

17.17*   The disk D of weight W and radius r is released from rest to press on the belt, which moves with a constant velocity $\mathbf{v}$ as shown. If the coefficient of kinetic friction between the disk and the belt is μ_k, determine the number of revolutions executed by the disk before its slipping on the belt ceases. (*Hint*. The answer is a function of r, v, μ_k, θ, and g.)

17.18   Solve Prob. 17.17* if $r = 0.1$ m, $v = 10$ m/s, $\mu_k = 0.2$, $\theta = 20°$, and $g = 9.81$ m/s².

17.19   Solve Prob. 17.17* if the direction of $\mathbf{v}$ is reversed.

17.20   The driver and the disabled tractor have a total weight of 1500 lb, excluding the four wheels. Each of the two front wheels has a radius $r = 10$ in., a weight of 30 lb, and a central radius of gyration of 7 in. The two rear wheels are much larger; each of them has a radius $R = 20$ in., a weight of 120 lb, and a central radius of gyration of 15 in. The driver sets the tractor in neutral gear and lets it be towed from rest by a constant force $\mathbf{P} = 100$ lb ← on a level surface until the tractor reaches a speed of 30 mph. Neglecting effects of friction and assuming that the wheels roll without slipping, determine the distance through which the tractor is towed.

Fig. P17.20

17.21   The 15-lb wheel A of central radius of gyration $\bar{k} = 5$ in. is released from rest to press on the disk B as shown, where B is driven by a motor to rotate at a constant angular velocity $\omega = 600$ rpm ↺. If $\mu_k = 0.2$ between A and B, determine the number of revolutions executed by the wheel before its slipping on the disk ceases.

17.22* Solve Prob. 17.21 if ω = 600 rpm $\downdownarrows$.

17.23 The 100-kg wheel shown has a central radius of gyration $\bar{k}_O$ = 150 mm and is initially rotating with ω = 300 rpm $\downdownarrows$ when the hydraulic cylinder *CD* is activated to exert a constant force $\mathbf{F}_{CD}$ on the lever *AC* for braking at *B*, where μ_k = 0.25. If the wheel is to stop in 3 revolutions, determine the magnitude of $\mathbf{F}_{CD}$.

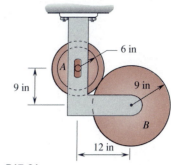

Fig. P17.21

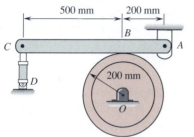

Fig. P17.23

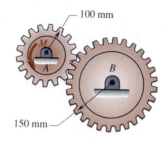

Fig. P17.24

17.24 The gears *A* and *B* shown have masses 5 kg and 12 kg, and central radii of gyration 80 mm and 120 mm, respectively. A constant torque $\mathbf{M}$ = 10 N·m $\downdownarrows$ is acting on the gear *A*. Determine the number of revolutions executed by the gear *A* during the interval in which the angular velocity of the gear *B* is increased from 300 rpm $\downdownarrows$ to 600 rpm $\downdownarrows$.

17.25* Solve Prob. 17.24 if the angular velocity of the gear *B* is increased from 150 rpm $\downdownarrows$ to 450 rpm $\downdownarrows$.

17.26 The flywheel at *A* weighs 40 lb and has a central radius of gyration of 1.2 ft. The system is released from rest in the position shown. After the 30-lb block *B* has moved 3 ft, the 15-lb collar *C* is lifted from the block *B* which continues to move. Determine the speed of the block *B* just before it strikes the bottom of the pit.

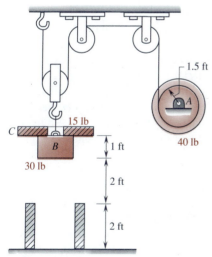

Fig. P17.26

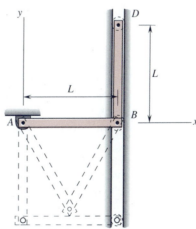

Fig. P17.27

17.27 Each of the two slender rods has a mass *m* and length *L*. If the system is released from rest in the position shown, determine the angular velocity of each rod as (a) the end *D* passes through the *x* axis, (b) the rod *AB* becomes vertical.

CONSERVATION OF ENERGY AND SPECIAL TOPIC

17.5   Conservation of Energy

We note that Eq. (13.44) developed in Sec. 13.12 is applicable to either a single particle or a system of particles. Along the same line of reasoning in Sec. 17.4, we readily see that Eq. (13.44) is also applicable to one or more rigid bodies. Thus, the *principle of conservation of energy* for any conservative system consisting of one or more rigid bodies is also of the form

$$T_1 + V_1 = T_2 + V_2 \qquad (17.17)$$

where T_1 and T_2 denote the total kinetic energies of the system in the positions 1 and 2, and V_1 and V_2 denote the total potential energies of the system in the positions 1 and 2, respectively. The sum of the kinetic energy T and the potential energy V of a rigid body is called the *total mechanical energy* of the rigid body. Naturally, Eq. (17.17) is a special case of Eq. (17.16) and is applicable to only *conservative systems*, where *displacement* and *velocity* (among others) are to be directly related. Note that bodies in conservative systems are subjected to only conservative forces and nonworking forces, if any.

The potential energies of a rigid body may include gravitational, elastic, and applied potential energies. For computing the gravitational potential energy of a rigid body, the weight force of the body is taken to act at the mass center of the body; i.e., the rigid body is taken as a particle located at the mass center of the body (cf. Sec. 13.9). For computing the elastic and applied potential energies of a rigid body, Eqs. (13.36), (13.38), (13.39), and (13.41) may appropriately be used (cf. Secs. 13.10 and 13.11).

EXAMPLE 17.6

Using the *principle of conservation of energy*, work Example 17.1.

Solution.   For convenience, the space diagram and the free-body diagram of the system are repeated as shown. Referring to Example 17.1, we note that the spring is unstretched in the initial position and the applied force $\mathbf{P} = 100$ N $\downarrow$ is a *constant force*. Therefore, the system is a *conservative system*; the initial position of the system may be chosen as the reference datum for the *gravitational*, the *elastic*, and the *applied* potential energies.[†] In the initial position, the flywheel is at rest; so, we have $T_1 = 0$ and $V_1 = (V_g)_1 + (V_e)_1 + (V_a)_1 = 0 + 0 + 0 = 0$. When the end A of the cord is displaced $\mathbf{q} = 0.6$ m $\downarrow$, the angular velocity of the flywheel is ω and the spring is stretched by the amount $x = 0.3$ m, as shown in the solution in Example 17.1. Thus, in the final position, we have $T_2 = \frac{1}{2}\bar{I}\omega^2 = \frac{1}{2}m\bar{k}_O\omega^2$ and $V_2 = (V_g)_2 + (V_e)_2 + (V_a)_2 = 0 + \frac{1}{2}kx^2 + (-Pq)$. Applying the *principle of conservation of energy*, $T_1 + V_1 = T_2 + V_2$, to the flywheel and noting that $m = 150$ kg, $\bar{k}_O = 0.3$ m, $P = 100$ N,

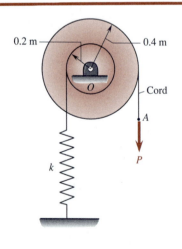

0.2 m — 0.4 m

O

Cord

A

P

k

[†]Cf. Secs. 13.9 through 13.11.

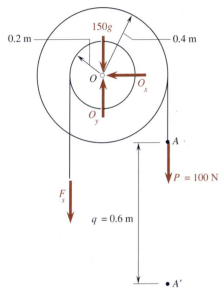

$q = 0.6$ m, $k = 1000$ N/m, and $x = 0.3$ m, we write

$$0 + 0 = \tfrac{1}{2}(150)(0.3)^2\omega^2$$

$$+ \left[0 + \tfrac{1}{2}(1000)(0.3)^2 - 100(0.6) \right]$$

$$\omega = 1.4907 \qquad \omega = 1.491 \text{ rad/s} \ \circlearrowright \ \blacktriangleleft$$

Note that the result obtained here is identical with that obtained in Example 17.1.

REMARK. We have here stressed that an *applied constant force* is a *conservative force*. As a matter of fact, any force **F** satisfying the condition $\nabla \times \mathbf{F} = \mathbf{0}$ (cf. Sec. 13.8) is a conservative force. This concept will help in setting the stage for more advanced studies.

Developmental Exercise

D17.16   Describe the *principle of conservation of energy* for a conservative system of rigid bodies.

EXAMPLE 17.7

A 5-lb disk A is connected to a 10-lb disk B by a cord. At the instant shown, the disk B is rotating at $(\omega_B)_1 = 15$ rad/s $\circlearrowright$. Determine how far the disk A will drop before the angular velocity of the disk B is increased to $(\omega_B)_2 = 30$ rad/s $\circlearrowright$.

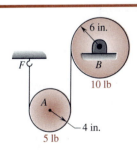

Solution.   This is a conservative system to which the *principle of conservation of energy* is applicable. Suppose that the disk A will drop a distance h and the reference datum is chosen as indicated in the sketch, where b is the initial distance between the two disks. As the disk A moves, it simply

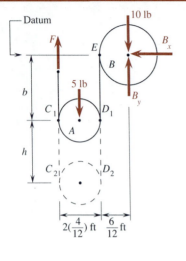

rolls without slipping down the left portion of the cord which serves as a stationary track; so, the velocity centers of the disk A in the positions 1 and 2 are at C_1 and C_2 as indicated. Furthermore, note that the extreme right point of the disk A and the extreme left point of the disk B must always have the same downward speed. We write

$$2r_A\omega_A = r_B\omega_B \qquad 2(\tfrac{4}{12})\omega_A = \tfrac{6}{12}\omega_B \qquad \omega_A = 0.75\omega_B$$

$$(\omega_B)_1 = 15 \text{ rad/s} \qquad (\omega_B)_2 = 30 \text{ rad/s}$$

$$(\omega_A)_1 = 0.75(15) \text{ rad/s} = 11.25 \text{ rad/s}$$

$$(\omega_A)_2 = 0.75(30) \text{ rad/s} = 22.5 \text{ rad/s}$$

In the position 1, the kinetic and potential energies are

$$T_1 = (T_1)_A + (T_1)_B = \frac{1}{2}(I_{C_1})_A(\omega_A)_1^2 + \frac{1}{2}\bar{I}_B(\omega_B)_1^2$$

$$= \frac{1}{2}\left[\frac{1}{2}\left(\frac{5}{32.2}\right)\left(\frac{4}{12}\right)^2 + \frac{5}{32.2}\left(\frac{4}{12}\right)^2\right](11.25)^2$$

$$+ \frac{1}{2}\left[\frac{1}{2}\left(\frac{10}{32.2}\right)\left(\frac{6}{12}\right)^2\right](15)^2$$

$$= 6.005 \text{ (ft·lb)}$$

$$V_1 = (V_1)_A + (V_1)_B = -5b + 0 = -5b$$

In the position 2, the kinetic and potential energies are

$$T_2 = (T_2)_A + (T_2)_B = \frac{1}{2}(I_{C_2})_A(\omega_A)_2^2 + \frac{1}{2}\bar{I}_B(\omega_B)_2^2$$

$$= \frac{1}{2}\left[\frac{1}{2}\left(\frac{5}{32.2}\right)\left(\frac{4}{12}\right)^2 + \frac{5}{32.2}\left(\frac{4}{12}\right)^2\right](22.5)^2$$

$$+ \frac{1}{2}\left[\frac{1}{2}\left(\frac{10}{32.2}\right)\left(\frac{6}{12}\right)^2\right](30)^2$$

$$= 24.020 \text{ (ft·lb)}$$

$$V_2 = (V_2)_A + (V_2)_B = -5(b + h) + 0 = -5(b + h)$$

Substituting the above expressions into $T_1 + V_1 = T_2 + V_2$, we write

$$6.005 - 5b = 24.020 - 5(b + h) \qquad h = 3.60 \text{ ft} \blacktriangleleft$$

Developmental Exercises

D17.17   Using the *principle of conservation of energy*, work Example 17.2.

D17.18   Using the *principle of conservation of energy*, work Example 17.3.

17.6 Power

The *power* input to a rigid body is defined as the sum of the time rates of doing work to the various particles of the rigid body at the instant considered. In the case of a force $\mathbf{F}$ acting on a rigid body, the power P input by $\mathbf{F}$ to the rigid body is

$$P = \mathbf{F} \cdot \mathbf{v} \qquad (17.18)$$

where $\mathbf{v}$ is the velocity *of the particle* (of the rigid body) acted on by $\mathbf{F}$ at the instant considered. In the case of a couple of moment $\mathbf{M}$ acting on a rigid body, the power P input by the couple to the rigid body is

$$P = \mathbf{M} \cdot \boldsymbol{\omega} \qquad (17.19)$$

where $\boldsymbol{\omega}$ is the angular velocity of the rigid body at the instant considered. In plane motion, we have $\mathbf{M} \parallel \boldsymbol{\omega}$. Thus, Eq. (17.19) gives

$$P = M\omega \qquad (17.20)$$

For units of power, refer to Sec. 13.2.

Developmental Exercises

D17.19 Refer to the system in Fig. 17.2. Assume that $N_B = 20$ N, $\mu_k = 0.3$, $r = 100$ mm, and the grinder is spinning at 600 rpm $\circlearrowright$. Determine the power input by (a) the friction force $\mathbf{F}$ at B to the grinder, (b) the reactive friction force $\mathbf{F}'$ at B to the rod.

D17.20 A steel shaft transmits 10 hp at 300 rpm. What is the torque transmitted by the shaft?

★17.7 Virtual Work in Kinetics: Force and Acceleration

We recall in Sec. 13.13 that the *principle of virtual work in kinetics* may be applied to solve a kinetic problem involving a one-degree-of-freedom system of particles where *force* and *acceleration* (among others) are to be directly related. Along the same line of reasoning in Sec. 17.4, it follows that *during a constrained virtual displacement of a set of connected rigid bodies with one degree of freedom, the total virtual work δU of all the external and internal force systems acting on the various rigid bodies in the set must be equal to the total virtual work δU_{eff} of all the effective force-moment systems on the various rigid bodies in the set*; i.e.,

$$\delta U = \delta U_{\text{eff}} \qquad (17.21)$$

If the set of rigid bodies under consideration is connected by *frictionless pins*, *inextensible cords*, or *links*, the total virtual work of the internal force system is zero, and δU in Eq. (17.21) reduces to the *total virtual work of the external force system only*. The property expressed in Eq. (17.21) is referred to as the *principle of virtual work in kinetics* for rigid bodies.[†] This principle complements the method of force and acceleration for rigid bodies in Sec. 16.7.

EXAMPLE 17.8

Using the *principle of virtual work in kinetics*, solve for α_{OA} in Example 16.18.

Solution.[†] For convenience, the space diagram of the system is repeated as shown, where each bar is of length $L = 5$ ft and weight $W = 10$ lb, and the system is released from rest. We are to determine α_{OA} of the bar OA immediately after release.

As explained in the solution in Example 16.18, we may, by symmetry, assume that $\alpha_{OA} = \alpha \circlearrowright$ and $\alpha_{AB} = \alpha \circlearrowleft$ and draw the vector-diagram equation consisting of the *free-body diagram* and the *effective-force diagram* for the entire system as shown. As the system is released and undergoes a *constrained virtual displacement*, the bar OA will have rotated about its acceleration center O to the position OA', while the bar AB will have rotated about its acceleration center Z to the position $A'B'$ as shown separately.[††] Since the triangles AOA' and AZA' are congruent, we see that the measures

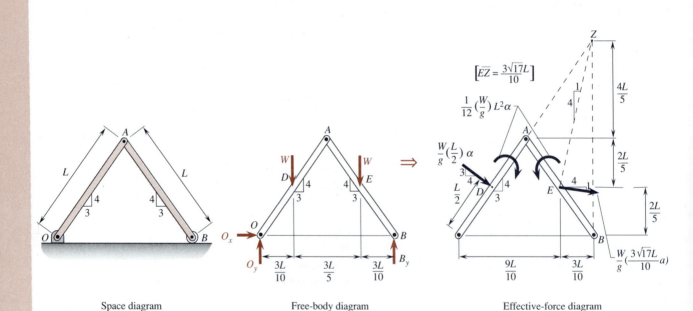

| Space diagram | Free-body diagram | Effective-force diagram |

[†]This principle and the solution in Example 17.8 were presented in the authors' paper, "Teaching the Concepts of Acceleration Center and Virtual Work in Dynamics," *Engineering Education*, vol. 79, no. 3, April 1989, pp. 441–443.

[††]The points O and Z are the displacement centers of the bars OA and OB, respectively. Cf. Sec. 10.6.

of the two infinitesimal rotations of *OA* and *AB* are equal in magnitude, which is denoted by $\delta\theta$. Since the set of rigid bodies is connected by frictionless pins, we see that δU in Eq. (17.21) is equal to the total virtual work of the external force system only. Referring to the diagrams shown and applying the *principle of virtual work in kinetics*, we write

$$\delta U: \; W\left(\frac{3L}{10}\,\delta\theta\right)(2) = \frac{1}{12}\left(\frac{W}{g}\right)L^2\alpha(\delta\theta)(2) + \frac{W}{g}\left(\frac{L}{2}\,\alpha\right)\left(\frac{L}{2}\,\delta\theta\right)$$

$$+ \frac{W}{g}\left(\frac{3\sqrt{17}\,L}{10}\,\alpha\right)\left(\frac{3\sqrt{17}\,L}{10}\,\delta\theta\right)$$

where the reaction components O_x, O_y, and B_y do no work, the weight forces do equal works, and the effective moments also do equal works. Solving for α, we write

$$\alpha = \frac{45g}{146L} = \frac{45(32.2)}{146(5)} \qquad \alpha_{OA} = 1.985 \text{ rad/s}^2 \; \text{↻} \quad \blacktriangleleft$$

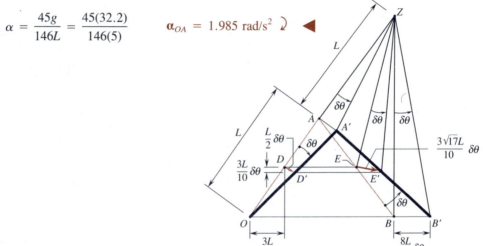

Constrained virtual displacement

REMARK. By comparing the solution in this example with that in Example 16.18, we readily note that, in order to find the magnitude α of the angular acceleration of the bar *OA* in the system, the *method of force and acceleration* requires the solution of at least *two* simultaneous equations, while the *principle of virtual work in kinetics* requires the solution of only *one* equation. The ability to *uncouple an unknown quantity* in some seemingly complex one-degree-of-freedom systems is, therefore, a salient feature of the principle of virtual work in kinetics. Clearly, the advantage of this principle resides in the fact that *the unknown constraining forces in the free-body diagram are nonworking forces during the constrained virtual displacement of the system.*

Developmental Exercise

D17.21   Describe the *principle of virtual work in kinetics* for rigid bodies.

EXAMPLE 17.9

Using the *principle of virtual work in kinetics*, solve for a_C in Example 16.19.

Solution.   For convenience, the space diagram of the system is repeated as shown, where a 40-kg block C is supported by a 20-kg cylinder A and a 30-kg cylinder B of the same radius $r = 100$ mm. We are to find a_C of the block C if a 50-N force **P** is applied to it and the cylinders roll without slipping.

As explained in the solution of Example 16.19, we may draw the vector-diagram equation consisting of the *free-body diagram* and the *effective-force diagram* for the entire system as shown. The constrained virtual displacement of the block C is $\overrightarrow{\delta x}_C = \delta x_C \leftarrow$, which is related to the constrained

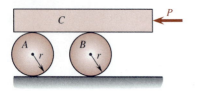

Space diagram

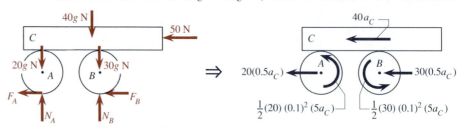

Free–body diagram   Effective–force diagram

angular virtual displacement $\overrightarrow{\delta\theta}$ of the cylinders A and B by $\delta x_C = 2r\,\delta\theta = 2(0.1)\,\delta\theta$ or $\delta\theta = 5\delta x_C$. The constrained virtual displacements of the centers of the cylinders are given by

$$\delta x_A = \delta x_B = r\delta\theta = 0.1(5\delta x_C) = 0.5\delta x_C$$

Thus, the *constrained virtual displacement* of the system may be indicated as shown. Referring to the diagrams shown and applying the *principle of virtual work in kinetics*, we write

$$\delta U: \quad 50\,\delta x_C = 40a_C(\delta x_C) + 20(0.5a_C)(0.5\,\delta x_C)$$

$$+\ 30(0.5a_C)(0.5\,\delta x_C)$$

$$+\ \tfrac{1}{2}\,(20)(0.1)^2(5a_C)(5\,\delta x_C)$$

$$+\ \tfrac{1}{2}\,(30)(0.1)^2(5a_C)(5\,\delta x_C)$$

where the weight forces and the reactions from the supports do no work during the constrained virtual displacement. Thus, we have

$$50 = 58.75a_C \qquad a_C = 0.851 \text{ m/s}^2 \leftarrow \quad \blacktriangleleft$$

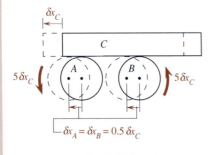

$\delta x_A = \delta x_B = 0.5\,\delta x_C$

Constrained
virtual displacement

REMARK. By comparing the solution in this example with that in Example 16.19, we readily note that, in order to find a_C of the block C in the system, the *method of force and acceleration* requires the solution of at least *three* simultaneous equations, while the *principle of virtual work in kinetics* requires the solution of only *one* equation. The ability to *uncouple an unknown quantity* in some seemingly complex one-degree-of-freedom systems is, again, demonstrated in this example by the principle of virtual work in kinetics.

Developmental Exercise

D17.22   Using the *principle of virtual work in kinetics*, solve for $\mathbf{a}_C$ in Example 16.19 if $r = 200$ mm. Note the change in $\mathbf{a}_C$.

PROBLEMS

17.28* through 17.45* Using the *principle of conservation of energy*, solve the following problems:

17.28* Prob. 17.1.[†]	17.29* Prob. 17.2.
17.30* Prob. 17.3*.	17.31* Prob. 17.4.
17.32* Prob. 17.5*.	17.33* Prob. 17.6.
17.34* Prob. 17.7*.	17.35* Prob. 17.8.
17.36* Prob. 17.9.	17.37* Prob. 17.10*.
17.38* Prob. 17.14.	17.39* Prob. 17.15*.
17.40* Prob. 17.16.	17.41* Prob. 17.20.
17.42* Prob. 17.24.	17.43* Frob. 17.25*.
17.44* Prob. 17.26.	17.45* Prob. 17.27.

17.46   A *cylinder* and a *sphere* of the same mass m and the same radius r are released from rest on an incline and roll without slipping to descend by $h = 1$ m in elevation as shown. Determine the velocity of the center O of each body after the descent.

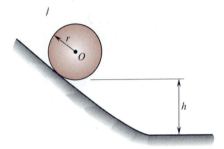

Fig. P17.46

17.47* Solve Prob. 17.46 if the bodies are replaced by a *pipe* and a *spherical shell*.

17.48   A 10-lb *cylinder* of radius 6 in. is released from rest in the position shown and rolls without slipping on the curved surface of radius 30 in. Determine (a) the speed of its center O when it reaches the bottom of the surface in terms of θ, (b) the value of θ for which the magnitude of the vertical reaction from the bottom of the surface is 20 lb.

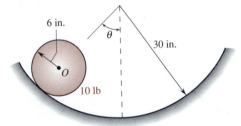

Fig. P17.48

17.49* Solve Prob. 17.48 if the cylinder is replaced by a *sphere* of the same radius and weight.

17.50   The 10-kg rod AB is released from rest when θ is essentially zero. If the modulus of the spring is $k = 1$ kN/m, determine $\mathbf{v}_A$ of the end A when (a) $\theta = 30°$, (b) $\theta = 70°$.

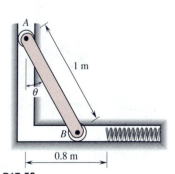

Fig. P17.50

[†]A couple of constant moment **M** is a conservative load. The corresponding potential energy is $V_a = -M\,\Delta\theta$, where $\Delta\theta$ is the angular displacement of the body.

17.51* The 10-kg rod AB is released from rest in the position shown, where the spring of modulus $k = 1$ kN/m has been compressed 0.4 m. Determine $\mathbf{v}_B$ of the end B when the rod passes through the vertical position.

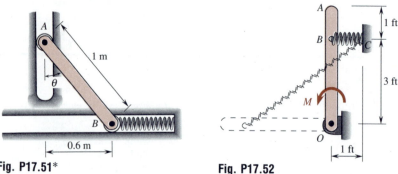

Fig. P17.51*   **Fig. P17.52**

17.52 A constant torque $\mathbf{M} = 10$ lb·ft $\circlearrowright$ is applied to the 20-lb rod OA which is initially at rest as shown. If the spring has a free length of 1 ft and a modulus $k = 2$ lb/ft, determine ω_{OA} as the rod OA passes through the horizontal position.[†]

17.53 The connecting rod OA weighs 10 lb and the gear C is stationary. The gear B weighs 20 lb and has a central radius of gyration of 6 in. If the system is released from rest in the position shown, determine ω_{OA} of the rod OA after it becomes vertical.

17.54* Solve Prob. 17.53 if the connecting rod OA and the gear B have the same weight of 20 lb.

17.55* The two slender bars OA and AB, each of mass 4 kg, are released from rest in the position shown, where $\theta = 60°$. If the modulus of the spring is $k = 400$ N/m, determine $\mathbf{v}_B$ of the end B as the bar OA passes through the position where (a) $\theta = 45°$, (b) $\theta = 30°$.

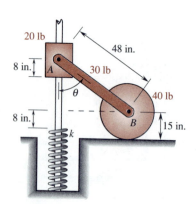

Fig. P17.53

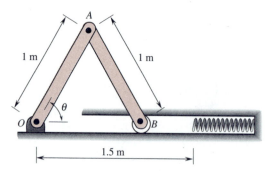

Fig. P17.55*

17.56 Solve Prob. 17.55* if the mass of each bar is 6 kg.

17.57 A 20-lb collar and a 40-lb disk are connected by a 30-lb rod AB as shown. It is known that the disk rolls without slipping and the modulus of the spring is 10 kips/ft. If the mechanism is released from rest when $\theta = 30°$, determine (a) the velocity of the collar when $\theta = 90°$, (b) the maximum deflection of the spring.

Fig. P17.57

[†]Cf. the footnote on p. 789.

17.58 A 40-kg sheave A of radius $r = 350$ mm and central radius of gyration $\bar{k} = 250$ mm carries a 160-kg load B and is suspended by a cable and a spring of modulus $k = 1.2$ kN/m. If the system is released from rest in the position shown, where the spring is stretched 100 mm, determine $\mathbf{v}_B$ of the load B after it has dropped 200 mm.

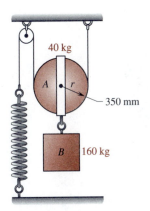

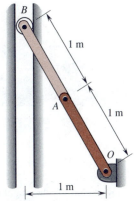

Fig. P17.58   **Fig. P17.59**

17.59 The two uniform rods OA and AB, each of mass 4 kg, are released from rest in the position shown. Determine $\mathbf{v}_B$ of the end B when (a) the rod OA is horizontal, (b) the ends B and O have the same elevation.

17.60 A 10-lb slender rod is released from rest when θ is slightly greater than zero. Determine the angle θ at which its bottom end A starts to lift off the ground.

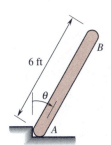

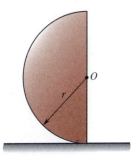

Fig. P17.60   **Fig. P17.61***

17.61* A *hemisphere* of mass m and radius r is released from rest in the position shown. If the hemisphere rolls without slipping, determine (a) its maximum angular speed, (b) the maximum reaction from the support.

17.62 Solve Prob. 17.61* if the hemisphere is replaced by a *semicircular disk* of mass m and radius r.

17.63* An oblong body of weight W is released from rest when θ is slightly greater than zero. Show that the angle θ at which its bottom end A starts to lift off the ground is given by the equation

$$2d^2 - (\bar{k}^2 + 3d^2) \cos\theta = 0$$

where $\bar{k}$ is the central radius of gyration of the body and d is the distance between its mass center G and its bottom end A.

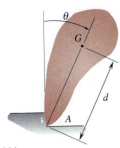

Fig. P17.63*

17.64   The 5-Mg dump of a dump truck is tilted at a constant rate of 5°/s about its hinge support at A by operating a hydraulic cylinder which acts at B as shown. If the gate is stuck during the tilting of the dump, determine for the hydraulic cylinder (a) the power required when $\theta = 30°$, (b) the maximum power required. (*Hint.* Does the solution depend on the values of h and L?)

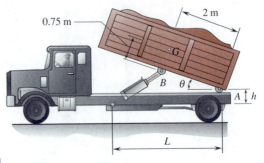

0.75 m   2 m   G   B θ $A \updownarrow h$ L

Fig. P17.64

17.65*   After power shutoff, the angular speed of the 1000-kg flywheel shown is reduced from 6000 rpm to 3000 rpm in a 2-min interval. If the central radius of gyration of the flywheel is 400 mm, determine the average power supplied by the flywheel to the system during that interval.

17.66   An electric motor is delivering 5 kW at 1800 rpm to the wheel shown. If the maximum tension in the belt is 160 N, determine the minimum value of μ_s between the belt and the drum of the wheel. (*Hint.* $T_2 = T_1 e^{\mu_s \beta}$.)

17.67   A 4-kg disk is supported by a cord which is connected to a spring as shown. The disk is released from the position where the spring is unstretched. If its center G attains a velocity of 1 m/s $\downarrow$ after descending 0.5 m, determine the modulus k of the spring.

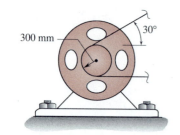

300 mm   30°

Fig. P17.65* and P17.66

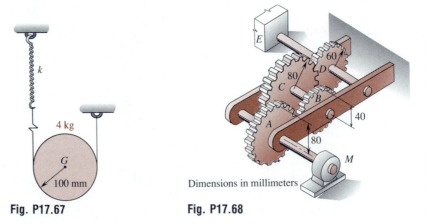

k   4 kg   G   100 mm

Fig. P17.67

E   60   80   D   C   B   A   40   80   M   Dimensions in millimeters

Fig. P17.68

17.68   A gear train is formed by four gears and three shafts as shown. If a power of 10 kW is steadily transmitted from the motor at M to the load at E and the motor run at 45 Hz, determine the magnitudes of the torques in the shafts MA, BC, and DE.

17.69   A section of *pipe* of radius r and mass m sits on a horizontal plate as shown. If the plate is pulled to the right from rest with a constant acceleration **a** and the pipe rolls on it without slipping, determine the total work done by the friction force to the pipe when the pipe has rotated one revolution. (*Hint.* Use the method of force and acceleration as well as the principle of work and energy.)

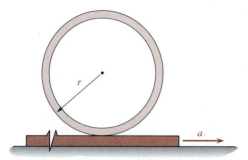

Fig. P17.69

17.70* Solve Prob. 17.69 if $\mathbf{a}$ = 0.5 m/s² → and the section of pipe is replaced by a 2-kg *solid cylinder* of radius 100 mm.

17.71* Solve Prob. 17.69 if $\mathbf{a}$ = 2 ft/s² → and the section of pipe is replaced by a 16.1-lb *sphere* of radius 6 in.

NOTE: Solve the following problems using the *principle of virtual work in kinetics* (cf. Sec. 17.7).

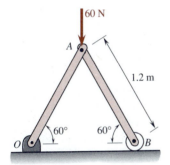

Fig. P17.72

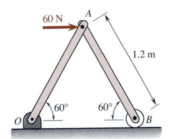

Fig. P17.73*

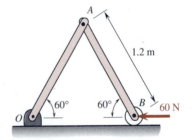

Fig. P17.74

17.72 through 17.74   Each of the two bars OA and AB has a mass of 4 kg. The system is released from rest in the position shown. Determine α_{AB} of the bar AB immediately after release.

17.75 and 17.76*   The 50-lb carriage D is supported by a 20-lb cylinder A, a 25-lb cylinder B, and a 30-lb cylinder C as shown, where the cylinders have the same radius of 4 in. If the cylinders roll without slipping, determine $\mathbf{a}_D$ of the carriage D when the 10-lb force acts on it.

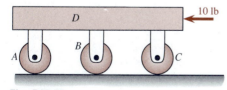

Fig. P17.75

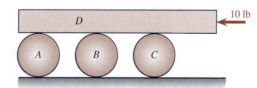

Fig. P17.76*

17.77* Solve Prob. 17.75 if all cylinders have the same weight of 20 lb.

17.78 Solve Prob. 17.76* if all cylinders have the same weight of 20 lb.

17.79 The 50-kg carriage E is supported by a 20-kg cylinder A, a 25-kg cylinder B, a 30-kg cylinder C, and a 35-kg cylinder D as shown, where the cylinders have the same radius of 100 mm. If the cylinders roll without slipping, determine $\mathbf{a}_E$ of the carriage E when the 20-N force acts on it.

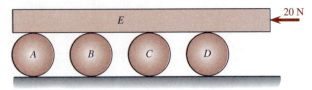

Fig. P17.79

17.80* Solve Prob. 17.79 if all cylinders have the same mass of 20 kg.

17.8 Concluding Remarks

This chapter is clearly a direct extension of Chap. 13. It covers three principles related to work and energy for studying the motion of rigid bodies. They are (1) the *principle of work and energy*, which is applicable to both conservative and nonconservative systems; (2) the *principle of conservation of energy*, which is applicable to conservative systems only; and (3) the *principle of virtual work in kinetics*, which complements the *method of force and acceleration* presented in Chap. 16.

We note that the principle of work and energy and the principle of conservation of energy are particularly well adapted to the study of the plane kinetics problems of rigid bodies where *velocity* and *displacement* (among others) are to be directly related without initially involving acceleration and time. On the other hand, the *principle of virtual work in kinetics* for rigid bodies is particularly well adapted to the study of the plane kinetics of systems of rigid bodies with one degree of freedom, where *force* and *acceleration* (among others) are to be directly related. As demonstrated in Examples 17.8 and 17.9, this principle provides a procedure which enables us to *uncouple an unknown quantity* in a seemingly complex one-degree-of-freedom system. Thus, the principle of virtual work in kinetics is an effective complement to the method of force and acceleration for one-degree-of-freedom systems of connected rigid bodies.

REVIEW PROBLEMS

17.81 The 20-lb rod shown is given an angular velocity of 3 rad/s ↻ in the vertical position where $\theta = 0$. Determine the angular velocity of the rod when (a) $\theta = 45°$, (b) $\theta = 90°$.

17.82* The ends of a 10-kg rod AB are constrained to move along the slots shown. Its end B is suspended by the spring BC which has a free length $L = 1$ m. The rod is released from rest when $\theta = 0$ and the maximum deflection of the end B is found to be $\delta = 0.5$ m. Determine the modulus k of the spring.

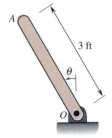

A

3 ft

θ

O

Fig. P17.81

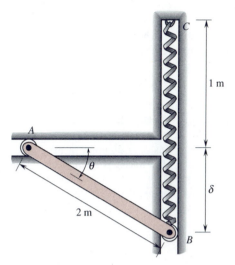

1 m

θ

2 m

δ

C

A

B

Fig. P17.82*

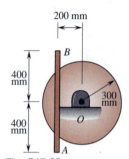

200 mm

B

400 mm

400 mm

O

300 mm

A

Fig. P17.83

17.83 A 2-kg rod AB is welded to a 4-kg disk as shown. If the system is released from rest, determine the maximum angular speed of the disk.

17.84 and 17.85* Four slender rods, each of length 500 mm and mass 2 kg, are welded together to form a square frame which is supported at its corner A as shown. If the frame is released from rest when θ is slightly less than 45°, determine $\mathbf{v}_B$ of its corner B when the rod AB becomes horizontal. (*Hint.* In Prob. 17.85*, $\overline{\mathbf{v}} = \overline{v} \downarrow$ when $\theta = 0$.)

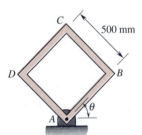

C

500 mm

D

B

θ

A

Fig. P17.84

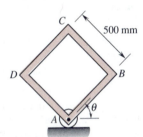

C

500 mm

D

B

θ

A

Fig. P17.85*

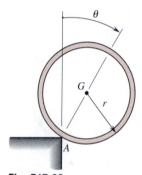

θ

G

r

A

Fig. P17.86

17.86 A section of *pipe* of mass m and radius r is placed at a sharp corner A and is released from rest when θ is slightly greater than zero. Assuming that no slipping occurs at A, determine the angle θ at which the pipe begins to leap off the corner A.

17.87 Solve Prob. 17.86 if the pipe is replaced by a *circular cylinder* of mass m and radius r.

17.88* Solve Prob. 17.86 if the pipe is replaced by a *sphere* of mass m and radius r.

17.89* An oblong body of weight W is placed at a sharp corner A and is released from rest when θ is slightly greater than zero. Assuming that no slipping occurs at A, show that the angle θ at which the body begins to leap off the corner A is given by

$$\cos\theta = \frac{2d^2}{\overline{k}^2 + 3d^2}$$

where $\overline{k}$ is the central radius of gyration of the body and d is the distance between its mass center G and its bottom end at A.

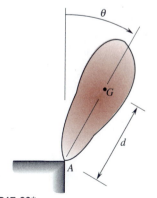

θ

$\cdot G$

d

A

Fig. P17.89*

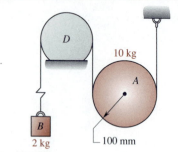

10 kg

D

A

B

2 kg

100 mm

Fig. P17.90

17.90 A 10-kg disk A is connected to a 2-kg block B by a cord which passes around a fixed drum D as shown, where $\mu_s = 0.25$ and $\mu_k = 0.2$ between the cord and the drum. If the system is released from rest, determine $\mathbf{v}_B$ of the block B after it has moved 1 m ↑.

17.91* Solve Prob. 17.90 if $\mu_k = 0.1$.

17.92 The driver and the disabled tractor have a total mass of 800 kg, excluding the four wheels. Each of the two front wheels has a radius $r = 250$ mm, a mass of 15 kg, and a central radius of gyration of 180 mm. The two rear wheels are much larger; each of them has a radius $R = 500$ mm, a mass of 60 kg, and a central radius of gyration of 380 mm. The tractor is set in neutral gear and is pulled from rest by a constant force $\mathbf{P} = 600$ N ←. Neglecting effects of friction and assuming that the wheels roll without slipping, determine the velocity of the tractor when it has been moved 10 m ←.

P

R

r

Fig. P17.92 and 17.93*

17.93* Refer to Prob. 17.92. Using the *principle of virtual work in kinetics*, determine the acceleration of the tractor when it is being pulled by the constant force $\mathbf{P} = 600$ N ←.

17.94* Each of the bars OA and AB weighs 20 lb and the disk at B weighs 30 lb. The system shown is released from rest in the position where $\theta = 60°$. If the disk rolls without slipping, determine ω_{OA} when $\theta = 30°$.

17.95 Each of the bars OA and AB weighs 20 lb and the disk at B weighs 30 lb. The system shown is released from rest in the position where $\theta = 60°$. The disk is known to roll without slipping. Using the *principle of virtual work in kinetics*, determine the angular acceleration of the bar OA immediately after release.

17.96* through **17.100*** Using the *principle of virtual work in kinetics*, solve the following problems:

17.96* Prob. 16.118.   17.97* Prob. 16.119*.
17.98* Prob. 16.132.   17.99* Prob. 16.133.
17.100* Prob. 16.137.

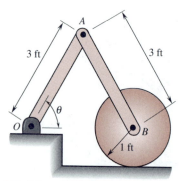

A

3 ft

3 ft

θ

O

B

1 ft

Fig. P17.94* and P17.95

Plane Kinetics of Rigid Bodies: Impulse and Momentum

Chapter 18

KEY CONCEPTS

- Linear impulse and angular impulse about a fixed point of a force acting on a rigid body, angular impulse of a couple, linear momentum and angular momentum of a rigid body in plane motion, and coefficient of restitution for eccentric impact.
- Principle of impulse and momentum for rigid bodies.
- Conservation of momentum for rigid bodies.
- Eccentric impact between rigid bodies.
- Principle of generalized virtual work for studying impulses and momenta in one-degree-of-freedom systems of rigid bodies.

It was shown in Chap. 14 that the principle of impulse and momentum was useful in studies involving velocity changes of particles resulting from the cumulative effect of forces which act on them over a specific interval of time. By this principle, it is possible to determine the change in velocity by analyzing the impulses exerted on the system and the change in momentum of the system, without first determining the acceleration and then integrating it over the time interval of motion. Such advantages still hold true when we extend the principle of impulse and momentum to rigid bodies in this chapter. Furthermore, the principle of generalized virtual work for particles in Sec. 14.10 is extended to rigid bodies in Sec. 18.7. It is shown in Examples 18.10 and 18.11 that the principle of generalized virtual work can effectively complement the principle of impulse and momentum for systems of rigid bodies with one degree of freedom.

IMPULSE AND MOMENTUM

18.1 Linear and Angular Impulses on a Rigid Body

The *linear impulse* $\mathbf{Imp}_{1\to2}$ of a force $\mathbf{F}$ acting on a rigid body during the time interval $t_1 \leq t \leq t_2$ is defined as the sum of the linear impulses exerted by **F** on the various particles of the rigid body during that interval.[†] We write

$$\mathbf{Imp}_{1\to2} = \int_{t_1}^{t_2} \mathbf{F}\, dt \tag{18.1}$$

The *linear impulse* of a force is often simply referred to as the *impulse* of the force. If the force **F** remains *constant* during the interval $\Delta t = t_2 - t_1$, we readily obtain

$$\mathbf{Imp}_{1\to2} = \mathbf{F}\,\Delta t \tag{18.2}$$

Similarly, the *angular impulse* $(\mathbf{Aimp}_A)_{1\to2}$ of a force **F** acting on a rigid body about a fixed point A during the time interval $t_1 \leq t \leq t_2$ is defined as the sum of the angular impulses exerted by the moment $\mathbf{M}_A$ of **F** about the point A during that interval. We write

$$(\mathbf{Aimp}_A)_{1\to2} = \int_{t_1}^{t_2} \mathbf{M}_A\, dt \tag{18.3}$$

If $\mathbf{M}_A$ remains *constant* during the interval $\Delta t = t_2 - t_1$, we get

$$(\mathbf{Aimp}_A)_{1\to2} = \mathbf{M}_A\,\Delta t \tag{18.4}$$

Note that the moment of a couple acting on a rigid body about a point is a free vector. The *angular impulse* $(\mathbf{Aimp}_A)_{1\to2}$ of a *couple* acting on a rigid body about a fixed point A during the time interval $t_1 \leq t \leq t_2$ is the same as that about any other point during that interval and is defined as a free vector equal to the integral of the torque **M** of the couple times the differential time dt, evaluated over that interval. We write

$$(\mathbf{Aimp}_A)_{1\to2} = \int_{t_1}^{t_2} \mathbf{M}\, dt \tag{18.5}$$

If **M** remains *constant* during the interval $\Delta t = t_2 - t_1$, we obtain

$$(\mathbf{Aimp}_A)_{1\to2} = \mathbf{M}\,\Delta t \tag{18.6}$$

Note that the angular impulse of a couple about an axis (instead of a point) is dependent on the direction of the axis.

Developmental Exercises

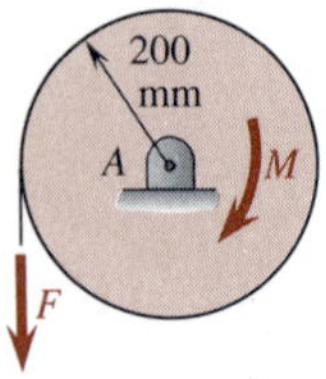

Fig. D18.1

D18.1 A 300-N force **F** pulls a cord which is wrapped around a disk mounted at A and, at the same time, a moment **M** of 50 N·m is applied to the disk as shown. If they act for an interval of 2 s, determine (a) the linear impulse of **F**, (b) the angular impulse of **F** about A, (c) the angular impulse of **M** about A.

D18.2 Do D18.1 if $F = 300t$ N and $M = 50t$ N·m, where t is the time in seconds.

D18.3 Define the *linear impulse* and the *angular impulse about a fixed point* of a force acting on a rigid body.

[†]Cf. the footnote on p. 595.

18.2 Momentum of a Rigid Body in Plane Motion

The vector sum of the momenta of the various particles of a rigid body is called the *resultant momentum of the rigid body*, which does not necessarily pass through the mass center G of the rigid body. However, by *Chasles' theorem* in Sec. 15.5, we may treat the *system of momenta* of the particles of a rigid body in plane motion as being equal to the sum of (1) the *translational system of momenta* corresponding to the translation of the rigid body with G whose velocity is $\bar{\mathbf{v}}$ in an inertial reference frame Oxy, and (2) the *rotational system of momenta* corresponding to the rotation of the rigid body about G with an angular velocity $\boldsymbol{\omega}$ in Oxy. Such a relation is depicted in Fig. 18.1, where $Gx'y'$ is a nonrotating reference frame attached to G, and r_i' is the radial distance from G to the ith particle P_i whose mass is Δm_i and whose velocity $\mathbf{v}_i$ in Oxy is given by

$$\mathbf{v}_i = \bar{\mathbf{v}} + (r_i'\omega)\mathbf{e}_t \qquad (18.7)$$

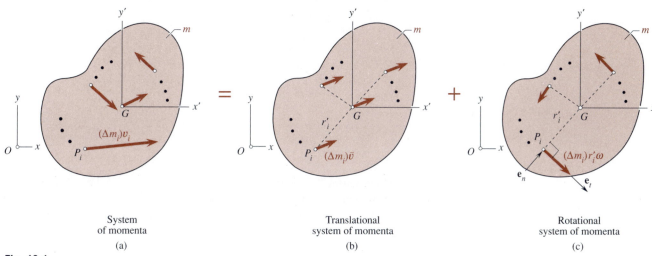

| System of momenta (a) | Translational system of momenta (b) | Rotational system of momenta (c) |

Fig. 18.1

Note that the vectors $\mathbf{e}_t$ and $\mathbf{e}_n$ in Eq. (18.7) and Fig. 18.1 are, respectively, the tangential and the normal unit vectors associated with the path of P_i rotating about G, and the total mass of the rigid body is m.

Letting $\mathbf{L}_{\text{trans}}$ be the *resultant linear momentum* of the *translational* system of momenta in Fig. 18.1(b), we write

$$\mathbf{L}_{\text{trans}} = \sum_{i=1}^{n} (\Delta m_i)\bar{\mathbf{v}} = \bar{\mathbf{v}} \sum_{i=1}^{n} (\Delta m_i) = \bar{\mathbf{v}}m$$

$$\mathbf{L}_{\text{trans}} = m\bar{\mathbf{v}} \qquad (18.8)$$

Letting $(\mathbf{H}_G)_{\text{trans}}$ be the *resultant angular momentum* about G of the *translational* system of momenta in Fig. 18.1(b) and applying the *principle of moments*, we write

$$(\mathbf{H}_G)_{\text{trans}} = \sum_{i=1}^{n} \mathbf{r}_i' \times (\Delta m_i)\bar{\mathbf{v}} = -\bar{\mathbf{v}} \times \sum_{i=1}^{n} \mathbf{r}_i' (\Delta m_i)$$

$$= -\bar{\mathbf{v}} \times \bar{\mathbf{r}}' m$$

where $\bar{\mathbf{r}}'$ is the position vector of G in $Gx'y'$. Clearly, $\bar{\mathbf{r}}' = \mathbf{0}$. Thus, we have

$$(\mathbf{H}_G)_{\text{trans}} = \mathbf{0} \tag{18.9}$$

From Eqs. (18.8) and (18.9), we find that *the translational system of momenta* in Fig. 18.1(b) *is equivalent to the resultant linear momentum* $\mathbf{L}_{\text{trans}}$ *acting through G* as illustrated in Fig. 18.2.

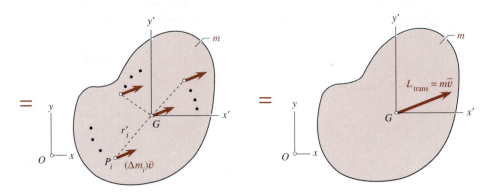

Fig. 18.2

Translational
system of momenta

(a)

Resultant of the
translational system

(b)

Now, letting $\mathbf{L}_{\text{rot}}$ be the *resultant linear momentum* of the *rotational* system of momenta in Fig. 18.1(c) and applying Eq. (18.7) as well as the *principle of moments*, we write

$$\mathbf{L}_{\text{rot}} = \sum_{i=1}^{n} (\Delta m_i)(r_i'\omega)\mathbf{e}_t = \sum_{i=1}^{n} (\Delta m_i)(\mathbf{v}_i - \bar{\mathbf{v}})$$

$$= \sum_{i=1}^{n} (\Delta m_i)\mathbf{v}_i - \bar{\mathbf{v}} \sum_{i=1}^{n} (\Delta m_i) = m\bar{\mathbf{v}} - \bar{\mathbf{v}}m$$

$$\mathbf{L}_{\text{rot}} = \mathbf{0} \tag{18.10}$$

Letting $(\mathbf{H}_G)_{\text{rot}}$ be the *resultant angular momentum* about G of the *rotational* system of momenta in Fig. 18.1(c) and noting that $\mathbf{e}_n \times \mathbf{e}_t = -\mathbf{k}$, we write

$$(\mathbf{H}_G)_{\text{rot}} = \sum_{i=1}^{n} (-r_i'\mathbf{e}_n) \times (\Delta m_i)(r_i'\omega)\mathbf{e}_t$$

$$= -\omega(\mathbf{e}_n \times \mathbf{e}_t) \sum_{i=1}^{n} r_i'^2 (\Delta m_i)$$

$$= (\omega\mathbf{k}) \int r'^2 \, dm = \omega\bar{I}$$

$$(\mathbf{H}_G)_{\text{rot}} = \bar{I}\boldsymbol{\omega} \tag{18.11}$$

where $\bar{I}$ is the moment of inertia of the rigid body about the central axis perpendicular to the plane of motion. From Eqs. (18.10) and (18.11), we find that the *rotational system of momenta* in Fig. 18.1(c) *is equivalent to the resultant angular momentum* $(\mathbf{H}_G)_{\text{rot}}$ *acting about G* as illustrated in Fig. 18.3.

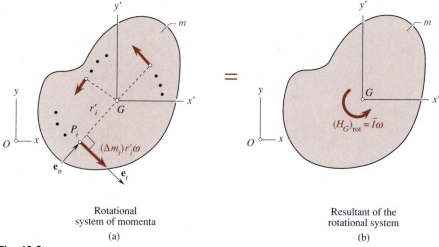

$=$

Rotational
system of momenta

(a)

Resultant of the
rotational system

(b)

Fig. 18.3

By *Chasles' theorem* and Eqs. (18.8) and (18.10), we conclude that the *resultant linear momentum* acts through G and is given by

$$\mathbf{L} = \mathbf{L}_{\text{trans}} + \mathbf{L}_{\text{rot}} = m\bar{\mathbf{v}} + \mathbf{0}$$

$$\mathbf{L} = m\bar{\mathbf{v}} \tag{18.12}$$

Moreover, by *Chasles' theorem* and Eqs. (18.9) and (18.11), we conclude that the coexisting *resultant angular momentum* acting about G is given by

$$\mathbf{H}_G = (\mathbf{H}_G)_{\text{trans}} + (\mathbf{H}_G)_{\text{rot}} = \mathbf{0} + \bar{I}\boldsymbol{\omega}$$

$$\mathbf{H}_G = \bar{I}\boldsymbol{\omega} \tag{18.13}$$

We see from Eqs. (18.12) and (18.13) that *the system of momenta of the particles of a rigid body in plane motion is equivalent to the resultant system of momenta consisting of the resultant linear momentum* $m\bar{\mathbf{v}}$ *acting through the mass center G and the resultant angular momentum* $\bar{I}\boldsymbol{\omega}$ *acting about G.* This important relation is illustrated in Fig. 18.4. Of course, $m\bar{\mathbf{v}}$ and $\bar{I}\boldsymbol{\omega}$ at G may be replaced by a *single equivalent linear momentum* $m\bar{\mathbf{v}}$ acting through an appropriate point.

In the case of a rigid body rotating in a plane about a fixed axis at O, the *resultant system of momenta* in Fig. 18.4(b) can be represented as shown in Fig. 18.5(a), where $m\bar{\mathbf{v}} \perp \overrightarrow{OG}$. Using the concept of equivalent systems, we can replace the system in Fig. 18.5(a) with an *equivalent single linear momentum* $m\bar{\mathbf{v}}$ acting through the point P as shown in Fig. 18.5(b). The distance d in this figure can readily be shown to be given by

$$d = \frac{\bar{k}^2}{\bar{r}} \tag{18.14}$$

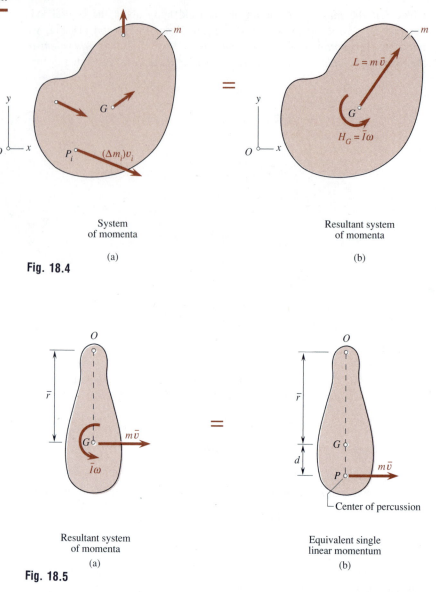

Fig. 18.4

System of momenta

(a)

Resultant system of momenta

(b)

Fig. 18.5

Resultant system of momenta

(a)

Equivalent single linear momentum

(b)

Center of percussion

where $\bar{k}$ is the central radius of gyration of the rigid body. Referring to Figs. 16.14 and 18.5 and noting that Eqs. (16.17) and (18.14) are identical, we see that the point P in Fig. 18.5(b) is coincident with the *center of percussion* of a rigid body rotating about a fixed pivot at O, as described in Sec. 16.6. Furthermore, it is well to note that $m\bar{\mathbf{v}}$ is a vector measured in units of kg·m/s or slug·ft/s, while $\bar{I}\omega$ is a vector measured in units of kg·m²/s or slug·ft²/s.[†]

[†]The resultant angular momentum $\bar{I}\omega$ has been called a *momentum couple* or *rotational momentum* by others. (Cf. the footnote on p. 90.)

Developmental Exercises

D18.4 A 10-kg disk of radius $r = 200$ mm rotates with $\boldsymbol{\omega} = 300$ rpm $\circlearrowright$ about a fixed axle through G as shown. Determine its (a) resultant linear momentum $\mathbf{L}$, (b) resultant angular momentum $\mathbf{H}_G$.

D18.5 The mass center G of a 20-lb cylinder of radius $r = 9$ in. moves with $\overline{\mathbf{v}} = 4$ ft/s $\rightarrow$. If the cylinder rolls without slipping, determine its (a) resultant linear momentum $\mathbf{L}$, (b) resultant angular momentum $\mathbf{H}_G$.

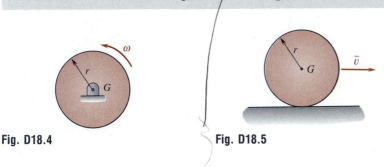

Fig. D18.4 **Fig. D18.5**

18.3 Principle of Impulse and Momentum

In Secs. 14.2 and 14.4, the principle of impulse and momentum was first developed for a particle and then for a system of particles. As a rigid body may be regarded as a system of a large number of particles which are at fixed distances from each other, we readily see that Eq. (14.12) for a system of particles is also applicable to a rigid body. Using the result derived in Sec. 18.2 and expressed in Fig. 18.4, we may state and illustrate the *principle of impulse and momentum* for a rigid body as follows:

The *resultant system of momenta* of a rigid body at the time t_1	+ The *linear and angular impulses* exerted on the rigid body from t_1 to t_2	= The *resultant system of momenta* of the rigid body at the time t_2	(18.15)

Note in Eq. (18.15) that the *resultant system of momenta* of a rigid body consists of the *resultant linear momentum* $m\overline{\mathbf{v}}$ acting through the mass center G and the *resultant angular momentum* $\overline{I}\boldsymbol{\omega}$ acting about G, as shown on the next page in Fig. 18.6. The *principle of impulse and momentum* for a rigid body, as expressed in Eq. (18.15) and illustrated in Fig. 18.6, is particularly well adapted to the solution of plane kinetic problems of rigid bodies where *time* and *velocity* (among others) are to be directly related without initially involving acceleration and displacement.

Developmental Exercise

D18.6 Describe the *principle of impulse and momentum* for a rigid body.

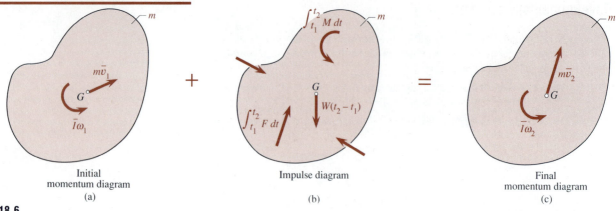

Fig. 18.6

Initial momentum diagram (a)   Impulse diagram (b)   Final momentum diagram (c)

EXAMPLE 18.1

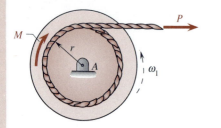

A flywheel with a drum of radius $r = 150$ mm has a mass of 10 kg and is rotating with $\omega_1 = 40$ rad/s ↻ at the time $t = 0$ when it is acted on by a moment $\mathbf{M} = 3$ N·m ↻ and a variable force $\mathbf{P} = 30t$ N → applied at the end of a cord which is wrapped around the drum as shown, where t is measured in seconds. If the central radius of gyration is $\bar{k}_A = 160$ mm, determine the angular velocity ω_2 of the flywheel at the time $t = 2$ s.

Solution. For this flywheel, we have $\mathbf{L} = m\bar{\mathbf{v}} = \mathbf{0}$ and $\mathbf{H}_G = \bar{I}\boldsymbol{\omega} = m\bar{k}_A^2\boldsymbol{\omega}$. Thus, we apply the *principle of impulse and momentum* to draw a vector-diagram equation involving momentum and impulse diagrams of the system as shown, where the impulses of $\mathbf{P}$ and the reaction component $\mathbf{A}_x$ are represented by *integrals* because both $\mathbf{P}$ and $\mathbf{A}_x$ are variable during $0 \leq t \leq 2$ s. We have

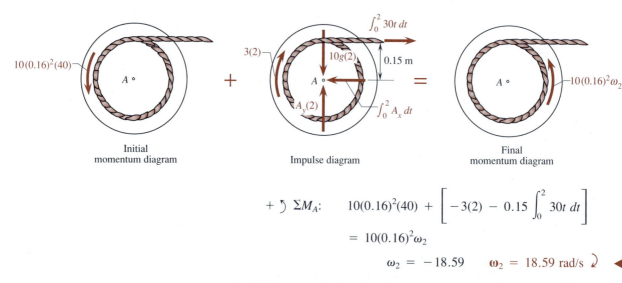

Initial momentum diagram   Impulse diagram   Final momentum diagram

$$+\circlearrowleft \ \Sigma M_A: \quad 10(0.16)^2(40) + \left[-3(2) - 0.15 \int_0^2 30t \, dt \right]$$

$$= 10(0.16)^2\omega_2$$

$$\omega_2 = -18.59 \qquad \boldsymbol{\omega}_2 = 18.59 \text{ rad/s} \ \circlearrowright \quad \blacktriangleleft$$

REMARK. The actual direction of $\boldsymbol{\omega}_2$ is opposite to that assumed in the final momentum diagram because ω_2 is found to be negative in the solution.

Developmental Exercise

D18.7   Refer to Example 18.1. If the wheel comes to a stop at $t = \tau$, determine τ.

EXAMPLE 18.2

A wheel of radius $r = 500$ mm and mass $m = 100$ kg has a pure rotation with an angular velocity $\omega_1 = 50$ rad/s $\circlearrowright$ at the time $t = 0$ when it is placed on an incline as shown, where $\theta = 15°$ and $\mu_k = 0.3$ between the wheel and the incline. If the central radius of gyration of the wheel is $\bar{k} = 400$ mm, determine (a) the time t_2 at which the wheel will start rolling without slipping, (b) the angular speed ω_2 of the wheel at the time t_2.

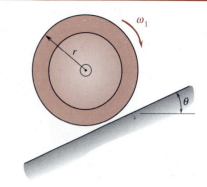

Solution.   We need to consider the motion of the wheel from the time $t = t_1 = 0$ when it is placed on the incline to the time $t = t_2$ when it starts rolling without slipping. Since *time* and *velocity* are to be directly related, we apply the *principle of impulse and momentum* to draw a vector-diagram equation for the system as shown, where **N** is the normal force and $\mathbf{F}_k = \mu_k N$ ↗ is the kinetic friction force.

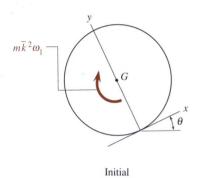

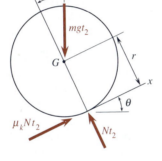

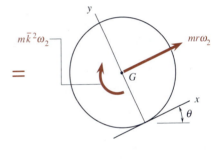

| Initial momentum diagram | Impulse diagram | Final momentum diagram |

Referring to the vector-diagram equation, we write

$$+\nwarrow \Sigma V_y: \qquad 0 + (Nt_2 - mgt_2 \cos\theta) = 0 \qquad (1)$$

$$+\nearrow \Sigma V_x: \qquad 0 + (\mu_k Nt_2 - mgt_2 \sin\theta) = mr\omega_2 \qquad (2)$$

$$+\circlearrowright \Sigma M_G: \qquad m\bar{k}^2\omega_1 - r\mu_k Nt_2 = m\bar{k}^2\omega_2 \qquad (3)$$

From the given data and equations above, we find that $N = mg \cos\theta$ and

$$t_2 = \frac{\bar{k}^2 r\omega_1}{g[\mu_k(r^2 + \bar{k}^2)\cos\theta - \bar{k}^2 \sin\theta]}$$

$$t_2 = 5.27 \text{ s} \quad \blacktriangleleft$$

$$\omega_2 = \frac{\bar{k}^2 (\mu_k \cos\theta - \sin\theta)\omega_1}{\mu_k (r^2 + \bar{k}^2)\cos\theta - \bar{k}^2 \sin\theta}$$

$$\omega_2 = 3.20 \text{ rad/s} \quad \blacktriangleleft$$

18.4   System of Rigid Bodies

For a *system of rigid bodies*, the principle of impulse and momentum may be applied to each rigid body separately. However, it is sometimes desirable to apply that principle to the system as a whole (cf. Examples 18.3 and 18.4). In the latter case, the momentum and impulse diagrams are drawn for the entire system of rigid bodies. The initial and final momentum diagrams should include the *resultant linear momentum* $m\bar{v}$ and the *resultant angular momentum* $\bar{I}\omega$, as appropriate, *for each component part of the system*. Linear and angular impulses of forces and moments *internal* to the system considered should be *omitted* in the impulse diagram, since they cancel out by occurring in collinear pairs equal in magnitude and opposite in sense.

In the present case, the *principle of impulse and momentum* may be stated as follows: *For a system of rigid bodies, the system of vectors in the initial momentum diagram plus the system of vectors in the impulse diagram during a time interval is equipollent to the system of vectors in the final momentum diagram.* In an abbreviated form, the *principle of impulse and momentum* for any case may, therefore, be written in terms of *equipollence*[†] as

$$\textbf{Syst Momenta}_1 + \textbf{Syst Ext Imp}_{1\to2} \Rightarrow \textbf{Syst Momenta}_2 \quad (18.16)^{\dagger\dagger}$$

In applying this principle, care should be taken *not* to add directly linear and angular momenta for the same reason that force and moment cannot be added directly. Furthermore, a *consistent system of kinetic units* (cf. Sec. 12.2) should be used in writing each term in an equation.

EXAMPLE 18.3

The flywheel of an engine has a drum of radius 0.4 m, a mass of 300 kg, and a central radius of gyration $\bar{k}_O = 0.5$ m. It is hoisting a 100-kg crate C as shown with an angular velocity $\omega_1 = 120$ rpm $\circlearrowright$ at the time $t = 0$ when the power is cut off. The system stops momentarily at the time $t = t_2$. Determine t_2.

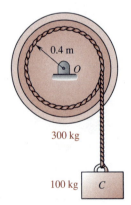

0.4 m

O

300 kg

100 kg C

[†]Note that if two systems of vectors are equivalent, they are also equipollent; but the converse is not always true.

[††]Recall that the symbol $\Rightarrow$ indicates *equipollence*.

Solution.   Let the motion of the flywheel and the crate as a system be considered from the time $t = t_1 = 0$ to the time $t = t_2$. Noting that $\omega_1 = 120$ rpm $\circlearrowright = 4\pi$ rad/s $\circlearrowright$, $\omega_2 = 0$, and $\bar{I} = m\bar{k}_O^2$, we apply the *principle of impulse and momentum* to draw a vector-diagram equation for the system as shown. Referring to the vector-diagram equation, we write

$$+\circlearrowright \ \Sigma M_O: \quad \{300(0.5)^2(4\pi) + 0.4(100)[0.4(4\pi)]\} - 0.4(100gt_2) = 0$$

$$t_2 = 9.1\pi/g \qquad\qquad t_2 = 2.91 \text{ s} \ \blacktriangleleft$$

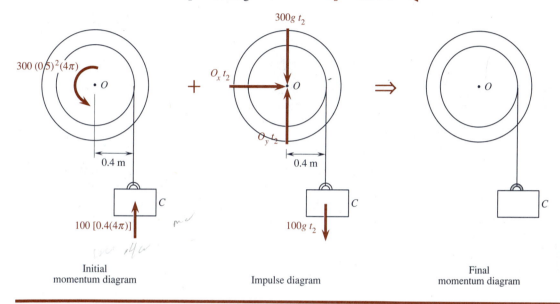

Initial momentum diagram	Impulse diagram	Final momentum diagram

Developmental Exercises

D18.10   Describe the *principle of impulse and momentum* for a system of rigid bodies.

D18.11   Refer to Example 18.3. Determine $\mathbf{v}_C$ of the crate C 2 s after the power is cut off.

EXAMPLE 18.4[†]

A 40-kg block C is supported by a 20-kg cylinder A and a 30-kg cylinder B as shown, where the cylinders have the same radius $r = 100$ mm. The system is at rest at the time $t = 0$ when a 50-N force $\mathbf{P}$ is applied to the block C. If the cylinders roll without slipping, determine the velocity $\mathbf{v}_C$ of the block C at the time $t = 2$ s.

Solution.   For the system as described, the cylinders must have the same angular velocity ω which is related to the velocity $\mathbf{v}_C$ of the block C by $v_C = 2r\omega = 2(0.1)\omega$ or $\omega = 5v_C$. Noting that the velocity of the center of

---

[†]For comparative study, the problem in this example is solved by using the *principle of generalized virtual work* in Example 18.10.

either cylinder is $\bar{v} = r\omega \leftarrow = 0.5v_C \leftarrow$, we apply the *principle of impulse and momentum* to draw the following vector-diagram equations:

Entire system

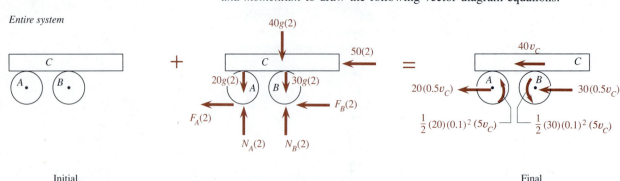

Initial
momentum diagram   Impulse diagram   Final
momentum diagram

Cylinder A

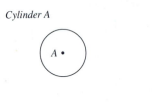   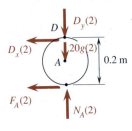   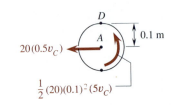

Initial
momentum diagram   Impulse diagram   Final
momentum diagram

Cylinder B

             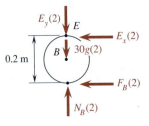   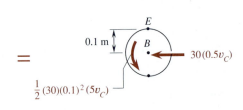

Initial
momentum diagram   Impulse diagram   Final
momentum diagram

Referring to the preceding vector-diagram equations, we write

Entire system

$$\xleftarrow{+} \Sigma V_x: \quad 0 + [50(2) + F_A(2) + F_B(2)]$$
$$= 40v_C + 20(0.5v_C) + 30(0.5v_C) \qquad (1)$$

Cylinder A

$$+\!\!\curvearrowright \Sigma M_D: \quad 0 + 0.2[F_A(2)] = 0.1[20(0.5v_C)]$$
$$- \tfrac{1}{2}(20)(0.1)^2(5v_C) \qquad (2)$$

Cylinder B

$$+\!\!\curvearrowright \Sigma M_E: \quad 0 + 0.2[F_B(2)] = 0.1[30(0.5v_C)]$$
$$- \tfrac{1}{2}(30)(0.1)^2(5v_C) \qquad (3)$$

Solution of Eqs. (1) through (3) yields

$$v_C = 1.702 \text{ m/s} \qquad F_A = 2.13 \text{ N} \qquad F_B = 3.19 \text{ N}$$

$$\mathbf{v}_C = 1.702 \text{ m/s} \leftarrow \quad \blacktriangleleft$$

REMARK: The acceleration of the block C under the same condition was determined in Example 16.19 to be $\mathbf{a}_C = 0.851 \text{ m/s}^2 \leftarrow$. Since $\mathbf{a}_C$ is constant we readily see in 2 s that $v_C = a_C t = 0.851(2) \text{ m/s} = 1.702 \text{ m/s}$, which is what we have obtained here.

Developmental Exercise

D18.12 Work Example 18.4 if the masses of the cylinders A and B are $m_A = m_B = 30 \text{ kg}$.

18.5 Conservation of Momentum

If the resultant force and the resultant moment of the external forces acting on a rigid body or a system of rigid bodies are equal to zero, Eq. (18.16) reduces to

$$\textbf{Syst Momenta}_1 \Rightarrow \textbf{Syst Momenta}_2 \qquad (18.17)$$

which expresses the *conservation of momentum*. In this case, *the system of the initial momenta is equipollent to the system of the final momenta*, and it follows that the linear momentum of the system in any direction as well as the angular momentum of the system about any point is conserved.

However, there are cases in which the resultant external force is not zero, yet in which *the resultant external moment about a given point or a given axis is zero*. In such cases, the resultant linear momentum is not conserved, yet *the resultant angular momentum about that given point or axis is conserved*. Such cases occur whenever the lines of action of the external forces (a) all pass through a given point or axis, (b) either intersect a given axis or are parallel to the given axis.

Problems involving *conservation of momentum* may be solved by applying the *principle of impulse and momentum* as before. It is usually beneficial to draw the vector-diagram equation for the impulses and momenta involved in the body or system of bodies during the interval under consideration. Scalar equations are written by appropriately summing and equating the components, or the moments, of the vector quantities in the vector-diagram equation. If the system is conservative, the principle of conservation of energy can also be applied to provide an additional scalar equation. The unknowns are then solved from the scalar equations.

Developmental Exercise

D18.13 Describe the cases in which the angular momentum of a system about a given point or a given axis is conserved, while the linear momentum of the system is not conserved.

EXAMPLE 18.5

The disk A with its attached shaft and motor armature have a mass of 20 kg and a central radius of gyration of 80 mm. The motor is mounted on the turntable B. The motor housing and the turntable B with its shaft have a mass of 30 kg and a radius of gyration of 240 mm about the y axis. Before the motor is turned on, the entire assembly is rotating with $\omega_1 = 5\mathbf{j}$ rad/s about the y axis. If the operating velocity of the motor is $\omega_{MA/B} = 600\mathbf{j}$ rpm relative to the turntable B and friction is negligible, determine the new angular velocity ω_2 of the turntable B after the motor is turned on.

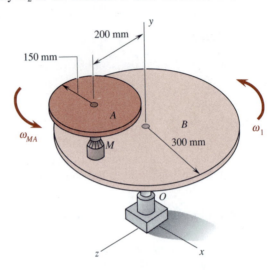

Solution.   Let the entire assembly be considered as a system. Then, the action and reaction in the motor are internal forces to the system. The weight forces are all parallel to the y axis. Thus, the angular momentum of the system about the y axis is conserved. Note that

$$\omega_{MA/B} = 600\mathbf{j} \text{ rpm} = 20\pi\mathbf{j}$$

rad/s and the final angular velocity of the motor armature and the disk A in the inertial reference frame $Oxyz$ is

$$\omega_{MA} = \omega_{MA/B} + \omega_B = (20\pi + \omega_2)\mathbf{j} \text{ rad/s}$$

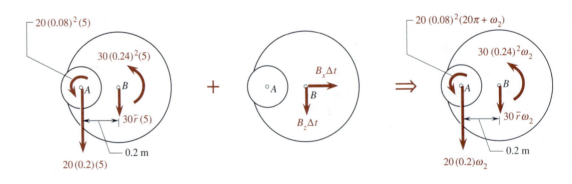

Initial
momentum diagram

Impulse diagram

Final
momentum diagram

Accordingly, we apply the *principle of impulse and momentum* to draw a vector-diagram equation as shown, where the impulses parallel to the y axis are not indicated and $\bar{r}$ represents the distance from the y axis to the center of mass of the composite body consisting of the motor housing and the turntable with its shaft. (We do not need to know $\bar{r}$ in the solution.) Referring to the vector-diagram equation, we write

$$+\,\circlearrowright\,\Sigma M_B\!: \quad 30(0.24)^2(5) + 20(0.08)^2(5)$$

$$+\ 0.2[20(0.2)(5)] + 0$$

$$=\ 30(0.24)^2\omega_2 + 20(0.08)^2(20\pi + \omega_2)$$

$$+\ 0.2[20(0.2)\omega_2]$$

$$\omega_2 = 1.972 \qquad \boldsymbol{\omega}_2 = 1.972\mathbf{j}\ \text{rad/s} \quad \blacktriangleleft$$

Developmental Exercises

D18.14 Refer to the vector-diagram equation in the solution in Example 18.5. Consider the composite body consisting of the motor housing and the turntable B with its shaft. Verify that the *resultant system of momenta* of this composite body in the initial momentum diagram is equivalent to a linear momentum acting through B and an angular momentum acting about B as indicated.

D18.15 Refer to Example 18.5. Determine the minimum operating velocity of the motor $\boldsymbol{\omega}_{MA/B}$ relative to the turntable B for which the turntable will reverse its direction of rotation after the motor is turned on.

EXAMPLE 18.6

A 15-lb solid sphere of radius 4 in. is mounted at B on the horizontal slender rod AC, which rotates freely about the vertical spindle with $\boldsymbol{\omega}_1 = 10\mathbf{j}$ rad/s. The cord holding the sphere is suddenly cut, and the sphere starts to slide outward. Neglecting the masses and the moments of inertia of the rod and spindle about the y axis, determine the angular velocity $\boldsymbol{\omega}_2$ of the rod when the sphere reaches the position C as shown.

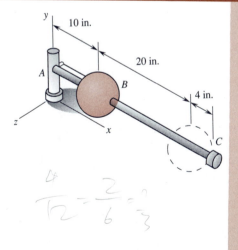

Solution. Since the spindle is free to rotate about the y axis, the reaction from the support to the spindle is composed of a force $\mathbf{A} = A_x\mathbf{i} + A_y\mathbf{j} + A_z\mathbf{k}$ and a moment $\mathbf{M} = M_x\mathbf{i} + M_z\mathbf{k}$. Thus, we apply the *principle of impulse and momentum* to draw a vector-diagram equation as shown on the next page, where the impulses parallel to the y axis are not indicated. Referring to the vector-diagram equation, we write

$$+\,\circlearrowright\,\Sigma M_y\!: \quad \left\{\frac{2}{5}\left(\frac{15}{32.2}\right)\left(\frac{4}{12}\right)^2(10) + \frac{10}{12}\left[\frac{15}{32.2}\left(\frac{10}{12}\right)(10)\right]\right\} + 0$$

$$=\ \frac{2}{5}\left(\frac{15}{32.2}\right)\left(\frac{4}{12}\right)^2\omega_2 + \frac{30}{12}\left[\frac{15}{32.2}\left(\frac{30}{12}\right)\omega_2\right]$$

$$\omega_2 = 1.174 \qquad \boldsymbol{\omega}_2 = 1.174\mathbf{j}\ \text{rad/s} \quad \blacktriangleleft$$

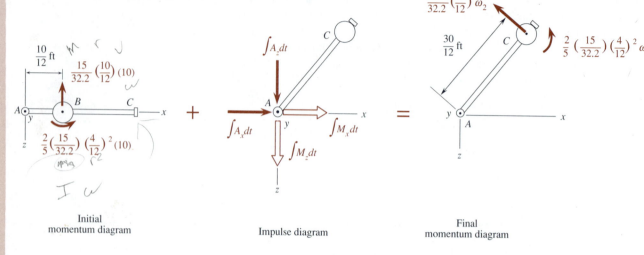

| Initial momentum diagram | Impulse diagram | Final momentum diagram |

REMARK. In the system considered, the angular impulse about the y axis is equal to zero at all times. Thus, the angular momentum of the system about the y axis is conserved. However, note that the direction of the linear momentum vector in the final momentum diagram is dependent on the orientation of the rod AC, which is unknown at t_2. This means that we are unable to sum and equate the components of the linear momentum vectors in the vector-diagram equation shown. We can do only what we have done with that vector-diagram equation in the above solution.

Developmental Exercise

D18.16   Work Example 18.6 if the rod weighs 3 lb and the moment of inertia of the spindle about the y axis is negligible.

PROBLEMS

18.1   A bolt is tightened by applying a couple as shown. The wheel has a mass of 20 kg and a central radius of gyration of 280 mm. If the wheel is rotated by the couple for 0.5 s and friction is negligible, determine the angular velocity of the wheel.

18.2 and 18.3*   The system shown is released from rest, where the 300-lb flywheel has a central radius of gyration $\bar{k}_O = 1.6$ ft. Determine the angular velocity of the flywheel 3 s after the release.

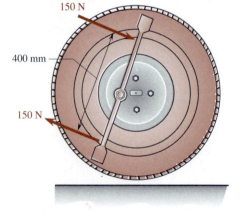

Fig. P18.1

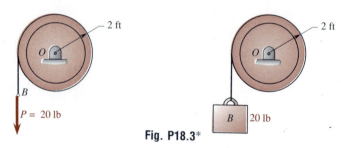

Fig. P18.2   **Fig. P18.3***

18.4   Solve Prob. 18.2 if $P = 20t$ lb, where t is the time measured in seconds.

18.5   A 20-kg cylinder of radius $r = 150$ mm is rotating with $\omega_1 = 12$ rad/s $\downcurvearrowright$ in a corner when a moment $\mathbf{M} = 10$ N·m $\downcurvearrowright$ is applied to it at the time $t = 0$. If $\mu_k = 0.1$ between all surfaces of contact, determine the angular velocity ω_2 of the cylinder when $t = 3$ s.

18.6*   Solve Prob. 18.5 if $\mathbf{M} = 10t$ N·m $\downcurvearrowright$, where t is the time measured in seconds.

18.7*   A bowling ball of mass m and radius r is released with a linear velocity $\mathbf{v}_1$ and no angular velocity as shown. Denoting the central radius of gyration of the ball by $\bar{k}$ and the coefficient of kinetic friction between the ball and the floor by μ_k, determine the time t_2 at which the ball begins rolling without slipping.

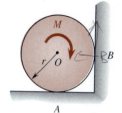

Fig. P18.5

$J = F \alpha t$

$H_1 + \int_{1 \to 2} = H_2$

$m \bar{r}^2 + I\omega_1 + B_y \mu(\frac{1}{2}A_x)\mu t_2 = I\omega_2$

$m k^2 \omega_1 - B_y \mu$

$v_2 = r\omega_2$ $\omega_1 = 0$

v_2 v_1

Fig. P18.7*

18.8   Solve Prob. 18.7* if $r = 110$ mm, $v_1 = 8$ m/s, $\bar{k} = 70$ mm, and $\mu_k = 0.2$.

18.9 and 18.10   A 50-kg gear of central radius of gyration $\bar{k}_O = 300$ mm is rolled up a rack from rest at the time $t = 0$ by a 200-N force $\mathbf{P}$ applied to a cord which is wrapped around the drum of the gear as shown, where $r_1 = 200$ mm and $r_2 = 400$ mm. Determine the speed of its center O when $t = 3$ s.

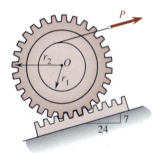

P

r_2 O r_1

24   7

Fig. P18.9 and P18.11

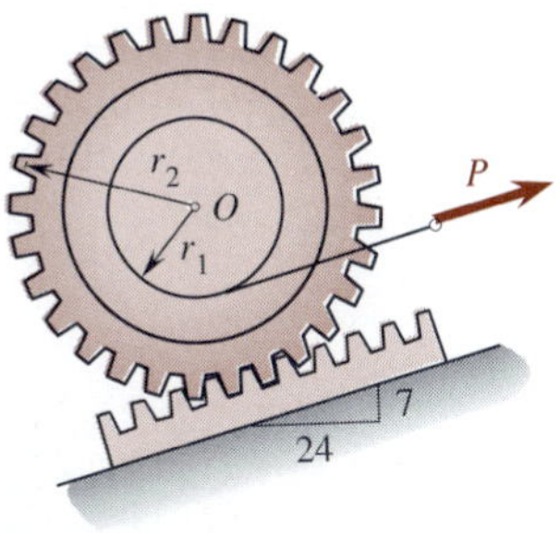

r_2 O r_1 P

24   7

Fig. P18.10 and P18.12*

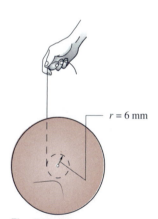

$r = 6$ mm

Fig. P18.13*

18.11 and 18.12*   A 100-lb gear of central radius of gyration $\bar{k}_O = 9$ in. is rolled up a rack from rest at the time $t = 0$ by a constant force $\mathbf{P}$ applied to a cord which is wrapped around the drum of the gear as shown, where $r_1 = 6$ in. and $r_2 = 12$ in. It is known that the angular velocity of the gear is 10 rad/s $\downcurvearrowright$ when $t = 3$ s. Determine the magnitude of $\mathbf{P}$.

18.13*   A 60-g yo-yo of central radius of gyration $\bar{k}_O = 18$ mm is released from rest at the time $t = 0$ as shown. Assume that the string is wound around the central peg such that the mean radius at which it unravels is $r = 6$ mm. Determine the time t_2 when it attains an angular speed of 600 rpm.

18.14   Constant tensions of 45 lb and 50 lb are applied to the two sides of the cable to hoist a load B of 55 lb as shown, where the pulley A weighs 30 lb and has a central radius of gyration of 7.5 in. At the time $t = 0$, the load B moves with $v_1 = 4$ ft/s $\downarrow$ and the pulley A rotates with $\omega_1 = 20$ rad/s $\downcurvearrowright$. Determine the velocity v_2 of the load and the angular velocity ω_2 of the pulley at the time $t = 3$ s.

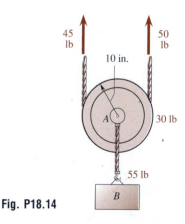

45 lb   50 lb

10 in.

A 30 lb

55 lb

B

Fig. P18.14

18.15   A cord is wrapped around the drum of a 60-kg wheel which has a central radius of gyration of 150 mm. If $\mu_s = 0.55$ and $\mu_k = 0.5$ between the wheel and the incline, as shown, and the system is released from rest at the time $t = 0$, determine the speed of the mass center G when $t = 4$ s.

18.16   The 100-kg wheel shown has a central radius of gyration $\bar{k}_O = 150$ mm and is rotating with $\omega = 300$ rpm ⟲ when the hydraulic cylinder CD is activated to exert a constant force $\mathbf{F}_{CD}$ on the lever AC for braking at B, where $\mu_k = 0.25$. If the wheel is to stop in 3 s, determine the magnitude of $\mathbf{F}_{CD}$.

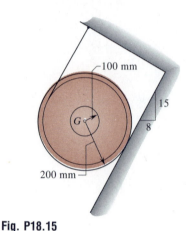

Fig. P18.15

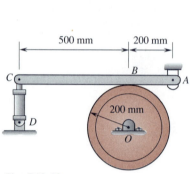

Fig. P18.16

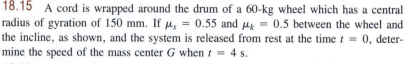

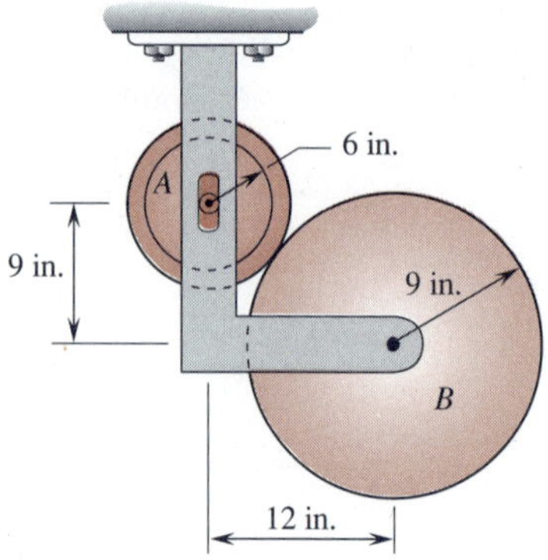

Fig. P18.17

18.17   The 15-lb wheel A has a central radius of gyration $\bar{k} = 5$ in. and is released from rest at the time $t = t_1 = 0$ to press on the disk B, which is driven by a motor to rotate at a constant angular velocity $\omega = 600$ rpm ⟲. If $\mu_k = 0.2$ between A and B, determine the time t_2 at which slipping between them ceases.

18.18*   Solve Prob. 18.17 if $\omega = 600$ rpm ⟳.

18.19*   The gears A and B shown have masses 5 kg and 12 kg and central radii of gyration 80 mm and 120 mm, respectively. A constant torque $\mathbf{M} = 10$ N·m ⟲ is acting on the gear A. Determine the time interval Δt during which the angular velocity of the gear B is increased from 300 rpm ⟳ to 600 rpm ⟳.

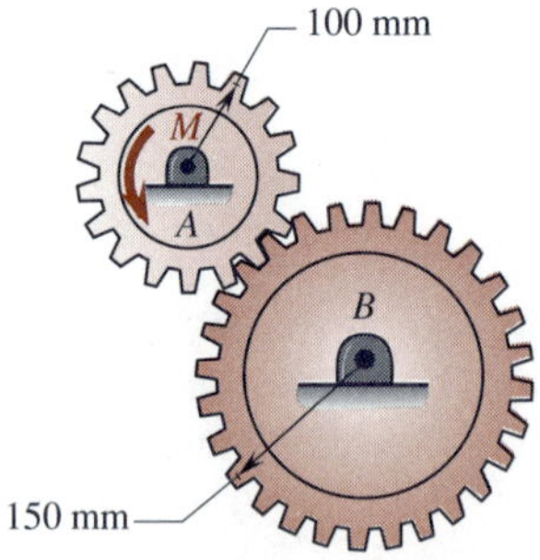

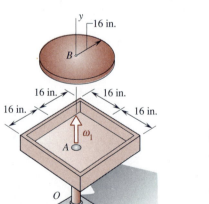

Fig. P18.21*

Fig. P18.19*

18.20   Solve Prob. 18.19* if the angular velocity of the gear B is increased from 150 rpm ⟳ to 450 rpm ⟳.

18.21*   The square tray A with its attached shaft weighs 20 lb and has a central radius of gyration of 14 in. The tray rotates freely with $\omega_1 = 180\mathbf{j}$ rpm about the y axis as shown. The 15-lb disk B is at rest when it is dropped into the tray. Determine (a) the final angular velocity ω_2 of the tray with the disk, (b) the percentage of loss of the kinetic energy of the system.

18.22   The 2-kg slender rod BC is hinged to the arm AB as shown. The arm with its attached shaft has a mass of 5 kg and a radius of gyration of 250 mm about the y axis. With the rod latched in the vertical position, the system rotates freely about the y axis with $\omega_1 = 120\mathbf{j}$ rpm. If the latch is released to let the rod lie in the horizontal position, determine the new angular velocity ω_2 of the system.

18.23   The arm AB with its attached shaft OA weighs 30 lb and has a radius of gyration of 8.6 in. about the y axis. The disk C with its attached shaft BC weighs 20 lb and has a radius of gyration of 8.4 in. about BC. Initially, the disk with its shaft is rotating with $\omega_1 = 300\mathbf{j}$ rpm while the arm with its shaft is at rest as shown. If the pin at D springs out to bear against the rotating disk to bring it to a stop relative to the arm AB and its shaft OA, determine the final angular velocity ω_2 with which the entire system will rotate freely about the y axis.

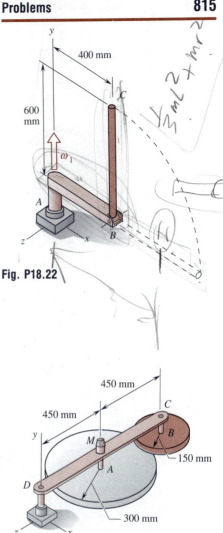

Fig. P18.22

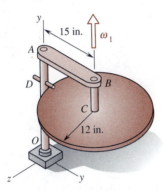

Fig. P18.23

18.24   In the assembly shown, the arm CD and the motor housing M have a mass of 30 kg and a radius of gyration of 500 mm about the y axis, the disk A and its attached shaft and motor armature have a mass of 25 kg and a radius of gyration of 200 mm about the axis of its shaft, and the 10-kg disk B has a radius of gyration of 100 mm about the axis of its shaft. The motor at M drives the disk A which turns the disk B without slipping. The entire assembly is at rest before the motor is turned on. Determine the angular speed of the arm CD when the motor is turned on and is operating at a speed of 600 rpm relative to the arm CD.

18.25*   A 10-kg block B is suspended by a cord which passes around a frictionless pulley of negligible mass at D and is wrapped around the drum of a 100-kg wheel as shown. The wheel with the drum has a central radius of gyration $\bar{k}_O = 0.5$ m and the system is released from rest. Knowing that the wheel rolls without slipping, determine $\mathbf{v}_B$ of the block B 2 s after the release.

Fig. P18.24

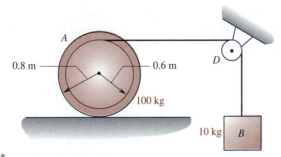

Fig. P18.25*

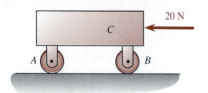

Fig. P18.26

18.26 The 30-kg carriage C is supported by the 25-kg and 20-kg cylindrical wheels A and B as shown, where the wheels have the same radius of 100 mm. If the system is pushed from rest by the 20-N force and the wheels roll without slipping, determine $\mathbf{v}_C$ of the carriage C 4 s later.

18.27* Solve Prob. 18.26 if both wheels have the same mass of 20 kg.

18.28 The double pulley shown has a mass $m = 10$ kg and a central radius of gyration $\bar{k}_O = 150$ mm. The inner and outer pulleys are rigidly attached and the center O is guided by a smooth pin at O to move along the inclined slot. The pulley is at rest when a 40-N force $\mathbf{P}$ is applied to the cord A at the time $t = 0$. Determine $\mathbf{v}_O$ of the center O when $t = 2$ s.

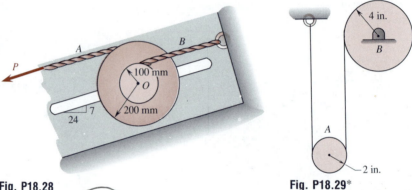

Fig. P18.28   **Fig. P18.29***

18.29* The 5-lb cylinder A and the 20-lb cylinder B are connected by a belt as shown. If the system is released from rest, determine (a) the angular velocity of the cylinder B after 4 s, (b) the tension in the portion of the belt connecting A and B.

18.30 Let m and k with appropriate subscripts denote the *mass* and the *radius of gyration* of a certain part in the gear train as shown. It is known that the gear A and the motor armature at M with their attached shaft have $m_{AM} = 18$ kg and $k_{AM} = 50$ mm, the gears B and C with their attached shaft have $m_{BC} = 20$ kg and $k_{BC} = 40$ mm, while the gear D and a working part at E with their attached shaft have $m_{DE} = 25$ kg and $k_{DE} = 45$ mm. If the motor M exerts a constant torque of 60 N·m on the shaft of the gear A for 5 s, determine the angular speed reached by the gear D.

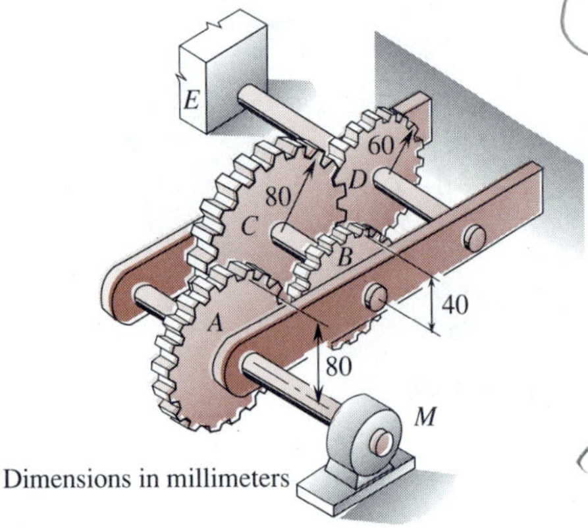

Dimensions in millimeters

Fig. P18.30

18.31 A 10-kg disk A is connected to a 2-kg block B as shown, where $\mu_k = 0.2$ between the cord and the drum at D. If the system is released from rest, determine $\mathbf{v}_B$ of the block B after 4 s.

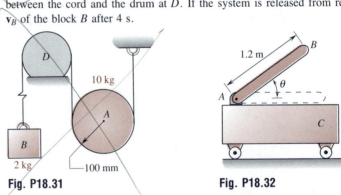

Fig. P18.31   **Fig. P18.32**

18.32 A system consists of a 10-kg rod AB and a 20-kg cart C as shown. The system is released from rest when $\theta = 60°$. Determine the velocity of the cart C when (a) $\theta = 30°$, (b) $\theta = 0$.

18.33 A 20-lb solid sphere B of radius 5 in. is mounted to slide on the 4-lb slender rod AC as shown. The sphere is held by a cord and is connected to the end C of the rod by a spring of modulus $k = 60$ lb/ft. At the instant shown, $r = 10$ in., the spring is neutral, and the system is rotating freely about the vertical spindle with $\omega_1 = 15\mathbf{j}$ rad/s. If the cord is suddenly cut and the moment of inertia of the spindle about the y axis is negligible, determine ω_2 of the system when $r = 20$ in.

18.34* Solve Prob. 18.33 if $\omega_1 = 30\mathbf{j}$ rad/s.

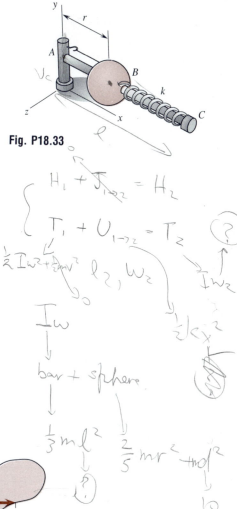

Fig. P18.33

IMPULSIVE MOTION AND SPECIAL TOPIC

18.6   Impulsive Motion of Rigid Bodies

We recall that an *impulsive motion* of a body is the motion of the body resulting from the action of an unbalanced *impulsive force*, which is a relatively large force acting on the body for a relatively short time interval and producing a significant change in the momentum of the body. Problems involving impulsive motion are generally amenable to solution by the *principle of impulse and momentum*. In such problems, the time interval considered is very short, *the bodies in the system may be assumed to occupy the same position during the time the impulsive forces act*. Furthermore, *nonimpulsive forces* (e.g., weight forces) *in juxtaposition with impulsive forces may usually be neglected*. Thus, the linear and angular impulses in problems involving impulsive motion of rigid bodies may readily be computed.

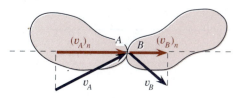

(a) Just before impact

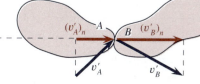

(b) Just after impact

Fig. 18.7

In impulsive motions resulting from *impact between two rigid bodies*, the *coefficient of restitution*, denoted by e, is often employed in the study if the bodies rebound after the impact. Let the points of contact between certain two colliding bodies be A and B as shown in Fig. 18.7, where the *line of impact* is a straight line normal to the contacting surfaces during the impact, $\mathbf{v}_A$ and $\mathbf{v}_B$ are the velocities of A and B just before the impact, $\mathbf{v}'_A$ and $\mathbf{v}'_B$ are the velocities of A and B just after the impact, and $(\mathbf{v}_A)_n$, $(\mathbf{v}_B)_n$, $(\mathbf{v}'_A)_n$, and $(\mathbf{v}'_B)_n$ are the normal components of $\mathbf{v}_A$, $\mathbf{v}_B$, $\mathbf{v}'_A$, and $\mathbf{v}'_B$ directed along the line of impact. These normal components are related to the *coefficient of restitution* e, as defined in Eq. (14.34), for the *eccentric impact* as follows:

$$(v'_B)_n - (v'_A)_n = e[(v_A)_n - (v_B)_n] \qquad (18.18)$$

The impact is called a *perfectly plastic impact* if $e = 0$, and a *perfectly elastic impact* if $e = 1$. When the line of impact passes through the centers of mass of the two colliding bodies, the impact becomes a *central impact*, which was studied in Secs. 14.8 and 14.9. Note that the comments on the values of e in the footnote on p. 622 apply to eccentric impacts between rigid bodies as well.

EXAMPLE 18.7

A 10-lb small sphere is dropped from a height $h = 5$ ft and strikes the 30-lb slender plank AB with a perfectly plastic impact at S as shown. Determine the angular velocity of the plank just after the impact.

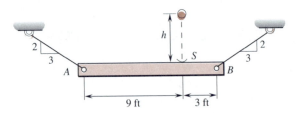

Solution. By the *principle of work and energy*, it can readily be determined that the velocity of the sphere just before it strikes the plank is $v_S = \sqrt{2gh} \downarrow = \sqrt{322}$ ft/s $\downarrow$. Immediately after a perfectly plastic im-

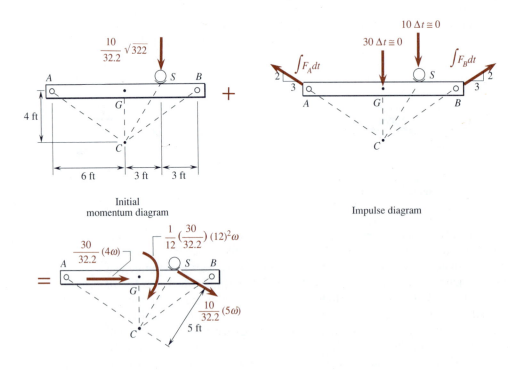

Initial
momentum diagram

Impulse diagram

Final
momentum diagram

pact, the plank and the sphere will rotate together as a rigid body with an angular velocity $\boldsymbol{\omega} = \omega \, \text{\Large ⤸}$ about their *velocity center* C. Note that the *weight forces* of the plank and the sphere in juxtaposition with the impulsive reactions $\mathbf{F}_A$ and $\mathbf{F}_B$ from the inextensible cords are *nonimpulsive forces*, which exert negligible impulses on the plank during the short time interval of impact. Thus, applying the *principle of impulse and momentum* to the sphere and the plank as a system, we have

$$+\text{\Large ⤸} \; \Sigma M_C: \quad 3\left(\frac{10}{32.2}\sqrt{322}\right) + 0 = \frac{1}{12}\left(\frac{30}{32.2}\right)(12)^2\omega$$

$$+ 4\left[\frac{30}{32.2}(4\omega)\right]$$

$$+ 5\left[\frac{10}{32.2}(5\omega)\right]$$

$$\omega = 0.494 \qquad \boldsymbol{\omega} = 0.494 \text{ rad/s} \; \text{\Large ⤸} \quad \blacktriangleleft$$

Developmental Exercises

D18.17   Refer to Example 18.7. Verify that $v_S = \sqrt{322}$ ft/s $\downarrow$.

D18.18   Refer to Example 18.7. Determine the loss of kinetic energy of the system during the impact.

EXAMPLE 18.8

The 20-kg wheel shown has a central radius of gyration $\bar{k} = 0.25$ m and rolls without slipping on the steps. If its angular velocity on the step AB is $\omega_1 = 10$ rad/s $\text{\Large ⤸}$ and its impact with the corner C is perfectly plastic, determine its angular velocity (a) immediately after the impact, (b) on the step CD.

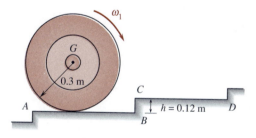

Solution.   Let the angular velocity of the wheel immediately after the impact be ω_C. The *weight force* of the wheel in juxtaposition with the impulsive reaction components $\mathbf{C}_x$ and $\mathbf{C}_y$ from the corner C is a *nonimpulsive force*, which exerts a negligible impulse on the wheel. Thus, applying the *principle of impulse and momentum* to the wheel, we have

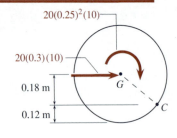

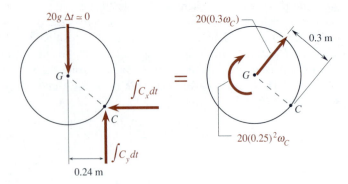

| Initial | Impulse diagram | Final |
| momentum diagram | | momentum diagram |

$$+ \, \circlearrowleft \, \Sigma M_C: \quad \{20(0.25)^2(10) + 0.18[20(0.3)(10)]\} + 0$$

$$= 20(0.25)^2\omega_C + 0.3[20(0.3\omega_C)]$$

$$\omega_C = 7.64 \qquad \mathbf{\omega_C = 7.64 \text{ rad/s} \, \circlearrowleft} \quad \blacktriangleleft$$

After impacting with the corner C, the wheel rises 0.12 m from the step AB to the step CD, while its angular velocity changes from $\boldsymbol{\omega}_C = 7.64$ rad/s $\circlearrowleft$ to $\boldsymbol{\omega}_2$. During this upward motion, the wheel simply rotates about C, its mass center moves from G to G' as shown, and the system is conservative. Note that the moment of inertia I_C of the wheel about C is

$$I_C = m\bar{k}^2 + md^2$$

$$= 20(0.25)^2 + 20(0.3)^2 \text{ kg}\cdot\text{m}^2$$

$$= 3.05 \text{ kg}\cdot\text{m}^2$$

Choosing the reference datum to be at G and applying the *principle of conservation of energy* we write

$$T_1 + V_1 = T_2 + V_2: \qquad \tfrac{1}{2}I_C\omega_C^2 + 0 = \tfrac{1}{2}I_C\omega_2^2 + mgh$$

$$\tfrac{1}{2}(3.05)(7.64)^2 + 0 = \tfrac{1}{2}(3.05)\omega_2^2 + 20(9.81)(0.12)$$

$$\omega_2 = 6.55 \qquad \mathbf{\omega_2 = 6.55 \text{ rad/s} \, \circlearrowleft} \quad \blacktriangleleft$$

Developmental Exercises

D18.19 Refer to Example 18.8. Determine the loss of kinetic energy of the wheel *during the impact*.

D18.20 Refer to Example 18.8. Determine the minimum magnitude of $\boldsymbol{\omega}_1$ with which the wheel will roll over the corner C to the step CD.

EXAMPLE 18.9†

Two slender rods of equal masses $m = 10$ kg may swing freely from the pivots at A and C as shown. The rod AB is released from rest in the horizontal position. When AB swings to the vertical position, its small knob at E hits the small knob at D on the rod CD which was at rest. If $d = 0.3$ m, $L = 1.2$ m, and the coefficient of restitution between the knobs at D and E on the rods is $e = 0.5$, determine the angular velocities ω'_{AB} and ω'_{CD} of the rods AB and CD immediately after the impact.

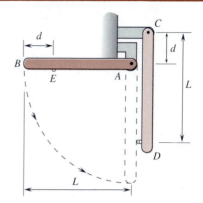

Solution. By the *principle of work and energy*, it can readily be determined that the angular velocity of the rod AB just before the impact takes place is $\omega_{AB} = \sqrt{3g/L} \; \circlearrowright = 4.952$ rad/s $\circlearrowright$. For simplicity in the solution, let the impulse of an impulsive force be equal to the product of the *average* impulsive force and the time interval Δt. Then, we apply the *principle of impulse and momentum* to draw the following vector-diagram equations:

Rod AB

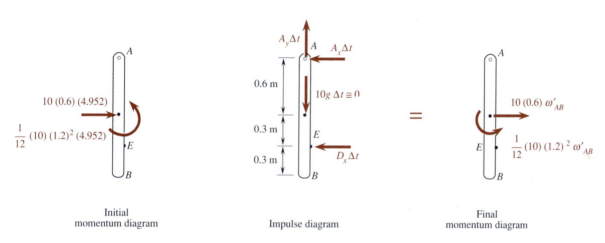

Initial momentum diagram   Impulse diagram   Final momentum diagram

Rod CD

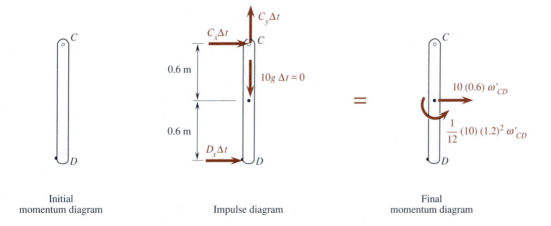

Initial momentum diagram   Impulse diagram   Final momentum diagram

---

†For comparative study, the problem in this example is solved by using the *principle of generalized virtual work* in Example 18.11.

Referring to the vector-diagram equation for the rod AB, we write

$$+ \circlearrowright \Sigma M_A: \qquad \tfrac{1}{12}(10)(1.2)^2(4.952) + 0.6[10(0.6)(4.952)]$$

$$- 0.9(D_x \, \Delta t)$$

$$= \tfrac{1}{12}(10)(1.2)^2 \omega'_{AB} + 0.6[10(0.6)\omega'_{AB}] \qquad (1)$$

Referring to the vector-diagram equation for the rod CD, we write

$$+ \circlearrowright \Sigma M_C: \qquad 0 + 1.2(D_x \, \Delta t)$$

$$= \tfrac{1}{12}(10)(1.2)^2 \omega'_{CD} + 0.6[10(0.6)\omega'_{CD}] \qquad (2)$$

The velocities of the points of contact E and D just before and just after the impact, in the direction of the line of impact, are $(\mathbf{v}_E)_n = 0.9(4.952)$ m/s $\rightarrow$, $(\mathbf{v}_D)_n = \mathbf{0}$, $(\mathbf{v}'_E)_n = 0.9\omega'_{AB} \rightarrow$, and $(\mathbf{v}'_D)_n = 1.2\omega'_{CD} \rightarrow$. Applying Eq. (18.18), which defines the *coefficient of restitution e* for the *eccentric impact*, we write

$$(v'_D)_n - (v'_E)_n = e[(v_E)_n - (v_D)_n]$$

$$1.2\omega'_{CD} - 0.9\omega'_{AB} = 0.5[0.9(4.952) - 0] \qquad (3)$$

Solution of Eqs. (1), (2), and (3) yields

$$\omega'_{AB} = 2.28 \qquad \omega'_{CD} = 3.57 \qquad D_x \, \Delta t = 14.26$$

$$\boldsymbol{\omega'_{AB} = 2.28 \text{ rad/s } \circlearrowright} \qquad \boldsymbol{\omega'_{CD} = 3.57 \text{ rad/s } \circlearrowright} \qquad \blacktriangleleft$$

Developmental Exercises

D18.21   Refer to Example 18.9. Verify that $\omega_{AB} = 4.952$ rad/s $\circlearrowright$.

D18.22   Work Example 18.9 if $d = 0.6$ m.

D18.23   Work Example 18.9 if the impact is *perfectly elastic*.

★18.7   Generalized Virtual Work: Impulse and Momentum

We recall from Sec. 14.10 that the *generalized virtual work* $\delta U'$ *of a linear vector* $\mathbf{L}$ (e.g., a linear impulse or linear momentum vector) acting on or associated with a particle during a virtual displacement $\delta \mathbf{r}$ of the particle is defined as

$$\delta U' = \mathbf{L} \cdot \delta \mathbf{r} \qquad (18.19)$$

In a similar manner, the *generalized virtual work* $\delta U'$ *of an angular vector* $\mathbf{A}$ (e.g., an angular impulse or angular momentum vector) acting on or associated with a rigid body during an angular virtual displacement $\overrightarrow{\delta \theta}$ of the rigid body is defined as

$$\delta U' = \mathbf{A} \cdot \overrightarrow{\delta \theta} \qquad (18.20)$$

By equivalence and equipollence in the *principle of impulse and momentum* in Secs. 18.3 and 18.4 and the nature of *constrained virtual displace-*

ment for a one-degree-of-freedom system in Sec. 13.13, it follows logically that, *during a constrained virtual displacement of a one-degree-of-freedom system of rigid bodies, the generalized virtual work of all the momentum vectors in the initial momentum diagram of the system plus the generalized virtual work of all the impulse vectors in the impulse diagram of the system must be equal to the generalized virtual work of all the momentum vectors in the final momentum diagram of the system.* This property is referred to as the *principle of generalized virtual work* for rigid bodies, which complements the traditional *principle of impulse and momentum* for rigid bodies.

EXAMPLE 18.10

Using the *principle of generalized virtual work*, do Example 18.4.

Solution. For convenience, the space diagram of the system is repeated as shown, where a 40-kg block C is supported by a 20-kg cylinder A and a 30-kg cylinder B of the same radius $r = 100$ mm. The system is at rest at the time $t = 0$ when a 50-N force **P** is applied to the block C. The cylinders are assumed to roll without slipping, and we are to find $\mathbf{v}_C$ of the block C at $t = 2$ s.

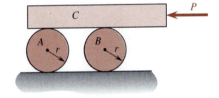

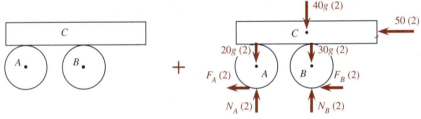

Initial
momentum diagram

Impulse diagram

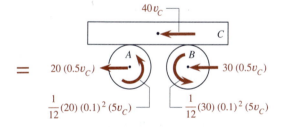

Final
momentum diagram

As explained in the solution in Example 18.4, we first draw the vector-diagram equation for the entire system as shown. The constrained virtual displacement of the block C is $\overrightarrow{\delta x_C} = \delta x_C \leftarrow$, which is related to the constrained angular virtual displacement $\delta\theta$ of the cylinders A and B by $\delta x_C = 2r\delta\theta = 2(0.1)\delta\theta$ or $\delta\theta = 5\delta x_C$. The constrained virtual displacements of the centers of the cylinders are given by $\delta x_A = \delta x_B = r\,\delta\theta = 0.1(5\delta x_C) = 0.5\,\delta x_C$. Thus, the constrained virtual displacement of the system may be indicated as shown. Referring to the diagrams shown and applying the *principle of generalized virtual work*, we

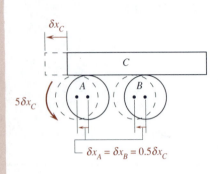

Constrained
virtual displacement

write

$$\delta U': \quad 0 + 50(2)\,\delta x_C = 40v_C(\delta x_C) + 20(0.5v_C)(0.5\,\delta x_C)$$

$$+ 30(0.5v_C)(0.5\,\delta x_C)$$

$$+ \tfrac{1}{2}(20)(0.1)^2(5v_C)(5\,\delta x_C)$$

$$+ \tfrac{1}{2}(30)(0.1)^2(5v_C)(5\,\delta x_C)$$

where the impulses of the weight forces and the reactions from the support to the cylinders do no generalized virtual work during the constrained virtual displacement. Thus, we have

$$100 = 58.75v_C \qquad v_C = 1.702 \text{ m/s} \leftarrow \blacktriangleleft$$

REMARK. By comparing the solution in this example with that in Example 18.4, we readily note that, in order to find v_C of the block C in the system, the traditional *principle of impulse and momentum* requires the solution of at least *three* simultaneous equations, while the *principle of generalized virtual work* requires the solution of only *one* equation. The ability to *uncouple an unknown quantity* in some seemingly complex one-degree-of-freedom systems is, therefore, a salient feature of the principle of generalized virtual work. Clearly, the advantage of this principle resides in the fact that *the impulses of the unknown constraining forces in the impulse diagram do not do any generalized virtual work during the constrained virtual displacement of the system.*

Developmental Exercise

D18.24   Describe the *principle of generalized virtual work* for rigid bodies.

EXAMPLE 18.11

Using the *principle of generalized virtual work*, work Example 18.9.

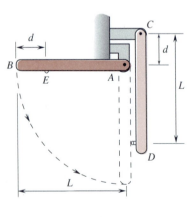

Solution.   For convenience, the space diagram of the system is repeated here as shown, where the 10-kg rod AB is released from rest in the horizontal position to hit the 10-kg rod CD which is initially at rest. It is given that $d = 0.3$ m, $L = 1.2$ m, and $e = 0.5$ between the knobs at D and E. We are to find ω'_{AB} and ω'_{CD} of the rods AB and CD just after the impact.

As indicated in the solution in Example 18.9, the angular velocity of the rod AB just before the impact takes place is $\omega_{AB} = 4.952$ rad/s $\downcurvearrowright$. Thus, we draw the vector-diagram equation for the entire system as shown. During the impact, the system behaves as a one-degree-of-freedom system, whose constrained virtual displacement may be drawn as indicated. Referring to the diagrams drawn and applying the *principle of generalized virtual work*, we write

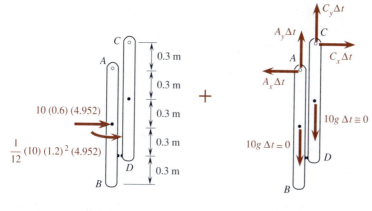

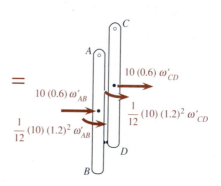

Initial momentum diagram	Impulse diagram	Final momentum diagram

$\delta U'$: $[10(0.6)(4.952)(0.8\,\delta\theta)$

$+ \frac{1}{12}(10)(1.2)^2(4.952)(\frac{4}{3}\,\delta\theta)] + 0$

$= 10(0.6)\omega'_{AB}(0.8\,\delta\theta) + \frac{1}{12}(10)(1.2)^2\,\omega'_{AB}(\frac{4}{3}\,\delta\theta)$

$+ 10(0.6)\omega'_{CD}(0.6\,\delta\theta) + \frac{1}{12}(10)(1.2)^2\omega'_{CD}(\delta\theta)$

which can be simplified and written as

$$4\omega'_{AB} + 3\omega'_{CD} = 4(4.952) \qquad (1)$$

Applying Eq. (18.18), which defines e, we write

$$(v'_D)_n - (v'_E)_n = e[(v_E)_n - (v_D)_n]$$

$$1.2\omega'_{CD} - 0.9\omega'_{AB} = 0.5[0.9(4.952) - 0] \qquad (2)$$

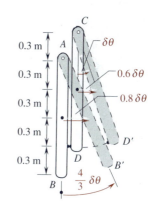

Constrained virtual displacement

Solution of Eqs. (1) and (2) yields

$$\omega'_{AB} = 2.28 \qquad \omega'_{CD} = 3.57$$

$$\boldsymbol{\omega'_{AB} = 2.28 \text{ rad/s} \quad \omega'_{CD} = 3.57 \text{ rad/s}} \quad \blacktriangleleft$$

REMARK. By comparing the solution in this example with that in Example 18.9, we readily note that Eq. (2) here and Eq. (3) in Example 18.9 are identical. Furthermore, by eliminating $D_x \Delta t$ from Eqs. (1) and (2) in Example 18.9 and simplifying, we obtain Eq. (1) in this example. Thus, the *principle of generalized virtual work* reduces the number of equations by *one* in solving the same problem in Example 18.9.

PROBLEMS

18.35 and 18.36   A 30-g bullet is fired with a velocity of 500 m/s into a 10-kg wooden rod AB suspended as shown. If $h = 0.75$ m and the bullet becomes embedded in 1 ms, determine (a) the angular velocity of the rod immediately after the impact, (b) the average impulsive reaction from the support at A.

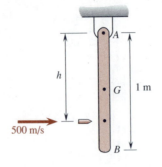

Fig. P18.35

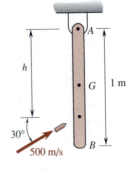

Fig. P18.36

18.37* Refer to Prob. 18.35. Determine (a) the distance h for which the average impulsive reaction from the support at A is zero, (b) the corresponding angular velocity of the rod immediately after the impact.

18.38* Solve Prob. 18.36 if $h = 0.5$ m.

18.39 A hammer exerts an impulse of 100 lb·s on a 300-lb sphere pivoted as shown. If $\theta = 60°$ and the impact lasts for 1.5 ms, determine (a) the angular velocity of the sphere immediately after the impact, (b) the average impulsive reaction from the support at A.

18.40 Solve Prob. 18.39 if $\theta = 120°$.

18.41 A player hits a tennis ball in a special way so that the ball has a velocity $\mathbf{v}$ and an angular velocity $\boldsymbol{\omega}$ just before it lands on the surface of the court as shown, where $\theta = 75°$ and $v = 10$ m/s. Assuming that the tennis ball can be taken as a spherical shell of diameter 65 mm and the ball does not skid during the impact, determine the minimum magnitude of $\boldsymbol{\omega}$ for which the ball will *not* rebound forward to the left.

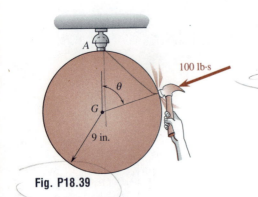

Fig. P18.39

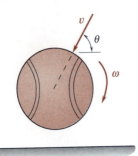

Fig. P18.41

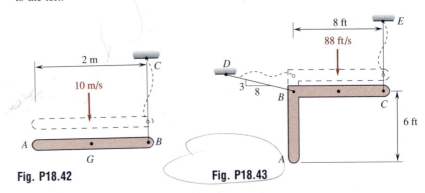

Fig. P18.42   **Fig. P18.43**

18.42 The rod AB is falling with a velocity of 10 m/s ↓ when the wire BC becomes taut. If the impact is perfectly plastic, determine the angular velocity of the rod and the velocity of its mass center G immediately after the impact.

18.43 A 6-lb rod AB and an 8-lb rod BC are welded together to form a single 14-lb rigid body which is falling with a velocity of 88 ft/s ↓ when the wires BD and CE simultaneously become taut. If the impact is perfectly plastic and is completed in 1 ms, determine (a) the angular velocity of the rigid body immediately after the impact, (b) the average impulsive reaction from each wire.

18.44 A 0.1-kg ball B is thrown with a velocity of 30 m/s at a 10-kg rod AC as shown. If the coefficient of restitution between the ball and the rod is $e = 0.8$, determine the velocity of the ball immediately after the impact.

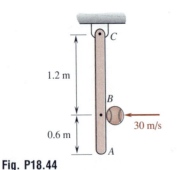

Fig. P18.44

18.45* A 4-oz ball B is thrown with a velocity of 100 ft/s at a 20-lb rod AC as shown. If the coefficient of restitution between the ball and the rod is $e = 0.7$, determine the angular velocity of the rod immediately after the impact.

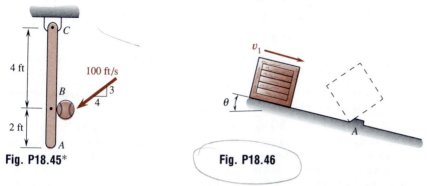

Fig. P18.45*　　　　　Fig. P18.46

18.46 A uniformly loaded square crate slides with a velocity v_1 when it strikes a small rigid obstruction at A as shown, where $\theta = 20°$. If the impact at A is perfectly plastic, determine the minimum magnitude of v_1 for which the crate will rotate over the obstruction.

18.47* Solve Prob. 18.46 if $\theta = 10°$.

18.48 Solve Prob. 18.46 if $\theta = 0$.

18.49 A section of pipe rolls without slipping down an incline which has an obstruction at A as shown, where $\theta = 20°$. If the velocity of the center of the pipe just before the impact at A is v_1 and the impact is perfectly plastic, determine the minimum magnitude of v_1 for which the pipe will roll over the obstruction.

18.50* Solve Prob. 18.49 if $\theta = 10°$.

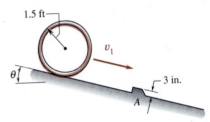

Fig. P18.49

18.51 A uniform slender rod is cut into four pieces which are then welded to form a rigid rectangular frame $ABCD$. The frame is released from rest in the position shown, and the floor is sufficiently rough to prevent the frame from slipping at the corner A. If the impact between the corner D and the floor is perfectly plastic, determine the smallest ratio of b/h for which the frame will not rebound in any way. (*Hint.* The answer is independent of the angle ϕ_0.)[†]

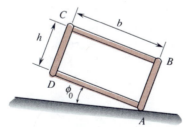

Fig. P18.51 and P18.52*

18.52* The rigid frame $ABCD$ is released from rest in the position shown, where the floor is sufficiently rough to prevent the frame from slipping at the corner A and the frame strikes the floor with an angular velocity ω_1 ↻. If $b = h$ and the impact between the corner D and the floor is perfectly plastic, determine (a) the angular velocity ω_2 of the frame immediately after the impact, (b) the percentage of the kinetic energy of the frame lost during the impact.

[†]The solution of this problem leads to a cubic equation. (Cf. App. E for the solution of a nonlinear equation.)

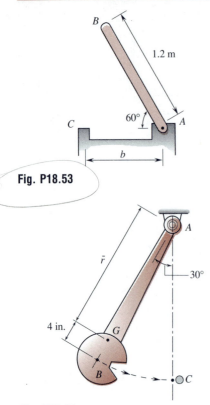

Fig. P18.53

Fig. P18.58

18.53 The 10-kg slender rod AB is released from rest in the position shown and hits the edge C, where $b = 0.6$ m. If the impact at C is perfectly elastic and is completed in 1 ms, determine (a) the angular velocity of the rod immediately after the impact, (b) the average impulsive reactions from the edge C and the hinge at A.

18.54* Solve Prob. 18.53 if $b = 0.75$ m.

18.55* Solve Prob. 18.53 if $b = 0.9$ m.

18.56 Solve Prob. 18.53 if the coefficient of restitution between the rod and the edge at C is $e = 0.5$.

18.57* Refer to Prob. 18.53. Determine the length b for which the average impulsive reaction from the hinge at A is equal to zero.

18.58 The pendulum AB of an impact testing machine weighs 80 lb and has a radius of gyration about its mass center G of 10 in. The pendulum is designed so that the horizontal component of the reaction from the pivot at A is essentially zero during its impact from the specimen at C as shown. Determine (a) the length $\bar{r}$, (b) the reaction from the pivot at A during the impact if the pendulum is released in the position shown.

18.59 A 5-kg sphere S is dropped from a height $h = 2$ m onto the end A of the 10-kg rod AB as shown, where $b = 1.8$ m. If the impact at A is perfectly elastic, determine immediately after the impact (a) the angular velocity of the rod, (b) the velocity of the sphere.

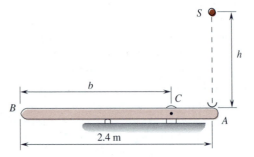

Fig. P18.59

18.60* Solve Prob. 18.59 if $b = 1.2$ m.

18.61* Solve Prob. 18.59 if $h = 4$ m.

18.62 A 10-lb sphere S is dropped from a height $h = 5$ ft onto the end A of the 20-lb rod AB as shown, where $b = 3$ ft and the 5-lb block D is not tied to the end B of the rod. If the impact at A is perfectly plastic, determine the height to which the block D will rise.

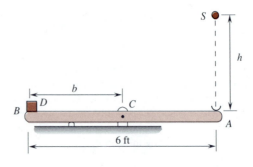

Fig. P18.62

18.63* Solve Prob. 18.62 if $b = 4$ ft.

18.64 A 45-g bullet is fired with a velocity of 650 m/s into a 12-kg wooden plank *AB* suspended by wires as shown. Determine the velocity of the bullet immediately after it becomes embedded in the plank.

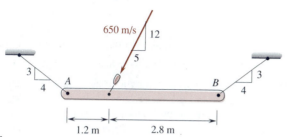

Fig. P18.64

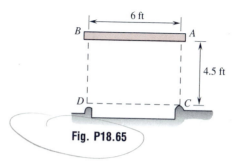

Fig. P18.65

18.65 The slender rod *AB* is dropped from rest in the position shown. The end *A* hits the corner *C* before the end *B* hits the corner *D* which is slightly lower than the corner *C*. If the impacts at *C* and *D* are perfectly plastic, determine the angular velocity of the rod immediately after it hits (a) the support *C*, (b) the support *D*.

18.66* A 20-Mg lunar landing module with mass center at *G* has a downward velocity of **v** = 2.5 m/s ↓ when one of its four legs becomes the first part of the module to hit the surface of the moon with a perfectly plastic impact. The module has a radius of gyration of 2 m about *G*, which is 3 m directly above the center of the square formed by the four feet as corners. If the diagonal of that square is 8 m, determine the angular speed of the module immediately after the first impact.

18.67 Solve Prob. 18.66* if **v** = 1.5 m/s ↓.

18.68* and 18.69 The rods *AB* and *CD* have equal weights *W* = 20 lb, and the rod *AB* is released from rest to hit the rod *CD* which is in equilibrium. If the coefficient of restitution between them is *e* = 0.6, determine their angular velocities immediately after the impact.

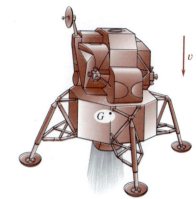

Fig. P18.66*

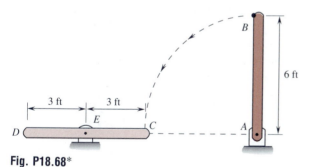

Fig. P18.68*

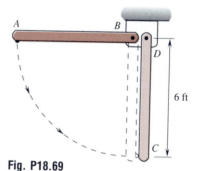

Fig. P18.69

18.70 Solve Prob. 18.68* if *e* = 0.8.

18.71* Solve Prob. 18.69 if *e* = 0.8.

18.72 and 18.73* The 50-lb carriage *D* is supported by a 20-lb cylinder *A*, a 25-lb cylinder *B*, and a 30-lb cylinder *C* as shown, where the cylinders have the same radius of 4 in. If the cylinders roll without slipping, determine the velocity **v**$_D$ attained by the carriage *D* after the 10-lb force has been applied to move it from rest for 1.2 s.

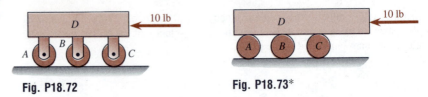

Fig. P18.72   **Fig. P18.73***

18.74*   Solve Prob. 18.72 if all cylinders have the same weight of 20 lb.

18.75   Solve Prob. 18.73* if all cylinders have the same weight of 20 lb.

18.76   The 50-kg carriage E is supported by a 20-kg cylinder A, a 25-kg cylinder B, a 30-kg cylinder C, and a 35-kg cylinder D as shown, where the cylinders have the same radius of 100 mm. If the cylinders roll without slipping, determine the velocity $\mathbf{v}_E$ attained by the carriage E after the 20-N force has been applied to move it from rest for 1.2 s.

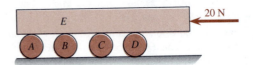

Fig. P18.76

18.77*   Solve Prob. 18.76 if all cylinders have the same mass of 20 kg.

18.78* through 18.83*   Using the *principle of generalized virtual work*, solve the following problems:

18.78* Prob. 18.72.	**18.79*** Prob. 18.73*.
18.80* Prob. 18.74*.	**18.81*** Prob. 18.75.
18.82* Prob. 18.76.	**18.83*** Prob. 18.77*.

18.8   Concluding Remarks

Clearly, this chapter is a direct extension of Chap. 14. We recognize that, unlike a particle which only translates in space, a rigid body may rotate about its central axis and translate in space. By Chasles' theorem, the *momentum* of a rigid body in plane motion may be described by the *linear momentum* $m\mathbf{v}$ acting through the mass center G of the rigid body and the *angular momentum* $\bar{I}\omega$ acting about G. The abilities to properly represent the angular impulse of a moment and the momentum of a rigid body in plane motion are the additional important skills to be developed when one progresses from Chap. 14 to Chap. 18.

     The *principle of impulse and momentum* for rigid bodies is very useful in attacking (1) kinetic problems of rigid bodies in plane motion where *time* and *velocity* (among others) are to be directly related, and (2) problems of *eccentric impact* between rigid bodies. The *drawing* of appropriate *vector-diagram equations* for the momenta and impulses in the system during the time interval considered is recommended because it provides valuable geometric aid in the solutions of such problems.

     The *principle of generalized virtual work* for rigid bodies is especially well adapted to the study of the plane kinetics of systems of rigid bodies with one degree of freedom, where impulse and momentum are involved.

As demonstrated in Example 18.10, this principle provides a procedure which enables us to *uncouple an unknown quantity* in a seemingly complex one-degree-of-freedom system. In problems of *eccentric impact*, this principle may help reduce the number of coupled simultaneous equations. Therefore, it is an effective complement to the principle of impulse and momentum.

REVIEW PROBLEMS

18.84 A 10-kg cylinder B is suspended by a cord which is wrapped around the drum of a 100-kg flywheel A as shown. The flywheel has a central radius of gyration $\bar{k}_O = 0.5$ m and the system is released from rest at the time $t = 0$. Neglecting effects of friction, determine the velocity of the cylinder when $t = 4$ s.

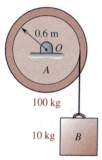

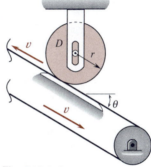

Fig. P18.84   Fig. P18.85*   Fig. P18.86*

18.85* A 100-lb gear of central radius of gyration $\bar{k}_O = 0.75$ ft is rolled up a rack from rest at the time $t = 0$ by the force $\mathbf{P} = 2(t + 15)$ lb $\nearrow$ which is applied to a cord wrapped around the drum of the gear and remains parallel to the rack as shown, where t is measured in seconds, $r_1 = 0.5$ ft, and $r_2 = 1$ ft. Determine the angular velocity of the gear when $t = 5$ s.

18.86* The disk D of radius r is released from rest at the time $t = 0$ to press on the belt, which moves with a constant velocity $\mathbf{v}$ as shown. If the coefficient of kinetic friction between the disk and the belt is μ_k, determine the time t_2 at which slipping of the disk on the belt ceases. (*Hint.* The answer is a function of v, μ_k, θ, and g.)

18.87 Solve Prob. 18.86* if $v = 10$ m/s, $\mu_k = 0.2$, and $\theta = 20°$.

18.88 The gears A and B shown have masses 5 kg and 12 kg, and central radii of gyration 80 mm and 120 mm, respectively. The gear A is rotated from rest at the time $t = 0$ by a moment $\mathbf{M} = 12t$ N·m $\circlearrowleft$, where t is measured in seconds. Determine the time t at which the angular velocity of the gear B is 300 rpm $\circlearrowright$.

18.89* Solve Prob. 18.88 if $\mathbf{M} = 12t^2$ N·m $\circlearrowleft$.

18.90 A solid sphere rolls without slipping with a horizontal velocity $\mathbf{v}_1 = 15$ m/s $\rightarrow$ when it hits the incline shown, where $\theta = 25°$. If the sphere does not rebound, determine (a) the new velocity $\mathbf{v}_2$ of the sphere as it starts up the incline, (b) the percentage of loss of kinetic energy of the sphere due to impact.

18.91* Solve Prob. 18.90 if $\mathbf{v}_1 = 30$ ft/s $\rightarrow$ and $\theta = 15°$.

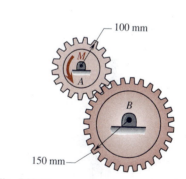

Fig. P18.88

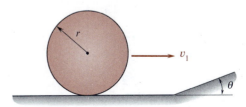

Fig. P18.90

18.92   A 120-lb acrobat stands at the end A of a 100-lb rigid plank ACB pivoted at C when a 150-lb weight W is dropped from a height $h_B = 10$ ft to hit the end B of the plank. If the impact at B is perfectly plastic and the acrobat stands completely rigid, determine the height h_A to which the acrobat at A will rise.

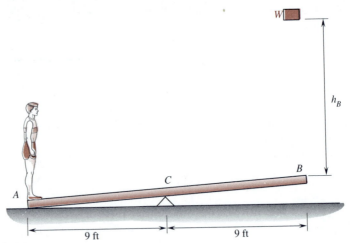

Fig. P18.92

18.93* Solve Prob. 18.92 if $h_B = 15$ ft.

18.94   A slender rod BC is connected by a cord to the support at A and is released from rest in the position shown. If the impact between the end C and the corner D is perfectly plastic, determine the angular velocity of the rod immediately after the impact.

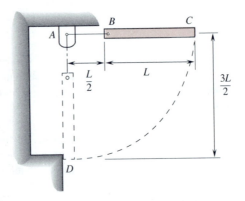

Fig. P18.94

18.95* Solve Prob. 18.94 if the impact is perfectly elastic.

18.96   Solve Prob. 18.94 if the coefficient of restitution between the end C and the corner D is $e = 0.6$.

18.97   The 10-kg rod AB rotates with $\omega = 4$ rad/s $\circlearrowright$ when it strikes the corner C as shown. If the coefficient of restitution between the rod and the corner C is $e = 0.5$, determine the maximum angular displacement of the rod after the impact.

18.98* Solve Prob. 18.97 if $e = 0.8$.

18.99   The 20-lb slender rod is released from rest in the position shown, where friction between the knobs A and B of the rod and their supports is negligible during

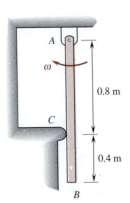

Fig. P18.97

the sliding. If impacts between the floor and the knobs of the rod are perfectly plastic and are completed in 0.1 s, determine the average impulsive reactions from the floor to the knobs of the rod.

18.100 A section of pipe weighing 64.4 lb rolls without slipping with $\omega_1 = 10$ rad/s $\circlearrowright$ before it crosses a 14-in. gap as shown. Assuming the impact to be perfectly plastic, determine the new angular velocity ω_2 of the pipe after it has crossed the gap to the other side.

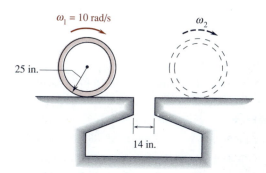

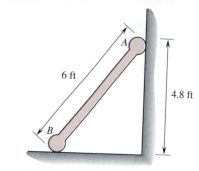

Fig. P18.99

Fig. P18.100

18.101* Refer to Prob. 18.100. Determine the minimum magnitude of ω_1 with which the pipe will cross the gap to the other side.

18.102* The 50-kg carriage C is supported by a 20-kg disk A and a 15-kg disk B as shown. Knowing that the disks roll without slipping, determine the velocity of the carriage 5 s after it is released from rest.

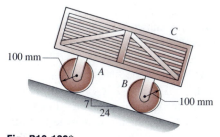

Fig. P18.102*

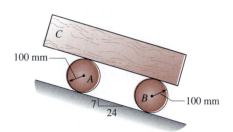

Fig. P18.104

18.103 Solve Prob. 18.102* if both disks A and B have the same mass of 20 kg.

18.104 A 50-kg wooden block C is supported by a 20-kg cylinder A and a 15-kg cylinder B as shown. Knowing that the cylinders roll without slipping, determine the velocity of the block 1.2 s after the system is released from rest.

18.105* Solve Prob. 18.104 if both cylinders A and B have the same mass of 20 kg.

Chapter 19

Motion of Rigid Bodies in Three Dimensions

KEY CONCEPTS

- Relation between the time derivatives of a vector taken in two different reference frames.
- Velocities and accelerations in different reference frames.
- Addition theorems for angular velocities and accelerations.
- Momentum and kinetic energy of a rigid body in general motion.
- Equations of motion for a rigid body in three-dimensional space, Euler's equations of motion, and Eulerian angles.
- Gyroscope, spin, nutation, precession, and steady precession.

The motion of rigid bodies in three-dimensional space is treated in this chapter. The fundamental relation between the time derivatives of a vector taken in two different reference frames is emphasized early in the chapter. The addition theorem for angular accelerations is shown to be different in form from that for angular velocities in general motion. Formulas for linear velocities and accelerations in different reference frames are given. The study of the kinetics of rigid bodies in space is based on the fundamental equations $\Sigma \mathbf{F} = \dot{\mathbf{L}}$ for translational motion, $\Sigma \mathbf{M}_G = \dot{\mathbf{H}}_G$ and $\Sigma \mathbf{M}_O = \dot{\mathbf{H}}_O$ for rotational motion, and their specialized forms for specific situations. In particular, Euler's equations of motion are derived, and the usual steps to be taken prior to applying Euler's equations of motion are outlined at the end of Sec. 19.8. Finally, we cover the gyroscopic motion and the cases of steady precession of axisymmetrical bodies either under a torque or in torque-free systems.

KINEMATICS OF RIGID BODIES IN SPACE

★19.1 Time Derivatives of a Vector in Two Reference Frames

Let $A\xi\eta\zeta$ be a reference frame *translating* in the fixed reference frame $OXYZ$, where the axes of $A\xi\eta\zeta$ remain parallel to the corresponding axes of $OXYZ$, as shown in Fig. 19.1. Then, the unit vectors $\mathbf{i}$, $\mathbf{j}$, $\mathbf{k}$ along the axes of $A\xi\eta\zeta$ are respectively equal to the unit vectors $\mathbf{I}$, $\mathbf{J}$, $\mathbf{K}$ along the axes of

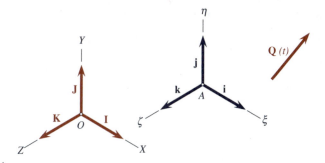

Fig. 19.1

$OXYZ$, and the scalar components (Q_ξ, Q_η, Q_ζ) of any vector $\mathbf{Q}$ in $A\xi\eta\zeta$ are correspondingly equal to its scalar components (Q_X, Q_Y, Q_Z) in $OXYZ$ at any given instant. This means that the mathematical descriptions of $\mathbf{Q}$ in the *fixed reference frame OXYZ* and in the *translating reference frame A$\xi\eta\zeta$* are identical, and the time derivatives of $\mathbf{Q}$ in $OXYZ$ and in $A\xi\eta\zeta$ must be equal; i.e.,

$$\left(\frac{d\mathbf{Q}}{dt}\right)_{OXYZ} = \left(\frac{d\mathbf{Q}}{dt}\right)_{A\xi\eta\zeta} \tag{19.1}$$†

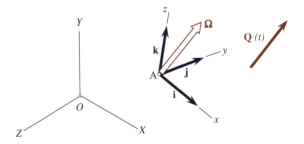

Fig. 19.2

Suppose that $Axyz$ is a reference frame moving arbitrarily (e.g., both translating and rotating) in the fixed reference frame $OXYZ$ as shown in Fig. 19.2, where $\boldsymbol{\Omega}$ is the *angular velocity* of $Axyz$ in $OXYZ$. As $Axyz$ moves in $OXYZ$, the time rates of change of the unit vectors $\mathbf{i}$, $\mathbf{j}$, $\mathbf{k}$ in $Axyz$ are zero because they are constant in both magnitude and direction in $Axyz$; i.e.,

$$(\dot{\mathbf{i}})_{Axyz} = (\dot{\mathbf{j}})_{Axyz} = (\dot{\mathbf{k}})_{Axyz} = 0 \tag{19.2}$$

On the other hand, the time rates of change of the unit vectors $\mathbf{i}$, $\mathbf{j}$, $\mathbf{k}$ in $OXYZ$ are generally not zero because their directions change from time to time as $Axyz$ moves in $OXYZ$, even though each of them has a constant magnitude of 1 unit. As shown in Fig. 19.3, the unit vector $\mathbf{i}$ is changed to $\mathbf{i}'$ during the time interval Δt, and $\Delta\mathbf{i}$ represents the change in $\mathbf{i}$ due to the rotation $\boldsymbol{\Omega}\,\Delta t$ of $Axyz$. For $\Delta t \approx 0$, we note from Fig. 19.3 that the magnitude and direction of $\Delta\mathbf{i}$ are given by the *cross product*

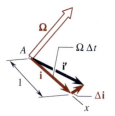

Fig. 19.3

$$\Delta\mathbf{i} = (\boldsymbol{\Omega}\,\Delta t) \times \mathbf{i} \tag{19.3}$$

---

†Vectors in kinematics are *free vectors*. A free vector is characterized only by its *magnitude* and its *direction*, but not its line of action.

Thus, the time derivative of **i** in *OXYZ* is

$$(\dot{\mathbf{i}})_{OXYZ} = \lim_{\Delta t \to 0} \frac{\Delta \mathbf{i}}{\Delta t} = \boldsymbol{\Omega} \times \mathbf{i}$$

NOTE: A quantity, or the derivative of a quantity, without the name of a specific reference frame as subscripts is to be understood as one measured, or taken, in the fixed reference frame, such as *OXYZ*.

We write

$$\dot{\mathbf{i}} = \boldsymbol{\Omega} \times \mathbf{i} \tag{19.4}$$

Similarly, we can show that

$$\dot{\mathbf{j}} = \boldsymbol{\Omega} \times \mathbf{j} \qquad \dot{\mathbf{k}} = \boldsymbol{\Omega} \times \mathbf{k} \tag{19.5}$$

For a vector **Q** in *Axyz* as shown in Fig. 19.2, we write

$$\mathbf{Q} = Q_x \mathbf{i} + Q_y \mathbf{j} + Q_z \mathbf{k} \tag{19.6}$$

Taking the time derivative in *Axyz* on both sides of Eq. (19.6) and applying Eqs. (19.2) and (15.41), we write

$$(\dot{\mathbf{Q}})_{Axyz} = (\dot{Q}_x)_{Axyz} \mathbf{i} + (\dot{Q}_y)_{Axyz} \mathbf{j} + (\dot{Q}_z)_{Axyz} \mathbf{k}$$

$$(\dot{Q}_x)_{Axyz} = (\dot{Q}_x)_{OXYZ} = \dot{Q}_x$$

$$(\dot{Q}_y)_{Axyz} = (\dot{Q}_y)_{OXYZ} = \dot{Q}_y$$

$$(\dot{Q}_z)_{Axyz} = (\dot{Q}_z)_{OXYZ} = \dot{Q}_z$$

$$(\dot{\mathbf{Q}})_{Axyz} = \dot{Q}_x \mathbf{i} + \dot{Q}_y \mathbf{j} + \dot{Q}_z \mathbf{k} \tag{19.7}$$

Taking the time derivative in *OXYZ* on both sides of Eq. (19.6) and applying Eqs. (15.41), (19.4), (19.5), and (19.7), we write

$$\dot{\mathbf{Q}} = (\dot{Q}_x \mathbf{i} + \dot{Q}_y \mathbf{j} + \dot{Q}_z \mathbf{k}) + Q_x \dot{\mathbf{i}} + Q_y \dot{\mathbf{j}} + Q_z \dot{\mathbf{k}}$$

$$= (\dot{\mathbf{Q}})_{Axyz} + Q_x (\boldsymbol{\Omega} \times \mathbf{i}) + Q_y (\boldsymbol{\Omega} \times \mathbf{j}) + Q_z (\boldsymbol{\Omega} \times \mathbf{k})$$

$$= (\dot{\mathbf{Q}})_{Axyz} + \boldsymbol{\Omega} \times (Q_x \mathbf{i} + Q_y \mathbf{j} + Q_z \mathbf{k})$$

$$\dot{\mathbf{Q}} = (\dot{\mathbf{Q}})_{Axyz} + \boldsymbol{\Omega} \times \mathbf{Q} \tag{19.8}[†]$$

Equation (19.8) is an important equation in this chapter; it permits us easily to calculate the time derivative of a vector in one reference frame if that vector is expressed in another. We see from Eq. (19.8) that the time derivative of any vector **Q** taken in the *fixed reference frame OXYZ* is equal to the sum of the following two parts: (a) the time derivative of **Q** taken in the *moving reference frame Axyz*, and (b) the cross product $\boldsymbol{\Omega} \times \mathbf{Q}$, where $\boldsymbol{\Omega}$ is the *angular velocity* of *Axyz* in *OXYZ*.

[†]Cf. Eq. (15.44).

Developmental Exercises

> D19.1 In the notation adopted, is the time derivative $\dot{\mathbf{Q}}$ taken in the *moving reference frame Axyz* or the *fixed reference frame OXYZ?*
>
> D19.2 Suppose that $\mathbf{Q}$ is a *constant* vector in *Axyz*, whose angular velocity in *OXYZ* is $\boldsymbol{\Omega}$. Express the first time derivative $\dot{\mathbf{Q}}$ and the second time derivative $\ddot{\mathbf{Q}}$ taken in *OXYZ*.

★19.2 Velocities in Different Reference Frames

In certain practical situations, a particle may move relative to two reference frames (or bodies) of interest to us. For example, a pin may be confined to move in a slot of a crank which is itself in motion relative to the ground. The general situation may be represented in Fig. 19.4, where a particle B moves along a path C in the reference frame *Axyz*, which is itself moving (i.e., translating and rotating) in the fixed reference frame *OXYZ*. For our study, we let the angular velocity of *Axyz* in *OXYZ* be $\boldsymbol{\Omega}$. The positions of A and B in *OXYZ* are specified by the position vectors $\mathbf{r}_A$ and $\mathbf{r}_B$, while the position of B relative to A is specified by the position vector $\mathbf{r}_{B/A}$.

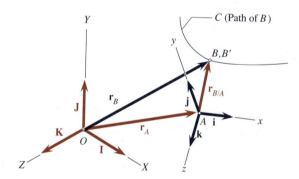

Fig. 19.4

I. LINEAR VELOCITIES IN *OXYZ* AND *Axyz*.

The analytical expression of $\mathbf{r}_{B/A}$ may be written in terms of the unit vectors $\mathbf{i}$, $\mathbf{j}$, and $\mathbf{k}$ in *Axyz*, or the unit vectors $\mathbf{I}$, $\mathbf{J}$, and $\mathbf{K}$ in *OXYZ*, as shown in Fig. 19.4. Letting (x_B, y_B, z_B) be the coordinates of B in *Axyz*, we write

$$\mathbf{r}_{B/A} = x_B\mathbf{i} + y_B\mathbf{j} + z_B\mathbf{k} \tag{19.9}$$

$$\mathbf{v}_{B/Axyz} = (\dot{\mathbf{r}}_{B/A})_{Axyz} = \dot{x}_B\mathbf{i} + \dot{y}_B\mathbf{j} + \dot{z}_B\mathbf{k} \tag{19.10}$$

Referring to Fig. 19.4, we write

$$\mathbf{r}_B = \mathbf{r}_A + \mathbf{r}_{B/A}$$

$$\mathbf{r}_B = \mathbf{r}_{B/A} + \mathbf{r}_A \tag{19.11}$$

Taking the time derivative in *OXYZ* on both sides of Eq. (19.11) and applying Eq. (19.8), we write

$$\dot{\mathbf{r}}_B = \dot{\mathbf{r}}_{B/A} + \dot{\mathbf{r}}_A = [(\dot{\mathbf{r}}_{B/A})_{Axyz} + \boldsymbol{\Omega} \times \mathbf{r}_{B/A}] + \dot{\mathbf{r}}_A$$

$$\mathbf{v}_B = \mathbf{v}_{B/Axyz} + \boldsymbol{\Omega} \times \mathbf{r}_{B/A} + \mathbf{v}_A \qquad (19.12)$$

where

$$\mathbf{r}_{B/A} = \overrightarrow{AB} = \text{position vector of } B \text{ relative to } A$$
$$\mathbf{v}_A, \mathbf{v}_B = \text{velocities of } A \text{ and } B \text{ in the } \textit{fixed} \text{ reference frame } OXYZ$$
$$\mathbf{v}_{B/Axyz} = \text{velocity of } B \text{ in the } \textit{moving} \text{ reference frame } Axyz$$
$$\boldsymbol{\Omega} = \text{angular velocity of } Axyz \text{ in } OXYZ$$

Suppose that B' is a point *of* the moving reference frame $Axyz$, and B' *coincides* with the moving particle B at the instant considered, as shown in Fig. 19.4. Following similar steps as those used in deriving Eqs. (15.50) and (15.52), we can show that the velocity of B' in $OXYZ$ is

$$\mathbf{v}_{B'} = \boldsymbol{\Omega} \times \mathbf{r}_{B/A} + \mathbf{v}_A \qquad (19.13)$$

and the velocity of the particle B in $OXYZ$ can be expressed as

$$\mathbf{v}_B = \mathbf{v}_{B/Axyz} + \mathbf{v}_{B'} \qquad (19.14)$$

II. ANGULAR VELOCITY $\boldsymbol{\Omega}$ OF *Axyz* IN *OXYZ*.

For defining the orientation of $Axyz$ in $OXYZ$, as shown in Fig. 19.4, let (a_{11}, a_{12}, a_{13}), (a_{21}, a_{22}, a_{23}), and (a_{31}, a_{32}, a_{33}) be the sets of *direction cosines* of the unit vectors $\mathbf{i}$, $\mathbf{j}$, and $\mathbf{k}$ with respect to the fixed reference frame $OXYZ$, respectively. Then, we have

$$\mathbf{i} = a_{11}\mathbf{I} + a_{12}\mathbf{J} + a_{13}\mathbf{K} \qquad (19.15)$$

$$\mathbf{j} = a_{21}\mathbf{I} + a_{22}\mathbf{J} + a_{23}\mathbf{K} \qquad (19.16)$$

$$\mathbf{k} = a_{31}\mathbf{I} + a_{32}\mathbf{J} + a_{33}\mathbf{K} \qquad (19.17)$$

where $\mathbf{I}$, $\mathbf{J}$, and $\mathbf{K}$ are the unit vectors directed along the axes of $OXYZ$. The direction cosines a_{mn} (m, $n = 1, 2, 3$) are functions of the time t as $Axyz$ moves in $OXYZ$. Therefore, the time derivatives of $\mathbf{i}$, $\mathbf{j}$, and $\mathbf{k}$ taken in $OXYZ$ are

$$\dot{\mathbf{i}} = \dot{a}_{11}\mathbf{I} + \dot{a}_{12}\mathbf{J} + \dot{a}_{13}\mathbf{K} \qquad (19.18)$$

$$\dot{\mathbf{j}} = \dot{a}_{21}\mathbf{I} + \dot{a}_{22}\mathbf{J} + \dot{a}_{23}\mathbf{K} \qquad (19.19)$$

$$\dot{\mathbf{k}} = \dot{a}_{31}\mathbf{I} + \dot{a}_{32}\mathbf{J} + \dot{a}_{33}\mathbf{K} \qquad (19.20)$$

Next, let the *angular velocity* $\boldsymbol{\Omega}$ of $Axyz$ in $OXYZ$ be written as

$$\boldsymbol{\Omega} = \Omega_x\mathbf{i} + \Omega_y\mathbf{j} + \Omega_z\mathbf{k} \qquad (19.21)$$

Then, Eqs. (19.6), (19.8), and (19.21) yield

$$\dot{\mathbf{Q}} = (\dot{\mathbf{Q}})_{Axyz} + (\Omega_x\mathbf{i} + \Omega_y\mathbf{j} + \Omega_z\mathbf{k}) \times (Q_x\mathbf{i} + Q_y\mathbf{j} + Q_z\mathbf{k})$$

$$\dot{\mathbf{Q}} = (\dot{\mathbf{Q}})_{Axyz} + [Q_x(-\Omega_y\mathbf{k} + \Omega_z\mathbf{j}) + Q_y(-\Omega_z\mathbf{i} + \Omega_x\mathbf{k})$$
$$+ Q_z(-\Omega_x\mathbf{j} + \Omega_y\mathbf{i})] \qquad (19.22)$$

Taking the time derivative on both sides of Eq. (19.6) in *OXYZ* and using Eq. (19.7), we write

$$\dot{\mathbf{Q}} = (\dot{Q}_x\mathbf{i} + \dot{Q}_y\mathbf{j} + \dot{Q}_z\mathbf{k}) + (Q_x\dot{\mathbf{i}} + Q_y\dot{\mathbf{j}} + Q_z\dot{\mathbf{k}})$$

$$\dot{\mathbf{Q}} = (\dot{\mathbf{Q}})_{Axyz} + (Q_x\dot{\mathbf{i}} + Q_y\dot{\mathbf{j}} + Q_z\dot{\mathbf{k}}) \qquad (19.23)$$

Comparing Eq. (19.23) with Eq. (19.22), we see that

$$\dot{\mathbf{i}} = -\Omega_y\mathbf{k} + \Omega_z\mathbf{j} \qquad (19.24)$$

$$\dot{\mathbf{j}} = -\Omega_z\mathbf{i} + \Omega_x\mathbf{k} \qquad (19.25)$$

$$\dot{\mathbf{k}} = -\Omega_x\mathbf{j} + \Omega_y\mathbf{i} \qquad (19.26)$$

By dot products, we find from Eqs. (19.24) through (19.26) that

$$\Omega_x = \mathbf{j}\cdot\mathbf{k} = -\dot{\mathbf{k}}\cdot\mathbf{j} \qquad (19.27)$$

$$\Omega_y = \dot{\mathbf{k}}\cdot\mathbf{i} = -\dot{\mathbf{i}}\cdot\mathbf{k} \qquad (19.28)$$

$$\Omega_z = \dot{\mathbf{i}}\cdot\mathbf{j} = -\dot{\mathbf{j}}\cdot\mathbf{i} \qquad (19.29)$$

Substituting Eqs. (19.27) through (19.29) into Eq. (19.21), we arrive at an equation defining the *angular velocity* $\boldsymbol{\Omega}$ of *Axyz* in *OXYZ* as follows:

$$\boldsymbol{\Omega} = (\dot{\mathbf{j}}\cdot\mathbf{k})\mathbf{i} + (\dot{\mathbf{k}}\cdot\mathbf{i})\mathbf{j} + (\dot{\mathbf{i}}\cdot\mathbf{j})\mathbf{k} \qquad (19.30)^\dagger$$

Thus, $\boldsymbol{\Omega}$ can be found when the time-dependent direction cosines a_{mn} (m, $n = 1, 2, 3$) in Eqs. (19.15) through (19.17) are known.

III. ADDITION THEOREM FOR ANGULAR VELOCITIES.
For a system consisting of several movable rigid bodies, it can be advantageous to use multiple reference frames, such as those shown in Fig. 19.5, where *OXYZ* continues to represent the *fixed reference frame*, and *Axyz*, *Bx'y'z'*, and *Cx"y"z"* are reference frames embedded in rigid bodies *A*, *B*, and *C*, respectively. The relations among the angular velocities of the various rigid bodies

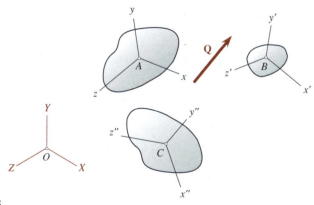

Fig. 19.5

†Cf. the footnote on p. 696.

with respect to each other are frequently utilized in the analysis of systems of rigid bodies moving in three dimensions. Suppose that we let $\omega_{A/B}$ denote *the angular velocity of the rigid body A* (and hence the reference frame *Axyz*) *with respect to the rigid body B* (and hence the reference frame *Bx'y'z'*), or more briefly *the angular velocity of A in B*. Similarly, let $\omega_{B/A}$, $\omega_{C/A}$, and $\omega_{C/B}$ denote the angular velocities of *B* in *A*, *C* in *A*, and *C* in *B*, respectively. Then, for any vector **Q** as shown in Fig. 19.5, we apply Eq. (19.8) to write

$$(\dot{\mathbf{Q}})_{Bx'y'z'} = (\dot{\mathbf{Q}})_{Axyz} + \omega_{A/B} \times \mathbf{Q} \qquad (19.31)$$

$$(\dot{\mathbf{Q}})_{Axyz} = (\dot{\mathbf{Q}})_{Bx'y'z'} + \omega_{B/A} \times \mathbf{Q} \qquad (19.32)$$

Adding Eqs. (19.31) and (19.32) and simplifying, we obtain

$$(\omega_{A/B} + \omega_{B/A}) \times \mathbf{Q} = \mathbf{0} \qquad (19.33)$$

Since Eq. (19.33) must hold true for any vector **Q**, we conclude that

$$\omega_{A/B} = -\omega_{B/A} \qquad (19.34)$$

Furthermore, referring to Fig. 19.5 and applying Eq. (19.8), we write

$$(\dot{\mathbf{Q}})_{Bx'y'z'} = (\dot{\mathbf{Q}})_{Cx''y''z''} + \omega_{C/B} \times \mathbf{Q} \qquad (19.35)$$

$$(\dot{\mathbf{Q}})_{Cx''y''z''} = (\dot{\mathbf{Q}})_{Axyz} + \omega_{A/C} \times \mathbf{Q} \qquad (19.36)$$

Subtraction of Eqs. (19.35) and (19.36) from Eq. (19.31) leads to

$$(\omega_{A/B} - \omega_{A/C} - \omega_{C/B}) \times \mathbf{Q} = \mathbf{0} \qquad (19.37)$$

Since Eq. (19.37) must hold true for any vector **Q**, we conclude that

$$\omega_{A/B} = \omega_{A/C} + \omega_{C/B} \qquad (19.38)$$

This property can be extended to *any number* of rigid bodies (or reference frames). For instance, we write

$$\omega_{A/O} = \omega_{A/B} + \omega_{B/O} = \omega_{A/B} + (\omega_{B/C} + \omega_{C/O})$$

$$\omega_{A/O} = \omega_{A/B} + \omega_{B/C} + \omega_{C/O} \qquad (19.39)$$

where the subscript *O* refers to the fixed reference frame *OXYZ*. Following our adopted convention of notations for quantities measured in the fixed reference frame, we may write Eq. (19.39) as

$$\omega_A = \omega_{A/B} + \omega_{B/C} + \omega_C \qquad (19.40)$$

The relation expressed in Eq. (19.38) or (19.40) is called the *addition theorem for angular velocities*. The validity of this theorem may seem intuitively obvious to the reader, but it will be shown in Sec. 19.3 that a similar relation does *not* hold true for angular accelerations in *general motion.*[†]

[†]Cf. Eqs. (19.48) and (19.50).

EXAMPLE 19.1

A connecting rod AB is attached to a disk by a ball-and-socket joint at A and to a collar at B by a clevis as shown. The disk is mounted on a vertical shaft at O and rotates in a horizontal plane at a constant rate of $\boldsymbol{\omega}_1 = 29\mathbf{J}$ rad/s. The collar at B slides freely on the vertical rod CD, which is fixed at C. For the position shown, determine (a) the velocity of the collar at B, (b) the angular velocity of the rod AB.

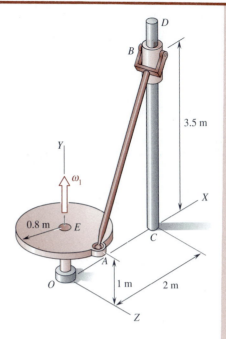

Solution.   We first let the reference frame $Axyz$ be embedded in the rod AB as indicated in the separate sketch, where the reference frame $OXYZ$ is fixed on the ground. Since the point A is attached to the disk and the collar at B slides freely on the vertical rod CD, we have

$$\mathbf{v}_A = \boldsymbol{\omega}_1 \times \mathbf{r}_{A/E} = 29\mathbf{J} \times 0.8\mathbf{K} = 23.2\mathbf{I}$$

$$\mathbf{v}_B = v_B\mathbf{J} \qquad \mathbf{r}_{B/A} = 2\mathbf{I} + 2.5\mathbf{J} - 0.8\mathbf{K}$$

The angular velocity $\boldsymbol{\omega}_{AB}$ of the rod AB is the same as the angular velocity $\boldsymbol{\Omega}$ of $Axyz$ in $OXYZ$. Expressing in rectangular components, we write

$$\boldsymbol{\omega}_{AB} = \boldsymbol{\Omega} = \Omega_X\mathbf{I} + \Omega_Y\mathbf{J} + \Omega_Z\mathbf{K}$$

Since B does not move in $Axyz$, we see that $\mathbf{v}_{B/Axyz} = \mathbf{0}$. Applying Eq. (19.12), we write

$$\mathbf{v}_B = \mathbf{v}_{B/Axyz} + \boldsymbol{\Omega} \times \mathbf{r}_{B/A} + \mathbf{v}_A$$

$$v_B\mathbf{J} = \mathbf{0} + \begin{vmatrix} \mathbf{I} & \mathbf{J} & \mathbf{K} \\ \Omega_X & \Omega_Y & \Omega_Z \\ 2 & 2.5 & -0.8 \end{vmatrix} + 23.2\mathbf{I}$$

$$v_B\mathbf{J} = (-0.8\Omega_Y - 2.5\Omega_Z + 23.2)\mathbf{I} + (0.8\Omega_X + 2\Omega_Z)\mathbf{J}$$
$$+ (2.5\Omega_X - 2\Omega_Y)\mathbf{K}$$

Equating the coefficients of the unit vectors, we obtain

$$0 = -0.8\Omega_Y - 2.5\Omega_Z + 23.2 \tag{1}$$

$$v_B = 0.8\Omega_X + 2\Omega_Z \tag{2}$$

$$0 = 2.5\Omega_X - 2\Omega_Y \tag{3}$$

The ball-and-socket joint at A and the collar-clevis connection at B allow the rod AB to rotate about the vertical rod CD and also about an axis perpendicular to the plane containing AB and CD. However, the rod AB *cannot* rotate about any axis (e.g., AF) which is perpendicular to CD and lies in the plane containing AB and CD. Thus, we see that $\boldsymbol{\Omega} \perp \overrightarrow{AF}$, or $\boldsymbol{\Omega} \cdot \overrightarrow{AF} = 0$. Since $\overrightarrow{AF} = 2\mathbf{I} - 0.8\mathbf{K}$, we write

$$(\Omega_X\mathbf{I} + \Omega_Y\mathbf{J} + \Omega_Z\mathbf{K}) \cdot (2\mathbf{I} - 0.8\mathbf{K}) = 0$$

$$2\Omega_X - 0.8\Omega_Z = 0 \tag{4}$$

Equations (1) through (4) yield

$$v_B = 18.56 \qquad \Omega_X = 3.2 \qquad \Omega_Y = 4 \qquad \Omega_Z = 8$$

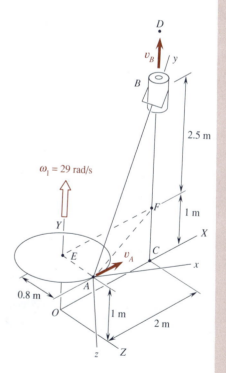

We obtain

$$\mathbf{v}_B = 18.56\mathbf{J} \text{ m/s} \blacktriangleleft$$

$$\boldsymbol{\omega}_{AB} = 3.2\mathbf{I} + 4\mathbf{J} + 8\mathbf{K} \text{ rad/s} \blacktriangleleft$$

REMARK. If the clevis at B is replaced with a ball-and-socket joint, then the rod AB is free to rotate about its axis AB. In this case, Eq. (4) becomes invalid and Eqs. (1) through (3) yield $\Omega_X = 0.8\Omega_Y$, $\Omega_Z = 9.28 - 0.32\Omega_Y$, and $v_B = 18.56$ m/s. Thus, v_B is determinate even though Ω_X, Ω_Y, and Ω_Z are indeterminate in the modified system.

Developmental Exercises

D19.3 Work Example 19.1 for the instant at which the joint A passes through the position $(0.8, 1, 0)$ m.

D19.4 Work Example 19.1 for the instant at which the joint A passes through the position $(0, 1, -0.8)$ m.

D19.5 Work Example 19.1 for the instant at which the joint A passes through the position $(-0.8, 1, 0)$ m.

EXAMPLE 19.2

The universal joint at U permits the transmission of power between the two shafts A and B which are out of alignment by an angle ψ as shown. The legs of the crosspiece C at U are perpendicular to each other, and the orientation of the yoke of A is defined by the angle θ which is measured from the vertical axis at U. If the angular speeds of the shafts A and B are ω_A and ω_B, respectively, determine the ratio ω_B/ω_A and interpret the result.

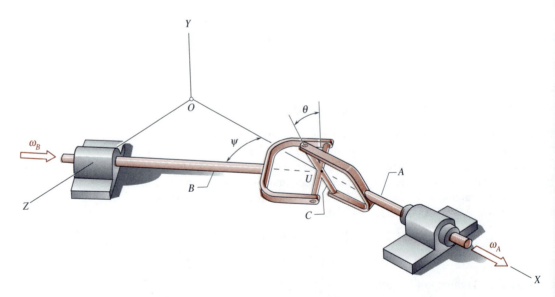

Solution. Let the unit vectors along the legs of the crosspiece C be $\mathbf{e}_1$ and $\mathbf{e}_2$, respectively, and the unit vector along the shaft B be $\boldsymbol{\lambda}$ as shown.

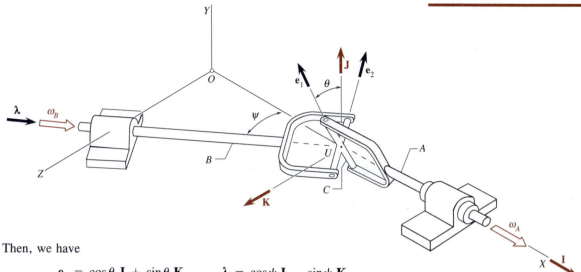

Then, we have

$$\mathbf{e}_1 = \cos\theta\,\mathbf{J} + \sin\theta\,\mathbf{K} \qquad \boldsymbol{\lambda} = \cos\psi\,\mathbf{I} - \sin\psi\,\mathbf{K}$$

Since $\mathbf{e}_2$ is perpendicular to both $\mathbf{e}_1$ and $\boldsymbol{\lambda}$, we have

$$\mathbf{e}_2 = \frac{\mathbf{e}_1 \times \boldsymbol{\lambda}}{|\mathbf{e}_1 \times \boldsymbol{\lambda}|}$$

$$= \frac{1}{G}\,(-\cos\theta\,\sin\psi\,\mathbf{I} + \sin\theta\,\cos\psi\,\mathbf{J} - \cos\theta\,\cos\psi\,\mathbf{K})$$

where

$$G = (\cos^2\theta + \sin^2\theta\,\cos^2\psi)^{1/2}$$

Applying the *addition theorem for angular velocities*, we write

$$\boldsymbol{\omega}_B = \boldsymbol{\omega}_{B/C} + \boldsymbol{\omega}_{C/A} + \boldsymbol{\omega}_A \tag{1}$$

Referring to the system shown, we write

$$\boldsymbol{\omega}_B = \omega_B\boldsymbol{\lambda} = \omega_B(\cos\psi\,\mathbf{I} - \sin\psi\,\mathbf{K})$$

$$\boldsymbol{\omega}_{B/C} = \omega_{B/C}\,\mathbf{e}_2 = \frac{\omega_{B/C}}{G}\,(-\cos\theta\,\sin\psi\,\mathbf{I} + \sin\theta\,\cos\psi\,\mathbf{J}$$

$$- \cos\theta\,\cos\psi\,\mathbf{K})$$

$$\boldsymbol{\omega}_{C/A} = \omega_{C/A}\mathbf{e}_1 = \omega_{C/A}(\cos\theta\,\mathbf{J} + \sin\theta\,\mathbf{K})$$

$$\boldsymbol{\omega}_A = \omega_A\mathbf{I}$$

Substituting the above expressions into Eq. (1) and equating the coefficients of the unit vectors, we have

$$\omega_B\cos\psi = -\frac{1}{G}\,\omega_{B/C}\cos\theta\,\sin\psi + \omega_A \tag{2}$$

$$0 = \frac{1}{G}\,\omega_{B/C}\sin\theta\,\cos\psi + \omega_{C/A}\cos\theta \tag{3}$$

$$-\omega_B\sin\psi = -\frac{1}{G}\,\omega_{B/C}\cos\theta\,\cos\psi + \omega_{C/A}\sin\theta \tag{4}$$

Eliminating $\omega_{C/A}$ in Eqs. (3) and (4), we get

$$\omega_{B/C} = G\omega_B \tan\psi \cos\theta \qquad (5)$$

Substituting Eq. (5) into Eq. (2) and simplifying, we obtain

$$\frac{\omega_B}{\omega_A} = \frac{\cos\psi}{1 - \sin^2\psi\,\sin^2\theta} \qquad \blacktriangleleft$$

When the misalignment angle ψ is equal to 0, the value of the ratio ω_B/ω_A becomes 1, or $\omega_B = \omega_A$, as expected. To aid in visualizing the result, we may make use of the digital root finder in App. E to plot the variation of this ratio versus θ for the case of $\psi = 60°$ as shown, where $Y = \omega_B/\omega_A$, $X = \theta$ (in degrees), and $0 \le \theta \le 180°$.

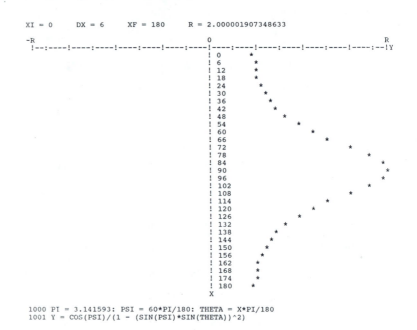

```
XI = 0        DX = 6        XF = 180        R = 2.000001907348633

-R                                    0                                    R
!--:----!----:----!----:----!----:----!----:----!----:----!----:----!----:--!Y
               ! 0                      *
               ! 6                      *
               ! 12                     *
               ! 18                     *
               ! 24                      *
               ! 30                       *
               ! 36                        *
               ! 42                         *
               ! 48                          *
               ! 54                           *
               ! 60                            *
               ! 66                              *
               ! 72                               *
               ! 78                                 *
               ! 84                                   *
               ! 90                                    *
               ! 96                                   *
               ! 102                                 *
               ! 108                               *
               ! 114                             *
               ! 120                           *
               ! 126                          *
               ! 132                        *
               ! 138                       *
               ! 144                      *
               ! 150                     *
               ! 156                    *
               ! 162                    *
               ! 168                    *
               ! 174                    *
               ! 180                    *
               X
```

```
1000 PI = 3.141593: PSI = 60*PI/180: THETA = X*PI/180
1001 Y = COS(PSI)/(1 - (SIN(PSI)*SIN(THETA))^2)
```

We can interpret from the same plot for the result that, for a given misalignment angle ψ in the range $0 < \psi < 90°$, the shaft whose yoke comes nearly or exactly perpendicular to the plane containing the axes of the two shafts must be turning more rapidly than the other shaft. However, the ratio ω_B/ω_A becomes zero and the system becomes inoperable when $\psi = 90°$. This is the *gimbal lock*.

Developmental Exercises

D19.6 Refer to Example 19.2. Making use of the digital root finder in App. E, plot the ratio ω_B/ω_A versus θ for $0 \le \theta \le 360°$ and (a) $\psi = 65°$, (b) $\psi = 70°$.

D19.7 Suppose that the time rates of change of the unit vectors **i**, **j**, and **k** along the axes of *Axyz*, as taken in *OXYZ*, are known. Define the angular velocity $\mathbf{\Omega}$ of *Axyz* in *OXYZ*.

D19.8 Describe the *addition theorem for angular velocities*.

★19.3 Accelerations in Different Reference Frames

I. LINEAR ACCELERATIONS IN *OXYZ* AND *Axyz*. We recall that the velocity $\mathbf{v}_B$ of a particle B, as shown in Fig. 19.4, is given by Eq. (19.12). Since the acceleration is the time derivative of the velocity, we take the time derivative in *OXYZ* on both sides of Eq. (19.12) and apply Eq. (19.8) to write

$$\dot{\mathbf{v}}_B = [(\dot{\mathbf{v}}_{B/Axyz})_{Axyz} + \mathbf{\Omega} \times \mathbf{v}_{B/Axyz}] + \dot{\mathbf{\Omega}} \times \mathbf{r}_{B/A}$$

$$+ \mathbf{\Omega} \times [(\dot{\mathbf{r}}_{B/A})_{Axyz} + \mathbf{\Omega} \times \mathbf{r}_{B/A}] + \dot{\mathbf{v}}_A$$

$$\mathbf{a}_B = \mathbf{a}_{B/Axyz} + \mathbf{\Omega} \times \mathbf{v}_{B/Axyz} + \dot{\mathbf{\Omega}} \times \mathbf{r}_{B/A}$$

$$+ \mathbf{\Omega} \times \mathbf{v}_{B/Axyz} + \mathbf{\Omega} \times (\mathbf{\Omega} \times \mathbf{r}_{B/A}) + \mathbf{a}_A$$

$$\mathbf{a}_B = \mathbf{a}_{B/Axyz} + \dot{\mathbf{\Omega}} \times \mathbf{r}_{B/A} + \mathbf{\Omega} \times (\mathbf{\Omega} \times \mathbf{r}_{B/A})$$

$$+ 2\mathbf{\Omega} \times \mathbf{v}_{B/Axyz} + \mathbf{a}_A \qquad (19.41)$$

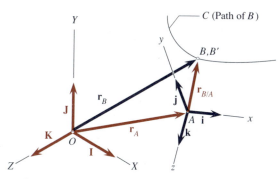

Fig. 19.4 (repeated)

Suppose that B' is a point *of* the moving reference frame *Axyz*, and B' *coincides* with the moving particle B at the instant considered, as shown in Fig. 19.4. Following similar steps as those used in deriving Eqs. (15.61) and (15.64), we can show that the acceleration of B' in *OXYZ* is

$$\mathbf{a}_{B'} = \dot{\mathbf{\Omega}} \times \mathbf{r}_{B/A} + \mathbf{\Omega} \times (\mathbf{\Omega} \times \mathbf{r}_{B/A}) + \mathbf{a}_A \qquad (19.42)$$

and the acceleration of the particle B in *OXYZ* can be expressed as

$$\mathbf{a}_B = \mathbf{a}_{B/Axyz} + \mathbf{a}_{B'} + 2\mathbf{\Omega} \times \mathbf{v}_{B/Axyz} \qquad (19.43)$$

Note in Eqs. (19.41) through (19.43) that $\mathbf{v}_{B/Axyz}$, $\mathbf{\Omega}$, and $\mathbf{r}_{B/A}$ are the same as those defined in Sec. 19.2; and

$\mathbf{a}_A$, $\mathbf{a}_B$ = accelerations of A and B in the *fixed* reference frame *OXYZ*

$\mathbf{a}_{B'}$ = acceleration of the *coinciding point B'* in *OXYZ*

$\mathbf{a}_{B/Axyz}$ = acceleration of B in the *moving* reference frame *Axyz*

$\dot{\mathbf{\Omega}}$ = angular acceleration of *Axyz* in *OXYZ*

Furthermore, note that $\boldsymbol{\Omega}$ is usually not perpendicular to $\mathbf{v}_{B/Axyz}$ and the magnitude of the Coriolis (or complementary) acceleration $2\boldsymbol{\Omega} \times \mathbf{v}_{B/Axyz}$ is generally *not* equal to $2\Omega v_{B/Axyz}$, as it was the case for the plane motion of a particle.

II. ADDITION THEOREM FOR ANGULAR ACCELERATIONS.

For a system consisting of several movable rigid bodies and reference frames as shown in Fig. 19.5, let $\boldsymbol{\alpha}_{A/B}$ denote *the angular acceleration of the rigid body A with respect to the rigid body B*, or more briefly *the angular acceleration of A in B*. Similarly, let $\boldsymbol{\alpha}_{A/C}$ and $\boldsymbol{\alpha}_{C/B}$ denote the angular accelerations of A in C, and C in B, respectively. Note in general motion that the *angular acceleration* of a rigid body in a reference frame is equal to the time derivative (taken in that reference frame) of the angular velocity (measured in that same reference frame) of the rigid body. Thus, we have

$$\boldsymbol{\alpha}_{A/B} = (\dot{\boldsymbol{\omega}}_{A/B})_{Bx'y'z'} \qquad (19.44)$$

$$\boldsymbol{\alpha}_{A/C} = (\dot{\boldsymbol{\omega}}_{A/C})_{Cx''y''z''} \qquad (19.45)$$

$$\boldsymbol{\alpha}_{C/B} = (\dot{\boldsymbol{\omega}}_{C/B})_{Bx'y'z'} \qquad (19.46)$$

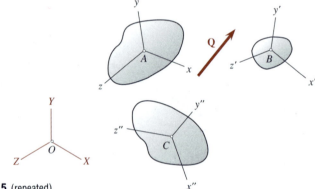

Fig. 19.5 (repeated)

Each term in the above equations involves *one* rigid body in *one* reference frame at a time and has a physical meaning. However, the quantity $(\dot{\boldsymbol{\omega}}_{A/C})_{Bx'y'z'}$ involves *one* rigid body in *two* reference frames and does not lend itself to a simple physical interpretation. By Eqs. (19.8) and (19.45) and the properties of cross products (cf. Sec. 4.8), we write

$$(\dot{\boldsymbol{\omega}}_{A/C})_{Bx'y'z'} = (\dot{\boldsymbol{\omega}}_{A/C})_{Cx''y''z''} + \boldsymbol{\omega}_{C/B} \times \boldsymbol{\omega}_{A/C}$$

$$(\dot{\boldsymbol{\omega}}_{A/C})_{Bx'y'z'} = \boldsymbol{\alpha}_{A/C} + \boldsymbol{\omega}_{C/B} \times \boldsymbol{\omega}_{A/C} \qquad (19.47)$$

Taking the time derivative in $Bx'y'z'$ on both sides of Eq. (19.38) and applying Eqs. (19.44) through (19.47), we write

$$(\dot{\boldsymbol{\omega}}_{A/B})_{Bx'y'z'} = (\dot{\boldsymbol{\omega}}_{A/C})_{Bx'y'z'} + (\dot{\boldsymbol{\omega}}_{C/B})_{Bx'y'z'}$$

$$\boldsymbol{\alpha}_{A/B} = (\boldsymbol{\alpha}_{A/C} + \boldsymbol{\omega}_{C/B} \times \boldsymbol{\omega}_{A/C}) + \boldsymbol{\alpha}_{C/B}$$

$$\boxed{\boldsymbol{\alpha}_{A/B} = \boldsymbol{\alpha}_{A/C} + \boldsymbol{\alpha}_{C/B} + \boldsymbol{\omega}_{C/B} \times \boldsymbol{\omega}_{A/C}} \qquad (19.48)^{\dagger}$$

[†]Cf. Eq. (19.38).

The property expressed in Eq. (19.48) can be extended to *any number* of rigid bodies (or reference frames). For instance, we write

$$\alpha_{A/O} = \alpha_{A/B} + \alpha_{B/O} + \omega_{B/O} \times \omega_{A/B}$$

$$= \alpha_{A/B} + (\alpha_{B/C} + \alpha_{C/O} + \omega_{C/O} \times \omega_{B/C})$$

$$+ \omega_{B/O} \times \omega_{A/B}$$

$$\alpha_{A/O} = \alpha_{A/B} + \alpha_{B/C} + \alpha_{C/O} + \omega_{B/O} \times \omega_{A/B}$$

$$+ \omega_{C/O} \times \omega_{B/C} \tag{19.49}$$

where the subscript O refers to the fixed reference frame $OXYZ$. Following our adopted convention of notations for quantities measured in the fixed reference frame, we may write Eq. (19.49) as

$$\alpha_A = \alpha_{A/B} + \alpha_{B/C} + \alpha_C + \omega_B \times \omega_{A/B} + \omega_C \times \omega_{B/C} \tag{19.50}^\dagger$$

The relation expressed in Eq. (19.48) or (19.50) is called the *addition theorem for angular accelerations*. Note that the cross-product terms in Eqs. (19.48) and (19.50) will vanish for plane motion, where the angular velocities are parallel to each other. Thus, the addition theorems for angular velocities and angular accelerations are of *different* form in general motion, but they become of the *same* form in plane motion.

EXAMPLE 19.3

At the instant shown, the bead B slides down a semicircular bent rod R with a constant speed $v_{B/R} = 10$ ft/s relative to the rod, while the rod is rotated about the X axis at the rate $\omega_R = 3\mathbf{I}$ rad/s and $\alpha_R = 4\mathbf{I}$ rad/s². Determine the velocity and acceleration of the bead relative to the ground.

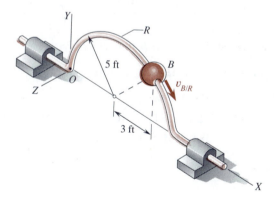

Solution. We first let the reference frame $Axyz$ be embedded in the bent rod R as shown, where the reference frame $OXYZ$ is fixed on the ground. Then, the angular velocity and angular acceleration of $Axyz$ in $OXYZ$, as shown in Fig. (a), are

$$\Omega = \omega_R = 3\mathbf{I} \text{ rad/s} \qquad \dot{\Omega} = \alpha_R = 4\mathbf{I} \text{ rad/s}^2$$

---

†Cf. Eq. (19.40).

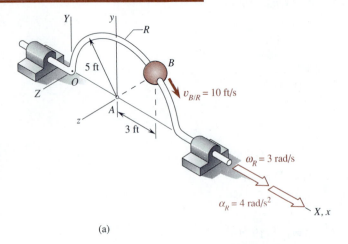

(a)

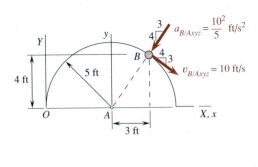

(b)

At the instant shown, we see from Fig. (b) that

$$\mathbf{r}_{B/A} = 3\mathbf{I} + 4\mathbf{J} \text{ ft}$$

$$\mathbf{v}_{B/Axyz} = \mathbf{v}_{B/R} = 10\left[\frac{1}{5}(4\mathbf{I} - 3\mathbf{J})\right] \text{ ft/s}$$

$$= 8\mathbf{I} - 6\mathbf{J} \text{ ft/s}$$

$$\mathbf{a}_{B/Axyz} = \mathbf{a}_n = \frac{10^2}{5}\left[\frac{1}{5}(-3\mathbf{I} - 4\mathbf{J})\right] \text{ ft/s}^2$$

$$= -12\mathbf{I} - 16\mathbf{J} \text{ ft/s}^2$$

Since the point A does not move, we have $\mathbf{v}_A = \mathbf{a}_A = \mathbf{0}$. By Eq. (19.12), the velocity of B in $OXYZ$ is

$$\mathbf{v}_B = \mathbf{v}_{B/Axyz} + \mathbf{\Omega} \times \mathbf{r}_{B/A} + \mathbf{v}_A$$

$$= (8\mathbf{I} - 6\mathbf{J}) + 3\mathbf{I} \times (3\mathbf{I} + 4\mathbf{J}) + \mathbf{0}$$

$$\mathbf{v}_B = 8\mathbf{I} - 6\mathbf{J} + 12\mathbf{K} \text{ ft/s} \quad \blacktriangleleft$$

By Eq. (19.41), the acceleration of B in $OXYZ$ is

$$\mathbf{a}_B = \mathbf{a}_{B/Axyz} + \dot{\mathbf{\Omega}} \times \mathbf{r}_{B/A} + \mathbf{\Omega} \times (\mathbf{\Omega} \times \mathbf{r}_{B/A})$$

$$+ 2\mathbf{\Omega} \times \mathbf{v}_{B/Axyz} + \mathbf{a}_A$$

$$= (-12\mathbf{I} - 16\mathbf{J}) + 4\mathbf{I} \times (3\mathbf{I} + 4\mathbf{J})$$

$$+ 3\mathbf{I} \times [3\mathbf{I} \times (3\mathbf{I} + 4\mathbf{J})]$$

$$+ 2(3\mathbf{I}) \times (8\mathbf{I} - 6\mathbf{J}) + \mathbf{0}$$

$$\mathbf{a}_B = -12\mathbf{I} - 52\mathbf{J} - 20\mathbf{K} \text{ ft/s}^2 \quad \blacktriangleleft$$

Developmental Exercises

D19.9 Refer to Example 19.3. Suppose that B' is a point *of* the bent rod, and B' *coincides* with the moving bead B at the instant considered. Determine the acceleration of B' in $OXYZ$ for the instant shown.

D19.10 Refer to Example 19.3. Determine the Coriolis acceleration of the bead B for the instant shown.

EXAMPLE 19.4

At the instant shown, the disk D spins in the yoke about the pin at A with an angular velocity $\boldsymbol{\omega}_1$ and an angular acceleration $\boldsymbol{\alpha}_1$ relative to the arm C, while the arm C rotates about the vertical shaft with an angular velocity $\boldsymbol{\omega}_2$ and an angular acceleration $\boldsymbol{\alpha}_2$ relative to the ground. For this instant, determine (a) the velocity $\mathbf{v}_B$ and acceleration $\mathbf{a}_B$ of the point B of the disk, (b) the angular velocity $\boldsymbol{\omega}_D$ and angular acceleration $\boldsymbol{\alpha}_D$ of the disk.

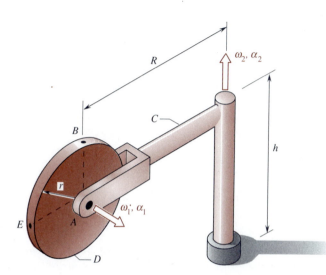

Solution.   We first let the reference frame $Axyz$ be embedded in the arm C (not the disk D) as shown, where the reference frame $OXYZ$ is fixed on the ground. Then, the angular velocity and angular acceleration of $Axyz$ in $OXYZ$ are

$$\boldsymbol{\Omega} = \omega_2 \mathbf{J} \qquad \dot{\boldsymbol{\Omega}} = \alpha_2 \mathbf{J}$$

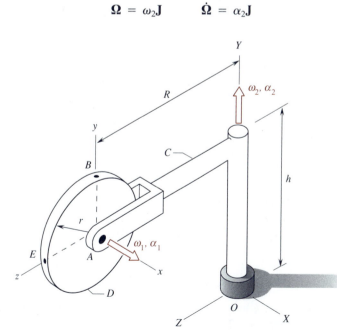

Since the point B describes a circular motion as the disk D spins about A in $Axyz$, we readily find for the instant shown that

$$\mathbf{r}_{B/A} = r\mathbf{J} \qquad \mathbf{v}_{B/Axyz} = r\omega_1\mathbf{K}$$

$$\mathbf{a}_{B/Axyz} = -r\omega_1^2\mathbf{J} + r\alpha_1\mathbf{K}$$

$$\mathbf{v}_A = R\omega_2\mathbf{I} \qquad \mathbf{a}_A = R\alpha_2\mathbf{I} - R\omega_2^2\mathbf{K}$$

By Eq. (19.12), the velocity of B in $OXYZ$ is

$$\mathbf{v}_B = \mathbf{v}_{B/Axyz} + \mathbf{\Omega} \times \mathbf{r}_{B/A} + \mathbf{v}_A$$

$$= r\omega_1\mathbf{K} + \omega_2\mathbf{J} \times r\mathbf{J} + R\omega_2\mathbf{I}$$

$$\mathbf{v}_B = R\omega_2\mathbf{I} + r\omega_1\mathbf{K} \quad \blacktriangleleft$$

By Eq. (19.41), the acceleration of B in $OXYZ$ is

$$\mathbf{a}_B = \mathbf{a}_{B/Axyz} + \dot{\mathbf{\Omega}} \times \mathbf{r}_{B/A} + \mathbf{\Omega} \times (\mathbf{\Omega} \times \mathbf{r}_{B/A})$$

$$+ 2\mathbf{\Omega} \times \mathbf{v}_{B/Axyz} + \mathbf{a}_A$$

$$= (-r\omega_1^2\mathbf{J} + r\alpha_1\mathbf{K}) + \alpha_2\mathbf{J} \times r\mathbf{J} + \omega_2\mathbf{J} \times (\omega_2\mathbf{J} \times r\mathbf{J})$$

$$+ 2\omega_2\mathbf{J} \times r\omega_1\mathbf{K} + (R\alpha_2\mathbf{I} - R\omega_2^2\mathbf{K})$$

$$\mathbf{a}_B = (2r\omega_1\omega_2 + R\alpha_2)\mathbf{I} - r\omega_1^2\mathbf{J} + (r\alpha_1 - R\omega_2^2)\mathbf{K} \quad \blacktriangleleft$$

The angular velocity of the disk D relative to the arm C is $\mathbf{\omega}_{D/C} = \omega_1\mathbf{I}$, while the angular velocity of the arm C in $OXYZ$ is $\mathbf{\omega}_C = \omega_2\mathbf{J}$. By the *addition theorem for angular velocities* [cf. Eq. (19.38)], the angular velocity of the disk D in $OXYZ$ is

$$\mathbf{\omega}_D = \mathbf{\omega}_{D/C} + \mathbf{\omega}_C \qquad \mathbf{\omega}_D = \omega_1\mathbf{I} + \omega_2\mathbf{J} \quad \blacktriangleleft$$

The angular acceleration of the disk D relative to the arm C is $\mathbf{\alpha}_{D/C} = \alpha_1\mathbf{I}$, while the angular acceleration of the arm C in $OXYZ$ is $\mathbf{\alpha}_C = \alpha_2\mathbf{J}$. By the *addition theorem for angular accelerations* [cf. Eq. (19.48)], the angular acceleration of the disk D in $OXYZ$ is

$$\mathbf{\alpha}_D = \mathbf{\alpha}_{D/C} + \mathbf{\alpha}_C + \mathbf{\omega}_C \times \mathbf{\omega}_{D/C}$$

$$= \alpha_1\mathbf{I} + \alpha_2\mathbf{J} + \omega_2\mathbf{J} \times \omega_1\mathbf{I}$$

$$\mathbf{\alpha}_D = \alpha_1\mathbf{I} + \alpha_2\mathbf{J} - \omega_1\omega_2\mathbf{K} \quad \blacktriangleleft$$

Developmental Exercises

D19.11 Refer to Example 19.4. Determine the velocity $\mathbf{v}_E$ and the acceleration $\mathbf{a}_E$ of the point E on the disk in $OXYZ$ for the instant shown.

D19.12 Describe the *addition theorem for angular accelerations*.

EXAMPLE 19.5

A right circular cone C rolls without slipping on the ZX plane as shown, where $r = 7$ in. and $L = 25$ in. At the instant shown, the axis OB of the cone rotates about the Y axis at the rate $\boldsymbol{\omega}_1 = 3\mathbf{J}$ rad/s and $\boldsymbol{\alpha}_1 = 4\mathbf{J}$ rad/s². Determine (a) the rate of spin of the cone about its axis BO and the angular velocity of the cone in $OXYZ$, (b) the angular acceleration of the cone in $OXYZ$.

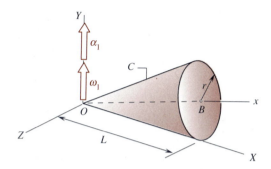

Solution. We first let $Axyz$ be a *moving reference frame* whose x axis is always collinear with the axis BO of the rolling cone as shown, and whose angular velocity and angular acceleration in the *fixed reference frame* $OXYZ$ are

$$\boldsymbol{\Omega} = \boldsymbol{\omega}_A = \boldsymbol{\omega}_1 = \omega_1\mathbf{J} \qquad \omega_1 = 3 \text{ rad/s}$$

$$\dot{\boldsymbol{\Omega}} = \boldsymbol{\alpha}_A = \boldsymbol{\alpha}_1 = \alpha_1\mathbf{J} \qquad \alpha_1 = 4 \text{ rad/s}^2$$

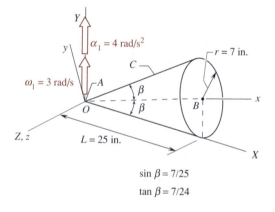

$$\sin \beta = 7/25$$
$$\tan \beta = 7/24$$

Since the cone C rolls without slipping, its angular velocity $\boldsymbol{\omega}_C$ in $OXYZ$ must be directed along its line of contact with the ZX plane. By inspection, we have

$$\boldsymbol{\omega}_C = -\omega_C\mathbf{I} \qquad (1)$$

The *rate of spin* of the cone C about its axis BO is equivalent to the angular velocity of the cone C in $Axyz$ and may be denoted by $\boldsymbol{\omega}_{C/Axyz}$, or more briefly $\boldsymbol{\omega}_{C/A}$. Noting that $\boldsymbol{\omega}_{C/A}$ is directed along $\overrightarrow{BO}$, we write

$$\boldsymbol{\omega}_{C/A} = \omega_{C/A}\boldsymbol{\lambda}_{BO} = \omega_{C/A}(-\cos\beta\,\mathbf{I} - \sin\beta\,\mathbf{J})$$

By the *addition theorem for angular velocities* [cf. Eq. (19.38)], the angular velocity of the cone C in $OXYZ$ is

$$\boldsymbol{\omega}_C = \boldsymbol{\omega}_{C/A} + \boldsymbol{\omega}_A = \omega_{C/A}(-\cos\beta\,\mathbf{I} - \sin\beta\,\mathbf{J}) + \omega_1\mathbf{J}$$

$$\boldsymbol{\omega}_C = -\omega_{C/A}\cos\beta\,\mathbf{I} + (\omega_1 - \omega_{C/A}\sin\beta)\mathbf{J} \qquad (2)$$

Comparing Eqs. (1) and (2), we find that

$$\omega_C = \omega_{C/A} \cos\beta$$

$$0 = \omega_1 - \omega_{C/A} \sin\beta$$

These two scalar equations and the given data yield

$$\omega_{C/A} = \frac{\omega_1}{\sin\beta} = \frac{3}{7/25} \qquad \omega_{C/A} = 10.71\boldsymbol{\lambda}_{BO} \text{ rad/s} \blacktriangleleft$$

$$\omega_C = \frac{\omega_1}{\tan\beta} = \frac{3}{7/24} \qquad \omega_C = -10.29\mathbf{I} \text{ rad/s} \blacktriangleleft$$

The angular acceleration of the cone C in $Axyz$ is $\boldsymbol{\alpha}_{C/A}$, which is directed along $\overrightarrow{BO}$. We write

$$\boldsymbol{\alpha}_{C/A} = \alpha_{C/A}(-\cos\beta\,\mathbf{I} - \sin\beta\,\mathbf{J})$$

By the *addition theorem for angular accelerations* [cf. Eq. (19.48)], the angular acceleration of the cone C in $OXYZ$ is

$$\boldsymbol{\alpha}_C = \boldsymbol{\alpha}_{C/A} + \boldsymbol{\alpha}_A + \boldsymbol{\omega}_A \times \boldsymbol{\omega}_{C/A}$$

$$= \alpha_{C/A}(-\cos\beta\,\mathbf{I} - \sin\beta\,\mathbf{J}) + \alpha_1\mathbf{J}$$

$$+ \omega_1\mathbf{J} \times \frac{\omega_1}{\sin\beta}(-\cos\beta\,\mathbf{I} - \sin\beta\,\mathbf{J})$$

$$= -\alpha_{C/A}\cos\beta\,\mathbf{I} + (\alpha_1 - \alpha_{C/A}\sin\beta)\mathbf{J} + \frac{\omega_1^2}{\tan\beta}\mathbf{K}$$

Since $\boldsymbol{\omega}_C$ always lies in the ZX plane, $\boldsymbol{\alpha}_C$ must also lie in the ZX plane. Thus, we see that $\boldsymbol{\alpha}_C \perp \mathbf{J}$, or $\boldsymbol{\alpha}_C \cdot \mathbf{J} = 0$, which leads to

$$\alpha_1 - \alpha_{C/A}\sin\beta = 0 \qquad \alpha_{C/A} = \frac{\alpha_1}{\sin\beta}$$

$$\boldsymbol{\alpha}_C = -\frac{\alpha_1}{\tan\beta}\mathbf{I} + \frac{\omega_1^2}{\tan\beta}\mathbf{K} = -\frac{4}{7/24}\mathbf{I} + \frac{3^2}{7/24}\mathbf{K}$$

$$\boldsymbol{\alpha}_C = -13.71\mathbf{I} + 30.9\mathbf{K} \text{ rad/s}^2 \blacktriangleleft$$

EXAMPLE 19.6

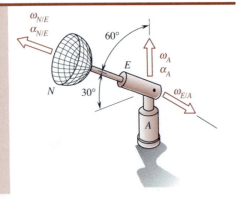

At the instant shown, the antenna N executes a *polarization* rotation about its axis in the body E at the rate of $\omega_{N/E} = 1.5$ rad/s and $\alpha_{N/E} = 2$ rad/s^2, the body E rotates about the pin in the body A to adjust the *elevation* at a constant rate $\omega_{E/A} = 1$ rad/s, while the body A rotates on the ground about the vertical axis to set the *azimuth* at the rate $\omega_A = 3$ rad/s and $\alpha_A = 2.5$ rad/s^2. For this instant, determine the angular velocity $\boldsymbol{\omega}_N$ and the angular acceleration $\boldsymbol{\alpha}_N$ of the antenna N.

Solution. We first let the reference frame $OXYZ$ be fixed on the ground as shown, where the rigid bodies E and A will serve as frames of reference.

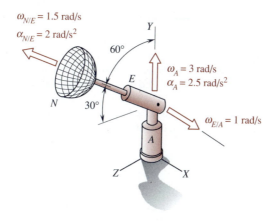

Referring to the sketch and the given data, we write

$$\boldsymbol{\omega}_{N/E} = 1.5(\cos 60° \, \mathbf{J} + \sin 60° \, \mathbf{K}) \text{ rad/s}$$

$$\boldsymbol{\alpha}_{N/E} = 2(\cos 60° \, \mathbf{J} + \sin 60° \, \mathbf{K}) \text{ rad/s}^2$$

$$\boldsymbol{\omega}_{E/A} = 1\mathbf{I} \text{ rad/s} \qquad \boldsymbol{\alpha}_{E/A} = \mathbf{0}$$

$$\boldsymbol{\omega}_A = 3\mathbf{J} \text{ rad/s} \qquad \boldsymbol{\alpha}_A = 2.5\mathbf{J} \text{ rad/s}^2$$

By the *addition theorem for angular velocities* [cf. Eq. (19.40)], the angular velocity $\boldsymbol{\omega}_N$ of the antenna N in $OXYZ$ is

$$\boldsymbol{\omega}_N = \boldsymbol{\omega}_{N/E} + \boldsymbol{\omega}_{E/A} + \boldsymbol{\omega}_A$$

$$\boldsymbol{\omega}_N = 1.5(\cos 60° \, \mathbf{J} + \sin 60° \, \mathbf{K}) + 1\mathbf{I} + 3\mathbf{J}$$

$$\boldsymbol{\omega}_N = 1\mathbf{I} + 3.75\mathbf{J} + 1.299\mathbf{K} \text{ rad/s} \quad \blacktriangleleft$$

Meanwhile, the angular velocity $\boldsymbol{\omega}_E$ of the body E in $OXYZ$ is

$$\boldsymbol{\omega}_E = \boldsymbol{\omega}_{E/A} + \boldsymbol{\omega}_A = 1\mathbf{I} + 3\mathbf{J} \text{ rad/s}$$

By the *addition theorem for angular accelerations* [cf. Eq. (19.50)], the angular acceleration $\boldsymbol{\alpha}_N$ of the antenna N in $OXYZ$ is

$$\boldsymbol{\alpha}_N = \boldsymbol{\alpha}_{N/E} + \boldsymbol{\alpha}_{E/A} + \boldsymbol{\alpha}_A + \boldsymbol{\omega}_E \times \boldsymbol{\omega}_{N/E} + \boldsymbol{\omega}_A \times \boldsymbol{\omega}_{E/A}$$

$$= 2(\cos 60° \, \mathbf{J} + \sin 60° \, \mathbf{K}) + \mathbf{0} + 2.5\mathbf{J}$$

$$+ (1\mathbf{I} + 3\mathbf{J}) \times 1.5(\cos 60° \, \mathbf{J} + \sin 60° \, \mathbf{K})$$

$$+ 3\mathbf{J} \times 1\mathbf{I}$$

$$\boldsymbol{\alpha}_N = 3.90\mathbf{I} + 2.20\mathbf{J} - 0.518\mathbf{K} \text{ rad/s}^2 \quad \blacktriangleleft$$

Developmental Exercises

D19.13 Refer to Example 19.6. For the instant shown, determine the angular velocity and angular acceleration of the antenna N relative to the body A.

D19.14 Refer to Example 19.6. For the instant shown, determine the angular velocity and angular acceleration of the body E relative to the ground.

PROBLEMS

19.1 At the instant shown, the crane turns about the vertical with a constant rate $\omega_1 = 0.25\mathbf{J}$ rad/s, while its boom, pivoted at O, is being raised with a constant rate $\omega_2 = 0.4\mathbf{K}$ rad/s relative to the cab. The length of the boom OP is 15 m. For this instant, determine (a) the velocity and acceleration of the tip P of the boom, (b) the angular velocity and angular acceleration of the boom.

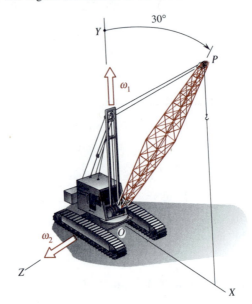

Fig. P19.1

19.2 Solve Prob. 19.1 if, at the instant shown, the crane turns about the vertical at the rate $\omega_1 = 0.25\mathbf{J}$ rad/s and $\alpha_1 = 0.3\mathbf{J}$ rad/s^2, while its boom is being raised at the rate $\omega_2 = 0.4\mathbf{K}$ rad/s and $\alpha_2 = 0.5\mathbf{K}$ rad/s^2 relative to the cab.

19.3* The bent rod rotates about the X axis with a constant rate $\omega = \pi\mathbf{I}$ rad/s. At the instant shown, the collar B moves toward C with a constant speed of 2 m/s relative to the rod. For this instant, determine the velocity and acceleration of the collar.

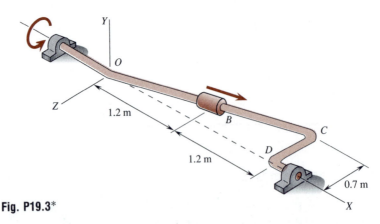

Fig. P19.3*

19.4 Solve Prob. 19.3* if, at the instant shown, the bent rod rotates at the rate $\omega = \pi\mathbf{I}$ rad/s and $\alpha = -(\pi/2)\,\mathbf{I}$ rad/s^2, while the collar B moves toward C with a speed of 2 m/s and an acceleration of 3 m/s^2, both relative to the rod.

19.5   A collar C oscillates along a rod AB as shown. At the instant considered, the turntable rotates at the rate $\omega = 2\mathbf{J}$ rad/s and $\alpha = 3\mathbf{J}$ rad/s^2, while the collar C is moving downward with a speed of 4 ft/s and an acceleration of 5 ft/s^2, both relative to the rod. For this instant, determine the velocity and acceleration of the collar.

19.6   The rod AB is connected by a ball-and-socket joint to the collar A and by a clevis to the collar B as shown. If $y_A = 0.032t^3$ ft, where t is the time measured in seconds, determine the velocity of the collar B and the angular velocity of the rod AB at the instant shown.

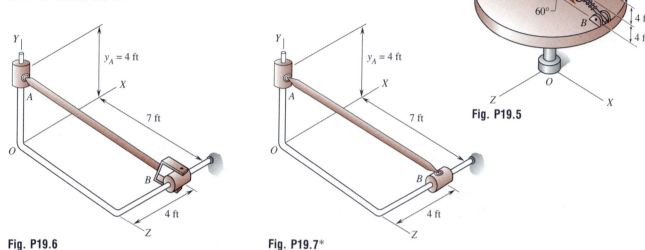

Fig. P19.5

Fig. P19.6   Fig. P19.7*

19.7*   The rod AB is connected by ball-and-socket joints to the collars A and B, which slide on the bent rod. If $y_B = 0.032t^3$ ft, where t is the time measured in seconds, determine the velocity and acceleration of the collar A at the instant shown.

19.8*   The disk D rotates about the pin at A at the constant rate $\omega_1 = -4\mathbf{I}$ rad/s relative to the frame F, while the frame F rotates about the fixed shaft at the constant rate $\omega_2 = 8\mathbf{K}$ rad/s. For the instant shown, determine the angular velocity and the angular acceleration of the disk.

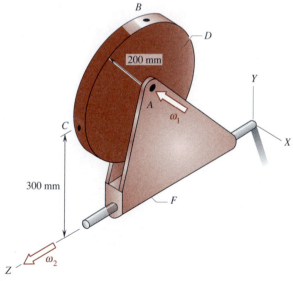

Fig. P19.8*

19.9 Refer to Prob. 19.8*. Determine the velocities and accelerations of the points B and C of the disk at the instant shown.

19.10 The disk D rotates about the pin at A at the constant rate $\omega_1 = 6\mathbf{I}$ rad/s relative to the arm F. At the instant shown, the arm F rotates at the rate $\omega_2 = 4\mathbf{J}$ rad/s and $\alpha_2 = -2\mathbf{J}$ rad/s^2 relative to the ground. For this instant, determine the angular acceleration of the disk D and the accelerations of the points B and C of the disk.

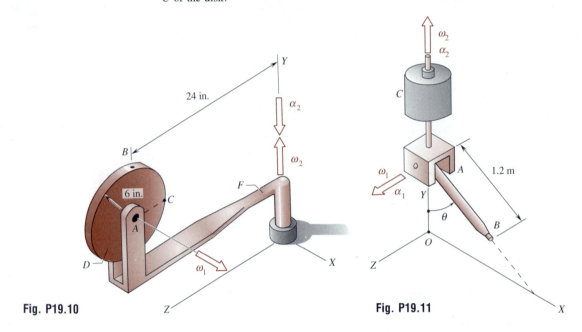

Fig. P19.10   **Fig. P19.11**

19.11 A 1.2-m arm AB for a remote-control mechanism, as shown, is rotated about the pin at A at the rate $\omega_1 = 4\mathbf{K}$ rad/s and $\alpha_1 = 2\mathbf{K}$ rad/s^2 relative to the clevis, while the clevis is turned at the rate $\omega_2 = 5\mathbf{J}$ rad/s and $\alpha_2 = 3\mathbf{J}$ rad/s^2 relative to the ground, where $\theta = 60°$. For this instant, determine the angular acceleration of the arm AB and the velocity and acceleration of the point B.

19.12 A small cone C rolls without slipping inside a fixed conical cavity F as shown. The cone C moves with a constant rate of spin $\omega_1 = 2$ rad/s about its axis BO. For the instant shown, determine (a) the angular velocity of its axis BO about the Y axis, (b) the angular velocity and angular acceleration of the cone C.

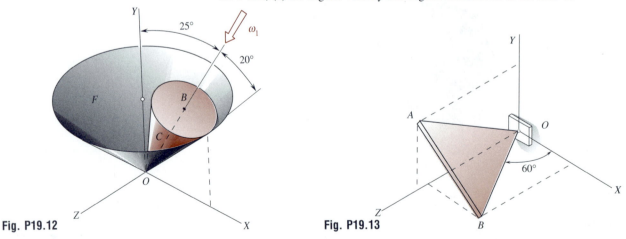

Fig. P19.12   **Fig. P19.13**

19.13 A triangular plate OAB is attached to a ball-and-socket at O as shown, where $\overline{OA} = 5$ ft, $\overline{OB} = 6$ ft, and $\overline{AB} = 7$ ft. The plate is moved in such a way that the edges OA and OB always lie in the YZ and ZX planes, respectively. If the edge OB is rotated at a constant rate $\omega_{OB} = 2\mathbf{J}$ rad/s, determine the angular velocity of the plate OAB and the velocity of the corner A at the instant shown.

19.14* Refer to Prob. 19.13. If the edge OA is rotated at a constant rate $\omega_{OA} = 3\mathbf{I}$ rad/s, determine the angular velocity of the plate OAB and the velocity of the corner B at the instant shown.

19.15 At the instant shown, the disk D rotates about its axis relative to the motor M at the rate $\omega_{D/M} = 2$ rad/s and $\alpha_{D/M} = 3$ rad/s², the motor M rotates about the mounting pin relative to the bracket B at the rate $\omega_{M/B} = 1.5$ rad/s and $\alpha_{M/B} = 2.5$ rad/s², while the bracket B rotates about the vertical shaft relative to the ground at the rate $\omega_B = 4$ rad/s and $\alpha_B = 5$ rad/s². For this instant, determine the angular velocity ω_D and the angular acceleration α_D of the disk D.

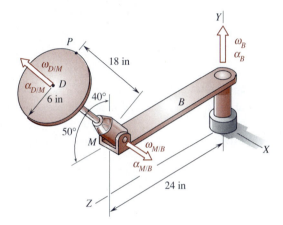

Fig. P19.15

19.16* Refer to Prob. 19.15. For the instant shown, determine the velocity and acceleration of the point P at the top of the disk.

KINETICS OF RIGID BODIES IN SPACE

★19.4   Moments and Products of Inertia of a Mass

The moment of inertia of a mass about a coordinate axis and the associated transfer formula as given by the parallel-axis theorem were presented in Secs. 16.1 and 16.3. To recapitulate the main points in those two sections, we refer to Fig. 16.5 and write

$$I_{xx} = \int (y^2 + z^2)\, dm$$
$$I_{yy} = \int (z^2 + x^2)\, dm \qquad (19.51)$$
$$I_{zz} = \int (x^2 + y^2)\, dm$$

$$I_{xx} = \bar{I}_{x'x'} + m(\bar{y}^2 + \bar{z}^2)$$
$$I_{yy} = \bar{I}_{y'y'} + m(\bar{z}^2 + \bar{x}^2) \qquad (19.52)$$
$$I_{zz} = \bar{I}_{z'z'} + m(\bar{x}^2 + \bar{y}^2)$$

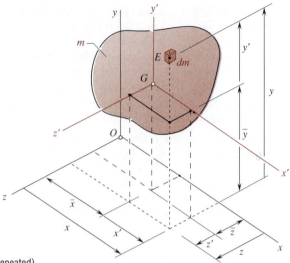

Fig. 16.5 (repeated)

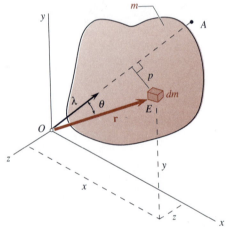

Fig. 19.6

where the mass moments of inertia about the coordinate axes x, y, z, and their parallel central axes x', y', z' are, for present purposes, denoted with double subscripts by I_{xx}, I_{yy}, I_{zz}, $\bar{I}_{x'x'}$, $\bar{I}_{y'y'}$, and $\bar{I}_{z'z'}$, respectively. The relations expressed in Eqs. (19.52) are called the *parallel-axis theorem* for mass moments of inertia.

Suppose that it is desired to determine the moment of inertia of a body about an arbitrary axis OA through the origin O as shown in Fig. 19.6. By definition, this quantity is given by

$$I_{OA} = \int p^2 \, dm \tag{19.53}$$

where p denotes the perpendicular distance from the differential mass element dm at $E(x, y, z)$ to the axis OA. We see that $p = r \sin\theta = 1 \cdot r \sin\theta = |\boldsymbol{\lambda} \times \mathbf{r}|$, where $\mathbf{r}$ is the position vector of E and $\boldsymbol{\lambda}$ is a unit vector directed along the axis OA. Note that

$$\mathbf{r} = x\mathbf{i} + y\mathbf{j} + z\mathbf{k}$$
$$\boldsymbol{\lambda} = \lambda_x\mathbf{i} + \lambda_y\mathbf{j} + \lambda_z\mathbf{k} \tag{19.54}$$

$$p^2 = |\boldsymbol{\lambda} \times \mathbf{r}|^2 = (\boldsymbol{\lambda} \times \mathbf{r}) \cdot (\boldsymbol{\lambda} \times \mathbf{r}) \tag{19.55}$$

where the scalar components λ_x, λ_y, and λ_z of the unit vector $\boldsymbol{\lambda}$ represent the direction cosines of the axis OA in $Oxyz$. It can be shown that Eqs. (19.53) through (19.55) yield

$$I_{OA} = \lambda_x^2 \int (y^2 + z^2) \, dm + \lambda_y^2 \int (z^2 + x^2) \, dm$$
$$+ \lambda_z^2 \int (x^2 + y^2) \, dm - 2\lambda_x\lambda_y \int xy \, dm$$
$$- 2\lambda_y\lambda_z \int yz \, dm - 2\lambda_z\lambda_x \int zx \, dm$$

$$I_{OA} = I_{xx}\lambda_x^2 + I_{yy}\lambda_y^2 + I_{zz}\lambda_z^2 - 2I_{xy}\lambda_x\lambda_y - 2I_{yz}\lambda_y\lambda_z - 2I_{zx}\lambda_z\lambda_x \tag{19.56}$$

where

$$I_{xy} = \int xy \, dm \qquad I_{yz} = \int yz \, dm \qquad I_{zx} = \int zx \, dm \tag{19.57}$$

The quantities I_{xy}, I_{yz}, and I_{zx} defined in Eqs. (19.57) are called the *products of inertia* of the body with respect to the x and y axes, the y and z axes, and the z and x axes, respectively. Clearly, we have

$$I_{xy} = I_{yx} \qquad I_{yz} = I_{zy} \qquad I_{zx} = I_{xz} \qquad (19.58)$$

Unlike the mass moment of inertia about an axis, which is always positive, the mass products of inertia may be positive, negative, or zero. *If the mass of a body is symmetrical with respect to a coordinate plane*, such as the xy plane in Fig. 19.7(a), *then the products of inertia of the body with respect to one of the two coordinate axes in the plane of symmetry and the third coordinate axis, such as I_{yz} or I_{zx}, must be equal to zero.* Furthermore, *if the mass of a body is symmetrical about two coordinate planes,* such as the xy and yz planes in Fig. 19.7(b), *then the product of inertia of the body with respect to any two of the coordinate axes must vanish.* In other words, we have $I_{xy} = I_{yz} = I_{zx} = 0$ for the homogeneous body in Fig. 19.7(b).

We note from Fig. 16.5 that

$$x = x' + \bar{x} \qquad y = y' + \bar{y} \qquad z = z' + \bar{z} \qquad (19.59)$$

Substituting Eqs. (19.59) into Eqs. (19.57) and applying the *principle of moments* as well as the definitions of mass products of inertia, we can verify that

$$I_{xy} = \bar{I}_{x'y'} + m\bar{x}\bar{y}$$

$$I_{yz} = \bar{I}_{y'z'} + m\bar{y}\bar{z} \qquad (19.60)$$

$$I_{zx} = \bar{I}_{z'x'} + m\bar{z}\bar{x}$$

The relations expressed in Eqs. (19.60) are called the *parallel-axis theorem* for mass products of inertia.

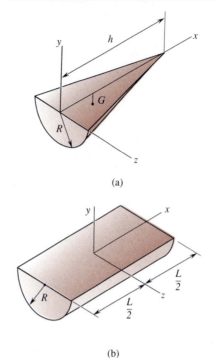

(a)

(b)

Fig. 19.7

EXAMPLE 19.7

The slender bent rod *OABC* weighs 9.66 lb/ft. Determine the moment of inertia I_{OC} of the bent rod about the axis *OC* as shown.

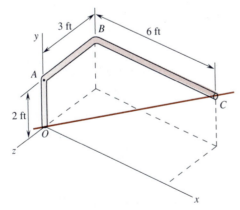

Solution. The mass density of the bent rod is

$$\rho_L = \frac{9.66}{32.2} \text{ slug/ft} = 0.3 \text{ slug/ft}$$

Thus, the masses of its three segments are

$$m_{OA} = 0.6 \text{ slug} \qquad m_{AB} = 0.9 \text{ slug}$$

$$m_{BC} = 1.8 \text{ slugs}$$

Referring to the bent rod as shown and applying the formula $\bar{I} = \frac{1}{12} mL^2$ (cf. Fig. 16.7 or Example 16.1) for a slender rod of mass m and length L as well as the *parallel-axis theorems*, we write

$$I_{xx} = [\tfrac{1}{12}(0.6)(2)^2 + 0.6(1)^2]$$

$$+ \{\tfrac{1}{12}(0.9)(3)^2 + 0.9[2^2 + (-1.5)^2]\}$$

$$+ \{0 + 1.8[2^2 + (-3)^2]\} \text{ slug·ft}^2$$

$$= 30.5 \text{ slug·ft}^2$$

$$I_{yy} = (0 + 0) + [\tfrac{1}{12}(0.9)(3)^2 + 0.9(1.5)^2]$$

$$+ [\tfrac{1}{12}(1.8)(6)^2 + 1.8(3^2 + 3^2)] \text{ slug·ft}^2$$

$$= 40.5 \text{ slug·ft}^2$$

$$I_{zz} = [\tfrac{1}{12}(0.6)(2)^2 + 0.6(1)^2] + [0 + 0.9(2)^2]$$

$$+ [\tfrac{1}{12}(1.8)(6)^2 + 1.8(3^2 + 2^2)] \text{ slug·ft}^2$$

$$= 33.2 \text{ slug·ft}^2$$

$$I_{xy} = (0 + 0) + (0 + 0) + [0 + 1.8(3)(2)] \text{ slug·ft}^2$$

$$= 10.8 \text{ slug·ft}^2$$

$$I_{yz} = (0 + 0) + [0 + 0.9(2)(-1.5)]$$

$$+ [0 + 1.8(2)(-3)] \text{ slug·ft}^2$$

$$= -13.5 \text{ slug·ft}^2$$

$$I_{zx} = (0 + 0) + (0 + 0)$$

$$+ [0 + 1.8(-3)(3)] \text{ slug·ft}^2$$

$$= -16.2 \text{ slug·ft}^2$$

For the axis OC, we write

$$\overrightarrow{OC} = 6\mathbf{i} + 2\mathbf{j} - 3\mathbf{k}$$

$$\boldsymbol{\lambda}_{OC} = \boldsymbol{\lambda} = \tfrac{1}{7}(6\mathbf{i} + 2\mathbf{j} - 3\mathbf{k})$$

$$\lambda_x = \tfrac{6}{7} \qquad \lambda_y = \tfrac{2}{7} \qquad \lambda_z = -\tfrac{3}{7}$$

Using the foregoing computed quantities and applying Eq. (19.56), we write

$$I_{OC} = 30.5(\tfrac{6}{7})^2 + 40.5(\tfrac{2}{7})^2 + 33.2(-\tfrac{3}{7})^2$$

$$- 2(10.8)(\tfrac{6}{7})(\tfrac{2}{7}) - 2(-13.5)(\tfrac{2}{7})(-\tfrac{3}{7})$$

$$- 2(-16.2)(-\tfrac{3}{7})(\tfrac{6}{7}) \text{ slug·ft}^2$$

$$= \frac{554.4}{49} \text{ slug·ft}^2$$

$$I_{OC} = 11.31 \text{ slug·ft}^2 \quad \blacktriangleleft$$

Developmental Exercises

D19.15   Refer to Example 19.7. Determine the moment of inertia of the bent rod about the axis (a) OB, (b) AC.

D19.16   Name the mass products of inertia which will vanish if the mass of the body is symmetrical with respect to (a) the xy plane, (b) the yz plane (c) the zx plane.

D19.17   Describe the condition under which the mass products of inertia will *all* vanish.

D19.18   Describe the *parallel-axis theorem* for mass products of inertia.

★19.5   Rotation of Axes: Principal Axes of Inertia

Let the coordinate system $Oxyz$ be rotated about its origin O so that a new coordinate system $Ox'y'z'$ is obtained as shown in Fig. 19.8, where $\boldsymbol{\lambda}_1$, $\boldsymbol{\lambda}_2$, and $\boldsymbol{\lambda}_3$ are the unit vectors directed along the x', y', and z' axes, respectively. In terms of the cartesian unit vectors $\mathbf{i}$, $\mathbf{j}$, and $\mathbf{k}$ in $Oxyz$, we may write

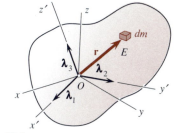

Fig. 19.8

$$\boldsymbol{\lambda}_1 = \lambda_{11}\mathbf{i} + \lambda_{12}\mathbf{j} + \lambda_{13}\mathbf{k}$$

$$\boldsymbol{\lambda}_2 = \lambda_{21}\mathbf{i} + \lambda_{22}\mathbf{j} + \lambda_{23}\mathbf{k} \qquad (19.61)$$

$$\boldsymbol{\lambda}_3 = \lambda_{31}\mathbf{i} + \lambda_{32}\mathbf{j} + \lambda_{33}\mathbf{k}$$

In Fig. 19.8, the coordinates of the point E, where the differential mass element dm is located, are (x, y, z) in $Oxyz$, but are (x', y', z') in $Ox'y'z'$. Since the position vector of E is $\mathbf{r} = x\mathbf{i} + y\mathbf{j} + z\mathbf{k}$ and

$$x' = \boldsymbol{\lambda}_1 \cdot \mathbf{r} \qquad y' = \boldsymbol{\lambda}_2 \cdot \mathbf{r} \qquad z' = \boldsymbol{\lambda}_3 \cdot \mathbf{r} \qquad (19.62)$$

we readily find that

$$x' = \lambda_{11}x + \lambda_{12}y + \lambda_{13}z$$

$$y' = \lambda_{21}x + \lambda_{22}y + \lambda_{23}z \qquad (19.63)$$

$$z' = \lambda_{31}x + \lambda_{32}y + \lambda_{33}z$$

Clearly, the orientation of $Ox'y'z'$ in $Oxyz$ is completely defined by the array (or matrix) of λ_{ij} $(i, j = 1, 2, 3)$, which can be represented by

$$\mathbf{R} = \begin{bmatrix} \lambda_{11} & \lambda_{12} & \lambda_{13} \\ \lambda_{21} & \lambda_{22} & \lambda_{23} \\ \lambda_{31} & \lambda_{32} & \lambda_{33} \end{bmatrix} \qquad (19.64\text{a})$$

$$\mathbf{R}^T = \begin{bmatrix} \lambda_{11} & \lambda_{21} & \lambda_{31} \\ \lambda_{12} & \lambda_{22} & \lambda_{32} \\ \lambda_{13} & \lambda_{23} & \lambda_{33} \end{bmatrix} \qquad (19.64\text{b})$$

where $\mathbf{R}$ is called the *rotation matrix*, and $\mathbf{R}^T$ is called the *transpose* of $\mathbf{R}$. It is well to recognize that the sets of elements in the rows of $\mathbf{R}$ represent the sets of direction cosines of x', y', and z' axes in $Oxyz$, respectively; while the sets of elements in the rows of $\mathbf{R}^T$ represent the sets of direction cosines of x, y, and z axes in $Ox'y'z'$, respectively. Of course, the direction cosines λ_{ij} $(i, j = 1, 2, 3)$ of $\boldsymbol{\lambda}_i$ $(i = 1, 2, 3)$ must be subject to the constraints that $\boldsymbol{\lambda}_i$ ($i = 1, 2, 3$) are unit vectors which are orthogonal to each other.

By using Eqs. (19.63) and the definitions for moments and products of inertia of a mass in Sec. 19.4, one can find $I_{x'x'}$, $I_{y'y'}$, $I_{z'z'}$, $I_{x'y'}$, $I_{y'z'}$, and $I_{z'x'}$ in terms of I_{xx}, I_{yy}, I_{zz}, I_{xy}, I_{yz}, I_{zx}, and the direction cosines λ_{ij} (i, $j = 1, 2, 3$) in the rotation matrix **R**. Nevertheless, the transformation of the moments and products of inertia of a body associated with $Oxyz$ into those associated with $Ox'y'z'$, as shown in Fig. 19.8, is best studied using *matrix algebra* or *tensor analysis*. A detailed coverage of such a study is beyond the scope of this chapter. Without dwelling on the details of matrix algebra or tensor analysis, we shall here simply point out that the arrays (or matrices) of inertia quantities

$$\mathbf{I} = \begin{bmatrix} I_{xx} & -I_{xy} & -I_{xz} \\ -I_{yx} & I_{yy} & -I_{yz} \\ -I_{zx} & -I_{zy} & I_{zz} \end{bmatrix} \tag{19.65a}$$

$$\mathbf{I}' = \begin{bmatrix} I_{x'x'} & -I_{x'y'} & -I_{x'z'} \\ -I_{y'x'} & I_{y'y'} & -I_{y'z'} \\ -I_{z'x'} & -I_{z'y'} & I_{z'z'} \end{bmatrix} \tag{19.65b}$$

are called the *inertia tensors* (or *inertia matrices*) at the origin O for a body. These two inertia tensors are second-order tensors and can be expressed in terms of each other by the laws of transformation for second-order tensors or the laws of rotation of axes for matrices. In terms of products of matrices, it can be shown that[†]

$$\boxed{\mathbf{I}' = \mathbf{R}\mathbf{I}\mathbf{R}^T \qquad \mathbf{I} = \mathbf{R}^T\mathbf{I}'\mathbf{R}} \tag{19.66}$$

where **R** and $\mathbf{R}^T$ are defined in Eqs. (19.64). It should be pointed out that the *negative* signs associated with the products of inertia in Eqs. (19.65) are needed when the products of inertia themselves are defined without a negative sign as those in Eqs. (19.57); otherwise, the laws relating $\mathbf{I}'$ to $\mathbf{I}$ in Eqs. (19.66) would be inapplicable.[††] Since tensors must obey the laws for their transformation under rotation of axes, we *cannot* call an arbitrary array of inertia quantities an "inertia tensor."

To explore the scalar equations implied in the first of Eqs. (19.66), we may first carry out the multiplications of the matrices in $\mathbf{R}\mathbf{I}\mathbf{R}^T$ and apply Eqs. (19.58) to simplify the result. Then we equate the corresponding elements in $\mathbf{I}'$ and the resulting product of $\mathbf{R}\mathbf{I}\mathbf{R}^T$; this will yield *six* independent scalar equations because the inertia tensor is symmetric. For example, in equating the element in the first row and first column of $\mathbf{I}'$ to the element in the first row and first column of the resulting product of $\mathbf{R}\mathbf{I}\mathbf{R}^T$, we find that

$$I_{x'x'} = \lambda_{11}^2 I_{xx} + \lambda_{12}^2 I_{yy} + \lambda_{13}^2 I_{zz}$$
$$- 2\lambda_{11}\lambda_{12}I_{xy} - 2\lambda_{12}\lambda_{13}I_{yz} - 2\lambda_{13}\lambda_{11}I_{zx} \tag{19.67}$$

As a check of Eq. (19.67) against Eq. (19.56), we may let the x' axis in

---

[†]See Gere, J. M. and W. Weaver, Jr., *Matrix Algebra for Engineers*, 2nd ed., Brooks/Cole, Monterey, CA, 1983. These laws of transformation may be *skipped* without prejudice to the understanding of the rest of the text.

[††]Cf. the footnote on p. 237.

Fig. 19.8 be the axis OA in Fig. 19.6. In this case, we have $\lambda_1 = \lambda$ and

$$\lambda_{11} = \lambda_x \qquad \lambda_{12} = \lambda_y \qquad \lambda_{13} = \lambda_z \qquad (19.68)$$

Substituting Eqs. (19.68) into Eq. (19.67) and comparing the result with Eq. (19.56), we see that the expressions for $I_{x'x'}$ and I_{OA} are identical (Q.E.D.). Thus, Eq. (19.56) is simply a subcase of Eqs. (19.66).

Note that the quantities in each array in Eqs. (19.65) are called the *components* of the inertia tensor at O. If the coordinate axes are oriented in such a way that the products of inertia of the body *all* vanish, then the coordinate axes are called the *principal axes of inertia* at O, and the moments of inertia about these principal axes are called the *principal moments of inertia* at O. When this is the case, the inertia tensor is "diagonalized" and may be written in the simplified form

$$\mathbf{I} = \begin{bmatrix} I_x & 0 & 0 \\ 0 & I_y & 0 \\ 0 & 0 & I_z \end{bmatrix} \qquad (19.69)$$

where $I_x = I_{xx}$, $I_y = I_{yy}$, and $I_z = I_{zz}$. Of these three principal moments of inertia at O, one will be the *maximum* and another a *minimum* of the body's moments of inertia about axes through the origin O.

Developmental Exercises

D19.19   Refer to Fig. 19.8 as well as Eqs. (19.61) and (19.64). (a) What do the sets of elements in the rows of the rotation matrix $\mathbf{R}$ represent? (b) What do the sets of elements in the rows of $\mathbf{R}^T$ represent? (c) Is the determinant of $\mathbf{R}$ given by $|\mathbf{R}| = \lambda_1 \cdot (\lambda_2 \times \lambda_3) = 1$? (d) Is the determinant of $\mathbf{R}^T$ given by $|\mathbf{R}^T| = \mathbf{i} \cdot (\mathbf{j} \times \mathbf{k}) = 1$?

D19.20   Define the principal axes and principal moments of inertia at a point for a body.

PROBLEMS

NOTE: Unless otherwise specified, the formulas given in Fig. 16.7 as well as the parallel-axis theorems may be applied in solving the following problems of mass moments and products of inertia.

19.17 and 19.18* Determine the products of inertia I_{xy}, I_{yz}, and I_{zx} of the body made of slender rods as shown.

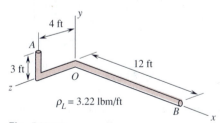

Fig. P19.17

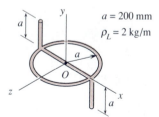

Fig. P19.18*

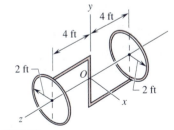

Fig. P19.19

19.19   A slender rod weighing 1.61 lb/ft is used to form a composite body as shown. Determine I_{xy}, I_{yz}, and I_{zx} of the body.

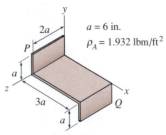

Fig. P19.20

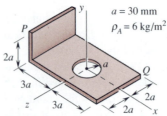

Fig. P19.21

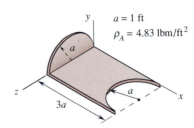

Fig. P19.22*

19.20 through 19.22* Determine I_{xy}, I_{yz}, and I_{zx} of the sheet metal part shown.

19.23 and 19.24* A piece of sheet metal is cut and bent into a machine component as shown. If its mass density is $\rho_A = 16$ kg/m², determine I_{xy}, I_{yz}, and I_{zx} of the component.

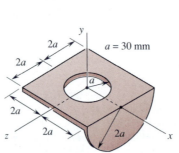

Fig. P19.23

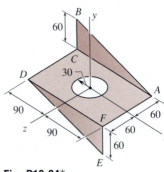

Dimensions in mm

Fig. P19.24*

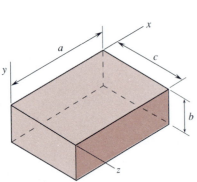

Fig. P19.25

19.25 A rectangular parallelepiped of mass m is shown. Determine I_{xy}, I_{yz}, and I_{zx} of the parallelepiped.

19.26 through 19.28 The mass density of the steel machine element is $\rho = 7.85$ Mg/m³. Determine I_{xy}, I_{yz}, and I_{zx} of the machine element.

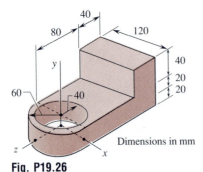

Dimensions in mm

Fig. P19.26

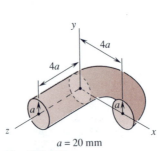

$a = 20$ mm

Fig. P19.27*

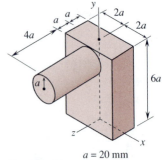

$a = 20$ mm

Fig. P19.28

19.29 and 19.30* Determine by integration I_{xy} of the body, as shown, whose mass is m. (*Hint.* Choose dm to be a first-order element parallel to the yz plane.)

19.31 A tetrahedron of mass density $\rho = 15$ slugs/ft³ is bounded by the coordinate planes and the plane $3x + 6y + 4z = 12$, where the lengths are in feet. Determine I_{xy} of the tetrahedron.

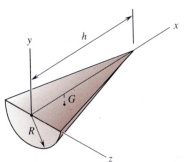

Fig. P19.29

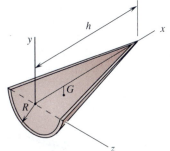

Fig. P19.30

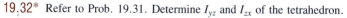

Fig. P19.31

19.32* Refer to Prob. 19.31. Determine I_{yz} and I_{zx} of the tetrahedron.

19.33 Refer to Prob. 19.17. Determine the moment of inertia of the body about the axis joining the points O and A.

19.34 Refer to Prob. 19.17. Determine the moment of inertia of the body about the axis joining the points A and B.

19.35 The slender bent rod weighs 3.22 lb/ft. Determine the moment of inertia of the bent rod about the axis joining the points O and C.

19.36 Refer to Prob. 19.35. Determine the moment of inertia of the bent rod about the axis joining the points A and C.

19.37 The solid right circular cone as shown has a mass m. Determine its moment of inertia about the axis OA which is the line of intersection of the surface of the cone and the xy plane.

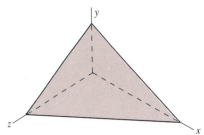

Fig. P19.35

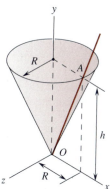

Fig. P19.37

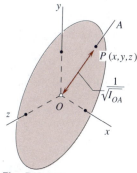

Fig. P19.38*

19.38* Suppose that the components of the inertia tensor at the origin O as well as I_{OA}, the moment of inertia about the axis OA, for a body are known. Furthermore, a point $P(x, y, z)$ is plotted on the axis OA at a distance $\overline{OP} = 1/\sqrt{I_{OA}}$ from O. Show that the locus of the point $P(x, y, z)$ is the surface of a *quadric* defined by

$$I_{xx}x^2 + I_{yy}y^2 + I_{zz}z^2 - 2I_{xy}xy - 2I_{yz}yz - 2I_{zx}zx = 1$$

which is known as the *ellipsoid of inertia*.

19.39* Refer to the sheet metal part in Fig. P19.20. Determine its moment of inertia about the axis joining the points P and Q.

19.40 Refer to the sheet metal part in Fig. P19.21. Determine its moment of inertia about the axis joining the points P and Q.

★19.6   Momentum of a Rigid Body

Suppose that a rigid body of mass m rotates with an angular velocity $\boldsymbol{\omega}$ and its mass center G moves with a velocity $\bar{\mathbf{v}}$ in the *fixed* reference frame $OXYZ$ as shown in Fig. 19.9, where $Gxyz$ is a *nonrotating* reference frame moving with G. Referring to this figure, we see that

$$\mathbf{r} = \bar{\mathbf{r}} + \mathbf{r}' \tag{19.70}$$

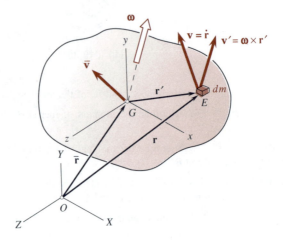

Fig. 19.9

Thus, the velocity $\mathbf{v}$ of the differential mass element dm at E in $OXYZ$ is given by

$$\mathbf{v} = \dot{\mathbf{r}} = \dot{\bar{\mathbf{r}}} + \dot{\mathbf{r}}' = \bar{\mathbf{v}} + [(\dot{\mathbf{r}}')_{Gxyz} + \mathbf{0} \times \mathbf{r}']$$

$$= \bar{\mathbf{v}} + [\boldsymbol{\omega} \times \mathbf{r}' + \mathbf{0}]$$

$$\mathbf{v} = \bar{\mathbf{v}} + \boldsymbol{\omega} \times \mathbf{r}' \tag{19.71}$$

Applying Eq. (19.71) and the *principle of moments*, we compute the *linear momentum* $\mathbf{L}$ of the rigid body in $OXYZ$ as follows:

$$\mathbf{L} = \int \mathbf{v} \, dm = \int (\bar{\mathbf{v}} + \boldsymbol{\omega} \times \mathbf{r}') \, dm$$

$$= \bar{\mathbf{v}} \int dm + \boldsymbol{\omega} \times \int \mathbf{r}' \, dm = \bar{\mathbf{v}}m + \boldsymbol{\omega} \times \mathbf{0}$$

$$\boxed{\mathbf{L} = m\bar{\mathbf{v}}} \tag{19.72}$$

Furthermore, we refer to Fig. 19.9 and apply Eq. (19.71) as well as the *principle of moments* to compute the angular momentum $\mathbf{H}_G$ of the rigid body about its mass center G in $OXYZ$ as follows:

$$\mathbf{H}_G = \int \mathbf{r}' \times \mathbf{v} \, dm = \int \mathbf{r}' \times (\bar{\mathbf{v}} + \boldsymbol{\omega} \times \mathbf{r}') \, dm$$

$$= \left(\int \mathbf{r}' \, dm \right) \times \bar{\mathbf{v}} + \int \mathbf{r}' \times (\boldsymbol{\omega} \times \mathbf{r}') \, dm$$

$$= \mathbf{0} \times \bar{\mathbf{v}} + \int \mathbf{r}' \times (\boldsymbol{\omega} \times \mathbf{r}') \, dm$$

$$\boxed{\mathbf{H}_G = \int \mathbf{r}' \times (\boldsymbol{\omega} \times \mathbf{r}') \, dm} \tag{19.73}$$

We see in Fig. 19.9 that the velocity of dm at E in $Gxyz$ is

$$\mathbf{v}' = \boldsymbol{\omega} \times \mathbf{r}' \tag{19.74}$$

Substituting Eq. (19.74) into Eq. (19.73), we find that

$$\mathbf{H}_G = \int \mathbf{r'} \times \mathbf{v'} \, dm = \mathbf{H}'_G \qquad (19.75)$$

which reveals that the angular momentum of a rigid body about its mass center G computed in the fixed reference frame $OXYZ$ and that computed in the nonrotating central reference frame $Gxyz$ are equal.

Noting that $Gxyz$ does not rotate, we may express $\mathbf{H}_G$, $\mathbf{r'}$, and $\boldsymbol{\omega}$ in rectangular components in $Gxyz$ as follows:

$$\mathbf{H}_G = H_x \mathbf{i} + H_y \mathbf{j} + H_z \mathbf{k} \qquad (19.76)$$

$$\mathbf{r'} = x\mathbf{i} + y\mathbf{j} + z\mathbf{k} \qquad (19.77)$$

$$\boldsymbol{\omega} = \omega_x \mathbf{i} + \omega_y \mathbf{j} + \omega_z \mathbf{k} \qquad (19.78)$$

Substituting Eqs. (19.76) through (19.78) into Eq. (19.73), carrying out the cross products, and combining terms, we get

$$H_x \mathbf{i} + H_y \mathbf{j} + H_z \mathbf{k}$$

$$= [\omega_x \int (y^2 + z^2) \, dm - \omega_y \int xy \, dm - \omega_z \int xz \, dm]\mathbf{i}$$
$$+ [-\omega_x \int yx \, dm + \omega_y \int (z^2 + x^2) \, dm - \omega_z \int yz \, dm]\mathbf{j}$$
$$+ [-\omega_x \int zx \, dm - \omega_y \int zy \, dm + \omega_z \int (x^2 + y^2) \, dm]\mathbf{k}$$

Recognizing that the integrals in this equation represent the central moments and products of inertia, we equate the coefficients of $\mathbf{i}$, $\mathbf{j}$, and $\mathbf{k}$ to obtain the following three equations:

$$H_x = \bar{I}_{xx}\omega_x - \bar{I}_{xy}\omega_y - \bar{I}_{xz}\omega_z$$

$$H_y = -\bar{I}_{yx}\omega_x + \bar{I}_{yy}\omega_y - \bar{I}_{yz}\omega_z \qquad (19.79)$$

$$H_z = -\bar{I}_{zx}\omega_x - \bar{I}_{zy}\omega_y + \bar{I}_{zz}\omega_z$$

These three equations expressing the components of $\mathbf{H}_G$ may be written in terms of arrays (or matrices) as follows:

$$\begin{bmatrix} H_x \\ H_y \\ H_z \end{bmatrix} = \begin{bmatrix} \bar{I}_{xx} & -\bar{I}_{xy} & -\bar{I}_{xz} \\ -\bar{I}_{yx} & \bar{I}_{yy} & -\bar{I}_{yz} \\ -\bar{I}_{zx} & -\bar{I}_{zy} & \bar{I}_{zz} \end{bmatrix} \begin{bmatrix} \omega_x \\ \omega_y \\ \omega_z \end{bmatrix} \qquad (19.80)$$

Using symbolic representation, we may write Eq. (19.80) as

$$\mathbf{H}_G = \bar{\mathbf{I}}\boldsymbol{\omega} \qquad (19.81)$$

where $\mathbf{H}_G$, $\bar{\mathbf{I}}$, and $\boldsymbol{\omega}$ are, respectively, the *angular momentum matrix*, the *inertia matrix*, and the *angular velocity matrix*, which are defined to represent the following arrays:

$$\mathbf{H}_G = \begin{bmatrix} H_x \\ H_y \\ H_z \end{bmatrix}$$

$$\bar{\mathbf{I}} = \begin{bmatrix} \bar{I}_{xx} & -\bar{I}_{xy} & -\bar{I}_{xz} \\ -\bar{I}_{yx} & \bar{I}_{yy} & -\bar{I}_{yz} \\ -\bar{I}_{zx} & -\bar{I}_{zy} & \bar{I}_{zz} \end{bmatrix} \qquad (19.82)$$

$$\boldsymbol{\omega} = \begin{bmatrix} \omega_x \\ \omega_y \\ \omega_z \end{bmatrix}$$

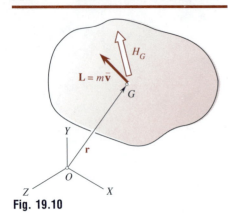

Fig. 19.10

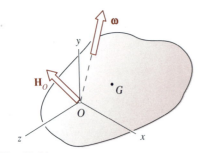

Fig. 19.11

By comparison, we clearly see that Eq. (19.81) resembles Eq. (18.13), and Eqs. (19.81) and (19.82) together are equivalent to Eqs. (19.79).

The foregoing analysis leads to the fact that the system of momenta of the particles of a rigid body in general motion is equivalent to a *momentum system* consisting of the linear momentum $\mathbf{L} = m\bar{\mathbf{v}}$ at the mass center G and the angular momentum $\mathbf{H}_G$ about G, as shown in Fig. 19.10, where the components of $\mathbf{H}_G$ are given by Eqs. (19.79). Once the momentum system at G has been determined, it can be replaced by an *equivalent momentum system* at another point.

For instance, if the momentum system at G in Fig. 19.10 were to be replaced by an equivalent momentum system at the fixed point O, the replacement would consist of the same linear momentum $\mathbf{L} = m\bar{\mathbf{v}}$ at O and a new angular momentum $\mathbf{H}_O$ given by

$$\mathbf{H}_O = \mathbf{H}_G + \bar{\mathbf{r}} \times m\bar{\mathbf{v}} \qquad (19.83)$$

On the other hand, it can be shown that the angular momentum $\mathbf{H}_O$ of a rigid body rotating with $\boldsymbol{\omega} = \omega_x\mathbf{i} + \omega_y\mathbf{j} + \omega_z\mathbf{k}$ about the origin O of a fixed reference frame $Oxyz$, as shown in Fig. 19.11, is given by

$$\mathbf{H}_O = H_x\mathbf{i} + H_y\mathbf{j} + H_z\mathbf{k} \qquad (19.84)$$

where

$$
\begin{aligned}
H_x &= I_{xx}\omega_x - I_{xy}\omega_y - I_{xz}\omega_z \\
H_y &= -I_{yx}\omega_x + I_{yy}\omega_y - I_{yz}\omega_z \\
H_z &= -I_{zx}\omega_x - I_{zy}\omega_y + I_{zz}\omega_z
\end{aligned}
\qquad (19.85)
$$

and the moments of inertia I_{xx}, I_{yy}, I_{zz} and the products of inertia I_{xy}, I_{yz}, I_{zx} are computed with respect to the fixed reference frame $Oxyz$.

Knowing how to compute the momentum system for a rigid body at its mass center G as shown in Fig. 19.10, we may solve appropriate problems by the *principle of impulse and momentum* in Sec. 18.3, which states that

The *resultant system of momenta* of a rigid body at the time t_1	+	The *linear* and *angular impulses* exerted on the rigid body from t_1 to t_2	=	The *resultant system of momenta* of the rigid body at the time t_2	(18.15) (repeated)

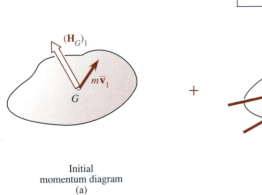

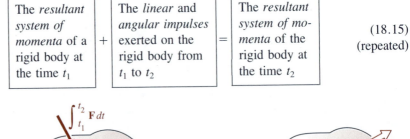

Initial momentum diagram (a)   Impulse diagram (b)   Final momentum diagram (c)

Fig. 19.12

Note that Fig. 19.12 is a graphical representation of Eq. (18.15) for a rigid body in general motion.

EXAMPLE 19.8

A 96.6-lb rectangular plate is falling with a downward velocity of 7 ft/s and zero angular velocity when its corner C strikes the corner O of a post as shown. If the impact is perfectly plastic, determine immediately after the impact (a) the angular velocity $\boldsymbol{\omega}$ of the plate, (b) the velocity $\bar{\mathbf{v}}$ of its mass center G.

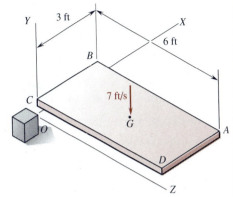

Solution. The mass of the plate is $m = 96.6/32.2$ slugs $= 3$ slugs. Letting the axes of the nonrotating central reference frame $Gxyz$ be parallel to the axes of the fixed reference frame $OXYZ$ and noting that the plate is symmetrical with respect to the xy and yz planes, we see that the x, y, z axes are the principal axes and we have, from Fig. 16.7,

$$\bar{I}_{xx} = \tfrac{1}{12} mc^2 = \tfrac{1}{12}(3)(6)^2 = 9 \text{ (slug·ft}^2)$$

$$\bar{I}_{yy} = \tfrac{1}{12} m(c^2 + a^2) = \tfrac{1}{12}(3)(6^2 + 3^2)$$

$$= 11.25 \text{ (slug·ft}^2)$$

$$\bar{I}_{zz} = \tfrac{1}{12} ma^2 = \tfrac{1}{12}(3)(3)^2 = 2.25 \text{ (slug·ft}^2)$$

$$\bar{I}_{xy} = \bar{I}_{yz} = \bar{I}_{zx} = 0$$

The components of the angular momentum about G are, from Eqs. (19.79),

$$H_x = 9\omega_x - 0 - 0 = 9\omega_x$$

$$H_y = -0 + 11.25\omega_y - 0 = 11.25\omega_y$$

$$H_z = -0 - 0 + 2.25\omega_z = 2.25\omega_z$$

Therefore, applying the *principle of impulse and momentum*, we have

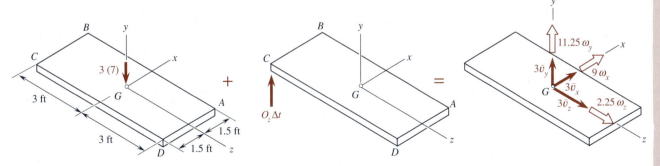

Initial
momentum diagram

Impulse diagram

Final
momentum diagram

Referring to the vector-diagram equation, we write

$$\Sigma V_x: \qquad 0 + 0 = 3\bar{v}_x \qquad \bar{v}_x = 0$$

$$\Sigma V_z: \qquad 0 + 0 = 3\bar{v}_z \qquad \bar{v}_z = 0$$

$$\Sigma M_C: \qquad (1.5\mathbf{i} + 3\mathbf{k}) \times [-3(7)\mathbf{j}] + \mathbf{0}$$

$$= (1.5\mathbf{i} + 3\mathbf{k}) \times (0\mathbf{i} + 3\bar{v}_y\mathbf{j} + 0\mathbf{k})$$

$$+ (9\omega_x\mathbf{i} + 11.25\omega_y\mathbf{j} + 2.25\omega_z\mathbf{k})$$

Carrying out the cross products in this equation and then equating the coefficients of the unit vectors, we get

$$63 = -9\bar{v}_y + 9\omega_x \tag{1}$$

$$0 = 11.25\omega_y \qquad \omega_y = 0$$

$$-31.5 = 4.5\bar{v}_y + 2.25\omega_z \tag{2}$$

Because of perfectly plastic impact at C, we have $\mathbf{v}_C = \mathbf{0}$ and

$$\mathbf{v}_G = 0\mathbf{i} + \bar{v}_y\mathbf{j} + 0\mathbf{k}$$

$$= (\omega_x\mathbf{i} + 0\mathbf{j} + \omega_z\mathbf{k}) \times (1.5\mathbf{i} + 3\mathbf{k})$$

which yields a nontrivial equation

$$\bar{v}_y = -3\omega_x + 1.5\omega_z \tag{3}$$

Solving Eqs. (1), (2), and (3) simultaneously, we obtain

$$\bar{v}_y = -6 \qquad \omega_x = 1 \qquad \omega_z = -2$$

Thus, we have

$$\boldsymbol{\omega} = \mathbf{i} - 2\mathbf{k} \text{ rad/s} \qquad \bar{\mathbf{v}} = -6\mathbf{j} \text{ ft/s} \blacktriangleleft$$

Developmental Exercises

D19.21 Refer to Example 19.8. Determine the velocities of the corners A, B, and D of the plate immediately after the impact.

D19.22 Refer to Fig. 19.9. If the mass of the body is symmetrical with respect to the xy and yz planes, write the components of the angular momentum $\mathbf{H}_G$ of the body about G in $OXYZ$.

EXAMPLE 19.9

A thin disk of mass $m = 45$ kg and radius $r = 300$ mm is mounted on a horizontal shaft with which the disk makes an angle of $\phi = 60°$ as shown. If the shaft rotates with $\boldsymbol{\omega} = 10\mathbf{I}$ rad/s, determine (a) the angular momentum $\mathbf{H}_G$ of the disk about its mass center G, (b) the angle θ between $\mathbf{H}_G$ and $\boldsymbol{\omega}$.

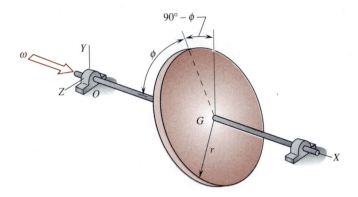

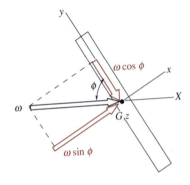

Solution.   Let *Gxyz* be a nonrotating reference frame oriented as shown.
Then, we write

$$\omega_x = \omega \sin\phi \qquad \omega_y = -\omega \cos\phi \qquad \omega_z = 0$$

$$\bar{I}_{xx} = \tfrac{1}{2} mr^2 \qquad \bar{I}_{yy} = \bar{I}_{zz} = \tfrac{1}{4} mr^2$$

$$\bar{I}_{xy} = \bar{I}_{yz} = \bar{I}_{zx} = 0$$

Applying Eqs. (19.79) and (19.76), we get

$$H_x = \tfrac{1}{2} mr^2 (\omega \sin\phi) - 0 - 0 = \tfrac{1}{2} mr^2 \omega \sin\phi$$

$$H_y = -0 + \tfrac{1}{4} mr^2 (-\omega \cos\phi) - 0 = -\tfrac{1}{4} mr^2 \omega \cos\phi$$

$$H_z = -0 - 0 + 0 = 0$$

$$\mathbf{H}_G = H_x \mathbf{i} + H_y \mathbf{j} + H_z \mathbf{k} = \tfrac{1}{4} mr^2 \omega (2 \sin\phi\, \mathbf{i} - \cos\phi\, \mathbf{j})$$

Expressing in terms of the unit vectors in *OXYZ*, we write

$$\mathbf{i} = \sin\phi\, \mathbf{I} + \cos\phi\, \mathbf{J} \qquad \mathbf{j} = -\cos\phi\, \mathbf{I} + \sin\phi\, \mathbf{J}$$

$$\mathbf{H}_G = \tfrac{1}{4} mr^2 \omega [2 \sin\phi\, (\sin\phi\, \mathbf{I} + \cos\phi\, \mathbf{J})$$

$$- \cos\phi\, (-\cos\phi\, \mathbf{I} + \sin\phi\, \mathbf{J})]$$

$$= \tfrac{1}{4} mr^2 \omega [(2 \sin^2\phi + \cos^2\phi)\mathbf{I} + \sin\phi \cos\phi\, \mathbf{J}]$$

Substituting *m* = 4 kg, *r* = 0.3 m, ω = 10 rad/s, and ϕ = 60° into the
expression for $\mathbf{H}_G$, we obtain

$$\mathbf{H}_G = 1.575\mathbf{I} + 0.390\mathbf{J} \text{ kg·m}^2\text{/s} \blacktriangleleft$$

The angle θ between $\mathbf{H}_G$ and $\boldsymbol{\omega}$ is determined as follows:

$$\cos\theta = \frac{\boldsymbol{\omega} \cdot \mathbf{H}_G}{\omega H_G} = \frac{15.75}{16.225}$$

$$\theta = 13.90° \blacktriangleleft$$

REMARK. The result shows that $\mathbf{H}_G$ and $\boldsymbol{\omega}$ for a rigid body in general motion
are generally *not* in the same direction.

Developmental Exercise

D19.23 Refer to Example 19.9. Determine (a) the value of the angle ϕ for which the angle θ between $\mathbf{H}_G$ and $\boldsymbol{\omega}$ is maximum, (b) the corresponding maximum value of θ.

★19.7 Kinetic Energy of a Rigid Body

Let us consider again the case in which a rigid body of mass m rotates with an angular velocity $\boldsymbol{\omega}$ and its mass center G moves with a velocity $\bar{\mathbf{v}}$ in the *fixed* reference frame *OXYZ* as shown in Fig. 19.9, where *Gxyz* is a *nonrotating* reference frame moving with G. The differential mass dm, as shown in this figure, has a velocity $\mathbf{v}$ in *OXYZ*, which is given by Eq. (19.71). Applying Eq. (19.71) and the vector identity $\mathbf{A} \cdot \mathbf{P} \times \mathbf{Q} = \mathbf{A} \times \mathbf{P} \cdot \mathbf{Q}$, we write[†]

$$v^2 = \mathbf{v} \cdot \mathbf{v} = (\bar{\mathbf{v}} + \boldsymbol{\omega} \times \mathbf{r}') \cdot (\bar{\mathbf{v}} + \boldsymbol{\omega} \times \mathbf{r}')$$

$$= \bar{v}^2 + 2\bar{\mathbf{v}} \cdot (\boldsymbol{\omega} \times \mathbf{r}') + (\boldsymbol{\omega} \times \mathbf{r}') \cdot (\boldsymbol{\omega} \times \mathbf{r}')$$

$$v^2 = \bar{v}^2 + 2\bar{\mathbf{v}} \cdot (\boldsymbol{\omega} \times \mathbf{r}') + \boldsymbol{\omega} \cdot \mathbf{r}' \times (\boldsymbol{\omega} \times \mathbf{r}') \qquad (19.86)$$

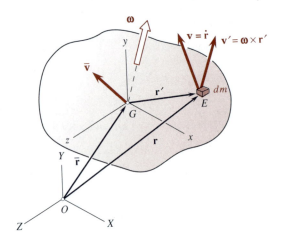

Fig. 19.9 (repeated)

Applying Eqs. (19.86) and (19.73) as well as the *principle of moments*, we compute the kinetic energy T of the rigid body in *OXYZ* as follows:

$$T = \tfrac{1}{2} \int v^2 \, dm$$

$$= \tfrac{1}{2} \bar{v}^2 \int dm + \bar{\mathbf{v}} \cdot \boldsymbol{\omega} \times \int \mathbf{r}' \, dm$$

$$+ \tfrac{1}{2} \boldsymbol{\omega} \cdot \int \mathbf{r}' \times (\boldsymbol{\omega} \times \mathbf{r}') \, dm$$

$$= \tfrac{1}{2} \bar{v}^2 m + \bar{\mathbf{v}} \cdot \boldsymbol{\omega} \times \mathbf{0} + \tfrac{1}{2} \boldsymbol{\omega} \cdot \mathbf{H}_G$$

$$T = \tfrac{1}{2} m\bar{v}^2 + \tfrac{1}{2} \boldsymbol{\omega} \cdot \mathbf{H}_G \qquad (19.87)$$

[†]Here we let $\mathbf{A} = \boldsymbol{\omega}$, $\mathbf{P} = \mathbf{r}'$, and $\mathbf{Q} = \boldsymbol{\omega} \times \mathbf{r}'$. Cf. Eq. (4.32).

where $\mathbf{H}_G$ is the angular momentum of the rigid body about G. Since $\frac{1}{2}m\bar{v}^2 = \frac{1}{2}\bar{\mathbf{v}} \cdot (m\bar{\mathbf{v}})$, the kinetic energy in Eq. (19.87) may be written in terms of dot products as

$$T = \tfrac{1}{2}\bar{\mathbf{v}} \cdot \mathbf{L} + \tfrac{1}{2}\boldsymbol{\omega} \cdot \mathbf{H}_G \qquad (19.88)$$

where $\mathbf{L}$ is the linear momentum defined in Eq. (19.72). Substituting Eqs. (19.76), (19.78), and (19.79) into Eq. (19.87), we obtain the explicit form of the kinetic energy T of a rigid body in $OXYZ$ as

$$\begin{aligned} T = \tfrac{1}{2}m\bar{v}^2 + \tfrac{1}{2}(\bar{I}_{xx}\omega_x^2 + \bar{I}_{yy}\omega_y^2 + \bar{I}_{zz}\omega_z^2 \\ - 2\bar{I}_{xy}\omega_x\omega_y - 2\bar{I}_{yz}\omega_y\omega_z - 2\bar{I}_{zx}\omega_z\omega_x) \end{aligned} \qquad (19.89)$$

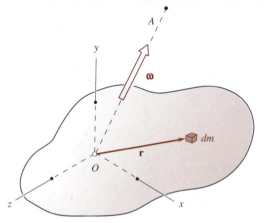

Fig. 19.13

In the particular case of a rigid body rotating with $\boldsymbol{\omega} = \omega_x\mathbf{i} + \omega_y\mathbf{j} + \omega_z\mathbf{k}$ about the origin O of a fixed reference frame $Oxyz$, as shown in Fig. 19.13, it can similarly be shown that the kinetic energy T of the rigid body is given by

$$\begin{aligned} T = \tfrac{1}{2}(I_{xx}\omega_x^2 + I_{yy}\omega_y^2 + I_{zz}\omega_z^2 - 2I_{xy}\omega_x\omega_y \\ - 2I_{yz}\omega_y\omega_z - 2I_{zx}\omega_z\omega_x) \end{aligned} \qquad (19.90)$$

where the moments of inertia I_{xx}, I_{yy}, I_{zz} and the products of inertia I_{xy}, I_{yz}, I_{zx} are computed with respect to the fixed reference frame $Oxyz$. Furthermore, suppose that OA is the axis of rotation or the *velocity axis* (i.e., the instantaneous axis of zero velocity) of the rigid body, as indicated in Fig. 19.13, at the instant considered. Then, $\boldsymbol{\omega}$ is directed along the axis OA, and the velocities of all particles of the rigid body are perpendicular to the radii of their circular arcs of paths centered on the axis OA. In this case, the kinetic energy T of the rigid body can readily be shown to be given by

$$T = \tfrac{1}{2}I_{OA}\omega^2 \qquad (19.91)$$

where I_{OA} is the moment of inertia of the body about its *velocity axis OA*.

Knowing how to compute the kinetic energy of a rigid body in general motion, we may solve appropriate problems by the *principle of work and energy*, which is represented by the equation

$$T_1 + U_{1 \to 2} = T_2 \qquad (19.92)$$

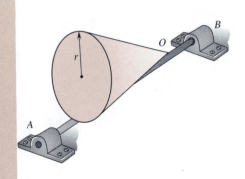

where T_1 and T_2 are the kinetic energies of the rigid body in the positions 1 and 2, respectively, and $U_{1\rightarrow2}$ is the work of the force system acting on the rigid body during its motion from the position 1 to the position 2. Note that the kinetic energy of a body has different values in different inertial reference frames moving with different constant velocities. It is important to use the *same* inertial reference frame in computing the kinetic energy and the work of a force or moment in a given kinetic problem.

EXAMPLE 19.10

A solid circular cone has a base of radius $r = 200$ mm, a height of $h = 480$ mm, and a mass of $m = 160$ kg. The cone is welded to a horizontal shaft AB which can rotate freely in its bearings as shown. If the cone is at rest in the upright position when it is given a slight push, determine its angular speed ω after it has rotated through 180°.

Solution. The slant height of the cone is $L = (r^2 + h^2)^{1/2}$. We may draw the side view of the cone as shown, where G is its mass center. Referring

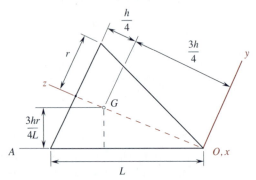

to this drawing and the formulas in Fig. 16.7, we write

$$I_{xx} = I_{yy} = \tfrac{3}{20} m(r^2 + 4h^2) \qquad I_{zz} = \tfrac{3}{10} mr^2$$

$$I_{xy} = I_{yz} = I_{zx} = 0$$

$$\lambda_{OA} = -\frac{r}{L}\mathbf{j} + \frac{h}{L}\mathbf{k}$$

$$\lambda_x = 0 \qquad \lambda_y = -\frac{r}{L} \qquad \lambda_z = \frac{h}{L}$$

Applying Eq. (19.56) to compute the moment of inertia, we write

$$I_{OA} = 0 + \frac{3}{20} m(r^2 + 4h^2)\left(-\frac{r}{L}\right)^2$$

$$+ \frac{3}{10} mr^2\left(\frac{h}{L}\right)^2 - 0 - 0 - 0$$

$$I_{OA} = \frac{3mr^2}{20L^2}(r^2 + 6h^2)$$

Since the problem directly involves displacement and velocity, it may be solved by using the *principle of work and energy*. As the cone rotates from rest through $180°$ to its lowest position, its mass center G will have descended a distance of $2(3hr)/(4L) = 3hr/(2L)$ and only its weight force $\mathbf{W} = mg \downarrow$ will do work during the rotation. Thus, we have $U_{1\to2} = 3mghr/(2L)$. We note that the shaft is also the *velocity axis* of the cone. Therefore, we apply Eq. (19.91) to write

$$T_1 = 0 \qquad T_2 = \tfrac{1}{2} I_{OA}\omega^2 = \frac{3mr^2}{40L^2}(r^2 + 6h^2)\omega^2$$

Applying the *principle of work and energy*, $T_1 + U_{1\to2} = T_2$, we write

$$0 + \frac{3mghr}{2L} = \frac{3mr^2}{40L^2}(r^2 + 6h^2)\omega^2$$

$$\omega^2 = \frac{20ghL}{r(r^2 + 6h^2)} = \frac{20gh(r^2 + h^2)^{1/2}}{r(r^2 + 6h^2)}$$

Substituting $g = 9.81 \text{ m/s}^2$, $h = 0.48$ m, and $r = 0.2$ m into the expression for ω^2, we obtain

$$\omega^2 = 172.14 \qquad \omega = 13.12 \text{ rad/s} \blacktriangleleft$$

Developmental Exercises

D19.24 Refer to Example 19.10. Determine the kinetic energy of the cone when it reaches its lowest position.

D19.25 Refer to the solid circular cone in Example 19.10. Determine the angular speed ω of the cone after it has rotated through $90°$.

D19.26 Write the kinetic energy of a rigid body in general motion in terms of dot products.

PROBLEMS

19.41 The bent rod $OABC$ has a total mass m and revolves about the Z axis with an angular velocity $\boldsymbol{\omega}$. For the position shown, determine (a) the angular momentum of the rod about the origin O, (b) the kinetic energy of the rod.

19.42 A 2.4-m slender rod AB of mass 4 kg is welded at its midpoint G to a 2.4-m horizontal shaft OC. In the position shown, the shaft rotates with an angular velocity $\boldsymbol{\omega} = 120\mathbf{I}$ rpm and the rod lies in the ZX plane. Determine (a) the angular momentum of the rod about G, (b) the kinetic energy of the rod.

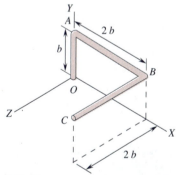

Fig. P19.41

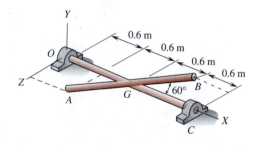

Fig. P19.42

19.43* A thin rectangular plate weighing 10 lb is mounted on a horizontal shaft with which the plate makes an angle of 50° as shown. If the shaft rotates with $\boldsymbol{\omega} = 6\mathbf{I}$ rad/s, determine (a) the angular momentum $\mathbf{H}_G$ of the plate about G, (b) the angle between $\mathbf{H}_G$ and $\boldsymbol{\omega}$, (c) the kinetic energy of the plate.

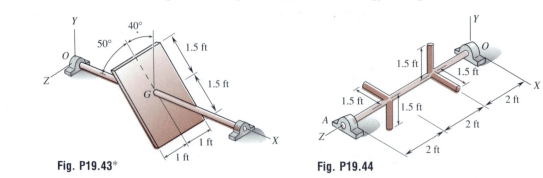

Fig. P19.43*   Fig. P19.44

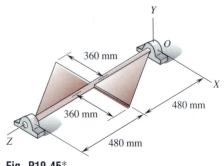

Fig. P19.45*

19.44 Four slender rods, each weighing 3 lb, are welded to a horizontal shaft as shown. If the shaft rotates at $\boldsymbol{\omega} = 90\mathbf{K}$ rpm, determine (a) the angular momentum $\mathbf{H}_O$ of the rods about the origin O, (b) the angle between $\mathbf{H}_O$ and $\boldsymbol{\omega}$, (c) the kinetic energy of the rods.

19.45* Two triangular plates, each of mass 5 kg, are welded to a horizontal shaft as shown. If the shaft rotates at $\boldsymbol{\omega} = 4\mathbf{K}$ rad/s in the position shown, determine (a) the angular momentum of the plates about the origin O, (b) the kinetic energy of the plates.

19.46 The 5-kg disk D rotates about the pin at A at the constant rate $\boldsymbol{\omega}_1 = -4\mathbf{I}$ rad/s relative to the frame F, while the frame F rotates about the fixed shaft at the constant rate $\boldsymbol{\omega}_2 = 8\mathbf{K}$ rad/s. For the instant shown, determine (a) the angular momentum of the disk about A, (b) the kinetic energy of the disk.

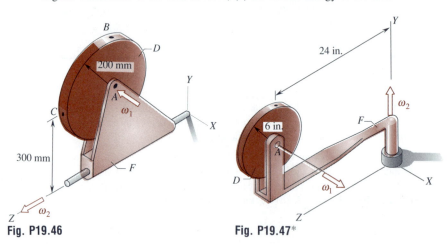

Fig. P19.46   Fig. P19.47*

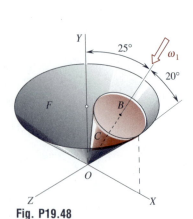

Fig. P19.48

19.47* The 12-lb disk D rotates about the pin at A at the constant rate $\boldsymbol{\omega}_1 = 6\mathbf{I}$ rad/s relative to the arm F, while the arm F rotates at the constant rate $\boldsymbol{\omega}_2 = 4\mathbf{J}$ rad/s relative to the ground. For the instant shown, determine (a) the angular momentum of the disk about A, (b) the kinetic energy of the disk.

19.48 A 6-kg cone C rolls without slipping inside a fixed conical cavity F as shown, where $\overline{OB} = 200$ mm. The cone C moves with a constant rate of spin $\boldsymbol{\omega}_1 = 2$ rad/s about its axis BO. For the instant shown, determine (a) the angular momentum of the cone about the origin O, (b) the kinetic energy of the cone.

19.49 At the instant shown, the 20-lb disk D rotates about its axis relative to the motor M at the rate $\omega_{D/M} = 2$ rad/s, the motor M rotates about the mounting pin relative to the bracket B at the rate $\omega_{M/B} = 1.5$ rad/s, while the bracket B rotates about the vertical shaft relative to the ground at the rate $\omega_B = 4$ rad/s. For this instant, determine (a) the angular momentum of the disk about its mass center, (b) the kinetic energy of the disk.

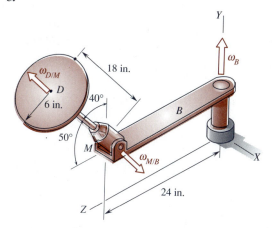

Fig. P19.49

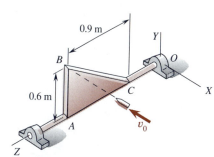

Fig. P19.50

19.50 The 8-kg triangular plate shown is in equilibrium when a 4-g bullet is fired into it with a velocity $\mathbf{v}_0 = -1800\mathbf{I}$ m/s and becomes embedded at B. Determine the angular velocity of the plate (a) immediately after the plate is hit by the bullet, (b) at the instant the plate has rotated 180°.

19.51 A 20-lb plate is suspended by a wire fastened at its mass center G. The plate is originally at rest. If an impulse of $\mathbf{I} = -12\mathbf{J}$ lb·s is momentarily applied at the corner A, determine the angular velocity of the plate immediately after the impact.

19.52* A 5-Mg space vehicle is translating at the velocity $\overline{\mathbf{v}} = 800\mathbf{J}$ m/s when it is struck at the point $A(1, 3, -2)$ m on its surface by a meteoroid of 0.8 kg which is traveling with a velocity $\mathbf{v}_0 = -200\mathbf{I} - 600\mathbf{J} + 300\mathbf{K}$ m/s. If the radii of gyration of the space vehicle are $k_X = k_Z = 1.2$ m and $k_Y = 0.6$ m, determine the angular velocity of the space vehicle after the meteoroid becomes embedded in it.

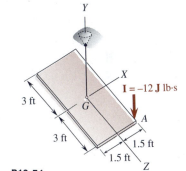

Fig. P19.51

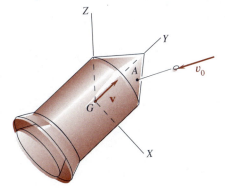

Fig. P19.52*

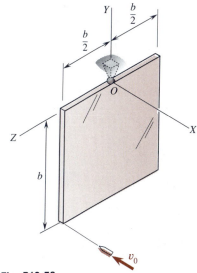

19.53 A heavy square plate of mass m is suspended from a ball-and-socket joint at O as shown. If a bullet of mass m_0 strikes the corner A of the plate with a velocity $\mathbf{v}_0 = -v_0\mathbf{I}$ and becomes embedded in the plate, determine the angular velocity of the plate immediately after the impact.

Fig. P19.53

★19.8   Equations of Motion for a Rigid Body

We know that a rigid body may be viewed as a special system of particles which keep fixed distances from each other at any time during the motion. In other words, the equations of motion expressed in terms of the time derivatives of the momenta of a system of particles in Sec. 14.4 are also applicable to a rigid body in general motion. Thus, for a rigid body moving in space, the motion must satisfy the following fundamental equations of motion

$$\Sigma \mathbf{F} = \dot{\mathbf{L}} \tag{19.93}$$

$$\Sigma \mathbf{M}_G = \dot{\mathbf{H}}_G \tag{19.94}$$

$$\Sigma \mathbf{M}_O = \dot{\mathbf{H}}_O \tag{19.95}$$

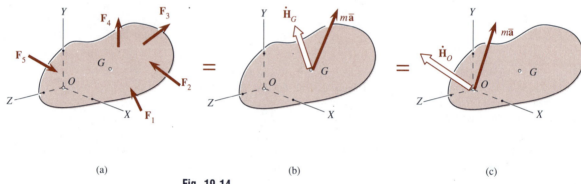

(a)   (b)   (c)

Fig. 19.14

I. TRANSLATIONAL MOTION.   Since $\dot{\mathbf{L}} = m\bar{\mathbf{a}}$ for a rigid body of mass m whose mass center G is moving with an acceleration $\bar{\mathbf{a}}$, we see that the force system in Fig. 19.14(a) is equivalent to the effective force-moment system consisting of $m\bar{\mathbf{a}}$ at G and $\dot{\mathbf{H}}_G$ about G as shown in Fig. 19.14(b), or the effective force-moment system consisting of $m\bar{\mathbf{a}}$ at O and $\dot{\mathbf{H}}_O$ about O as shown in Fig. 19.14(c). In any case, the *translational motion* of the rigid body is governed by Eq. (19.93) or the vector equation

$$\Sigma \mathbf{F} = m\bar{\mathbf{a}} \tag{19.96}$$

which entails the following three scalar equations:

$$\Sigma F_x = m\bar{a}_x \qquad \Sigma F_y = m\bar{a}_y \qquad \Sigma F_z = m\bar{a}_z \tag{19.97}$$

where $\Sigma \mathbf{F} = \Sigma F_x \mathbf{i} + \Sigma F_y \mathbf{j} + \Sigma F_z \mathbf{k}$ is the resultant force, and $\bar{\mathbf{a}} = \bar{a}_x \mathbf{i} + \bar{a}_y \mathbf{j} + \bar{a}_z \mathbf{k}$ is the acceleration of the mass center G.

II. ROTATIONAL MOTION ABOUT THE MASS CENTER G.   We recognize that $\dot{\mathbf{H}}_G$, the time derivative of the angular momentum about G, is simply equal to $\bar{I}\alpha$ in plane motion, where $\bar{I}$ is the moment of inertia about the central axis perpendicular to the plane of motion and α is the angular acceleration of the rigid body. In general motion, $\dot{\mathbf{H}}_G$ is much more involved. We may employ a *moving reference frame Gxyz*, which is not embedded in the rigid body but whose origin moves with the mass center G of the rigid body as shown in Fig. 19.15. Applying Eq. (19.8), we write

$$\dot{\mathbf{H}}_G = (\dot{\mathbf{H}}_G)_{Gxyz} + \mathbf{\Omega} \times \mathbf{H}_G \qquad (19.98)$$

where $\mathbf{\Omega}$ is the angular velocity of *Gxyz* in *OXYZ*. Substituting Eq. (19.98) into Eq. (19.94), we see that the *rotational motion* of a rigid body must satisfy the vector equation

$$\Sigma\mathbf{M}_G = (\dot{\mathbf{H}}_G)_{Gxyz} + \mathbf{\Omega} \times \mathbf{H}_G \qquad (19.99)$$

If we let *Gxyz* in Fig. 19.15 be *embedded* in the rigid body whose angular velocity is $\boldsymbol{\omega}$ in the fixed reference frame *OXYZ*, then the moments and products of inertia in *Gxyz* become constant and $\mathbf{\Omega}$ and $\boldsymbol{\omega}$ are equal. We have

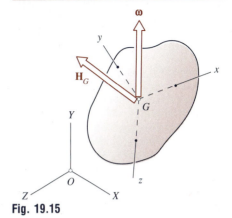

Fig. 19.15

$$\mathbf{\Omega} = \boldsymbol{\omega} = \omega_x\mathbf{i} + \omega_y\mathbf{j} + \omega_z\mathbf{k} \qquad (19.100)$$

$$\mathbf{H}_G = H_x\mathbf{i} + H_y\mathbf{j} + H_z\mathbf{k} \qquad \begin{array}{c}(19.76)\\ \text{(repeated)}\end{array}$$

$$(\dot{\mathbf{H}}_G)_{Gxyz} = \dot{H}_x\mathbf{i} + \dot{H}_y\mathbf{j} + \dot{H}_z\mathbf{k} \qquad (19.101)$$

where, by virtue of Eq. (19.79),

$$\dot{H}_x = \bar{I}_{xx}\dot{\omega}_x - \bar{I}_{xy}\dot{\omega}_y - \bar{I}_{xz}\dot{\omega}_z \qquad (19.102a)$$

$$\dot{H}_y = -\bar{I}_{yx}\dot{\omega}_x + \bar{I}_{yy}\dot{\omega}_y - \bar{I}_{yz}\dot{\omega}_z \qquad (19.102b)$$

$$\dot{H}_z = -\bar{I}_{zx}\dot{\omega}_x - \bar{I}_{zy}\dot{\omega}_y + \bar{I}_{zz}\dot{\omega}_z \qquad (19.102c)$$

Expressing the resultant moment $\Sigma\mathbf{M}_G$ about G in *Gxyz*, we write

$$\Sigma\mathbf{M}_G = \Sigma M_x\mathbf{i} + \Sigma M_y\mathbf{j} + \Sigma M_z\mathbf{k} \qquad (19.103)$$

Substituting Eqs. (19.100), (19.76), and (19.101) through (19.103) into Eq. (19.99), carrying out the cross product, and equating coefficients of the unit vectors $\mathbf{i}$, $\mathbf{j}$, and $\mathbf{k}$, we obtain the following three scalar equations:

$$\Sigma M_x = \bar{I}_{xx}\dot{\omega}_x - (\bar{I}_{yy} - \bar{I}_{zz})\omega_y\omega_z - \bar{I}_{xy}(\dot{\omega}_y - \omega_z\omega_x)$$
$$- \bar{I}_{yz}(\omega_y^2 - \omega_z^2) - \bar{I}_{zx}(\dot{\omega}_z + \omega_x\omega_y) \qquad (19.104a)$$

$$\Sigma M_y = \bar{I}_{yy}\dot{\omega}_y - (\bar{I}_{zz} - \bar{I}_{xx})\omega_z\omega_x - \bar{I}_{yz}(\dot{\omega}_z - \omega_x\omega_y)$$
$$- \bar{I}_{zx}(\omega_z^2 - \omega_x^2) - \bar{I}_{xy}(\dot{\omega}_x + \omega_y\omega_z) \qquad (19.104b)$$

$$\Sigma M_z = \bar{I}_{zz}\dot{\omega}_z - (\bar{I}_{xx} - \bar{I}_{yy})\omega_x\omega_y - \bar{I}_{zx}(\dot{\omega}_x - \omega_y\omega_z)$$
$$- \bar{I}_{xy}(\omega_x^2 - \omega_y^2) - \bar{I}_{yz}(\dot{\omega}_y + \omega_z\omega_x) \qquad (19.104c)$$

These are the *general rotational equations of motion* in terms of a body-axis coordinate system at the mass center G.

In the case the x, y, and z body axes in Fig. 19.15 are chosen to coincide with the *principal axes of inertia* at G of the rigid body, Eqs. (19.104) reduce to

$$\Sigma M_x = \bar{I}_x \dot{\omega}_x - (\bar{I}_y - \bar{I}_z)\omega_y\omega_z$$

$$\Sigma M_y = \bar{I}_y \dot{\omega}_y - (\bar{I}_z - \bar{I}_x)\omega_z\omega_x \qquad (19.105)$$

$$\Sigma M_z = \bar{I}_z \dot{\omega}_z - (\bar{I}_x - \bar{I}_y)\omega_x\omega_y$$

where $\bar{I}_x = \bar{I}_{xx}$, $\bar{I}_y = \bar{I}_{yy}$, and $\bar{I}_z = \bar{I}_{zz}$ are the principal moments of inertia at G. Equations (19.105) are called *Euler's equations of motion* for $\boldsymbol{\Omega} = \boldsymbol{\omega}$ for a rigid body about its mass center, after the Swiss mathematician Leonhard Euler (1707–1783).[†] In the present case, we do have $\boldsymbol{\Omega} = \boldsymbol{\omega}$ and

$$\dot{\boldsymbol{\omega}} = (\dot{\boldsymbol{\omega}})_{OXYZ} = (\dot{\boldsymbol{\omega}})_{Gxyz} + \boldsymbol{\Omega} \times \boldsymbol{\omega}$$

$$= (\dot{\boldsymbol{\omega}})_{Gxyz} + \boldsymbol{\omega} \times \boldsymbol{\omega} = (\dot{\boldsymbol{\omega}})_{Gxyz}$$

$$= \dot{\omega}_x \mathbf{i} + \dot{\omega}_y \mathbf{j} + \dot{\omega}_z \mathbf{k}$$

This special relation means that, when $\boldsymbol{\Omega} = \boldsymbol{\omega}$, the quantities $\dot{\omega}_x$, $\dot{\omega}_y$, and $\dot{\omega}_z$, which are the rectangular components of $(\dot{\boldsymbol{\omega}})_{Gxyz}$, can be obtained by first finding the angular acceleration $\dot{\boldsymbol{\omega}}$ of the rigid body in the fixed reference frame $OXYZ$, via Eq. (19.50), and then resolving it into components in the x, y, and z directions of the body-axis coordinate system $Gxyz$. Note that $(\dot{\boldsymbol{\omega}})_{Gxyz}$ is the time derivative, taken in the moving reference frame $Gxyz$, of the angular velocity $\boldsymbol{\omega}$ of the body in the fixed reference frame $OXYZ$. Do *not* assume that $\dot{\omega}_x = 0$ when $\omega_x = 0$ in $Gxyz$, etc. Moreover, note that Euler's equations of motion apply *only* for moments summed about the principal axes of inertia.

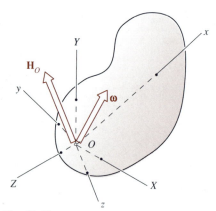

Fig. 19.16

III. ROTATIONAL MOTION ABOUT A FIXED POINT O.

When a rigid body is constrained to rotate about a fixed point, we may first let the fixed point be the origin of the fixed reference frame $OXYZ$ and then let the angular velocity of the rigid body in $OXYZ$ be $\boldsymbol{\omega}$, as shown in Fig. 19.16. Furthermore, we may employ a rotating reference frame $Oxyz$ whose angular velocity in $OXYZ$ is $\boldsymbol{\Omega}$. In general, $\boldsymbol{\Omega}$ is not equal to $\boldsymbol{\omega}$. Using Eq. (19.95) and applying Eq. (19.8), we readily see that the *rotational motion* of a rigid body must also satisfy the vector equation

$$\Sigma \mathbf{M}_O = (\dot{\mathbf{H}}_O)_{Oxyz} + \boldsymbol{\Omega} \times \mathbf{H}_O \qquad (19.106)$$

If we let $Oxyz$ in Fig. 19.16 be *embedded* in the rigid body, then the moments and products of inertia in $Oxyz$ become constant and $\boldsymbol{\Omega}$ and $\boldsymbol{\omega}$ are equal. In this case, Eq. (19.106) can similarly be shown to entail the following three scalar equations:

[†]Leonhard Euler left his birthplace Basel, Switzerland, for St. Petersburg, Russia, in 1727. From then on, his life and scientific work were closely associated with the St. Petersburg Academy and with Russia.

$$\Sigma M_x = I_{xx}\dot{\omega}_x - (I_{yy} - I_{zz})\omega_y\omega_z - I_{xy}(\dot{\omega}_y - \omega_z\omega_x)$$
$$- I_{yz}(\omega_y^2 - \omega_z^2) - I_{zx}(\dot{\omega}_z + \omega_x\omega_y) \qquad (19.107a)$$

$$\Sigma M_y = I_{yy}\dot{\omega}_y - (I_{zz} - I_{xx})\omega_z\omega_x - I_{yz}(\dot{\omega}_z - \omega_x\omega_y)$$
$$- I_{zx}(\omega_z^2 - \omega_x^2) - I_{xy}(\dot{\omega}_x + \omega_y\omega_z) \qquad (19.107b)$$

$$\Sigma M_z = I_{zz}\dot{\omega}_z - (I_{xx} - I_{yy})\omega_x\omega_y - I_{zx}(\dot{\omega}_x - \omega_y\omega_z)$$
$$- I_{xy}(\omega_x^2 - \omega_y^2) - I_{yz}(\dot{\omega}_y + \omega_z\omega_x) \qquad (19.107c)$$

These are the *general rotational equations of motion* in terms of a body-axis coordinate system at the fixed point O as shown in Fig. 19.16. If the x, y, and z body axes are chosen to coincide with the *principal axes of inertia* at O for the rigid body, Eqs. (19.107) reduce to

$$\Sigma M_x = I_x\dot{\omega}_x - (I_y - I_z)\omega_y\omega_z$$

$$\Sigma M_y = I_y\dot{\omega}_y - (I_z - I_x)\omega_z\omega_x \qquad (19.108)$$

$$\Sigma M_z = I_z\dot{\omega}_z - (I_x - I_y)\omega_x\omega_y$$

where $I_x = I_{xx}$, $I_y = I_{yy}$, $I_z = I_{zz}$ are the principal moments of inertia at the fixed origin O. Equations (19.108) are called *Euler's equations of motion* for $\boldsymbol{\Omega} = \boldsymbol{\omega}$ for a rigid body about a fixed point.

IV. EULER'S EQUATIONS OF MOTION FOR $\boldsymbol{\Omega} \neq \boldsymbol{\omega}$.

At times, a body of revolution (such as a spinning top or gyroscope) may be made to rotate about a fixed point O on its axis of symmetry, as shown in Fig. 19.17, where the z axis of the moving reference frame $Oxyz$ is chosen to be *always* directed along the axis of symmetry of the body. If $Oxyz$ is *not embedded* in the body, the angular velocity $\boldsymbol{\Omega}$ of $Oxyz$ in the fixed reference frame $OXYZ$ may be different from the angular velocity $\boldsymbol{\omega}$ of the body in $OXYZ$; i.e., $\boldsymbol{\Omega} \neq \boldsymbol{\omega}$. Because of symmetry, the x, y, and z axes are principal axes of inertia at O and the principal moments of inertia I_x, I_y, and I_z all remain *constant* during the motion of the body. In this case, we have

$$\boldsymbol{\omega} = \omega_x\mathbf{i} + \omega_y\mathbf{j} + \omega_z\mathbf{k}$$

$$\boldsymbol{\Omega} = \Omega_x\mathbf{i} + \Omega_y\mathbf{j} + \Omega_z\mathbf{k}$$

$$\mathbf{H}_O = I_x\omega_x\mathbf{i} + I_y\omega_y\mathbf{j} + I_z\omega_z\mathbf{k}$$

Substituting the above equations into Eq. (19.106), carrying out the indicated differentiation and the cross product, and equating the coefficients of the unit vectors, we obtain

$$\Sigma M_x = I_x\dot{\omega}_x - I_y\Omega_z\omega_y + I_z\Omega_y\omega_z$$

$$\Sigma M_y = I_y\dot{\omega}_y - I_z\Omega_x\omega_z + I_x\Omega_z\omega_x \qquad (19.109)$$

$$\Sigma M_z = I_z\dot{\omega}_z - I_x\Omega_y\omega_x + I_y\Omega_x\omega_y$$

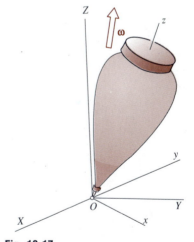

Fig. 19.17

where $\dot{\omega}_x$, $\dot{\omega}_y$, and $\dot{\omega}_z$ are the rectangular components of $(\dot{\boldsymbol{\omega}})_{Oxyz}$. Note that $(\dot{\boldsymbol{\omega}})_{Oxyz}$ is the time derivative, taken in the moving reference frame $Oxyz$, of the angular velocity $\boldsymbol{\omega}$ of the body in the fixed reference frame $OXYZ$. *In this particular case*, we have $\boldsymbol{\Omega} \neq \boldsymbol{\omega}$ and

$$\dot{\boldsymbol{\omega}} = (\dot{\boldsymbol{\omega}})_{Oxyz} + \boldsymbol{\Omega} \times \boldsymbol{\omega} \neq (\dot{\boldsymbol{\omega}})_{Oxyz}$$

Equations (19.109) are called *Euler's equations of motion* for $\boldsymbol{\Omega} \neq \boldsymbol{\omega}$ for a body of revolution.

Note that the usual steps to be taken prior to applying *Euler's equations of motion* in Eqs. (19.105), (19.108), or (19.109) are as follows:

a. Embed the x, y, and z axes in the rigid body (or a reference frame) so that they are, and remain to be, the principal axes of inertia at a fixed point O or the mass center G of the rigid body during the motion.
b. Compute $\boldsymbol{\omega}$ in $OXYZ$ then find ω_x, ω_y, and ω_z. [Cf. Eq. (19.40).]
c. If $\boldsymbol{\Omega} = \boldsymbol{\omega}$, compute $\dot{\boldsymbol{\omega}}$ in $OXYZ$ then find $\dot{\omega}_x$, $\dot{\omega}_y$, and $\dot{\omega}_z$. [Cf. Eq. (19.50).]
d. If $\boldsymbol{\Omega} \neq \boldsymbol{\omega}$, compute $(\dot{\boldsymbol{\omega}})_{Oxyz}$ or $(\dot{\boldsymbol{\omega}})_{Gxyz}$, *not* $\dot{\boldsymbol{\omega}}$, then find $\dot{\omega}_x$, $\dot{\omega}_y$, and $\dot{\omega}_z$.
e. Compute the principal moments of inertia at O or G, as appropriate.
f. Draw the free-body diagram of the rigid body.

If the principal axes of inertia at a fixed point O or the mass center G are not readily identifiable for the rigid body, its rotational motion may be studied by applying, instead, the general equations such as Eqs. (19.99) and (19.106), or (19.104) and (19.107).

EXAMPLE 19.11

A slender rod OA of length $L = 0.5$ m and mass $m = 2$ kg can rotate freely about the pin in a clevis which remains at the position O and is rotated with the shaft about the vertical axis with a constant angular velocity $\boldsymbol{\omega}$ as shown. If the rod OA and the vertical axis form a constant angle $\theta = 30°$, determine the magnitude of $\boldsymbol{\omega}$.

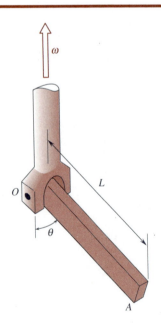

Solution.   At the instant considered, the rod OA may be assumed to lie in the XY plane of the fixed reference frame $OXYZ$ as shown, where G is the mass center of the rod. We first let the reference frame $Oxyz$ be *embedded in the rod*, where the x, y, and z axes are the principal axes of inertia. The angular velocity of the rod can readily be written as

$$\boldsymbol{\omega} = \omega \sin\theta \, \mathbf{i} + \omega \cos\theta \, \mathbf{j}$$

which is constant in $OXYZ$. Clearly, the angular acceleration of the rod in $OXYZ$ is $\dot{\boldsymbol{\omega}} = \mathbf{0}$. Thus, we have

$$\omega_x = \omega \sin\theta \qquad \omega_y = \omega \cos\theta$$

$$\omega_z = \dot{\omega}_x = \dot{\omega}_y = \dot{\omega}_z = 0$$

For the rod and the axes shown, we have

$$I_x = I_z = \tfrac{1}{3} mL^2$$

$$I_y = I_{xy} = I_{yz} = I_{zx} = 0$$

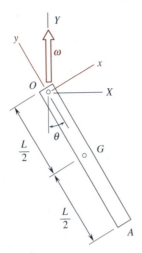

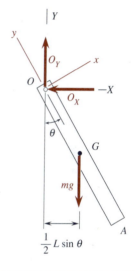

Referring to the free-body diagram drawn and applying *Euler's equations of motion* in Eqs. (19.108), we write

$$\Sigma M_z = I_z \dot{\omega}_z - (I_x - I_y)\omega_x \omega_y:$$

$$-(\tfrac{1}{2} L \sin\theta)(mg) = 0 - (\tfrac{1}{3} mL^2 - 0)(\omega \sin\theta)(\omega \cos\theta)$$

which simplifies to

$$\tfrac{1}{6} mL \sin\theta \, (3g - 2L\omega^2 \cos\theta) = 0$$

Since $\sin\theta = \sin 30° \neq 0$, we obtain

$$\omega^2 = \frac{3g}{2L \cos\theta} = \frac{3(9.81)}{2(0.5) \cos 30°}$$

$$\omega = 5.83 \text{ rad/s} \qquad \blacktriangleleft$$

Developmental Exercises

D19.27 Refer to Example 19.11. Determine (a) the acceleration $\bar{\mathbf{a}}$ of the mass center G, (b) the reaction components $\mathbf{O}_X$ and $\mathbf{O}_Y$.

D19.28 Refer to Example 19.11. Determine the maximum value of ω for which the rod will remain vertical (i.e., $\theta = 0$).

D19.29 Refer to Example 19.11. Determine the kinetic energy of the rod OA in $OXYZ$.

EXAMPLE 19.12

A thin disk having a radius $r = 1$ ft and weighing 16.1 lb rotates with a constant angular speed $\omega_1 = 10$ rad/s relative to the bent axle OAG, which has a negligible weight and is itself rotating with a constant angular velocity $\boldsymbol{\omega}_2 = 5\mathbf{J}$ rad/s about the Y axis of the fixed reference frame $OXYZ$ as shown. Determine the force-moment system, $\mathbf{O}$ and $\mathbf{M}_O$, representing the dynamic reaction from the support at O.

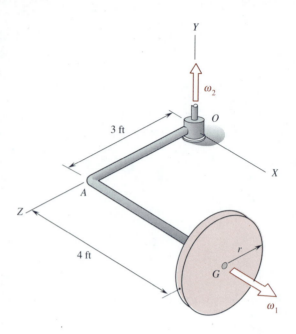

Solution. We first let the reference frame $Gxyz$ be *embedded in the disk* as shown (below left), where the x, y, and z axes are the principal axes at the mass center G of the disk. By the addition theorem for angular velocities [cf. Eq. (19.38)], the angular velocity of the disk in $OXYZ$ is

$$\boldsymbol{\omega} = \boldsymbol{\omega}_1 + \boldsymbol{\omega}_2 = \omega_1\mathbf{i} + \omega_2\mathbf{j}$$

By the addition theorem for angular accelerations [cf. Eq. (19.48)], we find that the angular acceleration of the disk in $OXYZ$ is

$$\dot{\boldsymbol{\omega}} = \boldsymbol{\alpha}_{\text{disk/axle}} + \boldsymbol{\alpha}_{\text{axle}} + \boldsymbol{\omega}_{\text{axle}} \times \boldsymbol{\omega}_{\text{disk/axle}}$$

$$= \mathbf{0} + \mathbf{0} + \boldsymbol{\omega}_2 \times \boldsymbol{\omega}_1 = \omega_2\mathbf{j} \times \omega_1\mathbf{i} = -\omega_1\omega_2\mathbf{k}$$

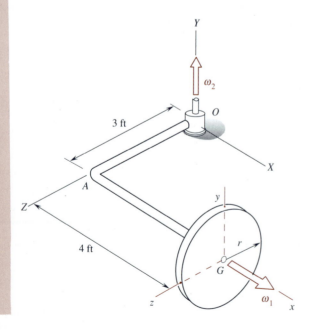

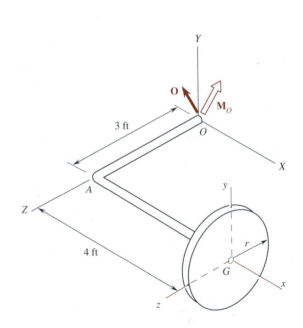

Thus, we have

$$\omega_x = \omega_1 \qquad \omega_y = \omega_2$$

$$\omega_z = \dot{\omega}_x = \dot{\omega}_y = 0 \qquad \dot{\omega}_z = -\omega_1\omega_2$$

For the disk and the axles shown, we write

$$\bar{I}_x = \tfrac{1}{2} mr^2 \qquad \bar{I}_y = \bar{I}_z = \tfrac{1}{4} mr^2$$

$$\bar{I}_{xy} = \bar{I}_{yz} = \bar{I}_{zx} = 0$$

Since the problem is mainly concerned with the dynamic reaction, the free-body diagram (bottom right p. 884) of the disk with its connecting bent axle may be drawn as shown, where the weight force **W** of the disk is a static force and is therefore neglected. We note that $\boldsymbol{\omega}_2$ is constant and the mass center G of the disk moves along a circular path of radius $\overline{OG} = 5$ ft at a constant speed. Thus, the acceleration $\bar{\mathbf{a}}$ of G is

$$\bar{\mathbf{a}} = \overline{OG}\, \omega_2^2\, \boldsymbol{\lambda}_{GO} = \omega_2^2\, \overrightarrow{GO} = -\omega_2^2(4\mathbf{I} + 3\mathbf{K})$$

Referring to the free-body diagram and applying the governing equation for the *translational motion*, we write

$$\Sigma \mathbf{F} = m\bar{\mathbf{a}}: \qquad \mathbf{O} = m\bar{\mathbf{a}} = -m\omega_2^2(4\mathbf{I} + 3\mathbf{K})$$

$$\mathbf{O} = -\frac{16.1}{32.2}\,(5)^2(4\mathbf{I} + 3\mathbf{K})$$

$$\mathbf{O} = -50\mathbf{I} - 37.5\mathbf{K} \text{ lb} \qquad \blacktriangleleft$$

Applying *Euler's equations of motion* in Eqs. (19.105), we write

$$\Sigma M_x = 0 - 0 = 0 \qquad \Sigma M_y = 0 - 0 = 0$$

$$\Sigma M_z = \bar{I}_z \dot{\omega}_z - (\bar{I}_x - \bar{I}_y)\omega_x\omega_y$$

$$\Sigma M_z = \tfrac{1}{4} mr^2\,(-\omega_1\omega_2) - (\tfrac{1}{2} mr^2 - \tfrac{1}{4} mr^2)\omega_1\omega_2$$

$$= -\tfrac{1}{2} mr^2\omega_1\omega_2$$

$$\Sigma \mathbf{M}_G = \Sigma M_x\mathbf{i} + \Sigma M_y\mathbf{j} + \Sigma M_z\mathbf{k} = -\tfrac{1}{2} mr^2\omega_1\omega_2\mathbf{k} \qquad (1)$$

Referring to the free-body diagram, we write

$$\Sigma \mathbf{M}_G = \overrightarrow{GO} \times \mathbf{O} + \mathbf{M}_O = \mathbf{0} + \mathbf{M}_O = \mathbf{M}_O \qquad (2)$$

Comparing Eq. (1) with Eq. (2), we see that

$$\mathbf{M}_O = -\tfrac{1}{2} mr^2\omega_1\omega_2\mathbf{k} = -\frac{1}{2}\left(\frac{16.1}{32.2}\right)(1)^2(10)(5)\mathbf{k}$$

$$= -12.5\mathbf{k}$$

Since $\mathbf{k} = \mathbf{K}$, we obtain

$$\mathbf{M}_O = -12.5\mathbf{K} \text{ lb·ft} \qquad \blacktriangleleft$$

Developmental Exercises

EXAMPLE 19.13

Two slender arms *BC* and *DEF* are welded to a horizontal shaft *OA* as shown, where $b = 0.3$ m and $L = 1.2$ m. The segments *BC*, *DE*, and *EF* have equal masses of $m = 2$ kg. If a torque $\mathbf{M}_O = 18\mathbf{K}$ N·m acts on the shaft and the shaft is rotating at $\boldsymbol{\omega} = 10\mathbf{K}$ rad/s in the position shown, determine the angular acceleration $\boldsymbol{\alpha}$ of the shaft and the dynamic reaction from the bearing at *A*.

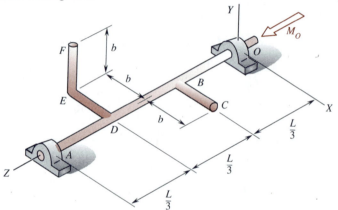

Solution.   Since the principal axes of inertia at either the fixed point *O* or the mass center of the system are not readily known, we simply let the

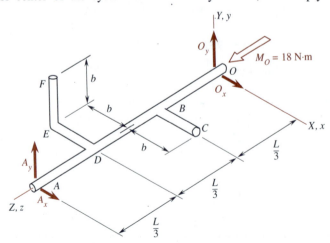

reference frame $Oxyz$ be *embedded in the shaft* as shown. We have $\mathbf{\Omega} = \mathbf{\omega}$
and

$$\mathbf{\omega} = 10\mathbf{K} = 10\mathbf{k} = \omega_z\mathbf{k} \qquad \omega_x = \omega_y = 0 \qquad \omega_z = 10$$

$$\dot{\mathbf{\omega}} = \alpha\mathbf{K} = \alpha\mathbf{k} = \dot{\omega}_z\mathbf{k} \qquad \dot{\omega}_x = \dot{\omega}_y = 0 \qquad \dot{\omega}_z = \alpha$$

Applying Eqs. (19.85) and (19.106), we write

$$H_x = -I_{xz}\omega_z \qquad H_y = -I_{yz}\omega_z \qquad H_z = I_{zz}\omega_z$$

$$\mathbf{H}_O = (-I_{xz}\mathbf{i} - I_{yz}\mathbf{j} + I_{zz}\mathbf{k})\omega_z$$

$$\Sigma\mathbf{M}_O = (\dot{\mathbf{H}}_O)_{Oxyz} + \mathbf{\Omega} \times \mathbf{H}_O$$

$$= (-I_{xz}\mathbf{i} - I_{yz}\mathbf{j} + I_{zz}\mathbf{k})\dot{\omega}_z$$

$$+ \omega_z\mathbf{k} \times (-I_{xz}\mathbf{i} - I_{yz}\mathbf{j} + I_{zz}\mathbf{k})\omega_z$$

$$\Sigma\mathbf{M}_O = (-I_{xz}\dot{\omega}_z + I_{yz}\omega_z^2)\mathbf{i}$$

$$+ (-I_{yz}\dot{\omega}_z - I_{xz}\omega_z^2)\mathbf{j} + I_{zz}\dot{\omega}_z\mathbf{k} \qquad (1)$$

Note that Eq. (1) would have been obtained if we directly applied Eqs.
(19.107). Referring to the free-body diagram, where weight forces are neglected for the purpose of determining dynamic reactions, we write

$$\Sigma\mathbf{M}_O = M_O\mathbf{k} + L\mathbf{k} \times (A_x\mathbf{i} + A_y\mathbf{j})$$

$$\Sigma\mathbf{M}_O = -LA_y\mathbf{i} + LA_x\mathbf{j} + M_O\mathbf{k} \qquad (2)$$

Comparing Eq. (1) with Eq. (2), we see that

$$-LA_y = -I_{xz}\dot{\omega}_z + I_{yz}\omega_z^2 \qquad (3)$$

$$LA_x = -I_{yz}\dot{\omega}_z - I_{xz}\omega_z^2 \qquad (4)$$

$$M_O = I_{zz}\dot{\omega}_z \qquad (5)$$

Noting that $m_{BC} = m_{DE} = m_{EF} = m$ and applying the parallel-axis theorem, we write

$$I_{xz} = \left[0 + m\left(\frac{b}{2}\right)\left(\frac{L}{3}\right)\right] + \left[0 + m\left(-\frac{b}{2}\right)\left(\frac{2L}{3}\right)\right]$$

$$+ \left[0 + m(-b)\left(\frac{2L}{3}\right)\right]$$

$$I_{yz} = 0 + 0 + \left[0 + m\left(\frac{b}{2}\right)\left(\frac{2L}{3}\right)\right]$$

$$I_{zz} = \tfrac{1}{3} mb^2 + \tfrac{1}{3} mb^2 + [\tfrac{1}{12} mb^2 + m(b^2 + \tfrac{1}{4} b^2)]$$

$$I_{xz} = -\tfrac{5}{6} mbL \qquad I_{yz} = \tfrac{1}{3} mbL \qquad I_{zz} = 2mb^2 \qquad (6)$$

Substituting Eqs. (6) into Eqs. (3) through (5) and noting that $m = 2$ kg,
$b = 0.3$ m, $L = 1.2$ m, $\omega_z = 10$ rad/s, and $M_O = 18$ N·m, we write

$$\alpha = \dot{\omega}_z = \frac{M_O}{2mb^2} = \frac{18}{2(2)(0.3)^2} = 50$$

$$A_x = -\frac{M_O}{6b} + \frac{5}{6}mb\omega_z^2 = -\frac{18}{6(0.3)} + \frac{5}{6}(2)(0.3)(10)^2 = 40$$

$$A_y = -\frac{5M_O}{12b} - \frac{1}{3}mb\omega_z^2 = -\frac{5(18)}{12(0.3)} - \frac{1}{3}(2)(0.3)(10)^2$$

$$= -45$$

$$\alpha = 50\mathbf{k} \ \text{rad/s}^2 \qquad \mathbf{A} = 40\mathbf{i} - 45\mathbf{j} \ \text{N} \blacktriangleleft$$

Developmental Exercise

D19.34   Refer to Example 19.13. Determine the dynamic reaction from the support at O. (*Hint.* Consider $\Sigma\mathbf{M}_A$.)

★19.9   Gyroscopic Motion: Steady Precession

I. MOTION OF A GYROSCOPE.   The usual concept of a *gyroscope* is an axisymmetrical body (i.e., a body of revolution) that spins rapidly about its geometric axis of symmetry. When mounted in a Cardan's suspension, as shown in Fig. 19.18, a gyroscope can spin freely about its geometric axis and may assume any orientation; however, its mass center always remains at a fixed point in space. For convenience, we let the reference frame $Oxyz$ be *embedded in the inner gimbal*, with the z axis directed along the axis of spin of the rotor and the y axis directed along the axis joining the supporting

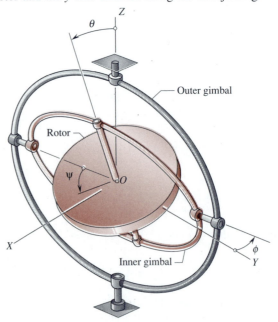

Fig. 19.18

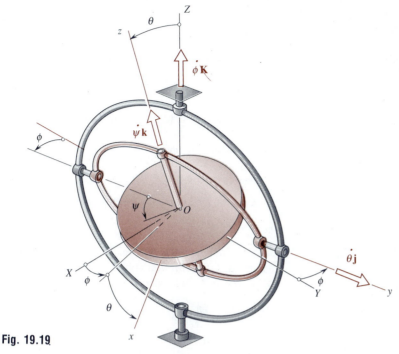

Fig. 19.19

bearings of the inner gimbal, so that the Z, z, and x axes are always co-planar. The position of the gyroscope in the fixed reference frame $OXYZ$ is completely characterized by the following three angles: (1) the angle ϕ specifying the rotation of the *outer gimbal* about the Z axis, (2) the angle θ specifying the rotation of the *inner gimbal* about the y axis which always lies in the XY plane of $OXYZ$, (3) the angle ψ specifying the rotation of the *rotor* about the z axis. The angles ϕ, θ, and ψ, which completely define the position of the gyroscope at any given time, are called the *Eulerian angles*. The Eulerian angles ϕ, θ, and ψ are also called the *precession angle*, the *nutation angle*, and the *spin angle* of the gyroscope, respectively.[†] Their time derivatives $\dot{\phi}$, $\dot{\theta}$, and $\dot{\psi}$, as indicated in Fig. 19.19, define, respectively, the *rate of precession*, the *rate of nutation*, and the *rate of spin* of the gyroscope; and the Z, y, and z axes are, respectively, called the *precession axis*, the *nutation axis*, and the *spin axis* of the gyroscope.

Clearly, the x, y, and z axes in Fig. 19.19 are the principal axes of inertia at the fixed point O of the rotor. Letting i.g. and o.g. denote the inner and outer gimbals, respectively, and applying Eqs. (19.38) and (19.40), we find that the angular velocity $\boldsymbol{\Omega}$ of $Oxyz$ (embedded in the inner gimbal) and the angular velocity $\boldsymbol{\omega}$ of the rotor in $OXYZ$ are

$$\boldsymbol{\Omega} = \boldsymbol{\Omega}_{\text{i.g./o.g.}} + \boldsymbol{\Omega}_{\text{o.g.}}$$

$$\boldsymbol{\Omega} = \dot{\theta}\mathbf{j} + \dot{\phi}\mathbf{K} \qquad (19.110)$$

$$\boldsymbol{\omega} = \boldsymbol{\omega}_{\text{rotor/i.g.}} + \boldsymbol{\omega}_{\text{i.g./o.g.}} + \boldsymbol{\omega}_{\text{o.g.}}$$

$$\boldsymbol{\omega} = \dot{\psi}\mathbf{k} + \dot{\theta}\mathbf{j} + \dot{\phi}\mathbf{K} \qquad (19.111)$$

[†]Unfortunately, the Eulerian angles (ϕ, θ, ψ) have *not* been represented by the same symbols in all books. It is important to choose a set to work with and be consistent.

Expressing the unit vector **K** in $Oxyz$, we have

$$\mathbf{K} = -\sin\theta\,\mathbf{i} + \cos\theta\,\mathbf{k} \tag{19.112}$$

Substituting Eq. (19.112) into Eqs. (19.110) and (19.111), we write

$$\mathbf{\Omega} = -\dot{\phi}\sin\theta\,\mathbf{i} + \dot{\theta}\mathbf{j} + \dot{\phi}\cos\theta\,\mathbf{k} \tag{19.113}$$

$$\mathbf{\omega} = -\dot{\phi}\sin\theta\,\mathbf{i} + \dot{\theta}\,\mathbf{j} + (\dot{\psi} + \dot{\phi}\cos\theta)\mathbf{k} \tag{19.114}$$

$$(\dot{\mathbf{\omega}})_{Oxyz} = -(\ddot{\phi}\sin\theta + \dot{\phi}\dot{\theta}\cos\theta)\mathbf{i} + \ddot{\theta}\mathbf{j}$$
$$+ (\ddot{\psi} + \ddot{\phi}\cos\theta - \dot{\phi}\dot{\theta}\sin\theta)\mathbf{k} \tag{19.115}$$

The principal moments of inertia of the rotor may be designated

$$I_x = I_y = I' \qquad I_z = I \tag{19.116}$$

Extracting Ω_x, Ω_y, Ω_z, ω_x, ω_y, ω_z, $\dot{\omega}_x$, $\dot{\omega}_y$, and $\dot{\omega}_z$ from Eqs. (19.113), (19.114), and (19.115), using Eq. (19.116), and applying *Euler's equations of motion* for $\mathbf{\Omega} \neq \mathbf{\omega}$ in Eqs. (19.109), we write

$$\Sigma M_x = I'\,(-\ddot{\phi}\sin\theta - \dot{\phi}\dot{\theta}\cos\theta) - I'\,(\dot{\phi}\cos\theta)(\dot{\theta})$$
$$+ I\dot{\theta}(\dot{\psi} + \dot{\phi}\cos\theta)$$

$$\Sigma M_y = I'\ddot{\theta} - I(-\dot{\phi}\sin\theta)(\dot{\psi} + \dot{\phi}\cos\theta)$$
$$+ I'\,(\dot{\phi}\cos\theta)(-\dot{\phi}\sin\theta)$$

$$\Sigma M_z = I(\ddot{\psi} + \ddot{\phi}\cos\theta - \dot{\phi}\dot{\theta}\sin\theta) - I'\dot{\theta}(-\dot{\phi}\sin\theta)$$
$$+ I'(-\dot{\phi}\sin\theta)(\dot{\theta})$$

These three equations may be simplified and written as

$$\Sigma M_x = -I'\,(\ddot{\phi}\sin\theta + 2\dot{\phi}\dot{\theta}\cos\theta) + I\dot{\theta}(\dot{\psi} + \dot{\phi}\cos\theta)$$

$$\Sigma M_y = I'\,(\ddot{\theta} - \dot{\phi}^2\sin\theta\cos\theta) + I\dot{\phi}\sin\theta(\dot{\psi} + \dot{\phi}\cos\theta) \tag{19.117}$$

$$\Sigma M_z = I\frac{d}{dt}\,(\dot{\psi} + \dot{\phi}\cos\theta)$$

Thus, the motion of a gyroscope subjected to an external force system is defined by Eqs. (19.117) when the mass of its gimbals is neglected.

Note that Eqs. (19.117) are useful in studying the motion of a *body of revolution* about its mass center or about a fixed point which lies on its axis of symmetry. For instance, a top spinning about its axis of symmetry with a high angular speed $\dot{\psi}$ and supported at the fixed point O may be depicted as shown in Fig. 19.20. In this figure, $OXYZ$ is the fixed reference frame, the z axis of the moving reference frame $Oxyz$ is chosen to be *always* directed along the axis of the top, the Z axis is the precession axis, the y axis is the nutation axis, the z axis is the spin axis, and the x, z, and Z axes are always coplanar. Thus, $Oxyz$ follows the motion of the top only in nutation and precession. Clearly, the position of the top is defined by the *Eulerian angles*: (a) the angle ϕ specifying its rotation about the precession axis, (b) the angle θ specifying its rotation about the nutation axis, (c) the angle ψ

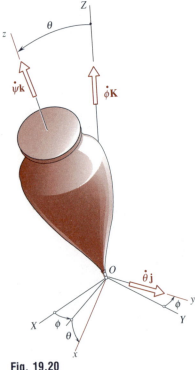

Fig. 19.20

specifying its rotation about the spin axis. A similar derivation will show that the motion of the top in Fig. 19.20 is also defined by Eqs. (19.117).

II. STEADY PRECESSION.

The *steady precession* of a gyroscope is a special case of gyroscopic motion in which the nutation angle θ, the rate of precession $\dot{\phi}$, and the rate of spin $\dot{\psi}$ all remain constant; i.e.,

$$\dot{\theta} = \ddot{\theta} = \ddot{\phi} = \ddot{\psi} = 0 \qquad (19.118)$$

By Eqs. (19.110), (19.113), (19.114), and (19.118), we write

$$\boldsymbol{\Omega} = \dot{\phi}\mathbf{K} = -\dot{\phi}\sin\theta\,\mathbf{i} + \dot{\phi}\cos\theta\,\mathbf{k} \qquad (19.119)$$

$$\boldsymbol{\omega} = -\dot{\phi}\sin\theta\,\mathbf{i} + \omega_z\mathbf{k} \qquad (19.120)$$

where $\omega_z = \dot{\psi} + \dot{\phi}\cos\theta$. Substituting Eqs. (19.118) into Eqs. (19.117), we get

$$\Sigma M_x = \Sigma M_z = 0$$

$$\Sigma M_y = [I(\dot{\psi} + \dot{\phi}\cos\theta) - I'\dot{\phi}\cos\theta]\dot{\phi}\sin\theta$$

which are equivalent to the vector equation

$$\Sigma\mathbf{M}_O = [I(\dot{\psi} + \dot{\phi}\cos\theta) - I'\dot{\phi}\cos\theta]\dot{\phi}\sin\theta\,\mathbf{j} \qquad (19.121)$$

or

$$\Sigma\mathbf{M}_O = (I\omega_z - I'\dot{\phi}\cos\theta)\dot{\phi}\sin\theta\,\mathbf{j} \qquad (19.122)$$

Since the mass center of the gyroscope is fixed in space, we have, by Eq. (19.96), $\Sigma\mathbf{F} = \mathbf{0}$; therefore, the force system to be applied to the gyroscope to maintain a steady precession is equivalent to a *torque*, which is equal to the moment given by the right-hand member of Eq. (19.121) or (19.122). Note that the torque required to maintain a (forced) steady precession must act on the rotor about the nutation axis which is perpendicular to the plane containing the precession axis and the spin axis, as illustrated in Fig. 19.21.

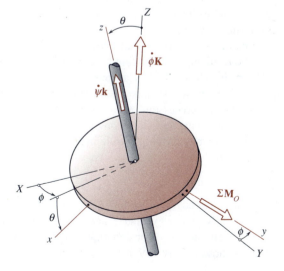

Fig. 19.21

In the particular case where the spin axis is perpendicular to the precession axis, we have $\theta = 90°$ and the required torque in Eq. (19.121) for steady precession reduces to

$$\Sigma \mathbf{M}_O = I \dot{\phi} \dot{\psi} \mathbf{j} \qquad (19.123)$$

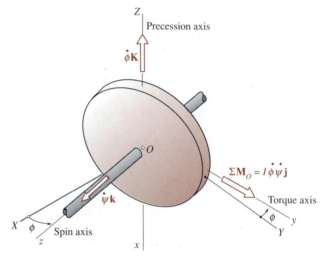

Fig. 19.22

In this case, the spin axis, the torque axis, and the precession axis form a right-handed trihedral as shown in Fig. 19.22. We can see from Eq. (19.123) that a large torque would be required to change the orientation of the axle of a rotor spinning at a high speed about its axis. This gyroscopic effect has many applications. For example, gyroscopes are used to stabilize ships and torpedos and to retain a fixed direction in inertial guidance systems.

III. TORQUE-FREE STEADY PRECESSION.   When the external force system exerts no moment about the mass center of a body, the motion of the body is referred to as a *torque-free motion*. We shall here consider the torque-free motion of an axisymmetrical body. Examples of this type of motion are furnished by artificial satellites and space vehicles, which may spin and precess during free flight.

A torque-free motion of an axisymmetrical body may be represented in Fig. 19.23. Since the resultant moment about the mass center G is zero, Eq. (19.94) yields $\dot{\mathbf{H}}_G = 0$. Thus, $\mathbf{H}_G$ has a constant magnitude and a fixed direction in space. We may first let this fixed direction be the direction of the Z axis, or the axis of precession, then let the z axis of the moving reference frame $Gxyz$ be always directed along the axis of symmetry of the body and the x axis be always in the plane defined by the z and Z axes, as shown in Fig. 19.23. We have

$$H_x = -H_G \sin\theta \qquad H_y = 0 \qquad H_z = H_G \cos\theta \qquad (19.124)$$

where θ is the angle between the z and Z axes, and H_G is the constant magnitude of $\mathbf{H}_G$. Clearly, the x, y, and z axes are the principal axes of inertia of the axisymmetrical body. Designating the principal moments of inertia of the body as those in Eqs. (19.116), we write

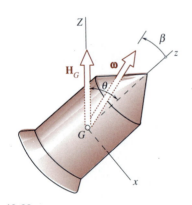

Fig. 19.23

$$H_x = I'\omega_x \qquad H_y = I'\omega_y \qquad H_z = I\omega_z \qquad (19.125)$$

It follows from Eqs. (19.124) and (19.125) that

$$\omega_x = -\frac{H_G \sin\theta}{I'} \qquad \omega_y = 0 \qquad \omega_z = \frac{H_G \cos\theta}{I} \qquad (19.126)$$

We see from Eqs. (19.111) and (19.126) that

$$. \dot\theta = \omega_y = 0 \qquad (19.127)$$

Thus, the angle θ between the z and Z axes is a constant and *the body is in steady precession about the Z axis* under no torque.

Substitution of $\dot\theta = 0$ into Eqs. (19.110) and (19.111) yields

$$\boldsymbol{\Omega} = \dot\phi\mathbf{K} \qquad \boldsymbol{\omega} = \dot\psi\mathbf{k} + \dot\phi\mathbf{K} \qquad (19.128)$$

The angle β in Fig. 19.23 is defined by

$$\tan\beta = -\frac{\omega_x}{\omega_z} \qquad (19.129)$$

Thus, dividing the first by the third in Eqs. (19.126), we obtain

$$\tan\beta = \left(\frac{I}{I'}\right)\tan\theta \qquad (19.130)$$

For torque-free motion, we set $\Sigma\mathbf{M}_O$ to zero in Eq. (19.121) to yield

$$\dot\phi = \frac{I\dot\psi}{(I' - I)\cos\theta} \qquad (19.131)$$

It is clear from Eqs. (19.129) through (19.131) that the directions of the spin and of the steady precession of an axisymmetrical body in torque-free motion depend on the given values of ω_x and ω_z as well as the relative magnitudes of I and I'. If the body is set to spin about its axis of symmetry, we have $\omega_x = 0$, $\omega_z \neq 0$, and $\beta = \theta = 0$; and the motion involves *no* precession. If the body is set to spin about a transverse axis, we have $\omega_x \neq 0$, $\omega_z = 0$, $\beta = \theta = 90°$; and the motion again involves *no* precession. However, there are two general cases which may be observed as follows:

1. If $I < I'$, then $\beta < \theta$, as shown in Fig. 19.24. This is the case of a body elongated in the direction of the spin axis (e.g., a space vehicle). The vector $\boldsymbol{\omega}$ is directed along the *velocity axis* (cf. Sec. 15.7) of the body and lies inside the angle ZGz, and the precession is said to be *direct*. Such a motion may be modeled by a *body cone* rolling around the outer surface of a fixed *space cone* without slipping, where the line of tangency of the two cones is the velocity axis.

2. If $I > I'$, then $\beta > \theta$, as shown in Fig. 19.25. This is the case of a body flattened in the direction of the spin axis (e.g., a satellite). The vector $\boldsymbol{\omega}$ is directed along the *velocity axis* of the body and lies outside the angle ZGz, and the precession is said to be *retrograde*, where the sense of the spin points in the negative direction of the z axis. The motion may be modeled by a *body cone* whose inside surface rolls on the outside surface of a fixed *space cone* without slipping, where the line of tangency of the two cones is the velocity axis.

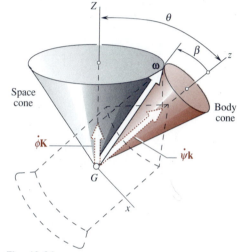

Fig. 19.24

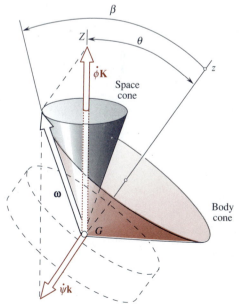

Fig. 19.25

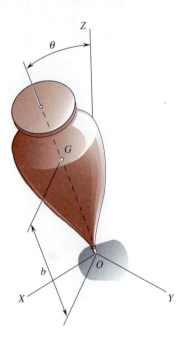

EXAMPLE 19.14

A spinning top of mass $m = 0.4$ kg is precessing about the vertical axis at a constant rate and is supported at the fixed point O as shown, where $b = 60$ mm and $\theta = 30°$. The radii of gyration of the top about its axis of symmetry and about a transverse axis through O are $k = 25$ mm and $k' = 45$ mm, respectively. If the top spins at the rate $\dot{\psi} = 120$ rad/s, determine its rate of precession.

Solution. The moments of inertia of the top are

$$I = mk^2 = 0.4(0.025)^2$$
$$I = 250 \times 10^{-6} \text{ kg·m}^2$$
$$I' = mk'^2 = 0.4(0.045)^2$$
$$I' = 810 \times 10^{-6} \text{ kg·m}^2$$

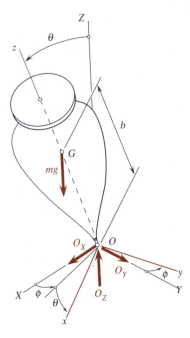

Since the motion is a *steady precession*, we use the reference frames as shown and apply Eq. (19.121) to write

$$\Sigma\mathbf{M}_O = [I(\dot{\psi} + \dot{\phi}\cos\theta) - I'\dot{\phi}\cos\theta]\dot{\phi}\sin\theta\,\mathbf{j} \qquad (1)$$

Referring to the free-body diagram shown, we write

$$\Sigma\mathbf{M}_O = (b\sin\theta)mg\mathbf{j} \qquad (2)$$

Comparing Eq. (1) with Eq. (2), we see that

$$[I(\dot{\psi} + \dot{\phi}\cos\theta) - I'\dot{\phi}\cos\theta]\dot{\phi} = mgb$$

or

$$(I' - I)\cos\theta\,\dot{\phi}^2 - I\dot{\psi}\dot{\phi} + mgb = 0 \qquad (3)$$

Substituting $I' = 810 \times 10^{-6}$ kg·m², $I = 250 \times 10^{-6}$ kg·m², $\theta = 30°$, $\dot{\psi} = 120$ rad/s, $m = 0.4$ kg, $b = 0.06$ m, and $g = 9.81$ m/s² into Eq. (3), we get

$$0.00048497\dot{\phi}^2 - 0.03\dot{\phi} + 0.23544 = 0$$

whose two roots are 9.22 and 52.6. Thus, we obtain two possible rates of precession:

$$\dot{\phi} = 9.22 \text{ rad/s} \qquad \dot{\phi} = 52.6 \text{ rad/s} \blacktriangleleft$$

REMARK. The lower rate of precession is more likely to be observed in reality because of the lower kinetic energy required of the top.

Developmental Exercises

D19.35 Refer to Example 19.14. Determine the two possible rates of precession of the top if the nutation angle is $\theta = 60°$.

D19.36 Refer to the gyroscope in Fig. 19.19. Suppose that the reference frame *Oxyz* is *embedded in the rotor* and has an orientation as shown at the instant considered. (a) Applying Eq. (19.50), determine the angular acceleration $\dot{\omega}$ of the rotor in *OXYZ*. (b) Derive Eqs. (19.117) by applying Euler's equations of motion for $\mathbf{\Omega} = \mathbf{\omega}$ in Eq. (19.108).

D19.37 Refer to the gyroscope in Fig. 19.19. Suppose that the reference frame *Oxyz* is embedded in the inner gimbal. (a) Determine $\mathbf{\omega}_{\text{rotor}/Oxyz}$ which is the angular velocity of the rotor in *Oxyz*. (b) Determine $(\dot{\mathbf{\omega}}_{\text{rotor}/Oxyz})_{Oxyz}$ which is the angular acceleration of the rotor in *Oxyz*. (c) Does the quantity $(\dot{\mathbf{\omega}})_{Oxyz}$, as given by Eq. (19.115), represent the angular acceleration of the rotor in *Oxyz*?

D19.38 Describe (a) direct precession, (b) retrograde precession.

PROBLEMS

19.54 A uniform slender rod is bent into the shape of a square and is attached by a collar at C to a vertical shaft which rotates with a constant angular velocity $\mathbf{\omega}$ as shown. Determine (a) the constant angle θ that the plane of the square forms with the vertical, (b) the maximum value of ω for which the square will remain vertical (i.e., $\theta = 0$).

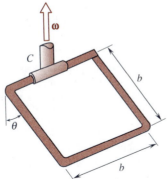

Fig. P19.54

19.55 The 20-kg slender rod *AB* is supported by a wire at *C* and pinned at *A* to a vertical axle *AD* which rotates with a constant angular velocity of 20 rad/s. Determine the tension in the wire and the magnitude of the reaction from the pin at *A*.

19.56* A thin rectangular plate weighing 10 lb is mounted on a horizontal shaft with which the plate makes an angle of 50° as shown. If the shaft rotates at the constant rate $\omega = 6\mathbf{I}$ rad/s, determine the moment exerted at *G* by the plate *on* the shaft due to the wobble of the plate.

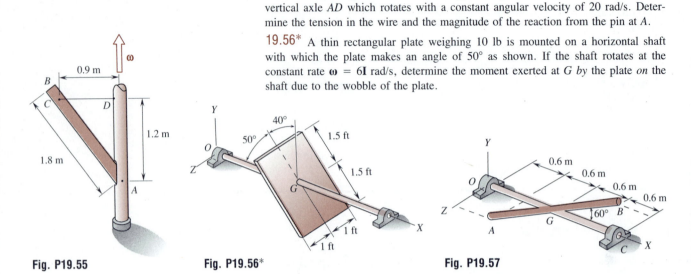

Fig. P19.55

Fig. P19.56*

Fig. P19.57

19.57 A 2.4-m slender rod *AB* of mass 4 kg is welded at its midpoint *G* to a 2.4-m horizontal shaft *OC*. If a torque $\mathbf{M}_O = 20\mathbf{I}$ N·m acts on the shaft and the shaft is rotating at $\omega = 120\mathbf{I}$ rpm in the position shown, determine the angular acceleration of the shaft and the dynamic reaction from the bearing at *C*.

19.58 Two triangular plates, each of mass 5 kg, are welded to a horizontal shaft as shown. If the shaft is rotating with $\omega = 5\mathbf{K}$ rad/s and $\alpha = 10\mathbf{K}$ rad/s^2 in the position shown, determine the torque acting on the shaft.

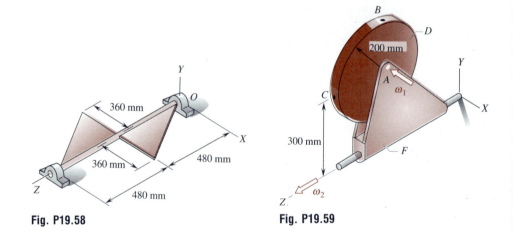

Fig. P19.58

Fig. P19.59

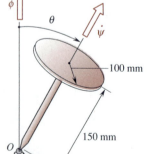

Fig. P19.60

19.59 The 5-kg disk *D* rotates about the pin at the constant rate $\omega_1 = -4\mathbf{I}$ rad/s relative to the frame *F*, while the frame *F* rotates about the fixed shaft at the constant rate $\omega_2 = 8\mathbf{K}$ rad/s. For the instant shown, determine the moment exerted by the disk on the pin at *A*.

19.60 The top consists of a thin disk of mass 3 kg and a slender rod of length 150 mm and negligible mass as shown, where $\theta = 30°$. If the top spins with $\dot{\psi} = 900$ rpm and precesses at a constant rate, determine the rate of precession.

19.61* Solve Prob. 19.60 if $\theta = 60°$.

19.62   A 10-lb sphere of radius 1 ft is attached to the end of a 3-ft rod of negligible mass which is supported by a ball-and-socket joint at the fixed position O. If the sphere precesses at the constant rate $\dot{\phi}$ = 15 rpm, determine its rate of spin $\dot{\psi}$.

19.63*   Solve Prob. 19.62 if $\dot{\phi}$ = 30 rpm.

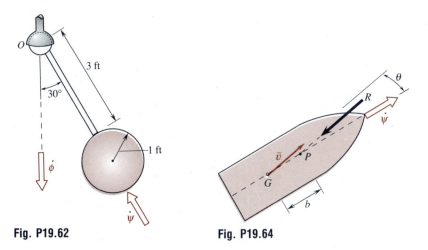

Fig. P19.62   **Fig. P19.64**

19.64   A 40-lb projectile is fired to move with a velocity $\bar{v}$ of 2000 ft/s which makes an angle $\theta = 5°$ with its axis of symmetry as shown. The projectile has a rate of spin $\dot{\psi}$ = 500 rad/s, and the air drag is equivalent to a force $\mathbf{R}$ of 30 lb that is parallel to $\bar{v}$ and passes through the point P which is a distance b = 5 in. from the mass center G. The radii of gyration of the projectile about its axis of symmetry and about a transverse axis through G are k = 1.5 in. and k' = 4.5 in., respectively. Determine its rate of steady precession.

19.65   Since the earth is an oblate spheroid, the gravitational system acting on the earth is equivalent to a force $\mathbf{F}$ and a torque $\mathbf{M}$ as shown in Fig. (a). The effect of the torque $\mathbf{M}$ is to cause the earth to have a forced precession at the rate $\dot{\phi}$ = 1 revolution per 25,800 years. The spin axis of the earth is now pointing to the proximity of Polaris, the North Star, as shown in Fig. (b). Thus, the spin axis of the earth will be pointing to near Vega some 12,900 years from now, as shown in Fig. (c). Determine the average magnitude of the torque $\mathbf{M}$. [*Hint*. Use Eqs. (12.29) and (12.59) and assume that $\bar{I} = 2mR^2/5$.]

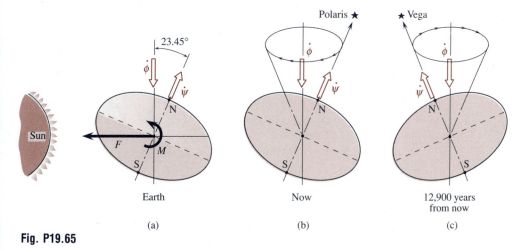

Earth   Now   12,900 years
                                                from now

(a)   (b)   (c)

Fig. P19.65

★19.10   Concluding Remarks

The concept of reference frames in connection with the time derivatives of vectors is very important. In general motion, the velocity of a body in a reference frame is the time derivative of the position in that reference frame, of the body, taken (or measured) in that same reference frame. Similarly, the acceleration of a body in a reference frame is the time derivative of the velocity in that reference frame, of the body, taken in that same reference frame. For studying rigid bodies in motion, it is exceedingly important to know the reference frame in which the time derivative of any vector is taken. For instance, note that

$$\dot{\boldsymbol{\omega}} = (\dot{\boldsymbol{\omega}})_{OXYZ} \qquad \mathbf{v}_{B/Axyz} = (\dot{\mathbf{r}}_{B/A})_{Axyz} \neq (\dot{\mathbf{r}}_{B/A})_{OXYZ}$$

$$\mathbf{a}_{B/Axyz} = (\dot{\mathbf{v}}_{B/Axyz})_{Axyz} \neq (\dot{\mathbf{v}}_{B/Axyz})_{OXYZ}$$

where $Axyz$ and $OXYZ$ denote moving and fixed reference frames, respectively. In particular, note that the vector $(\dot{\mathbf{r}}_{B/A})_{OXYZ}$ represents simply the time derivative of $\mathbf{r}_{B/A}$ in $OXYZ$ and carries the units of a velocity, but it is *not the velocity* of B in $Axyz$ nor $OXYZ$; the vector $(\dot{\mathbf{v}}_{B/Axyz})_{OXYZ}$ represents simply the time derivative of $\mathbf{v}_{B/Axyz}$ in $OXYZ$ and carries the units of an acceleration, but it is *not the acceleration* of B in $Axyz$ nor $OXYZ$. However, $(\dot{\mathbf{r}}_{B/Axyz})_{Axyz}$ is the velocity of B in $Axyz$, and $(\dot{\mathbf{v}}_{B/Axyz})_{Axyz}$ is the acceleration of B in $Axyz$.

In Eq. (19.30), we have shown that the angular velocity of a moving reference frame in a fixed reference frame is definable in terms of the time rates of change of the cartesian unit vectors embedded in the moving reference frame as measured in the fixed reference frame. It represents a *general* definition for the angular velocity of a reference frame or a rigid body. The addition theorems for angular velocities and angular accelerations are particularly useful in studying the motion of a system which consists of two or more movable rigid parts. These two theorems become identical in form only in plane motion where angular velocities of all rigid bodies in the system are parallel to each other.

After studying Sec. 19.4 and 19.5, one should be able to apply the parallel-axes theorems to compute the moments and products of inertia of a body. Besides, it is important to remember the following property: *If the mass of a body is symmetrical about two coordinate planes, then all products of inertia of the body vanish and the coordinate axes are principal axes of inertia at the origin.* Concepts and skills related to mass moments and products of inertia are needed in studying the kinetics of rigid bodies in space.

Since the translational motion of a rigid body is similar to the motion of a particle, the main effort in studying the kinetics of rigid bodies in space lies in the study of their rotational motion. One can always apply the concise equations for rotational motion in Eqs. (19.99) and (19.106). If appropriate, Euler's equation of motion in one of the three special situations in Eqs. (19.105), (19.108), and (19.109) may be applied. For ease in getting started, one may find it useful to review the usual steps to be taken prior to applying Euler's equations of motion as outlined at the end of Sec. 19.8.

The motion of a gyroscope is an important type of the general motion of a rigid body. Unfortunately, the equations governing such a motion are nonlinear simultaneous differential equations whose solutions may require the use of numerical methods. In studying gyroscopic motion, we focus our attention mainly on problems related to steady precession, which is a special physical phenomenon exhibited by some dynamical systems.

REVIEW PROBLEMS

19.66   At the instant considered, the disk C rotates relative to the arm at the rate $\omega_1 = 5\mathbf{I}$ rad/s and $\alpha_1 = 2\mathbf{I}$ rad/s^2, while the arm rotates relative to the ground at the rate $\omega_2 = 4\mathbf{J}$ rad/s and $\alpha_2 = 3\mathbf{J}$ rad/s^2. Determine (a) the angular velocity and angular acceleration of the disk C, (b) the velocity and acceleration of the point P at the top of the disk C.

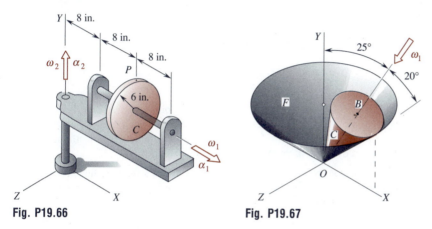

Fig. P19.66   Fig. P19.67

19.67   The right circular cone C rolls at a constant rate on the inside of the fixed right circular conic cavity F without slipping as shown. If the cone C makes one complete trip around the conic cavity F every 5 s, determine the angular velocity and angular acceleration of the cone C.

19.68*   Two slender rods, each of length L and mass m, are welded together to form a T bar. The T bar is attached to the pin of a clevis which rotates with the vertical shaft at a constant angular velocity ω as shown. Determine (a) the angle θ that the plane of the T bar forms with the vertical, (b) the maximum value of ω for which the T bar will remain vertical.

19.69   A thin disk of mass 45 kg and radius 300 mm is mounted on a horizontal shaft with which the disk makes an angle of 60° as shown. If the shaft rotates at the constant rate $\omega = 10\mathbf{I}$ rad/s, determine the moment exerted at G by the disk *on* the shaft due to the wobble of the disk.

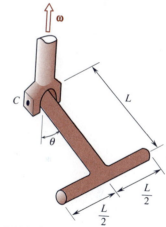

Fig. P19.68*

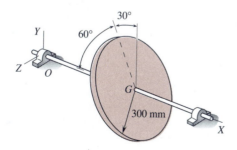

Fig. P19.69

19.70 Four slender rods, each weighing 3 lb, are welded to a horizontal shaft as shown. If a torque $M_O = 15\mathbf{K}$ lb·ft acts on the shaft and the shaft is rotating at $\omega = 90\mathbf{K}$ rpm in the position shown, determine the angular acceleration of the shaft and the dynamic reactions from the bearings at O and A.

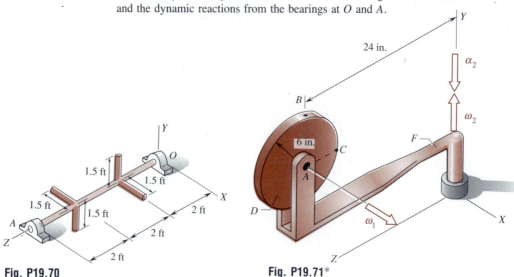

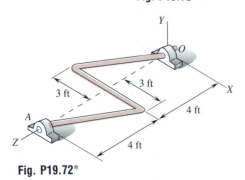

Fig. P19.70

Fig. P19.71*

19.71* The 12-lb disk D rotates about the pin at A at the constant rate $\omega_1 = 6\mathbf{I}$ rad/s relative to the arm F. At the instant shown, the arm F rotates at the rate $\omega_2 = 4\mathbf{J}$ rad/s and $\alpha_2 = -2\mathbf{J}$ rad/s² relative to the ground. Determine the moment exerted by the disk on the pin at A.

19.72* A slender rod weighing 2 lb/ft is bent to form a shaft as shown. If the shaft rotates at a constant rate of $\omega = 10\mathbf{K}$ rad/s, determine the dynamic reactions from the supports at O and A.

Fig. P19.72*

19.73 A cylinder is connected by a wire OA to a support as shown, where $h > 2r$, the wire has a steady precession about the vertical at the rate $\dot\phi$, and the cylinder spins at the constant rate $\dot\psi$ about its axis of symmetry. Derive the expression for the angle θ which the cylinder makes with the vertical if $\gamma = \theta$.

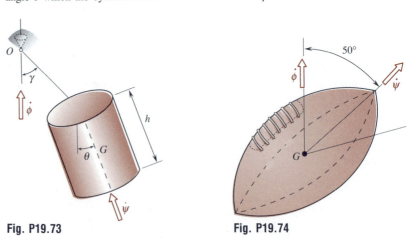

Fig. P19.73

Fig. P19.74

19.74 The football shown spins at the rate $\dot\psi = 6$ rad/s and has a steady precession about the vertical. If air resistance is negligible and the ratio of the axial to transverse moments of inertia is $I/I' = 1/3$, determine (a) the rate of precession $\dot\phi$, (b) the angular velocity ω of the football.

Vibrations

Chapter 20

KEY CONCEPTS

- Free and forced vibrations, damping, undamped and damped vibrations, amplitude, and simple harmonic motion.
- Natural frequency, natural circular frequency, natural cyclic frequency, and period of vibration.
- Viscous damper, dashpot, damped natural frequency, and damped period of vibration.
- Overdamping, critical damping, underdamping, damping factor, and logarithmic decrement.
- Forcing frequency, steady-state amplitude, phase angle, magnification factor, resonance, and transmissibility.

The study of vibratory motion of a body about a certain position of equilibrium is an important and special area of dynamics. Examples include studies of the response of a structure to earthquake, the vibration of a rotating machine, the flutter of aircraft wings, and the vibration of a musical instrument. We can see that the analysis of vibrations is a very extensive and well-developed subject. In this chapter, we limit our study to mainly the fundamentals of vibrations of single-degree-of-freedom systems of particles and rigid bodies. A free vibration is an oscillatory motion without the action of a recurrent excitation. Free vibrations are studied in the first part of the chapter using the method of force and acceleration as well as the energy method. A forced vibration is an oscillatory motion under the action of a recurrent excitation; forced vibrations are studied in the latter part of the chapter.

FREE VIBRATIONS

★20.1   Undamped Free Vibrations

The *vibration* of a body is an oscillatory motion of the body about a certain position of equilibrium. The swinging of a pendulum and the flutter of the wings of an aircraft are common examples of vibration. The subject of vibration deals with the study of oscillatory motions of bodies and the forces associated with them.

Vibration can be dichotomized in several ways; e.g., free and forced vibrations, undamped and damped vibrations, linear and nonlinear vibrations, and deterministic and random vibrations. The *free vibration* of a body is an oscillatory motion, caused by an initial disturbance to the body, without the action of a recurrent excitation. If a recurrent excitation acts on a body, the resulting vibration is called a *forced vibration*. The phenomenon of gradual loss or dissipation of mechanical energy due to friction or rolling resistance during oscillation is known as *damping*. An *undamped vibration* is an oscillatory motion of an idealized system experiencing no damping.[†] If damping in the system is taken into account, the resulting vibration is called a *damped vibration*. A body is said to be in *linear vibration* if the differential equation governing its oscillation is linear; otherwise, it is in *nonlinear vibration*. If a vibratory system is subjected to an excitation which is known at any given time, the resulting vibration is called a *deterministic vibration*. If the excitation is random or nondeterministic, the resulting vibration is called *random vibration*.

Let us first consider the undamped free vibration of a spring-mass system as shown in Fig. 20.1, where a body with mass m is suspended by a linear spring having a modulus k and a free length L. The spring is taken to have negligible mass compared with the mass m. In part (a) of Fig. 20.1, the spring is unloaded. It is seen in part (b) that the lower end of the spring moves from P to O to undergo a static deflection δ_{st} when it is loaded to hold the body in equilibrium. Clearly, the spring force of magnitude $k\delta_{st}$ is balanced by the weight force of magnitude mg; i.e.,

$$mg = k\delta_{st} \tag{20.1}$$

The coordinate x defining the position of the displaced body as measured from its equilibrium position is called the *position coordinate* of the body.

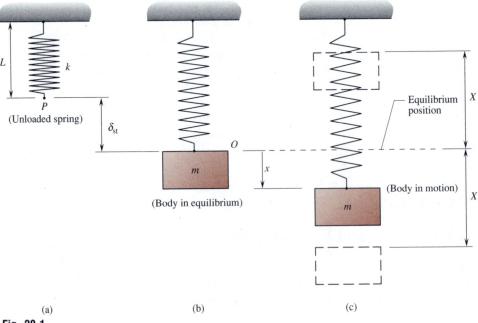

(a)   (b)   (c)

Fig. 20.1

[†]Unless otherwise indicated, damping is generally assumed to be negligible in the study.

If the body is displaced to the position $x = X$ and then released with no initial velocity, the body will move back and forth through the equilibrium position O. This generates a vibration of the body with a maximum displacement X from O as shown in part (c) of Fig. 20.1. The maximum displacement of a vibrating body from its equilibrium position is called the *amplitude*. Of course, the vibration can also be generated by imparting the body an initial velocity beside an initial displacement.

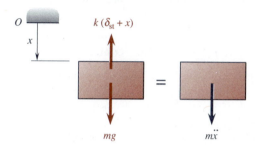

Fig. 20.2

The vibratory behavior of the body can be predicted by examining the dynamic condition of the body at a general time t in a general position x as shown in Fig. 20.2. Note that the unknown effective force is mathematically assumed to point in the positive direction of the position coordinate x. Referring to this figure (or applying Newton's second law), we write

$$+\downarrow \Sigma F: \qquad mg - k(\delta_{st} + x) = m\ddot{x}$$

Substituting Eq. (20.1) into this equation and rearranging, we get

$$m\ddot{x} + kx = 0 \qquad (20.2)$$

which is the differential equation of motion of the body. Setting

$$\frac{k}{m} = \omega_n^2 \qquad \text{or} \qquad \omega_n = \sqrt{k/m} \qquad (20.3)$$

we can write Eq. (20.2) in the form

$$\ddot{x} + \omega_n^2 x = 0 \qquad (20.4)$$

The physical significance of ω_n as defined will be clarified shortly.

We note that Eq. (20.4) is similar to Eq. (12.42); therefore, we can similarly assume that the general solution of Eq. (20.4) is

$$x = A \cos\omega_n t + B \sin\omega_n t \qquad (20.5)$$

where A and B are two arbitrary constants to be determined from the initial conditions of the body. Taking the time derivative of Eq. (20.5), we write

$$\dot{x} = -A\omega_n \sin\omega_n t + B\omega_n \cos\omega_n t \qquad (20.6)$$

Suppose that the body is situated at the initial position x_0 and has an initial velocity $\dot{x}_0$ at the time $t = 0$. Imposing these two initial conditions on Eqs. (20.5) and (20.6), we find that the values of the arbitrary constants are

$$A = x_0 \qquad (20.7)$$

$$B = \dot{x}_0/\omega_n \qquad (20.8)$$

Substituting these values of A and B into Eq. (20.5), we get

$$x = x_0 \cos\omega_n t + (\dot{x}_0/\omega_n) \sin\omega_n t \qquad (20.9)$$

which is the general solution expressed in terms of the initial conditions. Moreover, it can be verified that Eq. (20.9) is equivalent to the equation

$$x = X \sin(\omega_n t + \phi) \qquad (20.10)$$

where X is the *amplitude* and ϕ is the *phase angle*. Their values are

$$X = \sqrt{x_0^2 + (\dot{x}_0/\omega_n)^2} \qquad (20.11)$$

$$\phi = \tan^{-1}(x_0\omega_n/\dot{x}_0) \qquad (20.12)$$

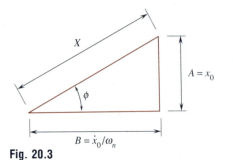

Fig. 20.3

The parameters X, ϕ, A, B, x_0, $\dot{x}_0$, and ω_n may geometrically be related to each other by a right triangle as shown in Fig. 20.3. If a pen is attached to the lower end of the spring, which suspends the body, and a strip of paper can be moved to the left at a constant speed v near the body to continuously record the position of the body, a sinusoidal curve given by Eq. (20.10) will be generated on the paper as shown in Fig. 20.4.

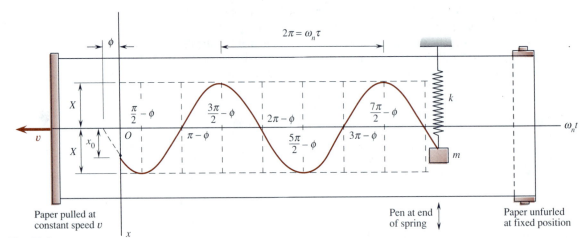

Fig. 20.4

For an additional geometrical representation, let a radius vector with a length equal to the amplitude X be drawn from the equilibrium position O of the body as shown in Fig. 20.5(a). If this radius vector is made to rotate clockwise about O at a constant angular velocity ω_n, the projection of this radius vector onto a vertical axis becomes identically equal to the position coordinate x of the body as given by Eq. (20.10). In Fig. 20.5(a), the magnitudes of the component vectors $\mathbf{A}$ and $\mathbf{B}$ are given by Eqs. (20.7) and (20.8). Clearly, the projection of the *tip* of the radius vector $\mathbf{X}$ onto a vertical axis corresponds to the position of the vibrating body. When this projection of the tip is plotted against the time t or the $\omega_n t$ axis, a sinusoidal curve is again generated as shown in Fig. 20.5(b).

The undamped free vibration described by Eq. (20.9) or (20.10) is often called *simple harmonic motion* and is characterized by the fact that the ac-

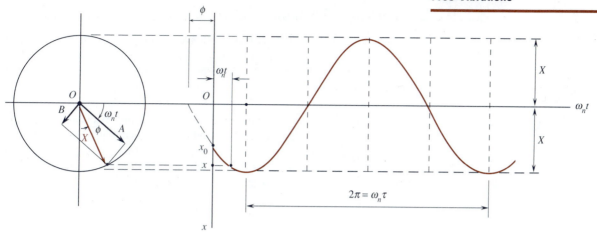

Fig. 20.5

celeration of the body is proportional to its displacement and of opposite
direction; i.e., $\ddot{x} = -\omega_n^2 x$. It is readily seen from Figs. 20.4 and 20.5 that
the *phase angle* ϕ represents the amount by which the sinusoidal curve is
displaced from the origin on the $\omega_n t$ axis. The constant angular velocity ω_n
with which the radius vector **X** rotates about the equilibrium position O is
called the *natural circular frequency*. Usually, ω_n is expressed in radians
per second (rad/s). The time required for the body to complete one cycle of
vibration is denoted by τ and is called the *period* of vibration. We see from
Figs. 20.4 and 20.5 that

$$2\pi = \omega_n \tau \qquad (20.13)$$

$$\tau = \frac{2\pi}{\omega_n} \qquad (20.14)$$

The number of cycles of vibration per unit of time is denoted by f_n and is
called the *natural cyclic frequency*. We write

$$f_n = \frac{1}{\tau} = \frac{\omega_n}{2\pi} \qquad (20.15)$$

Since ω_n is proportional to f_n, the words *circular* and *cyclic* in the names
for ω_n and f_n are often *omitted* for simplicity. In other words, the term
natural frequency in vibrations may refer to either ω_n or f_n as appropriate.
The period τ is usually expressed in seconds (s). However, the units for the
natural frequency f_n are given the special name *hertz* (Hz), after the German
physicist Heinrich Rudolf Hertz (1857–1894). Note that 1 Hz = 1 cycle per
second.[†]

[†]In 1965, the Institute of Electrical and Electronics Engineers (IEEE) adopted the unit hertz
(Hz), which is now also used in vibration studies.

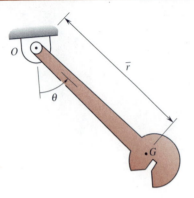

EXAMPLE 20.1

A pendulum used in impact tests is shown, where G is its mass center and $\bar{r} = 0.8$ m. If the period of small-amplitude vibration of the pendulum about the pivot O is $\tau = 2$ s, determine the radius of gyration $\bar{k}_G$ of the pendulum about G.

Solution. Since no damping is stated, the pendulum is taken to have un-damped free vibration. Denoting the mass of the pendulum by m, we draw the free-body diagram and the effective-force diagram of the pendulum in a general angular position θ as shown, where $\dot{\theta}$ and $\ddot{\theta}$ are the angular velocity and angular acceleration of the pendulum, respectively.

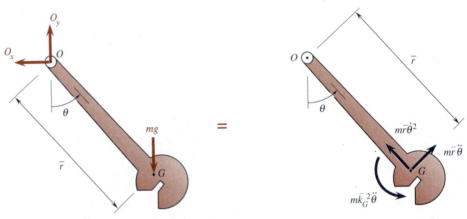

We write

$$+\circlearrowleft\ \Sigma M_O: \qquad -(\bar{r}\sin\theta)mg = \bar{r}(m\bar{r}\ddot{\theta}) + m\bar{k}_G^2\ddot{\theta}$$

$$(\bar{r}^2 + \bar{k}_G^2)\ddot{\theta} + \bar{r}g\sin\theta = 0$$

which is a nonlinear differential equation of motion. This means that the swinging of a pendulum is inherently a *nonlinear vibration*. However, $\sin\theta \approx \theta$ in small-amplitude vibration and the equation may be linearized. We write

$$(\bar{r}^2 + \bar{k}_G^2)\ddot{\theta} + \bar{r}g\theta = 0$$

$$\ddot{\theta} + \frac{\bar{r}g}{\bar{r}^2 + \bar{k}_G^2}\,\theta = 0 \qquad (1)$$

Comparing Eq. (1) with Eq. (20.4), we readily see that the natural frequency ω_n of the pendulum is given by

$$\omega_n^2 = \frac{\bar{r}g}{\bar{r}^2 + \bar{k}_G^2} \qquad (2)$$

By Eq. (20.14), we have

$$\omega_n^2 = \frac{4\pi^2}{\tau^2} \qquad (3)$$

For $\tau = 2$ s, $\bar{r} = 0.8$ m, and $g = 9.81$ m/s^2, we write, from Eqs. (2) and (3),

$$\frac{4\pi^2}{\tau^2} = \frac{\bar{r}g}{\bar{r}^2 + \bar{k}_G^2}$$

$$\frac{4\pi^2}{(2)^2} = \frac{0.8(9.81)}{(0.8)^2 + \bar{k}_G^2}$$

$$\bar{k}_G = 0.394 \text{ m} \qquad \bar{k}_G = 394 \text{ mm} \blacktriangleleft$$

Developmental Exercises

D20.1 Refer to Example 20.1. (a) Determine the natural circular frequency ω_n and the natural cyclic frequency f_n of the pendulum. (b) Does the term *natural frequency* in vibrations refer to either ω_n or f_n as appropriate?

D20.2 A slender rod OA oscillates about the pivot at O as shown. If $L = 1$ m, determine the natural frequency ω_n and the period τ of the rod.

D20.3 Solve D20.2 if $L = 1$ ft.

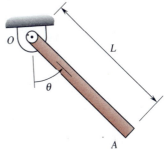

Fig. D20.2

EXAMPLE 20.2

A circular cylinder of weight $W = 40$ lb and radius $r = 6$ in. is suspended from a slender rod OA as shown. The cylinder is rotated, thus twisting the rod, through 90° and then released. If the observed period of torsional vibration is $\tau = 1.5$ s, determine (a) the torsional spring modulus K of the rod, (b) the maximum angular velocity reached by the cylinder.

Solution. Let θ denote the angular displacement of the cylinder from its equilibrium position. We note that the weight force of the cylinder is balanced by the tensile force in the rod all the time. The unbalanced action on the cylinder during its torsional vibration is the restoring moment **M** exerted

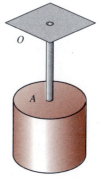

by the rod on the cylinder, where $M = K\theta$. Since the mass center of the cylinder does not move, the effective force on the cylinder is zero. However, the effective moment on the cylinder is $\bar{I}\alpha = \frac{1}{2}(W/g) r^2\ddot{\theta}\ \curvearrowright$. Thus, the free-body diagram and the effective-force diagram of the cylinder may be drawn as shown. We write

$$+ \circlearrowleft \ \Sigma M_A: \qquad\qquad -K\theta = \frac{1}{2} \frac{W}{g} r^2 \ddot{\theta}$$

$$\ddot{\theta} + \frac{2gK}{Wr^2} \theta = 0 \qquad\qquad (1)$$

Comparing Eq. (1) with Eq. (20.4), we readily see that the natural frequency ω_n of the cylinder is given by

$$\omega_n^2 = \frac{2gK}{Wr^2} \qquad\qquad (2)$$

By Eq. (20.14), we have

$$\omega_n^2 = \frac{4\pi^2}{\tau^2} \qquad\qquad (3)$$

For $\tau = 1.5$ s, $r = 6$ in. $= 0.5$ ft, $W = 40$ lb, and $g = 32.2$ ft/s^2, we write, from Eqs. (2) and (3),

$$\frac{4\pi^2}{\tau^2} = \frac{2gK}{Wr^2}$$

$$\frac{4\pi^2}{(1.5)^2} = \frac{2(32.2)K}{40(0.5)^2}$$

$$K = 2.72 \qquad\qquad K = 2.72 \ \text{lb·ft/rad} \ \blacktriangleleft$$

According to Eqs. (20.10) through (20.12), the general solution of Eq. (1) is given by

$$\theta = \Theta \sin(\omega_n t + \phi) \qquad\qquad (4)$$

$$\Theta = \sqrt{\theta_0^2 + (\dot{\theta}_0/\omega_n)^2} \qquad\qquad (5)$$

$$\phi = \tan^{-1}(\theta_0 \omega_n / \dot{\theta}_0) \qquad\qquad (6)$$

where θ_0 and $\dot{\theta}_0$ are the initial angular position and initial angular velocity of the cylinder. For the cylinder as given, we have $\theta_0 = 90° = \pi/2$ rad, $\dot{\theta}_0 = 0$, and $\omega_n = 2\pi/\tau = 2\pi/1.5$ rad/s. Taking the time derivative of Eq. (4), we write

$$\dot{\theta} = \Theta \omega_n \cos(\omega_n t + \phi)$$

Thus, the maximum angular velocity reached by the cylinder is

$$\dot{\theta}_{max} = \Theta \omega_n = \sqrt{(\pi/2)^2 + 0} \ (2\pi/1.5) = 6.58$$

$$\dot{\theta}_{max} = 6.58 \ \text{rad/s} \ \blacktriangleleft$$

Developmental Exercise

D20.4 Refer to Example 20.2. Suppose that the vibration is generated by imparting the cylinder an initial angular velocity of $\dot{\theta}_0 = 0.5$ rad/s beside the initial angular displacement of $\theta_0 = 90°$. Determine the maximum angular velocity reached by the cylinder.

EXAMPLE 20.3

Suppose that the cylinder of weight $W = 40$ lb and radius $r = 6$ in. in Example 20.2 is detached from the rod and placed on a cylindrical surface of radius $R = 2$ ft as shown. If the cylinder rolls without slipping on the surface, determine its period τ of small-amplitude vibration.

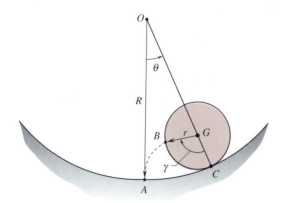

Solution.   When the cylinder rolls without slipping, the point B (of the cylinder) which was coincident with the point A (of the surface) must have moved in such a way that the arcs of contact AC and CB are equal; i.e., $R\theta = r\gamma$. However, the angular displacement (or the change of orientation) of the cylinder in the inertial reference frame (or with respect to the ground) is β, where $\gamma = \theta + \beta$. We write

$$R\theta = r(\theta + \beta) \qquad \beta = \frac{1}{r}(R - r)\theta$$

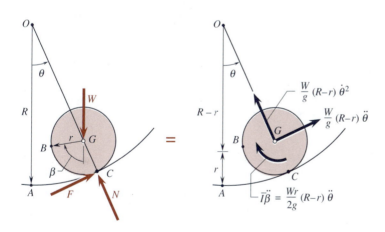

The angular acceleration of the cylinder in the inertial reference frame is $\ddot{\beta}$, not $\ddot{\gamma}$. The effective moment on the cylinder is given by

$$\bar{I}\ddot{\beta} = \frac{1}{2}\frac{W}{g}r^2 \cdot \frac{1}{r}(R - r)\ddot{\theta} = \frac{Wr}{2g}(R - r)\ddot{\theta}$$

Moreover, the mass center G of the cylinder is constrained to describe a circular path of radius $R - r$ about the center O. Its tangential and normal components of acceleration are

$$a_t = (R - r)\ddot{\theta} \qquad a_n = (R - r)\dot{\theta}^2$$

Therefore, the free-body diagram and the effective-force diagram of the cylinder may be drawn as shown. We write

$$+\circlearrowleft \Sigma M_C: \qquad -(r \sin\theta)W = r\left[\frac{W}{g}(R - r)\ddot{\theta}\right] + \frac{Wr}{2g}(R - r)\ddot{\theta}$$

$$\frac{3(R - r)}{2g}\ddot{\theta} + \sin\theta = 0$$

In small-amplitude vibration, we have $\sin\theta \approx \theta$. Thus, we write

$$\ddot{\theta} + \frac{2g}{3(R - r)}\theta = 0$$

which, as before, indicates that the natural frequency ω_n is given by

$$\omega_n^2 = \frac{2g}{3(R - r)}$$

For the given system, we have $R = 2$ ft, $r = 6$ in. $= 0.5$ ft, and $g = 32.2$ ft/s^2. The period τ of small-amplitude vibration of the cylinder is

$$\tau = \frac{2\pi}{\omega_n} = 2\pi\sqrt{\frac{3(R - r)}{2g}} = 2\pi\sqrt{\frac{3(2 - 0.5)}{2(32.2)}}$$

$$\tau = 1.661 \text{ s} \blacktriangleleft$$

REMARK. An alternative solution for the problem in this example is presented in Example 20.4.

Developmental Exercise

D20.5 Work Example 20.3 if the cylinder is replaced by a sphere of the same radius and the same weight.

★20.2   Energy Method

We note in plane motion that the kinetic energy T of a rigid body of mass m is given by

$$T = \tfrac{1}{2}m\bar{v}^2 + \tfrac{1}{2}\bar{I}\omega^2 \qquad\qquad (17.13)$$
$$\text{(repeated)}$$

or

$$T = \tfrac{1}{2}I_C\omega^2 \qquad\qquad (17.14)$$
$$\text{(repeated)}$$

where $\bar{v}$ is the speed of the mass center, ω is the angular speed, $\bar{I}$ is the central mass moment of inertia, and I_C is the mass moment of inertia about the velocity center C of the rigid body. For a particle of mass m moving with a speed v, its kinetic energy T is simply given by

$$T = \tfrac{1}{2} mv^2$$

(13.22)
(repeated)

We know from Sec. 13.8 that the potential energy of a body in a conservative force field relative to an appropriate reference datum is defined as *the amount of work which the body will receive from the conservative force if the body travels from its present position to the reference datum.* Thus, the *gravitational potential energy* V_g of a body with a weight W near the surface of the earth at a position of altitude y above the reference datum is given by

$$V_g = Wy$$

(13.32)
(repeated)

If the reference datum in an elastic force field is chosen to be the position of the body where the spring attached to the body is neutral, the elastic potential energy V_e of the body is given by

$$V_e = \tfrac{1}{2} kx^2$$

(13.36)
(repeated)

where k is the modulus and x is the change in length of the spring.

Note that the *total mechanical energy* E_{total} of a body is the sum of the kinetic energy T and the potential energy V of the body. In a conservative system, as defined in Sec. 13.8, the total mechanical energy of the body remains constant during its motion from one position to another; i.e.,

$$E_{total} = T + V = \text{constant}$$

(20.16)

This property is referred to as the *principle of conservation of energy* in Secs. 13.12 and 17.5. The *energy method* is based on the principle of conservation of energy.

In Eq. (20.16), observe that the kinetic energy T is nonnegative, but, depending on the choice of reference datum, the potential energy V can be positive, zero, or negative during vibration; i.e., the minimum value of T is zero, but *the minimum value of V may be less than zero.* Thus, for a conservative vibratory system with $E_{total} = T + V$, we have

$$T_{min} = 0$$

(20.17)

$$T_{max} = E_{total} - V_{min}$$

(20.18)

$$V_{max} = E_{total}$$

(20.19)

$$T_{max} \neq V_{max} \quad \text{if} \quad V_{min} \neq 0$$

(20.20)

$$T_{max} = V_{max} \quad \text{if} \quad V_{min} = 0$$

(20.21)

However, the value of E_{total} may vary from one choice of reference datum to another.

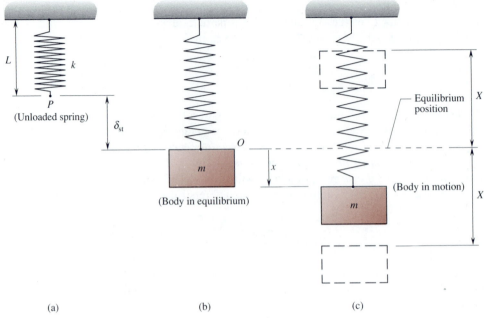

(a) (b) (c)

Fig. 20.1 (repeated)

I. REFERENCE DATUM *P*.

The spring-mass system in Fig. 20.1 is a conservative system. Let us choose the *unloaded position P* of the lower end of the spring as the *reference datum* for both the gravitational and the elastic potential energies of the body. When the body is displaced to the position x, its kinetic energy T and potential energy V are, by definition,

$$T = \tfrac{1}{2} m\dot{x}^2 \tag{20.22}$$
$$V = V_g + V_e = -mg(\delta_{st} + x) + \tfrac{1}{2} k(\delta_{st} + x)^2$$

Since $mg = k\delta_{st}$ from Eq. (20.1), we may simplify V to the form

$$V = \tfrac{1}{2} k(x^2 - \delta_{st}^2) \tag{20.23}$$

Clearly, it demonstrates that $V < 0$ whenever $x^2 < \delta_{st}^2$. By the energy method, the total mechanical energy E_{total} is

$$E_{total} = T + V = \tfrac{1}{2} m\dot{x}^2 + \tfrac{1}{2} k(x^2 - \delta_{st}^2)$$
$$= \text{constant}$$

Taking the time derivative to eliminate the constant, we write

$$\tfrac{1}{2} m(2\dot{x})\ddot{x} + \tfrac{1}{2} k(2x)\dot{x} = 0$$

$$\dot{x}(m\ddot{x} + kx) = 0 \tag{20.24}$$

Since the velocity $\dot{x}$ of the body is not *always* zero during vibration, Eq. (20.24) yields the governing equation of motion

$$m\ddot{x} + kx = 0 \tag{20.25}$$

This result agrees with Eq. (20.2).

II. REFERENCE DATUM 0.

If we choose the *equilibrium position O* as the *reference datum* for both the gravitational and the elastic potential energies of the body in Fig 20.1, we write, by definition,

$$T = \tfrac{1}{2} m\dot{x}^2 \tag{20.26}$$

$$V = V_g + V_e = -mgx + \int_x^0 k(\delta_{st} + x)\,(-dx)$$

$$= -mgx + (k\delta_{st}x + \tfrac{1}{2} kx^2)$$

$$= (-mg + k\delta_{st})x + \tfrac{1}{2} kx^2$$

Substituting Eq. (20.1) into the above expression for V, we get

$$V = \tfrac{1}{2} kx^2 \tag{20.27}$$

This shows that the choice of the equilibrium position O as the reference datum does simplify the final expression of V. By the energy method, the total mechanical energy E_{total} is

$$E_{total} = T + V = \tfrac{1}{2} m\dot{x}^2 + \tfrac{1}{2} kx^2$$
$$= \text{constant}$$

Taking the time derivative to eliminate the constant, we write

$$\tfrac{1}{2} m(2\dot{x})\ddot{x} + \tfrac{1}{2} k(2x)\dot{x} = 0$$

$$\dot{x}(m\ddot{x} + kx) = 0 \tag{20.28}$$

Equation (20.28) agrees with Eq. (20.24) and will similarly yield

$$m\ddot{x} + kx = 0 \tag{20.29}$$

We see that different choices of the reference datum for the potential energies in the energy method do not affect the final outcome for the equation of motion of a conservative system.

EXAMPLE 20.4

Using the *energy method*, solve for τ in Example 20.3.

Solution. Since the mass center G of the cylinder is constrained to describe a circular path of radius $R - r$ about the center O, its speed is

$$\bar{v} = (R - r)\dot{\theta}$$

As discussed in the solution in Example 20.3, the angular displacement of the cylinder is

$$\beta = \frac{1}{r}(R - r)\theta$$

Thus, the angular velocity of the cylinder is

$$\dot{\beta} = \frac{1}{r}(R - r)\dot{\theta}$$

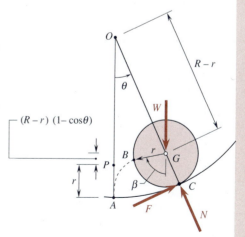

By Eq. (17.13), the kinetic energy T of this cylinder is

$$T = \frac{1}{2}\frac{W}{g}(R-r)^2\dot{\theta}^2 + \frac{1}{2}\left(\frac{1}{2}\frac{W}{g}r^2\right)\cdot\frac{1}{r^2}(R-r)^2\dot{\theta}^2$$

$$= \frac{3}{4}\frac{W}{g}(R-r)^2\dot{\theta}^2$$

Next, let the position of the mass center at its equilibrium position P, as shown, be chosen as the reference datum for the gravitational potential energy V_g. Noting that the reactions $\mathbf{N}$ and $\mathbf{F}$ from the support are nonworking forces, we find that the potential energy V of the conservative system is simply

$$V = V_g = W(R-r)(1-\cos\theta)$$

For this conservative system, we set $E_{\text{total}} = T + V = \text{constant}$; i.e.,

$$\frac{3}{4}\frac{W}{g}(R-r)^2\dot{\theta}^2 + W(R-r)(1-\cos\theta) = \text{constant}$$

Taking the time derivative to eliminate the constant, we write

$$\frac{3}{4}\frac{W}{g}(R-r)^2(2\dot{\theta})\ddot{\theta} + W(R-r)(\sin\theta)\dot{\theta} = 0$$

$$\frac{3W}{2g}(R-r)^2\dot{\theta}\left[\ddot{\theta} + \frac{2g}{3(R-r)}\sin\theta\right] = 0$$

Since $\dot{\theta}$ is not always zero during vibration and $\sin\theta \approx \theta$ for small-amplitude vibration, we find that

$$\ddot{\theta} + \frac{2g}{3(R-r)}\theta = 0$$

$$\omega_n^2 = \frac{2g}{3(R-r)} \qquad \tau = \frac{2\pi}{\omega_n}$$

For the given system in Example 20.3, we have $R = 2$ ft, $r = 0.5$ ft, and $g = 32.2$ ft/s^2. Thus, the period is

$$\tau = 2\pi\sqrt{\frac{3(R-r)}{2g}} = 2\pi\sqrt{\frac{3(2-0.5)}{2(32.2)}}$$

$$\tau = 1.661 \text{ s} \blacktriangleleft$$

Developmental Exercises

D20.6 Verify that the same result for the kinetic energy T in the solution in Example 20.4 can be obtained by applying Eq. (17.14).

D20.7 Using *energy method*, solve D20.2.

PROBLEMS

20.1 A particle in simple harmonic motion on the x axis has an amplitude of 2 in. and a period of 0.5 s. Determine its maximum velocity $\dot{x}_{\text{max}}$ and maximum acceleration $\ddot{x}_{\text{max}}$.

20.2   The forces in the springs are zero when the block of mass $m = 40$ kg is in equilibrium as shown. If the moduli of the two connecting springs are $k_1 = 640$ N/m and $k_2 = 800$ N/m, determine the natural frequency f_n of vibration of the block.

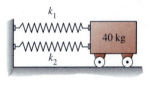

Fig. P20.2

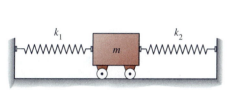

Fig. P20.3* and P20.4

20.3*   Determine the period of vibration of the 100-kg block B which is connected by a spring of modulus 2.5 kN/m as shown.

20.4   The 100-kg block B is connected by a spring of modulus 2.5 kN/m as shown. If the block is given an initial displacement of 20 mm $\searrow$ from its equilibrium position and an initial velocity of 75 mm/s $\searrow$, determine the amplitude and the phase angle of vibration of the block.

20.5   The spring supporting the slender rod AB, as shown, has a modulus of 600 lb/ft. If the period of vibration is 0.5 s, determine the weight of the rod.

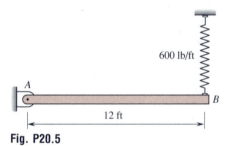

Fig. P20.5

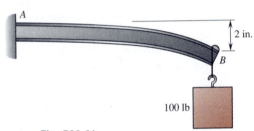

Fig. P20.6*

20.6*   A 100-lb weight causes a cantilever beam to have a static deflection of 2 in. as shown, where the mass of the cantilever can be taken as negligible. Determine the natural frequency f_n of vibration of the weight.

20.7   Two springs of moduli $k_1 = 2$ kN/m and $k_2 = 3$ kN/m are used to connect a 40-kg block as shown. The block is given an initial displacement of 30 mm $\rightarrow$ from its equilibrium position and released. Determine the period and the maximum velocity of vibration if the springs are arranged (a) in parallel, (b) in series.

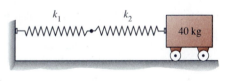

(a)   (b)

Fig. P20.7   Fig. P20.8*

20.8*   Three springs of moduli $k_1 = 20$ lb/in. and $k_2 = k_3 = 40$ lb/in. are arranged to suspend a cylinder of weight $W = 32$ lb as shown. If the amplitude of vibration of the cylinder is 6 in., determine the maximum velocity and maximum acceleration of the cylinder.

20.9 Two slender shafts in series with torsional moduli K_1 = 30 N·m/rad and K_2 = 20 N·m/rad are used to suspend an automobile wheel as shown. When the wheel is given an initial angular displacement and released, it makes 10 oscillations in 12 s. Determine the moment of inertia of the wheel about its central axis along the shafts.

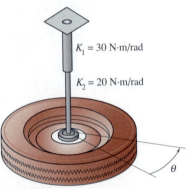

Fig. P20.9

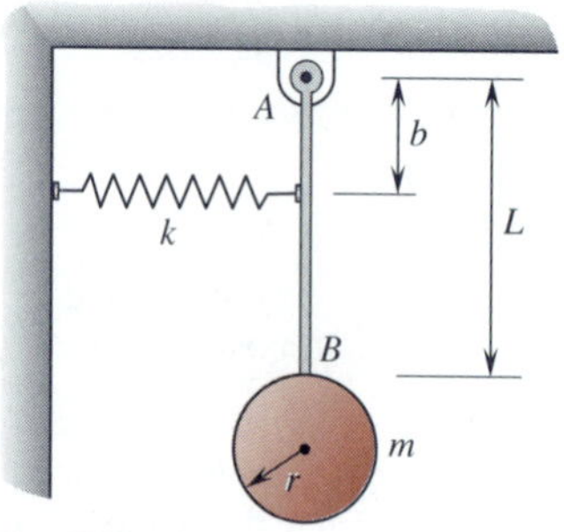

Fig. P20.10

20.10 A sphere of radius r and mass m is welded to a slender rod of negligible mass as shown, where k is the modulus of the connecting spring. Derive an expression for the natural frequency ω_n of small-amplitude vibration.

20.11 A homogeneous slender rod is bent into the shape of a square as shown, where b = 1 ft. Determine the period of small-amplitude vibration.

20.12* Solve Prob. 20.11 if b = 1 m.

20.13 A homogeneous slender rod is bent into the shape of an angle as shown, where L = 600 mm. Determine the natural frequency f_n of small-amplitude vibration.

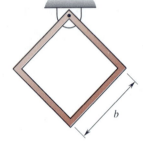

Fig. P20.11

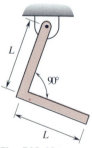

Fig. P20.13

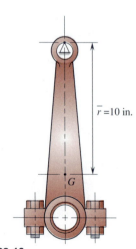

Fig. P20.16

Fig. P20.14

20.14 A homogeneous semicircular disk is suspended as shown. Determine the radius r for which the natural frequency of the disk is 1 Hz.

20.15* Solve Prob. 20.14 if the semicircular disk is replaced by a hemisphere of radius r.

20.16 A connecting rod weighing 10 lb is suspended as shown. If the period of small-amplitude oscillation is observed to be 1.5 s, determine the moment of inertia of the connecting rod about its mass center G.

20.17 A U-tube with uniform cross section contains a column of mercury as shown. Determine the length L of the mercury column for which the period of vibration is 2 s.

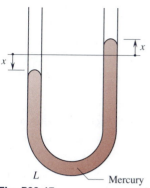

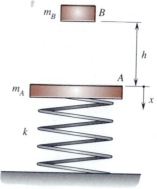

Fig. P20.17

Fig. P20.18

20.18 A spring of modulus k supports a disk A of mass m_A in equilibrium as shown. If another disk B of mass m_B is dropped from a height h and the impact is perfectly plastic, determine in terms of parameters of the system the subsequent motion of the disk since impact as defined by the position coordinate x which is measured from the original equilibrium position of the disk A.

20.19 An equilateral triangular plate ABC of mass m is suspended by three vertical wires as shown. If the plate is slightly rotated about its vertical central axis and released, determine the length L of the wires for which the period of vibration is 2 s.

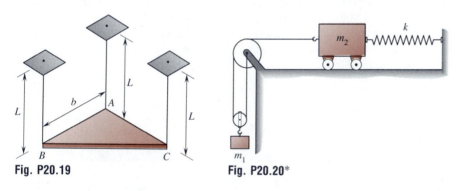

Fig. P20.19

Fig. P20.20*

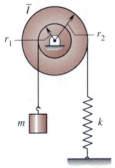

20.20* A cylinder of mass m_1 and a block of mass m_2 are connected as shown, where k is the modulus of the connecting spring. Neglecting masses of the pulleys, derive an expression for the period of vibration of the system.

20.21 A cylinder of mass m and a double pulley are connected as shown, where k is the modulus of the connecting spring. If the moment of inertia of the double pulley about its axis of support is $\bar{I}$, derive an expression for the natural frequency f_n of vibration of the system.

Fig. P20.21

★20.3 Damped Free Vibrations

In reality, every vibratory mechanical system experiences some degree of kinetic friction and suffers some dissipation of mechanical energy. The kinetic friction forces are the *damping forces*. These forces may be caused by *dry friction* between sliding surfaces, by *viscous friction* when a body moves

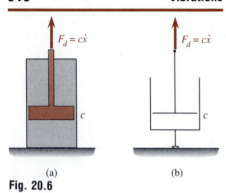

(a)   (b)

Fig. 20.6

on a lubricated surface or in a fluid, or by *internal friction* between mole-
cules or elements of a deformable body.

Although the precise mechanism of damping is rather complex, the effects of damping on many vibratory systems can be simulated by the effect
of a *viscous damper*, which is often referred to as a *dashpot*. A dashpot
consists of a cylinder filled with a viscous fluid and a plunger with some
tolerance or holes for the fluid to flow from one side to the other. A simple
dashpot is illustrated in Fig. 20.6(a), which is usually schematically represented as shown in Fig. 20.6(b). The magnitude F_d of the damping force is
proportional to the speed $\dot{x}$ of the plunger relative to the cylinder. We write

$$F_d = c\dot{x} \tag{20.30}$$

where the constant of proportionality c is known as the *coefficient of viscous
damping* and has units of N·s/m or lb·s/ft. When a dashpot is connected to
a body, the damping force exerted by the dashpot on the body is always
directed opposite to the velocity of the body relative to the dashpot. In application, the mass of a dashpot is usually assumed to be negligible compared with the mass of the body to which it is connected.

To be specific, let us consider the damped free vibration of a body with
mass m, which is suspended by a spring of modulus k and a simple dashpot
having a coefficient of viscous damping c as shown in Fig. 20.7. In part (a)
of Fig. 20.7, the spring and the dashpot are unloaded.[†] It is seen in part (b)
that the lower ends of the spring and the dashpot move from P to O to
undergo a static deflection δ_{st} when they are loaded to hold the body in
equilibrium. Since the dashpot exerts no force on a body in equilibrium, the
weight force of magnitude mg is balanced solely by the spring force of
magnitude $k\delta_{st}$; i.e.,

$$mg = k\delta_{st} \tag{20.31}$$

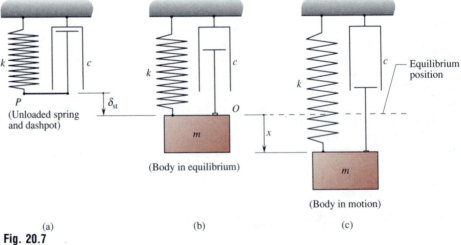

(a)   (b)   (c)

Fig. 20.7

[†]A dissipative restoring mechanism consisting of a linear spring in parallel with a simple dashpot, as shown in Fig. 20.7(a), is known as a *Voigt model* in viscoelasticity.

which is identical with Eq. (20.1). As before, we let the coordinate x defining the position of the displaced body be measured from its equilibrium position O, as shown in Fig. 20.7(b) and (c).

Suppose that the body is set in motion by certain initial conditions. The dynamic behavior of the body can be predicted by examining the condition of the body at a general time t in a general position x as shown in Fig. 20.8. Referring to this figure (or applying Newton's second law), we write

$$+\downarrow \Sigma F: \qquad mg - k(\delta_{st} + x) - c\dot{x} = m\ddot{x}$$

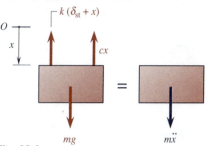

Fig. 20.8

Substituting Eq. (20.31) into this equation and rearranging, we get

$$m\ddot{x} + c\dot{x} + kx = 0 \tag{20.32}$$

which is the differential equation of motion of the body. Using Eq. (20.3) and setting

$$\frac{c}{m} = 2\zeta\omega_n \qquad \text{or} \qquad \zeta = \frac{c}{2\sqrt{km}} \tag{20.33}$$

we can write Eq. (20.32) in the form

$$\ddot{x} + 2\zeta\omega_n\dot{x} + \omega_n^2 x = 0 \tag{20.34}$$

The parameter ζ, as defined in Eq. (20.33), is called the *damping factor* or *damping ratio*. It can be shown that ζ is nondimensional and is a measure of the severity of damping.

Substituting a solution of the form $x = Ae^{\lambda t}$ into Eq. (20.34) and recognizing that $Ae^{\lambda t} \neq 0$, we obtain the *characteristic equation*

$$\lambda^2 + 2\zeta\omega_n\lambda + \omega_n^2 = 0 \tag{20.35}$$

whose roots are

$$\lambda_{1,2} = \left(-\zeta \pm \sqrt{\zeta^2 - 1}\right)\omega_n \tag{20.36}$$

By superposition, the general solution for Eq. (20.34) is

$$x = A_1 e^{\lambda_1 t} + A_2 e^{\lambda_2 t} \tag{20.37}$$

where A_1 and A_2 are arbitrary constants to be determined from the initial conditions $x(0) = x_0$ and $\dot{x}(0) = \dot{x}_0$. The characteristics of the damped motion are highly dependent on whether the radicand $\zeta^2 - 1$ in Eqs. (20.36) is positive, zero, or negative.

1. CASE OF $\zeta > 1$ (*overdamping*).   Both roots λ_1 and λ_2 in Eqs. (20.36) are negative and have distinct values. Thus, Eq. (20.37) may be written as

$$x = A_1 e^{-(\zeta+\sqrt{\zeta^2-1})\omega_n t} + A_2 e^{-(\zeta-\sqrt{\zeta^2-1})\omega_n t} \tag{20.38}$$

It can be shown by imposing the initial conditions on Eq. (20.38) that

$$A_1 = \frac{\dot{x}_0 + \left(\zeta + \sqrt{\zeta^2 - 1}\right)\omega_n x_0}{2\omega_n\sqrt{\zeta^2 - 1}} \tag{20.39}$$

$$A_2 = -\frac{\dot{x}_0 + \left(\zeta - \sqrt{\zeta^2 - 1}\right)\omega_n x_0}{2\omega_n\sqrt{\zeta^2 - 1}} \tag{20.40}$$

The motion as given by Eq. (20.38) diminishes exponentially as the time t increases. The motion is nonvibratory regardless of the initial conditions imposed on the system and is referred to as *aperiodic*.

2. CASE OF $\zeta = 1$ (*critical damping*).

Both roots λ_1 and λ_2 in Eqs. (20.36) become equal to $-\omega_n$. The general solution for the special case of equal roots is given by

$$x = (A_1 + A_2 t)e^{-\omega_n t} \tag{20.41}$$

Imposition of initial conditions on Eq. (20.41) yields

$$A_1 = x_0 \tag{20.42}$$

$$A_2 = \dot{x}_0 + \omega_n x_0 \tag{20.43}$$

Again, the motion as given by Eq. (20.41) is *aperiodic* and diminishes exponentially as the time t increases. Nevertheless, a critically damped system regains its equilibrium position faster than an otherwise overdamped system, as illustrated in Fig. 20.9. Note that, for certain initial conditions (e.g., $x_0 > 0$ and $\dot{x}_0 < 0$), the body in the damped system with $\zeta \geq 1$ may cross the equilibrium position at most once.

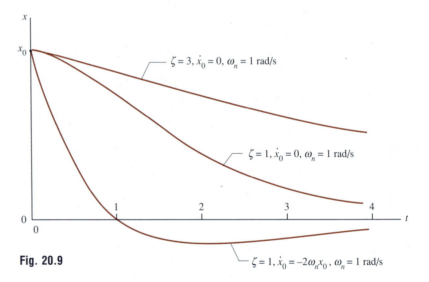

Fig. 20.9

3. CASE OF $\zeta < 1$ (*underdamping*).

The radicand $\zeta^2 - 1$ in Eqs. (20.36) is negative, and Eq. (20.37) may be written as

$$x = e^{-\zeta\omega_n t}(A_1 e^{i\omega_d t} + A_2 e^{-i\omega_d t}) \tag{20.44}$$

where $i = \sqrt{-1}$ and

$$\omega_d = \sqrt{1 - \zeta^2}\,\omega_n \tag{20.45}$$

Applying Euler's formula $e^{\pm i\theta} = \cos\theta \pm i\sin\theta$, we can rewrite Eq. (20.44) as

$$x = e^{-\zeta\omega_n t}(C_1 \cos\omega_d t + C_2 \sin\omega_d t) \tag{20.46}$$

where $C_1 = A_1 + A_2$ and $C_2 = i(A_1 - A_2)$. By imposing the initial conditions on Eq. (20.46), we get

$$C_1 = x_0 \qquad C_2 = (\dot{x}_0 + \zeta\omega_n x_0)/\omega_d \qquad (20.47)$$

It can be verified that Eq. (20.46) is equivalent to the equation

$$x = X e^{-\zeta\omega_n t} \sin(\omega_d t + \phi) \qquad (20.48)$$

where

$$X = \sqrt{x_0^2 + [(\dot{x}_0 + \zeta\omega_n x_0)/\omega_d]^2} \qquad (20.49)$$

$$\phi = \tan^{-1}[x_0\omega_d/(\dot{x}_0 + \zeta\omega_n x_0)] \qquad (20.50)$$

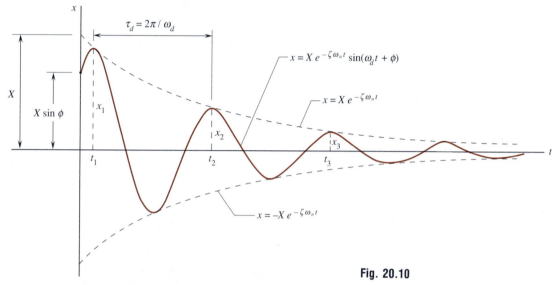

Fig. 20.10

We note that Eqs. (20.48) through (20.50) degenerate into Eqs. (20.10) through (20.12), respectively, if the damping factor $\zeta = 0$. Unlike preceding cases of $\zeta \geq 1$, the motion as given by Eq. (20.48) is *vibratory* with exponentially decaying amplitudes as shown in Fig. 20.10. The dashed curves defined by

$$x = \pm X e^{-\zeta\omega_n t}$$

are the envelopes of the motion curve. Although this motion does not exactly repeat itself, the parameter ω_d defined in Eq. (20.45) is called the *damped natural frequency* and the time

$$\tau_d = \frac{2\pi}{\omega_d} = \frac{2\pi}{\sqrt{1 - \zeta^2}\,\omega_n} \qquad (20.51)$$

is called the *damped period*.

Developmental Exercises

D20.8 What is the value or range of values of the *damping factor* ζ if a system is (a) critically damped, (b) overdamped, (c) underdamped?

D20.9 If the system in Fig. 20.7 is critically damped, show that the coefficient of viscous damping has the value

$$c = c_c = 2\sqrt{km}$$

*20.4 Logarithmic Decrement

We saw in Sec. 20.3 that a damped system may have free oscillations only when it is *underdamped* (i.e., $\zeta < 1$), and its amplitudes decay exponentially as shown in Fig. 20.10. The larger the damping, the greater will be the rate of decay of the amplitudes.

Suppose that x_1 is the first peak value of x occurring at $t = t_1$ and x_2 is the second peak value of x occurring at $t = t_2 = t_1 + \tau_d$ as indicated in Fig. 20.10. These peak values x_1 and x_2 are the amplitudes of the body in the first and second cycles of damped vibration. Referring to Fig. 20.10 and applying Eqs. (20.48) and (20.51), we write

$$x_1 = X e^{-\zeta\omega_n t_1} \sin(\omega_d t_1 + \phi) \tag{20.52}$$

$$
\begin{aligned}
x_2 &= X e^{-\zeta\omega_n(t_1+\tau_d)} \sin[\omega_d(t_1 + \tau_d) + \phi] \\
&= e^{-\zeta\omega_n\tau_d} \{X e^{-\zeta\omega_n t_1} \sin[2\pi + (\omega_d t_1 + \phi)]\} \\
&= e^{-\zeta\omega_n\tau_d} \{X e^{-\zeta\omega_n t_1} \sin(\omega_d t_1 + \phi)\} \\
&= e^{-\zeta\omega_n\tau_d} \{x_1\}
\end{aligned}
$$

$$x_2 = x_1 e^{-\zeta\omega_n\tau_d} \tag{20.53}$$

By Eqs. (20.52) and (20.53), we write

$$\frac{x_1}{x_2} = e^{\zeta\omega_n\tau_d}$$

which can be shown to be also the ratio of the nth and $(n+1)$th amplitudes x_n and x_{n+1}. The *logarithmic decrement* δ is defined as the natural logarithm of the ratio of any two successive amplitudes of a body in damped vibration; i.e.,

$$\delta = \ln\left(\frac{x_n}{x_{n+1}}\right) \tag{20.54}$$

For $n = 1$, we have

$$\delta = \ln\left(\frac{x_1}{x_2}\right) = \ln(e^{\zeta\omega_n\tau_d}) = \zeta\omega_n\tau_d$$

Applying Eq. (20.51), we write

$$\delta = \frac{2\pi\zeta}{\sqrt{1 - \zeta^2}} \tag{20.55}$$

which can be solved to yield

$$\zeta = \frac{\delta}{\sqrt{4\pi^2 + \delta^2}} \tag{20.56}$$

This means that the damping factor ζ, which reflects the amount of damping in a system, can conveniently be determined by measuring the rate of decay of free oscillations of the damped system. If a system is slightly damped with $\zeta \ll 1$, then $\delta \approx 2\pi\zeta$ or $\zeta \approx \delta/(2\pi)$.

EXAMPLE 20.5

A 10-kg mass is connected by a spring of modulus 160 N/m and a dashpot as shown. It is observed that the ratio of any two successive amplitudes in free vibration is 1:0.6. Determine (a) the logarithmic decrement, (b) the damping factor, (c) the damped natural frequency of the system, (d) the coefficient of viscous damping.

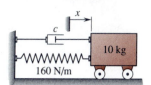

Solution. We know from the given data that $m = 10$ kg, $k = 160$ N/m, and the motion is governed by Eq. (20.32). The logarithmic decrement is

$$\delta = \ln(x_n/x_{n+1}) = \ln(1/0.6) = 0.5108$$

$$\delta = 0.511 \blacktriangleleft$$

By Eq. (20.56), we find that the damping factor is

$$\zeta = \frac{\delta}{\sqrt{4\pi^2 + \delta^2}} = \frac{0.5108}{\sqrt{4\pi^2 + (0.5108)^2}} = 0.08103$$

$$\zeta = 0.0810 \blacktriangleleft$$

The undamped natural frequency is

$$\omega_n = \sqrt{\frac{k}{m}} = \sqrt{\frac{160}{10}} \qquad \omega_n = 4 \text{ rad/s}$$

Therefore, the damped natural frequency is

$$\omega_d = \sqrt{1 - \zeta^2}\, \omega_n = \sqrt{1 - (0.08103)^2}\,(4) = 3.987$$

$$\omega_d = 3.99 \text{ rad/s} \blacktriangleleft$$

By Eq. (20.33), we find that the coefficient of viscous damping is

$$c = 2\zeta\sqrt{km} = 2(0.08103)\sqrt{160(10)}$$

$$c = 6.48 \text{ N·s/m} \blacktriangleleft$$

Developmental Exercises

D20.10 Using Eqs. (20.48) and (20.51), show that the ratio of the amplitudes x_n and x_{n+1} in the nth and $(n+1)$th cycles is

$$\frac{x_n}{x_{n+1}} = e^{\zeta \omega_n \tau_d}$$

D20.11 According to Eq. (20.54), the ratio of any two successive amplitudes must have the same value of e^δ, where δ is the *logarithmic decrement*; i.e.,

$$\frac{x_1}{x_2} = \frac{x_2}{x_3} = \frac{x_3}{x_4} = \cdots = \frac{x_n}{x_{n+1}} = e^\delta$$

Using this fact, show that

$$\delta = \frac{1}{n} \ln\left(\frac{x_1}{x_{n+1}}\right)$$

PROBLEMS

20.22 A 10-kg block is connected by a spring and a dashpot as shown. The block vibrates with a damped natural frequency of 2 Hz. If the dashpot is removed, the block is observed to vibrate with a natural frequency of 4 Hz. Determine (a) the modulus k of the spring, (b) the coefficient of viscous damping c of the dashpot.

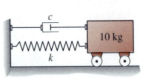

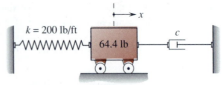

Fig. P20.22   **Fig. P20.23 and P20.26**

20.23 The 64.4-lb block in the spring-mass-dashpot system, as shown, is given an initial displacement of $x_0 = 3$ in. at the time $t = 0$ and then released. Determine the damping factor ζ and the position x of the block as a function of t if the dashpot has a coefficient of viscous damping $c = 60$ lb·s/ft.

20.24* Solve Prob. 20.23 if $c = 40$ lb·s/ft.

20.25 Solve Prob. 20.23 if $c = 20$ lb·s/ft.

20.26 An impulse of 3 lb·s → is exerted at the time $t = 0$ on the 64.4-lb block in the spring-mass-dashpot system, which is at rest as shown. If the dashpot has a coefficient of viscous damping $c = 30$ lb·s/ft, determine the position x of the block as a function of t.

20.27* A critically damped oscillator of natural frequency ω_n is set in motion with the initial conditions $x_0 > 0$ and $\dot{x}_0 > 0$ at the time $t = 0$. Determine the time t_m at which the displacement of the oscillator is maximum.

20.28 The oscillator shown in part (a) has a damped period of $\tau_d = 0.5$ s. If two successive amplitudes a full cycle apart are measured to be 17 mm and 12 mm as shown in part (b), determine the modulus k of the spring and the coefficient of viscous damping c of the dashpot.

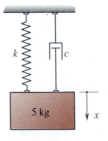

(a)
Fig. P20.28

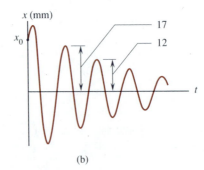

(b)

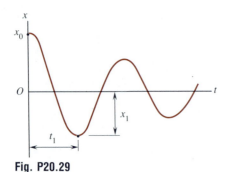

Fig. P20.29

20.29 A spring-mass-dashpot system has a modulus $k = 160$ N/m, a mass $m = 2.5$ kg, and a coefficient of viscous damping $c = 7$ N·s/m. The system is released from rest from an initial position x_0 as indicated. Determine (a) the time t_1 at which the overshoot distance x_1 is greatest, (b) the ratio x_1/x_0.

20.30   The slender rod AB of mass m is supported by a spring of modulus k and a dashpot with coefficient of viscous damping c. Determine the damping factor ζ and the damped period τ_d in terms of the parameters of the system.

20.31*   A spring-mass-dashpot system is arranged as shown. Neglecting the mass of the crank AOB, derive the equation of motion of the block in terms of the parameters of the system.

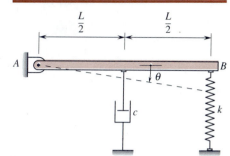

Fig. P20.30

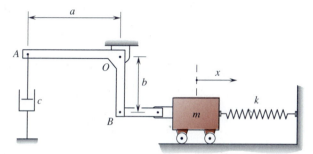

Fig. P20.31*

FORCED VIBRATIONS

★20.5   Harmonic Excitation

We saw in preceding sections that a mechanical system is set in free vibration by a certain initial disturbance. If a recurrent excitation acts on a body, the resulting vibration is a *forced vibration*.

The recurrent excitation may be externally applied or may be generated within the system by a rotating unbalance. Moreover, the recurrent excitation may also be applied to the support or foundation of the system. For fundamental study of forced vibrations, let us consider the motion of a spring-mass-dashpot system acted on by a harmonic force $F_0 \sin \omega t$ applied as shown in Fig. 20.11(a), where ω is the frequency of excitation and is

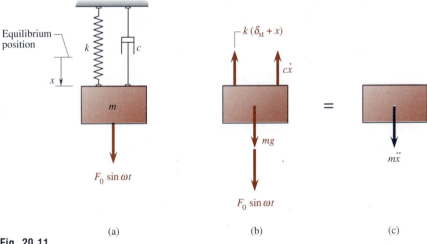

Fig. 20.11

called the *forcing frequency*. The free-body and effective-force diagrams of the body are shown in parts (b) and (c) of Fig. 20.11, where δ_{st} is the static deflection of the spring due to the weight mg. Referring to parts (b) and (c) of the figure, we write

$$+\downarrow \ \Sigma F: \qquad F_0 \sin \omega t + mg - k(\delta_{st} + x) - c\dot{x} = m\ddot{x}$$

Substituting Eq. (20.31) into this equation and rearranging, we get

$$m\ddot{x} + c\dot{x} + kx = F_0 \sin \omega t \qquad (20.57)$$

which is the differential equation of motion of the body. Using Eqs. (20.3) and (20.33), we can write Eq. (20.57) in the form

$$\ddot{x} + 2\zeta\omega_n\dot{x} + \omega_n^2 x = \frac{F_0}{m} \sin \omega t \qquad (20.58)$$

which is a nonhomogeneous second-order differential equation with constant coefficients.

The general solution of Eq. (20.58) is obtained by adding a particular solution x_p of Eq. (20.58) to the complementary solution x_c, which is equivalent to the general solution of the homogeneous equation in Eq. (20.34); i.e.,

$$x = x_c + x_p \qquad (20.59)$$

The complementary solution x_c is given by one of Eqs. (20.38), (20.41), and (20.48), depending on the case of damping. If the system is underdamped (as in most vibratory systems), we write

$$x_c = X_1 e^{-\zeta\omega_n t} \sin(\omega_d t + \phi_1) \qquad (20.60)$$

where X_1 and ϕ_1 are arbitrary constants, which are generally different from the values of X and ϕ given in Eqs. (20.49) and (20.50). In any case, the complementary solution x_c decays exponentially and will eventually be *damped out*. Thus, the complementary solution x_c for a damped system is significant only when a *transient solution* is being sought.

The particular solution x_p is *any* solution of the nonhomogeneous equation in Eq. (20.58). In the present case, x_p represents a *steady-state solution* which may be assumed to resemble the excitation in frequency but differ from it in phase; i.e.,

$$x_p = X \sin(\omega t - \varphi) \qquad (20.61)$$

where the *steady-state amplitude* X and the *phase angle* φ are to be determined in terms of parameters in the system. Substituting x_p for x in Eq. (20.58) and rearranging, we obtain

$$(\omega_n^2 - \omega^2)X \sin(\omega t - \varphi) + 2\zeta\omega_n\omega X \cos(\omega t - \varphi)$$

$$= \frac{F_0}{m} \sin \omega t \qquad (20.62)$$

Letting $\omega t - \varphi = \pi/2$ (or $\omega t = \pi/2 + \varphi$) in Eq. (20.62), we get

$$(\omega_n^2 - \omega^2)X = \frac{F_0}{m} \cos\varphi \qquad (20.63)$$

Letting $\omega t - \varphi = 0$ (or $\omega t = \varphi$) in Eq. (20.62), we get

$$2\zeta\omega_n\omega X = \frac{F_0}{m}\sin\varphi \qquad (20.64)$$

Squaring both members of Eqs. (20.63) and (20.64) and adding, we have

$$[(\omega_n^2 - \omega^2)^2 + (2\zeta\omega_n\omega)^2]X^2 = (F_0/m)^2 \qquad (20.65)$$

Dividing Eqs. (20.64) and (20.63) member by member, we have

$$\frac{2\zeta\omega_n\omega}{\omega_n^2 - \omega^2} = \tan\varphi \qquad (20.66)$$

Using Eq. (20.3) to substitute k/ω_n^2 for m, we can write Eqs. (20.65) and (20.66) in nondimensional forms as follows:

$$\frac{kX}{F_0} = \frac{1}{\sqrt{[1 - (\omega/\omega_n)^2]^2 + [2\zeta(\omega/\omega_n)]^2}} \qquad (20.67)$$

$$\tan\varphi = \frac{2\zeta(\omega/\omega_n)}{1 - (\omega/\omega_n)^2} \qquad (20.68)$$

Since F_0/k is the deflection of the spring by a static force F_0, the nondimensional ratio $kX/F_0 = X/(F_0/k)$ is known as the *amplitude ratio* or *magnification factor*. Equations (20.67) and (20.68) are plotted for several values of ζ in Figs. 20.12 and 20.13. Results pertaining to the undamped forced vibration can be obtained by simply setting $\zeta = 0$.

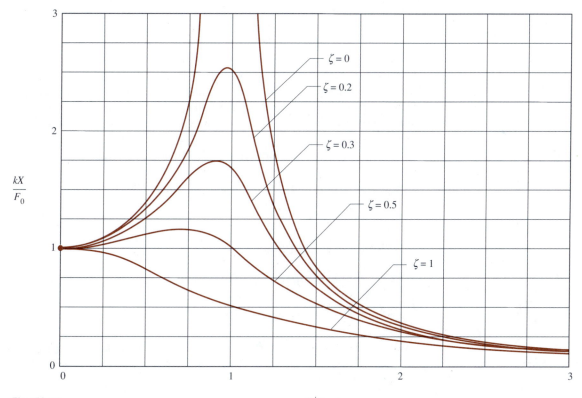

Fig. 20.12

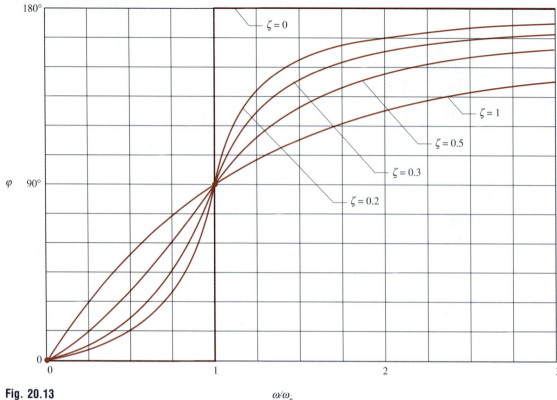

Fig. 20.13

ω/ω_n

Note in Fig. 20.12 that the *amplitude* of the steady-state response has a peak value when damping is small and the forcing frequency ω is near the value of the natural frequency ω_n. The phenomenon of exhibiting a peak amplitude is referred to as *resonance*. We see that the resonant peak is higher for smaller values of ζ; the resonant peak approaches infinity when ζ approaches zero. Thus, damping is beneficial in limiting the resonant peak. We also see that the steady-state amplitude is smaller when ω is much larger than ω_n.

The *phase angle* φ as shown in Fig. 20.13 ranges from 0 to 180°. It varies with the forcing frequency and the damping in the system. Without damping, the phase angle is zero if $\omega < \omega_n$ and is 180° if $\omega > \omega_n$. At $\omega = \omega_n$, the phase angle for all cases (with or without damping) is equal to 90°.

Developmental Exercises

D20.12 Define the terms: (a) forcing frequency, (b) magnification factor, (c) resonance.

D20.13 Verify that the abscissa and ordinate of the resonant peak in Fig. 20.12 are, respectively, $\sqrt{1 - 2\zeta^2}$ and $1/(2\zeta\sqrt{1 - \zeta^2})$.

D20.14 Refer to D20.13. Determine the threshold value of the damping factor ζ below which the magnification factor may exceed 1 for a certain range of the frequency ratio ω/ω_n.

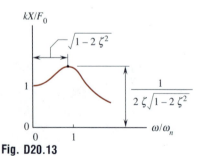

Fig. D20.13

*20.6 Transmissibility

Since $mg = k\delta_{st}$ as given in Eq. (20.31), the free-body and effective-force diagrams in parts (b) and (c) of Fig. 20.11 may be simplified as shown in Fig. 20.14. Referring to this figure, we write

$$+\downarrow \Sigma F: \qquad F_0 \sin\omega t + (-kx) + (-c\dot{x}) = m\ddot{x} \qquad (20.69)$$

In steady-state vibration, x_c in Eq. (20.59) will be damped out; i.e., $x_c \to 0$. By Eqs. (20.59) and (20.61), we have

$$x = X \sin(\omega t - \varphi)$$

$$\dot{x} = \omega X \cos(\omega t - \varphi)$$

$$\ddot{x} = -\omega^2 X \sin(\omega t - \varphi)$$

Using these equations, we write

$$-kx = -kX \sin(\omega t - \varphi) = kX \sin[\pi + (\omega t - \varphi)]$$

$$-c\dot{x} = -c\omega X \cos(\omega t - \varphi) = c\omega X \sin\left[\frac{3\pi}{2} + (\omega t - \varphi)\right]$$

$$m\ddot{x} = -m\omega^2 X \sin(\omega t - \varphi) = m\omega^2 X \sin[\pi + (\omega t - \varphi)]$$

Substituting the above equations into Eq. (20.69), we obtain

$$F_0 \sin\omega t + kX \sin[\pi + (\omega t - \varphi)]$$

$$+ c\omega X \sin\left[\frac{3\pi}{2} + (\omega t - \varphi)\right]$$

$$= m\omega^2 X \sin[\pi + (\omega t - \varphi)] \qquad (20.70)$$

According to Eq. (20.70) and noting the phase differences among the applied force F_0, the spring force kX, the damping force $c\omega X$, and the effective force $m\omega^2 X$ on the body, we can depict the relationship among the vectors for these forces in steady-state forced vibration as shown in Fig. 20.15.

The force transmitted to the support of the body, as shown in Fig. 20.11(a), is clearly the vector sum of the spring force and the damping force. This transmitted force is denoted by the dashed vector $\mathbf{F}_T$ in Fig. 20.15. Referring to this figure and applying Pythagorean theorem and Eqs. (20.3), (20.33), and (20.67), we write

$$F_T^2 = (kX)^2 + (c\omega X)^2$$

$$= (kX)^2\left[1 + \left(\frac{c\omega}{m}\frac{m}{k}\right)^2\right]$$

$$= (kX)^2[1 + (2\zeta\omega_n\omega/\omega_n^2)^2]$$

$$= \frac{F_0^2}{[1 - (\omega/\omega_n)^2]^2 + [2\zeta(\omega/\omega_n)]^2}\{1 + [2\zeta(\omega/\omega_n)]^2\}$$

$$\frac{F_T}{F_0} = \sqrt{\frac{1 + [2\zeta(\omega/\omega_n)]^2}{[1 - (\omega/\omega_n)^2]^2 + [2\zeta(\omega/\omega_n)]^2}} \qquad (20.71)$$

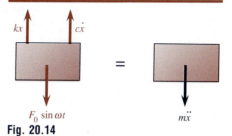

$F_0 \sin\omega t$

Fig. 20.14

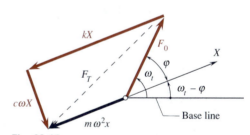

Fig. 20.15

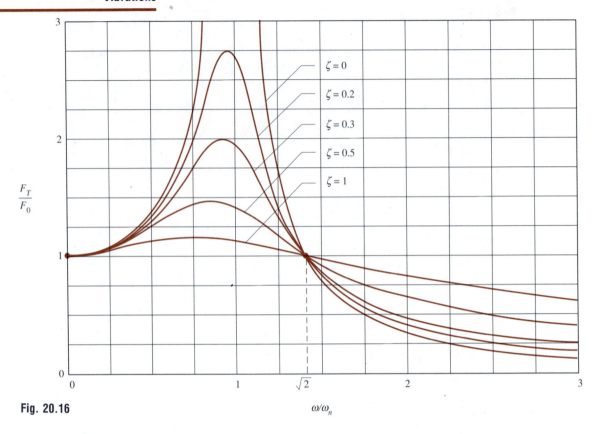

Fig. 20.16

ω/ω_n

The ratio of the maximum value F_T of the transmitted force to the maximum value F_0 of the disturbing force is known as the *transmissibility*.[†] Equation (20.71) is plotted for several values of ζ in Fig. 20.16. Results pertaining to the undamped forced vibration can be obtained by simply setting $\zeta = 0$. In particular, note in Fig. 20.16 that the transmissibility is less than 1 only when the ratio ω/ω_n is greater than $\sqrt{2}$. If $\omega/\omega_n < \sqrt{2}$, more damping results in less transmissibility; however, more damping results in more transmissibility if $\omega/\omega_n > \sqrt{2}$. As expected, the curves show that transmissibility is greatest at resonance.

Besides avoiding resonance, the transmission of vibratory forces generated by machines and other causes can be minimized by a proper isolator design. For instance, the ratio ω/ω_n can be increased to reduce transmissibility by using a softer spring or a larger mass in the system for a given forcing frequency ω.

[†]This is also known as the force transmissibility. A displacement transmissibility is similarly defined in vibration analysis. Cf. Prob. 20.36*.

EXAMPLE 20.6

A piece of equipment with a mass $m = 40$ kg is supported by springs with a total modulus $k = 100$ kN/m and a dashpot as shown. The equipment is acted on by harmonic excitations ranging in frequencies from 30 Hz to 60 Hz. If the isolation of force transmission is known to be 80%, determine the damping factor ζ and the corresponding coefficient of viscous damping c of the dashpot.

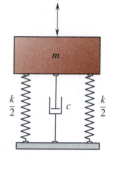

Solution. We know from the given data that $m = 40$ kg and $k = 100$ kN/m $= 10^5$ N/m. Thus, we have

$$\omega_n^2 = \frac{k}{m} = \frac{10^5}{40} \qquad \omega_n = 50 \text{ rad/s}$$

Since the smaller forcing frequency is more likely to incur more transmissibility, we choose to use the 30-Hz frequency to write

$$\omega = 2\pi f = 2\pi(30) \qquad \omega = 60\pi \text{ rad/s}$$

$$\frac{\omega}{\omega_n} = \frac{60\pi}{50} = \frac{6\pi}{5} > \sqrt{2}$$

For 80% isolation, the transmissibility is $1 - 0.8 = 0.2$. Applying Eq. (20.71), we write

$$0.2 = \sqrt{\frac{1 + [2\zeta(6\pi/5)]^2}{[1 - (6\pi/5)^2]^2 + [2\zeta(6\pi/5)]^2}}$$

which can be solved to yield

$$\zeta = 0.331 \quad \blacktriangleleft$$

By Eq. (20.33), we write

$$c = 2\zeta\sqrt{km} = 2(0.331)\sqrt{10^5(40)} = 1324$$

$$c = 1.324 \text{ kN·s/m} \quad \blacktriangleleft$$

Developmental Exercises

D20.15 Define *transmissibility* in forced vibration.

D20.16 Refer to Example 20.6. Determine (a) whether the isolation is increased or decreased if the dashpot is removed, (b) the isolation when the dashpot is removed.

PROBLEMS

20.32 For the system shown, determine the amplitude of the steady-state response if (a) $c = 10$ lb·s/ft, (b) $c = 0$.

20.33 For the system shown, determine the magnification factor if (a) $c = 5$ lb·s/ft, (b) $c = 0$.

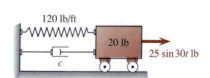

Fig. P20.32 and P20.33

20.34 A harmonic force $F_0 \sin \omega t$ acts on a given spring-mass-dashpot system, where F_0 is constant. When the ratio ω/ω_n is increased from 1 to 2, the amplitude of the steady-state response is observed to decrease by 80%. Determine the damping factor ζ of the system.

20.35* Verify that the abscissa and ordinate of the resonant peak in Fig. 20.12 are, respectively,

$$[(1 + 8\zeta^2)^{1/2} - 1]^{1/2}/(2\zeta) \quad \text{and}$$

$$2\sqrt{2}\zeta^2/[(1 + 8\zeta^2)^{1/2} - 1 - 4\zeta^2 + 8\zeta^4]^{1/2}$$

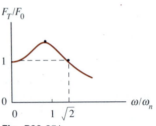

Fig. P20.35*

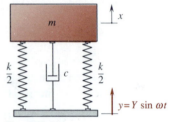

Fig. P20.36*

20.36* The base of a viscously damped spring-mass system is given a harmonic excitation of $y = Y \sin \omega t$, and the displacement x of the mass m is measured from an equilibrium position. Show that (a) the equation of motion of the mass is given by

$$m\ddot{x} + c\dot{x} + kx = Y\sqrt{k^2 + c^2\omega^2}\,\sin(\omega t + \beta)$$

where $\beta = \tan^{-1}(c\omega/k)$, (b) the ratio of the steady-state amplitude X of the mass to the amplitude Y of the excitation, called the *displacement transmissibility,* is identical with the expression for the transmissibility F_T/F_0 in Eq. (20.71); i.e.,

$$\frac{X}{Y} = \sqrt{\frac{1 + [2\zeta(\omega/\omega_n)]^2}{[1 - (\omega/\omega_n)^2]^2 + [2\zeta(\omega/\omega_n)]^2}}$$

20.37 A spring-dashpot-supported vehicle travels at $v = 60$ mi/h over a sinusoidal road surface as shown, where $W = 2000$ lb, $k = 1000$ lb/in., and it is critically damped. Derive the equation of motion of W and determine the steady-state amplitude X of W.

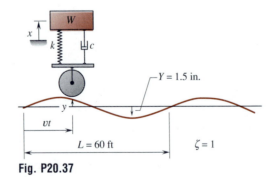

Fig. P20.37

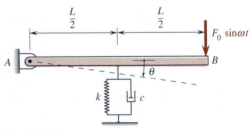

Fig. P20.38

20.38 A slender rod of mass m is subjected to a harmonic excitation as shown. Assuming small-amplitude vibration, determine the steady-state angular amplitude and phase angle of the rod in terms of the parameters of the system.

*20.7 Concluding Remarks

Except in finding the *period* of orbit in central-force motion in the latter part of Chap. 12, we were mostly concerned with the motion and the force system associated with the motion of particles and rigid bodies *at a given instant*, rather than over a period of time, in preceding chapters. In this chapter, we turn our attention to the study of the vibratory motion of a system *over a period of time*, rather than just at a given instant.

The analysis of vibrations involves, among other things, the derivation of the equation of motion valid for anytime during the period of motion of the system. There are several methods which can be used to accomplish this task. For fundamental study, two methods are used; they are (a) the *method of force and acceleration* for both conservative and nonconservative systems, and (b) the *energy method* for conservative systems. In either method, an inertial reference frame must be used in writing the velocity and acceleration of the body. In the energy method, special care should be exercised in computing the potential energy with respect to a properly chosen reference datum. The working definition for the potential energy as given in the text should facilitate the computation. Once the equation of motion for a particular system is established and written in the "standard" form, the natural frequency and other characteristics of the steady-state vibration can usually be·ascertained from the coefficients of that equation.

Excessive vibrations and damping in mechanical systems are generally undesirable because of the noise and energy dissipation which accompany them. With an understanding of the vibratory behavior of the system, engineers can eliminate or reduce them by appropriately designing the parameters of the system. The increasing use of high-speed machines and high-strength structures in modern engineering certainly makes the analysis of vibrations increasingly important.

REVIEW PROBLEMS

20.39   The 20-kg flywheel shown has a radius of gyration of 300 mm about its mass center G. Determine its period of small-amplitude vibration.

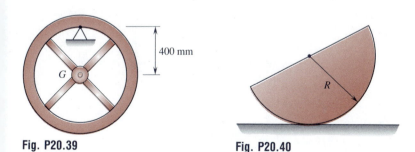

Fig. P20.39   **Fig. P20.40**

20.40   Determine the frequency of small-amplitude vibration of the hemisphere shown, where $R = 300$ mm.

20.41   A section of semicircular shell of mass m and radius r rests at the bottom of a cylindrical surface of radius R as shown. If the shell rocks without slipping on the surface, determine its period of small-amplitude vibration.

20.42   Determine the weight W of a cylinder, as shown, whose period of vibration is 0.2 s.

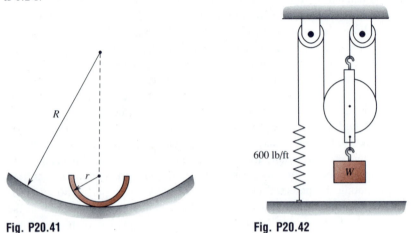

Fig. P20.41   **Fig. P20.42**

20.43   The right end of the connecting spring in the system shown is given a harmonic motion $x_2 = X_2 \sin\omega t$, and the position of the mass m is defined by the coordinate x_1. Derive the equation of motion and determine the steady-state amplitude and the phase angle of the mass in terms of the parameters of the system.

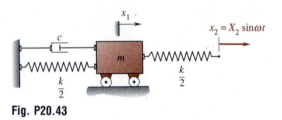

Fig. P20.43

20.44   A slender bar of mass m is subjected to a harmonic excitation as shown. Assuming small-amplitude vibration, determine the steady-state angular amplitude and phase angle of the bar in terms of the parameters of the system.

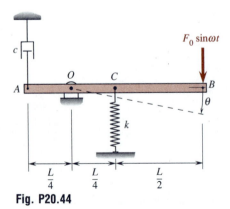

Fig. P20.44

Prior Basic Mathematics   Appendix A

Prior mathematical skills needed for studying Chaps. 1 through 5 involve an ability to perform calculation with a calculator and a proficiency in basic algebra, geometry, and trigonometry. However, some prior background in basic analytic geometry and elementary calculus will be needed starting with Chap. 6. The following problems are designed to reflect the typical prior skills required. Answers to these problems are provided at the end of the text. Some review materials on the aforementioned basic mathematics are presented in App. B.

Choose the correct or best item to complete the sentence or answer the question in each of the following:

1. The solution for x in the equation $3x + 3 = -2$ is
   (a) 3/5.   (b) $-3/5$.   (c) 5/3.
   (d) $-5/3$.   (e) -1.

2. The simultaneous equations $2x + 5y = 21$ and $-3x + y = 11$ have a solution given by
   (a) $(-2, 5)$.   (b) $(-5, -2)$.   (c) $(2, 5)$.
   (d) $(-5, 2)$.   (e) $(-2, -5)$.

3. The roots of the quadratic equation $x^2 + x - 12 = 0$ are
   (a) -1 and 12.   (b) 1 and -12.   (c) -2 and 6.
   (d) 3 and -4.   (e) 3 and 4.

4. The roots of the quadratic equation $x^2 + x - 9 = 0$ are
   (a) -1 and 9.   (b) 1 and -9.   (c) 2.54 and -3.54.
   (d) 5.08 and -7.08.   (e) -5.08 and 7.08.

5. Expansion of $(a + b)^3$ yields
   (a) $a^3 + b^3$.   (b) $a^3 - 2a^2b + 2ab^2 - b^3$.
   (c) $a^3 + 2a^2b + 2ab^2 + b^3$.   (d) $a^3 - 3a^2b + 3ab^2 - b^3$.
   (e) $a^3 + 3a^2b + 3ab^2 + b^3$.

6. The term $(x^2)^3$ is equivalent to
   (a) x^5.   (b) x^2.   (c) x^6.
   (d) x^3.   (e) $x^{2.5}$.

7. The term $(a/b)^{-3/2}$ is equivalent to
   (a) $(b/a)^{3/2}$.   (b) $-a^3/b^2$.   (c) $a^{3/2}/b^{3/2}$.
   (d) $\sqrt{(a/b)^3}$.   (e) $-\sqrt{(a/b)^3}$.

8. The term $\frac{1}{2} \ln (a^{1/3})$ is equivalent to
   (a) $\ln (a^{5/6})$.   (b) $\ln (a^{6/5})$.   (c) $\frac{1}{3} \ln (a^{1/2})$.
   (d) $\ln (a^{2/3})$.   (e) $\ln (a^{3/2})$.

9. The term $\frac{1}{3}\ln(a/b)$ is equivalent to
   (a) $\ln(a^{1/3}b)$. (b) $\ln(a^{1/3}) + \ln(b^{1/3})$. (c) $\frac{1}{3}(\ln a + \ln b)$.
   (d) $\frac{1}{3}(\ln a)(\ln b)$. (e) $\frac{1}{3}(\ln a - \ln b)$.

10. If $3.45^x = 8$, then the value of x is
    (a) 1.679. (b) 1.583. (c) 1.402. (d) 1.338.
    (e) 1.238.

11. The value of the determinant $\begin{vmatrix} 5 & -2 \\ 8 & -2 \end{vmatrix}$ is

    (a) 26. (b) -26. (c) 9. (d) 6. (e) -6.

12. The value of the determinant $\begin{vmatrix} 5 & -7 & 9 \\ 2 & 0 & 0 \\ 6 & -4 & 1 \end{vmatrix}$ is

    (a) -14. (b) -54. (c) 54. (d) 58. (e) -58.

13. The value of $[6.44 + (4.75)^{1/3}(e^{0.4\pi})\ln 3.65 - 3\sin 35° \cot 215°]$ is
    (a) 7.30. (b) 8.90. (c) 11.63 (d) 12.88. (e) 16.54.

Fig. P14

14. The shortest distance from the point A to the line which is tangent to the circle at the point B, as shown, is
    (a) 9.5 m. (b) 9 m. (c) 8.5 m. (d) 8 m. (e) 7.5 m.

Fig. P15

15. The length of the line segment $\overline{AB}$, as shown, is
    (a) 5.29 ft. (b) 5.66 ft. (c) 6.29 ft. (d) 6.66 ft.
    (e) 6.92 ft.

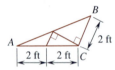
Fig. P16

16. The area of the triangle ABC shown is
    (a) 9 m^2. (b) 12 m^2. (c) 18 m^2. (d) 21 m^2. (e) 24 m^2.

17. The area of the trapezoid shown is
    (a) 28 in^2. (b) 42 in^2. (c) 48 in^2. (c) 63 in^2. (e) 84 in^2.

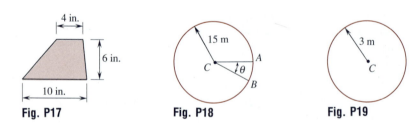

Fig. P17 **Fig. P18** **Fig. P19**

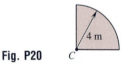
Fig. P20

18. If $\theta = 3/5$ rad, the length of the arc AB, as shown, is
    (a) 0.6 m. (b) 1.667 m. (c) 9 m. (d) 12 m. (e) 8 m.

19. The circumference of the circle shown is
    (a) 3π m^2. (b) 6π m^2. (c) 9π m^2. (d) 3π m. (e) 6π m.

20. The area of the quarter circular sector shown is
    (a) 2π m^2. (b) 4π m^2. (c) 8π m^2. (d) 2π m. (e) 4π m.

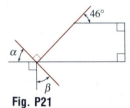
Fig. P21

21. In the figure shown, the angle α is
    (a) 44°. (b) 45°. (c) 46°. (d) 47°. (e) 60°.

22. The length of the side $\overline{BC}$ of the right triangle shown is
    (a) 8 ft.   (b) 6 ft.   (c) 7.5 ft.   (d) 4.12 ft.   (e) 8.5 ft.

23. In the triangle shown, $\tan\theta$ is equal to
    (a) 3/5.   (b) 5/3.   (c) 3/4.   (d) 4/3.   (e) 4/5.

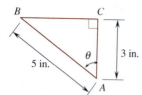

Fig. P23

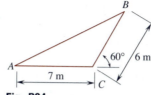

Fig. P24

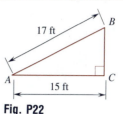

Fig. P22

24. The length of the side $\overline{AB}$ of the triangle shown is
    (a) 13 m.   (b) 11.27 m.   (c) 9.22 m.   (d) 8.95 m.
    (e) 6.48 m.

25. The length of the side $\overline{AB}$ of the right triangle shown is
    (a) 12.6 m.   (b) 11.8 m.   (c) 10.5 m.   (d) 9.24 m.
    (e) 8.51 m.

26. The angle θ of the right triangle shown is
    (a) 35.0°.   (b) 36.9°.   (c) 44.4°.   (d) 45.6°.   (e) 53.1°.

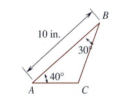

Fig. P25

Fig. P26

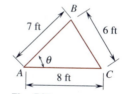

Fig. P27

27. The angle θ of the triangle shown is
    (a) 0.785 rad.   (b) 0.813 rad.   (c) 1.05 rad.
    (d) 1.57 rad.   (e) 3.14 rad.

28. The length of the side $\overline{AC}$ of the triangle shown is
    (a) 5.00 in.   (b) 5.08 in.   (c) 5.32 in.   (d) 5.53 in.
    (e) 5.77 in.

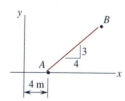

Fig. P28

29. The term $\cos^2\theta$ is equivalent to
    (a) $(1 - \sin 2\theta)/2$.   (b) $(1 - \cos 2\theta)/2$.   (c) $(1 + \sin 2\theta)/2$.
    (d) $(1 + \cos 2\theta)/2$.   (e) $-\sin 2\theta$.

30. The expression $1 - \sin^2\theta$ is equivalent to
    (a) $\tan^2\theta$.   (b) $\sec^2\theta$.   (c) $1 + \cot^2\theta$.   (d) $\cos^2\theta - 1$.
    (e) $\cos^2\theta$.

31. The equation of the line AB shown is
    (a) $3x - 4y = 0$.   (b) $4x - 3y = 0$.
    (c) $3x - 4y + 12 = 0$.   (d) $3x - 4y - 12 = 0$.
    (e) $4x - 3y + 12 = 0$.

32. The length of the line segment $\overline{AB}$ shown is
    (a) 22 m.   (b) 24 m.   (c) 26 m.   (d) 28 m.   (e) 38 m.

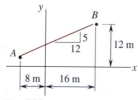

Fig. P31

Fig. P32

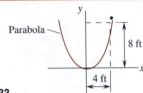

Fig. P33

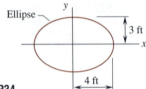

Fig. P34

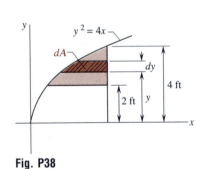

Fig. P38

33. The equation of the parabola shown is
   (a) $y^2 = 16x$. (b) $x^2 = 2y$. (c) $x^2 = y + 8$.
   (d) $y^2 = x + 60$. (e) $4y = 8x$.

34. The equation of the ellipse shown is
   (a) $\dfrac{x}{4} + \dfrac{y}{3} = 1$. (b) $\dfrac{x}{4} - \dfrac{y}{3} = 1$. (c) $\dfrac{x^2}{4} + \dfrac{y^2}{3} = 1$.
   (d) $\dfrac{x^2}{16} - \dfrac{y^2}{9} = 1$. (e) $\dfrac{x^2}{16} + \dfrac{y^2}{9} = 1$.

35. Which of the following represents an ellipse in polar coordinates?
   (a) $\dfrac{1}{r} = 3\cos\theta$. (b) $\dfrac{1}{r} = 3\theta$. (c) $\dfrac{1}{r} = 3 + 3\cos\theta$.
   (d) $\dfrac{1}{r} = 3 + 4\cos\theta$. (e) $\dfrac{1}{r} = 4 + 3\cos\theta$.

36. The differential area element dA in polar coordinates, as shown, is
   (a) $u\,du\,d\theta$. (b) $\theta\,du$. (c) $du\,d\theta$. (d) $u\,d\theta$. (e) $dx\,dy$.

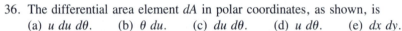

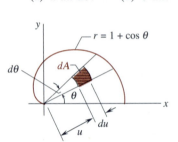

Fig. P36

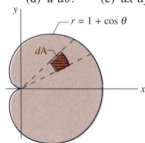

Fig. P37

37. The area enclosed by the cardioid shown is given by
   (a) $\displaystyle\int_0^{2\pi} (1 + \cos\theta)\,d\theta$. (b) $\displaystyle 2\int_{\theta=0}^{\pi}\int_{u=0}^{1+\cos\theta} du\,d\theta$.
   (c) $\displaystyle\int_{\theta=0}^{2\pi}\int_{u=0}^{1+\cos\theta} du\,d\theta$. (d) $\displaystyle 2\int_{\theta=0}^{\pi}\int_{u=0}^{1+\cos\theta} u\,du\,d\theta$.
   (e) $\displaystyle\int_{\theta=0}^{2\pi}\int_{u=0}^{1} (1 + \cos\theta)\,du\,d\theta$.

38. The differential area element dA, as shown, is
   (a) $(y^2/4)\,dy$. (b) $2x^{1/2}\,dx$. (c) $[(y^2 - 16)/4]\,dy$.
   (d) $[(16 - y^2)/4]\,dy$. (e) $(4 - y)\,dy$.

39. If $\dfrac{dx}{dt} > 0$, which of the following statements is *always* true?
   (a) The magnitude of x is increasing.
   (b) The algebraic value of x is increasing.
   (c) The magnitude of x is decreasing.
   (d) The algebraic value of x is decreasing.
   (e) The value of x is positive.

40. The slope of the normal to the curve $2x^2 + 9y = 0$ at the point $(3, -2)$ is
   (a) 3/4. (b) 4/3. (c) −4/3. (d) −3/4. (e) 3/5.

41. The radius of curvature of the curve $2x^2 + 9y = 0$ at the point $(3, -2)$ is
    (a) 0.096.   (b) 1.25.   (c) 3.75.   (d) 10.42.   (e) 15.52.

42. The differential area element dA, as shown, is
    (a) $2x^{1/2}\, dx$.   (b) $\frac{1}{4} y^2\, dy$.   (c) $(2x^{1/2} - 2)\, dx$.
    (d) $(4 - 2x^{1/2})\, dx$.   (e) $(2x^{1/2} - 4)\, dx$.

43. The area of the shaded region shown is given by
    (a) $\int_1^3 2y^{1/2}\, dy$.   (b) $\int_1^4 2y^{1/2}\, dy$.   (c) $\int_0^4 \frac{x^2}{4}\, dx$.
    (d) $\int_2^4 \frac{x^2}{4}\, dx$.   (e) $\int_1^3 (4 - y)\, dy$.

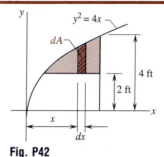

Fig. P42

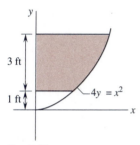

Fig. P43

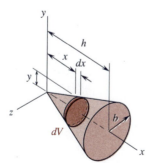

Fig. P44

44. The area of the shaded region shown is given by
    (a) $\int_0^4 \left(\frac{1}{4} x^2 - 2x^{1/2} \right) dx$.   (b) $\int_0^4 \frac{1}{4} x^2\, dx$.

    (c) $\int_0^4 \left(2y^{1/2} - \frac{1}{4} y^2 \right) dy$.   (d) $\int_0^4 2y^{1/2}\, dy$.   (e) $\int_0^4 x\, dy$.

45. The volume of the right circular cone shown is given by
    (a) $\int_0^h (2\pi b/h)x\, dx$.   (b) $\int_0^h 2\pi x\, dx$.   (c) $\int_0^h \pi x^2\, dx$.
    (d) $\int_0^h (\pi h^2/b^2)x^2\, dx$.   (e) $\int_0^h (\pi b^2/h^2)x^2\, dx$.

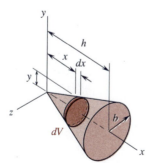

Fig. P45

46. The value of $0.234 \int_1^9 (2x - x^{-1} - x^{-1/2})\, dx$ is
    (a) 76.0.   (b) 73.8.   (c) 17.78.   (d) 17.27.   (e) -2.20.

47. Which of the following is *false*?
    (a) $\sin^2 x + \cos^2 x = 1$.   (b) $\sinh^2 x + \cosh^2 x = 1$.
    (c) $\sec^2 x - \tan^2 x = 1$.   (d) $\text{sech}^2 x + \tanh^2 x = 1$.
    (e) $\csc^2 x - \cot^2 x = 1$.

48. Which of the following represents the shape of the curve defined by $y = \cosh x$:

(a)   (b)   (c)   (d)   (e)

Fig. P48

49. If $e^x = y + (1 + y^2)^{1/2}$, then y is equal to
    (a) $e^x + e^{-x}$.   (b) $e^x - e^{-x}$.   (c) $\tanh x$.
    (d) $\cosh x$.   (e) $\sinh x$.

50. The value of $\displaystyle\int_0^1 \sinh x\, dx$ is
    (a) -1.543.   (b) -0.1752.   (c) 0.1752.   (d) 0.543.
    (e) 1.175.

Review of Basic Mathematics

B1   ALGEBRA

B1.1   Fundamental Principles

The following principles allow us to solve a great many equations:

1. *Addition principle*. If the equation $a = b$ is true, then $a + c = b + c$ is true, for any number or expression c.
2. *Multiplication principle*. If the equation $a = b$ is true, then $ac = bc$ is true for any number or expression c.
3. *Principle of zero products*. For any numbers or expressions a and b, if $ab = 0$, then $a = 0$ or $b = 0$, and if $a = 0$ or $b = 0$, then $ab = 0$.
4. *Principle of powers*. If the equation $a = b$ is true, then $a^n = b^n$ is true for any number n.

B1.2   Roots of a Quadratic Equation

If $ax^2 + bx + c = 0$ and $a \neq 0$, then

$$x = \frac{-b \pm \sqrt{b^2 - 4ac}}{2a}$$

B1.3   Solution of n Linear Simultaneous Equations

A set of n linear simultaneous equations in n unknowns $x_1, x_2, \ldots, x_n$ may be written in the form

$$
\begin{aligned}
a_{11}x_1 + a_{12}x_2 + \cdots + a_{1n}x_n &= b_1 \quad &(1) \\
a_{21}x_1 + a_{22}x_2 + \cdots + a_{2n}x_n &= b_2 \quad &(2) \\
\cdots \qquad \cdots \qquad \cdots \qquad \cdots \qquad \cdots & &\cdots \\
a_{n1}x_1 + a_{n2}x_2 + \cdots + a_{nn}x_n &= b_n \quad &(n)
\end{aligned}
$$

If these n equations are independent, nonhomogeneous, and consistent, then they have a unique nontrivial solution. Such a solution may be obtained by the following steps:

1. From one of the n given equations, express the unknown x_1 in terms of the remaining $n - 1$ unknowns $x_2, x_3, \ldots, x_n$.
2. Substitute the expression of x_1 obtained in step (1) into the remaining $n - 1$ equations to obtain a set of $n - 1$ linear simultaneous equations in $n - 1$ unknowns $x_2, x_3, \ldots, x_n$.
3. From one of the $n - 1$ equations obtained in step (2), express the unknown x_2 in terms of the remaining $n - 2$ unknowns $x_3, x_4, \ldots, x_n$.
4. Substitute the expression of x_2 obtained in step (3) into the remaining $n - 2$ equations to obtain a set of $n - 2$ linear simultaneous equations in $n - 2$ unknowns $x_3, x_4, \ldots, x_n$.
5. Continue the reduction process similar to the above four steps until the size of the set is reduced to one linear equation in one unknown.
6. Solve for the single unknown in the last equation and perform the back substitutions of determined values into the expressions of other unknowns to obtain the values of all other unknowns.

B1.4   Second- and Third-Order Determinants

Second- and third-order determinants can be expanded by special schemes as follows:

$$= aei + bfg + cdh - ceg - bdi - afh$$

Note that the sign to be used with each individual product is indicated at the tail of each individual arrow. Furthermore, the above schematic methods of expansion *cannot* be used with determinants of order higher than three. However, any determinant may be expanded by using the cofactors of its row elements or column elements. Using the cofactors of the elements of the first row of a third-order determinant, we write

$$\begin{vmatrix} a & b & c \\ d & e & f \\ g & h & i \end{vmatrix} = a\begin{vmatrix} e & f \\ h & i \end{vmatrix} - b\begin{vmatrix} d & f \\ g & i \end{vmatrix} + c\begin{vmatrix} d & e \\ g & h \end{vmatrix}$$

$$= a(ei - fh) - b(di - fg) + c(dh - eg)$$

If *any* two rows or columns of a determinant are interchanged, the value of the determinant will change its sign. For example, we write

$$\begin{vmatrix} a & b & c \\ d & e & f \\ g & h & i \end{vmatrix} = -\begin{vmatrix} d & e & f \\ a & b & c \\ g & h & i \end{vmatrix} = \begin{vmatrix} d & e & f \\ g & h & i \\ a & b & c \end{vmatrix} = -\begin{vmatrix} g & h & i \\ d & e & f \\ a & b & c \end{vmatrix}$$

Thus, by interchanging the rows twice, we write

$$\begin{vmatrix} a & b & c \\ d & e & f \\ g & h & i \end{vmatrix} = \begin{vmatrix} d & e & f \\ g & h & i \\ a & b & c \end{vmatrix} = \begin{vmatrix} g & h & i \\ a & b & c \\ d & e & f \end{vmatrix}$$

This property is useful in studying the volume of a parallelepiped using the scalar triple product in Sec. 4.15.

B1.5 Binomials

(a) $a^2 - b^2 = (a + b)(a - b)$
(b) $a^3 - b^3 = (a - b)(a^2 + ab + b^2)$
(c) $(a + b)^2 = a^2 + 2ab + b^2$
(d) $(a - b)^2 = a^2 - 2ab^2 + b^2$
(e) $(a + b)^3 = a^3 + 3a^2b + 3ab^2 + b^3$
(f) $(a - b)^3 = a^3 - 3a^2b + 3ab^2 - b^3$
(g) $(a + b)^n = a^n + na^{n-1}b + \dfrac{n(n - 1)}{2!}a^{n-2}b^2$

$$+ \frac{n(n - 1)(n - 2)}{3!}a^{n-3}b^3 + \cdots + b^n$$

(binomial theorem)

B1.6 Exponents

(a) $A^a A^b = A^{a+b}$
(b) $(A^a)^b = A^{ab}$
(c) $A^{-a} = \dfrac{1}{A^a}$
(d) $A^{a/b} = (A^a)^{1/b} = \sqrt[b]{A^a}$

B1.7   Logarithms

The logarithm of a number N to a base b is the exponent x indicating the power to which b must be raised to produce the number N. Thus, we write $\log_b N = x$ if $b^x = N$. The base of the common logarithm is 10, while that of the natural logarithm is e, where

$$e = 1 + \frac{1}{1!} + \frac{1}{2!} + \frac{1}{3!} + \frac{1}{4!} + \cdots = 2.718281828 \cdots$$

The natural logarithm is usually abbreviated as ln. We have

(a) $\ln (AB) = \ln A + \ln B$
(b) $\ln (A/B) = \ln A - \ln B$
(c) $\ln (A^a) = a \ln A$
(d) $A^a = e^{a \ln A}$

B2   GEOMETRY

B2.1   Useful Postulates and Theorems

1. *Vertical angles*: Vertical angles formed by two intersecting lines are congruent. In Fig. [a], we have $\theta_1 = \theta_3$ and $\theta_2 = \theta_4$.
2. *Corresponding angles*: When a transversal intersects two parallel lines, corresponding angles are congruent, and conversely. In Fig. [b], we have $\theta_1 = \theta_3$ and $\theta_2 = \theta_4$.
3. *Angle sum for triangles*: The sum of the measures of the angles of a triangle is 180°. In Fig. [c], we have $\theta_1 + \theta_2 + \theta_3 = 180°$.
4. *Pythagorean theorem*: In a right triangle, the square of the length of its hypotenuse is equal to the sum of the squares of the lengths of its legs.[†] In Fig. [d], we have $c^2 = a^2 + b^2$.
5. *Congruent angles*: If the two lines including an angle are respectively perpendicular to the two lines including another angle, then these two angles must be either congruent angles or supplementary angles. In Fig. [e], we have $\theta_1 = \theta_2$ and $\theta_1 + \theta_3 = 180°$.

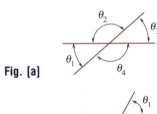

Fig. [a]

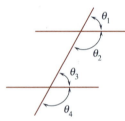

Fig. [b]

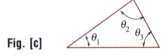

Fig. [c]

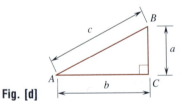

Fig. [d]

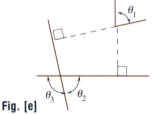

Fig. [e]

Fig. [f]

Center of curve

6. *Center of curve*: A line drawn perpendicular to the tangent to a curve and passing through the point of tangency will also pass through the

---

[†]The general formula $[\frac{1}{2}(m^2 + 1)]^2 = m^2 + [\frac{1}{2}(m^2 - 1)]^2$, where m is any odd integer, is attributed to the Greek philosopher and mathematician Pythagoras (ca. 580–500 B.C.).

center of the curve at that point. In Fig. [f], the point of tangency is P and $\overline{CP} \perp \overline{AB}$.

7. *Shortest distance*: The shortest distance from a point to a line is equal to the length of the perpendicular segment drawn from the point to the line. In Fig. [g], d_s is the shortest distance from the point P to the line AB.

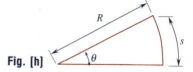

Fig. [g]

B2.2 Radian Measure

One radian is defined as the angle which subtends a circular arc whose length is equal to the length of the radius of the circular arc. As shown in Fig. [h], we have

$$\theta = \frac{s}{R} \qquad s = R\theta$$

where θ is measured in radians.

Fig. [h]

B2.3 Mensuration Formulas for Some Common Shapes

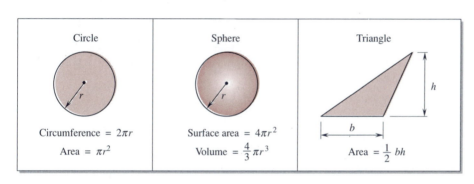

Circle	Sphere	Triangle
Circumference $= 2\pi r$	Surface area $= 4\pi r^2$	
Area $= \pi r^2$	Volume $= \frac{4}{3}\pi r^3$	Area $= \frac{1}{2}bh$

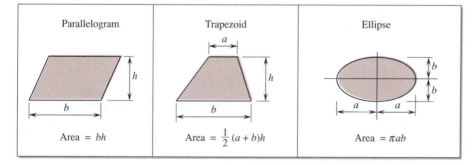

Parallelogram	Trapezoid	Ellipse
Area $= bh$	Area $= \frac{1}{2}(a+b)h$	Area $= \pi ab$

B3 TRIGONOMETRY

B3.1 Definitions with Mnemonics

Let a right triangle be oriented as shown in Fig. [a], where the angle θ is on the left and the right angle is on the right. Then, the values of $\sin\theta$,

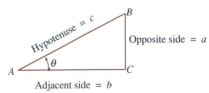

Fig. [a]

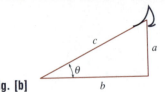

Fig. [b]

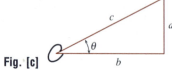

Fig. [c]

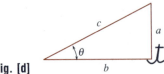

Fig. [d]

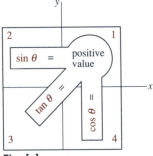

Fig. [e]

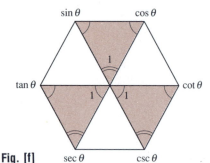

Fig. [f]

$\cos\theta$, and $\tan\theta$ are, respectively, equal to those obtained by *dividing the first side touched into the second side touched* when the letters *s*, *c*, and *t* are written in *script* on the triangle as shown in Figs. [b], [c], and [d]. We have

$$\sin\theta = \frac{a}{c} = \frac{\text{Opposite side}}{\text{Hypotenuse}} = \theta - \frac{\theta^3}{3!} + \frac{\theta^5}{5!} - \frac{\theta^7}{7!} + \cdots \ (\theta \text{ in rad})$$

$$\cos\theta = \frac{b}{c} = \frac{\text{Adjacent side}}{\text{Hypotenuse}} = 1 - \frac{\theta^2}{2!} + \frac{\theta^4}{4!} - \frac{\theta^6}{6!} + \cdots \ (\theta \text{ in rad})$$

$$\tan\theta = \frac{a}{b} = \frac{\text{Opposite side}}{\text{Adjacent side}} = \theta + \frac{\theta^3}{3} + \frac{2\theta^5}{15} + \frac{17\theta^7}{315} + \cdots \ (\theta \text{ in rad})$$

B3.2 Some Useful Properties with Mnemonics

The quadrants in which the trigonometric functions have positive values may be indicated as shown in Fig. [e]. Furthermore, several trigonometric identities may readily be written from the hexagon shown in Fig. [f]. The mnemonic rules associated with Fig. [f] are as follows:[†]

1. The product of the functions at the ends of each diagonal is equal to 1. We have

$$\sin\theta \csc\theta = 1 \qquad \cos\theta \sec\theta = 1 \qquad \tan\theta \cot\theta = 1$$

2. Any function is equal to the product of its two neighboring functions. We have

$$\sin\theta = \cos\theta \tan\theta \qquad\qquad \csc\theta = \sec\theta \cot\theta$$

$$\cos\theta = \sin\theta \cot\theta \qquad\qquad \sec\theta = \tan\theta \csc\theta$$

$$\tan\theta = \sin\theta \sec\theta \qquad\qquad \cot\theta = \cos\theta \csc\theta$$

3. In each of the three shaded triangles, the square of the quantity at the lower angle is equal to the sum of the squares of the quantities at the upper two angles. We have

$$\sin^2\theta + \cos^2\theta = 1$$

$$\tan^2\theta + 1 = \sec^2\theta \qquad 1 + \cot^2\theta = \csc^2\theta$$

B3.3 Other Useful Trigonometric Identities

(a) $\sin(\alpha \pm \beta) = \sin\alpha \cos\beta \pm \cos\alpha \sin\beta$

(b) $\cos(\alpha \pm \beta) = \cos\alpha \cos\beta \mp \sin\alpha \sin\beta$

(c) $\tan(\alpha \pm \beta) = \dfrac{\tan\alpha \pm \tan\beta}{1 \mp \tan\alpha \tan\beta}$

(d) $\sin^2\theta = \frac{1}{2}(1 - \cos 2\theta)$

[†]Note that the order in which the trigonometric functions appear on the hexagon is that $\sin\theta$ and $\cos\theta$ are in the first row, $\tan\theta$ and $\cot\theta$ are in the second row, and $\sec\theta$ and $\csc\theta$ are in the third row as shown.

(e) $\cos^2\theta = \frac{1}{2}(1 + \cos 2\theta)$

(f) $\sin\alpha \pm \sin\beta = 2 \sin\frac{1}{2}(\alpha \pm \beta) \cos\frac{1}{2}(\alpha \mp \beta)$

(g) $\cos\alpha + \cos\beta = 2 \cos\frac{1}{2}(\alpha + \beta) \cos\frac{1}{2}(\alpha - \beta)$

(h) $\cos\alpha - \cos\beta = -2 \sin\frac{1}{2}(\alpha + \beta) \sin\frac{1}{2}(\alpha - \beta)$

(i) $\sin\alpha \cos\beta = \frac{1}{2}[\sin(\alpha + \beta) + \sin(\alpha - \beta)]$

(j) $\sin\alpha \sin\beta = \frac{1}{2}[\cos(\alpha - \beta) - \cos(\alpha + \beta)]$

(k) $\cos\alpha \cos\beta = \frac{1}{2}[\cos(\alpha + \beta) + \cos(\alpha - \beta)]$

B3.4   Law of Sines

For a triangle with angles and sides as shown in Fig. [g], we may write the law of sines as follows:

$$\frac{a}{\sin\alpha} = \frac{b}{\sin\beta} = \frac{c}{\sin\gamma}$$

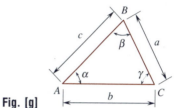

Fig. [g]

B3.5   Law of Cosines

For a triangle with angles and sides as shown in Fig. [g], we may write the law of cosines as follows:

$$a^2 = b^2 + c^2 - 2bc\,\cos\alpha \qquad b^2 = c^2 + a^2 - 2ca\,\cos\beta$$

$$c^2 = a^2 + b^2 - 2ab\,\cos\gamma$$

B3.6   Inverse Functions

(a) $\alpha = \sin^{-1} x$ if $x = \sin\alpha$

(b) $\theta = \csc^{-1} u$ if $u = \csc\theta$

(c) $\beta = \cos^{-1} y$ if $y = \cos\beta$

(d) $\phi = \sec^{-1} v$ if $v = \sec\phi$

(e) $\gamma = \tan^{-1} z$ if $z = \tan\gamma$

(f) $\psi = \cot^{-1} w$ if $w = \cot\psi$

B4   ANALYTIC GEOMETRY

B4.1   Equation of a Straight Line

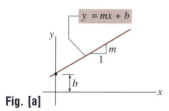

Fig. [a]

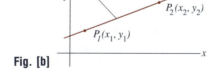

Fig. [b]

B4.2   Equations of Conic Sections

A *conic section* is any curve which is the locus of a point that moves in such a way that the ratio of its distance from a fixed point to its distance from a fixed line is constant. This ratio is called the *eccentricity*, the fixed point the *focus*, and the fixed line the *directrix* of the curve. These are illustrated as shown, where ε = *eccentricity*.

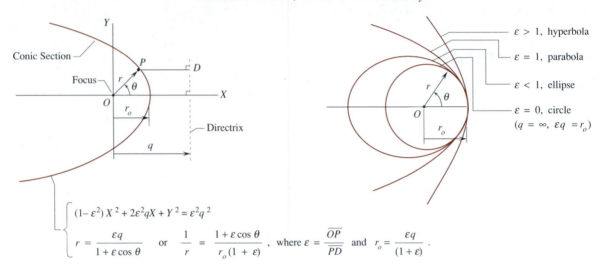

$\varepsilon > 1$, hyperbola

$\varepsilon = 1$, parabola

$\varepsilon < 1$, ellipse

$\varepsilon = 0$, circle
$(q = \infty,\ \varepsilon q = r_o)$

$$(1 - \varepsilon^2)\,X^2 + 2\varepsilon^2 qX + Y^2 = \varepsilon^2 q^2$$

$$r = \frac{\varepsilon q}{1 + \varepsilon \cos \theta} \quad \text{or} \quad \frac{1}{r} = \frac{1 + \varepsilon \cos \theta}{r_o(1 + \varepsilon)}, \quad \text{where } \varepsilon = \frac{\overline{OP}}{\overline{PD}} \ \text{ and } \ r_o = \frac{\varepsilon q}{(1 + \varepsilon)}.$$

B4.3   Equation of a Circle

Note that $\varepsilon = 0$, $q = \infty$, and $\varepsilon q = R$.

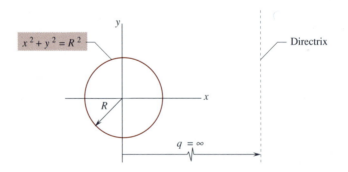

$x^2 + y^2 = R^2$

Directrix

R

$q = \infty$

B4.4   Equation of an Ellipse[†]

Note that $a^2 - b^2 = c^2$, $\varepsilon = \dfrac{c}{a}$, $b = a(1 - \varepsilon^2)^{1/2}$, and $q = \pm\dfrac{b^2}{c}$.

[†]If a conic section has two foci, the geometric center of the conic section does not lie between a focus and its associated directrix.

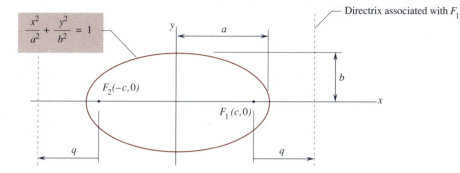

Directrix associated with F_1

$$\frac{x^2}{a^2} + \frac{y^2}{b^2} = 1$$

$F_2(-c,0)$

$F_1(c,0)$

a

b

q

q

B4.5   Equation of a Parabola

Note that $\varepsilon = 1$ and $q = -2c$.

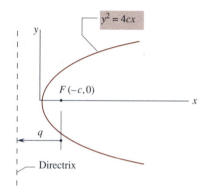

$y^2 = 4cx$

$F(-c,0)$

q

Directrix

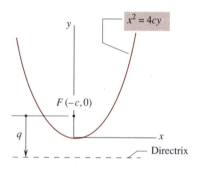

$x^2 = 4cy$

$F(-c,0)$

q

Directrix

B4.6   Equation of a Hyperbola

Note that $a^2 + b^2 = c^2$, $\varepsilon = \dfrac{c}{a}$, $b = a(\varepsilon^2 - 1)^{1/2}$, and $q = \pm\dfrac{b^2}{c}$.

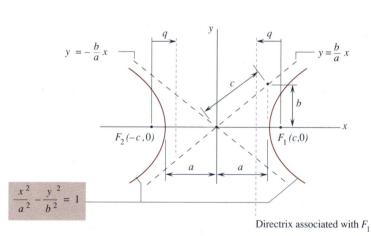

$y = -\dfrac{b}{a}x$

$y = \dfrac{b}{a}x$

q

q

c

b

$F_2(-c,0)$

$F_1(c,0)$

a

a

$$\frac{x^2}{a^2} - \frac{y^2}{b^2} = 1$$

Directrix associated with F_1

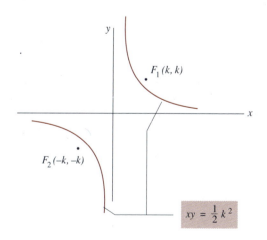

$F_1(k,k)$

$F_2(-k,-k)$

$xy = \dfrac{1}{2}k^2$

B5   CALCULUS

B5.1   Derivatives

(a) $\dfrac{dC}{dx} = 0$ (C is a constant.)

(b) $\dfrac{dx}{dx} = 1$

(c) $\dfrac{dy}{dt} = \dfrac{dy}{dx}\dfrac{dx}{dt}$

(d) $\dfrac{dx^n}{dx} = n\,x^{n-1}$

(e) $\dfrac{de^x}{dx} = e^x$

(f) $\dfrac{da^x}{dx} = (\ln a)\,a^x$

(g) $\dfrac{d}{dx}(\ln x) = \dfrac{1}{x}$

(h) $\dfrac{d}{dx}(\sin x) = \cos x$

(i) $\dfrac{d}{dx}(\cos x) = -\sin x$

(j) $\dfrac{d}{dx}(\tan x) = \sec^2 x$

(k) $\dfrac{d}{dx}(\cot x) = -\csc^2 x$

(l) $\dfrac{d}{dx}(\sec x) = \sec x \tan x$

(m) $\dfrac{d}{dx}(\csc x) = -\csc x \cot x$

(n) $\dfrac{d}{dx}(uv) = \dfrac{du}{dx}v + u\dfrac{dv}{dx}$

(o) $\dfrac{d}{dx}\left(\dfrac{u}{v}\right) = \dfrac{v\dfrac{du}{dx} - u\dfrac{dv}{dx}}{v^2}$

B5.2   Integrals

(a) $\int u\,dv = uv - \int v\,du$ (integration by parts)

(b) $\int x^n\,dx = \dfrac{1}{n+1}x^{n+1}$ $(n \neq -1)$

(c) $\int \dfrac{dx}{x} = \ln x$

(d) $\int \sin x\,dx = -\cos x$

(e) $\int \cos x\,dx = \sin x$

(f) $\int \tan x \, dx = -\ln(\cos x)$

(g) $\int \cot x \, dx = \ln(\sin x)$

(h) $\int \sec x \, dx = \ln(\sec x \tan x)$

(i) $\int \csc x \, dx = -\ln(\csc x + \cot x)$

(j) $\int \sin^n x \, dx = -\frac{1}{n} \sin^{n-1} x \cos x + \frac{1}{n}(n-1) \int \sin^{n-2} x \, dx$

(k) $\int \cos^n x \, dx = \frac{1}{n} \cos^{n-1} x \sin x + \frac{1}{n}(n-1) \int \cos^{n-2} x \, dx$

(l) $\int \sec^n x \, dx = \frac{1}{n-1} \sin x \sec^{n-1} x + \frac{n-2}{n-1} \int \sec^{n-2} x \, dx$

(m) $\int \csc^n x \, dx = -\frac{1}{n-1} \cos x \csc^{n-1} x + \frac{n-2}{n-1} \int \csc^{n-2} x \, dx$

(n) $\int e^x \, dx = e^x$

(o) $\int (a^2 - x^2)^{3/2} \, dx = \frac{1}{8}\left[2x(a^2 - x^2)^{3/2} \right.$
$$\left. + 3a^2 x(a^2 - x^2)^{1/2} + 3a^4 \sin^{-1} \frac{x}{a} \right]$$

(p) $\int (a^2 - x^2)^{-1/2} \, dx = \sin^{-1} \frac{x}{a}$

(q) $\int (x^2 \pm a^2)^{-1/2} \, dx = \ln[x + (x^2 \pm a^2)^{1/2}]$

(r) $\int x^{-1}(x^2 - a^2)^{-1/2} \, dx = \frac{1}{a} \cos^{-1} \frac{a}{x}$

(s) $\int x^2 (a^2 - x^2)^{-1/2} \, dx = \frac{1}{2}\left[-x(a^2 - x^2)^{1/2} + a^2 \sin^{-1} \frac{x}{a} \right]$

(t) $\int (x^2 \pm a^2)^{1/2} \, dx = \frac{1}{2}\{x(x^2 \pm a^2)^{1/2} \pm a^2 \ln[x + (x^2 \pm a^2)^{1/2}]\}$

(u) $\int x^2 (x^2 \pm a^2)^{1/2} \, dx = \frac{1}{8}\{2x(x^2 \pm a^2)^{3/2} \mp a^2 x(x^2 \pm a^2)^{1/2}$
$$- a^4 \ln[x + (x^2 \pm a^2)^{1/2}]\}$$

(v) $\int x^3 (a^2 \pm x^2)^{1/2} \, dx = \frac{1}{15}(\pm 3x^2 - 2a^2)(a^2 \pm x^2)^{3/2}$

(w) $\int (a^2 - x^2)^{1/2} \, dx = \frac{1}{2}\left[x(a^2 - x^2)^{1/2} + a^2 \sin^{-1} \frac{x}{a} \right]$

(x) $\int x(a^2 - x^2)^{1/2} \, dx = -\frac{1}{3}(a^2 - x^2)^{3/2}$

(y) $\int x^2 (a^2 - x^2)^{1/2} \, dx = \frac{1}{8}\left\{ -2x(a^2 - x^2)^{3/2} \right.$
$$\left. + a^2 \left[x(a^2 - x^2)^{1/2} + a^2 \sin^{-1} \frac{x}{a} \right] \right\}$$

(z) $\int x^3 (a^2 - x^2)^{1/2} \, dx = \frac{1}{15}(3x^2 + 2a^2)(a^2 - x^2)^{3/2}$

B5.3   Differential of a Function

For the purpose of this text, the first-order differential (or variation) δy of a function $y = y(x)$ may be obtained from the relationship

$$\delta y = \frac{dy}{dx} \delta x$$

For example, $\delta(x^2) = 2x \, \delta x$, $\delta(\sin x) = (\cos x) \, \delta x$, etc.

B5.4 Curvature κ and Radius of Curvature ρ

(a) In rectangular coordinates:

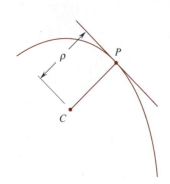

$$\kappa = \frac{\left|\dfrac{d^2y}{dx^2}\right|}{\left[1 + \left(\dfrac{dy}{dx}\right)^2\right]^{3/2}} \qquad \rho = \frac{1}{\kappa}$$

(b) In polar coordinates:

$$\kappa = \frac{\left|r^2 + 2\left(\dfrac{dr}{d\theta}\right)^2 - r\dfrac{d^2r}{d\theta^2}\right|}{\left[r^2 + \left(\dfrac{dr}{d\theta}\right)^2\right]^{3/2}} \qquad \rho = \frac{1}{\kappa}$$

B5.5 Expansion of $f(x)$ into Series

(a) Taylor's series:

$$f(x) = f(a) + (x - a)\,f'(a) + \frac{(x - a)^2}{2!}f''(a) + \cdots$$

$$= \sum_{n=0}^{\infty} \frac{(x - a)^n}{n!}f^{(n)}(a)$$

(b) Maclaurin's series:

$$f(x) = f(0) + x\,f'(0) + \frac{x^2}{2!}f''(0) + \cdots = \sum_{n=0}^{\infty} \frac{x^n}{n!}f^{(n)}(0)$$

B5.6 Hyperbolic Sine and Cosine Functions

(a) $\sinh x = \frac{1}{2}(e^x - e^{-x}) = x + \dfrac{x^3}{3!} + \dfrac{x^5}{5!} + \dfrac{x^7}{7!} + \cdots$

(b) $\cosh x = \frac{1}{2}(e^x + e^{-x}) = 1 + \dfrac{x^2}{2!} + \dfrac{x^4}{4!} + \dfrac{x^6}{6!} + \cdots$

(c) $\sinh^{-1} x = \ln\left(x + \sqrt{x^2 + 1}\right)$

(d) $\cosh^{-1} x = \ln\left(x + \sqrt{x^2 - 1}\right)$

(e) $\dfrac{d}{dx}(\sinh x) = \cosh x$

(f) $\dfrac{d}{dx}(\cosh x) = \sinh x$

(g) $\int \sinh x\, dx = \cosh x$

(h) $\int \cosh x\, dx = \sinh x$

B5.7   Some Hyperbolic Identities with Mnemonics

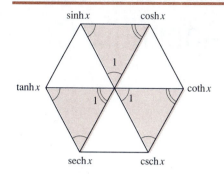

The mnemonics associated with the hexagon for hyperbolic functions shown consists of the following basic rules:

1. The product of the functions at the ends of each diagonal is equal to 1. We have

$$\sinh x \ \operatorname{csch} x = 1 \qquad \cosh x \ \operatorname{sech} x = 1 \qquad \tanh x \ \coth x = 1$$

2. Any function is equal to the product of its two neighboring functions.

$$\sinh x = \cosh x \ \tanh x \qquad \operatorname{csch} x = \operatorname{sech} x \ \coth x$$
$$\cosh x = \sinh x \ \coth x \qquad \operatorname{sech} x = \tanh x \ \operatorname{csch} x$$
$$\tanh x = \sinh x \ \operatorname{sech} x \qquad \coth x = \cosh x \ \operatorname{csch} x$$

3. In each of the three shaded triangles, the square of the quantity at the upper right side angle is equal to the sum of the squares of the quantities at the other two angles. We have

$$\cosh^2 x = 1 + \sinh^2 x \qquad 1 = \tanh^2 x + \operatorname{sech}^2 x$$
$$\coth^2 x = 1 + \operatorname{csch}^2 x$$

B5.8   Newton-Raphson Method

To extract a root of the equation $y(x) = 0$ by the Newton-Raphson method, we first assume a trial root $x = x_1$. Then the next improved trial root $x = x_2$ is given by $x_2 = x_1 - y(x_1)/y'(x_1)$ where $y(x_1) = y|_{x=x_1}$, and $y'(x_1) = dy/dx|_{x=x_1}$. The procedure is repeated until the correction term $y(x_i)/y'(x_i)$ for the $(i + 1)$th trial is a negligible fraction of the trial root $x = x_i$.

B5.9   Leibnitz's Rule

The taking of a derivative of a definite integral must be consistent with the rule that

$$\frac{d}{dx} \int_\alpha^\beta f(x,t) \, dt = \int_\alpha^\beta \frac{\partial}{\partial x} f(x,t) \, dt + f(x,\beta) \frac{d\beta}{dx} - f(x,\alpha) \frac{d\alpha}{dx}$$

where $\alpha = \alpha(x)$ and $\beta = \beta(x)$.

Appendix C

Infinitesimal Angular Displacements

Finite angular displacements (or finite rotations) are known to disobey the commutative law for the addition of vectors.[†] Thus, *finite* angular displacements do *not* qualify as vectors, despite the fact that they have magnitudes and directions given by the right-hand rule. However, we shall here show that *infinitesimal angular displacements are vectors*.

Let **r** be the position of the particle P of a rigid body, which initially occupies a region indicated by the dashed curve as shown in Fig. C.1. If the body undergoes successively an infinitesimal rotation $d\theta_1$ about the axis OA_1 and then an infinitesimal rotation $d\theta_2$ about the axis OA_2, the particle at P will go first to the point Q and then to the point P' as indicated, where $\overline{PQ} \approx \overline{C_1P}\, d\theta_1$ and $\overline{QP'} \approx \overline{C_2Q}\, d\theta_2$. If such successive infinitesimal rotations took place in reverse order, the particle at P would have gone first to the point R and then to the point P'' as indicated.

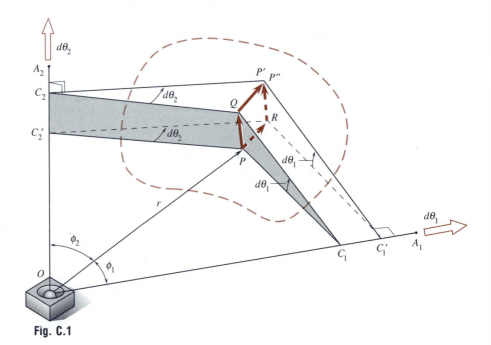

Fig. C.1

[†]Cf. D1.3.

Since the rotations $d\theta_1$ and $d\theta_2$ are infinitesimal, we infer that, within the framework of *first-order accuracy*, P'' must coincide with P', and

$$\overline{PQ} = \overline{RP'} = (r \sin\phi_1)d\theta_1$$

$$\overline{PR} = \overline{QP'} = (r \sin\phi_2)d\theta_2$$

Recalling that a rotation has a magnitude and a direction consistent with the right-hand rule, we may use the "mechanics" of cross products to write the differential displacements

$$\overrightarrow{PQ} = \overrightarrow{RP'} = \overrightarrow{d\theta}_1 \times \mathbf{r} \qquad\qquad (C.1)$$

$$\overrightarrow{PR} = \overrightarrow{QP'} = \overrightarrow{d\theta}_2 \times \mathbf{r} \qquad\qquad (C.2)$$

As the quadrilateral $PQP'R$ is a parallelogram, we have

$$\overrightarrow{PQ} + \overrightarrow{QP'} = \overrightarrow{PR} + \overrightarrow{RP'} \qquad\qquad (C.3)$$

Substituting Eqs. (C.1) and (C.2) into Eq. (C.3), we write

$$\overrightarrow{d\theta}_1 \times \mathbf{r} + \overrightarrow{d\theta}_2 \times \mathbf{r} = \overrightarrow{d\theta}_2 \times \mathbf{r} + \overrightarrow{d\theta}_1 \times \mathbf{r}$$

$$(\overrightarrow{d\theta}_1 + \overrightarrow{d\theta}_2) \times \mathbf{r} = (\overrightarrow{d\theta}_2 + \overrightarrow{d\theta}_1) \times \mathbf{r} \qquad\qquad (C.4)$$

Since Eq. (C.4) holds true for the position vector $\mathbf{r}$ of any point P, we must have

$$\overrightarrow{d\theta}_1 + \overrightarrow{d\theta}_2 = \overrightarrow{d\theta}_2 + \overrightarrow{d\theta}_1 \qquad\qquad (C.5)$$

Thus, *infinitesimal angular displacements* (or infinitesimal rotations, such as virtual angular displacements) *are vectors* because they obey the commutative law for the addition of vectors as shown in Eq. (C.5) and are consistent with the parallelogram law. Since the time variable t is a scalar, we readily see that *angular velocities* $\boldsymbol{\omega}_1 = \overrightarrow{d\theta}_1/dt$, $\boldsymbol{\omega}_2 = \overrightarrow{d\theta}_2/dt$ *and angular accelerations* $\boldsymbol{\alpha}_1 = d\boldsymbol{\omega}_1/dt$, $\boldsymbol{\alpha}_2 = d\boldsymbol{\omega}_2/dt$ *are also vectors.*

Appendix D

Simultaneous Equations Solver

```
10 REM   SOLUTION OF SIMULTANEOUS EQUATIONS   (File name: SOSE)
12 REM
14 REM   A program to solve N linear simultaneous equations
16 REM   in N unknowns, [A][X] = [B], via matrix inversion
18 REM
20 REM     This is a BASIC program which runs on IBM PC and
22 REM   compatibles.  With 60K or more bytes free for BASIC,
24 REM   it can accommodate the solution and checking of at
26 REM   least 45 equations in 45 unknowns if running in double
28 REM   precision, and 65 equations in 65 unknowns if running
30 REM   in single precision.  The program automatically prints
32 REM   the DETERMINANT of the coefficient matrix [A] and does
34 REM   the back substitution of the solution for [X] into the
36 REM   original equations to see how well the product [A][X]
38 REM   CHECKs the constant vector [B].
40 REM
42 REM              D I R E C T I O N S :
44 REM   (a) If it is desired to divert the output to a
46 REM       printer, use, instead, the program SOSEP,
48 REM       which is also stored on the diskette.
50 REM   (b) Enter data as illustrated in Lines 1000-4000,
52 REM       then run the program.
54 REM   -------------------------------------------------------
56 REM   |            C A U T I O N !                          |
58 REM   |    This program is intended to find unique solu-    |
60 REM   | tions.  Sometimes, the rounding-off process with-   |
62 REM   | in the computer may prevent the computer from       |
64 REM   | obtaining an exact value of zero for the determi-   |
66 REM   | nant of a singular matrix.  Thus, in the output     |
68 REM   | of this program, if the product [A][X] in the       |
70 REM   | checking by back substitution differs greatly       |
72 REM   | from the original [B], then the solution for [X]    |
74 REM   | should be discarded.  In this case, the matrix      |
76 REM   | [A] is possibly singular or the set of simulta-     |
78 REM   | neous equations is ill-conditioned.                 |
80 REM   -------------------------------------------------------
82 REM     Programmed by: I. C. Jong, University of Arkansas
84 REM
300 DEFDBL A-H,O-Z: REM Omit this line for single precision.
310 CLS: PRINT "There are N equations in [A][X] = [B]."
320 B$ = "Rowwise, the matrix [A] =": G1$ = "Subroutine "
330 C$ = "The constant vector [B] =": G2$="execution time ="
340 PRINT "N =";: READ N: D$ = "The determinant of [A] = "
350 E$ = "In checking, [A][X] gives [B] =": J$ = " = "
360 F$ = "There is no unique solution for [A][X] = [B]."
370 IF N=0 OR N <> ABS(INT(N)) THEN PRINT " Illegal N!": END
380 PRINT N: ND=N+1: IF N>1 THEN ND=N: P$ = "*IN PROGRESS..."
390 DIM  A(ND,ND), AP(N,N), U(ND,ND), X(N), B(N): PRINT B$
400 FOR I=1 TO N: FOR J=1 TO N: READ A(I,J): AP(I,J)=A(I,J)
410 PRINT A(I,J);: NEXT J: PRINT: NEXT I: PRINT C$
420 FOR I = 1 TO N: READ B(I): PRINT B(I);: NEXT I: PRINT
430 PRINT: PRINT "*PRESS SPACE BAR TO COMPUTE..."
440 SP$ = INKEY$: ON -(SP$ <> CHR$(32)) GOTO 440: PRINT P$
450 PRINT D$;: GOSUB 900: C = T: GOSUB 700: GOSUB 900
460 E = T: PRINT D: IF D = 0 THEN PRINT F$: GOTO 540
470 PRINT "The solution yields [X] =": FOR I=1 TO N: X(I)=0
480 FOR J = 1 TO N: X(I) = X(I) + U(I,J)*B(J): NEXT J
490 PRINT SPC(8);"X ("; I; J$; X(I): NEXT I
500 ON -(N<10) GOTO 520: PRINT "*PRESS SPACE BAR TO CHECK.";
510 SP$ = INKEY$: ON -(SP$ <> CHR$(32)) GOTO 510: PRINT
520 PRINT E$: FOR I = 1 TO N: B(I) = 0: FOR J = 1 TO N
530 B(I)=B(I)+AP(I,J)*X(J):NEXT J:PRINT SPC(11);B(I):NEXT I
540 I$ = " second.": K$ = " seconds.": ET$ = " Less than one"
550 IF E >= C+1 THEN ET$ = STR$(E-C): IF E >= C+2 THEN I$=K$
```

```
560 PRINT G1$; G2$; ET$; I$: END
570 REM
700 REM ......... SUBROUTINE  DANDI ......... (10-27-86)
702 REM 1. To let this subroutine use maximum pivots in find-
704 REM    ing the Determinant and Inverse of [A], do the
706 REM    following: (a) omit Lines 740 and 750, and (b)
708 REM    delete the word REM in each of Lines 745 and 746.
710 REM 2. Upon return: D = Determinant, [U] = Inverse of [A]
720 FOR I=1 TO N:FOR J=1 TO N:U(I,J)=0:IF I=J THEN U(I,J)=1
730 NEXT J: NEXT I: D = 1: N1 = N - 1: FOR I = 1 TO N1
740 ON -(A(I,I) <> 0) GOTO 790: K = I: FOR J = I TO N1
745 REM    BG = ABS(A(I,I)): K = I: FOR J = I TO N1: JP=J+1
746 REM    IF BG < ABS(A(JP,I)) THEN K=JP: BG = ABS(A(K,I))
750 JP = J + 1: IF A(JP,I) <> 0 THEN K = JP: GOTO 770
760 NEXT J: IF A(K,I) = 0 THEN D = 0: RETURN
770 ON -(K=I) GOTO 790: FOR J=1 TO N:P=A(I,J):A(I,J)=A(K,J)
780 A(K,J)=P:P=U(I,J):U(I,J)=U(K,J):U(K,J)=P:NEXT J: D = -D
790 AI = A(I,I): D = D*AI: FOR J=1 TO N: A(I,J) = A(I,J)/AI
800 U(I,J)=U(I,J)/AI: NEXT J: NI=N-I: FOR J=1 TO NI: IJ=I+J
810 AJ = A(IJ,I): FOR K=1 TO N: A(IJ,K) = A(IJ,K)-AJ*A(I,K)
820 U(IJ,K) = U(IJ,K) - AJ*U(I,K): NEXT K: NEXT J: NEXT I
830 D = D*A(N,N): IF D = 0 THEN RETURN
840 FOR I=1 TO N: U(N,I) = U(N,I)/A(N,N): NEXT I: A(N,N)=1
850 FOR I=1 TO N1: NI=N-I: FOR J=1 TO NI: IJ=I+J:AK=A(I,IJ)
860 FOR K = 1 TO N: A(I,K) = A(I,K) - AK*A(IJ,K)
870 U(I,K)=U(I,K)-AK*U(IJ,K): NEXT K: NEXT J: NEXT I:RETURN
900 REM .............. SUBROUTINE  TIME ..............
910 T$=TIME$:H$=MID$(T$,1,2):M$=MID$(T$,4,2):S$=MID$(T$,7,2)
920 T = VAL(H$)*3600 + VAL(M$)*60 + VAL(S$): RETURN
930 REM
1000 DATA 6:     REM Enter N, the number of equations, here.
2000 REM Enter, rowwise, the coefficient matrix [A] below:
2010 DATA .6, .96, -1, 0, 0, 0
2020 DATA .8, -.28, 0, 0, 0, 0
2030 DATA -.6, 0, 0, .384615, 0, 0
2040 DATA -.8, 0, 0, -.923077, 1, 0
2050 DATA 0, -.96, 0, -.384615, 0, .882353
2060 DATA 0, .28, 0, .923077, 0, .470588
3000 REM Enter the constant vector [B] below:
4000 DATA 0, 0, 0, 0, 0, 98.1
```

The preceding program, named SOSE, is to be run under BASIC and is stored on the computer diskette labeled TUTORIAL, as described in App. F. Following the DIRECTIONS contained in the first REMark segment of the program, we show two sample runs as follows:[†]

☐ SOLUTION OF SIMULTANEOUS EQUATIONS IN EXAMPLE 3.8

```
There are N equations in [A][X] = [B].
N = 6
Rowwise, the matrix [A] =
 .6   .96  -1    0    0    0
 .8  -.28   0    0    0    0
-.6   0    0    .384615   0    0
-.8   0    0   -.923077   1    0
 0   -.96  0   -.384615   0   .882353
 0   .28   0    .923077   0   .470588
The constant vector [B] =
 0   0    0    0    0   98.1

The determinant of [A] =  .382262138896008
The solution yields [X] =
      X ( 1 ) =  24.38565007683738
      X ( 2 ) =  69.6732859338211
      X ( 3 ) =  81.51774454257069
      X ( 4 ) =  38.04165216151848
      X ( 5 ) =  54.6238942137679
      X ( 6 ) =  92.38677098912871
In checking, [A][X] gives [B] =
           1.77635683940025D-15
          -4.440892098500626D-16
          -1.332267629550188D-15
           0
          -1.77635683940025D-15
           98.1
Subroutine execution time = 3 seconds.
```

[†]Notice the first item in the DIRECTIONS in the program.

☐ CASE OF SINGULAR COEFFICIENT MATRIX [A]

```
There are N equations in [A][X] = [B].
N = 3
Rowwise, the matrix [A] =
   1   2   3
   4   5   6
   7   8   9
The constant vector [B] =
   1   2   3

The determinant of [A] =  0
There is no unique solution for [A][X] = [B].
Subroutine execution time = Less than one second.
```

Digital Root Finder[†]

```
10 REM      DIGITAL ROOT FINDER AND PLOTTER     (File name: DIGITAL)
12 REM
14 REM      A program to find the real roots of Y(X) = 0 to any
16 REM        desired digits of accuracy by plotting Y versus X
18 REM
20 REM        This is a BASIC program which runs on IBM PC and
22 REM    compatibles.  It plots in 80 columns from X = XI, with
24 REM    increment DX, to X = XF.
26 REM
28 REM                  D I R E C T I O N S:
30 REM    ------------------------------------------------------------
32 REM    1. If it is desired to divert the output to a printer,
34 REM       use, instead, the program DIGITALP, which is also
36 REM       stored on the diskette.
38 REM    2. To find a root of a given nonlinear equation defined
40 REM       by Y(X) = 0, do the following: (a) define Y = Y(X) in
42 REM       BASIC in Lines 1000-1999; (b) run the program to ex-
44 REM       plore the locations of the roots by making trial runs
46 REM       until the graph crosses the X axis; (c) read the two
48 REM       neighboring digital values of X which bound the root
50 REM       at the crossing of the graph over the X axis; (d) use
52 REM       a smaller increment (e.g., one-tenth of the preceding
54 REM       increment) to plot between those two values of X; (e)
56 REM       repeat steps (c) and (d) until the root is determined
58 REM       to as many digits of accuracy as desired or permitted.
60 REM    ------------------------------------------------------------
62 REM       Programmed by: I. C. Jong, University of Arkansas
64 REM
100 DEFDBL A-H, O-Z: REM Omit this line for single precision.
110 INPUT "XI = "; XI: INPUT "DX = "; DX: INPUT "XF = "; XF
120 N% = INT(ABS((XF - XI)/DX) + 1.5): DIM Y(N%): PRINT
130 PRINT "*IN PROGRESS": R=0: X = XI: FOR I=1 TO N%:   GOSUB 1000
140 Y(I) = Y: X = XI + I*DX: AY = ABS(Y(I)): IF AY>R THEN R = AY
150 NEXT I: SF=R/38: X = XI: FOR I = 1 TO 7: Y$ = Y$+":----!----"
160 NEXT I: Y$ = " !--" + Y$ + ":--!Y": I = 1: PRINT: PRINT
170 PRINT "XI =";XI; "     DX =";DX; "     XF =";XF; "     R =";R
180 PRINT: PRINT "-R"; SPC(37); "0"; SPC(37); "R": PRINT Y$
190 P% = INT(Y(I)/SF + 39 + .5): X% = P% - 40: LT = LEN(STR$(X))
200 Y% = INT(Y(I)/SF-LT-1.5): Z% = 38 - P%: IF P% >= 39 THEN 220
210 PRINT SPC(P%); "*"; SPC(Z%); "!"; X: GOTO 260
220 IF P% = 39 THEN PRINT SPC(P%); "*"; X: GOTO 260
230 IF 44>=P% AND P%>39 THEN PRINT SPC(39)"!"SPC(X%)"*":GOTO 260
240 IF Y% < 0 THEN PRINT SPC(39); "!"; X: GOTO 260
250 PRINT SPC(39); "!"; X; SPC(Y%); "*"
260 X = XI + I*DX: I = I + 1: IF X < XF + DX/3 THEN 190
270 PRINT SPC(39); "X": PRINT: LIST 1000-1999: END
1000 Y = 20 - X*((EXP(50/X)+EXP(-50/X))/2 - 1)◄───────────  Y = Y(X)
2000 RETURN
```

[†]This program is provided for the convenience of those who may find it helpful in finding the root or roots of any nonlinear equation $y(x) = 0$, as well as plotting curves, with a computer. In the plot, the y axis is proportionally calibrated with the computed value of R; however, the x axis is digitally calibrated with any increment chosen by the user.

SOLUTION OF A TRANSCENDENTAL EQUATION (CF. EXAMPLE 8.24.)

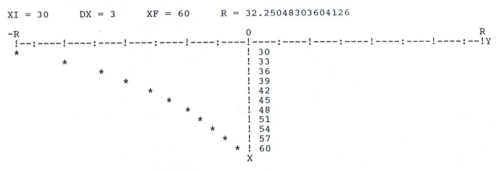

```
XI = 30     DX = 3     XF = 60     R = 32.25048303604126

-R                                              0                                    R
 !--:----!----:----!----:----!----:----!----:----!----:----!----:--!Y
   *                                            ! 30
       *                                        ! 33
           *                                    ! 36
             *                                  ! 39
               *                                ! 42
                 *                              ! 45
                   *                            ! 48
                     *                          ! 51
                       *                        ! 54
                         *                      ! 57
                           * !                    60
                                                X

1000 Y = 20 - X*((EXP(50/X)+EXP(-50/X))/2 - 1)
```

(a) A trial run which indicates that the root is located at X > 60

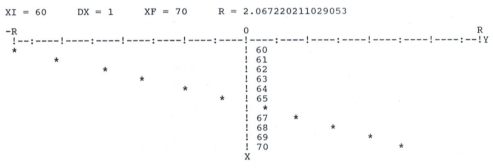

```
XI = 60     DX = 1     XF = 70     R = 2.067220211029053

-R                                              0                                    R
 !--:----!----:----!----:----!----:----!----:----!----:----!----:--!Y
   *                                            ! 60
        *                                       ! 61
            *                                   ! 62
              *                                 ! 63
                 *                              ! 64
                   *                            ! 65
                                                !  *
                                                ! 67      *
                                                ! 68         *
                                                ! 69           *
                                                ! 70              *
                                                X

1000 Y = 20 - X*((EXP(50/X)+EXP(-50/X))/2 - 1)
```

(b) Next trial run

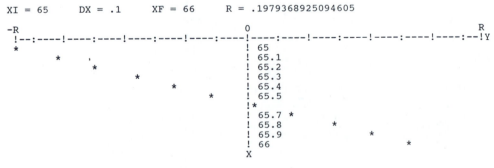

```
XI = 65     DX = .1     XF = 66     R = .1979368925094605

-R                                              0                                    R
 !--:----!----:----!----:----!----:----!----:----!----:----!----:--!Y
   *                                            ! 65
        *       ,                               ! 65.1
            *                                   ! 65.2
               *                                ! 65.3
                 *                              ! 65.4
                    *                           ! 65.5
                                                !*
                                                ! 65.7 *
                                                ! 65.8      *
                                                ! 65.9         *
                                                ! 66             *
                                                X

1000 Y = 20 - X*((EXP(50/X)+EXP(-50/X))/2 - 1)
```

(c) Next run[†]

[†]Additional runs and outputs show that $Y = 0$ when $X \approx 65.586$. Since $\cosh(50/x) = [\exp(50/x) + \exp(-50/x)]/2$, we may take $c = 65.59$ as the root of the equation $20 = c(\cosh(50/c) - 1)$ encountered in Example 8.24.

Supplementary Software

Appendix F

As alluded to in the Preface, diskettes containing supplementary software for this text may be obtained from bookstores or the publisher. The software is designed for an IBM PC® or compatible computer with at least 384K of random access memory, a CGA or compatible graphics display card, and DOS 2.0 or higher.

The BASIC programs CROSSPD, STRIPPD, CENTL, CENTA, AMOI, PROJECT, SPACFT, SOSE, and DIGITAL have been documented in Secs. 4.9, 4.15, 6.4, 6.8, 7.10, 11.12, 12.11, and Apps. D and E of the text. These nine BASIC programs plus a 40-column version of DIGITAL, named DIGIT40, and a compiled program named ISE40.EXE (which is an interactive version of SOSE for $N \leq 40$), as well as the tutorial modules for all chapters, are stored in the Mech * E TUTORIAL diskette. Of course, the BASIC programs are to be run under a BASIC interpreter (e.g., BASICA.COM or GWBASIC.EXE). They all output to the screen of the monitor. If it is desired to output to a printer which is properly connected, the user may use, instead, the corresponding programs ending with a P (e.g., CROSSPDP, STRIPPDP, CENTLP, CENTAP, AMOIP, PROJECTP, SPACFTP, SOSEP, and DIGITALP) which are also stored on that diskette. To run the tutorial modules, the user may do the following: (a) set the computer in DOS, (b) insert the TUTORIAL diskette in drive A, (c) key in TUTOR at the prompt from the DOS (e.g., A:>TUTOR), and (d) make a selection from the menu as directed on the screen.

The eight programs UNIT, VECTOR, AREA, TRUSS, TRAJECT, ORBIT, LINKAGE, and VIBRATE are compiled programs designed to complement selected sections in the text. These compiled programs are stored on the Mech * E UTILITY diskette. They are described in this appendix and may be run by doing the following: (a) set the computer in DOS, (b) insert the UTILITY diskette in drive A, (c) key in DYNASOFT at the prompt from the DOS (e.g., A:>DYNASOFT), and (d) make a selection from the menu as directed on the screen.

Naturally, the principal sources of guidance in learning a subject are the *instructor* and the *textbook*. Students are advised to first study the text as assigned by the instructor. Then, as needed, they may utilize the software to strengthen their understanding of the concepts as covered in the following programs:

TUTOR: A Multiple-Choice Tutorial

This program contains twenty modules covering Chaps. 1 through 20. It is designed to help students strengthen as well as assess their understanding of basic concepts in engineering mechanics by providing an interactive tutorial consisting of a series of multiple-choice questions related to one chapter at a time. As a user, you are to read each question carefully, go over the prospective answers thoughtfully, and key in the letter (e.g., A, B, C, D, or E) corresponding to the correct or best answer and then press the RE-TURN key. After your response, a window will appear to provide a comment or a hint. This process will continue until the correct answer is entered.

You receive a full credit for a question if you answer it correctly on the first try, half a credit if you answer it correctly on the second try, and no credit thereafter. Your score is continually updated on the screen. At the end of the quiz, a summary of credits earned for the individual questions, along with the total score, is shown. If you did not make 100% on the quiz, you may elect to retry those questions you missed on the first try. Note that the order of the prospective answers has been randomized and may not be the same when the question is posed again.

In responding to any question, you may enter END to terminate the quiz or SAVE to save the record of the quiz up to the current question for future restart. Entering HELP will recall the foregoing information.

UNITS: A Program to Convert Units

This program complements the material covered in Secs. 1.7 and 1.8 of the text. It is designed to convert one set of units to another. When it is run, a window appears on the top half of the screen giving a list of recognized units where upper and lower case modes are significant. Compound units may be formed by using the symbols * for multiplication, / for division, and $\wedge$ for raising a unit to a power. Any prefixed or compound unit may be converted to any compatible unit(s). In handling a compound unit, note that this program regards all units to the right side of / as being divided into the units written to its left side. Furthermore, units rasied to powers are assumed to have integer exponents.

At the prompt to enter the unit(s), one has several options. For example, one may enter a number with a unit or combination of units. If the number is omitted, it is defaulted to 1. If a question mark (?) is entered at the prompt, it will highlight a unit in the window, and a brief explanation of its meaning will appear. The left and right cursor keys may be used to move the highlight around the window. To leave the window and return to the prompt, press the ESC key. Entering nothing at the prompt (by simply pressing the RETURN key) will terminate the program.

VECTOR: A Program to Introduce Operations of Vectors in a Plane

This program complements the material covered in Secs. 2.1 through 2.5, 4.7 through 4.9, and 4.14 of the text. It is designed to introduce and demonstrate certain algebraic operations of vectors. It presents the addition, subtraction, dot product, and cross product of vectors in the xy plane. Initially, the screen shows a set of rectangular axes and prompts the user to interactively generate a vector **A** in either cartesian or polar form. By using the cursor keys, the initial vector shown on the screen can be incrementally elongated, shortened, or rotated to the liking of the user. This vector is then saved by pressing the RETURN key. A prompt to generate a second vector **B** will follow in a similar manner. Meanwhile, pressing the F1 key during the input process will call up a HELP screen.

After both vectors **A** and **B** have been saved into memory, the program will proceed to show a table which contains blanks to ber completed by the user. However, the user may let the computer fill a blank by entering a question mark (?) in the blank. After the table is completed, the user will be asked to enter a choice which will demonstrate one of the algebraic operations as listed. Note that pressing the F2 kwy will recall the graphical display of the vectors **A** and **B** which were generated earlier.

AREA: A Program to Evaluate Areal Properties

This program complements the material covered in Secs. 6.7, 6.8, 7.2 through 7.8, and 7.10 of the text. It is designed to evaluate the areal properties (e.g., area, centroid, moments and products of inertia, principal moments and axes of inertia, and radii of gyration) of any plane area whose boundary is defined (or approximated) by N line segments connecting N nodes (or points), where $N \leq 500$. The program allows the following three ways to enter the data:

1. *Line input*: Enter interactively the numerical values of the coordinates of each node from the keyboard as prompted by the program.
2. *Graphical input*: Use the cursor keys to move the cross hair on the screen of the monitor to a desired position and then press the RETURN key to graphically enter the coordinates of a node. Repeat until the coordinates of all nodes on the boundary are entered.
3. *File input*: If the data for the coordinates of the various nodes on the boundary of an area have previously been saved by this program in a data file, those data may be recalled by·choosing the file input mode and specifying the name of that file (e.g., ZBAR.DAT).

Once an area is defined, its properties may be avaluated and displayed. If desired, the area can be modified, translated, or rotated by using the Text Editor and/or the Graphics Editor in the program. Besides, the data can be saved to a disk file and recalled later. Thus, this program can give the user an experience in computer aided design (CAD).

TRUSS: A Program to Evaluate Forces in Truss Members

This program complements the material covered in Secs. 8.1 through 8.9 of the text. It is designed to evaluate the axial forces in the members of any statically determinate plane truss under the action of any forces at the joints.

The truss may contain up to 26 joints (designated by the letters of the alphabet, A through Z), and it may be supported by an appropriate number of hinges, rollers, or links at its joints, as long as Eq. (8.2) is satisfied. The program allows the following two ways to enter the data:

1. *Line input*: Enter interactively the answers to a series of questions about the truss as prompted by the computer. A user is expected to specify the following: a name for identifying the given truss and for saving the data if desired, the names of the joints (A through Z) in the truss, the coordinates of the joints, the supports (e.g., hinge, roller, or link) existing at the joints, and the *x* and *y* components of the forces acting on the joints.
2. *File input*: If the data for the truss have previously been saved by this program in a data file, those data may be recalled by choosing the file input mode and specifying the name of that file (e.g., EX8P2.DAT).

Upon completing either the line input or the file input, the truss with its loads will first be graphically displayed. The user may then follow the prompt and the menu on the screen for calculating the axial forces and other editing and recalculating opportunities. Naturally, this program can give the user an experience in computer aided engineering (CAE).

TRAJECT: A Program to Demonstrate Trajectories of a Projectile

This program complements the material covered in Sec. 11.12 of the text. It is designed to demonstrate the free flights of a spherical projectile through a vacuum (as an idealized case) and through a fluid (e.g., air or water). The user can specify the launch parameters (e.g., the initial position and the initial velocity) and the medium in which the projectile will travel. The initial velocity vector of the projectile may be entered in either cartesian or polar form. If the spherical projectile is to travel through a fluid, the diameter of the sphere as well as other data should be entered at the prompt from the computer. The program plots the trajectory starting from an initial position and ending in a position which is established by specifying the *x* coordinate, *y* coordinate, or time. Then, a menu is presented for selections to plot multiple trajectories, change the launch parameters or medium, or exit the program. To access the help screen, press the F1 key.

ORBIT: A Program to Simulate Space Flight

This program complements the material covered in Secs. 12.8 through 12.12. It is designed to provide the user with a computer simulation of launching, controlling, and landing a spacecraft flying in the earth-moon system. The spacecraft with a given amount of fuel flies from the earth. The user controls the forward or reverse thrust engines by using the up, down, left, and right cursor keys. The spacecraft can be placed in orbit around the earth, flown to the moon, orbited around the moon or landed on it, and returned to the earth. A ''time accelerator'' is on board with which the user can make time pass up to 2000 times faster. For example, a 90-minute orbit around the earth would take only 2.7 seconds, and a 2-week lunar excursion can be done in about 10 minutes. The time accelerator is controlled by using the Pg Up and Pg Dn keys; it can be toggled on and off by using the END key.

The program constantly displays a control panel showing various flight parameters such as the settings of the thrusters, the elapsed time since launch, its speed and acceleration, the apogee and perigee in an orbit, the period of orbit, the current distance from the earth and moon surfaces, and the remaining fuel supply. Pressing the F1 key at any time brings up a help screen, and pressing the ESCAPE key exits the program.

LINKAGE: A Program to Demonstrate Motion of a Fourbar Linkage

This program complements the material covered in Secs. 15.5 through 15.10. It is designed to allow a user to create, modify, and analyze the plane motion of fourbar linkages. A fourbar linkage consists of a crank, a coupler, a follower, and the ground link. The user is asked to specify the initial coordinates of the four connecting pins in the system. The path, velocity, and acceleration of a chosen point on the coupler can be continuously monitored. After a pause to compute the kinematic quantities involved in the plane motion of the linkage, the program graphically displays the linkage and animates its motion as the crank rotates. If the mechanism locks up, the rotation of the crank is confined to the operable range. The user has several options while the mechanism is being displayed. For example, pressing the + or − keys speeds or slows the rotation of the crank, pressing the S key stops the motion and displays a kinematic analysis of the mechanism at the selected position, pressing the R key resumes the motion, and pressing the T key toggles a trace of the path of the chosen point on the coupler. The user may press the F1 key for help in entering any data or command, and press the ESCAPE key to exit the program.

VIBRATE: A Program to Plot Vibrations of a SDOF System

This program complements the material covered in Secs. 20.1 through 20.6. It is designed to allow the user to design and analyze the free and forced vibration of a single-degree-of-freedom (SDOF) system and to plot response curves. The user is asked to specify the mass, the modulus of the spring, the viscous damping coefficient, the initial displacement and velocity of the mass, and the parameters A, B, C, and D in the forcing function $F = A * \sin(B * t) + C * \cos(D * t)$. The program then calculates the damping factor and other characteristics associated with the vibration.

The program animates the motion of the mass and plots the response curves. By selecting from the menu, the user can compare the responses of previously defined systems by showing them on the same plot, redefine the current system, or exit the program. A help screen is available at any time by pressing the F1 key.

Answers to Selected Developmental Exercises and Problems

Chapter 1

Developmental Exercises

D1.3 (i) No, (ii) no. **D1.4** No. **D1.5** F, a. **D1.6** (a) Magnitude, (b) orientation, (c) sense. **D1.9** a, b, d. **D1.10** (a) (a, b), (b) ∡ θ, (c) no. **D1.17** a, c, d. **D1.20** Yes. **D1.23** Seven. **D1.28** (a) kg, m, s; (b) lb, ft, s. **D1.29** (a) 10 lb·s²/ft, (b) 5.24 kg·m/s², (c) 1.452 kg, (d) 30×10^{-6} s. **D1.30** (a) 5, (b) 6, (c) 4, (d) 5, (e) 4, (f) 7, (g) 4. **D1.31** (a) 41.5, (b) 41.5, (c) 42.5, (d) 42.5, (e) 0.00624, (f) 1.36×10^6, (g) 5.22×10^6. **D1.32** (a) 42, (b) 42, (c) 42, (d) 43, (e) 0.0062, (f) 1.4×10^6, (g) 5.2×10^6. **D1.33** (a) 33.4, (b) 31.0, (c) 16.80, (d) 0.499, (e) 26.7×10^3, (f) 12.17×10^{-6}, (g) 5.67, (h) 2.28, (i) 1.264, (j) 0.814, (k) −0.581. **D1.34** No. **D1.35** (a) 0.0508 m, (b) 5.49 m, (c) 28.3 m³. **D1.37** Yes. **D1.39** (a) No, (b) no, (c) yes, (d) no. **D1.41** b. **D1.43** (i) a, c; (ii) b, d. **D1.44** 34.4×10^{-9} ft³/(slug·s²). **D1.45** c.

Problems

1.1 (a) 55.9 mi/h, (b) 82.0 ft/s. **1.2** 238 L/s. **1.4** 76.0 ft/s. **1.5** 3.32 kg. **1.6** 17.39 psi. **1.7** 137.9 MPa. **1.9** (a) 53.5 lb, (b) 238 N. **1.10** 1.030 in. **1.12** 7.71 kg. **1.13** 24.1×10^3 mi/h. **1.15** b. **1.16** c. **1.17** b. **1.18** d. **1.19** a. **1.20** a. **1.21** b. **1.22** e. **1.23** e. **1.24** d.

Chapter 2

Developmental Exercises

D2.2
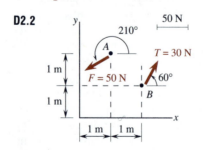

D2.3
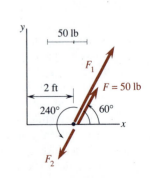

D2.4 $\theta_P = 36.9°$, $\theta_Q = 157.4°$.   **D2.5** c.   **D2.6** d.   **D2.8**
Parallelogram law.   **D2.9** $P = 73.2$ lb, $Q = 51.8$ lb.   **D2.10** $Q =$
29.8 lb, $\theta_Q = 320.9°$.   **D2.13** (a) $G_x = 80i$ N, $G_y = 60j$ N, $H_x =$
$-120i$ N, $H_y = 50j$ N; (b) $G_x = 80$ N, $G_y = 60$ N, $H_x = -120$ N, H_y
$= 50$ N; (c) 120 N.   **D2.14** (a) $-240i + 70j$ N, (b) 163.7°.
D2.15 $50i - 480j$ lb.   **D2.16** 71.8°.   **D2.17** $x = -2$, $y = 41$.
D2.18 (a) $F_x = 3$ kN, $F_y = -4$ kN, $F_z = -12$ kN; (b) $r_x = 40$ m, r_y
$= 0$, $r_z = -9$ m.   **D2.19** 9 kips.   **D2.20** (a) θ_y, θ_z; (b) no; (c) no;
(d) yes, $\cos\theta_x$.   **D2.21** (a) $F_x = 0$, $F_y = 8$ lb, $F_z = -15$ lb; (b) $\theta_x =$
90°, $\theta_y = 61.9°$, $\theta_z = 151.9°$.   **D2.22** (a) $6i - 4j + 12k$ m, (b)
14 m.   **D2.23** (a) $-6i + 2j + 3k$ m, (b) 7 m, (c) $\frac{1}{7}(-6i + 2j +$
$3k$).   **D2.24** (a) $0.707i - 0.5j + 0.5k$, (b) $70.7i - 50j + 50k$ lb.
D2.25 (a) $\frac{1}{9}(4i + 7j - 4k)$, (b) $60i - 240j + 80k$ lb.   **D2.26** $460i +$
$460j - 320k$ lb at O.

Problems

2.1 $C = 184.4$ lb, $\theta_C = 192.5°$.   **2.3** $P = 544$ N, $Q = 254$ N, $\theta_P =$
143.1°, $\theta_Q = 53.1°$.   **2.4** $D = 680$ lb, $\theta_D = 298.5°$.   **2.6** $P =$
135.9 N, $Q = 63.4$ N, $\theta_P = 165°$, $\theta_Q = 255°$.   **2.7** $\theta_A = 223.9°$,
$\theta_B = 115.9°$.   **2.9** $A = 273$ N, $B = 245$ N, $\theta_B = 315°$.   **2.11** 6.64
kips.   **2.12** 34.4°.   **2.14** $B = 1050$ N, $\theta_B = 150.9°$.   **2.16** $\theta_2 =$
27.3°, $R = 1544$ lb.   **2.17** $F = 170$ N, $S = 410$ N, $\theta_F = 61.9°$, θ_S
$= 282.7°$.   **2.19** $P = 1.44$ kips, $Q = 3.08$ kips.   **2.20** $18.5i -$
$1.5j$ kN.   **2.22** $R = -3.5i + 4.73j$ kN.   **2.24** $200i + 347j$ N.
2.25 390 N.   **2.27** 20 lb.   **2.28** $F_{AC} = 250$ N, $F_{BC} = 520$ N.
2.43 (a) $F_x = 3$ kN, $F_y = -12$ kN, $F_z = 4$ kN; (b) 5 kN; (c) $\frac{1}{13}(3i -$
$12j + 4k)$; (d) $\theta_x = 76.7°$, $\theta_y = 157.4°$, $\theta_z = 72.1°$.   **2.46** (a)
$-0.388i + 0.423j + 0.819k$, (b) 0.906.   **2.47** (a) $0.322i + 0.928j$
$- 0.1857k$, (b) 0.371.   **2.49** (a) $0.1664i - 0.766j + 0.621k$, (b)
0.643.   **2.51** $-35i + 360j - 40k$ lb.   **2.52** $F_{PA} = 125$ lb, $F_{PB} =$
90 lb, $F_{PC} = 195$ lb.   **2.53** $-16.8j$ kN.   **2.54** $P = 10.77$ N, $Q =$
6.32 N, $D = 17.09$ N, $\theta_P = 158.2°$, $\theta_Q = 341.6°$, $\theta_D = 159.4°$.
2.55 $\theta_A = 64.2°$, $\theta_B = 161.4°$.   **2.56** $S = 621$ N, $T = 312$ N.
2.57 2.12 kN.   **2.58** $-50i + 320j - 100k$ lb.   **2.59** $F_{DA} = 130$
N, $F_{DB} = 125$ N, $F_{DC} = 175$ N.   **2.61** $T_{DA} = 9.9$ kN, $T_{DB} = 1.2$
kN, $T_{DC} = 15.6$ kN.

Chapter 3

Developmental Exercises

D3.3 True: a, b, d, false: c, e.   **D3.4** (a) Internal, (b) external.
D3.7 (a) From B to A, (b) from A to B.   **D3.8** Yes: a, b, c; no: d.
D3.9 a, b.   **D3.10** (a) 200 N, (b) 400 N.   **D3.14** As a body with a
shape.   **D3.15** No.   **D3.16** b.

D3.17

D3.18

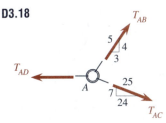

D3.19 431 N.　　**D3.20** Yes.　　**D3.23** 3.82 MN/m.　　**D3.24** 162.6 lb
D3.25 Two.　　**D3.29** $-600\mathbf{i} + 1800\mathbf{k}$ lb.

Problems

P3.1

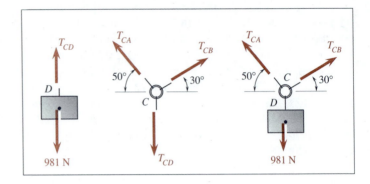

P3.4

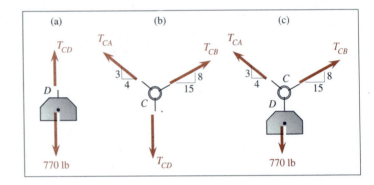

P3.9

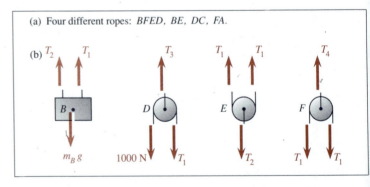

P3.12

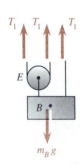

3.17 $T_{CA} = 863$ N, $T_{CB} = 640$ N.　　**3.18** $T_{CA} = 1.766$ kN, $T_{CB} =$
2.35 kN.　　**3.20** $T_{CA} = 750$ lb, $T_{CB} = 680$ lb.　　**3.21** $T_{CA} = 1436$ N,
$T_{CB} = 1759$ N.　　**3.23** 25 lb.　　**3.25** 306 kg.　　**3.26** 6 slugs.
3.28 $P = 577$ lb, $Q = 911$ lb.　　**3.29** $S_1 = S_2 = 11.33$ N.　　**3.30**
1.057 kN.　　**3.31** 49.9°.　　**3.32** (a) 294 N, (b) 196.2 N, (c) 196.2 N,
(d) 147.2 N.　　**3.34** 2.45 kN.　　**3.35** $P = 1.218$ kN, $\theta_P = 97.5°$.
3.37 $T_{CA} = 100$ lb, $T_{CB} = 340$ lb.　　**3.38** 207 lb $< F <$ 430 lb.
3.39 26.5 kg.　　**3.40** 312 kg.　　**3.42** $m_A = 176.7$ kg, $m_B = 413$
kg.　　**3.43** $T_{CG} = 29.4$ N, $T_{AB} = 20.6$ N, $T_{AC} = T_{BC} = 44.1$ N, T_{AD}
$= 58.9$ N, $T_{BE} = 58.9$ N, $T_{CF} = 76.5$ N.　　**3.45** $T_{AB} = 4.47$ lb, T_{AD}
$= 11.31$ lb, $T_{AE} = 15.68$ lb, $T_{AJ} = 9$ lb, $T_{BC} = 3.48$ lb, $T_{BF} = 3.22$ lb,

$T_{CD} = 3.14$ lb, $T_{CG} = 3.78$ lb, $T_{DH} = 13.35$ lb, $T_{DI} = 12$ lb.   **3.46** $(-\frac{2}{3}T_{AB} + \frac{3}{5}T_{AD})\mathbf{i} + (\frac{2}{3}T_{AB} + \frac{4}{5}T_{AC} + \frac{4}{5}T_{AD} - 450g)\mathbf{j} + (-\frac{1}{3}T_{AB} + \frac{3}{5}T_{AC})\mathbf{k}$ N.   **3.48** $(-\frac{3}{7}T_{BA} + \frac{3}{5}T_{CA} - \frac{3}{7}T_{DA})\mathbf{i} + (\frac{6}{7}T_{BA} + \frac{4}{5}T_{CA} + \frac{6}{7}T_{DA} - 300)\mathbf{j} + (-\frac{2}{7}T_{BA} + \frac{2}{7}T_{DA})\mathbf{k}$ lb.   **3.50** 2.21 kN.   **3.51** 353 N.   **3.52** 105 lb.   **3.54** $50\mathbf{i} + 27.7\mathbf{j} - 15\mathbf{k}$ N.   **3.55** 1099 N.   **3.56** 2.75 kN.   **3.57** $T_{AP} = 88.3$ N, $T_{BQ} = 78.5$ N, $T_{OR} = 98.1$ N, $T_{AB} = 33.7$ N, $T_{AC} = 66.4$ N, $T_{AD} = 52.5$ N, $T_{AO} = 26.2$ N, $T_{BG} = 57.2$ N, $T_{BH} = 72.8$ N, $T_{BO} = 47.0$ N, $T_{OE} = 58.8$ N, $T_{OF} = 68.1$ N.   **3.58** 196.2 N.   **3.59** 578 lb.   **3.60** (a) 156 lb/ft, (b) 1.675 ft.   **3.63** $T_{DA} = 125$ N, $T_{DB} = 260$ N, $T_{DC} = 325$ N.   **3.65** 12 lb.   **3.66** 6.87 kN.

Chapter 4

Developmental Exercises

D4.4 (a) Newton-meters (N·m), (b) pound-feet (lb·ft).   **D4.6** (a) Up.   **D4.8** (a) 25 kN ↰ 16.26°, (b) 361 m ↲ 56.3°, (c) 4 kN·m ↩, (d) 700 lb·ft ↪.   **D4.9** 1540 lb·ft ↩.   **D4.10** 560 N·m ↪.   **D4.14** (a) $M_A - 600$ N·m ↩, (b) $M_A - 4A_y + 1000$ N·m ↩.   **D4.15** $M_x = -320$ lb·ft, $M_y = -800$ lb·ft, $M_z = 640$ lb·ft.   **D4.16** $M_{AB} = M_{Az} - \frac{18}{13}T_{CD} + 15$ kN·m, $M_{CB} = 12A_y - 30$ kN·m.   **D4.18** a, b, d.   **D4.19** 0.   **D4.20** (a) 150°, (b) 750 lb·ft, (c) $\mathbf{k}$, (d) $750\mathbf{k}$ lb·ft.   **D4.23** $-3\mathbf{i} + 4\mathbf{j} - 12\mathbf{k}$.   **D4.24** (a) $-\mathbf{k}$, (b) $\mathbf{0}$.   **D4.26** (a) $-3\mathbf{i} - 6\mathbf{k}$, (b) $-3\mathbf{i} - 12\mathbf{j}$, (c) $\mathbf{i} + \mathbf{j} + \mathbf{k}$.   **D4.27** (a) 78.2 ft², (b) 42.0°, (c) $-0.0767\mathbf{i} + 0.383\mathbf{j} + 0.920\mathbf{k}$.   **D4.29** (a) $\mathbf{P} = -140\mathbf{i} + 20\mathbf{j} - 50\mathbf{k}$ N; (b) $\mathbf{r}_{AB} \times \mathbf{P}$; (c) $-10\mathbf{i} - 70\mathbf{j}$ N·m; (d) yes; (e) 0.471 m.   **D4.30** $-10\mathbf{i} + 70\mathbf{j}$ N·m.   **D4.32** $2\mathbf{i} - 6\mathbf{j} - 24\mathbf{k}$ kN·m.   **D4.34** $400\mathbf{i} + 800\mathbf{j}$ lb·ft for (a) through (d).   **D4.35** (a) $-8\mathbf{i}$ kN·m, (b) $16\mathbf{i} - 12\mathbf{k}$ kN·m, (c) $8\mathbf{i} - 12\mathbf{k}$ kN·m.   **D4.36** (a) -4, (b) -4, (c) -10.39, (d) 0, (e) 16, (f) 0.   **D4.37** (a) 173.2, (b) 8.66, (c) 17.32.   **D4.38** 7.   **D4.39** (a) $\frac{1}{5}(4\mathbf{i} + 3\mathbf{k})$, (b) $\frac{1}{7}(6\mathbf{i} + 3\mathbf{j} + 2\mathbf{k})$, (c) $\frac{6}{7}$, (d) $\frac{6}{7}$, (e) 31.0°, (f) 6 ft, (g) 6 ft.   **D4.40** 125.7°.   **D4.41** (a) 6 ft³, (b) 6 ft³, (c) no.   **D4.43** 5 ft.   **D4.44** $450\mathbf{i} - 700\mathbf{j} + 500\mathbf{k}$ lb·ft.   **D4.45** (a) $\frac{1}{7}(2\mathbf{i} - 3\mathbf{j} + 6\mathbf{k})$, (b) 2100 N·m, (c) $600\mathbf{i} - 900\mathbf{j} + 1800\mathbf{k}$ N·m.   **D4.46** (a) $-420\mathbf{i} + 140\mathbf{k}$ N, (b) 443N, (c) 443 N, (d) 4.74 m.   **D4.47** a, b, c, e.

Problems

4.1 20.1 N·m ↪.   **4.2** 620 lb·ft ↪.   **4.4** 8.11 lb·ft ↩.   **4.5** 40 lb·ft ↩.   **4.6** 53.6 N·m ↪.   **4.8** $3P - 30$ kN·m ↩.   **4.10** $8B_y - 2000$ lb·ft ↩.   **4.11** $24 - 6A_y$ kN·m ↩.   **4.12** $3A_x - 2A_y - 8$ kN·m ↩.   **4.14** (a) 66.6° or 246.6°, (b) 349 N·m.   **4.16** 62.4° or 185.0°.   **4.17** 220 lb·ft ↩.   **4.18** 300 N·m ↪.   **4.20** $200 - \frac{2}{5}P$ N·m ↩.   **4.21** $240 - \frac{6}{5}P$ N·m ↩.   **4.24** $2A_x - 3A_y - 240$ N·m ↩.   **4.25** $380 - \frac{38}{13}R$ lb·ft ↩.   **4.26** $M_x = -160$ N·m, $M_y = -220$ N·m, $M_z = 120$ N·m.   **4.28** $M_x = -1200$ lb·ft, $M_y = 1800$ lb·ft, $M_z = 1200$ lb·ft.   **4.29** $M_x^R = -4P + 8$ kN·m, $M_y^R = -6A_z + M_{Ay} - 3P + 2$ kN·m, $M_z^R = 6A_y - M_{Az}$ kN·m.   **4.30** $M_x^R = -6A_y + 36$ kN·m, $M_y^R = 6A_x + \frac{6}{13}M_A - \frac{12}{7}T_{BC} - 11$ kN·m, $M_z^R =$

$\frac{12}{13} M_A + \frac{6}{7} T_{BC} - 24$ kN·m. **4.32** $M_x^R = 870 - 9C_y$ lb·ft, $M_y^R = 4T_{EF} + 9B_x - 12C_z + M_{Cy}$ lb·ft, $M_z^R = 2T_{EF} + 12C_y + M_{Cz} - 1800$ lb·ft. **4.33** (a) $5\mathbf{i} + \mathbf{j} - 3\mathbf{k}$, (b) $-5\mathbf{i} - \mathbf{j} + 3\mathbf{k}$, (c) $-5\mathbf{i} - \mathbf{j} + 3\mathbf{k}$, (d) 5.92, (e) 40.2°. **4.35** 65.7 m². **4.36** 122.2 ft². **4.38** 42.4°. **4.39** 53.0°. **4.40** 19.46°. **4.41** (a) 51.2 ft², (b) 34.7°. **4.42** (a) 163.2 m², (b) 40.4°. **4.44** $6.5\mathbf{k}$ N·m. **4.45** $-680\mathbf{k}$ lb·ft. **4.46** $-800\mathbf{i} - 170\mathbf{j} - 200\mathbf{k}$ N·m. **4.47** (a) $-640\mathbf{i} - 800\mathbf{j} - 320\mathbf{k}$ N·m, (b) $-740\mathbf{i} - 1000\mathbf{j} - 1120\mathbf{k}$ N·m. **4.49** $(\frac{24}{7} T_{BF} - \frac{24}{7} T_{BG} + 300)\mathbf{i} + (\frac{36}{7} T_{BF} + \frac{36}{7} T_{BG} - 1200)\mathbf{k}$ kN·m. **4.50** $(M_C + B_y + C_y + 110)\mathbf{i} + (-C_x - 2C_z - 20)\mathbf{j} + (2C_y + 100)\mathbf{k}$ lb·ft. **4.51** $(-12A_z + 300)\mathbf{i} + (12A_x - 1200)\mathbf{k}$ kN·m. **4.53** (a) $\mathbf{M}_1 = -6\mathbf{i} - 8\mathbf{k}$ kN·m, $\mathbf{M}_2 = -8\mathbf{k}$ kN·m; (b) and (c) $-6\mathbf{i} - 16\mathbf{k}$ kN·m. **4.54** (a) $\mathbf{M}_1 = 6\mathbf{j} + 8\mathbf{k}$ kN·m, $\mathbf{M}_2 = 20\mathbf{k}$ kN·m; (b) and (c) $6\mathbf{j} + 28\mathbf{k}$ kN·m. **4.55** (a) $\mathbf{M}_1 = 400\mathbf{i} + 600\mathbf{k}$ lb·ft, $\mathbf{M}_2 = -800\mathbf{i} + 1200\mathbf{k}$ lb·ft, $\mathbf{M}_3 = 2000\mathbf{j}$ lb·ft; (b) and (c) $-400\mathbf{i} + 2000\mathbf{j} + 1800\mathbf{k}$ lb·ft. **4.57** (a) 76.8°, (b) 105.0°. **4.60** (a) $-600\mathbf{i} + 400\mathbf{j} - 300\mathbf{k}$ N·m, (b) 500 N·m, (c) 500 N·m, (d) $400\mathbf{j} - 300\mathbf{k}$ N·m. **4.61** 500 N·m. **4.62** 62.4 N. **4.64** (a) -18 N·m, (b) tighten. **4.65** 600 mm. **4.67** 196.1 m. **4.68** $8B_y - 540$ lb·ft $\circlearrowright$. **4.69** (a) $-126\mathbf{i} - 108\mathbf{j} - 36\mathbf{k}$ N·m, (b) -24 N·m, (c) tighten, (d) 629 mm, (e) 141.4 mm. **4.70** (a) $1000\mathbf{i} + 400\mathbf{j} + 300\mathbf{k}$ N·m, (b) 840 N·m. **4.71** (a) $(-4B_y + 2B_z)\mathbf{i} + (\frac{4}{3} T_{DE} + 4B_x - 4B_z + 1800)\mathbf{j} + (\frac{4}{3} T_{DE} + 2B_x - 4B_y)\mathbf{k}$ N·m, (b) $600 - \frac{4}{9} T_{DE}$ N·m.

Chapter 5

Developmental Exercises

D5.1 $\mathbf{M}_P' = \overrightarrow{PB} \times \mathbf{F} + \mathbf{M}_B = \overrightarrow{PB} \times \mathbf{F} + \overrightarrow{BA} \times \mathbf{F} = (\overrightarrow{PB} + \overrightarrow{BA}) \times \mathbf{F} = \overrightarrow{PA} \times \mathbf{F} = \mathbf{M}_P$. Thus, $\mathbf{M}_P = \mathbf{M}_P'$ is true. **D5.2** (a) $\mathbf{F}_O = 100\mathbf{j}$ N, $\mathbf{M}_O = 1200\mathbf{k}$ N·m; (b) $\mathbf{F}_B = 100\mathbf{j}$ N, $\mathbf{M}_B = 500\mathbf{i} + 1200\mathbf{k}$ N·m; (c) yes. **D5.3** (a) $4\mathbf{i} + 3\mathbf{j}$ kN, (b) $3\mathbf{i} + 6\mathbf{k}$ kN·m, (c) $4\mathbf{j}$ kN·m, (d) $3\mathbf{i} + 4\mathbf{j} + 6\mathbf{k}$ kN·m. **D5.5** $-10\mathbf{i} + 5\mathbf{j} + 10\mathbf{k}$ lb·ft. **D5.7** a, c, and f; d and e. **D5.8** (a) $10\mathbf{j}$ kN, (b) $-20\mathbf{j} + 10\mathbf{k}$ kN·m, (c) $\mathbf{j}$, (d) $-20\mathbf{j}$ kN·m, (e) $10\mathbf{k}$ kN·m, (f) $-\mathbf{i}$ m, (g) (1, 0, 0) m.

D5.12

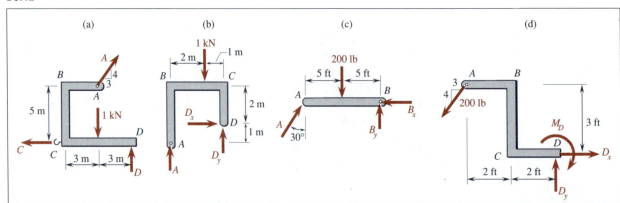

D5.15 b, c.   **D5.16** (a) Equilibrium problem of a *particle* in a plane, (b) equilibrium problem of a *rigid body* in a plane.   **D5.17** (a) $F_A = 80$ lb, $F_B = 60$ lb; (b) $F_A = 64$ lb, $F_B = 36$ lb.   **D5.18** (a) Unstable, (b) neutral, (c) stable.   **D5.20** Two-force: a, b; three-force: c.   **D5.21** Two-force: (a) *ABC*; three force: (a) *CDE*, (b) *CDE*, (c) *ABC*, *CDE*.   **D5.22** (a) Third degree, (b) first degree, (c) second degree.

D5.23

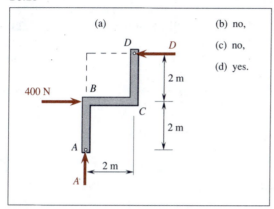

(a)

(b) no,

(c) no,

(d) yes.

D5.24 Partially constrained: three prospective independent equilibrium equations to be applied, $\Sigma F_y \neq 0$, two unknowns.

D5.25

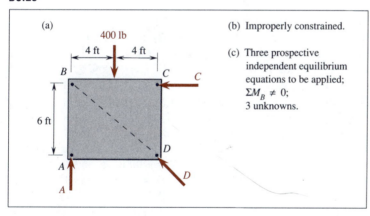

(a)

(b) Improperly constrained.

(c) Three prospective independent equilibrium equations to be applied; $\Sigma M_B \neq 0$; 3 unknowns.

D5.26 (a) 4, (b) 3, (c) B_y, (d) A_y, (e) no, (f) no, (g) no, (h) no.   **D5.27** $B_x = 980$ lb ←, $B_y = 2.35$ kips ↑.   **D5.29** $A_x = 50$ lb ←, $A_y = 30$ lb ↑.   **D5.32** No.   **D5.33** Wide.   **D5.34** (a) (i) 6, (ii) 6, (iii) no; (b) (i) 9, (ii) 6, (iii) statically indeterminate to the third degree.

Problems

5.1 $\mathbf{F}_O = 100$ N ⟋ 53.1°, $\mathbf{M}_O = 176$ N·m ⟳.   **5.2** $\mathbf{F}_O = -42\mathbf{i} + 24\mathbf{j} + 24\mathbf{k}$ lb, $\mathbf{M}_O = -160\mathbf{i} - 280\mathbf{j}$ lb·ft.   **5.3** $\mathbf{F}_C = 100$ N ⟋ 53.1°, $\mathbf{M}_C = 112$ N·m ⟳.   **5.5** $\mathbf{F}_O = -50\mathbf{i} - 50\mathbf{j}$ lb, $\mathbf{M}_O = -240\mathbf{k}$ lb·ft.   **5.6** $\mathbf{F}_O = 30$ kN ↑, $\mathbf{M}_O = 1440$ kN·m ⟲.   **5.7**

$F_O = -500\mathbf{j}$ lb, $M_O = 2.8\mathbf{i} - 4.5\mathbf{k}$ kip·ft. **5.9** $F_A = -50\mathbf{i} - 50\mathbf{j}$ lb, $M_A = -140\mathbf{k}$ lb·ft. **5.10** $F_A = 30$ kN $\uparrow$, $M_A = 1560$ kN·m $\downarrow$. **5.11** $F_A = 500$ lb $\downarrow$, $M_A = 2.8\mathbf{i} + 3\mathbf{k}$ kip·ft. **5.13** (b) $P = 200$ N, $Q = 200$ N, $M = 100$ N·m; (c) $P = 100$ N, $Q = 100$ N, $M = 400$ N·m; (d) $P = 300$ N, $Q = 100$ N, $M = 200$ N·m. **5.15** $P = 80$ lb, $Q = 35$ lb, $M = 17.5$ lb·ft. **5.16** $P = 50$ N, $Q = 16$ N, $d = 0.2$ m. **5.17** $A_x = 100$ N, $B = 50$ N, $P_y = 150$ N, $P_z = 225$ N, $Q = 100$ N, $S = 200$ N. **5.19** (a) $P = 40$ N, $Q = 230$ N, $S = 200$ N; (b) $R = 40\mathbf{i} - 30\mathbf{j}$ N at A, $M_\| = 400\mathbf{i} - 300\mathbf{j}$ N·m. **5.21** $-50\mathbf{i} - 50\mathbf{j}$ lb at (4.8, 0, 0) ft. **5.22** $30\mathbf{j}$ kN at (48, 0, 0) m. **5.23** $-500\mathbf{j}$ lb at (9, 0, 5.6) ft.

P5.25

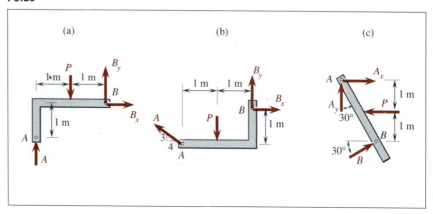

5.28 $A = 110$ lb $\searrow$ 53.1°, $B = 66\mathbf{i} + 32\mathbf{j}$ lb. **5.29** $A = 160$ lb $\searrow$ 53.1°, $B = 216\mathbf{i} - 128\mathbf{j}$ lb. **5.31** $A = 6$ kN $\measuredangle$ 53.1°, $C = -3.6\mathbf{i} + 3.2\mathbf{j}$ kN. **5.32** $A = 6.4\mathbf{i} + 3.2\mathbf{j}$ kN, $C = 8$ kN $\searrow$ 36.9°. **5.34** (a) $A = 3$ kN $\uparrow$, $B = 3$ kN $\uparrow$; (b) $A = 3$ kN $\searrow$ 36.9°, $B = 2.4\mathbf{i} + 4.2\mathbf{j}$ kN; (c) $A = 3.75\mathbf{i} - 1.299\mathbf{j}$ kN, $B = 2.60$ kN $\measuredangle$ 30°. **5.35** (a) $A = 8$ kN $\measuredangle$ 53.1°, $B = -4.8\mathbf{i} - 0.4\mathbf{j}$ kN; (b) $A = 6$ kN $\uparrow$, $M_A = 18$ kN·m $\circlearrowright$; (c) $A = 1.6\mathbf{i} + 4.8\mathbf{j}$ kN, $B = 2$ kN $\searrow$ 36.9°. **5.36** (a) $A = 500$ lb $\searrow$ 53.1°, $B = 300\mathbf{i} + 200\mathbf{j}$ lb; (b) $A = 2.6$ kips $\measuredangle$ 53.1°, $B = -1.56\mathbf{i} - 1.48\mathbf{j}$ kips; (c) $A = 200$ lb $\measuredangle$ 36.9°, $B = -160\mathbf{i} + 480\mathbf{j}$ lb. **5.38** $A = 48\mathbf{i} + 86\mathbf{j}$ lb, $D = 80$ lb $\searrow$ 53.1°. **5.39** $A = 58.9\mathbf{i} + 1109\mathbf{j}$ N, $D = 490$ N $\measuredangle$ 53.1°. **5.40** 74.8 kN $\measuredangle$ 87.7°. **5.41** (a) 20.6 kN; (b) 5.21 kN $\measuredangle$ 65.8°. **5.42** 4 Mg. **5.43** $A = 900\mathbf{i} - 900\mathbf{j}$ lb. **5.45** (a) 100.3 N; (b) $T_{AB} = 56.6$ N, $T_{AC} = 88.2$ N, $T_{AD} = 126.7$ N, $T_{BC} = 18.37$ N, $T_{BE} = 63.6$ N, $T_{CG} = 98.1$ N. **5.46** (a) 32.5°, (b) 63.8 lb. **5.47** (a) 4.71 m, (b) 72.5 N. **5.48** Partially constrained. **5.50** Improperly constrained. **5.52** W. **5.54** $T_A = 490$ N, $T_B = 981$ N, $T_C = 490$ N. **5.56** $B = 1600\mathbf{i} + 98.1\mathbf{j}$ N, $C = -1200\mathbf{i} - 49.0\mathbf{j}$ N. **5.57** $B = 1500\mathbf{i} + 122.6\mathbf{j}$ N, $C = -1200\mathbf{i} - 73.6\mathbf{j}$ N. **5.58** $Wr^2\theta/L$. **5.60** $T_{BF} = 85.8$ kN, $T_{BG} = 188.8$ kN, $A = 117.7\mathbf{i} + 353\mathbf{j} + 29.4\mathbf{k}$ kN. **5.61** $A = 24.5\mathbf{i} + 98.1\mathbf{j}$ N, $B = 24.5\mathbf{i}$ N, $C = -49.0\mathbf{i}$ N. **5.62** 4 kN. **5.63** 1120 lb. **5.64** 108 kN. **5.65** 84 kN. **5.68** $A = 2.4\mathbf{i} + 8.8\mathbf{j}$ kN, $M_A = 12\mathbf{j} - 27.2\mathbf{k}$ kN·m. **5.69** $A = 400\mathbf{j} + 360\mathbf{k}$ lb, $M_A = -4.16\mathbf{k}$

kip·ft.   **5.70 A** $= 120\mathbf{i} - 12\mathbf{j} - 8\mathbf{k}$ kN.   **5.71 A** $= 21\mathbf{i} + 36\mathbf{j} - 75\mathbf{k}$ kN, $\mathbf{M}_A = 30\mathbf{j} + 72\mathbf{k}$ kN·m.   **5.74** $T_{DF} = 1$ kN, $\mathbf{A} = 6\mathbf{i} + 2\mathbf{j} + 6\mathbf{k}$ kN, $\mathbf{B} = -3\mathbf{j} - 10\mathbf{k}$ kN.   **5.75** $T_{EF} = 1800$ lb, $\mathbf{A} = 900\mathbf{i} - 450\mathbf{j}$ lb, $\mathbf{B} = -300\mathbf{i} - 750\mathbf{j} + 1500\mathbf{k}$ lb.   **5.78 A** $= 500\mathbf{i} + 600\mathbf{k}$ N, $\mathbf{M}_A = 900\mathbf{j} + 1200\mathbf{k}$ N·m, $\mathbf{B} = -500\mathbf{i} + 600\mathbf{j}$ N.   **5.80 A** $= 200\mathbf{i} + 150\mathbf{k}$ lb, $\mathbf{C} = -200\mathbf{i} + 300\mathbf{j}$ lb, $\mathbf{D} = 300\mathbf{j} - 150\mathbf{k}$ lb.   **5.82 B** $= -25\mathbf{j} - 30\mathbf{k}$ N, $\mathbf{C} = 125\mathbf{j} + 30\mathbf{k}$ N, $\mathbf{D} = -60\mathbf{i} - 100\mathbf{j}$ N.   **5.83 A** $= -175\mathbf{i} + 100\mathbf{j} + 100\mathbf{k}$ lb, $\mathbf{B} = -75\mathbf{i} + 200\mathbf{j} - 100\mathbf{k}$ lb, $\mathbf{D} = 250\mathbf{i}$ lb.   **5.84** (a) (3.5, 6, 0) m; (b) $(\mathbf{j} + \mathbf{k})/\sqrt{2}$; (c) $\mathbf{A} = -128.8\mathbf{i} + 196.2\mathbf{j} - 98.1\mathbf{k}$ N, $\mathbf{B} = 128.8\mathbf{i}$ N, $\mathbf{E} = 98.1\mathbf{j} + 98.1\mathbf{k}$ N.   **5.85** (a) 8, (b) 6, (c) yes, (d) $\mathbf{C}_y = 360\mathbf{j}$ N.   **5.86** (a) 7, (b) 6, (c) yes, (d) $\mathbf{B}_y = 450\mathbf{j}$ lb.   **5.87** (a) 8; (b) 6; (c) yes; (d) $T_{EF} = 1080$ N, $\mathbf{C}_y = 360\mathbf{j}$ N, $\mathbf{M}_{Cz} = -720\mathbf{k}$ N·m.   **5.89** $T_{AD} = 54.6$ N, $T_{AE} = 38.7$ N, $T_{BF} = 78.6$ N, $T_{BG} = 48.0$ N, $T_{CI} = 34.0$ N, $T_{CH} = 75.1$ N.   **5.90** (a) (i) A_x, A_y, M_A, and C; (ii) $\Sigma F_x = 0$, $\Sigma F_y = 0$, $\Sigma M_A = 0$; (iii) none; (iv) statically indeterminate. (b) (i) A, C, T_{EF}; (ii) $\Sigma F_x = 0$, $\Sigma F_y = 0$, $\Sigma M_A = 0$; (iii) $\Sigma F_x = 0$; (iv) improperly constrained. (c) (i) A_x, A_y, A_z, M_{Ay}, M_{Az}, T_{EF}; (ii) all 6 eqs., (iii) none; (iv) statically determinate.   **5.93** (a) $\mathbf{A} = -2\mathbf{i} + \mathbf{j}$ kN, $\mathbf{M}_A = 13$ kN·m $\circlearrowleft$; (b) $\mathbf{A} = 275$ lb $\measuredangle 60°$; (c) $\mathbf{A} = 12\mathbf{i} + 37\mathbf{j}$ lb.   **5.97** 18 kN.   **5.98 A** $= -20\mathbf{i} + 18\mathbf{j} - 10\mathbf{k}$ kN, $\mathbf{M}_A = -20\mathbf{j} - 90\mathbf{k}$ kN·m, $\mathbf{C} = 12\mathbf{j}$ kN.   **5.99** (a) 7 unknowns (A:4, B:1, D:1, F:1), 6 equations; (b) $\mathbf{A}_x = 3\mathbf{i}$ kips, $\mathbf{A}_z = 6\mathbf{k}$ kips, $\mathbf{D} = 1\mathbf{j}$ kip, $T_{FG} = 18$ kips.

Chapter 6

Developmental Exercises

D6.1 (a) $\frac{1}{4}$ kN/m^3, (b) $\frac{1}{4}$ kN/m.   **D6.2** 1 ft.   **D6.3** 2 ft.   **D6.4** (a) $(-4.25, 8.65)$ km; (b) $(-1, 5.6)$ km.   **D6.7** (a) $d\theta$, (b) $\pi/2$ ft, (c) $\frac{1}{2}$ ft^2, (d) $1/\pi$ ft, (e) $\pi/4$ ft^2, (f) $\frac{1}{2}$ ft.   **D6.8** (0.652, 2.52) ft.   **D6.9** (2.49, -0.1195, -0.448) m.   **D6.10** $T_A = 10$ lb, $T_B = 15$ lb, $T_C = 75$ lb.   **D6.12** (a) $(4 - \frac{1}{4}y^2)\,dy$, (b) $2y^{1/2}\,dy$.   **D6.13** (a) $4x^{1/2}\,dx$ (b) $(4 - \frac{1}{4}x^2)\,dx$.   **D6.15** (a) $(2\pi R/h^2)(R^2 + h^2)^{1/2}\,x\,dx$, (b) $\pi R(R^2 + h^2)^{1/2}$, (c) $\frac{2}{3}\pi R h(R^2 + h^2)^{1/2}$, (d) $\frac{2}{3}h$.   **D6.16** $\bar{x} = [4/(3\pi)](a + b)$, $\bar{y}$, $= 4b/(3\pi)$.   **D6.17** $T_A = 610$ N, $T_B = 809$ N.   **D6.18** 6.38 m.   **D6.19** 1 kip ↑.   **D6.20 B** $= 2$ kN ↑, $\mathbf{M}_B = 1$ kN·m $\circlearrowleft$.   **D6.22** (a) $R/2$, (b) πR, (c) $\pi R L$.   **D6.24** (a) $R/3$, (b) $\frac{2}{3}\pi R$, (c) $\frac{1}{3}\pi R^2 h$.   **D6.25** (a) $(\pi/h^2)(R^2 x^2\,dx)$, (b) $\frac{1}{3}\pi R^2 h$, (c) $\frac{1}{4}\pi R^2 h^2$, (d) $\frac{3}{4}h$.   **D6.26** No.   **D6.27** 9.42 kN ←.   **D6.28** $(\frac{11}{8}a, 0, 0)$.   **D6.29** (a) (2, 1, 0.5) m; (b) (1.6, 1.6, 0.4) m.   **D6.30** (0.568, 0, 0) m.

Problems

6.1 (a) $c = 16$ m, $k = -64$ m^2; (b) $c = 3$ ft, $k = 2$.   **6.2** (a) 0.25 kN/m, (b) 0.707 kN/m.   **6.3** $\bar{x} = 1$ m.   **6.4** $\bar{x} = 1.85$ m, $\bar{z} = 0.6$ m.   **6.5** $\bar{x} = 1.5$ ft, $\bar{y} = 2.5$ ft.   **6.6** $1/\pi$ m.   **6.7** $R\,(\sin\alpha)/\alpha$.   **6.8** 2.82 ft.   **6.9** $\bar{x} = 0.5$ m, $\bar{y} = 3$ m.   **6.10** $\bar{x} = 0.569$ m, $\bar{y} =$

1.395 m. **6.12** $\bar{x} = -0.0981$ m, $\bar{y} = 1.465$ m. **6.13** $\bar{x} = -0.200a$, $\bar{y} = 0.280a$. **6.15** $\bar{x} = 2.55$ m, $\bar{y} = 0.8$ m, $\bar{z} = 0.75$ m. **6.16** $\bar{x} = 0.553$ ft, $\bar{y} = 0.246$ ft, $\bar{z} = 1.510$ ft. **6.18** 1.075 m. **6.19** 0.931m. **6.21** A = 71.9N ↑, B = 87.0 N ↑. **6.22** $T_A = 22.9$ N, $T_D = 128.4$ N. **6.24** $B_x = 35.7$ lb ←, $B_y = 186.7$ lb ↑, D = 59.5 lb ∠ 53.1°. **6.25** $T_B = 134.3$ lb, $T_C = 51.3$ lb, $T_D = 80$ lb. **6.27** $\bar{x} = 3$ m, $\bar{y} = 1.2$ m. **6.28** $\bar{x} = 1.5$ ft, $\bar{y} = 1.2$ ft. **6.30** $\bar{x} = 1.75$ ft, $\bar{y} = 1.7$ ft. **6.31** $\bar{x} = 4a/(3\pi)$, $\bar{y} = 4b/(3\pi)$. **6.33** $\bar{x} = 0.934$ ft, $\bar{y} = 0.75$ ft. **6.34** $\bar{x} = 1.098$ m, $\bar{y} = 0.224$ m. **6.36** $\bar{x} = 36$ mm, $\bar{y} = 14$ mm. **6.37** $\bar{x} = 23.5$ in., $\bar{y} = 60.8$ in. **6.40** $\bar{x} = 17.42$ in., $\bar{y} = 12.79$ in. **6.42** $\bar{x} = 6.49$ in., $\bar{y} = 5.35$ in. **6.43** $\bar{x} = 100$ mm, $\bar{y} = 65$ mm. **6.45** $T_A = 14.72$ N, $T_B = 8.83$ N. **6.46** $T_A = 20.4$ N, $T_B = 21.2$ N. **6.48** $\bar{x} = 2.8$ ft, $\bar{y} = \bar{z} = 0$. **6.49** $\bar{x} = 0.456$ m, $\bar{y} = \bar{z} = 0$. **6.50** $\bar{x} = \pi R/4$, $\bar{y} = \pi h/8$, $\bar{z} = 0$. **6.52** $\bar{x} = 0.446$ m, $\bar{y} = -0.1377$ m, $\bar{z} = 0.895$ m. **6.53** $\bar{x} = 0.394$ m, $\bar{y} = -1.873$ m, $\bar{z} = -0.986$ m. **6.54** A = 1.2 kN ↑, B = 1.5 kN ↑. **6.55** A = 9 kips ↑, $M_A = 27$ kip·ft ↺. **6.56** A = 2.08i + 4.86j kips, B = 3.47 kips ∠ 53.1°. **6.57** A = 1100 N ↑, B = 1450 N ↑. **6.58** B = 27.2 kN ↑, C = 42.8 kN ↑. **6.60** 143.6 in². **6.61** $(\pi/2)(8 + \pi)a^2$. **6.63** 118.4 in³. **6.64** $(\pi/12)(4 + 3\pi)a^3$. **6.66** 4.95 kg. **6.67** 39.0 L. **6.68** $\bar{x} = \bar{y} = 0$, $\bar{z} = 1.5$ m. **6.69** $\bar{x} = \bar{y} = 0$, $\bar{z} = 3h/8$. **6.70** $\bar{x} = 0$, $\bar{y} = 2.5$ ft, $\bar{z} = -1$ ft. **6.71** $\bar{x} = 0$, $\bar{y} = a(8R - 3a)/[2(3\pi R - 4a)]$, $\bar{z} = -(12\pi R^2 - 32Ra + 3\pi a^2)/[2\pi(3\pi R - 4a)]$. **6.72** $\bar{x} = \bar{y} = 0$, $\bar{z} = 9$ in. **6.74** 4. **6.75** $\bar{x} = \bar{z} = 0$, $\bar{y} = 5.60$ ft. **6.76** $\bar{x} = 5.93$ in., $\bar{y} = 0.414$ in., $\bar{z} = 0$. **6.77** 3.07 m. **6.78** 11.51 ft. **6.80** 749 lb. **6.82** 70.9 kN. **6.83** 9.6 ft. **6.84** 981 N. **6.86** (a) $\pi R^3 \rho g$, (b) $3R/4$. **6.87** $T_A = 988$ N, $T_B = 2.45$ kN. **6.89** $\bar{x} = 4a/5$, $\bar{y} = 3b/4$, $\bar{z} = 2c/3$. **6.90** $\bar{x} = 2.4$ ft, $\bar{y} = 1.2$ ft, $\bar{z} = 0.8$ ft. **6.91** $\bar{x} = 20$ ft. **6.92** D = 288 lb ∠ 67.4°, A = −110.7i − 8.30j lb. **6.93** (a) 140 mm, (b) 138.3 mm. **6.95** A = 9.68 kN ↑, B = 7.32 kN ↑. **6.97** $\bar{x} = \bar{y} = 0$, $\bar{z} = 6$ in. **6.99** $\bar{x} = h/4$, $\bar{y} = R (\sin\alpha)/(2\alpha)$, $\bar{z} = 0$. **6.100** $T_A = 0$, $T_B = 9.07$ kN. **6.102** $\bar{x} = 2.79$ in., $\bar{y} = \bar{z} = 0$.

Chapter 7

Developmental Exercises

D7.4 $(bh/12)(b^2 + h^2)$. **D7.5** $\frac{1}{4}\pi R^4$. **D7.6** $I_x = \frac{1}{12} bh^3$, $I_y = \frac{1}{12} b^3 h$. **D7.7** No. **D7.8** No. **D7.9** through **D7.11** No. The integrals do not conform to the definition in Sec. 7.2. **D7.13** $\bar{k}_x = (\sqrt{3}/6)h$, $\bar{k}_y = (\sqrt{3}/6) b$. **D7.14** $(\sqrt{2}/2) R$. **D7.17** (a) $bh^3/3$, (b) $7b^3h/3$, (c) $(bh/3)(h^2 + 7b^2)$. **D7.18** 50 ft⁴. **D7.20** (a) 0, (b) 0, (c) 0, (d) 0. **D7.21** $\frac{1}{24} b^2 h^2$. **D7.23** (a) $\frac{1}{3} x^2$ dx, (b) $\frac{1}{18} x^5$ dx, (c) 6.75 ft⁴. **D7.24** (a) 4.32×10^3 mm⁴, (b) 5.48 mm. **D7.25** (a) 27.6×10^3 mm⁴, (b) 13.86 mm. **D7.26** (a) 779 mm⁴, (b) 4.27 mm. **D7.27** $I_x = 114.6$ in⁴, $I_y = 427$ in⁴, $I_{xy} = -173.2$ in⁴. **D7.30** Yes. **D7.31** 300 mm⁴. **D7.32** X: (I_x, I_{xy}); Y: $(I_y, -I_{xy})$. **D7.34** $(I_{avg}, 0)$.

D7.35

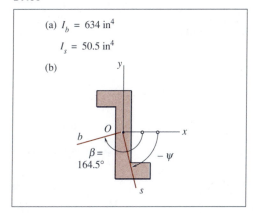

(a) $I_b = 634$ in^4

$I_s = 50.5$ in^4

(b)

D7.39 $\frac{1}{12} mc^2$.   **D7.40** $\frac{1}{2} mr^2$.   **D7.41** No.   **D7.42** 1088 slug·ft^2.   **D7.44** $0.548r$.   **D7.45** 8 kg·m^2.   **D7.46** No.
D7.49 $(m/20)(3r^2 + 2h)$.   **D7.50** $I_x = 1.89$ kg·m^2, $k_x = 410$ mm.
D7.51 $I_z = 8.2$ kg·m^2, $k_z = 854$ mm.   **D7.52** $I_x = 0.310$ slug·ft^2, $k_x = 7.89$ in.   **D7.53** $I_z = 0.487$ slug·ft^2, $k_z = 9.89$ in.   **D7.54** $I_x = 9.31 \times 10^{-3}$ kg·m^2, $k_x = 56.6$ mm.   **D7.55** $I_y = 4.09 \times 10^{-3}$ kg·m^2, $k_y = 37.5$ mm.

Problems

7.1 $\frac{1}{12} bh^3$.   **7.2** 3.15×10^3 mm^4.   **7.3** 6.75×10^3 mm^4.   **7.4** $\frac{3}{35} a^4$.   **7.5** 0.305 ft^4.   **7.6** $(\pi/16) ab^3$.   **7.8** 1470 mm^4.   **7.9** 686 mm^4.   **7.10** $\frac{3}{35} a^4$.   **7.11** 1.6 ft^4.   **7.13** $\frac{1}{12} bh(3b^2 + h^2)$.
7.14 4.62×10^3 mm^4.   **7.16** $\frac{6}{35} a^4$.   **7.17** 1.905 ft^4.   **7.25** $I_x = 32.4$ m^4, $k_x = 1.342$ m.   **7.26** $I_x = 43.2$ ft^4, $k_x = 1.897$ ft.   **7.28** $I_y = 12.15$ m^4, $k_y = 1.643$ m.   **7.29** $I_y = 10.8$ m^4, $k_y = 1.342$ m.
7.31 $I_x = \frac{1}{8}(2\alpha - \sin 2\alpha)R^4$, $k_x = [(2\alpha - \sin 2\alpha)/(8\alpha)]^{1/2}R$.   **7.32** $I_x = 6.96 \times 10^{-3} a^4$, $k_x = 0.1332a$.   **7.34** $I_y = \frac{1}{8}(2\alpha + \sin 2\alpha)R^4$, $k_y = [(2\alpha + \sin 2\alpha)/(8\alpha)]^{1/2}R$.   **7.35** $I_y = 0.1403a^4$, $k_y = 0.598a$.   **7.37** $J_O = \frac{1}{2}\alpha R^4$, $k_O = (\sqrt{2}/2) R$.   **7.38** $J_O = 0.1473a^4$, $k_O = 0.612a$.
7.40 19.50 m^4.   **7.41** 0.914 ft^4.   **7.43** 0.305 m^4.   **7.46** 5.33 m^4.   **7.47** 0.583 ft^4.   **7.49** -0.533 m^4.   **7.50** -1.333 m^4.
7.52 $\bar{I}_x = 1.102$ m^4, $\bar{k}_x = 0.714$ m.   **7.53** $\bar{I}_x = 355$ in^4, $\bar{k}_x = 3.33$ in.   **7.55** $\bar{I}_x = 13.6 \times 10^9$ mm^4, $\bar{k}_x = 238$ mm.   **7.56** $\bar{I}_x = 35.1 \times 10^6$ mm^4, $\bar{k}_x = 62.4$ mm.   **7.58** $\bar{I}_y = 0.324$ m^4, $\bar{k}_y = 0.387$ m.
7.59 $\bar{I}_y = 50.7$ in^4, $\bar{k}_y = 1.258$ in.   **7.61** $\bar{I}_y = 6.4 \times 10^9$ mm^4, $\bar{k}_y = 163.3$ mm.   **7.62** $\bar{I}_y = 8.78 \times 10^6$ mm^4, $\bar{k}_y = 31.2$ mm.   **7.64** -4.8×10^9 mm^4.   **7.65** -12.15×10^6 mm^4.   **7.67** 83.5×10^3 in^4.   **7.69** 1619 in^4.   **7.71** 1.541×10^9 mm^4.   **7.74** 1.019×10^9 mm^4.   **7.75** 1670 in^4.   **7.76** 237×10^3 in^4.   **7.79** $\bar{I}_x = 127.5$ in^4, $\bar{I}_y = 27.2$ in^4, $\bar{J}_C = 154.7$ in^4.   **7.80** $\bar{I}_x = 92.8$ in^4, $\bar{I}_y = 497$ in^4, $\bar{J}_C = 590$ in^4.   **7.81** $\bar{I}_x = 267$ in^4, $\bar{I}_y = 305$ in^4, $\bar{J}_C = 572$ in^4.
7.82 9.25 in.   **7.84** $\bar{I}_u = 279$ in^4, $\bar{I}_v = 126.7$ in^4, $\bar{I}_{uv} = -131.6$ in^4.
7.85 $\bar{I}_u = 4.56 \times 10^6$ mm^4, $\bar{I}_v = 2.09 \times 10^6$ mm^4, $\bar{I}_{uv} = -1.036 \times 10^6$ mm^4.   **7.86** $\bar{I}_u = 12.36 \times 10^9$ mm^4, $\bar{I}_v = 7.64 \times 10^9$ mm^4, $\bar{I}_{uv} = 5.52 \times 10^9$ mm^4.   **7.88** $\bar{I}_u = 270$ in^4, $\bar{I}_v = 178.4$ in^4, $\bar{I}_{uv} = -90.3$

in^4.   **7.98** $\bar{I}_u = 7.69 \times 10^6$ mm^4, $\bar{I}_v = 36.2 \times 10^6$ mm^4, $\bar{I}_{uv} = -10.85 \times 10^6$ mm^4.   **7.99** $\bar{I}_u = 9.31 \times 10^6$ mm^4, $\bar{I}_v = 8.08 \times 10^6$ mm^4, $\bar{I}_{uv} = 4.82 \times 10^6$ mm^4.

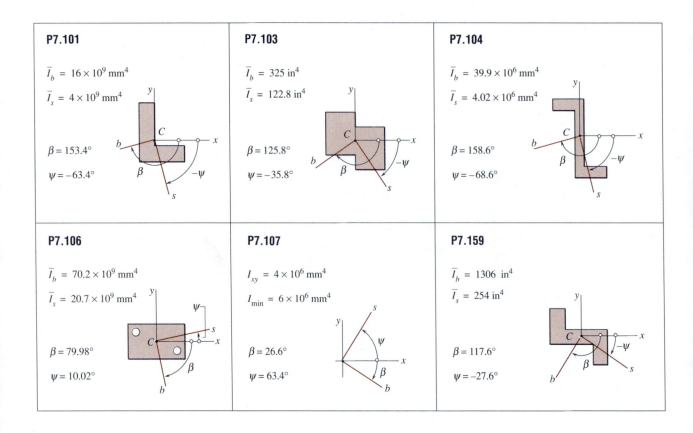

P7.101

$\bar{I}_b = 16 \times 10^9$ mm^4

$\bar{I}_s = 4 \times 10^9$ mm^4

$\beta = 153.4°$

$\psi = -63.4°$

P7.103

$\bar{I}_b = 325$ in^4

$\bar{I}_s = 122.8$ in^4

$\beta = 125.8°$

$\psi = -35.8°$

P7.104

$\bar{I}_b = 39.9 \times 10^6$ mm^4

$\bar{I}_s = 4.02 \times 10^6$ mm^4

$\beta = 158.6°$

$\psi = -68.6°$

P7.106

$\bar{I}_b = 70.2 \times 10^9$ mm^4

$\bar{I}_s = 20.7 \times 10^9$ mm^4

$\beta = 79.98°$

$\psi = 10.02°$

P7.107

$I_{xy} = 4 \times 10^6$ mm^4

$I_{min} = 6 \times 10^6$ mm^4

$\beta = 26.6°$

$\psi = 63.4°$

P7.159

$\bar{I}_b = 1306$ in^4

$\bar{I}_s = 254$ in^4

$\beta = 117.6°$

$\psi = -27.6°$

7.110 $I_x = \frac{1}{3} mb^2$, $I_y = \frac{1}{3} ma^2$, $I_z = \frac{1}{3} m(a^2 + b^2)$.   **7.117** $I_x = 21.5 \times 10^{-3}$ slug·ft^2, $k_x = 2.31$ in.   **7.118** $I_x = 326$ kg·mm^2, $k_x = 18.97$ mm.   **7.120** $\frac{1}{12} m(4b^2 + c^2)$.   **7.122** $I_y = \frac{1}{12} m(4a^2 + c^2)$, $k_y = (\sqrt{3}/6)(4a^2 + c^2)^{1/2}$.   **7.123** $I_y = \frac{1}{6} m(a^2 + c^2)$, $k_y = (\sqrt{6}/6)(a^2 + c^2)^{1/2}$.   **7.124** $I_x = 19.44$ kg·m^2, $k_x = 367$ mm.   **7.126** $I_x = 0.944$ kg·m^2, $I_y = 0.512$ kg·m^2.   **7.128** $I_x = 60.9 \times 10^{-3}$ kg·m^2, $I_y = 143.2 \times 10^{-3}$ kg·m^2.   **7.130** $I_z = 2.4$ slug·ft^2, $k_z = 1.414$ ft.   **7.131** $I_z = 103.6 \times 10^{-3}$ kg·m^2, $k_z = 158.7$ mm.   **7.132** $I_x = 55 \times 10^{-3}$ slug·ft^2, $I_y = 185 \times 10^{-3}$ slug·ft^2.   **7.134** $I_x = 1.259$ slug·ft^2, $I_y = 2.32$ slug·ft^2.   **7.136** $I_z = 748$ kg·mm^2, $k_z = 69.3$ mm.   **7.137** $I_z = 1.179$ slug·ft^2, $k_z = 1.145$ ft.   **7.138** $I_y = 110.8$ kg·mm^2, $k_y = 35.4$ mm.   **7.140** $I_y = 419$ kg·mm^2, $k_y = 47.1$ mm.   **7.142** $I_x = 0.933$ slug·ft^2, $k_x = 11.04$ in.   **7.143** $I_x = 681$ kg·mm^2, $k_x = 60.1$ mm.   **7.144** $I_y = 1.570 \times 10^{-3}$ kg·m^2, $k_y = 25.2$ mm.   **7.146** $I_y = 7.46 \times 10^{-3}$ kg·m^2, $k_y = 34.5$ mm.   **7.148** $I_x = 39.5$ slug·ft^2, $k_x = 10.36$ in.   **7.149** $I_x = 20.2 \times 10^{-3}$ kg·m^2, $k_x = 56.7$ mm.   **7.150** $I_z = 21.3 \times 10^{-3}$ slug·ft^2, $k_z = 2.46$ in.   **7.151** $I_x = 4.25$ kg·m^2, $k_x = 204$ mm.   **7.152** $I_y = 4.66 \times 10^{-3}$ kg·m^2, $I_z = 2.63 \times 10^{-3}$ kg·m^2.   **7.156** $\bar{I}_x = 136$ in^4, $\bar{I}_y = 40$ in^4, $\bar{k}_x = 2.38$ in.,

$\bar{k}_y = 1.291$ in.   **7.157** $\bar{I}_x = 480$ in^4, $\bar{I}_y = 1080$ in^4, $\bar{I}_{xy} = -432$ in^4, $\bar{I}_u = 1004$ in^4, $\bar{I}_v = 556$ in^4, $\bar{I}_{uv} = -476$ in^4.   **7.159** $\bar{I}_b = 1306$ in^4, $\bar{I}_s = 254$ in^4, $\beta = 117.6°$, $\psi = -27.6°$.   **7.161** $I_x = 1.660 \times 10^9$ mm^4, $I_y = 2.31 \times 10^9$ mm^4, $I_{xy} = 1.720 \times 10^9$ mm^4.   **7.162** 2.60 N.   **7.166** $I_x = 26.6$ slug·ft^2, $k_x = 3.79$ ft.   **7.168** $I_x = 1.130 \times 10^{-3}$ kg·m^2, $k_x = 48.9$ mm.   **7.169** $I_y = 2.18 \times 10^{-3}$ kg·m^2, $k_y = 67.9$ mm.   **7.171** $I_y = 46.1 \times 10^{-3}$ kg·m^2, $k_y = 86.1$ mm.

Chapter 8

Developmental Exercises

D8.6 9 joints.   **D8.7** 6 members.   **D8.8** No.   **D8.10** (a) *ACF* and *BDE*; (b) *AB*, *CD*, and *EF*; (c) compound truss.   **D8.12** F_{CD}.   **D8.13** Yes.   **D8.15** *AB*, *BC*, *EF*, *EG*.   **D8.16** $F_{AB} = 6$ kips *C*, $F_{AD} = 4.8$ kips *T*, $F_{BC} = 8$ kips *C*, $F_{BD} = 10$ kips *T*, $F_{CD} = 4.8$ kips *T*.   **D8.17** $F_{CF} = -\frac{10}{13} F_{EF}$, $F_{CD} = \frac{25}{39} F_{EF}$.   **D8.18** 10 kips *T*.   **D8.19** $F_{EK} = 46.9$ kips *T*, $F_{DE} = 30$ kips *T*, $F_{HO} = 30$ kips *C*.   **D8.20** 15 kips *C*.   **D8.21** c.   **D8.22** 17 kN *T*.   **D8.23** 3.9 kN *C*.   **D8.24** 5.2 kN *T*.   **D8.25** 28 kN *T*.   **D8.26** 17 kN *C*.   **D8.28** 24 members.   **D8.29** (a) $\overline{AM} = \frac{1}{2}$ m, $\overline{OA} = (\sqrt{3}/3)$ m, $\overline{OD} = (\sqrt{6}/3)$ m; (b) $\frac{1}{3}$ kN *C*; (c) $F_{AB} = F_{AC} = F_{BC} = 136.1$ N *T*.   **D8.32** (a) A solid circular dot, (b) a small circular curve.

D8.33

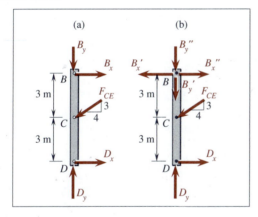

D8.34

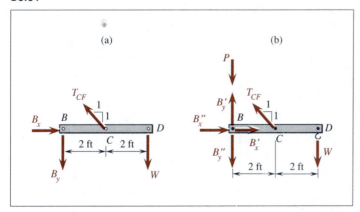

D8.52

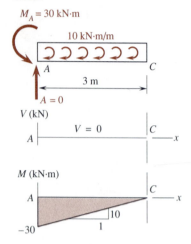

$M_A = 30$ kN·m

10 kN·m/m

A C

3 m

$A = 0$

V (kN)

$V = 0$ C

A ————————— x

M (kN·m)

A C

          10
          1
-30

D8.36 $A = -360i$ lb, $B = 480j$ lb, $C' = 360i - 480j$ lb.   **D8.37** $A = -960i - 1220j$ lb, $B' = 960i + 1340j$ lb.   **D8.38** 11.70 kips.
D8.39 (a) $(\sqrt{3}i - j)/2$, (b) 16.35j N, (c) $-7.08i + 4.09j$ N, (d) $7.08i + 12.26j$ N, (e) $13.83k$ N·m, (f) **0**, (g) $13.83k$ N·m.   **D8.40** Horizontal.   **D8.46** True.   **D8.47** Yes.   **D8.49** (a) -12 kN/m, (b) -18 kN·m/m, (c) 16 kN·m/m.   **D9.50** Yes.   **D8.51** (a) ↑, (b) ↓.   **D8.54** 1082 lb.   **D8.55** $T_{BC} = 1082$ lb, $T_{CD} = 949$ lb.
D8.56 d.   **D8.57** No.   **D8.58** Horizontal.   **D8.60** 5.64 kN.
D8.61 (a) 30.7 m, (b) 51.9 m.   **D8.62** (a) 126.6 m, (b) 1340 N.
D8.63 109.6 m.

Problems

8.1 $F_{AB} = 12.75$ kN T, $F_{AC} = 15.70$ kN T, $F_{BC} = 19.62$ kN C.   **8.2** $F_{AB} = 3.2$ kN C, $F_{AC} = 0$, $F_{BC} = 12.6$ kN C.   **8.4** $F_{AB} = 260$ lb T, $F_{AC} = 170$ lb T, $F_{AD} = 0$, $F_{BC} = 240$ lb C, $F_{CD} = 320$ lb C.   **8.5** $F_{AB} = 5.6$ kN T, $F_{AH} = 6$ kN T, $F_{BC} = 0$, $F_{BD} = 3.4$ kN T, $F_{BH} = 5$ kN C, $F_{CD} = 0$, $F_{DE} = 6$ kN T, $F_{DF} = 3.4$ kN C, $F_{DH} = 0$, $F_{EF} = 0$, $F_{FG} = 5.6$ kN C, $F_{FH} = 5$ kN T, $F_{GH} = 0$.   **8.7** $F_{AB} = 75$ lb T, $F_{AF} = 45$ lb C, $F_{BC} = 90$ lb T, $F_{BE} = 75$ lb C, $F_{BF} = 0$, $F_{CD} = 90$ lb T, $F_{CE} = 0$, $F_{DE} = 150$ lb C, $F_{EF} = 45$ lb C.   **8.8** $F_{AB} = 15$ kN C, $F_{AH} = 12$ kN T, $F_{BC} = 5$ kN C, $F_{BG} = 10$ kN C, $F_{BH} = 12$ kN T, $F_{CD} = 5$ kN C, $F_{CG} = 6$ kN T, $F_{DE} = 5$ kN C, $F_{DF} = 0$, $F_{DG} = 0$, $F_{EF} = 4$ kN T, $F_{FG} = 4$ kN T, $F_{GH} = 12$ kN T.   **8.10** $F_{AB} = 157.0$ kN T, $F_{BC} = 78.5$ kN T, $F_{BE} = 58.9$ kN T, $F_{BF} = 98.1$ kN C, $F_{CD} = 58.9$ kN T, $F_{CE} = 98.1$ kN C, $F_{DE} = 0$, $F_{EF} = 78.5$ kN C.   **8.11** $F_{AB} = 52$ kN C, $F_{AF} = 16$ kN C, $F_{BC} = 39$ kN C, $F_{BE} = 15$ kN C, $F_{BF} = 15.62$ kN T, $F_{CD} = 30$ kN C, $F_{CE} = 39$ kN T, $F_{DE} = 0$, $F_{EF} = 36$ kN T.   **8.14** Simple truss.   **8.15** Compound truss (LM's: *EI, EM, HI*).   **8.18** *AN, DL, DM, EL, FK, FL, GJ, GL, IJ, JK, KL, LM, MN*.   **8.21** $F_{AB} = 8$ kN C, $F_{AD} = 12$ kN C, $F_{AE} = 8$ kN C, $F_{BC} = 6$ kN C, $F_{BD} = 10$ kN T, $F_{CD} = 10$ kN T, $F_{CH} = 8$ kN T, $F_{DE} = 20$ kN T, $F_{EF} = 24$ kN T, $F_{EI} = 20$ kN C, $F_{FG} = 8$ kN T, $F_{FI} = 12$ kN T, $F_{GH} = 6$ kN T, $F_{GI} = 10$ kN C, $F_{HI} = 10$ kN C.   **8.22** $F_{AB} = 3$ kN T, $F_{AD} = 12.5$ kN T, $F_{BC} = 2.5$ kN T, $F_{BE} = 2.5$ kN C, $F_{CD} = 2.5$ kN C, $F_{DE} = 9$ kN C.
8.24 $F_{AB} = 425$ lb T, $F_{AC} = 225$ lb C, $F_{AD} = 380$ lb T, $F_{BC} = 375$ lb C, $F_{BF} = 200$ lb C, $F_{CD} = 300$ lb C, $F_{DE} = 20$ lb T, $F_{DG} = 300$ lb T, $F_{EF} = 425$ lb C, $F_{EG} = 225$ lb T, $F_{FG} = 375$ lb T.   **8.26** $F_{BK} = 40$ kN T, $F_{CJ} = 20$ kN C.   **8.27** $F_{EI} = 6$ kN C, $F_{DJ} = 72$ kN T.
8.29 $F_{AF} = 2.55$ kips T, $F_{BE} = 1$ kip T.   **8.30** $F_{BL} = 1.5$ kN C, $F_{CL} = 22.4$ kN T.   **8.32** $F_{CO} = 27.5$ kips T, $F_{EP} = 22.5$ kips T.   **8.34** $F_{AB} = 39$ kN C, $F_{FG} = 42$ kN C.   **8.35** $F_{GI} = 0$, $F_{DL} = 5$ kN T.
8.37 $F_{FG} = 2.5$ kN C, $F_{CJ} = 12.5$ kN C.   **8.38** $F_{LM} = 1250$ lb T, $F_{EN} = 966$ lb T, $F_{GK} = 0$.   **8.39** $F_{BC} = 150$ kN T, $F_{BI} = 120$ kN C, $F_{DH} = 30$ kN C.   **8.41** $F_{AB} = P/3$ C, $F_{CD} = P$ C, $F_{EF} = 2P/3$ C.
8.42 $F_{AB} = 0.471P$ C, $F_{CD} = 0.770P$ C, $F_{EF} = 0.345P$ T.   **8.44** $F_{BH} = 89.3$ kN T, $F_{AI} = 233$ kN C.   **8.46** $F_{IJ} = 50$ kN C, $F_{DG} = 25$ kN T.   **8.48** *AF, BF, EF, FO, FG*.   **8.49** $F_{BA} = 5$ kN C, $F_{BC} = 43.2$ kN C, $F_{BD} = 33.8$ kN T, $F_{BE} = 13.93$ kN T.

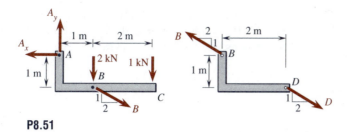

P8.51

8.55 $\mathbf{A} = -10\mathbf{i} + 8\mathbf{j}$ kN, $\mathbf{B} = 10\mathbf{i} - 7\mathbf{j}$ kN.   **8.57** $\mathbf{A} = 12\mathbf{i} - 6\mathbf{j}$ kN, $\mathbf{B} = -21\mathbf{i} + 8\mathbf{j}$ kN, $\mathbf{C} = 9\mathbf{i} - 2\mathbf{j}$ kN.   **8.58** $\mathbf{A} = 400\mathbf{i} + 1000\mathbf{j}$ lb, $\mathbf{B} = -2.4\mathbf{i} - \mathbf{j}$ kips, $\mathbf{C} = 2\mathbf{i}$ kips.   **8.59** $\mathbf{A} = -900\mathbf{i} - 435\mathbf{j}$ N, $\mathbf{F} = 900\mathbf{i} + 1155\mathbf{j}$ N.   **8.60** $\mathbf{A} = -200\mathbf{i} - 1200\mathbf{j}$ lb, $\mathbf{B} = 400\mathbf{i} + 600\mathbf{j}$ lb, $\mathbf{C} = -200\mathbf{i} + 600\mathbf{j}$ lb.   **8.61** $\mathbf{A} = -16\mathbf{i} + 3.5\mathbf{j}$ kN, $M_A = 10.5$ kN·m ↻, $\mathbf{G} = 8\mathbf{j}$ kN.   **8.64** $\mathbf{A} = 57.7$ lb →, $\mathbf{B} = 200$ lb ↑, $\mathbf{C} = 200$ lb ↑.   **8.65** 300 N.   **8.67** $\mathbf{C} = 171$ lb ←, $\mathbf{D} = 75\mathbf{i} - 40\mathbf{j}$ lb.   **8.68** (a) 500 N·m; (b) $\mathbf{B} = 375\mathbf{j}$ N, $\mathbf{C} = -375\mathbf{j}$ N, $\mathbf{E} = 0$.
8.70 8.51 lb.   **8.71** 173.2 N·m   **8.72** 720 lb.   **8.74** 260 N.
8.75 5.5 kN.   **8.77** 26P.   **8.78** 506 N.   **8.79** $F_{DE} = 10.61$ kN C, $F_{CF} = 17.30$ kN C.   **8.80** (a) $F_{BJ} = 493$ N C, $F_{CE} = 4$ kN C, $F_{FG} = 4.68$ kN C; (b) $\mathbf{A} = -1132\mathbf{i} - 40.5\mathbf{j}$ N, $\mathbf{D} = 4.28\mathbf{i} + 1.95\mathbf{j}$ kN.
8.81 (a) $-14\mathbf{k}$ kN, (b) $8\mathbf{i} - 8\mathbf{j}$ kN, (c) $-24\mathbf{k}$ kN·m, (d) $56\mathbf{i} + 98\mathbf{j}$ kN·m.

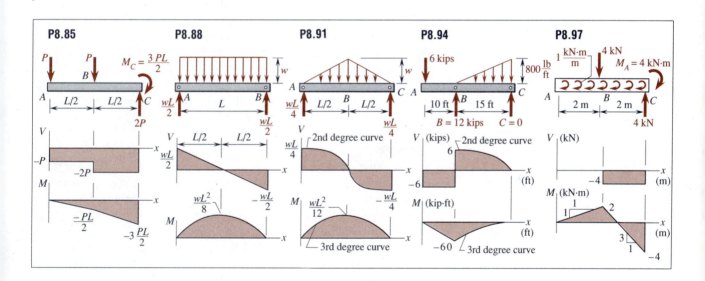

8.99 $V = \dfrac{w_o L}{12} - \dfrac{w_o}{L^2}\left(\dfrac{x^2 L}{2} - \dfrac{x^3}{3}\right)$, $M = \dfrac{w_o Lx}{12} - \dfrac{w_o}{L^2}\left(\dfrac{x^3 L}{6} - \dfrac{x^4}{12}\right)$.

8.102 $h_B = 15$ ft, $h_C = 20$ ft, $h_D = 12$ ft.   **8.103** (a) 108 N, (b) 1020 N.   **8.104** 5.3 m.   **8.106** 1345 m.   **8.108** $T_{min} = 2.5$ kips, $T_{max} = 2.69$ kips.   **8.109** $T_{min} = 9.91$ kN, $T_{max} = 11.29$ kN.   **8.110** 102.6 ft.   **8.111** 104.0 m.   **8.113** 4.73 m.   **8.115** 22.6 ft.
8.116 (a) 11.35°, (b) 16.88°, (c) 2.53 kips.   **8.118** 22.5 ft.   **8.119** (a) 11.31°, (b) 16.70°, (c) 2.53 kips.   **8.123** $F_{GH} = 62.5$ kN C, $F_{EI} =$

18.23 kN T, F_{BK} = 42.4 kN T, F_{KJ} = 73.7 kN C, F_{CL} = 36.1 kN T.
8.124 B = $-7.5\mathbf{i} - 0.4\mathbf{j}$ kips, **E** = $7.5\mathbf{i} + 1.2\mathbf{j}$ kips.   **8.125** 2.51
kN.   **8.127** (a) 8.00 kips, (b) 54.8 ft.   **8.128** 2.37 m.

Chapter 9

Developmental Exercises

D9.3

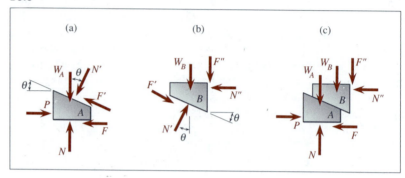

D9.5 No.   **D9.6** (a) 0.4, (b) 0.35, (c) 25 lb →, (d) 28 lb →.   **D9.9**
(a) 20°, (b) 0.364.   **D9.10** (a) 30°, (b) 0.577.   **D9.12** (a) 110 lb, (b)
24 lb, (c) 24.2 lb, (d) not move.   **D9.13** (a) Yes (sliding), (b) no, (c)
yes (tipping).   **D9.14** 346 lb.   **D9.15** (a) 14.60°, (b) 942 N.
D9.16 290 lb.   **D9.17** 21.4 N.   **D9.18** (a) 25.2 lb, (b) through the
corner at D, (c) 25 lb, (d) 25 lb.   **D9.19** 0.268.   **D9.22** P = 5.62
kN, M' = 5.04 N · m.   **D9.23** 51 lb ↑.   **D9.24** (a) 1.875 lb · ft, (b)
0.075 in.   **D9.27** (a) 2π rad, (b) 0.366.   **D9.28** $2\pi/3$ rad, (b) P_{min}
= 58.1 N, P_{max} = 165.6 N.   **D9.30** (a) 19.92, (b) 12.99.

Problems

9.1 $\frac{1}{2}$.   **9.2** $4/(3\pi)$.   **9.4** 30°.   **9.6** 26.1°.   **9.7** Yes.   **9.8**
Yes.   **9.10** No.   **9.11** Yes.   **9.13** $\frac{1}{2}$.   **9.15** 0.686.   **9.16**
0.646.   **9.18** 0.611.   **9.20** (a) $\frac{1}{3}$, (b) 20 lb.   **9.21** 17.27 N.
9.23 (a) Crates will move, (b) F_A = 193.6 N, F_B = 92.2 N.   **9.25**
73.3°.   **9.26** 132 lb.   **9.27** (a) Yes, (b) no, (c) no.   **9.28** 28.4
lb.   **9.29** 32.9 N $\leq P \leq$ 1.265 kN.   **9.31** 26.6 mm.   **9.32**
36.9°.   **9.34** 0.48.   **9.35** 0.204.   **9.37** 0.478.   **9.39** No.
9.41 5.66 kg $\leq m_A$ = 9.43 kg.   **9.42** 65.4° $\leq \theta \leq$ 69.7°.   **9.43** 9.5
kg $\leq m_B \leq$ 188 kg.   **9.45** 10 lb.   **9.46** 10.8 lb.   **9.48** 10.44
lb.   **9.49** 10.8 lb.   **9.51** 56.3°.   **9.52** 17.40 lb.   **9.53** 59.5
N.   **9.54** 172.4 N.   **9.56** $\frac{1}{3}$.   **9.57** 36.9°.   **9.60** 126.1 N.
9.61 157.6 N.   **9.62** 53.6 lb.   **9.64** 8.81 lb · ft.   **9.65** 28.1.
9.67 9.03 N · m.   **9.69** (a) 20.4 N · m, (b) 11.74 N · m.   **9.70** (a)
114.3 lb, (b) 109.9 lb.   **9.71** 536 N.   **9.73** 0.215.   **9.75** (a)
$3P/(2\pi R^2)$, (b) $\mu_s PR$.   **9.76** 15.18 N.   **9.77** 114.8 N.   **9.79** 4 full
turns.   **9.80** 3.62 N.   **9.82** 0.350.   **9.83** 0.1166.   **9.85**
0.348.   **9.87** 0.348.   **9.88** 336 N.   **9.89** 1.033 kN.   **9.91** 68.5
lb.   **9.93** 306 N.   **9.94** 139.7 N.   **9.95** 86.0 lb · ft ↻ .

9.97 285 lb.   **9.98** 31.9 N.   **9.100** 26.3 kg.   **9.102** 0.221.
9.104 58.9 N·m $\curvearrowright$.   **9.106** 1.382 kip·ft.   **9.108** 13 lb.   **9.109**
74.3 N (rotation of A and tipping of B).   **9.110** (a) 496 N, (b) 0.497
N·m.   **9.113** 216 mm.   **9.115** 3.42 kg $\leq m_B \leq$ 181.5 kg.
9.116 θ_A = 49.1°, μ_s = 0.235.   **9.117** 5.53.

Chapter 10

Developmental Exercises

D10.2 (a) 3 ft →, (b) 5 ft.   **D10.3** (a) $\pi/4$ rad $\curvearrowright$, (b) $5\pi/4$ rad.
D10.5 (a) -200 J, (b) 200 J, (c) -100 J.   **D10.6** (a) Equal to, (b) less
than.   **D10.7** (a) 98.1 J, (b) through (e) 0.   **D10.8** No.   **D10.10**
$(\frac{4}{5} F - A_x)\, dx$.   **D10.11** $(\frac{3}{5} F - A_y)\, dy$.   **D10.13** (a) $L(\delta\theta)^2$, (b) yes,
(c) $L\,\delta\theta$ ↑.   **D10.14** $\overrightarrow{\delta y_C}$ = 4 $\delta\theta$ ft ↓, $\overrightarrow{\delta x_D}$ = 10 $\delta\theta$ ft →, $\overrightarrow{\delta y_E}$ =
8 $\delta\theta$ ft ↓, $\overrightarrow{\delta x_A}$ = **0**, $\overrightarrow{\delta y_A}$ = 4 $\delta\theta$ ft ↓.   **D10.15** $\overrightarrow{\delta x_E}$ = 6 $\delta\theta$ ft →,
$\overrightarrow{\delta y_E}$ = 4 $\delta\theta$ ft ↓.   **D10.16** (1) $\sin\theta$ = 1.7 $\sin\phi$; (2) (a) 0.4 $\delta\theta$ $\curvearrowright$,
(b) $\overrightarrow{\delta x_E}$ = 1.12 $\delta\theta$ m ←, $\overrightarrow{\delta y_C}$ = 2.1 $\delta\theta$ m ↑.   **D10.19** (a) **P** =
$(P/5)(-3\mathbf{i} + 4\mathbf{j})$, **Q** = $(Q/17)(15\mathbf{i} + 8\mathbf{j})$; (b) $(\frac{84}{85} Q - 168)\delta r_1$; (c) 170 N;
(d) 250 N.   **D10.20** (a) δx_C = 3 $\delta\theta$ m →, $\overrightarrow{\delta y_C}$ = 2 $\delta\theta$ m ↑, $\delta\phi$ =
$\delta\theta$ $\curvearrowright$; (b) 4 kN·m $\curvearrowright$.   **D10.21** 400 lb →.   **D10.22** $\mathbf{D}_x$ = 1050 lb
→, $\mathbf{D}_y$ = 1100 lb ↑.   **D10.23 G** = 7 kN ↑, **K** = 1 kN ↑.
D10.24 9 kN ←.   **D10.25** 200 N←.   **D10.26** 60 lb.   **D10.33**
No.   **D10.35** (a) $-(WL/2)(4\cos\theta + \sin\theta)$, (b) 14.04°.   **D10.37**
Stable.   **D10.38** Unstable.   **D10.39** Unstable.

Problems

10.1 (a) 240 ft·lb, (b) 240 ft·lb.   **10.2** (a) 288 ft·lb, (b) 100.8
ft·lb.   **10.3** 90 ft·lb.   **10.5** 95.1 J.   **10.6** 185.7 J.   **10.7**
2.5 J.   **10.9** 28.0 kg.   **10.10** 1.453 slugs.   **10.12** $(A_y - 300)\, dy$
ft·lb.   **10.15** 300 N ↑.   **10.16** 400 lb ↑.   **10.18** 4.2 ft.
10.19 510 N.   **10.21** 3 kN.   **10.22** 1.2 kips.   **10.24** 0.477 m and
1.566 m.   **10.25** 33.4° and 81.2°.   **10.27 A** = 2 kN ↑, **H** = 13
kN ↑.   **10.29** $\mathbf{A}_x$ = 900 lb ←, $\mathbf{A}_y$ = 500 lb ↑, $\mathbf{M}_A$ = 2000 lb·ft $\curvearrowright$;
G = 650 lb ↑.   **10.31** $\mathbf{A}_x$ = 9.5 kN ←, $\mathbf{A}_y$ = 47 kN ↑, $\mathbf{B}_x$ = 9.5
kN →, $\mathbf{B}_y$ = 27 kN ↓.   **10.32** 119.6 kN.   **10.33** 111.4 lb.
10.35 1.6 kips.   **10.37** 34.2 kN.   **10.38** 20 lb.   **10.45** Neutral.
10.46 Unstable.   **10.48** Unstable.   **10.50** Unstable.   **10.54** $a/b <$
$4/(3\pi - 4)$.   **10.56** $a/b < \sqrt{2}/2$.   **10.57** $a/b < 1$.   **10.58** $h <$
1.019 m.   **10.60** $k > mg/h$.   **10.61** $\theta = \tan^{-1} mg/(2kL)$, $dV/d\theta =$
0, $d^2V/d\theta^2 = L[(kL)^2 + (\frac{1}{2} mg)^2]^{1/2} > 0$.   **10.62** $\theta = 0$, unstable; $\theta =$
75.5°, stable; $\theta = 180°$, unstable.   **10.64** $\theta = 0$, unstable; $\theta = 180°$,
stable.   **10.65** $\theta = 0$, unstable; $\theta = 77.8°$, stable.   **10.66** 8 kg.
10.68 $m > 8$ kg.   **10.69** $\theta = 30°$, stable; $\theta = 90°$, unstable.   **10.71**
Unstable.   **10.72** 90 lb, not dependent on a and b.   **10.73 A** = $-6\mathbf{i}$
+ 4.5**j** kN, $\mathbf{M}_A$ = 13.5 kN·m $\curvearrowright$.   **10.74 A** = 250**i** + 120**j** lb, $\mathbf{M}_A$
= 900 lb·ft $\curvearrowright$.   **10.75** 464 lb·ft $\curvearrowright$.   **10.79 A** = 22**i** + 6**j** kN,
$\mathbf{M}_A$ = 66**k** kN·m; **B** = $-2\mathbf{i} + 6\mathbf{j}$ kN.   **10.80** $P < ka^2/(a + b)$.
10.81 (a) a/b, (b) $Q > WLa/(2b^2)$.   **10.82** $P < 0.293\ K/L$.

Chapter 11

Developmental Exercises

D11.5 (a) -48 ft/s, (b) 4 s, (c) 30 ft, (d) 124 ft.   **D11.6** $\Delta x = 20$ m, $x_T = 50$ m, $v_{avg} = 4$ m/s, $|v|_{avg} = 10$ m/s.   **D11.7** (a) $a_0 = 6$ ft/s^2, $a_6 = -30$ ft/s^2; (b) -12 ft/s^2.   **D11.9** $v = v_0 + (A/\omega) \sin \omega t$, $x = x_0 + v_0 t - (A/\omega^2) \cos \omega t$.   **D11.10** $v = -(A/\omega) \cos \omega t$, $x = -(A/\omega^2) \sin \omega t$.   **D11.12** $x = (v_0/\omega) \sin \omega t$.   **D11.13** $v = v_0/(1 + bv_0 t)$, $x = (1/b) \ln (1 + bv_0 t)$.   **D11.14** (a) -40 ft/s, (b) 2.48 s.   **D11.15** (a) $x_A < x_B$, (b) $v_C > v_D$, (c) $a_D < a_C$.   **D11.16** 2.   **D11.17** 2 m/s ↑.   **D11.18** $\Delta x = 7.5$ ft, $x_T = 107.5$ ft.   **D11.19** $\tau = 3.46$ s, $t_2 = 11.46$ s.   **D11.20** (a) 62.5 m, (b) 77.5 m.   **D11.25** a, d, e.   **D11.26** (a) $2\mathbf{i} + 3\mathbf{j} - 3\mathbf{k}$, (b) -6, (c) $6\mathbf{i} - 9\mathbf{j} - 5\mathbf{k}$.   **D11.27** $2t^3\mathbf{i}$.   **D11.28** $v = 12.81$ m/s, $\theta_v = 128.7°$, $a = 10.77$ m/s^2, $\theta_a = 248.2°$.   **D11.29** 3.1 ft/s^2 ↓.   **D11.31** (a) 55.7 s, (b) 15.08 s.   **D11.32** (a) 29.7 s, (b) 14.30 s.   **D11.33** (a) $\theta_1 = 55.0°$, $\theta_2 = 49.0°$; (b) the projectile cannot reach the target.   **D11.36** b.   **D11.37** 30.6 m.   **D11.38** 4.5 m/s^2.   **D11.39** (a) No, (b) no.   **D11.40** $\mathbf{v} = 4.44$ ft/s ⟋ 45°, $\mathbf{a} = 22.1$ ft/s^2 ⟍ 26.6°.   **D11.41** Yes.   **D11.42** (a) $4\pi\mathbf{i} - 80\pi^2\mathbf{j} + (200\pi^3 - \pi)\mathbf{k}$ m^2/s^3, (b) $0.00201\mathbf{i} - 0.1264\mathbf{j} + 0.992\mathbf{k}$, (c) 7.26°, (d) 7.26°.

Problems

11.1 $t_1 = 2$ s, $x_1 = 10$ m, $a_1 = -6$ m/s^2, $t_2 = 4$ s, $x_2 = 6$ m, $a_2 = 6$ m/s^2.   **11.3** $t = 3$ s, $x = 8$ m, $v = -3$ m/s.   **11.4** $t = 4$ s, $x = 57$ mm, $v = -3$ mm/s.   **11.6** (a) 3 s or 6 s, (b) 297 in.   **11.7** $x = t^4 - 32t + 48$ m.   **11.8** (a) 6 s, (b) 608 in.   **11.11** (a) ± 4 s^{-1}, (b) $\pi/8$ s.   **11.12** (a) ± 6 m/s, (b) 3.46 m, (c) 2 m, (d) 8 m/s.   **11.14** (a) 12.77 km, (b) 1.593 Mm, (c) 25.5 Mm.   **11.16** (a) 11.18 km/s, (b) 6.95 mi/s.   **11.17** (a) 1.195×10^{-3} ft^{-1}, (b) 31.7 s.   **11.19** (a) 1910 ft, (b) 21.7 s.   **11.20** (a) 99.9 ft, (b) 81.6 ft/s.   **11.21** 107.3 m.   **11.24** (a) 11.25 s, (b) 309 m.   **11.26** (a) -2.5 m/s^2, (b) 10 s.   **11.27** 1.874 mi.   **11.28** $v = [(b - c)g - bk]e^{-bt}/[b(b - c)] + ke^{-ct}/(b - c) - g/b$, $x = [(b - c)g - bk](1 - e^{-bt})/[b^2(b - c)] + k(1 - e^{-ct})/[c(b - c)] - gt/b$.   **11.29** $\mathbf{v}_B = 1$ ft/s ↓, $\mathbf{a}_B = 1$ ft/s^2 ↑.   **11.30** $\mathbf{v}_B = 0.75$ ft/s ↑, $\mathbf{a}_B = 0.25$ ft/s^2 ↓.   **11.32** (a) 1.6 m/s ←; (b) $\mathbf{v}_C = 0.8$ m/s ←, $\mathbf{v}_D = 2.4$ m/s ←; (c) 0.4 m/s →.   **11.34** 100 mm/s →.   **11.35** (a) 8 m, (b) 8 m/s ↑.   **11.36** $\mathbf{v}_D = 0.4$ m/s ↑, $\mathbf{a}_D = 9$ mm/s^2 ↑.

11.38 $t = 4.5$ s.

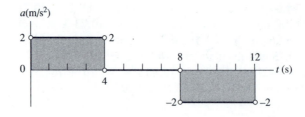

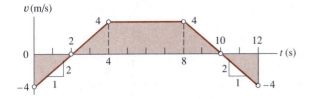

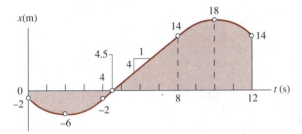

11.41 (a) $\Delta x = -17$ ft, $x_T = 47$ ft; (b) $t_1 = 4$ s, $t_2 = 12.17$ s.

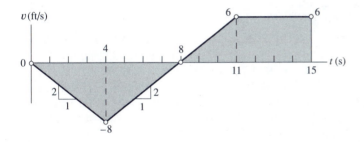

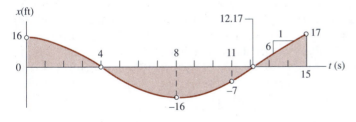

11.44 (a) 3.08 ft/s², (b) 378 ft.   **11.45** (a) 0.5 m/s², (b) 20 m/s.
11.46 (a) To avoid collision, $L_{min} = 227$ m $> L = 210$ m; (b) 9.54 m/s
$\rightarrow$; (c) $(x_T)_A = 121.9$ m, $(x_T)_B = 88.1$ m.   **11.47** (a) $a_A = -1.667$
m/s², $a_B = -1.25$ m/s² ; (b) $(x_T)_A = 120$ m, $(x_T)_B = 90$ m.   **11.48**
14.25 s.   **11.49** 6 s.   **11.51** $\mathbf{v} = 50$ ft/s $\measuredangle$ 53.1°, $\mathbf{a} = 49.5$ ft/s² $\measuredangle$
76.0°.   **11.52** $\mathbf{v} = -12\mathbf{i} + 3\mathbf{j}$ m/s, $\mathbf{a} = -12\mathbf{i}$ m/s².   **11.55** $v =$
41 m/s, $a = 40$ m/s².   **11.56** 48 mm/s $\rightarrow$.   **11.57** 30.6 mm/s² $\rightarrow$.
11.59 $-1.358\mathbf{i} - 1.019\mathbf{j}$ mm/s².   **11.60** 6.26 m/s $< v_0 < 9.40$ m/s.

11.61 1.193 m $< L <$ 3.193 m.   **11.62** 7.38 ft/s $< v_0 <$ 16.82 ft/s.
11.63 (a) $v_0^2(\sin 2\theta)/g$, (b) 45°, (c) v_0^2/g, (d) $v_0^2/(4g)$.   **11.65** (a) 45°,
(b) $4h$.   **11.66** (a) 45°, (b) 14.68 m.   **11.67** (a) 41.3°, (b) 22.7 m.
11.69 23.9°.   **11.70** 40.9°.   **11.72** 6.63 ft.   **11.73** (a) $\theta_1 =$
55.0°, $\theta_2 = 31.4°$; (b) $t_1 = 69.7$ s, $t_2 = 46.9$ s.   **11.74** 1.55 h.
11.76 (a) 15.17 min, (b) 29.6°.   **11.77** (a) 82.5 km/h ⦨ 49.1°,
(b) 45.8 m ⦨ 49.1°, (c) 90.1 m.   **11.79** 71.6° ↘.   **11.80** 79.7 km/h.
11.82 2.81 m/s².   **11.83** $v_B = 198.1$ m/s, $\theta = 60°$.   **11.84** (a)
4.77 mi/s, (b) 1.542 h.   **11.86** 250 m.   **11.87** 44.8 ft/s.   **11.88**
$\mathbf{a}_t = 1.732$ mm/s² ⦨ 30°, $\mathbf{a}_n = 1$ mm/s² ⦨ 60°.   **11.89** $\mathbf{v} = 400$ m/s
↑, $\mathbf{a} = 18.48$ m/s² ↑.   **11.90** $\mathbf{v} = L\dot{\theta}/\cos^2\theta$ ↑.   **11.92** $\mathbf{v} = \pi/3$
m/s ↓, $\mathbf{a} = \pi^2/12$ m/s² →.   **11.93** $\mathbf{v} = 1.888$ m/s ⦨ 33.7°, $\mathbf{a} =$
1.371 m/s² ⦨ 36.9°.   **11.95** $v = 1.047$ ft/s, $a = 2.21$ ft/s².   **11.96**
$v = 4.56$ ft/s, $a = 6.61$ ft/s².   **11.98** $\dot{r} = 0.528$ m/s, $\ddot{r} = -0.1114$
m/s², $\dot{\theta} = -0.0616$ rad/s, $\ddot{\theta} = -0.1398$ rad/s².   **11.99** $\dot{r} = 30$ mm/s,
$\ddot{r} = 2.76$ mm/s², $\dot{\theta} = -0.0530$ rad/s, $\ddot{\theta} = 0.00325$ rad/s².   **11.101**
4.55°.   **11.103** $v = 2.96$ m/s, $a = 0.956$ m/s².   **11.104** $v = 2.47$
ft/s, $a = 2.84$ ft/s².   **11.105** (a) ± 3 s^{-1}, (b) $\pi/6$ s.   **11.106** 0.5 m/s
↑.   **11.107** $\Delta x = 6$ ft, $x_T = 66$ ft, $t_1 = 5.16$ s, $t_2 = 8.75$ s.   **11.108**
$\theta_1 = 43.3°$, $\theta_2 = 58.0°$.   **11.109** 21.0 Mm.   **11.110** $\mathbf{v} = 231$ m/s
↑, $\mathbf{a} = 52.9$ m/s² ↑.

Chapter 12

Developmental Exercises

D12.3 (a) m, kg, s, N; (b) ft, lb, s, slug.   **D12.4** (a) 98.1 N, (b) 98.1
$\times 10^3$ N, (c) 98.1 $\times 10^{-6}$ N, (d) 98.1 $\times 10^{-3}$ N, (e) 322 lb, (f) 10
lb.   **D12.6** 0.1147.   **D12.7** 8.61 ft/s² ↓.   **D12.8** 17.21 kg.
D12.9 (a) 24$\mathbf{i}$ m/s², (b) 48$\mathbf{i}$ m/s, (c) 64 m.   **D12.10** (a) $-1280\mathbf{k}$ m/s²,
(b) $-1280\mathbf{k}$ m/s, (c) -1024 m.   **D12.11** 68.0 km/h.   **D12.12** 5.11
m/s.   **D12.13** (a) 0, (b) 0.   **D12.14** (a) $(40/41)\mathbf{e}_\theta + (9/41)\mathbf{e}_z$, (b)
12.68°, (c) 42.0 m.   **D12.15** No.   **D12.16** (a) $-15\mathbf{i} - 314\mathbf{j} -$
157.6$\mathbf{k}$ lb·ft, (b) no.   **D12.18** 5.97 $\times 10^{24}$ kg.   **D12.19** (a) Yes, (b)
no.   **D12.22** (a) 10.75 $\times 10^3$ mi²/s, (b) 4.96 $\times 10^3$ mi, (c) 800 mi, (d)
4.90 $\times 10^3$ mi, (e) 76.3 $\times 10^6$ mi², (f) 1.970 h.   **D12.23** Yes.
D12.24 (a) Hyperbola, (b) parabola, (c) ellipse, (d) circle.   **D12.28** (a)
4.34 $\times 10^3$ mi, (b) 4.33 $\times 10^3$ mi, (c) 59.0 $\times 10^6$ mi², (d) 10.18 $\times 10^3$
mi²/s, (e) 1.611 h.   **D12.29** (a) 1.981 h, (b) 7.82 h.   **D12.31** $r_P =$
28.6 $\times 10^6$ mi, $r_Q = 43.4 \times 10^6$ mi.

Problems

12.2 (a) 6.44 ft/s² ↓, (b) 2.5 lb.   **12.3** 0.2 N/m.   **12.4** (a) 32.2 ft/s²
↑, (b) 10.73 ft/s² ↑, (c) 1.4 ft/s² ↑.   **12.6** (a) 23.6 ft/s² ↗, (b) 14.19
ft/s² ↓, (c) 11.19 lb.   **12.9** (a) 7.36 m/s² →, (b) 490 N.   **12.10**
42.7 m.   **12.12** $a_A = 3.21$ m/s², $a_B = 1.507$ m/s².   **12.13** $a_A =$
1.512 m/s², $a_B = 0$.   **12.14** (a) 39.1 lb, (b) 35.5 lb.   **12.15** (a) 2.92
m/s² ↘, (b) 5.66 N.   **12.17** (a) 7.51 ft/s², (b) 66.7 lb ←, (c) 100 lb →.
12.18 (a) $\mathbf{a}_A = 2.03$ m/s² →, $\mathbf{a}_B = 2.71$ m/s² →; (b) 30.5 N.   **12.21**

(a) $\mathbf{a}_A = 3.41$ ft/s$^2 \rightarrow$, $\mathbf{a}_B = 6.81$ ft/s$^2 \rightarrow$; (b) 4.97 lb.   **12.22** (a) 47.4 lb; (b) $T_1 = 83$ lb, $T_2 = 41.5$ lb.   **12.24** (a) 150.7 lb; (b) $T_1 = 88$ lb, $T_2 = 44$ lb.   **12.25** (a) 170.8 kg; (b) $T_1 = 889$ N, $T_2 = 445$ N.
12.26 (a) 9.33 kg; (b) $T_1 = 97.1$ N, $T_2 = 194.2$ N.   **12.28** $\mathbf{a}_A = 0.892$ m/s$^2 \uparrow$, $\mathbf{a}_B = 0.892$ m/s$^2 \uparrow$, $\mathbf{a}_C = 2.68$ m/s$^2 \downarrow$, $T_1 = 428$ N, $T_2 = 214$ N.   **12.29** $\mathbf{a}_A = 6.99$ ft/s$^2 \uparrow$, $\mathbf{a}_B = 2.81$ ft/s$^2 \downarrow$, $\mathbf{a}_C = 12.61$ ft/s$^2 \downarrow$, $\mathbf{a}_D = 26.6$ ft/s$^2 \uparrow$, $\mathbf{a}_E = 26.3$ ft/s$^2 \downarrow$, $\mathbf{a}_F = 17.51$ ft/s$^2 \downarrow$, $T_1 = 182.5$ lb, $T_2 = 91.3$ lb, $T_3 = 45.6$ lb.   **12.32** $x = 424$ m, $y = -912$ m, $z = 1578$ m.   **12.34** $440\mathbf{i} + 1824\mathbf{j} - 1946\mathbf{k}$ ft.   **12.36** 132.5 N.   **12.37** (a) 11.55 lb, (b) 7.47 ft/s.   **12.39** $T_{AB} = 35.0$ N, $T_{BC} = 84.2$ N.   **12.40** $k_A = 0.5$, $k_B = 0.704$, $k_C = 0.296$, $k_D = 0.2$.   **12.42** (a) 49.0 N; (b) $T_{AB} = 62.8$ N, $\mathbf{a} = 5.89$ m/s^2 ⟋ 36.9°.
12.43 (a) 4.57 mi, (b) 2.09° s^{-1}.   **12.44** 195.3 km/h.   **12.46** 1.115 rev/s.   **12.47** 0.657.   **12.49** 0.408.   **12.50** $\rho = 1569$ mi, $a_t = 28.8$ ft/s^2.   **12.52** $\mathbf{F}_f = 19.29$ N ⟋, $\mathbf{F}_s = 36.1$ N ⟋.   **12.54** (a) 1.5 m, (b) 2 N.   **12.56** (a) 3.25 lb, (b) 2.5 lb.   **12.58** (a) $-54.4\mathbf{i} - 55.4\mathbf{j} - 7.5\mathbf{k}$ N, (b) 68.7 N C.   **12.59** (a) $-49.0\mathbf{i} - 199.9\mathbf{j} - 3\mathbf{k}$ lb, (b) 195.8 lb C.   **12.61** (a) $-1.2\mathbf{i} - 0.8\mathbf{j} + 0.9\mathbf{k}$ m, (b) $1.6\mathbf{i} + 0.7\mathbf{j} - 2.9\mathbf{k}$ m/s, (c) $-\mathbf{i} - 10\mathbf{j} + 2.5\mathbf{k}$ m/s^2.   **12.63** $\mathbf{a}_{1/G} = \mathbf{i} + 12.5\mathbf{k}$ m/s^2, $\mathbf{a}_{2/G} = -4\mathbf{i} - 6.5\mathbf{k}$ m/s^2, $\mathbf{a}_{3/G} = 8.5\mathbf{i} - 2.5\mathbf{k}$ m/s^2.   **12.65** $\mathbf{a}_{1/G} = 15.9\mathbf{i} + 2\mathbf{j} - 26\mathbf{k}$ ft/s^2, $\mathbf{a}_{2/G} = 2.4\mathbf{i} - 10\mathbf{j} + 4\mathbf{k}$ ft/s^2, $\mathbf{a}_{3/G} = -14.1\mathbf{i} + 10\mathbf{j} + 14\mathbf{k}$ ft/s^2.   **12.66** (a) 0.1380, (b) 6.95 km/s.
12.69 (a) 3.38 mi/s, (b) 11.77×10^3 mi^2/s, (c) 110.0×10^6 mi^2, (d) 2.60 h.   **12.70** (a) 563 mi, (b) 4.60 mi/s.   **12.71** 22.2×10^3 mi or 35.8 Mm.   **12.73** 10.59×10^3 mi.   **12.76** (a) 2.50 Gm; (b) $r_P = 147.0$ Gm, $r_Q = 152.0$ Gm; (c) 146.3 Gm.   **12.77** (a) 2.22×10^3 Mm^2/s, (b) 333×10^3.   **12.78** (a) 6.37 km/s, (b) 3.95 km/s, (c) 6.32 km/s.   **12.80** (a) 1.706 km/s, (b) 1.579 km/s, (c) 31.3 m/s.   **12.81** (a) 2.05 h, (b) 0.967 h.   **12.82** 1.487 km/s.   **12.84** 72.8°.   **12.86** (a) $\frac{1}{2} R \sqrt{(1 + \varepsilon)gr_P}$, (b) $R[(g/r_P)(1 + \varepsilon^2 + 2\varepsilon \cos \theta)/(1 + \varepsilon)]^{1/2}$, (c) $\tan^{-1}[(1 + \varepsilon \cos \theta)/(\varepsilon \sin \theta)]$.   **12.87** $[(2r_P^3/g)^{1/2}/(3R)][3 + \tan^2 (\theta/2)]$ $\tan (\theta/2)$.   **12.89** 2.17 h.   **12.90** (a) 19.26 kg, (b) 339 N.   **12.91** (a) 64 kN, (b) 32 kN T, (c) 14 kN T.   **12.92** 97.5 km/h.   **12.93** (a) 4.34 Mm, (b) 5.50 km/s, (c) 0.1849, (d) 29.5×10^3 km^2/s, (e) 2.38 h.
12.94 (a) 1.412 mi/s, (b) 0.998 mi/s, (c) 2.06 h.   **12.96** 30.3 km/s.

Chapter 13

Developmental Exercises

D13.2 (a) -200 J, (b) 200 J, (c) -100 J.   **D13.3** 1000 ft·lb.
D13.4 (a) dx, (b) $6x\,dx$, (c) 3 J.   **D13.5** (a) $2(1 - x)\,dx$, (b) $12\,x(1 - x)\,dx$, (c) 2 J; no, $\mathbf{F}$ is not a conservative force.   **D13.7** No.   **D13.8** (a) Zero, (b) negative, (c) positive.   **D13.9** (a) 769 W, (b) 0.961.
D13.10 (a) Yes, (b) yes, (c) no, (d) no.   **D13.11** -53.1 MJ.
D13.12 $U_W = 117.7$ J, $U_s = 225$ J.   **D13.15** $-mv^2/2$.   **D13.16** a, c, d.   **D13.18** 6.56 Mm.   **D13.19** 56.4°.   **D13.20** 1.835 m $\downarrow$.
D13.21 (a) 16 ft·lb, (b) yes.   **D13.22** 25.2 ft·lb.   **D13.26** At infinity.   **D13.30** Suddenly.   **D13.33** No.   **D13.35** 8.02 ft/s.
D13.38 1.635 m/s^2 $\leftarrow$.

Problems

13.2 (a) $T = 16.18$ J, $v = 2.54$ m/s, (b) $T = 98.1$ J, $v = 6.26$ m/s.
13.4 (a) 10.80 m, (b) 4.32 m/s.   **13.5** 42.1 ft/s.   **13.6** (a) 13.09 ft/s,
(b) 28.0 lb.   **13.8** 2.68 m/s.   **13.10** 6.33 mi/s.   **13.11** 4.54 mi/s.
13.14 6.02 Mm.   **13.15** 4.35×10^3 mi.   **13.16** 6.66 Mm.
13.17 8.88 m/s.   **13.19** 1.449 m/s.   **13.20** (a) 1 ft, (b) 4.01 ft/s.
13.21 (a) 3.61 m/s, (b) 0.986 m.   **13.24** (a) 13.19 in., (b) 4.60 ft/s.
13.25 4.79 ft/s.   **13.27** 2.74 ft/s.   **13.28** 5.49 m/s.   **13.29** 4.15 m/s.
13.31 2.47 m/s.   **13.33** 0.1462.   **13.34** 2.94 m/s^2 $\measuredangle$ 28.1°.
13.35 39.5 mi/h.   **13.36** 38.6 mi/h.   **13.38** 0.114 m/s^2 $\uparrow$.
13.51 (a) $\sqrt{2gR}$, (b) no.   **13.53** (a) $gmR^2(r_Q - r_P)/[2r_P(r_P + r_Q)]$,
(b) $gmR^2(r_Q - r_P)/[2r_Q(r_P + r_Q)]$, (c) $gmR^2(r_Q - r_P)/(2r_Pr_Q)$.   **13.55**
(a) 9.77 m/s, (b) 17.70 ft/s.   **13.57** 1.274 m.   **13.58** (a) 0.709 m,
(b) 2.81 m/s.   **13.59** 1.2 m.   **13.60** 1.414 m/s.   **13.62** 6.50 in.
13.63 6.61 in.   **13.64** 2.02 m/s.   **13.65** (a) 89.2 N, (b) no.
13.67 18.39 ft/s.   **13.78** 0.25.   **13.79** (a) 1.5 ft, (b) 4.91 ft/s.
13.81 9.00 m/s.   **13.84** 37.4 hp.

Chapter 14

Developmental Exercises

D14.2 $9\mathbf{i} - 6\mathbf{j} - 27\mathbf{k}$ N·s.   **D14.3** (a) $12.42\mathbf{i} - 15.53\mathbf{j} + 18.63\mathbf{k}$
lb·s, (b) $45\mathbf{j} - 24\mathbf{k}$ N·s.   **D14.5** 2.73 s.   **D14.8** $-155.2\mathbf{i}$ kg·m^2/s.
D14.9 40 kN.   **D14.10** 2.68 m/s.   **D14.11** $F_1 = 235$ N, $F_2 =$
117.7 N, $\mathbf{v}_A = 4.89$ m/s $\uparrow$, $\mathbf{v}_B = 7.89$ m/s $\downarrow$, $\mathbf{v}_C = 3.77$ m/s $\downarrow$.
D14.12 $t_1 = 0.5$ s, $x_1 = 378$ m.   **D14.14** 98.3 lb $\measuredangle$ 33.9°.
D14.15 668×10^9 lb·ft·s.   **D14.16** $v_{max} = 5.53$ mi/s, $v_{min} = 2.45$ mi/s.
D14.17 8.66 kg·m^2/s.   **D14.18** $v_{max} = 12.25$ m/s, $v_{min} = 7.07$ m/s.
D14.21 34.9°.   **D14.22** 0.5.   **D14.24** 18.44%.   **D14.25** 0.577.
D14.27 3.6 m/s $\uparrow$.   **D14.31** 160 N $\measuredangle$ 60°.   **D14.32** 6.37×10^3
hp.   **D14.33** $\mathbf{B}_x = 12.44$ lb $\rightarrow$, $\mathbf{B}_y = 68.7$ lb $\uparrow$.   **D14.34** (a)
$\frac{1}{2}\rho_L g L^2 + \rho_L L v_0^2$, (b) $\frac{1}{2}\rho_L L v_0^2$, (c) $\frac{1}{2}\rho_L g L^2$, (d) $\frac{1}{2}\rho_L L v_0^2$, (e) $\rho_L g(L - y)$.
D14.36 30 kN.

Problems

14.1 615 lb.   **14.3** 29.6 km/h.   **14.4** 7.95 s.   **14.6** (a) 4 s, (b) 7
s, (c) 11.27 ft/s, (d) 9.65 s.   **14.7** (a) $-12\mathbf{i} + 6\mathbf{j} - 4\mathbf{k}$ N, (b) 7 m/s^2,
(c) 18 m/s.   **14.9** (a) $v_y = 2$, $v_z = -8$, (b) $84\mathbf{j}$ kg·m^2/s.   **14.11**
$-231\mathbf{i} + 93\mathbf{j} + 48\mathbf{k}$ kg·m^2/s.   **14.12** $\theta_A = 52.0°$, $\theta_B = 47.7°$.
14.13 0.4 m/s.   **14.15** (a) 4.75 ft/s, (b) 1.475 lb, (c) 151.1 lb.
14.16 (a) 0.8 m/s, (b) 15 kN, (c) 8.57%.   **14.17** $v_B = 3.55$ m/s, $v_C =$
1.383 m/s.   **14.19** 1498 ft/s.   **14.20** 43.0 m/s.   **14.23** 14.71 kips
$\triangledown$ 84.0°.   **14.24** 31.7 N $\measuredangle$ 10.89°.   **14.25** 3.75 m/s.   **14.28**
17.04 ft/s.   **14.29** 29.4 lb.   **14.31** 89.3 lb.   **14.32** 107.2 kg.
14.33 18.62 kg.   **14.35** $T_1 = 215$ N, $T_2 = 431$ N, $\mathbf{v}_A = 5.25$ m/s $\downarrow$,
$\mathbf{v}_B = 2.38$ m/s $\downarrow$, $\mathbf{v}_C = 5.01$ m/s $\uparrow$.   **14.36** $T_1 = 19.19$ lb, $T_2 =$
38.4 lb, $T_3 = 19.19$ lb, $\mathbf{v}_A = 9.77$ ft/s $\uparrow$, $\mathbf{v}_B = 18.01$ ft/s $\uparrow$, $\mathbf{v}_C =$
2.59 ft/s $\downarrow$, $\mathbf{v}_D = 6.24$ ft/s $\uparrow$, $\mathbf{v}_E = 33.5$ ft/s $\downarrow$, $\mathbf{v}_F = 23.2$ ft/s $\downarrow$.

14.37 $v_A = 112.3$ ft/s, $v_B = 122.0$ ft/s, $v_C = 146.3$ ft/s.   **14.39** (a) $30\mathbf{i}$ m/s; (b) $111\mathbf{i} + 45\mathbf{j}$ m; (c) 1.5 s; (d) $(v_A)_y = -34$, $(v_B)_y = 16$, $(v_C)_y = 15$.   **14.42** (a) 76.5°, (b) 5.07 km/s.   **14.43** (a) 449 m, (b) $82.8° < \phi_S < 97.2°$.   **14.44** (a) 80.7° or 99.3°, (b) 0.1822, (c) 2.01 h.   **14.46** 4.88 km/s.   **14.47** (a) 14 m/s, (b) 0.742 m.   **14.49** 12.15 m/s.   **14.50** (a) 13.65 ft/s, (b) 8.89 in.   **14.52** (a) 0.949 m, (b) 1.896 m/s.   **14.53** 6.39 m/s.   **14.55** (a) $v'_A = 6.2$ ft/s ←, $v'_B = 5.8$ ft/s →, (b) 35.3%.   **14.56** (a) 0.904 m/s ←, (b) 1.021 m →.   **14.58** (a) 0.633, (b) 2.55 ft/s →.   **14.59** $v'_A = 6.5$ m/s ∡ 60°, $v'_B = 3.9$ m/s ←.   **14.60** $v'_A = 8.76$ m/s ⦨ 51.2°, $v'_B = 2.20$ m/s →.   **14.61** $v'_A = 3.99$ m/s ⦡ 11.23°, $v'_B = 0.978$ m/s →.   **14.63** 1.091 m.   **14.65** 53.8 kN.   **14.76** 475 lb.   **14.78** 52.8 lb.   **14.79** 1.8 kN.   **14.80** 1 kN.   **14.83** $\mathbf{B} = -160\mathbf{i} - 162\mathbf{j}$ N, $\mathbf{C} = 214\mathbf{i}$ N.   **14.84** $\mathbf{B} = -300\mathbf{i} - 1106\mathbf{j}$ N, $\mathbf{C} = 1.344\mathbf{j}$ kN.   **14.85** 2.03 in.   **14.87** (a) 900 kW, (b) 657 kW, (c) 73.0%.   **14.88** 15.80 kN.   **14.90** 771 km/h.   **14.91** (i) 55.8 kN, (ii) 828 km/h.   **14.92** $3\rho_L g(L - y)$.   **14.93** (a) $\sqrt{2gy/3}$, (b) $gt^2/6$, (c) $\rho_L g L^2/6$.   **14.94** $3\rho_L g y$.   **14.96** 3.58 tons/s.   **14.97** 11.65 kg/s.   **14.99** (a) 1.141 km/s, (b) 6.54 km/s.   **14.100** (a) 10.74 m/s →, (b) 10.96 s.   **14.101** 702 m/s.   **14.102** (a) 0.270 s, (b) 0.289 m →, (c) 0.504 m →.   **14.103** (a) 0.324 s, (b) 0.295 m →, (c) 0.605 m →.   **14.105** 3.01 s.   **14.106** $\mathbf{r}_A = 99\mathbf{i} + 18\mathbf{j}$ ft, $\mathbf{v}_A = 23\mathbf{i} + 6\mathbf{j}$ ft/s.   **14.107** 1.25 rad/s.   **14.108** $a_{\max} = 9.48 \times 10^3$ mi, $a_{\min} = 520$ mi.   **14.109** 0.371 kg.   **14.110** (a) 2.5 s, (b) -0.2 ft/s².   **14.111** 0.841.

Chapter 15

Developmental Exercises

D15.2 (a) Rotation, (b) translation, (c) general plane motion, (d) translation.   **D15.3** No.   **D15.4** (a) 375 rev, (b) 20 s.   **D15.5** $\Delta\theta_A = 75$ rad ↻ , $\Delta\theta_B = 25$ rad ↺ .   **D15.6** 0.942 m/s.   **D15.7** $\omega = 4$ rad/s ↺ , $\alpha = 30$ rad/s² ↺ , $a_P = 102$ in./s² ⦨ 61.9°.   **D15.11** The equation linking the motions of the key points of the system.   **D15.12** (a) Zero, (b) $\boldsymbol{\omega}_{AB} \times \overrightarrow{AB}$.   **D15.13** 90 rpm ↻ .   **D15.16** Yes.   **D15.17** Generally, $\mathbf{a}_C \neq \mathbf{0}$.   **D15.18** 0.433 m/s ∡ 33.7°.   **D15.19** $\alpha = 0.32$ rad/s² ↺ , $\mathbf{a}_A = 1.55$ m/s² ↓ .   **D15.20** 302 ft/s² ∡ 83.2°.   **D15.21** 156.1 ft/s² ⦡ 64.7°.   **D15.22** 1.505 m/s² ∡ 23.5°.   **D15.23** 0.918 m/s² ⦨ 78.7°.   **D15.26** 32 ft/s² →.   **D15.27** No.   **D15.28** (a) Z at C; (b) $\mathbf{a}_B = 0.48$ m/s² ←, $\mathbf{a}_D = 0.339$ m/s² ⦡ 45°.   **D15.29** $\omega = 0.3$ rad/s ↺ , $\alpha = 0.32$ rad/s² ↺ , $v_A = 1.2$ m/s ↓ , $a_A = 1.55$ m/s² ↓ .   **D15.30** (a) $20 \cos\theta - 15 \sin\theta$; (b) $\omega = 0.24$ rad/s ↺ , $\alpha = 0.0168$ rad/s² ↺ .   **D15.31** $\alpha_{BD} = \alpha_{DE} = 0$.   **D15.32** (a) A fixed reference frame, (b) a moving reference frame, (c) the angular velocity of $Axyz$ in $OXYZ$, (d) the time derivative of $\mathbf{i}$ taken in $OXYZ$, (e) the time derivative of $\mathbf{i}$ taken in $Axyz$, (f) the same as $(\dot{\mathbf{i}})_{OXYZ}$.   **D15.33** a, c.   **D15.34** $\boldsymbol{\Omega} \times \mathbf{Q}$.   **D15.37** $\omega_{OD} = 0.56$ rad/s ↺ , $v_{B/OD} = 0.96$ m/s ∡ 36.9°.   **D15.38** (a) Yes, (b) no.   **D15.42** $\alpha_{OD} = 5.99$ rad/s² ↻ , $a_{B/OD} = 0.717$ m/s² ∡ 36.9°.   **D15.43** $\mathbf{v}_{A/B} = 5\mathbf{j}$ m/s, $\mathbf{a}_{A/B} = -4\mathbf{i} - 20\mathbf{j}$ m/s².   **D15.44** $\mathbf{v}_{B/Axyz} = -5\mathbf{j}$ m/s, $\mathbf{a}_{B/Axyz} = 4\mathbf{i} + 20\mathbf{j}$ m/s².

Problems

15.1 (a) 4 s, (b) 45 s.   **15.2** $\omega_{max} = 6.58$ rad/s, $\alpha_{max} = 41.3$ rad/s^2.
15.4 $\alpha_c = \pi$ rad/s^2 ↻, $\Delta t = 20$ s.   **15.5** $\alpha = 15.59$ rad/s^2 ↺, $a_Q =$
3.6 m/s^2.   **15.7** 0.478 m.   **15.8** $\omega_B = 15\mathbf{i}$ rad/s, $\alpha_B = 7.5\mathbf{i}$ rad/s^2.
15.10 Estimate by applying Eq. (15.8); thus, (a) 3.0 rad/s^2, (b) 2.5
rad/s^2.   **15.12** 7.68 ft/s^2.   **15.14** $h(\omega_B)_0^2(r_A^2 + r_B^2)/(2\pi r_A^3)$.   **15.15**
$\mathbf{v}_P = -24\mathbf{i} + 66\mathbf{j} - 10\mathbf{k}$ m/s, $\mathbf{a}_P = -792\mathbf{i} - 338\mathbf{j} - 330\mathbf{k}$ m/s^2.
15.16 (a) 5.46 m; (b) $v = 70.9$ m/s, $a_t = 0$, $a_n = 922$ m/s^2, $a = 922$
m/s^2.   **15.17** $\omega = 2$ rad/s ↻, $\mathbf{v}_A = 2.6$ m/s ⦨ 67.4°.   **15.19** $\mathbf{v}_A =$
0.6 m/s →, $\mathbf{v}_B = 0.849$ m/s ⦨ 45°.   **15.20** $\omega = 1.5$ rad/s ↻, $(v_Q)_y =$
3.9.   **15.22** $\omega_{AB} = 21.9$ rad/s ↻, $\mathbf{v}_P = 42.6$ ft/s ↑.   **15.23** $\omega_P =$
3150 rpm ↻, $\omega_C = 1260$ rpm ↻.   **15.25** $\omega_{AB} = 0.196$ rad/s ↻, ω_O
$= 1.054$ rad/s ↺.   **15.26** (a) $\omega_{AB} = 1.075$ rad/s ↺, $\mathbf{v}_B = 0.215$ m/s
↓; (b) $\omega_{AB} = 0.866$ rad/s ↺, $\mathbf{v}_B = 0.310$ m/s ↓.   **15.28** (a) Yes, (b)
4.2 m/s →.   **15.29** (a) Yes, (b) 2.1 m/s ⦨ 36.9°.   **15.31** $\omega_{BD} = 3.75$
rad/s ↺, $\omega_{DE} = 7.5$ rad/s ↻.   **15.32** $\mathbf{v}_B = 210$ mm/s ←, $\omega_{CD} =$
1.125 rad/s ↺.   **15.34** 28 in./s →.   **15.36** 2.56 m/s ⦨ 69.4°.
15.51 (a) 3.10 in. above B, (b) 5.53 mi/h ←.   **15.53** (a) 150 mm, (b)
510 mm/s ⦨ 61.9°, (c) 300 mm/s ⦨ 36.9°.   **15.54** (a) 70 mm, (b) 2
rad/s ↻, (c) 278 mm/s ⦨ 59.7°.   **15.55** $\omega = 0.6$ rad/s ↺, $\mathbf{v}_B = 50.9$
mm/s ⦨ 45°.   **15.57** $\omega_{AB} = 0.2$ rad/s ↻, $\mathbf{v}_B = 0.52$ m/s ⦨ 59.5°.
15.58 (a) 0.12 rad/s ↻, (b) 0.9 ft/s ⦨ 53.1°, (c) 1.02 ft/s ⦨ 81.2°.
15.60 $\omega_{AB} = 0.8$ rad/s ↺, $\mathbf{v}_B = 0.532$ m/s ⦨ 74.3°.   **15.61** $\mathbf{v}_B = 56$
in./s ↑, $v_M = 50.3$ in./s ⦨ 72.6°.   **15.63** $\omega_A = \omega_B = 20$ rad/s ↻.
15.65 $\omega_{DE} = 1.6$ rad/s ↻, $\mathbf{v}_F = 488$ mm/s ⦨ 10.39°.   **15.66** (a) 1.8
rad/s, (b) 366 mm/s.   **15.67** $\alpha = 3$ rad/s^2 ↻, $\mathbf{a}_A = 7.8$ m/s^2 ⦨
67.4°.   **15.68** $\mathbf{a}_B = 4.45$ m/s^2 ⦨ 57.4°, $\mathbf{a}_C = 3.75$ m/s^2 ↑, $\mathbf{a}_E =$
2.82 m/s^2 ⦨ 25.2°.   **15.70** (a) $\omega = 2$ rad/s ↻, $\alpha = 5$ rad/s^2 ↻; (b)
9.06 m/s^2 ⦨ 83.7°.   **15.71** $x = 4$ m, $y = 1$ m.   **15.73** $\alpha_P = 35$
rad/s^2 ↻, $\alpha_C = 14$ rad/s^2 ↻, $\mathbf{a}_T = 27.2$ m/s^2 ⦨ 84.1°.   **15.74** $\alpha_{AB} =$
0.336 rad/s^2 ↻, $\alpha_O = 1.939$ rad/s^2 ↺.   **15.75** $\alpha_{AB} = 9$ rad/s^2 ↻, $\mathbf{a}_B$
$= 660$ mm/s^2 ↑.   **15.77** $\alpha_{BD} = 81$ rad/s^2 ↻, $\mathbf{a}_D = 63$ m/s^2 ⦨
36.9°.   **15.78** $\alpha_{BD} = 4.2$ rad/s^2 ↻, $\mathbf{a}_D = 70.6$ in./s^2 ⦨ 87.4°.
15.80 $\alpha_{BD} = 3$ rad/s^2 ↺, $\mathbf{a}_D = 63$ in./s^2 →.   **15.82** (a) $\mathbf{a}_B = 372$
mm/s^2 ←; (b) $\alpha_{AD} = 1.282$ rad/s^2 ↺, $\alpha_{BD} = 11.23$ rad/s^2 ↺, $\alpha_{CD} =$
8.37 rad/s^2 ↺.   **15.83** (a) $\mathbf{a}_B = 0.42$ m/s^2 ←, $\mathbf{a}_E = 1.449$ m/s^2 ⦨
14.38°; (b) $\alpha_{AD} = 7.2$ rad/s^2 ↻, $\alpha_{BD} = 4$ rad/s^2 ↺, $\alpha_{CD} = 2.25$ rad/s^2
↺.   **15.85** (a) $\alpha_{AB} = 3$ rad/s^2 ↺, $\alpha_{BC} = 4$ rad/s^2 ↺; (b) 59.1 in./s^2 ⦨
66.0°.   **15.86** (a) 22 rad/s^2 ↺, (b) 48 rad/s^2 ↻, (c) 56.3 m/s^2 ⦨
7.66°.   **15.87** (a) 0.5 rad/s^2 ↻; (b) $\mathbf{a}_B = 1.3$ m/s^2 ⦨ 59.5°, $\mathbf{a}_D = 1.2$
m/s^2 ⦨ 36.9°.   **15.88** (a) 0.2 rad/s^2 ↻; (b) $\mathbf{a}_B = 1.7$ ft/s^2 ⦨ 81.2°, $\mathbf{a}_D$
$= 1.5$ ft/s^2 ⦨ 53.1°.   **15.89** (a) 0.25 rad/s^2 ↺; (b) $\mathbf{a}_B = 134.2$ mm/s^2
⦨ 63.4°, $\mathbf{a}_D = 75$ mm/s^2 ↑.   **15.90** $9\mathbf{i} + 4\mathbf{j}$ in.   **15.91** (a) 75 mm,
(b) 0.8 rad/s^2 ↻, (c) 156 mm/s^2 ⦨ 67.4°.   **15.97** $\alpha_{AB} = 1.8$ rad/s^2 ↻,
$\mathbf{a}_E = 207$ mm/s^2 ⦨ 29.2°.   **15.98** $\mathbf{a}_D = 721\alpha_{OB}$ ⦨ 19.44°, $\mathbf{a}_E =$
$300\alpha_{OB}$ mm/s^2 ⦨ 53.1°.   **15.99** $\mathbf{a}_G = 1.3\alpha$ m/s^2 ⦨ 22.6°, $\mathbf{a}_H = 0.7\alpha$
m/s^2 ←.   **15.101** $bv_A/(x^2 + b^2)$ ↻.   **15.102** $v_A/(b^2 - x^2)^{1/2}$ ↺.
15.103 $bv_A^2(2x^2 - b^2)/[x^2(x^2 - b^2)^{3/2}]$ ↺.   **15.105** $v_A^2 x/(b^2 - x^2)^{3/2}$ ↺.
15.107 $\omega_{BD} = 0.580$ rad/s ↺ and $\mathbf{v}_D = 6.32$ ft/s ↓   (or $\omega_{BD} = 1.380$
rad/s ↻ and $\mathbf{v}_D = 1.519$ ft/s ↑).   **15.108** $\omega_{AD} = 3$ rad/s ↺, $\mathbf{v}_{B/AD} =$
8 ft/s ←.   **15.109** $\omega_{AD} = 0.4$ rad/s ↺, $\mathbf{v}_{B/AD} = 1.35$ m/s →.

15.111 $\alpha_{AD} = 41.5$ rad/s^2 ↷, $\mathbf{a}_{B/AD} = 38$ ft/s^2 ←.   **15.113** $\alpha_{AD} = 3.94$ rad/s^2 ↶, $\mathbf{a}_{B/AD} = 1.286$ m/s^2 ∡ 67.4°.   **15.115** $\alpha_{AE} = 0.0745$ rad/s^2 ↶, $\mathbf{a}_{B/AE} = 16.22$ mm/s^2 ⦨ 16.26°.   **15.116** $\omega_D = 0.25$ rad/s ↶, $\alpha_D = 0.01$ rad/s^2 ↷.   **15.117** $\omega = 0.75$ rad/s ↷, $\alpha = 0.9$ rad/s^2 ↶.   **15.118** (a) 0.714 rad/s^2 ↷, (b) 12.18 in./s^2 ⦨ 62.4°.   **15.120** (a) 2.04 rad/s ↶, (b) 28.6 in./s ⦨ 47.6°.   **15.122** (a) 1.25 rad/s ^{2} ↷, (b) 3 ft/s.   **15.124** $\alpha_{AB} = 0.0135$ rad/s^2 ↷, $\alpha_{BD} = 0.024$ rad/s^2 ↶.   **15.125** $\mathbf{v}_B = 600$ mm/s ⦨ 53.1°, $\mathbf{a}_B = 1200$ mm/s^2 ∡ 53.1°.   **15.126** $\mathbf{a}_E = 0.6$ ft/s^2 ↓, $\alpha_{AB} = 1.31$ rad/s^2 ↶.   **15.128** (a) $\alpha_{BD} = 2$ rad/s^2 ↶, $\alpha_{DE} = 3.5$ rad/s^2 ↶; (b) 2.82 m/s^2 ⦨ 16.50°.   **15.129** $\mathbf{a}_P = (2u\omega - 2b\omega^2 + u^2/b)\mathbf{i}$, $\mathbf{a}_Q = -4b\omega^2\mathbf{i} + (b\omega^2 - 2u\omega)\mathbf{k}$, $\mathbf{a}_R = -(5b\omega^2 + 2u\omega)\mathbf{i}$, $\mathbf{a}_S = -4b\omega^2\mathbf{i} + (2u\omega - b\omega^2)\mathbf{k}$.   **15.131** 3.47 rad/s^2 ↷.   **15.133** $\alpha_{CD} = 17$ rad/s^2 ↶, $\mathbf{a}_{A/CD} = 2$ m/s^2 ←.

Chapter 16

Developmental Exercises

D16.3 $\frac{1}{12}mc^2$.   **D16.4** $\frac{1}{2}mr^2$.   **D16.5** No.   **D16.6** 1088 slug·ft^2.   **D16.8** 0.548r.   **D16.9** 8 kg·m^2.   **D16.10** No.   **D16.13** $\frac{1}{20}m(3r^2 + 2h^2)$.   **D16.14** $I_x = 1.89$ kg·m^2, $k_x = 410$ mm.   **D16.15** $I_z = 8.2$ kg·m^2, $k_z = 854$ mm.   **D16.16** $I_x = 0.310$ slug·ft^2, $k_x = 7.89$ in.   **D16.17** $I_z = 0.487$ slug·ft^2, $k_z = 9.89$ in.   **D16.18** $I_x = 9.31 \times 10^{-3}$ kg·m^2, $k_x = 56.6$ mm.   **D16.19** $I_y = 4.09 \times 10^{-3}$ kg·m^2, $k_y = 37.5$ mm.   **D16.20** $m\bar{\mathbf{a}}$ and $\bar{I}\alpha$ at G.   **D16.22** 2.45 in.   **D16.24** 818 mm.   **D16.25** (a) No, (b) yes.   **D16.26** $-40.1\mathbf{i} + 205\mathbf{j}$ N.   **D16.27** 270 N.   **D16.29** 116.1 N ↑.   **D16.30** (a) $\mathbf{F} = 24$ lb →, $\mathbf{N} = 55.6$ lb ↑; (b) 0.432.   **D16.31** $\mathbf{O}_x = 4.93$ lb →, $\mathbf{O}_y = 10.72$ lb ↑.   **D16.32** N_A, N_B, D_y, E_y.   **D16.33** 0.909 m/s^2 ←.

Problems

16.3 $I_x = \frac{1}{3}mb^2$, $I_y = \frac{1}{3}ma^2$, $I_z = \frac{1}{3}m(a^2 + b^2)$.   **16.10** $I_x = 21.5 \times 10^{-3}$ slug·ft^2, $k_x = 2.31$ in.   **16.11** $I_x = 326$ kg·mm^2, $k_x = 18.97$ mm.   **16.13** $\frac{1}{12}m(4b^2 + c^2)$.   **16.15** $I_y = \frac{1}{12}m(4a^2 + c^2)$, $k_y = (\sqrt{3}/6)(4a^2 + c^2)^{1/2}$.   **16.16** $I_y = \frac{1}{6}m(a^2 + c^2)$, $k_y = (\sqrt{6}/6)(a^2 + c^2)^{1/2}$.   **16.17** $I_x = 19.44$ kg·m^2, $k_x = 367$ mm.   **16.19** $I_x = 0.944$ kg·m^2, $I_y = 0.512$ kg·m^2.   **16.21** $I_x = 60.9 \times 10^{-3}$ kg·m^2, $I_y = 143.2 \times 10^{-3}$ kg·m^2.   **16.23** $I_z = 2.4$ slug·ft^2, $k_z = 1.414$ ft.   **16.24** $I_z = 103.6 \times 10^{-3}$ kg·m^2, $k_z = 158.7$ mm.   **16.25** $I_x = 55 \times 10^{-3}$ slug·ft^2, $I_y = 185 \times 10^{-3}$ slug·ft^2.   **16.27** $I_x = 1.259$ slug·ft^2, $I_y = 2.32$ slug·ft^2.   **16.29** $I_z = 748$ kg·mm^2, $k_z = 69.3$ mm.   **16.30** $I_z = 1.179$ slug·ft^2, $k_z = 1.145$ ft.   **16.31** $I_y = 110.8$ kg·mm^2, $k_y = 35.4$ mm.   **16.33** $I_y = 419$ kg·mm^2, $k_y = 47.1$ mm.   **16.35** $I_x = 0.933$ slug·ft^2, $k_x = 11.04$ in.   **16.36** $I_x = 681$ kg·mm^2, $k_x = 60.1$ mm.   **16.37** $I_y = 1.570 \times 10^{-3}$ kg·m^2, $k_y = 25.2$ mm.   **16.39** $I_y = 7.46 \times 10^{-3}$ kg·m^2, $k_y = 34.5$ mm.   **16.41** $I_x = 39.5$ slug·ft^2, $k_x = 10.36$ in.   **16.42** $I_x = 20.2 \times 10^{-3}$ kg·m^2, $k_x = 56.7$ mm.   **16.43** $I_z = $

21.3×10^{-3} slug·ft^2, $k_z = 2.46$ in. **16.44** $I_x = 4.25$ kg·m^2, $k_x = 204$ mm. **16.45** $I_y = 4.66 \times 10^{-3}$ kg·m^2, $I_z = 2.63 \times 10^{-3}$ kg·m^2. **16.49** 2.86 m/s^2. **16.50** 24.2 ft/s^2. **16.53** 3.68 m/s^2. **16.54** 13.42 ft/s^2. **16.56** $a_B = 25.8$ ft/s^2 ⟋ 53.1°, $F_{BE} = 22.5$ lb T, $F_{CD} = 1.5$ lb T. **16.57** (a) $(40 - 19.62 \sin \theta)$ rad/s^2 ↺, (b) 1.043 kN T. **16.58** (a) 7.74 s, (b) 7.49 s, (c) 5.40 s. **16.60** 1.150 rad/s. **16.61** $\mathbf{A} = 6.41$ kN ↑, $\mathbf{B} = 7.32$ kN ↑. **16.63** (a) -37.2%, (b) -52.9%. **16.64** (a) 4.8 rad/s^2 ↻, (b) 4.32 rad/s^2↻. **16.65** (a) $\alpha = 3g/(4b)$ ↺ , $\mathbf{A} = \frac{3}{8}W\mathbf{i} + \frac{5}{8}W\mathbf{j}$; (b) $\alpha = 3g/(5b)$ ↺ , $\mathbf{A} = \frac{3}{10}W\mathbf{i} + \frac{7}{10}W\mathbf{j}$; (c) $\alpha = 2g/(3r)$ ↺ , $\mathbf{A} = \frac{1}{3}W\mathbf{j}$. **16.66** (a) 5.25 lb, (b) 7.97 lb. **16.68** $-466\mathbf{i} + 12.81\mathbf{j}$ lb. **16.70** 0.931 N·m ↺ . **16.72** $-45.6\mathbf{i} + 70.8\mathbf{j}$ N. **16.73** 21.8 kN ↑. **16.74** 4.04 rad/s^2 ↻ . **16.75** (a) 89.2 rad/s^2 ↻ , (b) 1.1 s. **16.78** 208 N. **16.79** 14.17 rev. **16.81** 2.41 rad/s^2 ↺ . **16.83** (a) 0.8 m, (b) 10 m/s^2 ↗. **16.84** $T_A = 1.949$ kN, $T_B = 2.91$ kN. **16.87** (a) 1.6 rad/s^2 ↻, (b) $\mathbf{a}_A = 4.06$ m/s^2 ↑, $\mathbf{a}_B = 2.34$ m/s^2 ↓. **16.88** $1.9W$ ⟍ 76.8°. **16.90** 2.55 rad/s^2 ↻ . **16.91** $0.787W$ ↑. **16.93** 0.663. **16.95** 1181 N ↑. **16.97** 2.39 s. **16.98** 0.903 s. **16.99** $\frac{2}{5}r$. **16.100** 3.11 s. **16.102** (a) 4.02 ft/s^2 ←, (b) 0.236. **16.105** (a) No slipping, (b) 1.28 m/s^2 →. **16.106** 2.10 rad/s ↺ . **16.108** $\alpha = 15.32$ rad/s^2↺ , $\mathbf{C} = -15.64\mathbf{i} + 88.9\mathbf{j}$ N. **16.110** (i) (a) $\frac{1}{3}a$ →, (b) $1.5b$; (ii) (a) $\frac{2}{7}a$ →, (b) $1.4b$. **16.111** 162.6 N. **16.113** $\alpha = 8.06$ rad/s^2 ↻ , $\mathbf{A} = 33.6$ N ↑, $\mathbf{B} = 0$. **16.115** 23.3 ft/s^2 ↓. **16.117** 19.95 N. **16.118** 2.28 m/s^2 ⟍ 16.26°. **16.120** (a) 41.5 rad/s^2 ↻, (b) 5.40 ft/s^2 ←. **16.122** 6.85 rad/s^2 ↺ , (b) 6.33 m/s^2 ⟋ 30°. **16.123** $I_x = 26.6$ slug·ft^2, $k_x = 3.79$ ft. **16.125** $I_x = 1.130 \times 10^{-3}$ kg·m^2, $k_x = 48.9$ mm. **16.126** $I_y = 2.18 \times 10^{-3}$ kg·m^2, $k_y = 67.9$ mm. **16.128** $I_y = 46.1 \times 10^{-3}$ kg·m^2, $k_y = 86.1$ mm. **16.129** $\alpha = 1.391$ rad/s^2 ↻ , $T_{AD} = 3.23$ lb, $T_{BF} = 12.65$ lb. **16.130** 1.951 m/s^2 ↓. **16.133** 2.96 m/s^2 ⟍ 16.26°. **16.134** $\alpha_{OA} = 3.86$ rad/s^2 ↻ , $\alpha_{AB} = 3.86$ rad/s^2 ↺ , $\mathbf{B} = 2.44\mathbf{i}$ lb, $\mathbf{O} = -6.33\mathbf{i} + 7.31\mathbf{j}$ lb. **16.135** $0 < P \le 62.8$ N. **16.136** 32 N. **16.137** 4.12 ft/s^2 ←. **16.138** 8.99 lb.

Chapter 17

Developmental Exercises

D17.2 (a) No, (b) no. **D17.3** (a) Yes, (b) no. **D17.5** (a) 15 J, (b) -15 J, (c) -5.24 J. **D17.9** (a) 0.8 m, (b) 2.96 rad/s^2 ↺ . **D17.10** 16.38 in. **D17.12** 2.68 m/s ↓. **D17.13** 1.143 m →. **D17.14** 80.4 N. **D17.15** 4.87 rad/s ↺ . **D17.19** (a) -37.7 J, (b) 0. **D17.20** 175.1 lb·ft. **D17.22** 0.851 m/s^2 ←, no change in $\mathbf{a}_C$.

Problems

17.1 10.96 rad/s ↻ . **17.2** 4.59 rad/s ↺ . **17.4** 4.66 rev. **17.5** Mass of gear should be 30 kg, not 50 kg; thus, $v_O = 1.225$ m/s. **17.8** 63.9 km/h. **17.9** 10.90 rad/s ↻ . **17.11** 0.309. **17.13** 20.3 N·m. **17.14** 15.53 kN/m. **17.16** 0.669 m/s. **17.18** 20.4 rev.

17.19 17.67 rev.   **17.20** 591 ft.   **17.21** 16.77 rev.   **17.23** 337. N.
17.24 5.77 rev.   **17.26** 8.58 ft/s.   **17.27** (a) $\omega_{AB} = \omega_{BD} = 1.530$
$\sqrt{g/L}$ ↻ ; (b) $\omega_{AB} = 0$, $\omega_{BD} = 3.46 \sqrt{g/L}$ ↻ .   **17.46** $\bar{v}_{cylinder} = 3.62$
m/s →, $\bar{v}_{sphere} = 3.74$ m/s →.   **17.48** (a) $9.27\sqrt{1 - \cos\theta}$ ft/s, (b)
75.5°.   **17.50** (a) 0.993 m/s ↓, (b) 3.45 m/s ↓.   **17.52** 4.90 rad/s
↻.   **17.53** 4.82 rad/s ↻.   **17.56** (a) 1.934 m/s →, (b) 1.755 m/s
→.   **17.57** (a) 16.13 ft/s ↓, (b) 1.911 in.   **17.58** 1.501 m/s ↓.
17.59 (a) 4.26 m/s ↓, (b) 6.61 m/s ↓.   **17.60** 53.1°.   **17.62** (a)
1.142 $\sqrt{g/r}$, (b) 1.553mg ↑.   **17.64** (a) 5.81 kW, (b) 8.56 kW.
17.66 Maximum belt tension should be 160 N, not 60 N; thus, $(\mu_s)_{min} =$
0.307.   **17.67** 33.2 N/m.   **17.68** $M_{MA} = 35.4$ N·m, $M_{BC} = 17.68$
N·m, $M_{DE} = 13.26$ N·m.   **17.69** $2\pi rma$.   **17.72** 4.77 rad/s^2 ↻.
17.74 8.11 rad/s^2 ↺.   **17.75** 1.982 ft/s^2 ←.   **17.78** 4.44 ft/s^2 ←.
17.79 0.219 m/s^2 ←.   **17.81** (a) 4.29 rad/s ↻, (b) 6.42 rad/s ↻.
17.83 4.63 rad/s.   **17.84** 1.561 m/s ↓.   **17.86** 60°.   **17.87**
55.2°.   **17.90** 1.809 m/s ↑.   **17.92** *Mass*, not weight, of each rear
wheel is 60 kg; thus, $v = 3.41$ m/s ←.   **17.95** 0.602 rad/s^2 ↺.

Chapter 18

Developmental Exercises

D18.1 (a) 600 N·s ↓, (b) 120 N·m·s ↻, (c) 100 N·m·s ↺.   **D18.2** (a)
600 N·s ↓, (b) 120 N·m·s ↻, (c) 100 N·m·s ↺.   **D18.4** (a) **0**, (b) 2π
N·m·s ↻.   **D18.5** (a) 2.48 lb·s →, (b) 0.932 lb·ft·s ↺.   **D18.7**
1.568 s.   **D18.8** (a) No, (b) yes.   **D18.9** (a) 4.19 s, (b) 6.52 rad/s.
D18.11 1.577 m/s ↑.   **D18.12** 1.6 m/s ←.   **D18.15** 991**j** rpm.
D18.16 1.866**j** rad/s.   **D18.18** 45.9 ft·lb.   **D18.19** 63.5 J.
D18.20 5.14 rad/s.   **D18.22** $\omega'_{AB} = 3.47$ rad/s ↻, $\omega'_{CD} = 2.97$
rad/s ↻.   **D18.23** $\omega'_{AB} = 3.17$ rad/s ↻, $\omega'_{CD} = 2.38$ rad/s ↻.

Problems

18.1 19.13 rad/s ↺.   **18.2** 5.03 rad/s ↻.   **18.4** 7.55 rad/s ↻.
18.5 102.6 rad/s ↺.   **18.8** 1.175 s.   **18.9** Mass of gear should be 30
kg, not 50 kg; thus, $v_O = 13.93$ m/s.   **18.10** Mass of gear should be
30 kg, not 50 kg; thus, $v_O = 1.126$ m/s.   **18.11** 29.4 lb.   **18.14** v_2
$= 7.36$ ft/s ↑, $\omega_2 = 14.35$ rad/s ↻.   **18.15** The cord should be
wrapped around a drum of radius 100 mm; thus, $v_G = 3.26$ m/s.
18.16 134.6 N.   **18.17** 2.24 s.   **18.20** 0.513 s.   **18.22** 56.1**j**
rpm.   **18.23** 52.1**j** rpm.   **18.24** The rims of the disks A and B should
be in contact, not overlap; thus, $\omega_{CD} = 22.1$ rpm.   **18.26** 0.821 m/s
←.   **18.28** 0.771 m/s ∠ 16.26°.   **18.30** 1501 rad/s.   **18.31** 6.55
m/s ↑.   **18.32** (a) 0.309 m/s ←, (b) **0**.   **18.33** Rod AC should have
a dimension of 30 in.; thus, $\omega_2 = 5.43$**j** rad/s and $v_r = 10.12$ ft/s.
18.35 (a) 3.36 rad/s ↻, (b) 1.866 kN →.   **18.36** (a) 2.91 rad/s ↻, (b)
1.616**i** − 7.50**j** kN.   **18.39** (a) 8.85 rad/s ↺, (b) 16.50**i** + 33.3**j** kips.
18.40 (a) 8.85 rad/s ↺, (b) 16.50**i** − 33.3**j** kips.   **18.41** 119.5 rad/s.
18.42 $\omega = 7.5$ rad/s ↻, $v_G = 7.5$ m/s ↓.   **18.43** (a) 10.92 rad/s ↻;
(b) $F_B = 8.69$ kips ∠ 20.6°, $F_C = 8.08$ kips ↑.   **18.44** 23.3 m/s →.
18.46 0.594$\sqrt{gh}$, $h =$ height of crate.   **18.48** 1.105$\sqrt{gh}$, $h =$ height
of crate.   **18.49** 1.844 ft/s.   **18.51** 2.   **18.53** (a) 4.61 rad/s ↺;

(b) $\mathbf{C}$ = 73.7 kN ↑ , $\mathbf{A}$ = 18.43 kN ↓ .   **18.56** (a) 2.30 rad/s ↻ ; (b) $\mathbf{C}$ = 55.3 kN ↑ , $\mathbf{A}$ = 13.83 kN ↓ .   **18.58** (a) 25 in., (b) 98.5 lb ↑ ≥ $\mathbf{A}$ ≥ 80 lb ↑.   **18.59** (a) 3.68 rad/s ↻ , (b) 4.05 m/s ↑ .   **18.62** 1.065 ft.   **18.64** 1.517 m/s ⦨ 28.1°.   **18.65** (a) 4.26 rad/s ↺ , (b) **0**.   **18.67** 0.207 rad/s.   **18.69** ω'_{AB} = 0.802 rad/s ↺ , ω'_{CD} = 3.21 rad/s ↺ .   **18.70** ω'_{AB} = 0.401 rad/s ↺ , ω'_{CD} = 7.22 rad/s ↻ .   **18.72** 2.38 ft/s ←.   **18.75** 5.33 ft/s ←.   **18.76** 0.263 m/s ←.   **18.84** 4.94 m/s ↓ .   **18.87** 2.57 s.   **18.88** 0.924 s.   **18.90** (a) 14.00 m/s ⦨25°, (b) 12.94%.   **18.92** 2.45 ft.   **18.94** $\sqrt{75g/(26L)}$ ↺ .   **18.96** $\sqrt{312g/(25L)}$ ↺ .   **18.97** 33.2° ↺ .   **18.99** $\mathbf{A}$ = 44.6 lb ↑ , $\mathbf{B}$ = 22.3 lb ↑ .   **18.100** 9.21 rad/s ↻ .   **18.103** 11.24 m/s ⦨16.26°.   **18.104** 3.52 m/s ⦨16.26°.

Chapter 19

Developmental Exercises

D19.1 $\dot{\mathbf{Q}}$ is taken in $OXYZ$.   **D19.2** $\dot{\mathbf{Q}} = \mathbf{\Omega} \times \mathbf{Q}$, $\ddot{\mathbf{Q}} = \dot{\mathbf{\Omega}} \times \mathbf{Q} + \mathbf{\Omega} \times (\mathbf{\Omega} \times \mathbf{Q})$.   **D19.3** (a) **0**, (b) $-19.33\mathbf{J}$ rad/s.   **D19.4** (a) $-18.56\mathbf{J}$ m/s, (b) $3.2\mathbf{I} + 4\mathbf{J} - 8\mathbf{K}$ rad/s.   **D19.5** (a) **0**, (b) $8.29\mathbf{J}$ rad/s.   **D19.7** $\mathbf{\Omega} = (\mathbf{j} \cdot \mathbf{k})\mathbf{i} + (\dot{\mathbf{k}} \cdot \mathbf{i})\mathbf{j} + (\mathbf{i} \cdot \mathbf{j})\mathbf{k}$.   **D19.9** $-36\mathbf{J} + 16\mathbf{K}$ ft/s².   **D19.10** $-36\mathbf{K}$ ft/s².   **D19.11** $\mathbf{v}_E = (r + R)\omega_2\mathbf{I} - r\omega_1\mathbf{J}$, $\mathbf{a}_E = (r + R)\alpha_2\mathbf{I} - r\alpha_1\mathbf{J} - [r\omega_1^2 + (r + R)\omega_2^2]\mathbf{K}$.   **D19.13** $\mathbf{\omega}_{N/A} = \mathbf{I} + 0.75\mathbf{J} + 1.299\mathbf{K}$ rad/s, $\mathbf{\alpha}_{N/A} = -0.299\mathbf{J} + 2.48\mathbf{K}$ rad/s².   **D19.14** $\mathbf{\omega}_E = \mathbf{I} + 3\mathbf{J}$ rad/s, $\mathbf{\alpha}_E = 2.5\mathbf{J} - 3\mathbf{K}$ rad/s².   **D19.15** (a) 23.0 slug·ft², (b) 18.08 slug·ft².   **D19.16** (a) I_{yz} and I_{zx}, (b) I_{xy} and I_{zx}, (c) I_{xy} and I_{yz}.   **D19.19** (a) The sets of direction cosines of x', y', and z' in $Oxyz$; (b) the sets of direction cosines of x, y, z in $Ox'y'z'$; (c) yes; (d) yes.   **D19.21** $\mathbf{v}_A = -12\mathbf{j}$ ft/s, $\mathbf{v}_B = -6\mathbf{j}$ ft/s, $\mathbf{v}_D = -6\mathbf{j}$ ft/s.   **D19.22** $H_x = \bar{I}_{xx}\omega_x$, $H_y = \bar{I}_{yy}\omega_y$, $H_z = \bar{I}_{zz}\omega_z$.   **D19.23** (a) 35.3°, (b) 19.47°.   **D19.24** 435 J.   **D19.25** 9.28 rad/s.   **D19.26** $T = \frac{1}{2}\bar{\mathbf{v}} \cdot \mathbf{L} + \frac{1}{2}\mathbf{\omega} \cdot \mathbf{H}_G$.   **D19.27** (a) $-4.25\mathbf{I}$ m/s²; (b) $\mathbf{O}_X = -8.50\mathbf{I}$ N, $\mathbf{O}_Y = 19.62\mathbf{J}$ N.   **D19.28** 5.42 rad/s.   **D19.29** 0.708 J.   **D19.31** $\mathbf{O} = -50\mathbf{I} + 16.1\mathbf{J} - 37.5\mathbf{K}$ lb, $\mathbf{M}_O = -48.3\mathbf{I} + 51.9\mathbf{K}$ lb·ft.   **D19.32** 170.3 ft·lb.   **D19.34** $5\mathbf{i} - 15\mathbf{j}$ N.   **D19.35** 8.53 rad/s or 98.6 rad/s.   **D19.37** (a) $\dot{\psi}\mathbf{k}$, (b) $\ddot{\psi}\mathbf{k}$, (c) no.

Problems

19.1 (a) $\mathbf{v}_P = -5.20\mathbf{I} + 3\mathbf{J} - 1.875\mathbf{K}$ m/s, $\mathbf{a}_P = -1.669\mathbf{I} - 2.08\mathbf{J} + 2.60\mathbf{K}$ m/s²; (b) $\mathbf{\omega} = 0.25\mathbf{J} + 0.4\mathbf{K}$ rad/s, $\mathbf{\alpha} = 0.1\mathbf{I}$ rad/s².   **19.2** (a) $\mathbf{v}_P = -5.20\mathbf{I} + 3\mathbf{J} - 1.875\mathbf{K}$ m/s, $\mathbf{a}_P = -8.16\mathbf{I} + 1.672\mathbf{J} + 0.348\mathbf{K}$ m/s²; (b) $\mathbf{\omega} = 0.25\mathbf{J} + 0.4\mathbf{K}$ rad/s, $\mathbf{\alpha} = 0.1\mathbf{I} + 0.3\mathbf{J} + 0.5\mathbf{K}$ rad/s².   **19.4** $\mathbf{v}_B = 1.92\mathbf{I} + 1.100\mathbf{J} - 0.56\mathbf{K}$ m/s, $\mathbf{a}_B = 2.88\mathbf{I} + 2.97\mathbf{J} + 2.61\mathbf{K}$ m/s².   **19.5** $\mathbf{v}_C = 3.46\mathbf{I} - 2\mathbf{J} - 13.86\mathbf{K}$ ft/s, $\mathbf{a}_C = -23.4\mathbf{I} - 2.5\mathbf{J} - 34.6\mathbf{K}$ ft/s².   **19.6** $\mathbf{v}_B = -2.4\mathbf{I}$ ft/s, $\mathbf{\omega}_{AB} = 0.258\mathbf{I} - 0.258\mathbf{J} - 0.1477\mathbf{K}$ rad/s.   **19.9** $\mathbf{v}_B = -4\mathbf{I} - 0.8\mathbf{K}$ m/s, $\mathbf{v}_C = -2.4\mathbf{I} + 0.8\mathbf{J}$ m/s, $\mathbf{a}_B = -35.2\mathbf{J}$ m/s², $\mathbf{a}_C = -12.8\mathbf{I} - 19.2\mathbf{J} - 3.2\mathbf{K}$ m/s².   **19.10** $\mathbf{\alpha}_D = -2\mathbf{J} - 24\mathbf{K}$ rad/s², $\mathbf{a}_B = 20\mathbf{I} - 18\mathbf{J} - 32\mathbf{K}$ ft/s², $\mathbf{a}_C =$

$-3\mathbf{I} - 6\mathbf{K}$ ft/s^2.   **19.11** $\boldsymbol{\alpha}_{AB} = 20\mathbf{I} + 3\mathbf{J} + 2\mathbf{K}$ rad/s^2, $\mathbf{v}_B = 2.4\mathbf{I} + 4.16\mathbf{J} - 5.20\mathbf{K}$ m/s, $\mathbf{a}_B = -41.4\mathbf{I} + 11.68\mathbf{J} - 27.1\mathbf{K}$ m/s^2.   **19.12** (a) $0.967\mathbf{J}$ rad/s; (b) $\boldsymbol{\omega}_C = 0.845(-\mathbf{I} - \mathbf{J})$ rad/s, $\boldsymbol{\alpha}_C = 0.818\mathbf{K}$ rad/s^2. **19.13** $\boldsymbol{\omega} = 0.274\mathbf{I} + 2\mathbf{J} + 0.475\mathbf{K}$ rad/s, $\mathbf{v}_A = -0.316\mathbf{J} + 1.333\mathbf{K}$ ft/s. **19.15** $\boldsymbol{\omega}_D = 1.5\mathbf{I} + 5.53\mathbf{J} + 1.286\mathbf{K}$ rad/s, $\boldsymbol{\alpha}_D = 7.64\mathbf{I} + 5.37\mathbf{J} - 1.774\mathbf{K}$ rad/s^2.   **19.17** $I_{xy} = I_{zx} = 0$, $I_{yz} = 1.8$ slug·ft^2.   **19.19** $I_{xy} = I_{zx} = 0$, $I_{yz} = 1.6$ slug·ft^2.   **19.20** $I_{xy} = -0.1125$ slug·ft^2, $I_{yz} = 0$, $I_{zx} = 0.0562$ slug·ft^2.   **19.21** $I_{xy} = -116.6 \times 10^{-6}$ kg·m^2, $I_{yz} = I_{zx} = 0$.   **19.23** $I_{xy} = -138.2 \times 10^{-6}$ kg·m^2, $I_{yz} = I_{zx} = 0$. **19.25** $I_{xy} = -\frac{1}{4}mab$, $I_{yz} = -\frac{1}{4}mbc$, $I_{zx} = \frac{1}{4}mca$.   **19.26** $I_{xy} = I_{zx} = 0$, $I_{yz} = -6.03 \times 10^{-3}$ kg·m^2.   **19.28** $I_{xy} = I_{zx} = 0$, $I_{yz} = 4.74 \times 10^{-3}$ kg·m^2.   **19.29** $-mRh/(5\pi)$.   **19.31** 24 slug·ft^2.   **19.33** 58.9 slug·ft^2.   **19.34** 15.39 slug·ft^2.   **19.35** 10.15 slug·ft^2.   **19.36** 24.2 slug·ft^2.   **19.37** $(3/20)mR^2(R^2 + 6h^2)/(R^2 + h^2)$.   **19.40** 392×10^{-6} kg·m^2.   **19.41** (a) $mb^2\omega(-0.8\mathbf{I} - 0.4\mathbf{J} + 3\mathbf{K})$, (b) $1.5mb^2\omega^2$. **19.42** (a) $18.10\mathbf{I} + 10.45\mathbf{K}$ kg·m^2/s, (b) 113.7 J.   **19.44** (a) $1.317\mathbf{I} + 1.317\mathbf{J} + 2.63\mathbf{K}$ slug·ft^2/s, (b) $35.3°$, (c) 12.41 ft·lb.   **19.46** (a) $-0.4\mathbf{I} + 0.4\mathbf{K}$ kg·m^2/s, (b) 16.8 J.   **19.48** (a) $-0.0597\mathbf{I} + 0.01599\mathbf{J}$ kg·m^2/s, (b) 18.45 mJ.   **19.49** (a) $0.0582\mathbf{I} + 0.365\mathbf{J} + 0.1763\mathbf{K}$ slug·ft^2/s, (b) 46.4 ft·lb.   **19.50** (a) $6\mathbf{K}$ rad/s, (b) $11.10\mathbf{K}$ rad/s.   **19.51** $19.32\mathbf{I} - 38.6\mathbf{K}$ rad/s.   **19.53** $-[3m_0v_c/(mb)](2\mathbf{J} + \mathbf{K})$.   **19.54** (a) $\cos^{-1}6g/(5b\omega^2)$ if $5b\omega^2 > 6g$, (b) $\sqrt{6g/(5b)}$.   **19.55** $T = 3.54$ kN, $A = 800$ N.   **19.57** $\alpha = 13.89\mathbf{I}$ rad/s^2, $C = 4.81\mathbf{J} + 54.7\mathbf{K}$ N.   **19.58** $2.16\mathbf{K}$ N·m.   **19.59** $3.2\mathbf{J}$ N·m.   **19.60** 3.60 rad/s or 23.6 rad/s. **19.62** 150.6 rad/s.   **19.64** 1.316 rad/s or 61.4 rad/s.   **19.65** 21.7×10^{21} N·m.   **19.66** (a) $\boldsymbol{\omega}_C = 5\mathbf{I} + 4\mathbf{J}$ rad/s, $\boldsymbol{\alpha}_C = 2\mathbf{I} + 3\mathbf{J} - 20\mathbf{K}$ rad/s^2; (b) $\mathbf{v}_P = -34\mathbf{K}$ in./s, $\mathbf{a}_P = -16\mathbf{I} - 150\mathbf{J} - 36\mathbf{K}$ in./s^2. **19.67** $\boldsymbol{\omega}_C = 1.098(-\mathbf{I} - \mathbf{J})$ rad/s, $\boldsymbol{\alpha}_C = 1.380\mathbf{K}$ rad/s^2.   **19.69** $-43.8\mathbf{K}$ N·m.   **19.70** $\alpha = 53.7\mathbf{K}$ rad/s^2, $\mathbf{O} = -3.32\mathbf{I} - 0.819\mathbf{J}$ N, $\mathbf{A} = 3.32\mathbf{I} + 0.819\mathbf{J}$ N.   **19.73** $\theta = \cos^{-1}6r^2\dot{\psi}/[(h^2 - 3r^2)\dot{\phi}]$. **19.74** (a) 4.67 rad/s, (b) $-3.58\mathbf{i} + 9\mathbf{k}$ rad/s.

Chapter 20

Developmental Exercises

D20.1 (a) $\omega_n = \pi$ rad/s, $f_n = 0.5$ Hz; (b) yes.   **D20.2** $\omega_n = 3.84$ rad/s, $\tau = 1.638$ s.   **D20.3** $\omega_n = 6.95$ rad/s, $\tau = 0.904$ s.   **D20.4** 6.60 rad/s.   **D20.5** 1.605 s.   **D20.7** $\omega_n = 3.84$ rad/s, $\tau = 1.638$ s.   **D20.8** (a) $\zeta = 1$, (b) $\zeta > 1$, (c) $\zeta < 1$.   **D20.14** $\sqrt{2}/2$. **D20.16** (a) Increased, (b) 92.4%.

Problems

20.1 $\dot{x}_{max} = 2.09$ ft/s, $\ddot{x}_{max} = 26.3$ ft/s^2.   **20.2** 0.955 Hz.   **20.4** $X = 25$ mm. $\phi = 53.1°$.   **20.5** 367 lb.   **20.7** (a) $\tau = 0.562$ s, $\dot{x}_{max} = 335$ mm/s; (b) $\tau = 1.147$ s, $\dot{x}_{max} = 164.3$ mm/s.   **20.9** 0.438 kg·m^2.   **20.10** $\{(5/m)[kb^2 + (L + r)mg]/[5(L + r)^2 + 2r^2]\}^{1/2}$. **20.11** 1.422 s.   **20.13** 0.627 Hz.   **20.14** 180.0 mm.   **20.16** 0.259

slug·ft^2.   **20.17** 1.988 m.   **20.18** $x = (m_B g/k)(1 - \cos \omega_n t) + [m_B\sqrt{2gh}/\sqrt{k(m_A + m_B)}] \sin \omega_n t$, where $\omega_n = \sqrt{k/(m_A + m_B)}$.   **20.19** 3.98 m.   **20.21** $[kr_2^2/(I + mr_1^2)]^{1/2}/(2\pi)$.   **20.22** (a) 6.32 kN/m, (b) 435 N·s/m.   **20.23** $\zeta = 1.5$, $x = 3.512 e^{-26.18t} - 0.512 e^{-3.82t}$ in.   **20.25** $\zeta = 0.5$, $x = 2\sqrt{3} e^{-5t} \sin (5\sqrt{3}t + \pi/3)$ in.   **20.26** $x = (3\sqrt{7}/35) e^{-7.5t} \sin 2.5\sqrt{7}t$ ft.   **20.28** $k = 792$ N/m, $c = 6.97$ N·s/m.   **20.29** (a) 0.399 s, (b) -0.572.   **20.30** $\zeta = \sqrt{3}c/(8\sqrt{km})$, $\tau_d = 16\pi m/\sqrt{3(64km - 3c^2)}$.   **20.32** (a) 0.564 in., (b) 0.683 in.   **20.33** (a) 0.259, (b) 0.273.   **20.34** 0.327.   **20.37** $(W/g)\ddot{x} + c\dot{x} + kx = ky + c\dot{y}$, $X = 1.731$ in.   **20.38** $\Theta = (12F_0/L)/\sqrt{(3k - m\omega^2)^2 + (3c\omega)^2}$, $\varphi = \tan^{-1}[3c\omega/(3k - 4m\omega^2)]$.   **20.39** 1.586 s.   **20.40** 0.691 Hz.   **20.41** $2\pi[2(\pi - 2)r(R - r)]^{1/2}/\{[\pi r + 2(R - r)]g\}^{1/2}$.   **20.42** 176.2 lb.   **20.43** $X_1 = \frac{1}{2} kX_2/\sqrt{(k - m\omega^2)^2 + (c\omega)^2}$, $\varphi = \tan^{-1}c\omega/(k - m\omega^2)$.   **20.44** $\Theta = (36F_0/L)/[(3k - 7m\omega^2)^2 + (3c\omega)^2]^{1/2}$, $\varphi = \tan^{-1}3c\omega/(3k - 7m\omega^2)$.

Appendix A

1. d   **2.** a   **3.** d   **4.** c   **5.** e   **6.** c   **7.** a   **8.** c
9. e.   **10.** a   **11.** d   **12.** e   **13.** c   **14.** b   **15.** a
16. b   **17.** b   **18.** c   **19.** e   **20.** b   **21.** a   **22.** a
23. d   **24.** b   **25.** e   **26.** c   **27.** b   **28.** c   **29.** d
30. e   **31.** d   **32.** c   **33.** b   **34.** e   **35.** e   **36.** a
37. d   **38.** d   **39.** b   **40.** a   **41.** d   **42.** c   **43.** b
44. c   **45.** e   **46.** d   **47.** b   **48.** a   **49.** e   **50.** d

INDEX

Absolute motion, 452
Absolute system (of units), 17
Acceleration, 14, 455, 479
  angular, 652, 846
  average, 455
  center, 683
  components of:
    cylindrical, 498, 521
    normal, 492–495, 520, 654
    radial, 496–497, 520
    rectangular, 482–483, 520
    tangential, 492–495, 520, 654
    transverse, 496–497, 520
  constant, 461, 486
  Coriolis, 704, 846
    interpretations for, 705
  curve, 470
  gravitational, 17
  instantaneous, 455
  of gravity, 486, 510
  resultant, 654
  rotational, 716
  translational, 716
Accelerations:
  in different reference frames, 702, 845
  in relative motion, 676
Accuracy, 11
  of virtual displacements, 409
Action and reaction, 14, 53
Addition:
  of forces (or vectors), 15, 25, 35, 44
  of moments, 108
Addition theorem:
  for angular accelerations, 846
  for angular velocities, 839
American Standard:
  channels, 243
  shapes, 243
Amplitude, 903
Amplitude ratio, 927
Analytical:
  expression, 34, 41, 44, 101
  vector method, 69
Analytic force field, 430, 576
Angle, 243
  direction, 41
  directional, 24
  lead, 375
  nutation, 889
  of contact, 387
  of kinetic friction, 355
  of repose, 356
  of static friction, 355
  phase, 905, 928
  precession, 889
  spin, 889
Angular:
  acceleration, 652, 846

impulse, 599, 798
momentum, 600, 801, 866
  conservation of, 536n, 603, 616
  of a particle, 600
  of a rigid body, 801, 867
  speed, 652
  velocity, 652, 696, 839
Aperiodic, 920
Apogee, 541
Applied potential energy, 434, 581
Area moments of inertia, 224
  Mohr's circle for, 251
  of common shapes, 244
  of composite areas, 241
  of rolled-steel structural sections, 242
  parallel-axis theorem for, 234
  polar, 226
  principal, 254
  situations involving, 224
Areal speed, 536, 616
Areal velocity, 536
Associative property, 25, 100
Astronomical unit of distance, 546
Auxiliary constant system, 631, 636
Average:
  acceleration, 455
  speed, 453
  velocity, 453, 477
Axes:
  central, 722
  centroidal, 722
  of inertia, principal, 863, 879
  rotation of, 861
Axis:
  acceleration, 683
  moment, 83
  nutation, 889
  of a wrench, 127
  of revolution, 204
  of rotation, 650
  of zero acceleration, 683
  of zero velocity, 670
  precession, 889
  spin, 889
  velocity, 670, 769
Axle friction, 379

Ball and socket, 154, 299
Ball support, 154
Barrel (bbl), 13
Beam, 202, 325
Bearing, 154, 379, 380
Belt friction, 386
Bending moment, 122, 324, 325
Bending moment diagram, 328
Binomials, A-9
Binormal unit vector, 495

Body centrode, 671
Body cone, 893
Body of revolution, 204
Bollard, 393
Boundary lubrication, 352, 353
Box product, 115
Built-in support, 136

Cable, 6, 54, 154, 336
  governing differential equation for, 338
  length of, 341, 344
  parabolic, 340
  sag of, 337, 340
  span of, 337, 340
Calculator, 12
Cant hook, 365
Cantilever beam, 136
Cardan's suspension, 888
Cartesian, 33, 41, 99
Catenary, 343
Center:
  acceleration, 683
  geographic, 178
  of force, 535
  of gravity, 184, 197, 212
  of mass, 212, 526, 569, 603, 722
  of percussion, 735, 802
  of pressure, 211, 225
  of zero acceleration, instantaneous, 683
  of zero velocity, instantaneous, 669
  population, 178
  velocity, 670, 769
Central:
  axes, 266, 722
  moments of inertia, 267
Central force, 535
Central-force motion, 535, 616
  differential equation for, 539
Central impact, 620, 818
Centrode, 671
Centroid, 722n
Centroidal:
  axes, 722n
    principal, 258
  moments of inertia, 224, 227
Centroids, 179
  of areas, 189, 195
  of lines, 179, 183
  of volumes, 208, 210
Chasles' theorem, 660, 731, 734, 799
Circle of friction, 379
Circular frequency, natural, 905
Circular orbit, 541, 543
Clamped support, 136
Coefficient:
  matrix, 68
  of kinetic friction, 354

of restitution, 622, 817
of rolling resistance, 397
of static friction, 354
of viscous damping, 918
Collar, 76, 135, 426
Collinear force system, 52
Commutative property, 25, 99, 112
Compatible virtual displacement, 587
Complementary acceleration, 699n, 704
Complementary solution, 540
Complex truss, 283
Components of a force, 27, 86
Components of a moment, 116
Composite:
   area moments and products of inertia, 241
   areas, 194
   bodies, 724–726
   lines, 182
   volumes, 209
Compound truss, 283
Compressive force, 284
Computer, use of, 1, 68, 73, 79, 80, 102,
   115, 174, 183, 196, 255, 291, 345,
   347, 398, 428, 430, A-22, A-25,
   A-27
Concurrent force system, 52
Conservation of:
   areal speed, 536
   energy, 583, 782, 911
   momentum, 603, 809
Conservative:
   force, 402, 431, 553, 561, 577, 766, 911
   force field, 576
   system, 430, 577
Constant of gravitation, 15, 534
Constrained motion, 750
Constrained virtual displacement, 587, 626,
   785, 822
Constraining forces and moments, 134, 154
Constraint condition, 467
Constraint equation, 42, 467, 512, 515, 606
Continuum, 6
Conversion of units, 11
Coordinate system:
   cylindrical, 498, 521
   polar, 496, 520
   radial and transverse, 496, 520
   rectangular, 482, 520
   tangential and normal, 492, 520
Coplanar force system, 52, 135
Cordlike elements, 6, 54, 135
Coriolis acceleration, 704–706
Coulomb friction, 352
Counters, 307
Couple, 24, 89, 107, 108
   angular impulse of a, 798
   momentum, 802n
   work of a, 767
Coupler link, 663n
Crank, 663n
Critical damping, 920
Critical load for buckling, 224, 256
Cross product, 98, 101, 412
   derivative of a, 482
Curvature, A-18
Curve:
   acceleration, 470
   position, 470
   velocity, 470

Curved arrow, 83, 90, 125, 653
Curvilinear motion, 451, 520
Curvilinear translation, 650
Customary units, U.S., 7, 17
Cylindrical components, 498, 521

d'Alembert's principle, 734, 738
Damped:
   free vibrations, 917
   natural frequency, 921
   period, 921
   vibration, 902
Damping, 902
   factor, 919
   forces, 917
   ratio, 919
Dashpot, 918
Datum, reference, 431, 433
Deceleration, 456
Deformation, period of, 621
Degrees of freedom, 436
Density:
   mass: of bodies, 212
      of laminas, 197
      of slender members, 184
   weight: of laminas, 197
      of slender members, 184
Derivatives:
   of a vector in two reference frames, 696,
      834
   of functions, A-16
   of vector functions, 481
Detached members, 309
Determinant, 102, 114, A-8
Determinate truss, 283
Deterministic vibration, 902
Diagram:
   effective-force, 508, 512, 737
   equivalent-force, 508n
   impulse, 597, 803–808
   kinetic, 508n
   momentum, 597, 803–808
   resultant-force, 508n
   second free-body, 508n
   spring-force, 562
Differential calculus, 413
Differential elements:
   of a line, 179
   of a mass, 212
   of an area, 188, 226
   of a volume, 208
   of a weight, 213
Differential of a function, A-17
Digital root finder, 349, A-25
Dimensional homogeneity, 176
Dimensions, 7
Direct central impact, 621
Direct impact, 620
Direct precession, 893
Direction:
   cosines, 42
   of a force, 3
   of a moment vector, 83, 85
   of a vector, 2, 3
Disk:
   brakes, 380
   clutches, 380
   friction, 380

Displacement, 400, 453, 477
   center, 413
   compatible virtual, 409
   constrained virtual, 410
Distributed loads, 6
   on beams, 202, 325
   on cables, 337, 340, 343
   on (submerged) surfaces, 210
Distributive property, 100, 113
Dot product, 112
   derivative of a, 482
Driver, 663n
Dry friction, 917
Dummy variables, 189
Dynamic equilibrium, 511
Dynamics, 5, 451

Eccentric impact, 620, 817
Eccentricity of conic section, 541
Eccentricity of orbit, 542
Effective force, 508, 526, 731
   diagram, 508, 512, 737
Effective force-moment system, 731–734
Efficiency, 557
Elastic central force motion, 535
Elastic force field, 580
Elastic potential energy, 433, 580
Embedded reference frame, 701, 707, 708,
   841, 847, 849, 879, 880, 882, 884,
   887, 888
Elliptic orbit, 541
Energy:
   conservation of, 583, 782, 911
   criterion, 447
   kinetic, 564, 768, 872, 911
   method, 910
   potential, 577, 583, 911
Engineering notation, 10
Equations:
   of circle, ellipse, hyperbola, parabola,
      straight line, A-13, A-14, A-15
   of equilibrium, 63, 74, 138, 146, 157,
      282, 300
   of motion, 508
      Euler's, 880–883
      for a rigid body, 878–883
      general rotational, 879
      scalar, 508
Equilibrium, 5
   stability of, 139, 439
Equipollent, 527
   systems of forces, 129, 447
Equivalent:
   couples, 108
   cross products, 100
   force, 512
   momentum system, 868
   system of forces, 128, 446
Escape velocity, 543
Eulerian angles, 889
Euler's equations of motion, 880–883
Euler's first law, 528n
Exponents, A-9
External force, 54
Eyebolt, 154

Finite elements, 192, 246

First moments:
   of a line, 180
   of a surface, 190
   of a volume, 208
Fixed:
   reference frame, 452, 694, 834
   support, 136, 155
Flat belt, 386
Follower link, 663*n*
Foot (ft), 8, 12
Force, 3
   applied conservative, 581
   applied constant, 581
   axial, 284
   body, 3, 17
   center of, 535
   central, 535
   characteristics of a, 3
   concentrated, 6
   conservative, 553, 561, 577, 766, 911
   contact (or surface), 3
   effective, 508, 526, 731
   equilibrium equations, 146, 147, 157, 160
   equivalent, 512
   gravitational, 534, 560
   impulsive, 610
   in a cordlike element, 54
   in a plane, 23
   in a truss member, 284
   in space, 41, 44
   inertia, 511
   nonimpulsive, 610
   nonworking, 577
   resultant, 508
   spring, 561
   systems, 52, 125, 126, 128, 130
Force and acceleration:
   method of, 507, 737
   using virtual work, 785
Forced vibration, 902, 925
Force-moment system, 125, 126
Forcing frequency, 926
Four-bar linkage, 663*n*
Frame, inertial reference, 452*n*
Frame link, 663*n*
Frames, 278, 307
   rotating reference, 694
Free-body diagram, 57, 58, 59, 79, 508,
       512, 734, 750
Free flight, 485, 539
Free length of a spring, 56
Free vector, 4, 89, 107
Free vibration, 901, 902
   damped, 917
Frequency:
   forcing, 926
   natural, 905
   natural circular, 905
   natural cyclic, 905
   natural damped, 921
   ratio, 930
Friction, 351
   angles of, 355
   axle, 379
   belt, 386, 390
   circle of, 379
   coefficient of, 354
   Coulomb, 352
   disk, 380

dry, 352
   force, 56, 352
      sense of, 352
   laws of dry, 353
   problems, types of, 356
Frictionless (surface), 54
Full film lubrication, 352
Fundamental laws, 13

Gallon (gal), 13
Generalized coordinates, 436
Generalized virtual work, 626, 822
General plane motion, 651, 660, 750, 769
General rotational equations of motion, 879
General solution, 540
Generating:
   area, 204
   curve, 204
Geometric:
   interpretation, 99, 103, 115
   notations, 87
Gimbals, inner and outer, 888
Governing differential equation for:
   a cable, 338
   central-force motion, 539
Graphical representation, 23
Gravitational:
   acceleration, 17
   central-force motion, 535, 539, 578, 617
   force, 534, 560
   force field, 577
   potential energy, 431, 578
   system (of units), 17
   weight, 17
Ground link, 663*n*
Guides for drawing a good free-body
       diagram, 59
Gyration, radius of, 720
Gyroscope, 888
Gyroscopic motion, 888

Hertz (Hz), 905
Hinge, 135, 155, 282
Hollow arrows, 83, 90, 125, 600*n*, 653
Horsepower, 557
Hydrostatics, 210
Hyperbolic:
   functions, 344, A-18
   trajectory, 541

Idealizations, 5
Idealized truss, 281
Impact, 620
   central, 620, 818
   direct central, 621
   eccentric, 620, 817
   line of, 620
   oblique, 620
   oblique central, 624
   perfectly elastic, 623, 818
   perfectly plastic, 623, 818
Improperly constrained, 143
Impulse:
   angular, 599, 798
   of a force, 595
   linear, 594, 797

Impulse and momentum:
   principle of, 596, 602, 626, 803, 817, 868
   using generalized virtual work, 626, 822
Impulsive force, 610, 817
Impulsive motion, 610, 817
Inch (in.), 9, 12
Incline, 53, 57
Inertia, 8, 14, 16:
   ellipsoid of, 865
   force, 511
   matrix, 862, 867
   moments of, 716
   products of, 859
   rotational, 715
   tensors, 862
   translational, 715
   vector, 511
Inertial reference frame, 14, 51, 452*n*
Infinitesimal:
   angular displacement, 401, A-20
   areal patch, 189
   areal strip, 188
Initial conditions, 457
Inner product, 112
Instability, 139
Instantaneous:
   acceleration, 455
   axis of zero acceleration, 683
   axis of zero velocity, 670
   center of zero acceleration, 683
   center of zero velocity, 669
   velocity, 453
Integrals, table of, A-16
Intermediate computations, 11
Internal force, 54, 323
Internal friction, 918
International gravity formula, 17*n*
Inverse functions, A-13
Isolator design, 930

Jack, scissors, 376
Jerk, 456
Jet engine, 634, 641
Joint of a structure, 278
Joule (J), 9, 402, 554
Journal bearing, 379

Kepler's laws, 546
Kilogram (kg), 8, 9, 10
Kinematic quantity, 452
Kinematics, 5, 451
   of particles, 451
   of rigid bodies, 649, 834
Kinetic:
   diagram, 508*n*
   energy, 564, 768, 872, 911
   friction force, 352
   quantity, 509
   units, 509
   work, 403
Kinetics, 5, 451
   of particles, 507, 552, 594
   of rigid bodies, 715, 765, 797, 857
   virtual work in, 587, 785
Kip, 12
Knife edge, 135

Laminas, centers of gravity of, 197
Laws of dry friction, 353
Leibnitz's rule, A-19
Length, 8, 9, 20
Line of action, 4
Line of impact, 620, 817
Linear:
    accelerations in different reference frames, 703, 845
    impulse, 594, 797
    momentum, 594, 866
        resultant, 801
    spring, 55
    velocities in different reference frames, 699, 837
    vibration, 902
Link, 135, 154, 282
    coupler, 663n
    follower link, 663n
    frame, 663n
    ground, 663n
    output link, 663n
Linkage equation, 663
    for accelerations, 678
    for velocities, 664
Linkage, four-bar, 663n
Linking joint or member, 283
Liter (L), 9, 13
Load, 202, 325
Loading diagram, 202
Logarithmic decrement, 922
Logarithms, 387, A-10

Machines, 308
Maclaurin's series, 410, A-18
Magnification factor, 927
Magnitude:
    of a force (or vector), 3, 34, 41
    of the moment of a couple, 89
    of the moment of a force, 83, 85, 103
Mass, 8, 9, 16, 20, 258
    center of, 526, 569, 603, 722
    density, 184, 197, 212
    flow, 630, 631
    moment of inertia, 258, 716, 857
        interpretations for, 743
        of bodies of common shapes, 269, 270
        of composite bodies, 268
        parallel-axis theorem for, 266
    products of inertia, 859
    relativistic, 596n
    rest, 596n
    variable, 630, 636
Mathematically homogeneous, 177
Matrix:
    angular momentum, 867
    angular velocity, 867
    inertia, 862, 867
Maximum number:
    of independent scalar equilibrium equations, 146n, 311
    of statically determinate unknowns, 157
Mechanical energy, total, 584, 782, 911
Mechanical power, 557
Mechanics, 4
Mensuration formulas, A-11
Meter (m), 8, 9

Method:
    of joints, 286, 300
    of sections, 291
Method of solution:
    analytical vector, 69
    for equilibrium in a plane:
        frames and machines, 311
        particles, 64, 69
        rigid bodies, 147
    for equilibrium in space:
        particles, 76
        rigid bodies, 159
    for friction problems:
        type I, 357
        type II, 359
        type III, 368
    semigraphical, 69
    scalar, 64
Metric units (see SI)
Mixed triple product, 115
Mnemonics, 99, A-12, A-19
Modulus of a spring, 562
Mohr's circle, 251
Moment:
    approach, 291
    arm, 83
    bending, 122, 325, 328
    center, 83
    equilibrium equations, 146, 147, 157, 160
    of a couple, 89, 107
    of a force:
        about a point, 83, 103
        about an axis, 83, 84, 91, 115
    of a quantity, 177
Moment of momentum, 601
Moments, principle of, 177
Moments and products of inertia of a mass, 857
Moments of inertia (see Area and mass moments of inertia)
Momentum, 14
    angular, 600, 801, 866
    conservation of, 603, 809
    couple, 802n
    linear, 594, 866
    moment of, 601
    resultant angular, 801
    resultant linear, 801
    resultant system of, 801, 802, 868
    rotational, 802n
Motion, 451
    central-force, 535, 616
    curvilinear, 451, 520
    dependent rectilinear, 467
    equations of, 508, 878
        Euler's, 880–883
        general rotational, 879
    general plane, 651, 660, 750, 769
    gravitational central-force, 578
    gyroscopic, 888
    impulsive, 610, 817
    of a rigid body, 649, 834
    of rockets, 636
    of the mass center, principle of, 528
    plane, 451
    rectilinear, 451
    relative, 452
    relative rectilinear, 465

    rotational, 878
    simple harmonic, 904
    translational, 878
    torque-free, 892
Moving reference frame, 694, 836
Multiforce members, 307

Natural:
    circular frequency, 905
    cyclic frequency, 905
    frequency, 905
    frequency, damped, 921
Necessary and sufficient conditions for:
    equilibrium of a particle, 63
    equivalent systems of forces, 129
Necessary conditions for:
    equilibrium of a rigid body, 138
    equilibrium of any body, 145
    statically determinate trusses, 283, 300
Negative vector, 24
Neutral equilibrium, 139, 439
Newton (N), 9, 17, 509
Newton-Raphson method, A-19
Newtonian:
    mechanics, 4
    reference frame, 52
Newton's:
    first law, 13, 596
    law of gravitation, 15, 534, 542
    second law, 14, 17, 508, 595
    third law, 14, 20, 53, 309, 323, 326, 352, 417, 511, 526
Nonimpulsive force, 610, 817
Nonlinear:
    spring, 56
    vibration, 902
Nonrigid truss, 281
Nonrotating reference frame, 651
Nonworking force, 431, 577
Normal component, 492–495, 520, 654
Normal force (or reaction), 56, 324, 352
Normal unit vector, 493
Numerical accuracy, 11
Nutation angle, 889
Nutation axis, 889

Oblique central impact, 624
Oblique impact, 620
Orbit, 541
    eccentricity of, 542
    period of, 542
Orders of differential area elements, 188
Orientation, 2, 3, 24, 104
Orthographic projections, 390
Osculating plane, 495
Output link, 663n
Overdamping, 919

Pappus-Guldinus, theorems of, 204
Parabolic:
    cable, 340
    trajectory, 486, 541
Parallel-axis theorem, 234, 239, 266, 723, 857, 859
Parallel component, 118, 324, 403
Parallelogram law, 15, 25, 27

Parametric method, 689
Partially constrained, 143
Particle, 5, 56, 63, 64, 73, 282, 451*n*
Particles, systems of, 525, 569, 602
Particular solution, 540
Pascal (Pa), 9, 19
Percussion, center of, 735, 802
Perfectly elastic impact, 623, 818
Perfectly plastic impact, 623, 818
Perigee, 541
Period of:
　deformation, 622
　orbit, 542
　restitution, 622
　vibration, 905
　　damped, 921
　　undamped, 905
Perpendicular component, 99, 118, 324
Phase angle, 904, 926
Pin-and-bracket support, 155
Pin-connected truss, 279
Pins, 135, 136, 308
Pitch of a screw, 375
Pitch of a wrench, 127
Pivot, 135
Plane truss, 281
Pliers, 127*n*
Point:
　of application, 3, 25, 766*n*
　of inflection, 330
Polar moment of inertia, 226
Polygon rule, 25, 69
Position, 453
　coordinate, 453, 902
　curve, 470
　vector, 42, 477
Possible sets of independent scalar
　　equilibrium equations, 145
Potential energy, 430, 436, 577, 583, 782,
　　911
　applied, 434, 581
　elastic, 433, 580
　gravitational, 431, 578
　principle of, 436
Pound (lb), 9, 17
Pound-mass (lbm), 8, 9, 17
Power, 552, 557
Powered flight, 485, 539
Precession:
　angle, 889
　axis, 889
　direct, 893
　retrograde, 893
　steady, 891, 892
Pressure, 210, 225
　diagram, 211
Primary unit, 9
Principal moments and axes of inertia, 254,
　　863, 879
Principle:
　of conservation of energy, 584, 782, 911
　of generalized virtual work, 823
　of impulse and momentum, 596, 602,
　　626, 803, 817, 868
　of moments, 177, 528, 569, 603, 732,
　　799, 859, 866, 872
　of motion of the mass center, 528
　of potential energy, 436

　of transmissibility, 16, 26, 293
　of virtual work, 417, 418, 447
　of virtual work in kinetics, 587, 785
　of work and energy, 552, 565, 770, 873
　rigid-body, 16
Product:
　of a scalar and a vector, 24
　of inertia of a mass, 857
　of inertia of an area, 236, 239
Projectile, 485
Projection, 42, 112, 113, 115
Pulleys (or sheaves), 54

Quadratic equation, A-7
Quantum mechanics, 5

Radial components, 496, 497, 520
Radian (rad), 9, 12, A-11
Radius:
　of curvature, A-18
　of gyration, 231, 264, 720
Random vibration, 902
Ratio, amplitude, 927
Ratio, damping, 919
Reaction, 14, 53, 512
Reactions, 56, 57, 134, 153
Rectangular:
　components, 33, 41, 116, 412, 482, 483,
　　520
　coordinate system, 52
　moments of inertia of an area, 226
Rectilinear motion, 451
　of a particle, 511, 512
Rectilinear translation, 650
Reduction of a system of forces, 126, 130
Reference datum, 577
　for elastic potential energy, 580, 912, 913
　for gravitational potential energy, 578,
　　579
Reference frame, 452*n*
Reference frames:
　accelerations in different, 702, 845
　time derivatives of a vector in two, 696
　use of rotating, 694
　velocities in different, 689, 837
Relations between shear and bending
　　moment, 330
Relative motion, 452, 465
Relativistic:
　mass, 596*n*
　mechanics, 5, 20
Replacement, 128
Reporting of answers, 11
Resolution of forces, 27, 35, 44, 125, 323
Resonance, 928
Rest mass, 596*n*
Restitution:
　coefficient of, 622, 817
　period of, 622
Resultant, 15
　force, 25, 105, 126
　moment, 104, 108, 126
Retrograde precession, 893
Right-hand rule, 84
Right-handed:
　rectangular coordinate system, 52
　trihedral, 97, 663, 892

Rigid body, 6, 137
Rigid-body principle, 15, 125
Rigidly connected truss, 279
Rigid truss, 281
Rockets, motion of, 636
Roller, 56, 135, 154
Rolling resistance, 396
Rotating reference frames, 694
Rotating unit vectors, 694
Rotation, 650, 652
　of axes, 250, 861
　matrix, 861
Rotational:
　inertia, 715
　kinetic energy, 769
　momentum, 802*n*
　motion (or effect), 4, 16, 89, 401, 878,
　　880
　system of momenta, 799
Rough surface, 135, 154, 351
Rounding rule, 10

Sag of a cable, 337, 340
Scalar, 2
　components of a vector, 34, 41
　equations of motion, 508
　method of solution, 64
　product (*see* Dot product)
　triple product, 114
Scientific notation, 10
Screwdriver, 127*n*
Screws, 374, 375
Second (s), 8, 9
Secondary units, 9
Second moments (*see* Area and mass
　　moments of inertia)
Sections, rolled-steel structural, 242, 243
Semigraphical method, 69
Sense, 2, 3
Shear, 324
　approach, 294
　diagrams, 328
　in cross sections of beams, 325
Sheaves (*see* Pulleys)
Shortest distance, 83, 104, 118
SI, 7–10, 17
Sign convention, 83, 325
Significant digits (or figures), 10
Simple harmonic motion, 904
Simple truss, 281
Simultaneous equations, A-7
　solver, A-22
Slender members, 184
Slider, 135, 154
Slope triangle, 24, 87
Slug, 8
Smooth support (or surface), 54, 56, 135,
　　154
Space:
　centrode, 671
　cone, 893
　diagram, 57
　truss, simple, 299
Specific weight, 211
Speed, 453
　angular, 652
　areal, 536, 616
　average, 453

Spin angle, 889
Spin axis, 889
Spring, 6
    force, 56, 405, 433
    free length of, 56
    modulus (or constant), 56, 562
Spring-force diagram, 562
Stability criteria, 440, 441, 447
Stable equilibrium, 139, 439
Standard:
    deviation, 232n
    gravitational acceleration, 17
Static friction force, 352
Statical indeterminacy, 143
Statically determinate, 143
    problems, 143
    trusses, 283, 300
    unknowns, 143, 157
Statics, 5
Steady precession, 891, 892
Steady-state amplitude, 926
Structure, 278
Submerged surface, 210, 211, 225
Subtraction of forces (or vectors), 26
Successive method, 297
Supports, 56, 134, 154
Surface of revolution, 204
Suspension bridges, 337, 340
System:
    auxiliary constant, 630, 636
    with steady mass flow, 630, 631
    with variable mass, 630, 636
System of:
    effective forces, 527, 731
    external forces, 527
    linked masses, 734
    momenta, 799, 802
    particles, 525, 569, 602
Systems of units, 7, 9

T joint, 285
Tangential component, 492–495, 520, 654
Tangential unit vector, 493
Taylor's series, 440, A-18
Tensile force, 54, 284
Tensor, 2, 237n
Tensors, inertia, 862
Tetrahedron truss, 299
Theorem:
    of acceleration center, 684
    of velocity center, 670
    of work, 554n
    Varignon's, 554
Theory of relativity:
    general, 20
    special, 20
Three-force body (or member), 142
Thrust bearing, 380
Time, 8, 9, 20
Time derivatives of a vector in two reference
    frames, 696, 834
Torque, 89, 122, 324, 375
Torque-free motion, 892
Torsional spring, 433
Total mechanical energy, 584, 782, 911
Trajectory:
    of particle, 486
    of spacecraft, 539

Transcendental equations, 174, 344, 345,
    347, 398
Transfer formula (see Parallel-axis theorem)
Transient solution, 926
Translating reference frame, 834
Translation, 650, 651, 731
    curvilinear, 650
    rectilinear, 650
Translational:
    inertia, 715
    kinetic energy, 769
    motion (or effect), 5, 16, 89, 400, 406,
        651, 878
    system of momenta, 799
Transmissibility, 929
    principle of, 16, 26, 293
Transpose, 861
Transverse components, 496, 497, 520
Triangle rule, 25
Triangular finite elements, 192, 200, 246,
    249
Trihedral, 97
Truss, 278
    joint, 278
    member, 279, 280, 282
    triangular, 281
Tutorials, A-27
Two-force body (or member), 141, 281,
    284, 299
Types of equilibrium, 139, 439

Undamped vibration, 902
Underdamping, 920
Unit, astronomical, 546
Units, 8
    kinetic, 509
Unit vector:
    axial, 498
    binormal, 495
    cartesian, 33, 41, 99
    general, 33, 43, 113
    normal, 493
    principal normal, 495
    radial, 496, 498
    rotating, 694
    tangential, 493
    transverse, 496, 498
Universal joint, 155
Unstable equilibrium, 139, 439
U.S. customary units, 7, 17

V belt, 390
V joint, 285
Variable systems of particles, 630
Varignon's theorem, 86, 105, 178, 554
Vector, 2
    addition, 25, 35, 44
    algebra, 35, 48
    components, 34, 41
    inertia, 511
    position, 477
    product (see Cross product)
    representation of a, 3, 33, 41
Vector-diagram equation, 508, 751, 804,
    809
Vector function, derivative of a, 481

Velocities:
    in different reference frames, 698, 837
    in relative motion, 662
Velocity, 453, 477
    angular, 652, 696, 839
    areal, 536
    average, 453, 477
    axis, 670, 769
    center, 670, 769
    curve, 470
    escape, 543
    instantaneous, 453
Vertical jumps, 330
Vibration, 901
    damped, 902, 917
    deterministic, 902
    forced, 902, 925
    free, 901, 902
    linear, 902
    nonlinear, 902
    period of, 905
        damped, 921
    random, 902
    undamped, 902
Virtual displacement, 409
    compatible, 409, 587
    constrained, 410, 587, 822
Virtual work, 417
    applications, 418
    generalized, 626, 822
    in kinetics, 587, 785
    principle of, 417, 418, 447
Viscous damper, 918
Viscous friction, 917

Watt (W), 557
Wedge, 353, 361
Weight, 17
    apparent, 17
    density, 184, 197
    force, 17, 53
    gravitational, 17
Wide-flange shapes (or sections), 243
Work:
    kinetic, 403
    of a couple, 406, 767
    of a force, 401, 552, 553n
        acting on a rigid body, 765, 766n
    of a gravitational force, 431
    of a spring force, 405, 433
    virtual, 417
Work and energy, principle of, 552, 565,
    770, 873
Worm gear, 382
Wrench, 126, 380

X joint, 297

Yard (yd), 12

Zero-force member, 285
Zero (or null) vector, 24

CENTROIDS OF LINES AND AREAS OF COMMON SHAPES

Quarter-circular arc	Semicircular arc	Segment of circular arc
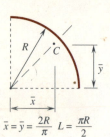 $\bar{x} = \bar{y} = \dfrac{2R}{\pi}$ $L = \dfrac{\pi R}{2}$	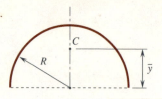 $\bar{y} = \dfrac{2R}{\pi}$ $L = \pi R$	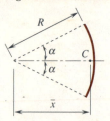 $\bar{x} = \dfrac{R \sin \alpha}{\alpha}$ $L = 2\alpha R$

Triangular area	Circular sector	Quarter-circular area
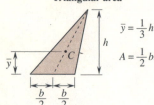 $\bar{y} = \dfrac{1}{3} h$ $A = \dfrac{1}{2} bh$	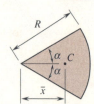 $\bar{x} = \dfrac{2R \sin \alpha}{3\alpha}$ $A = \alpha R^2$	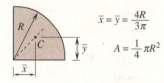 $\bar{x} = \bar{y} = \dfrac{4R}{3\pi}$ $A = \dfrac{1}{4} \pi R^2$

Semicircular area	Parabolic area	Semiparabolic area
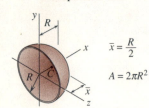 $\bar{y} = \dfrac{4R}{3\pi}$ $A = \dfrac{1}{2} \pi R^2$	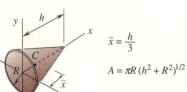 $\bar{y} = \dfrac{3}{5} h$ $A = \dfrac{4}{3} ah$	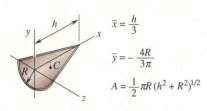 $\bar{x} = \dfrac{3}{8} a$ $\bar{y} = \dfrac{3}{5} h$ $A = \dfrac{2}{3} ah$

Hemispherical shell	Circular conic shell	Half conic shell
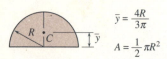 $\bar{x} = \dfrac{R}{2}$ $A = 2\pi R^2$	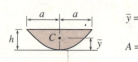 $\bar{x} = \dfrac{h}{3}$ $A = \pi R (h^2 + R^2)^{1/2}$	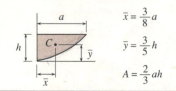 $\bar{x} = \dfrac{h}{3}$ $\bar{y} = -\dfrac{4R}{3\pi}$ $A = \dfrac{1}{2} \pi R (h^2 + R^2)^{1/2}$

CENTROIDS OF VOLUMES OF COMMON SHAPES

Hemisphere	Paraboloid of revolution	Semiellipsoid of revolution
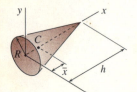 $\bar{x} = \dfrac{3}{8} R$ $V = \dfrac{2}{3} \pi R^3$	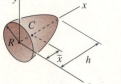 $\bar{x} = \dfrac{h}{3}$ $V = \dfrac{1}{2} \pi R^2 h$	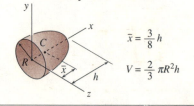 $\bar{x} = \dfrac{3}{8} h$ $V = \dfrac{2}{3} \pi R^2 h$

Circular cone	Half circular cone	Pyramid
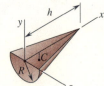 $\bar{x} = \dfrac{h}{4}$ $V = \dfrac{1}{3} \pi R^2 h$	$\bar{x} = \dfrac{h}{4}$ $\bar{y} = -\dfrac{R}{\pi}$ $V = \dfrac{1}{6} \pi R^2 h$	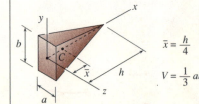 $\bar{x} = \dfrac{h}{4}$ $V = \dfrac{1}{3} abh$